Biochemistry of Foods

Third Edition

Biochemistry of Foods

Third Edition

Edited by

N. A. Michael Eskin and Fereidoon Shahidi

AMSTERDAM • BOSTON • HEIDELBERG • LONDON • NEW YORK • OXFORD • PARIS
SAN DIEGO • SAN FRANCISCO • SINGAPORE • SYDNEY • TOKYO

Academic Press is an Imprint of Elsevier

Academic Press is an imprint of Elsevier
32 Jamestown Road, London NW1 7BY, UK
225 Wyman Street, Waltham, MA 02451, USA
525 B Street, Suite 1800, San Diego, CA 92101-4495, USA

First edition 1971
Second edition 1990
Third edition 2013

Permissions may be sought directly from Elsevier's Science & Technology Rights Department in Oxford, UK: phone +44 (0) 1865 843830; fax +44 (0) 1865 853333; email: permissions@elsevier.com. Alternatively, visit the Science and Technology Books website at www.elsevierdirect.com/rights for further information

Notice
No responsibility is assumed by the publisher for any injury and/or damage to persons or property as a matter of products liability, negligence or otherwise, or from any use or operation of any methods, products, instructions or ideas contained in the material herein. Because of rapid advances in the medical sciences, in particular, independent verification of diagnoses and drug dosages should be made

British Library Cataloguing-in-Publication Data
A catalogue record for this book is available from the British Library

Library of Congress Cataloging-in-Publication Data
A catalog record for this book is available from the Library of Congress

ISBN : 978-0-12-242352-9

For information on all Academic Press publications
visit our website at www.store.elsevier.com

Typeset by TNQ Books and Journals

Printed and bound in United States of America

12 13 14 15   10 9 8 7 6 5 4 3 2 1

Working together to grow
libraries in developing countries

www.elsevier.com | www.bookaid.org | www.sabre.org

ELSEVIER   BOOK AID International   Sabre Foundation

We would like to dedicate this book to:
Dr. H. Michael Henderson, a dear friend and
colleague with whom I wrote the first edition of
BIOCHEMISTRY OF FOODS who passed away in 2009.

Contents

Part I
Biochemical Changes in Raw Foods

Part II
Biochemistry of Food Processing

6. Browning Reactions in Foods 245

N. A. Michael Eskin, Chi-Tang Ho and Fereidoon Shahidi

Part III
Biochemistry of Food Spoilage

The advances made since the publication of the first edition of *Biochemistry of Foods* in 1971 have been extraordinary. This was evident by the more comprehensive approach taken in preparing the second edition in 1990. In the third edition we have tried to keep true to the second edition by significantly updating certain chapters, adding several new chapters, and replacing the chapter on food enzymes with recombinant DNA technologies. It is a real privilege and pleasure to have written an important textbook that spans almost my entire career as a food biochemist. I was particularly delighted that Dr. Fereidoon Shahidi agreed to co-edit this edition and together we have tried to present a book that stands out as an authoritative textbook for teachers, students, and researchers in this very important and dynamic field in food science.

The book is organized into four major sections. Part I deals with important biochemical changes occurring in raw foods that affect quality. In addition to discussing the biochemistry of cereal development in Chapter 1, a section has been added on legumes. Chapter 2 covers postharvest changes in fruits and vegetables with a more extensive discussion of flavor and storage. Chapter 3 provides an in-depth discussion of postmortem changes responsible for converting muscle into edible meat and fish. Chapter 4 covers the latest information on the complex chemical changes involved in the biosynthesis of milk. The last chapter in this section, Chapter 5,

presents important information on the biochemical changes associated with the development of eggs. Part II focuses on the biochemical changes occurring during processing. Chapter 6 presents an extensive coverage of non-enzymatic browning reactions in foods during heating and storage. Chapter 7 details the biochemistry of brewing, and Chapter 8 provides a detailed discussion of the biochemical processes involved in producing cheese and yogurt. Chapter 9 reviews oil processing and fat modification. Part III deals with selected areas associated with food spoilage. Chapter 10 presents a detailed discussion of enzymatic browning, while Chapter 11 provides a comprehensive review of lipid oxidation. The final chapter in this section, Chapter 12, provides an updated and revised coverage of off-flavors in milk. The final section, Part IV, on Biotechnology, provides an extensive review of recombinant DNA technologies in Chapter 13.

In this edition we are particularly grateful to colleagues from around the world for their important contributions to many of the chapters in this book. We are also appreciative of our wives for allowing us the many hours needed to pull this book together. Finally, we would like to acknowledge the outstanding editorial assistance provided by the staff of Elsevier that made the completion of this book a labor of love.

N.A. Michael Eskin and Fereidoon Shahidi

Michel Aliani, Department of Human Nutritional Sciences, University of Manitoba, Winnipeg, Manitoba, Canada

Peter Eck, Department of Human Nutritional Sciences, University of Manitoba, Winnipeg Manitoba, Canada

N. A. Michael Eskin, Department of Human Nutritional Sciences, University of Manitoba, Winnipeg, Manitoba, Canada

H. Douglas Goff, Department of Food Science, University of Guelph, Ontario, Canada

Juan He, Department of Food Science and Technology, Oregon State University, Corvallis, Oregon, USA

Arthur R. Hill, Department of Food Science, University of Guelph, Guelph, Ontario, Canada

Chi-Tang Ho, Food Science Department, Cook College, Rutgers University, New Brunswick, New Jersey, USA

Ernst Hoehn, Swiss Federal Research Station, Switzerland

Prashanti Kethireddipalli, Department of Food Science, University of Guelph, Guelph, Ontario, Canada

Herman Lutterodt, Department of Nutrition and Food Science, University of Maryland, College Park, Maryland, USA

Yoshinori Mine, Department of Food Science, University of Guelph, Guelph, Ontario, Canada

Michael C. Qian, Department of Food Science and Technology, Oregon, Oregon State University, Corvallis, Oregon, USA

Christiane Queiroz, Departamento de Nutrição Básica e Experimental, Instituto de Nutrição Josué de Castro, Universidade Federal do Rio de Janeiro, Brazil

Karen M. Schaich, Department of Food Science, Rutgers University, New Brunswick, New Jersey, USA

Fereidoon Shahidi, Department of Biochemistry, Memorial University of Newfoundland, St. John's, Newfoundland, Canada

Margaret Slavin, Department of Nutrition and Food Studies, George Mason University, Fairfax, Virginia, USA

Graham G. Stewart, GGStewart Associates, 13 Heol Nant Castan, Rhiwbina, Cardiff, UK

Vera Lúcia Valente Mesquita, Departamento de Nutrição Básica e Experimental, Instituto de Nutrição Josué de Castro, Universidade Federal do Rio de Janeiro, Brazil

Pedro Vazquez-Landaverde, CICATA-IPN Unidad Queretaro, Colonia Colinas del Cimatario, Queretaro, Mexico

Monica Whent, Department of Nutrition and Food Science, University of Maryland, College Park, Maryland, USA

Liangli Yu, Department of Nutrition and Food Science, University of Maryland, College Park, Maryland, USA

Hua Zhang, Department of Food Science, University of Guelph, Guelph, Ontario, Canada

Ying Zhong, Department of Biochemistry, Memorial University of Newfoundland, St. John's, Newfoundland, Canada

Kequan Zhou, Department of Nutrition and Food Science, Wayne State University, Detroit, Michigan, USA

Biochemical Changes in Raw Foods

Chapter 1

Cereals and Legumes

Kequan Zhou,* Margaret Slavin,[†] Herman Lutterodt,** Monica Whent,** N. A. Michael Eskin[‡] and Liangli Yu**

*Department of Nutrition and Food Science, Wayne State University, Detroit, Michigan, USA, [†]Department of Nutrition and Food Studies, George Mason University, Fairfax, Virginia, USA, **Department of Nutrition and Food Science, University of Maryland, College Park, Maryland, USA, [‡]Department of Human Nutritional Science, University of Manitoba, Winnipeg, Manitoba, Canada

Chapter Outline

Biochemistry of Foods. DOI: http://dx.doi.org/10.1016/B978-0-12-242352-9.00001-1

PART I: CEREALS

I. INTRODUCTION

Cereals are members of the large monocotyledonous grass family, the Gramineae, which mainly consists of wheat, maize, barley, oats, rice, and sorghum (Anderson *et al.*, 2000). Cereal-based foods have been a staple dietary source for the world's population for centuries. Cereal grains contain the macronutrients (protein, fat, and carbohydrate) required by humans for growth and maintenance, contributing approximately 70% and 50% of the total calories and protein, respectively (Topping, 2007). Cereal grains also supply important minerals, vitamins, and other micronutrients essential for optimal health. They still provide 20% of magnesium and zinc, 30–40% of carbohydrate and iron, 20–30% of riboflavin and niacin, and over 40% of thiamine in the diet (Marston and Welsh, 1980). Global cereal production per capita fluctuated around 280 kg per year during the first half of the twentieth century, as seen in Figure 1.1 (Gilland, 2002). The world production of cereal is projected to be 3555 Mt, with per capita production of 378 kg (Gilland, 1998). Cereal-based food, especially whole grains, have shown the potential for health promotion, linking to reduced risk of several chronic diseases such as coronary heart disease, type 2 diabetes, and certain types of cancer (Truswell, 2002; Montonen *et al.*, 2003; Slavin, 2000; Slavin *et al.*, 1999). These beneficial effects are attributed to numerous phytochemicals contained in the grains. Therefore, this chapter will discuss those biochemical changes taking place during development, germination, and storage of cereal grains, with particular attention to wheat.

II. CEREAL GRAIN STRUCTURE

The different tissues constituting the cereal seed are generally described in terms of their embryogenic origin and structure (Evers *et al.*, 1999). The cereal seed is composed of three main tissues: the embryo, the endosperm, and the aleurone layer surrounding the storage endosperm. This is illustrated for wheat in Figure 1.2, in which the endosperm comprises over 80% of the grain weight while the aleurone cells and the germ tissue containing the embryo embedded in the surrounding scutellum account for 15% and 3%, respectively. The peripheral tissues of the grain overlying the starchy endosperm are made up successively from the outer to the inner surface: the outer pericarp, the inner pericarp, seed coat, hyaline layer, and the aleurone layer (Barrona *et al.*, 2007). The germ comprises the embryonic axis and the scutellum. The anatomical structure of all cereal grains is essentially similar with some minor

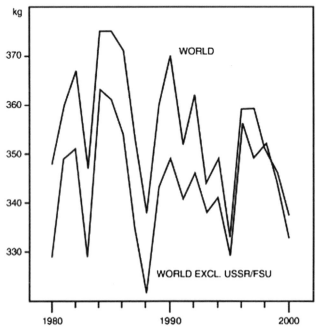

FIGURE 1.1 Cereal production per capita, 1980–2000. *(Sources: FAOSTAT; US Bureau of the Census; adapted from Gilland, 2002.)*

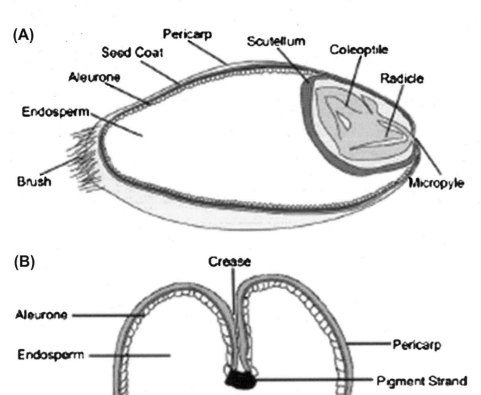

FIGURE 1.2 Diagram of wheat grain showing major structures in (A) longitudinal and (B) transverse sections *(Rathjen et al., 2009 with permission).*

differences. For instance, wheat and maize are surrounded by a fruit coat or pericarp and seed coat or testa, together referred to as a naked caryopsis. In the case of barley, oats, and rice, however, an additional husk is found surrounding the caryopsis or kernel of the grain.

III. CEREAL GRAIN COMPOSITION

Cereal grains are highly nutritious, and the major components of the grain are proteins (approximately 10−15%) and starch (approximately 60−70% of grain), with non-starch polysaccharides derived from the cell walls accounting for about 3−8% of the total (Saulniera *et al.*, 2007). The composition of cereals varies and highly depends on grain variety, growing conditions, husbandry, and infection (Tester, 1995). The starchy endosperm constitutes the major portion of the cereal seed and provides the nutrients necessary for embryo development during germination. The nutrients are made available by the release of enzymes from the aleurone layer and embryo, which hydrolyze the endosperm reserves. These reserves are contained in discrete storage bodies identified as starch granules and protein bodies. These components have major effects on the use of cereal grain due to their physiochemical properties during milling and food processing. It is worth noting that the non-digestible carbohydrates in cereal grains have received increased attention as a significant source of dietary fiber, and are purported to have an impact on the nutritional quality and health-enhancing effects of cereal foods (Salmeron *et al.*, 1997). In addition, polysaccharides in cereals are associated with many other substances, mostly with proteins, polyphenols, and phytate, which can modify mineral binding by dietary fiber (Vitali *et al.*, 2008).

A. Amyloplasts

Amyloplasts are plastids or organelles responsible for the storage of starch granules. The rate of starch synthesis in cereal grains is one of the factors affecting both grain size and yield (Kumar and Singh, 1980). In the mature endosperm of wheat, barley, and rye, starch is found as two distinct fractions based on the size of the granules. The primary or A-type starch granules range in size from 20 to 45 μm, while the secondary or B-type granules rarely exceed 10 μm in diameter (Evers, 1973). Examination of the particle size distribution in wheat endosperm starch by Evers and Lindley (1977) showed that those starch granules less than 10 μm in diameter accounted for approximately one-third of the total weight of starch. The presence of these two starch granule types in wheat kernels was confirmed in studies by Baruch *et al.* (1979). They found that the size of the starch was affected by seasonal changes in much the same way as grain yield and protein content. The starch granule occupies only a very small part of the total plastid during initial kernel development but accounts for close to 93% at maturity (Briarty *et al.*, 1979). In the mature endosperm, A-type starch granules account for only 3% of the total number of granules although they represent 50–70% of the total weight, owing to their larger size (Evers and Lindley, 1977). The smaller B-type granules, however, make up 97% of the total number of starch granules but account for only 25–50% of the overall weight.

Isolated starch granules also contain protein, most of which can be removed by washing repeatedly with water. A small part of the protein, however, remains strongly associated with the granule itself. Lowy *et al.* (1981) found that this protein fraction is readily extracted with salt solution and suggested that it is associated with the starch granule surface. This extractable fraction accounts for 8% of the total protein in the starch granule. The major protein fraction has a molecular weight of around 30,000 and is associated with both A- and B-type starch granules. Based on amino acid analysis, this protein is quite different from wheat gluten. An additional protein fraction was extracted from A-type starch granules but only following gelatinization in the presence of sodium dodecyl sulfate. This fraction was quite different, and based on electrophoresis, was thought to be part of the internal granule components.

B. The Starch Granule

Starch granule shape can be characteristic of a genus and species (Ellis, 1998). The shape and size of the starch granule vary with the different cereals (Table 1.1). The size distribution of the starch granules in the amyloplasts and the composition of starch granules and their properties change during granule development. The large A-type starch granules of wheat, barley, and rye are lenticular, while the smaller B-type starch granules are spherical or polyhedral. The starch granules of rice, oats, and maize are irregular and polyhedral in shape, those of rice being comparable in size to the B-type starch granules of wheat and barley, while those of maize are larger (Ellis, 1998). Starch is composed of amylose and amylopectin, with the level of amylose ranging from 20% to 30% for most cereal starches (Katz *et al.*, 1993). In the case of certain varieties of maize, barley, and rice, the starch is composed almost exclusively

TABLE 1.1 Structure and Amylose Content of Some Whole Granular Cereal Starches

Source	Granule Shape	Granule Size (nm)	Amylose Content (%)
Wheat	Lenticular or round	20–25	22
Maize	Round or polyhedral	15	28
Waxy maize	Round	15 (5–15)	1
High-amylose	Round or irregular sausage-shaped	25	52
Barley	Round or elliptical	20–25	22
Rice	Polygonal	3–8	17–19[a] 21–23[b]
Oats	Polyhedral	3–10	23–24

[a]Japonica
[b]Indica
Adapted from Lineback (1984).

of amylopectin; these are referred to as 'waxy'. High-amylose starches are also found, for example, in the case of amylomaize.

A portion of amylose in the starch granule is complexed with lysophospholipids as a function of the stage of development of the endosperm, at which the amylose is formed (Morrison, 1993). B-type granules are initiated later in grain development than A-type granules. The proportions of amylose and of lysophospholipids rose in both A-type and B-type starch granules during grain development of wheat and barley (Morrison and Gadan, 1987; McDonald, 1991).

C. Biosynthesis of Starch

Sucrose is believed to be the main source of carbon for starch synthesis in the cereal endosperms, and is converted to starch via a series of enzyme-catalyzed reactions (Duffus, 1993). Starch synthesis is achieved through the action of starch synthase, which can utilize either adenosine diphosphoglucose (ADPG) or uridine diphosphoglucose (UDPG) as a substrate (Recondo and Leloir, 1961). ADPG appears to be the more active glucosyl donor and is formed by the action of ADPG-pyrophosphorylase (Preiss and Levi, 1979).

$$ATP + \alpha\text{-Glucose-1-P} \rightarrow ADPG + PP_i$$

The amount of inorganic pyrophosphate (PP_i) in developing grains is controlled by the enzyme alkaline inorganic pyrophosphatase (EC 3.6.1.1). This enzyme limits the accumulation of PP_i and was thought to be the controlling factor in starch synthesis, as PP_i inhibited ADPG-pyrophosphorylase in sweet corn (Amir and Cherry, 1972). The activity of both ADPG-pyrophosphorylase and alkaline pyrophosphatase was studied by Kumar and Singh (1983) during the development of wheat grain. Their results, shown in Figure 1.3, indicate that both enzymes increased steadily, reaching a maximum 28 days after anthesis, but then declined with maturity. The rapid increase in alkaline pyrophosphatase activity 14 days after anthesis corresponded with the period of rapid starch synthesis. The inability of the intermediate metabolites of sucrose—starch conversion to inhibit the activity of alkaline pyrophosphatase eliminated any possible regulatory role for this enzyme in starch biosynthesis.

D. Sucrose Starch Conversion in Developing Grains

The amount of free sugars formed during the development of wheat was examined by Kumar and Singh (1981) in relation to grain size and starch content. Their results, summarized in Figure 1.4, indicate that the non-reducing sucrose reached a maximum level 14 days after anthesis, then declined and leveled off after 28 days. Starch synthesis

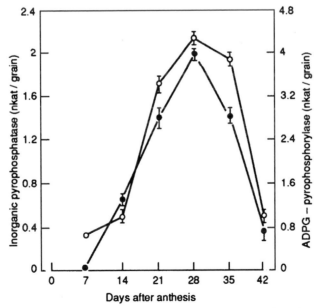

FIGURE 1.3 Activity of alkaline inorganic pyrophosphatase (○) and ADPG-pyrophosphorylase (●) during wheat grain development (Kumar and Singh, 1983).

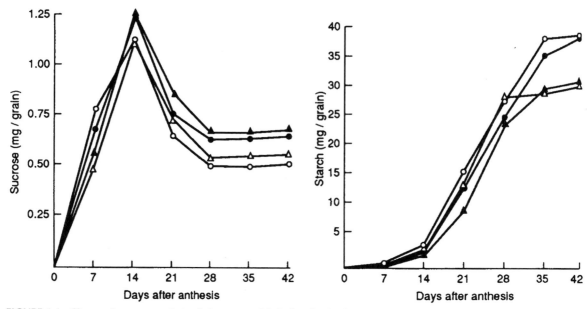

FIGURE 1.4   Changes in sucrose and starch (mg per grain) during the development of four wheat grains *(Kumar and Singh, 1981).*

was negligible after 7 days but increased markedly after 14 days, then continued until 35 days after anthesis. The rapid decline in sucrose and reducing sugars once starch synthesis commenced suggested the involvement of hydrolytic enzymes, including invertase. The activity of this enzyme was found by Kumar and Singh (1980) to decrease to negligible levels after 21 days compared to the rather rapid rise in sucrose—UDP glucosyl transferase activity (Figure 1.5). The latter enzyme, also referred to as sucrose synthetase, catalyzes the first step in the formation of starch from sucrose, as discussed later in this section. The parallel activities of sucrose—UDP glucosyl transferase and starch synthesis suggested that this enzyme played a major role in the hydrolysis of sucrose. Kumar and Singh (1984) suggested that the initial role of invertase was to provide substrates for energy-liberating respiratory enzymes needed for sustaining active cell division.

Chevalier and Lingle (1983) reported that insoluble invertase was located mainly in the outer pericarp, with only slight activity in the endosperm (Figure 1.6). These researchers monitored sucrose synthetase activity, which was

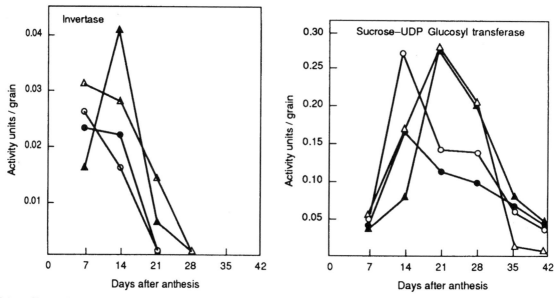

FIGURE 1.5   Changes in invertase and sucrose-UDP glucosyl transferase activities during the development of wheat grains *(Kumar and Singh, 1980).*

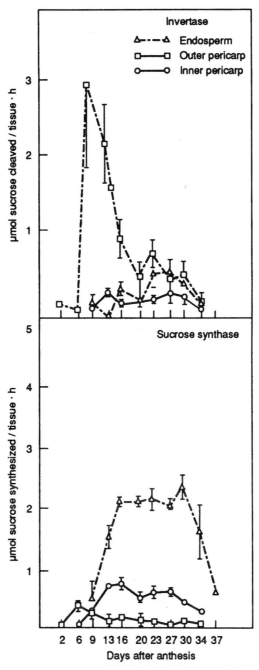

FIGURE 1.6 Distribution of invertase and sucrose synthetase activities in the endosperm and pericarp of developing wheat grains. *(Reproduced from Chevalier and Lingle, 1983.)*

found predominantly in the endosperm. Using whole wheat kernels, Kumar and Singh (1980) monitored invertase activity during the early stages of grain development and that of sucrose synthetase as the grain matured. Chevalier and Lingle (1983) found an increase in free sucrose in mature wheat and barley kernels, which was consistent with earlier research with wheat (Cerning and Guilbot, 1973), barley (Laberge *et al.*, 1973), and rice (Singh and Juliano, 1977). The marked rise in sucrose observed by Lingle and Chevalier (1980) in the endosperm fraction was accompanied by decreasing sucrose synthetase activity. This decline in synthetase activity was considered to be an important factor in controlling grain filling. It appeared to be responsible for accumulation of sucrose in extracellular spaces such as the endosperm cavity since the endosperm was now unable to utilize any incoming sucrose. The overall effect was to prevent any more sucrose from entering the kernel.

Kumar and Singh (1984) confirmed the accumulation of sucrose up to 14 days after anthesis, which represented rapid translocation from photosynthetic parts to the wheat endosperm followed by active starch synthesis. Previous work by Chevalier and Lingle (1983) demonstrated the movement of sucrose from the phloem to the endosperm in developing wheat and barley kernels. Using wheat endosperm slices, Rijven and Gifford (1983) also found sucrose to be the preferred substrate for starch synthesis as it was not hydrolyzed before its uptake by the endosperm.

E. Starch Synthesis

The *in vivo* synthesis of starch involves phosphorylase or synthetase leading to the formation of the linear polymer amylose. Once sucrose enters the endosperm it becomes the starting point for amylose synthesis. The first step involves its conversion to UDPG by sucrose synthetase (sucrose–UDP glucosyl transferase):

$$\text{Sucrose} + \text{UDP} \xrightarrow{\text{Sucrose–UDPG glucosyl transferase}} \text{UDPG} + \text{Fructose-1-P}$$

Following this, fructose-1-P is converted to glucose-1-P by phosphoglucoisomerase, hexokinase, and phosphoglucomutase. Glucose-1-P is metabolized to ADPG by ADPG-pyrophosphorylase:

$$\text{Glucose-1-P} + \text{ATP} \xrightarrow{\text{ADPG-phosphorylase}} \text{ADPG} + \text{PP}_i$$

The absence of any detectable PP_i suggested that it is rapidly hydrolyzed by pyrophosphatase since, as discussed earlier, it is a potent inhibitor of ADPG-phosphorylase (Amir and Cherry, 1971). Amylose synthesis, as discussed earlier, can be mediated directly by starch synthetase involving UDPG or indirectly by ADPG-starch synthetase to ADPG via glucose-1-P:

$$\text{ADPG} + \text{Primer (Gn)} \xrightarrow{\text{ADPG-starch synthetase}} \text{Glycosyl primer} \rightarrow \text{(Gn} + 1) + \text{ADP}$$

It appeared, however, that the ADPG reaction was the preferred one for starch biosynthesis in developing wheat grains (Kumar and Singh, 1984). In this reaction glucose is repeatedly transferred from ADPG to a small glucan primer until the elongated starch chain is formed. The extremely small amount of glucose-1-P in the developing grain suggested that it was rapidly utilized, pointing to a possible regulatory role for phosphoglucomutase, the enzyme responsible for its formation, in starch biosynthesis. Kumar and Singh (1984) proved conclusively that the termination of starch accumulation in mature wheat grains was due to the loss in synthetic capacity of the endosperm and not due to the unavailability of sucrose.

Joshi *et al.* (1980) attempted to explain the regulation of starch biosynthesis in normal and Opaque-2 maize during development of the endosperm. Opaque-2 maize was nutritionally superior although it had decreased grain yield and a lower protein and starch content. These researchers monitored the activities of sucrose–UDP glucosyl transferase, glucose-6-phosphate ketoisomerase, and soluble and bound ADPG-starch glucosyl transferase in the developing endosperm for 30 days following pollination. Except for sucrose-UDP glucosyl transferase, all the other enzymes were much lower in Opaque-2 maize, compared to the normal maize during the latter stages of endosperm development. The lower activity of these enzymes was responsible for the reduced amount of starch in Opaque-2 maize, which had 15% less starch content per endosperm. This was accompanied by a decreased protein synthesis in the Opaque-2 endosperm, which explained the reduced enzyme synthesis during the later stages of endosperm development.

F. Starch Synthesis: Amylopectin

Biosynthesis of the branched chain amylopectin requires the formation of the amylose via phosphorylase or synthase as described in the previous section. The branch points (α-(1,6)-D-glucosidic linkage) required for amylopectin are introduced by the branching enzyme Q-enzyme (EC 2.4.1.18). Borovsky *et al.* (1979) concluded that the introduction of 1,6-branch points is a random process in which the Q-enzyme interacts with two 1,4-glucan chains held together in a possible double-helix arrangement.

Amylose and amylopectin are synthesized concurrently in the ratio of 1:4 for ordinary starches (Robyt, 1984). Several hypotheses have been developed to explain the side-by-side occurrence of amylose and amylopectin in the starch granule, although our understanding of starch biosynthesis remains incomplete (Erlander, 1958; Geddes and Greenwood, 1969; Marshal and Whelan, 1970). One such hypothesis suggested that some mechanism was operating which protected the linear polymer from the branching enzyme (Whelan, 1958, 1963). The participation of phospholipids in the regulation of amylopectin was proposed by Vieweg and De Fekete (1976), since phospholipids inhibit the action of the branching enzyme. Thus, only amylose without attached phospholipids could theoretically be converted, although this remains to be verified. Another hypothesis, discussed earlier, is the possible specificity of the branching enzyme for a double-helix arrangement involving the shorter amylopectin chains (Borovsky *et al.*, 1979; Robyt, 1984).

G. Protein Bodies

Protein bodies are membrane-bound cellular organelles containing storage proteins located in the starchy endosperm of cereals (Pernollet, 1978, 1982). They are also found in the aleurone layer, although these differ in composition, structure, and function. While the protein bodies in the endosperm have only a storage function, those in the aleurone layer possess both synthetic and secretory functions (Simmonds and O'Brien, 1981). Protein bodies in the aleurone layer are 2–4 μm in diameter with globoid and crystalline inclusions, while those in the starchy endosperm have a homogeneous granular structure devoid of inclusions. These differences have been confirmed in wheat, barley, maize, and rice by examination of their ultrastructural differences, as indicated in Table 1.2.

In members of the *Triticum* species, these protein bodies vanish as the grain matures, as observed for wheat seeds (Simmonds, 1972; Pernollet and Mossé, 1983) and rye seeds (Parker, 1981). This results in the conversion of the spherical protein granules into irregularly shaped protein masses which eventually become the matrix protein, which is no longer bound by a membrane between the starch granules.

H. Origin of Protein Bodies

The origin of protein bodies in the endosperm is still unclear. Most researchers support their synthesis on the rough endoplasmic reticulum (RER) (Campbell *et al.*, 1981; Miflin *et al.*, 1981; Miflin and Burgess, 1982; Parker and Hawes, 1982), although Bechtel *et al.* (1982a, b) favored secretion of the wheat storage proteins. Irrespective of the mechanism proposed, initiation and formation of the protein bodies involve the active participation of the Golgi apparatus. Pernollet and Camilleri (1983) examined protein body formation and development in wheat endosperm and found the polypeptides stored in all protein bodies to be similar. Earlier work by Tanaka *et al.* (1980) suggested that only one kind of protein was stored in wheat endosperm. The presence of all cell storage proteins in the protein bodies, however, pointed to a common synthetic pathway operating in wheat seeds. The polypeptides in the protein bodies were similar to those in the endoplasmic reticulum. This suggested that the storage proteins were secretory proteins discharged into the endoplasmic reticulum before being translocated to the protein bodies. This model conflicted with the soluble mode of gliadin synthesis proposed by Bechtel *et al.* (1982a, b), but was in agreement with studies carried out by Greene (1981) and Donovan *et al.* (1982). These researchers reported that messenger RNAs encoding gliadin molecules were translated on polysomes bound to the endoplasmic reticulum.

TABLE 1.2 Ultrastructural Differences Between Aleurone Layer and Starchy Endosperm Protein Bodies

| Species | Aleurone Layers | | Endosperm | |
	Diameter (μm)	Structure	Diameter (μm)	Structure
Wheat	2–3	Two kinds of inclusion	0.1–8	No inclusion; granular structure
	4–5	One globoid and one crystalloid	1–2	No inclusion; lamellar structure
Barley	2–3	Two kinds of inclusion	2	No inclusion; lamellar structure
	4–5	One globoid and one crystalloid	1–2	No inclusion; lamellar structure
Rice	1.5–4	Globoid	2–5	No inclusion; homogeneous
	1–3	Globoid	2–5	No inclusion; homogeneous
Maize			1–2	No inclusion; homogeneous

Adapted from Pernollet (1978). Reprinted with permission. Copyright © Pergamon Press.

FIGURE 1.7 Schematic diagram of wheat endosperm protein body formation and evolution *(Pernollet and Camilleri, 1983).*

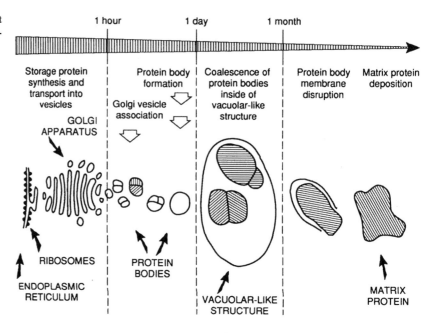

Three distinct stages were noted by Pernollet and Camilleri (1983) during the development of wheat protein bodies. The initial stage involved protein synthesis of storage proteins and their association as small vesicles into bodies of 5—10 μm in the first month after anthesis. During the next stage, the formation of the small protein bodies slowed down and instead they coalesced into much larger bodies (50—100 μm). The instability of the membrane of these large protein bodies and the mechanical pressure of the developing starch granules resulted in disruption of the membrane with the release of the protein bodies into the matrix protein. This loss of the protein bodies and the matrix protein formation is characteristic of the final stage of development of the mature wheat endosperm. The model proposed by Pernollet and Camilleri (1983) in Figure 1.7 summarizes the sequence of events leading to the formation of protein bodies in wheat and their eventual disruption.

The protein bodies of barley are similar to those of wheat, but differ quiet markedly from those of maize. In maize the membrane is derived from the endoplasmic reticulum, which completely encloses the protein bodies. This differs from wheat and barley, where the endoplasmic reticulum is disrupted by wheat and barley protein body aggregates which are not completely surrounded by this membrane. Oparka and Harris (1982) reported that rice protein bodies were surrounded by a membrane derived from the endoplasmic reticulum.

I. Classification of Plant Proteins

Plant proteins were first classified by Osborne (1895) as albumin, globulin, prolamins, and glutelins on the basis of solubility in different solvents as summarized in Scheme 1.1.

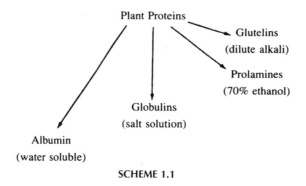

SCHEME 1.1

Several modifications have since been introduced to improve extraction of these fractions. Current practice extracts a combined albumin–globulin fraction as salt-soluble protein while the prolamines are extracted with aqueous propan-1-ol or propan-2-ol plus a reducing agent (Shewry *et al.*, 1980). This method is appropriate for the study of the basic genetic products but quite inappropriate from a technological point of view, as reducing agents result in the re-establishment of new disulfide bonds that change the solubility of the fractions. To prevent denaturation of the glutelin fraction by alkali extraction, alternative extractants such as buffers containing the detergent sodium dodecyl sulfate (SDS) at pH 10 are used (Moreaux and Landry, 1968). The relative proportion of the Osborne protein fractions in the seeds of wheat, barley, maize, and rye are summarized in Table 1.3.

J. Prolamins

The major storage proteins present in the starchy endosperm of wheat, barley, and maize are the alcohol-soluble proteins, the prolamins. These account for 30–60% of the total grain nitrogen depending on species, nutritional status, and genotype of the plant (Bright and Shewry, 1983; Shewry *et al.*, 1981). The prolamin fractions identified for different cereal species are listed in Table 1.4.

Prolamins derive their name from their unusually high content of proline and amide nitrogen (glutamine). This protein fraction is deficient in the essential amino acid lysine. Oats and rice differ substantially from other cereals in containing very little prolamin (5–10%), with the major storage proteins being globulin and a glutelin-like compound, respectively. Thus, these cereals have much more lysine, making them nutritionally superior. Electrophoretic separation of the different prolamin fractions on the basis of molecular size is accomplished by polyacrylamide gel electrophoresis in the presence of sodium dodecyl sulfate (SDS–PAGE). This permits identification of different polypeptide patterns in prolamins, which vary considerably among different cultivars of the same species. PAGE is a widely used technique for varietal identification of single seeds of wheat and barley. When there are only minor differences, two-dimensional isoelectric focusing (IEF) and PAGE can be effectively applied. Using

TABLE 1.3 Relative Proportions (%) of the Osborne Protein Fractions in Cereal Seeds

Cereal	Non-protein N	Albumins	Globulins	Prolamins	Glutelins	Residues
Barley[a]	11.6		15.6	45.2	18.0	5.0
Wheat[b]		33.1		60.7		6.2
Maize[a]	4.4	0.9	1.5	55.4	22.9	–
Rice[c]		15.7		6.7	61.5	15.4
Oats[d]	11		56	9	23	–

[a] % Total seed N (%)
[b] % Recovered seed N (%)
[c] % Total protein (%)
[d] % Recovered protein (%)
From Bright and Shewry (1983) with permission.

TABLE 1.4 Prolamin Fractions of Cereal Grain

Species	Trivial name
Wheat	Gliadin
Maize	Zein
Barley	Hordein
Oats	Avenin

TABLE 1.5 Prolamin Fractions of Wheat, Barley, and Maize

Wheat	MW	Barley	MW	Maize	MW
α-Gliadin	32,000	B-hordein	35,000—46,000	20K	20,000—21,000
β-Gliadin	40,000	C-hordein	45,000—72,000	22K	22,000—23,000
ω-Gliadin	40,000—72,000	D-hordein	100,000	9K	9,000—10,000
HMW subunits	95,000—136,000			14K	13,000—14,000

MW: molecular weight; HMW: high molecular weight.

these procedures the polypeptides identified for prolamin fractions in wheat, barley, and maize are summarized in Table 1.5.

The wheat gliadins are classified into two groups based on their electrophoretic mobility at low pH. The first group includes the fastest fraction, α-gliadin, followed by β-, γ-, and ω-gliadins, while the second group, with a much higher apparent molecular weight (95,136,000), is referred to as high-molecular-weight units (HMU). All the gliadin fractions are deficient in lysine and threonine. Three groups of hordein protein were separated from barley by SDS—PAGE and referred to as B, C, and D. They differed from each other in apparent molecular weights and amino acid composition (Miflin and Shewry, 1977). The C fraction had only trace amounts of sulfur amino acids while the D fraction was rich in glycine (13%). Lysine was particularly low in all the hordein protein fractions ($< 1\%$) while B and C hordeins were also deficient in threonine.

The zein component of maize protein, although not well defined, was composed of two major and two minor fractions. The two major fractions had apparent molecular weights of 20,000—21,000 and 22,000—23,000, while minor fractions were 9,000—10,000 and 13,000—14,000, referred to as 22K, 9K, and 14K zein, respectively. All of these fractions were deficient in lysine. Unlike the other cereal grains, the major storage proteins of oats were 12S and 7S globulins as prolamins accounted for less than 15% of the total grain nitrogen (Peterson and Smith, 1976). Burgess and Miflin (1985) showed that 7S globulin was located mainly in the embryo while 12S globulin, the larger fraction, was predominant in the endosperm. Based on SDS—PAGE it appeared that the globulin and prolamin fractions were localized in different protein bodies.

K. Protein Synthesis

The development of cereal seed protein is associated with at least three stages. The first stage is characterized by rapid cell division in which protein synthesis remains quite low. When cell division ceases this is followed by an increase in the RER and accumulation of soluble nucleotides (Briarty *et al.*, 1979; Jenner, 1968). This results in a rapid synthesis of storage proteins which is related to initiation and synthesis of messenger RNA (mRNA) as well as the efficiency of mRNA translation. The accumulation of mRNA in developing wheat seeds was correlated with protein synthesis by Greene (1983). Using labeled [5-^{3}H] uridine and L-[^{3}H] leucine, he studied the synthesis, functioning, and stability of storage protein RNAs. Three developmental stages were apparent:

1. A change from seed protein synthesis of non-storage to storage protein.
2. An increase in the rate of accumulation of poly(A)+RNA.
3. An increase in the level of transcription mRNA.

A direct relationship between mRNA levels and the rate of protein synthesis is shown in Figure 1.8. Synthesis of the gliadin peptide was predominant from 15 to 25 days following flowering and paralleled the increase in poly(A)+ RNA. Thus, the storage protein gene expression in wheat endosperm is an mRNA-limiting process based on the amount of storage protein that the mRNA synthesized near the end of endosperm cell division. Okita and Greene (1982) previously identified mRNAs as the major messenger species in Cheyenne responsible for gliadin synthesis 20—25 days after anthesis. For a more detailed review of cereal proteins the article by Laszity (1984) is recommended.

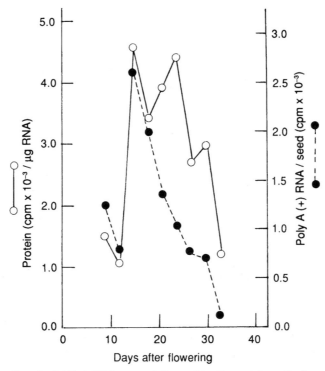

FIGURE 1.8 **Development profiles of poly(A) + RNA accumulation and** *in vitro* **protein synthesis capacity in wheat** *(Greene, 1983).*

TABLE 1.6 Lipid Content of Whole-grain Cereals

Cereal	Crude Fat (%)
Wheat	1.8
Maize kernel	0.4−1.7
Barley	3.3−4.6
Oats	5.4
Rice	1.9−3.1

L. Lipids

Lipids are distributed throughout the cereal grain as part of the intracellular membranes and spherosomes. They are stored as triglyceride-rich droplets in the spherosomes of the aleurone layer which are found clustered around the aleurone grains or with the plasmalemma (Buttrose, 1971; Chamura, 1975; Morrison *et al.*, 1975; Morrison, 1978). Spherosomes are also present in the embryo, scutellum, and coleoptile (Buttrose and Soeffly, 1973; Jelseman *et al.*, 1974). Lipids are also found in starch, primarily as monoacyl lysophosphatidyl ethanolamine, and lysophosphatidylcholine, and as inclusion complexes with amylose inside the starch granule. There appears to be a correlation between the amylose and lipid content of cereals, for example, waxy maize has little lipid while high-amylose or amylomaize starch has a higher lipid content than normal maize starch (Acker and Becker, 1971). The distribution of lipids in mature cereal kernels is shown in Table 1.6.

The major fatty acids present in grain lipids are linoleic, oleic, palmitic, and linolenic acids, in order of decreasing amounts (Price and Parsons, 1975). The cereal lipids can be separated into polar and non-polar lipids by solvent fraction. For example, in the case of a hard red spring wheat, Waldron, the polar and non-polar lipids accounted for

TABLE 1.7 Distribution of Wheat Lipids within Wheat Tissues[a,b]

Total Lipids							
				Endosperm (44.8%)			
Germ (30.4%)		Aleurone Layer (24.8%)		Non-starch (29.2%)		Starch (15.6%)	
Non-polar lipids (24.1%)	Polar lipids (6.3%)	Non-polar lipids (17.9%)	Polar lipids (6.9%)	Non-polar lipids (9.7%)	Polar lipids (19.5%)	Non-polar lipids (0.7%)	Polar lipids (14%)

[a]*Adapted from Hargin and Morrison (1980)*
[b]*calculated and adapted from data by Hargin and Morrison (1980)*
Data are expressed as a percentage of total lipids.

49.6% and 50.4% of the total lipids, respectively (Hargin and Morrison, 1980). The distribution of these lipid fractions within the wheat tissues is shown in Table 1.7.

The germ contains one-third of the total wheat lipids, of which 80% are neutral triglycerides. Aleurone lipids account for one-quarter of the total lipids, with 80% being non-polar in nature. The endosperm, however, accounts for almost half of the whole kernel lipids. The endosperm starch is associated with 15.6% of the total lipids, of which 96% are phospholipids. The predominant phospholipid in starch endosperm is lysophosphatidyl-choline (Hargin and Morrison, 1980).

The biosynthesis of lipids begins with the formation of fatty acids by a multistep process involving a multienzyme complex, the acyl protein carrier (ACP) fatty acid synthetase. Once formed, they are esterified with glycerol to triglycerides, which serve as an important source of energy during germination of cereals. They are responsible for maintaining the embryo and aleurone layer during the initial stages of germination until sugars are provided from the starchy endosperm.

IV. GERMINATION OF CEREALS

Germination of cereals is important in the malting industry, which depends on a certain degree of starch degradation. In the production of baked products, however, it is important that most of the starch granules remain intact. Thus germination or sprouting of cereal grains affects the grading of wheat and cereal grains as a result of the damage it causes. According to the *Grain Primer* (US Department of Agriculture, 1957), sprouting of wheat is defined as 'kernels which have the germ end broken open from germination, and kernels from which sprouts have broken off'. This is prevalent during wet weather when the moisture content is increased. The preharvest germination of the wheat reduces grain yield, flour yield, and flour quality. This has an adverse effect on the breadmaking properties of the flour because of the enhanced hydrolysis of the dough starch by α-amylase (Buchanan and Nicholas, 1980). If the activity of α-amylase is excessive it produces a bread product with a wet, sticky crumb.

A. Mobilization of Cereal Starches by α-Amylase

The native starch granule in wheat is attacked by certain α-amylase isoenzymes. Two groups were separated by Sargeant (1979) during germination of wheat, one of which hydrolyzed the starch granules. Halmer (1985) pointed out that since starch hydrolysis is normally carried out with soluble and not native granular starch granules, it is difficult to relate total amylolytic activity, as measured in the laboratory, to the granule-degrading activity of the cereal grain *in vivo*. The hydrolysis of starch by α-amylases is characterized by the endocleavage of amylose and amylopectin (Abbott and Matheson, 1972).

B. Biosynthesis of α-Amylase during Germination

The importance of α-amylase activity in baking and brewing has focused considerable attention on its secretion during germination. A major controversy has centered on whether the site of α-amylase biosynthesis is in the scutellum or aleurone layer (Akazawa and Hara-Nishimura, 1985). In the case of barley grains, α-amylase formation *de novo* has been reported in both the scutellum and aleurone layer (Briggs, 1963, 1964; Chrispeels and Varner,

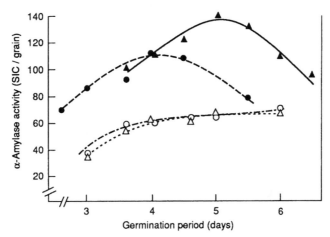

FIGURE 1.9 α-Amylase activity in decorticated barley germinated in the K₂SO₄ and GA₃. No additives (○); K₂SO₄ (50 mM) (Δ); GA₃ (50 μg/ml) (●); GA₃ (50 μg/ml) and K₂SO₄ (50 mM) (▲) *(Raynes and Briggs, 1985).*

1967). The biosynthesis and secretion of this enzyme appear to involve the plant hormone gibberellin GA₃. This hormone is produced by the embryo, and triggers the production of α-amylase as well as other hydrolytic enzymes in the aleurone layer (Briggs *et al.*, 1981). The increase in enzyme activity was attributed to an increase in the level of α-amylase mRNA (Bernal-Lugo *et al.*, 1981; Higgins *et al.*, 1976). In the case of wheat, the aleurone layer also becomes the target of hormone-induced enzymes, including increased synthesis of α-amylase (Filmer and Varner, 1967; Melcher and Varner, 1971). Varty *et al.* (1982) found that the plant hormone abscisic acid inhibited both transcription and translation of α-amylase mRNA in isolated wheat aleurone tissue. This explained the ability of abscisic acid to inhibit the induction of α-amylase by GA₃ (Chrispeels and Varner, 1967). Studies by Raynes and Briggs (1985) showed increased α-amylase production in decorticated barley grains germinated with or without gibberellic acid. Their results, shown in Figure 1.9, indicate that the onset and amount of enzyme activity were affected by GA₃ and K₂SO₄. The presence of K₂SO₄ appeared to delay the destruction of α-amylase (Briggs, 1968). Based on studies with rice scutellum, calcium also appears to play a role in the biosynthesis and secretion of α-amylase with the possible involvement of calmodulin (Mitsui *et al.*, 1984).

A number of researchers reported that the major isoenzyme form of α-amylase in germinated mature grain or aleurone tissue incubated with GA₃ was α-AMY1 (MacGregor, 1983; Marchylo *et al.*, 1981; Sargeant, 1979, 1980). This differed from α-amylase production in pre-mature excised embryo/scutellar tissue, where α-AMY2 was the predominant isoenzyme formed even in the presence of GA₃. While this tissue normally produces little α-amylase activity in pre-mature wheat grain, once removed from the caryopsis it starts synthesizing α-amylase, resulting in the characteristic cytological changes associated with germination. Cornford *et al.* (1987) further examined the production of α-amylase in embryo/scutellar tissue from pre-mature wheat and found that it was influenced by embryo age. While both α-AMY1 and α-AMY2 forms were detected by rocket-line immuno-electrophoresis in the presence of GA₃, it was the production of α-AMY2 that was stimulated by the addition of this growth substance. Abscisic acid inhibited the production of α-AMY1 and several α-AMY2 bands, although four active α-AMY2 bands were still detected. This switching of developmental to germinative mode by the excised embryos, in terms of α-amylase production, may be due to the loss of abscisic acid from the embryo (Triplett and Quatrano, 1982).

MacGregor and Matsuo (1982) conducted a detailed study on initial starch degradation during germination in endosperms of barley and wheat kernels. The kernels examined were all carefully split longitudinally through the crease without distorting any of the structural features (Figure 1.10). Using scanning electron microscopy, similar physical changes were evident in both barley and wheat kernels during the initial stages. Starch degradation appeared to commence at the endosperm—embryo junction, then moved along the junction to the dorsal edge of the kernel. This effect was only observed once extensive degradation of the cell wall material and protein matrix in the endosperm had occurred. These results were consistent with earlier work showing that α-amylase synthesis during germination commenced in the embryo (Gibbons, 1979, 1980; Okamoto *et al.*, 1980). Irrespective of where α-amylase is synthesized, it is ultimately discharged into the endosperm, where starch hydrolysis takes place.

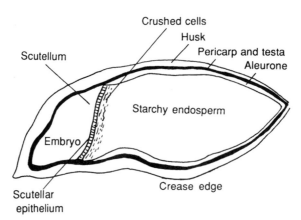

FIGURE 1.10   Longitudinal section of barley kernel cracked open through the crease edge *(MacGregor and Matsuo, 1982).*

C. α-Amylase Activity in Germinated Cereals

During the course of germination, starch is degraded by α-amylases and simple sugars are released (Kruger, 1972a,b). Lineback and Ponpipom (1977) monitored the degradation of starch during the germination of cereals, including wheat and oats. They found that an increase in α-amylase activity was accompanied by a rise in free sugars in all cereals examined. The amount produced reflected the degree of damaged starch in the flour milled from the germinated seed. Although the highest α-amylase activity was associated with germinated wheat, the degradation of starch was less than in the other cereals. Starch degradation in wheat was evident by erosion of the granule surface and the equatorial groove. The starch granule from oats was much more resistant to enzyme attack, however, with little damaged starch in the milled flour.

Germination studies conducted on five wheat cultivators by Reddy *et al.* (1984a) showed α-amylase development to be temperature dependent. Wheat kernels germinated in growth chambers at 15.5°C developed the highest enzyme activity compared to 20°C for the field-grown kernels. The activity did not increase significantly until the third day of germination and then rose markedly after 6 days.

D. Effect of Germination on Flour Quality

Lukow and Bushuk (1984) examined the effect of germination on wheat flour quality. Using flours from two cultivars of Canadian hard red spring wheat they found that α-amylase activity was quite low but increased 1600- and 3000-fold during germination. This marked increase in enzyme activity was accompanied by a rise in reducing sugars, which explained the inferior baking characteristics of the germinated flours. The major effect of α-amylase activity was to reduce the water-binding properties of the flour by degradation of the gelatinized starch. The overall result was the production of bread in which the crumbs were damp and sticky (Jongh, 1967; Thomas and Lukow, 1969).

Kruger and Matsuo (1982) studied the effect of preharvest sprouting on the pasta-making quality of durum wheat. α-Amylase activity increased 155- and 320-fold when germinated for 72 and 120 hours, respectively. While cooling during semolina and spaghetti production decreased α-amylase activity, it did not destroy the enzyme immediately. These researchers noted that α-amylase was still active during the first 6 minutes of cooking spaghetti and accounted for the production of reducing sugars, the substantial loss of solids, and the detrimental effect on spaghetti quality.

E. Treatment of Sprouted Grain: Reduction of α-Amylase

Germination of wheat grains commences at harvest time with an adverse effect on quality (Meredith and Pomeranz, 1985). The major culprit is α-amylase activity, which increases during germination, while β-amylase activity remains unchanged. Various methods have been examined to improve the properties of the sprouted grain. Since the starch fraction of sprouted wheat was of good quality, efforts focused on inhibiting α-amylase activity using heat or chemical agents (Bean *et al.*, 1974; Cawley and Mitchell, 1968; McDermott and Elton, 1971; Westermarck-Rosendahl *et al.*, 1979). Early research by Schultz and Stephan (1960), for example, reported an improvement in structure when wheat was treated with acids. Fuller *et al.* (1970) used hydrochloric acid followed by neutralization with ammonia to reduce α-amylase activity, but their method proved impractical. Several α-amylase inhibitors were

TABLE 1.8 Effect of Sodium Polyphosphate on the Falling Number of Sprout-damaged Wheat

Chemical Agent	Concentration[a] (%)	Falling Number
Sodium polyphosphate	0.1	147[b]
	0.5	175[c]
	1.0	250[c]

[a]Based on meal weight (moisture content 15.0%)
[b]Difference significant at 5%
[c]Difference significant at 1%
Adapted from Westermarck-Rosendahl et al. (1979).

examined by Westermarck-Rosendahl et al. (1979) to improve the baking qualities of sprouted wheat. The most promising agents were trisodium phosphate, disodium phosphate, sodium polyphosphate, SDS, calcium steoryl lactylate, and citric acid. Evaluations were based on the falling number test values for grain samples in which the optimum for baking flour was around 200 seconds (Greenaway, 1969). These α-amylase inhibitors caused an increase in falling number values well above 200 seconds, as shown in Table 1.8 for sodium polyphosphate.

The falling number test measures the time it takes for a plunger to fall freely through a suspension of flour in water and the effect of starch amylolytic degradation on the viscosity of the flour/water paste. The faster the decrease in viscosity of the flour paste, the lower the falling number value. Further research by Westermarck-Rosendahl et al. (1980) showed that the most promising of the 23 enzyme inhibitors examined were trisodium phosphate and disodium hydrogen phosphate. These were particularly effective in reducing the stickiness problem associated with flours from sprouted wheat as well as improving crumb characteristics. Alternative solutions discussed by Meredith and Pomeranz (1985) included the elimination of sprout-susceptible lines through breeding and selection programs.

F. Mobilization of Proteins during Germination

Essential amino acids increase during germination or sprouting of cereal grains (Dalby and Tsai, 1976; Tsai et al., 1975). For example, both lysine and tryptophan increased during the germination of wheat, barley, oats, and rice. The extent of the increase was directly related to the decrease in prolamin content of the grain. A substantial increase in lysine of 50% was noted for wheat, compared to only a slight increase in oats (Figure 1.11). The level of prolamin in oats, however, was much lower than that in wheat. Jones and Tsai (1977) reported an increase in the lysine and tryptophan content of the embryo of normal maize and a corresponding decrease in the endosperm. The higher level of lysine is required for embryo growth and development, as observed previously by Singh and Axtell (1973) in studies on barley embryo and endosperm proteins. The precursors for lysine biosynthesis in maize may be provided by mobilization of zein reserves in the endosperm.

The release of amino acids during germination of wheat was investigated by Tkachuk (1979). After 122 hours of germination at 16.5°C the proline and glutamine content increased 100- and 80-fold, respectively, while lysine increased only 12-fold (Table 1.9). These results represent the changes taking place in the whole wheat kernels and may not reflect changes occurring in the embryo or aleurone layer. Nevertheless, they do illustrate that considerable proteolysis occurs during germination, which could be a method for assessing the extent of germination.

Kruger (1984), using high-performance liquid chromatography in the gel permeation mode, monitored the molecular weight profiles of buffer-soluble (0.5 M sodium phosphate buffer, pH 7.0, containing 0.5 M sodium chloride) proteins in both sound and germinated wheat kernels. Of the molecular weight protein group examined, the low-molecular-weight peptides and amino acids exhibited the largest change. This was further evidence for the increased solubilized amino nitrogen, particularly amino acids, during germination. Little change occurred during the first 2 days of germination compared to after 6 days. Further studies by Kruger and Marchylo (1985) examined mobilization of protein during the germination of five wheat cultivars. Six major protein components were eluted, of which only the low-molecular-weight species underwent major changes during germination. These results confirmed earlier work by Kruger (1984) and Lukow and Bushuk (1984) which showed that a very rapid hydrolysis of wheat endosperm proteins occurs following limited endopeptidase activity during the initial period of germination.

FIGURE 1.11 Changes in protein and prolamin content during germination of wheat and oats. *(Adapted from Dalby and Tsai, 1976.)*

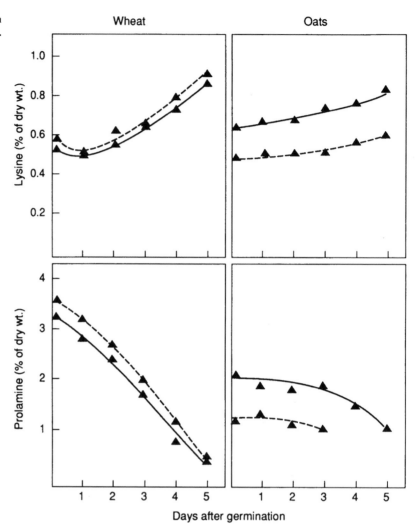

Increased release of free amino acids during germination suggests extensive mobilization of the storage proteins during this period. The mechanism controlling this process remains poorly understood. A number of proteases has been found in wheat grain, including endopeptidases, carboxypeptidases, and aminopeptidases (Grant and Wang, 1972; Kruger, 1973; Preston and Kruger, 1976a, b, 1977; Kruger and Preston, 1978). Of these, carboxypeptidase is prominent in the endosperm, where it represents one-quarter of the total endopeptidase activity (Preston and Kruger, 1976a). These enzymes have a negligible effect on the endosperm reserves during the first 2 days of germination, possibly because of their compartmentalization, the presence of protease inhibitors, or insolubilization of the substrate. During the course of germination there is limited endopeptidase activity resulting in the formation of intermediate products which are then degraded by carboxypeptidase to amino acids (Kruger and Marchylo, 1985). Only a fraction of the storage proteins is affected at any time, which explains the similarity in protein patterns for sprouted and mature seeds.

G. Lipid Mobilization during Germination

Germination and sprouting of cereal grains are accompanied by an increase in total lipid content (Lorenz, 1980; Rahnotra *et al.*, 1977). The presence of lipase in ungerminated wheat and barley seeds is extremely low but develops as soon as germination commences (Huang and Moreau, 1978; Taverner and Laidman, 1972). In sharp contrast to these cereals, oats are rich in lipase activity (Matlashewski *et al.*, 1982). Lipase (triacylglycerol lipase, EC 3.1.1.30) hydrolyzes triacyiglycerols, diacyiglycerols, and possibly monoacylglycerols, producing fatty acids. The major

TABLE 1.9 Effect of Germination at 16.5°C on the Production of Selected Free Amino acids in Wheat cv. 'Neepawa'

Amino Acid (μmole/g N)	Germination Period (Hours)	
	0	122
Tryptophan	47	50
Lysine	5.7	63
Histidine	2.2	72
Glutamic acid	64	95
Methionine	2.4	27
Isoleucine	5.1	140
Leucine	6.0	170
Tyrosine	4.5	72
Phenylalanine	4.2	150
Proline	7.8	790
Glutamine	12	920

Adapted from Tkachuk (1979).

difficulty involved in measuring lipase activity is due to the insolubility of the substrate in aqueous solution. This difficulty has been partially overcome using water-soluble substrates such as *p*-nitrophenyl (Pnp) acetate, or butyrate, or by forming a stable emulsion with olive oil. A specific method for assaying lipase was developed by Matlashewski *et al.* (1982) using radioactive triacylglycerols in which the fatty acid moiety was labeled. Using this method, Baxter (1984) examined lipase activity in both germinated and ungerminated barley. The results obtained in Figure 1.12 show that lipase activity increased slowly during the initial 2 days of germination but then rose sharply after 3 days. Two distinct lipase fractions were separated with similar molecular weights (400,000 range) but different ionic properties. The major fraction (I) was associated with the embryo, while the smaller lipase fraction (II) was located in the endosperm. Taverner and Laidman (1972) identified lipase in wheat embryo and endosperm, each induced by different factors. Urquardt *et al.* (1984) separated oat embryos from the rest of the kernel and monitored changes in lipase activity during germination. The initial increase in lipase activity appeared to be primarily in the bran layer,

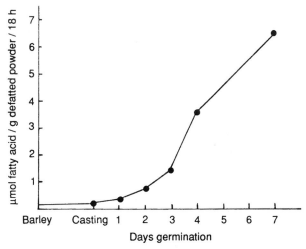

FIGURE 1.12 **Lipase activity in aqueous extracts of barley (variety Sonja) during germination.** *(Baxter, 1984.)*

with little or no activity in the endosperm (Urquardt *et al.*, 1983). While the primary role of lipase is hydrolysis of the storage triacylglycerols, its physiological role remains obscure.

V. STORAGE OF GRAINS

Following harvest, cereal grains such as wheat are stored in either sacks or bulk silos. These grains are traditionally recognized for their keeping quality, which is affected by moisture, temperature, and invasion by rodents, insects, bacteria, and fungi. Postharvest grain losses worldwide appear to be around 3–10%, sometimes up to 15%, depending on local conditions and resources (Harris, 1984). This section will focus on the effects of moisture and temperature on the quality of grains.

A. Respiration

When cereal grains are dry very little respiration occurs. If the moisture content of the seeds rises above 14%, respiration increases until a critical moisture level is attained. At this point respiration accelerates rapidly with the subsequent heating of the grain. This marked rise in respiration is attributed in part to germination and growth of molds such as *Aspergillus* and *Penicillium*. Grain respiration is affected by moisture, temperature, and oxygen tension, although the moisture content is of paramount importance in the commercial storage of cereal grains.

1. Effect of Moisture Content

Exposure of grain will result in the uptake of moisture until equilibrium is reached with the water vapor in the atmosphere. Thus, the moisture content of grain is controlled by the relative humidity in the environment, which in terms of grain storage is the nature of the interstitial atmosphere. When exposed to an atmosphere of uniform relative humidity at a constant temperature, the relative humidity of the stored grain reaches an equilibrium referred to as the equilibrium relative humidity (ERH). The relationship between relative humidity and moisture content is defined by the sorption isotherm, the shape of which is sigmoid. This is due to the larger equilibrium moisture content during desorption compared to adsorption at a given ERH. Figure 1.13 shows the moisture isotherm obtained at 30°C for maize with the characteristic sigmoid curve resulting from the greater water content of the desorption isotherm (Denloye and Ade-John, 1985).

The equilibrium moisture content is quite low in grains and only after the isotherm reaches 80% relative humidity does the moisture content rise exponentially with relative humidity (Oxley, 1948). The moisture content that is regarded as safe for grain is that in equilibrium with 70% relative humidity (Pixton and Warburton, 1971). Microbial growth will only occur above 75% relative humidity, resulting in extensive deterioration of the grain.

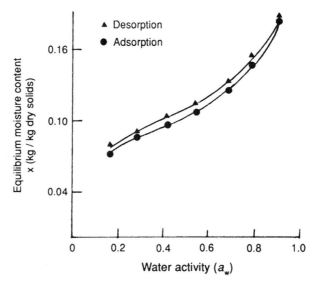

FIGURE 1.13   Moisture sorption isotherms for maize (30°C). *(Reprinted with permission from Denloye and Ade-John, 1985. © Pergamon Press.)*

Under extremely wet conditions the grain may be harvested at a moisture content that is too high for safe storage. This necessitates the use of drying to reduce the moisture content of the grain, which can then be stored with minimal loss in seed viability, nutritive value, and breadmaking properties (Bushuk, 1978). Spillane and Pelhate (1982) attempted to bypass the drying step by storing barley harvested with a high moisture content (> 30%) under ventilated conditions. Unless the rise in grain temperature, due to respiration, could be controlled, an explosive growth of yeasts and bacteria would take place. This was prevented by continuously ventilating the silo for a month, which removed a good portion of the heat generated by respiration, and so reduced the final grain temperature to below the critical point of 16°C. The moisture content of the grain was reduced to 16% and the relative humidity of the grain environment to around 80% at the end of the storage period. Under these conditions the growth of yeasts and bacteria was suppressed while quality factors remained intact.

2. Effect of Temperature

The ERH is affected only slightly by changes in temperature. Ayerst (1965) reported that a rise or fall of 10°C resulted in a 3% change in ERH over a relative humidity range of 40–90%. At higher relative humidities the change never exceeded 1% (Pixton and Warburton, 1975). Using Manitoba wheat, Pixton (1968) showed that at 10% moisture content the ERH increased by 6% when heated at 70°C compared to only 2% when the moisture content was 14%. Prolonging the heating for more than an hour produced further change. Denloye and Ade-John (1985) noted a decrease in the equilibrium moisture content for maize kept at a constant relative humidity as temperature was changed from 30°C to 50°C (Figure 1.14).

Since grain is stored in bulk, the movement of heat and moisture in the stored grain is extremely important. Anderson et al. (1943) first showed that movement of moisture occurred over a temperature gradient from high to low. This process was extremely slow and involved diffusion with some convection currents. The main effect of heating appears to be related to the translocation of moisture brought about by the temperature gradients in the grain.

B. Prolonged Storage of Grains and Flour

Pixton et al. (1975) monitored the changes in quality of wheat stored for 16 years under conditions of low temperature (4.5–0.5°C) and low oxygen concentrations (< 2% by volume). Two different pest-free dry wheats, Manitoba and Cappelle, with respective moisture contents of 11.9% and 12.6%, were placed in bins in 1-ton lots. The moisture content did not change significantly over this period. The crude protein and salt-soluble protein content remained unchanged for both wheats irrespective of the storage conditions. A slight increase in the total fat of 0.5% was observed for both wheat varieties, which was attributed to carbohydrate metabolism during the long storage period. This was based on the slight reduction in total sugars observed in these wheat samples by Pixton and Hill

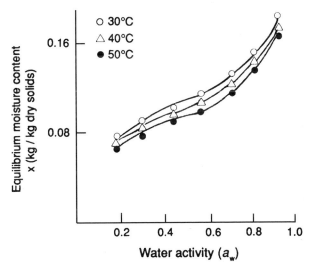

FIGURE 1.14 **Desorption moisture isotherm for maize at different temperatures.** *(Reprinted with permission from Denloye and Ade-John, 1985.* © *Pergamon Press.)*

(1967) after 8 years, although maltose and sucrose changed very little during the subsequent storage period. These researchers also monitored vitamin B, which remained unchanged throughout the storage period.

A high viability was observed for wheats stored at 4.5°C (96%) compared to only one-third viability when held at ambient temperature for the same period. As long as the wheat was protected from atmospheric moisture, rapid temperature changes, and insects, the baking quality remained intact, although some supplementation with fungal α-amylase was required.

PART II: LEGUMES

I. INTRODUCTION

The term legume encompasses more than 13,000 different species, all of the family Leguminosae. Legumes play a dominant role in the diets of humans across the globe. Of the thousands of species, however, only relatively few are widely grown commercially: soybeans, peanuts, dry beans, peas, broadbeans, chickpeas, and lentils. Of these seven, soybean is by far the most widely produced. Many other species of legumes play an important role in local food production in various corners of the world, but they are too numerous to be discussed in this chapter. Legumes are perhaps best known for their high plant protein content, due to nitrogen fixation allowed by the symbiotic relationship with bacteria. Table 1.10 shows the Food and Agriculture Organization (FAO) estimations for the world 2007 production of the major food legumes. This portion of the chapter will discuss the composition of legume seeds and the biochemical changes that occur during seed development, germination, storage, and fermentation.

II. LEGUME SEED STRUCTURE

Despite great variation in the macronutrient composition of legumes, their basic seed structure is the same. Mature seeds contain three major components: the seed coat (testa), the embryo, and the endosperm. Most legume seeds, however, have very little endosperm at maturity, as the cotyledons of the embryo make up a majority of the seed weight and contain the necessary stores for growth. Thus, the cotyledons provide the great majority of the nutritional components of interest to food value, with the exception of fiber and calcium, of which a significant portion is found in the seed coat (Kadam *et al.*, 1989). The structure of a typical soybean seed is shown in Figures 1.15 and 1.16. Size, shape, color, and thickness of the seed coat vary among the different legumes, although the basic structure prevails.

III. LEGUME SEED COMPOSITION

A. Proximate Composition

Food legumes vary greatly in their nutrient composition, depending on the type and variety of seed, soil conditions, and environmental factors. The proximate compositions of some major food legumes grown in the USA are presented

TABLE 1.10 World Production of Legumes in 2007

Legume	Million Tonnes
Beans, dry	19.3
Beans, green	6.4
Broad beans, dry	4.9
Peanuts, with shells	34.9
Lentils	3.9
Peas, dry	10.1
Peas, green	8.3
Soybeans	216.1

Data from the Food and Agriculture Organization (2008).

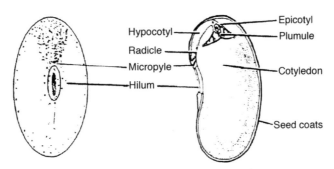

FIGURE 1.15   Structure of a soybean seed. *(From Liu, 1997, p. 4. With kind permission from Springer Science + Business Media BV.)*

in Table 1.11. The protein content of the selected legumes ranges from 19.30% to 26.12% of the edible portion, although crude protein content has been reported to range between 15% and 45% (Kadam *et al.*, 1989), with some soybean varieties containing as much as 50% protein (Vaidehi and Kadam, 1989). Carbohydrate content ranges from 24% to 68% (Reddy *et al.*, 1984b), and appears to be inversely related to the lipid content. Legume seeds high in carbohydrates have low lipid content, and vice versa. A classical example is peanut, which has a very high lipid content (49.24%) and relatively low carbohydrate content (16.13%) (Table 1.11). Potassium is by far the most abundant mineral in most food legumes (Iqbal *et al.*, 2006; USDA 2008), with soybeans containing as much as 1.80 g/100 g edible portion (Table 1.11). Phosphorus, copper, iron, calcium, and magnesium are some of the important minerals found in significant amounts in legumes. Niacin and pantothenic acid account for the most quantitatively important vitamins in legumes, and most are also a good source of folate.

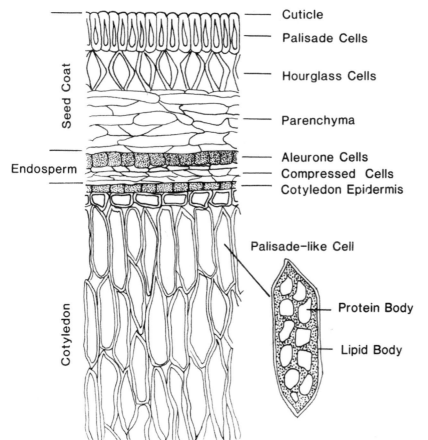

FIGURE 1.16   Cross-section of soybean seed coat. *(From Bair, 1979. © Craig Bair.)*

TABLE 1.11 Proximate Composition of Some Important Food Legumes

Nutrient	Glycine max L.	Cicer arietinum	Arachis hypogaea	Pisum sativum	Vicia faba	Lens culinaris	Phaseolus vulgaris
Water (g)	8.54	11.53	6.50	11.27	10.98	10.40	11.02
Protein (g)	36.49	19.30	25.80	24.55	26.12	25.80	21.60
Total lipid (g)	19.94	6.04	49.54	1.16	1.53	1.06	1.42
Ash (g)	4.87	2.48	2.33	2.65	3.08	2.67	3.60
Carbohydrate, by difference (g)	30.16	60.65	16.13	60.37	58.29	60.08	62.36
Fiber, total dietary (g)	9.30	17.4	8.50	25.5	25	30.50	15.2
Calcium (mg)	277	105	92	55	103	56	123
Iron (mg)	15.70	6.24	4.58	4.43	6.70	7.54	5.02
Magnesium (mg)	280	115	168	115	192	122	171
Phosphorus (mg)	704	366	376	366	421	451	352
Potassium (mg)	1797	875	705	981	1062	955	1483
Sodium (mg)	2	24	18	15	13	6	5
Zinc (mg)	4.89	3.43	3.27	3.01	3.14	4.78	3.65
Copper (mg)	1.658	0.847	1.144	0.866	0.824	0.519	0.841
Manganese (mg)	2.517	2.204	1.934	1.391	1.626	1.33	1.06
Selenium (mg)	17.8	8.2	7.2	1.60	8.2	8.30	3.2

Nutrient values are per 100 g edible portion.
Data from USDA (2008).

B. Protein

1. Nitrogen Fixation

Biological nitrogen fixation (BNF) is achieved by diazotrophs, microorganisms that possess the enzyme nitrogenase, which converts atmospheric dinitrogen (N_2) into organic nitrogen (usually ammonia). Legumes benefit from BNF by forming symbiotic associations with some nitrogen-fixing bacteria. The soil bacteria *Azorhizobium*, *Bradyrhizobium*, and *Rhizobium*, in association with legumes, are responsible for most of the nitrogen fixed biologically (Freiberg *et al.*, 1997). The bacteria inhabit specialized organs called nodules on the roots of the legumes, which is the site where nitrogen fixation takes place. The process of legume nodulation is a complex one, and is controlled by several genetic and environmental factors (Hirsch, 1992; Schubert, 1995).

Nodulation may be seen to proceed through three stages: preinfection, nodule initiation, and differentiation, with the flavonoids in the seed coat thought to serve as chemoattractants that induce *Rhizobium* nod genes (see review by Hirsch, 1992). Nitrogenase is sensitive to oxygen concentration, requiring very low partial pressure of oxygen in order to fix atmospheric nitrogen. Legume root nodules, as part of the symbiotic relationship with the bacteria, synthesize leghemoglobin in response to being infected or inoculated by the bacteria. Leghemoglobin is an oxygen-binding protein that maintains a low enough oxygen tension to protect the oxygen-labile nitrogenase enzyme from inactivation but high enough to make bacterial respiration possible. It has been reported that the apoprotein portion of leghemoglobin is synthesized by the plant, while the bacteria contribute the heme (iron complexed with a porphyrin ring) (O'Brian *et al.*, 1987). Another study, however, suggests that both apoprotein and heme components of leghemoglobin are synthesized by the plant (Santana *et al.*, 1998). The entire process of atmospheric nitrogen fixation can be reduced to the following chemical equation:

$$N_2 + 8H^+ + 8e^- + 16\ ATP \rightarrow 2NH_3 + H_2 + 16ADP + 16\ P_i$$

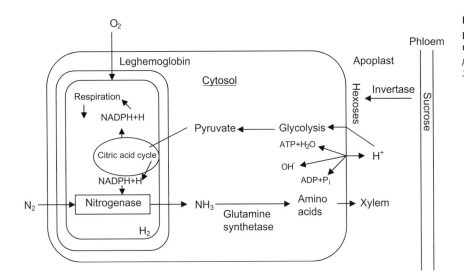

FIGURE 1.17 Model metabolic and transport pathways in infected cells of legume nodules. *(Redrawn from Schubert, 1995, p. 102. With kind permission from Springer Science + Business Media BV.)*

The energy (ATP) required to drive this process, as well as other metabolites necessary for the survival of the symbiotic bacteria, is supplied by the host plant in the form of sucrose through the vascular bundles in the inner cortex of the nodule (Serraj *et al.*, 1999). Soil acidity, drought, and soil mineral nitrogen content are a few of the environmental factors that influence the rate of nitrogen fixation (Schubert, 1995). The ammonia formed, present in its ammonium ion form, is released into the cytosol across a diffusion gradient (Schubert, 1995). Figure 1.17 presents a good schematic summary of the metabolic and transport pathways in the nodule.

2. Classification

Most of the protein in legumes is located in the cotyledons and embryonic axis, with the seed coat containing very little protein (Singh *et al.*, 1968). Legume seed proteins can be classified based on their functionality into structural and storage proteins. Structural proteins, sometimes referred to as enzymatic or catalytic proteins, are made up of protease inhibitors, lectins, lipoxygenases, and amylase inhibitors. Together they make up a small percentage of the total protein in the seeds, are found in the cotyledon, and are responsible for cell metabolism (Duranti and Gius, 1997). These structural proteins are albumins, soluble in water, and influence the postharvest taste and digestibility of food legumes. Specific examples and their effects will be examined in a later section. The storage proteins, which make up the bulk of legume seed proteins, are insoluble in water but soluble in salt solution and belong to the globulin class of proteins. Found primarily in the parenchyma cells of the cotyledons, storage proteins are contained in small membrane-bound organelles called protein bodies (Tombs, 1967; Duranti and Gius, 1997; Herman and Larkins, 1999), and range in size from 2 to 20 μm in diameter (Vaidehi and Kadam, 1989). Storage proteins provide the carbon and nitrogen building blocks necessary for seed growth during germination. They are further classified based on their sedimentation coefficients into four main fractions: 2S, 7S, 11S, and 15S. Fractions with higher sedimentation coefficients (up to 18S) have been reported in some soybean strains (Duranti and Gius, 1997). None of the fractions are, however, homogeneous. The 2S and 15S fractions are made up mainly of enzyme inhibitors and allergenic factors (Vaidehi and Kadam, 1989). The 7S globulins make up the bulk of the 7S fraction, while 11S globulin is the only protein in the 11S fraction. The 7S and 11S globulins together account for 50% of proteins in some soybeans (Vaidehi and Kadam, 1989), but usually contribute more than 70% of total proteins in most soybean and legume seeds (Kimura *et al.*, 2008; Natarajan *et al.*, 2006). Some legumes, e.g. French bean and cowpea, have a predominance of 7S globulins as the storage protein (Kimura *et al.*, 2008).

3. Protein Structure and Properties

The 7S and 11S globulins, because they make up the bulk of legume seed proteins, have been extensively studied. These globulins may be structurally similar, but differences exist from one legume to another in their subunit profiles and amino acid sequences, that give rise to differences in overall protein functionality.

The 7S globulins in different species of legumes are referred to by different names. The most abundant of the 7S globulins in soybeans is β-conglycinin. It is a trimeric glycoprotein made up of three types of subunit, α, α′, and β (Natarajan *et al.*, 2006; Rickert *et al.*, 2004). Different combinations of the subunits give rise to heterogeneous fractions with different functional properties among varieties of soybean (Rickert *et al.*, 2004). The α and α′ subunits are each composed of a core region with 418 amino acid residues, and extension regions with 125 and 141 residues, respectively (Maruyama *et al.*, 1999, 2002). The β subunit has only a core region made up of 416 amino acid residues (Maruyama *et al.*, 1999). In the study by Maruyama *et al.* (1999), the relationship between structure and physico-chemical properties of these β-conglycinin subunits was studied. Their results indicated that the subunits differed in their thermal stabilities, solubilities, emulsifying abilities, surface hydrophobicities, and heat-induced associations. They also found that these properties varied with changes in conditions such as pH and ionic strength. They concluded that the core regions of the subunits were responsible for determining surface hydrophobicity and thermal stability, while solubility, heat-induced association, and emulsifying ability depended on the extension regions, the carbohydrate moieties, and the core regions.

The predominant 7S protein in common beans (*Phaseolus vulgaris* L.) is known as phaseolin. It has 420 amino acid residues at synthesis, but loses 21 residues during maturation (Slightom *et al.*, 1983). Like the soybean 7S protein, phaseolin is a trimeric protein, with α, β, and γ subunits (Blagrove *et al.*, 1983; Slightom *et al.*, 1983). In general, legume 7S proteins show pH and ionic strength-dependent association and dissociation equilibria (Duranti and Gius, 1997).

Glycinin, the 11S globulin in soybean seeds, is a hexamer consisting of five types of subunit, G1, G2, G3, G4, and G5, with G1 and G2 being allergens (Natarajan *et al.*, 2006). Each subunit is composed of an acidic and a basic polypeptide linked by a single disulfide bridge (Staswic *et al.*, 1981). The amino acid sequences of the polypeptides differ between and within species, giving rise to heterogeneous fractions with different functional properties. For example, differences in gel strength of different glycinin fractions have been found to be dependent on the amino acid sequence of the acidic

TABLE 1.12 Amino Acid Content of Some Important Food Legumes

Amino Acid	Glycine max L.	Cicer arietinum	Arachis hypogaea	Pisum sativum	Vicia faba	Lens culinaris	Phaseolus vulgaris
Tryptophan (g)	0.591	0.185	0.250	0.275	0.247	0.223	0.256
Threonine (g)	1.766	0.716	0.883	0.872	0.928	0.895	0.909
Isoleucine (g)	1.971	0.828	0.907	1.014	1.053	1.078	0.954
Leucine (g)	3.309	1.374	1.672	1.760	1.964	1.809	1.725
Lysine (g)	2.706	1.291	0.926	1.772	1.671	1.740	1.483
Methionine (g)	0.547	0.253	0.317	0.251	0.213	0.212	0.325
Cystine (g)	0.655	0.259	0.331	0.373	0.334	0.327	0.235
Phenylalanine (g)	2.122	1.034	1.337	1.132	1.103	1.230	1.168
Tyrosine (g)	1.539	0.479	1.049	0.711	0.827	0.667	0.608
Valine (g)	2.029	0.809	1.082	1.159	1.161	1.238	1.130
Arginine (g)	3.153	1.819	3.085	2.188	2.411	1.928	1.337
Histidine (g)	1.097	0.531	0.652	0.597	0.664	0.702	0.601
Alanine (g)	1.915	0.828	1.025	1.080	1.070	1.042	0.905
Aspartic acid (g)	5.112	2.270	3.146	2.896	2.916	2.758	2.613
Glutamic acid (g)	7.874	3.375	5.390	4.196	4.437	3.868	3.294
Glycine (g)	1.880	0.803	1.554	1.092	1.095	1.014	0.843
Praline (g)	2.379	0.797	1.138	1.014	1.099	1.042	0.916
Serine (g)	2.357	0.973	1.271	1.080	1.195	1.150	1.175

Amino acid values are per 100 g edible portion.
Data from USDA (2008).

polypeptide chain (Nakamura *et al.*, 1984). Several studies have been conducted into the structure−function relationship of glycinin (Mori *et al.*, 1981; Nakamura *et al.*, 1984; Riblett *et al.*, 2001; Khatib *et al.*, 2002).

4. Protein Quality

Protein quality is generally defined by its amino acid composition, digestibility, and bioavailability. It is well known that legume proteins are low in the essential sulfur-containing amino acid methionine, while being especially rich in lysine. The second limiting amino acid in legume protein is tryptophan, but in a few legumes (cowpeas, lentils, and greenpeas) it was the most limiting amino acid (Iqbal *et al.*, 2006). The effect of these deficiencies is observed more markedly on growth than on protein requirements for maintenance (Patwardhan, 1962). The amino acid profiles of some important food legumes are presented in Table 1.12. The nutritional value of the legume proteins can be assessed by a variety of methods. The most commonly used include the essential amino score with reference to the FAO/World Health Organization (WHO) standard amino acid profile and protein efficiency ratio (based on growth response in experimental animals, usually rats). The digestibility coefficient of legume proteins varies greatly, from 51% to 92% (Patwardhan, 1962), and is influenced by the presence of anti-nutritional factors (Duranti and Gius, 1997).

C. Carbohydrates

1. Overview

The total carbohydrate of dry pulses varies greatly, ranging from 24% in winged bean seeds to 68% found in cowpea seeds (Table 1.13). In general, the total carbohydrate may consist of soluble and insoluble fractions. Soluble

TABLE 1.13 Total Carbohydrate and Starch Contents of Food Legumes

Legume	Total Carbohydrate	Starch	Amylose	Gelatinization Temperature (°C)
Winged bean seed	24.0−42.2	−	−	−
Smooth pea	56.6	36.9−48.6	5.3−8.7	65−69
Wrinkled pea	−	24.0−36.6	10.2−15.1	>99
Great bean	61.2−61.5	44.0	9.9	−
California small white bean	−	57.8	7.7	−
Red kidney bean	56.3−60.5	31.9−47.0	17.5−37.2	64−68
Navy bean	58.4	27.0−52.7	22.1−36.0	68−74
Pinto bean	−	51.0−56.5	25.8	−
Pink bean	−	42.3	14.9−35.3	−
Black eye bean	−	41.2	15.8−38.3	−
Black gram	56.5−63.7	32.2−47.9	43.9	71.5−74
Bengal gram	60.1−61.2	37.2−50.0	31.8−45.8	−
Mung bean	53.3−61.2	37.0−53.6	13.8−35.0	63−69
Red gram	57.3−58.7	40.4−48.2	38.6	−
Soybean	25.2−33.5	0.2−0.9	15.0−20.0	73−81
Broad bean	57.3	41.2−52.7	22.0−35.0	
Lentil	59.7	34.7−52.8	20.7−45.5	58−61
Cowpea	56.0−68.0	31.5−48.0	−	−
Lupine seed	−	0.3−3.5	−	−

Values are reported in g/100 g on a dry weight basis.
Data from Reddy *et al.* (1984b).

carbohydrate may include monosaccharides and oligosaccharides, whereas the insoluble fraction may include starch and dietary fiber and other polysaccharides. These carbohydrate components differ in their functionality and impact on human health. Individual pulse seeds may differ in their total carbohydrate content and carbohydrate composition, which may lead to their different nutritional values and food utilizations.

2. Insoluble Carbohydrate

a. Starch

Starch is the primary component of legume carbohydrates. Legume seeds vary greatly in their starch content and composition. As shown in Table 1.13, California small white beans contain 57.8% of starch, while soybeans contain as little as 0.2% of starch. In general, soybean, lupine, and winged beans have a lower starch level. The legume seed starches may have a high amylose concentration and differ significantly in their amylose and amylopectin ratios. The amylose content of legumes varies from 5.3% in smooth peas to 43.9% in black gram, as shown in Table 1.13 (Reddy *et al.*, 1984b).

The gelatinization temperature of various legume starches generally ranges from 60°C to 90°C (Table 1.13), which is comparable to the gelatinization temperature of corn starch and higher than that of waxy maize starch. Gelatinization temperature is determined by the structure and composition of the starch: a substantial amount of amylopectin promotes the gelatinization process while the degree of amylopectin branching in the starch varies the gelatinization temperature. Other factors that may alter the gelatinization temperature include the presence of bound lipids, protein and phosphate, and starch granule size.

It has recently been realized that starch may not be completely hydrolyzed and absorbed after digestion, although legume starch may contribute significantly to total energy intake. The starch components that are not hydrolyzed in the human gastrointestinal tract are classified as resistant starch (RS). Resistant starch may have an improved glycemic index (GI), an indicator of the effect of carbohydrates on blood glucose level. Many factors may alter starch digestibility and the formation of RS. These may include the inherent properties of starch such as granular structure and ratio of amylose and amylopectin, presence or treatment of heat, moisture content, interaction of starch with other chemicals, and storage and processing conditions (Sajilata *et al.*, 2006; Siddhuraju and Becker, 2005; Bravo *et al.*, 1998; Tovar and Melito, 1996). For instance, thermal treatment decreased RS concentration from 2.4% to 1.9% in the field pea, from 3.3% to 2.5% in lentil beans, and from 3.4% to 2.3% in chickpea (Rochfort and Panozzo, 2007). In contrast, steam-heating increased RS concentration from 1.9% to 6.0% in black beans, and from 0.8% to 4.0% in lima beans, whereas dry pressure cooking increased RS from 0.8% to 2.1% in lima beans, but had no effect on that in black beans, suggesting the effect of inherent properties and treatment of heat and moisture on the formation of RS (Tovar and Melito, 1996).

The effect of gelatinization on digestibility was reported by Sandhu and Lim (2007), who compared the pasting temperature of several legumes: mung bean, chickpea, field pea, lentil, black gram, and pigeon pea. Mung bean yielded the lowest pasting temperature at 50.2°C, compared to 51.4°C for chickpea, suggesting higher digestibility. Sandhu and Lim confirmed mung bean's high digestibility, as the level of RS was low at 50.3%. In comparison, pigeon pea starch, with 78.9% RS, has a low digestibility and GI. As Sandhu and Lim concluded, mung bean with high digestibility is suitable for malnourished patients, while pigeon pea is more preferable for diabetic patients.

Soaking and autoclaving also altered the digestibility of starch in mucuna beans, which has 28% starch composition (Siddhuraju and Becker, 2005). After soaking in water, the percentage of digestible starch increased from 67.4% to 87.2%. This was accompanied by a significant decrease in the amount of RS from 88.3 g/kg in raw seeds to 48.1 g/kg after soaking. Earlier work by Chau and Cheung (1997) found that soaking two Chinese legume seeds increased starch digestibility by 36.4—98.2%, while heat treatments increased starch digestibility by 6—7-fold, and germination increased starch digestibility by 1—2-fold. The content of RS in raw and cooked legume seeds is listed in Table 1.14.

b. Dietary Fiber

Legume seeds are excellent sources of dietary fibers. Their total fiber concentration ranges from about 1.2% (w/w) in black gram, Bengal gram, red gram, and mung beans, to 25.6% (w/w) in Bengal gram beans (Table 1.15). Total fiber concentration and composition may vary greatly in the same type of legume bean. For instance, total fiber concentration varied from 1.2% to 25.6% and cellulose content ranged from 1.1% to 13.7% in Bengal gram beans

TABLE 1.14 Resistant Starch (RS) Content in Raw and Cooked Legume Seeds

Legume	Raw (% RS)	Cooked (% RS)
Field pea[a]	2.4	1.9
Lentil[a,b]	3.3—6.53	2.5
Chickpea[a,b]	3.4—5.7	2.3
Black bean[c]	1.9	6.0
Red bean[c]	0.8	NA
Lima bean[c]	2.0	4.0
Kidney bean[b]	4.6—6.6	NA
Soybean[b]	0.2—0.3	NA
Soybean faba bean[b]	0.3—5.6	NA
Pea — smooth[b]	5.6—7.1	NA
Pea — wrinkled[b]	9.6—10.3	NA
Wheat bran	0.4	NA

NA: not available
[a]Data from Rochfort and Panozzo (2007)
[b]Mikulikova et al. (2008)
[c]Tovar and Melito (1996)

TABLE 1.15 Fiber Compositions of Food Legumes

Legume	Total Fiber	Cellulose	Lignin	Hemicellulose
Winged bean seed	3.4—12.5	—	0.7—1.0	1.36
Smooth pea	4.6—7.0	0.9—4.9	0.5—0.9	1.0—5.1
Wrinkled pea	7.6	1.2—4.2	0.3—1.0	0.9—6.6
Great bean	4.5—6.7	—	—	—
Red kidney bean	3.7	2.5—5.9	2.7—3.1	0.3
Navy bean	3.4—6.6	3.2	0.1	0.5—4.9
Pinto bean	4.3—7.2	9.0	1.8—3.0	4.0
Pink bean	—	6.0	0.2	—
Black eye bean	3.1	4.9	0.1	—
Black gram	1.2—19.5	5.0	3.8	10.7
Bengal gram	1.2—25.6	1.1—13.7	2.9—7.1	0.6—9.1
Mung bean	1.2—12.8	2.5—4.6	2.2—7.2	0.3—9.1
Red gram	1.2—20.3	7.3	2.9	10.1
Soybean	2.4—5.5	—	—	7.6
Broad bean	8.0	1.0	0.7—1.1	4.0—4.6
Lentil	2.6	4.1	2.6	6.0
Cowpea	1.7—4.0	—	0.6—1.8	—
Lupine seed	3.0	—	0.7—0.8	9.3—9.9

Values are reported in g/100 g on a dry weight basis.
Data from Reddy et al. (1984b).

(Salunkhe *et al.*, 1985). Legume bean fiber may contain significant levels of cellulose, hemicellulose, and lignins, along with minor amounts of pectic substance, arabinogalactan, and xyloglucan (Sathe *et al.*, 1984).

Growing evidence indicates the potential health beneficial effects of legume bean fiber and other dietary fibers. These beneficial effects may include, but are not limited to, an increase in fecal bulk and fecal moisture, reduction of plasma cholesterol level, improved GI, and reduced risk of colon cancer (Nwokolo, 1996). Dietary fiber molecules cannot be digested by the human small intestine but are fermented by the microorganisms in the colon to short-chain fatty acids. These reduce local pH, enhance intestinal content passage, and lead to enhanced elimination of bile acids. The reduced risk of colon cancer may be attributed to the production of short-chain fatty acids. Dietary fiber may absorb and trap bile acids and enhance their elimination in feces. This effect may stimulate the conversion of cholesterol to bile acids in liver and reduce the plasma cholesterol level.

3. Soluble Carbohydrate

Legume seeds contain significant levels of water soluble carbohydrates, including trace amounts of monosaccharides, such as glucose and arabinose in soybeans, and measurable concentrations of disaccharides and oligosaccharides, often including sucrose, raffinose, stachyose, verbascose, and ajugose (Sathe *et al.*, 1984; Nwokolo, 1996). Food legume seeds differ significantly in their total soluble sugar content and composition (Table 1.16). Wrinkled peas may have about 10.2–15.1% total soluble sugar, whereas soybean may contain 5% total soluble sugar (Reddy *et al.*, 1984b). The individual oligosaccharides are not evenly distributed in the different fractions of legume seeds (Vaidehi and Kadam, 1989), which may demand consideration when producing or using legume-based food ingredients. As

TABLE 1.16 Soluble Carbohydrate Compositions of Food Legumes

Legume	Sucrose	Raffinose	Stachyose	Verbascose	Ajugose	Total Soluble
Winged bean seed	0.3–8.2	0.2–2.0	0.1–3.6	0.04–0.9	—	3.4
Smooth pea	2.3–2.4	0.3–0.9	2.2–2.9	1.7–2.3	0.06	5.3–8.7
Wrinkled pea	2.3–4.2	1.2–1.6	2.9–5.5	2.2–4.2	0.13	10.2–15.1
Great bean	2.0–3.8	0.3–0.7	2.3–3.8	—	—	9.9
California small white bean	3.0	0.3–0.7	2.9–3.7	0.1	—	7.7
Red kidney bean	1.6	0.3–0.9	2.4–4.0	0.1–0.5	—	8.0
Navy bean	2.2–3.5	0.4–0.7	2.6–3.5	0.1–0.4	—	5.6–6.2
Pinto bean	2.	0.4–0.6	2.9–3.0	0.1–0.2	—	6.7
Pink bean	1.4	0.2–0.4	0.2–0.4	—	—	—
Black eye bean	2.6	0.4–1.0	0.4–0.9	—	—	—
Black gram	0.7–1.5	0–1.3	0.9–3.0	3.4–3.5	—	3.0–7.1
Bengal gram	0.7–2.9	0.7–2.4	2.1–2.6	0.4–4.5	—	3.5–9.0
Mung bean	0.3–2.0	0.3–2.6	1.2–2.8	1.7–3.8	—	3.9–7.2
Red gram	2.7	1.0–1.1	2.7–3.0	4.0–4.1	—	3.5–10.2
Soybean	—	0.7–1.3	2.2–4.2	0–0.3	—	5.3
Broad bean	1.4–2.7	0.1–0.5	0.5–2.4	1.6–2.1	—	3.1–7.1
Lentil	1.8–2.5	0.4–1.0	1.9–2.7	1.0–3.1	—	4.2–6.1
Cowpea	1.8–3.1	0.4–1.2	2.0–3.6	0.6–3.1	—	6.0–13.0
Lupine seed	1.0–2.6	0.5–1.1	0.9–7.1	0.6–3.4	0.3–2.0	7.4–9.5

Values are reported in g/100 g on a dry weight basis.
Data from Reddy *et al.* (1984b); Kamath and Belavady (1980).

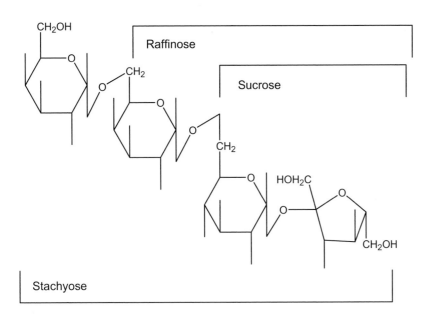

FIGURE 1.18 Structure of legume soluble carbohydrates stachyose, raffinose, and sucrose.

shown in Figure 1.18, these oligosaccharides are non-reducing sugars derived from sucrose by adding one or more galactose units via α (1→6) linkages to the glucose moiety of the sucrose unit. They are commonly known as the raffinose family or galacto-oligosaccharides. Raffinose oligosaccharides are not digested in the human small intestine owing to the absence of α-1,6-galactosidase in human intestinal mucosa, although sucrose is hydrolyzed and absorbed.

The raffinose family of oligosaccharides has been associated with flatulence and abdominal discomfort after legume consumption. These oligosaccharides tend to draw fluid into the lumen by osmosis, and lead to possible abdominal distension, cramps, and diarrhea. In the large intestine, these oligosaccharides may be broken down to monosaccharides by enzymes produced by local microorganisms. This increases the local osmolality and leads to greater water retention. The monosaccharides may also be further utilized by the microorganisms, which may produce significant amounts of gas and small molecular weight acids. The acids lower local pH, which may irritate the lining of the colon, and increase the movement of the intestinal contents. Both the fluid retention and increased movement of intestinal contents may cause diarrhea. The gas products, including carbon dioxide, hydrogen, and methane, may lead to bloating and cause problems in individuals with colonic pathologies, including irritable bowel syndrome.

The level of oligosaccharides in food may be reduced by improving food processing conditions, such as soaking the seeds or meal in water (Reddy *et al.*, 1984b; Vaidehi and Kadam, 1989; Martin-Carrejas *et al.*, 2006). Breeding efforts and genetic modification have been explored to reduce these oligosaccharides in legumes. In addition, seed germination, fermentation and α-galactosidase treatment, and irradiation are possible approaches to eliminate raffinose family oligosaccharides from legumes (Reddy *et al.*, 1984b; Rochfort and Panozzo, 2007). Completely removing the raffinose family of oligosaccharides may not eliminate flatulence from legume consumption, as dietary fiber components also contribute to flatulence. Interestingly, further studies have suggested that these oligosaccharides may provide potential health beneficial effects, such as immunomodulation and altering digestive passage speed (Parsons *et al.*, 2000; Rochfort and Panozzo, 2007).

4. Conclusion

Carbohydrates, one of the major constituents of legume seeds, may serve as an excellent source of dietary fiber and provide health benefits to consumers. Individual fractions of legume carbohydrate may contribute differently in human nutrition and food safety, and food functionality. A number of postharvest treatments, including those involved in food preparation, may affect the digestibility of legume carbohydrate and alter their nutritional values and beneficial properties, as well as undesirable properties such as flatulence production.

D. Lipids

Plant seeds store lipids in small spherical organelles with diameters ranging from 0.5 to 2.5 µm, called oil bodies (Tzen and Huang, 1992). Theses organelles, also called spherosomes, are located in the cotyledons and contain mostly triglycerides. Lipid synthesis in soybean seeds is controlled through regulation of the levels of fatty acid biosynthetic proteins, the presence of which depends on the developmental stage of the seed (Ohlrogge and Kuo, 1984).

Legume seed oils are generally good sources of polyunsaturated fatty acids, especially the essential fatty acids omega-6 linoleic acid and omega-3 linolenic acid. Soybean has by far the highest amount of linoleic acid, and good amounts of oleic and palmitic acids (Table 1.17). Peanut (*Arachis hypogaea*) has the highest monounsaturated fatty acid content on a per-weight basis, almost entirely composed of oleic acid.

Phytosterols are gaining prominence as cholesterol-lowering nutraceutical agents. They are structurally similar to cholesterol, and are found in the cell membranes of plant cells as rigidifying components. From Table 1.17, peanut has the highest amount of phytosterols per seed weight, followed by soybeans, peas, and broadbeans.

Phospholipids are another class of lipids found in legume seeds. They are a large family of polar lipids, the most prominent among which is lecithin from soybeans. Soy lecithin has many health benefits and has been utilized in many functional foods (Wang *et al.*, 2006). The major phospholipids in broad beans include phosphatidylcholine, phosphatidylehtanolamine, and phosphatidylinositol (Yoshida *et al.*, 2009). Legume seed phospholipid content diminishes as the seed matures owing to a decrease in the membrane component proportions in the developing seed (Wang *et al.*, 2006).

E. Other Components of Interest

1. Enzyme Inhibitors

Legume seeds contain low-molecular-weight proteins (Clemente and Domoney, 2006) that inhibit the activity of hydrolases such as proteases, amylases, and lipases (Lajolo and Genovese, 2002). Of these, protease inhibitors are the most important and extensively studied.

Soybean protease inhibitors are classified into two families, Bowman—Birk and Kunitz, which differ in their structure, weight, and activity. Bowman—Birk inhibitors have a molecular mass of between 6 and 10 kDa, with seven disulfide bridges, and can act against trypsin and chymotrypsin simultaneously at independent binding sites (Lajolo and Genovese, 2002; Becker-Ritt *et al.*, 2004). They are made up of two discrete polypeptide binding loops that hold a binding site each, making it possible for them to inhibit two molecules of enzymes at the same time (Clemente and Domoney, 2006). The Kunitz inhibitors are smaller, with only one polypeptide chain of molecular mass about 2 kDa (Becker-Ritt *et al.*, 2004). They have two disulfide bridges and act specifically against trypsin (Lajolo and Genovese, 2002).

TABLE 1.17 Fatty Acid and Phytosterol Composition of Some Major Food Legumes

Legume	Fatty Acids of Interest (g)							Phytosterol (mg)
	16:0	18:1	18:2	18:3	Saturated	Monounsaturated	Polyunsaturated	
Glycine max	2.116	4.348	9.925	1.330	2.884	4.404	11.255	161
Cicer arietinum	0.501	1.346	2.593	0.101	0.626	1.358	2.694	35
Arachis hypogaea	5.154	23.756	15.555	0.003	6.834	24.429	15.559	220
Pisum sativum	0.125	0.232	0.411	0.084	0.161	0.242	0.495	135
Vicia faba	0.204	0.297	0.581	0.046	0.254	0.303	0.627	124
Lens culinaris	0.133	0.180	0.404	0.109	0.156	0.189	0.516	—
Phaseolus vulgaris	0.343	0.123	0.332	0.278	0.366	0.123	0.610	—

Amino acid values are per 100 g edible portion.
Data from USDA (2008).

The effect of these protease inhibitors on digestion and metabolism is well documented. In a study by Grant *et al.* (1995), the contents and effects of long-term dietary exposure to protease inhibitors and lectins in four major food legumes (soybean, cowpea, kidney bean, and lupin) in rats were investigated. In terms of content, soybean had the highest amount of 24.6 g trypsin inhibited/kg and 12.0 g chymotrypsin inhibited/kg, while lupin had the lowest of 1.1 g trypsin inhibited/kg and 1.4 g chymotrypsin inhibited/kg. Consumption of the soybean diet, which was high in protease inhibitors and low in lectins, over a 700-day period, resulted in extensive enlargement of the pancreas in the rats, with some exhibiting macroscopic nodules on the pancreas. In an earlier study by Jaffe and Lette (1968), growth was severely hampered in rats fed beans (*P. vulgaris*) low in hemagglutinin and high in trypsin inhibitor activity.

It has been shown that aqueous heat treatment of soybean at 80°C for 40 minutes significantly reduced protease inhibitor activity and improved growth in rats (Armour *et al.*, 1998). Not withstanding the established anti-nutritive activities of legume protease inhibitors, recent studies suggest possible health beneficial applications for these seed components. Soybean protease inhibitors, especially the Bowman—Birk family, have been shown to possess anticarcinogenic properties (see review by Kennedy, 1998). A study by Lin and Ng (2008) revealed a dimeric Kunitz-type trypsin inhibitor in black soybean (*Glycine max* cv. 'Dull Black') that stimulates nitric oxide production by macrophages and inhibits human immunodeficiency virus-1 (HIV-1) reverse transcriptase. It also showed promise as an inhibitor of cell proliferation in liver and breast cancer cells. More research is needed to confirm and elucidate the mechanisms behind these novel biological activities of legume protease inhibitors.

α-Amylase inhibitors are characterized as oligomeric proteins made up of glycopeptide subunits (Berre-Anton *et al.*, 1997). They are synthesized on the endoplasmic reticulum as preproteins, glycosylated, and transported to the protein storage vacuoles (Pueyo *et al.*, 1993). Two isoforms, α-AI1 and α-AI1′, have been isolated from kidney beans (*P. vulgaris*), with both showing inhibition of mammalian α-amylase by a mixed non-competitive inhibition mechanism (Berre-Anton *et al.*, 1997). The same study found that the α-AI1 form has a low optimum activity pH of 4.5 at 30°C, suggesting that this enzyme may exhibit specificity for pancreatic α-amylase, whose optimum activity pH falls within a similar range. In the common bean, molecular level control of α-amylase inhibitors appears to be exerted by a gene that closely resembles the one responsible for coding lectins in both structure and amino acid composition (Moreno and Chrispeels, 1989). Inhibition of starch digestion by α-amylase inhibitor has been shown to decrease dietary protein and fat utilization in experimental rats, resulting in retarded growth (Pusztai *et al.*, 1995).

2. Lectins

Legume lectins are a group of homologous glycoproteins found mostly in the seeds (Loris *et al.*, 1998). The protein's affinity for sugar moieties is specific and reversible (Lajolo and Genovese, 2002; Hamelryck *et al.*, 1996), and this specificity has been used to explain their role in rhizobia—host plant recognition during the initiation of nitrogen fixation (Bohlool and Schmidt, 1974). In the experiment by Bohlool and Schmidt (1974), soybean lectin, labeled with fluorescein isothiocyanate, selectively bound to only soybean nodulating strains of *Rhizobium japonicum*. They concluded that the soybean lectins must be interacting with specific polysaccharides on the surface of the appropriate *Rhizobium* cell, leading to initiation of nodulation. Legume lectins show considerable similarity in their primary, secondary, and tertiary structures, but small variations in amino acid sequence result in vast differences in quaternary structure (Srinivas *et al.*, 2001) and carbohydrate specificities (Loris *et al.*, 1998).

Lectins are agglutinating by binding to sugar moieties in cell membranes, disrupting membrane structure. This activity, when it takes place in the gut, interferes with digestion and absorption, making lectins anti-nutritional factors. Lajolo and Genovese (2002) reported that purified soybean lectins impaired growth, induced small intestine enlargement and damage, and stimulated hypertrophy and hyperplasia of the pancreas in experimental rats. Phytohemagglutinin, the seed lectin from *P. vulgaris*, is made up of two polypeptide subunits (E and L), which have been found to be erythroagglutinating and leucoagglutinating, respectively (Hamelryck *et al.*, 1996). Legume lectins are usually inactivated by thermal processing, as cooking at atmospheric pressure for 15 minutes was found to be adequate to curtail their anti-nutritive effects.

Despite their well-known anti-nutritive properties, research is beginning to show possible health beneficial effects of some legume lectins. Raw kidney beans high in lectins were found to reduce lipid accumulation in obese rats (Pusztai *et al.*, 1998). From that study, Pusztai and co-workers concluded that it may be possible to develop dietary adjunct or therapeutic agent from the bean lectin to stimulate gut function and reduce obesity.

3. Lipoxygenase

Lipoxygenase is a non-heme, iron-bearing, monomeric enzyme that catalyzes the dioxygenation of fatty acids containing (1Z,4Z)-pentadiene systems (Schilstra *et al.*, 1994). In legumes, lipoxygenases catalyze the hydroperoxidation of lipids, leading to the development of off-flavors. This usually takes place in the seeds postharvest. The physiological role of lipoxygenases in the plant is still not clear, but they have been found to be capable of oxidizing plant pigments (chlorophyll and carotenoids) and cholesterol (Sessa, 1979). It appears that the activity of the enzyme is influenced by the presence of its hydroperoxide products (Smith and Lands, 1972). Enzyme activity also increases during processing when the cell walls are ruptured and cellular control mechanisms are no longer effective (Sessa, 1979). Off-flavor development depends on the fatty acid composition, and results from the formation of volatile short-chain aldehydes, ketones, and alcohols (Kobayashi *et al.*, 1995; Yuan and Chang, 2007).

IV. EFFECTS OF GERMINATION

Germination has long been seen as a possible method of enhancing the nutritive value and esthetic qualities of legume seeds (Chen *et al.*, 1975). The uses of germinated seeds vary: they can be dried and ground into flour for uses similar to those of ungerminated seed flours. The fresh germinated seed is eaten as a vegetable, particularly in Eastern cultures, while a growing awareness has made these 'sprouts' a more common diet component in Western culture as well. Conversely, germination can also be a hindrance to legume purveyors. Whereas the shelf-life of dried, mature seeds is lengthy under proper storage conditions, fresh sprouts are relatively perishable, requiring stricter temperature and humidity control. Changes to the seed composition during germination are discussed here, but it should be noted that comparison of germination studies is difficult owing to protocol variations in relative humidity, temperature, seed age, seed species, and time of germination, among others. In addition, publications analyzing germinated seeds that do not provide results on a dry weight basis make it difficult to draw conclusions: the increased water weight of the seedling inherently gives the appearance of falling levels of the macronutrients and kilocalories on a per-weight basis. The body of research has become too complex to thoroughly discuss each of these variables here. Thus, an overall discussion of the major trends of germination is presented. Please refer to the literature for specific details.

A. Carbohydrates

Of particular interest when discussing legume carbohydrates in germination are raffinose and stachyose, two α-galactosides that are major offenders in the production of flatulence. Aman (1979) notes that these molecules generally decrease during the early days of germination, coinciding with an increase in fructose (Viana *et al.*, 2005). Other research supports this assertion in various legumes, including most notably soybeans, black beans, and chickpeas [El-Adawy, 2002 (chickpeas); Martin-Cabrejas *et al.*, 2008 (soybeans); Donangelo *et al.*, 1995]. Donangelo *et al.* (1995) suggests that α-galactosidase activity during germination may be responsible for such changes. Viana *et al.* (2005) confirmed this assertion both by chemical substrate reactions and by partially purifying the enzyme. The carbohydrate composition of germinating soybean seeds is shown in Figure 1.19.

The magnitude of the decrease in α-galactosides varies across species of legumes, as do the amount and type of fiber present in the seeds (Donangelo *et al.*, 1995; Vanderstoep, 1981; Martin-Cabrejas *et al.*, 2008). Of these works, Martin-Cabrejas *et al.*, (2008) and one other (Bau *et al.*, 1997) report a decrease in total dietary fiber in soybean germination, whereas Donangelo *et al.*, (1995) report a slight increase. Aman (1979) found no difference in mung beans and a small decrease in fiber in chickpeas, while El-Adawy (2002) reported a small increase in fiber when chickpeas germinated. Nonetheless, differences in fiber content of chickpeas are in the range of $\pm 2\%$ weight, indicating little dietary significance. Germinated soybeans, however, lost fiber equivalent to 10% of the dry matter in 24 hours (Martin-Cabrejas *et al.*, 2008).

B. Lipids

It is well established that lipids decrease in legume seeds during germination, on a dry weight basis [Bau *et al.*, 1997 (triglycerides); Chen *et al.*, 1975 (triglycerides); El-Adawy, 2002 (fat); Mostafa and Rahma, 1987 (oil content)], while the extent of the decrease varied, the apparent decrease may be due to an increase in other dry matter in the

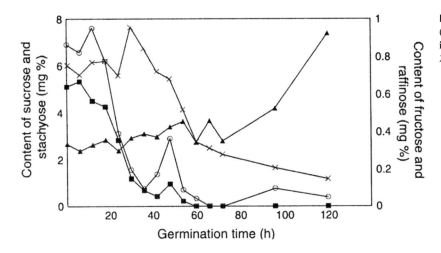

FIGURE 1.19 Contents (mg %) of raffinose (○), stachyose (■), fructose (▲), and sucrose (×) in germinating soybean seeds. *(From Viana* et al., *2005. © Elsevier.)*

seedling (Liu, 1997). It appears that the degradation of lipids, in addition to carbohydrates, is used to fuel the growth of the seedling and the accompanying processes (Bau *et al.*, 1997).

C. Proteins

Many studies have shown an increase in total protein contents in various legumes by dry weight after germination (El-Adawy, 2002; Bates *et al.*, 1977; Kakade and Evans, 1966; Martin-Cabrejas *et al.*, 2008; Palmer *et al.*, 1973). This may be attributable to (1) the use of carbohydrates and lipids as energy sources for the germinating seeds, thus allowing for protein to contribute a greater proportion of the remaining weight, and (2) the production of enzymes. Fewer studies have shown only slight increases in protein or no increase during germination (Ahmad and Pathak, 2000).

Several groups have studied the impact of germination on the composition of free amino acids, including non-protein amino acids. Kuo *et al.* (2004) reported that germination affected the free protein and non-protein amino acids of beans, peas, and lentils differently, with no specific trend across all three legumes. They also noted that light during germination affected the amino acid contents of the species differently. Rodriguez *et al.* (2008) and Urbano *et al.* (2005) both showed a decrease in protein nitrogen as germination progressed in beans, peas, and lentils, coinciding with a similar magnitude increase in non-protein nitrogen. Changes in individual amino acids were again species specific. A more complete discussion of amino acids and protein turnover in soybeans specifically can be found in Bau *et al.* (1997).

D. Vitamins and Minerals

While there is much variation in legume seed compositional changes during germination, the least disputed topic may be that of vitamin C. The contents of vitamin C, or ascorbic acid, increased during germination in soybeans (Ahmad and Pathak, 2000; Bau *et al.*, 1997) and various peas and beans (Sangronis and Machado, 2007; Chen *et al.*, 1975; Vanderstoep, 1981; Khattak *et al.*, 2007; Fernandez-Orozco, 2008). The respiration process is said to be triggered by ascorbic acid (Sangronis and Machado, 2007).

The B vitamins are next, although there is a less clear trend. Riboflavin has been shown to increase in soybeans under some conditions (Ahmad and Pathak, 2000; Bau *et al.*, 1995), as well as in chickpeas (El-Adawy, 2002) and peas (Urbano, 2005). Minor increases in thiamine have also been shown in germinated soybeans and white, black, and pigeon beans (Ahmad and Pathak, 2000; Sangronis and Machado, 2007), while a decrease (El-Adawy, 2002; Urbano *et al.*, 2005) or no change (Vanderstoep, 1981) was reported in chickpeas and peas. Results on niacin are similarly mixed and more limited (Bau *et al.*, 1997; El-Adawy, 2002).

The data on Vitamins E and A are limited (Bau *et al.*, 1997). Chen *et al.* (1975) reported low values in peas and beans, although their values for germinated seeds increased over their dry seeds. Fernandez-Orozco *et al.* (2008) showed differences in tocopherol content of two varieties of soybeans, although both showed an overall increase in all four tocopherol isomers and vitamin E activity after germination. Mung bean seeds, in contrast, showed an overall decrease in tocopherol contents and vitamin E activity, despite a slight increase in the α-tocopherol isomer.

The effect of germination on minerals is also varied in the literature. Calcium showed increases in soybeans (Bau *et al.*, 1997; Donangelo *et al.*, 1995) and black, white, and pigeon peas (Sangronis and Machado, 2007). Iron content appeared to decrease across various legumes (Chen *et al.*, 1975; Donangelo *et al.*, 1995; Sangronis and Machado, 2007; Vanderstoep, 1981). Although iron content drops, the increased vitamin C in germinated seeds may make it more bioavailable. Other minerals do not appear to have as clear trends in germination or as much evidence. Worthy of note, it has been suggested that the mineral content of the water used for germination has the potential to significantly affect the mineral content of the germinated legumes.

E. Anti-Nutritional Factors

A reduction in anti-nutritional factors of legumes has been one of the driving forces behind legume germination research, as destruction of these compounds without the use of heat processing may afford an easier solution to their removal. Anti-nutritional factors were discussed earlier. The majority of research has been focused on the effect of germination on trypsin inhibitors and phytate.

1. Trypsin Inhibitor Activity

Trypsin inhibitors prevent the action of trypsin in the gut, which breaks down proteins into absorbable amino acids and small peptides. Thus, a reduction in trypsin inhibitors would theoretically improve protein absorption. There are many reports of a reduction in trypsin inhibitor activity (TIA) during germination, including peas (Urbano *et al.*, 2005), various beans and soybean (Sangronis and Machado, 2007; Donangelo *et al.*, 1995; Mostafa and Rahma, 1987), and chickpeas (El-Adawy, 2002). However, some researchers have reported little or no change in TIA (Vanderstoep, 1981; Kakade and Evans, 1966) while others found an increase in TIA (Palmer *et al.*, 1973). Kakade and Evans (1966) showed that soaking the seeds for longer periods decreased TIA, which may help to explain differences among studies if the germination protocols were not standardized with respect to their soaking/rinsing methods.

Despite accurate measurements of TIA, the ultimate measure of interest is still the utilization of protein by live animals, for which feeding studies and protein efficiency ratio (PER) are often the instruments of choice. Germination for 2 and 4 days with and without light showed increased PER values for peas, but decreased after 6 days (Urbano *et al.*, 2005). A rat feeding study showed improved nutritive value of kidney beans after germination, despite increases in TIA mentioned above (Palmer *et al.*, 1973). *In vitro* protein digestibility tests have also indicated better digestibility in chickpeas (El-Adaway, 2002), various beans (Sangronis and Machado, 2007), and soybeans (Mostafa and Rahma, 1987).

2. Phytic Acid

Phytates, although possessing antioxidant capacity, are still often viewed as anti-nutritional factors owing to their ability to bind and prevent absorption of minerals. Therefore, decreases in phytate, or phytic acid, are considered a desirable outcome of processing. Khattak *et al.* (2007) reported that germination of chickpeas under blue light produced the greatest decrease in phytic acid, while Urbano *et al.* (2005) showed a similar decrease in phytates in peas regardless of the presence of light during germination. Decreases were also seen in various beans (Sangronis and Machado, 2007) and chickpeas (El-Adawy, 1997), whereas no decrease in phytic acid activity was seen in soybeans and black beans after 2 days' germination (Donangelo *et al.*, 1995). The breakdown of phytates during germination is thought to occur because of a surge in phytase activity (Bau *et al.*, 1997).

F. Nutraceutical Components

Research on the antioxidant components in germinated legumes is recent and growing. Isoflavones in soy are of particular interest because of the connection between their consumption and decreased incidence of chronic diseases. Therefore, it was questioned whether germination would positively or negatively affect the contents and composition of isoflavones in soy seeds. Zhu *et al.* (2005) studied the isoflavone contents of two soybean varieties grown to

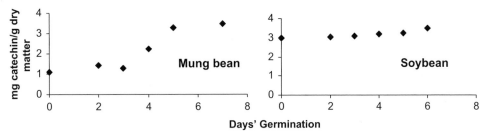

FIGURE 1.20   Total phenolic contents of germinated soybean and mung bean seeds. *(Adapted from Fernandez-Orozco et al., 2008.)*

specific hypocotyl lengths. Although there were differences between the two varieties, they concluded that total isoflavone concentration in mg/g dry ground seed increased at 1 day of germination, then decreased until the end of the study. However, this increase and decrease coincided with an increase and decrease in the malonyl derivatives of isoflavones, which have a significantly higher molecular weight than the aglycone forms, though similar antioxidant activity on a molar basis. For this reason, it is difficult to interpret the antioxidant effect of these changes and is recommended that isoflavone values be reported on a molar basis.

Fernando-Orozco *et al.*, (2008) studied both the phenolic contents and various measures of antioxidant capacity. Soybean seeds and mung beans both increased in total phenolic contents during several days' germination time. The increase in phenolics was accompanied by increases in antioxidant capacity of the germinated seeds, as increases were also seen in Trolox equivalent antioxidant capacity, peroxyl radical-trapping capacity, and inhibition of lipid peroxidation. Total phenolic content results can be seen in Figure 1.20.

G. Esthetic Food Quality

Germinated legume seeds are commonly thought to have improved organoleptic qualities. Ahmad and Pathak (2000) showed a decrease in 'beany flavor' and overall improved sensory scores after 3 days' germination of soybeans. Germination also enhanced the sensory scores of a soy-breadfruit product in Nigeria (Ariahu *et al.*, 1999). Chen *et al.*, (1975) investigated the acceptability of bean and pea sprouts, although not specifically in comparison to their dry seeds, and all varieties were found to be acceptable to the sensory panel.

V. EFFECTS OF FERMENTATION

Fermentation of legumes has been used by humans for millennia as a method to preserve food, introduce variation into the diet, and decrease cooking times. The detailed history of fermentation is well covered elsewhere (Hesseltine, 1965; Deshpande *et al.*, 2000). In particular, the FAO Agricultural Services Bulletin 142 from the year 2000 is an excellent resource on both the scientific and societal rationale for fermentation (Deshpande *et al.*, 2000). It addresses major crop legumes in addition to some that are more locally produced, including the biochemical changes that occur in fermentation using different methods and a variety of microbial cultures. Thus, only a short review of fermentation is given here.

The function of fermentation depends somewhat on the locale where it is being performed. In remote or developing areas, where refrigeration is not readily available, fermentation mainly serves as a food preservation technique through the generation of products such as alcohol and lactic and acetic acids. Food preservation is less of a concern in modern countries, and the reasons for choosing to ferment a food shift towards achieving desirable flavor and consistency, among other esthetic qualities. Fermentation also reduces cooking time in legumes by enzymatically breaking down structural components, so products require less energy input to be cooked, a desirable outcome in both developing and modern parts of the world.

The above benefits are seemingly obvious to the cook and perhaps explain why early humans chose to ferment foods, but there are other functions of fermenting foods that have only become apparent through careful observation and science. The anti-nutrients in legumes discussed previously, including trypsin inhibitors, phytates, and gas-producing oligosaccharides, are reduced, often to insignificant levels. This reduction is seen both because of enzymatic action, but also because soaking and boiling (often steps in the fermentation process) leach or destroy these components. Partly because of this reduction of anti-nutrients and partly because the microbial enzymes 'predigest' carbohydrates, proteins and triglycerides, nutrient availability in the products increases. The presence of the microbes may also add protein, amino acids, and/or vitamins to the fermented food.

Concern over aflatoxins arises when discussing legumes and fungi, but the FAO reports that they have generally not been of large concern in fermentation products, particularly in modern processing, but also largely in traditional processing (Deshpande *et al.*, 2000).

Since the FAO report, several articles have been published on the effect of fermentation on the isoflavone content and antioxidant properties of legumes. Fermentation of soybean with *Rhizopus oligosporus* showed increases in total phenolic contents and scavenging activity against the 1,1-diphenyl-2-picrylhydrazyl (DPPH) radical (McCue and Shetty, 2003). Fermentation of black bean with various filamentous fungi showed similar increases in total phenolic contents (Lee *et al.*, 2008). Pyo *et al.*, (2005) showed an increase in antioxidant activity against both 2,2'-azino-bis-3-ethylbenzthiazoline-6-sulfonic acid (ABTS) and DPPH radicals after fermentation, and also the bioconversion of the glycoside forms of isoflavones to their aglycone form. Other groups have also shown similar results: the sugar side-chains are cleaved off the glycoside isoflavones, leaving behind aglycone forms in higher amounts (Ikeda *et al.*, 1995; Chun *et al.*, 2007).

VI. STORAGE

The storage of legumes has received considerable attention, particularly because of the potential for immense loss of food through biological aging in addition to microbial, insect, and rodent infestation. Physical factors such as moisture, temperature, and oxygen concentration affect the rates of deterioration and infestation. The amount of water in the seeds and the environment seems to play the dominant role in determining the rate of seed deterioration during storage (Liu, 1997). An older, but still relevant, discussion of the topic can be found in Salunkhe, Kadam, and Chavan's *Postharvest Biotechnology of Food Legumes* (1983). A more recent discussion of soybean seed aging during storage can be found in Liu's *Soybeans: Chemistry, Technology, and Utilization* (1997).

A. Respiration, Moisture, and Temperature

Respiration plays a major role in storage stability of legumes, as discussed previously for cereals. The rate of respiration is higher in oilseeds than in cereals (Salunkhe *et al.*, 1985). In addition, seeds of higher moisture content have a higher rate of respiration. Respiration of the stored seeds increases moisture, thus encouraging the growth of mold. It is obvious that it becomes critical to dry seeds before storage to prevent the cycle from beginning. Temperature also affects the rate of respiration of legume seeds, halving it for every decrease of 10°C (Salunkhe *et al.*, 1985). For soybeans, a moisture content of 13% or below has been suggested to be adequate for ambient storage in the USA (Liu, 1997), although lower temperature and moisture contents down to 5°C and 11% have been suggested to be 'ideal' (Liu, 1997).

Ambient conditions also need to be considered in storage of legumes. A similar moisture equilibrium exists in legumes, as was discussed previously for cereals. Legume seeds stored at a relative humidity of 70% will equilibrate to about 14% moisture content (Dobie, 1982). *Codex Alimentarius* (1995) standards recommend two levels of seed moisture content for storage based on climate and length of storage. Selected values are shown in Table 1.18. For those seeds stored without a seed coat, the moisture content should be 2% below what is shown.

B. Seed Aging and Food Quality

Of most interest is not only *that* seeds age, but *what* happens in the process and how it affects the seeds' intended uses. A decrease in protein extractability or solubility with increased storage time has been shown in multiple studies (Saio *et al.*, 1980, 1982; Liu *et al.*, 2008; Narayan *et al.*, 1988a). Extractability of the 11S protein decreased more rapidly than that of the 7S (Saio *et al.*, 1982). Decreases were exacerbated by adverse storage conditions of ≥ 30°C, $\geq 80\%$ relative humidity, and/or excessive storage times. Saio and Baba (1980) showed that poor storage of soybeans resulted in deformations of protein bodies and loss of starch granules. Whole beans have been shown to be more resistant to degradation than milled flours (Saio *et al.*, 1982). The color of soybeans changed from an initial creamy yellow to light brown after 9 years of ambient storage (Narayan *et al.*, 1988a).

Degradation of the protein in legumes, particularly soybeans, is of interest because of their food uses. Improperly stored soybeans produce a lower yield of tofu (Narayan *et al.*, 1988b; Hou and Chang, 2004), softer tofu, and fewer

TABLE 1.18 Selected Suggested Moisture Contents of Legume Seeds

Legume	Recommended Moisture Content (%)	
	Tropical Climate, Long term	Moderate Climate, Short Term
Bean	15	19
Lentil	15	16
Chickpea	14	16
Pea	15	18

Data from *Codex Alimentarius* (1995).

solids in soybean milk (Saio *et al.*, 1980). Hou and Chang (2004) suggested that assessing the color change of soybeans in storage may be a rapid method for predicting the quality of tofu made from those soybeans. Organoleptic scores for soymilk, tofu, and soynuts decreased significantly with an increase in storage time of the beans up to 9 years (Narayan *et al.*, 1988b). Improper storage also resulted in decreases in other functional properties, including emulsifying activity, emulsifying stability, thermal stability, and the protein disperse index (Liu *et al.*, 2008).

Other changes have been noted in soybeans during storage, including hydrolysis of neutral fats to free fatty acids (Yanagi *et al.*, 1985); a decrease in free sugar (Hou and Chang, 2004), available lysine, trypsin inhibitor activity, and lipoxygenase activity; and an increase in non-protein nitrogen and peroxide values (Narayan *et al.*, 1988a).

C. Effect on Isoflavones

Because of the ability of isoflavones to mitigate chronic disease, research on the effects of storage has expanded. Several groups have shown that isoflavones in stored soybeans (Hou and Chang, 2002; Lee *et al.*, 2003), soy protein isolate and soy flour (Pinto *et al.*, 2005) saw reductions in the malonylglucoside forms and increases in the glucoside and aglycone forms. Total isoflavones tended to remain at about the same level during study periods of 1 and 3 years (Lee *et al.*, 2003; Pinto *et al.*, 2005). The rate of conversion was enhanced by increased temperature and humidity and decreased to non-significant levels with refrigeration (Hou and Chang, 2002; Pinto *et al.*, 2005).

Further discussion of isoflavones as affected by soybean storage and processing, including a discussion of the kinetics of isoflavone degradation, can be found in a review article by Shimoni (2004).

REFERENCES

Abbot, I.R., Matheson, N.K., 1972. Starch depletion in germinating wheat, wrinkled seed peas and senescing tobacco leaves. Phytochemistry 11, 1261–1272.

Acker, L., Becker, G., 1971. Recent studies on the lipids of cereal starches. II. Lipids of various types of starch and their binding to amylose. Staerke 23, 419–424.

Ahmad, S., Pathak, D.K., 2000. Nutritional changes in soybean during germination. J. Food Sci. Technol. 37, 665.

Akazawa, T., Hara-Nishimura, I., 1985. Topographic aspects of biosynthesis extracellular secretion, and intracellular storage of proteins in plant cells. Annu. Rev. Physiol. 36, 441–472.

Aman, P., 1979. Carbohydrates in raw and germinated seeds from mung bean and chick pea. J. Sci. Food Agric. 30, 869–875.

Amir, J., Cherry, J.H., 1971. Chemical control of sucrose conversion to polysaccharides in sweet corn after harvest. J. Agric. Food Chem. 19, 954–957.

Amir, J., Cherry, J.H., 1972. Purification and properties of adenosine diphosphoglucose pyrophosphorylase from sweet corn. Plant Physiol. 49, 893–897.

Anderson, J.A., Babbitt, J.D., Meredith, W.O.S., 1943. The effect of temperature differential on the moisture content of wheat. Can. J. Res. Sect. C 21, 297–306.

Anderson, J.W., Hanna, T.J., Peng, X., Kryscio, R.J., 2000. Whole grain foods and heart disease risk. J. Am. Coll. Nutr. 19, 291S–299S.

Ariahu, C.C., Ukpabi, U., Mbajunwa, K.O., 1999. Production of African breadfruit (*Treculia africana*) and soybean (*Glycine max*) seed based food formulations. 1: Effects of germination and fermentation on nutritional and organoleptic quality. Plant Foods Hum. Nutr. 54, 193–206.

Armour, J.C., Perera, R.L.C., Buchan, W.C., Grant, G., 1998. Protease inhibitors and lectins in soya beans and effects of aqueous heat-treatment. J. Sci. Food Agric. 78, 225–231.

Ayerst, G., 1965. Determination of water activity of some hygroscopic food materials by a dew-point method. J. Sci. Food Agric. 16, 71–78.

Bair, C.W., 1979. Microscopy of soybean seeds: cellular and subcellular structure during germination, development, and processing with emphasis on lipid bodies. Ph.D. Thesis, Iowa State University, Ames.

Barrona, C., Surgeta, A., Rouau, X., 2007. Relative amounts of tissues in mature wheat (*Triticum aestivum* L.) grain and their carbohydrate and phenolic acid composition. J. Cereal Sci. 45, 88–96.

Baruch, D.W., Meredith, P., Jenkins, L.D., Sinimonds, L.D., 1979. Starch granules of developing wheat kernels. Cereal Chem. 56, 554–558.

Bates, R.P., Knapp, F.W., Araujo, P.E., 1977. Protein quality of green-mature, dry mature and sprouted soybeans. J. Food Sci. 42, 271–272.

Bau, H.-M., Villaume, C., Nicolas, J.-P., Mejean, L., 1997. Effect of germination on chemical composition, biochemical constituents and antinutritional factors of soya bean (*Glycine max*) seeds. J. Sci. Food Agric. 73, 1–9.

Baxter, D.E., 1984. Recognition of two lipases from barley and green malt. J. Inst. Brew. 90, 277–281.

Bean, M.M., Nimmo, C.C., Fullington, J.G., Keagy, P.M., Mecham, U.K., 1974. Dried Japanese noodles. II. Effect of amylase, protease, salts, and pH on noodle doughs. Cereal Chem. 51, 427–433.

Bechtel, D.B., Gaines, R.L., Pomeranz, Y., 1982a. Protein secretion in wheat endosperm. Formation of the protein matrix. Cereal Chem. 59, 336–343.

Bechtel, D.B., Gaines, R.L., Pomeranz, Y., 1982b. Early stages in wheat endosperm formation and protein body initiation. Ann. Bot. (London) 50, 507–518.

Becker-Ritt, A.B., Mulinari, F., Vasconcelos, I.M., Carlini, C.R., 2004. Antinutritional and/or toxic factors in soybean (*Glycine max* (L.) Merril) seeds: comparison of different cultivars adapted to the southern region of Brazil. J. Sci. Food Agric. 84, 263–270.

Bernal-Lugo, L., Beachy, R.N., Varner, J.E., 1981. The response of barley aleurone layers to giberellic acid includes the transcription of new sequences. Biochim. Biophys. Acta. 102, 617.

Berre-Anton, V., Bompard-Gilles, C., Payan, F., Rouge, P., 1997. Characterization and functional properties of the α-amylase inhibitor (α-AI) from kidney bean (*Phaseolus vulgaris*) seeds. Biochim. Biophys. Acta. 1343, 31–40.

Blagrove, R.J., Colman, P.M., Lilley, G.G., Donkelaar, A.V., Suzuki, E., 1983. Physicochemical and structural studies of phaseolin from French bean seed. Plant Food Hum. Nutr. 33, 227–229.

Bohlool, B.B., Schmidt, E.L., 1974. Lectins: a possible basis for specificity in the *Rhizobium*–legume root nodule symbiosis. Science New Ser. 185, 269–271.

Borovsky, D., Smith, E.E., Whelan, W.J., French, D., Kikumoto, S., 1979. The mechanism of Q-enzyme action and its influence on the structure of amylopectin. Arch. Biochem. Biophys. 198, 627–631.

Bravo, L., Siddhuraju, P., Saura-Calixto, F., 1998. Effect of various processing methods on the *in vitro* starch digestibility and resistant starch content of Indian pulses. J. Agric. Food Chem. 46, 4667–4674.

Briarty, L.G., Hughes, C.E., Evers, A.D., 1979. The developing endosperm of wheat. A stereological analysis. Ann. Bot. (London) 44, 641–658.

Briggs, D.E., 1963. Biochemistry of barley germination: action of gibberellic acid on barley endosperm. J. Inst. Brew. 69, 13–19.

Briggs, D.E., 1964. Origin and distribution of α-amylase in malt. J. Inst. Brew. 70, 14–24.

Briggs, D.E., 1968. α-Amylase in germinating, decorticated barley. III. Effects of adding CCC and other chemical substances. Phytochemistry 7, 539–554.

Briggs, D.E., Hough, J.S., Stevens, R., Young, T.W., 1981. Malting and Brewing Science I. Malt and Sweet Wort, second ed. Chapman & Hall, London.

Bright, S.W.J., Shewry, P.R., 1983. Improvement of protein quality in cereals. CRC Crit. Rev. Plant Sci. 1, 49–93.

Buchanan, A.M., Nicholas, E.M., 1980. Sprouting, alpha-amylase, and breadmaking quality. Cereal Res. Commun. 8, 23–28.

Burgess, S.R., Miflin, B.J., 1985. The localization of oat (*Avena sativa* L.) seed globulins in protein bodies. J. Exp. Bot. 36, 945–954.

Bushuk, W., 1978. Biochemical changes in edible plant tissue during maturation and storage. In: Hultin, H.O., Miller, M. (Eds.), Post-harvest Biology and Biotechnology. Foods and Nutrition Press, Westport, CT, Chapter I.

Buttrose, M.S., 1971. Ultrastructure of barley aleurone cells as shown by freeze etching. Planta 96, 13–26.

Buttrose, M.S., Soeffly, A., 1973. Ultrastructure of lipid deposits and other contents in freeze-etched coleoptile cells of ungerminated rice grains. Aust. J. Biol. Sci. 26, 357.

Campbell, W.P., Lee, J.W., O'Brien, T.P., Smart, M.G., 1981. Endosperm morphology and protein body formation in developing wheat grain. Aust. J. Plant Physiol. 8, 5–19.

Cawley, J.E., Mitchell, T.A., 1968. Inhibition of wheat α-amylase by bran phytic acid. J. Sci. Food Agric. 19, 106–108.

Cerning, J., Guilbot, A., 1973. Changes in the carbohydrate composition during development and maturation of the wheat and barley kernel. Cereal Chem. 50, 220–232.

Chamura, S., 1975. Histochemical investigation of the accumulation of phosphorised lipid in aleurone cells of rice kernels. Nippon Saku-motsu Gakkai Kiji 44, 243.

Chau, C.-F., Cheung, P.C.-K., 1997. Effect of various processing methods on antinutrients and *in vitro* digestibility of protein and starch of two Chinese indigenous legume seeds. J. Agric. Food Chem. 45, 4773–4776.

Chen, L.H., Wells, C.E., Fordham, J.R., 1975. Germinated seeds for human consumption. J. Food Sci. 40, 1290–1294.

Chevalier, P., Lingle, S.E., 1983. Sugar metabolism in developing kernels of wheat and barley. Crop. Sci. 23, 272–277.

Chrispeels, M.J., Varner, J.E., 1967. Hormonal control of enzyme synthesis and mode of action of gibberellic acid and abscisin in aleurone layer of barley. Plant Physiol. 42, 1008–1016.

Chun, J., Kim, G.M., Lee, K.W., Choi, I.D., Kwon, G.-H., Park, J.-Y., et al., 2007. Conversion of isoflavone glucosides to aglycones in soymilk by fermentation with lactic acid bacteria. J. Food Sci. 72 (2), M39–M44.

Clemente, A., Domoney, C., 2006. Biological significance of polymorphism in legume protease inhibitors from the Bowman–Birk family. Curr. Prot. Pept. Sci. 7, 210–216.

Codex, Alimentarius, 1995. Codex Standard for Certain Pulses. Codex Standard, 171–1989 (Rev. 1 1995) 3.2.1.1. Food and Agriculture Organization of the United Nations and the World Health Organization.

Cornford, C.A., Black, M., Daussant, J., Murdoch, K.M., 1987. α-Amylase production by pre-mature wheat (*Triticum aestivum* L.) embryos. J. Exp. Bot. 38, 277–285.

Dalby, A., Tsai, C.Y., 1976. Lysine and tryptophan increases during germination of cereal grains. Cereal Chem. 53, 222–226.

Denloye, A.O., Ade-John, A.O., 1985. Moisture sorption isotherms of some Nigerian food grains. J. Stored Prod. Res. 21, 53–58.

Deshpande, S.S., Salunkhe, D.K., Oyewole, O.B., Azam-Ali, S., Battcock, M., Bressani, R., 2000. Fermented grain legumes, seeds and nuts — a global perspective. FAO Agricultural Services Bulletin, 142. Food and Agriculture Organization of the United Nations, Rome.

Dobie, P., 1982. Storage losses in legumes. J. Nutr. Soc. 41, 75−79.

Donangelo, C.M., Trugo, L.C., Trugo, N.M.F., Eggum, B.O., 1995. Effect of germination of legume seeds on chemical composition and on protein and energy utilization in rats. Food Chem. 53, 23−27.

Donovan, G.R., Lee, J.W., Longhurst, T.J., 1982. Cell-free synthesis of wheat prolamins Aust. J. Plant Physiol. 9, 59−68.

Duffus, C.M., 1993. Starch synthesis and deposition in developing cereal endosperms. In: Shewry, P.R., Stobart, A.K. (Eds.), Seed Storage Compounds: Biosynthesis, Interactions and Manipulation. Clarendon Press, Oxford, pp. 191−209.

Duranti, M., Gius, C., 1997. Legume seeds: protein content and nutritional value. Field Crop. Res. 53, 31−45.

El-Adawy, T., 2002. Nutritional composition and antinutritional factors of chickpeas (Cicer arietinum L.) undergoing different cooking methods and germination. Plant Food Hum. Nutr. 57, 83−97.

Ellis, R.P., Cochrane, M.P., Dale, M.F.B., Duffus, C.M., Lynn, A., Morrison, I.M., et al., 1998. Starch production and industrial use. J. Sci. Food Agric. 77, 289−311.

Erlander, S.P., 1958. Proposed mechanism for the synthesis of starch from glycogen. Enzymologia 19, 273−283.

Evers, A.D., 1973. The size distribution among starch granules in wheat endosperm. Staerke 25, 303−304.

Evers, A.D., Lindley, J., 1977. The particle size distribution in wheat endosperm starch. J. Sci. Food Agric. 28, 98−102.

Evers, A.D., Blakeney, A.B., O'Brien, L., 1999. Cereal structure and composition. Aust. J. Agric. Res. 50, 629−650.

Fernandez-Orozco, R., Frias, J., Zielinski, H., Piskula, M.K., Kozlowska, H., Vidal-Valverde, C., 2008. Kinetic study of the antioxidant compounds and antioxidant capacity during germination of Vigna radiate cv. emerald, Glycine max cv. jutro and Glycine max cv. merit. Food Chem. 111, 622−630.

Filmer, P., Varner, J.E., 1967. A test for de novo 18 synthesis of enzymes, density labeling with H_2O^{18} of barley α-amylase induced by gibberellic acid. Proc. Natl. Acad. Sci. U.S.A. 58, 1520−1526.

Food and Agriculture Organization of the United Nations. 2008. Available at http://faostat.fao.org

Freiberg, C., Fellay, R., Bairoch, A., Broughton, W.J., Rosenthal, A., Perret, X., 1997. Molecular basis of symbiosis between Rhizobium and legumes. Nature 387, 394−401.

Fuller, P., Hutchinson, J.B., McDermott, E.E., Stewart, B.A., 1970. Inactivation of α-amylase in wheat and flour with acid. J. Sci. Food Agric. 21, 27−31.

Geddes, R., Greenwood, C.T., 1969. Biosynthesis of starch granules. IV. Observations on the biosynthesis of the starch granule. Staerke 21, 148−153.

Gibbons, G.C., 1979. On the localisation and transport of α-amylase during germination and early seedling growth of. Hordeum vulgare. Carlsberg Res. Commun. 44, 353−366.

Gibbons, G.C., 1980. On the sequential determination of α-amylase transport and cell wall breakdown in germinating seeds of. Hordeum vulgare. Carlsberg Res. Commun. 45, 177−184.

Gilland, B., 1998. Plants and population: is there time? NAS Colloquium, 1−2.

Gilland, B., 2002. World population and food supply: can food production keep pace with population growth in the next half-century? Food Policy 27, 47−63.

Grant, D.R., Wang, C.C., 1972. Dialyzable components resulting in proteolytic activity in extracts of wheat flour. Cereal Chem. 49, 201−207.

Grant, G., Dorward, P.M., Buchan, W.C., Armour, J.C., Pusztai, A., 1995. Consumption of diets containing raw soya beans (Glycine max), kidney beans (Phaseolus vulgaris), cowpeas (Vigna unguiculata), or lupin seeds (Lupinus angustifolius) by rats for up to 700 days: effects on body composition and organ weights. Br. J. Nutr. 73, 17−29.

Greenaway, W.T., 1969. The sprouted wheat problem: the search for a solution. Cereal Sci. Today 14 (390), 393−395, 406.

Greene, F.C., 1981. In vitro synthesis of wheat (Triticum aestivum L.) storage proteins. Plant Physiol. 68, 778−783.

Greene, F.C., 1983. Expression of storage protein genes in developing wheat (Triticum aestivum L.) seeds. Correlation of RNA accumulation and protein synthesis. Plant Physiol. 71, 40−46.

Halmer, P., 1985. The mobilization of storage carbohydrates in germinated seeds. Physiol. Veg. 23, 107−125.

Hamelryck, T.W., Do-Thi, M., Poortmans, F., Chripeels, M.J., Wyns, L., Loris, R., 1996. The crystallographic structure of phytohemagglutinin-L. J. Biol. Chem. 271, 20479−20485.

Hargin, K.D., Morrison, W.R., 1980. The distribution of acyl lipids in the germ, aleurone, starch and non-starch endosperm of four wheat varieties. J. Sci. Food Agric. 31, 877−888.

Harris, L., 1984. Postharvest grain losses in the developing world. Cereal Foods World 29, 456.

Herman, E.M., Larkins, B.A., 1999. Protein storage bodies and vacuoles. Plant Cell. 11, 601−613.

Hesseltine, C.W., 1965. A millennium of fungi, food and fermentation. Mycol. 7 (2), 149−197.

Higgins, T.J.V., Zwar, J.A., Jacobsen, J.V., 1976. Gibberellic acid enhances the level of translatable mRNA for α-amylase in barley aleurone layers. Nature (London) 260, 166−169.

Hirsch, A.M., 1992. Developmental biology of legume nodulation. New Phytol. 122, 211−237.

Hou, H.J., Chang, K.C., 2002. Interconversions of isoflavones in soybeans as affected by storage. J. Food Sci. 67, 2083−2089.

Hou, H.J., Chang, K.C., 2004. Storage conditions affect soybean color, chemical composition and tofu qualities. J. Food Process. Pres. 28, 473−488.

Huang, A.A.C., Moreau, R.A., 1978. Lipases in the storage tissue of peanut and other oilseeds during germination. Planta. 141, 111−116.

Ikeda, R., Ohta, N., Watanabe, T., 1995. Changes of isoflavones at various stages of fermentation in defatted soybeans [abstract]. J. Jpn. Soc. Food Sci. 42, 322−327.

Iqbal, A., Khalil, I.A., Ateeq, N., Khan, M.S., 2006. Nutritional quality of important food legumes. Food Chem. 97, 331−335.

Jaffe, W.G., Lette, C.L.V., 1968. Heat-labile growth-inhibiting factors in beans (Phaseolus vulgaris). J. Nutr. 94, 203−210.

Jelseman, C.L., Morre, D.J., Ruddat, M., 1974. Isolation and characterization of spherosomes from aleurone layers of wheat. Proc. Indiana Acad. Sci. 84, 166−178.

Jenner, C.F., 1968. The composition of soluble nucleotides in the developing wheat grain. Plant Physiol. 43, 41−49.

Jones, R.A., Tsai, C.Y., 1977. Changes in lysine and tryptophan content during germination of normal and mutant maize seed. Cereal Chem. 54, 565−571.

Jongh, G., 1967. Amylase determination. Getreide Mehl. 17, 1−4.

Joshi, S., Lodha, M.L., Mehta, S.L., 1980. Regulation of starch biosynthesis in normal and opaque-2 maize during endosperm development. Phytochemistry 19, 2305−2309.

Kadam, S.S., Deshpande, S.S., Jambhale, N.D., 1989. Seed structure. In: Salunkhe, D.K., Kadam, S.S. (Eds.), CRC Handbook of World Food Legumes: Nutritional Chemistry, Processing Technology, and Utilization, Vol. I. CRC Press, Boca Raton, FL, pp. 23–50.

Kakade, M.L., Evans, R.J., 1966. Effect of soaking and germinating on the nutritive value of navy beans. J. Food Sci. 31, 781–783.

Kamath, M.V., Belavady, B., 1980. Unavailable carbohydrates of commonly consumed Indian food. J. Sci. Food Agric. 31, 194–202.

Katz, F.R., Furcsik, S.L., Tenbarge, F.L., Hauber, R.J., Friedman, R.B., 1993. Behaviour of starches derived from varieties of maize containing different genetic mutations: effects of starch genotype on granular morphology. Carbohydr. Polym. 21, 133–136.

Kennedy, A.R., 1998. The Bowman–Birk inhibitor from soybeans as an anticarcinogenic agent. Am. J. Clin. Nutr. 68 (Suppl), 1406S–1412S.

Khatib, K.A., Herald, T.J., Aramouni, F.M., MacRitchie, F., Schapaugh, W.T., 2002. Characterization and functional properties of soy β-conglycinin and glycinin of selected genotypes. J. Food Sci. 67, 2923–2929.

Khattak, A.B., Zeb, A., Bibi, N., Khalil, S.A., Khattak, M.S., 2007. Influence of germination techniques on sprout yield, biosynthesis of ascorbic acid and cooking ability, in chickpea (Cicer arietinum L. Food Chem. 103, 115–120.

Kimura, A., Fukuda, T., Zhang, M., Motoyama, S., Maruyama, N., Utsumi, S., 2008. Comparison of physicochemical properties of 7S and 11S globulins from peas, fava bean, cowpea, and French bean with those of soybean—French bean 7S globulin exhibits excellent properties. J. Agric. Food Chem. 56, 10273–10279.

Kobayashi, A., Tsuda, Y., Hirata, N., Kubota, K., Kitamura, K., 1995. Aroma constituents of soybean [Glycine max (L.) Merril] milk lacking lipoxygenase isozymes. J. Agric. Food Chem. 43, 2449–2452.

Kruger, J.E., 1972a. Changes in the amylases of hard red spring wheat during growth and maturation. Cereal Chem. 49, 379–390.

Kruger, J.E., 1972b. Changes in the amylase of hard spring wheat during germination. Cereal Chem. 49, 391–398.

Kruger, J.E., 1973. Changes in the levels of proteolytic enzymes from red spring wheat during growth and maturation. Cereal Chem. 50, 122–132.

Kruger, J.E., 1984. Rapid analysis of changes in the molecular weight distribution of buffer-soluble proteins during germination of wheat. Cereal Chem. 61, 205–208.

Kruger, J.E., Marchyto, B.A., 1985. Examination of the mobilization of storage proteins of wheat kernels during germination by high-performance reversed-phase and gel permeation chromatography. Cereal Chem. 62, 1–5.

Kruger, J.E., Matsuo, R.R., 1982. Comparison of alpha-amylase and simple sugar levels in sound and germinated durum wheat during pasta processing and spaghetti cooking. Cereal Chem. 59, 26–31.

Kruger, J.E., Preston, K.R., 1978. Changes in aminopeptidases of wheat kernels during growth and maturation. Cereal Chem. 55, 360–372.

Kumar, R., Singh, R., 1980. The relationship of starch metabolism to grain size in wheat. Phytochemistry 19, 2299–2303.

Kumar, R., Singh, R., 1981. Free sugars and their relationship with grain size and starch content in developing wheat grains. J. Sci. Food Agric. 32, 229–234.

Kumar, R., Singh, R., 1983. Alkaline inorganic pyrophosphatase from immature wheat grains. Phytochemistry 22, 2405–2407.

Kumar, R., Singh, R., 1984. Levels of free sugars, intermediate metabolites and enzymes of sucrose-starch conversion in developing wheat grains. J. Agric. Food Chem. 32, 806–808.

Kuo, Y.-H., Rozan, P., Lambein, F., Frias, J., Vidal-Valverde, C., 2004. Effects of different germination conditions on the contents of free protein and non-protein amino acids in commercial legumes. Food Chem. 86, 537–545.

Laberge, D.E., MacGregor, A.W., Meredith, W.O.S., 1973. Changes in the free sugar content of barley kernels during maturation. J. Inst. Brew. 79, 471–477.

Lajolo, F.M., Genovese, M.I., 2002. Nutritional significance of lectins and enzyme inhibitors from legumes. J. Agric. Food Chem. 50, 6592–6598.

Lasztity, R., 1984. The Chemistry of Cereal Proteins. CRC Press, Boca Raton, FL.

Lee, I.H., Hung, Y.H., Chou, C.C., 2008. Solid-state fermentation with fungi to enhance the antioxidative activity, total phenolic and anthocyanin contents of black bean [abstract]. Int. J. Food Microbiol. 121, 150–156.

Lee, S.J., Ahn, J.K., Kim, S.H., Kim, J.T., Han, S.J., Jung, M.Y., Chung, I.M., 2003. Variation in isoflavone of soybean cultivars with location and storage duration. J. Agric. Food Chem. 51, 3382–3389.

Lin, P., Ng, T.B., 2008. A stable trypsin inhibitor from Chinese dull black soybeans with potentially exploitable activities. Process Biochem. 43, 992–998.

Lineback, D.R., 1984. The starch granule; organization and properties. Bakers Dig. 58 (3), 16–21.

Lineback, D.R., Ponpipom, S., 1977. Effects of germination of wheat, oats, and pearl millet on alpha-amylase activity and starch degradation. Staerke 29, 52–60.

Lingle, S.E., Chevalier, P., 1980. Vascularization of developing barley kernels. Plant Physiol. 65 (Suppl), 105.

Liu, K., 1997. Soybeans: Chemistry, Technology and Utilization. Chapman and Hall, New York.

Liu, C., Wang, X., Ma, H., Zhang, Z., Gao, W., Xiao, L., 2008. Functional properties of protein isolates from soybeans stored under various conditions. Food Chem. 111, 29–37.

Lorenz, K., 1980. Cereal sprouts: composition, nutritive value, food applications. CRC Crit. Rev. Food Sci. Nutr. 13, 353–385.

Loris, R., Hamelryck, T., Wyns, L., 1998. Legume lectin structure. Biochem. Biophys. Acta. 1383, 9–36.

Lowy, G.D.A., Sargeant, J.G., Schofield, J.D., 1981. Wheat starch granule protein: the isolation and characterization of a salt-extractable protein from starch granules. J. Sci. Food Agric. 32, 371–377.

Lukow, O.M., Bushuk, W., 1984. Influence of germination on wheat quality. II. Modification of endosperm protein. Cereal Chem. 61, 340–344.

McCue, P., Shetty, K., 2003. Role of carbohydrate-cleaving enzymes in phenolic antioxidant mobilization from whole soybean fermented with Rhizopus oligosporus [abstract]. Food Biotechnol. 17 (1), 27–37.

McDermott, E.E., Elton, G.A., 1971. Effect of surfactants on the α-amylase activity of wheat flour. J. Sci. Food Agric. 2, 131–135.

McDonald, A.M.L., Stark, J.R., Morrison, W.R., Ellis, R.P., 1991. The composition of starch granules from developing barley genotypes. J. Cereal Sci. 13, 93–112.

MacGregor, A.W., 1983. Cereal α-amylases: synthesis and action pattern. In: Daussant, J., Mosse, J., Vaughan, J. (Eds.), Seed Proteins. Academic Press, New York, pp. 1–34.

MacGregor, A.W., Matsuo, R.R., 1982. Starch degradation in endosperms of barley and wheat kernels during initial stages of germination. Cereal Chem. 59, 210–216.

Marchylo, B.A., Lacroix, L.J., Kruger, J.E., 1981. α-Amylase synthesis in wheat kernels as influenced by seed coat. Plant Physiol. 67, 89−91.

Marshall, J.J., Whelan, J.J., 1970. Incomplete conversion of glycogen and starch by crystalline amyloglucosidase and its importance in the determination of amylaceous polymers. FEBS. Lett. 9, 85−88.

Marston, R.M., Welsh, S.O., 1980. Nutrient content of the national food supply. Nat. Food Rev. Winter ed.

Martin-Cabrejas, M.A., Aguilera, Y., Benitez, V., Molla, E., Lopez-Andreu, F.J., Esteban, R.M., 2006. Effect of industrial dehydration on the soluble carbohydrates and dietary fiber fractions in legumes. J. Agric. Food Chem. 54, 7652−7657.

Martin-Cabrejas, M.A., Diaz, M.F., Aguilera, Y., Benitez, V., Molla, E., Esteban, R.M., 2008. Influence of germination on the soluble carbohydrates and dietary fibre fractions in non-conventional legumes. Food Chem. 107, 1045−1052.

Maruyama, N., Sato, R., Wada, Y., Matsumura, Y., Goto, H., Okuda, E., Nakagawa, S., Utsumi, S., 1999. Structure−physiochemical function relationships of soybean β-conglycinin constituent subunits. J. Agric. Food Chem. 47, 5278−5284.

Maruyama, N., Salleh, M.R.M., Takahashi, K., Yagasaki, K., Goto, H., Hontani, N., Nakagawa, S., Utsumi, S., 2002. Structure−physiochemical function relationships of soybean β-conglycinin heterotrimers. J. Agric. Food Chem. 50, 4323−4326.

Matlashewski, G.J., Urquhart, A.A., Sahasrabudhe, M.R., Altosaar, I., 1982. Lipase activity in oat flour suspensions and soluble extracts. Cereal Chem. 59, 418−422.

Melcher, V., Varner, J.E., 1971. Protein release of barley aleurone layers. J. Inst. Brew. 77, 456−461.

Meredith, P., Pomeranz, Y., 1985. Sprouted grain. Adv. Cereal Sci. Technol. 7, 239−320.

Miflin, B.S., Burgess, S.R., 1982. Protein bodies from developing wheat and peas, the effects of protease treatment. J. Exp. Bot. 33, 251−260.

Miflin, B.J., Shewry, P.R., 1977. An introduction to the extraction and characterization of barley and maize prolamins. In: Miflin, B.J., Shewry, P.R. (Eds.), Techniques for the Separation of Barley and Maize Seed Proteins. Academic Press, New York, p. 13.

Miflin, B.J., Burgess, S.R., Shewry, P.R., 1981. The development of protein bodies in the storage tissues of seeds, subcellular separations of homogenates of barley, maize and wheat endosperms of pea cotyledons. J. Exp. Bot. 32, 199−219.

Mikulikova, D., Masar, S., Kraic, J., 2008. Biodiversity of legume health-promoting starch. Starch 60, 426−432.

Mitsui, T., Christeller, J.T., Hara-Nishimura, I., Akazawa, T., 1984. Possible roles of Ca^{2+} and calmoldulin in the biosynthesis and secretion of α-amylase in rice seed scutellar epithelium. Plant Physiol. 75, 21−25.

Montonen, J., Knekt, P., Jarvinen, R., Aromaa, A., Reunanen, A., 2003. Whole-grain and fiber intake and the incidence of type 2 diabetes. Am. J. Clin. Nutr. 77, 62−92.

Moreaux, T., Landry, J., 1968. Extractin selective des proteines du grain de mais et en particulier de la fraction 'glutelines'. C. R. Hebd. Seances Acad. Sci. 266, 2302.

Moreno, J., Chrispeels, M.J., 1989. A lectin gene encodes the α-amylase inhibitor of the common bean. Proc. Natl. Acad. Sci. U.S.A. 86, 7885−7889.

Mori, T., Utsumi, S., Inaba, H., Kitamura, K., Harada, K., 1981. Differences in subunit composition of glycinin among soybean cultivars. J. Agric. Food Chem. 29, 20−23.

Morrison, W.R., 1978. Cereal lipids, Chapter 4. In: Pomeranz (Ed.), Advances in Cereal Science and Technology, Vol. 2. American Association of Cereal Chemistry, St. Paul, MN.

Morrison, I.N., Kuo, J., O'Brien, T.P., 1975. Histochemistry and fine structure of developing wheat aleurone cells. Planta. 123, 105−116.

Morrison, W.R., 1993. Cereal starch granule development and composition. In: Shewry, P.R., Stobart, K. (Eds.), Seed Storage Compounds. Oxford Science Publications, Oxford, UK, pp. 175−206.

Morrison, W.R., Gadan, H., 1987. The amylose and lipid contents of starch granules in developing wheat endosperm. J. Cereal Sci. 5, 263−275.

Mostafa, M.M., Rahma, E.H., 1987. Chemical and Nutritional changes in soybean during germination. Food Chem. 23, 257−275.

Nakamura, T., Utsumi, S., Kitamura, K., Harada, K., Mori, T., 1984. Cultivar differences in gelling characteristics of soybean glycinin. J. Agric. Food Chem. 32, 647−651.

Narayan, R., Chauhan, G.S., Verma, N.S., 1988a. Changes in the quality of soybean during storage. Part 1 − Effect of storage on some physico-chemical properties of soybean. Food Chem. 27, 13−23.

Narayan, R., Chauhan, G.S., Verma, N.S., 1988b. Changes in the quality of soybean during storage. Part 2 − Effect of Soybean storage on the sensory qualities of the products made therefrom. Food Chem. 30, 181.

Natarajan, S.S., Xu, C., Bae, H., Caperna, T.J., Garrett, W.M., 2006. Characterization of storage proteins in wild (Glycine soja) and cultivated (Glycine max) soybean seeds using proteomic analysis. J. Agric. Food Chem. 54, 3114−3120.

Nwokolo, E., 1996. The need to increase consumption of pulses in the developing world. In: Nwokolo, E., Smartt, J. (Eds.), Food and Feed from Legumes and Oilseeds. Chapman and Hall, London, pp. 3−11.

O'Brian, M.R., Kirshbom, P.M., Maier, R.J., 1987. Bacterial heme synthesis is required for the expression of the leghemoglobin but not the apoprotein in soybean root nodules. Proc. Natl. Acad. Sci. U.S.A. 84, 8390−8393.

Ohlrogge, J.B., Kuo, T.-M., 1984. Control of lipid synthesis during soybean seed development: enzymic and immunochemical assay of acyl carrier protein. Plant Physiol. 74, 622−625.

Okamoto, K., Kitano, H., Akazawa, T., 1980. Biosynthesis and excretion of hydrolases in germinating cereal seeds. Plant Cell Physiol. 21, 201−204.

Okita, T.W., Greene, F.C., 1982. The wheat storage proteins: isolation and characterization of gliadin messenger RNAs. Plant Physiol. 69, 834−839.

Oparka, K.J., Harris, N., 1982. Rice protein-body formation, all types are initiated by dilation of the endoplasmic reticulum. Planta. 154, 184−188.

Osborne, T.B., 1895. The proteins of barley. J. Am. Chem. Soc. 17, 539−567.

Oxley, T.A., 1948. The scientific principles of grain storage. Northern Publishing, Liverpool.

Palmer, R., McIntosh, A., Pusztai, A., 1973. The nutritional evaluation of kidney beans (Phaseolus vulgaris): the effect on nutritional value of seed germination and changes in trypsin inhibitor content. J. Sci. Food Agric. 24, 937−944.

Parker, M.L., 1981. The structure of mature rye endosperm. Ann. Bot. (London) 47, 181−186.

Parker, M.L., Hawes, C.R., 1982. The Golgi apparatus in developing endosperm of wheat (Triticum aestivum L.). Planta. 154, 277−283.

Parsons, C.M., Zhang, Y., Araba, M., 2000. Nutritional evaluation of soybean meals varying in oligosaccharide content. Poultry Sci. 79, 1127–1131.

Patwardhan, V.N., 1962. Pulses and beans in human nutrition. Am. J. Clin. Nutr. 11, 12–30.

Pernollet, J.-C., 1978. Protein bodies of seeds, ultrastructure, biochemistry, and degradation. Physiol. Veg. 20, 259–276.

Pernollet, J.-C., 1982. Les corpuscles protéiques des graines, stade transitoire de vacuoles spécialisees. Physiol. Veg. 20, 259.

Pernollet, J.-C., Camilleri, C., 1983. Formation and development of protein bodies in the wheat endosperm. Physiol. Veg. 21, 1093–1103.

Pernollet, J.-C., Mossé, J., 1983. Structure and location of legume or cereal seed storage proteins. In: Daussant, J., Mosse, J., Vaughan, J. (Eds.), Seed Proteins. Academic Press, New York, pp. 155–191.

Peterson, D.M., Smith, D., 1976. Changes in nitrogen and carbohydrate fractions in developing oat groats. Crop. Sci. 16, 67–71.

Pinto, M.D.S., Lajolo, F.M., Genovese, M.I., 2005. Effect of storage temperature and water activity on the content and profile of isoflavones, antioxidant activity, and in vitro protein digestibility of soy protein isolates and defatted soy flours. J. Agric. Food Chem. 53, 6340–6346.

Pixton, S.W., 1968. The effect of heat treatment on the moisture content/relative humidity equilibrium relationship of Manitoba wheat. J. Stored Prod. Res. 4, 267–270.

Pixton, S.W., Hill, S.T., 1967. Long-term storage of wheat. II. J. Sci. Food Agric. 18, 94–98.

Pixton, S.W., Warburton, S., 1971. Moisture content relative humidity equilibrium of some cereal grains at different temperatures. J. Stored Prod. Res. 6, 283–293.

Pixton, S.W., Warburton, S., 1975. The moisture content equilibrium relative humidity relationship of rice bran at different temperatures. J. Stored Prod. Res. 11, 1–8.

Pixton, S.W., Warburton, S., Hill, S.T., 1975. Longterm storage of wheat. III. Some changes in the quality of wheat observed during 16 years of storage. J. Stored Prod. Res. 11, 177–185.

Preiss, J., Levi, C., 1979. Metabolism of starch in leaves. Encycl. Plant Physiol. New Ser. 6, 282–312.

Preston, K.R., Kruger, J.E., 1976a. Location and activity of proteolytic enzymes in developing wheat kernels. Can. J. Plant Sci. 56, 217–223.

Preston, K.R., Kruger, J.E., 1976b. Purification and properties of two proteolytic enzymes with carboxypeptidase activity in germinated wheat. Plant Physiol. 58, 516–520.

Preston, K.R., Kruger, J.E., 1977. Specificity of two isolated wheat carboxypeptidases. Phytochemistry 16, 525–528.

Price, P.B., Parsons, J.G., 1975. Lipids of seven cereal grains. J. Am. Oil Chem. Soc. 52, 490–493.

Pueyo, J.J., Hunt, D.C., Chrispeels, M.J., 1993. Activation of bean (Phaseolus vulgaris) α-amylase inhibitor requires proteolytic processing of the proprotein. Plant Physiol. 101, 1341–1348.

Pusztai, A., Grant, G., Duguid, T., Brown, D.S., Peumans, W.J., Van Damme, E.J.M., Bardocz, S., 1995. Inhibition of starch digestion by α-amylase inhibitor reduces the efficiency of utilization of dietary proteins and lipids and retards the growth of rats. J. Nutr. 125, 1554–1562.

Pusztai, A., Grant, G., Buchan, W.C., Bardocz, S., de Carvalho, A.F.F.U., Ewen, S.W.B., 1998. Lipid accumulation in obese Zucker rats is reduced by inclusion of raw kidney bean (Phaseolus vulgaris) in the diet. Br. J. Nutr. 79, 213–221.

Pyo, Y.-H., Lee, T.-C., Lee, Y.-C., 2005. Effect of lactic acid fermentation on enrichment of antioxidant properties and bioactive isoflavones in soybean. J. Food Sci. 70 (3), S215–S220.

Rahnotra, G.S., Loewe, R.J., Lehmann, T.A., 1977. Breadmaking quality and nutritive of sprouted wheat. J. Food Sci. 42, 1373–1375.

Rathjen, J.R., Strounina, E.V., Mares, D.J., 2009. Water movement into dormant and non-dormant wheat (Triticum aestivum L.) grains. J. Expt. Bot. 60 (6), 1619–1631.

Raynes, J.G., Briggs, D.E., 1985. Genotype and the production of α-amylase in barley grains germinated in the presence and absence of gibberellic acid. J. Cereal Sci. 3, 55–65.

Recondo, E., Leloir, L.F., 1961. Adenosine diphosphate glucose and starch synthesis. Biochem. Biophys. Res. Commun. 6, 85–88.

Reddy, L.V., Cling, T.M., Metzer, R.J., 1984a. Alpha–amylase activity in wheat kernels matured and germinated under different temperature conditions. Cereal Chem. 61, 228–231.

Reddy, N.R., Pierson, M.D., Sathe, S.K., Salunkhe, D.K., 1984b. Chemical, nutritional and physiological aspects of dry bean carbohydrates — a review. Food Chem. 13, 25–68.

Riblett, A.L., Herald, T.J., Schmidt, K.A., Tilley, K.A., 2001. Characterization of β-conglycinin and glycinin soy protein fractions from four selected soybean genotypes. J. Agric. Food Chem. 49, 4983–4989.

Rickert, D.A., Johnson, L.A., Murphy, P.A., 2004. Functional properties of improved glycinin and β-conglycinin fractions. J. Food Sci. 69, 303–311.

Rijven, A.H.G.C., Gifford, R.M., 1983. Accumulation and conversion of sugars developing wheat grains. 3. Non-diffusional uptake of sucrose, the substrate preferred by endosperm slices. Plant Cell Environ. 6, 417–425.

Robyt, J.F., 1984. Enzymes in the hydrolysis and synthesis of starch. In: Whistler, R.L., BeMiller, J.N., Paschall, E.F. (Eds.), Starch: Chemistry Technology, second ed. Academic Press, New York.

Rochfort, S., Panozzo, J., 2007. Phytochemicals for health, the role of pulses. J. Agric. Food Chem. 55, 7981–7994.

Rodriguez, C., Frias, J., Vidal-Valverde, C., Hernandez, A., 2008. Correlations between some nitrogen fractions, lysine, histidine, tyrosine, and ornithine contents during the germination of peas, beans, and lentils. Food Chemistry 108 (1), 245–252.

Saio, K., Baba, K., 1980. Microscopic observation on soybean structural changes in storage [abstract]. Nippon Shokuhin Kogyo Gakkaishi 27, 343.

Saio, K., Nikkuni, I., Ando, Y., Otsuru, M., Terauchi, Y., Kito, M., 1980. Soybean quality changes during model storage studies. Cereal Chem. 57, 77–82.

Saio, K., Kobayakawa, K., Kito, M., 1982. Protein denaturation during model storage studies of soybeans and meals. Cereal Chem. 59, 408–412.

Sajilata, M.G., Singhal, R.S., Kulkarni, P.R., 2006. Resistant starch — a review. Compr. Rev. Food Sci. F. 5, 1–17.

Salmeron, J., Ascherio, A., Rimm, E.B., Colditz, G.A., Spiegelman, D., Jenkins, D.J., Stampfer, M.J., Wing, A.L., Willett, W.C., 1997. Dietary fiber, glycemic load, and risk of NIDDM in men. Diabetes Care 20, 545–550.

Salunkhe, D.K., Kadam, S.S., Chavan, J.K., 1985. Postharvest Biotechnology of Food Legumes. CRC Press, Boca Raton, FL. 29–52.

Sandhu, S.S., Lim, S.-T., 2007. Digestibility of legume starches as influenced by their physical and structural properties. Carbohydr. Polym. 71, 245–252.

Sangronis, E., Machado, C.J., 2007. Influence of germination on the nutritional quality of *Phaseolus vulgaris* and *Cajunis cajan*. *LWT J. Food Sci. Technol.* 40, 116—120.

Santana, M.A., Pihakaski-Maunsbach, K., Sandal, N., Marcker, K.A., Smith, A.G., 1998. Evidence that the plant host synthesizes the heme moiety of leghemoglobin in root nodules. Plant Physiol. 116, 1259—1269.

Sargeant, J.G., 1979. The α-amylase isoenzymes of developing and germinating wheat grain. In: Laidman, D.L., Wyn Jones, R.G. (Eds.), Recent Advances in the Biochemistry of Cereals. Academic Press, New York, pp. 339—343.

Sargeant, J.G., 1980. α-Amylase isoenzymes and starch degradation. Cereal Res. Commun. 8, 77—85.

Sathe, S.K., Deshpande, S.S., Salunkhe, D.K., 1984. Dry beans of *Phaseolus*. A review. Part 2. Chemical composition: carbohydrates, fiber, minerals, vitamins, and lipids. Crit. Rev. Food Sci. 21 (1), 41—93.

Saulniera, L., Sadoa, P.-E., Branlardb, G., Gilles, C., Guillona, F., 2007. Wheat arabinoxylans: exploiting variation in amount and composition to develop enhanced varieties. J. Cereal Sci. 46, 261—281.

Schilstra, M.J., Veldink, G.A., Vliegenhart, J.F.G., 1994. The dioxygenation rate in lipoxygenase catalysis is determined by the amount of iron(III) lipoxygenase in solution. Biochemistry 33, 3974—3979.

Schubert, S., 1995. Nitrogen assimilation by legumes — process and ecological limitations. Fert. Res. 42, 99—107.

Schulz, A., Stephan, H., 1960. Untersuchungen unber eine zweckmassige Verarbeitung auswuchsgeschadigter Roggenmehle. Brot Gebaeck 14, 240.

Serraj, R., Sinclair, T.R., Purcell, L.C., 1999. Symbiotic N_2 fixation response to drought. J. Exp. Bot. 50, 143—155.

Sessa, D.J., 1979. Biochemical aspects of lipid-derived flavors in legumes. J. Agric. Food Chem. 27, 234—239.

Shewry, P.R., Field, J.M., Kirkman, M.A., Faulks, A.J., Miflin, B.J., 1980. The extraction, solubility and characterization of two groups of barley storage polypeptides. J. Exp. Bot. 31, 393—407.

Shewry, P.R., Miflin, B.J., Forde, B.C., Bright, S.W.J., 1981. Conventional and novel approaches to the improvement of the nutritional quality of cereal and legume seeds. Sci. Prog. (Oxf.) 67, 575—600.

Shimoni, E., 2004. Stability and shelf life of bioactive compounds during food processing and storage: soy isoflavones. J. Food Sci. 69 (6), R160—R166.

Siddhuraju, P., Becker, K., 2005. Nutritional and antinutritional composition, *in vitro* amino acid availability, starch digestibility and predicted glycemic index of differentially processed mucuna beans (*Mucuna Pruriens* var. *utilis*): an under-utilised legume. Food Chem. 91, 275—286.

Simmonds, D.H., 1972. The ultrastructure of the mature wheat endosperm. Cereal Chem. 49, 212—222.

Simmonds, D.H., O'Brien, T.P., 1981. Morphological and biochemical development of the wheat endosperm. Adv. Cereal Sci. Technol. 4, 5—70.

Singh, R., Axtell, J.D., 1973. High lysine mutant gene (hI) that improves protein quality and biological value of grain sorghum. Crop. Sci. 13, 535—539.

Singh, R., Juliano, B.O., 1977. Free sugars in relation to starch accumulation in developing rice grain. Plant Physiol. 59, 417—421.

Singh, S., Singh, H.D., Sikka, K.C., 1968. Distribution of nutrients in the anatomical parts of common Indian pulses. Cereal Chem. 45, 13—18.

Slavin, J.L., 2000. Mechanisms for the impact of whole grain foods on cancer risk. J. Am. Coll. Nutr. 19, 300S—307S.

Slavin, J.L., Martini, M.C., Jacobs Jr., D.R., Marquart, L., 1999. Plausible mechanisms for the protectiveness of whole grains. Am. J. Clin. Nutr. 70, 459S—463S.

Slightom, J.L., Sun, S.M., Hall, T.C., 1983. Complete nucleotide sequence of a French bean storage protein gene: phaseolin. Proc. Natl. Acad. Sci. U.S.A. 80, 1897—1901.

Smith, W.L., Lands, W.E.M., 1972. Oxygenation of unsaturated fatty acids by soybean lipoxygenase. J. Biol. Chem. 247, 1038—1047.

Spillane, P.A., Pelhate, J., 1982. Changes in quality and in microbiological activity during extended storage of high-moisture grain. Cereal Foods World 27, 107—111.

Srinivas, V.R., Reddy, G.B., Ahmad, N., Swaminathan, C.P., Mitra, N., Surolia, A., 2001. Legume lectin family, the 'natural mutants of the quaternary state', provide insights into the relationship between protein stability and oligomerization. Biochem. Biophys. Acta. 1527, 102—111.

Staswick, P.E., Hermodson, M.A., Nielsen, N.C., 1981. Identification of the acidic and basic subunit complexes of glycinin. J. Biol. Chem. 256, 8752—8755.

Tanaka, K., Sugimoto, T., Ogawa, M., Kasai, Z., 1980. Isolation and characterization of two types of protein bodies in rice endosperm. Agric. Biol. Chem. 44, 1633—1639.

Taverner, R.J.A., Laidman, D.L., 1972. The induction of lipase activity in the germinating wheat grain. Phytochemistry 11, 989—997.

Tester, R.F., Morrison, W.R., Ellis, R.H., Piggot, J.R., Batts, G.R., Wheeler, T.R., Morison, J.I.L., Hadley, P., Ledward, D.A., 1995. Effects of elevated growth temperature and carbon dioxide levels on some physicochemical properties of wheat starch. J. Cereal Sci. 22, 63—71.

Thomas, B., Lukow, G., 1969. Starch degradation during dough raising with regard to α-amylase. Brot Gebaeck 23, 24.

Tkachuk, R., 1979. Free amino acids in germinated wheat. J. Sci. Food Agric. 30, 53—58.

Tombs, M.P., 1967. Protein bodies of the soybean. Plant Physiol. 42, 797—813.

Topping, D., 2007. Cereal complex carbohydrates and their contribution to human health. J. Cereal Sci. 46, 220—229.

Tovar, J., Melito, C., 1996. Steam-cooking and dry-heating produce resistant starch in legumes. J. Agric. Food Chem. 44, 2642—2645.

Triplett, B.A., Quatrano, R.S., 1982. Timing localization and control of wheat germ agglutin synthesis in developing wheat embryos. Dev. Biol. 91, 491—496.

Truswell, A.S., 2002. Cereal grains and coronary heart disease. Eur. J. Clin. Nutr. 56, 1—14.

Tsai, C.Y., Dalby, A., Jones, R.A., 1975. Lysine and tryptophan increases during germination of maize seed. Cereal Chem. 52, 356—360.

Tzen, J.T.C., Huang, A.H.C., 1992. Surface structure and properties of plant seed oil bodies. J. Cell Biol. 117, 327—335.

Urbano, G., Aranda, P., Vilchez, A., Aranda, C., Cabrera, L., Porres, J.M., Lopez-Jurado, M., 2005. Effects of germination on the composition and nutritive value of proteins in *Pisum sativum*, L. Food Chem. 93, 671—679.

Urquardt, A.A., Altosaar, I., Matlashewski, G.J., 1983. Localization and lipase activity in oat grains and milled oat fractions. Cereal Chem. 60, 181—183.

Urquardt, A.A., Brumell, C.A., Altosaar, I., Matlashewski, G.J., Sahasrabudhe, M.R., 1984. Lipase activity in oats during grain maturation and germination. Cereal Chem. 61, 105—108.

US Department of Agriculture. 1957. *Grain Grading Primer*. Misc. Publ. No. 740. USDA, Washington, DC, p. 32.

US Department of Agriculture, Agricultural Research Service. 2008. USDA National Nutrient Database for Standard Reference, Release 21. Nutrient Laboratory Home Page. Available at http://www.ars.usda.gov/ba/bhnrc/ndl

Vaidehi, M.P., Kadam, S.S., 1989. Soybean. In: Salunkhe, D.K., Kadam, S.S. (Eds.), CRC Handbook of World Food Legumes: Nutritional Chemistry, Processing Technology, and Utilization, vol. 3. CRC Press, Boca Raton, FL, pp. 1–31.

Vanderstoep, J., 1981. Effect of germination on the nutritive value of legumes. Food Technology 35 (3), 83–85.

Varty, K., Arreguin, B., Gomez, M.T., Lopez, P.J.T., Gomez, M.A.L., 1982. Effects of abscisic acid and ethylene on the gibberellic acid-induced synthesis of α-amylase by isolated wheat aleurone layers. Plant Physiol. 73, 692–697.

Viana, S.F., Guimaraes, V.M., Jose, I.C., Goreti de Amneida e Oliveira, M., Brunoro Costa, N.M., Goncalves de Barros, E., Moreira, M.A., Tavares de Rezende, S., 2005. Hydrolysis of oligosaccharides in soybean flour by soybean α-galactosidase. Food Chem. 93, 665–670.

Vieweg, G.H., De Fekete, M.A.R., 1976. The effect of phospholipids on starch metabolism. Planta. 129, 155–222.

Vitali, D., Dragojevića Vedrina, I., Šebečića, B., 2008. Bioaccessibility of Ca, Mg, Mn and Cu from whole grain tea-biscuits: impact of proteins, phytic acid and polyphenols. Food Chem. 110, 62–68.

Wang, L., Wang, T., Fehr, W.R., 2006. Effect of seed development stage on sphingolipid and phospholipid contents in soybean seeds. J. Agric. Food Chem. 54, 7812–7816.

Westermarck-Rosendahl, C., Junnila, L., Koivistionen, P., 1979. Efforts to improve the baking properties of sprout-damaged wheat by reagents reducing α-amylase activity. I. Screening tests by the falling number method. Lebensm. Wiss. Technol. 12, 321–324.

Westermarck-Rosendahl, C., Junnila, L., Koivistoinen, P., 1980. Efforts to improve the baking properties of sprout-damaged wheat by reagents reducing α-amylase activity. III. Effects on technological properties of flour. Lebensm. Wiss. Technol. 13, 193.

Whelan, W.J., 1958. In: Ruhiand, W. (Ed.), Handbuch der Pflanzenphysiologie, vol. 6. Springer, Berlin, p. 154.

Whelan, W.J., 1963. Recent advances in starch metabolism. Staerke 15, 247–251.

Yanagi, S.O., Galeazzi, M.A.M., Saio, K., 1985. Properties of soybean in model storage studies. Agric. Biol. Chem. 49, 525–528.

Yoshida, H., Saiki, M., Yoshida, N., Tomiyama, Y., Mizushina, Y., 2009. Fatty acid distribution in triacylglycerols and phospholipids of broad beans (*Vicia faba*). Food Chem. 112, 924–928.

Yuan, S., Chang, S.K., 2007. Selected odor compounds in soymilk as affected by chemical composition and lipoxygenases in five soybean materials. J. Agric. Food Chem. 55, 426–431.

Zhu, D., Hettiarachchy, N.S., Horax, R., Chen, P., 2005. Isoflavone contents in germinated soybean seeds. Plant Food Hum. Nutr. 60, 147–151.

Chapter 2

Fruits and Vegetables

N. A. Michael Eskin* and Ernst Hoehn[†]

*Department of Human Nutritional Sciences, Faculty of Human Ecology, University of Manitoba, Winnipeg, Manitoba, Canada, [†]Swiss Federal Research Station, Switzerland

Chapter Outline

I. INTRODUCTION

Characteristics of fruits and vegetables such as flavor, color, size, shape, and absence of external defects ultimately determine their acceptance by consumers. The development of these characteristics is the result of many chemical and biochemical changes that occur following harvesting and storage. Since harvesting fruits and vegetables at their correct stage of maturity is critical for the development of a highly acceptable product for the fresh market, or for processing, it is important to understand more fully what changes are taking place. This chapter will highlight those changes occurring within fruits and vegetables during the postharvest period. It is during this period that fruits and

Biochemistry of Foods. DOI: http://dx.doi.org/10.1016/B978-0-12-242352-9.00002-3

vegetables show a gradual reduction in quality concurrent with transpiration and respiration, as well as with other biochemical and physiological changes. Ultimately the plant material deteriorates because of the undesirable enzyme activity and spoilage microorganisms.

The growth and maturation of fruits and vegetables are dependent on photosynthesis and absorption of water and minerals by the parent plant. Once detached, however, they are independent units in which respiratory processes play a major role. This chapter will focus on those changes in postharvest fruits and vegetables that affect quality.

II. RESPIRATION

Respiration is the fundamental process whereby living organisms carry out the exothermic conversion of potential energy into kinetic energy. In higher plants the major storage products are sucrose and starch. These are completely oxidized in the presence of oxygen to carbon dioxide and water, with the production of adenosine triphosphate (ATP):

$$C_6H_{12}O_6 + 6O_2 \rightarrow 6CO_2 + 6H_2O + \text{energy (heat and ATP)}$$

The latter is the form in which energy is stored within the cell. The contribution of proteins and lipids to plant respiration is difficult to assess but can occur via the formation of acetyl-coenzyme A (CoA). In the absence of oxygen, anaerobic respiration occurs, resulting in only a partial degradation of carbohydrates and a lower ATP production.

The metabolic pathways involved in the respiration of plant tissue result in the conversion of starch or sucrose to glucose-6-P. The latter is then oxidized by glycolysis (Embden−Meyerhoff pathway) or the pentose phosphate pathway to triose phosphate, which enters the tricarboxylic acid cycle by way of pyruvate (Scheme 2.1) (ap Rees, 1977). Finally, in a third stage, oxidative phosphorylation converts NADH and FADH$_2$ into chemical energy in the form of ATP (Browse et al., 2006).

The contribution of these two major pathways of carbohydrate oxidation to plant respiration remains unresolved. Difficulties were encountered with the experimental techniques used in assessing the relative roles of these pathways based on the production of $^{14}CO_2$ or labeled intermediates from labeled hexoses (ap Rees, 1980). Evidence shows that both pathways exist in plant tissues (ap Rees, 1974) and that they change considerably during plant development (ap Rees, 1977). Current evidence supports the glycolytic pathway as the predominant one operating, while the maximum contribution of the pentose phosphate pathway may not exceed 30% of the total (ap Rees, 1980). The relative importance of these pathways probably depends on the particular plant, the organ, and the state of maturity.

Respiration rates of fruits and vegetables are affected by many environmental factors. In cases where this leads to negative effects on plant tissue it is defined as stress. During the storage of fruits and vegetables (Section VIII) effects of low temperatures, reduction in oxygen (O_2) concentration and increase in carbon dioxide (CO_2) concentration in the storage atmosphere are utilized to extend the storage life of produce. However, maintaining an adequate energy status is required to prevent browning or senescence of harvested fruits and vegetables (Saquet et al., 2000, 2003a; Xuan et al., 2005; Song et al., 2006; Jiang et al., 2007). It is well established that lowering of O_2 concentrations during controlled atmosphere storage reduces respiration rates and energy supply and that severe limitations of O_2 induce fermentative (anaerobic) respiration and metabolism in stored produce (Scheme 2.1). The net yield of ATP during anaerobic respiration is only 2 moles of ATP of hexose sugar compared with 36 moles of ATP per mole of hexose in aerobic respiration. Hence, energy status may be insufficient and provoke storage disorders (Jiang et al., 2007).

A. Fruits

A large number of fruits exhibit a sudden sharp rise in respiratory activity following harvesting, referred to as the climacteric rise in respiration. This phenomenon was first noted by Kidd and West (1922, 1930a) as an upsurge in carbon dioxide gas at the end of the maturation phase of apples. Since then there have been numerous reports on this phenomenon in a wide range of fruits. The appropriateness of the term climacteric was questioned by Rhodes (1970), who suggested that it should be all inclusive and describe the 'whole of the control phase in the life of fruit triggered by ethylene and the concomitant changes occurring'. McGlasson et al. (1978), however, suggested that respiratory climacteric was the more appropriate term to describe this gaseous phenomenon. Biale and Young (1981) nevertheless still preferred the more inclusive description in which climacteric defined those physical, chemical, physiological, and metabolic changes associated with the increased rate of respiration covering the transition phase from growth and maturation to the final stages of senescence. Essentially, climacteric defines the last stages of the fruit at the cellular level, which determine the quality of the fruit that is shipped to the consumer.

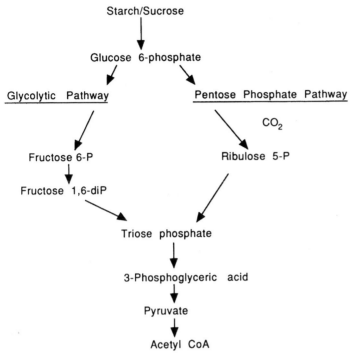

SCHEME 2.1 Glycolytic and pentose phosphate pathways.

Biale (1960a, b) tentatively classified fruits as either climacteric or non-climacteric according to their respiratory rates. A later review by Biale and Young (1981), however, suggested a more extensive list of fruits from both groups as shown in Table 2.1. Eventually fruits such as cantaloupe, honeydew melon, and figs were included, all of which are considered climacteric (Lyons *et al.*, 1962; Pratt and Groeschel, 1968; Marei and Crane, 1971). A few rare fruits were also added, namely, breadfruit (Biale and Barcus, 1970), guavas, and mammee apples (Akamine and Goo, 1978, 1979a, b; Saltveit, 2004).

The period immediately prior to the climacteric rise, when the respiratory level is at a minimum, is known as the preclimacteric. Following the completion of the climacteric rise is the postclimacteric phase, in which a decline in the respiratory rate occurs. Unlike the sudden rise in respiratory activity which characterizes climacteric fruits, non-climacteric fruits exhibit a steady fall in respiratory activity. This downward trend in the respiratory activity was originally observed for lemons stored at 15°C by Biale and Young (1947) and later in oranges (Biale, 1960a, b). Figure 2.1 illustrates the difference in respiratory activity between climacteric and non-climacteric fruits, for example, avocados and lemons.

In the original classification by Biale (1960a, b), oranges were classified as non-climacteric fruits since they had a low rate of respiration. It was soon evident that some members of the citrus fruit family had respiratory activities similar to those of climacteric fruits. In spite of the higher respiratory rates, a downward trend was observed for both 'Valencia' and 'Washington Navel' cultivars. This decline in respiration was observed by Bain (1958) in 'Valencia' oranges from fruit set to maturity. Nevertheless, Trout *et al.* (1960) reported a typical respiratory rise in oranges stored at 4.3−10°C. Their results were attributed by Biale and Young (1981) to possible chilling injury at the lower storage temperatures. Only half of those oranges kept at 10°C exhibited a rise in respiration following harvesting, which could be due to the immaturity of some of these fruits. Aharoni (1968) also used the term climacteric to describe the increase in respiratory rate in postharvest, young, unripe 'Washington Navel', 'Shamouti', and 'Valencia' oranges, as well as in Marsh's seedless grapefruit stored at 16°C and 20°C. In contrast, the full-sized and mature fruit did not exhibit any rise in respiratory activity. Eaks (1970) examined the respiratory patterns of several species of citrus fruit throughout ontogeny. The small and immature oranges and grapefruit exhibited a rise in respiration and ethylene activity when stored at 20°C for several days after harvest. As the weight of the fruit increased, characteristic of maturation, the level of carbon dioxide and ethylene production decreased until full maturity was approached or attained, at which point no change in respiration was noted. Based on this study it was

TABLE 2.1 Respiratory Activity of Selected Fruit

Climacteric	Non-climacteric
Apple	Blueberry
Apricot	Grape
Avocado	Grapefruit
Banana	Java plum
Breadfruit	Lemon
Fig	Olive
Guava	Orange
Mammee apple	Pineapple
Muskmelon cantaloupe	Strawberry
	Honeydew melon

Adapted from Biale and Young (1981).

evident that citrus fruits were correctly classified as non-climacteric. Rhodes (1970) noted that had the climacteric been defined as the period of enhanced metabolic activity during the transition from the growth phase to senescence in fruits, the confusion with citrus fruits could have been avoided. A similar situation was observed in grapes, in which a respiratory rise was reported by Peynaud and Riberau-Gayon (1971) to accompany rapid growth, which was referred to as 'rudimentary climacteric'. However, this was resolved when the postmaturation change in respiratory activity showed a typical non-climacteric pattern consistent with earlier work by Geisler and Radler (1963). A similar controversy arose with respect to the pineapple and while Dull *et al.* (1967) found a slight upward respiratory trend it was not typical of climacteric fruit. The identification of several tomato mutants by Herner and Sink (1973) without any climacteric pattern was later reviewed by Tigchelaar *et al.* (1978a, b). These mutants were unable to produce ethylene and had low levels of carotenoids. A particular feature was the extremely low levels of polygalacturonase, which accounted for their prolonged firmness.

B. Vegetables

Once the vegetable becomes detached from the parent plant, metabolism continues to take place, although it is the catabolic reactions that soon become dominant. The climacteric rise in respiration characteristic of certain fruits such

FIGURE 2.1 Respiratory trends in climacteric fruit, exemplified by the avocado, compared with non-climacteric fruit, depicted by the lemon (Biale *et al.*, 1954). *[Reprinted with permission of the copyright owner, American Society of Plant Physiology (ASPP).]*

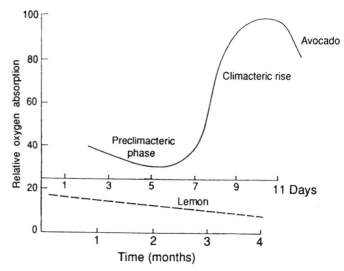

as apples and avocados is not apparent in vegetables, where there is no clear-cut division between maturation and breakdown.

The intensity and rate of respiration vary with the particular plant, the degree of maturity, and whether the vegetable is actively growing at the time of harvest or functioning as a storage organ. For example, McKenzie (1932) reported higher respiration intensity in freshly harvested immature lettuce (*Lactuca sativa*) during the first 12 hours, which then fell to the same level as that of mature lettuce. The deterioration of several vegetables was examined by Platenius (1942), who found that the initial respiratory rate for asparagus (*Asparagus officinalis*) at 24°C was almost 50 times greater than for potatoes. The respiration rate declined for all vegetables during 60 days' storage irrespective of the temperature. The initial rates of respiration appeared to be a useful indicator of the potential storage life of the crop during precooling and early storage. A high respiratory rate, however, was indicative of a short storage life, while the reverse was true for crops with a low respiratory rate. This is illustrated for a number of vegetables in Figure 2.2.

Vegetables and fruits can be classified as extremely high, very high, high, moderate, low, or very low according to the respiration rates (Saltveit, 2004). For example, young tissues, such as growing parts of asparagus or developing seeds of green peas, have high respiration rates, while low rates are evident in storage organs such as stems (potatoes) and bulbs (onions). Leafy vegetables appear to be moderate, while some vegetables such as cabbage can be stored at a low temperature for considerable periods. Other vegetables, including cucumbers (*Cucumis sativa*), are particularly susceptible to chilling injury if stored over a temperature range of 0−10°C (Eaks and Morris, 1956). Many other commodities, in particular those originating in the tropics or subtropics, are prone to chilling injuries when their temperature falls below 10−12°C (Wang, 1989; Saltveit, 2004). Chilling injury is diagnosed by an increase in respiration in which a plateau is reached corresponding to chilling, after which there is a decline in respiration.

1. Control of the Climacteric Rise

The dramatic upsurge in respiratory activity associated with the climacteric has been attributed to a number of different factors. One theory proposed was the breakdown in 'cell membrane permeability or organization resistance' (Solomos and Laties, 1973). While such changes are evident during ripening, the question remains as to whether they are the cause or the consequence of the ripening process (Theologis and Laties, 1978). The second theory focused on increased protein synthesis as a necessary prerequisite for the ripening process (Brady *et al.*, 1976; Richmond and Biale, 1966) or enhanced ATP turnover with respiratory stimulation (Biale, 1960b). Subsequent research, however, showed there was no difference in the respiratory capacity of mitochondria obtained from preclimacteric and climacteric avocado tissue (Biale, 1969). In addition, the respiratory rate of uncoupled preclimacteric avocado slices

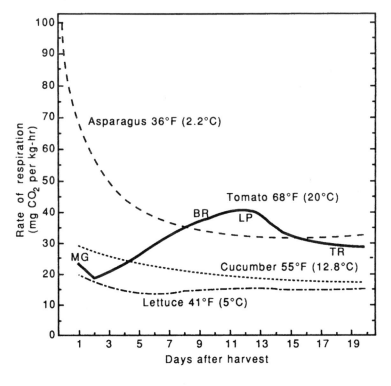

FIGURE 2.2 Rate of respiration of a shoot (asparagus), a leafy vegetable (head lettuce), a non-ripening fruit (cucumber), and a ripening fruit (tomato) at temperatures commonly encountered during their marketing (for tomato: MG = mature green; BR = breaker; LP = light pink; TR = table red) (data taken from Lipton, 1977; Pratt *et al.*, 1954; Workman *et al.*, 1957) (Ryall and Lipton, 1979).

was quite sufficient to facilitate the climacteric rise (Millerd *et al.*, 1953). This was confirmed in studies using isolated mitochondria from avocado (Biale *et al.*, 1957). Further research by Lance *et al.* (1965) and Hobson *et al.* (1966), using improved techniques, isolated mitochondria from avocado throughout all stages of the climacteric and found that the oxidative and phosphorylating activities remained unchanged provided all cofactors were present. The stimulation of glycolysis in avocado fruit under anaerobic conditions also demonstrated considerable latent glycolytic capacity (Solomos and Laties, 1974). These studies indicated the adequacy of the enzymes in preclimacteric fruits to sustain the respiratory climacteric. This was consistent with studies by Frenkel *et al.* (1968) and McGlasson *et al.* (1971), who found that while inhibiting protein synthesis prevented the ripening of intact peas and banana slices, it did not decrease the upsurge in respiration.

The dramatic burst in respiratory activity that accompanies the ripening of climacteric-type fruits appears to be due to some change in the mitochondrial respiratory function *in vivo* (Biale, 1960a, b). A possible explanation for this was attributed to an increase in cyanide-resistant respiration (Solomos and Laties, 1976; Solomos, 1977). Solomos and Laties (1974, 1976) observed that cyanide initiated identical physiological and biochemical changes in avocado and potato tubers. Cyanide is a known inhibitor of cytochrome oxidase, the terminal oxidase in the electron transport system. Thus, there is present a cyanide-insensitive pathway which permits the aerobic oxidation of respiratory substrates in the presence of cyanide (Bendall and Bonner, 1971). This cyanide-resistant pathway or alternative pathway was reported to be present in ethylene-responsive fruits (Solomon and Laties, 1974, 1976). Further research by Theologis and Laties (1978) led to an examination of this pathway in the respiration of ripening avocados and bananas. These researchers found that the surge in respiration during the climacteric in the intact fruit was cytochrome mediated. The preclimacteric fruit had the capacity to sustain electron transport through the cytochrome pathway, although it remained unexpressed. During the climacteric rise, the alternative pathway appears to remain at a low level of activity and may be involved in the generation of peroxide (Rich *et al.*, 1976). The enzyme responsible for the cyanide-resistant respiration is called the alternative oxidase (Vanlerbergh and McIntosh, 1997). The function of the alternative pathway is not completely resolved, but it may diminish detrimental effects of stress on respiration (Tucker, 1993; Wagner and Krab, 1995; Møller, 2001). A discussion of the possible regulatory role of cyanide in ethylene biosynthesis can be found in Section IV.

2. Enzymatic Control

The possibility of enzymatic activity being the controlling factor in the climacteric has been suggested. Tager and Biale (1957) noted a rise in carboxylase and aldolase activity during the ripening of bananas accompanied by a shift from the pentose phosphate to the glycolytic pathway. This may occur during the transition period from the preclimacteric to the postclimacteric phase in fruit ripening.

a. Malic Enzyme

Hulme *et al.* (1963) reported a sharp increase in the activities of malic enzyme and pyruvate carboxylase during the ripening of apples. This explained the slight uptake of oxygen during ripening of apples compared to the marked increase in carbon dioxide evolution.

$$
\begin{array}{ccc}
\text{COOH} & & \text{CH}_3 \\
| & \text{NADP}^+ \quad \text{NADPH} + \text{H}^+ & | \\
\text{CH}_2 & \xrightarrow{\quad\quad\quad\quad} & \text{C}{=}\text{O} + \text{CO}_2 \\
| & \text{Malic dehydrogenase} & | \\
\text{CHOH} & & \text{COOH} \\
| & & \text{Pyruvate} \\
\text{COOH} & & \\
\text{Malic acid} & &
\end{array}
$$

Several studies showed that malic enzyme exhibited varying degrees of cyanide insensitivity depending on the activity of the enzyme (Lance *et al.*, 1967; Macrae, 1971; Coleman and Palmer, 1972; Neuburger and Douce, 1980). Moreau and Romani (1982) examined the oxidation of malate during the climacteric rise in avocado mitochondria with particular focus on the cyanide-insensitive alternative pathway. The increase in malic enzyme activity paralleled the increase in malate oxidation as ripening advanced through the climacteric. Malate is oxidized by malic dehydrogenase through the cytochrome pathway. It can also be oxidized by the malic enzyme, which involves the alternative pathway via a rotenone-insensitive NADH dehydrogenase located in the inner layer of the mitochondrial membrane (Palmer, 1976; Marx and Brinkmann, 1978; Rustin *et al.*, 1980). These researchers concluded that the malic enzyme and alternative oxidase pathway probably function under conditions of relatively low ATP demands and high energy change characteristics at the later stages of the climacteric. While the regulation of electron transport via cytochrome and alternative pathways remains to be clarified in avocado mitochondria, the involvement of the alternative pathway cannot be totally ruled out.

b. Phosphofructokinase and Pyrophosphate: Fructose-6-Phosphate Phosphotransferase

Salimen and Young (1975) examined the possibility that the climacteric was regulated by enzyme activation involving phosphofructokinase (PFK) (ATP:D-fructose-6-phosphate-1-phosphotransferase, EC 2.7.1.11). This was based on research by Barker and Solomos (1962), who observed an increase in fructose 1,6-diphosphate during the ripening of bananas and in tomatoes by Chalmers and Rowan (1971). This increase in fructose 1,6-diphosphate was attributed to activation of PFK. Salimen and Young (1975) reported that activation of this enzyme accounted for a 20-fold increase in fructose 1,6-diphosphate during the ripening process. Electrophoretic separation of PFK showed that no new species of the enzyme were produced during the climacteric, with the enzyme remaining in the oligomeric form. Rhodes (1971) reported that PFK was present in the oligomeric weight form species in tomato fruits up to the climacteric phase while both the oligomeric and low-molecular-weight species were isolated in the postclimacteric phase. Isaac and Rhodes (1982) later found that PFK existed in the oligomeric form at the 'breaker' stage during tomato ripening. Using gel-permeation chromatography Isaac and Rhodes (1987) identified a single peak corresponding to the oligomeric form of PFK at the green and breaker stages. Two peaks were separated, however, at the orange and red stages of tomato ripening, which corresponded to oligomeric and monomeric forms of the enzyme (Figure 2.3). To explain the behavior of PFK, these researchers proposed that stimulation of the enzyme occurred because of leakage of inorganic phosphate (P_i) from the vacuole as a consequence of permeability changes in the membrane during initiation of the climacteric. The continued leakage of P_i and citrate affected the enzyme at the molecular level by dissociating the oligomeric form of the enzyme into monomeric subunits during the later stages of ripening.

A study by Bennett *et al.* (1987) examined the role of glycolytic regulation of the climacteric in avocado fruit. They used *in vivo* ^{31}P nuclear magnetic resonance spectroscopy to monitor the levels of phosphorylated nucleotides. They focused particular attention on pyrophosphate:fructose-6-phosphate phosphotransferase (PFP), an alternative enzyme identified in pineapple by Carnal and Black (1979). This enzyme catalyzes the identical reaction as PFK, utilizing PP_i instead of ATP as a phosphate donor, and is activated by fructose 2,6-biphosphate. An increase in the amount of fructose 2,6-phosphate concurrently with the rise in respiration suggested to these researchers that PFP may also be involved in the regulation of ripening in avocado fruit.

III. INITIATION OF RIPENING

Ethylene is one of many volatile substances emanating from fruits and vegetables which was subsequently identified by Gane (1934) as the active component for the stimulation of ripening. The application of minute quantities of ethylene, on the order of 1 ppm, stimulates respiratory activity, induces ripening, and hastens the onset of the climacteric. Thus, ethylene was soon recognized as a plant hormone that initiates the ripening process as well as regulating many aspects of plant growth, development, pathogen and wounding responses, senescence, and both abiotic and biotic stress responses (Abeles, 1973; Gazzarrini and McCourt, 2001; Alexander and Grierson, 2002). Based on respiration, fruits can be grouped as climacteric and non-climacteric, which was described in the previous part (Lelievre *et al.*, 1997). Ripening in climacteric fruits (Table 2.1) manifests itself by an upsurge in respiration and a concomitant burst of ethylene. In non-climacteric fruits respiration shows no change and ethylene production remains at a very low level during ripening. In addition to increased respiratory metabolism, ethylene stimulates its own biosynthesis in ripening climacteric fruits (Burg and Burg, 1965). The application of increased levels of ethylene to climacteric fruits hastens the onset of the climacteric rise, accompanied by an increase in oxygen uptake (Figure 2.4). Hence, ethylene is required for ripening because the ripening process can be inhibited by genetic control (Oeller *et al.*, 1991; Theologis *et al.*, 1993; Picton *et al.*, 1993; Ayub *et al.*, 1996; Brummell, 2005) or by ethylene action inhibitors such as 1-methyl-cyclopropene (1-MCP) (Sisler and Serek, 1997). With respect to non-climacteric fruits, an increase in oxygen absorption accompanies the application of ethylene. In the case of climacteric fruits, once ethylene exerts the respiratory rise the process cannot be reversed. This is in sharp contrast to non-climacteric

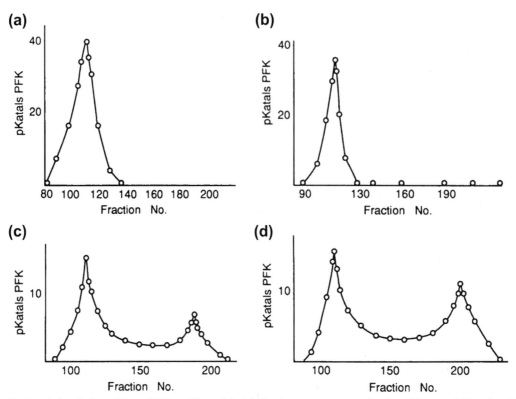

FIGURE 2.3   Elution of phosphofructokinase (PFK) on Ultrogel AcA 34 for the enzyme preparation from **(a) green, (b) breaker, (c) orange, and (d) red tomatoes (Isaac and Rhodes, 1987).** *(Reprinted with permission. Copyright © by Pergamon Press.)*

fruits, in which the respiratory activity returns to the level of the control once ethylene treatment is terminated (Vendrell *et al.*, 2001).

At one time ethylene was considered a by-product rather than a ripening hormone as the amount present during the preclimacteric phase in many fruits was insufficient to stimulate ripening (Biale *et al.*, 1954). This conclusion was based on the amount of ethylene emanating from the fruit rather than the intracellular concentrations and was measured using manometric techniques that were far too insensitive. Subsequent research using gas chromatography provided ample evidence for the presence of ethylene in the intracellular spaces (Burg and Burg, 1965). It is generally accepted that the ethylene levels required to stimulate ripening fall within 0.1–1.0 ppm, completely beyond the range of normal manometric techniques. Table 2.2 summarizes the change in internal ethylene levels during the ripening of some climacteric and non-climacteric fruits. In the case of avocado, mango, and pears the ethylene level before the climacteric rise was lower than the accepted threshold of 0.1 ppm. Biale and Young (1971) noted that the rapid initiation of ripening of avocado required levels of ethylene greater than 1 ppm. These researchers pointed out some 10 years later (Biale and Young, 1981) that it was difficult to generalize about the minimum levels of ethylene needed to induce the climacteric rise because of the scant data available. Peacock (1972) suggested that the effectiveness of ethylene was a function of the log of its concentration, length of exposure, and time of application after harvest. As the fruit approached maturity it was evident that there was a decrease in sensitivity to ethylene.

However, studies by Vendrell and McGlasson (1971) on preclimacteric banana fruit tissue and syca-more figs (Zeroni *et al.*, 1976) reported that exogenous ethylene exerted a negative feedback regulation on ethyl-ene production. Autoinhibition of ethylene production was also reported for non-climacteric fruit such as wounded flavedo tissues of citrus (Riov and Yang, 1982a). On the basis of such observations, McMurchie *et al.* (1972) proposed the concept of system 1 and system 2 ethylene. System 1 is responsible for the low-rate production of ethylene in preclimacteric fruit and for most, if not all, ethylene produced by vegetative tissues. System 2 is accountable for the high-rate production of ethylene observed during the climacteric and is autocatalytically induced by ethylene. The transition to the system 2 is presumably due to the transcriptional activation of different ACC synthase (ACS) and ACC oxidase (ACO) genes (Scheme 2.2) (Barry *et al.*, 2000; Alexander and Grierson, 2002).

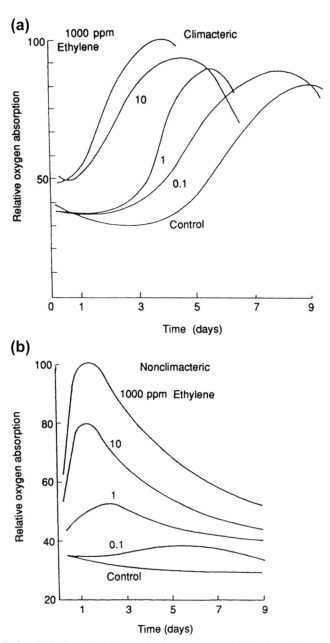

FIGURE 2.4 Oxygen uptake by fruits which show the climacteric phenomenon and by fruits which do not, in relation to concentration of external ethylene (Biale, 1964).

It is evident that ethylene plays a crucial and complex role in the regulation of climacteric fruit ripening. Ethylene affects encoding of many specific enzymes required for ripening, while some other compound(s), as yet unidentified, may induce this in non-climacteric fruits. Hence, it appears that both ethylene-dependent and -independent ripening pathways coexist in climacteric and non-climacteric fruits. Isolation of genes associated with the ripening processes in non-climacteric fruit may lead to progress in elucidating the regulation of fruit ripening (Aharoni *et al.*, 2000; Flores *et al.*, 2002; Li *et al.*, 2011).

IV. BIOSYNTHESIS OF ETHYLENE

Various precursors of ethylene have been proposed but it is well established that methionine appears to be the main one in higher plants (Kende, 1993; Alexander and Grierson, 2002).

TABLE 2.2 Internal Ethylene Content (ppm) in some Climacteric and Non-climacteric Fruit

Fruit	Variety	Preclimacteric	Onset	Climacteric Peak
Climacteric				
Avocado[a]	Fuerte	0.03	0.09	25
Banana[b]	Gros Michel	0.1	1.5	40
Mango[b]	Kent Haden	0.01	0.08	3
Pear[c]	Anjou	0.09	0.4	40
Non-climacteric			Steady state	
Lemon[a]			0.1–0.2	
Orange[a]			0.1–0.2	
Lime[a]			0.3–2.0	

[a]From Akamine and Goo (1979b)
[b]from Burg and Burg (1962)
[c]from Kosiyachinda and Young (1975)

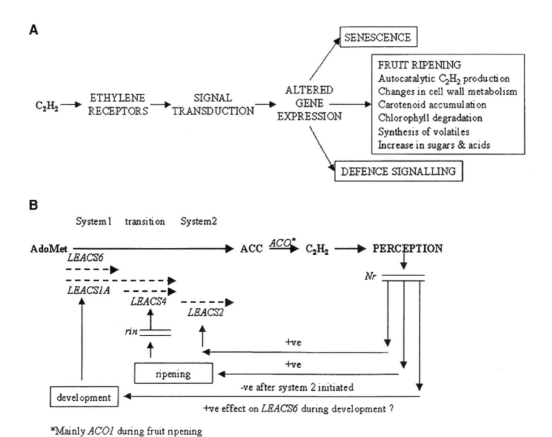

SCHEME 2.2 (A) Schematic representation of the role that ethylene plays in fruit ripening. (B) Model proposing the differential regulation of ACS gene during the transition from system 1 to system 2 ethylene synthesis. The symbols −ve (negative) and +ve (positive) refer to the action of ethylene on signaling pathways resulting in repression (−ve) or stimulation (+ve) of gene expression (Barry *et al.*, 2000; Alexander and Grierson, 2002).

A. Methionine as a Precursor of Ethylene

Lieberman and Mapson (1964) initially examined the production of hydrocarbons, including ethane and ethylene, in model systems containing peroxidized linoleic acid, Cu^{2+}, and ascorbic acid. To test whether ethylene production from linoleic acid involved free radicals they added a free radical quencher, methionine. Instead of methionine inhibiting the reaction they found that the production of ethylene was greatly enhanced. Further work showed that ethylene could be produced in the absence of peroxidized lipids as long as methionine–Cu^{2+}-ascorbate was present (Lieberman et al., 1965). It was soon shown that methionine was in fact the biological precursor of ethylene in plants (Lieberman et al., 1965; Yang, 1974). Using ^{14}C-labeled methionine, Lieberman et al. (1966) demonstrated its conversion to ethylene in apple fruit tissue. The fact that the C1 from methionine yielded carbon dioxide, and C3 and C4 yielded ethylene in both chemical systems and plant tissue, suggested that a common mechanism was involved. The two systems were quite different, however, as methionine was converted via methional with the methyl sulfide group yielding volatile dimethyl sulfide in the model systems. This differed in plant tissue, as methionine is limiting so that the sulfur group is recycled for resynthesis of methionine.

1. The Yang Cycle, Recycling of Methionine

The inhibition of ethylene production from methionine in the presence of DNP, an uncoupler of oxidative phosphorylation, suggested the formation of S-adenosyl-L-methionine (SAM) as an intermediate in this process (Burg, 1973; Murr and Yang, 1975). Using labeled methionine, Adams and Yang (1977) reported that the CH_3-S group of methionine was released as 5-methylthioadenosine (MTA) during ethylene synthesis in apple slices. MTA could only be formed as a degradation product if ethylene was synthesized from SAM. In addition to MTA, these researchers detected 5-methylthioribose (MTR), a degradation product of MTA in apple tissue. This suggested that the CH_3-S unit of MTR combined with a four-carbon receptor, such as homoserine, to form methionine, while the ribose group split off. It was subsequently found that the ribose unit of MTA/MTR was directly incorporated into methionine along with the CH_3-S group. Yung and Yang (1980) demonstrated that three MTR molecules were involved in methionine formation, with the ribose moiety modified to form the 2,3-aminobutyrate portion of methionine while the CH_3-S unit remained intact:

$$
\begin{array}{ccc}
\text{CH}_3 & \text{TPP Mg}^{2+} & \text{CH}_3 \\
| & & | \quad + \text{CO}_2 \\
\text{C}=\text{O} & \xrightarrow{\hspace{2cm}} & \text{CHO} \\
| & \text{Pyruvate carboxylase} & \\
\text{COOH} & &
\end{array}
$$

This pathway explains how methionine is recycled and maintained within plants. The overall pathway involved in the resynthesis of methionine from MTA is shown in Scheme 2.3. MTR-1 phosphate is converted to 2-oxo-4-methylthiobutanoic acid, from which methionine is re-formed. Miyazaki and Yang (1987) examined the methionine cycle enzymes in a number of fruits and showed that the conversion of MTR to methionine in ripening apples was not a limiting factor in the formation of ethylene. This cycle has been named the Yang cycle in plant biochemistry texts (Bradford, 2008).

2. Methionine and Ethylene Biosynthesis

Early studies by Hansen (1942) and Burg and Thimann (1959) showed that ethylene production ceased when apples and pears were stored in an atmosphere of nitrogen. On re-exposure to oxygen, however, the production of ethylene was restored. The rapid production of ethylene suggested the accumulation of an intermediate compound during anaerobic storage. Adams and Yang (1979), using L[U-^{14}C]methionine, identified 1-aminocyclopropane-1-carboxylic acid (ACC) as the intermediate formed in apple fruit stored under nitrogen. It would appear therefore that methionine is first converted to S-adenosylmethionine, which then undergoes fragmentation to ACC and MTA. These researchers also found that labeled ACC was converted to ethylene when the apple tissue was incubated in air, which suggested the following sequence:

$$\text{Methionine} \rightarrow \text{SAM} \rightarrow \text{ACC} \rightarrow \text{Ethylene}$$

The conversion of methionine to SAM involves methionine adenosyl-transferase (ATP:methionine S-adenosyl-transferase, EC 2.5.1.6). This enzyme was reported in plant tissues by Konze and Kende (1979) in relation to ethylene

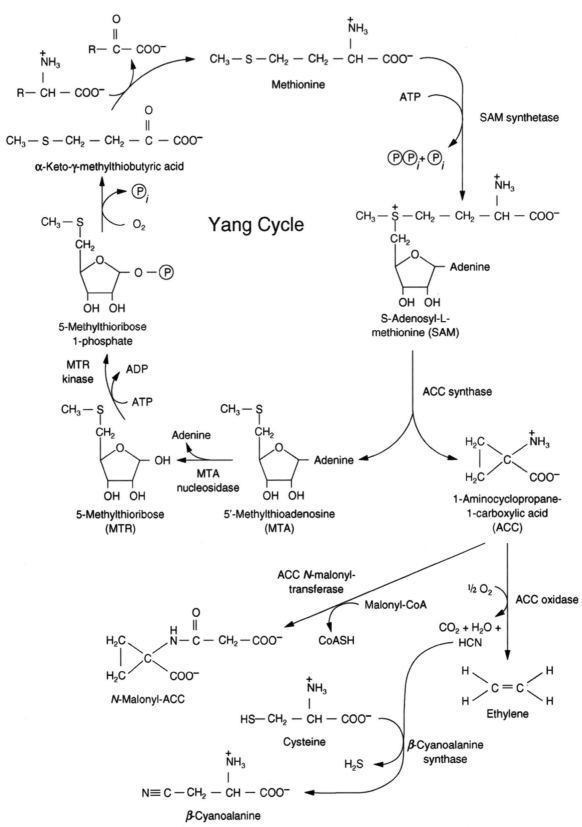

SCHEME 2.3   The Yang cycle and formation of ethylene and other products from ACC (Bradford, 2008).

production. The addition of aminoethoxylvinylglycine (AVG), an inhibitor of pyridoxal phosphate-mediated enzyme reactions (Rando, 1974), was subsequently shown to inhibit ethylene production from methionine. The part of the reaction sequence affected was SAM to ACC, which involved the participation of pyridoxal phosphate (Adams and Yang, 1979). The enzyme involved, ACC synthase, was identified in tomato preparations and shown to be activated by pyridoxal phosphate (Boller *et al.*, 1979; Yu *et al.*, 1979). ACC synthase was later identified and studied in apples (Bufler and Bangerth, 1983; Bufler, 1984), tomatoes (Acaster and Kende, 1983), cantaloupe (Hoffman and Yang, 1980), and citrus peel (Riov and Yang, 1982b).

The application of ACC to plant organs was shown by Lurssen *et al.* (1979) to enhance ethylene production. These researchers speculated that ACC was derived from methionine via SAM or ACC. The enzyme system involved in the formation of ethylene from ACC appeared to be associated with cellular particles (Mattoo and Lieberman, 1977; Imaseki and Watanabe, 1978). Disruption of the cellular membrane by either treatment with lipophilic compounds or osmotic shock reduced ethylene production in plant tissues (Odawara *et al.*, 1977; Imaseki and Watanabe, 1978). The particular step inhibited was identified as ACC to ethylene (Apelbaum *et al.*, 1981). An enzyme extract capable of converting ACC to ethylene was reported in pea seedlings by Konze and Kende (1979). Similar systems have been reported in a carnation microsomal system (Mayak *et al.*, 1981) and a pea microsomal system (McRae *et al.*, 1982). The search for the enzyme system that is responsible for the conversion of ACC to ethylene was finally comprehensively described by Kende (1993). It was long assumed that the ethylene-forming system was membrane associated and was referred to as the ethylene-forming enzyme (EFE). Yang and Hoffman (1984) suggested that ACC might be oxidized by an enzyme, ACC hydroxylase, to *N*-hydroxy-ACC, which is then broken down to ethylene and cyanoformic acid. The latter is extremely labile and spontaneously fragments to carbon dioxide and hydrogen cyanide (HCN). Support for this was based on studies by Peiser *et al.* (1983), who reported incorporation of [1-^{14}C] ACC into [4-^{14}C] asparagine in mung bean hypocotyls at levels similar to the production of ethylene. These findings, together with work by Miller and Conn (1980), who demonstrated incorporation of Na—CN into asparagine in mung bean, suggested the following pathway:

(MTR)   (Methionine)

The identification of EFE was eventually accomplished based on molecular cloning of tomatoes (Slater *et al.*, 1983). The first ACO gene was discovered through antisense expression of a clone, pTOM13 (Holdsworth *et al.*, 1987). The role of this enzyme in ethylene synthesis was then confirmed by expression of pTOM13 in yeast and *Xenopus* oocytes (Hamilton *et al.*, 1990; Spanu *et al.*, 1991). Further ACO genes have been identified in tomato and other climacteric fruit such as apples, avocado, bananas, and melons (Holdsworth *et al.*, 1988; Barry *et al.*, 1996; Blume and Grierson, 1997; Nakatsuka *et al.*, 1998; Jiang and Fu, 2000; Llop-Tous *et al.*, 2000). Ververidis and John (1991) extracted EFE activity from melon fruits under N_2 gas and addition of Fe^{2+} and ascorbate. These conditions enabled full recovery of EFE activity and upon fractionation by centrifugation, the enzyme was recovered in the soluble fraction. Ververidis and John (1991) proposed that EFE be referred to as ACC oxidase (ACO) since it seemed to be related to 2-oxoglutarate-dependent dioxygenases which require Fe^{2+} and ascorbate for *in vitro* activity. ACO has since been extracted from apple (Dong *et al.*, 1992; Fernandes-Maculet and Yang, 1992; Kuai and Dilley, 1992) and from avocado (McGarvey and Christoffersen, 1992). Dong *et al.* (1992) stated that ACO activity was enhanced *in vivo* by carbon dioxide and *in vitro* it was entirely dependent on the presence of carbon dioxide. Furthermore, they observed that ACO exhibited an absolute requirement for Fe^{2+} and ascorbate but not for 2-oxoglutarate. Based on these observations they postulated the following stoichiometry for the conversion of ACC to ethylene:

$$\text{ACC} + O_2 + \text{ascorbate} \xrightarrow{Fe^{2+},\ CO_2} C_2H_4 + CO_2 + \text{HCN} + \text{dehydroascorbate} + 2H_2O$$

The role of the Fe(II) ion is to bind ACC and O_2 simultaneously and promote electron transfer, which initiates catalysis of ACC to ethylene (Pirrung, 1999; Rocklin *et al.*, 1999).

B. Regulation of Ethylene in Ripening Fruits

1. ACC Synthase and ACC Oxidase

The climacteric rise in fruits is associated with enhanced ethylene production at the onset of ripening. Positive feedback regulation of ethylene biosynthesis is a characteristic feature of ripening fruits. Exposure to exogenous ethylene triggers a large increase in ethylene production due to the induction of both key enzymes, ACS and ACO (Chang and Bleecker, 2004; Génard and Gouble, 2005; Li *et al.*, 2011). The changes in the internal level of ACC were examined by Hoffman and Yang (1980) during the ripening of avocados as well as the effect of exogenous ACC on ethylene synthesis in the preclimacteric fruit. Their results in Figure 2.5 show that ACC was present at extremely low levels in the preclimacteric fruit (< 0.1 nmole/g) but increased dramatically just before the onset of ethylene, decreasing to 5 nmole/g in the overripe fruit. The low level of ACC in the preclimacteric fruit was attributed to the inability to convert SAM to ACC. Addition of exogenous ACC to the preclimacteric tissue did increase ethylene production but only to a limited degree (Adams and Yang, 1977, 1979). Thus, the formation of ACC from SAM appeared to be the rate-controlling step in the biosynthesis of ethylene (Yang, 1980). Further confirmation was provided by Liu *et al.* (1985), who studied the effect of ethylene treatment on the production of ethylene in climacteric tomato and cantaloupe fruits. They found that when exposed to exogenous ethylene the increased activity of the ethylene-forming enzyme (ACO) preceded any increase in ACC synthase in these preclimacteric fruits. Examination of *ACO*mRNA expression patterns in different tissues and at different developmental stages confirmed the regulatory role of ACO in the ethylene production during ripening (Holdsworth *et al.*, 1987; Hamilton *et al.*, 1990; Balague *et al.*, 1993; Barry *et al.*, 1996; Alexander and Grierson, 2002).

A variety of stresses, such as wounding, hypoxia, chilling, freezing, or drought, can induce ethylene biosynthesis as a result of the increase in ACC synthase activity (Wang *et al.*, 2002). Morin *et al.* (1985) found that cold storage of 'Passé-Crassane' pears was required to initiate ethylene ripening and induce the synthesis of free or conjugated ACC. During cold storage (0°C) both free and conjugated ACC increased together with ribosomes and mRNA. When these pears were then transferred to 15°C, a marked rise in ethylene occurred, followed by the climacteric period (Hartmann *et al.*, 1987).

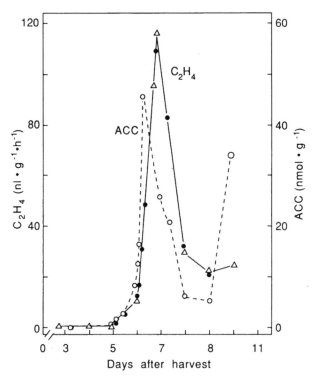

FIGURE 2.5   Changes in ACC content of avocado fruit at various stages of ripening. Each ACC value is from a single fruit which has been monitored for ethylene production and assigned an arbitrary stage of ripeness by comparison with the established climacteric patterns of ethylene production (Hoffman and Yang, 1980).

2. Cyanide

The production of cyanide in the biosynthesis of ethylene from ACC was shown in studies by Peiser *et al.* (1984) and Pirrung (1985). Pirrung and Brauman (1987) suggested that cyanide might regulate ethylene formation during the climacteric period. They proposed that in ethylene biosynthesis the cytochrome and cyanide-resistant respiratory chains were connected via cytochrome *c* oxidase. The inhibition of cytochrome *c* oxidase by cyanide during ethylene biosynthesis favored the alternative pathway which, in turn, led to ACC synthesis. Gene expression for ACC synthase in response to the alternative respiratory pathway could explain differences between climacteric and non-climacteric fruit. However, there is evidence that plant tissues possess ample capacity to detoxify HCN formed during ethylene biosynthesis and that the concentration of HCN is kept at a low level (Yip and Yang, 1988). The key enzyme to detoxify HCN is β-cyanoalanine synthease (EC 4.4.1.9). The β-cyanoalanine is further metabolized to asparagine or to γ-glutamyl-β-cyanoalanine.

3. Organic Acids

De Pooter *et al.* (1982) observed an increase in carbon dioxide production and premature ripening when intact 'Golden Delicious' apples were treated with propionic and butyric acids. This change was identical to apples treated with ethylene and suggested a role for these acids in ethylene production. It appeared feasible that ethylene could be produced from these carboxylic acids according to system 1, which then triggered normal ethylene production via system 2. Further work by these researchers (De Pooter *et al.*, 1984) confirmed the premature ripening of the intact 'Golden Delicious' apples when treated with acetic or propionic acid vapors. A small part of labeled [2-^{14}C]propionic acid was transformed into [^{14}C]ethylene which then acted as a trigger for ripening (system 2). The degree of fruit maturity was thought to be a major factor, which suggested that in the unripe apples the small amount of ethylene produced was probably derived from simple organic acids. Thus, the concentration at which ethylene can trigger ripening depends on the availability of simple organic acids. The ability of carbon dioxide to delay the onset of ripening in fruits was shown by Bufler (1984) to be due, in part, to inhibition of ACC synthase development (Figure 2.6).

4. Lipid Peroxidation: Lipoxygenase

The production of ethylene has been correlated with changes in hydroperoxide levels, peroxidase activity, as well as increases in lipoxygenase activity in ripening fruits (Meigh *et al.*, 1967; Frenkel and Eskin, 1977; Frenkel, 1979; Marcelle, 1991; Sheng *et al.*, 2000; Liu *et al.*, 2008). Studies by Adams and Yang (1979) and Konze *et al.* (1980) both suggested peroxidation as one mechanism for the formation of ethylene from ACC; however, the involvement of peroxidase was discounted by later researchers (Machackova and Zmrhal, 1981; Rohwer and Mader, 1981). The enzymatic conversion of ACC to ethylene in a cell-free system was found to be sensitive to catalase and inhibited by hydrogen peroxide (Konze and Kende, 1979; McRae *et al.*, 1982). The effect of hydroperoxides on the enzymatic conversion of ACC to ethylene was investigated by Legge and Thompson (1983) using a model system composed of microsomal membranes from etiolated peas. Addition of hydroperoxides stimulated ethylene production in model systems containing ACC, as shown in Figure 2.7. Hydrogen peroxide, a known inhibitor of lipoxygenase, was found to inhibit ethylene formation. Lipoxygenase forms lipid hydroperoxides from linoleic acid (Eskin *et al.*, 1977). A 1.5-fold increase in ethylene production occurred following the addition of linoleic acid to model systems containing this enzyme. There appeared to be an interaction between lipoxygenase activity, a hydroperoxide derivative, and the ethylene-forming enzyme (ACO). McRae *et al.* (1982) provided evidence, based on spin-trapping data, that oxygen was involved in the formation of ethylene from ACC by pea microsomal membranes. Since hydroperoxides facilitate oxygen activation, the promotion of oxygen via this mechanism could lead to the formation of ethylene. Legge *et al.* (1982) detected free radical formation using a diagnostic spin trap 4-MePyBN which required ACC, oxygen, and hydroperoxides. Their results suggested that free radicals were derived from ACC in the microsomal system producing ethylene. Conversion of ACC to ethylene by pea microsomal membranes is mediated via a free radical intermediate requiring hydroperoxides and oxygen. The increase in free radical formation was attributed by Kacperska and Kubacka-Zabalska (1984) to lipoxygenase-mediated oxidation of polyunsaturated fatty acids. This was confirmed with *in vitro* and *in vivo* studies by Kacperska and Kubacka-Zabalska (1985), who found that an increase in ethylene from ACC in winter rape leaf disks resulted from lipoxygenase activity. Similar observations were reported by Bousquet and Thimann (1984) using oat leaf segments. In more recent studies

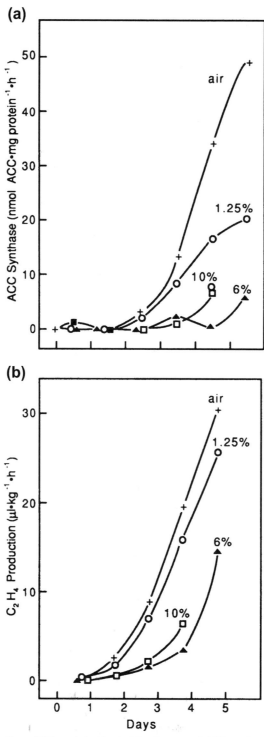

FIGURE 2.6   Effect of different concentrations of CO$_2$ on induction and development of ACC synthase activity (a) and ethylene production (b) in preclimacteric treated apples (Bufler, 1984). Apples were transferred from hypobaric storage to normal pressure and 25°C and immediately treated with air (+), 1.25% (○), 6% (▲), or 10% (□) CO$_2$ (Bufler, 1984). *(Reprinted with permission of copyright owner, American Society of Plant Physiology (ASPP)).*

relationships between lipoxygenase (LOX) activity and ethylene biosynthesis have been reported for kiwi fruit (Xu *et al.*, 2003; Zhang *et al.*, 2006), tomato (Sheng *et al.*, 2000), and oriental pear cultivars (Xu *et al.*, 2008). However, LOX activity may be involved in ethylene biosynthesis as well as several aspects of fruit ripening affecting flavor development and response to low temperatures.

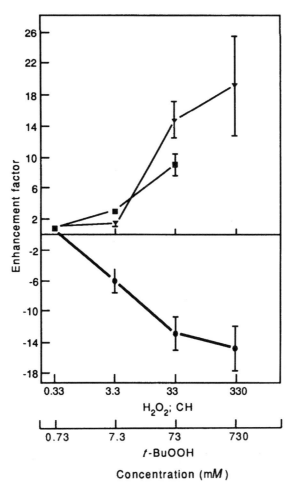

FIGURE 2.7   **Effects of hydrogen peroxide (●), *t*-butylhydroperoxide (▲), and cumene hydroperoxide (■) on the conversion of ACC to ethylene by pea microsomal membranes.** Final concentrations of hydroperoxide added to the basic reaction mixture are indicated along the abscissa. The enhancement factor is the ratio of ethylene produced in the presence of added hydroperoxide relative to that produced in its absence. Values represent the mean ± SE (*n* = 3) (Legge and Thompson, 1983). *(Reprinted with permission. Copyright © by Pergamon Press.)*

5. Galactose

Studies by Gross (1985) and later by Kim *et al.* (1987) showed that ethylene production was stimulated by galactose during the ripening of tomato fruit. Galactose is a product of reduced cell wall synthesis of galactan and increased activity of cell wall galactosyl residues by β-galactosidase (Lackey *et al.*, 1980; Pressey, 1983). Stimulation of ACC synthase activity by exogenous galactose to the pericarp tissue of green tomatoes suggested a relationship between cell wall turnover and ethylene biosynthesis in ripening tomato fruit (Kim *et al.*, 1987, 1991). Other studies confirmed that cell wall fragments stimulated or induced ethylene production (Tong *et al.*, 1986; Brecht and Huber, 1988). Later studies indicated relationships between short-chain oliogalacturonides and induction and expression of a gene encoding ACO in tomato plants and the regulation mode of transcription factors (Simpson *et al.*, 1998; Li *et al.*, 2011). Similarly, relationships between PG gene expression and softening and ethylene production were reported for kiwi fruit (Wang *et al.*, 2000).

V. COLOR CHANGES

One of the first changes observed during the ripening of many fruits is the loss of green color. This is followed by the development of red colors in some fruits and vegetables due to the formation of anthocyanins. The color changes in some fruits are summarized in Table 2.3. These changes take place immediately following the climacteric rise in respiration and are accompanied by textural changes in the fruit. In the case of leafy vegetables such as cabbage, lettuce, and Brussels sprouts, the loss of chlorophyll is also responsible for the symptom of yellowing during senescence (Lipton, 1987; Lipton and Ryder, 1989).

A. Chlorophyll Changes during Ripening

Ethylene has been reported to promote the degradation of chlorophyll during fruit ripening (Burg and Burg, 1965). A study by Hardy *et al.* (1971) noted the stimulation of chlorophyll biosynthesis in excised cotyledons of cucumber seeds (*Cucumis sativus*) by ethylene and light. Further work by Alscher and Castelfranco (1972) found that stimulation of chlorophyll synthesis occurred only in the dark as exposure to light inhibited chlorophyll synthesis. Little reference can be found on the stimulation of chlorophyll synthesis by ethylene, although the cucumber has provided an excellent system for studying chlorophyll production over the past decade (Pardo *et al.*, 1980; Chereskin *et al.*, 1982; Fuesler *et al.*, 1982; Hanamoto and Castelfranco, 1983). In contrast to stimulation of chlorophyll synthesis by ethylene in the dark, the biosynthesis of anthocyanins in red cabbage was only stimulated by ethylene when exposed to light.

B. Chlorophyll Biosynthesis

Chlorophylls, the major plant tetrapyrroles, are essential to human and animal life via their role in photosynthesis. The biosynthesis of chlorophyll is a highly coordinated process that involves numerous enzymes (Beale, 1999). By applying molecular genetic techniques it has been possible to identify and characterize those genes encoding these enzymes (Beale, 2005) (Table 2.4). The biosynthesis of chlorophyll appears to involve three distinct phases (Tanaka and Tanaka, 2006). The first phase is the biosynthesis of chlorophyll *a* from glutamate (Vavilin and Vermass, 2002; Willows, 2003; Eckhardt and Grimm, 2004; Grossman *et al.*, 2004). The second phase, also known as the chlorophyll cycle, involves the interconversion of chlorophyll *a* to chlorophyll *b* (Ruediger, 2002). The third and final phase of chlorophyll metabolism concerns itself with degradation of chlorophyll *a* (Pinta *et al.*, 2000; Eckhardt and Grimm, 2004).

1. Phase 1: Glutamate to Chlorophyll a

Shemin and Russell (1953) first demonstrated the role of δ-aminolevulinic acid (ALA) in the biosynthesis of the tetra-pyrrole nucleus of chlorophyll. The biosynthesis of ALA involves two distinct pathways (von Wettstein *et al.*, 1995). One pathway, used by animals (for heme), yeasts, and a number of bacteria, involves condensation of succinyl-CoA and glycine by the pyridoxal-P-containing enzyme 5-aminolevulinic acid synthase (EC 2.3.1.37) (Gibson *et al.*, 1958; Kikuchi *et al.*, 1958). The second and characteristic pathway in higher plants and algae, known as the C_5 pathway, is a three-step pathway in which glutamate is subsequently converted to 5-aminolevulinate. Using ^{14}C-labeled glutamate the intact five-carbon skeleton was shown to be directly incorporated into 5-aminolevulinate (Beale *et al.*, 1975; Meller *et al.*, 1975; Porra, 1986). The conversion to ALA requires activation of glutamate at the α-carbonyl by ligation to $tRNA^{Glu}$ (Kannangara *et al.*, 1984). ALA formation, the first step in the biosynthesis of chlorophyll, was thought to regulate the total amount of tetrapyrroles formed (Beale, 1999). Subsequent research suggested that ALA synthesis is subjected to feedback regulation by heme, and inhibited by FU, a regulatory protein, in the dark (Meskauskiene *et al.*, 2001; Goslings *et al.*, 2004). The initial reactions responsible for the synthesis of ALA in higher plants are shown in Scheme 2.4.

Activation of glutamate to glutamyl tRNA by glutamyl-tRNA synthetase (EC 6.1.1.17) involves ligation of $tRNA^{Glu}$, a process normally associated with plastid protein synthesis. The activated carboxyl group of glutamyl

---

TABLE 2.3 Color Changes Occurring in Some Fruits During Ripening

	Ripening Stage	
Fruit	Immature	Ripe
Apple	Green	Yellow/red[a]
Banana	Green	Yellow
Pear	Green	Yellow
Strawberry	Green	Red

[a]*Depending on which variety*

TABLE 2.4 Genes Encoding the Enzymes in the Chlorophyll Biosynthetic Pathway in Angiosperms (Beale, 2005)

Step[a]	Enzyme Name	Gene Name(s)[b]
1	Glutamyl-tRNA reductase	HEMA1
		HEMA2
		HEMA3
2	Glutamate 1-semialdehyde aminotransferase (glutamate 1-semialdehyde aminomutase)	GSA1 (HEM1)
		GSA2 (HEM2)
3	Porphobilinogen synthase (5-aminolevulinate dehydratase)	HEMB1
		HEMB2
4	Hydroxymethylbilane synthase (porphobilinogen deaminase)	HEMC
5	Uroporphyrinogen III synthase (uroporphyrinogen III co-synthase)	HEMD
6	Uroporphyrinogen decarboxylase	HEME1
		HEME2
7	Coproporphyrinogen oxidative decarboxylase	HEMF1
		HEMF2
8	Protoporphyrinogen oxidase	HEMG1
		HEMG2
9	Mg chelatase D subunit	CHLD
	Mg chelatase H subunit	CHLH
	Mg chelatase I subunit	CHLI1
		CHLI2
10	Mg-protoporphyrinogen IX methyltransferase	CHLM
11	Mg-protoporphyrinogen IX monomethylester cyclase	CRD1 (ACSF)
12	Divinyl reductase	DVR
13	NADPH:protochlorophyllide oxidoreductase	PORA
		PORB
		PORC
14	Chlorophyll synthase	CHLG
15	Chlorophyllide a oxygenase	CAO (CHL)

[a]The step numbers correspond to the numbers in Scheme 2.4.
[b]The gene names are those for Arabidopsis thaliana. Alternative names for the enzymes and genes are indicated in parentheses. Multiple genes for a given enzyme are indicated by numerical suffixes, except for POR genes (step 13), for which letter suffixes are used.

tRNA is then reduced to a formyl group by glutamyl-tRNA reductase (GluTR), resulting in the formation of glutamate-1-semialdehyde (GSA). This is followed by intermolecular amino-exchange reactions in which GSA is converted to ALA (Tanaka and Tanaka, 2007). Glutamate 1-semialdehyde aminotransferase (EC 5.4.3.8) catalyzes this reaction with pyridoxal-P or pyridoxamine-P as cofactor. The mechanism appears to involve the catalysis of two successive transformation reactions (Mau and Wang, 1988; Mayer *et al.*, 1993). A diamino intermediate, 4,5-diaminovaleric acid, is initially formed by transfer of an amino group from pyridoxamine-P to GSA, forming pyridoxal-P. In the second reaction, the amino group is transferred back to pyridoxal-P,

SCHEME 2.4 Biosynthesis of chlorophyll in higher plants. Numbered arrows refer to the enzymes listed in Table 2.4. Reactions 12 and 13 can occur in either order, depending on the availability of substrates. Reaction 14 can use either of the substrates indicated. The position numbers of the two vinyl groups are indicated for 3.8-divinyl protochlorophyllide (Beale, 2005).

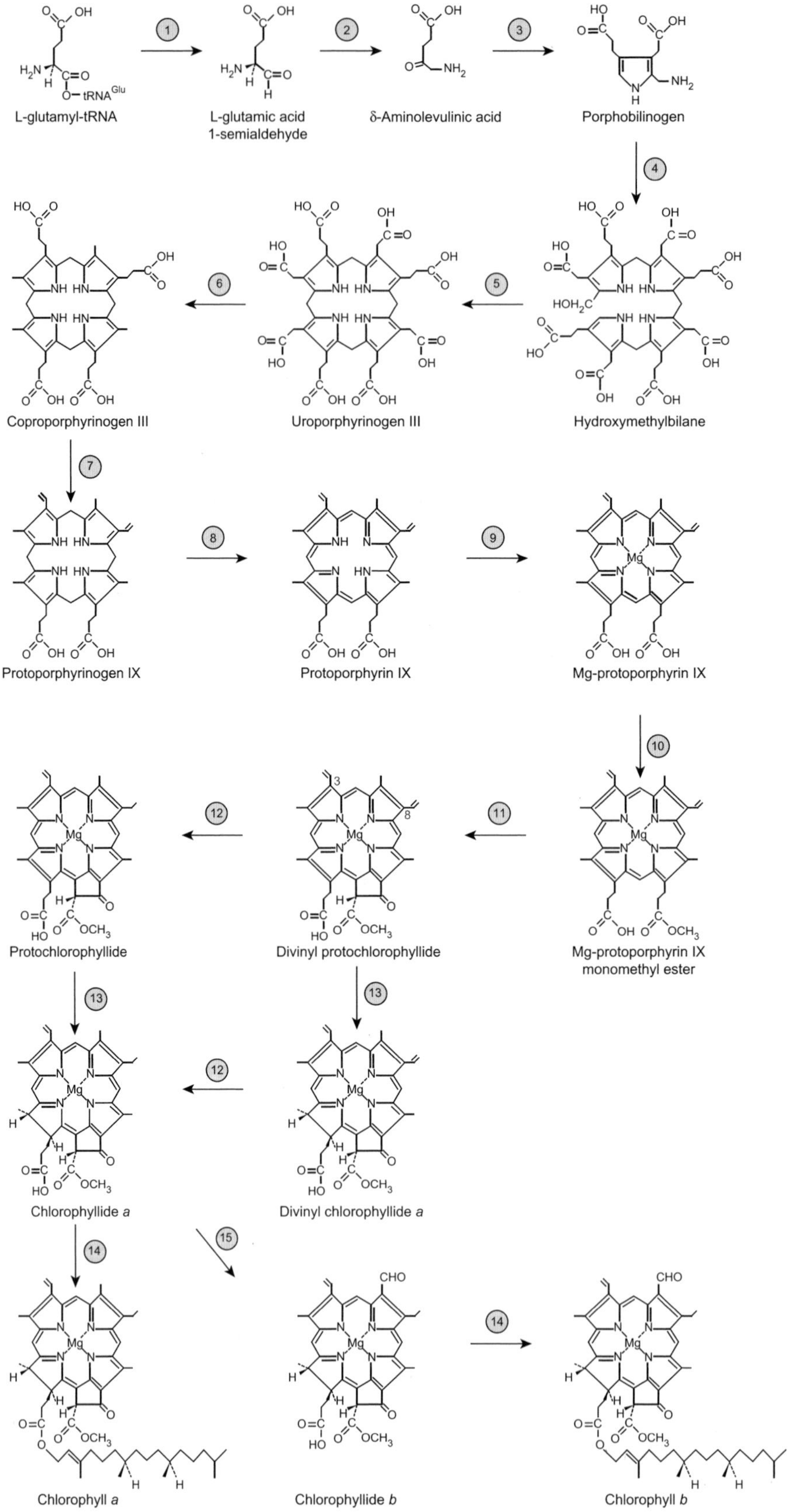

regenerating pyridoxamine-P and forming ALA. This is supported by kinetic studies based on time-resolved spectrophometric measurements (Smith *et al.*, 1998). Once ALA is formed two molecules condense to form porphobilinogen (PBG), catalyzed by ALA dehydratase [5-aminolevulinate hydrolase (ALAD), EC 4.2.1.24] (Dresel and Falk, 1953; Schmid and Shemin, 1955). It is during this step that an aliphatic compound is converted into an aromatic one.

ALAD has been studied extensively in animal tissue and photosynthetic bacteria and in a few plants including wheat (Nandi and Waygood, 1967), soybean tissue culture (Tigier *et al.*, 1968, 1970), mung bean (Prasad and Prasad, 1987), and spinach (Liedgens *et al.*, 1983). ALAD resides in the chloroplasts, where it appears in the soluble form in the plastid stroma or loosely bound to lamellae.

The first tetrapyrrole intermediate, a linear hydroxymethylbilane porphyrin precursor, was identified by Battersby *et al.* (1979) and Jordan and Seehra (1979). This results from the head-to-tail condensation of four molecules of PBG catalyzed by PBG deaminase. This linear molecule is enzymatically closed by uroporphyrinogen III synthase to form the first cyclic tetrapyrrole, uroporphyrinogen III.

(2) 5-Aminolevulinic acid   Porphobilinogen

The steps leading to the formation of protoporphyrin IX, outlined in Scheme 2.5, will only be discussed briefly, as a detailed discussion can be found in several excellent reviews (Avissar and Moberg, 1995; Jaffe, 2003, 2004; Shoolingin-Jordan, 2003; Beale, 2005). Uroporphyrinogen III is converted to coproporphyrinogen III by uroporphyrinogen III decarboxylase, which decarboxylates the acetic acid groups on the pyrrole rings A, B, C, and D (Jackson *et al.*, 1976). This is followed by oxidative decarboxylation of the propionic acid groups on pyrrole groups A and B by the oxygen-dependent co-proporphyrinogen III oxidase (CPOX), forming protoporphyrinogen IX (Games *et al.*, 1976). The final step is the formation of protoporphyrin, during which six electrons are removed from protoporphyrinogen IX by the FAD-containing enzyme protoporphyrinogen IX oxidase (Poulson and Polglase, 1975).

Chelation of protoporphyrin IX is mediated by Mg chelatase and requires a high concentration of ATP (Pardo *et al.*, 1980). This enzyme is composed of three subunits, Ch1H, Ch1I, and Ch1D, with average molecular weights of 140, 40, and 7 kDa, respectively. The catalytic site is on Ch1H, which is activated by Ch1I and Ch1D binding together (Davison *et al.*, 2005). The enzyme, Mg-protoporphyrin IX methyltransferase (MgMT), then transfers a methyl group from *S*-adenosyl-L-methionine to the carboxyl group on 13-proprionate to form Mg-protoporphyrin-*N*-monomethyl ester (Fuesler *et al.*, 1982). Fuesler and co-workers (1982) demonstrated the following reaction sequence in which metal chelation preceded methylation. Using a high-performance liquid chromatography (HPLC) procedure to separate Mg-protoporphyrin and Mg-protoporphyrin-Me ester, Fuesler *et al.* (1982) demonstrated the following reaction sequence:

$$\text{Protoporphyrin IX} \xrightarrow{\text{Mg}^{2+}\text{ATP}} \text{Mg-protoporphyrin} \xrightarrow{\text{SAM}} \text{Mg-protoporphyrin-Me}$$

Mg-protoporphyrin-Me ester is then converted by Mg-protoporphyrin IX monomethyl cyclase (MgCy) to 3,8-divinyl-protochlorophyllide by incorporating an atomic oxygen. Protochlorophyllide oxidoreductase (POR) then reduces the D ring of 3,8-divinyl protochlorophyllide to 3,8-divinyl chlorophyllide. The 8-vinyl substituent in the side-chain of the pyrrole ring B is then reduced by divinylchlorophyllide reductase (DVR) to 3-vinyl chlorophyllide *a* (monovinyl chlorophyllide *a*). DVR can also reduce 3,8-divinyl protochlorophyllide but the efficiency of this substrate was reported by Tanaka and Tanaka (2007) to be substantially lower than 3,8-divinyl chlorophyllide. Based on this, they revised the conventional reaction steps by placing the DVR reaction after the POR reaction. The final step is the biosynthesis of chlorophyll *a* from 3-vinyl chlorophyllide. Chlorophyll synthase esterifies the 17-proprionate substituent on pyrrole ring D of monovinyl chlorophyllide *a* with geranyl geraniol, which is reduced to phytol.

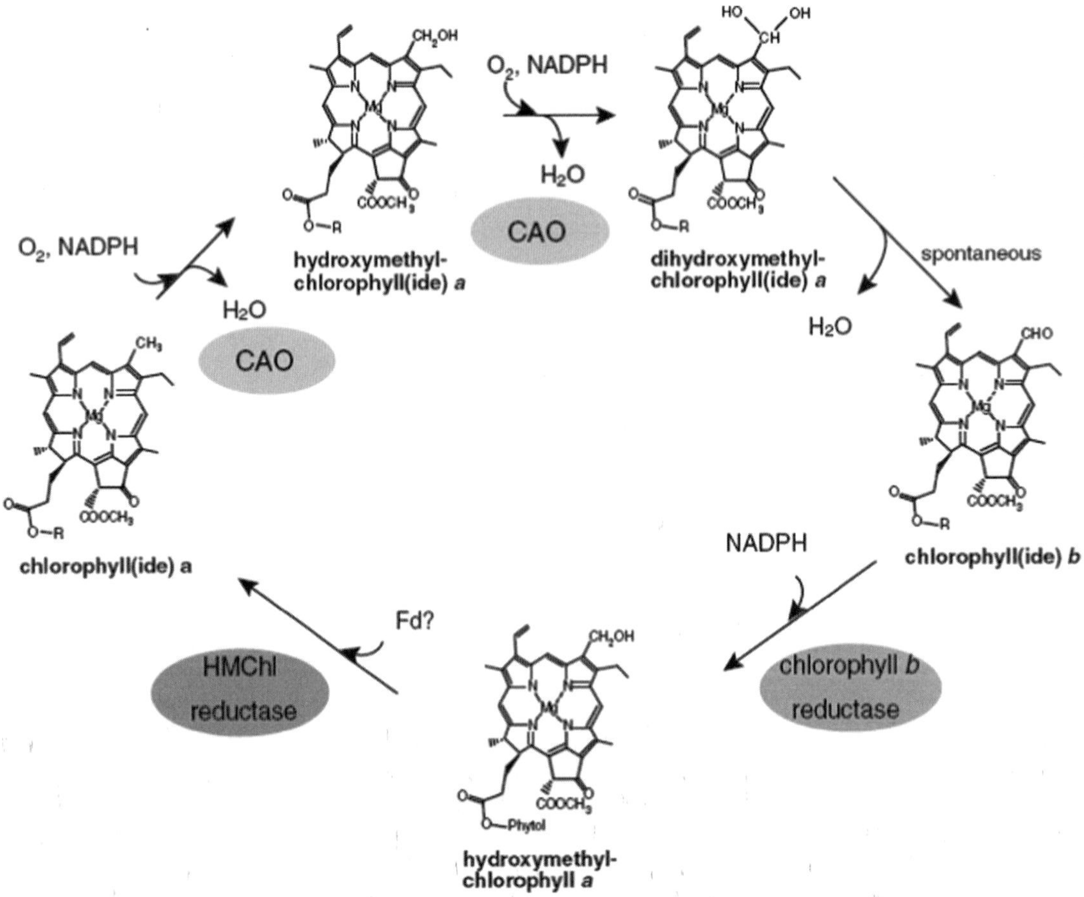

SCHEME 2.5 Proposed biosynthetic route of the chlorophyll cycle. The forward reaction may occur with chlorophyll or chlorophyllide. The chlorophyll to chlorophyllide conversion and reverse reaction are catalyzed by chlorophyllase and chlorophyll synthase, respectively, neither of which are depicted in the diagram. R = either proton or phytol; Fd = ferrdoxin (Tanaka *et al.*, 2011).

2. Phase 2: The Chlorophyll Cycle

In this phase a portion of the chlorophyll *a* pool is converted to chlorophyll *b* (Scheme 2.5). This is a two-step process catalyzed by a single polypeptide. The enzyme involved, a Rieske-type monooxygenase, chlorophyllide *a* oxygenase (CAO), was first characterized by Tanaka and co-workers (Tanaka *et al.*, 1998; Espineda *et al.*, 1999). Cloning and functional expression of the CAO gene subsequently showed that CAO catalyzed two oxygenation steps in which chlorophyllide *a* was converted to chlorophyllide *b* (Oster *et al.*, 2000). The first oxygenation step by CAO produced the intermediate 7-hydroxymethylchlorophyllide *a*, which was then oxygenated to chlorophyllide *b*. The latter is then phytylated by chlorophyll synthase to chlorophyll *b*. A recent review by Tanaka and Tanaka (2011) includes a discussion of how plants use the chlorophyll cycle to regulate the synthesis and destruction of a specific subset of light-harvesting complexes.

C. Regulation of Chlorophyll Biosynthesis

Of the three steps involved in ALA synthesis, glutamyl-tRNA reduction appears to be the limiting one in chlorophyll biosynthesis (Tanaka and Tanaka, 2007). Current evidence suggests that glutamyl tRNA reductase (GluTR) activity is controlled through feedback regulation by the end products (Meskauskiene *et al.*, 2001). A second negative regulator of tetrapyrrole biosynthesis was later discovered, the FLU protein, which appeared to control chlorophyll biosynthesis by directly interacting with GluTR (Goslings *et al.*, 2004). Other possible regulators include CAO's conversion of chlorophyll *a* to chlorophyll *b* (Tanaka *et al.*, 2001; Tanaka and Tanaka, 2005; Pattanayak *et al.*, 2005). For a more detailed discussion of these regulators the paper by Tanaka and Tanaka (2007) is recommended.

D. Mechanism of Chlorophyll Degradation

The third and final phase of chlorophyll metabolism involves the degradation of chlorophyll (Takamiya *et al.*, 2000; Eckhardt and Grimm, 2004). This phase is central to leaf senescence and fruit ripening, with the reaction mechanism having been well established since the last edition of this book (Matile *et al.*, 1999; Hortensteiner, 1999; Krautler, 2003; Eckhardt *et al.*, 2004). The mechanism of chlorophyll degradation involves its conversion to colorless non-fluorescent catabolites (NCCs) as shown Scheme 2.6.

The breakdown of chlorophyll *a* in plants involves four consecutive steps catalyzed by chlorophyllase, Mg-dechelatase, pheophorbide *a* oxygenase and red chlorophyll catabolite reductase, respectively (Harpaz-Saad *et al.*, 2007). The first step is the removal of phytol by the enzyme chlorophyllase (chlorophyll chlorophyllidihydrolase, EC 3.1.1.14), an intrinsic membrane glycoprotein located in the lipid envelope of the thylakoid membranes (Bacon and Holden, 1970; Schoch and Vielwerth, 1983). This enzyme was first discovered a century ago by Willstater and Stoll (1913). Since then, there have been many studies trying to elucidate the mechanisms of reaction. The inhibition of *Euglena gracilis* chlorophyllase by *p*-chloromercuribenzoic acid (PCMB) suggested the possible involvement of cysteine. However, PCMB proved ineffective

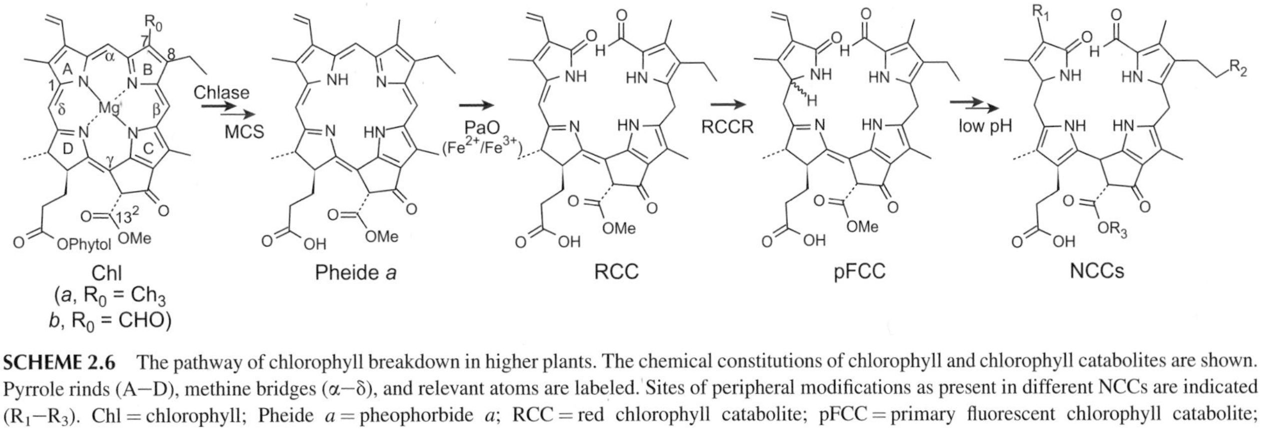

SCHEME 2.6 The pathway of chlorophyll breakdown in higher plants. The chemical constitutions of chlorophyll and chlorophyll catabolites are shown. Pyrrole rinds (A–D), methine bridges (α–δ), and relevant atoms are labeled. Sites of peripheral modifications as present in different NCCs are indicated (R_1–R_3). Chl = chlorophyll; Pheide *a* = pheophorbide *a*; RCC = red chlorophyll catabolite; pFCC = primary fluorescent chlorophyll catabolite; NCC = non-fluorescent chlorophyll catabolite (Pruzinska *et al.*, 2005).

against *Phaeodactylum tricornium* chlorophyllase (Terpstra, 1977). Later work by Khalyfa *et al.* (1995) found that diisopropyl fluorophosphate, a serine hydrolase inhibitor, inhibited *Paseodactylim* chlorophyllase. These inhibition results suggested that the amino acid residues involved in the chlorophyllase reaction varied with the source of the enzyme. Cloning the chlorophyllase gene from the higher plant *Chenopodium album*, Tsuchiya *et al.* (2003) found that it contained a lipase motif with an active serine residue. Identification of serine, histidine and aspartic acid at the active site of chlorophyllase indicated that it was similar to serine hydrolase. Based on these results the following mechanism for chlorophyllase was proposed (Scheme 2.7). Azoulay-Shemer *et al.* (2011) found that dual N- and C-terminal processing was involved in the maturation of citrus chlorophyllase.

The enzyme involved in removal of Mg from chlorophyllide *a* with the formation of pheophorbide *a* has been referred to as Mg-dechelatase (Owens and Falkowski, 1988; Ziegler *et al.*, 1988; Shimokawa *et al.*, 1990; Shioi *et al.*, 1991). Using chlorophyllin, Mg-dechelatase was assayed in rape cotyledons by Vicentini *et al.* (1995). The enzyme appeared to be located in a latent form in the thylakoids but present at high levels in presenescent chloroplasts. While there have been a number of studies on Mg-dechelation, none of them has yet identified Mg-dechelatase (Shioi *et al.*, 1996; Tan *et al.*, 2000; Costa *et al.*, 2002; Suzuki and Shioi, 2002). Using chlorophyllide *a*, Shioi *et al.* (1996) found that the Mg-dechelation reaction required a small heat-stable substance, referred to as a metal-chelating substance (MCS). Suzuki and Shioi (2002) used an artificial substrate, chlorophyllin *a*, and showed that a purified Mg-releasing protein from *Chenopodium album* had weak peroxidase activity. This was consistent with an earlier study by Azuma *et al.* (1999), who showed that horseradish peroxidase exhibited Mg-dechelating activity. Subsequent research using extracts of *C. album* by Kunieda *et al.* (2005), however, found that peroxidase and glutathione *S*-transferase could only release magnesium from the artificial substrate chlorophyllin *a*. This eliminated their possible role in the degradation of chlorophyll, as only the purified low-molecular-mass protein exhibited Mg-dechelating activity with the native substrate chlorophyllide *a*.

Pheophorbide *a* oxygenase (PAO), identified in the soluble protein fraction of *C. alba* by Shioi *et al.* (1995), catalyzes the conversion of pheophorbide *a* to pyropheophorbide *a*. PAO, a Rieske-type iron–sulfur oxygenase (Pruzinska *et al.*, 2003, 2005), is the key enzyme responsible for opening the chlorine macrocycle of pheophorbide *a* (Hortensteiner and Krautler, 2011). Unsuccessful attempts to biochemically characterize PAO in isolated chloroplast membranes led to the identification of a second enzyme, red chlorophyll catabolite reductase (RCCR). Only after the addition of stromal proteins, which contained RCCR, could PAO activity be determined. Together these two enzymes catalyzed the formation of a non-polar fluorescent chlorophyll catabolite (FCC) as primary FCC (pFCC). The required opening of the macrocycle of pheophorbide *a* by the PAO oxygenase formed the elusive red chlorophyll catabolite (RCC), which was then reduced by RCCR to FCC. pFCC was then further modified by unidentified hydroxylating enzymes to a variety of colorless end products (Scheme 2.8).

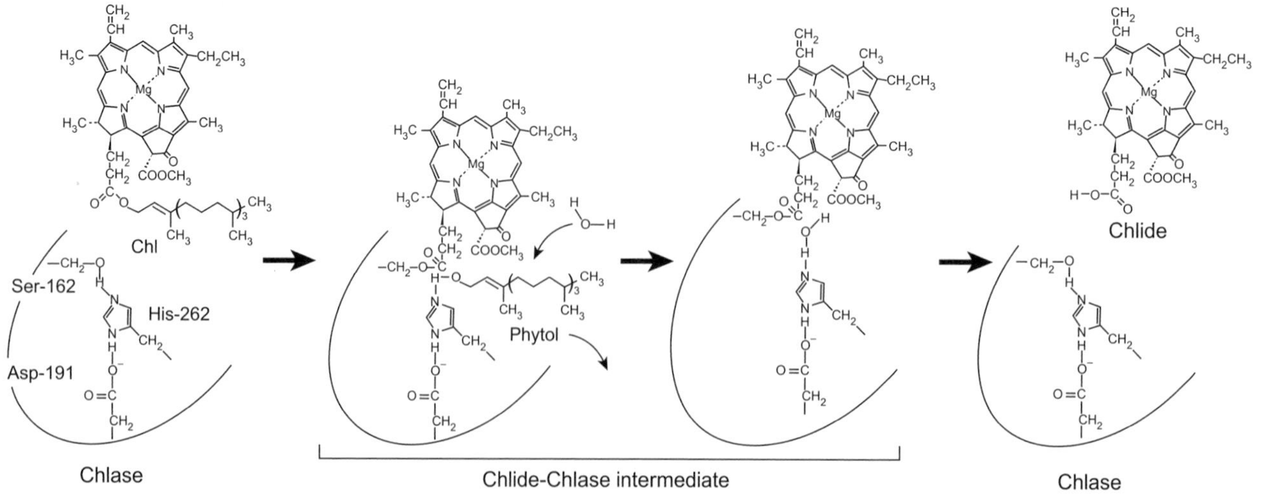

SCHEME 2.7 A possible model for the catalytic mechanism of the recombinant *Chenopodium album* chlorophyllase (CaCLH) (Tsuchiya *et al.*, 2003).

1. Chlorophyll Degradation: Processing and Storage

The retention of chlorophyll is generally used as a measure of quality in green vegetables as chlorophyll degradation occurs in damaged tissue during blanching and processing (Sweeney and Martin, 1961; Heaton and Marangoni, 1996; Tijkens et al., 2001). Several pathways have been proposed for the loss of chlorophyll during processing and storage of fruits and vegetables. In addition to enzymes, weak acids, oxygen, light, and heat can all result in the breakdown of chlorophyll. This loss of green color can be undesirable and such changes need to be minimized. One of the main reactions is replacement of the Mg^{2+} atom in chlorophyll by hydrogen under acidic conditions with the formation of pheophytin (Minguez-Mosquera et al., 1989). The latter pigment is associated with a color change from a bright green to a dull olive green. This reaction was first recognized by Campbell (1937) to cause discoloration in stored frozen peas. The influence of pH on the conversion of chlorophyll to pheophytin is generally considered to be the main cause of discoloration of foods during processing. Thus, vegetables turn a dull olive green when heated or placed in an acidic environment (Gold and Weckel, 1959; Gunawan and Barringer, 2000). As a consequence, efforts to maintain chlorophylls during heat processing included pH control, high-temperature short time processing, or a combination (Gupte and Francis, 1964; Buckle and Edwards, 1970; Schwartz and Lorenzo, 1991).

The rate of conversion to pheophytin was shown to be first order with respect to the acid concentration (Joslyn and Mackinney, 1938). The formation of pheophytin has been the subject of a large number of studies (Gupte et al., 1964; Hermann, 1970; LaJollo et al., 1971; Robertson and Swinburne, 1981). A linear relationship was reported by Walker (1964) between the appearance and pheophytin formation for frozen beans stored for up to 1 year. LaJollo et al. (1971) noted that the formation of pheophytin was the predominant reaction at water activity (a_w) levels greater than 0.32 in freeze-dried, blanched spinach purée stored at 37°C and 55°C under nitrogen and air. Chlorophyll a was degraded far more rapidly than chlorophyll b, by a factor of 2.5–3.0, consistent with earlier reports (Schanderl et al., 1962; Gupte et al., 1964). LaJollo et al. (1971) reported a linear relationship between a_w and log time for a 20% loss of chlorophyll (Figure 2.8). These early studies primarily used spectrophotometric and colorimetric methods to establish the kinetics of chlorophyll degradation. Later, researchers used HPLC to measure chlorophylls a and b (Steet and Tong, 1996; Mangos and Berger, 1997; Weemaes et al., 1999). Using HPLC, Koca et al. (2006) confirmed the first order degradation of chlorophyll a and b in blanched green peas over a temperature range of 70–100°C in buffered solutions at pH 5.5, 6.5, and 7.5. Using the CIE-L*a*b* color

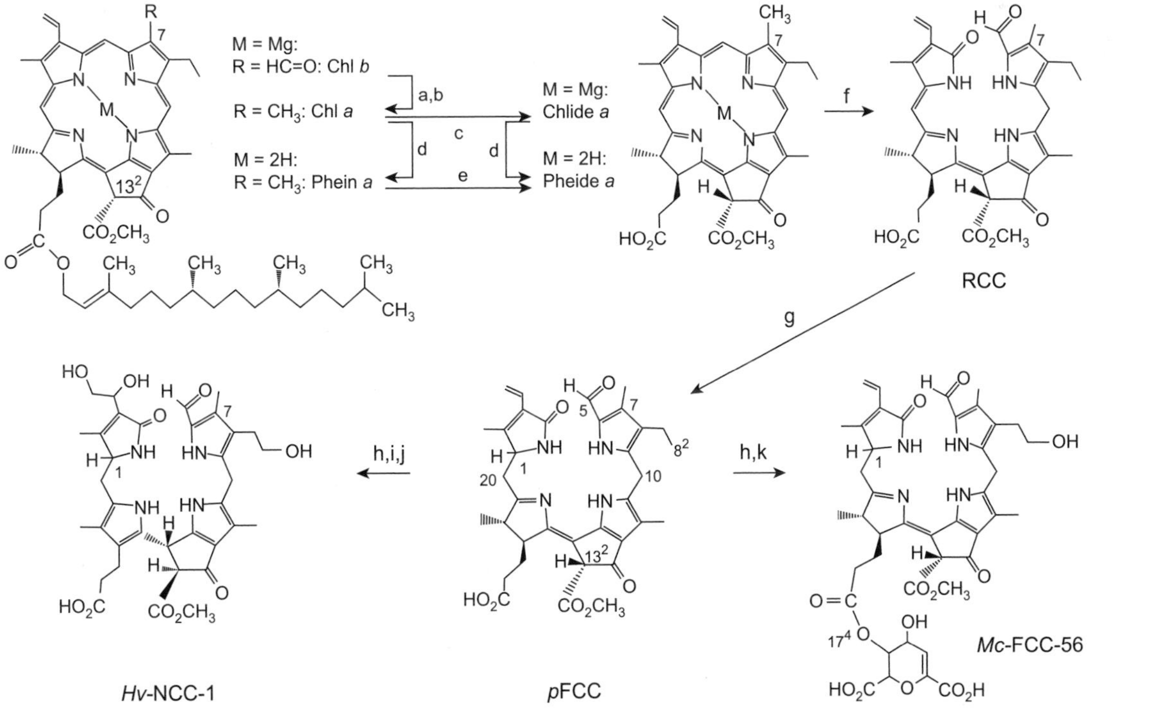

SCHEME 2.8 Representative structural outline of major catabolites delineating the major paths of chlorophyll breakdown in higher plants (Krautler and Hortensteiner, 2006; Moser et al., 2009). (Hortensteiner and Krautler, 2011).

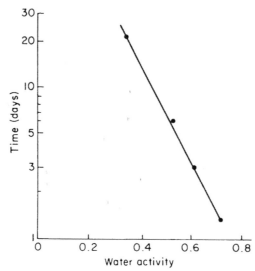

FIGURE 2.8 **Time required for 20% loss of chlorophyll in spinach at different water activities (37°C in air) (La Jollo et al., 1971).** *(Copyright © by Institute of Food Technologists.)*

measuring system, they found a significant correlation between the change in visual color parameters ($-a$, $-b$, and h values) and chlorophyll degradation.

Besides pheophytinization, chlorophyllase converts chlorophylls to chlorophyllides with the loss of the phytol group. The combined action of chlorophyllase and acid results in the loss of Mg^{2+} and phytol group with the formation of pheophorbides (White *et al.*, 1963). Pheophorbide was also found to be the major degradation product during the brining of cucumbers (Jones *et al.*, 1961, 1963). Several new products were identified by Schwartz *et al.* (1981) in heated spinach purée, including pyropheophytins *a* and *b*. These were formed from the corresponding pheophytins as a result of the loss of the carbomethoxy group and separated by HPLC. Schwartz and von Elbe (1983) examined the formation of pyropheophytins in heated spinach. Pheophytins *a* and *b* were both detected after heating at 121°C for up to 15 minutes, but then a decline in these pigments occurred. Further heating resulted in the formation of pyropheophytin with *a* detected after 4 minutes while *b* was not observed until after 15 minutes. A sample of blanched spinach purée heated for almost 2 hours at 126°C produced only pyropheophytins *a* and *b*. No further changes were noted, indicating that these were the final products of degradation. The olive-green color associated with canned vegetables was attributed to pyropheophytins. These pigment derivatives probably accounted for the unidentified pigments reported by Buckle and Edwards (1969) and LaJollo *et al.* (1971) to represent 20–30% of the total pigments. The inability of these researchers to detect pyropheophytins was due to the absence of high-resolution chromatographic techniques required to separate these derivatives. Schwartz and von Elbe (1983) proposed the following kinetic sequence to explain the formation of pyropheophytins:

$$\text{Chlorophyll} \rightarrow \text{Pheophytin} \rightarrow \text{Pyropheophytin}$$

A later study by Teng and Chen (1999) examined the changes in chlorophylls and their derivatives during the heating of spinach leaves using HPLC with a photodiode array or positive ion fast atom bombardment mass spectrometry (FAB-MS). They showed the first order degradation of both chlorophyll *a* and *b* and that the rate constant was greater for microwave cooking and blanching than for steaming and baking. Chlorophyll epimers and pheophytins were the major degradation products formed during baking and blanching while pyrochlorophylls *a* and *b* were detected after steaming for 30 minutes or microwave cooking for 1 minute. Microwave cooking appeared to favor the formation of pyrochlorophylls *a* and *b*, while steaming favored the formation of pyropheophytins *a* and *b*. The formation of these chlorophyll derivatives during the heating of spinach leaves was described by these researchers in Scheme 2.9.

To preserve the desirable fresh green color of vegetables, the potential use of metallochlorophyll complexes has been examined. A regreening phenomenon was observed involving the formation of such complexes resulting from the introduction of copper (Cu) and zinc (Zn) in the chlorophyll pyrrole ring. These appear to form a strong bond that is more resistant to acid and heat than the normal magnesium (Mg^{2+})-containing chlorophyll (Humphrey, 1980). Pheophytin *a*, pyropheophytin *a*, and phephorbide *a* were found to be much more reactive with zinc than were

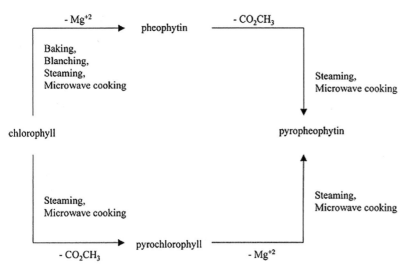

SCHEME 2.9 Formation pathway of pyro-chlorophylls and their derivatives in spinach leaves during heating (Teng and Chan, 1999).

the corresponding *b* forms (von Elbe *et al.*, 1986; LaBorde and von Elbe, 1990). Several patents were issued for improving the color of green vegetables by blanching or packing green vegetables in zinc or copper solutions (Leake and Kirk, 1992; LaBorde and von Elbe, 1996). A study by Canjura *et al.* (1999) found an improvement in the green color of aseptically processed fresh and frozen green peas when blanched in a solution of $ZnCl_2$ that may prove to be viable for preserving the green color of vegetables.

Holden (1965) also attributed decoloration of chlorophyll in legume seeds to the oxidation of fatty acids by lipoxygenase (LOX). This reaction was coupled with the degradation of fatty acid hydroperoxides and required a thermolabile factor to be present in the crude extract. Subsequent research by Zimmerman and Vick (1970) identified an enzyme, linoleate hydroperoxide isomerase, as the heat-labile factor. The bleaching effect on chlorophyll was attributed to an oxidation–reduction reaction in which the ketohydroxy fatty acid, the isomerized product formed by hydroperoxide isomerase, and a portion of the conjugated double-bond system of chlorophyll were involved. Imamura and Shimizu (1974) disproved the involvement of hydroperoxide isomerase and confirmed the role of LOX in chlorophyll bleaching. Later researchers confirmed that different LOX isoenzymes played a role in bleaching chlorophyll and carotenes (Grosch *et al.*, 1976; Ramadoss *et al.*, 1978). The participation of the different LOX isoenzymes in carotene and chlorophyll bleaching has been the subject of a number of investigations (Hilderbrand and Hymovitz, 1982; Reynolds and Klein, 1982; Cohen *et al.*, 1984; King and Klein, 1987). The effect of soybean LOX-1 on wheat chloroplasts investigated by Kockritz *et al.* (1985) suggested that this enzyme, which selectively attacks free fatty acids, may be involved in senescence and chloroplast breakdown. LOX activity has been reported previously in chloroplasts of peas (Borisova and Budnitskaya, 1975; Douillard and Bergeron, 1978). Because of its role in the deterioration of vegetable quality, LOX was considered to be a possible indicator enzyme for optimizing blanching prior to freezing. However, Gokmen *et al.* (2005) clearly showed that peroxidase (POD) was more heat stable than LOX, so the inactivation of POD was a better indicator for the adequacy of blanching.

E. Carotenoids

During the maturation of many fruits there is a change in color from green to orange or red. This is due to the loss in chlorophyll and the unmasking and synthesis of carotenoids (MacKinney, 1961). Such pigmentation changes are accompanied by structural changes in the chloroplasts. The granal–integranal network, in particular, becomes disorganized, resulting in the formation of chromoplasts (Thomson, 1966; Spurr and Harris, 1968; Camara and Brangeon, 1981). The chromoplasts no longer contain chlorophyll or photosynthetic pigments but become the major site for carotenoid biosynthesis (Camara and Brangeon, 1981).

Carotenoids are C_{40} isoprenoid compounds composed of isoprene units joined head to tail to form a system of conjugated double bonds (Eskin, 1979). They are classified into two groups, carotenes and xanthophylls. The carotenes are structurally related to hydrocarbons while the xanthophylls include the corresponding oxidized derivatives (hydroxy, epoxy, and oxy compounds) and are frequently esterified. Examples of carotenes are α- and β-carotenes in carrots and lycopene in tomatoes, while xanthophylls include capsanthin and capsorubin found in red pepper.

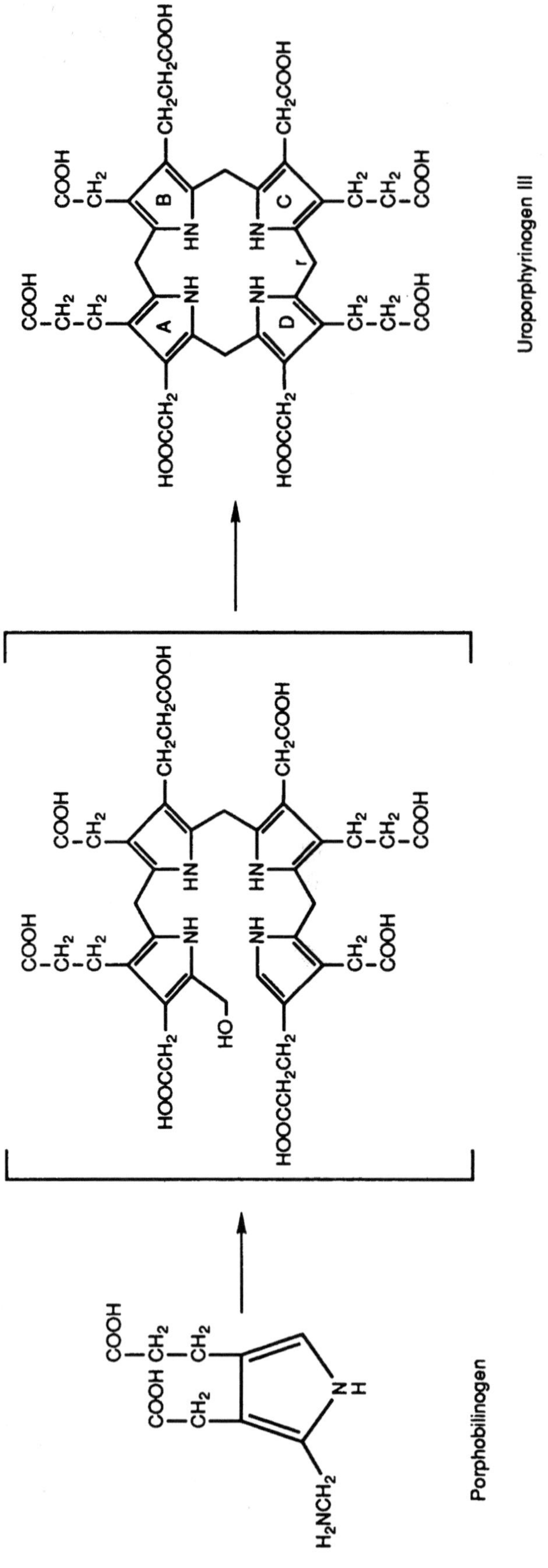

Porphobilinogen

Uroporphyrinogen III

Carotenoids are synthesized within the chromoplast from isopentyl pyrophosphate (IPP). At one time IPP biosynthesis was thought to occur via a single pathway involving acetate/mevalonate (Britton, 1982). However, a second pathway was discovered in aubacteria, green algae, and higher plants in which IPP was formed from glyceraldehyde 3-phosphate and pyruvate (Lichtenthaler *et al.*, 1997a, b). Both pathways are present in higher plants but differ in their localization. The mevalonate pathway enzymes are found in the cytosol, while the glyceraldehyde 3-phosphate pathway, referred to as the 1-deoxy-D-xylulose-5-phosphate pathway (DXP), operates in the plastid (Scheme 2.10).

It remained somewhat unclear whether acetyl-CoA was synthesized within the chloroplast or was of extraplastidic origin. Grumbach and Forn (1980) showed quite clearly that acetyl-CoA was formed within the chloroplast and could synthesize carotenoids autonomously. Chromoplasts from both red peppers and red daffodils exhibited an ability to synthesize carotenoids from isopentyl phosphates (Beyer *et al.*, 1980; Camara *et al.*, 1982). The presence of enzymes capable of synthesizing acetyl-CoA suggested that a similar autonomy existed in the chromoplasts capable of synthesizing carotenoids.

The formation of geranylgeranyl pyrophosphate from mevalonic acid (MVA) involves phosphorylation by mevalonate kinase (ATP:mevalonate 5-phosphotransferase, EC 27.1.3.6). This enzyme has been identified in many plants, including pumpkin seedlings (Loomis and Battaille, 1963), green leaves and etiolated cotyledons of French beans (*Phaseolus vulgaris*) (Rogers *et al.*, 1966; Gray and Keckwick, 1969, 1973), orange juice vesicles (Potty and Breumer, 1970), and melon cotyledons (*Cucumis mello*) (Gray and Keckwick, 1972). Phosphorylation of MVA-5P to MVA-5 pyrophosphate (MVA-5PP) is then catalyzed by 5-phosphomevalonate kinase (ATP:phosphomevalonate phosphotransferase, EC 2.7.4.2). MVA-5 PP is then decarboxylated by pyrophosphomevalonate decarboxylase [ATP:5-pyrophosphomevalonate carboxylase (dehydrating), EC 4.1.1.33]. This enzyme catalyzes a bimolecular reaction in which ATP and 5-pyrophosphomevalonate are converted to isopentyl pyrophosphate, ADP, phosphate, and carbon dioxide (Scheme 2.11).

Isomerization of isopentyl pyrophosphate to dimethylallyl pyrophosphate is catalyzed by isopentyl pyrophosphate isomerase (EC 5.3.3.2). The double bond is isomerized from position 3 in isopentyl pyrophosphate to position 2 in dimethylallyl pyrophosphate. This enzyme was isolated from pumpkin fruit by Ogura *et al.* (1968). One molecule of dimethylallyl pyrophosphate then condenses with one, two, or three molecules of isopentyl pyrophosphate, leading to the formation of geranylgeranyl pyrophosphate (Scheme 2.12). The last reactions are catalyzed by a group of

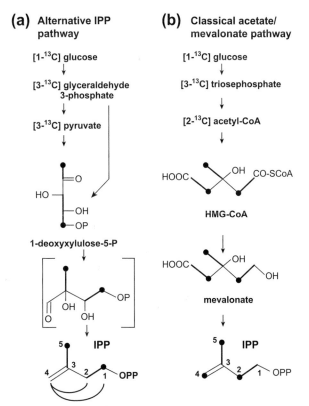

SCHEME 2.10 Labeling of isopentyl diphosphate (IPP) from [^{13}C]glucose via (a) a novel alternative and (b) the classical acetate/mevalonate pathway of IDPP biosynthesis (Lichtenthaler *et al.*, 1997b).

SCHEME 2.11   Formation of IPP from mevalonic acid.

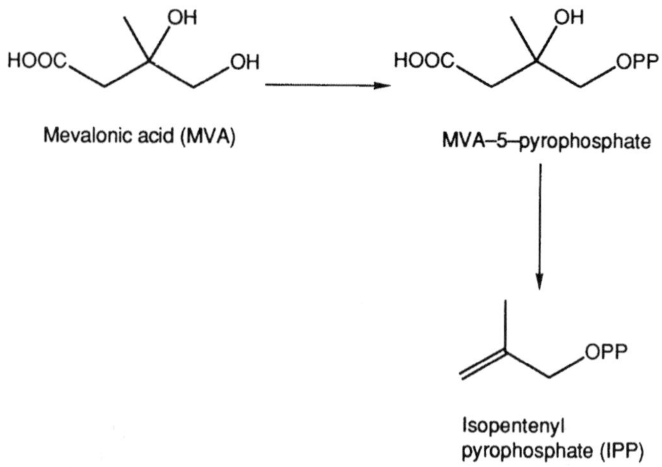

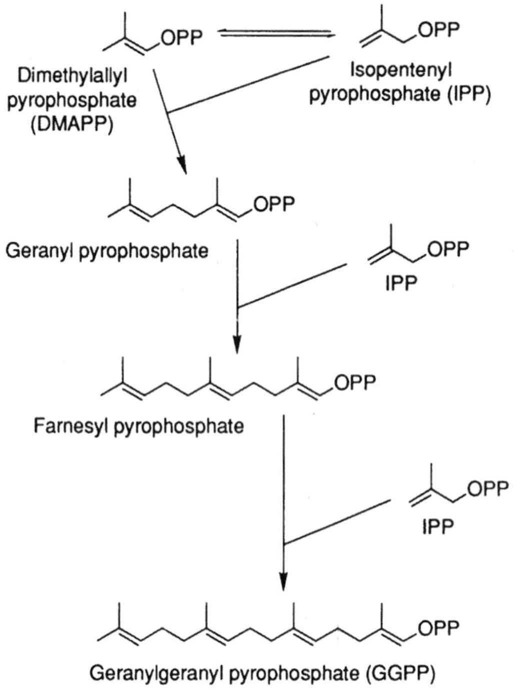

Mevalonic acid (MVA)

MVA–5–pyrophosphate

Isopentenyl
pyrophosphate (IPP)

Dimethylallyl
pyrophosphate
(DMAPP)

Isopentenyl
pyrophosphate (IPP)

Geranyl pyrophosphate

IPP

Farnesyl pyrophosphate

IPP

Geranylgeranyl pyrophosphate (GGPP)

SCHEME 2.12   Geranylgeranyl pyrophosphate biosynthesis from IPP.

enzymes referred to as prenyl transferases. The enzyme responsible for the synthesis of farnesyl pyrophosphate was partially purified from pumpkin seed (Eberhardt and Rilling, 1975).

The formation of the first C_{40} hydrocarbon 15-*cis*-phytoene results from condensation of two molecules of geranylgeranyl pyrophosphate (C_{20}) (Scheme 2.13). The intermediate in this reaction, prephytoene pyrophosphate, loses a proton, which results in a double bond at the C15 position. Maudinas *et al.* (1975) identified a soluble enzyme system from tomato fruit plastids which was capable of synthesizing *cis*-phytoene from isopentyl pyrophosphate. Phytoene is quite colorless and is converted to colored carotenoids by a series of desaturation steps which produce a conjugated double-bond system. Unlike phytoene, which is a 15-*cis* isomer, the colored carotenoids are all *trans* so that isomerization to the *trans* form must take place during the desaturation process. The mechanism involves the loss of hydrogen by trans-elimination and may be mediated by an enzyme complex in the membrane, possibly involving metal ions or cytochromes in a simple electron transfer system (Britton, 1979). The sequential desaturation of phytoene to lycopene was proposed by Porter and Lincoln (1950) and is shown in Scheme 2.14. A similar enzyme

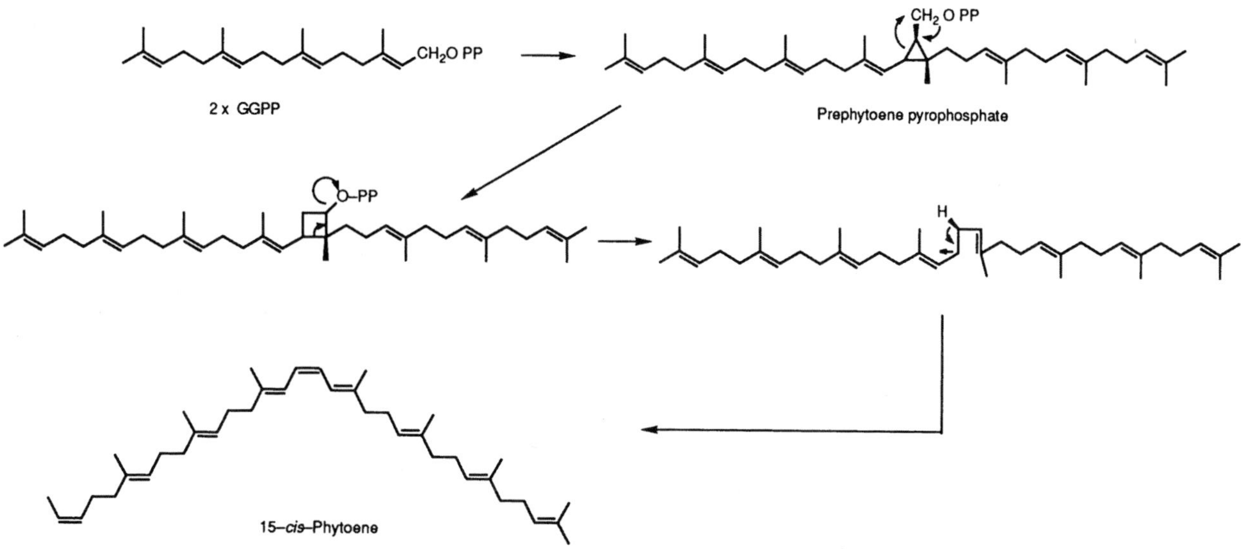

SCHEME 2.13 Mechanism of phytoene biosynthesis (Britton, 1982).

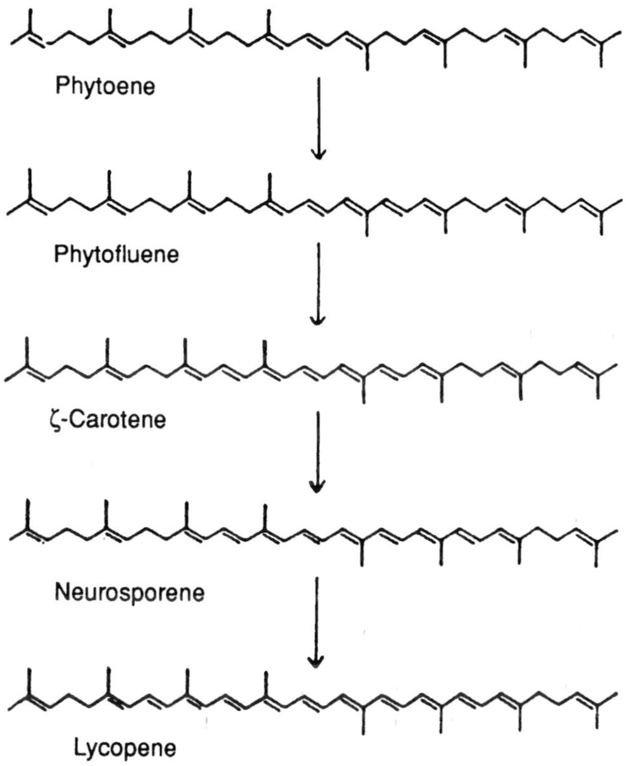

SCHEME 2.14 Sequence of desaturation reactions leading from phytoene to lycopene (Britton, 1982).

system was reported by Qureshi *et al.* (1974) in tangerine mutant tomatoes capable of converting *cis*-β-carotene into all-*trans* carotene. The conversion of neurosporene into lycopene has been found in fungal systems only (Davies, 1973; Bramley *et al.*, 1977).

The final step in carotenoid biosynthesis is cyclization with the formation of at least one or two cyclic end groups in the carotenoids (Scheme 2.15). The conversion of lycopene to α-, β, and γ-carotenes was demonstrated in the presence of soluble enzymes from tomato fruit plastids and spinach chloroplasts by Kuwasha *et al.* (1969). Cyclization was shown to be inhibited by nicotine and CPTA [2-(4-chlorophenyltrio)triethyl ammonium chloride]

SCHEME 2.15 Overall scheme for the biosynthesis of bicyclic carotenes from neurosporene (Britton, 1982).

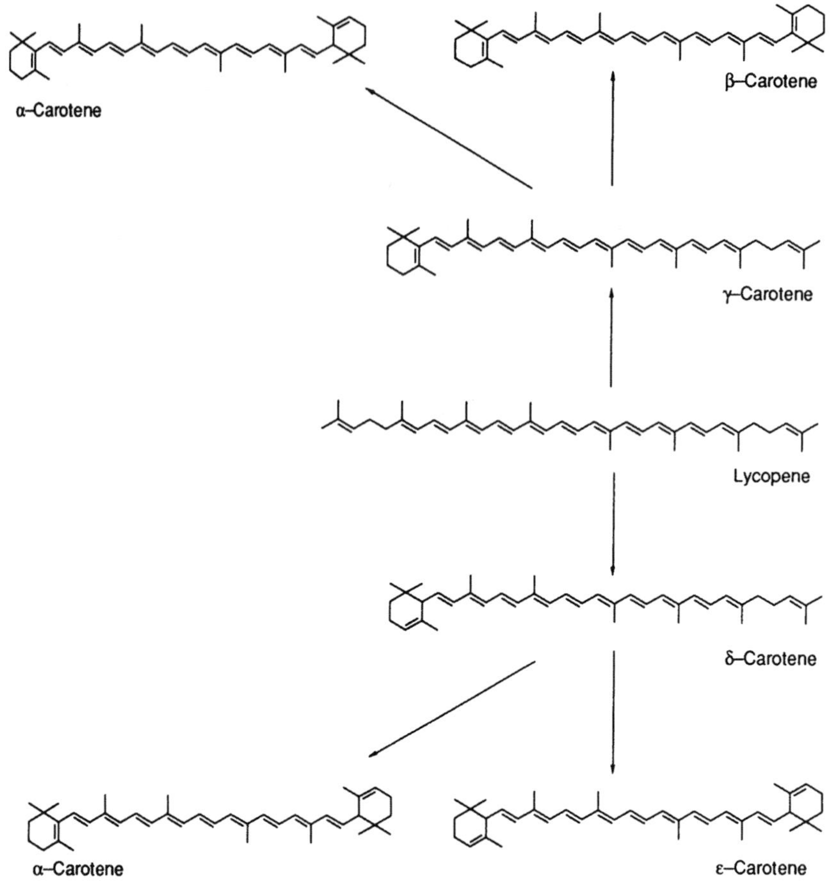

α–Carotene

β–Carotene

γ–Carotene

Lycopene

δ–Carotene

α–Carotene

ε–Carotene

with the accumulation of lycopene in citrus fruits treated with these inhibitors (Britton, 1982). Following cyclization, oxygen is incorporated as a hydroxyl group at C3 or an epoxide at the 5,6 position (Takeguchi and Yamamoto, 1968; Britton, 1976). The latter involves a series of reactions referred to as the xanthophyll cycle.

Camara *et al.* (1982) examined the site of carotenoid biosynthesis in chromoplasts of semi-ripened pepper (*Capsicum annium* L.) fruits. Incubation of [1-^{14}C] isopentyl diphosphate with different chromoplast fractions showed that the membrane was incapable of synthesizing carotenoids, while the stroma synthesized the first colorless carotenoid, phytoene (Table 2.4). Thus, phytoene synthetase, the enzyme responsible, could be a useful indicator of the chromoplast stroma. Increased incorporation of the labeled substrate in the presence of chromoplast membranes was attributed to their desaturation and cyclization systems, which produced colored carotenoids. This study showed that the enzymes involved in carotenoid biosynthesis were compartmentalized in the chromoplasts. The stroma synthesized phytoene synthesis, which underwent desaturation and cyclization to the colored carotenoids in the membrane. These researchers proposed that a protein carrier transferred phytoene to the chromoplast membrane, or alternatively that phytoene synthetase itself was bound to the membrane, where it discharged phytoene for further reactions.

1. Carotenoid Changes during Ripening

Ebert and Gross (1985) examined the carotenoid pigments in the peel of ripening persimmon (*Diospyros kaki* cv. 'Triumph'). A steady decline in chlorophyll (*a* and *b*) was observed during the course of ripening which disappeared in the harvest-ripe fruit (Table 2.5). The chloroplast carotenoids (α- and β-carotenes, lutein, violaxanthin, and neoxanthin) decreased, followed by the gradual synthesis of the chromoplast carotenoids (cryptoxanthin, antheraxanthin, and zeaxanthin). Ikemefuna and Adamson (1985) monitored changes in the chlorophyll and carotenoid pigments of ripening palm fruit (*Elaeis quineeris* 'Palmal') and noted a similar degeneration of chloroplasts and

TABLE 2.5 Carotenoid Changes in the Peel of Persimmon (*Dlospyros kaki* cv. 'Triumph') during Postharvest Ripening

	Ripening Stage		
	Harvest-ripe	Intermediate	Fully Ripe
Total carotenoids (µg/g fr. wt)	128.0	366.0	491.0
Carotenoid pattern (% of total carotenoids)			
Phytofluene	—	—	0.4
α-Carotene	1.6	1.2	1.0
β-Carotene	9.4	7.6	6.7
Mutatochrome	—	0.7	—
γ-Carotene	—	0.4	—
Lycopene	1.1	0.5	8.2
β-Cryptoxanthin	29.2	50.0	48.2
Cryptoxanthin 5,6-epoxide	0.9	1.2	1.9
Cryptoflavin	0.7	2.1	2.9
Lutein	12.4	5.5	4.1
Zeaxanthin	9.3	9.7	5.9
Mutatoxanthin	0.8	4.7	1.8
Isolutein	0.5	—	0.3
trans-Antheraxanthin	5.4	2.0	4.8
cis-Antheraxanthin	6.2	2.2	2.3
Luteoxanthin	1.7	1.8	1.9
trans-Violaxanthin	6.9	3.7	3.8
cis-Violaxanthin	6.7	1.5	2.0
Neoxanthin	7.2	5.2	3.8

From Ebert and Gross (1985). Reprinted with permission. Copyright © by Pergamon Press.

formation of chromoplasts. These changes were accompanied by increase in carotenogenesis with the formation of α- and β-carotenes as the major pigments in the ripe fruit.

Farin *et al.* (1983) examined the change in carotenoids during the ripening of the Israeli mandarin hybrid 'Michal' (*Citrus reticulata*). This particular fruit is the most highly colored of the citrus fruits with a bright-reddish color. Total chlorophyll decreased rapidly in the peel and completely disappeared at the ripening stage (Table 2.6). The total carotenoids decreased at the color break stage due to a decline of the chloroplast carotenoids, δ-carotene, lutein, violaxanthin, and neoxanthin. A marked rise in carotenogenesis followed, with an increase in the chromoplast carotenoids, cryptoxanthin and C_{30} apocarotenoids. The two C_{30} apocarotenoids, β-citraurin and β-citraurinene, accounted for 26.1% and 9.9%, respectively, of the total carotenoids in the ripe fruit. Their formation appeared to require asymmetric degradation of a C_{40} fragment from the side of the C_1 carotenoid, cryptoxanthin.

Gross *et al.* (1983) studied changes in pigments and ultrastructure during the ripening of pummelo (*Citrus grandis* Osbech). Chlorophyll decreased from 90 µg/g at the unripe (green) stage to 11 µg/g in the ripe pale-yellow fruit, totally disappearing by the fully ripe stage. The chloroplast carotenoids, β-carotene, lutein, violaxanthin, and neoxanthin, all decreased during ripening, with the β-carotene totally disappearing. Accumulation of phytofluene was evident at the color break, reaching 67% of the total carotenoids in the ripe fruit. Besides phytofluene, other chromoplast carotenoids were detected, including δ-carotene, neurosporene cryptoxanthin, and cryptoflavin.

TABLE 2.6 Pigment Distribution in the Flavedo of a Mandarin Hybrid (*Citrus reticulate cv.* 'Michal') during Ripening

	Peel		
	Green	Color Break	Ripe
Fruit diameter (cm)	4.70	4.85	5.10
Chlorophyll *a* (µg/g)	240.0	52.6	—
Chlorophyll *b* (fr. wt)	86.0	15.8	—
Total carotenoids	143.4	51.0	174.1
Carotenoid pattern (% of total carotenoids)			
Phytofluene	—	5.2	3.1
α-Carotene	9.7	2.4	0.2
β-Carotene	6.9	2.5	0.3
ζ-Carotene	—	—	0.4
δ-Carotene	—	—	0.1
Mutatochrome	—	—	0.5
Lycopene	—	—	—
β-Apo-8′-carotenal	0.7	0.7	1.3
α-Cryptoxanthin	1.9	—	—
β-Cryptoxanthin	—	3.1	6.4
Cryptoxanthin 5,6-epoxide	—	—	0.4
Cryptoxanthin 5′,6′-epoxide	—	—	0.3
β-Citraurinene	—	9.5	9.9
β-Citraurin	—	12.3	26.1
Lutein	23.5	12.8	2.6
Zeaxanthin	3.9	1.6	1.0
Mutatoxanthin	1.5	0.2	0.2
trans-Antheraxanthin	3.6	4.3	1.8
cis-Antheraxanthin	—	—	2.5
Luteoxanthin	5.8	5.6	9.1
trans-Violaxanthin	14.0	11.7	9.8
cis-Violaxanthin	11.0	18.2	19.8
trans-Neoxanthin	11.7	6.6	4.2
cis-Neoxanthin	3.6	3.3	—
Neochrome	—	—	—
trans-Trollixanthin	—	—	—
Trollichrome	—	—	—
Unknown	2.2	—	—

Adapted from Farin *et al.* (1983). Reprinted with permission. Copyright © by Pergamon Press.

These studies showed that compositional changes in carotenoids during ripening reflect the transformation of chloroplasts to chromoplasts. Earlier work by Eilati *et al.* (1975) on ripening Shamouti 'orange peel' showed that parallel transformations occurred irrespective of whether the fruit was attached to or detached from the tree. The transformation of chloroplasts to chromoplasts during degreening of citrus fruits was reported by Huff (1984) to be regulated by the accumulation of sugar in the epicarp. The reverse transformation associated with the regreening of certain citrus species was found to be accompanied by the disappearance of sugars. This regreening phenomenon of *Citrus sinensis* fruit epicarp observed by Thomson *et al.* (1967) was attributed to the reversion of chromoplasts to chloroplasts. Thus, sucrose promoted the formation of chromoplasts while nitrogen stabilized the chloroplasts by retarding the degreening process.

2. Carotenoid Degradation: Processing and Storage

Carotenoids impart most of the yellow and orange colors to fruits and vegetables such as pineapples and carrots. The unsaturated nature of carotenoids renders them particularly susceptible to isomerization and oxidation, resulting in a loss of color which is most pronounced following oxidation. The latter can be brought about by the action of lipoxygenase, which can bleach carotenoids (Eskin *et al.*, 1977). From recent research, it appears that differences in the abilities of lipoxygenases from fruits and vegetables to oxidize carotenoids are due to the isoenzymes present. Lipoxygenase isoenzymes are classified as either type 1 or type 2 depending on their pH optimum and product specificity. Of these isoenzymes, lipoxygenase-2 exhibits a more acidic pH optimum and is involved in cooxidation reactions leading to pigment discoloration (Klein and Grossman, 1985). For example, pea lipoxygenase-2 was shown by Arens *et al.* (1973) to be an effective oxidizer of carotenoids. Chepurenko *et al.* (1978) attributed carotenoid bleaching to the combined effect of the lipoxygenase isoenzymes in peas and not just to lipoxygenase-2. Yoon and Klein (1979) showed definite differences between the rate of carotenoid oxidation for these two pea lipoxygenase isoenzymes. These enzymes were also involved in the biosynthesis of traumatic acid, a wound-healing hormone, as well as the plant growth regulator jasmonic acid (Zimmerman and Coudron, 1979; Vick and Zimmerman, 1983).

Carotenoids are extremely susceptible to non-enzymatic oxidation in dehydrated fruits and vegetables. For example, powdered, dehydrated carrots were reported by MacKinney *et al.* (1958) to lose 21% of their carotenoids when stored in air. Of particular importance to their stability is the amount of moisture present in dehydrated products. The effect of a_w on the degradation of β-carotene in model systems was studied by Chou and Breene (1972). These researchers showed that at a_w of 0.44 the oxidative decoloration of β-carotene was reduced compared to the corresponding dry system with or without the presence of the antioxidant butylated hydroxytoluene (BHT) (Figure 2.9). It is clear from this study that water acts as a barrier to oxygen diffusion.

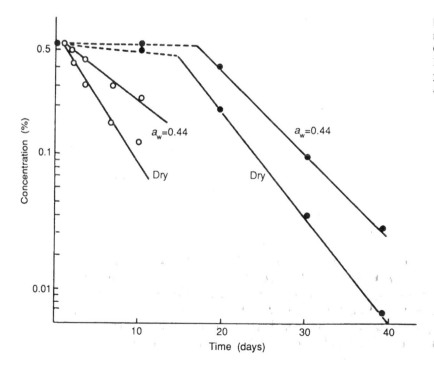

FIGURE 2.9 Decoloration of β-carotene in model systems. Dry refers to systems held over solid CaCl$_2$; $a_w = 0.44$ to systems held over saturated K$_2$CO$_3$ (O: control; ●: BHT-butylated hydroxytoluene) (Chou and Breene, 1972). *(Copyright © by Institute of Food Technologists.)*

When carotenoids are heated in the absence of air some of the *trans* double bonds undergo isomerization to the corresponding *cis* isomers. Acids also catalyze isomerization from the all-*trans* form to the corresponding *cis* isomer. The change in shape associated with the *cis* isomers reduces the resonance in the molecule as well as the color intensity. These reactions were responsible for the difference in quality between canned and fresh pineapple (Singleton *et al.*, 1961). A change in the spectrum of the extracted carotenoid pigments was reported for canned pineapples, including loss of a peak at 466 nm together with peaks of shorter wavelengths. This spectral shift, although too subtle for the human eye, caused a slight color shift from orange—yellow in the fresh pineapple to a more lemon—yellow in the canned product. This shift was attributed to isomerization of carotenoids with 5,6-epoxide groups to the corresponding 5,6-furanoid oxides. In the intact fruit the natural cell vacuole acids have a catalytic effect by coming into contact with the carotenoid-bearing plastids when cell membranes are disrupted during handling or processing. Several studies examined the effect of cooking carrots on the formation of *cis* isomers. Klaui (1973), however, suggested that the effects of cooking on color and carotene content were insignificant.

F. Anthocyanins

Anthocyanins are responsible for the attractive pink, red, purple, and blue colors of flowers, leaves, fruits, and vegetables (Harborne, 1967). These are water-soluble pigments that accumulate in the epidermal cells of fruits as well as in roots and leaves (Timberlake and Bridle, 1975, 1982; Harborne, 1976). These pigments are formed as the fruit matures and ripens, as is evident for strawberries (Fuleki, 1969), certain varieties of cherries, raspberries, cranberries, and apples (Zapsalis and Francis, 1965; Proctor and Creasy, 1971; Cansfield and Francis, 1970; Craker and Wetherbee, 1973; Bishop and Klein, 1975), black grapes (Liao and Luh, 1970), and blueberries (Suomalainen and Keranen, 1961), as well as in such vegetables as red cabbage, russet potato, radishes, and red onion (Fuleki, 1971; Small and Pecket, 1982). The primary role of anthocyanins is that of an insect or bird attractant for pollination of flowers and for fruit seed dissemination by animals (Brouillard, 1983).

Anthocyanins are flavonoid pigments whose structure is based on the phenylpropanoid nucleus. They occur in nature as glycosides in which the aglycone forms, or anthocyanidins, are substituted flavylium salts. The structural formula of the flavylium cation is as follows:

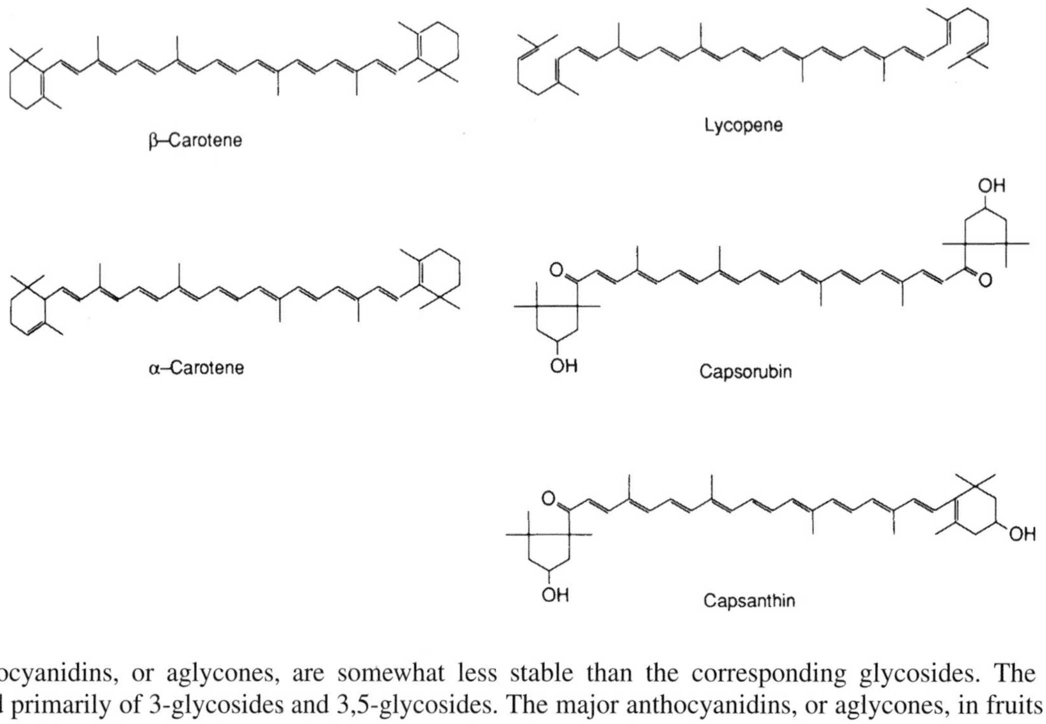

β—Carotene

Lycopene

α—Carotene

Capsorubin

Capsanthin

The anthocyanidins, or aglycones, are somewhat less stable than the corresponding glycosides. The latter are composed primarily of 3-glycosides and 3,5-glycosides. The major anthocyanidins, or aglycones, in fruits are listed in Scheme 2.16. D-Glucose, D-lactose, l-rhamnose, and D-xylose are the main sugars linked to the anthocyanidin at the 3-position. The intensity of the color is determined by the nature of the pigment, its concentration, and pH, as well as the presence of pigment mixtures, copigments, and certain metallic ions (Brouillard, 1983). Anthocyanins are dissolved in the aqueous vascular sap in the plant cell, which is slightly acidic. These pigments tend to be more stable

SCHEME 2.16 Major food anthocyanins.

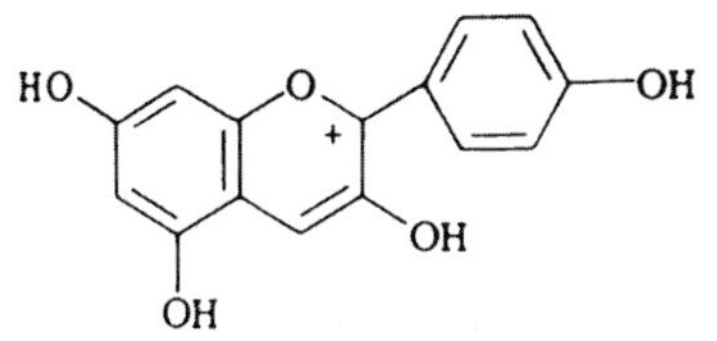

Pelargonidin (3, 5, 7, 4´-
tetrahydroxyflavylium cation)

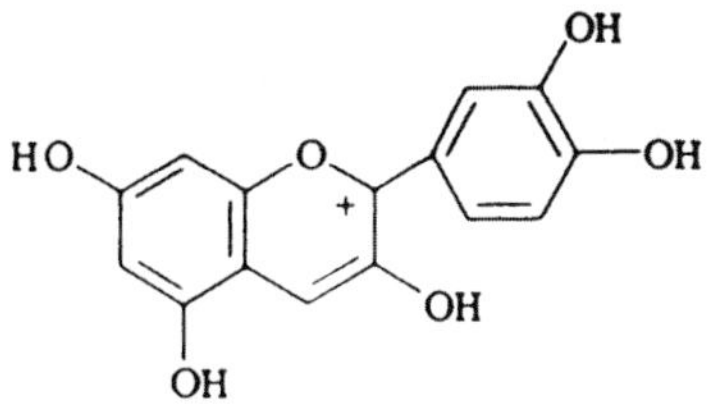

Cyanidin (3, 5, 7, 3´, 4´-
pentahydroxyflavylium cation)

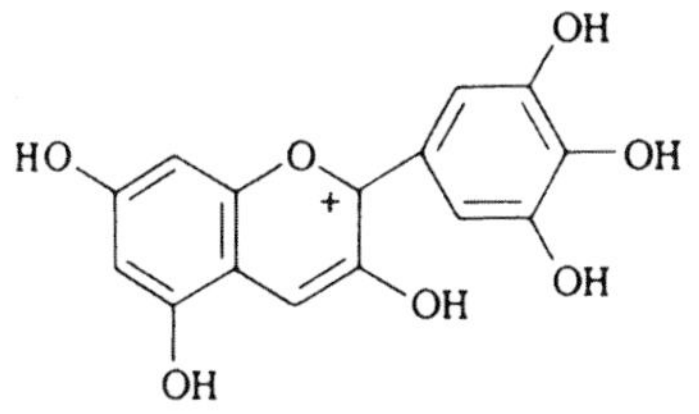

Delphinidin (3, 5, 7, 3´, 4´, 5´-
hexahydroxyflavylium cation)

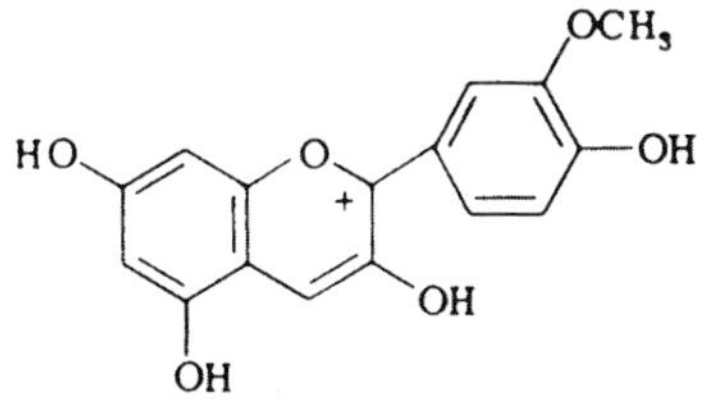

Peonidin (3, 5, 7, 4´-tetrahydroxy-
3´-methoxyflavylium cation)

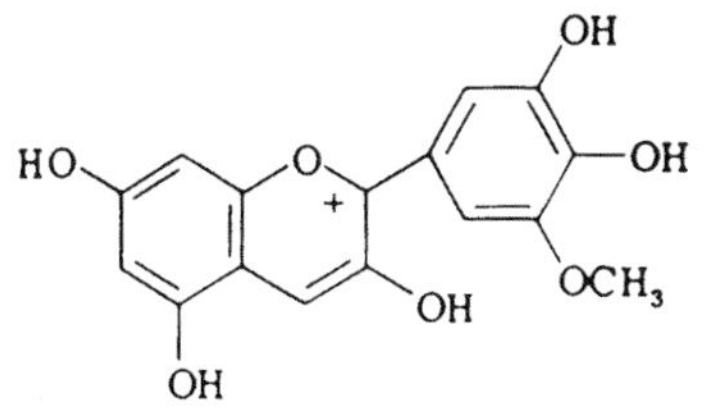

Petunidin (3, 5, 7, 3´, 4´-pentahydroxy-
5´-methoxyflavylium cation)

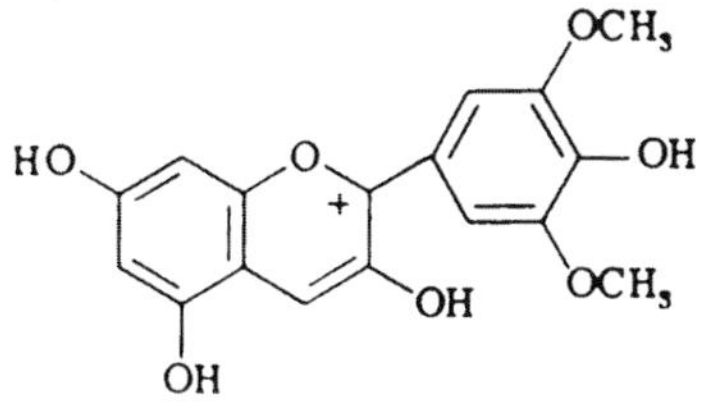

Malvidin (3, 5, 7, 4´-tetrahydroxy-
3´, 5´-dimethoxyflavylium cation)

under these acidic conditions. The different shades of colors of anthocyanidins reflect the nature of their hydroxylation and methoxylation patterns. An increase in hydroxylation is accompanied by an increase in the blue hue, whereas methoxylation enhances the red color (Braverman, 1963). The distribution of anthocyanidins in some edible fruits is shown in Table 2.7.

1. Biosynthesis of Anthocyanins

The biosynthesis of anthocyanins is well established (Mol *et al.*, 1989; Forkmann, 1991), with a generalized scheme shown in Scheme 2.17 (Holton and Cornish, 1995). The synthesis of aromatic amino acids is achieved in plants via the shikimic acid pathway. The synthesis of all flavonoids, including anthocyanins, involves three key enzymes,

TABLE 2.7 Distribution of Anthocyanins in Some Edible Fruit

Anthocyanidin	Fruit
Cyanidin	Blackberry, rhubarb
Cyanidin, delphinidin	Blackcurrant
Cyanidin, peonidin	Cherry
Cyanidin, pelargonidin	Strawberry

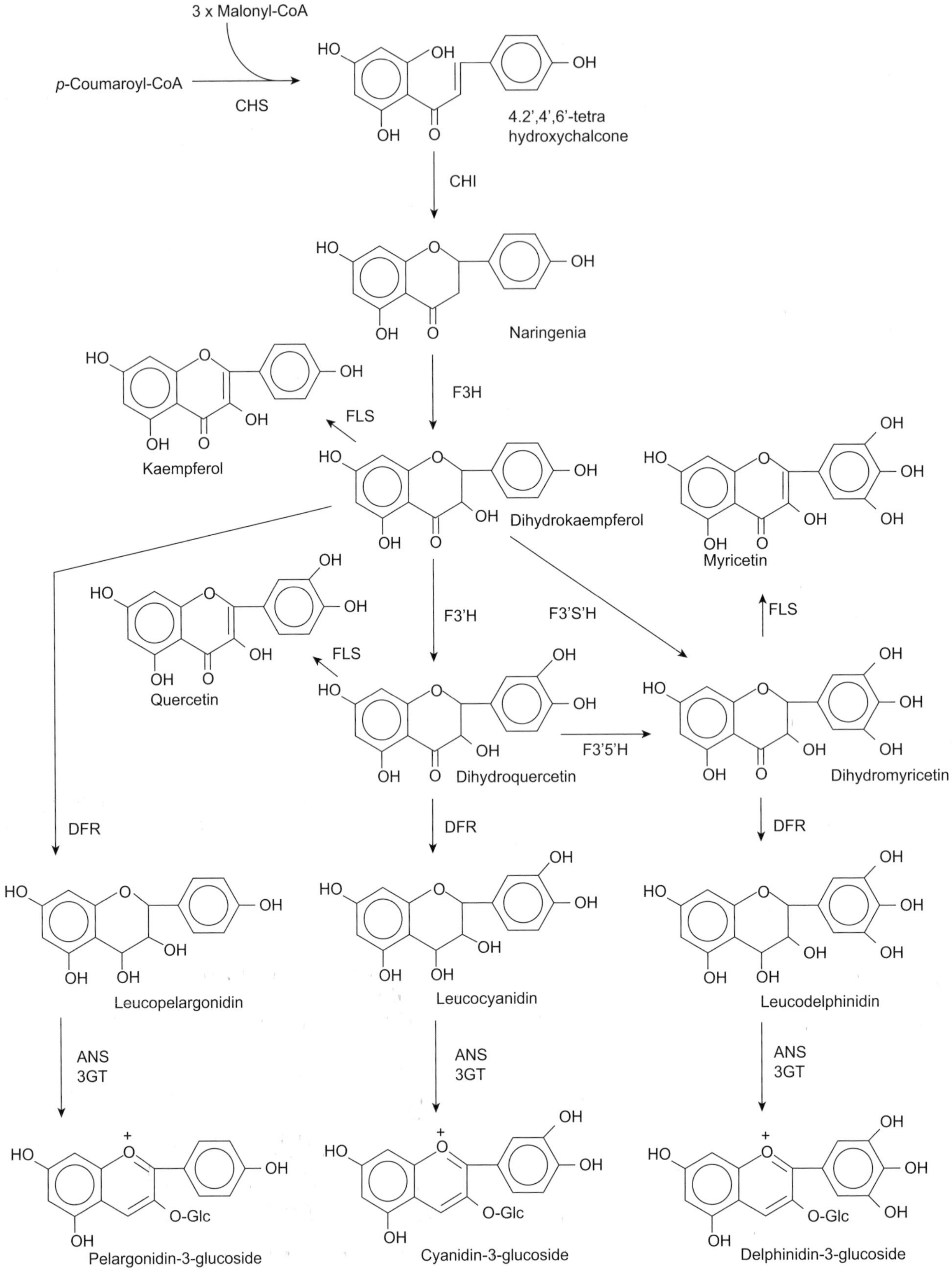

SCHEME 2.17   Anthocyanin and flavonoid biosynthetic pathway (Holton and Cornish, 1995).

chalcone synthase (CHS), chalcone isomerase (CHI), and flavonoid 3-hydroxylase (F_3H) (Gou *et al.*, 2011). These enzymes appear to be present as multiprotein enzyme complexes (referred to as metabolons) channeling the substrate through specific subpathways (Winkel, 2004; Jorgensen *et al.*, 2005). For example CHS, CHI, F_3H, and dihidro-flavonoid-4-reductase (DFR) are all located in the endoplasmic reticulum in *Arabidopsis* (Lepiniec *et al.*, 2006). CHS is responsible for the stepwise condensation of the phenylpropanoid precursor, *p*-coumaryl-CoA, with three molecules of malonyl CoA to form tetrahydroxychalcone. The next enzyme, CHI, isomerizes the yellow-colored chalcone to the colorless naringenin. The third step, the formation of dihydrokaemferol, is catalyzed by F_3H. CHI is responsible for the formation of the six-membered heterocyclic ring C of the flavonone which was first isolated by Moustafa and Wong (1967). Three other enzymes then catalyze the conversion of colorless dihydroflavonols to anthocyanins. DFR reduces dihydroflavols to flavan-3,4-ci-diols (leucoanthicyanidins) which then undergo oxidation, dehydration, and glycosylation to form the various anthocyanins, brick-red perlogonidin, red cyanidin, and blue delphinidin pigments (Holton and Cornish, 1995).

The first gene involved in flavonoid biosynthesis was the CHS gene isolated from parsley (Kreuzaler *et al.*, 1983). This led to the isolation of clones of two different CHS genes from petunia (Reif *et al.*, 1985). Of 12 different CHS genes later identified in petunia, however, only four appear to be expressed (Koes *et al.*, 1989). Over the past two decades the transcriptional control of flavonoid biosynthesis has been extensively studied (Broun, 2005). Genetic screens have identified several classes of transcriptional regulators (Rubin *et al.*, 2009). For a more detailed discussion of the regulators of anthocyanin biosynthesis the paper by Gou *et al.* (2011) is recommended.

2. Anthocyanins: Effect of Processing

Anthocyanins are generally unstable during processing with a net loss of color during canning, bottling, and other thermal processing operations. Fruits and vegetables contain many enzymes capable of decolorizing anthocyanins; however, these can be inactivated by blanching. Such enzymes include polyphenol oxidase, anthocyanase, and peroxidase (Grommeck and Markakis, 1964; Peng and Markakis, 1963; Sakamura *et al.*, 1966). In addition to enzymes, thermal stresses, pH, oxygen, and light have all been reported to accelerate anthocyanin degradation (Eskin, 1979; Brouillard, 1983; Hubbermann, 2005; Bordignon *et al.*, 2007; Queiroz *et al.*, 2009; Patras *et al.*, 2010). A recent paper by Hillman *et al.* (2011) showed that the degradation of anthocyanins in grape juice and concentrate increased with increasing temperature and juice concentration. However, of these, pH is the most important factor affecting the stability of the anthocyanin (Mazza and Brouillard, 1987).

Anthocyanins are stable under acidic conditions although model systems show, in most cases, that they are in a colorless form. The ability to retain their color in plants is attributed to the formation of complexes with other phenolics, nucleic acids, sugars, and amino acids, as well as metallic ions such as calcium, magnesium, and potassium (Brouillard, 1983). Several acylated anthocyanins have been identified in flowers which exhibit remarkable stability in neutral or weakly acidic solutions. One such pigment isolated from the petals of Chinese bellflower (*Platycodon grandiflorium*) is platyconin (Saito *et al.*, 1971). Its structure was subsequently confirmed by ^{1}H-nuclear magnetic resonance (NMR) spectroscopy (Goto *et al.*, 1983). The two acyl groups in this anthocyanin, one located above the pyrilium ring and the other below it, result in its stability in neutral solutions. The presence of two or more acyl residues linked to sugars appears to provide excellent color stability in neutral conditions (Mazza and Brouillard, 1987). Acylated anthocyanins have since been identified in many fruits and vegetables including red onion, red cabbage, red potatoes, broccoli, black carrots, radishes, purple sweet potatoes, and grapes (Giusti and Wrolstadt, 2003; Stintzing and Carle, 2004; Pliszka *et al.*, 2009). Their enhanced stability makes them very attractive as food colorants. Terahara *et al.* (2004) identified four acylated anthocyanins in the callus induced from the storage of purple-fleshed sweet potato (*Ipomoea batatas* L.). They all exhibited higher stability in neutral pH as well as greater DPPH radical scavenging properties compared to the corresponding non-acylated anthocyanins.

The effect of pH on the color of anthocyanins is well established. Harper (1968) showed that perlargonidin chloride was quite stable in a pH range of 1−3 and existed as a deep-red oxonium ion (I). As the pH increased towards neutrality, a loss of red color was observed because of the formation of a colorless pseudobase (II), which was in equilibrium with its keto form (III). The latter then underwent ring opening to the α-diketone form (IV), which was present at pH 3−7. Above pH 7, the anhydro base (V) predominated, producing a purple color which rapidly faded below pH 7 owing to the formation of the pseudobase (II) and α-diketone. Increasing the pH to more alkaline conditions, the purple color changed from mauve to blue because of the formation of the ionized anhydro base (VI). The formation of a brown precipitate was due to degradation of the ketone, which was responsible for the decoloration in anthocyanin-containing fruits during prolonged storage (Scheme 2.18).

SCHEME 2.18 Changes in the molecular structure of perlargonidin chloride with pH (Harper, 1968).

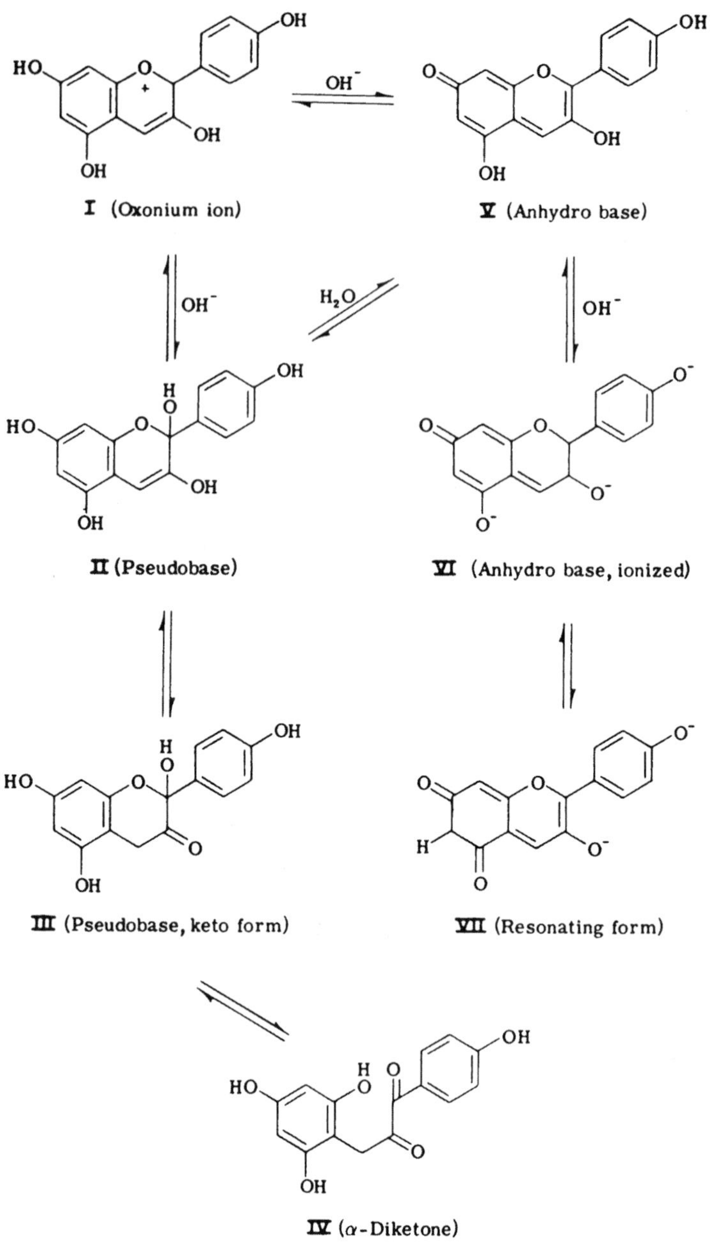

I (Oxonium ion)   V (Anhydro base)

II (Pseudobase)   VI (Anhydro base, ionized)

III (Pseudobase, keto form)   VII (Resonating form)

IV (α-Diketone)

However, Gonnet (1998) showed that spectral variations to describe the color changes in anthocyanin solutions, as affected by pH, should include the entire spectral curve and not just their visible lambda max. The color description should also include tristimulus color attributes of hue, saturation, and lightness. While the CIELAB scale has been reported in a wide range of colors, juices, extracts, and flowers, and some pure anthocyanins, Torskangerpoll and Anderson (2005) noted that no work has been conducted on color variation during the prolonged storage of anthocyanins over the entire pH range. These researchers examined the effect that anthocyanin structure, such as 5-glycosidic substitution and aromatic acylation, has on color stability over 98 days' storage at 10°C over a pH range of 1.1–10.5. Under these conditions, three anthocyanin pigments were examined: cyanidin 3-glucoside, cyanidin-3-(2′-glucosylglucoside)-5-glucoside, and cyanidin-3-(2′-(2′-sinapoylglucosyl)-6′-sinapoylglucoside)-5-glucoside. They confirmed the importance of structure in determining the properties of anthocyanins, with the acylated anthocyanins exhibiting greater stability at all pH values except at pH 1.1 and at the more alkaline conditions. At pH 1.1, all pigments retained their color during storage, whereas they all underwent dramatic color changes under the more alkaline conditions.

VI. TEXTURE

The texture of fruits and vegetables is related to the structure and organization of the plant cell walls and intercellular cementing substances (Eskin, 1979). The architecture of the cell wall has been the subject of numerous studies suggesting that it is composed of cellulose fibrils located in a matrix of pectic substances, hemicellulose, proteins, lignin, low-molecular-weight solutes, and water (Van Buren, 1979). In edible plants, the primary cell wall is of central interest as secondary cell walls are virtually absent in mature fruit (Nelmes and Preston, 1968). In fact, the primary cell wall provides most of the dietary fiber in our diet (Jarvis, 2011). During the ripening of fruits, a loss of texture results from the deterioration of the primary cell wall constituents. This is in sharp contrast to the maturation of vegetables, which is accompanied by a toughening of the texture. The latter is due to the development of the secondary cell walls in vegetables, which contribute to the tough and fibrous texture, as a result of lignin deposition.

A. Cell Wall

The cell walls of plants are composed primarily of complex carbohydrates (Northcote, 1963; Jarvis, 2011). The presence of a protein fraction was later established in the primary cell wall of plants (Lamport, 1965). Because of the complexity of the cell wall, numerous models have been proposed, including the first complete model by Keegstra *et al.* (1973). These models tended to raise more questions than answers, however, as none described the cell wall structure adequately. The general approach has been to obtain hydrolyzed fragments, characterize them, and reconstruct the individual fragments. Early research suggested that the hydroxyproline-rich proteins are covalently linked to the pectic substances through the hemicellulose fraction xyloglucan (Hayashi, 1989). Since then, three major families comprising the hydroxyproline-rich glycoproteins (HRGPs) have been identified (Kieliszewski and Shpak, 2001). Textural changes occurring during ripening of fruit are due primarily to enzymatic changes in cell wall architecture. Most affected is the middle lamella, an intercellular cement between the primary cell walls of adjacent cells, which is particularly rich in pectic substances.

1. Cell Wall Constituents

a. Polysaccharides

An important constituent of the primary wall of higher plants is cellulose. It is present as linear aggregates or microfibrils of around 36 β-1,4-glucan chains about 3 nm in diameter (Somerville, 2006). The individual chains are held together by hydrogen bonds through the hydroxyl groups at carbon-6 in one cellulose chain with glycosidic oxygens of the adjacent chains (Northcote, 1972). Within the microfibril are highly organized or crystalline regions as well as some amorphous regions. Cellulose serves as a scaffold that binds other cell wall components (Lerouxel *et al.*, 2006). For a thorough discussion of cellulose structure and biosynthesis, the following reviews are recommended: Doblin *et al.* (2003), Hayashi *et al.*, (2005), Saxena and Brown (2005), and Joshi and Mansfield (2007).

Flavylium cation

b. Pectic Substances

Pectic substances comprise one-third of the dry substance of the primary cell walls of fruits and vegetables (Van Buren, 1979). They are complex polysaccharides that include domains of homogalacturonan (HG) and rhamnogalacturonan I (RG-I) with minor amounts of rhamnogalacturonan II (RG-II) (Lerouxel *et al.*, 2006). The basic structure of HG, which accounts for over 60% of the cell wall pectins, consists of a polymer of 1,4-linked α-D-galacturonic acid in which some of the carboxyl groups are methyl esterified (Scheme 2.19) (Ridley *et al.*, 2001). In some plants HG may be partially O-acetylated at O_2 and O_3 (Ishii, 1995, 1997). HG is covalently linked to RG-I and RG-II and was hypothesized to be covalently cross-linked to the hemicelluloses fraction, xyloglucan (XG)

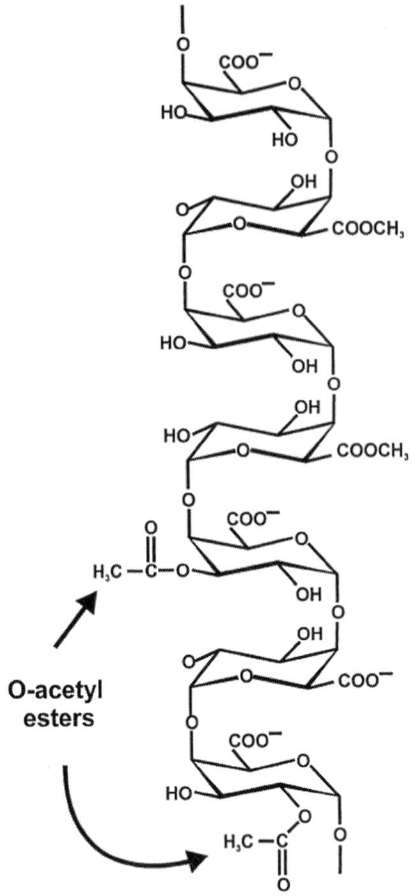

SCHEME 2.19 The primary structure of homogalacturonan (Ridley *et al.*, 2001).

(Popper and Fry, 2008). The potato cell wall is rich in HG, which is particularly dense in the middle lamellae (Bush, 2001). The HG in the cell walls of tomatoes and mangoes was reported to contain about 35% and 52% uronic acid, respectively (Muda *et al.*, 1995). RG-1 is composed of a backbone of the repeating disaccharide, in which the GalpA residues may be acetylated on C2 and/or C3 (Komalavilas and Mort, 1989) (Scheme 2.20).

SCHEME 2.20 A model showing the structural features of rhamnogalacturonan I (Ridley *et al.*, 2001).

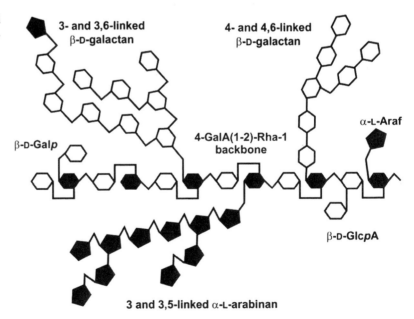

RG-II is structurally quite different, with a backbone of [→4)- α-D-GalpA-(1→2)-α-L-Rhap-(1→] (O'Neill *et al.*, 1990) (Scheme 2.21). For a more detailed description of pectins, the reviews by Ridley *et al.* (2001) and Caffall and Mohnen (2009) are recommended.

c. Hemicellulose

Hemicelluloses, the third and heterogeneous group of polysaccharides in plant cell walls, all have β-(1→4)-linked backbones with an equatorial configuration (Scheller and Ulvskov, 2010). The main groups present in all terrestrial plants include xyloglucan, xylans, mannans, glucomannans, and β-(1→3,1→4)-glucans. At one time galactans, arabinans, and arabinogalactans were also considered to be hemicelluloses. However, Scheller and Ulvskov (2010) pointed out that they appear to be a part of pectin molecules and do not have the equatorial β-(1→4)-linked backbone structure. Xyloglucan, the major hemicellulose in the primary cell walls, is a neutral polysaccharide composed of a β-(1→4)-D-glucose backbone with side-chains rich in α-D-xylose, β-D-gulactose and occasionally some α-fucose residues (Popper and Fry, 2008). For a more detailed discussion of hemicelluloses the review article by Scheller and Uvskov (2010) is strongly recommended.

d. Proteins

Proteins identified in the primary cell walls of plants include those rich in hydroxyproline. This family of hydroxyproline-rich glycoproteins (HRGPs) includes three major groups, the highly periodic and lightly arabinosylated repetitive proline-rich proteins (PRPs), the periodic and highly arabinosylated extensins, and the most highly glycosylated and least periodic arabinogalactan-proteins (Kieliszewski and Shpak, 2001). In addition to extensin and hydroxyl-proline-rich proteins, a third group of proteins has been identified in plant cell walls. These are glycine-rich proteins which are also part of structural components of the plant cell wall (Ringli *et al.*, 2001). A proteomic approach to plant cell wall proteins was presented by Jamet *et al.* (2006).

e. Lignin

Lignin includes a large group of aromatic polymers produced from the oxidative coupling of 4-hydroxyphenylpropanoids ($C_6–C_3$) (Boerjan *et al.*, 2003; Ralph *et al.*, 2004). These polymers are primarily deposited in the secondary cell walls and provide mechanical strength to the plant cell walls. They are also responsible for the

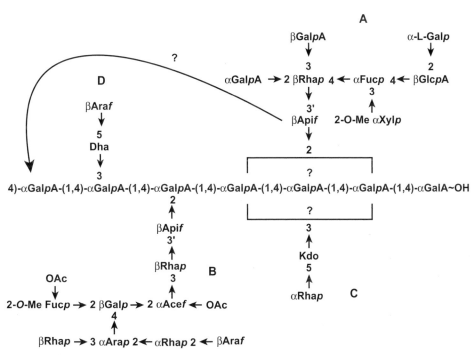

SCHEME 2.21 The primary structure of rhamnogalacturonan II (Ridley *et al.*, 2001).

toughness of such vegetables as celery and asparagus (Segerlind and Herner, 1972; Herner, 1973). For a more detailed discussion of lignin a recent review by Vanholme *et al.* (2010) is recommended.

f. Cell Wall Biosynthesis

Considerable progress has been made over the past two decades on the biosynthesis of the complex polysaccharides and their deposition in the cell wall (Lerouxel *et al.*, 2006; Joshi and Mansfield, 2007; Caffall and Mohnen, 2009). While there are still gaps in our knowledge, we have a much better understanding of its regulation (Zhong and Ye, 2007).

Pectin biosynthesis occurs in the Golgi apparatus, a complex organelle composed of stacks of flattened vesicles containing proteins. The current model for pectin biosynthesis suggests that nucleotide sugars are synthesized on the cytosolic side of the Golgi and transported into the Golgi lumen by membrane-spanning protein transporters, or alternatively synthesized in the Golgi lumen (Scheller *et al.*, 1999; Scheible and Pauly, 2004; Mohnen, 2008; Mohnen *et al.*, 2008). Multiple activities are required for the complete biosynthesis of HG, RG-I and RG-II, with many glucosyltransferases (GT) involved (Scheller *et al.*, 1999; Moyne *et al.*, 1999, 2008). For example, the complete synthesis of HG requires several specific catalysts, α1,4-GalATs (HG:GalATs), to transfer D-GalA from UDP-D-GalA to the growing HG polymer (Caffall and Mohnen, 2010). HG:GalAT has since been characterized in tomato and turnip (Lin *et al.*, 1966; Bolwell and Northcote, 1983), peas (Sterling *et al.*, 2001), and pumpkin (Ishii, 2002).

Once formed, GHHG can be modified by the addition of methyl groups at the C6 carboxyl group or acetyl groups at O2 and O3 of the GalA residues (Willats *et al.*, 2001). Pectin methyltransferases (PMTs) then act specifically on the pectic polysaccharides in the Golgi apparatus by transferring a methyl group from the donor *S*-adenosyl-methionine (SAM) (Villemez *et al.*, 1966; Kauss *et al.*, 1967; Bruyant-Vannier *et al.*, 1996), releasing them in a highly methylesterified form (Li *et al.*, 1997; Lennon and Lord, 2001). Multiple PMT enzymes are needed for the synthesis of pectin, as different isoforms were discovered each exhibiting their own pH optima and substrate preference (Krupkova *et al.*, 2007). For a discussion of the biosynthesis of RG-I and RG-II, the article by Caffall and Mohnen (2009) is recommended.

The biosynthesis of a cellulose-like ($1 \rightarrow 4$) glucan involves several cytosolic and membrane-bound enzymes that primarily produce pools of UDP-glucose. The latter provides the primary substrate for cellulose synthase (Carpita and McCann, 2000). Two pathways were found to lead directly to the production of UDP-glucose (Joshi and Mansfield, 2007). In the first pathway, sucrose is cleaved by sucrose synthase, releasing UDP-glucose and fructose. The second pathway is much more dependent on the hexose phosphate pool, with UDP-glucose pyrophosphorylase (UGPase) phosphorylating glucose-1-phosphate. An alternative, and indirect pathway for producing UDP-glucose, is through the hydrolysis of sucrose by invertase. Irrespective of how UDP-glucose is produced, it is then catalyzed by glycosyltransferase to form the cellulose polymer. This enzyme is referred to as inverting, as the glucosyl residue in UDP-glucose is in the α-configuration which must be converted to the β-linkage, exclusive to cellulose (Guerriero *et al.*, 2010). A major breakthrough in understanding cell wall biosynthesis was identification of genes encoding cellulose synthase (*Ces*) in cotton fiber (Pear *et al.*, 1996; Delmer, 1999). Cellulase synthase (*Ces*) has since been identified in other higher plants. For a more detailed review of cellulose biosynthesis including the formation of microfibrils, the following reviews are recommended: Lerouxel *et al.* (2006), Joshi and Mansfield (2007), and Guerriero *et al.* (2010).

2. Cell Wall Degradation

The softening of fruits during ripening is attributed to changes in pectin (Kertesz, 1951; Hobson, 1967; Tavakoli and Wiley, 1968; Pressey *et al.*, 1971). This is characterized by a decrease in the level of insoluble pectin (protopectin) with a concomitant increase in soluble pectic substances (Ben-Arie *et al.*, 1979). Protopectin is the generic name ascribed to the water-insoluble high-molecular-weight parent pectin substances present in unripe fruit. Little is known about this polymer other than that its hydrolysis during ripening yields soluble polyuronides, pectin, and pectinic acids (Worth, 1967; Eskin, 1979; John and Dey, 1986). The latter are polygalacturonic acid polymers varying in the degree of methylation. The pectic material in the middle lamella is quite different from that found in the primary cell walls and consists of pectic substances in the free state or as the corresponding calcium pectate (Eskin, 1979). The degradation of the loosely bound soluble pectin is responsible for the softening of the texture during ripening (Doreyappa *et al.*, 2001). These pectin-degrading enzymes are classified according to their mode of action on pectin and pectic substances and include polygalacturonases (PG) (EC 3.2.1.15), pectin methylesterases (PM) (EC 3.1.1.11), and pectate and pectin lyases (PL) (EC 4.2.2.2) (Sakai *et al.*, 1993; Wong, 1995; Chauhan, 2001).

a. Polygalacturonases

Polygalacturonase (PG), an important pectolytic glycanase, has been primarily implicated in the softening of fruit during ripening (Poovaiah and Nukuya, 1979). It is considered to be primarily responsible for the dissolution of the middle lamella during ripening (Jackman and Stanley, 1995; Voragen *et al.*, 1995). Since the preferred substrate of these enzymes is D-galacturonans, Rexova-Benkova and Markovic (1976) referred to them as D-galacturonases. These enzymes have been reported in many fruits, including peaches (Pressey and Avants, 1973a), pears (McCready and McComb, 1954; Pressey and Avants, 1976; Bartley and Knee, 1982), and tomatoes (Foda, 1957; Patel and Phaff, 1960a, b; Hobson, 1964). The activity of these enzymes increases during the ripening process, when it hydrolyzes pectic material in the middle lamellae and cell walls (Hobson, 1965; Pressey, 1977). The change in polygalacturonase activity during ripening is illustrated in Figure 2.10 for peaches. Pressey *et al.* (1971) found that the increase in enzyme activity was accompanied by an increase in water-soluble pectin and fruit softening.

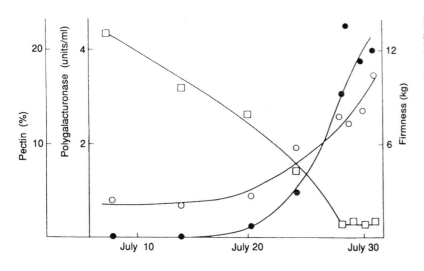

Cellobiose

Two types of polygalacturonases or D-galacturonases have been identified, endo and exo. The former randomly hydrolyzes the glycosidic bonds in the pectin molecule while the exo-enzyme acts from the terminal end of the pectin molecule (Scheme 2.22). In the presence of endopolygalacturonases, the pectin molecules are rapidly degraded into smaller units, accompanied by a marked decrease in viscosity. Both forms of the enzyme are found in pears (Pressey and Avants, 1976; Bartley and Knee, 1982) and peaches (Pressey and Avants, 1978). However, some fruit species lack any detectable endo-PG, including apples (Brackmann *et al.*, 1996; Goulao *et al.*, 2007), strawberries (Abeles and Takeda, 1990), and grapes (Nunan *et al.*, 2001). The greater degree of softening reported by Pressey and Avants (1978) for 'freestone' peaches compared to 'clingstone' peaches was attributed to the absence of endopolygalacturonase in the clingstone fruit. The random degradation of pectin by

FIGURE 2.10 Changes in firmness (□), PGA activity (●), and water-soluble pectin (○) in Elberta peaches during ripening (Pressey *et al.*, 1971). *(Copyright © by Institute of Food Technologists.)*

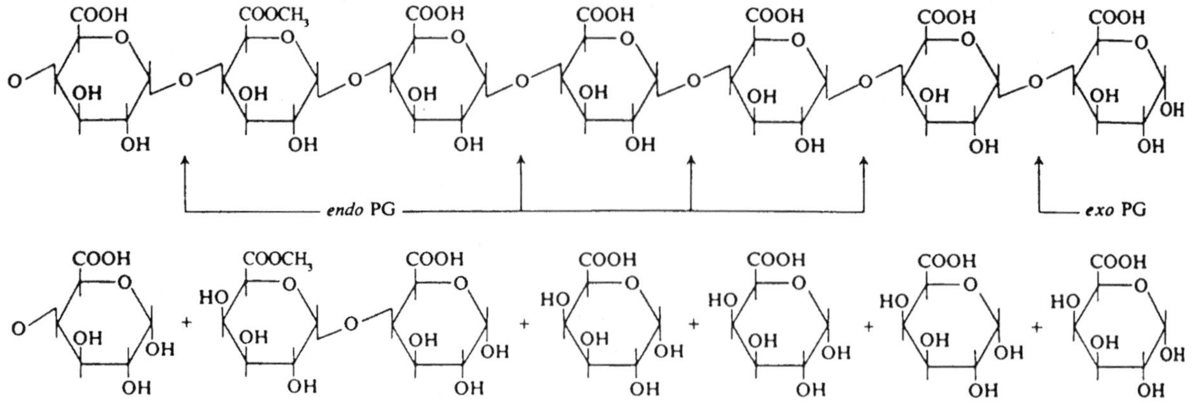

SCHEME 2.22   Action of exo- and endo-polygalacturonases.

endopolygalacturonase together with exopolygalacturonase rapidly solubilized the pectin in freestone peaches. The absence of endopolygalacturonase in clingstone peaches was evident by the greater retention of protopectin during ripening.

Exopolygalacturonases (EC 3.2.1.67) have been identified in peaches (Pressey and Avants, 1973a), pears (Pressey and Avants, 1976), cucumbers (Pressey and Avants, 1975; McFeeters et al., 1980), and bananas (Markovic et al., 1975). Exopolygalacturonase is the only D-galacturonase found in apples and is responsible for the release of galacturonic acid and polyuronides (Bartley, 1978). Studies using cell wall isolates from tomatoes (Wallner and Bloom, 1977; Gross and Wallner, 1979; Themmen et al., 1982) and pears (Ahmed and Labavitch, 1980) demonstrated that endopolygalacturonases played the major role in pectin degradation during ripening. The release of a water-soluble polymer (WSP) of molecular weight 20,000 containing galacturonic acid and rhamnose by endopolygalacturonase from the cell walls of red tomatoes was reported by Gross and Wallner (1979) to be identical with the polymer released by the same enzyme from the cell walls of mature green tomatoes. While the activity of polygalacturonases in ripe red tomatoes is predominantly the endo form, it is the exo form that is mainly present in the corresponding green fruit (Pressey and Avants, 1973b; Tucker et al., 1980). Although exopolyga- lacturonase represents only a small fraction of polygalacturonase activity, it was present throughout the ripening of tomatoes at fairly constant levels. Pressey (1987) suggested, therefore, that it was unlikely that exopolygalactur- onase played an important part in pectin degradation, but they might have some role in the growth and development of the tomato fruit.

Isoenzyme forms of PG have been reported in a wide range of fruits including bananas (Pathak and Sanwal, 1998), strawberries (Nogota et al., 1993), and peaches (Pressey and Avants, 1973a). Two isoforms of endo-PG, PG_1 and PG_2, were reported in tomatoes (Pressey and Avants, 1973b). Subsequent studies by Tucker et al. (1980) suggested that PG_1 was a dimer of PG_2. Further work by Pressey (1986a, b) suggested that PG_1 was a combination of PG_2 and a β-subunit.

b. Pectin Methyl Esterase: Pectinesterase

Pectin methyl esterases (PMs), or pectinesterases, are widely distributed in many fruits, including bananas (Buescher and Tigchelaar, 1975), peaches (Nagel and Patterson, 1967), and strawberries (Barnes and Patchett, 1976). Considerable confusion surrounds the early studies on PM during ripening. For example, Hultin and Levine (1965) noted a rise in PM activity during the ripening of bananas which was not observed by De Swardt and Maxie (1967) when they used polyvinylpyrrolidine (PVP) to remove polyphenols. Brady (1976) also found very little change in PM activity in banana extracts following the addition of 2-mercaptobenzothiazole, a potent inhibitor of banana polyphenol oxidase. PVP was later found by Awad and Young (1980) to suppress PM activity, although no inhibitory effects were exerted by the endogenous phenols. PM does not appear to have a major role in fruit softening, as it is present at high levels in underdeveloped fruit, such as tomatoes and bananas, prior to ripening (Barnes and Patchett, 1976; Brady, 1976; Pressey and Avants, 1982a; Tucker et al., 1982). In the case of avocados, however, there is a dramatic drop in PME activity just before ripening, which was reported to be a useful index of fruit maturity (Zauberman and Schiffman-Nadel, 1972). A decrease in PM activity in avocados by as much as 50% prior to the climacteric proved to be a reliable indicator of softening time when stored in controlled atmospheres (Barmore and Rouse, 1976). A drop of 80% in PME activity was also reported in avocados by Awad and Young (1980).

The traditional view of PM is of de-esterification of the cell wall galacturonans followed by polygalacturonase action. This de-esterification of galacturonans was found by several researchers (Dahodwala *et al.*, 1974; Rexova-Benkova and Markovic, 1976) to enhance the activities of both endo- and exopolygalacturonases. Stimulation of tomato endopolygalacturonases by PM was reported by Pressey and Avants (1982b), although these enzymes were capable of hydrolyzing highly esterified substrates at pH 3.5. The presence of a high degree of pectin methylation during the ripening of avocados (Dolendo *et al.*, 1966), apples (Knee, 1978), and peaches (Shewfelt *et al.*, 1971) pointed to a rather limited role for PM in fruit softening. Ben-Arie and Sonego (1980) attributed the development of woolly breakdown of peach flesh during cold storage to the inhibition of polygalacturonase activity and enhancement of PM. This phenomenon was attributed to the inability of peaches to undergo the desirable textural changes associated with normal development. Recent studies by von Mollendorff and De Villiers (1988), however, showed that the primary cause of woolliness in peaches was the sudden rise in the level of polygalacturonase, while the role of PME was far less clear. Nevertheless, the cause of woolliness in peaches remains confusing as some studies have associated woolliness in peaches with an increase in PM activity (Brummell *et al.*, 2004; Girardi *et al.*, 2005), a decrease in PM activity (Buescher and Furmanski, 1978; Ben-Arie and Sonego, 1980), or a lack of expression of PM (Obehland and Carroll, 2000; Zhou *et al.*, 2000).

It is evident from these discussions that fruit softening is due to compositional changes in the cell walls of fruit mediated by the combined activity of polygalacturonases and PME. This results in the release of soluble polyuronide with a corresponding decrease in the molecular weight of the polyuronide polymer (Gross and Wallner, 1979; Huber, 1983; Seymour *et al.*, 1987a). The action of polygalacturonase, as discussed previously, may be limited to the demethylated regions of the polygalacturonan, which is brought about by the action of PM. The enhanced production of polyuronides was reported by Pressey and Avants (1982b) in isolated cell walls of mature green tomato fruit in the presence of PME. Seymour *et al.* (1987a), using enzyme-inactivated cell wall preparations to eliminate the effect of any endogenous enzymes, found that polyuronide breakdown was much lower in *in vivo* than in *in vitro* studies.

The fact that pectin was not completely de-esterified, in spite of high levels of PM, suggested that this enzyme may be restricted *in vivo*. Further studies by Seymour *et al.* (1987b) on tomatoes also demonstrated the lower solubilization of polyuronides *in vivo*, which was attributed to the restriction of PM action. The release of two discrete-sized oligomers, together with galacturonic acid, suggested that the combined pectolytic action was not completely random. These researchers indicated the importance of identifying these oligomers *in vivo* in light of the recent discovery of cell wall elicitors.

c. Pectic Lyase

Pectic lyase (PL) [poly(1,4-α-D galacturonide) lyase, EC 4.2.2.1] cleaves the de-esterified or esterified galacturonate units of pectin by trans-β-elimination of the hydrogen on the C4 and C5 positions of galacturonic acid (Prasanna *et al.*, 2007). While there are exo and endo forms of PL, most of those studied are endo type, which act randomly (Wong, 1995).

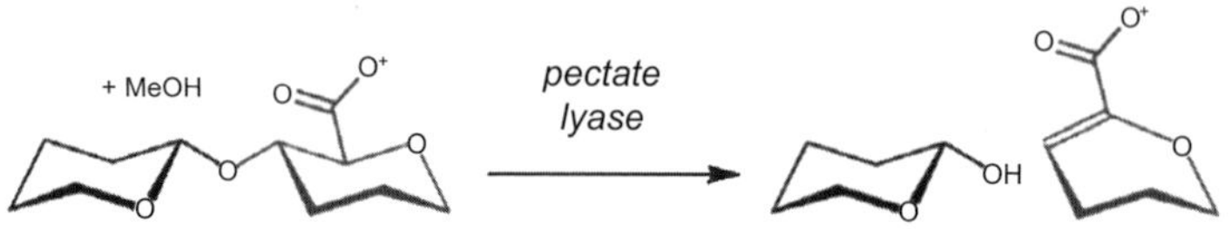

Adapted from Gilbert *et al.* (2008)

A pectate lyase gene was isolated from ripe strawberries and proposed as a candidate for pectin degradation and loss of fruit firmness (Medina-Escobar *et al.*, 1997). Using transgenic strawberries, Jimenez-Bermudez *et al.* (2002) manipulated softening by antisense expression of a pectate lyase gene. Their results indicated that the rapid softening of strawberries could be slowed down by the pectate lyase gene. More recent research by Payasi and Sanwal (2003) demonstrated pectate lyase activity in ripening bananas. Further purification of pectate lyase by Payasi *et al.* (2008) showed that it required Mg^{2+}, in addition to Ca^{2+}, for maximum activity.

d. Cellulase

Cellulose degradation also occurs during the ripening of tomatoes (Babbitt *et al.*, 1973; Pharr and Dickinson, 1973; Sobotka and Stelzig, 1974), strawberries (Barnes and Patchett, 1976), avocados (Pesis *et al.*, 1978; Awad and Young, 1980), and Japanese pear fruit (Yamaki and Kakiuchi, 1979). The enzyme involved, cellulase, is composed of several distinct enzymes referred to as the 'cellulase complex' (King and Vessal, 1969). These include C_1-cellulase, C_x-cellulase, cellobiase, and exocellulase, which together catalyze the degradation of cellulose as follows:

Insoluble cellulose

↓ C_1-cellulase

Soluble cellulose derivatives

↓ C_x-cellulase

Cellobiose

↓ cellobiase (β-1,4-glucosidase)

Glucose

The degradation of insoluble cellulose to soluble derivatives is poorly understood but appears to involve C_1-cellulase. The breakdown of soluble cellulose is mediated by C_x-cellulase, also referred to by its systematic name 1,4-glucan-4-glucanohydrolase, which randomly cleaves the internal linkages in the cellulose chain. Conflicting reports in the literature suggest that cellulase activity was absent or present at very low levels in unripe fruit, whereas others found cellulase activity in immature tomato fruit (Hobson, 1968; Babbitt *et al.*, 1973). Information on the cellulase complex was derived almost exclusively from studies on microbial cellulase. The first study to identify a similar complex in plants was by Sobotka and Stelzig (1974), who partially purified four cellulase fractions from tomato using ammonium sulfate fractionation. These researchers identified C_1-cellulase, C_1-cellulase, β-glucosidase, and cellobiase as the first cellulase complex capable of completely degrading insoluble cellulose in plants. Pharr and Dickinson (1973) were unable to identify an enzyme in tomato fruit that could degrade insoluble cellulose, although they did report the presence of C_x-cellulase (EC 3.2.1.4) and cellobiase (EC 3.2.1.21). The presence of a cellulase complex in plants still remains to be established.

The role of cellulase in fruit softening is still somewhat speculative. The only direct evidence with respect to the involvement of cellulase was that reported by Babbitt et al. (1973). These researchers investigated the effect of the growth regulators ethephon and gibberellic acid on cellulase and polygalacturonase activities in ripening tomatoes. In the presence of ethephon, cellulase activity increased initially and then declined after 6 days, while polygalacturonase activity increased (Figure 2.11). This contrasted with the almost complete inhibition of polygalacturonase activity by gibberellic acid while cellulase activity continued to increase. The decrease in overall firmness of the tomato fruit in the presence of gibberellic acid pointed to a definite role for cellulase in fruit softening. These researchers proposed that cellulase initiated fruit softening by degrading the cellulose fibrils in the cell walls, which permitted pectic enzymes to penetrate the middle lamella. This could explain the observation by Awad and Young (1979), who found

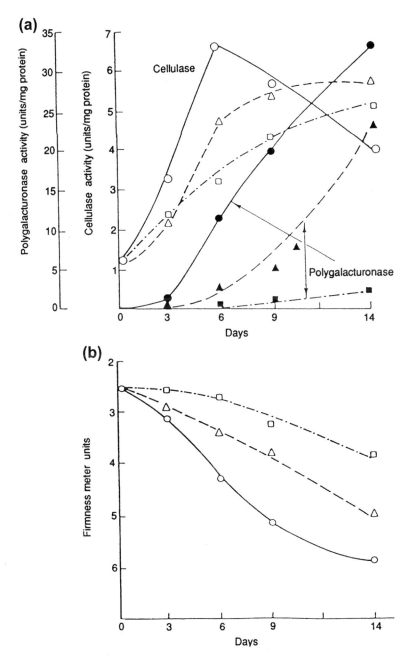

FIGURE 2.11   Effect of ethephon (◯) and gibberellic acid (☐) on enzyme activity (a) and firmness (b) of tomatoes (△, control) (Babbitt et al., 1973).

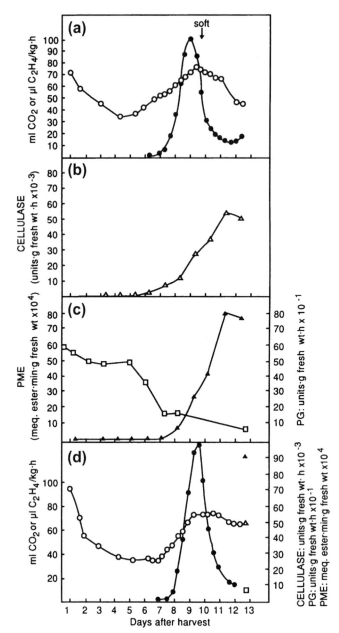

FIGURE 2.12   Postharvest trends in cellulase (△), PG (▲), and PME (□) activity and in CO₂ (○) and C₂H₄ (●) production in an individual Fuerte avocado fruit. Fruit in (a) was edibly soft after 9.5 days and in (d) after 10.5 days (Awad and Young, 1979).

that an increase in cellulase activity preceded increases in poylgalacturonase and ethylene production in ripening avocado fruit and subsequent softening (Figure 2.12).

Using isoelectric focusing (IEF), Kanellis and Kalaitzis (1992) isolated multiple active forms of cellulase in the mesocarp of ripe avocados. The cellulases were separated into 11 distinct forms capable of degrading carboxymethylcellulose (CMC) on CMC-agarose gels. Sexton *et al.* (1997) localized cellulase in the regions of red raspberry fruit associated with abscission, suggesting that it may be involved in fruit separation as well as softening.

e. β-Galactosidase

The loss of galactose from the cell walls of apples, strawberries, and tomatoes during ripening is the result of the action by β-galactosidase (EC 3.2.1.23). The increased activity of this enzyme was correlated with the loss of firmness during the ripening and storage of apples (Bartley, 1974, 1978; Wallner, 1978; Berard *et al.*, 1982). Evidence for this is provided by a decrease in the galactose content of apple cortex cell walls (Knee, 1973), an increase in the soluble polyuronide content (Knee, 1975), the ability of β-galactosidase to break down β-(1→4)-linked galactan (Bartley, 1974), and the release of galactose from cell wall preparations (Bartley, 1978). Dick *et al.* (1984) provided preliminary evidence for the regulation of β-galactosidase activity in 'McIntosh' apples by the presence of an endogenous inhibitor. Unlike β-galactosidase in apples, that found in tomatoes did not appear to be involved in cell wall hydrolysis of galactans (Gross and Wallner, 1979). In fact, these researchers were unable to detect any β-galactosidase activity during the ripening of tomatoes. This contrasted with a later study by Pressey (1983), who isolated three enzymes responsible for β-galactosidase activity in tomatoes. One of these enzymes hydrolyzed tomato galactans and increased in activity during the ripening process. This suggested a possible role for β-galactosidase in fruit softening. Since the galactan polysaccharide in tomato fruit is (1→4)-linked, the β-galactosidase involved must be β-1,4-galactosidase. The inability of Gross and Wallner (1979) to detect any galactanase activity was attributed to the preparation of their extracts from frozen fruit. As pointed out by Pressey (1983), the yield of β-galactosidase in frozen fruit is very low compared to that in the corresponding fresh tomato extracts. Ranwala *et al.* (1992) monitored the changes in the soluble and wall-bound β-galactosidases during the ripening of muskmelon fruit. They demonstrated the involvement of β-galactosidases in modifying the pectic and hemicelluloses components, indicating a possible role in the softening of muskmelon.

VII. FLAVOR

The flavor of fruits and vegetables is a complex interaction between aroma and taste. Aroma is produced by the volatiles synthesized during fruit ripening, which include aldehydes, ketones, alcohols, esters, lactones, terpenes, and sulfur compounds (Baldwin, 2004; Pérez and Sanz, 2008). Vegetables, with a few exceptions, tend to be more bland in flavor. Their aroma is determined by secondary metabolites such as terpenes, phenolics, glucosinolates, and cysteine sulfoxides (Jones, 2008). Taste is provided by many non-volatile components, including mainly sugars and acids, as well as compounds with bitter taste or astringency present in the fruit flesh or vegetable tissue.

A. Aroma

Volatiles responsible for aroma originate from proteins, carbohydrates, lipids, and vitamins, as shown in Scheme 2.23. The aroma characteristics of individual fruit and vegetable crops are determined by genetics as well as

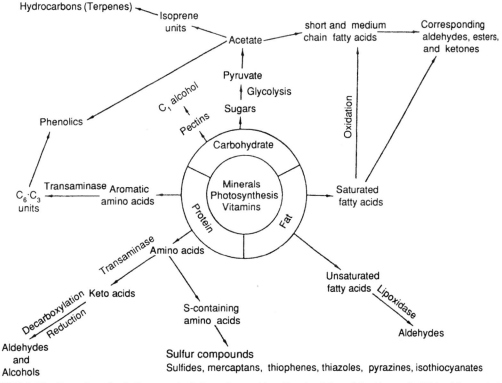

SCHEME 2.23 Formation of volatile aroma in fruits and vegetables. (Reprinted from Salunkhe *et al.*, 1976, with permission.)

preharvest, harvesting, and postharvest factors, and usually develop during ripening and maturation (Kader, 2008). Many volatiles have been identified in the literature and were reviewed by Salunkhe and Do as early as 1976. Since then, studies on a wide range of produce have been undertaken. A well-researched fruit is the apple, in which the impact of preharvest and postharvest factors has been extensively studied. Over 300 volatile compounds are produced by apple fruits; the most abundant group comprises esters (Yahia, 1994; Dixon and Hewett, 2000). Over 400 volatile compounds are generated by tomato fruit (Hobson and Grierson, 1993) and differences in flavor between varieties are due to variation in aroma volatile production (Brauss *et al.*, 1998). However, qualitative and quantitative differences between research groups may be caused by the use of a vast range of extraction and analytical methods (Saevels *et al.*, 2004). Furthermore, instrumental analysis of volatiles must be combined with sensory analysis to provide insight into the impact of volatile compounds on flavor perception (Baldwin *et al.*, 2007; Song, 2007). A new approach to elucidate responses of human odorant receptors to aroma compounds includes molecular sensory science (Greger and Schieberle, 2007; Schmiech *et al.*, 2008). In view of optimizing and improving the flavor quality of fruits and vegetables and gaining a better understanding of flavor volatile production and regulation, current research has been directed towards the isolation of genes encoding key enzymes involved in the biosynthesis of aroma volatiles (Song, 2007; Pech *et al.*, 2008) based on research in recent decades focusing on volatile identification and biosynthesis. Control of formation of volatiles depends on many factors such as the genes involved and hence on their expression and transcription, the functionality of enzymes catalyzing modifications of substrates to volatiles, and the availability of substrates or precursors, as well as the energy status of cells (Saquet *et al.*, 2003b; Souleyre *et al.*, 2005; Song, 2007; Schwab, 2008; Wyllie, 2008). This section will focus briefly on a few of the biogenic pathways involved in volatile formation during ripening.

1. Aldehydes, Alcohols, and Esters

Short-chain unsaturated aldehydes and alcohols and straight chain and branched chain esters are important contributors to the aroma volatiles of fruits. These are formed during the short ripening period associated with the climacteric rise in respiration (Paillard, 1968; Romani and Ku, 1968; Tressl *et al.*, 1970; Song and Bangerth, 2003; Lara *et al.*, 2003; Souleyre *et al.*, 2005). Studies on apples, bananas, pears, and tomatoes have shown them to be synthesized from amino acids or fatty acids (Eskin *et al.*, 1977; Eskin, 1979; Song and Bangerth, 2003; Rapparini and Predieri, 2003).

a. Amino Acids as Precursors

An increase in 3-methyl-1-butanol, isopentyl acetate, isopentyl butyrate, and isovalerate volatiles was reported by Dalal (1965) during tomato ripening. At the same time, 3-methyl-1-butanal increased up to the breaker stage and then decreased. The similarity between the alcohol portion of these esters and the carbon skeleton of 3-methyl-1-butanal suggested to Yu *et al.* (1968a) that they were synthesized from this aldehyde. Since leucine had an identical carbon skeleton to that of 3-methyl-1-butanal, the possible role of this amino acid in the synthesis of this aldehyde was investigated by Yu *et al.* (1968c). On the basis of their work with L-[^{14}C]leucine, the following pathway was proposed:

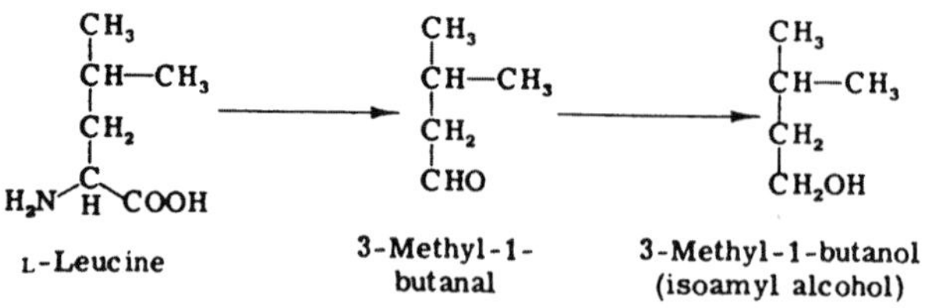

In addition to leucine, aspartic acid, alanine, and valine were shown to produce carbonyl compounds when added to tomato extracts (Yu *et al.*, 1968c; Buttery and Ling, 1993). The decrease in the levels of these amino acids during ripening pointed to the presence of enzymes capable of utilizing them. The soluble fraction obtained by centrifugation of tomato extracts was particularly active on leucine, while aspartic acid and alanine were metabolized by the mitochondrial fraction. Based on the detection of large amounts of glutamic acid in tomato extracts, Freeman and Woodbridge (1960) and Yu *et al.* (1967) both pointed to the presence of active transaminases. This was confirmed in a subsequent study by Yu *et al.* (1968b), who found a marked production of glutamic acid when extracts of field-grown tomatoes at the green and ripe stages were incubated with these amino acids. Yu and Spencer (1969) incubated L-leucine with fresh tomato extracts and isolated α-keto-isocaproic acid among the products formed. Confirmation with labeled [^{14}C]leucine established the presence of L-leucine:2 keto-glutarate amino transferase in tomatoes. Scheme 2.24 summarizes the reactions leading to 3-methyl-1-butanol from L-leucine. Studies with deuterium-labeled substrates in apples corroborated these findings (Rowan *et al.*, 1996, 1999). Finally, aromatic amino acids can be converted to esters through the same pathways in different fruits (Tressl and Albrecht, 1986; Tikunov *et al.*, 2005).

b. Fatty Acids as Precursors

The volatile carbonyls responsible for the aroma of tomatoes and bananas as well as other commodities are synthesized from unsaturated fatty acids (Goldstein and Wick, 1969; Jadhav *et al.*, 1972; Song and Bangerth, 2003). β-Oxidation and the lipoxygenase pathway are the main metabolic pathways generating aroma in fruits (Baker *et al.*, 2006). The major unsaturated fatty acids in the pericarp of tomatoes were shown to be oleic, linoleic, and linolenic acids (Kapp, 1966). Jadhav *et al.* (1972) reported a decrease in the levels of both linoleic and linolenic acids as the tomato fruit ripened. A marked decrease in linoleic acid was also observed by Goldstein and Wick (1969) in ripe banana pulp, which suggests a possible relationship between this fatty acid and the production of volatiles. Similarly, Song and Bangerth (2003) reported that fatty acids served as precursors of aroma volatiles in preclimacteric and climacteric apple fruits.

Incorporation of ^{14}C-labeled linoleic and linolenic acids into hexanal using tissue slices or cell-free tomato extracts suggested to Jadhav *et al.* (1972) the involvement of lipoxygenase. This was confirmed by the total inhibition of carbonyls in the presence of hydrogen peroxide, a recognized inhibitor of lipoxygenase. Consequently, this enzyme was monitored during tomato ripening and found to increase in activity at the onset of the climacteric. This appeared to explain the increase in volatiles that accompanied maturation of the tomato fruit (Dalal *et al.*, 1968). Kazeniac and Hall (1970) reported the presence of higher levels *cis*-3-hexenal, *trans*-2-hexenal, and *n*-hexanol in fully ripened tomato fruit. The formation of *trans*-2-hexenal resulted from the instability of *cis*-3-hexenal to the acidic pulp and juice of the tomato, with isomerization to the *trans* isomer. Stone *et al.* (1975) showed that *cis*-3-hexenal was the major volatile of tomato distillates, while Jadhav *et al.* (1972) found *n*-hexanol to be the major volatile formed. This discrepancy suggested to Stone *et al.* (1975) that *cis*-3-hexenal was a precursor of *n*-hexanol, but this was discounted when only 2% of *cis*-3-[^{14}C]hexenal was incorporated into the alcohol form. Jadhav *et al.* (1972) attributed the presence of hexanol, propanol, 2,4-decadienal, 2,6-heptadiene, and *cis*-3-hexenal in tomato volatiles to the formation of 9-, 12-, and 16-hydroperoxides by the action of lipoxygenase on linoleic and linolenic acids. Gaillard and Matthew (1977), however, reported that the major fatty acid hydroperoxides formed from linoleic and linolenic acids were 9- and 13-hydroperoxides in a ratio of 95:5. Of these, only the 13-hydroperoxide was cleaved to form the non-volatile compound 12-oxo-dodec-*cis*-9-enoic acid, together with hexanal and *cis*-3-hexenal from linoleic and linolenic acids, respectively (Scheme 2.25). Zamora *et al.* (1987) characterized lipoxygenase from tomato fruit and confirmed the 9-hydroperoxide isomer to be the major one formed from linoleic acid. The ratio of 9- to 13-hydroperoxide isomers produced from linoleic acid was found to be 24:1, in close agreement with that reported previously by Gaillard and Matthew (1977).

Buttery and co-workers (1987) developed improved trapping methods for the quantitative analysis of the major C_4–C_6 volatiles in tomato fruit. In addition to inactivating tomato enzymes which affected the volatiles during

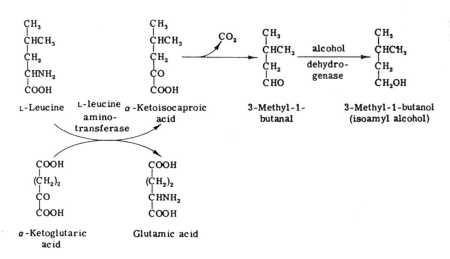

SCHEME 2.24 Biogenesis of isoamyl alcohol from L-leucine in tomato fruit.

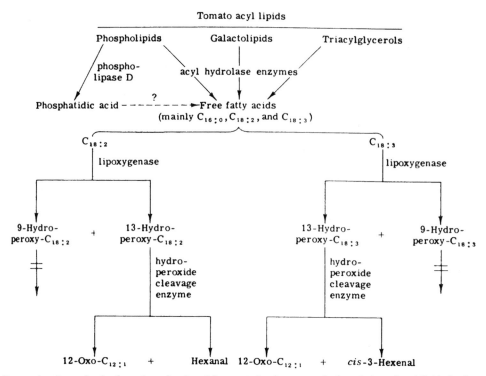

SCHEME 2.25 Proposed pathway for the formation of carbonyl fragments by the enzymatic degradation of acyl lipids in disrupted tomato fruits (Galliard and Matthew, 1977).

isolation, they overcame the problem of isomerization of *cis*-3-hexenal to *trans*-2-hexenal reported by Kazeniac and Hall (1970). Using their procedure, which involved Tenax trapping and CaCl₂ enzyme deactivation, they identified *cis*-3-hexenal among the major volatiles present. These researchers attributed the lack of flavor in tomatoes purchased in the supermarket to the lower levels of *cis*-3-hexenal present compared to the higher levels present in vine-ripened tomatoes. In addition, they reported that storing tomatoes in the refrigerator caused further loss of flavor in part because of lower levels of *cis*-3-hexenal. This effect of cold storage on tomato flavor was in agreement with previous work by Lammers (1981). In addition to *cis*-3-hexenal, other important flavor volatiles were β-ionone, 1-penten-3-one, hexanal, *cis*-3-hexanol, *trans*-2-hexanal, 2- and 3-methylbutanol, 2-iso-butylthiazole, and 6-methyl-5-hepten-2-one.

Trans-2-hexenal was also identified among the volatiles of 'Gros Michel' bananas by Issenberg and Wick (1963). Using a volatile enrichment technique, Tressl and Jennings (1972) confirmed the presence of *trans*-2-hexenal in the headspace of ripening bananas. Separation of volatile fractions from bananas by Palmer (1971) showed the presence of *cis*-3-hexenal, *trans*-2-hexenal, and *n*-hexanal among the aldehydes formed. Tressl and Drawert (1973) found that green banana homogenates produced *trans*-2, *cis*-6-nonadienal, *trans*-2-nonenal, and 9-oxanonanoic acid, similar to that reported in cucumbers (Fleming *et al.*, 1968). Tressl and Drawert (1973) detected hexanal, *trans*-2-hexenal, and 12-oxo-*trans*-10-dodecenoic acid when green bananas were exposed to ethylene and stored for 4 days at 15°C. Incorporation of ¹⁴C-labeled linoleic and linolenic acids into these volatiles pointed to the involvement of lipoxygenase. Labeled 13- and 9-hydroperoxyoctadecadienoic acids incubated with crude banana extracts were converted to C₆—C₉ aldehydes as shown in Scheme 2.26. Aldehyde lyase, the enzyme responsible for the breakdown of the hydroperoxy derivatives, was also identified in germinating watermelon seedlings by Vick and Zimmerman (1976). This enzyme catalyzed the formation of 12-oxo-*trans*-10-dodecenoic acid and hexanal from 13-hydroperoxy-*cis*-9,*trans*-11-octadecanoic acid. This enzyme differed from hydroperoxide cleavage enzyme in tomatoes by producing *trans*-2-enals as the primary products.

Studies on apple volatiles by Flath *et al.* (1967) identified 20 compounds from 'Delicious' apple essence to be 'character impact' compounds, including hexanol, *trans*-2-hexenal, and ethyl-2-methyl butoneate. Dürr and Röthlin (1981) showed that a mixture including these and some additional components, amounting to a total of 11 components, was sufficient to compose a typical apple-like odor. Kim and Grosch (1979) partially purified

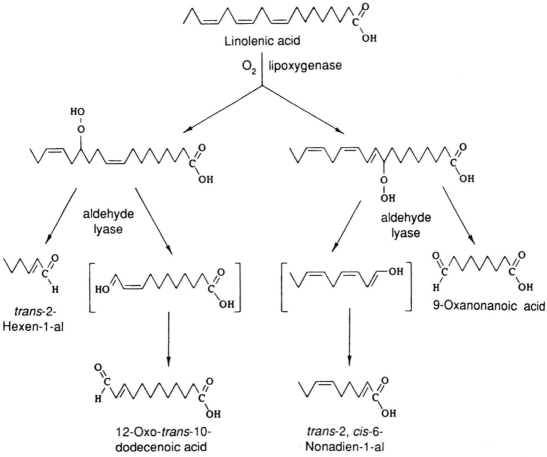

SCHEME 2.26 Reaction scheme for enzymatic splitting of linolenic acid into aldehydes and oxo acids. (Reprinted with permission from Tressl and Drawert, 1973. Copyright © by the American Chemical Society.)

lipoxygenase from apple homogenates, which produced 13-hydroperoxyoctadeca-9,11-dienoic acid from linoleic acid. The latter was converted to 2-hexenal and hexanol in a similar manner to that described previously.

B. Taste

The characteristic taste of fruits is determined by the content of sugars and organic acids. The ratio of sugar to acid is particularly useful as an index of ripeness for many fruits. Thus, studies concerning fruit quality often rely on measurements of sugars, acids, total soluble solids (°Brix), and titratable acidity (Harker *et al.*, 2002; Hoehn *et al.*, 2003). The Thiault quality index for 'Golden Delicious' apples, for example, is calculated based on total sugar content plus 10 times the acidity (Thiault, 1970). The level of sucrose and L-malic acid together with protein patterns was suggested by Gorin (1973) as parameters for assessing the quality of 'Golden Delicious' apples. Hammett *et al.* (1977) found a high correlation between the ratio of soluble solids to acid content with days from full bloom (DFFB) for 'Golden Delicious' apples. The ratio of sugars or soluble solids and acids indicates the quality of citrus and other fruits (Vangdal, 1985; Fellers, 1991; Mitchell *et al.*, 1991). Sugars and acids not only contribute to sweetness and sourness in tomatoes but are responsible for the overall flavor intensity (Kader *et al.*, 1977; Stevens *et al.*, 1979; Jones and Scott, 1983). Sweetness and hence sugars contribute to consumer liking of carrots. Studies on carrots indicated that sugars, in addition to eliciting sweetness, suppressed the bitterness exerted by 6-methoxymellein (isocumarin) (Seljasen *et al.*, 2001a, b; Höhn *et al.*, 2003; Varming *et al.*, 2004; Kreutzmann *et al.*, 2007).

In addition to these components, the presence of tannins, phenolic compounds classified into hydrolyzable and non-hydrolyzable, affects taste (Lesschaeve and Noble, 2005). Unlike the hydrolyzable tannins, which yield gallic acid and glucose by enzymatic hydrolysis, the non-hydrolyzable tannins are resistant to enzymatic hydrolysis. The latter appears to be responsible for astringency in many under-ripe fruits. The loss of astringency in persimmon was

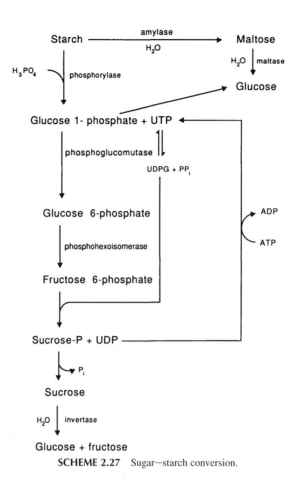

SCHEME 2.27 Sugar–starch conversion.

attributed by Matsuo and Itoo (1982) to immobilization of tannin with acetaldehyde formed during ripening. No substantial changes in the composition or amount of polyphenols in the fruit of *Rubus* sp. were detected during ripening by several researchers (Haslam *et al.*, 1982; Okuda *et al.*, 1982a, b). On the basis of studies with model systems, Ozawa *et al.* (1987) proposed that the loss of astringency during fruit ripening was due, in part, to the possible interaction between polyphenols and proteins in the fruit.

1. Starch–Sugar Conversion

Sugars and transient starch are synthesized in the growing plant by photosynthesis. This is translocated, mainly in the form of sucrose, from the chloroplasts via the phloem to the growing cells in the plant, where it is resynthesized into starch. This sucrose–starch conversion appears to involve the sequence of reactions shown in Scheme 2.27.

During the postharvest period, starch is transformed into sucrose, glucose, and fructose. This is affected by the physiological condition of the fruit and vegetables as well as storage temperature and time. Starch hydrolysis is among the more conspicuous changes accompanying ripening of many climacteric fruit. For example, a drop in starch content from 22% to 1% was reported by Palmer (1971) to accompany bananas as they passed from the preclimacteric to the climacteric phase. In other fruit such as apples and pears the progress of starch degradation is used to determine ripeness stage and harvest date (Smith *et al.*, 1979; De Jager *et al.*, 1996). However, the starch content in apples and pears is less than in bananas and hence the increase in sugars due to starch conversion is usually moderate (Murayama *et al.*, 2002; Travers *et al.*, 2002). The presence of phosphorylase and amylase enzymes was observed in the storage tissues of many fruits, although their respective roles in the ripening process remained unclear. Young *et al.* (1974) observed an increase in α-amylase activity during the ripening of bananas but were unable to confirm its involvement in starch hydrolysis. β-Amylase and phosphorylase were also found during ripening but the presence of enzyme inhibitors impeded their assay during the preclimacteric phase. Yang and Ho (1958) suggested that phosphorylase had a role in starch degradation during the climacteric. This was later confirmed in gamma-irradiated 'Cavendish' bananas by Surendranathan and Nair (1973). Three phosphorylases were partially purified from ripe bananas by Singh and Sanwall (1973, 1975, 1976), each exhibiting different biochemical properties.

Starch hydrolysis was examined by Chitarra and LaJollo (1981) during the ripening of hybrid 'Marmello' bananas. This hybrid exhibited abnormal behavior as the peel, color, aroma, and texture did not undergo changes normally associated with ripening. In addition, the starch content dropped to 5% during ripening because of the possible presence of inhibitors. Raising the storage temperature from 20 to 25°C accelerated the climacteric and reduced the time required to stabilize the starch level at 5% from 24.5 to 8 days. The starch level in bananas could be further reduced to 3.3% by storing at 30°C; however, the resulting fruit was overripe and deteriorated rapidly. This contrasted with 'Dwarf Cavendish' bananas, in which complete starch degradation (97.7%) occurred during the climacteric. The lower sugar content in the 'Marmello' banana was responsible for the difference in taste compared to the corresponding 'Cavendish' variety. Phosphorylase activity in 'Marmello' bananas was reported to remain constant during the preclimacteric although it paralleled the change in starch content. This differed from 'Cavendish' fruit, in which a 50% increase in phosphorylase activity was reported prior to the climacteric and preceded any starch degradation. These results pointed to the possible involvement of phosphorylase in starch–sucrose transformations as well as sucrose synthetase (Areas and LaJollo, 1981). The exclusive involvement of α-amylase in starch degradation was seriously questioned since it occurred before any increase in α-amylase activity. The inhibition of α-amylase synthesis by cycloheximide during banana ripening did not block starch degradation, which supported a role for phosphorylase.

Phosphorylase activity has been associated with the cold storage of potato tubers at 4°C (Hyde and Morrison, 1964). This enzyme is responsible for the initial reaction in the cold-induced sweetening of potatoes (Isherwood, 1976). The overall result is a marked increase in the sugar content of the potato, resulting in the production of chips with unacceptable dark coloration (Talburt and Smith, 1975). To prevent this, potatoes are generally stored at or above 10°C, referred to as 'conditioning'. The biochemical mechanism involved in these starch/sugar transformations remains unclear, although the effect of storage temperature on a number of enzymes has been reported (Kennedy and Isherwood, 1975; Pollock and ap Rees, 1975; Isherwood, 1976; Dixon and ap Rees, 1980). Increase in sugar phosphates and sucrose was attributed by Pollock and ap Rees (1975) to the cold lability of some of the glycolytic enzymes. One of these, phosphofructokinase, was shown to be a major cause of low-temperature sweetening of potato tubers (Dixon et al., 1981). The cold lability of this enzyme appeared to be due to denaturation of its oligomeric enzyme complex and its dissociation into subunits. This resulted in the inability of the enzyme to oxidize hexose phosphates, with their subsequent accumulation and conversion to sucrose.

2. Organic Acids

Fruit ripening is accompanied by changes in organic acids. These reach a maximum during the growth and development of the fruit on the tree, but decrease during storage, as well as being highly dependent on temperature. The Krebs cycle is active in the cells of higher plants in generating a variety of organic acids, including citric, malic, and succinic acids. Citric and malic acids are important constituents of most fruits, with oranges, lemons, and strawberries being high in citric acid, and apples, pears, and plums high in malic acid. During ripening, these organic acids decrease as they are used as substrates for respiration. Depletion of acids is accompanied by a decrease in starch content and an increase in sugars responsible for fruit sweetness and diminishing of acid taste (Schifferstein and Frijters, 1990; Ackermann et al., 1992; Harker et al., 2002). Akhavan and Wrolstadt (1980) reported that ripening 'Bartlett' pears attained a maximum sugar content of 13.5% (Figure 2.13), and a total acidity of 6 milliequivalents and flesh firmness of 6 lbs on the fourth day of ripening. Malic and citric acids accounted for the major changes in total acidity both before and after ripening (Figure 2.14). Similar evolvement of sugars and acids accompanied by an increase in pear aroma was observed in 'Conference' pears (Höhn et al., 1996). These organic acids contribute to the pH in tomatoes, which is particularly important in processing (Davies and Hobson, 1981). Picha (1987) reported citric and malic acids to be the major organic acids in cherry tomato fruit. During ripening, changes in these organic acids were evident, with citric acid increasing from the immature green to mature green stage while malic acid decreased from the mature green to the table-ripe stage. Furthermore, acid and sugar levels affect the flavor quality of tomatoes and other fruits (Malundo et al., 1995).

In addition to these organic acids, ascorbic acid is prominent in fruits. It is present in plant tissues, mainly in the reduced form, but can be oxidized to dehydroascorbic acid by the action of ascorbic acid oxidase (see Chapter 10).

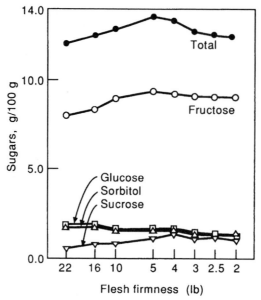

FIGURE 2.13 **Changes in sugars during the ripening of Bartlett pears (Akhavan and Wrolstad, 1980).** *(Copyright © by Institute of Food Technologists.)*

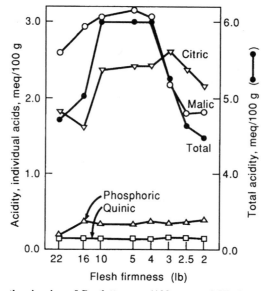

FIGURE 2.14 **Changes in acids during the ripening of Bartlett pears (Akhavan and Wrolstad, 1980).** *(Copyright © by Institute of Food Technologists.)*

The presence of L-quinic acid as a minor organic acid has also been reported in a number of fruits, including pears (Akhavan and Wrolstad, 1980). Other organic acids reported include oxalic and citramalic acids.

VIII. STORAGE

To ensure that an adequate year-round supply of fruits and vegetables is available for the consumer, as well as the food industry, various storage methods have been developed. The oldest of these involves cold or refrigerated storage of the perishable produce, while the others are modified atmosphere (MA) or modified atmosphere packaging (MAP) and controlled atmosphere (CA) storage. The latter technology is expensive in terms of storage facilities, equipment, and maintenance. Common to all these technologies is that fruits and vegetables remain alive and carry on processes of all living tissues, in contrast to other preservation techniques such as heat processing, dehydration, or freezing, and others which arrest respiration and other metabolic processes. Commercial interest in the development and application of MA and CA for transportation and storage of fruits and vegetables has triggered extensive research in these fields in recent decades and has led to continuous development and refining, resulting in continuing and increasing application of these technologies worldwide.

A. Cold Storage

The oldest and most popular method for prolonging the shelf-life of perishable produce is cold storage. This is based on the fact that biological reactions, such as respiration and other metabolic processes, as well as decay, depend on temperature. The Van't Hoff rule indicates that the velocity of a reaction decreases two- to three-fold for every 10°C decrease in temperature. Thus, storage of fruits and vegetables at the lowest, but above freezing, temperatures would seem most beneficial in terms of maximal storage duration. However, some commodities show chilling injuries when exposed to low but not freezing temperatures. Most tropical or subtropical, and some crops of temperate zones, are sensitive to chilling injury. Thus, physiological and biochemical alteration may occur in sensitive species in response to low-temperature exposure and result in the development of chilling injury symptoms, such as surface pitting, discoloration, internal breakdown and flesh browning, failure to ripen, and loss of flavor, as well as decay (Ontario Ministry of Agriculture, 1998; Perez-Tello et al., 2009; Schotsmans et al., 2009). The critical temperature for chilling injury varies with the commodity and cultivar, and is thus genetically determined. Furthermore, the sensitivity of crops may be affected by preharvest conditions such as climatic factors and cultural practices. Officially recommended storage conditions for fruits and vegetables take this into account and, accordingly, storage temperatures recommended do not underrun critical temperatures (USDA, 2004). However, there are other factors that affect the storage life of the different commodities, including adequate relative humidity and the presence of ethylene in the storage atmosphere. Thus, for most vegetables, a high humidity must be maintained to prevent wilting or shriveling. On the other hand, a temperature range between $-1°C$ and $+3°C$ together with 70–75% relative humidity and good ventilation has generally been recommended for extending the storage life of onions over the winter period.

Good ventilation in a cold storeroom prevents the accumulation of ethylene or other volatile components in the store atmosphere. Ethylene increases the respiratory activity of most fruits and vegetables and hence shortens storage life. In addition, it has other beneficial as well as detrimental effects (Saltveit, 2005). Some of the detrimental effects, such as accelerated senescence, chlorophyll loss, and excessive softening, may cause unacceptable losses. Ethylene in the storage atmosphere may alter the flavor of commodities. A well-documented case describes bitterness in carrots induced by traces of ethylene (Seljasen, 2001a). Storability of commodities may be linked to the concentration of some constituents in the harvested produce; thus, Hanaoka and Ito (1957) reported that a high soluble sugar content in onion bulbs was a good indicator of their storage potential. Later researchers also found that the metabolism of carbohydrates in onions was closely linked to their storage performance (Kato, 1966; Toul and Popsilova, 1966). A more detailed study on carbohydrate metabolism in stored onions by Rutherford and Whittle (1982) showed that increased fructose resulted from the hydrolysis of the storage oligosaccharides. A low fructose content in freshly harvested onion bulbs indicated poor storage properties, with the reverse being true for onion bulbs high in fructose. Measurement of alkaline invertase in onions by Rutherford and Whittle (1984) showed that its activity reflected the level of fructose present. Monitoring fructose content and invertase activity in harvested onions provided a useful indicator for assessing their stability during cold storage.

B. Controlled Atmosphere Storage

Controlled atmosphere (CA) storage refers to storage in atmospheres different from normal air (20–21% O_2, about 0.03% CO_2, 78–79% N_2, and trace quantities of other gases) and its strict control during the entire storage time. MA and MAP rely also on storage or transportation of produce in atmospheres different from normal air, but in contrast to CA, the atmosphere is not controlled after its establishment. Thus, strict temperature control management is mandatory to effect a positive impact of MA and MAP on the quality and shelf-life of produce. Modified or controlled atmosphere conditions reduce or retard biochemical processes such as respiration, ripening, and yellowing of fruits and vegetables. The modified environment contains typically low levels of oxygen and high levels of carbon dioxide, which slow down the catabolic processes, in particular respiration, as well as aging, thereby extending the storage life of fruits and vegetables. Worth mentioning is the effect on ethylene: oxygen is required for its synthesis and both oxygen and carbon dioxide affect its biological activity. Hence, controlled or modified atmospheres slow down the synthesis and activity of ethylene during storage.

In the 1960s it became evident that maintaining as low levels of oxygen as tolerated by produce was advantageous. This was reflected in the terminology used for different storage regimes, such as LO (low-oxygen) storage or ULO (ultra-low-oxygen) storage (Hoehn *et al.*, 2009). In general, the lowest possible level of oxygen corresponds to the anaerobic compensation point (ACP) (Figure 2.15). The ACP is the oxygen concentration at which carbon dioxide production is at a minimum (Boersig *et al.*, 1988).

The overall benefit of CA storage is that produce can retain its freshness and eating quality for much longer periods. Not all fruits and vegetables, however, can be stored under these conditions. Considerable research has been conducted on a wide range of produce, but apples remain worldwide the major fruit handled commercially under CA storage conditions. Other commodities stored or shipped commercially under controlled atmospheres include pears, kiwis, berries, cabbage, onions, lettuce, avocados, asparagus, and bananas. One of the major problems associated with CA storage is that different conditions are required for the storage of different cultivars. For example, apples have been extensively studied, with lists of recommended conditions for specific cultivars available from different regions of the USA, England, and Europe (Stoll, 1970, 1973; Blanpied, 1977; Fidler *et al.*, 1977; Porritt, 1977; Kader, 2003; Kupferman, 2003; Saltveit, 2003b; Brecht, 2006; Erkan and Wang, 2006). This makes CA storage very expensive as conditions have to be tailor-made for the different cultivars of the same crop as well as for the different fruits and vegetables.

C. New Developments in Postharvest Storage

The establishment of commercial applications of CA storage technology was based on research on the effects of oxygen, carbon dioxide, and ethylene (C_2H_2) on respiration and ripening in pome fruits and berries by Kidd and West (1927, 1930b, 1937, 1950). Since then, this technology has been continuously developed and refined (Dilley, 2006). In the new millennium two new developments were implemented. The first was the application of 1-methyl-cyclopropene (1-MCP) as a postharvest chemical (Sisler and Blankenship, 1996). 1-MCP is a low-boiling compound. For its commercial use it is complexed with α-cyclodextrin to produce a water-soluble powder (Daly and Kourelis, 2000). For application, the soluble powder is mixed with water, thereby releasing 1-MCP as a gas and then dispersing it around the produce. The application rates recommended are in the range of 0.2−1 ppm (Kostansek and Pereira, 2003). Treatment durations of 12−24 hours are generally sufficient to achieve a full response. Since 2002 1-MCP has been approved for postharvest application of apples in many countries (Prange and DeLong, 2003). In some countries application for bananas has been approved, while for other commodities governmental approval is pending. 1-MCP binds to the ethylene receptors and inhibits the action of both internal and external ethylene, resulting in a considerable retarding of ripening processes. However, individual commodities respond differently to 1-MCP (Watkins, 2002; Blankenship and Dole, 2003) and more research remains to be done on 1-MCP effects on both postharvest and non-postharvest processes occurring in plants.

FIGURE 2.15 Effect of oxygen concentration on respiration and aerobic compensation point (ACP). The ACP is the O_2 concentration at which respiration (CO_2 production) is at a minimum. *(From Gasser et al., 2008. With permission.)*

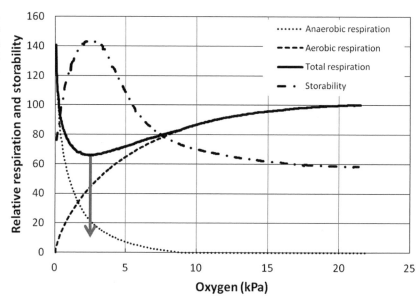

A second new development in postharvest techniques encompasses dynamic or adaptive control of controlled atmosphere storage (DCA). Recommendations for CA and MA of fruits and vegetables are usually based on numerous empirical storage trials to find the optimal CA conditions. The conditions yielding the best storage results are normally recommended and maintained from the beginning to the end of storage (Prange *et al.*, 2005; Gasser *et al.*, 2008). Since stored fruits and vegetables are living tissues, it is reasonable to expect their metabolism to be dynamic in nature during storage. Thus, static CA conditions are not likely to be optimal and may even be incorrect at some point during the storage period. Saltveit (2003b) pointed out this aspect and advocated the need for dynamic CA. A promising technique to dynamically control CA appears to be chlorophyll fluorescence measurements (Prange *et al.*, 2003, 2005; Gasser *et al.*, 2005). Chlorophyll fluorescence is affected by low oxygen and high carbon dioxide (DeEll *et al.*, 1995, 1998) and research indicates that the technique detects the lowest level of oxygen that is tolerated by stored fruits and vegetables. This oxygen concentration would correspond to the ACP (Gasser *et al.*, 2005, 2007). Based on the principle of chlorophyll fluorescence, the Harvest Watch™ system has been developed and subsequently patented (Prange *et al.*, 2007). The first commercial applications of DCA for apples were undertaken in 2004 in Italy and the USA and, since then, its application for storage of apples has increased steadily worldwide. Storing apples just above oxygen threshold concentrations (ACP) resulted in improved quality retention and it has been reported that this technology maintains apple quality after storage equal to, or better than, 1-MCP (Zanella *et al.*, 2005). Adaptive control of CA conditions entails monitoring the response of horticultural produce to changes in storage parameters or stressful conditions. Another indicator represents the fermentation threshold. It can be monitored and deduced from increases of acetaldehyde or ethanol in the storage atmosphere. Devices to monitor ethanol have been developed and were applied in the CA storage of 'Elstar' apples in the Netherlands (Veltman *et al.*, 2003c). Other techniques may be incorporated into CA systems in the future, for example, control of storage rots by natural antimicrobial compounds delivered as gas at critical periods during storage, or the addition of novel gases (NO, N_2O) to increase the benefits of current CA (Hoehn *et al.*, 2009).

REFERENCES

Abeles, F.B., 1973. Ethylene in Plant Biology. Academic Press, New York.

Abeles, F.B., Takeda, F., 1990. Cellulase activity and ethylene in ripening strawberry and apple fruit. Sci. Hortic 47, 77–87.

Acaster, M.A., Kende, H., 1983. Properties and partial purification of ACC synthetase. Plant Physiol. 72, 139–145.

Ackermann, J., Fischer, M., Amado, R., 1992. Changes in sugars, acids, and amino acids during ripening and storage of apples (cv. Glockenapfel). J. Agric. Food Chem. 40, 1131–1134.

Adams, D.O., Yang, S.F., 1977. Methionine metabolism in apple tissue: implication of *S*-adenosylmethionine as an intermediate in the conversion of methionine to ethylene. Plant Physiol. 60, 892–896.

Adams, D.O., Yang, S.F., 1979. Ethylene biosynthesis. Identification of 1-aminocyclopropane-1-carboxylic acid as an intermediate in the conversion of methionine to ethylene. Proc. Natl. Acad. Sci. U.S.A. 76, 170–174.

Aharoni, Y., 1968. Respiration of oranges and grapefruits harvested at different stages of development. Plant Physiol. 43, 99–102.

Aharoni, A., Keizer, L.C.P., Bouwmeester, H.J., 2000. Identification of the SAAT gene involved in strawberry flavor biogenesis by use of DNA microarrays. Plant Cell 12, 647–661.

Ahmed, A.E., Labavitch, J.M., 1980. Cell wall metabolism in ripening fruit. II. Changes in carbohydrate-degrading enzymes in ripening 'Bartlett' pears. Plant Physiol. 65, 1014–1016.

Akamine, E.K., Goo, T., 1978. Respiration and ethylene production in Mammee apple (*Mammea americana* L). J. Am. Soc. Hortic. Sci. 103, 308–310.

Akamine, E.K., Goo, T., 1979a. Concentrations of carbon dioxide and ethylene in the cavity of attached papaya fruit. HortScience 14, 138–139.

Akamine, E.K., Goo, T., 1979b. Respiration and ethylene production in fruits of species and cultivars of *Pridium* and species of *Eugenia*. J. Am. Soc. Hortic. Sci. 104, 632–635.

Akhavan, I., Wrolstad, R.E., 1980. Variation of sugars and acids during ripening of pears and in the production and storage of pear concentrate. J. Food Sci. 45, 499–501.

Alexander, L., Grierson, D., 2002. Ethylene biosynthesis and action in tomato: a model for climacteric fruit ripening. J. Exp. Bot. 53, 2039–2055.

Alscher, R.G., Castelfranco, P.A., 1972. Stimulation by ethylene of chlorophyll biosynthesis in dark-grown cucumber cotyledons. Plant Physiol. 50, 400–403.

Apelbaum, A., Burgoon, A.C., Anderson, J.D., Solomos, T., Lieberman, M., 1981. Some characteristics of the system converting aminocyclopropane-1-carboxylic acid to ethylene. Plant Physiol. 67, 80–84.

Areas, J.A.C., LaJollo, F.M., 1981. Starch transformation during banana ripening: I – The phosphorylase phosphatise behaviour in *Musa acuminate*. J. Food Biochem. 5, 19–37.

Arens, D., Seilmeier, W., Weber, E., Kloos, G., Grosch, W., 1973. Purification and properties of a carotene cooxidizing lipoxygenase from peas. Biochim. Biophys. Acta 327, 295–305.

Avissar, Y.J., Moberg, P., 1995. The common origin of the pigments of life-early steps in chlorophyll biosynthesis. Plant Photosynth. Res. 44, 221–242.

Awad, M., Young, R.E., 1979. Postharvest variation in cellulose, polygalacturonase, and pectinmethylesterase in avocado (*Persea americana* Mill cv. Fuerte) fruits in relation to respiration and ethylene production. Plant Physiol. 64, 306–308.

Awad, M., Young, R.E., 1980. Avocado pectinmethylesterase activity in relation to temperature, ethylene and ripening. J. Am. Hortic. Sci. 105, 638–641.

Ayub, R., Guis, M., BenAmor, M., Gillot, L., Roustan, J.P., Latche, A., Bouzayen, M., Pech, J.C., 1996. Expression of ACC oxidase antisense gene inhibits ripening of cantaloupe melon fruits. Nat. Biochem. 212, 27–34.

Azoulay-Shemer, T., Harpaz-Saad, S., Cohen-Peer, R., Mett, A., Spicer, V., Lovat, N., Krokhin, O., Brand, A., Gidoi, D., Standing, K.G., Goldschmidt, E.E., Eyal, Y., 2010. Dual N- and C-terminal processing of citrus chlorophyllase precursor within the plastid membranes leads to the mature enzyme. Plant Cell Physiol. 52, 70–83.

Azuma, R., Takahashi, Y., Kurata, H., Kawano, T., Shimokwa, K., Adachi, M., 1999. Does peroxidase act as a 'Mg-dechetalase?'. Plant Peroxidase Newslett. 13, 145–151.

Babbitt, J.K., Powers, M.J., Patterson, M.E., 1973. Effects of growth regulators on cellulase, polygalacturonase, respiration, color and texture of ripening tomatoes. J. Am. Soc. Hortic. Sci. 98, 77–81.

Bacon, M.F., Holden, M., 1970. Chlorophyllase of sugar-beet leaves. Phytochemistry 9, 115–125.

Bain, J.A., 1958. Morphological and physiological changes in the developing fruit of Valencia orange (Citrus sinesis L. Osbeck. Aust. J. Bot. 6, 1–23.

Baker, A., Graham, I.A., Hodsworth, M., Smith, S.M., 2006. Chewing the fat: β-oxidation in signaling and development. Trends Plant Sci. 11, 124–132.

Balague, C., Watson, C.F., Turner, A.J., Rouge, P., Picton, S., Pech, J.C., Grierson, D., 1993. Isolation of a ripening and wound-induced cDNA from Cucumis melo L. encoding a protein with homology to the ethylene-forming enzyme. Eur. J. Biochem. 212, 27–34.

Baldwin, E.A., 2004. Flavor. In: The commercial storage of fruits, vegetables, and florist and nursery stocks. USDA, ARS. http://www.ba.ars.usda.gov/hb66/contents.html (18 March 2011).

Baldwin, E.A., Plotto, A., Goodner, K., 2007. Shelf-life versus flavor-life for fruits and vegetables: how to evaluate this complex trait. Stewart Postharvest Rev. 1, 1–10.

Barker, J., Solomos, T., 1962. Mechanisms of carbohydrate breakdown in plants. Nature (London) 196 189–189.

Barmore, C.R., Rouse, A.H., 1976. Pectinesterase activity in controlled atmosphere stored avocados. J. Am. Hortic. Soc. 101, 294–296.

Barnes, M.F., Patchett, B.J., 1976. Cell wall degrading enzymes, and the softening of senescent strawberry fruit. J. Food Sci. 41, 1392–1395.

Barry, C.S., Blume, B., Bouzayen, M., Cooper, W., Hamilton, A.J., Grierson, D., 1996. Differential expression of the 1-amino-cyclopropane-1-carboxylic acid oxidase gene family of tomato. Plant J. 9, 525–535.

Barry, C.S., Llop-Tous, M.I., Grierson, D., 2000. The regulation of 1-amincyclopropane-1-cyclic acid syntheas gene expression during transition from system-1 to system-2 ethylene synthesis in tomato. Plant Physiol. 123, 979–986.

Bartley, I.M., 1974. B-Galactosidase activity in ripening apples. Phytochemistry 13, 2107–2111.

Bartley, I.M., 1978. Exo-polygalacturonase of apple. Phytochemistry 17, 213–216.

Bartley, I.M., Knee, M., 1982. The chemistry of textural changes in fruit during storage. Food Chem. 9, 47–58.

Battersby, A.R., Fookes, C.J.R., Matcham, G.W.J., McDonald, E., Gustafson-Porter, K.E., 1979. Proof of synthesis that unrearranged hydroxymethylbilane is the product from deaminase and the substrate for cosynthetase in the biosynthesis of urogen III. J. Chem. Soc. Chem. Comm. 316, 155–158.

Beale, S.I., 1999. Enzymes of chlorophyll biosynthesis. Photosynth. Res. 60, 45–73.

Beale, S.I., 2005. Green genes gleaned. Trends Plant Sci. 10, 309–312.

Beale, S.I., Gough, S.P., Granick, S., 1975. Biosynthesis of δ-aminolevulinic acid from the intact carbon skeleton of glutamic acid in greening barley. Proc. Natl. Acad. Sci. U.S.A. 72, 2719–2723.

Ben-Arie, R., Sonego, L., 1980. Pectolytic enzyme involved in wooly breakdown of stored peaches. Phytochemistry 19, 2553–2555.

Ben-Arie, R., Sonego, L., Frenkel, C., 1979. Changes in pectic substances in ripening pears. J. Am. Soc. Hortic. Sci. 104, 500–505.

Bendall, D.S., Bonner Jr., D.W., 1971. Cyanide-insensitive respiration in plant mitochondria. Plant Physiol. 47, 236–245.

Bennett, A.B., Smith, G.M., Nichols, B.G., 1987. Regulation of climacteric respiration in ripening avocado. Plant Physiol. 83, 973–976.

Berard, L.S., Lougheed, E.C., Murr, D.P., 1982. β-Galactosidase activity of 'McIntosh' apples in storage. HortScience 17, 660–661.

Beyer, P.J., Kreuz, K., Kleinig, H., 1980. β-Carotene synthesis in isolated chromoplasts from. Narcissus pseudonarcissus. Planta 150, 435–438.

Biale, J.B., 1960a. The postharvest biochemistry of tropical and subtropical fruits. Adv. Food Res. 10, 293–354.

Biale, J.B., 1960b. Respiration of fruits. In: Ruhland, W. (Ed.), Handbuch der Pflanzenphysiologie, Vol. 12, Part 2. Springer, Berlin, p. 586.

Biale, J.B., 1964. Growth, maturation and senescence in fruits: recent knowledge on growth regulation and on biological oxidations has been applied to studies with fruits. Science 146, 880–888.

Biale, J.B., 1969. Metabolism at several levels of organization in the fruit of avocado, Pesea americana. Mill. Qual. Plant. Mater. Veg. 19, 141–153.

Biale, J.B., Barcus, D.E., 1970. Respiration patterns in tropical fruits of the Amazon basin. Trop. Sci. 12, 93–104.

Biale, J.B., Young, R.E., 1947. Critical oxygen concentrations for the respiration of lemons. Am. J. Bot. 34, 301–309.

Biale, J.B., Young, R.E., 1971. The avocado pear. In: Hulme, A.C. (Ed.), The Biochemistry of Fruits and their Products, Vol. 2. Academic Press, London, pp. 1–63.

Biale, J.B., Young, R.E., 1981. Respiration and ripening in fruits — retrospect and prospect. In: Friend, J., Rhodes, M.J.C. (Eds.), Recent Advances in the Biochemistry of Fruits and Vegetables. Academic Press, New York, pp. 1–39.

Biale, J.B., Young, R.E., Olmstead, A.J., 1954. Fruit respiration and ethylene production. Plant Physiol. 29, 168–174.

Biale, J.B., Young, R.E., Popper, C.S., Appleman, W.E., 1957. Metabolic processes in cytoplasmic particles of the avocado fruit. 1. Preparative procedure, cofactor requirements, and oxidative phosphorylation. Physiol. Plant. 10, 48–63.

Bishop, R.C., Klein, R.M., 1975. Photo-promotion of anthocyaninsynthesis in harvested apples. HortScience 10, 126–127.

Blankenship, S.M., Dole, J.M., 2003. 1-Methylcyclopropene: a review. Postharvest Biol. Technol. 28, 1–25.

Blanpied, G.D., 1977. Requirements and recommendations for eastern and midwestern apples. Mich. State Univ. Hortic. Rep. 28, 225–230.

Blume, B., Grierson, D., 1997. Expression of ACC oxidase promoter-GUS fusion in tomato and Nicotiana plumbaginifolia regulated by developmental and environmental stimuli. Plant J. 12, 731–746.

Boerjan, W., Ralph, J., Baucher, M., 2003. Lignin biosynthesis. Annu. Rev. Plant Biol. 54, 519–546.

Boersig, M.R., Kader, A.A., Romani, R.J., 1988. Aerobic—anaerobic respiration transition in pear fruit and cultured pear fruit cells. J. Am. Soc. Hortic. Sci. 113, 869—873.

Boller, T., Herner, R.C., Kende, H., 1979. An assay for the ethylene precursor 1-aminocyclo-propane-1-carboxylic acid and studies on its enzymatic formation. Planta 145, 293—303.

Bolwell, G.P., Northcote, D.H., 1983. Induction by growth factors of polysaccharide synthases in bean cell suspension cultures. Biochem. J 210, 509—515.

Bordignon, M.T.L., Gauche, C., Gris, E.F., Falcao, L.D., 2007. Colour stability of anthocyanins from Isabel grapes (*Vitis labrusca* L.) in model systems. LWT Food Sci. Technol. 40, 594—599.

Borisova, I.G., Budnitskaya, E.V., 1975. Lipoxygenase of chloroplasts. Dokl. Akad. Nauk SSSR 225, 439 (Cited Chem. Abstr. 84: 14788p, 1976).

Bousquet, J.F., Thimann, K.V., 1984. Lipid peroxidation forms ethylene from 1-aminocyclopropane-1-carboxylic acid and may operate in leaf senescence. Proc. Natl. Acad. Sci. U.S.A. 81, 1724—1727.

Brackmann, A., Streif, J., Bangerth, F., 1966. Relationship between reduced aroma production and lipid metabolism of apples after long-term controlled-atmosphere storage. J. Am. Soc. Hortic. Sci. 88, 98—104.

Bradford, K.J., 2008. Shang Fa Yang: pioneer in plant ethylene biochemistry. Plant Sci. 175, 2—7.

Brady, C.J., 1976. The pectinesterase of the pulp of banana fruit. Aust. J. Plant Physiol. 3, 163—172.

Brady, C.J., Palmer, J.K., O'Connell, P.B.H., Smillie, R.M., 1976. An increase in protein synthesising during ripening of the banana fruit. Phytochemistry 9, 1037—1047.

Bramley, P.M., Than, A., Davies, B.H., 1977. Alternative pathways of carotene cyclisation in *Phycomyces blackesleeanus*. Phytochemistry 16, 235—238.

Brauss, M.S., Linforth, R.S.T., Taylor, A.J., 1998. Effect of variety, time of eating, and fruit-to-fruit variation on volatile release during eating of tomato fruit (*Lycopersicon esculentum*). J. Agric. Food Chem. 46, 2287—2292.

Braverman, J.B.S., 1963. Introduction to the Biochemistry of Foods. Elsevier, Amsterdam.

Brecht, J.K., 2006. Controlled atmosphere, modified atmosphere and modified atmosphere packaging for vegetables. Stewart Postharvest Rev. 5 (5), 1—6.

Brecht, J.K., Huber, D.J., 1988. Products released from enzymically active cell wall stimulate ethylene production and ripening in pre-climacteric tomato (*Lycopersicon esculentur* Mill.) fruit. Plant Physiol. 88, 1037—1041.

Britton, G., 1976. Biosynthesis of carotenoids. In: Goodwin, T.W. (Ed.), Chemistry and Biochemistry of Plant Pigments, Vol. 1. Academic Press, New York, pp. 262—327.

Britton, G., 1979. Carotenoid biosynthesis — a target for herbicide activity. Z. Naturforsch. 34C, 979—985.

Britton, G., 1982. Biosynthesis of carotenoids. In: Goodwin, T.W. (Ed.), Chemistry and Biochemistry of Plant Pigments, Vol. 1. Academic Press, New York, pp. 262—327.

Brouillard, R., 1983. The *in vivo* expression of anthocyanin colour in plants. Phytochemistry 22, 1311—1323.

Broun, P., 2005. Transcriptional control of flavonoid biosynthesis, a complex network of conserved regulators involved in multiple aspects of differentiation in *Arabidopsis*. Curr. Opin. Plant Biol. 8, 272—279.

Browse, J., Møfler, I.M., Rasmusson, A.G., 2006. Respiration and lipid metabolism. In: Taiz, L., Zeiger, E. (Eds.), Plant Physiol.ogy. Sinauer Associates, Sunderland, MA, pp. 251—288. Chapter 11.

Brummell, D.A., 2005. Regulation and genetic manipulation of ripening in climacteric fruit. Stewart Postharvest Rev. 3, 1—19.

Brummell, D.A., Cin, V.D., Lurie, S., Crisoto, C.H., Labavitch, J.M., 2004. Cell wall metabolism during the development of chilling injury in cold-stored peach fruit: association of mealiness with assisted disassembly of cell wall pectins. J. Exp. Bot 55, 2041—2052.

Bruyant-Vannier, M.-P., Gaudinet-Schauman, A., Bourlard, T., Morvan, C., 1966. Solubilization and partial characterization of pectinmethyl-transferase from flax cells. Plant Physiol. Biochem 34, 489—499.

Buckle, K.A., Edwards, R.A., 1969. Chlorophyll degradation products from processed pea puree. Phytochemistry 8, 1901—1906.

Buckle, K.A., Edwards, R.A., 1970. Chlorophyll, colour and pH changes in HTST processed green pea puree. J. Food Technol. 5, 173—186.

Buescher, R.W., Furmanski, R.J., 1978. Role of pectinesterase and polygalacturonase in the formation of woolliness in peaches. J. Food Sci. 43, 264—266.

Buescher, R.W., Tigchelaar, E.C., 1975. Pectinesterase, polygalacturonase, Cx-cellulase activities and softening of x in the rin tomato mutant. HortSci 10, 4—25.

Bufler, G., 1984. Ethylene-enhanced 1-aminocyclopropane-1-carboxylic acid synthase activity in ripening apples. Plant Physiol. 75, 192—195.

Bufler, G., 1986. Ethylene-promoted conversion of 1-aminocyclopropane-1-carboxylic acid to ethylene in peel of apple at various stages of fruit development. Plant Physiol. 80 539—492.

Bufler, G., Bangerth, F., 1983. Effects of propylene and oxygen on the ethylene producing system of apples. Physiol. Plant. 58, 486—492.

Burg, S.P., 1973. Ethylene in plant growth. Proc. Natl. Acad. Sci. U.S.A. 70, 591—597.

Burg, S.P., Burg, E.A., 1962. Role of ethylene in fruit ripening. Plant Physiol. 37, 179.

Burg, S.P., Burg, E.A., 1965. Ethylene action and the ripening of fruits. Science 148, 1190—1196.

Burg, S.P., Thimann, K.V., 1959. The physiology of ethylene formation in apples. Proc. Natl. Acad. Sci. U.S.A. 45, 335—344.

Bush, M.S., Marry, M., Huxcham, I.M., Jarvis, M.C., McCann, M.C., 2001. Developmental regulation of pectic epitopes during potato tuberisation. Planta 213, 869—880.

Buttery, R.G., Ling, L.C., 1993. Volatile compounds of tomato fruit and plant parts: relationship and biogenesis. ACS Symp. Ser 525, 23—34.

Buttery, R.G., Teranishi, R., Ling, L.C., 1987. Fresh tomato aroma volatiles: a quantitative study. J. Agric. Food Chem. 35, 540—544.

Caffall, K.H., Mohnen, D., 2009. The structure, function, and biosynthesis of plant cell pectic polysaccharides. Carbohydr. Res. 344, 1879—1900.

Camara, B., Brangeon, J., 1981. Carotenoid metabolism during chloroplast to chromoplast transformation in *Capsicum annuum* fruit. Planta 151, 359.

Camara, B., Badat, F., Moneger, R., 1982. Sites of biosynthesis of carotenoids in *Capsicum* chromoplasts. Eur. J. Biochem 127, 255—258.

Campbell, H., 1937. Undesirable color changes in frozen peas stored at insufficiently low temperatures. J. Food Sci. 2, 55—57.

Canjura, F.L., Watkins, R.H., Schwartz, S.J., 1999. Color improvement and metallo- chlorophyll complexes in continuous flow aseptically processed peas. J. Food Sci. 64, 987—990.

Cansfield, P.E., Francis, F.J., 1970. Quantitative methods for anthocyanins. 5. Separation of cranberry phenolics by electrophoresis and chromatography. J. Food Sci. 35, 309−311.

Carnal, N., Black, C.C., 1978. Pyrophosphate-dependent 6-phosphofructokinase, a new glycolytic enzyme in pineapple leaves. Biochem. Biophys. Res. Commun. 86, 20−26.

Carpita, N., McCann, M., 2000. The cell wall. In: Buchanan, B., Gruissem, W., Jones, R. (Eds.), Biochemistry and Molecular Biology. American Society of Plant Physiologists, pp. 52−108.

Chalmers, D.J., Rowan, K.S., 1971. The climacteric in ripening tomato fruit. Plant Physiol. 48, 235−240.

Chang, C., Bleecker, A.B., 2004. Ethylene biology. More than a gas. Plant Physiol. 136, 2895−2899.

Chauhan, S.K., Tyagi, S.M., Singh, D., 2001. Pectinolytic liquefaction of apricot, plum, and mango pulps for fruit extraction. Int. J. Food Prop. 4, 103−109.

Chepurenko, N.V., Borisova, I.G., Budnitskaya, E.V., 1978. Isolation and characterization of isoenzymes of lipoxygenase from pea seeds. Biochem. (English Transl.) 43, 480−485.

Chereskin, B.M., Wong, Y.-S., Castelfranco, P.A., 1982. In vitro synthesis of the chlorophyll isocyclic ring. Transformation of magnesium−protoporphyrin IX and magnesium−protoporphyrin mono methyl ester into magnesium-2,4-divinyl pheoporphyrin a. Plant Physiol. 70, 987.

Chitarra, A.B., LaJollo, F.M., 1981. Phosphorylase, phosphatase, α-amylase activity and starch breakdown during ripening of Marmelo banana (Musa acuminata (olla) × musubalbisiani (olla) ABB group). Whole fruit and thin slices. J. Am. Soc. Hortic. Sci. 106, 579−584.

Chou, H., Breene, N.M., 1972. Oxidative discoloration of β-carotene in low moisture foods. J. Food Sci. 37, 66−68.

Cohen, B.-S., Grossman, S., Pinsky, A., Klein, B., 1984. Chlorophyll inhibition of lipoxygenase in growing plants. J. Agric. Food Chem. 32, 516−519.

Coleman, J.O.D., Palmer, J.M., 1972. The oxidation of malate by isolated plant mitochondria. Eur. J. Biochem. 26, 499−509.

Costa, M.L., Crevello, P.M., Chaves, A.R., Marines, G.A., 2002. Characterization of Mg-dechelatase activity obtained from Fragaria × ananassa. Plant Physiol. Biochem. 40, 111−118.

Craker, L.E., Wetherbee, P.J., 1973. Ethylene, carbon dioxide, and anthocyanin synthesis. Plant Physiol. 52, 177−179.

Dahodwala, S., Humphrey, A., Weibel, M., 1974. Pectic enzymes: individual and concerted kinetic behavior of pectinesterase and pectinase. J. Food Sci. 39, 920−926.

Dalal, K.B., 1965. Investigation into flavor chemistry with special reference to synthesis of volatiles in developing tomato fruit (Lycopersicon esculentum) under field and greenhouse growing conditions. PhD thesis. Utah State University, Logan, UT.

Dalal, K.B., Salunkhe, D.K., Olson, L.E., Do, J.Y., Yu, M.H., 1968. Volatile components of developing tomato fruit grown under field and greenhouse conditions. Plant Cell Physiol. 9, 389−400.

Daly, J., Kourelis, B., 2000. Synthesis methods, complexes and delivery methods for the safe and convenient storage, transport and application of compounds for inhibiting the ethylene response in plants. US Patent No. 6,017,849 25 January 2000.

Davies, B.H., 1973. Carotene biosynthesis in fungi. Pure Appl. Chem. 35, 1−28.

Davies, J.N., Hobson, G.E., 1981. The constituents of tomato fruit − the influence of environment, nutrition, and genotype. CRC Crit. Rev. Food Sci. Nutr. 15, 205−280.

Davison, P.A., Schubert, H.L., Reid, J.D., Iorg, C.D., Heroux, A., Hill, C.P., Hunter, C.N., 2005. Structural and biochemical characterization of Gun4 suggests a mechanism for its role in chlorophyll biosynthesis. Biochemistry 44, 7603−7612.

De Jager, A., Johnson, D., Höhn, E., 1996. Determination and prediction of optimum harvest date of apples and pears. ECSC-EC-EAEC, Brussels−Luxembourg.

De Pooter, H.L., Montens, J.P., Dirinck, P.J., Willaert, G.A., Schamp, N.M., 1982. Ripening induced in preclimacteric immature Golden Delicious apples by proprionic and butyric acids. Phytochemistry 21, 1015−1016.

De Pooter, H.L., D'Ydewalle, Y.E., Willaert, G.A., Dirink, P.J., Schamp, N.M., 1984. Acetic and propionic acids, inducer of ripening in preclimacteric Golden Delicious apples. Phytochemistry 23, 23−26.

De Swardt, G.H., Maxie, E.C., 1967. Pectin methylesterase in the ripening banana. S. Afr. J. Agric. Res. 10, 501−506.

DeEll, J.R., Prange, R.K., Murr, D.P., 1995. Chlorophyll fluorescence as a potential indicator of controlled-atmosphere disorders in 'Marshall' McIntosh apples. HortScience 30, 1084−1085.

DeEll, J.R., Prange, R.K., Murr, D.P., 1998. Chlorophyll fluorescence techniques to detect atmospheric stress in stored apples. Acta Hortic 464, 127−131.

Delmer, D.P., 1999. Cellulose biosynthesis: exciting times for a difficult field of study. Annu. Rev. Plant Physiol. Plant Mol. Biol. 50, 245−276.

Dick, A.J., Laskey, G., Lidster, P.D., 1984. Inhibition of β-galactosidase isolated from 'McIntosh' apples. HortScience 19, 552−553.

Dilley, D.R., 2006. Development of controlled atmosphere storage technologies. Stewart Postharvest Rev. 6 (5), 1−8.

Dixon, J., Hewett, E.W., 2000. Factors affecting apple aroma/flavor volatile concentration: a review. N.Z. J. Crop Hortic. Sci. 28, 155−173.

Dixon, W.L., ap Rees, T., 1980. Identification of the regulatory steps in glycolysis in potato tubers. Phytochemistry 19, 1297−1301.

Dixon, W.L., Franks, F., ap Rees, T., 1981. Cold lability of phosphofructokinase from potato tubers. Phytochemistry 20, 969−972.

Doblin, M.S., Vergara, C.E., Read, S.M., Newbigin, E., Bacic, A., 2003. Plant cell wall biosynthesis: making bricks. In: Rose, J.K.C. (Ed.), The Plant Cell Wall. Blackwell, Oxford, pp. 183−222.

Dolendo, A.L., Luh, B.S., Pratt, H.K., 1966. Relation of pectic and fatty acid changes to respiration rate during ripening of avocado fruit. J. Food Sci. 31, 332−336.

Dong, J.G., Fernandez-Maculet, J.C., Yang, S.F., 1992. Purification and characterization of 1-aminocyclopropane-1-carboxylate oxidase from apple fruit. Proc. Natl. Acad. Sci. U.S.A 89, 9789−9793.

Doreyappa Gowda, I.N., Huddar, A.G., 2001. Studies on ripening changes in mango (Mangifera indica L.) fruits. J. Food Sci. Technol. 38, 135−137.

Douillard, R., Bergeron, E., 1978. Activite lipoxyginasque de chloroplasts plantules de Ble. C. R. Hebd. Seances Acad. Sci., Ser. D 286, 753.

Dresel, E.J.B., Falk, J.E., 1953. Conversion of δ-aminolevulinic acid to porphobilinogen in a tissue system. Nature 172, 1185.

Dull, G.G., Young, R.E., Biale, J.B., 1967. Respiratory patterns in fruit of pineapple, Ananas comosus, detached at different stages of development. Physiol. Plant. 20, 1059−1065.

Dürr, P., Röthlin, M., 1981. Development of synthetic apple juice odour. Lebensm. Wiss. Technol. 14, 313−314.

Eaks, I.L., 1970. Respiratory response, ethylene production, and response to ethylene of citrus fruit during ontogeny. Plant Physiol. 45, 334–338.

Eaks, I.L., Morris, L., 1956. Respiration of cucumber fruits associated with physiological injury at chilling temperatures. Plant Physiol. 31, 308–314.

Ebert, G., Gross, J., 1985. Carotenoid changes in the peel of ripening persimmon (*Diospyros Kam*) Triumph. Phytochemistry 24, 29–32.

Eberhardt, N.L., Rilling, H.C., 1975. Phenyltransferase from *Saccharomyces cerevisiae*. Purification to homogeneity and molecular properties. J. Biol. Chem. 250, 863–866.

Eckhardt, U., Grimm, B., 2004. Recent advances in chlorophyll biosynthesis and breakdown in higher plants. Plant Mol. Biol. 56, 1–14.

Eckhardt, U., Grimm, B., Hortensteiner, S., 2004. Recent advances in chlorophyll biosynthesis and breakdown in higher plants. Plant Mol. Biol. 56, 1–14.

von Elbe, J.H., Huang, A.S., Attoe, E.L., Nank, W.K., 1986. Pigment composition and colour of conventional Veri-Green® canned beans. J. Agric. Food Chem. 34, 52–54.

Eiati, S.K., Budowski, P., Monselise, S.P., 1975. Carotenoid changes in the 'Shamouti' orange peel during chloroplast–chromoplast transformation on and off the tree. J. Exp. Bot. 26, 624–632.

Erkan, M., Wang, C.Y., 2006. Modified and controlled atmosphere storage of subtropical crops. Stewart Postharvest Rev. 5 (4), 1–8.

Eskin, N.A.M., 1979. Plant Pigments, Flavors and Textures: The Chemistry and Biochemistry of Selected Compounds. Academic Press, New York.

Eskin, N.A.M., Grossmann, S., Pinsky, A., 1977. The biochemistry of lipoxygenase in relation to food quality. CRC Crit. Rev. Food Sci. Nutr. 9, 1–40.

Espineda, C.E., Linford, A.S., Devine, D., Brusslan, J., Yamasato, A., 1999. The *AtCAO* gene, encoding chlorophyll *a* oxygenase, is required for chlorophyll *b* synthesis in *Arabidopsis thaliana*. Proc. Natl. Acad. Sci. U.S.A. 96, 10507–10511.

Farin, D., Ikan, R., Gross, J., 1983. The carotenoid pigments in the juice and flavedo of a mandarin hybrid (*Citrus reticulate*) cv. Michal during ripening. Phytochemistry 22, 403–408.

Fellers, P.J., 1991. The relationship between the ratio of degree Brix to percent acid and sensory flavor in grapefruit juice. Food Technol. 45 68, 70, 72–75.

Fernandez-Maculet, J.C., Yang, S.F., 1992. Extraction and partial characterization of the ethylene-forming enzyme from apple fruit. Plant Physiol. 99, 751–754.

Fidler, J.C., Wilkinson, B.B., Edney, K.L., Sharples, R.O., 1977. The biology of apple and pear storage. Res. Rev. – Commonw. Agric. Bur. Engl. 3, 235.

Flath, R.A., Black, D.R., Guadagni, D.G., McFadden, W.H., Schultz, T.H., 1967. Identification and organoleptic evaluation of compounds in Delicious apple essence. J. Agric. Food Chem. 15, 29–35.

Fleming, H.P., Cobb, W.Y., Etchells, J.L., Bell, T.A., 1968. The formation of carbonyl compounds in cucumbers. J. Food Sci. 33, 572–576.

Flores, F.B., El Yahauoui, F., de Billerbeck, G., Romojaro, F., Latche, A., Bouzayen, M., Pech, J.C., Ambid, C., 2002. Role of ethylene in the biosynthetic pathway of aliphatic ester aroma volatiles in Charantais Canteloupe melons. J. Exp. Bot. 53, 201–206.

Foda, Y.H., 1957. Pectic changes during ripening as related to flesh firmness in the tomato. PhD thesis. University of Illinois, Urbana.

Forkmann, G., 1991. Flavonoids as flower pigments. The formation of the natural spectrum and its extension by genetic engineering. Plant Breed 106, 1–26.

Freeman, J.A., Woodbridge, C.G., 1960. Effect of maturation, ripening and truss position on the free amino acid content in tomato fruits. Proc. Am. Soc. Hortic. Sci. 76, 515–523.

Frenkel, C., 1979. Role of oxidative metabolism in the onset of senescence in plant storage. Z. Ernaehrungswiss 18, 209–212.

Frenkel, C., Eskin, N.A.M., 1977. Ethylene evolution as related to changes in hydroperoxides in ripening tomato fruit. HortScience 12, 552–553.

Frenkel, C., Klein, I., Dilley, D.R., 1968. Protein synthesis in relation to ripening of pome fruits. Plant Physiol. 43, 1146–1153.

Fuesler, T.P., Hanamoto, C.M., Castelfranco, P.A., 1982. Separation of Mg-protoporphyrin IX and Mg-protoporphyrin IX monomethyl ester synthesized *de novo* by developing cucumber etioplasts. Plant Physiol. 69, 421–423.

Fuleki, T., 1969. The anthocyanins of strawberry, rhubarb, radish and onion. J. Food Sci. 34, 365–369.

Fuleki, T., 1971. Anthocyanins in red onion. Allium cepa. J. Food Sci. 36, 101–104.

Gaillard, T., Matthew, J.A., 1977. Lipoxygenase-mediated cleavage of fatty acids to carbonyl fragments in tomato fruits. Phytochemistry 16, 339–343.

Games, D.E., Jackson, A.H., Jackson, J.R., Belcher, R.V., Smith, S.G., 1976. Biosynthesis of protoporphyrin-IX from coproporphyrinogen III. J. Chem. Soc. Chem. Commun. 6, 187–189.

Gane, R., 1934. Production of ethylene by some ripening fruits. Nature (London) 134 1008–1008.

Gasser, F., Dättwyler, D., Schneider, K., Naunheim, W., Hoehn, E., 2005. Effects of decreasing oxygen levels in the storage atmosphere on the respiration and production of volatiles of 'Idared' apples. Acta Hortic 682, 1585–1592.

Gasser, F., Eppler, T., Naunheim, W., Gabioud, S., Hoehn, E., 2007. Control of critical oxygen level during dynamic CA storage of apples by monitoring respiration as well as chlorophyll fluorescence. Acta Hortic 796, 69–76.

Gasser, F., Eppler, T., Naunheim, W., Gabioud, S., Höhn, E., 2008. Control of critical oxygen level during dynamic CA storage of apples. Agrarforschung 15, 98–103.

Gazzarrini, S., McCourt, P., 2001. Genetic interactions between ABA, ethylene and sugar signaling pathways. Curr. Opin. Plant Biol. 4, 339–343.

Geisler, G., Radler, F., 1963. Developmental and ripening processes in grapes of *Vitis*. Ber Dtsch. Bot. Ges. 76, 112–119.

Génard, M., Gouble, B., 2005. ETHY. A theory of fruit climacteric ethylene emission. Plant Physiol. 139, 531–545.

Gibson, H.D., Laver, W.G., Neuberger, A., 1958. Initial stages in the biosynthesis of porphyrins. II. The formation of 5-aminolevulinic acid from glycine and succinyl-CoA by particles from chicken erythrocytes. Biochem. J. 70, 71–81.

Girardi, C.L., Corrent, A.R., Lucchetta, L., Zanuzo, M.R., da Costa, T.S., Brackmann, A., Twyman, R.M., Nora, F.R., Nora, L., Silva, J.A., Rombaldi, C.V., 2005. Effect of ethylene, intermittent warming and controlled atmosphere on postharvest quality and the occurrence of woolliness in peach (*Prunus persica* cv. Chiripa) during cold storage. Postharvest Biol. Technol. 38, 25–38.

Giusti, M.M., Wrolstad, R.E., 2003. Acylated anthocyanins from edible sources and their applications in food systems. Biochem. Eng. J. 14, 217–225.

Gokmen, V., Bahceci, S., Serpen, A., Acar, J., 2005. Study of lipoxygenase and peroxidase as blanching indicator enzymes in peas: change of enzymic activity, ascorbic acid and chlorophylls during frozen storage. LWT 38, 903—908.

Gold, H.J., Weckel, K.G., 1959. Degradation of chlorophyll to pheophytin during sterilization of canned green peas. Food Technol. 13, 281—286.

Goldstein, J.L., Wick, E.L., 1969. Lipids in ripening banana fruit. J. Food Sci. 34, 482—484.

Gonnet, J.F., 1998. Color effects of copigmentation of anthocyanins. Revisited − 1. A colorimetric definition using the CIELAB scale. Food Chem. 63, 409—415.

Gorin, N., 1973. Several compounds in Golden Delicious apples as possible parameters of acceptability. J. Agric. Food Chem. 21, 671—673.

Goslings, D., Meskauskiene, R., Kim, C., Lee, K.P., Nater, M., Apel, K., 2004. Concurrent interactions of heme and FLU with Glut RNA reductase (HEMA1), the target of metabolic feedback inhibition of tetrapyrrole biosynthesis, in dark- and light-grown Arabidopsis plants. Plant J. 40, 957—967.

Goto, T., Kondo, T., Tamura, H., Kawahori, K., Hatton, H., 1983. Structure of platyconin, a diacylated anthocyanin isolated from the Chinese bell-flower, Platycondon grandiflorum. Tetrahedr. Lett. 24, 2181—2184.

Gou, J.-Y., Felippes, F.F., Liu, C.-J., Weigel, D., Wang, J.-W., 2011. Negative regulation of znthocyanin biosynthesis in Arabidopsis by a miR156-targeted SPL transcription factor. Plant Cell 23, 1512—1522.

Goulao, L.F., Oliviera, C.M., 2007. Molecular identification of novel differentially expressed mRNAs up-regulated during ripening of apples. Plant Sci. 172, 306—318.

Gray, J.C., Keckwick, R.G.O., 1969. Mevalonate kinase from etiolated cotyledons of French beans. Biochem. J 113, 37.

Gray, J.C., Keckwick, R.G.O., 1972. The inhibition of plant mevalonate kinase preparation by phenyl pyrophosphates. Biochim. Biophys. Acta 279, 290—296.

Gray, J.C., Keckwick, R.G.O., 1973. Mevalonate kinase in green leaves and etiolated cotyledons of the French bean Phaseolus vulgaris. Biochem. J 133, 335—347.

Greger, V., Schieberle, P., 2007. Characterisation of the key aroma compounds in apricots (Prunus armeniaca) by application of molecular sensory science concepts. J. Agric. Food Chem. 55, 5221—5228.

Grommeck, R., Markakis, P., 1964. Effect of peroxidase on anthocyanin pigments. J. Food Sci. 29, 53—57.

Grosch, W., Laskaway, G., Weber, F., 1976. Formation of volatile carbonyl compounds and cooxidation of β-carotene by lipoxygenase from wheat, potato, flax and beans. J. Agric. Food Chem. 24, 456—459.

Gross, K.C., 1985. Promotion of ethylene evolution and ripening of tomato fruit by galactose. Plant Physiol. 79, 306—307.

Gross, K.C., Wallner, S.J., 1979. Degradation of cell wall polysaccharides during tomato fruit ripening. Plant Physiol. 63, 117—120.

Gross, J., Timberg, R., Graej, M., 1983. Pigment and ultrastructural changes in the developing Mammalo Citrus grandis 'Goliath'. Bot. Gaz. (Chicago) 144, 401—404.

Grossman, A.R., Lohr, M., Im, C.S., 2004. Chlamydomonas reinhardtii in the landscape of pigments. Annu. Rev. Genet. 38, 119—173.

Grumbach, K.H., Forn, B., 1980. Chloroplast autonomy in acetyl coenzyme a formation and terpenoid synthesis. Z. Nuturforsch 35C, 645—648.

Guerriero, G., Fugelstad, J., Bulone, V., 2010. What do we really know about cellulose biosynthesis in higher plants? J. Integrative Plant Biol. 52, 161—175.

Gunawan, M.I., Barringer, S.A., 2000. Green colour degradation of blanched broccoli (Brassica oleracea) due to acid and microbial growth. J. Food Proc. Pres. 24, 253—263.

Gupte, S.M., Francis, F.J., 1964. Effect of pH adjustment high-temperature short-time processing on colour and pigment retention in spinach puree. Food Technol. 18, 1645—1658.

Gupte, S.M., El-Bisi, H.M., Francis, F.J., 1964. Kinetics of thermal degradation of chlorophyll in spinach puree. J. Food Sci. 29, 379—382.

Hamilton, A., Lycett, G.W., Grierson, D., 1990. Antisense gene that inhibits synthesis of the hormone ethylene in transgenic plants. Nature 346, 284—287.

Hammett, L.K., Kirk, H.J., Todd, H.G., Hale, S.A., 1977. Association between soluble solids/acid content and days from full bloom of 'Golden Delicious' apple fruits. J. Am. Soc. Hortic. Sci. 102, 429—434.

Hanamoto, M., Castelfranco, P.A., 1983. Separation of monovinyl and divinyl protochlorophyllides and chlorophyllides from etiolated and phototransformed cucumber cotyledons. Plant Physiol. 73, 79.

Hanaoka, T., Ito, K., 1957. Studies on the keeping quality of onions. 1. Relation between the characters of bulbs and their sprouting during storage. J. Hortic. Assoc. Jpn 26, 129—136.

Hansen, E., 1942. Quantitative study of ethylene production in relation to respiration of pears. Bot. Gaz. (Chicago) 103, 543—558.

Harker, F.R., Marsh, K.B., Young, H., Murray, S.H., Gunson, F.A., Walker, S.B., 2002. Sensory interpretation of instrumental measurements 2: sweet and acid taste of apple fruit. Postharvest Biol. Technol. 24, 241—250.

Harborne, J.B., 1967. Comparative Biochemistry of the Bioflavonoids. Academic Press, New York.

Harborne, J.B., 1976. Functions of flavonoids in plants. In: Goodwin, T.W. (Ed.)Chemistry and Biochemistry of Plant Pigments, 2nd ed., Vol. 1. Academic Press, New York, p. 525.

Hardy, S.I., Castelfranco, P.A., Rebeiz, C.A., 1971. Effect of the hypoctyl hook on chlorophyll accumulation in excised cotyledons of Cucumis sativus L. Plant Physiol. 47, 705—708.

Harpaz-Saad, S., Azoulay, Y., Arazi, T., Ben-Yaakov, E., Mett, A., Shiboleth, Y.M., Hortensteiner, S., Gidoni, D., Gal-On, A., Goldschmidt, E.E., Eyal, Y., 2007. Chlorophyllase is a rate-limiting enzyme in chlorophyll catabolism and is posttranslationally regulated. Plant Cell 19, 1007—1022.

Harper, K.A., 1968. Structural changes of flavylium salts. IV. Polarographic and spectrometric examination of perlargonidin chloride. Aust. J. Chem. 21, 221—227.

Hartmann, C., Drouet, A., Morin, F., 1987. Ethylene and ripening of apple, pear and cherry fruit. Plant Physiol. 25, 505—512.

Haslam, E., Gupta, R.K., Al-Shafi, S.M.K., Layden, K., 1982. J. Chem. Soc., Perkin Trans. 1, 2525.

Hayashi, T., 1989. Xyloglucans in the primary cell wall. Annu. Rev. Plant Physiol. Plant Mol. Biol. 46, 139—168.

Hayashi, T., Yohida, K., Park, Y.W., Konishi, T., Baba, K., 2005. Cellulose metabolism in plants. Int. Rev. Cytol. 247, 1—34.

Heaton, J.W., Marangoni, A.G., 1996. Chlorophyll degradation in processed foods and senescent plant tissues. Trends Food Sci. Technol. 7, 8–15.

Hermann, J., 1970. Berechnungder chermischenj und seorischen Verandderungen unserer Lebensmittel bei Erhitzungs-und Lageruugsprozessen. Ernaehrungsforschung 15, 279.

Herner, R.C., 1973. Fiber determination. Progress report on asparagus research. Mich. Agric. Exp. Stn. Rep. 217, 11.

Herner, R.C., Sink, K.C., 1973. Ethylene production and respiratory behaviour of rin tomato mutant. Plant Physiol. 52, 38–42.

Hilderbrand, D.F., Hymovitz, T., 1982. Carotene and chlorophyll bleaching by soybeans with and without seed lipoxygenase. J. Agric. Food Chem. 30, 705–708.

Hillman, C.R., Burin, V.M., Bordignon-Luiz, M.T., 2011. Thermal degradation kinetics of anthocyanins in grape juice and concentrate. Int. J. Food Sci. Technol. 46, 1997–2000.

Hobson, G.E., 1965. The firmness of tomato fruit in relation to polygalacturonase activity. J. Hortic. Sci. 40, 66–72.

Hobson, G., 1967. The effects of alleles at the 'never ripe' locus on the ripening tomato fruit. Phytochemistry 6, 1337–1341.

Hobson, G., 1968. Cellulase activity during the maturation and ripening of tomato fruit. J. Food Sci. 33, 588–591.

Hobson, G.E., Grierson, D., 1993. Tomato. In: Seymour, G., Taylor, J., Tucker, G. (Eds.), Biochemistry of Fruit Ripening. Chapman and Hall, London, pp. 405–442.

Hobson, G.E., Lance, C., Young, R.E., Biale, J.B., 1966. Isolation of active subcellular particles from avocado fruit at various stages of ripeness. Nature (London) 209, 1242–1243.

Hoehn, E., Gasser, F., Guggenbühl, B., Künsch, U., 2003. Efficacy of instrumental measurements for determination of minimum requirements of firmness, soluble solids, and acidity of several apple varieties in comparison to consumer expectations. Postharvest Biol. Technol. 27, 27–37.

Hoehn, E., Prange, R.K., Vigneault, C., 2009. Storage technology and applications. In: Yahia, E.M. (Ed.), Modified and Controlled Atmospheres for the Storage, Transportation, and Packaging of Horticultural Commodities. CRC Press, Boca Raton, FL, pp. 160–192.

Hoffman, N.E., Yang, S.F., 1980. Changes of 1-aminocyclopropane-1-carboxylic acid content in ripening fruits in relation to their ethylene production rates. J. Am. Soc. Hortic. Sci. 105, 492–495.

Höhn, E., Gasser, F., Dätwyler, D., Jampen, M., 1996. Harvest, storage and ripening of pears. AgrarForschung 3, 369–372.

Höhn, E., Schärer, H., Künsch, U., 2003. Karottengeschmack — Beliebtheit, Süssigkeit und Bitterkeit. AgrarForschung 10, 144–149.

Holden, M., 1965. Chlorophyll bleaching by legume seeds. J. Sci. Food Agric. 16, 312–325.

Holdsworth, M.J., Bird, C.R., Ray, J., Schuch, W., Grierson, D., 1987. Structure and expression of an ethylene-related mRNA from tomato. Nucleic Acids Res. 15, 731–739.

Holdsworth, M.J., Schuch, W., Grierson, D., 1988. Organization and expression for a wound/ripening-related small multigene family from tomato. Plant Mol. Biol. 11, 81–88.

Holton, T.A., Cornish, E.C., 1995. Genetics and biochemistry of anthocyanin biosynthesis. Plant Cell 7, 1071–1083.

Hortensteiner, S., 1999. Chlorophyll breakdown in higher plants and algae. Cell. Mol. Life Sci. 56, 330–347.

Hortensteiner, S., Krautler, B., 2011. Chlorophyll breakdown in high plants. Biochim. Biophys. Acta 1807, 977–988.

Hubbermann, E.M., Heins, A., Stockmann, H., Schwartz, K., 2006. Influence of acids, salt, sugars and hydrocolloids on the colour stability of anthocyanin rich blackcurrant and elderberry concentrates. Eur. Food Res. Technol. 223, 83–90.

Huber, D.J., 1983. Poluronide degradation and hemicellulase modifications in ripening tomato fruit. J. Am. Hortic. Soc. 108, 405–409.

Huff, A., 1984. Sugar regulation of plastid interconversion in epicarp of citrus fruit. Plant Physiol. 76, 258–261.

Hulme, A.C., Jones, J.D., Woolworton, L.S.C., 1963. The respiration climacteric in apple fruits. Proc. R. Soc. Lond. Ser. B 158, 514–535.

Hultin, H.O., Levine, A.S., 1965. Pectinmethylesterase in the ripening banana. J. Food Sci. 30, 917.

Humphrey, A.M., 1980. Chlorophyll. Food Chem. 5, 57–67.

Hyde, R.B., Morrison, J.W., 1964. The effect of storage temperature on reducing sugars, pH and phosphorylase enzyme activity in potato tubers. Am. Potato J. 41, 163–168.

Ikemefuna, J., Adamson, J., 1985. Chlorophyll and carotenoid changes in ripening palm fruit, Elacis guimneesis. Phytochemistry 23, 1413–1415.

Imamura, M., Shimizu, S., 1974. Metabolism of chlorophyll in higher plants. IV. Relationship of fatty acid oxidation and chlorophyll bleaching in plant extracts. Plant Cell Physiol. 15, 187–190.

Imaseki, H., Watanabe, A., 1978. Inhibition of ethylene inhibition by osmotic shock. Further evidence for control of ethylene production. Plant Cell Physiol. 19, 345–348.

Isaac, J.E., Rhodes, M.J.C., 1982. Purification and properties of phosphofructokinase from fruits of Lycopersicon esculentum. Phytochemistry 21, 1553–1556.

Isaac, J.E., Rhodes, M.J.C., 1987. Phosphofructokinase and ripening in Lycopersicon esculentum fruits. Phytochemistry 26, 649–653.

Isherwood, F.A., 1976. Mechanisms of starch sugar interconversion in Solanum tuberosum. Phytochemistry 15, 33–41.

Ishii, T., 1995. Pectic polysaccharides from bamboo shoot cell walls. Mokuzai Gakkaishi 41, 669–674.

Ishii, T., 1997. O-acetylated oligosaccharides from pectins from potato tuber cell walls. Plant Physiol. 113, 1265–1272.

Ishii, T., 2002. A sensitive and rapid bioassay of homogalacturonan synthase using 2- Aminobenzamide-labeled oligogalacturonides. Plant Cell Physiol. 43, 1386–1389.

Issenberg, P., Wick, E.L., 1963. Volatile components of bananas. J. Agric. Food Chem. 11, 2–8.

Jackman, R.L., Stanley, D.W., 1995. Perspectives in the textural evaluation of plant foods. Trends Food Sci. Technol. 6, 187–194.

Jackson, A., Sancovich, H.A., Ferramola, A.M., Evans, N., Games, D.E., Matlin, S.A., Elder, G.E., Smith, S.G., 1976. Macrocyclic intermediates in the biosynthesis of porphyrins. Philos. Trans. R. Soc. Lond 273, 191–206.

Jadhav, S., Singh, B., Salunkhe, D.K., 1972. Metabolism of unsaturated fatty acids in tomato fruit. Linoleic and linolenic acid as precursors of hexanal. Plant Cell Physiol. 13, 449–459.

Jaffe, E.K., 2003. An unusual phylogenetic variation in the metal ion binding sites of porphobilinogen synthase. Chem. Biol. 10, 25–34.

Jaffe, E.K., 2004. The pophobilinogen synthase catalyzed reaction mechanism. Bioorg. Chem. 32, 316–325.

Jamet, E., Canut, H., Boudart, G., Pont-Lesica, R.F., 2006. Cell wall proteins: a new insight through proteomics. Trends Plant Sci. 11, 33–39.

Jarvis, M.C., 2011. Plant cell walls: supramolecular assemblies. Food Hydrocoll. 25, 257–262.

Jiang, Y.M., Fu, J.R., 2000. Ethylene regulation of fruit ripening: molecular aspects. Plant Growth Regul. 30, 193–200.

Jiang, Y., Jiang, Y., Qu, H., Duan, X., Luo, Y., Jang, W., 2007. Energy aspects in ripening and senescence of harvested horticultural crops. Stewart Postharvest Rev. 2, 1–5.

Jimenez-Bermudez, S., Redono-Nevado, J., Mufioz,-Blanco, J., Caballero, J.L., Lopez-Aranda, J.M., Valpuesta, V., Pliego-Alfaro, F., Queseda, M.A., Mercado, J.A., 2002. Manipulation of strawberry fruit softening by antisense expression of a pectate lyase gene. Plant Physiol. 128, 751–759.

John, M.A., Dey, P.M., 1986. Post harvest changes in fruit cell wall. Adv. Food Res. 30, 139–193.

Jones, I.D., White, R.C., Gibbs, E., 1961. The formation of pheophorbides during brine preservation of cucumber. Food Technol. 15, 172.

Jones, I.D., White, R.C., Gibbs, E., 1963. Influence of bleaching or brining treatments on the formation of chlorophyllides, pheophytins and pheophorbides in green plant tissue. J. Food Sci. 28, 437.

Jones, M.G., 2008. Formation of vegetable flavor. In: Brückner, B., Grant Wyllie, S. (Eds.), Fruit and Vegetable Flavor. Woodhead, Cambridge, pp. 71–102.

Jones, R.A., Scott, S.J., 1983. Genetic potential to improve tomato flavor in commercial F hybrids. J. Am. Soc. Hortic. Sci. 109, 318.

Jordan, P.M., Seehra, J.S., 1979. The biosynthesis of uroporphyrinogen III. Order of assembly of the four porphobilinogen molecules in the formation of the tetrapyrrole ring. FEBS Lett. 104, 364–366.

Jorgensen, K., Rasmussen, A.V., Morant, M., Nielsen, A.H., Bjarnholt, N., 2005. Metabolon formation and metabolic channelling in the biosynthesis of plant natural products. Curr. Opin. Plant Biol. 8, 280–291.

Joshi, P., Mansfield, S.D., 2007. The cellulose paradox – simple molecule, complex biosynthesis. Curr. Opin. Plant Biol. 10, 220–226.

Joslyn, M.A., Mackinney, G., 1938. The rate of conversion of chlorophyll to pheophytin. J. Am. Chem. Soc. 60, 1132–1136.

Kacperska, A., Kubacka-Zabalska, M., 1984. Is lipoxygenase involved in biosynthesis of wound ethylene? In: Siegenthaler, P.A., Eichenberger, W. (Eds.), Structure, Function and Metabolism of Plant Lipids. Elsevier, Amsterdam, pp. 521–522.

Kacperska, A., Kubacka-Zabalska, M., 1985. Is lipoxygenase involved in the formation of ethylene from ACC? Physiol. Plant 64, 333–338.

Kader, A.A., 2003. A summary of CA requirements and recommendations for fruits other than apples. Acta Hortic. 600, 737–740.

Kader, A.A., 2008. Perspective flavor quality of fruits and vegetables. J. Sci. Food Agric. 88, 1863–1868.

Kader, A.A., Stevens, M.A., Albright-Holton, M., Morris, C.C., Algazi, M., 1977. Effect of fruit ripeness when picked on flavor and composition in fresh market tomatoes. J. Am. Soc. Hortic. Sci. 102, 724–731.

Kanellis, A.K., Kalaitzis, P., 1992. Cellulase occurs in multiple active forms in ripe avocado mesocarp. Plant Physiol. 98, 530–534.

Kannangara, C.G., Gough, S.P., Oliver, R.P., Rasmussen, S.K., 1984. Biosynthesis of delta-aminolevulinate in greening barley leaves IV. Activation of glutamate by ligation to RNA. Carlsberg Res. Commun 43, 185–194.

Kapp, P.P., 1966. Some effects of variety, maturity, and storage on fatty acids in fruit pericarp of Lycopersicon esculentum Mill. Diss. Abstr. 27, 77B.

Kato, T., 1966. Physiological studies on the bulbing and dormancy of onion plant. VIII. Relations between dormancy and organic constituents of bulbs. J. Jpn. Soc. Hortic. Sci. 35, 142–151.

Kauss, H., Swanson, A.L., Hassid, W.Z., 1967. Biosynthesis of the methyl ester groups of pectin by transmethylation from S-adenosyl-L-methionine. Biochem. Biophys. Res. Commun. 26, 234–240.

Kazeniac, S.J., Hall, R.M., 1970. Flavor chemistry of tomato volatiles. J. Food Sci. 35, 519–530.

Keegstra, K., Talmadge, K.W., Bauer, W.D., Albershein, P., 1973. The structure of plant cell walls. III. A model of the walls of suspension-cultured cells based on the interconnections of the macromolecular components. Plant Physiol. 51, 188–197.

Kende, H., 1993. Ethylene biosynthesis. Annu. Rev. Plant Physiol. Plant Mol. Biol. 44, 283–307.

Kennedy, M.G.H., Isherwood, F.A., 1975. Activity of phosphorylase in Solanum tuberosum during low temperature storage. Phytochemistry 14, 667–670.

Kertesz, Z.I., 1951. The Pectic Substances. Wiley Interscience, New York.

Khalyfa, A., Kermasha, S., Marsot, P., Goetghebeur, M., 1995. Purification and characterization of chlorophyllase from alga Phaeodactylum tricornutim by preparative native electrophoresis. Appl. Biochem. Biotechnol 53, 11–27.

Kidd, F., West, C., 1922. 'Brown heart', a functional disease of apples and pears. Food Invest. Board Rep. 1921, 14.

Kidd, F., West, C., 1927. Gas storage of fruit. Great Britain Department of Scientific Industrial Research Food Investigation Board Report Vol. 30, 87.

Kidd, F., West, C., 1930a. Physiology of fruits: changes in the respiratory activity of apples during senescence at different temperatures. Proc. R. Soc. Lond. 106, 93–109.

Kidd, F., West, C., 1930b. The gas storage of fruit. II. Optimum temperatures and atmospheres. J. Pomol. Hortic. Sci. 8, 67–77.

Kidd, F., West, C., 1937. Action of carbon dioxide on the respiration activity of apples. Effect of ethylene on the respiration activity and climacteric of apples. Individual variation in apples. Great Britain Department of Scientific Industrial Research Food Investigation Board Report, 101–115.

Kidd, F., West, C., 1950. The refrigerated gas storage of apples. Variation in apples. Great Britain Department of Scientific Industrial Research Food Investigation Board Leaflet No. 6 (rev.), 16.

Kieliszewski, M., Shpak, E., 2001. Synthetic genes for the elucidation of glycosylation codes for arabinogalactan proteins and other hydroxyproline-rich glycoproteins. Cell Mol. Life Sci. 58, 1386–1398.

Kikuchi, G., Shemin, D., Bachmann, B.J., 1958. The enzymic synthesis of delta-aminolevulinic acid. Biochim. Biophys. Acta 28, 219–220.

Kim, I., Grosch, W., 1979. Partial purification of a lipoxygenase from apples. J. Agric. Food Chem. 27, 243–246.

Kim, J.K., Gross, K.C., Solomos, T., 1987. Characterization of the stimulation of ethylene production by galactose in tomato (Lycopersicon esculentum Mill.) fruit. Plant Physiol. 85, 804–807.

Kim, J., Gross, K.C., Solomos, T., 1991. Galactose metabolism and ethylene production during development and ripening of tomato fruit. Postharvest Biol. Technol. 1, 67–80.

King, D.L., Klein, B.P., 1987. Effect of flavonoids and related compounds on soybean lipoxygenase-1 activity. J. Food Sci. 52, 220–221.

King, K.J.N., Vessol, M.J., 1969. Enzymes in the cellulose complex. Adv. Chem. Ser. 95, 7–25.

Klaui, H., 1973. Carotinoide in Lebensmitteln. Func. Prop. Fats Foods, Adv. Study Course 1971.

Klein, B., Grossman, S., 1985. Co-oxidation reactions of lipoxygenase in plant systems. Adv. Free Rad. Biol. Med. 1, 309−343.

Knee, M., 1973. Polysaccharide changes in cell walls of ripening apples. Phytochemistry 12, 1543−1549.

Knee, M., 1978. Metabolism of polymethylgalacturonate in apple cortical tissue during ripening. Phytochemistry 17, 1261−1264.

Koca, N., Karadeniz, F., Burdulu, H.S., 2006. Effect of pH on chlorophyll degradation and colour loss in blanch green peas. Food Chem. 100, 609−615.

Kockritz, A., Schewe, T., Hieke, B., Hass, W., 1985. The effect of soybean lipoxygenase-1 on chloroplasts from wheat. Phytochemistry 24, 381−384.

Koes, R.E., Spelt, C.E., Mol, J.M., 1989. The chalcone synthase multigene family of Petunia hybrida (V30): differential, light-regulated expression during flower development and UV light induction. Plant Mol. Biol. 12, 213−225.

Komalavalis, P., Mort, A.J., 1989. The acetylation of O-3 of galacturonic acid in the rhamnose-rich region of pectins. Carbohydr. Res. 189, 261−272.

Kostansek, E., Pereira, W., 2003. Successful application of 1-MCP in commercial storage facilities. Acta Hortic 628, 213−219.

Konze, J.R., Kende, H., 1979. Ethylene formation from 1-amino-cyclo-propane-1-carboxylic acid in homogenates of etiolated pea seedlings. Planta 146, 293−301.

Konze, J.R., Jones, J.F., Boller, T., Kende, H., 1980. Effect of 1-amino-cyclopropane-1-carboxylic acid on the production of ethylene in senescing flowers of Ipomoea tricolor Cav. Plant Physiol. 66, 566−571.

Krautler, B., 2003. Chlorophyll breakdown and chlorophyll catabolites. In: Kadish, K.M., Smith, K., Guillard, R. (Eds.), The Porphyrin Handbook, Vol. 13. Elsevier, Oxford, pp. 183−209.

Krautler, B., Hortensteiner, S., 2006. Chlorophyll catabolites and the biochemistry of chlorophyll breakdown. In: Grimm, B., Porra, R., Rudiger, W., Scheer, H. (Eds.), Chlorophylls and Bacteriochlorophylls: Biophysics, Functions and Applications. Springer, Dordrecht, pp. 237−260.

Kreutzmann, S., Thybo, A.K., Bredie, W.L.P., 2007. Training of a sensory panel and profiling of winter hardy and coloured carrot genotypes. Food Qual. Pref. 18, 482−489.

Kreuzaler, F., Ragg, H., Fautz, E., Kuhn, D.N., Hahlbrock, K., 1983. UV-induction of chalcone synthase mRNA in cell suspension cultures of Petroselinum hortense. Proc. Natl. Acad. Sci. U.S.A. 80, 2591−2593.

Krupkova, E., Immerzeel, P., Pauly, M., Smulling, T., 2007. The tumorous shoot development 2 gene of Arabidopsis encoding a putative methyltransferase is required for cell adhesion and co-ordinated plant development. Plant J. 50, 735−750.

Kuai, J., Dilley, D.R., 1992. Extraction, partial purification, and characterization of 1-aminocyclopropane-1-carboxylic acid oxidase from apple fruit. Postharvest Biol. Technol. 1, 203−211.

Kuneida, t., Amano, T., Shioi, Y., 2005. Search for chlorophyll degradation enzyme, Mg-dechetalase, from extracts of Chenopodium album with native and artificial substrates. Plant Sci. 169, 177−183.

Kupferman, E., 2003. Controlled atmosphere storage of apples and pears. Acta Hortic 600, 729−735.

Kuwasha, S.C., Subbarayan, C., Beeler, D.A., Porter, J.W., 1969. The conversion of lycopenes-15,15′H to cyclic carotenes by soluble extracts of higher plant plastids. J. Biol. Chem. 244, 3635−3642.

LaBorde, L.F., von Elbe, J.H., 1990. Zinc complex formation in heated vegetable purees. J. Agric. Food Chem. 38, 484−487.

LaBorde, L.F., von Elbe, J.H., 1996. Method for improving the color of containerized green vegetables. US Patent No. 5,482,727.

Lackey, G.D., Gross, K.C., Wallner, S.J., 1980. Loss of tomato cell wall galactan may involve reduced rate synthesis. Plant Physiol. 66, 532−533.

LaJollo, F.M., Tannenbaum, S.R., Labuza, T.P., 1971. Reaction of limited water concentration. 2. Chlorophyll degradation. J. Food Sci. 36, 850−853.

Lammers, S.M., 1981. All About Tomatoes. Ortho Books. Chevron Chemical Co., San Francisco, CA, p. 93.

Lamport, D.T.A., 1965. The protein component of primary cell walls. Adv. Bot. Res. 2, 151−218.

Lance, C., Hobson, G.E., Young, R.E., Biale, J.B., 1965. Metabolic processes in cytoplasmic particles of the avocado fruit. VII. Oxidative and phosphorylative activities throughout the climacteric cycle. Plant Physiol. 40, 1116−1123.

Lance, C., Hobson, G.E., Young, J.B., Biale, J.B., 1967. Metabolic processes in cytoplasmic particles in avocado fruit. IX. The oxidation of pyruvate and malate during the climacteric cycle. Plant Physiol. 42, 471−478.

Lara, I., Mio, R.M., Fuentes, T., Sayez, G., Graell, J., Lopez, M.L., 2003. Biosynthesis of volatile aroma compounds in pear fruit stored under long-term controlled-atmosphere conditions. Postharvest Biol. Technol. 29, 29−39.

Leake, L.H., Kirk, L.K., 1992. Method for color preservation in canned green vegetables US Patent No. 5,114,724.

Legge, R.L., Thompson, J.E., 1983. Involvement of hydroperoxides and an ACC-derived free radical in the formation of ethylene. Phytochemistry 22, 2161−2166.

Legge, R.L., Thompson, J.E., Baker, J.E., 1982. Free radical-mediated formation of ethylene from 1-aminocyloprane-1-carboxylic acid, a spin-trap study. Plant Cell Physiol. 23, 171−177.

Lelievre, J.M., Latche, A., Jones, B., Bouzayen, M., Pech, J.C., 1997. Ethylene and fruit ripening. Physiol. Plant. 101, 727−739.

Lennon, K.A., Lord, E.M., 2001. In vivo pollen tube cell of Arabidopsis thaliana 1. Tube cell cytoplasm and wall. Protoplasma 214, 45−56.

Lepiniec, L., Debeaujon, I., Routaboul, J.-M., Baudry, A., Pourcel, L., Nesi, N., Caboche, M., 2006. Genetics and biochemistry of seed flavonoids. Annu. Rev. Plant Biol. 57, 405−430.

Lerouxel, O., Cavalier, D.M., Liepman, A.H., Keegstra, K., 2006. Biosynthesis of plant cell wall polysaccharides − a complex process. Curr. Opin. Plant Biol. 9, 621−630.

Lesschaeve, I., Noble, A.C., 2005. Polyphenols: factors influencing their sensory properties and their effects on food and beverage preferences. Am. J. Clin. Nutr. 81 (Suppl.), 330S−335S.

Li, L., Zhu, B., Yang, P., Fu, D., Zhu, Y., Luo, Y., 2011. The regulation mode of RIN transcription factor involved in ethylene biosynthesis in tomato fruit. J. Sci. Food Agric. 91, 1822−1828.

Li, Y.Q., Cai, G., Mascatelli, A., Cresti, M., 1997. Functional interaction among cytoskeleton, membranes and cell wall in the pollen tubes of flowering plants. Int. Rev. Cytol. 176, 133−199.

Liao, F.W.H., Luh, B.S., 1970. Anthocyanin pigments in Tinto cao grapes. J. Food Sci. 35, 41−46.

Lichtenthaler, H.K., Schwender, J., Disch, A., Rohmer, M., 1997a. Biosynthesis of isoprenoids in high plant chloroplasts proceeds via a mevalonate-independent pathway. FEBS Lett. 400, 271−274.

Lichtenthaler, H.K., Rohmer, M., Schwender, J., 1997b. Two independent biochemical pathways for isopentyl diphosphate and isoprenoid biosynthesis in higher plants. Physiol. Plant 101, 643−652.

Lieberman, M., Mapson, L., 1964. Genesis and biogenesis of ethylene. Nature (London) 204, 343−345.

Lieberman, M., Mapson, L.W., Kunishi, A.T., Wardale, D.A., 1965. Ethylene production from methionine. Biochem. J. 97, 449−459.

Lieberman, M., Kunishi, A.T., Mapson, L.W., Wardale, D.A., 1966. Stimulation of ethylene production in apple tissue slices by methionine. Plant Physiol. 41, 376−382.

Liedgens, W., Lutz, C., Scheider, H.A.N., 1983. Molecular properties of 5-aminolevulinic acid dehydratase from *Spinacia oleracea*. Eur. J. Biochem. 135, 35−39.

Lin, T.-Y., Elbein, A.D., Su, J.D., 1966. Substrate specificity in pectin synthesis. Biochem. Biophys. Res. Commun. 22, 650−657.

Lipton, W.J., 1977. Recommendations for CA storage of broccoli, brussel sprouts, cabbage, cauliflower, asparagus and potatoes. Mich. State Univ. Hortic. Rep. 28, 277−280.

Lipton, W.J., 1987. Senescence in leafy vegetables. HortScience 22, 854−859.

Lipton, W.J., Ryder, E.J., 1989. Lettuce. In: Eskin, N.A.M. (Ed.), Quality and Preservation of Vegetables. CRC Press, Boca Raton, FL, pp. 212−245.

Liu, X., Liao, M., Deng, G., Chen, S., Ren, Y., 2008. Changes in activity of PG, PE, Cx and LOX in pulp during fruit growth and development oft two different ripening-season pear cultivars. Amer. Euras. J. Agric. Environ. Sci. 3, 445−450.

Liu, Y., Hoffman, N.E., Yang, S.F., 1985. Promotion by ethylene of the capability to convert 1-aminocyclopropane-1-carboxylic acid to ethylene in preclimacteric tomato and cantaloupe fruits. Plant Physiol. 77, 407−411.

Llop-Tous, I., Barry, C.S., Grierson, D., 2000. Regulation of ethylene biosynthesis in response to pollination in tomato flowers. Plant Physiol. 123, 971−978.

Loomis, W.D., Battaile, J., 1963. Biosynthesis of terpenes. III. Mevalonic kinase from higher plants. Biochim. Biophys. Acta 67, 54−63.

Lurssen, K., Naumann, K., Schroder, R., 1979. 1-Aminocyclopropane-1-carboxylic acid − An intermediate in the ethylene biosynthesis in higher plants. Z. Pflanzenphysiol. 92, 285−294.

Lyons, J.M., McGlasson, W.B., Pratt, H.K., 1962. Ethylene production, respiration, and internal gas concentrations in cantaloupe fruits at various stages of maturity. Plant Physiol. 37, 31−36.

Machackova, I., Zmrhal, Z., 1981. Is peroxidase involved in ethylene biosynthesis? Physiol. Plant. 53, 479−482.

MacKinney, G., 1961. Coloring matters. In: Sinclair, W.B. (Ed.), The Orange: Its Biochemistry and Physiology. University of California, Berkley, CA, pp. 302−333.

MacKinney, G., Lukton, A., Greenbaum, A., 1958. Carotenoid stability in stored dehydrated carrots. Food Technol. 12, 164−166.

McCready, B.M., McComb, E.A., 1954. Pectic constituents in ripe and unripe fruit. Food Res. 19, 530−535.

McFeeters, R.F., Bell, T.A., Fleming, H.P., 1980. An endopolygalacturonase in cucumber fruit. J. Food Biochem. 4, 1.

McGlasson, W.B., Wade, N.L., Adato, I., 1978. In: Letham, D.S., Goodwin, P.B., Higgins, T.J. (Eds.), Phytohormones and Related Compounds: A Comprehensive Treatise, Vol. 1. Academic Press, London, pp. 475−519.

McGarvey, D.J., Christoffersen, R.E., 1992. Characterization and kinetic parameters of ethylene-forming enzyme from avocado fruit. J. Biol. Chem. 267, 5964−5967.

McGlasson, W.B., Palmer, J.K., Vendrell, M., Brady, C., 1971. Metabolic studies with banana fruits. II. Effect of inhibitors on respiration, ethylene production and ripening. Aust. J. Biol. Sci. 24, 1103−1114.

McKenzie, K.A., 1932. Respiration studies with lettuce. Proc. Am. Soc. Hortic. Sci. 28, 244−248.

McMurchie, E.J., McGlasson, W.B., Eaks, I.L., 1972. Treatment of fruit with propylene gives information about biogenesis of ethylene. Nature 237, 235−236.

Macrae, A.R., 1971. Malic enzyme activity of plant mitochondria. Phytochemistry 10, 2343−2347.

McRae, D.G., Baker, J.E., Thompson, J.E., 1982. Evidence for involvement of the superoxide radical in the conversion of 1-aminocyclopropane-1-carboxylic acid to ethylene by pea microsomal membranes. Plant Cell Physiol. 23, 375−383.

Malundo, T.M.N., Shewfelt, R.L., Scott, J.W., 1995. Flavor quality of fresh tomato (*Lycopersicon esculentum* Mill.) as affected by sugar and acid levels. Postharvest Biol. Technol. 6, 103−110.

Mangos, T.J., Berger, R.G., 1997. Determination of major chlorophyll degradation products. Z. Lebensm. Unters. Forsch. A 204, 345−350.

Marcelle, R.D., 1991. Relationship between mineral content, lipoxygenase activity, levels of 1-aminocyclopropane-1-carboxylic acid and ethylene emission in apple fruit flesh disks during storage. Postharvest Biol. Technol. 1, 101−109.

Marei, N., Crane, J.C., 1971. Growth and respiratory response of fig (*Ficus carica* L. cv. Mission) fruits to ethylene. Plant Physiol. 48, 249−254.

Markovic, O., Heinrichova, K., Lenkey, B., 1975. Pectolytic enzymes from banana. Collect. Czech. Chem. Commun. 40, 769−774.

Marx, R., Brinkmann, K., 1978. Characteristics of rotenone-insensitive oxidation of matrix-NADH broad bean mitochondria. Planta 142, 83−90.

Matile, P., Hortensteiner, S., Thomas, H., 1999. Chlorophyll degradation. Annu. Rev. Plant Mol. Biol. 50, 67−95.

Matsuo, T., Itoo, S., 1982. A model experiment for deastringency of persimmon fruit with high carbon dioxide treatment, *in vitro* gelation of kaki-tannin by reacting with acetaldehyde. Agric. Biol. Chem. 46, 683−689.

Mattoo, A.K., Lieberman, M., 1977. Localization of the ethylene-synthesising system in apple tissue. Plant Physiol. 60, 794−799.

Mau, Y.H., Wang, W.Y., 1988. Biosynthesis of delta-aminolevulinic acid in *Chlamydomonas reinhardtii*: study of the transamination mechanism using specifically labeled glutamate. Plant Physiol. 86, 793−797.

Maudinas, B., Bucholtz, M.L., Porter, J.W., 1975. The partial purification and properties of a phytoene synthase complex isolated from tomato fruit plastids. Abstr. 4th Int. Symp. Carotenoids, 41.

Mayak, S., Legge, R.L., Thompson, J.E., 1981. Ethylene formation from 1-aminocyclopropane-1-carboxylic acid by microsomal membranes from senescing carnation flowers. Planta 153, 49−55.

Mayer, S.M., Gawlita, E., Avuissar, Y.J., Anderson, V.E., Beale, S.I., 1993. Intermolecular nitrogen transfer in the enzymic conversion of glutamate to delta-aminolevulinic acid by extracts of *Chlorella vulgaris*. Plant Physiol. 10, 1029−1038.

Mazza, G., Brouillard, R., 1987. Recent developments in the stabilization of anthocyanins in food products. Food Chem. 25, 207−225.

Medina-Escobar, N., Cardenas, J., Moyano, E., Caballeo, J.L., Munozo-Blanco, J., 1997. Cloning, molecular characterization and expression pattern of a strawberry ripening-specific CDNA with sequence homology to pectate lyase from higher plants. Plant Mol. Biol. 34, 867–877.

Meigh, D.F., Jones, J.D., Hulme, A.C., 1967. The respiration climacteric in the apple. Production of ethylene and fatty acids in fruit attached and detached from the tree. Phytochemistry 6, 1507–1515.

Meller, E., Belkin, S., Harel, E., 1975. The biosynthesis δ-aminolevulinic acid in greening maize leaves. Phytochemistry 14, 2399–2402.

Meskauskiene, R., Nater, M., Goslings, D., Kessler, F., op den Camp, R., Apel, K., 2001. FLU: a negative regulator of chlorophyll biosynthesis in Arabidopsis thaliana. Proc. Natl. Acad. Sci. U.S.A. 98, 12826–12831.

Miller, C.O., Conn, E.E., 1980. Metabolism of hydrogen cyanide by higher plants. Plant Physiol. 65, 1199–1202.

Millerd, A., Bonner, J., Biale, J.B., 1953. The climacteric rise in fruit respiration as controlled by phosphorylative coupling. Plant Physiol. 28, 521–531.

Minguez-Mosquera, M.I., Garrido-Fernandez, J., Gandul-Rojas, B., 1989. Pigment changes in olives during fermentation and brine storage. J. Agric. Food Chem. 37, 8–11.

Mitchell, F.G., Mayer, G., Biasi, G., 1991. Effect of harvest maturity on storage performance of 'Hayward' kiwifruit. Acta Hortic 297, 617–625.

Miyazaki, J.H., Yang, S.F., 1987. Inhibition of the methionine cycle enzymes. Phytochemistry 26, 2655–2660.

Mohnen, D., 2008. Pectin structure and biosynthesis. Curr. Opin. Plant Biol. 11, 266–277.

Mohnen, D., Bar-Peled, M., Somerville, C., 2008. Biosynthesis of plant cell walls. In: Himmel, M.E. (Ed.), Biomass Recalcitrance. Blackwell, Oxford, pp. 94–187.

Mol, J., Stuitje, A., Gerats, A., van der Krol, A., Jorgensen, R., 1989. Saying it with genes: molecular flower breeding. Trends Biotechnol 7, 148–153.

von Mollendorff, L.J., De Villiers, O.T., 1988. Role of pectolytic enzymes in the development of wooliness in peaches. J. Hortic. Sci. Biotechnol. 63, 53–58.

Møller, I.M., 2001. Plant mitochondria and oxidative stress. Electron transport, NADPH turnover and metabolism of reactive oxygen species. Annu. Rev. Plant Physiol. Plant Mol. Biol. 52, 561–591.

Moreau, F., Romani, R., 1982. Malate oxidation and cyanide insensitive respiration in avocado mitochondria during the climacteric cycle. Plant Physiol. 70, 1385–1390.

Morin, F., Rigault, R., Hartmann, C., 1985. Conséquences d'un séjour au froid sur le métabolism de l'ethylene au cours de la maturation de la poire Passe-Crassane après recolte. Physiol. Veg. 23, 353–359.

Moser, S., Muller, T., Oberhuber, M., Krautler, B., 2009. Chlorophyll catabolites – chemical and structural footprints of a fascinating biological phenomenon. Eur. J. Org. Chem., 21–31.

Moustafa, E., Wong, E., 1967. Purification and properties of chalcone isomerase from soya bean seed. Phytochemistry 6, 625–632.

Muda, P., Seymour, G.B., Errington, N., Tucker, G.A., 1995. Carbohydr. Polym. 26, 255–260.

Murayama, H., Ikai, S., Fukushima, T., 2002. Changes in starch content in 'La France' pears during storage. Acta Hortic 596, 871–874.

Murr, D.P., Yang, S.F., 1975. Inhibition of in vivo conversion of methionine to ethylene by L-canaline and 2,4-dinitrophenol. Plant Physiol. 55, 79–82.

Nagel, C.W., Patterson, M.E., 1967. Pectic enzymes and development of the pear (Pyrus communis). J. Food Sci. 32, 294.

Nakatsuka, A., Murachi, S., Okunishi, H., Shiomi, S., Nakano, R., Kubo, Y., Inaba, A., 1998. Differential expression and internal feedback regulation of 1-amino-cyclopropane-1-carboxylate oxidase, and ethylene receptor genes in tomato fruit during development and ripening. Plant Physiol. 118, 1295–1305.

Nandi, D.L., Waygood, G.R., 1967. Biosynthesis of porphyrins in wheat leaves. Can. J. Biochem. 45, 322.

Nelmes, B.J., Preston, R.D., 1968. Wall development in apple fruit: a study of the life history of a parenchyma. J. Exp. Bot. 19, 496–518.

Neuburger, M., Douce, R., 1980. Effect of bicarbonate and oxaloacetate on malate oxidation by spinach leaf mitochondria. Biochim. Biophys. Acta 589, 176–189.

Nogota, Y., Ohta, H., Voragen, A.G.J., 1993. Polygalacturonase in strawberry fruit. Phytochemistry 34, 617–620.

Northcote, D.H., 1963. The biology and chemistry of the cell walls of higher plants, algae and fungi. Int. Rev. Cytol. 14, 223–265.

Northcote, D.H., 1972. Chemistry of the plant cell wall. Annu. Rev. Plant Physiol. 23, 113–122.

Nunan, K.J., Davies, C., Robinson, S.O., Fincher, G.M., 2001. Expression patterns of cell wall- modifying enzymes during grape-berry development. Planta 214, 257–264.

Obenland, D.M., Carroll, T.R., 2000. Mealiness and pectolytic activity in peaches and nectarines in response to heat treatment and cold storage. J. Am. Soc. Hortic. Sci. 125, 723–728.

Odawara, S., Watanabe, A., Imaseki, H., 1977. Involvement of cellular membranes in regulation of ethylene production. Plant Cell Physiol. 18, 569–575.

Oeller, P.W., Wong, L.M., Taylor, L.P., Pike, D.A., Theologis, A., 1991. Reversible inhibition of tomato fruit senescence by antisense RNA. Science 254, 275–286.

Ogura, K., Nishino, T., Seto, S., 1968. Purification of prenyltransferase and isopentyl pyrophosphate isomerase of pumpkin fruit and some of their properties. J. Biochem. (Tokyo) 64, 197–203.

Okuda, T., Hatano, T., Ogawa, N., 1982a. Rugosin D, E, F and G, dimeric and trimeric hydrolyzable tannins. Chem. Pharm. Bull. 30, 4234–4237.

Okuda, T., Yoshida, T., Kuwaharu, M., Memon, M., Shingu, T., 1982b. Agrimonin and potentillin, an ellagitannin dimer and monomer having an α-glucose core. J. Chem. Soc. Chem. Commun., 162–164.

O'Neill, M., Albersheim, P., Darville, A., 1990. The pectic polysaccharides of primary cell walls. In: Dey, D.M. (Ed.), Methods in Plant Biochemistry, Vol. 2. Academic Press, London, pp. 415–441.

Ontario Ministry of Agriculture Food & Rural Affairs, 1998. Chilling injury of horticultural crops, factsheet. http://www.omafra.gov.on.ca/english/crops/facts/98-021.htm (15 August 2010).

Oster, U., Tanaka, R., Tanaka, A., Rudiger, W., 2000. Cloning and functional expression of the gene encoding the key enzyme for chlorophyll b biosynthesis (CAO) from Arabidopsis thaliana. Plant J. 21, 305–310.

Owens, T.G., Falkowski, P.G., 1982. Enzymatic degradation of chlorophyll a by marine phytoplankton in vitro. Phytochemistry 21, 979–984.

Ozawa, T., Lilley, T.H., Haslam, E., 1987. Polyphenol interactions: astringency and the loss of astringency in ripening fruit. Phytochemistry 26, 2937–2942.

Paillard, N., 1968. Analyse de l'arome de pommes de la variété Calville blanc par chromatographic sur colonne capillaire. Fruits 23, 283.

Palmer, J.K., 1971. The banana. In: Hulme, A.C. (Ed.), The Biochemistry of Fruits and Their Products, Vol. 2. Academic Press, London, pp. 65–105.

Palmer, J.M., 1976. The organization and regulation of electron transport in plant mitochondria. Annu. Rev. Plant Physiol. 27, 133–157.

Pardo, A.D., Chereskin, B.M., Castelfranco, P.A., Franceschi, V.R., Wezelman, B.E., 1980. ATP requirement for higher plants. Annu. Rev. Plant Physiol. 24, 129.

Pathak, N., Sanwal, G.G., 1998. Multiple forms of polygalacturonase from banana fruit. Phytochemistry 48, 249–255.

Patras, A., Brunton, N.P., O'Donnell, C., Tiwari, B.K., 2010. Effect of thermal processing on anthocyanin stability in foods: mechanisms and kinetics of degradation. Trends Food Sci. Technol. 21, 3–11.

Pattanayak, G.K., Biswal, A.K., Reddy, V.S., Tripathy, V.C., 2005. Light-dependent regulation of chlorophyll *b* biosynthesis in chlorophyllide *a* oxygenase overexpressing tobacco plants. Biochim. Biophys. Res. Commun. 326, 466–471.

Payasi, A., Sanwal, G.G., 2003. Pectaste lyase during ripening of banana fruit. Phytochemistry 63, 243–248.

Payasi, A., Misra, P.C., Sanwal, G.G., 2008. Purification and characterization of pectate lyase from banana (*Musa acuminate*) fruits. Phytochemistry 67, 861–869.

Peacock, B.C., 1972. Role of ethylene in the initiation of fruit ripening. Queensl. J. Agric. Sci. 29, 137–145.

Pear, J.R., Kawagoe, Y., Shreckengost, W.E., Delmer, D.P., Stalker, D.M., 1996. Higher plants contain homologs of bacterial celA genes encoding the catalytic unit of cellulose synthase. Proc. Natl. Acad. Sci. U.S.A. 93, 12637–12642.

Pech, J.C., Latché, A., van der Resr, B., 2008. Genes involved in the biosynthesis of aroma volatiles and biotechnical applications. In: Brückner, B., Grant Wyllie, S. (Eds.), Fruit and Vegetable Flavour. Woodhead, Cambridge, pp. 254–271.

Peiser, G.D., Wang, T.-T., Hoffman, N.E., Yang, S.F., 1983. Evidence for CN formation from [^{1}C]ACC during *in vivo* conversion of ACC to ethylene [abstract]. Plant Physiol. Suppl. 72 (No. 203).

Peiser, G.D., Wang, T.-T., Hoffman, N.E., Yang, S.F., Liu, H.-W., Walsh, C.T., 1984. Formation of cyanide from carbon 1 of 1-aminocyclopropane-1-carboxylic acid during its conversion to ethylene. Proc. Natl. Acad. Sci. U.S.A. 81, 3059–3063.

Peng, C.Y., Markakis, P., 1963. Effect of phenolase on anthocyanins. Nature (London). 199, 597–598.

Pérez, A.G., Sanz, C., 2008. Formation of fruit flavor. In: Brückner, B., Grant Wyllie, S. (Eds.), Fruit and Vegetable Flavor. Woodhead, Cambridge, pp. 41–70.

Perez-Tello, G.O., Martinez-Telllez, M.A., Vargas-Arisouro, I., Gonzalez-Aguilar, G., 2009. Chilling injury in mamey sapote fruit (*Pouteria sapota*): biochemical and physiological responses. Am. J. Agric. Biol. Sci. 4, 137–145.

Pesis, E., Fuchs, Y., Zauberman, G., 1978. Cellulase activity in avocado. Plant Physiol. 61, 416–419.

Peynaud, E., Riberau-Gayon, P., 1971. The grape. In: Hulme, A.C. (Ed.), The Biochemistry of Fruits and their Products, Vol. 2. Academic Press, London, pp. 171–205.

Pharr, D.M., Dickinson, D.B., 1973. Partial characterization of Cx cellulase and cellobiase from ripening tomato x fruit. Plant Physiol. 51, 577–583.

Picha, D.H., 1987. Sugar and organic acid content of cherry tomato fruit at different ripening stages. HortScience 22, 94–96.

Picton, S., Barton, S.L., Bouzayen, M., Hamilton, A.J., Grierson, D., 1993. Altered fruit ripening and leaf senescence in tomatoes expressing an antisense ethylene-forming enzyme transgene. Plant J. 3, 469–481.

Pinta, V., Picaud, M., Reiss-Husson, F., Astier, C., 2002. Rubrivivax gelatinous acsF (previously orf358) codes for a conserved, putative binuclear-iron-cluster-containing protein involved in aerobic oxidative cyclization of Mg-protoporphyrin IX monomethylester. J. Bacteriol. 164, 746–753.

Pirrung, M.C., 1985. Ethylene biosynthesis. 3. Evidence concerning the fate of C1–N1 of ACC. Bioorg. Chem. 13, 219–226.

Pirrung, M.C., 1999. Ethylene biosynthesis from 1-aminocyclopropanecarboxylic acid. Acc. Chem. Res. 32, 711–718.

Pirrung, M.C., Brauman, J.I., 1987. Involvement of cyanide in the regulation of ethylene biosynthesis. Plant Physiol. Biochem. (Paris) 25, 55–61.

Platenius, H., 1942. Effect of temperature on the respiration rate and quotient of some vegetables. Plant Physiol. 17, 179–197.

Pliszka, B., Huscza-Ciolkowska, G., Mieleszko, E., Czaplicki, S., 2009. Stability and antioxidative properties of acylated anthocyanins in three cultivars of red cabbage (*Brassica oleracea* L. var. *capitata* L. f. rubra). J. Sci. Food Agric. 89, 1154–1158.

Pollock, C.J., ap Rees, T., 1975. Activities of enzymes of sugar metabolism in cold-stored tubers of *Solanum tuberosum*. Phytochemistry 14, 613–617.

Poovaiah, B.W., Nukuya, A., 1979. Polygalacturonase and cellulose enzymes in the normal Rutgen and Mutant rin tomato fruits and their relationship to the respiratory climacteric. Plant Physiol. 64, 534–537.

Popper, Z.A., Fry, S.C., 2008. Xyloglucan-pectin linkages are formed intra-protoplasmically, contribute to wall-assembly, and remain stable in the cell wall. Planta 227, 781–794.

Porra, R.J., 1986. Labelling of chlorophylls and precursors by [2-^{14}C] glycine and 2-[-1-^{14}C]oxoglutarate in *Rhodopseudomonas spheroides* and *Zea mays*. Eur. J. Biochem. 156, 111–121.

Porritt, S.W., 1977. Conditions and practices used in CA of apples in Western US and Canada. Mich. State Univ. Hortic. Rep. 28, 231–232.

Porter, J.W., Lincoln, P.E., 1950. *Lycopersicon* selection containing a high content of carotenes and colorless polyenes. II. The mechanism of carotene biosynthesis. Arch. Biochem. 27, 390–403.

Potty, V.H., Breumer, J., 1970. Formation of isoprenoid pyrophosphates from mevalonate by orange enzymes. Phytochemistry 9, 1229–1237.

Poulson, R., Polglase, W.J., 1975. The enzymatic conversion of protoporphyrinogen-IX to protoporphyrin-IX. Protoporphyrinogen oxidase activity in mitochondrial extracts of *Saccharomyces cerevisiae*. J. Biol. Chem. 250, 1269–1274.

Prange, R.K., DeLong, J.M., 2003. 1-Methylcyclopropene: the 'magic bullet' for horticultural products? Chron. Hortic. 43, 11–14.

Prange, R.K., DeLong, J.M., Daniels-Lake, B.J., Harrison, P.A., 2005. Quality management through respiration control: is there a relationship between lowest acceptable respiration, chlorophyll fluorescence and cytoplasmatic acidosis? Acta Hortic 682, 823–830.

Prange, R.K., DeLong, J.M., Harrison, P.A., Leyte, J., McLean, S.D., Scrutton, J.G.E., Cullen, J.J., 2007. Method and apparatus for monitoring a condition in chlorophyll containing matter. US Patent No. 7,199,376.

Prasad, D.D.K., Prasad, A.R.K., 1987. Effect of lead and mercury on chlorophyll synthesis in mungbean seedlings. Phytochemistry 26, 881–883.

Prasanna, V., Prabha, T.N., Tharanathan, R.N., 2007. Fruit ripening phenomena – an overview. CRC Crit. Rev. Food Sci. 47, 1–19.

Pratt, H.K., Groeschl, J.D., 1968. In: Wrightmann, F., Setterfield, G. (Eds.), Biochemistry and Physiology of Plant Growth Substances. Runge Press, Ottawa, pp. 1295–1302.

Pratt, H.K., Morrris, L.L., Tucker, C.L., 1954. Temperature and lettuce deterioration of broccoli varieties. In: Proceedings of Conference on Transportation of Perishables. University of California, Davis, pp. 77–83.

Pressey, R., 1977. Enzymes involved in fruit softening. ACS Symp. Ser. 47, 172–191.

Pressey, R., 1983. β-Galactosidases in ripening tomatoes. Plant Physiol. 71, 132–135.

Pressey, R., 1986a. Extraction and assay of tomato polygalacturonases. HortScience 21, 490–492.

Pressey, R., 1986b. Changes in polygalacturonases isoenzymes and converter in tomatoes during ripening. HortScience 21, 1183–1185.

Pressey, R., 1987. Exopolygalacturonase in tomato fruit. Phytochemistry 26, 1867–1870.

Pressy, R., Avants, J.K., 1973a. Separation and characterization of endopolygalacturonase and exopolygalagturonase from peaches. Plant Physiol. 52, 252–256.

Pressey, R., Avants, J.K., 1973b. Two forms of polygalacturonase in tomatoes. Biochim. Biophys. Acta 309, 363–369.

Pressey, R., Avants, J.K., 1975. Cucumber polygalacturonase. J. Food Sci. 40, 937–939.

Pressy, R., Avants, J.K., 1976. Pear polygalacturonases. Phytochemistry 15, 1349–1351.

Pressy, R., Avants, J.K., 1978. Difference in polygalacturonase composition of Clingstone and Freestone peaches. J. Food Sci. 43, 1415–1417.

Pressy, R., Avants, J.K., 1982a. Pectic enzymes in Long Keeper tomatoes. HortScience 17, 398–400.

Pressy, R., Avants, J.K., 1982b. Solubilization of cell walls by tomato polygalacturonases: effects of pectinesterase. J. Food Biochem. 6, 57–74.

Pressey, R., Hinton, D.M., Avants, J.K., 1971. Development of polygalacturonase activity and solubilisation of pectin in peaches during ripening. J. Food Sci. 36, 1070–1073.

Proctor, J.T., Creasy, L.L., 1971. Effects of supplementary light on anthocyanin synthesis in 'McIntosh' apples. J. Am. Hortic. Sci. 96, 523–526.

Pruzinska, A., Tanner, G., Anders, I., Roca, M., Hortensteiner, S., 2003. Chlorophyll breakdown: pheophorbide a oxygenase is a Rieske-type iron-sulfur protein, encoded by the accelerated cell death 1 gene. Proc. Natl. Acad. Sci. U.S.A 100, 15259–15264.

Pruzinska, A., Tanner, G., Aubry, S., Anders, I., Moser, S., Muller, T., Organia, K.-H., Ktautler, B., Youn, J.-Y., Liljegren, S.J., Hortensteiner, S., 2005. Chlorophyll breakdown in senescent Arabidopsis leaves. Characterization of chlorophyll catabolites and of chlorophyll catabolic enzymes involved in the degreening reaction. Plant Physiol. 139, 52–63.

Queiroz, F., Oliveira, C., Pinho, O., Ferreira, I.M., 2009. Degradation of anthocyanins and anthocyanidins in blackberry jams/stuffed fish. J. Agric. Food Chem. 57, 10712–10717.

Qureshi, A.A., Manok, K., Qureshi, N., Porter, J.W., 1974. The enzymatic conversion of cis-[C]phyto-14-fluene, trans[C]carotene to poly-cis acyclic carotenes by a cell-free preparation of tangerine tomato fruit plastids. Arch. Biochem. Biophys. 162, 108–116.

Ralph, J., Lundquist, K., Brunow, G., Lu, F., Kim, H., Schatz, P.F., Marita, J.M., Hatfield, R.D., Ralph, S.A., Christensen, J.H., Boerjan, W., 2004. Lignins: natural polymers from oxidative coupling of 4-hydroxyphenyl-propanoids. Phytochem. Rev. 3, 29–60.

Ramadoss, C.S., Pistorius, E.K., Axelrod, B., 1978. Coupled oxidation of carotene by lipoxygenase requires two isozymes isoenzyme. Arch. Biochem. Biophys. 190, 549–552.

Rando, R.R., 1974. βγ-Unsaturated amino acids as irreversible inhibitors. Nature (London) 250, 586–587.

Ranwala, P., Suematsu, C., Masuda, H., 1992. The role of β-galactosidases in the modification of cell wall components during muskmelon fruit ripening. Plant Physiol. 100, 1318–1325.

Rapparini, F., Predieri, S., 2003. Pear volatiles. Hortic. Rev. 28, 237–324.

ap Rees, T., 1974. Pathways of carbohydrate breakdown in higher plants. In: Northcote, D.H. (Ed.), Plant Biochemistry. MTP Int. Rev. Sci. Biochem. Ser. 1, Vol. 11. Butterworth, London, p. 51.

ap Rees, T., 1977. Conservation of carbohydrate by non-photosynthetic cells of higher plants. Symp. Soc. Exp. Biol. 31, 7–32.

ap Rees, T., 1980. Assessment of the contributions of metabolic pathways to plant respiration. In: Davies, D.D. (Ed.), The Biochemistry of Plants, Vol. 2. Academic Press, New York, pp. 1–29.

Reif, H.J., Niesbach, U., Salamini, F., Seedler, H., Rohde, W., 1985. Cloning and analysis of two chalcone genes for chalcone synthase from Petunia hydrida. Mol. Gen. Genet. 199, 208–215.

Rexova-Benkova, L., Markovic, O., 1976. Pectic enzymes. Adv. Carbohydr. Chem. Biochem. 33, 323–385.

Reynolds, P.A., Klein, B.P., 1982. Purification and characterization of a type-1 lipoxygenase from pea seeds. J. Agric. Food Chem. 30, 1157–1163.

Rhodes, M.J.C., 1970. The climacteric and ripening of fruits. In: Hulme, A.C. (Ed.), The Biochemistry of Fruits and their Products, Vol. 1. Academic Press, London, p. 524.

Rhodes, M.J.C., 1971. Respiration and senescence of plant organs. In: Hulme, A.C. (Ed.), The Biochemistry of Fruits and their Products, Vol. 2. Academic Press, New York.

Rich, P.R., Boveris, A., Bonner, W.D., Moore, A.L., 1976. Hydrogen peroxide generated by the alternative oxidase of higher plants. Biochem. Biophys. Res. Commun. 3, 695–703.

Richmond, A., Biale, J.B., 1966. Protein and nucleic acid metabolism in fruits. Studies of amino acid incorporation during the climacteric rise in respiration of the avocado. Plant Physiol. 41, 1247–1253.

Ridley, B.L., O'Neill, M.A., Mohnen, D., 2001. Pectins: structure, biosynthesis, and oligogalacturonide-related signalling. Phytochemistry 57, 929–967.

Ringli, C., Keller, B., Ryser, U., 2001. Glycine-rich proteins as structural components of plant cell walls. Cell Mol. Life Sci. 58, 1430–1441.

Riov, J., Yang, S.F., 1982a. Autoinhibition of ethylene production in citrus peel discs. Suppression of 1-aminocyclopropane-1-carboxylic acid synthesis. Plant Physiol. 69, 687–690.

Riov, J., Yang, S.F., 1982b. Effects of exogenous ethylene on ethylene production in citrus leaf tissue. Plant Physiol. 70, 136–141.

Robertson, G.L., Swinburne, D., 1981. Changes in chlorophyll and pectin after storage and canning of kiwifruit. J. Food Sci. 46, 1557–1559.

Rocklin, A.M., Tierney, D.L., Kofman, V., Brunhuber, N.M.W., Hoffman, B.M., Christoffersen, R.E., Reich, N.O., Lipscomb, J.D., Que, L., 1999. Role of the nonheme Fe(II) enter in the biosynthesis of the plant hormone ethylene. Proc. Natl. Acad. Sci. U.S.A 96, 7905–7909.

Rogers, L.J., Sha, S.P.J., Goodwin, T.W., 1966. Mevalonate-kinase isoenzymes in plant cells. Biochem. J. 100, 14c.

Rohwer, F., Mader, M., 1981. The role of peroxidase in ethylene formation from 1-aminocyclopropane-1-carboxylic acid. Z. Pflanzenphysiol. 104, 363–372.

Romani, R.J., Ku, L., 1968. Direct gas chromatographic analysis of volatiles produced by ripening fruit. J. Food Sci. 31, 558–560.

Rowan, D.D., Lane, H.P., Allen, J.M., Fielder, S., Hunt, M.B., 1996. Biosynthesis of 2-methylbutyl, 2-methyl-2-butenyl, and 2-methyl-butanoate esters in Red Delicious and Granny Smith apples using deuterium-labelled substrates. J. Agric. Food Chem. 44, 3276–3285.

Rowan, D.D., Allen, J.M., Fielder, S., Hunt, M.B., 1999. Biosynthesis of straight chain ester volatiles in red delicious and Granny Smith apples using deuterium-labelled precursors. J. Agric. Food Chem. 47, 2553–2562.

Rubin, R., Toghe, T., Matsuda, F., Saito, K., Scheible, W.-R., 2009. Members of the LBD family of transcriptional factors repress anthocyanin synthesis and affect additional nitrogen responses in Arabidopsis. Plant Cell. 25, 3567–3584.

Ruediger, W., 2002. Biosynthesis of chlorophyll b and the chlorophyll cycle. Photosynth. Res. 74, 187–193.

Rustin, P., Moreu, F., Lance, C., 1980. Malate oxidation in plant mitochondria via malic enzyme and the cyanide insensitive electron transport pathway. Plant Physiol. 66, 457–462.

Rutherford, P.P., Whittle, R., 1982. The carbohydrate composition of onions during long-term cold storage. J. Hortic. Sci. 57, 349–356.

Rutherford, P.P., Whittle, R., 1984. Methods of predicting the long-term storage of onions. J. Hortic. Sci. 59, 537.

Ryall, A.L., Lipton, W.L., 1979. Vegetable as living products – respiration and heat production. In: Handling, Transportation, and Storage of Fruits and Vegetables, 2nd ed., Vol. 1. Avi, Westport, CT. Chapter 1.

Saevels, S., Lammertyn, J., Berna, A.Z., Veraverbeke, E.A., Di Natale, C., Nicolai, B.M., 2004. An electronic nose and mass spectrometry-based electronic nose for assessing apple quality during shelf life. Postharvest Biol. Technol. 31, 9–19.

Saito, N., Osawa, Y., Hayashi, K., 1971. Platyconin, a new acylated anthocyanin in Chinese bell-flower, Platycodon grandiflorum. Phytochemistry 10, 445–447.

Sakai, T., Sakamoto, T., Hallaert, J., Vadamme, E.J., 1993. Pectin, pectinase and protopectinase: production, properties and application. Adv. Appl. Microbiol. 39, 213–294.

Sakamura, S., Shibusa, S., Obata, Y.T., 1966. Separation of a polyphenol oxidase activity for anthocyanin degradation in eggplant. J. Food Sci. 31, 317–319.

Salimen, S.O., Young, R.E., 1975. The control properties of phosphofructokinase in relation to the respiratory climacteric in banana fruit. Plant Physiol. 55, 45–50.

Saltveit, M.E., 2003a. Is it possible to find an optimal controlled atmosphere? Postharvest Biol. Technol. 27, 3–13.

Saltveit, M.E., 2003b. A summary of CA requirements and recommendations for vegetables. Acta Hortic 600, 723–727.

Saltveit, M.E., 2004. Respiratory metabolism. In: The commercial storage of fruits, vegetables, and florist and nursery stocks. USDA, ARS. http://www.ba.ars.usda.gov/hb66/contents.html (18 March 2011).

Saltveit, M.E., 2005. Ethylene effects. In: The commercial storage of fruits, vegetables, and florist and nursery stocks. USDA, ARS. http://www.ba.ars.usda.gov/hb66/contents.html (15 August 2010).

Salunkhe, D.K., Do, J.Y., Maga, J.A., 1976. Biogenesis of aroma constituents of fruits and vegetables. CRC Crit. Rev. Food Sci. Nutr. 8, 161–190.

Saquet, A.A., Streif, J., Bangerth, F., 2000. Changes in ATP, ADP and pyridine nucleotide levels related to the incidence of physiological disorders in 'Conference' pears and 'Jonagold' apples during controlled atmosphere storage. J. Hortic. Sci. Biotechnol. 75, 243–249.

Saquet, A.A., Sttreif, J., Bangerth, F., 2003a. Energy metabolism and membrane lipid alterations in relation to brown heart development in 'Conference' pears during delayed controlled atmosphere storage. Postharvest Biol. Technol. 30, 123–132.

Saquet, A.A., Streif, J., Bangerth, F., 2003b. Impaired aroma production of CA-stored 'Jonagold' apples as affected by adenin and pyridine nucleotide levels and fatty acid concentrations. J. Hortic. Sci. Biotechnol. 78, 695–705.

Saxena, I.M., Brown Jr., R.M., 2005. Cellulose biosynthesis: current views and evolving concepts. Ann. Bot. 96, 9–21.

Schanderl, S.H., Chichester, C.O., Marsh, B.L., 1962. Degradation of chlorophyll and several derivatives in acid solution. J. Org. Chem. 27, 3865–3868.

Scheible, W.-R., Pauly, M., 2004. Glycosyltransferases and cell wall biosynthesis: novel players and insights. Curr. Opin. Plant Biol. 7, 285–295.

Scheller, H.V., Ulvskov, P., 2010. Hemicelluloses. Annu. Rev. Plant Biol. 61, 263–289.

Scheller, H.V., Doong, R.L., Ridley, B.L., Mohnen, D., 1999. Pectin biosynthesis: a solubilised α1,4-galacturonosyltransferase from tobacco catalyzes the transfer of galacturonic acid from UDP-galacturonic acid onto the nonreducing end of hemogalacturonan. Planta 207, 512–517.

Schifferstein, H.N.J., Frijters, J.E.R., 1990. Sensory integration in citrate/sucrose mixtures. Chem. Senses 15, 87–109.

Schmid, R., Shemin, D., 1955. The enzymic formation of porphobilinogen from 5-aminolevulinic acid and its conversion to protoporphyrin. J. Am. Chem. Soc. 77, 506–508.

Schmiech, L., Uemeura, D., Hofman, T., 2008. Reinvestigation of the bitter compounds in carrots (Daucus carota L.) by using a molecular sensory science approach. J. Agric. Food Chem. 56, 10252–10260.

Schoch, S., Vielwerth, F.X., 1983. Chlorophyll degradation in senescent tobacco cell culture (Nicotiana tabacum var 'Samsun'). Z. Pflanzenphysiol. 110, 309–317.

Schotsmans, W.C., DeLong, J.M., Larrigaudière, C., Prange, R.K., 2009. Effects of physiological disorders. In: Yahia, E.M. (Ed.), Modified and Controlled Atmospheres for the Storage, Transportation, and Packaging of Horticultural Commodities. CRC Press, Boca Raton, FL, pp. 160–192.

Schwab, W., 2008. Role of metabolome diversity in fruit and vegetable quality: multifunctional enzymes and volatiles. In: Brückner, B., Grant Wyllie, S. (Eds.), Fruit and Vegetable Flavor. Woodhead, Cambridge, pp. 272–286.

Schwartz, S.J., von Elbe, J.H., 1983. Kinetics of chlorophyll degradation to pyropheophytins in vegetables. J. Food Sci. 48, 1303–1306.

Schwartz, S.J., Lorenzo, T.V., 1991. Chlorophyll stability during continuous aseptic processing and storage. J. Food Sci. 56, 1059−1062.

Schwartz, S.J., Wool, S.L., von Elbe, J.H., 1981. High performance liquid chromatography of chlorophylls and their derivatives in fresh and processed spinach. J. Agric. Food Chem. 29, 533−537.

Segerlind, L.J., Herner, R.C., 1972. On the fiber content problem of processed asparagus. CASAE Publ. 72, 882.

Seljasen, R., Hoftun, H., Bengtson, G.B., 2001a. Sensory quality of ethylene-exposed carrots (Daucus carota L., cv. 'Yukon') related to the contents of 6-methoxymellein, terpenes and sugars. J. Sci. Food Agric. 81, 54−61.

Seljasen, R., Bengtsson, G.B., Hoftun, H., Vogt, G., 2001b. Sensory and chemical changes in five varieties of carrot (Daucus carota L.) in response to mechanical stress at harvest and post-harvest. J. Sci. Food Agric. 81, 436−447.

Sexton, R., Palmer, J.M., Whyte, N.A., Littlejohns, S., 1997. Cellulase, fruit softening and abscission in red raspberry Rubus ideaus L. cv. Gen Clova. Plant Physiol. 80, 371−376.

Seymour, G.B., Harding, S.E., Taylor, A.J., Hobson, G.E., Tucker, G.A., 1987a. Poluronide solubilization during ripening of normal and mutant tomato fruit. Phytochemistry 26, 1871−1875.

Seymour, G.B., Lasslett, Y., Tucker, G.A., 1987b. Differential effects of pectolytic enzymes on tomato poluronides in vivo and in vitro. Phytochemistry 26, 3137−3139.

Shemin, D., Russell, R.S., 1953. δ-Aminolevulinic acid. Its role in the biosynthesis of porphyrins and purines. J. Am. Chem. Soc. 75, 4873−4874.

Sheng, J., Luo, Y., Wainwright, H., 2000. Studies on lipoxygenase and the formation of ethylene in tomato. J. Hortic. Sci. Biotechnol. 75, 69−71.

Shewfelt, A.L., Panter, V.A., Jen, J.J., 1971. Textural changes and molecular characteristics of pectin constituents in ripening peaches. J. Food Sci. 36, 573.

Shimokawa, K., Hashizume, A., Shioi, Y., 1990. Pyropheophorbide a, a catabolite of ethylene-induced chlorophyll a degradation. Phytochemistry 29, 2105−2106.

Shioi, Y., Watanabe, K., Takamiya, K., 1966. Enzymatic conversion of pheophorbide a in leaves of Chenopodium album. Plant Cell Physiol. 37, 1143−1149.

Shioi, Y., Tatsumi, Y., Shimokawa, K., 1991. Enzymatic degradation of chlorophyll in Chenopodium album. Plant Cell Physiol. 32, 87−93.

Shioi, Y., Masuda, T., Takamiya, K., Shimokawa, K., 1995. Breakdown of chlorophylls by soluble proteins extracted from leaves of Chenopodium album. J. Plant Physiol. 145, 416−421.

Shioi, Y., Tomita, N., Tsuchiya, T., Takamiya, K., 1996. Conversion of chlorophyllide to pheophorbide by Mg-dechelating substance in extracts if Chenopodium album. Plant Physiol. Biochem. 34, 41−47.

Shoolingin-Jordan, P.M., 2003. The biosynthesis of coproporphyrinogen III. In: Kadisk, K.M. (Ed.), The Porphyrin Handbook. Elsevier, Amsterdam, pp. 33−74.

Simpson, S.D., Ashford, D.A., Harvey, D.J., Bowles, D.J., 1998. Short chain oligogalactuuronides induce ethylene production and expression of the gene encoding aminocyclopropane 1-carboxylic acid oxidase in tomato plants. Glycobiology 8, 579−583.

Singh, S., Sanwall, G.G., 1973. An allosteric α-glucan phosphorylase from banana fruits. Biochim. Biophys. Acta 309, 280−288.

Singh, S., Sanwall, G.G., 1975. Characterization of multiple forms of α-glucan phosphorylase from Musa paradisiaca. Phytochemistry 14, 113−118.

Singh, S., Sanwall, G.G., 1976. Multiple forms of α-glucan phosphorylase in banana fruits: properties and kinetics. Phytochemistry 15, 1447−1451.

Singleton, V.L., Gortner, W.A., Yang, H.Y., 1961. Carotenoid pigments of pineapple fruit. 1. Acid-catalyzed isomerisation of the pigments. J. Food Sci. 26, 49−52.

Sisler, E.C., Blankenship, S.M., 1996. Methods of counteracting an ethylene response in plants. US Patent No. 5,518,988 21 May 1996.

Sisler, E.C., Serek, M., 1997. Inhibition of ethylene response in plants at the receptor level: recent development. Physiol. Plant. 101, 577−582.

Slater, A., Maunders, M.L.J., Edwards, K., Schuch, W., Grierson, D., 1983. Isolation and characterization of cDNA clones for tomato polygalacturonase and other ripening-related proteins. Plant Mol. Biol. 5, 137−147.

Small, C.J., Pecket, R.C., 1982. The ultrastructure of anthocyanoplasts in red-cabbage. Planta 154, 97−99.

Smith, M.A., King, P.T., Grimm, B., 1998. Transient-state kinetic analysis of Synochococcus glutamate 1-semialdehyde aminotransferase. Biochemistry 37, 319−329.

Smith, R.B., Lougheed, E.L., Franklin, E.W., McMillan, L., 1979. The starch iodine test for determining stage of maturation in apples. Can. J. Plant Sci. 59, 725−735.

Sobotka, F.E., Stelzig, A.A., 1974. An apparent cellulose complex in tomato (Lycopersicon esculentum L.) fruit. Plant Physiol. 53, 759−763.

Solomos, T., 1977. Cyanide-resistant respiration in higher plants. Plant Physiol. 28, 279−297.

Solomos, T., Laties, G.G., 1973. Cellular organization and fruit ripening. Nature (London) 245, 390−392.

Solomos, T., Laties, G.G., 1974. Similarities between ethylene and cyanide action in triggering the rise in respiration in potato slices. Plant Physiol. 54, 506−511.

Solomos, T., Laties, G.G., 1976. Induction by ethylene of cyanide-sensitive and cyanide-resistant plant tissues. Plant Physiol. 58, 47−50.

Somerville, C.R., 2006. Cellulose synthesis in higher plants. Annu. Rev. Cell Dev. Biol. 22, 53−78.

Song, J., 2007. Flavour volatile production and regulation in apple fruit. Stewart Postharvest Rev. 2, 2−8.

Song, J., Bangerth, F., 2003. Fatty acids as precursors for aroma volatile biosynthesis in pre-climacteric and climacteric apple fruit. Postharvest Biol. Technol. 30, 113−121.

Song, L.L., Jiang, Y.M., Gao, H.Y., Li, C.T., Liu, H., You, Y.L., Sun, J., 2006. Effects of adenosine triphosphate on browning and quality of harvested litchi fruit. Am. J. Food Technol. 1, 173−178.

Souleyre, J.F., Greenwood, D.R., Friel, E.N., Karunairetnam, S., 2005. An alcohol acyl transferase from apple (cv. Royal Gala), MpAAT1, produces esters involved in apple fruit flavor. FEBS J. 272, 3132−3144.

Spanu, P., Reinhardt, D., Boller, T., 1991. Analysis and cloning of the ethylene-forming enzymes from tomato by functional expression of its mRNA in Xenopus laevis oocytes. EMBO J. 10, 2007−2013.

Spurr, A.R., Harris, W.M., 1968. Ultrastructure of chloroplasts and chromoplasts in Capsicum anuum. I. Thylakoid membrane changes during fruit ripening. Am. J. Bot. 55, 1210−1224.

Steet, J.A., Tong, C.H., 1996. Degradation kinetics of green color and chlorophylls in peas by colorimetry and HPLC. J. Food Sci. 61, 924−928.

Sterling, J.D., Quigley, H.F., Orellana, A., Mohnen, D., 2001. The catalytic site of the pectin biosynthetic enzyme α-1,4-galacturonosyltransferase is located in the lumen of Golgi. Plant Physiol. 127, 360–371.

Stevens, M.A., Kader, A.A., Albright, M., 1979. Potential for increasing tomato flavor via increased sugar and acid content. J. Am. Soc. Hortic. Sci. 104, 40–52.

Stintzing, F.C., Carle, R., 2004. Functional properties of anthocyanins and betalains in plants, food, and in human nutrition. Trends Food Sci. Technol. 15, 19–38.

Stoll, K., 1970. Apfellagerung in kontrollierter Atmosphäre (CA) unter modifizierten Konditionen. Schweiz. Z. Obst. Weinbau 106, 334–337.

Stoll, K., 1973. Tables on the storage of fruits and vegetables in controlled atmospheres. Eidg. Forschungsanst. Obst. Wein Gartenbau, Wadenswil, Flugschr., 78.

Stone, E.J., Hall, R.M., Kazeniac, S.J., 1975. Formation of aldehydes and alcohols in tomato fruit from UC labelled linolenic and linoleic acids. J. Food Sci. 40, 1138–1141.

Suomalainen, H., Keranen, A.J.A., 1961. The first anthocyanins appearing during the ripening of blueberries. Nature (London) 191, 498–499.

Surendranathan, K.K., Nair, P.M., 1973. Alterations in carbohydrate metabolism of γ-irradiated Cavendish banana. Phytochemistry 12, 241–249.

Suzuki, T., Shioi, Y., 2002. Re-examination of Mg-dechelation reaction in the degradation of chlorophylls using chlorophyllin a as a substrate. Photosynth. Res. 74, 217–223.

Sweeney, J.P., Martin, M.E., 1961. Stability of chlorophyll in vegetables as affected by pH. Food Technol. 15, 263–266.

Tager, J.M., Biale, J.B., 1957. Carboxylase and aldolase activity in the ripening banana. Physiol. Plant. 10, 79–85.

Takamiya, K., Tsichiya, Y., Ohta, H., 2000. Degradation pathway(s) of chlorophyll: was gene cloning revealed? Trends Plant Sci. 5, 426–431.

Takeguchi, C.A., Yamamoto, H., 1968. Light-induced oxygen-18 uptake by epoxy xanthophylls in New Zealand spinach leaves (*Tetragonia expansa*). Biochim. Biophys. Acta 150, 459–465.

Talburt, W.F., Smith, O., 1975. Potato Processing, 3rd ed. Avi, Westport, CT.

Tan, L., Okazawa, A., Fukuaki, E., Kobayashi, A., 2000. Removal of magnesium by Mg-dechetalase is a major step in the chlorophyll-degrading pathway in *Ginkgo biloba* in the process of autumnal tints. Z. Naturforsch. 55c, 923–926.

Tanaka, A., Tanaka, R., 2005. Effects of chlorophyllide a oxygenase overexpression on light acclimation in *Arabidopsis thaliana*. Photosynth. Res. 85, 327–340.

Tanaka, A., Tanaka, R., 2006. Chlorophyll metabolism. Curr. Opin. Plant Biol. 9, 248–255.

Tanaka, A., Ito, H., Tanaka, R., Yoshida, K., Okada, K., 1998. Chlorophyll a oxygenase (CAO) is involved in chlorophyll b formation from chlorophyll *a*. Proc. Natl. Acad. Sci. U.S.A. 95, 12719–12723.

Tanaka, R., Tanaka, A., 2007. Tetrapyrrole biosynthesis in higher plants. Annu. Rev. Plant Biol. 58, 321–346.

Tanaka, R., Tanaka, A., 2011. Chlorophyll cycle regulates the construction and destruction of the light-harvesting complexes. Biochim. Biophys. Acta 1807, 968–976.

Tanaka, R., Koshino, Y., Sawa, S., Ishiguro, S., Okada, K., Tanaka, A., 2001. Overexpression of chlorophyllide a oxygenase (CAO) enlarges the antenna size of Photosystem II in *Arabidopsis thaliana*. Plant J. 26, 365–373.

Tavakoli, M., Wiley, R.C., 1968. Relation between trimethylsilyl derivatives of fruit tissue polysaccharides in apple texture. Proc. Am. Soc. Hortic. Sci. 92, 780.

Teng, S.S., Chen, B.H., 1999. Formation of pyrochlorophylls and their derivatives in spinach leaves during heating. Food Chem. 65, 367–373.

Terahara, N., Konczak, I., Ono, H., Yoshimoto, M., Yamakawa, O., 2004. Characterization of acylated anthocyanins in callus induced from storage root of purple-fleshed sweet potato, *Ipomoea batatas* L. J. Biomed. Biotechnol. 2004, 279–286.

Terpstra, W., 1977. A study of the properties and activity of chlorophyllase in photosynthetic membranes. Z. Planzenphysiol. 85, 139–146.

Themmen, A.P.N., Tucker, G.A., Grierson, D., 1982. Degradation of isolated tomato cell walls by purified polygalacturonase *in vitro*. Plant Physiol. 69, 249.

Theologis, A., Laties, G.G., 1978. Respiratory contribution of the alternate pathway during various stages of ripening in avocado and banana fruit. Plant Physiol. 62, 249–256.

Theologis, A., Oeller, P.W., Wong, L.M., Rottmann, W.H., Gantz, D.M., 1993. Use of a tomato mutant constructed with reverse genetics to study fruit ripening, a complex developmental process. Dev. Genet. 14, 282–295.

Thiault, J., 1970. Etude de criteres objectives de la qualite gustative de pommes Golden Delicious. Bull. Techn. Inf. Minist. Agric. Paris 248, 191–201.

Thomson, W.W., 1966. Ultrastructural development of chromoplasts in Valencia oranges. Bot. Gaz. (Chicago) 127, 133–139.

Thomson, W.W., Lewis, L.N., Coggins, C.W., 1967. The reversion of chromoplasts to chloroplasts in Valencia oranges. Cytology 32, 117–124.

Tigchelaar, E.C., McGlasson, W.B., Buescher, R.W., 1978a. Genetic regulation of tomato fruit ripening. HortScience 13, 508–513.

Tigchelaar, E.C., McGlasson, W.B., Franklin, M.J., 1978b. Natural and ethephon-stimulated ripening of F hybrids of the ripening inhibitor (*rin*) and non-ripening inhibitor (*nor*) mutants of tomato (*Lycopersicon esculentum* Mill.). Aust. J. Plant Physiol. 5, 449–456.

Tigier, H.A., Batile, A. M. del C., Locascio, G., 1968. Porphyrin biosynthesis in the soybean callus system. II. Improved purification and some properties of δ-aminolevulinic acid dehydratase. Enzymologia 38, 43–56.

Tigier, H.A., Batile, A. M. del C., Locascio, G., 1970. Porphyrin biosynthesis in soybean callus tissue system. Isolation, purification and general properties of δ-aminolevulinic acid dehydratase. Biochim. Biophys. Acta 151, 300–302.

Tijkens, L.M.M., Barringer, S.A., Biekman, E.S.A., 2001. Modelling the effect of pH on the colour degradation of blanched broccoli. Innovative Food Sci. Emerg. Technol. 2, 315–322.

Tikunov, Y., Lommen, A., Ric De Vos, C.H., Verhoeven, H.A., Bino, R.J., Hall, R.D., Bovy, A.G., 2005. A novel approach for non-targeted data analysis for metabolites. Large scale-profiling of tomato fruit volatiles. Plant Physiol. 39, 1125–1137.

Timberlake, C.F., Bridle, P., 1975. Anthocyanins. In: Harorne, J.B., Mabry, T.J., Mabry, H. (Eds.), The Flavonoids. Chapman and Hall, London, pp. 214–266.

Timberlake, C.F., Bridle, P., 1982. Distribution of anthocyanins in food plants. In: Markakis, P. (Ed.), Anthocyanins as Food Colors. Academic Press, New York, p. 125.

Tong, C.B., Labavitch, J.M., Yang, S.F., 1986. The induction of ethylene production from pear cell culture by cell wall fragments. Plant Physiol. 81, 929–930.

Torskangerpoll, K., Andersen, O.M., 2005. Colour stability of anthocyanins in aqueous solutions at various pH values. Food Chem. 89, 427–440.

Toul, V., Popsilova, J., 1966. Chemical composition of onion varieties (*Allium cepa* L. Bull. Hortic. Abstr. 38, 3385.

Travers, I., Jacquet, A., Brisset, A., Maite, C., 2002. Relationship between the enzymatic determination of starch and the starch iodine index in two varieties of cider apples. J. Sci. Food Agric. 82, 983–989.

Tressl, R., Albrecht, W., 1986. Biogenesis of aromas through acyl pathways. In: Parliment, T.H., Croteau, R. (Eds.), Biogeneration of Aromas. ACS, Washington DC, pp. 560–565.

Tressl, R., Drawert, F., 1973. Biogenesis of banana volatiles. J. Agric. Food Chem. 21, 560–565.

Tressl, R., Jennings, W.G., 1972. Production of volatile compounds in the ripening banana. J. Agric. Food Chem. 20, 189–192.

Tressl, R., Drawert, F., Heimann, W., 1970. Uber die Biogenese von Aromastoffen bei Pflanzen und Früchten. V. Anreicherung, *Trennung und Identifizierrung von Banaenaromastoffen*. Z. Lebensm. Unters. Forsch. 142, 249–263.

Trout, S.A., Huelin, F.E., Tindale, G.B., 1960. The respiration of Washington navel and Valencia oranges. Div. Food Preserv. Tech. Pap. (Aust., C.S.I.R.O.) 14, 11.

Tsuchiya, T., Suzuki, T., Yamada, T., Shimada, H., Masuda, T., Takamiya, K., 2003. Chlorophyllase as a serine hydrolase: identification of a putative catalytic triad. Plant Cell Physiol. 44, 96–101.

Tucker, G.A., 1993. Introduction. In: Seymour, G., Taylor, J., Tucker, G. (Eds.), Biochemistry of Fruit Ripening. Chapman & Hall, London, pp. 1–51.

Tucker, G.A., Robertson, N.G., Grierson, D., 1980. Changes in polygalacturonase isoenzymes during ripening of normal and mutant tomato fruit. Eur. J. Biochem. 112, 119–124.

Tucker, G.A., Robertson, N.G., Grierson, D., 1982. Purification and changes in activities of tomato pectin-esterase isoenzymes. J. Sci. Food Agric. 33, 396–400.

USDA, ARS, 2004. The commercial storage of fruits, vegetables, and florist and nursery stocks. http://www.ba.ars.usda.gov/hb66/contents.html (15 August 2010).

Vangdal, E., 1985. Quality criteria for fruit for fresh consumption. Acta Agric. Scand 35, 41–47.

Van Buren, J.P., 1979. The chemistry of texture in fruits and vegetables. J. Tex. Stud. 10, 1–23.

Vanholme, R., Demedts, B., Morreel, K., Ralph, J., Boerjan, W., 2010. Lignin biosynthesis and structure. Plant Physiol. 153, 895–905.

Vanlerberghe, G.C., McIntosh, L., 1997. Alternative oxidase: from gene to function. Annu. Rev. Plant Physiol. Plant Mol. Biol. 48, 703–734.

Varming, C., Jensen, K., Moller, S., Brockhoff, P.B., Christiansen, T., Edelenbos, M., Bjorn, G.K., Poll, L., 2004. Eating quality of raw carrots — correlation between flavour compounds, sensory profiling analysis and consumer liking test. Food Qual. Prefer. 15, 531–540.

Vavilin, D.V., Vermass, W.F., 2002. Regulation of tetrapyrrole biosynthetic pathway leading to heme and chlorophyll in plants and cyanobacteria. Physiol. Plant 115, 9–24.

Veltman, R.H., Verschoor, J.A., Ruijsch van Dungsteren, J.H., 2003. Dynamic control systems (DCS) for apples (*Malus domestica* Borkh.

('Elstar')): optimal quality through storage based on product response. Postharvest Biol. Technol. 27, 79–86.

Vendrell, M., McGlasson, W.B., 1971. Inhibition of ethylene production in banana fruit tissue by ethylene treatment. Aust. J. Biol. Sci. 24, 885–895.

Vendrell, M., Dominguez-Puigjaner, E., Llop-Tous, I., 2001. Climacteric versus non-climacteric physiology. Acta Hortic. 553, 345–349.

Ververidis, P., John, P., 1991. Complete recovery *in vitro* of ethylene-forming enzyme activity. Phytochemistry 30, 725–727.

Vicentini, F., Iten, F., Matile, P., 1995. Development of an assay for Mg-dechelatase. Physiol. Plant 94, 57–63.

Vick, B.A., Zimmerman, D.C., 1976. Lipoxygenase and hydroperoxide lyase in germinating watermelon seedlings. Plant Physiol. 57, 780–788.

Vick, B.A., Zimmerman, D.C., 1983. The biosynthesis of jasmonic acid: a physiological role for plant lipoxygenase. Biochem. Biophys. Res. Commun. 111, 470–477.

Villemez, C.L., Swanson, A.L., Hassid, W.Z., 1966. Properties of a polygalacturonic acid-synthesizing enzyme system from *Phaseolus aureus* seedlings. Arch. Biochem. Biophys. 116, 446–452.

Voragen, A.G.J., Pilnik, W., Thibault, J.-F., Axelos, M.A.V., Renard, C.M.C.G., 1995. Pectins. In: Stephen, A.M. (Ed.), Food Polysaccharides and their Applications. Marcel Dekker, New York, pp. 287–339.

Wagner, A.M., Krab, K., 1995. The alternative respiration pathway in plants: role and regulation. Physiol. Plant. 95, 318–325.

Wallner, S.J., 1978. Apple fruit β-galactosidase and softening in storage. J. Am. Soc. Hortic. Sci. 103, 364.

Wallner, S.J., Bloom, H.L., 1977. Characteristics of tomato cell wall degradation *in vitro*. Implications for the study of fruit softening. Plant Physiol. 55, 94–98.

Wang, C.Y., 1989. Chilling injury of fruits and vegetables. Food Rev. Int. 5, 209–236.

Wang, K.L.-C., Li, H., Ecker, J.R., 2002. Ethylene biosynthesis and signaling networks. Plant Cell 14 (Suppl. 1), S131–S151.

Wang, Z.Y., MacRay, E., Wright, M.A., Bolitho, K.M., Ross, G.S., Atkinson, R.G., 2000. Polygalacturonase gene expression in kiwi fruit: relationship to fruit softening and ethylene production. Plant Mol. Biol. 42, 317–328.

Watkins, C.B., 2002. Ethylene synthesis, mode of action, consequences and control. In: Knee, M. (Ed.), Fruit Quality and its Biological Basis. Sheffield Academic Press, Sheffield, pp. 180–224.

Weemaes, C., Ooms, V., Van Loey, A.M., Hendricks, M.E., 1999. Kinetics of chlorophyll degradation and color loss in heated broccoli juice. J. Agric. Food Chem. 47, 2404–2409.

von Wettstein, D., Gough, S., Kannangara, C.G., 1995. Chlorophyll biosynthesis. Plant Cell 7, 1039–1057.

White, R.C., Jones, I.D., Gibbs, E., 1963. Determination of chlorophylls, chlorophyllides and pheophorbides in plant material. J. Food Sci. 28, 431–436.

Willats, W.G.T., McCartney, L., Mackie, W., Knox, J.P., 2001. Pectin: cell biology and prospects for functional analysis. Plant Mol. Biol. 47, 9–27.

Willows, R.D., 2003. Biosynthesis of chlorophylls from protoporphyrin IX. Nat. Prod. Rep 20, 327–341.

Willstater, R., Stoll, A., 1913. Die wirkungen der chlorophyllase. In: Untersuchungen uber Chlorophyll. Springer, Berlin, pp. 172–187.

Winkle, B.S., 2004. Metabolic channeling in plants. Annu. Rev. Plant Biol. 55, 85–107.

Wong, D.W.S., 1995. Pectic enzymes. In: Food Enzymes: Structure and Mechanism. Chapman and Hall, New York, pp. 212–236.

Workman, M., Pratt, H.K., Morris, L., 1957. Studies on the physiology of tomato fruits. 1. Respiration and ripening behavior at 20°C as related to date of harvest. Proc. Am. Soc. Hortic. Sci. 69, 352–365.

Worth, H.G.J., 1967. The chemistry and biochemistry of pectic substances. Chem. Rev. 67, 465–473.

Wyllie, S.G., 2008. Flavour quality of fruit and vegetables: are we on the brink of major advances? In: Brückner, B., Grant Wyllie, S. (Eds.), Fruit and Vegetable Flavor. Woodhead, Cambridge, pp. 3–10.

Xu, L., Liao, M., Deng, G., Chen, S., Ren, Y., 2008. Changes in activity of PG, PE, Cx and LOX in pulp during fruit growth and development of two different ripening-season pear cultivars. Amer. Euras. J. Agric. Environ. Sci. 3, 445–450.

Xu, W.P., Chen, K.S., Xu, C.J., Zhang, S.L., 2003. Changes of lipoxygenase activity and ethylene biosynthesis in Actinidia fruit stored at 0°C and on shelf at 20°C after cold storage. Sci. Agric. Sin. 36, 1196–1201.

Xuan, H., Streif, J., Saquet, A., Romheld, V., Bangerth, F., 2005. Application of boron with calcium affects respiration and the ATP/ADP ratio in 'Conference' pears during controlled atmosphere storage. J. Hortic. Sci. Biotechnol. 80, 633–637.

Yahia, E.M., 1994. Apple flavor. Hortic. Rev. 16, 197–234.

Yamaki, S., Kakiuchi, N., 1979. Changes in hemi-cellulose degrading enzymes during development and ripening of Japanese pear fruit. Plant Cell Physiol. 20, 311–321.

Yang, S.F., 1974. Ethylene biosynthesis in fruit tissues. Colloq. Int. C.N.R.S. 238.

Yang, S.F., 1980. Regulation of ethylene biosynthesis. HortScience 15, 238–243.

Yang, S.F., Ho, H.K., 1958. Biochemical studies on postripening banana. J. Chin. Chem. Soc. (Taiwan) 5, 71–85.

Yang, S.F., Hoffman, N.E., 1984. Ethylene biosynthesis and its regulation in higher plants. Annu. Rev. Plant Physiol. 35, 155–189.

Yip, W.-K., Yang, S.F., 1988. Cyanide metabolism in relation to ethylene production in plant tissues. Plant Physiol. 88, 473–476.

Yoon, S., Klein, B., 1979. Some properties of pea lipoxygenase isoenzymes. J. Agric. Food Chem. 27, 955–962.

Young, R.E., Salimen, S., Sornrivichai, P., 1974. Enzyme regulation associated with ripening in banana fruit. Colloq. Int. C.N.R.S. 238, 271.

Yu, M.H., Spencer, M., 1969. Conversion of L-leucine to certain keto acids by a tomato enzyme preparation. Phytochemistry 8, 1173–1178.

Yu, M.H., Olson, D.E., Salunkhe, D.K., 1967. Precursors of volatile components in tomato fruit. I. Compositional changes during development. Phytochemistry 6, 1457–1465.

Yu, M.H., Salunkhe, D.K., Olson, L.E., 1968a. Production of 3-methylbutanol from l-leucine by tomato extract. Plant Cell Physiol. 9, 633–638.

Yu, M.H., Olson, D.E., Salunkhe, D.K., 1968b. Precursors of volatile components in tomato fruit. III. Enzymatic reaction products. Phytochemistry 7, 555–565.

Yu, M.H., Olson, L.E., Salunkhe, D.K., 1968c. Precursors of volatile components in tomato fruit. III. Enzymatic reaction products. Phytochemistry 7, 561–565.

Yu, U.B., Adams, D.O., Yang, S.F., 1979. 1-Aminocyclopropane-1-carboxylate synthase, a key enzyme in ethylene biosynthesis. Arch. Biochem. Biophys. 198, 280–286.

Yung, K.H., Yang, S.F., 1980. Biosynthesis of wound ethylene. Plant Physiol. 66, 281–285.

Zamora, R., Olias, J.M., Mesias, J.L., 1987. Purification and characterization of tomato lipoxygenase. Phytochemistry 26, 345–347.

Zanella, A., Cazazanelli, P., Panarese, A., Coser, A., Cecchinel, M., Rossi, O., 2005. Fruit fluorescence response to low oxygen stress: modern storage technologies compared to 1-MCP treatment of apple. Acta Hortic 682, 1535–1542.

Zapsalis, C., Francis, F.J., 1965. Cranberry anthocyanins. J. Food Sci. 30, 396–399.

Zauberman, A., Schiffman-Nadel, M., 1972. Pectin methylesterase and polygalacturonase in avocado fruit at various stages of development. Plant Physiol. 49, 864–865.

Zeroni, M., Galil, J., Ben-Yehoshua, S., 1976. Autoinhibition of ethylene formation in non ripening stages of the fruit of sycamore fig (*Ficus sycamores* L.). Plant Physiol. 57, 647–650.

Zhang, B., Chen, K., Bowen, J., Allan, A., Espley, R., Karunairetnam, S., Ferguson, I., 2006. Differential expression within the LOX gene family in ripening kiwifruit. J. Exp. Bot. 57, 3825–3836.

Zhong, R., Ye, Z.H., 2007. Regulation of cell wall biosynthesis. Curr. Opin. Plant Biol. 10, 564–572.

Zhou, H.-W., Sonego, L., Khalchitski, A., Ben-Arie, R., Lers, A., Lurie, S., 2000. Cell wall enzymes and cell wall changes in 'Flavortop' nectarines: mRNA abundance, enzyme activity, and changes in pectin and natural polymers during ripening and in wooly fruit. J. Am. Soc. Hortic. Sci. 125, 630–637.

Ziegler, R., Blaheta, A., Guha, N., Schonegge, B., 1988. Enzymatic formation of pheophorbide and pyropheophorbide during chlorophyll degradation in a mutant *Chlorella fusca* Shihira et Kraus. J. Plant Physiol. 132, 327–332.

Zimmerman, D.C., Coudron, C.A., 1979. Identification of traumatin, a wound hormone, as 12-oxo-*trans*-10-dodecendic acid. Plant Physiol. 63, 536–541.

Zimmerman, D.C., Vick, A.A., 1970. Hydroperoxide isomerase: a new enzyme of lipid metabolism. Plant Physiol. 46, 445–453.

Chapter 3

Meat and Fish

N. A. Michael Eskin,* Michel Aliani* and Fereidoon Shahidi†

*Department of Human Nutritional Sciences, Faculty of Human Ecology, University of Manitoba, Winnipeg, Canada, †Department of Biochemistry, Memorial University of Newfoundland, St. John's, Canada

Chapter Outline

Biochemistry of Foods. DOI: http://dx.doi.org/10.1016/B978-0-12-242352-9.00003-5

I. INTRODUCTION

Meat is defined as the flesh of animals used as food. A more precise definition is provided by the US Food and Drug Administration (Meyer, 1964): meat is that derived from the muscles of animals closely related to man biochemically and therefore of high nutritive value. The more conventional animal species include cattle, pig, sheep, and the avian species chicken and turkey. In fish, however, it is often the white muscle that provides the main nutritional source. The per capita consumption of muscle foods in the USA has changed over the past decade by chicken replacing pork as the second most preferred muscle (Table 3.1). Seafood, particularly fish, accounts for approximately 20% of all animal proteins consumed worldwide (Zabel *et al.*, 2003). In the developing continents, Africa, Asia, and Latin America, the consumption of meat and fish is still extremely low or non-existent, as evidenced by the increasing incidence of malnutrition. This lack of high-grade proteins and the accompanying deficiency in essential amino acids still remains the world's most urgent problem.

This chapter will discuss the dynamic changes involved in the conversion of muscle to meat or edible fish. Following the death of the animal or fish, many chemical, biochemical, and physical changes occur leading to the development of postmortem tenderness. A greater understanding of these changes should make an important contribution to the production of high-quality meat or fish products. While the degradation of myofibrillar and cytoskeleton proteins is desirable for postmortem tenderization of mammalian and avian muscles, such changes can

TABLE 3.1 US Per Capita Meat Consumption 1970-2008[a]

Year	Beef	Lamb	Pork	Veal	Chicken	Turkey	Fish	Total Meat
1970	51.8	1.2	28.1	1.0	15.8	4.2	2.8	**104.9**
1972	52.3	1.2	28.1	0.8	16.3	4.7	2.9	**106.3**
1974	52.5	0.8	27.4	0.8	15.6	4.5	2.7	**104.3**
1976	57.7	0.7	23.7	1.3	16.5	4.6	3.4	**107.9**
1978	53.5	0.6	24.7	1.0	17.5	4.5	3.5	**105.3**
1980	46.9	0.6	30.4	0.6	18.9	5.3	3.3	**108.0**
1982	47.2	0.6	26.2	0.7	19.5	5.5	3.1	**102.8**
1984	48.0	0.6	27.6	0.7	20.4	5.7	3.4	**106.4**
1986	48.4	0.6	26.4	0.8	21.3	6.7	3.7	**107.9**
1988	44.6	0.6	28.5	0.6	22.8	8.1	3.7	**108.9**
1990	41.6	0.6	27.1	0.4	24.4	9.1	3.7	**106.9**
1992	40.6	0.6	28.7	0.4	26.8	9.2	3.6	**109.9**
1994	40.9	0.5	28.6	0.4	28.1	9.1	3.9	**111.5**
1996	41.6	0.5	26.4	0.5	28.1	9.4	3.8	**110.3**
1998	41.3	0.5	28.1	0.3	29.1	9.1	3.5	**111.9**
2000	42.0	0.5	27.9	0.3	31.2	9.0	3.4	**114.3**
2002	42.0	0.5	28.1	0.2	32.7	9.2	3.7	**116.4**
2004	41.0	0.5	28.0	0.2	34.2	8.8	3.4	**116.1**
2006	40.8	0.4	26.8	0.2	35.1	8.7	4.0	**116.0**
2008	38.8	0.4	26.8	0.2	33.8	9.1	3.8	**112.9**

[a]*The data is adapted from the USDA/Economic Research Service, www.ers.usda.gov, Feb. 1, 2011. The per capita meat "consumption" is measured in lbs/year and reflects the total amount of meat supply in the United States adjusted for loss. The USDA calculates "consumption" by dividing the total annual supply of a specific meat by the annual population and adjusting that figure for loss by subtracting 1) the estimated amount of meat lost when converting it from the farm to a market ready product, 2) the estimated amount of meat lost at the retail level due to spoilage, and 3) the estimated amount of meat lost when consumers cook it at home. The fish category includes both fresh and frozen fish. http://vegetarian.procon.org/view.resource.php?resourceID=004716#V*

lead to unfavorable changes in fish muscle (Geesink *et al.*, 2000b). Texture remains one of the most important quality attributes affecting consumer acceptance of meat and fish products. In meat products, however, it is tenderness that is important to the consumer, while firmness is key to fish quality (Cheret *et al.*, 2007). Considerable advances have been made over the past two decades with a better understanding of the molecular mechanisms involved in the tenderization and textural changes in both meat and fish (Delbarre-Ladrat *et al.*, 2006; Terova *et al.*, 2011).

II. THE NATURE OF MUSCLE

While muscles are classified into several types, it is the striated or voluntary muscle which constitutes lean meat. The basic unit of the muscle is the fiber, a multinucleate, cylindrical cell bounded by an outer membrane, the sarcolemma. These fibers associate together into bundles, and are enclosed by a sheath of connective tissue, the perimysium. Fiber bundles are held together by connective tissue and covered by a connective tissue sheath, the epimysium. Connective tissues important to the texture and edibility of the meat and fish include fibrous proteins, collagen, reticulin, and elastin. Fish muscle has much less connective tissue, thus providing less of a problem in tenderization.

A. Structure

Individual muscle fibers are composed of myofibrils which are 1–2 μm thick and are the basic units of muscular contraction. The skeletal muscle of fish differs from that of mammals in that the fibers arranged between the sheets of connective tissue are much shorter. The connective tissue is present as short transverse sheets (myocommata) which divide the long fish muscles into segments (myotomes) corresponding in numbers to those of the vertebrae (Dunajski, 1979). The individual myofibrils are separated by a fine network of tubules, the sarcoplasmic reticulum. Within each fiber is a liquid matrix referred to as the sarcoplasm, which contains mitochondria, enzymes, glycogen, adenosine triphosphate (ATP), creatine, and myoglobin.

Examination of myofibrils under a phase-contrast light microscope shows them to be cross-striated due to the presence of alternating dark or A-bands and light or I-bands. These structures in the myofibrils appear to be very similar in both fish and meat. The A-band is traversed by a lighter band or H-zone, while the I-band has a dark line in the middle known as the Z-line. A further dark line, the M-line, is observed at the center of the H-zone. The basic unit of the myofibril is the sarcomere, defined as the unit between adjacent Z-lines as shown in Figure 3.1. Examination of the sarcomere by electron microscopy reveals two sets of filaments within the fibrils, a thick set consisting mainly of myosin, and a thin set containing primarily F-actin.

In addition to the paracrystalline arrangement of the thick and thin set of filaments, there appears to be a filamentous 'cytoskeletal structure' composed of connectin and desmin (Young *et al.*, 1980–1981). Connectin is now recognized as the major myofibrillar protein in the 'gap filaments' in muscle and is present throughout the sarcomere of

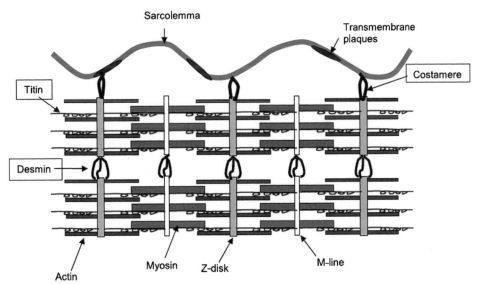

FIGURE 3.1 Schematic representation of muscle myofibrillar proteins showing the major components of the sarcomere. Boxes indicate the cytoskeleton structures and proteins susceptible to postmortem cleavage. (*Kemp* et al.*, 2010*)

skeletal muscle (Maruyama *et al.*, 1976a). These gap or G-filaments were reported by Locker and Leet (1976a, b) to span the region between the thick and thin filaments in fibers of overstretched beef muscle. Locker (1984) proposed that each gap filament formed a core to an A-band. Connectin was subsequently characterized as the protein titin, consisting of the three fractions titin-1, -2, and -3, reported by Wang *et al.* (1979) to account for 10–15% of the myofibrillar proteins in chicken breast. Titin-3 is now recognized as a distinct protein and referred to as nebulin (Wang and Williamson, 1980). Desmin, on the other hand, was reported by several researchers to be present in the periphery of each Z-disk in chicken skeletal muscle (Lazarides and Hubbard, 1976; Grainger and Lazarides, 1978). It may have a role in maintaining alignment of adjacent sarcomeres, which unifies the contractile process of the separate myofibrils.

B. Cytoskeleton

The cytoskeleton of muscle is composed of two elements, gap filaments and intermediate filaments (Stanley, 1983).

1. Gap Filaments

Gap filaments (G-filaments) were originally identified by Hanson and Huxley (1955) as extremely thin elastic 'S-filaments' responsible for keeping the actin filaments together. The model in Figure 3.2, proposed by Hoyle, showed that these filaments were located parallel to the fiber axis extending between the Z-disks, and were referred to as 'gap filaments' (Sjöstrand, 1962). These filaments were found by Maruyama *et al.* (1976b, 1977) to be composed of a rubbery, insoluble protein called 'connectin'. Wang *et al.* (1979) identified a high-molecular-weight protein which was referred to as titin. Subsequent research showed the high-molecular-weight components of connectin to be titin (Maruyama *et al.*, 1981). Titin appeared to be the major cytoskeleton protein in the sarcomere responsible for muscle cell integrity (Wang and Ramirez-Mitchell, 1979, 1983a, b). Locker and Leet (1976b) proposed that each gap filament formed a core within an A-band and linked the two thick filaments in adjacent sarcomeres through the Z-line. Wang and Ramirez-Mitchell (1984), using four distinct monoclonal antibodies to rabbit titin, showed that titin passed from the M-line through the A-band and into the I-band, thereby discounting the central core model.

Wang *et al.* (1979) also identified a large myofibrillar protein in vertebrate skeletal muscle referred to as nebulin. This protein was later isolated from the myofibrils of rabbit psoas and chicken breast muscles using immunological and electrophoretic techniques. It was found to be distinct from titin (Wang and Williamson, 1980; Murayama *et al.*, 1981; Ridpath *et al.*, 1982, 1984). The location of nebulin in the myofibril was at the N^2 line. Wang and Ramirez-Mitchell (1983b) presented an alternative model for the G-filaments consisting of an elastic filamentous matrix containing both titin and nebulin as additional sarcomere constituents (Locker, 1984).

Nebulin is a family of giant proteins (600–900 kDa) in skeletal muscle that forms a set of inextensible filaments anchored at the Z-line (Wang and Wright, 1988). For the past two decades it was considered by many investigators to be a molecular ruler that dictated the length of the actin thin filament. Several properties exhibited by nebulin allowed it to function as a molecular ruler (Horowitz, 2006). These included, first, its ability to extend the entire length of the thin filament with its C-terminus anchored in the Z-disc and its N-terminus towards the center of the sarcomere (Wright *et al.*, 1993). Secondly, the length of nebulin can vary by alternative splicing, enabling it to correlate with the thin filament lengths of the different muscles (Kruger *et al.*, 1991; Labeit *et al.*, 1991). Thirdly, nebulin has a unique protein structure that enables it to bind all the major components of the thin filament. The critical role of nebulin in maintaining proper skeletal function was shown in nebulin-deficient mice which died within 2 weeks of birth as a result of muscle weakness (Bang *et al.*, 2006; Witt *et al.*, 2006). In addition to its role as a molecular ruler, recent evidence suggests that nebulin regulates contraction by optimizing the actin–myosin interaction and controlling calcium in the sarcoplasmic reticulum (Root and Wang, 1994, 2001; Ottenheijm *et al.*, 2008; Bang *et al.*, 2009; Chandra *et al.*, 2009). The interaction of the C-terminus of nebulin appeared to be regulated by desmin, suggesting

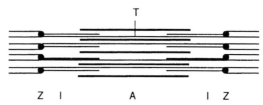

FIGURE 3.2   Diagram showing proposed model for muscles including a very thin elastic filament (T) extending between Z-disks (Z) and parallel to the A-band (A) and I-band (I). *(Hoyle, 1967)*

that it may be involved in maintaining the lateral alignment of myofibrils, an important property for coordinated and efficient contraction (Bang *et al.*, 2002; Conover *et al.*, 2009; Tonino *et al.*, 2010). Recently, Pappas *et al.* (2010) failed to find a role for nebulin in actin polymerization as it appeared to regulate thin actin filament lengths by a stabilization mechanism that prevented actin depolymerization. Defects in the nebulin gene are associated with congenital nemaline myopathy, a debilitating condition of muscle weakness (Gokhin *et al.*, 2009). The important role that nebulin plays in muscle health and disease was reviewed by Labeit *et al.* (2011).

2. Intermediate Filaments

These filaments linking the myofibrils laterally to the sarcolemma are intermediate in size (10 nm in diameter) between the actin (6 nm in diameter) and myosin (14–16 nm in diameter) filaments (Ishikawa *et al.*, 1968). The protein isolated from these filaments, desmin, also referred to as skeletin, is located in the periphery of the Z-disk in the filamentous form (Lazaride and Hubbard, 1976; Richardson *et al.*, 1981). The cytoskeleton role of desmin is to connect Z-lines of adjacent myofibrils (O'Shea *et al.*, 1981; Robson *et al.*, 1984).

C. Connective Tissue

The interstitial space in muscle cells is occupied by three proteins, collagen, reticulin, and elastin, together referred to as connective tissue. The endomysium layer surrounding the muscle fibers is composed of fine reticular and collagenous fibrils, while elastin is sparsely distributed in the muscle with the blood, capillary, and nervous systems (Asghar *et al.*, 1984). Bundles of these muscle fibers are surrounded by a thicker connective tissue, the perimysium. These connective tissues appear to unite at the ends of the muscle in the thick tendon fibers as shown in Figure 3.3 (Etherington and Sims, 1981). During muscle contraction, movement is transmitted via the tendon to the skeleton. The limited elasticity of collagen in the tendon permits the translation of muscle contraction into a high degree of movement.

1. Collagen

The major protein of connective tissue is collagen, a glycoprotein. It was originally thought to be composed of two polypeptide chains, α_1- and α_2-chains, which formed a triple helical structure. At least 10 different α-chains are now known which appear to be responsible for the different types of collagen so far identified. Of these, collagen I and III are the most abundant types in muscle, with minor quantities of IV, V, VI, XII, XIV, XV, and XIX. These differ from each other in their primary structure and amino acid composition (Asghar *et al.*, 1984). Morphologically, three discrete collagen depots, endomysium, perimysium, and epimysium, make up the three-dimensional network of collagen and elastin fibers embedded in a matrix of proteoglycans (Lepetit, 2008). The endomysium is the layer of connective tissue surrounding each muscle and overlying the basement membrane (McCormick, 1999). The epimysium includes the connective sheath surrounding individual muscles and continuous with the tendon joining other muscles (McCormick, 1999). It is extremely tough and resistant to shear and solubilization. The perimysium accounts for the bulk of the intramuscular connective tissue (IMCT) and plays a key role in determining meat texture differences (Lewis and Purlow, 1990). The subunit of the collagen fiber is the collagen monomer tropocollagen. This is composed of three polypeptide α-chains arranged in a pattern which allows the staggered overlap of one polypeptide chain over the other, as shown for the type I tropocollagen monomer in Figure 3.4 (Asghar and Henrickson, 1982).

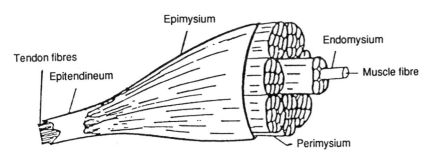

FIGURE 3.3 Connective tissues of muscle. *(Etherington and Sims, 1981)*

FIGURE 3.4   Schematic representation of the type I tropocollagen monomer, composed of two identical pro-α_1(I) chains (solid lines) and one pro-α_2(I) chain (dashed line). *(Asghar and Henrickson, 1982)*

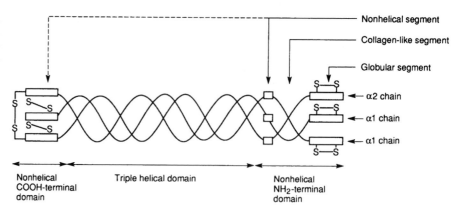

The major collagen component of both the epimysium and perimysium is type I collagen, while types III, IV, and V are located in the endomysium (Bailey and Peach, 1968; Bailey and Sims, 1977). Since collagen is the principal component of connective tissue, the texture of meat is greatly influenced by it. Bailey (1972) suggested that an acceptable meat texture requires a certain degree of cross-linkages in collagen. A lack or an overabundance of such linkages in collagen produces meat that is either too tender or too tough. The toughness associated with meat from older animals is attributed to the high degree of stable and mature cross-linkages in the collagen fibers (Eyre *et al.*, 1984; Reiser *et al.*, 1992). The relative amounts of these stable cross-links appear to determine the texture of cooked meat (Bailey and Light, 1989; Bailey, 1990).

a. Collagen and Meat Texture

Attempts to correlate total collagen of muscles with meat texture have resulted in conflicting reports. Dransfield (1977) found a definite relationship between total muscle collagen and toughness. Other studies showed that the qualitative nature of collagen, rather than quantity, ultimately affected texture (Bailey, 1972; Shinomokai *et al.*, 1972; Bailey *et al.*, 1979; Bailey and Sims, 1977). A later study by Light *et al.* (1985) examined the role of epimysial, perimysial, and endomysial collagen in the texture of six bovine muscles. These researchers reported a correlation between both collagen fiber diameter and collagen content of perimysial and endomysial connective tissue and meat toughness. A linear plot was obtained when the number of heat-stable cross-links was plotted against the compressive force (kg) (Figure 3.5) for six muscle perimysia cooked at 75°C using the data of Dransfield (1977). Although not as clear-cut, similar trends were observed for both epimysial and endomysial

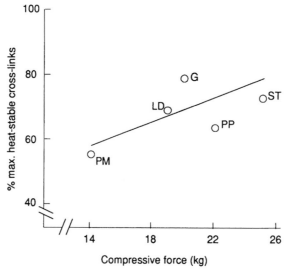

FIGURE 3.5   Plot of total heat-stable (keto-imine) cross-links in six perimysia versus compressive force estimated after cooking the muscles for 1 hour at 75°C (from data by Dransfield, 1977). PM: Psoas major; LD: longissimus dorsi; PP: pectoralis profundis; G: gastrocnemius; ST: semitendinosus. *(Light* et al., *1985)*

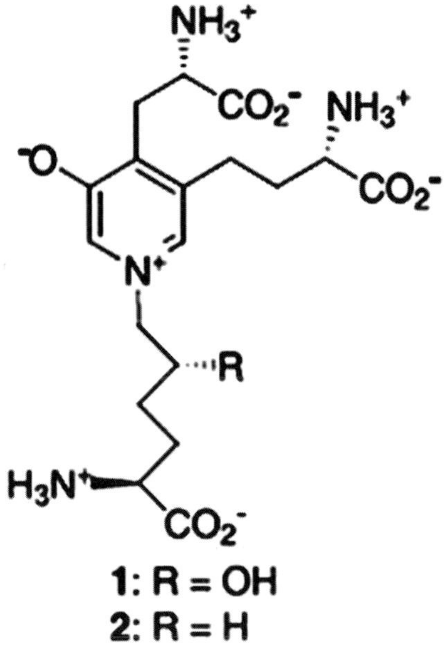

$$NH_3^+$$
$$CO_2^-$$
$$NH_3^+$$
$$CO_2^-$$
$$R$$
$$H_3N^+$$
$$CO_2^-$$

1: R = OH
2: R = H

FIGURE 3.6 **Pyridinoline** *(Allevi and Anastasia, 2003).*

muscle samples. Based on these results it was apparent that the cross-links have a crucial role in determining tenderness or toughness of meat. If the primary cause for meat fracture and breakdown, according to Purslow (1985), is via the perimysium or at the perimysial–endomysial junctions, the nature of the cross-links between these fibers could be extremely important.

The toughening of the meat in older animals was attributed to changes in collagen cross-links. These cross-links appeared to be mediated by changes in lysyl oxidase during advanced aging (Bailey and Shimokomaki, 1971; Robins *et al.*, 1973). The toughening of the meat during aging was related to a decrease in the immature collagen cross-link, hydroxylysinoketonorleucine (HLKNL) (Shimokomaki *et al.*, 1972). The main mature cross-link, pyridinoline (PYR) (Figure 3.6) was supposedly derived from two reducible HLKNL and stabilized collagen in the older animals (Fujimoto, 1977; Eyre and Oguchi, 1980; McCormick, 1999). The formation of PYR was found to be related to meat texture and formed bridges in the different types of collagen (Shimokomaki *et al.*, 1990; Nakano *et al.*, 1991; Bosselmann *et al.*, 1995). A study by Coro *et al.* (2002) reported an increase in collagen from 0.448 to 0.568% in Ross chicken breast muscle (pectoralis major) aged from 20 to 540 days. During this period, PYR also increased from 0.009 to 0.101 mol/mol collagen, and was inversely related to collagen solubility. Textural changes measured by shear values, however, showed that a decrease in tenderness was affected more by the formation of PYR cross-links than by an increase in collagen synthesis. A decrease in collagen solubility was also directly related to an increase in PYR concentration. Reviews by Lepetit (2007, 2008) discuss the role of collagen in meat tenderness or toughness.

b. Collagen and Fish Texture

Fish muscles generally contain only one-tenth of the collagen found in red meats. They are divided by thin membranes, myocommata, into segments or myotomes as shown in Figure 3.7. The myocommata are composed

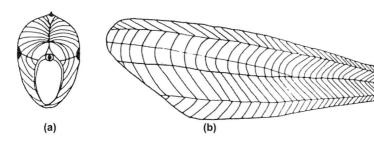

(a)   **(b)**

FIGURE 3.7 **The metameric structure of fish muscles.** The patternsof lines on the cross-section (a) and longitudinal section (b) represent the arrangement of sheets of connective tissue in the muscles. *(Dunajski, 1979)*

of connective tissue, with each muscle fiber surrounded by a cell wall or basement membrane containing thin collagen fibrils. The integrity of the fish muscles is maintained by the connective tissue of the myocommata and collagen fibers, which together form the endomysial reticulum. If the myotomes are not connected to the myocommata, slits and holes form in the flesh, which is characteristic of gaping. This results in the deterioration of fish quality as the fish fillets fall apart and become quite unacceptable. Love *et al.* (1972) attributed the development of this problem to the rupturing of the endomysial and myocommata connection brought about by rough handling or bending of the stiffened fish. The development of gaping, accompanied by softening of the tissue, is a consequence of deterioration during storage and is enhanced by rapid and strong muscular contractions during rigor (Love, 1988; Bremner, 1999; Taylor *et al.*, 2002). Examination of this phenomenon in cod and wolfish by Ofstad *et al.* (2006) attributed the increase in gaping during ice storage to the degradation of proteoglycans and glycoproteins, major constituents of the IMCT. They play an important role in the spatial organization of the collagen fibers by anchoring cells to the extracellular matrix in the collagen network. The greater propensity for gaping in cod, compared to wolfish, was attributed to the more rapid detachments occurring between myofibers prior to the myotomes to myocommata detachments (Ofstad *et al.*, 2006). Around 40% of the downgrading of Atlantic salmon (*Salmo salar* L.) during secondary processing was attributed to either soft flesh or gaping (Michie, 2001). In addition, preslaughter stress of salmon was reported to accelerate fish fillet softening, increasing gaping as well as fillet color and drip loss (Kiessling *et al.*, 2004; Erikson and Misimi, 2008; Morkore *et al.*, 2008). A novel tensile testing method, using a mechanical texture analyzer, was recently developed by Ashton *et al.* (2010) for assessing texture and gaping in salmon fillets. The contribution of connective tissue to the texture of cooked fish remains unclear, compared to its role in meat, as a consequence of the smaller amount present. For example, the high level of connective tissue in dogfish requires a cooking temperature of 45°C, which is still substantially lower than the cooking temperature of 92°C for 1 hour needed to obtain the same degree of tenderness for beef.

III. CONTRACTION OF MUSCLE

While the majority of studies have been conducted on mammalian muscle, it is apparent that similar changes occur in fish muscle. It is generally accepted that contraction and relaxation of striated muscle occurs by the sliding action of the thick filaments over the thin filaments, with the length of the filaments remaining the same (Rowe, 1974). Myosin possesses adenosine triphosphatase (ATPase) activity, which requires the presence of magnesium and calcium ions. It is the regulation of myofibrillar ATPase which determines the contractile response of the muscle. This enzyme catalyzes the hydrolytic cleavage of ATP, thereby providing the most immediate source of energy for muscular contraction:

$$ATP + H_2O \rightarrow ADP + H_3PO_4$$

ΔG_{298} (Standard free energy change at 25°C) $= -11.6$ kcal/mole

In resting muscle the activity of ATPase is very low, resulting in the slow release of adenosine diphosphate (ADP) and inorganic phosphorus at the active sites of myosin and actin. Once muscle stimulation occurs, the head of myosin, containing the actin-combining and enzymatic sites, interacts with actin with the rapid release of ADP and inorganic phosphate (P_i). The increase in ATP hydrolysis can be several hundred times that observed in the resting state (Perry, 1979). This is accompanied by a conformational change in the myosin-head, causing a change in the angle it makes with the actin filament. The overall result is that the actin monomer with the myosin head attached moves forward by approximately $5-10$ nm (Huxley, 1969). Once ADP and P_i are released from the myosin head, the actin monomer detaches itself to permit a fresh molecule of Mg-ATP^{2-} to be picked up by the enzymatic site on the myosin head and the enzyme—substrate complex is re-established. Muscular contraction is thus characterized by a rapid conversion of ATP to ADP and inorganic-phosphate, and on completion the muscle returns to its resting state. The latter is characterized by the capacity of the substrate—enzyme complex at enzymatic sites in the myosin head to be released once stimulation occurs. The hydrolysis of myosin in the presence of actin has been studied by a number of researchers, although detailed steps remain to be characterized (Chock *et al.*, 1976; Eccleston *et al.*, 1976).

A. Regulation of Muscle Contraction: Troponin and Tropomyosin

The regulation of muscular contraction appears to involve the release of calcium from the vesicles of the sarcoplasmic reticulum, located in the myofibrils. Calcium is released when the stimulus is received at the muscle fiber by way of the central nervous system. It stimulates myosin ATPase, thus releasing the energy necessary for muscle contraction to facilitate the sliding action of actin filaments which form contractile actomyosin (Huxley, 1964). This was confirmed in studies by Goodno *et al.* (1978), who reported up to 100-fold increase in ATPase activity of the myofibril when calcium ion (Ca^{2+}) levels were increased. Calcium regulates actin—myosin interaction by directly binding to the troponin C component of the I-filament.

Troponin is a myofibrillar protein associated with the thin filaments, which appears to control the interaction between actin and myosin. It is an elongated molecule of molecular weight 80 kDa, which is attached to tropomyosin, another myofibrillar protein. They both provide the regulatory system in muscular contraction (Ebashi, 1974). Tropomyosin, a long, coiled α-helix, is located in each of the two long-pitch helical grooves of the actin monomers on thin filaments (Seymour and O'Brien, 1980). Murray and Weber (1974) showed that troponin (Tn) and tropomyosin (TM) interacted with seven actin molecules (Figure 3.8). In contrast, troponin is found at discrete intervals of 38 nm along the thin actin filaments and is associated with the stripes on the I-band. Troponin is composed of three subunits, troponin C (TnC), troponin T (TnT), and troponin I (TnI). It appears to have multiple, compartmentalized functions, with each subunit exerting a particular role including binding tropomyosin or Ca^{2+} or inhibiting actomyosin ATPase (Greaser and Gergely, 1973). When the muscle is stimulated, TnC first binds Ca^{2+} then neutralizes TnI inhibition of actomyosin ATPase which is then translated into conformational changes in protein via tropomyosin. The role of TnT has remained obscure although it has since been reported to enhance actomyosin ATPase activity at high Ca^{2+} and may play a dual role in filament regulation (Dahiya *et al.*, 1994; Potter *et al.*, 1995; Tobacman, 1996). Recent research by Murakami *et al.* (2008) showed that head-to-tail polymerization of tropomyosin was important for binding to actin and for regulating actin-myosin contraction. The 2.1Å solution structure of crystals containing overlapping tropomyosin N and C termini is shown in Figure 3.8A. They emphasized the importance of the molecular swivel generated by TnT at the junction between successive tropomyosin for regulating striated muscle contraction.

Skeletal muscles include a variety of fibers, classified into slow and fast types, composed of complex multiple myofibril proteins. Schiaffino and Reggianni (1996) showed that the major proteins responsible for the different fiber types were myosin heavy-chain isoforms. The fiber types appeared to have different ATPase activities which may impact meat quality traits during the conversion of muscle to meat. Ouali (1992) suggested myofibrils might be the limiting step in the development of meat texture. Later researchers correlated muscle fiber types and size to meat quality traits in pigs (Chang *et al.*, 2003; Ryu and Kim, 2005) and in beef (Crouse *et al.*, 1991; Ozawa *et al.*, 2000; Maltin *et al.*, 2003). There are four calcium binding sites in TnC of fast muscles compared to three calcium binding sites in slow muscles. In the presence of low calcium levels the formation of cross-links is inhibited by the troponin—tropomyosin complex. An increase in calcium levels following stimulation of the muscle results in the binding of calcium to TnC and the formation of the actomyosin complex. Accumulation of calcium in the

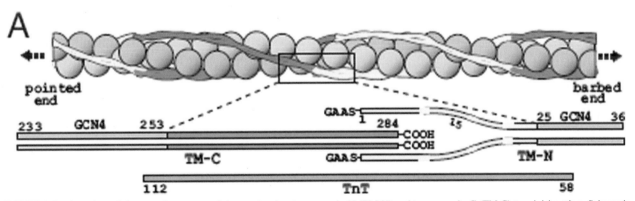

FIGURE 3.8 Overview of the crystal structure of the overlapping tropomyosin N (TM-N) and tropomyosin C (TM-C) termini junction. Schematic representation of the constructs used (O-actin). The C-terminal (TM-C) (residues 254—284, preceded by a 20 fragment of the GCN4 leucine zipper to stabilize dimerization) and N-terminal (TM-N) [residues 1—24, followed by a 12 residue fragment of the GCN4 leucine zipper, and preceded by N-terminal GlyAlaAlaSer-extension (Murakami *et al.*, 2008).

sarcoplasmic reticulum is achieved against a concentration gradient requiring an active transport pumping system involving ATP. This is hydrolyzed by ATPase present in the membranes of the sarcoplasmic reticulum (de Meis and Vianna, 1979).

B. Mechanism of Tropomyosin Action

Extensive studies using electron micrographs of a myosin subfragment (S-1) suggested that tropomyosin regulated muscle contraction by steric blocking and unblocking of the myosin interaction sites in the muscle thin filaments (Haselgrove, 1972; Huxley, 1972b; Parry and Squire, 1973). This theory was questioned by Seymour and O'Brien (1980), who proposed that tropomyosin was located on the opposite side of the thin filament helix axis from the binding sites of myosin S-1 (Moore *et al.*, 1970). Further studies by Taylor and Amos (1981), based on three-dimensional image reconstructions of electron micrographs of thin filaments decorated with myosin S-1, suggested that the location of the binding sites proposed by Moore *et al.* (1970) was incorrect. Taylor and Amos (1981) clearly demonstrated that tropomyosin was located on the same side of the actin helix. Later research by Mendelson (1982) using high-resolution reconstructions of X-ray scattering of myosin S-1 further supported the model for attachment of myosin S-1 to actin presented by Taylor and Amos (1981). Thus, regulation of muscle contraction appeared to occur by steric blocking of actin–S-1 myosin interaction. Amos *et al.* (1982) presented structural evidence that the head of myosin S-1 interacted with two sites on F-actin. In the active state tropomyosin was thought to occupy a position near the middle of the actin groove, while in the inhibited state it lay on the other side of the groove where it could interface with interactions between S-1 and actin at the first or second site. Based on the three dimensional image reconstructions of actin-myosin complexes, Taylor and Amos (1981) and Amos *et al.* (1982) suggested a globular shape, while research by Garrigos and co-workers (1992) supported a more anisomeric model, in agreement with other researchers (Milligan and Flicker, 1987; Karijama, 1988). For further information on the structural mechanism of ATP-induced dissociation of rigor myosin from actin, the paper by Kuhner and Fischer (2011) is recommended.

Tropomyosins (TPM) are present in a large number (> 20) of isoforms encoded by four genes, TPM1(α-TM, α-TM$_{fast}$), TPM2(β-TM), TPM3(α-TM,α-TM$_{slow}$), and TMP4 (Perry, 2001). The three major isoforms expressed in skeletal muscle are TPM1, TPM2, and TPM3. Oe *et al.* (2007) determined the cDNA sequences in TPM isoforms in bovine skeletal muscles from Holstein cows. Based on the amino acid sequence from two-dimensional gel electrophoresis, they identified three TPM isoforms expressed in five physiologically different bovine skeletal muscles (masseter, diaphragm, psoas major, longissimus thoracis and semitendinosis). The three isoforms identified, TPM1, TPM2, and TPM3, had 93.3%, 99.6%, and 100% amino acid homology to the human sequence, respectively. TPM2 accounted for approximately 50% of the TPM in all of the muscles examined; however, the presence of TPM2 and TPM3 depended on whether the muscle was a fast or slow type. The higher sequence homology between TPM1 and TPM3 suggested that the ratio of these TPMs may be an important factor in determining the contractile properties of skeletal muscles.

The structural basis for troponin and tropomyosin activation of muscle was recently reported by Lehman *et al.* (2009), in which troponin appeared to be involved in both activation and inhibition of muscular contraction. Key structural changes in troponin resulted in marked steric effects that control actin activation of myosin ATPase and its subsequent contraction.

C. ATP and the Lohmann Reaction

On cessation of the stimulus, calcium ions are removed by a 'relaxing factor' with the sarcoplasmic reticulum acting as a calcium pump (Newbold, 1966). While the primary source of ATP in muscle is derived from ADP by aerobic respiration, it can also be resynthesized from ADP and creatine phosphate (CP) by the Lohmann reaction:

$$\text{ADP} + \text{CP} \xrightarrow{\text{ATP:creatine phosphotransferase}} \text{ATP} + \text{creatine}$$

ATP:creatine phosphotransferase and creatine phosphate are both located in the sarcoplasm. This reaction is important in conditions leading to muscle fatigue, representing an immediate pathway for the resynthesis of ATP. Consequently, muscular activity can continue until adequate amounts of ATP are generated via carbohydrate degradation.

D. Fish Muscle Contraction

Fish muscle consists of two types, red and white muscle, in which rigor contractions have been shown to differ. While the proportions of these two muscles vary from one species to the next, red muscle never exceeds 10% of the total muscle for any species, for example, tuna. The dark and white muscle content of 16 species of fish was measured by Obatake and Heya (1985) using a rapid direct gravimetric method on the heated fish. With the exception of saury pike, the dark muscle of all the other species of fish never exceeded the 10% level. The ratio of dark muscle to whole muscle accounted for over 12% in the so-called red meat species (e.g. sardine, saury pike, frigate herring, and round herring) compared to less than 3% in the white meat species, such as yellow sea bream and silago. The dark or red muscle is characterized by a high myoglobin content as well as distinct proteins (Hamoir and Konosu, 1965). Obatake *et al.* (1985) found that the dark muscle of fish had higher levels of extractive nitrogen constituents, as well as creatine, compared to white muscles.

Rigor contraction has been shown to be far greater in red muscle of fish compared to white muscle (Figure 3.9) and corresponded more closely with that of mammalian muscle (Buttkus, 1963). The role of contraction, tension, and elasticity associated with the development of rigor mortis in postmortem fish muscles is still poorly understood. Bate-Smith and Bendall (1956), in studies on rabbit muscle, found considerable shortening during the rigor period which was not concomitant with stiffening and rarely occurred at room temperature. In contrast, the red muscle of lingcod and trout consistently produced a postmortem contraction at 20°C, the rate being indicative of the condition of the fish prior to death (Buttkus, 1963). White muscle is generally regarded as the nutritional flesh of fish, therefore most studies have been confined to this tissue. The importance of red muscle in the postmortem changes of fish cannot be ignored, however, since as yet there is no process capable of separating these two muscles in the fish processing industry.

Trucco *et al.* (1982) reported that the visual and tactile estimation of rigidity of fish during the prerigor, full rigor, and postrigor stages originally proposed by Cutting (1939) was still the most reproducible method. Their results, shown in Figure 3.10, indicate that for sea bream (*Sparus pagnes*) it took 10 hours for rigor mortis to develop and approximately 30 hours for its resolution, compared to 55 hours in anchovy. However, resolution of rigor mortis was not apparent until 80 hours. The development of the actomyosin complex during rigor mortis was monitored by the reduced viscosity of the high-strength muscle extract (Crupkin *et al.*, 1979). Thus, the course of rigor mortis could be followed by measuring the viscosity of the extract (Figure 3.11).

Bito *et al.* (1983) developed the rigidity/rigor index to determine the stage of rigor mortis in fish. The upper half of the whole fish was placed on a horizon table on one side, suspending the other half or tail part off the edge. The vertical distance (L) between the base of the caudal fin and the table surface was measured over time intervals, with rigor index calculated as:

$$\text{Rigor index } (\%) = [(L_0 - L)/L_0] \times 100\%$$

where L_0 is the value measured immediately after death, and L values are recorded at different time intervals after death.

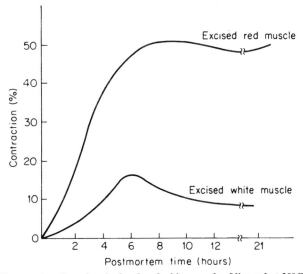

FIGURE 3.9 **Rigor contraction of excised red and white muscle of lingcod at 20°C.** *(Buttkus, 1963)*

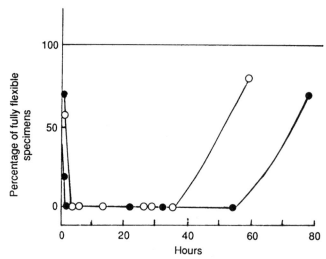

FIGURE 3.10   Evolution of rigor mortis in sea bream (○) and anchovy (●). *(Trucco* et al., *1982) © Elsevier*

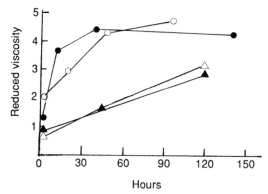

FIGURE 3.11   Reduced viscosity of the high ionic strength muscle extract during the development of rigor mortis in hake (○), Patagonian blue whiting (●), tail hake (▲), and grenadier (△). *(Trucco* et al., *1982) © Elsevier*

Using the rigor index, Wang *et al.* (1998) examined rigor mortis and ATP degradation in cultivated Atlantic salmon (*Salmo salar*). The data obtained followed a similar pattern to that observed by Crupkin *et al.* (1979) for sea bream using viscosity as an index of rigor. Wang *et al.* (1998) showed that rigor mortis started 8 hours after death, reaching a maximum between 24 and 30 hours (Figure 3.12). Complete dissolution of rigor mortis in Atlantic salmon occurred between 60 and 70 hours. The different stages of fish were classified using the rigor index as: no stiffening during prerigor (rigor index of $< 10\%$), fully stiffened in rigor mortis (rigor index $< 100\%$), and postrigor (rigor index $< 10\%$). The ATP content of the Atlantic salmon muscle decreased from 7.25 to 0.14 µmol/g in rigor mortis, consistent with a decrease in ATP of less than 1 µmol/g reported in plaice by Iwamoto *et al.* (1987) when a rigor index of 100% was also reached. The K value (see Section IV.B, below) for Atlantic salmon was also found to increase from 0.7 to 10.6% to 41.1% at prerigor, rigor, and postrigor stages. These values were within the range reported by Erikson *et al.* (1997) for unstressed farmed Atlantic salmon.

IV. CONVERSION OF MUSCLE TO MEAT AND EDIBLE FISH

A vast array of biochemical and physicochemical reactions takes place from the time that the animal or fish is killed until it is consumed as meat or edible fish. This period can be divided into three distinct stages:

1. The prerigor state, when the muscle tissue is soft and pliable. This is characterized biochemically by a fall in ATP and creatine phosphate levels as well as by active glycolysis. Postmortem glycolysis results in the conversion of glycogen to lactic acid, causing the pH to fall. The extent of pH change varies from one species to another as well as among

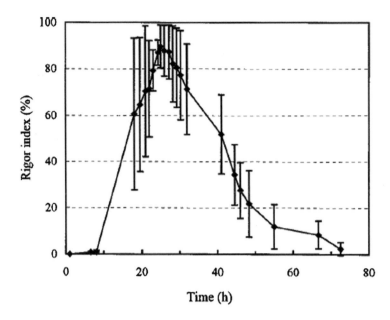

FIGURE 3.12 Rigor index histories of Atlantic salmon stored at 0°C. The curve shows the mean of four replicated tests and the error bars show standard deviation. *(Wang et al., 1998)*

different muscles. Nevertheless, in well-fed, rested animals the glycogen reserves are large so that in the postmortem state the meat produced has a lower pH than meat produced from animals exhausted at the time of slaughter.

2. The development of the stiff and rigid condition in the muscle known as rigor mortis. This occurs as the pH falls and is associated with the formation of actomyosin. The loss of extensibility associated with the formation of actomyosin proceeds slowly at first (the delay period) and then extremely rapidly (fast phase). The onset of rigor mortis normally occurs at 1—12 hours postmortem and may last for a further 15—20 hours in mammals, depending on a number of factors to be discussed later. Fish generally exhibit a shorter rigor mortis period commencing 1—7 hours after death, with many factors affecting its duration.

3. The postrigor state, during which the meat and fish muscles gradually tenderize, becoming organoleptically acceptable as aging progresses. Mammalian meat usually attains optimum acceptability when stored for 2—3 weeks at 2°C following dissolution of rigor.

The importance of rigor mortis in fish is recognized by the fishing industry, since in addition to retarding microbial spoilage, it affords a stiffness to the fish which is generally recognized by the consumer as a sign of good quality. The rigor period, however, is also a distinct disadvantage with respect to the filleting of fish, as it renders the fish too stiff to process. Thus, filleting is carried out following dissolution of rigor, or on trawlers, immediately before the development of rigor.

The principal changes following death are summarized in Figure 3.13. Following the death of the animal or fish, circulation of the blood ceases, which results in a complex series of changes in the muscular tissue. Since blood is an ideal medium for spoilage microorganisms, as much as possible is removed from the animal carcass to ensure that the edibility and keeping quality of the meat are maintained. With respect to fish, only some of the larger species are bled; the blood differs from that of warm-blooded animals and coagulates far more rapidly. The advent of the modern fish-processing trawler brought with it a number of problems, particularly surface discoloration in sea-frozen fillets. This discoloration in the prerigor processed cod fillets developed from surface contamination by the fish blood in the form of methemoglobin (Kelly and Little, 1966). This phenomenon can be prevented in prerigor fish fillets of ocean perch (*Sebastes marinus*) and cod (*Gadhus morhua*) by correct bleeding methods.

The most immediate effect of the stoppage of blood circulation and the removal of blood from the muscle tissue is depletion of the oxygen supply to the tissue and the subsequent fall in oxidation—reduction potential. This results in an inability to resynthesize ATP, as the electron transport chain and oxidative phosphorylation mechanisms are no longer operative.

A. ATP and Postmortem Changes

The major source of ATP supply to the muscle fibers is lost following the death of the animal or fish, since glycogen can no longer be oxidized to carbon dioxide and water. In its place, anaerobic metabolism takes over, resulting in the

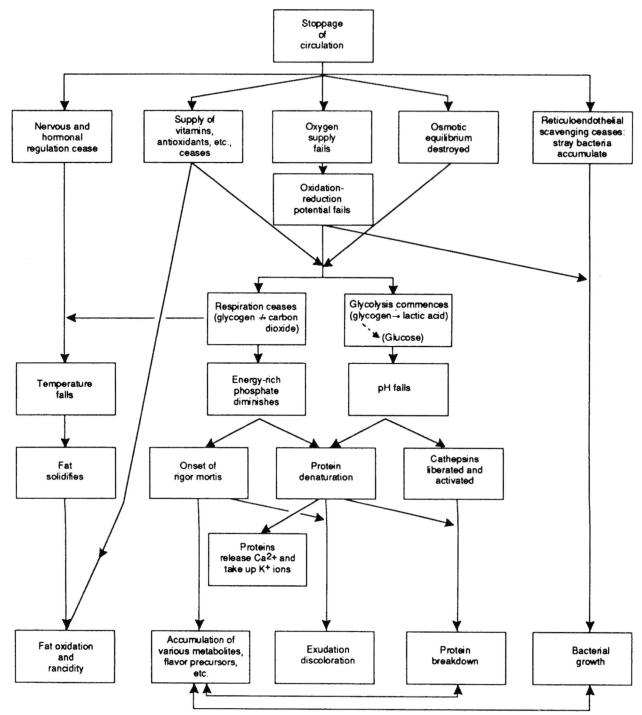

FIGURE 3.13   **The consequences of stoppage of the circulation in muscular tissue. (Lawrie, 1998)** *(Reprinted with permission. Copyright © by Wood Publishing Ltd.)*

conversion of glycogen to lactic acid. Under normal aerobic conditions 39 molecules of ATP are produced for each glucosyl unit of glycogen oxidized compared to only three molecules of ATP for each hexose unit broken down under anaerobic conditions. The time for the first phase development of rigor mortis is determined by the postmortem level of ATP. The level of ATP is also depleted by the non-contractile ATPase activity of myosin which maintains the temperature and structural integrity of the muscle cell (Bendall, 1973). This results in the production of inorganic phosphate, which stimulates the degradation of glycogen to lactic acid. Inorganic phosphate is essential for the phosphorolysis of glycogen to glucose l-phosphate by muscle phosphorylase, which is the initial step in the degradation of glycogen. In addition to the ATPase of myosin, the sarcoplasmic reticulum has ATPase activity.

The level of ATP is maintained in the muscles after death by an active creatine kinase which catalyzes the resynthesis of ATP from ADP and creatine phosphate (Lawrie, 1966; Newbold, 1966). Thus, in the early postmortem or prerigor period, the concentration of ATP remains relatively constant, whereas there is a rapid decline in creatine phosphate levels. In studies with rabbit muscle of well-rested and relaxed animals the creatine phosphate levels were high immediately postmortem but fell rapidly to one-third of the original before any detectable loss in ATP (Bendall, 1951). A rapid fall in creatine phosphate levels was also observed in poultry muscle accompanied by the liberation of free creatine (De Fremery, 1966). The transitory rise in pH in poultry muscle immediately postmortem was attributed to the presence of creatine. Studies by Hamm (1977) on ground beef muscle showed that creatine phosphate was totally degraded within 1–2 hours postmortem.

As discussed earlier, mammalian muscle is capable of maintaining its ATP level for as long as several hours postmortem, compared to fish skeletal muscle, which generally exhibits a rapid decline in ATP levels (Tomlinson and Geiger, 1962). Some species of fish, however, can maintain a constant ATP level but must be in an unexercised state prior to slaughter. The relationship between ATP and creatine phosphate levels in mammalian muscles appears to be similar to that reported in fish skeletal muscle (Partmann, 1965).

The continued activity of the several ATPases in the muscle cell, including in the sarcoplasmic reticulum, mitochondria, sarcolemma, and the myofibrils, presumably contributes to the depletion of ATP in the muscle. Of these, Hamm et al. (1973) concluded that it was the myofibrillar ATPase, rather than membrane or sarcoplasmic reticulum ATPase, which was probably responsible for the degradation of ATP in postmortem skeletal muscle. There is an overall decrease in the ATP level as a consequence of ATPase, activity, a decrease in creatine phosphate, and the inability of postmortem glycolysis to synthesize ATP at an effective rate.

The development of rigor mortis in fish is also related to the reduction in ATP. Depletion of creatine phosphate, adenosine monophosphate (AMP), and glycogen reserves and the subsequent inability to resynthesize ATP results in the formation of the actomyosin complex. This is accompanied by the muscle becoming tough and inextensible. Unlike land animals, rigor mortis in fish terminates far more rapidly. Jones and Murray (1961) reported that the onset of rigor mortis for cod occurred when ATP dropped to 5% of the original level in the rested fish. This was corroborated for a number of Indian fish by Nazir and Magar (1963), although some species were found to enter rigor at much higher phosphate levels, for example, *Mugul dussumieri* and *Harpodon nehereus*. Jones et al. (1965) followed the steady decline in ATP during the postmortem period for cod until the point at which rigor developed. The ATP levels for unexercised and exercised cod muscle were 2.35 and 0.82 μmole/g, respectively, which showed the effect of exercise in determining the ATP levels at which rigor was established. A reduction in the time required for rigor mortis development is also associated with excessive struggling by fish during capture, which has been correlated with a reduction in creatine phosphate. The depletion of ATP in cod prior to the onset of rigor was shown by Fraser et al. (1961) to be 1.25 μmole/g muscle. Iwamoto et al. (1987) found that in spiked plaice (*Paralichthys olivaceus*), the ATP level remained constant for a short time after death as creatine phosphate was degraded prior to ATP breakdown. Using the rigor index method developed by Bito et al. (1983), Cappelin and Jessen (2002) monitored the changes in ATP, IMP, and glycogen in 16 different muscle positions in cod. While the fish were considered in rigor, the dorsal and tail muscles were not in rigor because they were still high in ATP.

B. Postmortem Metabolism of ATP

The development of rigor mortis in animals or fish is a direct response to the decline of ATP. Bendall and Davey (1957) observed that the liberation of ammonia occurred when rabbit voluntary muscle was fatigued or passed into rigor. This was shown to arise from deamination of adenylic acid to inosine-5′-monophosphate (IMP). They postulated the direct deamination of ADP in which ammonia was produced in equimolar proportions to the disappearance of adenosine nucleotides, primarily AMP, during the development of rigor. Tsai et al. (1972) reported the presence of ATP, ADP, and IMP in prerigor porcine muscle and traces of AMP. The levels of ATP and ADP declined rapidly in the postmortem muscle, while the concentration of IMP, inosine, and hypoxanthine increased markedly.

Fraser et al. (1961) reported an increase in ammonia during the resolution of rigor in cod muscle. The postmortem degradation of ATP follows a similar pattern to that in mammalian muscles, in which ATP is rapidly degraded to ADP by the sarcoplasmic ATPase, and hydrolyzed by myokinase to AMP. AMP is then converted to IMP by deaminase action (Saito and Arai, 1958). Nucleotides, particularly IMP, are recognized as important contributors to the flavor of

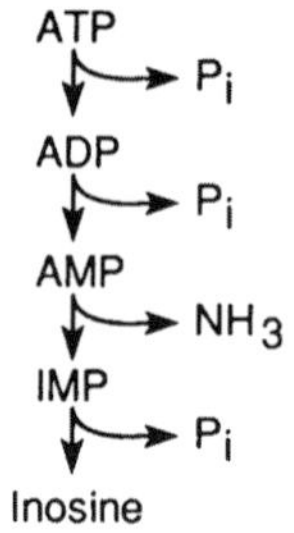

ATP
  └──▶ P$_i$
ADP
  └──▶ P$_i$
AMP
  └──▶ NH$_3$
IMP
  └──▶ P$_i$
Inosine

SCHEME 3.1 Postmortem degradation of ATP to inosine in meat and fish.

good-quality fish (Hashimoto, 1965). Scheme 3.1 summarizes the postmortem degradation of ATP in both fish and meat in which inosinic acid is dephosphorylated to inosine.

IMP and guanosine-5′-monophosphate (GMP) have been reported to enhance the meaty flavor and suppress sulfurous, fatty, burnt, starchy, bitter, and hydrolyzed vegetable type flavors (Wagner *et al.*, 1963; Kuninaka, 1967). Flavor enhancers, such as monosodium glutamate (MSG), IMP, and GMP are natural components of meat and are believed to make a major contribution to meat flavor (Farmer, 1999). They have been shown to improve flavor and have been used by the Japanese for many years to give 'umami' (Reineccius, 1994; Maga, 1994). When using added IMP, Kurtzman and Sjostrom (1964) concluded that canned chicken-containing noodle soup was not flavor enhanced. However, other products evaluated, including canned beef noodle soup, did show improvement with IMP addition.

At equimolar concentrations, IMP has most effect on aroma, but at concentrations related to those reported in the literature for red meat, ribose, IMP, and glucose-6-phosphate were all found to increase 'meaty' and 'roasted' aromas (Farmer *et al.*, 1996). Cambero *et al.* (2000) reported that flavor properties of beef broths were more influenced by the IMP concentration than GMP and AMP concentration and that small non-amino acid nitrogen compounds are more important than peptidic nitrogen with higher molecular weight than 600 Da. In addition to its flavor-potentiating action, IMP has been reported in early studies to be essential for development of meaty aroma during cooking (Batzer *et al.*, 1960; Koehler and Jacobson, 1967). Farmer *et al.* (1996) reported that at four times (340 mg/100 g) the reported concentration in beef and pork, IMP caused highly or very highly significant increases in both 'meaty' and 'roasted' aroma. They also reported that, in most cases, the addition of twice the reported concentration of IMP gave no significant effect. Mottram and Madruga (1994) observed an increase in many sulfur-containing furans, including 2-methyl-3-(methyldithio)furan, on addition of IMP to beef at 10 times its natural concentration. They affirmed that IMP was a precursor for 2-methyl-3-furanthiol and mercaptoketones, but 2-furanmethanethiol was not derived from IMP. IMP appears to act as a source of these thiols via the Maillard reaction and this reaction appears to involve the intermediate formation of 4-hydroxy-5-methyl-3(2H)-furanone and dicarbonyls, such as butanedione and pentane-dione, and their subsequent reaction with hydrogen sulfide or cysteine (Mottram and Madruga, 1994). IMP also appears to suppress the 'green' odor, probably caused by hexanal, and to reduce the formation of *n*-aldehydes and other lipid-oxidation products (Farmer *et al.*, 1996).

Nucleotides and nucleosides are potential precursors of free ribose and ribose phosphate, which have been implicated in Maillard reactions during processing and storage of flesh (Tarr, 1954; Macy *et al.*, 1964). During thermal processing of flesh foods, IMP is degraded into inosine and hypoxanthine, which have been reported to be either flavorless or bitter (Arya and Parhar, 1979).

Macy *et al.* (1970) reported that the concentration of IMP decreases, whereas that of free nucleosides and bases increases during roasting of beef, lamb, and pork. Piskarev *et al.* (1972) reported that both AMP and IMP increase during thermal sterilization of freshly slaughtered meat, but decrease in aged meat. Suryanarayana-Rao *et al.* (1969) reported that IMP, inosine, and hypoxanthine decrease during canning of shrimps, but Hughes and Jones (1966) and Mori *et al.* (1974) reported that hypoxanthine is stable at canning temperatures. Aliani and Farmer (2005a, b) reported IMP to be generally the most abundant nucleotide in commercially available chicken muscle, with average quantities of IMP in chicken breast and leg muscle of 83.7 and 44.6 mg/100 g, respectively. Inosine, an enzymatic breakdown product of IMP, was the second major compound in most chickens (36.2 and 28.5 mg/100 g in breast and leg, respectively), while GMP and hypoxanthine were detected at lower concentrations. AMP was least abundant in most of the chickens analyzed in this study.

The biochemical reactions involved in the formation of these nucleotides, nucleosides, and bases in meat have been studied for several decades.

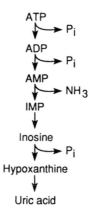

SCHEME 3.2   Degradation of ATP in fish muscle.

Early work by Bendall and Davey (1957) demonstrated that these reactions in meat were catalyzed by ATPase, myokinase, and deaminase at 37 and 17°C. In addition to the reactions shown in Scheme 3.1, the presence of ITP and IDP was reported and attributed to the following reaction:

$$ADP \rightarrow IDP + NH_3$$

$$2\,IDP \rightarrow ITP + IMP$$

Small amounts of inosine and hypoxanthine, degradation products of IMP, were also found in postrigor mammalian muscle. The conversion of ATP to IMP occurred by the time the ultimate pH was reached, while the degradation of IMP followed the establishment of the final pH (Lawrie, 1966).

The degradation of ATP in fish muscle also leads to the formation of IMP, which is subsequently hydrolyzed to uric acid (Saito et al., 1959; Kassemsarn et al., 1963; Tarr, 1966) (Scheme 3.2). The presence of 5'-nucleotidase activity was reported in carp muscle by Tomioka and Endo (1984, 1985).

The level of these nucleotides changes drastically following the death of the fish. Many estimates have been made of fish freshness based on the level of nucleotides (Saito et al., 1959), ammonia (Ota and Nakamura, 1952), amines (Karube et al., 1980), and volatile acids (Suzuki, 1953). Of these, the production of nucleotides from ATP appears to be the most reliable indicator in fish. While inosine is comparatively tasteless, its conversion to hypoxanthine gives rise to a bitter substance (Jones, 1965). Conflicting reports refuting this were subsequently presented by a number of Japanese workers, including Hashimoto (1965), who suggested that hypoxanthine was tasteless. The presence of hypoxanthine in fish muscles was proposed as a chemical index of freshness and quality in fish (Jones et al., 1964; Spinelli et al., 1964). Dugal (1967) suggested that an average rate of hypoxanthine formation could be obtained for a group of fish which would reflect the degree of freshness for a particular species of freshwater fish. Watanabe et al. (1984) estimated fish freshness by monitoring the level of IMP by an enzyme sensor. Karube et al. (1984) developed a multifunctional enzyme sensor system for assessing fish freshness based on measuring the levels of IMP, inosine, and hypoxanthine. This was based on the changes observed in ATP, ADP, and AMP levels in sea bass, saurel, mackerel, and yellowfish following death. Their results, shown in Figure 3.14, indicate a rapid decrease and reduction in ATP and ADP levels 24 hours after death as well as a drop in AMP to less than 1 µmole/g.

IMP increased sharply during the first 24 hours postmortem and then decreased gradually, accompanied by a rise in inosine and hypoxanthine. The changes in these nucleotides varied with the individual fish species. Since ATP, ADP, and AMP were still present in some of the fish varieties for up to 2 weeks they included these nucleotides with hypoxanthine, inosine, and IMP in the overall equation defining fish freshness.

Saito et al. (1959) first proposed the term 'K value' as an indicator of fresh fish defined as the ratio of inosine plus hypoxanthine to the total amount of ATP-related compounds. The K value has since been used to express freshness of marine products (Lee et al., 1982; Uchiyama and Kakuda, 1984). Ryder (1985) developed a rapid method for computing K values based on the quantitative measurement of ATP and its degradation products using high-performance liquid chromatography (HPLC). Surette et al. (1988) monitored the postmortem breakdown of ATP-related compounds in Atlantic cod (*Gadus morhua*) and reported that inosine hydrolysis and hypoxanthine formation resulted from both autolytic and bacterial enzyme activity. These nucleotide catabolites provide a useful index of quality as their presence is affected by spoilage bacteria and mechanical damage during handling. These researchers

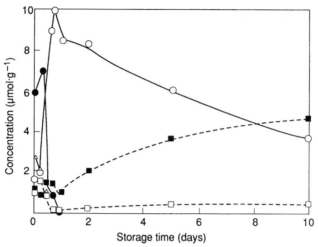

FIGURE 3.14   **Time-course of ATP decomposition and associated reactions in sea bass.** ●: ATP; △: ADP; □: AMP; ○: IMP; ■: HxR + Hx. *(Reprinted with permission from Karube* et al.*, 1984. Copyright © by the American Chemical Society.)*

suggested that a diagnostic kit for measuring catabolites, such as inosine monophosphate (IMP), inosine (INO), and hypoxanthine (Hx), could provide a useful tool for assessing the fresh quality of cod.

$$K = \frac{[\text{Inosine}] + [\text{Hypoxanthine}] \times 100}{[\text{ATP}] + [\text{ADP}] + [\text{AMP}] + [\text{IMP}] + \text{inosine} + \text{hypoxanthine}}$$

G and *P* values were derived from the K value. Burns *et al.* (1985) defined the *G* value as an index for evaluating the shelf-life of lean fish during ice storage, while a second quality indicator, *P*, was introduced to serve as an indicator of spoilage during the early stages of chilled storage.

$$P = \frac{\text{Inosine} + \text{Hypoxanthine}}{[\text{AMP}] + [\text{IMP}] + \text{inosine} + [\text{Hypoxanthine}]}$$

The choice of one indicator over another depends on the degradation patterns of the metabolites (Shahidi *et al.*, 1994).

The applicability of the *K* value for assessing the freshness of edible meat was reported by Nakatani *et al.* (1986). By monitoring the changes in ATP degradation products in beef and rabbit muscles during cold storage they proposed the following new index K_0, in which adenosine is represented by AD and xanthine by X:

$$\frac{\text{Inosine} + \text{hypoxanthine} + \text{xanthine} \times 100}{\text{ATP} + \text{ADP} + \text{AMP} + \text{IMP} + \text{AD} + \text{INO} + \text{Hx} + \text{X}}$$

Subsequent research by Fujita *et al.* (1988) found that this index could also be used to assess the freshness of both pork and chicken.

Most studies dedicated to understanding the ATP degradation pathway by analyzing the resulting products failed to report the concentrations of ribose and/or ribose-1-phosphate (R1P), which may result from the breakdown of inosine to hypoxanthine, and ribose-5-phosphate (R5P), which can be a product of a direct IMP breakdown to hypoxanthine (Lee and Newbold, 1963). These potential breakdown reactions of IMP are presented in Figure 3.15. It has been suggested that in beef, sugars may contribute to the sweet flavors and aromas upon heating (MacLeod, 1994).

The lack of information on natural concentrations of these key sugars in meat might be explained by the fact that most analytical methods used in recent years were adequate for nucleotides and not for sugars. In order to overcome this problem, a postcolumn derivatization method was employed (Aliani and Farmer, 2002) to determine sugar and sugar phosphates in chicken breast (24.7 mg) and leg (14.1 mg/100 g) from five different local commercial suppliers. Ribose and R5P have been shown to be important in aroma development in heated model systems and also contribute to the desirable browning reactions in meat. The addition of small quantities of ribose to raw meat has been shown to increase the quantities of key odor compounds as well as desirable roasty and meaty notes in cooked meat (Farmer *et al.*, 1999). Aliani *et al.* (2005a, b), using different sensory studies, reported the importance of ribose as a key flavor precursor in chicken meat and suggested that a two- to four-fold increase in the natural concentration of ribose (25 mg/100 g wet weight) may be sufficient to significantly improve the roasted and chicken odors in cooked chicken breast (Aliani and Farmer, 2005a, b).

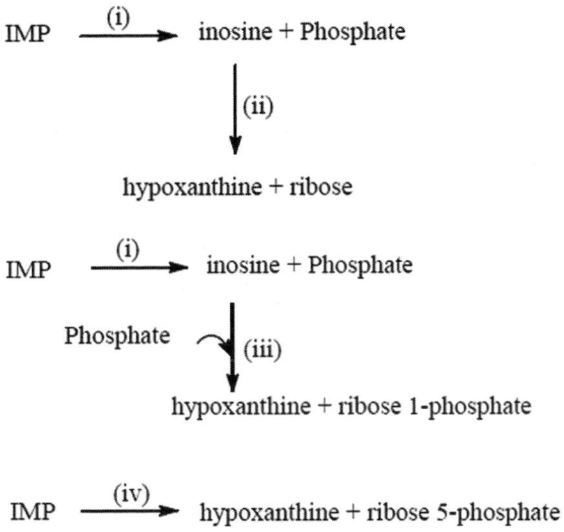

IMP $\xrightarrow{\text{(i)}}$ inosine + Phosphate

$\Big\downarrow$ (ii)

hypoxanthine + ribose

IMP $\xrightarrow{\text{(i)}}$ inosine + Phosphate

Phosphate $\rightharpoondown$ (iii)

hypoxanthine + ribose 1-phosphate

IMP $\xrightarrow{\text{(iv)}}$ hypoxanthine + ribose 5-phosphate

FIGURE 3.15   Degradation of IMP by three potential pathways. *(Adapted from Lee and Newbold, 1963)*

Several sugars and sugar phosphates, such as ribose (Aliani and Farmer, 2005b; Mottram and Nobrega, 1998), R5P (Mottram and Nobrega, 1998), glucose, glucose-6-phosphate (G6P) (Farmer *et al.*, 1999), mannose, and fructose (Aliani and Farmer, 2005b; Madruga *et al.*, 2010), are usually found in most types of meat such as chicken (Aliani and Farmer, 2005a), beef (Koutsidis *et al.*, 2008), pork (Meinert *et al.*, 2009), and goat (Madruga *et al.*, 2010), and are believed to display flavor-generating potential (Meinert *et al.*, 2009).

Postmortem conditioning contributed to an increase in ribose concentrations and is likely to have a major effect on flavor formation (Koutsidis *et al.*, 2008). Ribose significantly increased from conditioning day 1 to day 21 at 4°C in beef longissimus lumborum muscle. Ribose has been reported to be the most heat-labile sugar, while fructose is the most stable (Macy *et al.*, 1964). Therefore, ribose is likely to be altered or breakdown during high temperatures. Ribose can be formed from the degradation of R5P (Aliani *et al.*, 2008). R5P has been shown to undergo Maillard reactions significantly more rapidly than similar sugars and sugar phosphates (Sandwick *et al.*, 2005). Madruga *et al.* (2010) reported that all sugars present in goat meat decreased during cooking, with fructose exhibiting the highest percentage loss of 66%. This can be attributed to the involvement of sugars in the Maillard reaction, which produces several pentose and hexose degradation products containing carbonyl groups which are main reactants for the formation of important heterocyclic compounds such as pyrazines, thiazoles, and pyridines in meat volatiles (Madruga *et al.*, 2010). The odors obtained from these reactions are dependent on the amino acid; however, the nature of the sugar is said to dictate the rate of the reaction (Kiely *et al.*, 1960).

C. Adenosine Nucleotides and Protein Denaturation

During frozen storage, deteriorative changes in fish texture have been reported as a consequence of protein denaturation (Dyer, 1951; Andou *et al.*, 1979, 1980; Acton *et al.*, 1983). The possible effect of adenosine nucleotides on protein denaturation was recently investigated by Jiang *et al.* (1987). These researchers assessed protein denaturation in fish frozen at $-20°C$ by extractability of actomyosin (AM) and monitored the activities of Ca-ATPase and Mg(EGTA)-ATPase [EGTA = ethylene glycol *bis*(2-aminoethylether)tetraacetic acid] in AM. During frozen storage the molecular weight of the myosin heavy chain and actin decreased. The least stable muscle was associated with the lowest levels of ATP, ADP, AMP, and IMP and the highest levels of inosine and hypoxanthine. A correlation of -0.80 was obtained between inosine and hypoxanthine and the Ca-ATPase total activity of AM compared to $+0.78$ for ATP, ADP, AMP, and IMP (Figures 3.16 and 3.17). These results point to the possible involvement of adenosine nucleotides with protein denaturation.

The rate of ATP breakdown was also found to be much faster in muscles exhibiting pale, soft, and exudative (PSE) or dark, firm, and dry (DFD) conditions as they had much lower levels of ATP and higher levels of IMP compared to normal meats (Tsai *et al.*, 1972; Lundstrom *et al.*, 1977; Essen-Gustavson *et al.*, 1991). A method for the early postmortem detection of PSE and DFD in pork was developed by Honikel and Fischer (1977) using the *R* value (the ratio of the absorption at 250 and 200 nm). Batlle *et al.* (2000) detected the PSE condition in longissimus dorsi muscle

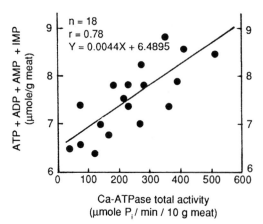

FIGURE 3.16   Relationship between the sum of the quantity of adenosine triphosphate (ATP), adenosine diphosphate (ADP), adenosine monophosphate (AMP), and inosine monophosphate (IMP) and Ca-ATPase total activity. *(Reprinted with permission from Jiang* et al., *1987. Copyright © by the American Chemical Society.)*

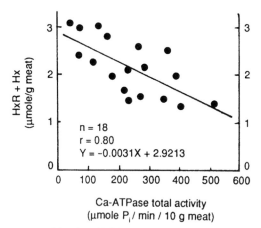

FIGURE 3.17   Relationship between the content of inosine (HxR) and hypoxanthine (Hx) and Ca-ATPase total activity. *(Reprinted with permission from Jiang* et al., *1987. Copyright © by the American Chemical Society.)*

of pork meats by measuring ATP-related compounds 2 hours postmortem. PSE meats were significantly lower ($p < 0.05$) in ATP, but higher in AMP, IMP, and hypoxanthine compared to normal meat. Measurement of K_0 values, R' values, and the ratios of IMP/ATP all proved useful indicators of PSE.

D. Postmortem Glycolysis

Once the supply of oxygen to the muscle tissue is depleted, glycogen, the main carbohydrate of animal and fish muscle, undergoes anaerobic glycolysis to lactic acid. Compared to mammalian muscle, the level of glycogen in fish muscle is reported to be much lower. Tomlinson and Geiger (1962), however, found a close similarity between the muscle glycogen levels for many species of both fish and warm-blooded animals. This was attributed to the excessive struggling normally associated with the capture of fish, resulting in depletion of the glycogen level compared to that in the rested fish.

Postmortem degradation of glycogen in fish muscle suggests that two possible pathways are involved: the hydrolytic or amylolytic pathway, and the phosphorolytic pathway. These are illustrated in Scheme 3.3.

The postmortem conversion of glucose-6-phosphate to glucose by phosphomonoesterase only occurs to a slight extent in fish muscle. Consequently, the hydrolytic pathway appears to be the main one operating in fish. This pathway was first postulated by Ghanekar *et al.* (1956) and has since been confirmed as the main degradative pathway of glycogen to glucose for most fish (Tarr, 1965; Burt, 1966; Nagayama, 1966). In mammalian muscle, however, it is

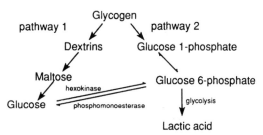

SCHEME 3.3 Postmortem degradation of glycogen.

the phosphorolytic pathway that is responsible for glycogen degradation. When stressed, catecholamines released from sympathetic neurons and adrenal medulla accelerate glycogen degradation through activation of glycogen phosphorylase. The activated phosphorylase rapidly removes glucose-1-P units from glycogen molecules (Poso and Puolanne, 2005).

Irrespective of which pathway is involved in the initial breakdown of glycogen, the final pathway of glycolysis is the same for either animal or fish muscle. The enzymes responsible have, in the main, been characterized and identified in mammalian muscle, with many since reported in many species of fish, including rainbow trout (MacLeod *et al.*, 1963; Tarr, 1968). The general reactions involved in the glycolytic pathway are outlined in Scheme 3.4.

The rate of postmortem glycolysis in muscles is affected by temperature, muscle fiber type, and hormone secretions, as well as the intensity of the nervous stimuli in the muscle before and during slaughter (Beecher *et al.*, 1965; Disney *et al.*, 1967; Tarrant *et al.*, 1972a, b; Bendall, 1973). The effect this has on the pH of the muscle will be discussed in the next section.

E. Postmortem pH

The production of lactic acid causes the pH of the muscle tissue to drop from the physiological pH of 7.2–7.4 in warm-blooded animals to the ultimate postmortem pH of around 5.3–5.5. A direct relationship was demonstrated by

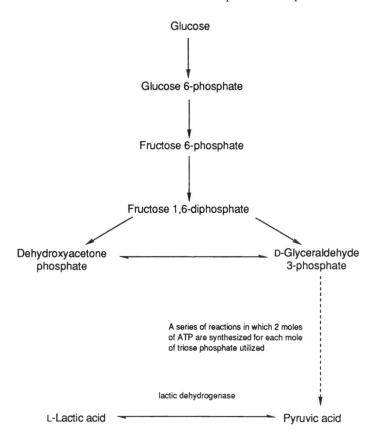

SCHEME 3.4 Metabolism of glucose to lactic acid.

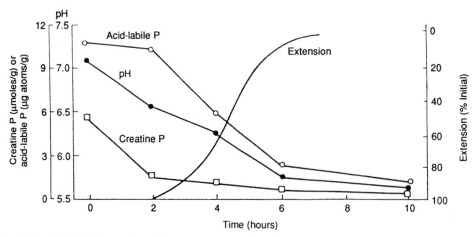

FIGURE 3.18   Chemical and physical changes in beef sternomandibularis muscle held at 37°C. Extension changes were recorded on an apparatus similar to that described by Bate-Smith and Bendall (1949) using a load of about 60 g/cm² and a loading–unloading cycle of 8 minutes on and 8 minutes off. Zero time: 1 hour 45 minutes postmortem. *(Newbold, 1966)*

Bate-Smith and Bendall (1949) between the rate of fall of pH in postmortem rabbit muscle and lactic acid production. It is particularly important to attain as low a pH as possible in the tissue, since in addition to retarding the growth of spoilage bacteria, it imparts a more desirable color to the meat. In the case of frozen fish, however, a higher pH is more desirable to prevent toughness. The final pH can be attained within the first 24-hour postmortem period, the glycolytic pathway being related to ATP production, the net fall of which is directly responsible for the development of rigor mortis. The interrelationship between the creatine phosphate disappearance, fall in the levels of ATP and pH, and the decrease in extensibility as a measure of rigor mortis is shown in Figure 3.18. ATP is the major source of acid-labile phosphorus, while the fall in pH is a measure of glycolysis.

A postmortem pH of 5.3–5.5 is attained in the muscles of well-rested animals fed just before slaughter, when glycogen is at a maximum level. Animals that undergo severe death struggling, however, are fatigued prior to slaughter and characterized by lower glycogen levels, as illustrated in Table 3.2 for chicken breast muscle.

The lower glycogen levels result in a higher final postmortem pH of around 6.0–6.5, producing a dark, dry, and close-textured meat that is much more susceptible to microbial spoilage (Cassens, 1966; Joseph, 1968a, b). This meat, referred to as DFD, represents a serious quality problem, particularly in beef (Tarrant, 1981), and is discussed further in Section K. The final postmortem pH in meat rarely falls below 5.3, although several exceptions have been reported. For example, in pig longissimus dorsi muscle, pH values ranging from 4.78 to 5.1 have been recorded (Lawrie *et al.*, 1958). Meat with a pH of 5.1–5.5 was found to be in an exudative condition with a whitish color and loose texture, while meat at pH 4.78 had abnormal muscle fibers. This is not unexpected, as the isoelectric point of the major meat proteins is around pH 5.5, which would lead to a loss of water-holding capacity (WHC).

The lactic acid concentration in fish is similarly dependent on the initial glycogen stores before death as well as on the treatment of the fish. Fish muscles have been reported to have lactic acid concentrations ranging from 0.29% in haddock (Ritchie, 1926) to 1.2–1.4% in tuna (Tomlinson and Geiger, 1962) during rigor mortis. Most fish exhibit a higher postmortem pH compared to warm-blooded animals of around 6.2–6.6 at full rigor. An exception is in the

TABLE 3.2 Effect of Slaughter Conditions on the Glycogen Concentration 3 Minutes Postmortem in Chicken Breast Muscle

Condition	Initial Glycogen Level (mg/g)
Anesthetized	8.3
Stunned	6.0
Struggling	3.4

Adapted from De Fremery (1966).

case of flatfish, where a final pH of 5.5, similar to that of mammals, has been reported. The struggling of fish during capture substantially depletes the glycogen stores, resulting in a high pH at rigor of around 7.0, giving rise to a condition known as 'alkaline rigor'. This condition was reported in cod as well as other fish species (Fraser *et al.*, 1961). A review by Wells (1987) noted that the capture, transportation, and handling of live fish is accompanied by substantial biochemical and physiological changes. The extent of such changes was dependent on species as well as environmental conditions. Eliminating stress in fish by allowing them to return to a resting condition during captivity prior to death should be the ultimate goal to avoid the abnormal pH changes incurred during struggling.

Low ultimate pH has also been associated with textural problems in fish such as halibut, Alaska pollack, and tuna (Patashnik and Groninger, 1964; Konagaya and Konagaya, 1979; Suzuki, 1981). Love (1975) noted that while low ultimate pH produced a tough texture, high ultimate pH resulted in a 'sloppy' soft texture in Atlantic cod (*Gadus morhua*). This condition rendered the fish unfit for filleting and produced a poor frozen product (Love *et al.*, 1982; MacCallum *et al.*, 1967). Postmortem changes in soft-textured cod caught off Newfoundland were examined by Ang and Haard (1985), who reported the lowest ultimate pH in cod that had been feeding heavily on capelin before capture. Atlantic cod do not feed during the months prior to spawning in May and June but feed intensely during the postspawning period. These researchers found that the muscles of cod caught during this intense feeding period were characterized by a persistent low ultimate pH up to 100 hours following rigor. Ang and Haard (1985) suggested that the altered metabolic state as a result of the heavy feeding caused the low and stable ultimate pH in the cod muscle, which was responsible for the soft texture in these fish.

F. Time-Course of Postmortem Glycolysis

The ultimate postmortem pH is dependent on the physiological state of the muscle, the type of muscle, and the species of animal or fish studied. The different rates of fall in pH with time are shown in Figure 3.19 for three species of animal.

Several postmortem changes observed visually in meat are related to the rate of decline in pH and temperature. For example, a rapid decline in pH in beef held at around body temperature resulted in changes in color, decreased WHC, and some muscle protein denaturation (Scopes, 1964; Chaudhry *et al.*, 1969; Lister, 1970; Follett *et al.*, 1974; Locker and Daines, 1975). Earlier work by Cassens (1966) and Briskey *et al.* (1966) studied the rate of glycolysis in porcine muscle by monitoring the decline in pH. In addition to the physiological state of the muscle they observed that certain pigs were predisposed to a rapid postmortem glycolysis. The properties of the meat associated with a fall in pH are summarized in Table 3.3.

Similar changes occur in fish muscle as the pH declines. A low postmortem pH was associated with poor fish texture, low WHC, and high drip loss (Kelly, 1969). A high water content in cod muscle was identified by Love (1975) in those fish with high postmortem pH, which correlated best with the texture of the cooked fish.

G. Effect of Temperature on Postmortem Glycolysis: Cold Shortening

The rate of postmortem glycolysis varies with temperature, as evident from differences in the final pH of mammalian muscles (Marsh, 1954; Cassens and Newbold, 1966; Newbold, 1966; Newbold and Scopes, 1967). These researchers

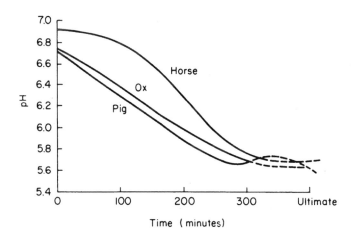

FIGURE 3.19 The effect of species in a given muscle (longissimus dorsi) and at a given temperature (37°C) on the progress of glycolysis measured by a fall in pH (Lawrie, 1966). *(Reprinted with permission. © Wood Publishing Ltd.)*

TABLE 3.3 Relationship Between Type of Fall in pH and Meat Properties

Final pH	Type of Decrease	Properties of Meat
6.0–6.5	Slow, gradual	Dark
5.7–6.0	Slow, gradual	Slightly dark
5.3–5.7	Gradual	Normal
5.3–5.6	Rapid	Normal to slightly dark
5.0	Rapid	Dark to pale but exudative
5.1–5.4 then up to 5.3–5.6	Rapid	Pale and exudative

Adapted from Cassens (1966).

all reported the hastening of rigor mortis as temperature is reduced from 5 to 1°C due to increased glycolytic activity and ATP hydrolysis. Stimulation of contractile actomyosin ATPase appeared to be potentiated by the release of Ca^{2+} ions. This phenomenon, known as 'cold shortening', results in toughening of cooked meat (Marsh and Leet, 1966). A 30–40-fold increase in the level of ionic calcium was reported by Davey and Gilbert (1974) in the myofibril region of beef muscle held at 0°C compared to 15°C. Jeacocke (1977) examined the relationship between temperature and postmortem pH decline in beef sternomandibularis muscle. The results illustrated in Figure 3.20 indicate a minimum fall in pH over 10–12°C which increased as the temperature dropped to 0°C, characteristic of cold shortening. This was attributed to an increase in glycolysis due to enhanced ATPase activity in contractile actomyosin. Cornforth *et al.* (1980) subsequently confirmed earlier research by Buege and Marsh (1975) that the mitochondrial content of the muscle was involved in cold shortening. These researchers also proposed a role for the sarcoplasmic reticulum in the reversibility of this phenomenon. This was attributed to the possible effect of temperature on the membrane of the sarcoplasmic reticulum and the subsequent release of Ca^{2+} ions.

Honikel *et al.* (1983) identified two types of shortening taking place in beef muscle. One which occurred above 20°C was referred to as 'rigor shortening' while the other taking place below 15°C was termed 'cold shortening'. In both cases muscle contraction was explained by the release of Ca^{2+} ions into the myofibrillar space in the presence of

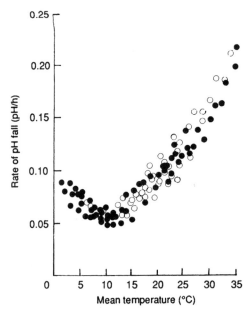

FIGURE 3.20 The rate of pH fall in beef sternomandibular muscle as a function of the mean temperature of the adjacent thermocouple junction. The results of 16 different temperature gradients are pooled. (●) Muscles vacuum-packed before insertion into the apparatus; (○) muscles not vacuum-packed. *(Jeacocke, 1977)*

adequate levels of ATP. The uptake of Ca^{2+} ions by the sarcoplasmic reticulum was particularly sensitive to both pH and temperature changes. For example, Honikel *et al.* (1983) found that rigor shortening commenced at pH 6.25 in the presence of 2.4 µmole ATP/g muscle. This represented optimum conditions for the uptake of Ca^{2+} ions by the sarcoplasmic reticulum as observed previously by Cornforth *et al.* (1980) and Whiting (1980). The myofibrillar Mg/Ca-ATPase activity was reported by Bendall (1969) to be independent of pH between 6 and 7. Both theories explain the development of rigor shortening at temperatures of 20°C and higher and at pH below 6.3. In sharp contrast, however, cold shortening developed at pH 7 in the presence of full ATP concentration (4 µmole/g) in the muscle. The occurrence of cold shortening was attributed by Cornforth *et al.* (1980) to the combined effect of the release of Ca^{2+} ions from the muscle mitochondria and the reduced uptake of Ca^{2+} ions by the sarcoplasmic reticulum.

Herring *et al.* (1965) first showed a relationship between cold shortening and sarcomere length. The potential for cold shortening varied with muscle types as red muscle was more susceptible to cold shortening (Bendall, 1973). As a result, pork is less susceptible than beef and lamb to cold shortening, as it is composed mostly of white fibers.

The development of cold shortening is highly undesirable and can be avoided by holding the meat at a minimum temperature of 15°C until the pH drops below 6.0. Lamb carcass, however, should be held for at least 16 hours to ensure that prerigor changes have been completed (McCrea *et al.*, 1971). This represents a delay for the meat-processing industry, which utilizes hot-deboning of beef carcasses which are cut and rapidly refrigerated below 15°C long before the pH falls below 6.0. One technique for rapidly reducing the pH of the carcasses to below 6.0 involves the use of electrical stimulation to accelerate postmortem glycolysis, as discussed in the next section.

H. Effect of Electrical Stimulation on Postmortem Glycolysis and Tenderness

Electrical stimulation of muscle has long been known to accelerate postmortem glycolysis and hasten the onset of rigor (Harsham and Detherage, 1951; De Fremery and Pool, 1960; Hallund and Bendall, 1965; Forrest and Briskey, 1967). Carse (1973) prevented cold shortening by subjecting freshly slaughtered lamb carcasses to 250 V pulses and attained pH 6.0 within 3 hours compared to 15.4 hours for the unstimulated carcass. This technique facilitated accelerated conditioning of lambs and is commercially used in New Zealand. Similar results were also reported in lamb by Bendall (1976) and Chrystall and Hagyard (1976). The latter researchers monitored the progress of glycolysis in stimulated and unstimulated lamb longissimus dorsi muscles by following the change in pH, as shown in Figure 3.21. Electrical stimulation substantially accelerated postmortem glycolysis, with the final pH 5.5 attained within 8 hours for the stimulated carcass compared to 24 hours for the control. Chrystall and Devine (1978) noted that during electrical stimulation, glycolysis was stimulated as much as 150-fold, which resulted in a marked drop in pH. Even following cessation of the stimulus, the rate of glycolysis can increase by as much as three-fold. The initial effect was attributed by Newbold and Small (1985) to activation of glycogen phosphorylase, which reached a peak following 30 seconds of electrical stimulation. The increased glycolysis and ATP turnover following cessation of electrical stimulation, however, remains unexplained.

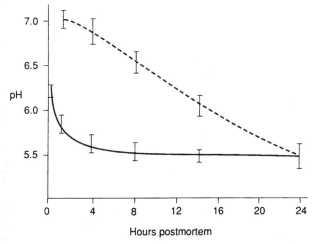

FIGURE 3.21   Time-course of pH fall in longissimus dorsi muscles of stimulated (− −) and unstimulated (——) animals. Standard deviation is shown by the vertical bar. *(Chrystall and Hagyard, 1976)*

Horgan and Kuypers (1985) examined postmortem glycolysis in rabbit longissimus dorsi muscles following high-voltage and low-voltage stimulation. As expected, there was an increase in the fall of pH as well as in, for example, phosphorylase activity following electrical stimulation. This was attributed to an increase in phosphorylase kinase activity and a substantial loss in phosphorylase and phosphatase activities. The yield of sarcoplasmic reticulum was reduced, suggesting an increase in the release of Ca^{2+} ions from the stimulated muscle. However, whether the results were due to physical disruption or to fast glycolysis was not determined, as later work by Hopkins and Thompson (2001) reported that free calcium concentration was a function of pH decline. The biochemical and physical effects of electrical stimulation on beef and sheep meat were reviewed by Hwang *et al.* (2003).

Electrical stimulation has been reported to improve tenderness in meat (Savell *et al.*, 1977; McKeith *et al.*, 1980). Its effect was attributed to several factors, including reduction of cold shortening capacity, myofibril fragmentation, and enhanced activity of acid proteases caused by lowering of the pH (Davey *et al.*, 1976; Chrystall and Hagyard, 1976; Savell *et al.*, 1977, 1978). Sonaiya *et al.* (1982) subjected one side of a cow carcass (semimembranosus, longissimus dorsi, and triceps brachii) to electrical stimulation and monitored the pH, temperature, and myofibril fragmentation index over a period of 1 week following treatment. A significant drop in pH occurred in the treated muscles compared to the corresponding untreated muscles. A higher myofibril fragmentation index (MFI) was observed for the electrically stimulated muscles, which is normally associated with the degradation of troponin T and the Z-disk by the calcium-activated factor. However, the appearance of the 30 kDa protein, normally associated with CAF action, did not reach a maximum level until 72 hours post-stimulation, by which time the MFI had decreased. It was apparent from this study that electrical stimulation did not enhance CAF activity as might have been expected. Since the degradation of troponin T was brought about primarily by the action of cathepsins, the effect of electrical stimulation was attributed to the enhanced activity of these proteolytic enzymes at the low pH.

Concern about installation costs and operator safety associated with high-voltage electrical stimulation of meat led to more frequent use of low-voltage electrical stimulation (LV-ES, voltage < 100 V) in many countries. Such treatment has been reported to improve tenderness in beef (Taylor and Marshall, 1980; Aalhus *et al.*, 1994). Chrystall *et al.* (1980), working with lambs, recommended that for efficiency, LV-ES treatment be carried out within a short time after slaughter. Mixed results reported on beef tenderness using LV-ES ranged from none to a slight improvement and even negative changes, including poorer quality (Unruh *et al.*, 1986; Redbottom *et al.*, 2001; Hwang *et al.*, 2003). Different muscles were reported to respond differently to stimulation on the same carcass (Devine *et al.*, 1984). A study by Rhee and Kim (2001) examined the effect of LV-ES and temperature conditioning at 2, 16 and 30°C for 3 hours on glycolysis and calpain activity in the loin (m. longissimus) obtained from the Korean native cattle (Hanwoo). LV-ES appeared more effective in accelerating the rate of glycolysis in longissimus compared to conditioning at 30°C, although both accelerated the decline in pH and depletion of ATP and glycogen. The activity of the calpains was found to decrease over 24 hours. The best treatment for accelerating glycolysis and improving calpain activity proved to be a combination of LV-ES and conditioning at 30°C. Hollung *et al.* (2007) later showed that LV-ES not only affected the muscles within the beef carcass differently, but also had a variable effect on Warner—Bratzler (WB) shear force and pH decline, in the same muscle from different animals. Some animals showed no effect by LV-ES on pH decline, while WB shear force values were improved. Thus, LV-ES may be responsible for other mechanisms, besides accelerated pH decline and cold shortening prevention, that must play an important role in meat tenderness.

A novel approach to enhancing meat quality in Australia is based on medium-voltage electrical stimulation (Pearce *et al.*, 2009). The alternating frequencies enhanced stimulation of the muscle, making it more tender than the untreated control after 30 days. No detrimental effects were observed for either drip loss or retail color display.

I. Prerigor Pressurization

Prerigor pressurization is another accelerated procedure for processing meat that has been developed (Macfarlane, 1973; Kennick *et al.*, 1980; Elkhalifa *et al.*, 1984a, b). Macfarlane (1973) was the first to tenderize prerigor meat using high-pressure treatment (103 MPa, 30—35°C for 1—4 minutes) and observed a rapid fall in pH as well as improved tenderness ratings. Since prerigor is rarely used, they examined the effect of high pressure on postrigor beef tenderness, which is discussed in Section P.

J. Glycolytic Enzymes

Investigations were conducted to determine the first glycolytic enzyme inhibited as the pH falls, as glycolysis ceases at pH values much higher than 5.3. Newbold and Lee (1965) found that phosphorylase was the limiting enzyme in minced sternomandibular muscle diluted with an equal volume of 0.16 M potassium chloride, which was consistent with earlier studies by Briskey and Lawrie (1961). Kastenschmidt *et al.* (1968) examined the metabolism of pig longissimus dorsi muscle and confirmed that phosphorylase was the primary control site in postmortem glycolysis. Phosphorylase is one of the key enzymes in glycolysis present in muscles (Scopes, 1970). In addition to this enzyme, Kastenschmidt *et al.* (1968) implicated phosphofructokinase and pyruvic kinase in glycolytic control. On the basis of these and related studies, the muscles in pigs were classified as 'fast' or 'slow' glycolyzing muscles (Briskey *et al.*, 1966; Kastenschmidt *et al.*, 1968).

K. Pale Soft Exudative and Dark Firm Dry Conditions

Meat quality defects such as PSE and DFD result in unattractive meats that are rejected by consumers (Viljoena *et al.*, 2002). The estimated percentage of PSE and DFD reported for meat in the USA was 16 and 10%, respectively (Cassens *et al.*, 1992). Other countries, including Canada, Portugal, and the UK, reported higher incidences (Fortin, 1989; Santos *et al.*, 1994). Surveys indicate that this problem is increasing rather than decreasing (Wariss, 2000). Associated with these phenomena are the considerable and continuing financial losses suffered by the meat industry in many countries (Adzitey and Nurul, 2011).

1. PSE

The PSE condition, mainly associated with porcine muscles, is due to the rapid drop in pH, to 5.3–5.8 within 1 hour following death, while the muscle temperature is still above 36°C. Although a considerable amount of research has been conducted on PSE, it remains a significant meat quality problem in the pork industry (Barbut *et al.*, 2008). The pale color, soft texture, and poor WHC of PSE meat has caused serious economic losses for the industry (Wang *et al.*, 2005). Early research showed that a combination of high temperature and low pH caused partial denaturation of muscle proteins (Bendall and Lawrie, 1964; Charpentier, 1969; Goutefongea, 1971). The effect of high temperature and low pH on the properties of phosphorylase was examined by Fischer *et al.* (1979); these conditions caused the denaturation of phosphorylase, with the resultant loss of activity, as well as decreased solubility in PSE muscles.

The development of PSE meats has been shown to be stress related. This is prevalent in pigs that are genetically sensitive to stress, which suffer acute preslaughter stessors, although it can occur in normal pigs as well (Topel *et al.*, 1969; Oliver *et al.*, 1988; Honkavaara, 1989; Kuchenmeister *et al.*, 2000). This phenomenon, once characterized as porcine stress syndrome, was shown to be linked to a single autosomal recessive gene. This caused aberrant calcium metabolism in postmortem muscle due to a mutation in the 615 amino acid (Arg615Cys) of the sarcoplasmic reticulum Ca^{2+} release channel. As a result, twice the amount of calcium is released from the sarcoplasmic reticulum compared to normal muscle (Mickelson and Louis, 1996). Pigs with this stress susceptibility were also reported by Kuchenmeister *et al.* (1999) to have a diminished uptake of Ca^{2+} in postmortem muscle. The increased release of sarcoplasmic Ca^{2+} was responsible for enhancing muscle metabolism by accelerating the production and accumulation of lactate in postmortem muscle. Gispert *et al.* (2000) conducted a complete survey of preslaughter conditions, including delivery from the farms to slaughter and frequency of halothane genotypes, on the development of PSE in five Spanish abattoirs. This led to a study by Guardia *et al.* (2004), who identified and assessed both environmental and genetic risk factors for PSE meat under commercial Spanish conditions. Using a polychotomous logistic regression model, they found that a decreased risk of PSE meat with transportation time depended on the stocking density. For example, the risk of PSE meat increased with stocking density for transportation times greater than 3 hours, while the reverse was true for shorter transits.

A similar condition was reported in beef muscle, described as pale, watery beef. Fischer and Hamm (1980) studied postmortem changes in fast-glycolyzing muscles of beef. Significant correlations were obtained in which a lower muscle pH was associated with lower WHC and lower glycogen and higher lactate levels. It was evident that phosphorylase activation occurred in fast-glycolyzing muscles, although the overall effect on beef quality was far less severe than in PSE pork.

Recent reports also identified the development of pale and exudative conditions in poultry meat around the world, characterized by a high L^* value and low WHC (Petracci *et al.*, 2009). Increasing evidence suggests that exposure of birds to thermal stress before slaughter was one factor responsible for PSE in broilers (Barbut, 1998; Mitchell and Kettlewell, 1998). The biological cause of broiler PSE is considered to be the excessive release of Ca^{2+} as a result of a genetic mutation of ryanodine receptors, a class of intracellular calcium channels located in the sarcoplasmic reticulum. Wilhelm *et al.* (2010) found that the 24-hour postmortem pectoralis major muscle classified as PSE had lower pH, WHC, and shear force values than the control samples. Higher protease activity in the PSE samples, particularly calpain activity, was also observed, which contributed to the eventual collapse of the sarcomere structure. Further work on pectoralis major muscles in turkey by Zhu *et al.* (2011) demonstrated the detrimental effects of elevated postmortem temperate (40°C) and glycolysis, which caused the denaturation of phosphorylase, leading to PSE turkey meat.

2. DFD

Improper handling of cattle before slaughter, causing physiological stress and exhaustion, is the major cause of DFD (Grandin, 1980; Tarrant, 1981). Excessive physical exercise in animals depletes the muscle glycogen, although other factors such as fasting, trauma, and psychological stress have also been implicated (Howard and Lawrie, 1956; Bergstrom and Hultman, 1966; Conlee *et al.*, 1976; Sugden *et al.*, 1976). Tarrant and Sherrington (1980) monitored the final postmortem pH of steer and heifer carcasses at a slaughtering plant in Ireland over a 3-year period. A seasonal effect was noted for the development of DFD with an average incidence of 3.2% in the carcasses examined. Measurement of the ultimate pH of the meat remained the best method for characterizing this phenomenon. The pH limits reported for the development of DFD ranged from 5.8–5.9 as the lower limit for normal meat to 6.2–6.3 as the upper limit for extreme DFD (Fjelkner-Modig and Ruderus, 1983; Tarrant, 1981). The high pH conditions result in negligible denaturation of proteins so that the water remains tightly bound with little or no exudate (Warriss, 2000). The incidence of DFD in Sweden was examined by Fabiansson *et al.* (1984), who classified beef carcasses with a pH less than 6.2 after 24 hours as DFD. The overall incidence of DFD in electrically stimulated and non-stimulated carcasses was 3.4 and 13.2%, respectively. Further studies continue to report on the development of DFD in beef (Viljoena *et al.*, 2002; Kannan *et al.*, 2002; Apple *et al.*, 2005; Zhang *et al.*, 2005).

L. Postmortem Changes in Meat and Fish Proteins

The decline in pH of the muscle to an acidic state, together with the various exothermic reactions, such as glycolysis, has a profound effect on the muscle proteins of both meat and fish. The retention of moisture by fresh meat is recognized as one the most important quality characteristics. The majority of the water is held within the myofibril, between the myofibrils and the cell membrane (sarcolemma), and between muscle cells or bundles (Offer and Trinick, 1983; Offer and Cousins, 1992). Consequently, any degradation of muscle proteins, particularly by proteases, could affect the WHC of the muscle. This section will discuss the phenomenon of protein lability within the muscle and its effect on meat and fish quality.

Shortly after death, the body temperature in cattle may rise from 37.6 to 39.5°C (Meyer, 1964). Even during refrigeration, meat cools slowly as a result of the various exothermic reactions taking place, such as glycolysis. This phenomenon is known as 'animal heat' and was recorded from ancient times. Consequently, postmortem changes in muscle proteins are affected very often by a combination of high temperatures and low pH. Such changes include loss of color and decrease in WHC (Cassens, 1966). In meat, it is the sarcoplasmic proteins that are denatured and become firmly attached to the surface of the myofilaments, causing the lightening of meat color (Bendall and Wismer-Pederson, 1962). In fish, sarcoplasmic proteins are generally more stable than myofibrillar proteins, and are unaffected by dehydration or prolonged cold storage.

1. Water-Holding Capacity

The major postmortem change in meat and fish muscle is the loss of water or exudation. In lean meat, which has around 75% water, the majority of the water is somewhat loosely bound by the meat proteins. In the prerigor state, meat has a high WHC, which falls within the first few hours following death to a minimal level coincident with the establishment of rigor mortis. This minimal level corresponds to the ultimate postmortem pH of 5.3–5.5, which is the isoelectric point of the major muscle proteins. A rapid fall in pH leads to a number of changes including some muscle protein denaturation (Scopes, 1964; Chaudhrey *et al.*, 1969) and loss of WHC (Disney *et al.*, 1967). Tarrant and Mothersill (1977) determined glycolytic rates at several locations in beef carcasses as they affect the properties of the muscles. The postmortem pH decreased with distance from the carcass surface, indicating an increased rate of glycolysis which was accompanied by a greater degree of protein denaturation and drip loss. Drip loss was measured by low centrifugation of intact muscle samples from several muscles, as shown in Table 3.4.

TABLE 3.4 Drip Losses in Four Muscles in the Round

Muscle	Depth (cm)	Percentage Drip $\pm$ SE (n)
M. biceps femoris	1.5	8.1 $\pm$ 0.7 (6)
	5	14.9 $\pm$ 0.8 (6)
	8	18.2 $\pm$ 0.7 (6)
M. semitendinosus	1.5	7.5 $\pm$ 1.0 (6)
	8	16.7 $\pm$ 1.2 (6)

SE: standard error; n: number of muscles examined.
Adapted from Tarrant and Mothersill (1977).

The muscles at a depth of 8 cm were paler, softer, and wetter than those at 5 cm or 1.5 cm after 2 days postmortem, which is somewhat characteristic of the PSE condition observed in pork, although not as pronounced. This increased drip loss was attributed to a decrease in the WHC of denatured muscle proteins and sarcolemma disruption. The smallest drip loss at 1.5 cm was attributed to the low temperatures ($< 15°C$) attained before the pH reached 6.0. This explained why rapid cooling of beef carcasses following slaughter minimized the amount of drip loss.

Evidence suggests that reduced degradation of key cytoskeleton proteins, such as the intermediate filament protein desmin, which ties the myofibril to the cell membrane, may allow shrinkage of the myofibril and hence shrinkage of the muscle. Such shrinkage, however, opens up drip channels and increases drip loss (Morrison *et al.*, 1998; Kristensen and Purlow, 2001; Rowe *et al.*, 2001). Consequently, preventing shrinkage, by increasing degradation of desmin, would allow more moisture to remain in the tissue. Melody *et al.* (2004) associated differences in drip loss in porcine longissimus dorsi muscle with decreased proteolysis of desmin (Table 3.5). Since desmin is degraded by μ-calpain (Huff-Lonergan *et al.*, 1996), autolysis and activation of this enzyme might explain the variation in desmin degradation and its influence on drip loss. Examination of porcine longissimus dorsi, semimembranosus, and psoas major muscles by Melody *et al.* (2004) showed that differences in μ-calpain activity, μ-calpain autolysis, and protein degradation could explain differences in both tenderness and WHC for these muscles. These results do not rule out a role for other intermediate filament proteins such as talion and vinculin. For a more detailed discussion on the role that biochemical and structural changes have on WHC, the review by Huff-Lonergan and Lonergan (2005) is recommended.

TABLE 3.5 Drip Loss Percentage and Desmin Degradation Product of the Porcine Longissimus Dorsi (LD), Semimembranosus (SM), and Psoas Major (PM)

Item	LD	SM	PM
Drip loss (%)			
24 hours of storage	1.57[x] $\pm$ 0.384	1.46[xy] $\pm$ 0.373	1.06[y] $\pm$ 0.250
96 hours of storage	2.95[b] $\pm$ 0.623	2.78[b] $\pm$ 0.551	1.10[c] $\pm$ 0.214
Desmin degradation product[a]			
45 minutes postmortem	0.126[c] $\pm$ 0.027	0.177[c] $\pm$ 0.036	0.286[b] $\pm$ 0.039
6 hours postmortem	0.165[c] $\pm$ 0.106	0.188[c] $\pm$ 0.056	0.415[b] $\pm$ 0.097
24 hours postmortem	0.306[xy] $\pm$ 0.081	0.233[y] $\pm$ 0.071	0.391[x] $\pm$ 0.057

[a]Ratios were calculated as the intensity of the desmin degradation product band in each sample over the intensity of the intact desmin degradation product band in the internal designated densitometry standard.
[b,c]Within a row, means without a common superscript letter differ ($p < 0.01$); [xy]within a row, means without a common superscript letter differ ($p < 0.05$).
From Melody *et al.* (2004).

In the case of fish muscle, the ultimate pH tends to be higher than meat, hardly falling below 6.0 even in full rigor. However, considerable water losses were reported from exercised fish muscle similar to those in mammalian skeletal muscle (Partmann, 1965). A rapid rise in expressible fluid was observed in cod stored in ice for a 168-hour period (Banks, 1955). A series of papers published by Tomlinson *et al.* (1965, 1966a, b) reported that a decrease in pH in pacific halibut (*Hippoglossus stenolepis*) caused protein insolubility, which resulted in a pale, soft, exudative condition resembling PSE in pork (Briskey, 1964). This condition, known as chalkiness in halibut, is a problem for the fishing industry in the Pacific northwest because of rejection by the consumer. A technical report on chalky halibut investigations from 1997 to 1999 by Kammer (2000) for the International Pacific Halibut Commission also identified water temperature, season, postcapture handling, and sexual maturity as factors affecting chalkiness in Pacific halibut. To alleviate this condition, fish are allowed to remain alive after capture to remove excess lactic acid, so that a normal postmortem pH is attained following death.

A study by Hagen (2008) confirmed the importance of pH to chalkiness in farmed Atlantic salmon (*Hippoglossus hippoglossus* L.). They also implicated the activity of cathepsins B and D in the development of chalkiness. A related study on commercially farmed Atlantic halibut by Hagen *et al.* (2008) also found that cathepsins had a significant impact on protein content and liquid loss. Of the cathepsins examined, cathepsin H was the largest contributor to liquid loss, accounting for approximately 48.8% of the total variance. To obtain the best quality Atlantic halibut, they strongly recommended harvesting in the fall (autumn) or early winter, when liquid loss and cathepsin activities were low.

2. Proteomics and Water-Holding Capacity

Using proteomics, Hwang *et al.* (2004) found a strong relationship between an increase in drip loss in pork longissimus muscle and the high rate of postmortem proteolysis. Three proteins, troponin T, adenylate kinase (the substrate of ATP-dependent proteinase SP-22) and D5-1 protein, were all found to be related to changes in drip loss. Using two-dimensional electrophoresis, Van de Wiel and Zhang (2007) isolated a number of protein markers that could serve as important to the drip loss. These included creatine phosphokinase M-type (CKP), desmin, and a transcription activator (SWI/SNF). A recent proteomic study by Yu *et al.* (2009) examined the relationship between meat quality and heat shock proteins (HSPs) in longissimus dorsi muscle of pigs. Using the enzyme-linked immunosorbent assay (ELISA) technique, Yu *et al.* (2009) observed a tendency for the four HSPs ($\alpha\beta$-crystalline, HSP27, HSP70, and HSP90) to decline as drip loss increased in the longissimum dorsi muscles, suggesting their close relationship with WHC in pork.

M. Postrigor Tenderness

The most widely used process for tenderizing meat involves postmortem aging of the carcass. An optimum aging period of 8−11 days for choice carcasses from beef was proposed by Smith *et al.* (1978), although longer periods are used by the meat industry. The aging (conditioning or ripening) process is accelerated by raising the temperature. For example, in the Tenderay process (American Meat Institute Foundation, 1960), beef is held at 15°C for a 3-day period in ultraviolet light to control the surface microbial spoilage. This contrasts with fresh pork, which is not aged because of the rapid onset of fat rancidity even at low temperatures. In addition to temperature, both electrical stimulation and pressurization accelerate postmortem aging, as discussed previously (Savell *et al.*, 1981; Koohmaraie *et al.*, 1984).

N. Mechanism of Postrigor Tenderization

Since the last edition of this book, there have been major advances in our understanding of the biochemistry of meat tenderness (Herrera-Mendez *et al.*, 2006; Luciano *et al.*, 2007; Kemp *et al.*, 2010; Huff-Lonergan *et al.*, 2010). Three proteolytic enzyme systems in muscle appear to play a role in postmortem tenderization. Listed in order of their discovery, they are the lysosomal cathepsins, calpains, and proteosomes or multicatalytic proteinase complexes (MCPs) (Koohmarie and Geesink, 2006). Cathepsins were the first enzymes discovered by De Duve *et al.* (1955) that appeared to be involved in meat tenderness. Calpains were identified later in rat brain by Guroff (1964) and received far greater attention because of their ability to alter Z-line intensity. More recent studies, however, suggested a possible role for proteosomes in tenderization, with some researchers concluding that 20S proteosome was a potential contributor to the tenderization of stored meat (Matsuishi and Okitani, 1997; Dutaud, 1998; Lamare *et al.*, 2002; Sentandreu *et al.*, 2002; Thomas *et al.*, 2004).

1. Calcium-Activated Factor: The Calpain Family

A calcium-dependent endogenous protease capable of degrading the Z-disk was first identified in skeletal muscle by Busch *et al.* (1972), and subsequently referred to as the calcium-activated factor (CAF). This factor was active at neutral pH in the presence of calcium ions (Dayton *et al.*, 1975, 1976a, b). It degraded tropomyosin, troponin T, troponin I, filamin, and C-protein, with no detectable effect on myosin, actin, α-actinin, or troponin C (Dayton *et al.*, 1975; Dayton and Schollmeyer, 1980). CAF was found by Olson *et al.* (1977) to cause major changes in the myofibrillar proteins. Using a combination of microscopic and sodium dodecyl sulfate—polyacrylamide gel electrophoresis (SDS—PAGE), they observed the degradation of the Z-line, fragmentation of the myofibril, and disappearance of troponin T. This was accompanied by the release of a 30 kDa myofibril component from aged beef longissimus dorsi muscle. The presence of a 3 kDa component had been reported earlier during the aging of chicken and beef muscles (Hay *et al.*, 1973; Penny, 1974; Samejima and Wolfe, 1976). Adding calcium ions to minced muscle samples, Cheng and Parrish (1977) increased the rate of troponin T degradation to the 30 kDa component. These results provided convincing evidence of the importance of CAF in the proteolysis and tenderization of postmortem muscle.

The role of CAF in meat tenderization was shown indirectly by Olson *et al.* (1976), who found that postmortem degradation of the Z-disk and myofibril fragmentation in bovine longissimus muscle correlated with tenderness scores as well as Warner—Bratzler shear values (Table 3.6). An objective method was subsequently developed for measuring myofibril fragmentation, referred to as the myofibril fragmentation index (MFI), which correlated with the tenderness scores of beef steaks (Olson and Parrish, 1977). Confirmation of the relationship was provided in later studies on beef loin tenderness and MFI by Culler *et al.* (1978), MacBride and Parrish (1977), and Parrish *et al.* (1979). MacBride and Parrish (1977) introduced the term 'myofibril fragmentation tenderness' to describe tenderness in conventionally aged beef carcasses.

TABLE 3.6 Effect of Postmortem Storage (2°C) on Myofibril Fragmentation Index (MFI) and Warner—Bratzler (W-B) Shear Force of Longissimus (L), Semitendinosus (ST), and Psoas Major (PM) Muscles

	Days of Postmortem Storage		
	1	3	6
MFI[a]			
L	49.6 ± 1.3	69.8 ± 1.1	76.3 ± 0.9
ST	48.8 ± 0.8	68.2 ± 1.1	77.6 ± 1.0
PM	47.1 ± 0.9	49.3 ± 1.1	54.7 ± 1.0
W-B[b]			
L	2.60 ± 0.20	2.23 ± 0.17	2.13 ± 0.12
ST	3.27 ± 0.11	2.72 ± 0.09	2.64 ± 0.11
PM	2.16 ± 0.12	1.94 ± 0.11	1.86 ± 0.17

Data are shown as mean ± standard error of five carcasses. Means not underscored by the same line are significantly different ($p < 0.05$).
[a]Absorbance per 0.5 mg myofibril protein × 200;
[b]kilograms of shear force per cm^2.
From Olson *et al.* (1976). Copyright © by Institute of Food Technologists.

These calcium-dependent proteases were subsequently referred to as calpains by Murachi (1985). Calpains are a large family of intracellular cysteine proteases of which 14 members have been identified so far (Goll *et al.*, 2003). The three main calpain isoforms found in skeletal muscle are μ-calpain, m-calpain, and skeletal specific calpain, p94 or calpain 3, together with calpastatin, an inhibitor of both μ- and m-calpains (Koohmaraie and Geesink, 2006). Calpain 3/p94 appeared to bind to titin at the N_2 line (Sorimachi *et al.*, 1995; Kinbara *et al.*, 1998), a site where proteolysis has been associated with tenderization (Taylor *et al.*, 1995). When not bound to titin, calpain 3 autolyzes rapidly even in the absence of Ca^{2+} (Kinbara *et al.*, 1998; Spencer *et al.*, 2002). Because of the difficulty in purifying active calpain 3, its involvement in tenderization was determined by correlating its expression or autolysis with proteolysis and tenderization. Parr *et al.* (1999a) were unable to find any association between the amount or autolysis of calpain 3 and changes in meat tenderness. In contrast, significant correlations were reported by Illian *et al.* (2004a, b) between the rate of calpain 3 autolysis with meat proteolysis and tenderization. However, further work by Geesink *et al.* (2011), using transgenic mice overexpressing calpastatin, found that activation of calpain 3 by autolysis produced negligible changes in postmortem proteolysis. Thus, calpain 3 appeared to play a minor role, if any, in postmortem proteolysis in muscle. The other two isoforms, however, differed with respect to the amount of Ca^{2+} needed for their activation. For example, μ-calpain required between 5 and 60 μM for half maximal activity compared to 300–1000 μM for m-calpain (Goll *et al.*, 1992). Of the two isomers, μ-calpain appeared to be the one with a major role in the postmortem degradation of myofibrillar proteins associated with tenderization (Geesink *et al.*, 2000a; Hopkins and Thompson, 2002; Koohmaraie *et al.*, 2002). μ-Calpain, because of its limited specificity, is unable to degrade proteins to their constituent amino acids, or degrade myosin or actin, the major myofibrillar proteins. A model proposed by Goll *et al.* (2003) and Neti *et al.* (2009) suggested that calpains release myofilaments from microfibrils, which are then completely degraded to amino acids by proteosome and lysosomes.

2. Calpastatin

Calpastatin, an endogenous inhibitor of calpain enzymes, is present in all tissues containing calpains. It is a highly polymorphic protein requiring calcium levels similar to those required to activate calpain (Goll *et al.*, 2003). Calpastatin is susceptible to proteolysis, although the resulting protein fragments retain inhibitory activity. Because of the ability of calpastatin to inhibit calpains, the presence of high levels would naturally prevent or reduce proteolysis of meat and tenderization. This is evident in Figure 3.22, as the random selection of commercial pigs with high levels of calpastatin produced a higher incidence of toughness a few hours following slaughter (Sensky *et al.*, 1998; Parr *et al.*, 1999b).

The genetic component of meat quality is extremely complex since such important traits such as color, flavor, juiciness, fat content, and tenderness are controlled by several different genes throughout the genome, referred to as quantitative trait loci (QTL). A QTL with large effects on pork tenderness was found at the same position on porcine chromosome 2 (SSC2q) (Stearns *et al.*, 2005). Among the genes in this region was calpastatin (CAST). Mutations in CAST, resulting in unregulated calpain activity and increased tenderness, were examined in a number of studies (Parr *et al.*, 1999b; Ciobanu *et al.*, 2004). Comparative analysis of SSC2q by Meyers *et al.* (2007) showed that it was orthologous to a segment of the human chromosome 5 (HSA5) and contained a strong positional candidate gene (CAST). Since some of the CAST polymorphs were associated with meat quality characteristics, it should provide a better understanding of the molecular basis of pork tenderness. Further work by Nonneman *et al.* (2011)

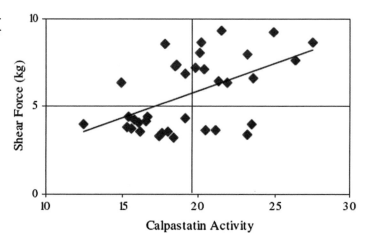

FIGURE 3.22   Correlation between slaughter (2 hour) calpastatin activity (× 10[7] fluorescence units/kg) and 8 day shear force in porcine longissimus dorsi (LD). *(Kemp* et al., *2010)*

resequenced the calpastatin regulatory and transcribed regions in pigs with divergently different shear force values. This enabled these researchers to identify possible mutations that could affect tenderness, with a total of 194 single nucleotide polymorphisms (SNPs) identified. Of these, four SNPs were consistently associated with pork tenderness in over 2826 pigs examined from four distinct populations. Nonneman *et al.* (2011) concluded that these calpastatin markers could be used to predict tenderness in pig populations. Markers are also available within calpastatin and μ-calpain genes for selecting beef cattle with the genetic propensity for producing tender meat (Casas *et al.*, 2006).

3. Lysosomal Proteases

At one time the lysosomal proteases, the cathepsins, were considered to be the main agents of postmortem tenderness in meat (Bate-Smith, 1948). However, their precise role in postmortem tenderization remains inconclusive, as there is limited degradation of actin and myosin, the primary substrates for cathepsins, during postmortem conditioning (Koohmaraie *et al.*, 1991). In addition, these enzymes must be released from lysosomes in order to access the myofibril proteins (Hopkins and Taylor, 2002). Disruption of the lysosomal membrane, however, can occur under conditions of low postmortem pH and high carcass temperature (O'Halloran *et al.*, 1997). Overall, there appeared to be little association between the activity of cathepsins and meat tenderness (Whipple *et al.*, 1990). Cathepsins include both exopeptidases and endopeptidases and are classified into cysteine (cathepsins B, H, I, and X), aspartic (cathepsins D and E), and serine (cathepsin G) peptidases (Sentandreuet *et al.*, 2002). They exhibit an optimum pH of around 5.5 and are active at 37°C. Schwartz and Bird (1977) showed that rat muscle cathepsins B and D were capable of hydrolyzing myosin. A cathepsin from rabbit muscle was also found to break down myosin, α-actinin, and actin (Okitani *et al.*, 1980). Cathepsin B was also reported to degrade troponin T (Noda *et al.*, 1981) and actin (Hirao *et al.*, 1984). Further research is necessary on the activities of cathepsins B and D, which have yet to be demonstrated in postmortem muscle (Parrish and Lusby, 1983). The combined effects of CAF and cathepsin D on the myofibrils of ovine longissimus dorsi muscles were examined by Elgasim *et al.* (1985). In both cases there was loss of the Z-line, but in the case of CAF it was totally destroyed. The Z-line proteins, α-actinin, desmin, and actin, were degraded or released by either of the enzymes. CAF degraded desmin and released α-actinin without affecting actin. Cathepsin D primarily affected α-actinin, with little or no effect on actin or desmin. Although inconclusive, it appears from this study that these two enzymes might act cooperatively in the degradation of the myofibrillar proteins.

A purified cathepsin L preparation from rabbit muscle was shown by Mikami *et al.* (1987) to degrade titin, nebulin, c-protein, α-actinin, myosin light chains, and tropomyosin in myofibrils prepared from rabbit, chicken and beef muscle. Catheptic damage was most rapid in chicken myofibrils and least rapid in beef myofibrils, consistent with the more rapid conditioning in chicken. Thomas *et al.* (2004) examined the role of proteasomes and cathepsins in ostrich meat tenderization. While cathepsin D remained very active throughout the 12-day postmortem aging period at 4°C, cathepsin B, L, and H activities were minimal during the first few hours postmortem but increased substantially after 12 days. Their results suggested that cathepsin D played a role in ostrich meat tenderization as B, L, and H were not active until day 1. Previous work on ostrich meat by Van Jaarsveld *et al.* (1997) found that cathepsins B, B + L and D activities all remained very stable after 12 days of storage at 4°C. In contrast, only cathepsin H activity was found to decrease significantly, by 40%, during that period. These results appeared to clarify the controversy surrounding the role of cathepsin in postmortem proteolysis discussed by Sentandreu *et al.* (2002).

The presence of cathepsin A in fish muscle was first reported by Makinodan and Ikeda (1976). This enzyme was subsequently purified by Toyohara *et al.* (1982) from carp muscle but did not appear to be directly involved in the postmortem proteolysis of the fish muscle. Later research by Aoki and Ueno (1997) attributed the postmortem degradation in the white muscle of aging mackerel at 0°C to the action of cathepsin L. While there have been a few reports of cathepsins in the muscles of some fish species (Wojtowicz and Odense, 1972; Matsumiya and Mochizuki, 1993; Porter *et al.*, 1996), a more comprehensive examination of their distribution in 24 fish species was carried out by Aoki *et al.* (2000). With the exception of cathepsin D, marine red-flesh fish were higher in cathepsins B, B-like, L, and latent L compared to marine white-flesh fish and freshwater fish. Since Yamashita and Konagay (1990b, 1991) reported that cathepsin L was the main cause of tissue degradation and softening in salmon, the large amounts present in the red flesh of marine fish could explain why they readily degrade and soften during storage.

4. Proteasomes

Proteasome, a multicatalytic protease complex, is the main enzyme responsible for protein degradation in the cytosol and nucleus. Proteasomes are involved in many biological processes and are particularly abundant in skeletal muscle

(Robert *et al.*, 1999). The most common form of proteasome is 26S proteasome, which contains a 19S regulatory subunit and a 20S catalytic core. The 20S proteasome, also known as the multicatalytic proteinase complex (MCP), is the catalytic core of protease complexes (Dahlmann *et al.*, 2001). It exists as a free entity or part of a core of larger particles with high proteolytic properties. Although initially, 20S-proteasome was thought unable to degrade proteins or polypeptides, later studies reported it exhibited both endopeptidase and exopeptidase activities (Forster and Hill, 2003; Liu *et al.*, 2003; Orlowski and Wilk, 2003). Several researchers also supported the ability of 20S-proteasome to degrade myofibrils and myofibrillar proteins *in vivo* (Taylor *et al.*, 1995; Matsuishi and Okitani, 1997). Dutaud *et al.* (2006) confirmed the ability of 20S-proteasome to degrade the Z-band in veal with a pH 5.8 or higher. The potential role of both proteasomes and cathepsins in the tenderization of ostrich meat was reported by Thomas *et al.* (2004). Koohmaraie and Gesink (2006) reported that the degradation pattern for myofibrillar proteins incubated with 20S-proteasome differed from that seen in postmortem muscle. The latter, however, does not completely exclude the proteasome from contributing to postmortem proteolysis (Houbak *et al.*, 2008).

5. Proteomics and Meat Tenderness

Using proteomics, Morzel *et al.* (2008) demonstrated the important contribution of sarcoplasmic proteins in beef longissimus thoracis muscles to meat tenderness and eating quality. Succinate dehydrogenase (SDH) was found to be the best common predictor of initial and overall tenderness, accounting for 67.6% and 57.8% of variation of these palatability traits. In addition, regression analysis showed that heat shock protein HSP27 in fresh meat and HSP27 in aged beef accounted for up to 91% variation in sensory scores. Based on data collected in biopsies and in samples from bovine longissimum thoracis muscles shortly after slaughter, Jia *et al.* (2009) identified peroxiredoxin-6, an antioxidant enzyme, as a potential protein marker for tenderness in beef.

O. Cytoskeleton and Meat Tenderness

1. G-Filaments

Davey and Graffhuis (1976) cooked beef neck muscle to maximum stretch and found that the G-filaments remained intact as long as the meat had not been aged. The tenderness of myofibrillar proteins involves weakening of the G-filaments. The rapid degradation of titin, a major protein component of the G-filaments, was reported by King (1984) to occur at elevated temperatures as well as at 0 and 15°C. Their results differed from those of Locker and Wild (1984), who found that titin was particularly resistant to aging for up to 20 days at 15°C. Nebulin, however, was found to disappear within 2 days, accompanied by an improvement in tenderness. No significant change in titin was reported, although a new B-band was detected by polyacrylamide electrophoresis (Locker and Wild, 1984). A later study by Lusby *et al.* (1983) examined the aging of beef longissimus dorsi muscle at three storage temperatures (2, 25, and 37°C) over a 7-day period. Nebulin was also found to be rapidly degraded, with the appearance of a new band corresponding to the B-band reported by Locker and Wild (1984). Lusby *et al.* (1983) claimed that titin-1 was converted to titin-2 as the time and temperature increased. However, Locker (1984) pointed out that the photographs of their gels were inconclusive because of a lack of resolution of their bands. Paterson and Parrish (1987) noted some minor differences in the titin content between tough and tender muscles at 7 days postmortem. Native titin T1 (3000 kDa), was later to be degraded early postmortem into a major degradation product, T2 (2000 kDa) (Wang *et al.*, 1979; Kurzban and Wang, 1988) and a subfragment (1200 kDa) (Matsuura *et al.*, 1988) by cleavage in the I-band region (Furst *et al.*, 1988). Using a monoclonal antibody method for titin (9 D 10) developed by Wang and Greaser (1985), Ringkob *et al.* (1988) showed that there was an alteration in the shape of titin within 2 days in postmortem beef muscle. Immunofluorescence microscopy of muscle sections and myofibrils enabled them to monitor changes in the I-band by showing transition from two- to four-band anti-titin staining patterns by the antibody per sarcomere in the muscle after 24–48 hours of storage. A similar observation for titin was made by Fritz and Greaser (1991) in postmortem beef muscle. This pointed to proteolysis of either titin or another protein to which titin may be attached.

A number of researchers used electron microscopy to show the occurrence of myofibril breakage near the Z-line during postmortem storage (Davey and Graafhuis, 1976; Taylor *et al.*, 1995). Boyer-Berri and Greaser (1998) determined whether changes in titin immunofluorescence patterns near the Z-line occurred postmortem and whether such changes were affected by muscle type. Using an antibody against a 56 kDa fragment (FE-RE) of titin located in the Z-line, they examined unaged and aged myofibrils obtained from four muscles (cutaneus trunct, rectus abdominis, psoas major, and masseter from five steers. Their results suggested that during postmortem aging the titin–actin interactions with the Z-line and N1—line might be weakened as some titin (FE RE) fragments are able to move away from the Z line. This was the first evidence for rearrangement of the Z-line region of titin during postmortem aging that could be related to the breakdown of the Z-line and increased fragility of the myofibril. Such titin changes could also be involved in the development of postmortem tenderness.

2. Intermediate Filaments

The degradation of desmin, which plays a role in linking adjacent myofibrils together, may cause physical changes to the postmortem muscle. Robson *et al.* (1984) monitored the changes in three major desmin fractions isolated from bovine skeletal muscle during postmortem aging for up to 7 days at 15°C by SDS—PAGE. Only a slight decrease in the amount of desmin was observed after 1 day, although after 7 days' storage a substantial loss of intact desmin was evident. This change was paralleled by a decline in troponin T (37 kDa band) which disappeared after 7 days' storage and was replaced by a band at around 30 kDa, as observed previously by Olson *et al.* (1977). The degradation of desmin may be carried out by CAF, since this has been shown by O'Shea *et al.* (1979) using purified desmin. Thus, the degradation of desmin and the connecting intermediate filaments may play an important role in meat tenderness during aging (Yamaguchi *et al.*, 1982, 1983a, b).

P. Effect of Pressurization on Tenderness

1. High-Pressure Treatment

Prerigor pressurization has been reported to increase meat tenderness (Macfarlane, 1973; Elgasim, 1977; Kennick *et al.*, 1980). As discussed in Section I, this approach was discontinued and replaced by high-pressure treatment of postrigor meat. However, no improvement in tenderness was initially reported by Bouton *et al.* (1977) in postrigor meat subjected to high-pressure treatment at low temperatures (< 30°C). By combining high pressure with heat, they found that an optimum temperature of 55—60°C improved the tenderness of the postrigor meat. Ma and Ledward (2004) treated postrigor beef longissimus dorsi to high pressures of up to 800 MPa over a temperature range of 20—70°C for 20 minutes. While no tenderizing effect was observed below 60°C, a significant decrease in hardness was found at 200 MPa and 60—70°C. A significant improvement in tenderness in postrigor beef sternomandibularis muscle was recently reported by Sikes *et al.* (2010) with high-pressure—high-temperature-treated compared to raw or heat treated beef (200 MPa at 60°C).

Since muscle enzymes important in meat tenderization are susceptible to high pressure, several studies have investigated the effects of such treatment. Elgasim (1977) showed that improvements in tenderness induced by prerigor pressurization could be related to lysosomal activity. An examination of bovine longissimus muscle exposed to different pressure levels by Elgasim *et al.* (1983) suggested the early release of lysosomal enzymes when conditions of pH and temperature were still conducive to rapid enzyme activity. This was based on the Z-line degradation observed in the pressurized muscle 24 hours postmortem. The pH of the muscle was less than 6.0 following pressurization, which was shown previously to facilitate the activity of cathepsin D, a lysosomal enzyme (Eino and Stanley, 1973; Robbins and Cohen, 1976). Since cathepsins are entrapped in lysosomal bodies, participation in meat aging and tenderization requires their release into the cytosol (Goll, 1983). Thus, improvement in tenderness associated with prerigor pressurization may be explained, in part, by the early release of lysosomal enzymes. This was evident in a study by Homma *et al.* (1994), who monitored acid phosphatase as an index of lysosomal membrane disruption. An increase in acid phosphatase activity was accompanied by increased cathepsin B, D, and L activities with pressures of up to 400 MPa. A further rise in pressure to 500 MPa, however, tended to decrease the activity of these cathepsins. Using electron microscopy to examine the membrane integrity of the lysosomes, Jung *et al.* (2000) established a correlation between lysosomal enzyme activities and lysosomal integrity in beef muscle subjected to high pressure. Acid phosphatase activities were always higher in the pressure-treated meat than in the corresponding control, and increased with pressure and time. In the case of cathepsin D, only the higher pressures caused a significant increase in activity. The highest values were observed for cathepsin D when beef muscle was subjected to a pressure of 500—600 MPa for short periods, less than 200 seconds. Using immunoelectron microscopy, Kubo *et al.* (2002) observed the release of cathepsin D from rabbit longissimus thoracis muscle pressurized at 100 MPa for 5 minutes followed by the absorption of cathepsin on the microfibrils. The brief exposure to high pressure appeared to have a far more drastic effect on the muscle than did normal conditioning at 2—4°C for 7 and 14 days.

Homma *et al.* (1995) found that calpain activity decreased in meat subjected to pressures above 100 MPa and was inactivated above 300 MPa. Meat pressurized at 200 MPa still had calpain activity while calpastatin was completely inactivated. The presence of calpain activity in pressurized meat below 300 MPa, however, suggested that it could still play a role in tenderization. Purification of proteasomes from rabbit muscle by Otsuka *et al.* (1998) reported an increase in activity at 150 MPa which decreased at higher pressures. Using electron microscopy, they attributed gaps between the filament structures, the complete loss of the M-line, and partial loss of the Z-line to proteasomes action.

2. High-Pressure Shockwaves

Several patents have reported high-pressure shockwaves as an alternative method for tenderizing beef (Godfrey, 1970; Long, 2000). This technology, referred to as hydrodynamic pressure (HDP) wave technology, involves the underwater detonation of explosives to generate shockwave pressure on meat submerged in a test chamber (Solomon *et al.*, 1997). The high-pressure shockwave generated by explosives resulted in dramatic improvements in the tenderness of beef, pork, lamb, and poultry (Solomon *et al.*, 1997, 1998; Meek *et al.*, 2000; Claus *et al.*, 2001a). The effect of the shockwaves was instantaneous, with improvements in tenderness ranging from 37% to 57%, equivalent to that obtained by aging (Solomon, 1998). As an alternative to explosives, high-efficiency compact sparkers were examined for generating shockwaves for meat tenderization (Schaefer 2004a–d, 2005). Sparkers have a similar effect to explosives, but differ by being electrically driven acoustic sources of high-pressure shockwaves. Some improvements in tenderness using sparkers were reported previously in poultry, beef, and pork (Claus *et al.*, 2001b; Claus, 2002; Sagili and Claus, 2003). A recent study by Bowker *et al.* (2011) showed favorable changes in the tenderness of beef loins using a high sparker efficiency consistent with previous studies using shockwaves.

Q. Lysosomal Enzymes and Collagen Degradation

The possible effects of lysosomal enzymes on the connective tissue of bovine muscles were examined by Wu *et al.* (1981). They showed that high-temperature conditioning (37°C) resulted in the release of lysosomal enzymes and enhancement of collagen solubilization. The increased breakdown of collagen fibers in the presence of lysosomal glycosidases was attributed to the degradation of the proteoglycan components which normally interfere with the degradation of collagen by collagenase (Eyre and Muir, 1974; Osebold and Pedrini, 1976).

V. MEAT PIGMENTS

Consumer acceptance of packaged fresh meat is primarily influenced by its color (Pirko and Ayres, 1957). The bright red color of fresh meat, caused by the pigment oxymyoglobin, is preferred by most consumers to meat of a darker or browner color. Metmyoglobin, the brown pigment responsible, develops during storage of the meat and is generally recognized by consumers to indicate lack of freshness. Consequently, a predominance of this pigment in packaged meat products results in their rejection. This discoloration of packaged meats, referred to by the industry as 'loss of bloom', is associated by the consumer with bacterial growth, although this may not always be the case. The annual consumption of meat, particularly beef, is well in excess of 100 lb per capita in North America, purchased mainly as raw meat. Any deleterious changes in the nature of these pigments is of great concern to the meat industry as it affects the consumer market. Approximately 15% of retail beef is discounted in price largely as a result of surface discoloration (Smith *et al.*, 2000), so it is economically important to maintain the bright red color associated with fresh meat.

The major pigment in meat is the purplish-red pigment myoglobin. Hemoglobin, the red blood pigment, was considered at one time to play only a minor role since blood is normally drained from the slaughtered carcass. Solberg (1968) stated, however, that the desirable color of meat is influenced by both myoglobin and hemoglobin. While the amount of hemoglobin reported in meat varies considerably in the literature, the relatively small amount may still be important in terms of color and stability. Myoglobin accounts for 10% of the total iron in an animal prior to slaughter, but in the case of a well-bled carcass it could account for as much as 95% of the total iron (Clydesdale and Francis, 1971). Since the majority of studies carried out on meat color have been concerned primarily with myoglobin, this section will confine its discussion to those changes associated with this pigment.

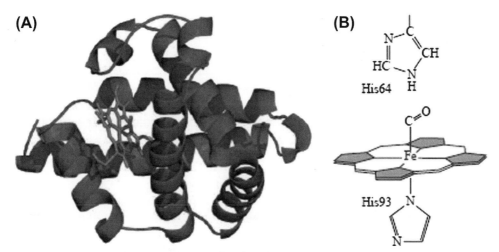

FIGURE 3.23 **Myoglobin consists of a backbone and heme-binding domain.** (A) The backbone of myoglobin consists of eight α -helices that wrap around a central pocket containing a heme group. (B) The protoheme group is bracketed or stabilized by histidine residues above (His64) and below (His93). *(Ordway and Garry, 2004.)*

A. Myoglobin

Myoglobin is a water-soluble protein composed of a single polypeptide chain, globin, containing eight α-helices. It is connected to a prosthetic heme group, an ion (II) protoporphyrin—IX complex (Pegg and Shahidi, 1997). The latter consists of four pyrrole groups containing a centrally located atom of iron (Figure 3.23). The molecular weights of mammalian myoglobins are generally larger than fish myoglobins, ranging from 14 to 18 kDa (Satterlee and Zecharia, 1972; Fosmire and Brown, 1976; Yamaguchi *et al.*, 1979; Ueki and Uchiai, 2004; Chaijan *et al.*, 2007). The primary factors responsible for the color of meat are the valence state of the iron atom and the ligand bond to the free binding site of the heme (Seideman *et al.*, 1984).

1. The Primary Structure of Myoglobins

Myoglobin stability varies among species as a result of differences in the amino acid sequences of their respective globins as well as secondary structures. Over the past few years the primary structures of myoglobin from goat, turkey, bison, tuna, and sardine have all been established (Chaijan *et al.*, 2007; Joseph *et al.*, 2010, 2011; Suman *et al.*, 2009). The complete primary structure of isolated goat myoglobin obtained by Edman degradation and tryptic and cyanogen bromide peptides was reported by Suman *et al.* (2009). They compared the amino acid sequence of goat myoglobin with homology of known ruminant myoglobin (sheep, cattle, buffalo, and deer) sequences. Goat myoglobin was similar to other livestock myoglobins by having 153 residues. However, the amino acid sequence in goat myoglobin was quite different from that in sheep and other ruminants. As a small ruminant meat animal, goat myoglobin was more similar to sheep than the larger ruminants. The two amino acid substitution differences between sheep and goat myoglobins were the smaller amino acids, threonine and glycine, at positions 8 and 52 in goat myoglobin, compared to the larger amino acids, glutamine and glutamic acid, at positions 8 and 52 in sheep myoglobin (THRgoat8GLNsheep and GLYgoat52GLU sheep) (Figure 3.24). Distal (HIS64) and proximal (HIS94) histidines, important for stabilizing the heme group and coordinating oxygen binding in other meat-producing livestock, were also conserved in goat myoglobin. Using the same techniques, Joseph *et al.* (2011) determined the complete amino acid sequence of turkey myoglobin. The percentage sequence similarities between turkey myoglobin and other myoglobins are summarized in Table 3.7. The amino acid sequence of turkey myoglobin was identical to that of chicken myoglobin, while it shared 92.5% homology with ostrich, 76.5% with pig, and less than 73% with ruminant myoglobins.

2. Myoglobin Changes in Raw Meat

Studies on fresh meat discoloration are concerned primarily with the formation of metmyoglobin in postrigor meat. The color of fresh meat is determined by the relative proportions of the three primary forms of myoglobin (Watts *et al.*, 1966). These include the three oxidative states of myoglobin: oxymyoglobin, deoxymyoglobin, and

```
Sequence No.        10        20        30        40        50
Goat        GLSDGEWTLV LNAWGKVEAD VAGHGQEVLI RLFTGHPETL EKEDKFKHLK
Sheep       GLSDGEWQLV LNAWGKVEAD VAGHGQEVLI RLFTGHPETL EKFDKFKHLK
Cattle      GLSDGEWQLV LNAWGKVEAD VAGHGQEVLI RLFTGHPETL EKFDKFKHLK
Buffalo     GLSDGEWQLV LNAWGKVETD VAGHGQEVLI RLFTGHPETL EKFDKFKHLK
Deer        GLSDGEWQLV LNAWGKVEAD VAGHGQEVLI RLFTGHPETL EKFDKFKHLK

Sequence No.        60        70        80        90        100
Goat        TGAEMKASED LKKHGNTVLT ALGGILKKKG HHEAEVKHLA ESHANKHKIP
Sheep       TEAEMKASED LKKHGNTVLT ALGGILKKKG HHEAEVKHLA ESHANKHKIP
Cattle      TEAEMKASED LKKHGNTVLT ALGGILKKKG HHEAEVKHLA ESHANKHKIP
Buffalo     TEAEMKASED LKKHGNTVLT ALGGILKKKG HHEAEVKHLA ESHANKHKIP
Deer        TEAEMKASED LKKHGNTVLT ALGGILKKKG HHEAEVKHLA ESHANKHKIP

Sequence No.        110       120       130       140       150
Goat        VKYLEFISDA IIHVLHAKHP SDFGADAQGA MSKALELFRN DMAAQYKVLG FQG
Sheep       VKYLEFISDA IIHVLHAKHP SDFGADAQGA MSKALELFRN DMAAQYKVLG FQG
Cattle      VKYLEFISDA IIHVLHAKHP SDFGADAQAA MSKALELFRN DMAAQYKVLG FHG
Buffalo     VKYLEFISDA IIHVLHDKHP SDFGADAQAA MSKALELFRN EMAAQYKVLG FHG
Deer        VKYLEFISDA IIHVLHAKHP SNFGADAQGA MSKALELFRN DMAAQYKVLG FQG
```

FIGURE 3.24   Comparison of amino acid sequences of goat myoglobin and other ruminant myoglobins. Differences between sheep and goat myoglobins at positions 8 and 52 are boxed. Goat = *Capra hircus*; sheep = *Ovis aries*; cattle = *Bos taurus*; buffalo = *Bubalus bubalis*; deer = *Cervus elaphus. (Suman* et al.*, 2009)*

metmyoglobin. The predominant form of myoglobin in brightly colored red meat is oxymyoglobin, in which myoglobin is bound to oxygen. Deoxymyoglobin, the reduced form of myoglobin, is a purplish pigment bound to water (or with oxygen removed) and is readily converted to oxymyoglobin. In contrast, the undesirable brownish color associated with discolored meats is metmyoglobin, in which the valency of the iron portion is converted to the ferric form.

TABLE 3.7 Percentage Sequence Similarities Among Turkey Myoglobin and Other Myoglobins

Species	Turkey	Chicken	Ostrich	Pig	Beef	Buffalo	Sheep	Goat
Turkey	100							
Chicken	100	100						
Ostrich	92.8	92.8	100					
Pig	76.5	76.5	74.5	100				
Beef	72.5	72.5	70.6	88.2	100			
Buffalo	71.2	71.2	69.9	86.9	98.0	100		
Sheep	72.5	72.5	70.6	89.5	98.7	96.7	100	
Goat	71.9	71.9	69.9	88.2	97.4	95.4	98.7	100

Turkey: *Meleagris gallopavo*; chicken: *Gallus gallus*; ostrich: *Struthio comelus*; pig: *Sus scrofa*; beef: *Bos taurus*; buffalo: *Bubalu bubalis*; sheep: *Ovis aries*; goat: *Capra hircus* (Joseph *et al.*, 2011).

TABLE 3.8 Comparison of Living Muscle Tissue and Meat

Condition	Living Muscle	Postrigor Muscle Meat
Myoglobin	Oxymyoglobin	Myoglobin
pH	7.35–7.43	5.3–5.5
Temperature (°C)	37.7–39.1	2–5
Osmotic pressure (% NaCl equiv.)	0.936	
Oxidation–reduction potential (mV)	+250	−50

From Solberg (1970).

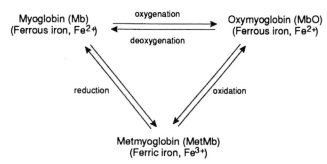

SCHEME 3.5 Myoglobin changes in fresh meat.

When the animal is slaughtered, oxygen is no longer available to the muscle tissue, resulting in the conversion of oxymyoglobin to deoxymyoglobin. Other changes during this period include pH, temperature, osmotic pressure, and oxidation–reduction potential (Table 3.8). Under these conditions the pigment changes occurring in meat are of great concern to the food technologist. Metmyoglobin, the undesirable brown pigment, is in equilibrium with the other pigment forms, as shown in Scheme 3.5. Further degradative reactions are mediated by bacterial action, causing irreversible damage to the porphyrin ring. This results in the formation of bile pigments, choleglobin, sulfmyoglobin, and oxysulfmyoglobin, as well as other nitrogenous compounds characteristic of meat spoilage.

3. Myoglobin Changes and Oxygen Tension

The formation of oxymyoglobin involves the complexing of oxygen with the heme group of myoglobin. This process, referred to as oxygenation, occurs under high oxygen tensions and favors the formation of the desirable bright cherry-red colored meat pigment. Under low oxygen tensions, however, metmyoglobin is formed (Taylor, 1972). In fresh meat both the reduced and oxygenated forms of myoglobin are present, with the predominating form determining the final color of the meat. The formation of oxymyoglobin involves the covalent binding of molecular oxygen to myoglobin (Clydesdale and Francis, 1971). Under conditions of low oxygen tension, oxygen dissociates from the heme to yield myoglobin. The latter, being unstable, is then oxidized to metmyoglobin (Pirko and Ayres, 1957). The formation of the metmyoglobin is accompanied by the loss of an electron in the iron molecule, resulting in a change from the ferrous (Fe^{2+}) to the ferric (Fe^{3+}) state (Giddings and Markarkis, 1973; Giddings, 1977a, b).

4. Effect of pH, Temperature, and Salt on the Formation of Metmyoglobin

The formation of metmyoglobin occurs under conditions of high temperature, low pH, ultraviolet light, and in the measure of salt and aerobic bacteria (Seideman *et al.*, 1984). Both high temperatures and low pH cause denaturation of the globin moiety, leaving the heme unprotected so that it undergoes rapid oxidation to metmyoglobin (Walters, 1975; Wallace *et al.*, 1982). The effect of salt is two-fold: it lowers the buffering capacity of the meat and promotes low oxygen tensions in meat. Either of these effects results in oxidation of myoglobin to metmyoglobin (Brooks, 1937; Seideman *et al.*, 1984).

5. Endogenous Meat Enzymes and Metmyoglobin Formation

Freshly stored meat metmyoglobin is formed by two opposing reactions, autoxidation and reduction:

$$\text{Metmyoglobin} \xrightarrow{\text{reduction}} \xleftarrow[\text{autooxidation}]{} \text{Myoglobin}$$

Myoglobin is now referred to as deoxymyoglobin, the reduced form of myoglobin (Fe^{2+}). These reactions, referred to as metmyoglobin-reducing activity (MRA), are responsible for the valence change in myoglobin from ferrous to ferric in the meat tissue. Differences in MRA between muscles contribute to the ability of some muscles to retain the bright red pigment for much longer periods. Stewart *et al.* (1965) separated these two reactions and showed that the reduction of metmyoglobin was carried out by enzymes. They reported considerable variation in the reducing activity (MRA) of their beef samples, which increased with a rise in pH and temperature. The name methemoglobin reductase is currently used to describe many reducing systems, including those involved with other heme proteins (Bekhit and Faustman, 2005).

Watts *et al.* (1966) concluded that the reduction of metmyoglobin and oxygen in meat was carried out via reduced nicotinamide adenine dinucleotide (NAD^+). Succinic dehydrogenase appeared to be one of the enzymes involved, as addition of succinate to meat increased oxygen utilization. Other intermediates of the citric acid cycle and the amino acid L-glutamate were found to stimulate the reduction of metmyoglobin (Saleh and Watts, 1968). The enzymes involved remained potentially active in the postmortem meat and were capable of resuming their activity in the presence of oxygen. This may occur if meat is ground or when cut surfaces are exposed to air, provided suitable hydrogen donors (e.g. $NADH^+$) are still available.

The reduction of metmyoglobin in postmortem meat is due mainly to enzymes in which the mitochondria act as a source of reducing equivalents for the reduction of pyridine nucleotides (MacDougall, 1982). One such enzyme, metmyoglobin reductase, was identified in bovine heart muscle (Hagler *et al.*, 1979). This enzyme required ferrocyanide and NADH and could reduce metmyoglobin back to myoglobin. Arihara *et al.* (1989) established the presence of NADH-cytochrome b_5 reductase in bovine skeletal muscle which had an absorption spectrum assembly identical to that of a previously purified erythrocyte cytochrome b_5 reductase from bovine red blood cells (Yubisui and Takeshita, 1980; Tamura *et al.*, 1983). Using electrophoretic immunoblotting techniques, Arihara *et al.* (1995) estimated that the amount of NADH-cytochrome b_5 reductase in bovine skeletal muscle was 13.8 ± 2.6 g/g tissue, ranging from 9.0 to 12.0 g/g tissue in biceps femoris muscles. Arihara *et al.* (1990) previously located the outer membrane of mitochondria cytochrome b on the surface of mitochondria in muscle tissue, and cytochrome b_5 in the microsomal fractions. Arihara *et al.* (1995) proposed the two pathways shown in Figure 3.25 to be involved in the

FIGURE 3.25 Metmyoglobin reduction pathways by NADH-cytochrome b_5 reductase: (a) situation at the mitochondria and (b) situation at the sarcoplasmic reticulum. *(Arihara et al., 1995)*

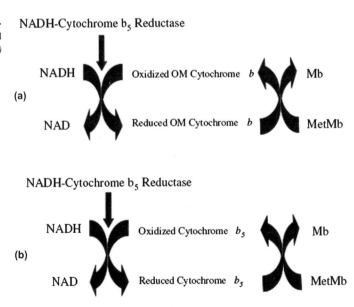

reduction of MetMb. The enzyme had a much higher affinity for cytochrome b_5 and could reduce both MetMb and MetHb. An excellent review of metmyoglobin reduction can be found in Bekhit and Faustman (2005).

B. Fish Pigments

The molecular weight of fish myoglobin is generally much smaller than that of mammalian myoglobin (Fosmire and Brown, 1976; Yamaguchi *et al.*, 1979; Chaijan *et al.*, 2007). The red color of tuna meat, a rich source of myoglobin, strongly influences acceptance by the consumer. Any discoloration during storage is associated with its conversion to metmyoglobin (Bito, 1965, 1976). The dark muscle of yellowfin tuna (*Thunnus albacores*), also rich in myoglobin, was reported to oxidize more rapidly than mammalian myoglobins (Brown and Mebine, 1969). Levy *et al.* (1985) isolated and characterized metmyoglobin reductase from the dark muscle of tuna which had a similar optimum pH to the corresponding bovine enzyme. The molecular weight of the tuna enzyme was similar to that of the dolphin reductase (30–40 kDa) but half that of bovine reductase (65 kDa). While the isoelectric pH values were different between tuna and bovine reductases, both enzymes were specific for NADH. Thiansilakul *et al.* (2011) recently isolated myoglobin from the dark muscle of eastern little tuna (*Euthynnus affinis*) and found that it was very prone to oxidation and denaturation at pH 3 at temperatures above 60°C. They recommended that processing be carried out under mild conditions to prevent undesirable changes in myoglobin from occurring.

Ochiai *et al.* (2009) showed that the structural stability of different tuna myoglobins (bigeye and bluefin) was affected by slight amino acid substitution. While highly conserved amino acid residues were responsible for the stability of myoglobin, a few non-conserved residues in the non-helical segments and heme pocket region greatly affected the thermal stability of myoglobin.

Earlier research by Miki and Nishimoto (1984) examined the relationship between loss of freshness and discoloration in several fish species, including skipjack, mackerel, and sea bream. In red-muscled fish, percentage metmyoglobin has provided a useful index of discoloration. Miki and Nishimoto (1984) plotted the log of (100 − metmyoglobin%), which represents the ratio of residual oxymyoglobin to total myoglobin against storage times at different temperatures. Their results for skipjack in Figure 3.26 clearly indicate a first order reaction corresponding to changes in freshness. These results are consistent with early work by Matsuura *et al.* (1962), who showed that the rate of autoxidation of isolated myoglobin from fish muscle also followed first order kinetics.

C. Preservation of Meat Pigments

The accelerated conversion of myoglobin to metmyoglobin under conditions of low oxygen partial pressure is extremely important when packaging fresh meat as color is the primary factor influencing meat purchasing decisions (Mancini and Hunt, 2005). For instance, meat containing 20% metmyoglobin is generally discriminated against by consumers (MacDougall, 1982) and if it exceeds 40% it is either downgraded or rejected (Greene *et al.*, 1971). George and Stratmann (1952) reported that the maximum rate of metmyoglobin formation occurred at around 1–1.4 mm partial pressures of oxygen, decreasing to a constant minimum rate above 30 mm. Consequently, the

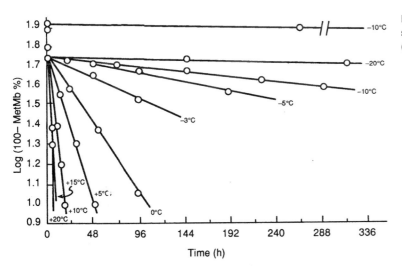

FIGURE 3.26 Changes in MetMb percentage of skipjack during storage at different temperatures. *(Miki and Nishomoto, 1984)*

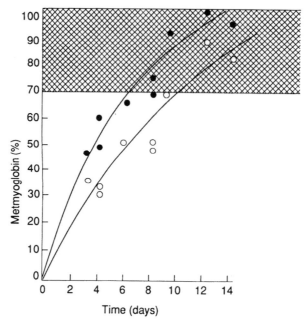

FIGURE 3.27   **Percentage metmyoglobin at the surface of bovine semimembranosus and adductor muscle slices stored at 4 ? 0.5°C in hermetically sealed semirigid polyvinyl-chloride packages containing an air or an enriched oxygen atmosphere.** ●: Air; ○: oxygen enriched; hatched area denotes unacceptable color (Daun *et al.*, 1971). *(Copyright © by Institute of Food Technologists.)*

oxygen permeability of the packaging film is of considerable importance when handling fresh meat products. Landrock and Wallace (1955) suggested that packaging should allow an oxygen penetration of 5 liters of $O_2/m^2/day/$ atmosphere to prevent the formation of metmyoglobin and subsequent browning.

Fellers *et al.* (1963), contrary to previous researchers, reported that oxygen levels higher than normally found in the atmosphere preserved oxymyoglobin. This was later confirmed by Bausch (1966). Thus, high oxygen partial pressures had considerable potential for enhancing the desirable color of meat. Very little deoxygenation takes place as the excess oxygen recombines with myoglobin to form oxymyoglobin. Daun *et al.* (1971) examined the effect of oxygen-enriched atmospheres on packaged fresh meat using a polyvinyl-chloride semirigid tray with a controlled headspace for storing fresh meat. Oxygen flushed into the headspace established an oxygen-enriched atmosphere and the formation of metmyoglobin on the surface of semimembranosus and adductor muscle slices was lower than meat stored under normal air atmosphere. Their results, shown in Figure 3.27, indicated that those samples stored in air were unacceptable, on the basis of color, after 6 days, while meat stored under oxygen-enriched atmospheres took 10 days to reach the same state. One of the major problems that has limited commercialization of this method was the development of fat rancidity under these conditions.

Of the many factors affecting the discoloration of packaged beef, differences between muscles was shown by Hood (1980) to be the most important single factor. Figure 3.28 shows decreased color stability for all four muscles as the temperature increased from 0 to 10°C. The rate of metmyoglobin accumulation was eight times faster in psoas major than in longissimus dorsi. Further work by O'Keeffe and Hood (1982) examined the biochemical factors affecting the rate of metmyoglobin discoloration in beef muscles. Psoas major was found to have a higher level of succinic dehydrogenase activity which, combined with its low myoglobin content, was responsible for a high oxygen consumption rate (OCR), resulting in low oxygen penetration and rapid formation of metmyoglobin compared to the situation in longissimus dorsi. Based on their research, enzymes responsible for metmyoglobin reduction were the main determinants of metmyoglobin formation in beef muscles, with psoas major being low in reducing activity and longissimus dorsi being high. The high-oxygen modified atmosphere packaging (MAP) saturates the meat pigment with oxygen, while carbon dioxide at a level above 20% inhibits bacterial growth (Enfors *et al.*, 1979; Nissen *et al.*, 1996; Sorheim and Dainty, 1996; Luno *et al.*, 2000). Thus, the strong relationship between color stability in the meat from different muscles is due to OCR and myoglobin and metmyoglobin reducing activities (Kropf, 1993; Sammel *et al.*, 2002; Seyfert *et al.*, 2006).

The interrelationship between lipid and pigment oxidation has been studied, since ferric pigments, such as metmyoglobin, are known to enhance lipid oxidation (Younathan and Watts, 1960; Brown *et al.*, 1963). Hutchins

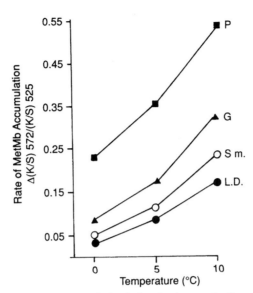

FIGURE 3.28   Effect of temperature on metmyoglobin accumulation in four bovine muscles from 10 experimental animals. P: m. psoas major; G: m. gluteus medius; S m.: m. semimembranosus; L.D.: m. longissimus dorsi. *(Hood, 1980)*

et al. (1967) reported good correlations between lipid oxidation and metmyoglobin formation in raw meat. Greene (1969) attempted to retard rancidity in refrigerated ground raw beef by adding several antioxidants, including propyl gallate (PG) and butylated hydroxyanisole (BHA). The desirable color of the ground beef was extended in the antioxidant-treated samples. Greene *et al.* (1971) treated samples of raw ground beef with a combination of PG or BHA and ascorbic acid (AA). Based on color scores using a trained panel, a consumer panel, and metmyoglobin measurements, the antioxidant-treated meat had a much longer shelf-life. Among the main limitations of high oxygen MAP are accelerated lipid oxidation and off-flavors (Jakobsen and Berrelsen, 2000; Jayasingh *et al.*, 2002) and premature browning during cooking (Torngren, 2003; John *et al.*, 2004, 2005).

Chapter 11, on lipid oxidation, discusses the myriad of aldehydes responsible for off-flavors and off-odors in food products, including meat (Pearson *et al.*, 1977). Studies showed that monounsaturated aldehydes incubated with oxymyoglobin significantly accelerated metmyoglobin formation (Yin and Faustman, 1993; Chan *et al.*, 1997; Faustman *et al.*, 1999). Among the many classes of aldehydes are the α,β-unsaturated aldehydes, which are extremely reactive with biomolecules (Witz, 1989). The most prominent and studied α,β-unsaturated aldehyde, 4-hydroxy-2-nonenal (HNE), is formed from linoleic acid oxidation in meat (Sakai *et al.*, 1998; Suhr and Kwon, 2005; Gasc *et al.*, 2007). Previous studies showed that a number of factors, including lipid oxidation products such as HNE, reduced the redox stability of oxymyoglobin with the formation of metmyoglobin (Faustman *et al.*, 1999; Alderton *et al.*, 2003). Suman *et al.* (2007) monitored the formation of metmyoglobin when HNE was incubated with porcine oxymyoglobin at pH 7.4 and 5.6, which represent normal physiological and postmortem conditions in the muscle. Their results, in Figure 3.29, show that metmyoglobin formation occurred more rapidly at pH 5.6 and was further enhanced in the presence of HNE. This contrasted with results obtained under alkaline conditions (pH 7.4) where metmyoglobin formation was much lower and not enhanced in the presence of HNE. While HNE can bind with cysteine, histidine, lysine, and arginine residues in proteins, HNE—myoglobin adducts are primarily formed at the histidine residues. Studies conducted separately with myoglobins from horse, cattle, pig, tuna, bison, chicken, and turkey all found that HNE promoted the oxidation of oxymyoglobin (Alderton *et al.*, 2003; Lee *et al.*, 2003a, b; Joseph *et al.*, 2010; Maheswarappa *et al.*, 2010). A recent study by Yin *et al.* (2011) examined the effect of HNE on the oxidation of chicken, turkey, pig, beef, sheep, horse, and deer oxymyoglobins. While HNE increased the oxidation of oxy-myoglobin in all species examined, the effect was far greater with beef, horse, sheep, and deer myoglobins than with chicken, pig, and turkey myoglobins. This was attributed to the presence of 13 or more histidine residues in beef, horse, sheep, and deer myoglobins compared with only nine histidine residues in chicken, pig, and turkey myoglobins. Using isotope-labeled phenyl isocyanate, Suman *et al.* (2007) attributed the lower susceptibility of porcine myoglobin to redox instability to the formation of fewer HNE adducts (three) compared to seven in bovine myoglobin. The histidine residues adducted by HNE in porcine myoglobin were HIS 24, 26, and 119, compared to HIS 24, 36, 81, 88, 93, 119, and 152 in bovine myoglobin.

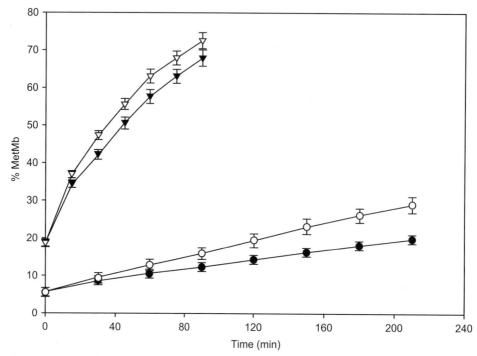

FIGURE 3.29 **MetMb formation during the reaction of porcine OxyMb (0.075 mM) with HNE (0.05 mM) at pH 7.4 or 5.6, at 37°C.** Standard error bars are indicated. τ: pH 5.6, control; ∇: pH 5.6, HNE; ●: pH 7.4, control; ○: pH 7.4, HNE. *(Suman et al., 2007)*

A combination of carbon dioxide and oxygen thus provides an alternative way of extending the shelf-life of meat. While carbon dioxide tended to inhibit spoilage bacteria, oxygen maintained the oxymyoglobin form of the pigment. Taylor and MacDougall (1973) reported the retention of the fresh color in beef for at least 1 week at 1°C in an atmosphere of 60% oxygen and 40% carbon dioxide. Gas packaging of retail cuts of fresh meat in oxygen-free nitrogen or carbon dioxide extended the storage of these products by minimizing oxidative changes and bacterial growth (Partmann and Frank, 1973; Huffman, 1974). O'Keeffe and Hood (1980−1981) examined the effect of anoxic gas packaging of fresh beef in flexible packages or desiccators incorporating a palladium catalyst as an oxygen-scavenging system. No discoloration was observed with meat stored in either carbon dioxide or nitrogen for up to 3 weeks at 0°C. Some differences observed between meat samples were attributed to differences between the animals.

Most fresh retail beef in the USA is displayed in Styrofoam™ trays covered with oxygen-permeable films to permit the rapid oxygenation of the surface myoglobins to form the desirable red color or bloom. Unfortunately, this only lasts for 1−7 days, depending on the muscle or cutting method, before browning or metmyoglobin formation occurs (Maudhavi and Carpenter, 1993). Extending this period for a further 3−4 days is possible using a modified atmosphere containing oxygen, carbon dioxide, and nitrogen (Manu-Tawiah *et al.*, 1991). However, a minimum of 21 days is required for color stability to meet the 7 days needed for packaging and distribution, 7 days for retail display, and 7 days in the home. This extension is possible by adding 0.5% carbon monoxide to the MAP in place of oxygen (Sorheim *et al.*, 1999). In the presence of carbon monoxide, myoglobin forms carboxymyoglobin, a very stable bright red pigment with an almost identical absorption spectrum to oxymyoglobin (El-Badawi *et al.*, 1964; Cornworth, 1994). The addition of low levels of carbon monoxide to MAP systems presents no danger to the consumer and this practice is widely used in Norway for packaging beef (Sorheim *et al.*, 1997). In 2001, a Scientific Committee on Food for the European Commission examined the use of carbon monoxide as a component of packing gases in MAP for fresh meat (SCF/CS/ADD/MSAd/204 Final). The committee concluded that there was no health concern from using 0.3−0.5% carbon monoxide in a gas mixture with carbon dioxide and nitrogen for fresh meat, as long as the storage temperature did not exceed 4°C. Care should be taken, however, to avoid inappropriate storage conditions as carbon monoxide could mask microbial spoilage. Cornforth and Hunt (2008) discussed the advantages and disadvantages of low-oxygen packaging of fresh meat with carbon monoxide and factors that should be taken into consideration when developing a packaging technology that protects both consumers and the meat industry.

REFERENCES

Aalhus, J.L., Jones, S.D.M., Lutz, S., Best, D.R., Robertson, W.M., 1994. The efficacy of high and low voltage electrical stimulation under different chilling regimes. Can. J. Anim. Sci. 74, 433–442.

Acton, J.C., Zieler, G.R., Burge Jr., D.L., 1983. Functionality of muscle constituents in the processing of comminuted meat products. CRC Crit. Rev. Food Sci. Nutr. 18, 99–121.

Adzitey, F., Nurul, H., 2011. Pale soft exudative (PSE) and dark, firm dry (DFD) meats: causes and measures to reduce these incidences – a mini review. Int. Food Sci. J. 18, 11–20.

Alderton, A.L., Faustman, C., Liebler, D.C., Hill, D.W., 2003. Induction of redox instability of bovine myoglobin by adduction with 4-hydroxy-2-nonenal. Biochemistry 42, 4398–4405.

Aliani, M., Farmer, L.J., 2002. Postcolumn derivatization method for determination of reducing and phosphorylated sugars in chicken by high performance liquid chromatography. J. Agric. Food Chem. 50, 2760–2766.

Aliani, M., Farmer, L.J., 2005a. Precursors of chicken flavor. I. Determination of some flavor precursors in chicken muscle. J. Agric. Food Chem. 53, 6067–6072.

Aliani, M., Farmer, L.J., 2005b. Precursors of chicken flavor. II. Identification of key flavor precursors using sensory methods. J. Agric. Food Chem. 53, 6455–6462.

Aliani, M., Kennedy, J.T., McRoberts, C.W., Farmer, L.J., 2008. Formation of flavor precursors by the AMP pathway in chicken meat. In: Interlaken, Switzerland, I., Wust, Blank, M., Yeretzian, C. (Eds.), Proceedings of the 12th Weurman Symposium. Institut fur Chemie und Bioloische, pp. 288–292.

American Meat Institute Foundation, 1960. The Science of Meat and Meat Products. Freeman, San Francisco, CA.

Amos, L.A., Huxley, H.E., Holmes, K.C., Goody, R.S., Taylor, K.A., 1982. Structural evidence that myosin heads may interact with two sites on F-actin. Nature (London) 299, 467–469.

Andou, S., Takama, K., Zama, K., 1979. Interaction between lipid and protein during frozen storage. I. Effect of oil dipping on rainbow trout muscle during frozen storage. Hokkaido Daigaku Kenkyu Iho 30, 282–287.

Andou, S., Takama, K., Zama, K., 1980. Interaction between lipid and protein during frozen storage. II. Effect off non-polar and polar lipids on rainbow trout myofibrils during frozen storage. Hokkaido Daigaku Kenkyu Iho 31, 201–209.

Ang, J.F., Haard, N.F., 1985. Chemical composition and postmortem changes in soft textured muscle from intensely feeding Atlantic cod (Gadus morhua L.). J. Food Biochem. 9, 49–64.

Aoka, T., Ueno, R., 1997. Involvement of cathepsins B and L in postmortem autolysis of mackerel muscle. Food Res. Int. 30, 585–591.

Aoka, T., Yamashita, T., Uneno, R., 2000. Distribution of cathepsins in red and white muscles among fish species. Fish. Sci. 66, 776–782.

Apple, J.K., Kegley, E.B., Galloway, D.L., Wistuba, T.J., Rakes, L.K., 2005. Duration of restraint and isolation stress as a model to study dark-cutting condition in cattle. J. Anim. Sci. 83, 1202–1214.

Arihari, K., Itoh, M., Kondo, Y., 1989. Identification of bovine skeletal muscle metmyoglobin reductase as an NADH-cytochrome b_5 reductase. J. Zootech. Sci. 6, 46–56.

Arihari, K., Itoh, M., Kondo, Y., 1990. Presence of cytochrome b5-like hemoprotein (OM cytochrome b) in rat muscles as a metmyoglobin reducing enzyme system component. Jpn. Soc. Scient. Fish 45, 1335–1339.

Arihari, K., Cassens, R.G., Greaser, M.L., Luchansky, J.B., Modziak, R.E., 1995. Localization of metmyoglobin-reducing enzyme (NADH-cytochrome b_5 reductase) system components in bovine skeletal muscle. Meat Sci. 39, 205–213.

Arya, S.S., Parhar, D.B., 1979. Changes in free nucleotides, nucleosides and bases during thermal processing of goat and sheep meats. Part I. Effect of temperature. Nahrung 23, 1–7.

Asghar, A., Henrickson, R.L., 1982. Chemical, biochemical, functional and nutritional characteristics of collagen in food systems. Adv. Food Res. 28, 231–372.

Asghar, A., Samejima, K., Yasui, T., 1984. Functionality of muscle proteins in gelation mechanisms of structured meat products. CRC Crit. Rev. Food Sci. Nutr. 22, 27–106.

Ashton, T.J., Michie, I., Johnston, I.A., 2010. A novel tensile test method to assess texture and gaping in salmon fillets. J. Food Sci. 74, S182–S190.

Bailey, A.J., 1972. The basis of meat texture. J. Sci. Food Agric. 23, 995–1007.

Bailey, A.J., 1990. The chemistry of collagen cross-links and their role in meat texture. In. Reciprocal Meat Conference Proceedings 23, 995–1007.

Bailey, A.J., Light, N.D., 1989. In: Connective Tissue in Meat and Meat Products. Elsevier Applied Science, London, pp. 1–355.

Bailey, A.J., Peach, C.M., 1968. Isolation and structural identification of labile intermolecular crosslink in collagen. Biochem. Biophys. Res. Commun. 33, 812–819.

Bailey, A.J., Shimokomaki, M., 1971. Age related changes in the reducible cross-links of collagen. FEBS Lett. 16, 86–88.

Bailey, A.J., Sims, T.J., 1977. Meat tenderness, distribution of molecular species of collagen in bovine muscle. J. Sci. Food Agric. 28, 565–570.

Bailey, A.J., Restall, D., Sims, T.J., Duance, V.C., 1979. Meat tenderness. Immunofluorescent localization of the isomorphic forms of collagen in bovine muscles of varying texture. J. Sci. Food Agric. 30, 203–210.

Bang, M.I., Gregorio, C., Labeit, S., 2002. Molecular dissection of the interaction of desmin with the C-terminal region of nebulin. J. Struct. Biol. 137, 119–127.

Bang, M.I., Li, X., Littlefield, R., Bremner, S., Thor, A., Knowlton, K.U., Lieber, R.I., Chen, J., 2006. Nebulin-deficient mice exhibit shorter thin filament lengths and reduced contractile function in skeletal muscle. J. Cell Biol. 173, 905–916.

Bang, M.I., Caremani, M., Brunello, F., Littlefield, R., Lieber, R.I., Chen, J., Lombardi, V., Linari, M., 2009. Nebulin plays a direct role in promoting strong actin–myosin interactions. FASEB J 23, 4117–4125.

Banks, A., 1955. The expressible fluid of fish fillets. IV. The expressible fluid of iced cod. J. Sci. Food Agric. 6, 584–587.

Barbut, S., 1998. Estimating the magnitude of the PSE problem in poultry. J. Muscle Foods 9, 35–49.

Barbut, S., Sosnicki, A.A., Lonergan, S.M., Knapp, T., Ciobanu, D.C., Gatcliffe, L.J., Huff-Lonergan, E., Wilson, E.W., 2008. Progress in reducing the pale, soft and exudative (PSE) problem in pork and poultry meat. Meat Sci. 79, 46–63.

Bate-Smith, E.C., 1948. The physiology and chemistry of rigor mortis, with special reference to the aging of beef. Adv. Food Res. 1, 1–38.

Bate-Smith, E.C., Bendall, J.R., 1949. Factors determining the time course of rigor mortis. J. Physiol. (London) 110, 47–65.

Bate-Smith, E.C., Bendall, J.R., 1956. Changes in muscle after death. Br. Med. Bull. 12, 230–235.

Batlle, N., Aristoy, M.-C., Toldra, E., 2000. Early postmortem detection of exudative pork meat based on nucleotide content. J. Food Sci. 65, 413–416.

Batzer, O.F., Santora, A.T., Tan, M.C., Landmann, W.A., Schweigert, B.S., 1960. Precursors of beef flavor. J. Agric. Food Chem. 8, 498–501.

Bausch, E.R., 1966. Color retention of red meat. Canadian Patent 742, 165.

Beecher, G.R., Briskey, E.J., Hockstra, W.G., 1965. Comparison of glycolysis and associated changes in light and dark portions of the porcine semitendinosus. J. Food Sci. 30, 477–486.

Bekhit, A.G.D., Faustman, C., 2005. Metmyoglobin reducing activity. Meat Sci. 71, 407–439.

Bendall, J.R., 1951. The shortening of rabbit muscles during rigor mortis: its relation to the breakdown of adenosine triphosphate and creatine phosphate and to muscular contraction. J. Physiol. (London) 114, 71–88.

Bendall, J.R., 1969. Muscles, Molecules and Movement. Heinemanns, London.

Bendall, J.R., 1973. Postmortem changes in muscle. In: Bourne, G.H. (Ed.)T he Structure and Function of Muscle, second ed., vol. 2. Academic Press, New York.

Bendall, J.R., 1976. Electrical stimulation of rabbit and lamb carcasses. J. Sci. Food Agric. 27, 819–826.

Bendall, J.R., Davey, C.L., 1957. Ammonia liberation during rigor mortis and its relation to changes in the adenine and inosine nucleotides of rabbit muscle. Biochim. Biophys. Acta. 26, 93–103.

Bendall, J.R., Lawrie, R.A., 1964. Watery pork. Anim. Breed. Abstr. 32, 1–8.

Bendall, J.R., Wismer-Pederson, J., 1962. Some properties of the fibrillar proteins of normal and watery pork muscle. J. Food Sci. 27, 144–159.

Bergstrom, J., Hultman, E., 1966. The effect of exercise on muscle glycogen and electrolytes. Scand. J. Clin. Lab. Invest. 18, 16–20.

Bito, M., 1965. Studies on the retention of meat color of frozen tuna. II. Effect of storage temperature on preventing discoloration of tuna meat during frozen storage. Nippon Suisan Gakkaishi 31, 534–539.

Bito, M., 1976. Studies on the retention of meat color of frozen tuna. Bull. Tokai Region. Fish. Res. Lab. 84, 51–113.

Bito, M., Yamada, K., Mikumo, Y., Amano, K., 1983. Studies on rigor mortis of fish. 1. Difference in the mode of rigor mortis among some varieties of fish by modified Cutting's method. Bull. Tokai Reg. Fish Res. Lab. 109, 89–96.

Bosselmann, A., Moller, C., Steinhart, H., Kirchgessner, M., Schwartz, F.J., 1995. Pyridinoline cross-links in bovine muscle collagen. J. Food Sci. 60, 953–958.

Bouton, P.E., Ford, A.L., Harns, P.V., Macfarlane, J.J., Oshea, J.M., 1977. Pressure–heat treatment of postrigor muscle. Effect on meat tenderness. J. Food Sci. 42, 132–135.

Bowker, B.C., Schaefer, R.B., Grapperhaus, M.J., Solomon, M.B., 2011. Tenderization of beef loins using a high efficiency sparker. Innov. Food Sci. Emerg. Technol. 12, 135–141.

Boyer-Berri, C., Greaser, M.L., 1998. Effect of postmortem storage on the Z-line region of titin in bovine muscle. J. Anim. Sci. 76, 1034–1044.

Bremner, H.A., 1999. Gaping in fish flesh. In: Satp, K., Sakaguchi, M., Bremner, H.A. (Eds.), Extracellular Matrix of Fish and Shellfish. Research Signpost, Trivandrum, India, pp. 81–94.

Briskey, E.J., 1964. Etiological status and associated studies of pale, exudative porcine musculature. Adv. Food Res. 13, 89–178.

Briskey, E.J., Lawrie, R.A., 1961. Comparative *in vitro* activities of phosphorylase b and cytochrome oxidase in preparations from two ox muscles. Nature (London) 192, 263–264.

Briskey, E.J., Kastenschmidt, L.L., Forrest, J.C., Beecher, G.R., Judge, M.D., Cassens, R.G., Hoekstra, W.G., 1966. Biochemical aspects of postmortem changes in porcine muscle. J. Agric. Food Chem. 14, 201–206.

Brooks, J., 1937. Color of meat. Food Ind. 9, 707.

Brown, W.D., Mebine, L.B., 1969. Autoxidations of oxymyoglobins. J. Biol. Chem. 244, 6696–6701.

Brown, W.D., Harris, L.D., Olcott, H.S., 1963. Catalysis of unsaturated lipid oxidation by iron protoporphyrin derivatives. Arch. Biochem. Biophys. 101, 14–20.

Buege, D.R., Marsh, B.B., 1975. Mitochondrial calcium and postmortem muscle shortening. Biochem. Biophys. Res. Commun. 65, 478–482.

Burns, G.B., Kee, P.J., Irvine, B.B., 1985. Objective procedure for fish freshness evaluation based on nucleotide changes using a HPLC system. Can. J. Fish Aquat. 22, 1373–1376.

Burt, J.R., 1966. Glycogenolytic enzymes of cod (*Gadus callarias*) muscle. J. Fish. Res. Board Can. 23, 527–538.

Busch, W.A., Stromer, M.H., Goll, D.E., Suzuki, A., 1972. Ca^{2+}-specific removal of Z lines from rabbit skeletal muscle. J. Cell Biol. 52, 367–381.

Buttkus, H., 1963. Red and white muscle of fish in relation to rigor mortis. J. Fish. Res. Board Can. 20, 45–58.

Cambero, M.I., Pereira-Lima, C.I., Ordonez, J.A., Garcia de Fernando, G.D., 2000. Beef broth flavour: relation of components with the flavour developed at different cooking temperatures. J. Sci. Food Agric. 80, 1519–1528.

Cappelin, G., Jessen, F., 2002. ATP, IMP and glycogen in cod muscles at onset and during development of rigor mortis depend on sample location. J. Food Sci. 67, 991–995.

Carse, W.A., 1973. Meat quality and the acceleration of postmortem glycolysis by electrical stimulation. J. Food Technol. 8, 163–166.

Casas, E., White, S.N., Wheeler, T.L., Shackelford, S.D., Koohmaraie, M., Riley, D.G., Chase Jr., C.C., Johnson, D.D., Smith, T.P., 2006. Effects of calastatin and μ-calpain markers in beef cattle on tenderness traits. J. Anim. Sci. 84, 520–525.

Cassens, R.G., 1966. General aspects of postmortem changes. In: Briskey, E.J., Cassens, R.G., Trautman, J.C. (Eds.), The Physiology and Biochemistry of Muscle as a Food. University of Wisconsin Press, Madison, WI, pp. 181–196.

Cassens, R.G., Newbold, R.P., 1966. Effects of temperature on postmortem metabolism in beef muscle. J. Sci. Food Agric. 17, 254–256.

Cassens, R.G., Kauffman, R.G., Scherer, A., Mecker, D.I., 1992. Variations in pork quality: a 1991 USA survey. In: Proceedings of the 38th International Congress of Meat Science and Technology. Clermont-Ferrand, France, pp. 237–240.

Chaijan, M., Benjakul, S., Visessanguan, W., Faustman, C., 2007. Characterisation of myoglobin from sardine (*Sardinella gibbosa*) dark muscle. Food Chem. 100, 156–164.

Chan, W.K.M., Decker, E.A., Yin, M., Faustman, C., 1997. Lipid oxidation induced by oxymyoglobin and metmyoglobin with involvement of H_2O_2 and superoxide anion. Meat Sci. 46, 181–190.

Chandra, M., Mamidi, R., Ford, S., Hidalgo, C., Witt, C., Ottenheijm, C., Labeit, S., Granzier, H., 2009. Nebulin alters cross-bridge cycling kinetics and increases the filament activation: a novel mechanism for increasing tension and reducing tension cost. J. Biol. Chem. 284, 30889–30896.

Chang, K.C., da Costa, N., Blackley, R., Southwood, O., Evans, G., Plastow, G., Wood, J.D., Richardson, R.I., 2003. Relationship of myosin heavy chain fibre types to meat quality traits in traditional and modern pigs. Meat Sci. 64, 93–103.

Charpentier, J., 1969. Postmortem biochemical characteristics of the sarcoplasmic reticulum of pig muscle. Ann. Biol. Anim. Biochim. Biophys. 9, 101–110.

Chaudhry, H.M., Parrish Jr., F.C., Goll, D.E., 1969. Molecular properties of postmortem muscle. VI. J. Food Sci. 34, 183–191.

Cheng, C.-S., Parrish Jr., F.C., 1977. Effect of Ca^{2+} on changes in myofibrillar proteins of bovine skeletal muscle. J. Food Sci. 42, 1621–1626.

Cheret, R., Delbarre-Ladrat, C., de Lamballerie-Anton, M., Verrez-Bagnis, V.V., 2007. Calpain and cathepsin activities in post mortem fish and meat muscles. Food Chem. 101, 1474–1479.

Chock, S.P., Chock, P.B., Eisenberg, E., 1976. Presteady-state kinetic evidence for a cyclic interaction of myosin subfragment one with actin during the hydrolysis of adenosine-5'-triphosphate. Biochemistry 15, 3244–3253.

Chrystall, B.B., Devine, C.E., 1978. Electrical stimulation, muscle tension, and glycolysis in bovine sternomandibularis. Meat Sci. 2, 49–58.

Chrystall, B.B., Hagyard, C.J., 1976. Electrical stimulation and lamb tenderness. N.Z.J. Agric. Res. 19, 7–11.

Chrystall, B.B., Devine, C.E., Davey, C.L., 1980. Studies in electrical stimulation: postmortem decline in nervous response in lambs. Meat Sci. 4, 69–76.

Ciobanu, D.C., Bastiaansen, J.W., Lonergan, S.M., Thomsen, H., Dekkers, J.C., Plastow, G.S., Rothchild, M.E., 2004. New alleles in calpastin gene are associated with meat quality traits in pigs. J. Anim. Sci. 82, 2829–2839.

Claus, R.R., 2002. Shock treatment: shock waves are an effective tool for tenderizing meat. Meat Poultry 48 (12), 61–63.

Claus, J.R., Schilling, M.B., Paroczay, E.W., Eastridge, J.S., Pursel, V.G., Mitchell, A.D., 2001a. Hydrodynamic shock wave tenderization effects using a cylinder processor on early deboned broiler breasts. Meat Sci. 58, 287–292.

Claus, J.R., Schilling, J.K., Marriott, N.G., Duncan, S.E., Solomon, M.B., Wang, H., 2001b. Tenderization of chicken and turkey meats with electrically produced hydrodynamic shockwaves. Meat Sci. 58, 283–286.

Clydesdale, F.M., Francis, F.J., 1971. The chemistry of meat color. Food Prod. Dev. 5 (3), 81. 87.

Cornforth, D.P., Hunt, M.C., 2008. Low-oxygen packaging of fresh meat with carbon monoxide: meat quality, microbiology, and safety. American Meat Science Association White Paper Series. 1–10.

Cornforth, D.P., Pearson, A.M., Merkel, R.A., 1980. Relationship of mitochondria and sarcoplasmic reticulum to cold shortening. Meat Sci. 4, 103–121.

Conlee, R.K., Renna, M.J., Winder, N.W., 1976. Skeletal muscle glycogen content, diurnal variation and effects of fasting. Am. J. Physiol. 231, 614–618.

Conover, G.M., Henderson, S.N., Gregorio, C.C., 2009. A myopathy-linked desmin mutation perturbs striated muscle actin filaments architecture. Mol. Biol. Cell. 20, 834–845.

Cornforth, D.P., 1994. Color – its basis and importance. In: Pearson, A.M., Dutson, T.R. (Eds.), Quality Attributes and their Measurements in Meat, Poultry and Fish Products. Blackie, London, pp. 34–78.

Coro, F.A.G., Youssef, E.Y., Shimokomaki, M., 2002. Age related changes in poultry breast meat collagen pyridinoline and texture. J. Food Biochem. 26, 533–541.

Crouse, J.D., Koohmaraie, M., Seideman, S.D., 1991. The relationship of muscle fibre size to tenderness of beef. Meat Sci. 30, 295–302.

Crupkin, M., Barassi, C.A., Martone, C.B., Trucco, R.E., 1979. Effect of storing hake (Merluccius merluccius hubbsi) on ice on the viscosity of the extract of soluble muscle protein. J. Sci. Food. Agric. 30, 911–913.

Culler, R.D., Parrish Jr., F.C., Smith, G.C., Cross, H.R., 1978. MFI relationship of myofibril fragmentation index to palatability and carcass attributes. J. Food Sci. 43, 1177–1180.

Cutting, C.L., 1939. Immediate post mortem changes in trawled fish. G.B. Dep. Sci. Ind. Res. Rep. Food Invest. Board, 39–40.

Dahiya, R., Butters, C.A., Tobacman, L.S., 1994. Equilibrium linkage analysis of cardiac thin filament assembly. Implications for the regulation of muscle contraction. J. Biol. Chem. 269, 29457–29461.

Dahlmann, B., Ruppert, T., Kloetzel, P.M., Kuehn, L., 2001. Subtypes of 20S proteasomes from skeletal muscle. Biochimie. 83, 295–299.

Daun, H., Solberg, M., Franke, W., Gilbert, S., 1971. Effect of oxygen-enriched atmospheres on storage quality of packaged fresh meat. J. Food Sci. 36, 1011–1014.

Davey, C.L., Gilbert, K.V., 1974. The mechanism of cold induced shortening in beef muscle. J. Food Technol. 9, 51–58.

Davey, C.L., Graafhuis, A.E., 1976. Structural changes in beef muscle during ageing. J. Sci. Food Agric. 27, 301–306.

Davey, C.L., Gilbert, K.V., Carse, W.A., 1976. Carcass electrical stimulation to prevent cold shortening toughness in beef. N.Z.J. Agric. Res. 19, 13–18.

Dayton, W.R., Schollmeyer, J.R., 1980. Isolation from papaine cardiac muscle of a calcium($2+$)-activated protease that partially degrades myofibrils. J. Mol. Cell. Cardiol. 12, 533–551.

Dayton, W.R., Goll, D.E., Stromer, M.H., Reville, W.J., Zeece, M.G., Robson, R.M., 1975. Some properties of a Ca^{2+}-activated protease that may be involved in myofibrillar protein activated protease that may be involved in myofibrillar protein turnover. In: Reich, E., Rifkin, D.B., Shaw, E. (Eds.), Proteases and Biological Control, vol. 2. Cold Spring Harbor Laboratory, Cold Spring Harbor, NY, pp. 551–577.

Dayton, W.R., Goll, D.E., Zeece, M.G., Robson, R.M., Reville, W.J., 1976a. A Ca^{2+}-activated protease possibly involved in myofibrillar protein turnover. Purification from porcine muscle. Biochemistry 15, 2150–2158.

Dayton, W.R., Reveille, W.J., Goll, D.E., Stromer, M.H., 1976b. A Ca^{2+}-activated protease possibly involved in myofibrillar protein turnover. Partial characterization of the purified enzyme. Biochemistry 15, 2159–2167.

De Duve, C., Pressman, B.C., Gianetto, R., Wattiaux, R., Appelman, F., 1955. Tissue fractionation studies 6. Intracellular distribution pattern of enzymes in rat liver tissue. Biochem. J. 60, 604–617.

De Fremery, D., 1966. Some aspects of post-mortem changes in poultry muscle. In: Briskey, E.J., Cassens, R.G., Trautman, J.C. (Eds.), The Physiology and Biochemistry of Muscle as a Food. University of Wisconsin Press, Madison, WI, pp. 205–212.

De Fremery, D., Pool, M.F., 1960. Biochemistry of chicken muscle as related to rigor mortis and tenderization. Food Res. 25, 73–87.

Delbarre-Ladrat, C., Cheret, R., Taylor, R., Verrez-Bagnis, V., 2006. Trends in postmortem aging in fish: understanding of proteolysis and disorganization of the myofibrillar structure. CRC Crit. Rev. Food Sci. Nutr. 46, 409–421.

Devine, C.E., Ellery, S., Wade, L., Chrystall, B.B., 1984. Differential effects of electrical stunning on the early post mortem glycolysis in sheep. Meat Sci. 11, 301–309.

Disney, J.G., Follett, M.J., Ratcliff, P.W., 1967. Biochemical changes in beef muscle postmortem. J. Sci. Food Agric. 18, 314–321.

Dransfield, E., 1977. Intramuscular composition and texture of beef muscles. J. Sci. Food Agric. 28, 833–842.

Dugal, L.C., 1967. Hypoxanthine in iced freshwater fish. J. Fish. Res. Board Can. 24, 2229–2239.

Dunajski, E., 1979. Texture of fish muscle. J. Texture Stud. 10, 301–318.

Dutand, D., 1998. Quantification and characterization of 20S proteasome in relation to meat tenderization. PhD thesis. Blaise Pascal University, Clermont-Ferrand, France.

Dutand, D., Aubry, L., Guignot, F., Monin, G., Oali, A., 2006. Bovine muscle 20S proteasome to meat tenderization as revealed by an ultrasonic approach. Meat Sci. 74, 337–344.

Dyer, W.J., 1951. Protein denaturation in frozen and stored fish. J. Fish. Res. Biol. 24, 2229.

Ebashi, S., 1974. Regulatory mechanism of muscle contraction with special reference to the catroponin–tropomyosin system. Essays Biochem. 10, 1–36.

Eccleston, J.F., Geeves, M.A., Trentham, D.R., Bagshaw, C.R., Mowa, V., 1976. The binding and cleavage of ATP in the myosin and actomyosin ATPase mechanisms. Colloq. Ges. Biol. Chem. 26, 42.

Eino, M.F., Stanley, D.W., 1973. Cathepic activity, textural properties and surface ultrastructure of postmortem beef muscle. J. Food Sci. 38, 45–50.

El-Badawi, A.A., Cain, R.F., Samuels, C.E., Anglemeir, A.F., 1964. Color and pigment stability of packaged refrigerated beef. Food Technol. 18, 154–163.

Elgasim, E.A., 1977. The effect of ultrahydrostatic pressure of prerigor muscle on characteristics of economic importance. MS thesis. Oregon State University, Corvallis.

Elgasim, E.A., Kennick, W.H., Anglemeir, A.F., Koohmaraie, M., Elkhalifa, E.A., 1983. Effect of pressurization on bovine lysosomal enzyme activity. Food Microstruct. 2, 91–98.

Elgasim, E.A., Koohmaraie, M., Anglemeir, A.F., Kennick, W.H., Elkhalifa, E.A., 1985. The combined effects of the calcium activated factor and cathepsin D on skeletal muscle. Food Microstruct. 4, 55–62.

Elkhalifa, E.A., Anglemeir, A.F., Kennick, W.H., Elgasim, E.A., 1984a. Effect of prerigor pressurization on post-mortem bovine muscle lactate dehydrogenase activity and glycogen degradation. J. Food Sci. 49, 593–594.

Elkhalifa, E.A., Anglemeir, A.F., Kennick, W.H., Elgasim, E.A., 1984b. Influence of prerigor pressurization on beef muscle creatine phosphokinase activity and degradation of creatine phosphate and adenosine triphosphate. J. Food Sci. 49, 595–597.

Enfors, S.O., Molin, G., Ternstrom, A., 1979. Effect of packaging under carbon dioxide, nitrogen, or air on the microflora of pork stored at 4°C. J. Appl. Bacteriol. 47, 197–208.

Erikson, U., Misimi, E., 2008. Atlantic salmon skin and fillet color changes effects by postmortem handling stress, rigor mortis, and ice storage. J. Food Sci. 73, C50–C59.

Erikson, U., Beyer, A.R., Sigholt, T., 1997. Muscle high-energy phosphates and stress affect K-values during ice-storage of Atlantic salmon (Salmo salar). J. Food Sci. 62, 43–47.

Essen-Gustavson, B., Karlstrom, K., Poso, R., 1991. Adenine nucleotide breakdown products in muscle at slaughter and their relation to meat quality in pigs with different halothane genotypes. Proceedings of the 37th International Congress on Meat Science Technology vol. 1 Kulmbach, Germany.

Etherington, D.J., Sims, T.J., 1981. Detection and estimation of collagen. J. Sci. Food Agric. 32, 539–546.

Eyre, D.R., Muir, G.C., 1974. Collagen polymorphism. Two molecular species in pig invertebral disc. FEBS Lett. 42, 192–196.

Eyre, D.R., Oguchi, H., 1980. The hydroxypyridinium crosslinks of skeletal collagens: their measurement, properties and a proposed pathway of formation. Biophys. Res. Commun. 92, 403–410.

Eyre, D.R., Paz, M.A., Gallop, P.M., 1984. Crosslinking in collagen and elastin. Ann. Rev. Biochem. 53, 717–748.

Fabiansson, S., Erichsen, I., Reutersward, M.L., 1984. The incidence of dark cutting beef in Sweden. Meat Sci. 10, 21–33.

Farmer, L.J., 1999. Poultry Meat Flavour. CABI, Bristol.

Farmer, L.J., Hagan, T.D.J., Paraskevas, O., 1996. A comparison of three sugars and inosine monophosphate as precursors of meat aroma. In: Taylor, A.J., Mottram, D.S. (Eds.), Flavour Science: Recent Developments. Royal Society of Chemistry, Cambridge, pp. 225–230.

Farmer, L.J., Hagan, T.D., Paraskevas, O., 1999. Role of selected precursors in meat flavour formation. In: Xiong, Y.L., Ho, C.T., Shahidi, F. (Eds.), Quality Attributes of Muscle Foods. Kluwer Academic, New York, pp. 159–172.

Faustman, C., Sun, Q., McClure, T.D., Liebler, D.C., 1999. Alpha, beta-unsaturated aldehydes accelerate oxymyoglobin oxidation. J. Agric. Food Chem. 47, 3140–3144.

Fellers, D.A., Wahba, I.J., Caldano, J.C., Ball, C.O., 1963. Factors affecting the color of packaged retail beef cuts. Origin of cuts, package type, and storage conditions. Food Technol. 17, 95.

Fischer, C., Hamm, R., 1980. Biochemical studies on fast glycolysing bovine muscles. Meat Sci. 4, 41–49.

Fischer, C., Hamm, R., Honikel, K.O., 1979. Changes in solubility and enzymic activity of muscle glycogen phosphorylase in PSE-muscles. Meat Sci. 3, 11–19.

Fjelkner-Modig, S., Ruderus, H., 1983. The influence of exhaustion and electrical stimulation on the meat quality of young bulls. Part 2. Physical and sensory properties. Meat Sci. 8, 203–220.

Follett, M.J., Norman, G.A., Ratcliff, P.W., 1974. The ante-rigor excision and air cooling of beef semimembranosus muscles at temperatures between −5°C and +15°C. J. Food Technol. 9, 509–523.

Forrest, J.C., Briskey, E.J., 1967. Response of striated muscle to electrical stimulation. J. Food Sci. 32, 483–488.

Forster, A., Hill, C.P., 2003. Proteasome degradation: enter the substrate. Trends Cell. Biol. 13, 550–553.

Fortin, A., 1989. Preslaughter management of pigs and its influence on the quality (PSE/DFD) of pork. Proceedings of the 35th International Congress of Meat Science and Technology. Roskilder. Danish Meat Science Research Institute. 981–986.

Fosmire, G.J., Brown, W.D., 1976. Yellowfin tuna (*Thunnus albares*) myoglobin: characterization and comparative stability. Comp. Biochem. Physiol. B. 55, 293–299.

Fraser, D.I., Punjamapirom, S., Dyer, W.J., 1961. Temperature and the biochemical processes occurring during rigor mortis in cod muscle. J. Fish Res. Board Can. 18, 641–644.

Fritz, J.D., Greaser, M.L., 1991. Changes in titin and nebulin in postmortem bovine revealed by gel electrophoresis, western blotting and immunofluorescence microscopy. J. Food Sci. 56, 607–610.

Fujita, T., Hori, Y., Otani, T., Kunita, Y., Sawa, S., Sakai, S., Tanaka, Y., Takagahara, I., Nakatani, Y., 1988. Applicability of the K_0 value as an index of freshness for porcine and chicken muscles. Agric. Biol. Chem. 52, 107–112.

Fujimoto, D., 1977. Isolation and characterization of a fluorescent material in bovine Achilles tendon collagen. Biochem. Biophys. Res. Commun. 76, 1124–1129.

Furst, D.O., Osborn, M., Nave, P., Weber, K., 1988. The organization of titin filaments in the half sarcomere revealed by monoclonal antibodies in immunoelectron microscopy: a map of ten nonrepetitive epitomes starting at the Z line extends close to the M line. J. Cell. Biol. 106, 1563–1572.

Gasc, N., Tache, S., Rathahao, E., Bertrand-Michel, J., Roques, V., Guerand, F., 2007. 4-Hydroxynonenal in foodstuffs: haem concentration, fatty acid composition and freeze-drying are determining factors. Redox. Rep. 12, 40–44.

Geesink, C., Ilian, M., Moton, J.D., Bickerstaffe, R., 2000a. Involvement of calpains in post-mortem tenderization: a review of recent research. Proc. N.Z. Soc. Anim. Prod. 60, 99–102.

Geesink, C., Morton, J., Kent, M., Bickerstaffe, R., 2000b. Partial purification and characterization of Chinook salmon (*Oncorhynchus tshawytscha*) calpains and an evaluation of their role in postmortem proteolysis. J. Food Sci. 65, 1318–1324.

Geesink, G.H., Taylor, R.G., Koohmaraie, M., 2005. Calpain 3/p94 is not involved in post-mortem proteolysis. J. Anim. Sci. 83, 1646–1652.

George, P., Stratmann, C.J., 1952. The oxidation of myoglobin to metmyoglobin by oxygen. 2. The relation between the first order rate constant and the partial pressure of oxygen. Biochem. J. 51, 418–425.

Ghanekar, D.S., Bal, D.V., Kamala, S., 1956. Enzymes of some elasmobranchs from Bombay. III. Amylases of Scoliodon sorrakowah and Sphyrna blochii. Proc. Indian Acad. Sci. Sect. B. 43, 134–148.

Giddings, G.G., 1977a. The basis of color in muscle foods. J. Food Sci. 42, 288–294.

Giddings, G.G., 1977b. The basis of color in muscle foods. CRC Crit. Rev. Food Sci. Nutr. 9, 81–144.

Giddings, G.G., Markarkis, P., 1973. On the interaction of myoglobin and hemoglobin with molecular oxygen and its lower oxidation states with cytochrome c. J. Food Sci. 38, 705–709.

Gispert, A., Faucitano, L., Oliver, M.A., Guardia, M.D., Coll, C., Siggens, K., Harvey, K., Diestre, A., 2000. A survey of slaughter conditions halothane gene frequency in five Spanish pig commercial abattoirs. Meat Sci. 55, 97–106.

Godfrey, J. 1970. Tenderizing meat. US Patent 3,492,688.

Gokhin, D.S., Bang, M.I., Zhang, J., Chen, J., Lieber, R.I., 2009. Reduced thin filament length in nebulin-knockout skeletal muscle alters isometric contractile properties. Am. J. Physiol. Cell. Physiol. 296, C1123–C1132.

Goll, D.E., Otsuka, Y., Nagainis, P.A., Shannon, J.D., Sathe, S.K., Muguruma, M., 1983. Role of muscle proteinases in maintenance of muscle integrity and mass. J. Food Biochem. 7, 137–177.

Goll, D.E., Thompson, V.F., Taylor, R.G., Christiansen, J.A., 1992. Role of the calpain system in muscle growth. Biochimie. 74, 225–237.

Goll, D.E., Thompson, V.F., Li, H., Wei, W., Cong, J., 2003. The calpain system. Physiol. Rev. 83, 731–801.

Goodno, C.C., Wall, C.M., Perry, S.V.P., 1978. Kinetics and regulation of the myofibrillar adenosine triphosphatase. Biochem. J. 175, 813–821.

Goutefongea, R., 1971. Influence du pH et de la température sur le solubilite des protéines musculaires du porc. Ann. Biol. Anim. Biochim. Biophys. 11, 233–244.

Grandin, T., 1980. The effect of stress on livestock and meat quality prior to and during slaughter. Int. J. Stud. Anim. Prob. 1, 313–337.

Granger, B.L., Lazarides, E., 1978. The existence of an insoluble Z-disc scaffold in chicken skeletal muscle. Cell (Cambridge Mass.) 15, 1253–1268.

Greaser, M.L., Gergely, J., 1973. Purification and properties of the components from troponin. J. Biol. Chem. 248, 2125–2133.

Greene, B.E., 1969. Lipid oxidation and pigment changes in raw beef. J. Food Sci. 34, 110–113.

Greene, B.E., Hsin, I., Zipser, M.W., 1971. Retardation of oxidative color changes in raw ground beef. J. Food Sci. 36, 940–942.

Guardia, M.D., Estany, J., Balasch, S., Oliver, M.A., Gispert, M., Dietstre, A., 2004. Risk assessment of PSE condition due to preslaughter conditions and RYRI gene in pigs. Meat Sci. 67, 471–478.

Guroff, G., 1964. A neutral calcium-activated proteinase from the soluble fraction of rat brain. J. Biol. Chem. 239, 149–155.

Hagen, O., 2008. Protease activity impacts flesh colour during post-mortem storage of farmed Atlantic halibut (*Hippoglossus hippoglossus*): a condition referred to as chalky halibut. Food Chem. 125, 1294–1298.

Hagen, O., Solberg, C., Johnston, I.A., 2008. Activity of aspargate (cathepsin D), cysteine proteases (cathepsins B, B + L, and H) and their influence on protein and water-holding capacity of muscle in commercially farmed Atlantic halibut (*Hippoglossus hippoglossus* L.). J. Agric. Food Chem. 56, 5953–5959.

Hagler, L., Coppes, R.I., Herman, R.H., 1979. Metmyoglobin reductase. Identification and purification of a reduced nicotinamide adenine dinucleotide-dependent enzyme from bovine heart which reduces metmyoglobin. J. Biol. Chem. 254, 6505–6514.

Hallund, O., Bendall, J.R., 1965. Long-term effect of electrical stimulation on the post-mortem fall of pH in the muscles of Landrace pigs. J. Food Sci. 30, 296–299.

Hamm, R., 1977. Postmortem breakdown of ATP and glycogen in ground muscle: a review. Meat Sci. 1, 15–39.

Hamm, R., Dalrymple, R.H., Honikel, K.O., 1973. Proceedings of the 19th Meeting of European Meat Research Workers vol. 1, 73.

Hamoir, G., Konosu, S., 1965. Carp myogens of white and red muscles. General composition and isolation of low molecular weight components of abnormal amino acid composition. Biochem. J. 96, 85–97.

Hanson, J., Huxley, H.E., 1955. The structural basis of contraction in striated muscle. Symp. Soc. Exp. Biol. 9, 228–264.

Harsham, A. and Detherage, F. E. 1951. Tenderization of meat. US Patent 2,544,681.

Haselgrove, J.C., 1972. X-ray evidence for a conformational change in the actin-containing filaments of vertebrate striated muscle. Cold Spring Harbor Symp. Quant. Biol. 37, 341–352.

Hashimoto, Y., 1965. Taste-producing substances in marine products. In: Kreuzer, R. (Ed.), The Technology of Fish Utilization. Fishing News (Books), London, p. 57.

Hay, J.D., Currie, R.W., Wolfe, F.H., Sanders, E.J., 1973. Effects of post-mortem aging on chicken muscle fibrils. J. Food Sci. 38, 981–986.

Herrera-Mendez, C.H., Becila, S., Boudjellal, A., Ouali, A., 2006. Meat aging: reconsideration of the current concept. Trends Food Sci. Technol. 17, 394–405.

Herring, H.K., Cassens, R.G., Briskey, E.J., 1965. Sarcomere length of free and restrained bovine muscles at low temperature as related to tenderness. J. Sci. Food Agric. 16, 379–384.

Hirao, T., Hara, T., Takahashi, K., 1984. Purification and characterization of cathepsin B from monkey skeletal muscle. J. Biochem. 95, 871–879.

Hollung, K., Veiseth, E., Frystein, T., Aass, L., Langsrud, O., Hildrum, K.I., 2007. Variation in the response to manipulation of post-mortem glycolysis in beef muscles by low-voltage electrical stimulation and conditioning temperature. Meat Sci. 77, 372–383.

Homma, N., Ikeuchi, Y., Suzuki, A., 1994. Effects of high pressure treatment on the proteolytic enzymes in beef. Meat Sci. 38, 219–228.

Homma, N., Ikeuchi, Y., Suzuki, A., 1995. Levels of calpain and calpastatin in meat subjected to high pressure. Meat Sci. 41, 251–260.

Honikel, K.O., Fischer, C., 1977. A rapid method for the detection of PSE and DFD porcine muscle. J. Food Sci. 42, 1633–1636.

Honikel, K.O., Roncales, P., Hamm, R., 1983. The influence of temperature on shortening and rigor onset in beef muscle. Meat Sci. 8, 221–241.

Honkavaara, M., 1989. Influence of selection phase, fasting and transport on porcine stress and on the development of PSE pork. J. Agric. Sci. Finl. 61, 415–423.

Hood, D.E., 1980. Factors affecting the rate of metmyoglobin accumulation in prepackaged beef. Meat Sci. 4, 247–265.

Hopkins, D.L., Taylor, R.G., 2002. Post-mortem muscle proteolysis and meat tenderization. In: te Pas, M., Everts, M., Haagsman, H. (Eds.), Muscle Development of Livestock Animals. CAB International, Cambridge, MA, pp. 363–389.

Hopkins, D.L., Thompson, J.M., 2001. Inhibition of protease activity. 2. Degradation of myofibrillar proteins, myofibril examination and determination of free calcium levels. Meat Sci. 59, 199–209.

Hopkins, D.L., Thompson, J.M., 2002. Factors contributing to proteolysis and disruption of myofibrillar proteins and the impact on tenderization in beef and sheep meat. Aust. J. Agric. Res. 53, 149–166.

Horgan, D.J., Kuypers, R., 1985. Post-mortem glycolysis in rabbit longissimus dorsi muscles following electrical stimulation. Meat Sci. 12, 225–241.

Horowitz, R., 2006. Nebulin regulation of actin filament lengths: new angles. Trends Cell. Biol. 16, 121–124.

Houbak, M.B., Ertbjerg, P., Therkildsen, M., 2008. In vitro studies to evaluate the degradation of bovine muscle proteins post-mortem by proteasome and microcalpain. Meat Sci. 79, 77–85.

Howard, A., Lawrie, R.A., 1956. Beef quality. II. Physiological and biological effects of various preslaughter treatments. Div. Food Preserv. Tech. Pap. (Aust. C.S.I.R.O.) 2, 18.

Hoyle, G., 1967. Diversity of striated muscle. Am. Zool. 7, 435–449.

Huff-Lonergan, E., Lonergan, S., 2005. Mechanisms of water-holding capacity of meat: the role of postmortem biochemical and structural changes. Meat Sci. 71, 194–204.

Huff-Lonergan, E., Mitsusashi, T., Beekman, D.D., Parrish Jr., F.C., Olson, D.G., Robson, R.M., 1996. Proteolysis of specific muscle structural proteins by μ-calpain at low pH and temperature is similar to degradation in postmortem bovine muscle. J. Anim. Sci. 74, 993–1008.

Huff-Lonergan, E., Zhang, W., Lonergan, S.M., 2010. Biochemistry of postmortem muscle — lessons on mechanism of meat tenderization. Meat Sci. 86, 184–195.

Huffman, D.L., 1974. Effect of gas atmospheres on microbial quality of pork. J. Food Sci. 39, 723–725.

Hughes, R.B., Jones, N.R., 1966. Measurement of hypoxanthine concentration in canned herring as an index of the freshness of the raw material, with a comment on flavour relations. J. Sci. Food Agric. 17, 434–436.

Hutchins, B.K., Liu, T.H.P., Watts, B.M., 1967. Effect of additives and refrigeration on the reducing activity, metmyoglobin and malonaldehyde of raw ground beef. J. Food Sci. 32, 214–217.

Huxley, H.E., 1964. Structural arrangements and the contraction mechanism in striated muscle. Proc. R. Soc. Lond. Ser. B. 160, 442–448.

Huxley, H.E., 1969. The mechanism of muscular contraction. Science 164, 1356–1365.

Huxley, H.A., 1972b. Structural changes in the actin and myosin-containing filaments [of muscle] during contraction. Cold Spring Harbor Symp. Quant. Biol. 37, 361–376.

Hwang, I.H., Devine, C.E., Hopkins, D.L., 2003. Review: The biochemical and physical aspects of electrical stimulation on beef and sheep tenderness. Meat Sci. 65, 677–691.

Hwang, I.H., Park, B.Y., Kim, S.H., Cho, J.M.L., 2004. Assessment of postmortem proteolysis by gel-based proteome analysis and its relationship to meat quality traits in pig longissimus. Meat Sci. 69, 79–91.

Ilian, M.A., Bekhit, A.E.D., Bickerstaffe, R., 2004a. The relationship between meat tenderization, myofibril fragmentation and autolysis of calpain 3 during post-mortem aging. Meat Sci. 66, 317–327.

Ilian, M.A., Bekhit, A.E.D., Bickerstaffe, R., 2004b. Does the newly discovered calpain 10 play a role in meat tenderization during post-mortem storage? Meat Sci. 66, 387–397.

Ishikawa, H., Bischoff, R., Holtzer, H., 1968. Mitosis and intermediate-sized filaments in developing skeletal muscle. J. Cell. Biol. 38, 538–555.

Iwamoto, M., Yamanaka, H., Watabe, S., Hashimoto, K., 1987. Effect of storage temperature on rigor mortis and ATP depletion in plaice (Paralichthys olivaceus) muscle. J. Food Sci. 52, 1514–1517.

Jakobsen, M., Bertelsen, G., 2000. Color stability and lipid oxidation of fresh beef. Development of a response surface model for predicting the effects of temperature, storage time and modified stress atmosphere composition. Meat Sci. 54, 49–57.

Jayasingh, P., Cornworth, D.P., Brennand, C.P., Carpenter, C.E., Whittier, D.R., 2002. Sensory evaluation of ground beef stored in high oxygen modified atmosphere. J. Food Sci. 67, 3493–3496.

Jeacocke, R.E., 1977. The temperature dependence of anaerobic glycolysis in beef muscle held in a linear temperature gradient. J. Sci. Food Agric. 28, 551–556.

Jia, X., Veiseth-Kent, E., Grove, H., Kuziora, P., Aass, L., Hildrum, K.I., Hollung, K., 2009. Peroxiredoxin-6 — a potential protein marker for meat tenderness in bovine longissimus thoracis muscle. J. Anim. Sci. 87, 2391–2399.

Jiang, S.-T., Hwang, B.-S., Tsao, C.-T., 1987. Protein denaturation and changes in nucleotides of fish muscle during frozen storage. J. Agric. Food Chem. 35, 22–27.

John, L., Cornforth, D.P., Carpenter, C.E., Sorheim, O., Pettee, B.C., Whittier, D.R., 2004. Comparison of odor and thiobarbituric acid values of cooked hamburger patties after storage of beef chubs in modified atmospheres. J. Food Sci. 69, 608–614.

John, L., Cornforth, D.P., Carpenter, C.E., Sorheim, O., Pettee, B.C., Whittier, D.R., 2005. Color and thiobarbituric acid values of cooked top sirloin steaks packaged in modified atmospheres of 80% oxygen, or 0.4% carbon monoxide, or vacuum. Meat Sci. 69, 441–449.

Jones, N.R., 1965. Interconversions of flavorous catabolites in chilled frozen fish. Prog. Refrig. Sci. Technol. Proc. 11th Int. Congr. Refrig. 1963, 917.

Jones, N.R., Murray, J., 1961. Nucleotide concentration in codling (*Gadus callarias*) muscle passing through rigor mortis at 0°C. Z. Vergl. Physiol. 44, 174–177.

Jones, N.R., Murray, J., Livingstone, E.I., 1964. Rapid estimations of hypoxanthine concentrations as indices of the freshness of chill-stored fish. J. Sci. Food Agric. 15, 763–774.

Jones, N.R., Burt, J.R., Murray, J., Stroud, G.D., 1965. Nucleotides and the analytical approach to the rigor mortis problem. In: Kreuzer, R. (Ed.), The Technology of Fish Utilization. Fishing News (Books), London, pp. 14–20.

Joseph, P., Suman, S.P., Li, S., Beach, C.M., Steinke, L., Fontaine, M., 2010. Characterization of bison (*Bison bison*) myoglobin. Meat Sci. 84, 71–78.

Joseph, P., Suman, S.P., Li, S., Claus, J.R., Fontaine, M., 2011. Primary structure of turkey myoglobin. Food Chem. 129, 175–178.

Joseph, R.L., 1968a. Biochemistry and quality in beef. Part I. Process Biochem. 3 (7), 20.

Joseph, R.L., 1968b. Biochemistry and quality in beef. Part II. Process Biochem. 3 (9), 32.

Jung, S., de Lamballerie-Anton, M., Taylor, R.G., Ghoul, M., 2000. High-pressure effects on lysosome integrity and lysosomal enzyme activity in bovine muscle. J. Agric. Food Chem. 48, 2467–2471.

Kammer, S.M., 2000. Chalky halibut investigations 1997–1999. Technical Report No. 4. International Pacific Halibut Commission, Washington, pp. 1–24.

Kannan, G., Chawan, C.B., Kouakou, B., Gelaye, B., 2002. Influence of packaging method and storage time on shear value and mechanical strength of intramuscular connective tissue of chevon. J. Anim. Sci. 80, 2383–2389.

Karijama, H., 1988. Shape of the myosin head in the rigor complex. Three-dimensional image reconstruction of the actin-tropomyosin heavy meromyosin complex. J. Mol. Biol. 204, 639–652.

Karube, I., Sato, I., Araki, Y., Suzuki, S., Hideaki, Y., 1980. Monoamine oxidase electrode in freshness testing of meat. Enzyme Microb. Technol. 2, 117–120.

Karube, I., Maatsuoka, H., Suzuki, S., Watanabe, E., Toyama, K., 1984. Determination of fish freshness with an enzyme sensor system. J. Agric. Food Chem. 32, 314–319.

Kassemsarn, B.O., Sang, P., Murray, J., Jones, N.R., 1963. Nucleotide degradation in the muscle of ice haddock, lemon sole and plaice. J. Food Sci. 28, 28–37.

Kastenschmidt, L.L., Hoekstra, W.G., Briskey, E.J., 1968. Glycolytic intermediates and co-factors in 'fast' and 'slow-glycolyzing' muscles of the pig. J. Food Sci. 33, 151–158.

Kelly, T.R., 1969. Quality in frozen cod and limiting factors on its shelf life. J. Food Technol. 4, 95–103.

Kelly, T.R., Little, W.T., 1966. Brown discolouration in prerigor cut fish fillets. J. Food Technol. 1, 121–129.

Kemp, C.M., Sensky, P.L., Bardsley, R.G., Buttery, P.I., Parr, T., 2010. Tenderness – an enzymatic view. Meat Sci. 84, 248–256.

Kennick, W.H., Elgasim, E.A., Holmes, Z.A., Meyer, P.F., 1980. The effects of ultra-hydrostatic pressurization of prerigor muscle on pre-rigor meat characteristics. Meat Sci. 4, 33–40.

Kiely, P.J., Nowlin, A.C., Moriarty, J.H., 1960. Bread aromatics from browning systems. Cereal Sci. Today 5, 273–274.

Kiessling, A., Espe, M., Roohonen, R., Morkore, T., 2004. Texture, gaping and colour of fresh and frozen Atlantic salmon flesh as affected by pre-slaughter isoeugenol or CO_2 anesthesia. Aquaculture 236, 645–657.

Kinbara, K., Ishiura, S., Tomioka, S., Sorimachi, H., Jeoong, S.Y., Amano, S., Kawasaki, H., Kilmerer, B., Kimuras, S., Labeit, S., Suzuki, K., 1998. Purification of p94, a muscle-specific calpain, and characterization of its autolysis. Biochem. J. 335, 589–596.

King, N.L., 1984. Breakdown of connectin during cooking of meat. Meat Sci. 11, 27.

Kinbara, K., Ishiura, S., Tomioka, S., Sarimachi, H., Jeong, S.Y., Amano, S., Kawasaki, H., Kilmerer, B., Kimuras, S., Labeit, S., Suzuki, K., 1998. Purification of native p94, a muscle calpain, and characterization of its autolysis. Biochem. J. 335, 589–596.

Koehler, H.H., Jacobson, M., 1967. Characteristics of chicken flavor-containing fraction extracted from raw muscle. J. Agric. Food Chem. 15, 707–712.

Konagaya, S., Konagaya, T., 1979. Acid denaturation of myofibrillar protein as the main cause of formation of 'yake niku', a spontaneously done meat, in red meat fish. Nippon Suisan Gakkaishi 45 (2), 245.

Koohmaraie, M., Geesink, G.H., 2006. Contribution of postmortem muscle biochemistry to the delivery of consistent meat quality and particular focus on the calpain system. Meat Sci. 74, 34–43.

Koohmaraie, M., Kennick, W.H., Elgasim, E.A., Anglemeir, A.F., 1984. Effect of prerigor pressurization on the activity of calcium-activated factor. J. Food Sci. 49, 680–684.

Koohmaraie, M., Whipple, G., Kretchmar, D.H., Crouse, D.H., Mermann, H.J., 1991. Postmortem proteolysis of longissimus muscle from beef, lamb and pork carcasses. J. Anim. Sci. 69, 617–624.

Koohmaraie, M., Kent, M.P., Shackelford, S.D., Veiseth, E., Wheeler, T.L., 2002. Meat tenderness and muscle growth: is there any relationship? Meat Sci. 62, 345–352.

Koutsidis, G., Elmore, J.S., Oruna-Concha, M.J., Campo, M.M., Wood, J.D., Mottram, D.S., 2008. Water-soluble precursors of beef flavour. II: Effect of post-mortem conditioning. Meat Sci. 79, 270–277.

Kristensen, L., Purlow, P.P., 2001. The effect of ageing on the water-holding capacity of pork: role of cytoskeleton proteins. Meat Sci. 58, 17–23.

Kropf, D.H., 1993. Color stability. Meat Focus Int. 2, 269–275.

Kruger, M., Wright, J., Wang, K., 1991. Nebulin as a tough regulator of thin filaments of vertebrate skeletal muscles: correlation of thin filament length. Nebulin size, and epitome profile. J. Cell Biol. 115, 97–107.

Kubo, T., Gerelt, B., Han, G.N., Sugiyama, T., Nishiumi, T., Suzuki, A., 2002. Changes in immunoelectron microscopic localization of cathepsin D in muscle induced conditioning or high-pressure treatment. Meat Sci. 61, 415–418.

Kuchenmeister, U., Kuhn, G., Wagner, J., Nurnberg, G., Ender, K., 1999. Postmortem changes in Ca^{2+} transporting proteins of sarcoplasmic reticulum in dependence on malignant hypothermia status in pigs. Mol. Cell. Biochem. 195, 37–46.

Kuchenmeister, U., Kuhn, G., Ender, K., 2000. Seasonal effects on Ca^{2+} transport of sarcoplasmic reticulum and on meat quality of pigs with different malignant hypertemia status. Meat Sci. 55, 239–245.

Kuhner, S., Fischer, S., 2011. Structural mechanism of the ATP-induced dissociation of rigor myosin from actin. PNAS. 108, 7793–7798.

Kuninaka, A., 1967. Flavour potentiators. In: Schultz, H.W., Day, E.A., Libbey, L.M. (Eds.), Chemistry and Physiology of Flavours. Avi, Westport, CT, pp. 515–535.

Kurtzman, C.H., Sjostrom, L.B., 1964. The flavor-modifying properties of disodium inosinate. Food Technol. 18, 221–225.

Kurzban, G.P., Wang, K., 1988. Grant polypeptides of skeletal muscle titin: sedimentation equilibrium in guanidine hydrochloride. Biochem. Biophys. Res. Commun. 150, 1155–1161.

Labeit, S., Gibson, T., Lakey, A., Leonard, K., Zeviani, M., Knight, P., Wardale, J., Trinick, J., 1991. Evidence that nebulin is a protein-ruler in muscle thin filaments. FEBS Lett. 282, 313–316.

Labeit, S., Ottenheijm, C.A.C., Granzier, H., 2011. Nebulin, a major player in muscle health and disease. FASEB J. 25, 822–829.

Lamare, M., Taylor, R.G., Farout, L., Briand, Y., Briand, M., 2002. Changes in proteosome activity during postmortem aging of bovine muscle. Meat Sci. 61, 199–204.

Landrock, A.H., Wallace, G.A., 1955. Discoloration of fresh red meat and its relationship to film oxygen permeability. Food Technol. 9, 194–196.

Lawrie, R.A., 1966. The effect of species in a given muscle (longossimus dorsi) and at a given temperature (37°C) on the progress of glycolysis measured by a fall in pH. Meat Science, First English edition. Woodhead Publishing Ltd., Cambridge, UK.

Lawrie, R.A., 1998. Meat Science, sixth ed. Woodhead Publishing Ltd., Cambridge, UK. p. 106.

Lawrie, R.A., Gatherum, D.P., Hale, H.P., 1958. Abnormally low ultimate pH in pig muscle. Nature (London) 182, 807–808.

Lazarides, E., Hubbard, B.D., 1976. Immunological characterization of the subunit of the 100 Å filaments from muscle cells. Proc. Natl. Acad. Sci. U.S.A. 73, 4344–4348.

Lee, C.A., Newbold, R.P., 1963. The pathway of degradation of inosinic acid in bovine skeletal muscle. Biochim. Biophys. Acta. 72, 349–352.

Lee, E.H., Oshima, T., Koizumi, C., 1982. High performance liquid chromatographic determination of K value as an index of freshness of fish. Bull. Jpn. Soc. Sci. Fish. 48, 255.

Lee, S., Faustman, C., Liebler, D.C., Phillips, A.L., 2003a. Porcine oxymyoglobin and lipid oxidation in vitro. Meat Sci. 63, 241–247.

Lee, S., Joo, S.T., Alderton, A.L., Hill, D.W., Faustman, C., 2003b. Porcine oxymyoglobin and lipid oxidation in yellowfin tuna (Thunnus albacores) loins. J. Food Sci. 68, 1664–1668.

Lehman, W., Galinska-Rakoczy, A., Hatch, V., Tobacman, L.S., Crig, R., 2009. Structural basis for the activation of muscle contraction by troponin and tropomyosin. J. Mol. Biol. 388, 673–681.

Lepetit, J., 2007. A theoretical approach of the relationships between collagen content, cross-links and meat tenderness. Meat Sci. 76, 147–159.

Lepetit, J., 2008. Collagen contribution to meat toughness: theoretical aspects. Meat Sci. 80, 960–967.

Levy, M.J., Livingston, D.J., Criddle, R.S., Brown, W.D., 1985. Isolation and characterization of metmyoglobin reductase from yellowfin tuna (Thunnus albacares). Comp. Biochem. Physiol. 81, 809–814.

Lewis, G.J., Purslow, P.P., 1990. Connective tissue differences in strength of cooked meat across the muscle fiber direction due to test specimen size. Meat Sci. 28, 183–194.

Light, N., Champion, A.E., Voyle, C., Bailey, A.J., 1985. The role of epimysial, perimysial and endomysial collagen in determining texture in six bovine muscles. Meat Sci. 13, 137–149.

Lister, D., 1970. The physiology of animals and the use of their muscle for food. In: Briskey, E.J., Locker, R.G., Marsh, B.B. (Eds.), The Physiology and Biochemistry of Muscle as a Food. University of Wisconsin Press, Madison, WI, p. 705.

Liu, M.J., Corboy, M.J., DeMartino, G.N., Thomas, P.J., 2003. Endoproteolytic activity of proteasome. Science 229, 408–411.

Locker, R.H., 1984. The role of gap filaments in muscle and meat. Food Microstruct. 3, 17–32.

Locker, R.H., Daines, G.J., 1975. Rigor mortis in beef sternomandibularis muscle at 37°C. J. Sci. Food Agric. 26, 1721–1733.

Locker, R.H., Leet, N.G., 1976a. Histology of highly stretched beef muscle. II. Further evidence on location and nature of gap filaments. J. Ultrastruct. Res. 55, 157–172.

Locker, R.H., Leet, N.G., 1976b. Histology of highly stretched beef muscle. IV. Evidence for movement of gap filaments through the Z-line, using the N-line and M-line as markers. J. Ultrastruct. Res. 56, 31–38.

Locker, R.H., Wild, D.J.C., 1984. The fate of the large proteins of the myofibril during tenderizing treatments. Meat Sci. 11, 89–108.

Long, J. B. 2000. Treatment of meat by capacitor discharge. US Patent 6,120,828.

Love, R.M., 1975. Variability in Atlantic cod (Gadus morhua) from the northeast Atlantic: a review of seasonal and environmental influences on various attributes of the flesh. J. Fish. Res. Board Can. 32, 2333–2342.

Love, R.M., 1988. Gaping. In: Love, R.M. (Ed.), The Food Fishes: Their Intrinsic Variation and Practical Implication. Farrand Press, London, pp. 161–180.

Love, R.M., Roberson, I., Smith, G.L., Whittle, K.J., 1972. The texture of cod muscle. J. Texture Stud. 5, 201–212.

Love, R.M., Lavety, J., Vellas, F., 1982. Unusual properties of the connective tissues of cod (Gadus murhua L. In: Martin, R.E., Flick, G.J., Hebard, C.E., Ward, D.R. (Eds.), Chemistry and Biochemistry of Marine Food Products. Avi, Westport, CT, pp. 67–73.

Luciano, F.B., Anton, A.A., Rosa, C.F., 2007. Biochemical aspects of meat tenderness: a brief review. Arch. Zootecnia. 56, 1–8.

Lundstrom, K., Nilsson, H., Malmfors, B., 1977. Interrelationships between meat quality characteristics in pigs. Proceedings of Muscle Function and Porcine Meat Quality, NJF Symposium. Acta Agric. Scand. (Suppl. 1), 71.

Luno, M., Roncales, P., Djenane, D., Beltran, J.A., 2000. Beef shelf life in low O_2 and high CO_2 atmospheres containing different low CO concentrations. Meat Sci. 55, 413–419.

Lusby, M.L., Ridpath, J.F., Parrish Jr., F.C., Robson, R.M., 1983. Effect of post mortem storage on the degradation of the myofibrillar protein titin in bovine longissimus muscle. J. Food Sci. 48, 1787–1790.

Ma, H.-J., Ledward, D.A., 2004. High pressure/thermal treatment effects on the texture of beef muscle. Meat Sci. 68, 347–355.

MacBride, M.A., Parrish Jr., F.C., 1977. The 30,000 dalton component of tender bovine longissimus muscle. J. Food Sci. 42, 1627–1629.

MacCallum, W.A., Jaffray, J.I., Churchill, D.N., Idler, D.R., Odense, P.H., 1967. Postmortem physicochemical changes in unfrozen Newfoundland trap-caught cod. J. Fish Res. Board Can. 24, 651–676.

McCormick, R.J., 1999. Extracellular modifications to muscle collagen: implications for meat quality. Poultry Sci. 78, 785–791.

McCrea, S.E., Secombe, C.G., Marsh, B.B., Carse, W.A., 1971. Studies in meat tenderness. 9. The tenderness of various lamb muscles in relation to their skeletal restraint and delay before freezing. J. Food Sci. 36, 566–570.

MacDougall, D.B., 1982. Changes in colour and opacity of meat. Food Chem. 9, 75—88.

Macfarlane, J.J., 1973. Prerigor pressurization of muscle, effect of pH, shear value and taste panel assessment. J. Food Sci. 38, 294—298.

McKeith, F.K., Smith, G.C., Dutson, T.R., Savell, J.W., Hostetler, R.L., Carpenter, Z.L., 1980. Electrical stimulation of intact or split steer and cow carcasses. J. Food Prot. 43, 795—798.

MacLeod, G., 1994. The flavour of beef. In: Shahidi, F. (Ed.), The Flavour of Meat and Meat Products. Blackie, Glasgow, pp. 4—37.

MacLeod, R.A., Jonas, R.E.E., Roberts, E., 1963. Glycolytic enzymes in the tissues of a salmonoid fish (*Salmo gairdnerii gaidnerii*). Can. J. Biochem. Physiol. 41, 1971—1981.

Macy, R.L., Neumann, H.D., Bailey, M.E., 1964. Water-soluble flavour and odour precursors of meat. I. Qualitative study of certain amino acids, carbohydrates, non amino-acid nitrogen compounds and phosphoric acid esters of beef, pork and lamb. J. Food Sci. 29, 136—141.

Macy, R.L., Naumann, H.D., Bailey, M.E., 1970. Water-soluble flavor and odor precursors of meat. 5. Influence of heating on acid extractable non-nucleotide chemical constituents of beef, lamb and pork. J. Food Sci. 35, 83—87.

Madruga, M.S., Elmore, J.S., Oruna-Concha, M.J., Balagiannis, D., Mottram, D.S., 2010. Determination of some water-soluble aroma precursors in goat meat and their enrolment on flavour profile of goat meat. Food Chem. 123, 513—520.

Maga, J.A., 1994. Umami flavor in meat. In: Shahidi, F. (Ed.), Flavor of Meat and Meat Products. Chapman and Hall, Glasgow, pp. 98—115.

Maheswarappa., N.B., Faustman, C., Tatiyaborworntham, N., Yin, S., Ramanathan, R., Mancini, R.A., 2010. Detection of 4-hydroxy-2-nonenal adducts of turkey and chicken myoglobins using mass spectrometry. Food Chem. 122, 836—840.

Makinodan, Y., Ikeda, S., 1976. Studies on fish muscle protease. VI. Separating carp muscle cathepsins A and D and some properties of carp muscle cathepsin A. Bull. Jpn. Soc. Sci. Fish. 42, 239—247.

Marsh, B.B., 1954. Rigor mortis in beef. J. Sci. Food Agric. 5, 70—75.

Marsh, B.B., Leet, N.G., 1966. Studies on meat tenderness. III. The effects of cold-shortening on tenderness. J. Food Sci. 31, 450—459.

Maruyama, K., Matsubara, S., Natori, R., Nonomura, Y., Kimura, S., Ohashi, K., Murakami, F., Handa, S., Eguchi, G., 1976a. Connectin, an elastic protein of muscle. Characterization and function. J. Biochem. (Tokyo) 82, 347—350.

Maruyama, K., Natori, R., Nonomura, Y., 1976b. New elastic protein from muscle. Nature (London) 262, 58—59.

Maruyama, K., Matsubara, S., Natori, R., Nonomura, Y., Kimura, S., Ohashi, K., Murakami, F., Harada, S., Eguchi, G., 1977. Connectin, an elastic protein of muscle. Characterization and function. J. Biochem. (Tokyo) 82, 317—337.

Maruyama, K., Kimura, M., Kimura, S., Ohashi, K., Suzuki, K., Katunuma, N., 1981. Connectin, an elastic protein muscle. Effects of proteolytic enzymes in situ. J. Biochem. (Tokyo) 89, 701.

Maltin, C., Balcerak, D., Tilley, R., Delday, M., 2003. Determinants of meat quality. Proc. Nutr. Soc. 62, 337—347.

Mancini, R.A., Hunt, M.C., 2005. Current research on meat color. Meat Sci. 71, 100—121.

Manu-Tawiah, W., Ammann, L.L., Sebranehh, J.G., Molins, A., 1991. Extending the color stability and shelf life of fresh meat. Food Technol. 45 (94), 96—98. 100—102.

Matsuishi, M., Okitani, A., 1997. Proteasome from rabbit skeletal muscle: some properties and effect on muscle proteins. Meat Sci. 45, 451—462.

Matsumiya, Y., Mochizuki, A., 1993. Existence of a new type of cysteine protease in ordinary and dark muscle of common mackerel. Bull. Coll. Agr. Vet. Med. Nihon Univ. 48, 112—118.

Matsuura, F., Hashimoto, K., Kikawada, S., Yamaguchi, K., 1962. Studies on the autoxidation velocity of fish myoglobin. Bull. Jpn. Soc. Sci. Fish. 28, 210—215.

Maudhavi, D.L., Carpenter, C.G., 1993. Aging and processing affect color, metmyoglobin reductase and oxygen consumption of beef muscles. J. Food Sci. 58, 939—942.

Meek, K.I., Claus, J.R., Duncan, S.E., Marriot, N.G., Solomon, M.B., Kathman, S.J., 2000. Quality and sensory characteristics of selected pre-rigor, early-deboned broiler breast meat tenderized using hydrodynamic shock waves. Poultry Sci. 79, 126—136.

Melody, J.L., Lonergan, S.M., Rowe, L.J., Huiatt, T.W., Mayes, M.S., Huff-Lonergan, E., 2004. Early postmortem biochemical factors influence tenderness and water-holding capacity of three porcine muscles. J. Anim. Sci. 82, 1195—1205.

Mendelson, R., 1982. X-ray scattering by myosin S-1, implications for the steric blocking model of muscle control. Nature (London) 298, 665.

Meinert, L., Schäfer, A., Bjergegaard, C., Aaslyng, M.D., Bredie, W.L.P., 2009. Comparison of glucose, glucose 6-phosphate, ribose, and mannose as flavour precursors in pork; the effect of monosaccharide addition on flavour generation. Meat Sci. 81, 419—425.

de Meis, L., Vianna, A.L., 1979. Energy interconversions by the Ca dependent ATPase of the sarcoplasmic reticulum. Annu. Rev. Biochem. 48, 275—289.

Meyer, L.H., 1964. Food Chemistry, third ed. Reinhold, New York.

Meyers, S.N., Rodroguez-Zas, S.L., Beever, J.E., 2007. Fine-mapping of QTL influencing pork tenderness on porcine chromosome 2. BMC Genomics. 8, 69.

Michie, I., 2001. Causes of downgrading in salmon farming industry. In: Kestin, S.C., Warriss, P.D. (Eds.), Farmed Fishing Quality. Blackwell Science, Oxford, pp. 129—136.

Mickelson, J.R., Louis, C.F., 1996. Malignant hyperthermia: excitation—contraction coupling, Ca^{2+} release channel and cell Ca^{2+} regulation defects. Physiol. Rev. 76, 537—592.

Mikami, M., Whiting, A.H., Taylor, M.A.J., Maciewicz, R.A., Etherington, D.J., 1987. Degradation of myofibrils from rabbit, chicken and beef by cathepsin 1 and lysomal lysates. Meat Sci. 21, 81—97.

Miki, H., Nishimoto, J., 1984. Kinetic parameters of freshness-lowering and discoloration based on temperature dependence in fish muscles. Bull. Jpn. Soc. Sci. Fish. 50, 281—285.

Milligan, R.A., Flicker, P.F., 1987. Structural relationships of actin, myosin, and tropomyosin revealed by cryo-electron miscroscopy. J. Cell Biol. 105, 29—39.

Mitchell, M.A., Kettlewell, P.J., 1998. Physiological stress and welfare of broiler chickens in transit: solutions not problems! Poultry Sci. 77, 1803—1814.

Moore, P.B., Huxley, H.E., De Roosier, D.J., 1970. Three-dimensional reconstruction of F-actin. Thin filaments and decorated filaments. J. Mol. Biol. 50, 279—295.

Mori, T., Hashida, W., Hirai, A., Kawaski, Y., 1974. Assessment of freshness of marine products based on composition in nucleic acid related substances in canned sea food. IV. Nucleic acid decomposing enzyme of oysters. Toyo Shokutin Kogyo Tanki Daigaku Toyo Shokutin Kenkyusho 11, 154—159.

Morkore, T., Mazo, T.P.L., Tahirovic, V., Einen, O., 2008. Impact of starvation and handling stress on rigor development and quality of Atlantic salmon (*Salmon salar* L.). Aquaculture 277, 231–238.

Morrison, E.H., Mielche, M.M., Porslow, P.P., 1998. Immunolocalisation of intermediate filament proteins in porcine meat. Fibre-type and muscle specific variations during conditioning. Meat Sci. 50, 91–104.

Morzel, M., Terlouw, C., Chambon, C., Micol, D., Picard, B., 2008. Muscle proteome and meat eating qualities of longissimus thoracis of Blonde d'Aquitaine young bulls. A central role of hsp7 isoforms. Meat Sci. 78, 297–304.

Mottram, D.S., Madruga, M.S., 1994. The role of inosine monophosphate as a flavour precursor in meat aroma. In: Maarse, H., van der Heij, D.G. (Eds.), Trends in Flavour Research. Elsevier, Amsterdam, pp. 339–344.

Mottram, D.S., Nobrega, I.C.C., 1998. Formation of volatile sulphur compounds in reaction mixtures containing cysteine and three different ribose compounds. In: Contis, E.T., Spanier, A.M., Shahidi, F. (Eds.), Food Flavors: Formation, Analysis and Packaging Influences. Proceedings of the 9th International Flavor Conference. Elsevier, Amsterdam, pp. 483–492.

Murachi, T., 1985. Calcium-dependent proteases and their specific inhibitors: calpain and calpastatin. Biochem. Soc. Symp. 49, 149–167.

Murakami, K., Stewart, M., Nozawa, K., Tomil, K., Kudou, N., Igarashi, N., Shirakihara, Y., Wakatsuki, S., Yasunaga, T., Wakabayashi, T., 2008. Structural basis for tropomyosin overlap in thin (actin) filaments and the generation of a molecular swivel by troponin T. PNAS. 105, 7200–7205.

Murray, J.M., Weber, A., 1974. The cooperative action of muscle protein. Nippon Suisan Gakkaishi. 50, 59–65.

Nagayama, F., 1966. Mechanisms of breakdown and synthesis of glycogen in tissues of marine animals. Nippon Suisan Gakkaishi. 32, 188.

Nakano, T., Thompson, J.R., Aherne, F.X., 1991. Concentration of the crosslink pyridinoline in porcine skeletal muscle epimysium. Can. J. Inst. Food Sci. Technol. 18, 100–102.

Nakatani, Y., Fujita, T., Sawa, S., Otani, T., Hori, Y., Takagahara, I., 1986. Changes in ATP-related compounds of beef and rabbit muscles and a new index of freshness of muscle. Agric. Biol. Chem. 50, 1751–1756.

Nazir, D.J., Magar, N.G., 1963. Biochemical changes in fish muscle during rigor mortis. J. Food Sci. 28, 1–7.

Neti, G., Novak, S.M., Thompson, V.F., Goll, D.E., 2009. Properties of easily releasable myofilaments: are they the first step in myofibrillar protein turnover? Am. J. Physiol. Cell Physiol. 296, C1383–C1390.

Newbold, R.P., 1966. Changes associated with rigor mortis. In: Briskey, E.J., Cassens, R.G., Trautman, J.C. (Eds.), The Physiology and Biochemistry of Muscle as Food. University of Wisconsin Press, Madison, WI, pp. 213–224.

Newbold, R.P., Lee, C.A., 1965. Post-mortem glycolysis in skeletal muscle. The extent of glycolysis in diluted preparation of mammalian muscle. Biochem. J. 97, 1–6.

Newbold, R.P., Scopes, R.K., 1967. Post-mortem glycolysis in ox skeletal muscle. Effects of temperature on the concentrations of glycolytic intermediates and cofactors. Biochem. J. 105, 127–136.

Newbold, R.P., Small, L.M., 1985. Electrical stimulation of post-mortem glycolysis in the semitendinosus muscle of sheep. Meat Sci. 12, 1–16.

Nissen, H., Sorheim, O., Dainty, R., 1996. Effects of vacuum, modified atmospheres and storage temperature on the microbial flora of packaged beef. Food Microbiol. 13, 183–191.

Noda, T., Isogai, L., Hayashi, H., Katunuma, N., 1981. Susceptibilities of various myofibrillar proteins to cathepsin B and morphological alteration of isolated myofibrils by this enzyme. J. Food Biochem. 90, 371–379.

Nonneman, D., Lindholm-Perry, A.K., Shackelford, S.D., King, D.A., Wheeler, T.L., Rohrer, G.A., Bierman, C.D., Schneider, J.F., Miller, R.K., Zerby, H., Moeller, S.J., 2011. Predictive markers for tenderness in commercial pig populations. J. Anim. Sci. 89, 2663–2672.

Obatake, A., Heya, H., 1985. A rapid method to measure dark content in fish. Bull. Jpn. Soc. Sci. Fish. 51, 1001–1004.

Obatake, A., Tsumiyama, S., Yamamoto, Y., 1985. Extractive nitrogenous constituents from the dark muscle of fish. Bull. Jpn. Soc. Sci. Fish. 5, 1461–1468.

Ochiai, Y., Ueki, N., Watabe, S., 2009. Effects of point mutations on the structural stability of tuna myoglobins. Comp. Biochem. Physiol. B. 153, 223–228.

Oe, M., Ohnishi-Kameyama, M., Nakajima, I., Muroya, S., Chikuni, K., 2007. Muscle type expression of tropomyosin isoforms in bovine skeletal muscles. Meat Sci. 75, 558–561.

Offer, G., Cousins, T., 1992. The mechanism of drip production: formation of two compartment of extracellular space in muscle post-mortem. J. Sci. Food Agric. 58, 107–116.

Offer, G., Trinick, J., 1983. On the mechanism of water holding in meat: the swelling and shrinking of myofibrils. Meat Sci. 8, 245–281.

Ofstad, R., Olsen, R.L., Taylor, R., Hannesson, K.O., 2006. Breakdown of intramuscular tissue of cod (*Gadus morhua* L.) and spotted wolfish (*Anarhichas minor* O.) related to gaping. LWT Food Sci. Technol. 39, 1143–1154.

O'Halloran, G.R., Troy, D.J., Buckley, D.J., Reville, W.J., 1997. The role of endogenous proteases in the tenderisation of fast glycolysing muscle. Meat Sci. 47, 187–210.

O'Keeffe, M., Hood, D.E., 1980–1981. Anoxic storage of fresh beef. 1. Nitrogen and carbon dioxide storage atmospheres. Meat Sci. 5, 27–39.

O'Keeffe, M., Hood, D.E., 1982. Biochemical factors influencing met-myoglobin formation on beef from muscles of differing colour stability. Meat Sci. 7, 209–228.

Okitani, A., Matsukura, U., Kato, H., Fujimaki, M., 1980. Purification and some properties of a myofibrillar protein-degrading protease, cathepsin rabbit skeletal muscle. J. Biochem. (Tokyo) 87, 1133–1143.

Oliver, M.A., Gispert, M., Diestre, A., 1988. Estudio del pH, de los musculos Longissimus dorsi y Semimembranosus en canals porcinis comerciales. Med. Vet. 1, 45–49.

Olson, D.G., Parrish Jr., F.C., 1977. Relationship of myofibril fragmentation index to measures of beefsteak tenderness. J. Food Sci. 42, 506–511.

Olson, D.G., Parrish Jr., F.C., Stromer, M.H., 1976. Myofibril fragmentation and shear resistance of three bovine muscles during post-mortem storage. J. Food Sci. 41, 1036–1041.

Olson, D.G., Parrish Jr., F.C., Dayton, W.R., Goll, D.E., 1977. Effect of postmortem storage and calcium activated factor on the myofibrillar proteins of bovine skeletal muscle. J. Food Sci. 42, 117–124.

Ordway, G.A., Garry, D.J., 2004. Myoglobin: an essential hemoprotein in striated muscle. J. Expt. Biol. 207, 3441–3446.

Orlowski, M., Wilk, S., 2003. Ubiquitin-independent proteolytic functions of the proteasome. Arch. Biochem. Biophys. 415, 1−5.

Osebold, W.R., Pedrini, V., 1976. Pepsin-solubilized collagen of human nucleus pulposus and annuls fibrosus. Biochim. Biophys. Acta. 434, 390−405.

O'Shea, J.M., Robson, R.M., Huiatt, T.W., Hartzer, M.K., Stromer, M.H., 1979. Purified desmin b from adult mammalian skeletal muscle: a peptide mapping comparison with desmin from adult mammalian and avian smooth muscle. Biochem. Biophys. Res. Commun. 89, 972−980.

O'Shea, J.M., Robson, R.M., Hartzer, M.K., Huiatt, T.W., Rathbun, W.E., Stromer, M.H., 1981. Purification of desmin from adult mammalian skeletal muscle. Biochem. J. 195, 345−356.

Ota, F., Nakamura, T., 1952. Change of ammonia content in fish meat by heating under pressure. Relation between increase of ammonia and the freshness of fish. Bull. Jpn. Soc. Sci. Fish. 18, 15.

Otsuka, Y., Homma, N., Shiga, K., Ushiki, J., Ikeuchi, Y., Suzuki, A., 1988. Purification and properties of rabbit muscle proteasome and its effect on myofibrillar structure. Meat Sci. 49, 365−378.

Ottenheijm, C.A., Fong, C., Vangheluwe, P., Wutack, F., Babu, G.J., Periasamy, M., Witt, C.C., Labeit, S., Granzier, H., 2008. Sarcoplasmic reticulum calcium uptake and speed of relaxation are depressed in nebulin-free skeletal muscle. FASEB J. 22, 2912−2917.

Ouali, A., 1992. Proteolytic and physicochemical mechanisms involved in meat texture development. Biochimie 74, 251−265.

Ozawa, S., Mitsuhashi, T., Mitsumoto, M., Matsumoto, S., Itoh, N., Itagaki, K., Kohno, Y., Dohgo, T., 2000. The characteristics of muscle fiber types of longissimus thoracis muscle and their influences on the quantity and quality of meat from Japanese black steers. Meat Sci. 54, 65−70.

Pappas, C.T., Krieg, P.A., Gregorio, C.C., 2010. Nebulin regulates actin filament lengths by a stabilization mechanism. J. Cell Biol. 189, 859−870.

Parr, T., Sensky, P.L., Scothern, G.P., Bardsley, R.G., Buttery, P.J., Wood, J.D., Warkup, C., 1999a. Relationship between skeletal muscle-specific calpain and tenderness of conditioned porcine longissimus muscle. J. Anim. Sci. 77, 661−668.

Parr, T., Sensky, P.L., Scothern, G.P., Bardsley, R.G., Buttery, P.J., Wood, J.D., Warkup, C., 1999b. Immunochemical study of the calpain system in porcine longissimus dorsi muscle with high and low shear force values. J. Anim. Sci. 77 (Suppl. 1), 164.

Parrish Jr., F.C., Lusby, M.L., 1983. An overview of a symposium on the fundamental properties of muscle proteins important in meat science. J. Food Biochem. 7, 125−135.

Parrish Jr., F.C., Vandell, C.J., Culler, R.D., 1979. Effect of maturity and marbling on the myofibril fragmentation index of bovine longissimus muscle. J. Food Sci. 44, 1668−1671.

Parry, D.A.D., Squire, J., 1973. Structural role of tropomyosin in muscle regulation, analysis of the X-ray diffraction patterns from relaxed and contracting muscles. J. Mol. Biol. 75, 33−55.

Partmann, W., 1965. Changes in proteins, nucleotides and carbohydrates during rigor mortis. In: Kreuzer, R. (Ed.), The Technology of Fish Utilization. Fishing News (Books), London, p. 4.

Partmann, W., Frank, H.K., 1973. Storage of meat in controlled gaseous atmospheres. Prog. Refrig. Sci. Technol., Proc. 13th Int. Conf. Refrig. 1971 (Vol. 3), 17.

Patashnik, M., Groninger, H.S., 1964. Observations on the milky condition in some Pacific coast fishes. J. Fish. Res. Board Can. 21, 335−346.

Paterson, B.C., Parrish Jr., F.C., 1987. SDS−PAGE conditions for detection of titin and nebulin in tender and tough bovine muscle. J. Food Sci. 52, 509−510.

Pearce, K.L., Hopkins, D.L., Williams, Jacob, R.H., Pethwick, D.W., Phillips, J.K., 2009. Alternating frequency to increase the response to stimulation from medium voltage electrical stimulation and the effect on objective meat quality. Meat Sci. 81, 188−195.

Pearson, A.M., Love, J.D., Shorland, F.B., 1977. Warmed over flavor in meat poultry and fish. Adv. Food Res. 23, 2−74.

Pegg, R.B., Shahidi, F., 1997. Unravelling the chemical identity of meat pigments. CRC Crit. Rev. Food Sci. Nutr. 37, 561−589.

Penny, I.F., 1974. The action of a muscle proteinase on the myofibrillar proteins of bovine muscle. J. Sci. Food Agric. 25, 1273−1284.

Perry, S.V., 1979. The regulation of contractile activity in muscle. Biochem. Soc. Trans. 7, 593−617.

Perry, S.V., 2001. Vertebrate tropomyosin: distribution, properties and function. J. Muscle Res. Cell Motil. 22, 5−49.

Petracci, M., Bianchi, M., Cavani, C., 2009. The European perspective on pale, soft, exudative conditions in poultry. Poultry Sci. 88, 1518−1523.

Pirko, P.C., Ayres, J.C., 1957. Pigment changes in packaged beef during storage. Food Technol. 11, 461−468.

Piskarev, A., Dibirasulaev, M., Korzhenko, V.P., 1972. Changes in the content of free amino acids and nucleotides in warm and matured meat during sterilization. Myasnaya Industriya USSR 43, 34−37.

Porter, R., Koury, B., Stone, F., 1996. Comparison of cathepsin B, D, H, and L activity in four species of Pacific fish. J. Food Biochem. 19, 429−442.

Poso, A., Puolanne, E., 2005. Carbohydrate metabolism in meat animals. Meat Sci. 70, 423−434.

Potter, J.D., Gergely, J., 1975. The calcium and magnesium binding sites of troponin and their role in the regulation of myofibrillar adenosine triphosphatase. J. Biol. Chem. 250 4628−4623.

Purslow, P.P., 1985. The physical basis of meat texture: observations on the fracture behaviour of cooked bovine M. semitendinosus. Meat Sci. 12, 39−60.

Redbottom, R., Lea, P., Hildrum, K.I., 2001. Relative influence of low-voltage electrical stimulation, chilling rate and ageing time on tenderness of M. longissimus dorsi of Norwegian cattle. Acta. Agric. Scand. A. Anim. Sci. 51, 184−191.

Reineccius, G., 1994. Source Book of Flavors. Chapman and Hall, New York.

Reiser, K., McCormick, R.J., Rucker, R.B., 1992. Enzymatic and nonenzymatic crosslinking of collagen and elastin. FASEB J. 6, 2439−2449.

Rhee, M.S., Kim, B.C., 2001. Effect of low voltage electrical stimulation and temperature conditioning on postmortem changes in glycolysis and calpains activities of Korean native cattle (Hanwoo). Meat Sci. 58, 231−237.

Richardson, F.L., Stromer, M.H., Huiatt, T.W., Robson, R.M., 1981. Immunoelectron and fluorescence microscope localisation of desmin in mature avian muscles. Eur. J. Cell Biol. 26, 91−101.

Ridpath, J.F., Robson, R.M., Huiatt, T.W., Trenkle, A.H., Lusby, M.L., 1982. Localization and rate of accumulation of nebulin in skeletal and cardiac muscle cell cultures. J. Cell Biol. 95, 361a.

Ringkob, T.P., Marsh, B.B., Greaser, M.L., 1988. Change in titin position in postmortem bovine muscle. J. Food Sci. 53, 276−277.

Ritchie, A.D., 1926. Lactic acid and rigor mortis. J. Physiol. (London) 6 (1), iv−v.

Robbins, F.M., Cohen, S.H., 1976. Effects of cathepsin enzymes from spleen on the microstructure of bovine semimembranous muscle. J. Texture Stud. 7, 137−142.

Robert, N., Briand, M., Taylor, R.G., Briand, Y., 1999. The effect of proteosome on myofibrillar structures in bovine skeletal muscle. Meat Sci. 51, 149−153.

Robins, S.P., Shimokomaki, M., Bailey, A.J., 1973. The chemistry of the collagen cross-links. Age-related changes in the reducible components of intact collagen fibres. Biochem. J. 131, 771−780.

Robins, S.P., 1983. Cross-linking of collagen. Isolation, structural characterization and glycosylation of pyridinoline. Biochem. J. 215, 167−173.

Robson, R.M., O'Shea, J.M., Hartzer, M.K., Rathbun, W.E., LaSalle, F., Schreiner, P.J., Kasang, L.E., Stromer, M.H., Lusby, M.L., Ridpath, J.F., Pang, Y.-Y., Evans, R.R., Zeece, M.G., Parrish, F.C., Huiatt, T.W., 1984. Role of new cytoskeletal elements in maintenance of muscle integrity. J. Food Biochem. 8, 1−24.

Root, D.D., Wang, K., 1994. Calmodulin sensitive interaction of human nebulin fragments with actin and myosin. Biochemistry 33, 12581−12591.

Root, D.D., Wang, K., 2001. High-affinity actin-binding nebulin fragments influence the actoS1 complex. Biochemistry 40, 1171−1186.

Rowe, L.J., Huff-Lonergan, E., Lonergan, S.M., 2001. Desmin degradation influences water-holding capacity and tenderness in fresh pork [abstract]. J. Anim. Sci. 20 (Suppl. 1).

Rowe, R.W.D., 1974. Collagen fiber arrangement in intramuscular connective tissue. Changes associated with muscle shortening and their possible relevance to raw meat toughness measurements. J. Food Technol. 9, 501−509.

Ryder, J.M., 1985. Determination of adenosine triphosphate and its breakdown products in fish muscle by high performance liquid chromatography. J. Agric. Food Chem. 33, 678−680.

Ryu, Y.C., Kim, B.C., 2005. The relationship between muscle fiber characteristics, postmortem metabolic rate, and meat quality of pig longissimus dorsi muscle. Meat Sci. 71 351−257.

Sagili, J.V.V., Claus, J.R., 2003. Tenderization effects of electrically produced hydrodynamic shock waves on top round and strip loin muscles of beef. Proceedings of the 56th Reciprocal Meat Conference. American Meat Science Association, Columbia, MO. 15−19 June.

Saito, T., Arai, K., 1958. Further studies of inosinic acid formation in carp muscle. Nippon Suisan Gakkaishi. 23, 579−580.

Saito, T., Arai, A., Matsuyoshi, M., 1959. A new method for estimating the freshness of fish. Bull. Jpn. Soc. Sci. Fish. 24, 749−750.

Sakai, T., Yamauchi, K., Kuwazuru, S., Gotoh, N., 1998. Relationships between 4-hydroxy-2-nonenal, 2-thiobarbituric acid reactive substances and n-6 polyunsaturated fatty acids in refrigerated and frozen pork. Biosci. Biotechnol. Biochem. 62, 2028−2029.

Saleh, B., Watts, B.M., 1968. Substrates and intermediates in the enzymatic reduction of metmyoglobin in ground beef. J. Food Sci. 33, 353−357.

Samejima, K., Wolfe, F.H., 1976. Degradation of myofibrillar protein components during postmortem aging of chicken muscle. J. Food Sci. 41, 250−254.

Sammel, L.M., Hunt, M.C., Kropf, D.H., Hachmeister, K.A., Johnston, D.E., 2002. Comparison of assays for metmyoglobin reducing ability in beef inside and outside semimembranosus muscle. J. Food Sci. 67, 978−984.

Sandwick, R., Johanson, M., Breuer, E., 2005. Maillard reaction of ribose-5-phosphate and amino acids. Ann. N.Y. Acad. Sci. 1043, 85−96.

Santos, C., Roseiro, L., Gonclaves, H., Melo, R.S., 1994. Incidence of different pork quality categories in a Portugal slaughterhouse: a survey. Meat Sci. 38, 279−287.

Saterlee, L.D., Zecharia, N.Y., 1872. Porcine and ovine myoglobin: isolation, purification, characterization and stability. J. Food Sci. 37, 909−912.

Savell, J.W., Smith, G.C., Dutson, T.R., Carpenter, Z.L., Suter, D.A., 1977. Effect of electrical stimulation on palatability of beef, lamb and goat meat. J. Food Sci. 42, 702−706.

Savell, J.W., Dutson, T.R., Smith, G.C., Carpenter, Z.L., 1978. Structural changes in electrically stimulated beef muscles. J. Food Sci. 43, 1606−1607.

Savell, J.W., McKeith, F.K., Smith, G.C., 1981. Reducing postmortem aging time of beef with electrical stimulation. J. Food Sci. 46, 1777−1781.

Schaefer, R.B., 2004a. Innovative water treatment technology for water treatment. BSF SBIR Phase 1 Final Report, Contract No. DMI-0318911, June 2004.

Schaefer, R.B., 2004b. Nontoxic sparker control of zebra mussels EPA SBIR Phase II, Final Report, Contract No. 68-D-02−059, June 2004.

Schaefer, R.B., 2004c. High efficiency long lifetime sparker sources US Patent No. 6,687,189, 3 February 2004.

Schaefer, R.B., 2004d. High efficiency and projection reflectors for light and sound US Patent No. 6,672,729, 6 January 2004.

Schaefer, R.B., 2005. High source level sparker for navy applications Phase II SBIR, Naval Air Warfare Center, Contract No. N00421-03-C-0570, September 2003−August 2005.

Schiaffino, S., Reggianni, C., 1996. Molecular diversity of myofibrillar proteins: gene regulation and functional significance. Physiol. Rev. 76, 371−423.

Schwartz, W.N., Bird, J.N.C., 1977. Degradation of myofibrillar proteins by cathepsins B and D. Biochem. J. 167, 811−820.

Scopes, R.K., 1964. The influence of post-mortem conditions on the solubilities of muscle proteins. Biochem. J. 91, 201−207.

Scopes, R.K., 1970. In: Briskey, E.J., Cassens, R.G., Marsh, B.B. (Eds.), The Physiology and Biochemistry of Muscle as a Food. Characterization and study of sarcoplasmic proteins, Vol. 2. University of Wisconsin Press, Madison, WI, pp. 471−492.

Seideman, S.C., Cross, H.R., Smith, G.C., Durland, P.R., 1984. Factors associated with fresh meat colour: a review. J. Food Qual. 6, 211−237.

Sensky, P.L., Parr, T., Bardsley, R.G., Buttery, P.J., Warkup, C.C., 1998. Differences in calpain enzyme system in tough and tender samples of porcine longissimus dorsi. Proc. Br. Soc. Anim. Sci. 16.

Sentandreu, M.A., Coulis, G., Ouali, A., 2002. Role of muscle endopeptidases and their inhibitors in meat tenderness. Trends Food Sci. Technol. 13, 398−419.

Seyfert, M., Mancini, R.A., Hunt, M.C., Tang, J., Faustman, C., Garcia, M., 2006. Colour stability, reducing activity and cytochrome c oxidase activity of five bovine muscles. J. Agric. Food Chem. 54, 8919−8925.

Seymour, J., O'Brien, E.J., 1980. The position of tropomyosin in muscle thin filaments. Nature (London) 283, 680−682.

Shahidi, F., Chong, X., Dunajski, E., 1994. Freshness quality in harp seal (*Phoca groenlandica*) meat. J. Agric. Food Chem. 42, 868−872.

Shinokomaki, M., Elsden, D.F., Bailey, A.J., 1972. Meat tenderness: age related changes in bovine intramuscular collagen. J. Food Sci. 37, 892–896.

Shinokomaki, M., Wright, D.W., Irwin, M.H., Van der Rest, M., Mayne, R., 1990. The structure and macromolecular organization of type IX collagen in cartilage. Ann. N.Y. Acad. Sci. 580, 1–7.

Sikes, A., Tomberg, E., Tume, R., 2010. A proposed method of tenderising post-rigor beef using high pressure heat treatment. Meat Sci. 84, 390–394.

Sjöstrand, F.S., 1962. The connections between A and I band filaments in striated frog muscle. J. Ultrastruct. Res. 7, 225–246.

Smith, G.C., Culp, G.R., Carpenter, Z.L., 1978. Postmortem aging of beef carcasses. J. Food Sci. 43, 823–826.

Smith, G.C., Belk, K.E., Sofos, J.N., Tatum, T.D., Williams, S.N., 2000. Economic implications of improved color stability in beef. In: Decker, E.A., Faustman, C., Lopez-Pote, C.S. (Eds.), Antioxidants in Muscle Foods: Nutritional Strategies to Improve Quality. Wiley Interscience, New York, pp. 397–426.

Solberg, M., 1968. Factors affecting fresh meat color. Proc. Meat Ind. Res. Conf., 32–40.

Solberg, M., 1970. The chemistry of color stability in meat: a review. Can. Inst. Food Technol. J. 3, 55–59.

Solomon, M.B., 1998. . The hydrodyne process for tenderizing meat. Proceedings of the 51st Reciprocal Meat Conference. 28 June–1 July 1998. American Meat Science Association, Storrs, CT, 171–176.

Solomon, M.B., Long, J.B., Eastridge, J.S., 1997. The Hydrodyne — a new process to improve meat tenderness. J. Anim. Sci. 75, 1534–1537.

Solomon, M.B., Long, J.B., Carpenter, C.E., Snowder, G.D., Cockett, N.E., 1998. Tenderizing callipyge lamb with the Hydrodyne process and electrical stimulation. J. Muscle Foods 9, 305–311.

Sonaiya, E.B., Stouffer, J.R., Beerman, D.H., 1982. Electrical stimulation of mature cow carcasses and its effects on tenderness, myofibril protein degradation and fragmentation. J. Food Sci. 47, 889–891.

Sorheim, N.H., Dainty, R.H., 1966. Effect of vacuum modified atmospheres and storage temperature on the microbial flora of packaged beef. Food Microbiol. 13, 183–191.

Sorheim, O., Aune, T., Nesbakken, T., 1997. Technological, hygienic and toxicological aspects of carbon monoxide used in modified-atmosphere packaging of meat. Trends Food Sci. Technol. 8, 307–312.

Sorheim, O., Nissen, H., Nesbakken, T., 1999. The storage life of beef and pork packaged in an atmosphere of low carbon monoxide and high carbon dioxide. Meat Sci. 52, 157–164.

Sorimachi, H., Toyama-Sorimachi, N., Saido, T.C., Kawasaki, H., Sugita, H., Miyasaka, M., Arahata, K., Ishiura, S., Suzuki, K., 1993. Muscle-specific calpain, p94, is degraded by autolysis immediately after translation, resulting in disappearance from muscle. J. Biol. Chem. 268, 10593–10605.

Spencer, M.J., Guyon, J.R., Sorimachi, H., Potts, A., Richards, I., Herasse, M., Chamberlain, J., Dalkilic, I., Kunkel, L.M., Beckman, J.S., 2002. Stable expression of calpain 3 from a muscle transgene in vivo: immature muscle in transgenic mice suggests a role for calpain 3 in muscle maturation. Proc. Natl. Acad. Sci. U.S.A. 99, 8874–8879.

Spinelli, J., Eklund, M., Miyauchi, D., 1964. Measurement of hypoxanthine in fish as a method of assessing freshness. J. Food Sci. 29, 710–714.

Stanley, D.W., 1983. A review of the muscle cell cytoskeleton and its possible relation to meat texture and sarcolemma emptying. Food Microstruct. 2, 99–109.

Stearns, T.M., Beaver, J.E., Southey, B.R., Ellis, M., Mckeith, F.K., Rodriguez-Zas, S.L., 2005. Evaluation of approaches to detect quantitative trait loci for growth, carcass, meat quality on swine chromosomes 2, 6, 13 and 18. II. Multivariate and principal component analyses. J. Anim. Sci. 83, 2472–2481.

Stewart, M.P., Hutchins, B.K., Zipser, M.W., Watts, B.M., 1965. Enzymatic reduction of metmyoglobin by ground beef. J. Food Sci. 30, 487–491.

Sugden, M.C., Sharples, S.C., Randle, J., 1976. Carcass glycogen as a potential source of glucose during short-term starvation. Biochem. J. 160, 817–819.

Suman, S.P., Faustman, C., Stamer, S.L., Liebler, D.C., 2007. Proteomics of lipid oxidation-induced oxidation of porcine and bovine oxymyoglobins. Proteomics 7, 628–640.

Suman, S.P., Joseph, P., Li, L., Steinke, L., Fontaine, M., 2009. Primary structure of goat myoglobin. Meat Sci. 82, 456–460.

Surette, M.E., Gill, T.A., LeBlanc, P.J., 1988. Biochemical basis of postmortem nucleotide catabolism in cod (Gadus morhua) and its relationship to spoilage. J. Agric. Food Chem. 36, 19–22.

Surh, J., Kwon, H., 2005. Estimation of daily exposure to 4-hydroxy-2-nonenals in Korean foods containing n-3 and n-6 polyunsaturated fatty acids. Food Addit. Contam. 22, 701–708.

Suryanarayana-Rao, S.V., Rangaswamy, J.R., Lahiry, N.L., 1969. Nucleotides and related compounds in canned shrimp. J. Fish. Res. Board Can. 26, 704–706.

Suzuki, T., 1953. Determination of volatile acids for judging the freshness of fish. Bull. Jpn. Soc. Sci. Fish. 19, 102–105.

Suzuki, T., 1981. Fish and Krill Protein Processing Technology. Applied Science Publishers, London, pp. 31–34.

Tamura, M., Yubisui, T., Takeshita, M., 1983. Microsomal NADH cytochrome b5 reductase of bovine brain purification and properties. J. Biochem. (Tokyo) 94, 1547–1552.

Tarr, H.L.A., 1954. The Maillard reaction in flesh foods. Food Technol. 8, 15–19.

Tarr, H.L.A., 1965. Pathways of glycogen breakdown. In: Kreuzer, R. (Ed.), The Technology of Fish Utilization. Fishing News (Books), London, p. 34.

Tarr, H.L.A., 1966. Post-mortem changes in glycogen, nucleotides, sugar phosphates and sugars in fish muscles. A review. J. Food Sci. 31, 846.

Tarr, H.L.A., 1968. Post-mortem degradation of glycogen and starch in fish muscle. J. Fish. Res. Board Can. 25, 1539–1559.

Tarrant, P.J.V., 1981. In: Hood, D.E., Tarrant, P.J.V. (Eds.), The Problem of Dark-Cutting in Beef. Martinus Nijhoff, The Hague, p. 462.

Tarrant, P.V., Mothersill, C., 1977. Glycolysis and associated changes in beef carcasses. J. Sci. Food Agric. 28, 739–749.

Tarrant, P.V., Sherrington, J., 1980. An investigation of ultimate pH in the muscles of commercial beef carcasses. Meat Sci. 4, 287–297.

Tarrant, P.J.V., McLoughlin, J.V., Harrington, M.G., 1972a. Anaerobic glycolysis in biopsy and post-mortem porcine longissimus dorsi muscle. Proc. R. Ir. Acad. Sect. B. 72B, 55–73.

Tarrant, P.J.V., Hegarty, P.V.J., McLoughlin, J.V., 1972b. High-energy phosphates and anaerobic glycolysis in the red and white fibers of porcine semitendinosus muscle. Proc. R. Ir. Acad. Sect. B. 72B, 229–251.

Taylor, A.A., 1972. Gases in fresh meat packaging. Meat World 5, 3–6.

Taylor, A.A., Amos, L.A., 1981. A new model for the geometry of the binding of myosin crosslinkages to muscle thin filaments. J. Mol. Biol. 147, 297.

Taylor, A.A., MacDougall, D.B., 1973. Fresh beef packed in mixtures of oxygen and carbon dioxide. J. Food Technol. 8, 453−461.

Taylor, D.G., Marshall, A.R., 1980. Low voltage electrical stimulation of beef carcasses. J. Food Sci. 45, 144−145.

Taylor, R.G., Geesink, G.H., Thompson, V.F., Koohmaraie, M., Goll, D.E., 1995. Is Z-disk degradation responsible for postmortem tenderization? J. Anim. Sci. 73, 1351−1367.

Taylor, R.G., Fjaera, S.O., Skjervold, P.O., 2002. Salmon fillet texture is determined by myofiber−myofiber and myofiber−myocommata attachment. J. Food Sci. 67, 1−5.

Terova, G., Preziosa, E., Marelli, S., Gornati, R., Bernadini, G., Saroglia, M., 2011. Applying transcriptomics to better understand the molecular mechanisms underlying fish filet quality. Food Chem. 124, 1268−1278.

Thiansilakul, Y., Benjakul, S., Richards, M.P., 2011. Effect of myoglobin from Eastern little tuna muscle on lipid oxidation of washed Asian seabass mince at different pH conditions. J. Food Sci. 76, C242−C249.

Thomas, A.R., Gondoza, H., Hoffman, L.C., Oosthuizen, V., Naude, R.J., 2004. The roles of proteosome, and cathepsins B, L, H and D in ostrich meat tenderization. Meat Sci. 67, 113−120.

Tobacman, L.S., 1996. Thin filament regulation of cardiac contraction. Annu. Rev. Physiol. 58, 447−481.

Tomioka, K., Endo, K., 1984. Purification of 5′-nucleotidase from carp muscle. Bull. Jpn. Soc. Sci. Fish. 50, 1077−1081.

Tomioka, K., Endo, K., 1985. Zn content and subunit structure of carp muscle 5′-nucleotidase. Bull. Jpn. Soc. Sci. Fish. 51, 857.

Tomlinson, N., Geiger, S.E., 1962. Glycogen concentration and postmortem loss of adenosine triphosphate in fish and mammalian muscle: a review. J. Fish. Res. Board Can. 19, 997−1003.

Tomlinson, N., Geiger, S.E., Dollinger, E., 1965. Chalkiness in halibut in relation to muscle pH and protein denaturation. J. Fish. Res. Board Can. 22, 653−663.

Tomlinson, N., Geiger, S.E., Dollinger, E., 1966a. Free drip, fresh pH and chalkiness in halibut. J. Fish. Res. Board Can. 23, 673−678.

Tomlinson, N., Geiger, S.E., Dollinger, E., 1966b. Influence of fishing methods on chalkiness in halibut. J. Fish. Res. Board Can 23, 925−928.

Tonino, P., Pappas, C.T., Hudson, B.D., Labeit, S., Gregorio, C.C., Granzier, H., 2010. Reduced myofibrillar connectivity and increased Z-disk width in bebulin-deficient skeletal muscle. J. Cell Sci. 123, 384−391.

Topel, D.G., Bicknell, E.J., Preston, K.S., Christian, L.L., Matsushima, L.Y., 1969. Porcine stress syndrome. Mod. Vet. Pract. 49, 40−60.

Torngren, M.A., 2003. Effect of packaging method on colour and eating quality of beef loin steaks. Brazil. In: Proceedings of the 49th International Congress of Meat Science and Technology, pp. 495−496.

Toyohara, H., Makinodan, Y., Ikeda, S., 1982. Purification and properties of carp muscle cathepsin. A. Bull. Jpn. Soc. Sci. Fish. 48, 1145.

Trucco, R.E., Lupin, H.M., Giannini, D.H., Crupkin, M., Boeri, R.L., Barassi, C.A., 1982. Study on the evolution of rigor mortis in batches of fish. Lebensm. Wiss. Technol. 15, 77−79.

Tsai, R., Cassens, R.G., Briskey, E.J., Greaser, M.L., 1972. Studies on nucleotide metabolism in porcine longissimus muscle postmortem. J. Food Sci. 37, 612−617.

Uchiyama, H., Kakuda, K., 1984. A simple and rapid method for measuring K value, a fish freshness index. Bull. Jpn. Soc. Sci. Fish. 50, 263−267.

Ueki, N., Ochiai, Y., 2004. Primary structure and thermostability of bigeye tuna myoglobin in relation to those of other scombridae fish. Fish. Sci. 70, 875−884.

Unruh, J.A., Kastner, C.L., Kropf, D.H., Dikeman, M.E., Hunt, M.C., 1986. Effects of low-voltage electrical-stimulation on meat quality and display color stability. Meat Sci. 18, 281−293.

Van de Wiel, D.F.M., Zhang, W.L., 2007. Identification of pork quality parameters by proteomics. Meat Sci. 77, 46−54.

Van Jaarsveld, F.P., Naude, R.J., Oelofsen, W., 1997. The effects of Ca ions, EGTA and storage time on myofibrillar protein degradation, levels of Ca^{2+}-dependent proteases and cathepsins B, H, L and D of ostrich skeletal muscle. Meat Sci. 45, 517−529.

Viljoena, H.F., de Kocka, H.L., Webb, E.C., 2002. Consumer acceptability of dark, firm and dry (DFD) and normal beef steaks. Meat Sci. 61, 181−185.

Wagner, J.R., Titus, D.S., Schade, J.E., 1963. New opportunities for flavor modification. Food Technol. 46, 730−736.

Wallace, W.J., Houtchens, R.A., Maxwell, J.C., Caughey, W.S., 1982. Mechanism of autooxidation for hemoglobins and myoglobins. Promotion of superoxide production by protons and anions. J. Biol. Chem. 257, 4966−4977.

Walters, C.L., 1975. In: Cole, D.J.A., Lawrie, R.A. (Eds.), Meat. Avi, Westport, CT, pp. 385−401.

Wang, D., Tang, J., Correia, L.R., Gill, T.A., 1998. Postmortem changes of cultivated Atlantic salmon and their effects on salt intake. J. Food Sci. 63, 634−637.

Wang, H., Pato, M.D., Shand, P.J., 2005. Biochemical properties of natural actomyosin extracted from normal and pale, soft, exudative pork loin after frozen storage. J. Food Sci. 70, 313−330.

Wang, K., 1983. Cytoskeletal matrix in striated muscle. The role of titin, nebulin and intermediate filaments. In: Cross-Bridge Mechanisms in Muscular and Cellular Control. Cold Spring Harbor Laboratory, Cold Spring Harbor, NY, pp. 439−452.

Wang, K., McClure, J., Tu, A., 1979. Titin: major myofibrillar components of striated muscle. Proc. Natl. Acad. Sci. U.S.A. 76, 3698−3702.

Wang, K., Ramirez-Mitchell, R., 1979. Titin: Possible candidate as putative longitudinal filaments in striated muscle. Proc. Natl. Acad. Sci. U.S.A. 76, 3698−3702.

Wang, K., Ramirez-Mitchell, R., 1983a. A network of transverse and longtitudinal intermediate filaments is associated with sarcomeres of adult vertebrate skeletal muscle. J. Cell Biol. 83, 389a.

Wang, K., Ramirez-Mitchell, R., 1983b. Ultrastructural morphology and epitope distribution of titin, a giant sarcomere-associated cytoskeletal protein. J. Cell Biol. 97, 257a.

Wang, K., Ramirez-Mitchell, R., 1984. Architecture of titin-containing cytoskeletal matrix in striated muscle. Mapping of distinct epitopes of titin specified by monoclonal antibodies. Biophys. J. 45, 392a.

Wang, K., Williamson, C.L., 1980. Identification of an N-line protein of striated muscle. Proc. Natl. Acad. Sci. U.S.A. 77, 3254−3258.

Wang, K., Wright, J., 1988. Sarcomere matrix of skeletal muscle: the role of thick filaments in the segmental extensibility of elastic titin filaments. Biophys. J. 53, 25a.

Wang, K., McClure, J., Tu, A., 1979. Titin: major myofibrillar component of striated muscle. Proc. Natl. Acad. Sci. U.S.A. 76, 3698−3702.

Wang, S.-M., Greaser, M.L., 1985. Immunocytochemical studies using a monoclonal antibody to bovine cardiac titin on intact and extracted myofibrils. J. Muscle Res. Cell Motil. 6, 293–312.

Warriss, P.D., 2000. Meat Science: An Introductory Text. CAB-International, Wallingford.

Watanabe, E., Toyama, K., Karube, I., Matsuoka, H., Suzuki, S., 1984. Determination of inosine-5-monophosphate in fish tissue with an enzyme sensor. J. Food Sci. 49, 114–116.

Watts, B.M., Kendrick, J., Zipser, M.W., Hutchins, B., Saleh, B., 1966. Enzymatic reducing pathways in meat. J. Food Sci. 31, 855–862.

Wells, R.M.G., 1987. Stress responses imposed by fish capture and handling: a physiological perspective. Food Technol. Aust. 39, 479–481.

Whipple, G., Koohmaraie, M., Dikeman, M.E., Crouse, J.D., Hunt, M.C., Klemm, R.D., 1990. Evaluation of attributes that affect longissimus dorsi tenderness in *Bos taurus* and *Bos indicus* cattle. J. Anim. Sci. 68, 2716–2728.

Whiting, R.C., 1980. Calcium uptake by bovine muscle mitochondria and sarcoplasmic reticulum. J. Food Sci. 45, 288–292.

Wilhelm, A.E., Maganhini, M.B., Hernandez-Blazquez, F.J., Ida, E.I., Shimokomaki, M., 2010. Protease activity and the ultrastructure of broiler chicken PSE (pale, soft, exudative) meat. Food Chem. 119, 1201–1204.

Witt, C.C., Burkart, C., Albeit, D., McNabb, M., Wu, Y., Granzier, H., Labeit, S., 2006. Nebulin regulates thin filament length, contractability, and Z-disk structure *in vivo*. EMBO J. 25, 3843–3855.

Witz, G., 1989. Biological interactions of alpha, beta-unsaturated aldehydes. Free Radic. Biol. Med. 7, 333–349.

Wojtowicz, M.B., Odense, P.H., 1972. Comparative study of the muscle castheptic activity of some marine species. J. Fish. Res. Board Can. 29, 85–90.

Wright, J., Huang, Q.Q., Wang, K., 1993. Nebulin in full-length template of actin filaments in the skeletal muscle sarcomere: an immunoelectron microscopic study of its orientation and span with site-specific monoclonal antibodies. J. Muscle Res. Cell Motil. 14, 476–483.

Wu, J.J., Dutson, T.P., Carpenter, Z.L., 1981. Effect of postmortem time and temperature on the release of lysosomal enzymes and their possible effect on bovine connective tissue components of muscle. J. Food Sci. 46, 1132–1135.

Yamaguchi, K., Takeda, N., Ogawa, K., Hashimoto, K., 1979. Properties of mackerel and sardine myoglobins. Bull. Jpn. Soc. Sci. Fish. 45, 1335–1339.

Yamaguchi, M., Robson, R.M., Stromer, M.H., Dahl, D.S., Oda, T., 1982. Nemaline rod bodies: structure and composition. J. Neurol. Sci. 56, 35–56.

Yamaguchi, M., Robson, R.M., Stromer, M.H., 1983a. Evidence for actin involvement in cardiac Z-line analogs. J. Cell Biol. 96, 435–442.

Yamaguchi, M., Robson, R.M., Stromer, M.H., Cholvin, N.R., Izumimoto, M., 1983b. Properties of soleus muscle Z-lines and induced Z-line analogs revealed by dissection with Ca-activated neutral protease. Anat. Rec. 206, 345–362.

Yamashita, M., Konagaya, S., 1990b. Purification and characterization of cathepsin L from the white muscle of chum salmon (*Oncorhynchus keta*). Comp. Biochem. Physiol. B. 96, 247–252.

Yamashita, M., Konagaya, S., 1991. Hydrolytic action of salmon cathepsins B and L to muscle structural proteins in respect of muscle softening. Nippon Suisan Gakkaisha 57, 1917–1922.

Yin, S., Faustman, C., 1993. Influence of temperature, pH and phospholipid composition upon the stability of myoglobin and phospholipid: a liposome model. J. Agric. Food Chem. 41, 853–857.

Yin, S., Fustman, C., Tatiyaboworntham, N., Ramathan, R., 2011. Species-specific myoglobin oxidation. J. Agric. Food Chem. 59, 12198–12203.

Younathan, M.T., Watts, B.M., 1960. Oxidation of tissue lipids in cooked pork. Food Res. 25, 538–543.

Young, O.A., Graafhuis, A.E., Davey, C. L, 1980–1981. Post-mortem changes in cytoskeletal proteins of muscle. Meat Sci. 5, 41–55.

Yu, J., Tang, S., Bao, E., Zhang, B., Hao, Q., Yue, Z., 2009. The effect of transportation on the expression of heat shock proteins and meat quality of M. longissimus dorsi in pigs. Meat Sci. 83, 474–478.

Yubisui, T., Takeshita, M., 1980. Fractionation and characterization of the purified NADH cytochrome b5 reductase of human erythrocytes as a FAD-containing enzyme. J. Biol. Chem. 255, 2454–2456.

Zabel, R.W., Harvey, E.J., Katz, S.L.G.T.P., Levin, P.S., 2003. Ecologically sustainable yield. Am. Sci. 91, 110–117.

Zhang, S.X., Farouk, K., Young, O.A., Wieliezko, K.J., Podmore, C., 2005. Functional stability of frozen normal and high pH beef. Meat Sci. 69, 346–765.

Zhu, X., Ruusunen, M., Gusella, M., Zhou, G., Puolanne, E., 2011. High post-mortem temperature combined with rapid glycolysis induces phosphorylase denaturation and produces pale and exudative characteristics in broiler pectoralis major muscles. Meat Sci. 89, 181–188.

Chapter 4

Milk

N. A. Michael Eskin* and H. Douglas Goff[†]

*Department of Human Nutritional Sciences, Faculty of Human Ecology, University of Manitoba, Winnipeg, Canada, [†]University of Guelph, Ontario, Canada

I. INTRODUCTION

Milk and dairy products are important sources of animal protein, vitamins, minerals, and essential fatty acids for infants and young adults (Walstra *et al.*, 2006). The major source of milk is obtained from Western breeds of dairy cattle (*Bos taurus*), although milk is available from other species of mammal in some countries. According to the Pasteurized Milk Ordinance of the United States Public Health Service (2011), milk is defined as the normal lacteal secretion, practically free from colostrum, obtained by the complete milking of one or more healthy cows, containing not less than 8.25% milk solids-not-fat (SNF) and not less than 3.25% of milk fat. The exclusion of colostrum, a fluid secreted immediately following parturition, is primarily for esthetic reasons. Over 70 billion gallons of milk are produced annually in the USA, with more than 30 billion gallons consumed as liquid milk (Goff and Griffiths, 2006).

Since the publication of the second edition there have been major advances in our understanding of the mechanism of milk secretion. These include the intracellular assembly of milk droplets and the nature and formation of the milk fat globule membrane (MFGM), as well as the regulation of the mammary synthesis of milk fat (Bauman *et al.*, 2006). Several comprehensive advanced dairy chemistry texts have also been published (Fox and McSweeney, 2003, 2006, 2009; Walstra *et al.*, 2006). This chapter will discuss the dynamic biochemical systems involved in the biosynthesis of milk components.

II. COMPOSITION OF MILK

The major constituents of milk are water (86−88%), milk fat (3−6%), protein (3−4%), lactose (5%), and minerals (ash) (0.7%), with total solids of 11−14%. The composition of milk is affected by a variety of factors, including breed, stage of lactation, nutritional and health status of the cow, season (which can be both temperature and stage of lactation effects, if calving is not even throughout the year), and genetic factors (Fox and McSweeney, 1998). Table 4.1 illustrates the influence of the four main breeds in North America on yield and milk composition and also

Biochemistry of Foods. DOI: http://dx.doi.org/10.1016/B978-0-12-242352-9.00004-7

TABLE 4.1 Changing Yield and Composition of Milk Over the Past Two Decades in Canada, as a Function of Breed

	All Breeds			Holstein			Jersey			Guernsey			Brown Swiss		
Year	Yield (kg)[a]	Fat (%)	Protein (%)	Yield (kg)	Fat (%)	Protein (%)	Yield (kg)	Fat (%)	Protein (%)	Yield (kg)	Fat (%)	Protein (%)	Yield (kg)	Fat (%)	Protein (%)
2009[b]	9592	3.77	3.22	9793	3.76	3.19	6371	4.87	3.81	6812	4.56	3.43	8128	4.05	3.48
2008	9642	3.78	3.23	9836	3.74	3.20	6435	4.84	3.81	6820	4.51	3.45	8366	4.04	3.48
2007	9538	3.77	3.22	9733	3.72	3.19	6412	4.82	3.78	6673	4.51	3.46	8159	4.06	3.48
2006	9481	3.75	3.21	9677	3.71	3.18	6331	4.83	3.77	6540	4.55	3.43	8064	4.06	3.46
2005	9422	3.76	3.21	9624	3.71	3.19	6279	4.85	3.77	6398	4.50	3.43	7792	4.12	3.48
2004	9458	3.72	3.22	9658	3.67	3.19	6291	4.85	3.77	6435	4.54	3.45	8048	4.07	3.47
2003	9519	3.73	3.23	9721	3.68	3.21	6344	4.87	3.81	6570	4.49	3.49	8038	4.04	3.49
2002	9511	3.72	3.25	9717	3.67	3.22	6407	4.86	3.84	6347	4.44	3.51	8215	4.03	3.50
2001	9242	3.72	3.24	9440	3.68	3.22	6186	4.87	3.83	6015	4.45	3.51	8020	4.02	3.50
2000	9152	3.70	3.23	9350	3.67	3.21	6203	4.90	3.83	5949	4.45	3.48	7920	3.96	3.48
1999	8960	3.69	3.24	9162	3.66	3.22	6072	4.89	3.85	5939	4.43	3.54	7585	4.03	3.54
1998	8738	3.70	3.24	8946	3.65	3.22	6002	4.88	3.84	5991	4.48	3.55	7105	4.04	3.50
1997	8427	3.72	3.24	8697	3.68	3.22	5753	4.90	3.86	5919	4.53	3.55	6818	4.05	3.52
1996	8424	3.76	3.25	8633	3.72	3.23	5720	4.93	3.86	5984	4.56	3.54	6910	4.08	3.53
1995	8251	3.74	3.24	8461	3.70	3.21	5620	4.89	3.86	5936	4.57	3.54	6709	4.05	3.49
1994	8103	3.74	3.26	8309	3.69	3.21	5501	4.94	3.94	5867	4.76	3.56	6718	4.06	3.51
1993	7988	3.75	3.25	8193	3.71	3.21	5408	4.94	3.88	5826	4.58	3.57	6639	4.07	3.52
1992	7807	3.73	3.24	8028	3.67	3.21	5244	4.92	3.91	5755	4.55	3.58	6483	4.03	3.48
1991	7523	3.70	3.21	7717	3.66	3.22	4992	4.90	3.92	5554	4.52	3.60	6176	4.01	3.54

[a]Based on a 305 day lactation period.
[b]In 2009, Holsteins accounted for 92.8% of the population, Jerseys 3.2%, Guernsey < 1%, and Brown Swiss < 1%.
Source: Canadian Dairy Information Centre, www.dairyinfo.gc.ca (2010).

shows the changes in yield and composition that have occurred over two decades owing to improved genetics and nutrition. Holsteins produce by far the majority of milk; although they are lower in fat and protein on a percentage weight basis, the yield is much higher, so they produce a higher total amount of fat and protein. Seasonal changes in fat, protein, and lactose and other solids can be seen in Table 4.2 from all milk produced in the Province of Ontario in 2006. Since calving is evenly distributed, these changes are due to temperature and feeding regimes. In countries where calving is more heavily concentrated in certain periods, such as spring, milk composition across the seasons can also be affected by stage of lactation.

Extensive studies on milk composition in North America were undertaken because of considerable interest in the percentage of SNF or protein in determining milk prices. This measurement was facilitated through the development of rapid, automated methods based on infrared analysis. Wilcox *et al.* (1971) completed a large study on milk composition in five breeds and reported that the effect of age on fat, SNF, protein, and total solids (TS) was as important as its effect on milk yield. This was followed by Norman *et al.* (1978), who examined the influence of age and month of calving on milk fat, SNF, and protein yields over a period of nine years. This was the largest study ever conducted and involved 106,411 cows from 2215 herds in 41 states. These researchers found that the levels of milk components were lower than those reported by Wilcox *et al.* (1971) but reflected more closely the average US yields for milk and fat. The average percentages for fat and SNF reported by Norman *et al.* (1978) were 3.8% and 8.6%, respectively. The effect of breed on milk composition was consistent with previous studies. A later study by Sommerfeldt and Baer (1986) monitored the variation in milk components from 1705 herds from eastern South

TABLE 4.2 Average Monthly Milk Composition (kg/hl) from the Province of Ontario, Canada

Month	Milk Fat	Milk Protein	Lactose and Other Solids
January	3.99	3.34	5.71
February	4.00	3.32	5.72
March	4.01	3.34	5.73
April	3.97	3.30	5.74
May	3.91	3.27	5.75
June	3.85	3.24	5.73
July	3.77	3.20	5.72
August	3.78	3.25	5.70
September	3.92	3.33	5.70
October	4.04	3.37	5.70
November	4.03	3.39	5.70
December	4.01	3.33	5.70

Data from University of Guelph Laboratory Services, http://www.guelphlabservices.com (2006).

Dakota, Minnesota, and northern Iowa. Milk was collected biweekly over a one-year period to evaluate the advantages and disadvantages of milk payment based on SNF. The greatest variability was observed for fat (8.4% coefficient of variation), followed by protein, TS, and SNF, with coefficients of variation of 6.3%, 4.1%, and 3.4%, respectively. Sommerfeldt and Baer (1986) did not find any consistent relationship between SNF and protein or fat in the milk examined and recommended that testing for SNF should also include the fat content and producer grade. The introduction of a milk component pricing plan was considered to be beneficial to both the producer and processor as milk high in protein would be particularly attractive to cheese manufacturers as this would lead to higher yields.

During the past decade, Stoop *et al.* (2006, 2008) showed the potential for selective breeding to alter milk composition as genetic effects showed a high heritability for lipid (0.52), protein (0.60), and lactose (0.64) contents, and moderate heritability for milk (0.44), lipid (0.37), and protein (0.34) yields. Using elite dairy cows (Swedish Red and Swedish Holstein breeds), classified by their high genetic merit, Glantz *et al.* (2009) reported higher yields for components of protein, lipids, carbohydrate profiles, and minerals. However, the content of milk components such as lipids and whey was found to decrease on average. Milk from these cows exhibited good gelation characteristics but was more susceptible to oxidation owing to a lower antioxidant capacity. This study showed that milk composition and processing characteristics could be used to adjust breeding practices for optimizing milk quality and stability.

A study in the Netherlands by Heck *et al.* (2009) examined the effect of seasonal variation on the detailed composition of milk from February 2005 to February 2006. Milk protein ranged from a low of 3.21 g/100 g in June to a high of 3.38 g/100 g in December, while milk fat increased from a low of 4.10 g/100 g in June to a high of 4.57 g/100 g in January. The largest seasonal changes were observed for fatty acid composition in which *trans* fatty acids, including conjugated linoleic acid, increased up to two-fold. In contrast, however, milk lactose remained unchanged throughout the seasons.

While environmental and physiological factors have been shown to affect milk fat, it is nutrition that appears to be the dominant factor (Bauman *et al.*, 2006). The most dramatic effect of nutrition is the low-fat milk syndrome, often referred to as milk fat depression (MFD). Reduction in milk fat was associated with diets containing fish and plant oil supplements as well as those high in concentrates and low in fiber.

III. MILK CONSTITUENTS

The overall composition of cow's milk provides useful information to the farmer and processor in determining quality and market value. This section will discuss the chemistry and biochemistry of major components in milk.

A. Lipids

The lipid component of cow's milk (3—5%) is secreted by the lactating cow as small fat globules ranging in size from < 1 to 15 μm in diameter. Each fat globule is surrounded by a special interfacial layer, or milk fat globule membrane (MFGM), composed of a lipid bilayer and protein (Spitsberg, 2005; Singh, 2006). This layer ensures that each lipid droplet is dispersed in the aqueous milk serum and is unable to aggregate with others. Kanno (1990) reported that the proteins and lipids in MFGM are present in a 1:1 weight ratio. Earlier work by Wooding and Kemp (1975) suggested that MFGM lipids contain a large amount of high-melting triacylglycerols, as indicated in an MFGM membrane model proposed by McPherson and Kitchen (1983). However, Walstra (1974, 1985) reported that this was an artifact caused by contamination of MFGM fragments with triacylglycerol crystals during the isolation procedure. MFGM lipids are now recognized to be predominantly polar, with a very small amount of neutral lipids including triacylglycerols, diacylglycerols, monoacylglycerols, and cholesterol and its esters. The overall gross composition of MFGM is shown in Table 4.3. A review of the literature shows that values vary considerably owing to differences in the isolation, purification, and analytical procedures used (Dewettinck et al., 2008). The MFGM layer also contains at least 50 polypeptides ranging in size from 10 to 300 kDa (Spitsberg and Gorwitt, 1998; Mather, 2000) as well as trace elements.

The structure of MFGM, from the lipid core outwards, was shown schematically by Dewettinck et al. (2008) and is reproduced in Figure 4.1. The inner core consists of a monolayer of polar lipids and proteins surrounding the intracellular fat droplet with an electron-dense proteinaceous coat on the inner face of the bilayer membrane with a true bilayer of polar lipids and proteins. The phospholipid bilayer appears to serve as the backbone of the membrane and exists in the fluid state. Choline-containing phospholipids together with glycolipids, cerebrosides, and gangliosides are found predominantly on the outside of the membrane, with phosphatidyl ethanolamine (PE), phosphatidylserine (PS), and sphingomyelin (SM) concentrated on the inner surface of the membrane. Partially embedded or loosely attached to the bilayer are peripheral membrane proteins. Of these, adipophilin (ADPH) is located in the inner polar layer, while xanthine dehydrogenase/oxidase (XDH/XO) is found between the two layers. Other major MFGM proteins, mucin 1 (MUCI), butyrophilin (BTN), cluster of differentiation 36 (CD36), and periodic acid/Schiff III (PAS III), are shown in the outer layer, with periodic acid/Schiff 6/7 (PAS 6/7) and proteose peptone 3 (PP3) loosely attached to the outside of the membrane. For a more detailed discussion of MFGM structure, the following papers are recommended: Dewettinck et al. (2008), Evers et al. (2008), Reinhardt and Lippolis (2008), Sanchez-Juanes et al. (2009), and Vanderghem et al. (2008).

The amphiphilic nature of the polar lipids (phospholipids and sphingolipids) in the MFGM layer has an important influence on the stability of the fat phase in milk and the changes that occur during the processing of milk and cream. McPherson and Kitchen (1983) summarized the factors affecting the composition of the MFGM layer as shown in Scheme 4.1. The effect of some of these compositional changes on the properties of the MFGM layer remains

TABLE 4.3 Estimated Average Composition of the Milk Fat Globule Membrane

Component	mg/100 g Fat Globules	g/100 g MFGM Dry Matter
Proteins	1800	70
Phospholipids	650	25
Cerebrosides	80	1
Cholesterol	40	2
Monoglycerides	+[a]	?
Water	+	—
Carotenoids + vitamin A	0.04	0.0
Iron	0.3	0.0
Copper	0.01	0.0
Total	> 2570	100

MFGM: milk fat globule membrane.
[a]Present, but quantity unknown.
From Dewettinck et al. (2008).

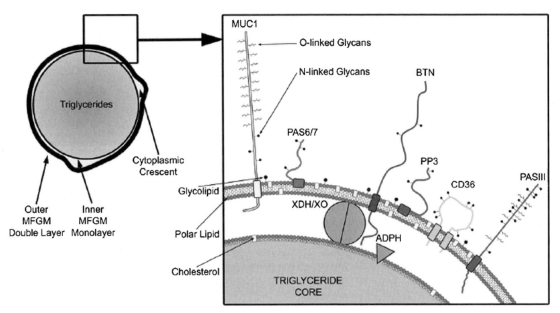

FIGURE 4.1   Structure of the fat globule with detailed arrangements of the main milk fat globule membrane (MFGM) proteins. The drawing is highly schematic and sizes are not proportional. A double layer of polar lipids is placed on the inner monolayer of polar lipids. Membrane-specific proteins are distributed along the membrane. Adipophilin (ADPH) is located in the inner polar lipid layer. Xanthine dehydrogenase/oxidase (XDH/XO) is located between the two layers. Mucin 1 (MUC1), butyrophilin (BTN), cluster of differentiation 36 (CD36), and periodic acid/Schiff III (PAS III) are located in the outer layer. Periodic acid/Schiff 6/7 (PAS 6/7) and proteose peptone 3 (PP3) are only loosely attached to the outside of the MFGM. The choline-containing phospholipids, PC and SM, and the glycolipids, cerebrosides and gangliosides, are largely located on the outside of the membrane, while phosphatidyl ethanolamine (PE), phosphatidylserine (PS), and phosphatidylinositol (PI) are mainly concentrated on the inner surface of the membrane *(Deeth, 1997; Mather and Keenan, 1998; Danthine et al., 2000; Mather, 2000; Harrison, 2002; Rasmussen et al., 2002; Evers, 2004; Dewettinck et al., 2008).*

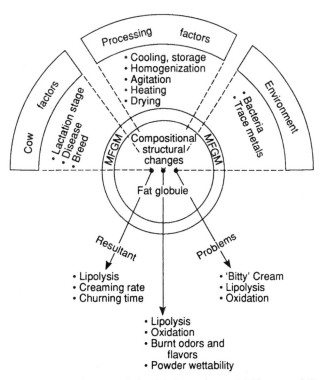

SCHEME 4.1   Factors affecting the milk fat globule membrane. *(McPherson and Kitchen, 1983)*

unclear. The outer surfaces of the MFGM are quite labile and can be removed by simple washing procedures and by temperature manipulation. Such losses of the outer surface have an impact on the processing and storage of milk.

Triacylglycerols account for 97–98% of the total milk fat. The fatty acid composition of the triacylglycerols is influenced by breed, management practices, and season, with saturated fats being lower in summer than in winter (Table 4.4). Milk fats are not markedly affected by changes in dietary lipids since some lipids are metabolized by the microbial activity in the cow's rumen (Garton, 1964). Nevertheless, modification of milk fat can be achieved by feeding the cow a triacylglycerol emulsion protected by a cross-linked protein membrane, to protect the fat from microbial metabolism in the rumen. This technique allows the fatty acid composition of the milk fat to be manipulated and the degree of unsaturation of fats to be modified. The fatty acid composition of bovine milk lipids is more varied than that of any other natural product, with 437 components listed by Patton and Jensen (1975). Of these,

TABLE 4.4 Fatty Acid Composition (% w/w) of Total Fatty Acids in Swedish Dairy Milk in 2001, given as Weighted Mean with Standard Deviation (SD); the Lowest and Highest Values Observed and p-Values for Geographical and Seasonal Variation are also given (Mansson, 2008)

Fatty Acid	Weighted Mean 2001	SD	Lowest Value Observed	Highest Value Observed	Seasonal Variation
4:0	4.4	0.1	4.0	5.1	n.s.
6:0	2.4	0.1	2.1	2.9	n.s.
8:0	1.4	0.1	1.2	1.9	n.s.
10:0	2.7	0.2	2.4	3.5	*
12:0	3.3	0.2	3.0	4.1	**
14:0	10.9	0.5	10.0	12.1	***
15:0	0.9	0.0	0.8	1.1	n.s.
16:0	30.6	0.9	28.7	34.1	**
17:0	0.4	0.0	0.4	0.5	**
18:0	12.2	0.4	10.3	13.3	n.s.
20:0	0.2	0.0	0.2	0.2	n.s.
Saturated fatty acids, total	69.4	1.7	67.1	74.4	***
10:1	0.3	0.0	0.2	0.4	n.s.
14:1	0.8	0.4	0.4	1.3	**
16:1	1.0	0.0	0.9	1.8	n.s.
17:1	0.1	0.0	<0.1	0.3	n.s.
18:1	22.8	1.0	19.7	24.7	***
Monounsaturated fatty acids, *cis*, total	25.0	1.0	22.2	26.7	**
18:2	1.6	0.1	1.4	1.8	n.s.
18:3	0.7	0.0	0.6	0.9	**
Polyunsaturated fatty acids, *cis*, total	2.3	0.1	2.0	2.5	n.s.
16:1t	0.4	0.1	0.3	0.4	***
18:1t	2.1	0.7	2.0	3.3	***
18:2t	0.2	0.0	0.1	0.5	n.s.
Trans fatty acids, total	2.7	0.7	0.6	3.9	***
CLA	0.4	0.1	0.3	0.5	***

n.s.: Not significant; *$p < 0.05$; **$p < 0.01$; ***$p < 0.001$.

saturated fatty acids account for 70%, monounsaturates 25%, *cis*-polyunsaturates 2.3%, and *trans* fatty acids 2.7%. Long-chain fatty acids (C14—C23) and short-chain fatty acids (C4—C6) account for 85% and 15% of the total fatty acids in the triacylglycerols of milk fat, respectively (Table 4.4).

1. Biosynthesis of Milk Fat

The synthesis of milk triacylglycerols relies on blood lipids and *de novo* synthesis in the mammary epithelial cells. The majority of C16—C18 fatty acids in bovine milk triglycerides are derived from dietary fat taken up from the blood, while the C4—C14 fatty acids appear to be synthesized *de novo* in the mammary gland. The dietary fatty acids are transported from the intestine to the mammary glands and other tissues by chylomicrons (Zinder *et al.*, 1974). These chylomicrons are formed in the intestinal epithelial cells from lipids absorbed in the intestine and then transported by the lymph in the intestinal lymphatic vessels to the thoracic duct via the bloodstream, to be taken up by various tissues of the body. Chylomicrons are spherical particles consisting of a triacylglycerol core with traces of cholesteryl esters. They are enclosed by a monolayer of surface film 25—30 Å in width composed primarily of phospholipids and diglycerides with small amounts of cholesterol, protein, monoglycerides, and fatty acids (Scow *et al.*, 1976). Evidence supports the hypothesis that triacylglycerols are taken up by the mammary gland from the chylomicrons, when present, and as low-density lipoprotein (LDL; $d < 1.05$ g/ml) (Moore and Christie, 1979). Mazur and Rayssiguier (1988) reported that dairy cows have characteristically low levels of triacylglycerols and very-low-density lipoprotein (VLDL; $d < 1.01$ g/ml). Most of the circulating lipids are related to high-density lipoprotein (HDL; $d = 1.06$-1.21 g/ml), which carries only small amounts of triacylglycerols. In other words, the triacylglycerols circulating in the blood as lipoprotein ($d > 1.05$) do not donate their fatty acids to the mammary gland for milk fat synthesis. Data presented by Brumby and Welch (1970) and Stead and Welch (1975) demonstrated quite clearly that serum lipoproteins from lactating cows contained negligible amounts of triacylglycerols. In spite of all these reports, Bickerstaffe (1971) still claimed that triacylglycerols were present in all lipoprotein fractions in the blood of lactating cows and were taken up by the mammary gland.

Under normal conditions, when a standard dietary regimen is fed to the cow, the main fatty acids in the blood triacylglycerols are C16:0, C18:0, and C18:1, which are taken up by the mammary gland. Modification of the diet, by feeding canola seed protected with formaldehyde to escape biohydrogenation in the rumen, resulted in an increase in C18:1, C18:2, and C18:3 content in the plasma of Holstein cows (Delbecchi *et al.*, 2001). This increase was reflected by a corresponding increase in the assimilation of C18:1, C18:2, and C18:3 by the mammary gland. It is clear, however, that the fatty acid composition in the plasma is not the sole determinant of fatty acids in the plasma triacylglycerols. Annison *et al.* (1967) and West *et al.* (1972) both observed that the total triacylglycerol in the arterial plasma was much higher in C18:0 and lower in C18:1. This selective assimilation was confirmed by Moore *et al.* (1969), who concluded that the mammary glands selectively assimilated a plasma triacylglycerol with low levels of C18:2 in the presence of high C18:2 plasma tri-acylglycerols for the synthesis of milk fat.

The distribution of triacylglycerol fatty acids among the different lipoprotein fractions was examined because the triacylglycerol in the chylomicrons and VLDL hydrolyzed during uptake into tissues by lipo-protein lipase. Goldberg *et al.* (2009) reviewed the regulation of fatty acid uptake by lipoprotein lipase. Lipoprotein lipase purified from cow's milk was shown by a number of researchers (Morley and Kuksis, 1972; Morley *et al.*, 1975; Paltauf *et al.*, 1974; Paltauf and Wagner, 1976) to have a high specificity for the acyl ester bond in position 1 of the triacylglycerol, hydrolyzing it to 2-monodactyl glycerol according to the following sequence:

Triglyceride
(TG)   2,3–Diacyl glycerol   2-Monoacyl glycerol

Studies on the uptake of blood triacylglycerol by the mammary gland and the role of lipoprotein lipase have been conducted primarily with non-ruminant animals. Nevertheless, the overall results indicate that chylomicrons and VLDL triacylglycerols of mammary tissue are the substrates and apoprotein and phospholipid cofactors for optimum lipoprotein lipase activity. The enzymatic hydrolysis of triacylglycerol occurs only in the chylomicron attached to the endothelium, possibly by hydrolysis of the acyl ester bond at the 1 position. Further hydrolysis of some of the remaining diglycerides at the lumenal surface may occur, and they are then taken up into the microvesicles and transported across the endothelial cells. The second acyl ester bond of the diglyceride is hydrolyzed as the micro-vesicle crosses the endothelium, where some monoglyceride is also hydrolyzed to free glycerol. Once the diglyc-erides enter the alveolar cells of the mammary gland they are resynthesized into milk triacylglycerol. Under normal dietary conditions, a fraction of plasma unesterified fatty acids equal to that released into the bloodstream by hydrolysis of chylomicron triacylglycerol is taken up for the synthesis of milk fat (Moore and Christie, 1979).

2. Fatty Acid Synthesis in the Mammary Gland

The synthesis of fatty acids in the mammary gland involves the stepwise condensation of C2 units by reversal of β-oxidation (Hele, 1954):

$$(1)\quad CH_3-\underset{\underset{O}{\|}}{C}-SCoA + HCO_3^- + ATP \longrightarrow HOOC-CH_2-\underset{\underset{O}{\|}}{C}-SCoA$$

$$+ADP + P_i$$

$$(11)\quad CH_3-\underset{\underset{O}{\|}}{C}-SCoA + (7)\ HOOC-CH_2-CH_2-SCoA + 14NADPH$$

$$\downarrow$$

$$CH_3-CH_2(CH_2{-}{-}{-}CH_2)_6CH_2COOH + 7CO_2 + 14NADP + 8CoASH$$

$$+6H_2O$$

Reaction (1) is catalyzed by acetyl-coenzyme A (acetyl-CoA) carboxylase forming malonyl-CoA, while the second reaction (11) is mediated by a group of enzymes known collectively as the fatty acid synthetase reaction sequence. Current evidence still indicates that acetate and β-hydroxybutyrate are the two main sources for the *de novo* synthesis of fatty acids in lactating ruminants. The fatty acid synthetase sequence consists of seven enzymes integrated into a multienzyme complex, which copurifies as a single protein (Smith, 1976). In animals, fatty acid synthetase requires NADPH and produces C16:0 as the main free fatty acid. This is illustrated in Scheme 4.2, in which each cycle is initiated by the transfer of a saturated acyl group from 4-phosphopantatheine thiol to the cysteine site (B2) with the simultaneous transfer of a malonyl group from the loading site (B1) to the 4-phosphopantatheine thiol. Subsequent reactions include condensation, ketoreduction, dehydration, and enoyl reduction, with the end product being a saturated acyl moiety. To synthesize palmityl this cycle has to be completed seven times. The termination of the acyl chain occurs through the action of a deacylase or thioesterase enzyme, which hydrolyzes the acyl-4-phosphopantatheine thioester bond.

Ruminant milk fat contains substantial amounts of short- and medium-chain fatty acids, which are synthesized in the mammary gland. The synthesis of fatty acids from [^{14}C]acetate by mammary tissue obtained by biopsy from cows before and after parturition was examined with tissue slices (Mellenberger *et al.*, 1973) and homogenates (Kinsella, 1975). These researchers noted that the synthesis of short- and medium-chain fatty acids varied with the physio-logical state of the mammary glands. Only trace amounts of C4:0 to C10:0 fatty acids were synthesized in mammary tissue 18 days before parturition, with C16:0, C14:0, C18:0, and C12:0 accounting for 60%, 30%, 5%, and 4% of the total fatty acids, respectively. This was in sharp contrast to 7 days before parturition, when C4:0−C10:0 fatty acids accounted for 40% of the total fatty acids while C16:0 represented only 30% of the total. A 30-fold increase in the rate of fatty acid synthesis was observed in both mammary slices and homogenates, which began 18 days before parturition and continued until 20 days after parturition. Based on these studies, it is clear that some hormonal control is involved in the synthesis of specific patterns of fatty acid by the mammary glands.

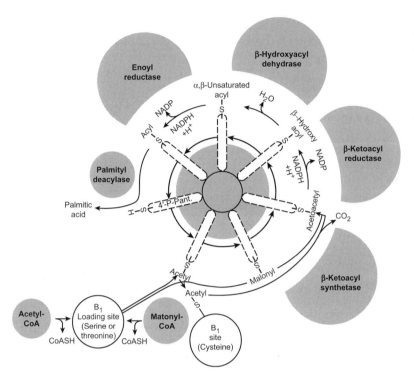

SCHEME 4.2 A mechanism of fatty acid synthesis by the mammalian prototype 1 system according to J. Porter. Note the flow of acetyl-CoA and malonyl-CoA to the loading site and the subsequent movement of the C2 and C3 units attached to the central ACP-like protein, which serves as the substrate component for the peripherally oriented enzyme systems. *(Adapted from Collomb et al., 2006.)*

The mechanism involved in chain length termination, particularly in the synthesis of short- and medium-chain fatty acids, remains unclear. Smith (1980) examined the mechanism of chain length termination in milk fat synthesis by selectively cleaving the thioesterase component from the fatty acid synthetase complex. This was based on previous studies which showed that limited proteolysis of the fatty acid synthetase complex from rat liver and mammary gland with chymotrypsin and trypsin produced defective chain termination (Agradi *et al.*, 1976). Trypsin was later shown to selectively cleave the thioesterase component of the fatty acid synthetase complex (Smith *et al.*, 1976; Dileepan *et al.*, 1978). The fatty acid synthetase complex appeared to be composed of two identical subunits, each susceptible to tryptic attack at three sites. Thioesterase I occupied a terminal locus at one end of the two polypeptide chains. It had a molecular weight of 25,000 with a trypsin-susceptible site near the center. The core of the fatty acid synthetase complex remaining after the addition of trypsin retained the 4'-phosphopantatheine moiety as well as the enzyme activities associated with it, with the exception of thioesterase. The complex was still capable of synthesizing long-chain fatty acids but now lacked the ability to terminate this process. Related studies using an inhibitor of thioesterase I activity also showed the ability of the modified multienzyme complex to synthesize a single acyl moiety. Unlike the dimer, fatty acid synthetase monomer catalyzed all of the reaction described previously with the exception of condensation, which is only carried out by the former (Kumar *et al.*, 1970). The separation of the components of fatty acid synthetase involved in chain growth and chain termination provided a unique opportunity to pinpoint chain length regulation. Since fatty acid synthetase from animal tissues preferentially synthesizes C16:0, a common mechanism must be responsible for the termination of the acyl chain. Thus, removal or inhibition of thioesterase I resulted in the modified multienzyme complex elongating the fatty acid chain to C22; however, the rate of elongation of C16:0 to C18:0 was much slower than the rate of formation of C16:0, with further chain elongation being even slower. If thioesterase I was involved in chain length determination, there would be little activity toward short- and medium-chain thioesters, with a sharp distinction between C14 and C16 thioesters. The enzymes from liver and mammary multienzymes were identical, with a preference for long-chain C16 and C18 acyl moieties.

Studies on the specificity of the condensing enzyme, the chain-elongating enzyme, and the chain-terminating enzyme (thioesterase I) pinpointed where chain lengthening was controlled. The inability of the condensing enzyme to elongate rapidly beyond C16, and of thioesterase I to hydrolyze acyl thioesters shorter than C16:0, ensured that C16:0 would be the main product synthesized. Examination of the great diversity in the fatty acid distribution between species shows that rats, rabbits, rodents, and mice all have large amounts of medium-chain fatty acids, while guinea-pigs have only long-chain fatty acids in milk fat. Smith (1980) demonstrated the presence of a tissue-specific thioesterase II in the mammary glands of rabbits, rats, and mice, which was responsible for modification of the product specificity of fatty acid synthetase. This enzyme was isolated and purified from rabbit and rat mammary glands and when added to a purified fatty acid synthetase it switched production from C16 to medium-chain fatty acids. It was intimately involved with the fatty acid synthetase complex in ruminant mammary glands.

3. Synthesis of Unsaturated Fatty Acids

It is well known that bovine mammary cells actively desaturate C18:0 to C18:1. Subcellular fractionation of lactating and non-lactating cow mammary tissues indicated that the desaturase activity occurred exclusively in the microsomes. Monounsaturated fatty acids, such as oleic acid (C18:1), account for 25% of the fatty acids in milk fat, while polyunsaturated fatty acids are around 2.4% by weight of the total fatty acids. Of these, linoleic acid (C18:2) and α-linolenic acid (C18:3) contributed 1.7% and 0.7%, by weight of the total fatty acids, respectively (Mansson, 2008). In addition, *trans* fatty acids with one or more *trans*-double bonds were reported to account for around 2.7% of the fatty acids in milk (Precht and Molkentin, 1995). These *trans* fatty acids, a mixture of positional and geometric isomers of octadecadinenoic acid containing conjugated double bonds, are referred to as conjugated linoleic acids (CLAs). The main CLA isomer in milk fat, however, is *cis*-9,*trans*-11, accounting for 80–90% of the total CLA (Parodi, 1977; Chin *et al.*, 1992; Sehat *et al.*, 1998). The biosynthesis of CLA in milk fat involves both isomerization and biohydrogenation of unsaturated fatty acids by rumen bacteria as well as Δ^9-desaturase in the mammary gland (Collomb *et al.*, 2006). The main isomer, *cis*-9,*trans*-11 CLA, is formed from linoleic and α-linolenic acids (Bauman *et al.*, 2003). The pathway involved in the biosynthesis of the *cis*-9,*trans*-11 CLA isomer is shown below in Scheme 4.3.

The scheme shows that linoleic acid is rapidly isomerized to CLA *cis*-9,*trans*-11 by *cis*-12,*trans*-11 isomerase and then hydrogenated by rumen bacteria, *Butyrivibrio fibrisolvens*, to vaccenic acid (VA, *trans*-11) (Kepler and Tove, 1967). Loor *et al.* (2002) found a strong correlation between the amount of C18:2 in the diet and the level of vaccenic acid in milk fat. Because VA is hydrogenated at a much slower rate by other microorganisms it accumulates in the rumen (Griinari *et al.*, 1997; Harfoot and Hazelwood, 1997). VA is then converted back to CLA *cis*-9,*trans*-11 by Δ^9-desaturase in the mammary gland (Griinari *et al.*, 2000; Piperova *et al.*, 2002; Mosley *et al.*, 2006). Kraft *et al.* (2003) proposed that CLA (*cis*-9,*trans*-11) is formed indirectly from α-linolenic acid (C18:3) by first being isomerized to a conjugated triene (*cis*-9,*trans*-11,*cis*-15, C18:3) by isomerase in the rumen and then reduced by biohydrogenation to *trans*-11,*cis*-5 C18:2,*trans*-11 18:1 (VA).

B. Milk Proteins

The proteins of milk are a heterogeneous mixture and include two main groups: caseins and whey or serum proteins. These are composed of six major proteins: α_{s1}-casein, α_{s2}-casein, β-casein, κ-casein, β-lactoglobulin, and α-lactalbumin. Other protein fractions are also present but at very low levels, including bovine serum albumin (BSA), immunoglobulins, lactoferritin (LF), PP3, and ceruloplasmin (Fox and Mulvihill, 1982). Caseins represent the major protein group and were defined by Jenness *et al.* (1956) as phosphoproteins precipitated from raw milk by acidification to pH 4.6 at 20°C. The residual proteins in the serum or whey after the removal of caseins were referred to as whey proteins. The nomenclature of proteins in cow's milk has undergone a number of revisions in response to the improved separation of the proteins by gel electrophoresis. The fifth revision of *Nomenclature of the proteins of cow's milk*, by Eigel *et al.* (1984), recommended that electrophoresis no longer be used as the basis for classification. Instead, they suggested that caseins be identified according to the homology of their primary amino acid sequences into four genetic families: α_{s1}-, α_{s2}-, β-, and κ-caseins. Gel electrophoresis was still useful, however, for identifying the different

Linoleic acid (C18:2 cis-9, cis-12)
↓   Rumen isomerase
CLA C18:2 *cis*-9,*trans*-11
Biohydrogenation   Δ 9-Desaturase in
in the rumen   ↓ ↑   the mammary gland
Vaccenic acid (VA) (C18:1 *trans*-11)
Biohydrogenation
in the rumen   ↓
Stearic acid (C18:0)

SCHEME 4.3 Biosynthesis of CLA isomer *cis*-9,*trans*-11 isomer in milk from C18:2. *(Adapted from Collomb et al., 2006)*

members of these families. Farrell *et al.* (2004), in the sixth revision of *Nomenclature of the proteins of cow's milk*, suggested that a flexible nomenclature system should be adopted that can incorporate new discoveries.

1. Caseins

Caseins account for approximately 80% of the total protein in cow's milk. They are present as macromolecular aggregates or 'casein micelles' ranging in size from 30 to 300 nm. These micelles scatter light and are responsible for the whitish opaque nature of skim milk. The major casein fractions, α_{s1}-, α_{s2}-, κ-, β-, and γ-caseins, account for 38%, 10%, 36%, 13%, and 3% of whole casein, respectively (Davies and Law, 1980). The individual casein fractions differ from each other in their behavior towards calcium ions. Waugh and von Hippel (1956) separated micellar casein into a calcium-sensitive and a calcium-insensitive fraction. This was based on the differential solubilities in the presence of specified amounts of calcium ions, resulting in the identification of two fractions referred to as α_s-casein and κ-casein. Aschaffenburg (1961), using alkaline urea paper electrophoresis, demonstrated genetic variants in cow's milk caseins, thus confirming genetic polymorphism. This was consistent with earlier studies by Aschaffenburg and Drewry (1955), who reported the presence of genetic variants in whey proteins. The three genetic variants for α_{s1}-casein were designated A, B, and C, based on their decreasing mobility during electrophoresis in urea-containing starch gels by Thompson *et al.* (1962). A fourth genetic variant, α_{s1}-D, was identified by Grosclaude *et al.* (1966). Genetic polymorphism occurs as a result of substitution of one or two amino acids in the same protein, although in the case of α_{s1}-D, eight amino acid residues are deleted. This phenomenon is directly related to breed and genus (Aschaffenburg, 1968; Bell *et al.*, 1981; Swaisgood, 1982). With the exception of α_{s1}-D and κ-caseins, differences between genetic variants, however, do not have any technological importance. Variants of β-casein were reported earlier by Aschaffenburg (1961) and designated A, B, and C. A fourth genetic variant, β-casein D, was identified in *Bos indicus* by Aschaffenburg *et al.* (1968).

The κ-casein fraction originally identified by von Hippel and Waugh (1955) and Waugh and von Hippel (1956) was considered to be identical to the protective colloid Z-casein described by Lindstrom-Lang and Kodoma (1925). It was later found that κ-casein was associated with α_s-casein, stabilizing it from being precipitated by calcium ions. This fraction appeared to stabilize the casein micelles as well as limit their size. κ-Casein was also composed of several genetic variants, including A and B (Neeling, 1964; Schmidt, 1964; Woychik, 1964).

2. Molecular and Structural Characteristics of Caseins

The complete amino acid sequence has been determined for the primary structure of all the major bovine caseins (Mercier *et al.*, 1971; Jollès *et al.*, 1972; Ribadeau-Dumas *et al.*, 1972; Grosclaude *et al.*, 1973), which permits their average hydrophobicity to be calculated.

a. α_{s1}-Casein

α-Casein is the largest fraction and includes those phosphoproteins precipitated at low calcium concentrations. The α_{s1}-casein (α_{s1}-CN) family makes up 40% of this fraction and contains 199 amino acid residues, of which 8.4% are prolyl residues evenly distributed throughout the polypeptide chain. A highly charged region is evident between residues 41 and 80, which contains eight very acidic phosphoseryl residues. The rest of the molecule has a net zero charge, although there are three strong hydrophobic regions at 1–40, 90–110, and 130–199 amino acid residues. α_{s1}-Casein is typical of caseins in being an amphiphilic molecule, dominated by an acidic peptide at one end, while the other end is capable of forming hydrophobic bonds. The five genetic variants designated for α_{s1}-casein are A, D, B, C, and E, in order of decreasing relative electrophoretic mobilities in alkaline gels containing urea. Three new genetic variants of α_{s1}-CN F have been reported since the last edition of this book. These include α_{s1}-CN F in German black and white cattle (Erhardt, 1993), α_{s1}-CN G in Italian brown cattle (Mariani *et al.*, 1995), and α_{s1}-CN H in African animals (Mahe *et al.*, 1999). The major genetic variant in *B. taurus* is the B variant. If such differences between genetic variants are due to phosphorylation, this is now designated by an Arabic number and a letter after the Latin letter which indicates the particular genetic variant (Eigel *et al.*, 1984). For example, the main genetic variant B for α_{s1}-casein in *B. taurus* with eight phosphoseryl residues is now designated α_{s1}-CN B-8P (Figure 4.2).

b. α_{s2}-Casein

α_{s2}-Casein (α_{s2}-CN) accounts for 10% of this fraction and is composed of 207 amino acid residues. It is the most hydrophilic of the caseins, with 10–13 phosphoseryl residues located in three charged regions, 8–16, 56–61, and

```
 1                                    10                                  20
H-Arg-Pro-Lys-His-Pro-Ile-Lys-His-Gln-Gly-Leu-Pro-Gln-Glu-Val-Leu-Asn-Glu-Asn-Leu-
21                                    30                                  40
 Leu-Arg-Phe-Phe-Val-Ala-Pro-Phe-Pro-Glu-Val-Phe-Gly-Lys-Glu-Lys-Val-Asn-Glu-Leu-
41                                    50                                  60
 Ser-Lys-Asp-Ile-Gly-SeP-Glu-SeP-Thr-Glu-Asp-Gln-Ala-Met-Glu-Asp-Ile-Lys-Gln-Met-
61                                    70                                  80
 Glu-Ala-Glu-SeP-Ile-SeP-SeP-SeP-Glu-Glu-Ile-Val-Pro-Asn-SeP-Val-Glu-Gln-Lys-His-
81                                    90                                  100
 Ile-Gln-Lys-Glu-Asp-Val-Pro-Ser-Glu-Arg-Tyr-Leu-Gly-Tyr-Leu-Glu-Gln-Leu-Leu-Arg-
101                                   110                                 120
 Leu-Lys-Lys-Tyr-Lys-Val-Pro-Gln-Leu-Glu-Ile-Val-Pro-Asn-SeP-Ala-Glu-Glu-Arg-Leu-
121                                   130                                 140
 His-Ser-Met-Lys-Glu-Gly-Ile-His-Ala-Gln-Gln-Lys-Glu-Pro-Met-Ile-Gly-Val-Asn-Gln-
141                                   150                                 160
 Glu-Leu-Ala-Tyr-Phe-Tyr-Pro-Glu-Leu-Phe-Arg-Gln-Phe-Tyr-Gln-Leu-Asp-Ala-Tyr-Pro-
161                                   170                                 180
 Ser-Gly-Ala-Trp-Tyr-Tyr-Val-Pro-Leu-Gly-Thr-Gln-Tyr-Thr-Asp-Ala-Pro-Ser-Phe-Ser-
181                                   190                                 199
 Asp-Ile-Pro-Asn-Pro-Ile-Gly-Ser-Glu-Asn-Ser-Glu-Lys-Thr-Thr-Met-Pro-Leu-Trp-OH
```

FIGURE 4.2 Primary structure of Bos α$_{s1}$-CN B-8P. *(Mercier et al., 1971; Grosclaude et al., 1973; Stewart et al., 1984; Nagao et al., 1984; Koczan et al., 1991; Farrell et al., 2004)*

129–133. It has a very strong hydrophobic C-terminal region (160–207) and a weak hydrophobic segment between 90 and 120 (Fox and Mulvihill, 1982). α$_{s1}$-Casein also contains cysteine residue(s), which may be involved in thermally induced thiol–disulfide interactions. Four genetic variants have been reported, designated A, B, C, and D. The milk components previously classified as α$_{s3}$-, α$_{s4}$-, α$_{s5}$-, and α$_{s6}$-casein (Annan and Manson, 1969; Rose *et al.*, 1970; Whitney *et al.*, 1976) were all components of the α$_{s2}$-CN family with a similar amino acid sequence differing only in the degree of phosphorylation. Eigel *et al.* (1984) recommended that a tentative nomenclature for the A variant be changed as indicated in Table 4.5. α$_{s5}$-Casein, however, is a dimer of the α$_{s3}$- and α$_{s4}$-caseins, now designated α$_{s2}$-CN A-12P and α$_{s2}$-CN A-11P, and joined by a disulfide bond (Hoagland *et al.*, 1971). As a result of cDNA sequencing (Stewart *et al.*, 1987) and genomic DNA sequencing (Groenen *et al.*, 1993), the primary structure of α$_{s2}$-CN A-11P, originally reported by Brignon *et al.* (1977), has been modified. This is shown in Figure 4.3, where Gln replaces Glu at position 87.

c. β-Caseins

β-Caseins (β-CN) comprise up to 45% of total caseins and are also phosphoproteins. They differ from α-caseins by their strong temperature-dependent association as well as the temperature dependency of their solubility in the presence of calcium ions. β-Casein has 209 amino acid residues, of which 16.7% are proline evenly distributed along the polypeptide, which limits the formation of an α-helix. Seven genetic variants are recognized, which separate differently by electrophoresis, depending on whether acidic or alkaline conditions are used. While A can be differentiated from B, C, and D by electrophoresis under alkaline conditions, the A variants can only be separated under acidic conditions. Details on the nature and nomenclature of these variants were reviewed by Eigel *et al.* (1984). Since

TABLE 4.5 Recommended Changes for Nomenclature of the α$_{s2}$-CN A Family

Former Nomenclature	Recommended Nomenclature
α$_{s2}$-CN A	α$_{s2}$-CN A-13
α$_{s3}$-CN	α$_{s2}$-CN A-12
α$_{s4}$-CN	α$_{s2}$-CN A-11
α$_{s6}$-CN	α$_{s2}$-CN A-10

```
1                           10                          20
H-Lys-Asn-Thr-Met-Glu-His-Val-SeP-SeP-SeP-Glu-Glu-Ser-Ile-Ile-SeP-Gln-Glu-Thr-Tyr-
21                          30                          40
  Lys-Gln-Glu-Lys-Asn-Met-Ala-Ile-Asn-Pro-SeP-Lys-Glu-Asn-Leu-Cys-Ser-Thr-Phe-Cys-
41                          50                          60
  Lys-Glu-Val-Val-Arg-Asn-Ala-Asn-Glu-Glu-Glu-Tyr-Ser-Ile-Gly-SeP-SeP-SeP-Glu-Glu-
61                          70                          80
  SeP-Ala-Glu-Val-Ala-Thr-Glu-Glu-Val-Lys-Ile-Thr-Val-Asp-Asp-Lys-His-Tyr-Gln-Lys-
81                          90                          100
  Ala-Leu-Asn-Glu-Ile-Asn-Gln-Phe-Tyr-Gln-Lys-Phe-Pro-Gln-Tyr-Leu-Gln-Tyr-Leu-Tyr-
101                         110                         120
  Gln-Gly-Pro-Ile-Val-Leu-Asn-Pro-Trp-Asp-Gln-Val-Lys-Arg-Asn-Ala-Val-Pro-Ile-Thr-
121                         130                         140
  Pro-Thr-Leu-Asn-Arg-Glu-Gln-Leu-SeP-Thr-SeP-Glu-Glu-Asn-Ser-Lys-Lys-Thr-Val-Asp-
141                         150                         160
  Met-Glu-SeP-Thr-Glu-Val-Phe-Thr-Lys-Lys-Thr-Lys-Leu-Thr-Glu-Glu-Glu-Lys-Asn-Arg-
161                         170                         180
  Leu-Asn-Phe-Leu-Lys-Lys-Ile-Ser-Gln-Arg-Tyr-Gln-Lys-Phe-Ala-Leu-Pro-Gln-Tyr-Leu-
181                         190                         200
  Lys-Thr-Val-Tyr-Gln-His-Gln-Lys-Ala-Met-Lys-Pro-Trp-Ile-Gln-Pro-Lys-Thr-Lys-Val-
201              207
  Ile-Pro-Tyr-Val-Arg-Tyr-Leu-OH
```

FIGURE 4.3   Primary structure of *Bos* α$_{s2}$-CN A-11P (Brignon *et al.*, 1977; Mahe and Grosclaude, 1982; Stewart *et al.*, 1987; Groenen *et al.*, 1993). Seryl residues (SeP) identified as phosphorylated are indicated in italic, bold type *(Farrell* et al., *2004).*

then, three new variants were identified by sequence as β-CN F, formerly known as β-CN X (Visser *et al.*, 1995), β-CN G (Dong and Ng-Kwai-Hang, 1998), and β-CN H (Han *et al.*, 2000).

The primary sequence for the most common A2 variant, β-CN A2-5P, is shown in Figure 4.4. β-Caseins are the most hydrophobic of the casein fractions because of the number of hydrophobic residues present, although they have a strongly charged N-terminal region. Techniques including circular dichroism and spherical rotary dispersion

```
1                           10                          20
H-Arg-Glu-Leu-Glu-Glu-Leu-Asn-Val-Pro-Gly-Glu-Ile-Val-Glu-SeP-Leu-SeP-SeP-SeP-Glu-
21                 ↓        30                          40
  Glu-Ser-Ile-Thr-Arg-Ile-Asn-Lys-Lys-Ile-Glu-Lys-Phe-Gln-SeP-Glu-Glu-Gln-Gln-Gln-
41                          50                          60
  Thr-Glu-Asp-Glu-Leu-Gln-Asp-Lys-Ile-His-Pro-Phe-Ala-Gln-Thr-Gln-Ser-Leu-Val-Tyr-
61                          70                          80
  Pro-Phe-Pro-Gly-Pro-Ile-Pro-Asn-Ser-Leu-Pro-Gln-Asn-Ile-Pro-Pro-Leu-Thr-Gln-Thr-
81                          90                          100
  Pro-Val-Val-Val-Pro-Pro-Phe-Leu-Gln-Pro-Glu-Val-Met-Gly-Val-Ser-Lys-Val-Lys-Glu-
101        ↓         ↓        110                         120
  Ala-Met-Ala-Pro-Lys-His-Lys-Glu-Met-Pro-Phe-Pro-Lys-Tyr-Pro-Val-Glu-Pro-Phe-Thr-
121                         130                         140
  Glu-Ser-Gln-Ser-Leu-Thr-Leu-Thr-Asp-Val-Glu-Asn-Leu-His-Leu-Pro-Leu-Pro-Leu-Leu-
141                         150                         160
  Gln-Ser-Trp-Met-His-Gln-Pro-His-Gln-Pro-Leu-Pro-Pro-Thr-Val-Met-Phe-Pro-Pro-Gln-
161                         170                         180
  Ser-Val-Leu-Ser-Leu-Ser-Gln-Ser-Lys-Val-Leu-Pro-Val-Pro-Gln-Lys-Ala-Val-Pro-Tyr-
181                         190                         200
  Pro-Gln-Arg-Asp-Met-Pro-Ile-Gln-Ala-Phe-Leu-Leu-Tyr-Gln-Glu-Pro-Val-Leu-Gly-Pro-
201              209
  Val-Arg-Gly-Pro-Phe-Pro-Ile-Ile-Val-OH
```

FIGURE 4.4   Primary structure of *Bos* β-CN A^2-5P. *(Ribadeau-Dumas* et al., *1972; Grosclaude* et al., *1973; Farrell* et al., *2004)*

```
*1                                    10                                    20
H-Glu-Glu-Gln-Asn-Gln-Glu-Gln-Pro-Ile-Arg-Cys-Glu-Lys-Asp-Glu-Arg-Phe-Phe-Ser-Asp-
   21                                 30                                    40
   Lys-Ile-Ala-Lys-Tyr-Ile-Pro-Ile-Gln-Tyr-Val-Leu-Ser-Arg-Tyr-Pro-Ser-Tyr-Gly-Leu-
   41                                 50                                    60
   Asn-Tyr-Tyr-Gln-Gln-Lys-Pro-Val-Ala-Leu-Ile-Asn-Asn-Gln-Phe-Leu-Pro-Tyr-Pro-Tyr-
   61                                 70                                    80
   Tyr-Ala-Lys-Pro-Ala-Ala-Val-Arg-Ser-Pro-Ala-Gln-Ile-Leu-Gln-Trp-Gln-Val-Leu-Ser-
   81                                 90                                    100
   Asn-Thr-Val-Pro-Ala-Lys-Ser-Cys-Gln-Ala-Gln-Pro-Thr-Thr-Met-Ala-Arg-His-Pro-His-
   101              ↓                  110                                   120
   Pro-His-Leu-Ser-Phe-Met-Ala-Ile-Pro-Pro-Lys-Lys-Asn-Gln-Asp-Lys-Thr-Glu-Ile-Pro-
   121                                130                                   140
   Thr-Ile-Asn-Thr-Ile-Ala-Ser-Gly-Glu-Pro-Thr-Ser-Thr-Pro-Thr-Thr-Glu-Ala-Val-Glu-
   141                                150                                   160
   Ser-Thr-Val-Ala-Thr-Leu-Glu-Asp-SeP-Pro-Glu-Val-Ile-Glu-Ser-Pro-Pro-Glu-Ile-Asn-
   161                       169
   Thr-Val-Gln-Val-Thr-Ser-Thr-Ala-Val-OH
```

FIGURE 4.5   Primary structure of *Bos* κ-CN B-1P (Mercier *et al.*, 1973). The arrow indicates the point of attack by chymosin (rennin). The asterisk (*) indicates pyroglutamate as the cyclized N-terminal. The site of post-translational phosphorylation (SeP) is indicated in italic, bold type; residues that may potentially be phosphorylated are underlined *(Farrell* et al.*, 2004).*

excluded the presence of secondary and tertiary structures, although Andrews *et al.* (1979) calculated the presence of 10% α-helix, 13% sheets, and 77% unordered structure in β-casein.

d. γ-Casein

At one time γ-casein was considered to be a distinct fraction accounting for 3% of whole casein. It was shown by electrophoresis to be identical to the C-terminal portion of β-casein (Gordon *et al.*, 1972; Groves *et al.*, 1973). Trieu-Cuot and Gripon (1981), using electrofocusing and two-dimensional electrophoresis of bovine caseins, obtained two-dimensional patterns by enzymatic hydrolysis of β-casein with bovine plasma similar to γ-casein. Studies conducted by a number of researchers suggested that β-casein was hydrolyzed by milk proteinase (plasmin) at three sites adjacent to lysyl residues 28, 104, and 106, producing six polypeptides including γ_1, γ_2, and γ_3-caseins, as well as protease peptones (heat-stable, acid-soluble phosphoproteins) found in milk (Groves *et al.*, 1973; Eigel, 1977; Andrews, 1979). Thus, γ-casein could arise by trypsin-like proteolysis of β-casein before or after milking. The fifth revision of the 'Nomenclature of the proteins of cow's milk' no longer categorized γ-caseins but considered them degradation products of β-casein (Eigel *et al.*, 1984). In the sixth edition, by Farrell *et al.* (2004), the plasmin cleavage products of β-CN were identified as γ_1, γ_2, and γ_3 CN variants and proteose peptones.

The open structure and hydrophobicity of caseins render them extremely susceptible to proteolysis, with a high propensity to formation of bitter peptides (Guigoz and Solms, 1976). Caseins are accessible to attack by the indigenous milk or psychotrophic proteinases, which do not affect the whey proteins (Fox, 1981).

e. κ-Casein

κ-Casein (κ-CN) contains 169 amino acid residues, of which 11.8% are proline (Figure 4.5). The major κ-CN component is carbohydrate free while the minor κ-CN, a glycoprotein, is thought to be glycosylated forms of the major κ-CN. The carbohydrate portion contains N-neuraminic acid (NANA), galactose (Gal), and *N*-acetylgalactosamine (NeuNAC), and is present as either a trisaccharide or a tetrasaccharide (Jollès and Fiat, 1979). The points of attachment between colostral κ-CN and the complex oligosaccharides were shown to be threonine (Thr) residues 131, 133, and 135. Saito and Itoh (1992) confirmed the structures of these carbohydrate moieties and identified three more. The lack of information regarding the structure of the minor κ-CN components made nomenclature of these casein fractions inconclusive in the fifth revision (Eigel *et al.*, 1984). In the sixth revision, Farrell *et al.* (2004) still precluded drawing up a precise nomenclature for these minor κ-casein fractions owing to their high degree of heterogeneity and the limited amounts of κ-CN.

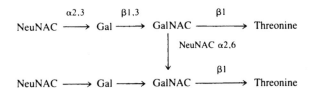

κ-Casein, an amphiphilic molecule with only one phosphoseryl residue, has charged trisaccharide or tetrasaccharide moieties located in the C-terminal segment. The rest of the molecule, however, is highly hydrophobic in character. This protein is unique in that it is soluble in calcium solutions that would normally precipitate the other casein fractions. κ-Casein exerts a stabilizing effect on the casein fractions by forming colloidal micelles. It is this protein fraction that is specifically hydrolyzed by rennin, which releases a macropeptide from the C-terminal region containing the carbohydrates. The specific bond hydrolyzed in κ-casein is 105–106, the Phe–Met linkage. The remaining product with the N-terminus and two-thirds of the original peptide chain is referred to as para-κ-casein. This reaction, discussed in Chapter 8, destabilizes the casein micelle, causing formation of the curd.

3. Whey Proteins

Whey, the major coproduct of the cheese and casein industries, was considered at one time to be a waste product by cheese makers and casein manufacturers (Smithers, 2008). However, it is now considered to be a functional food because of the bioactive properties that whey proteins and individual fractions exhibit (Kruger et al., 2005; Michaelidou and Steijns, 2006). It contains a group of proteins that remain soluble in the milk serum or whey following the precipitation and removal of the caseins at pH 4.6 and 20°C. The major whey proteins, β-lactoglobulin (β-LG), and α-lactalbumin (α-LA), are globular proteins that account for 70–80% of the total whey proteins (Chatterton et al., 2006). The other minor fractions include lactoferrin and lactoperoxidase.

a. β-Lactoglobulins

β-LGs normally provide immunological protection and, because of this and other properties, they are now highly valued as nutraceuticals. They are present as genetic variants, with A and B variants found most frequently in most breeds of cows. The A variant is expressed at a higher level than either the B (Aschaffenburg and Drewry, 1955) or C variant (Hill et al., 1996; Ng-Kwai-Hang and Grosclaude, 2003). Additional variants have also been reported, including H, I, J, and W. The primary sequence of β-LG B is shown in Figure 4.6 and contains 162 amino acids (Farrell et al., 2004).

b. α-Lactalbumin

α-LA normally accounts for 3–4% of the total protein in milk or 20% of the whey proteins (Swaisgood, 1995). It is a globular protein containing 123 amino acids (Brew et al., 1970) with a high degree of homology between the α-LA of other animals including human milk. Together with $β_{1,4}$-galactosyltransferase it forms the lactose synthase complex. As discussed later, it facilitates the formation of lactose from glucose and UDP galactose. Two genetic variants are predominant in the mature protein α-LA, A and B (Jenness, 1974), with a third variant C reported, but not yet confirmed, by DNA or protein sequencing (Bell et al., 1981). It is higher in human milk than cow's milk, suggesting that infant formula could be enriched with α-LA whey proteins (Chatterton et al., 2006). The primary structure for α-LA is shown in Figure 4.7.

c. Lactoferritin

LF, the major member of a family of proteins with specific iron-binding properties, is found in the milk of most species (Schanbacher et al., 1993). It occurs as a single polypeptide chain containing 689 amino acids with varying degrees of glycosylation. LF provides the first line of defense against infection and inflammation (Ward et al., 2002), and also displays antiviral activity (Van de Strate et al., 2001).

d. Bovine Serum Albumin

BSA accounts for around 8% of the total whey proteins in milk and is physically and immunologically identical to blood serum albumin (Coulson and Stevens, 1950; Polis et al., 1950). It contains 583 amino acids and 17 disulfide

```
        1                                    10                                      20
                                                   |<---helical--->|    |<----- β-A -----
     H-Leu-Ile-Val-Thr-Gln-Thr-Met-Lys-Gly-Leu-Asp-Ile-Gln-Lys-Val-Ala-Gly-Thr-Trp-Tyr-
       21                                    30                                      40
     ----- -----    β-A ---->|    |<-helical->|
     Ser-Leu-Ala-Met-Ala-Ala-Ser-Asp-Ile-Ser-Leu-Leu-Asp-Ala-Gln-Ser-Ala-Pro-Leu-Arg-
       41                                    50                                      60
         |<---- ---- β-B ---- ------>|                       |<--- --- --- β-C --- --- ---
     Val-Tyr-Val-Glu-Glu-Leu-Lys-Pro-Thr-Pro-Glu-Gly-Asp-Leu-Glu-Ile-Leu-Leu-Gln-Lys-
       61                                    70                                      80
     β-C-->|    |<--- ---- ---- ---- β-D ---- ---- ---- -->|
     Trp-Glu-Asn-Gly-Glu-Cys-Ala-Gln-Lys-Lys-Ile-Ile-Ala-Glu-Lys-Thr-Lys-Ile-Pro-Ala-
       81                                    90                                     100
     |<--- --- β-E --- --->|            |<---- ---- ---- β-F ---- ---- ---->|
     Val-Phe-Lys-Ile-Asp-Ala-Leu-Asn-Glu-Asn-Lys-Val-Leu-Val-Leu-Asp-Thr-Asp-Tyr-Lys-
       101                                   110                                    120
         |<---- ----- β-G ----- ---->|              |<-helical->|        |<---- β-H --
     Lys-Tyr-Leu-Leu-Phe-Cys-Met-Glu-Asn-Ser-Ala-Glu-Pro-Glu-Gln-Ser-Leu-Ala-Cys-Gln-
       121                                   130                                    140
     --- β-H -->|                   |<----- ----- ---- α-helix ---- ----- ------>|
     Cys-Leu-Val-Arg-Thr-Pro-Glu-Val-Asp-Asp-Glu-Ala-Leu-Glu-Lys-Phe-Asp-Lys-Ala-Leu-
       141                                   150                                    160
                     |<---- β-I ---->|              |<---helical--->|
     Lys-Ala-Leu-Pro-Met-His-Ile-Arg-Leu-Ser-Phe-Asn-Pro-Thr-Gln-Leu-Glu-Glu-Gln-Cys-
                     162
     His-Ile-OH
```

FIGURE 4.6   **Primary structure of** *Bos* **β-LG B (Eigel** *et al.*, **1984).** The free sulfhydryl groups are on Cys121 in the native form of the protein (Bewley *et al.*, 1997; Brittan *et al.*, 1997; Brownlow *et al.*, 1997; Quin *et al.*, 1998a, b, 1999). The sequence positions of the major secondary structural features, α-helix, helical regions, and β-strands (β-A to β-I) are shown above the main sequence (*Farrell* et al., *2004*).

bonds (Hirayama *et al.*, 1990). It plays an important role in the transport, metabolism, and distribution of ligands (Carter and Ho, 1994) as well as in free radical protection.

C. Casein Micelle

The casein proteins in milk are found in a colloidal particle called the casein micelle. These colloidal particles, along with the fat globules, comprise the dispersed phase in milk, surrounded by the serum phase, which consists of lactose, the whey proteins, and dissolved minerals and vitamins. Micelle sizes are polydisperse and can be fitted to a log normal size distribution ranging from 50 to 400 nm with the peak centered at 200 nm, as determined by sedimentation

```
        1                                    10                                      20
     H-Glu-Gln-Leu-Thr-Lys-Cys-Glu-Val-Phe-Arg-Glu-Leu-Lys-Asp-Leu-Lys-Gly-Tyr-Gly-Gly-
       21                                    30                                      40
     Val-Ser-Leu-Pro-Glu-Trp-Val-Cys-Thr-Thr-Phe-His-Thr-Ser-Gly-Tyr-Asp-Thr-Gln-Ala-
       41                                    50                                      60
     Ile-Val-Gln-Asn-Asn-Asp-Ser-Thr-Glu-Tyr-Gly-Leu-Phe-Gln-Ile-Asn-Asn-Lys-Ile-Trp-
       61                                    70                                      80
     Cys-Lys-Asp-Asp-Gln-Asn-Pro-His-Ser-Ser-Asn-Ile-Cys-Asn-Ile-Ser-Cys-Asp-Lys-Phe-
       81                                    90                                     100
     Leu-Asp-Asp-Asp-Leu-Thr-Asp-Asp-Ile-Met-Cys-Val-Lys-Lys-Ile-Leu-Asp-Lys-Val-Gly-
       101                                   110                                    120
     Ile-Asn-Tyr-Trp-Leu-Ala-His-Lys-Ala-Leu-Cys-Ser-Glu-Lys-Leu-Asp-Gln-Trp-Leu-Cys-
                     123
     Glu-Lys-Leu-OH
```

FIGURE 4.7   **Primary structure of** *Bos* **α-LA B (Brew** *et al.*, **1970; Vanaman** *et al.*, **1970).** The disulfide bridges in the molecule are between positions 6 and 120, 28 and 111, 61 and 77, and 73 and 91 (*Farrell* et al., *2004*).

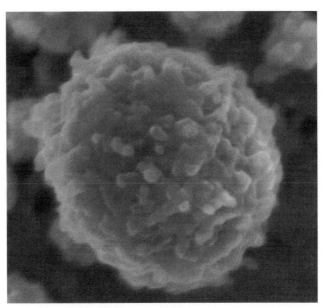

FIGURE 4.8 Image of a casein micelle from field-emission scanning electron microscopy. See Dalgleish *et al.* (2004) for methodology. Bar = 150 nm. *(Courtesy of Professor D. Goff, University of Guelph.)*

field flow fractionation and dynamic light scattering (de Kruif, 1998). Molecular weight determinations suggest that individual micelles could contain as many as 20,000 individual protein molecules (de Kruif and Holt, 2003). In addition to protein (94%), the micelles contain small ions such as calcium, phosphate, magnesium, and citrate, referred to as colloidal calcium phosphate (CaP). These micelles have an open, porous structure, as seen by electron microscopy (Figure 4.8). They are highly hydrated, with a voluminosity of 2−4 ml/g. The stability of the micelles is particularly important as they exert a great influence on the processing properties of the milk. This has resulted in a number of studies on the nature of protein−protein and protein−ion interactions within the micelle structure. The micelle is extremely stable under some conditions of processing, e.g. concentration, ultrafiltration, pelleting, and drying, while being very unstable under others, e.g. acidification and renneting (Horne, 1998). The intricacy of the micelle structure may be related to its biological function in milk, to carry a large amount of highly insoluble CaP to the mammalian young in liquid form, and to form a clot in the stomach for more efficient nutrition (de Kruif and Holt, 2003).

Numerous models of the surface and internal structure of casein micelles have been proposed over a period of more than 50 years, including the core coat, chain polymer, and submicelle models (Waugh and Noble, 1965; Payens, 1966, 1979; Garnier and Ribadeau-Dumas, 1970; Waugh *et al.*, 1970; Schmidt and Payens, 1976; Slattery, 1976; Schmidt, 1980; McMahon and Brown, 1984; Walstra, 1990, 1999). Current models include the CaP nanocluster model (Holt *et al.*, 1998, 2003; de Kruif and Holt, 2003), the dual-binding model (Horne, 1998, 2002, 2006), and the interlocking lattice model (McMahon and Oommen, 2008). With all of the available information and techniques that have been applied to studying casein micelles, it appears now that these models all differ only in fine detail but the essential elements of casein structure have been agreed upon. The reader is referred to a number of recent references for a complete discussion of casein micelle structures (de Kruif and Holt, 2003; Dalgleish *et al.*, 2004; Horne, 2006; McMahon and Oommen, 2008).

1. Casein Micelle Structure

About 75−80% of the proteins in milk are classed as casein protein, that which precipitates at pH 4.6. Most, but not all, of this casein protein exists in the casein micelle, which contains other components as well as casein, including calcium, phosphate, citrate, minor ions, lipase and plasmin enzymes, and entrapped milk serum. This particle is a casein protein:calcium-phosphate complex, and not a true micelle in the colloidal sense. The molar ratio of proteins within the micelle at the time of secretion is approximately α_{s1}: α_{s2}: β: $\kappa = 4:1:3.5:1.5$, or about 38% α_{s1}-, 10% α_{s2}-, 36% β-, and 13% κ-casein. α_s- and β-caseins are extensively phosphorylated on serine residues and will precipitate in the presence of calcium ions or CaP, whereas κ-casein is not sensitive to calcium precipitation. Owing to the high concentration of proline residues, α_s- and β-caseins have been referred to as rheomorphic, lacking a specific secondary structure. They also have high hydrophobicity (Walstra *et al.*, 2006).

The κ-casein content requires special consideration. It plays a unique role during gelation of milk by chymosin in that it is cleaved by chymosin at the 105−106 Phe−Met bond into para-κ-casein (residues 1−105) and the hydrophilic caseinomacropeptide (residues 106−169) fractions. An inverse relationship between κ-casein content and micelle size has been reported (Schmidt, 1979; McGann *et al.*, 1980; Donnelly *et al.*, 1984). Micelles in the 154 nm diameter range may contain as little as 4% κ-casein whereas those in the 62 nm diameter range contain up to 12% κ-casein (Donnelly *et al.*, 1984). Decreases in α_{s1}-, α_{s2}-, and β-casein were noted as κ-casein increased. Artificial micelles can be prepared if α_{s1}- or β-casein is absent, but not if κ-casein is absent (Schmidt, 1980). Thus, it plays an important role in casein micelle stabilization. The macropeptide portion, once cleaved by chymosin, is readily soluble and migrates away from the micelle. The para-casein micelle, devoid of the negatively charged caseinomacropeptide, coagulates readily with other para-casein micelles forming curd. Therefore, κ-casein must exist at or near the surface of the micelle, orientated within easy access of the caseinomacropeptide to cleavage by chymosin. This macropeptide is thought to exist in a hairy manner covering the outer surface of the micelle (Walstra, 1979; Walstra *et al.*, 1981; Holt and Dalgleish, 1986; Holt and Horne, 1996) and stabilizing it by a combination of electrostatic and steric mechanisms.

About 7% of the dry matter of the micelles consists of inorganic material, principally calcium and phosphate (Holt *et al.*, 1998). Milk contains approximately 117 mg calcium per 100 g of milk. Milk serum contains 40 mg per 100 g of serum, only 32% of the calcium content. The rest is associated with the casein micelle, approximately 31 mg per gram of dry casein. Likewise, of 203 mg of inorganic phosphate per 100 g of milk, only 53% is present in the serum and the rest, 37 mg per gram of dry casein, is associated with the micelle. The micelle also contains 5.6 mg citrate, 3.3 mg K, 1.5 mg Mg, and 0.9 mg Na per gram of dry casein (Walstra *et al.*, 2006).

The micelles may contain as much as 7.9 g of water (Farrell *et al.*, 1990), or milk serum depleted of large solute molecules such as globular proteins (Walstra *et al.*, 2006), per gram of protein. The micelle is more highly solvated than most globular proteins and therefore has a rather porous structure. The voluminosity of the micelles ranges from 2 to 4 ml per gram of casein, depending on how far one considers the outer hairy layers of the micelle to extend, representing 6−12% of the volume fraction of milk (Walstra, 1979).

Many studies on the size distribution of the micelles have been conducted using electron microscopy, light scattering, and controlled pore glass chromatography methods. The average diameter has been reported as 25−140 nm, and the volume surface average diameter as 86 nm (Schmidt and Payens, 1976). Micelle sizes can be fitted to a log normal size distribution ranging from 50 to 400 nm with the peak centered at 200 nm, as determined by sedimentation field flow fractionation and dynamic light scattering (de Kruif, 1998), although results vary according to the technique used. A small number of very large particles, up to 800 nm in diameter, and a large number of small particles have been reported (Walstra *et al.*, 2006).

The molar ratios of the casein fractions of the micelle given above refer to the micelle at the time of secretion. However, it has been recognized that β-casein is able to migrate out of the micelle to the serum phase in a reversible manner without causing collapse of the micellar structure (Creamer *et al.*, 1977; Ali *et al.*, 1980). This migration is temperature dependent. As much as 60% of the β-casein has been found in the serum phase after 48 hours at 4°C. This serum β-casein is free to interchange with micellar casein. These changes, however, are reversible upon rewarming to 37°C. This cold dissociation phenomena has a pedagogical impact on a model for the casein micelle, as it suggests the importance of hydrophobic interactions for intramolecular interactions in the interior of the micelle, and also has a technological impact, particularly on the cheese industry, since cheese-making parameters may be altered (Ali *et al.*, 1980) and enhanced proteolysis of the β-casein by plasmin and proteinases of the bacterial microflora may occur in the serum phase (Fox and Guiney, 1973; Creamer *et al.*, 1977).

2. Casein Micelle Models

All of the above considerations have led to numerous proposed models for the casein micelle, which have evolved and been continually refined with new experimental techniques and results. For many years, the most accepted model was the submicelle model elaborated by Schmidt (1980) and refined by Walstra (1990, 1999). The strongest evidence for submicelles of 8−20 nm in diameter came from the porous nature of the micelle, elaborated by early electron microscopy techniques as having a raspberry-like appearance (Schmidt and Buchheim, 1970, 1976). The submicelles were suggested to contain a hydrophobic core (Farrell *et al.*, 1990) covered by a hydrophilic coat much less dense than globular proteins (Farrell *et al.*, 1990), which is at least partly comprised of the polar moieties of κ-casein (Walstra, 1990). The κ-casein formed hair-like projections from the surface of the micelle, with an effective thickness of the hairy layer of least 7 nm (Walstra, 1990).

The submicelle model called for two types of micelles, those rich in κ-casein at the surface and those not. Submicelles were suggested to be aggregated by CaP until such time as the entire surface was coated with κ-casein, thus containing hundreds of submicelles held together by CaP. This, in fact, is the great difficulty with this model: there is no inherent reason why, during the synthesis of caseins and casein micelles in the secretory cell, there would be two types of submicelles of differing composition, nor have any such submicelles ever been found (McMahon and McManus, 1998; Holt *et al.*, 2003; Dalgleish *et al.*, 2004).

Association of casein proteins begins in the Golgi vesicles within the secretory cell. The fully phosphorylated but non-aggregated caseins leave the Golgi apparatus in these vesicles. Calcium and phosphate pass through the membrane into the Golgi vesicles, and when their concentration exceeds the solubility of CaP, micellar aggregation begins to occur. When the Golgi vesicles fuse to the apical plasma membrane, intact micelles are emptied into the alveolar lumen (Farrell *et al.*, 2006).

It must be remembered that one of the biological roles of the casein micelle is to carry much higher concentrations of bioavailable calcium to the infant mammalian than can be carried in solution. The model proposed by Holt *et al.* (1998, 2003) described casein micelles as a homogeneous network of casein polymers containing nanoclusters of CaP. In this model, the biological role of the micelle is addressed, as it has been shown that the phosphoserine residues of α_s- and β-caseins can prevent CaP from precipitating once the association of calcium and phosphate during biosynthesis exceeds solubility (Holt *et al.*, 1998). Hence, it is the biological role of the caseins to sequester CaP into a colloidal particle. The CaP–casein nanocluster, then, resembles to some extent the older casein submicelle, but it has been derived from a better understanding of the biosynthesis process. The dual-binding model of Horne (1998, 2002, 2006) emphasizes the dual abilities of the various casein proteins, except for κ-casein, either to self-associate or to associate with CaP. It is the self-association of caseins that begins to aggregate the nanoclusters. During biosynthesis, formation and aggregation of nanoclusters continue, but κ-casein association with either β- or α_s-casein acts as a chain terminator, which eventually leads to a surface coverage of κ-casein and cessation of aggregation. This evolves the role of κ-casein from the previous submicellar models, as it explains how κ-casein ends up forming the surface hairy layer and it eliminates the need for two different populations of submicelles, as discussed above. Marchin *et al.* (2007), using small-angle X-ray scattering analysis, also concluded that casein micelles were likely to consist of a complex network of protein chains. The only particulate substructures present were considered to be CaP nanoclusters.

McMahon and Oommen (2008) have brought the ideas of the nanocluster model and the dual-binding model together to elaborate what they see as an interlocked lattice type of supramolecular structure, which includes both protein chains and protein–CaP nanoclusters. In their study, they immobilized casein micelles on poly-L-lysine and parlodion-coated copper grid with proteins and CaP stained with uranyl oxalate. Their protocol included instantaneous freezing and drying under high vacuum to obtain stereo images by high-resolution transmission electron microscopy (TEM). These, together with the cross-sectional freeze-fracture replica images produced by TEM by Heertje *et al.* (1985) and Karlsson *et al.* (2007) and the field-emission scanning electron microscope images of Dalgleish *et al.* (2004), provide perhaps the best structural representations to date. Based on these images, McMahon and Oommen (2008) proposed the supramolecular structure shown in Figure 4.9, as an interlocked lattice. The proteins appear as spheres of 8 nm in diameter that both surround CaP nanoclusters and extend as short chains between the interlocking points and out from the periphery. The overall structure is irregular and can accommodate a large diversity of linkages among the proteins such as chain extenders (β-casein or α_{s1}-casein), chain branch points (α_{s1}-casein or α_{s2}-casein), chain terminators (κ-casein), and interlocking points (CaP nanoclusters). The supramolecular structure proposed allows for a very stable colloidal particle composed of thousands of protein molecules and hundreds of CaP nanoclusters. The distance between interlocking sites was in agreement with the 19 nm interval for CaP nanoclusters proposed by de Kruif and Holt (2003). The supramolecular model also allowed for the predominance of κ-casein as terminal molecules on the periphery with protuberances that extended into the surrounding environment. This description, in effect, reconciles the nanocluster and dual-binding models into what seems to be a coherent whole.

The surface of the micelle deserves a little more discussion. It is clear that κ-casein plays a vital role in stabilizing the micelle and exists with its macropeptide extending into the serum. However, the surface is penetrable by chymosin to the Phe–Met bond of κ-casein, by β-lactoglobulin during high-temperature heating as it is known to interact

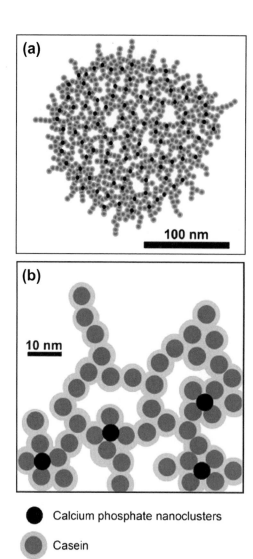

FIGURE 4.9 **Schematic diagram of an interlocking lattice model of the casein micelle with calcium phosphate aggregates throughout the entire supramolecule and chains of proteins extending between them.** Drawn as cross-sectional scaled views of (a) the complete supramolecule, and (b) a portion of the supramolecule periphery. Calcium phosphate nanoclusters are shown with a diameter of 4.8 nm and approximately 18 nm apart; the caseins are shown with a hydrodynamic diameter of 8 nm *(McMahon and Oommen, 2008).*

with κ-casein at the surface of the micelle (Diaz *et al.*, 1996), and also by large polymers such as negatively charged κ-carrageenan (Spagnuolo *et al.*, 2005). It must also allow for β-casein to dissociate reversibly during cold aging. In addition, calculations suggest that there is not sufficient κ-casein to cover the entire surface (Dalgleish, 1998). Hence, this implies that it is not a 'hairy tennis ball' (Holt and Dalgleish, 1986; de Kruif, 1998) type of surface structure but rather that it must be distributed unevenly, leaving large gaps or pores in the surface, perhaps at the end of micro-tubules as described by Dalgleish *et al.* (2004).

D. Biosynthesis of Milk Proteins

The major portion of milk proteins is synthesized by highly specialized mammary secretory cells under genetic control. These produce highly specific proteins that are unique to lactation. The starting materials are free amino acids absorbed from the bloodstream via the basal membrane by a process that involves active transport (Christensen, 1975). The possible role of the γ-glutamyl peptidase cycle in mammary amino uptake was suggested by Baumrucker and Pocius (1978). The enzyme involved, glutamyl transpeptidase (EC 2.3.2.2), catalyzes the transfer of the γ-glutamyl residue from glutathione and/or other γ-glutamyl components to amino acids or peptides (Meister *et al.*, 1976). This enzyme is thought to regulate cellular glutathione and amino acid transport via the γ-glutamyl cycle:

$$\text{glutathione} \quad + \quad \text{amino acid}$$
$$\downarrow$$
$$\gamma\text{-glutamyl-AA} \quad + \quad \text{cysteinyl glycine}$$

The enzyme γ-glutamyl transpeptidase is secreted into the milk where it associates with the milk membranes, including the MFGM or another membrane obtained from skim milk. The latter is derived from plasma membranes, Golgi apparatus, endoplasmic reticulum, and secretory vessels (Kitchen, 1974). Pocius *et al.* (1981) noted that the level of glutathione was extremely low in the plasma of lactating Holstein cows compared to the level in the blood, where it was 200-fold higher. From *in vitro* studies these researchers found that when arteriovenous differences for free amino acids in plasma were quantitated, there was an apparent shortage of cysteine for milk protein synthesis. The uptake of glutathione by the mammary gland, however, was more than sufficient to account for any cysteine secreted in milk. Pocius and Baumrucker (1980) studied the *in vitro* uptake pattern of nine essential amino acids by mammary slices compared to the known *in vivo* uptake pattern of the same amino acids in the cow's udder. A significant linear correlation ($r = 0.91$) was evident between these patterns of uptake in spite of the many assumptions made (Figure 4.10).

The biosynthesis of milk proteins is similar to other systems in which the genetic message is transmitted from DNA to messenger RNA and then translated at the ribosomal level into the amino acid sequence of polypeptides. Following synthesis, the export milk protein leaves the ribosomes on the outer surface of the rough endoplasmic reticulum and is transported to the Golgi apparatus, where alteration of some of the export protein and synthesis of other major constituents of milk occur.

E. Lactose

The major carbohydrate of milk is lactose, a disaccharide of galactose and glucose linked by an α-(1 → 4) glycosidic bond. It is sometimes referred to as the milk sugar and accounts for 2% of normal cow's milk. Its complete name is lactose α-(1 → 4)galactosylglucopyranose.

The biosynthesis of lactose is catalyzed by the enzyme lactase synthetase in which glucose acts as the galactosyl acceptor (Watkins and Hassid, 1962). This enzyme is located in the luminal face of the Golgi dicytosome membrane, where it receives both glucose and UDP-galactose from the cytosol. The reactions involved are shown in Figure 4.11, in which the uridine nucleotide cycle appears to functionally link the two regions (Kuhn *et al.*, 1980). The transfer of glucose, UDP-galactose, and UMP through the Golgi membrane is probably facilitated by a specific carrier in the membrane. The formation of UMP by the enzyme nucleotide diphosphatase permits the removal of UDP released in the

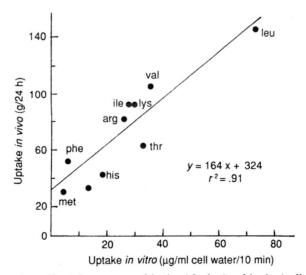

FIGURE 4.10   Relationship between amino acid uptake as measured *in vitro* (abscissa) and *in vivo* (ordinate) (Pocius and Baumrucker, 1980). *(Calculated from data by Clark* et al., *1975.)*

FIGURE 4.11 Uridine nucleotide cycle, supporting lactose synthesis and functionally linking the cytosol and Golgi lumen compartments of the mammary secretory cell. A = Galactosyl transferase; B = α-lactalbumin *(Kuhn* et al., *1980).*

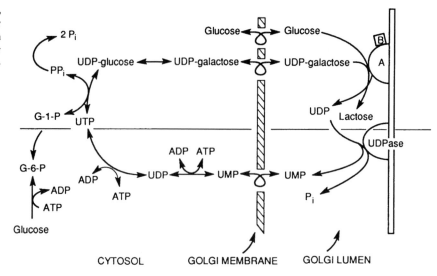

lactose synthetase reaction. This is important as UDP competitively inhibits the lactose synthetase enzyme to form UDP galactose (Kuhn and White, 1975, 1976, 1977). The major steps involved in the biosynthesis of lactose are as follows:

(1) $\text{UTP} + \text{glucose-1-P} \xrightarrow{\text{UDPG-pyrophosphorylase(EC 2.7.79.)}} \text{UDPG-glucose} + \text{PP}_i$

(2) $\text{UDP-glucose} \xrightarrow{\text{UDPG-galactose-4-epimerase(EC 5.1.3.2)}} \text{UDPG-galactose}$

(3) $\text{UDP-galactose} \xrightarrow{\text{Lactase synthetase(EC 2.6.1.22)}} \text{lactose} + \text{UD}$

Kuhn *et al.* (1980), from their studies on rat mammary glands, summarized the benefits accrued by the compartmentalization of the lactose synthetase system. The free energy involved could be used to synthesize lactose without creating osmotic problems for the rest of the cells. The concentration of UDP-glucose in the cytosol must not exceed that of UDP-galactose by a factor of three. UDP-glucose must itself be prevented from entering the Golgi lumen as it would inhibit lactose synthetase. The separation of nucleoside diphosphatase prevents hydrolysis of diphosphate in the cytosol and the subsequent depletion of phosphate energy. The role of α-lactalbumin (α-LA) in the biosynthesis of lactose was discovered by Brodbeck and Ebner (1966), who found that lactose synthetase was composed of two components, an 'A' protein and an 'R' protein. The A protein was identified by Brew *et al.* (1968) as galactosyltransferase, while the R protein was shown by Brodbeck *et al.* (1967) to be α-LA. Interaction between the two components of lactase synthase (EC 2.4.1.22), α-LA, and galactosyltransferase, is essential for the catalysis of lactose biosynthesis. Because α-LA is involved in the last step in the biosynthesis of lactose it is critical for controlling lactation and section of milk (Brew *et al.*, 1968; Lo *et al.*, 1998). While α-LA has no catalytic function, it acts as a specific carrier protein in facilitating the action of galactosyl transferase, also known as $\beta_{1,4}$-galactosyltransferase (β_4Gal-T-1, EC 2.4.1.38). In the absence of α-LA, βGal-T-1 transfers galactose from UDP-galactose to *N*-acetylglucosamine. During the synthesis of lactose, manganese glucose, UDP-galactose, and galactosyl transferase combine with α-LA to form a dimer, which accepts millimolar concentrations of glucose, forming lactose under physiological conditions. α-LA appears to be the major regulator during lactogenesis by altering not only the sugar specificity of β-Gal-T-1 but also its sugar donor specificity (Ramakrishnan *et al.*, 2001, 2002). Thus, in the presence of α-LA, β-Gal-T-1 transfers galactose to glucose, which is its lactose synthase activity. Other possible regulators identified were D-glucose, UDP-galactose, calcium ions, and protein generation within the Golgi lumen, as well as the rate-limiting properties of the Golgi membrane.

REFERENCES

Agradi, E., Libertini, L., Smith, S., 1976. Specific modification of fatty acid synthetase from lactating rat mammary gland by chymotrypsin and trypsin. Biochem. Biophys. Res. Commun. 68, 894−900.

Ali, A.E., Andrews, A.T., Cheeseman, G.C., 1980. Influence of storage of milk on casein distribution between the micellar and soluble phases and its relationship to cheesemaking parameters. J. Dairy Res. 47, 371−382.

Andrews, A.L., 1979. The formation and structure of some proteose—peptone components. J. Dairy Res. 46, 215—218.

Andrews, A.L., Atkinson, D., Evans, M.T.A., Finer, E.G., Green, J.P., Phillips, M.C., Roberston, R.N., 1979. The conformation and aggregation of bovine β-casein A. I. Molecular aspects of thermal aggregation. Biopolymers 18, 1105—1121.

Annan, W.D., Manson, W., 1969. A fractionation of the α-casein complex of bovine milk. J. Dairy Res. 36, 259—268.

Annison, E.F., Linzell, J.L., Fazakerley, S., Nichols, B.W., 1967. The oxidation and utilization of palmitate, stearate, oleate and acetate by the mammary gland of the fed goat in relation to their overall metabolism, and the role of plasma phospholipids and neutral lipids in milk-fat synthesis. Biochem. J. 102, 637—647.

Aschaffenburg, R., 1961. Inherited casein variants in cow's milk. Nature (Lond.) 176, 218.

Aschaffenburg, R., 1968. Reviews of the progress of dairy science. Section G. Genetics, genetic variants of milk proteins, their breed distribution. J. Dairy Res. 35, 447—460.

Aschaffenburg, R., Drewry, J., 1955. Occurrence of different beta-lacto-globulins in cow's milk. Nature (Lond.) 176, 218—219.

Aschaffenburg, R., Sen, A., Thompson, M.P., 1968. Genetic variants of casein in Indian and African zebu cattle. Comp. Biochem. Physiol. 25, 177—184.

Bauman, D.E., Cori, B.A., Peterson, G.P., 2003. The biology of conjugated linoleic acids in ruminants. In: Sebedio, J.-L., Christie, W.W., Adolf, R. (Eds.), Advances in Conjugated Linoleic Acid Research, Vol. 2. AOCS Press, Champaign, IL, pp. 267—282.

Bauman, D.E., Mather, I.H., Wall, R.J., Lock, A.L., 2006. Major advances associated with the biosynthesis of milk. J. Dairy Sci. 89, 1235—1243.

Baumrucker, C.R., Pocius, P.A., 1978. γ-Glutamyl transpeptidase in lactating mammary secretory tissue of cow and rat. J. Dairy Sci. 61, 309—314.

Bell, K., Hopper, K.E., McKenzie, H.A., 1981. Bovine α-lactalbumin C and α-, β- and κ-caseins of Bali (Banteng) cattle. Bos (Bibos) javanicus. Aust. J. Biol. Sci. 34, 149—159.

Bewley, M.C., Quin, B.Y., Jaeson, G.B., Sawyer, L., Baker, E.N., 1997. β-Lactoglobulin and its variants. A three dimensional structural perspective. In: Milk Protein Polymorphism. International Dairy Federation, Special Issue 9702. International Dairy Federation, Brussels, pp. 100—109.

Bickerstaffe, R., 1971. Uptake and metabolism of fat. In: Falconer, I.R. (Ed.), Lactation. Butterworth, London, pp. 317—332.

Brew, K., Vanaman, T.C., Hill, R.L., 1968. The role of α-lactalbumin and the A protein in lactose synthetase: a unique mechanism for the control of biological reaction. Proc. Natl. Acad. Sci. U.S.A. 59, 491—497.

Brew, K., Castellino, F.J., Vanaman, T.C., Hill, R.L., 1970. The complete amino acid sequence of bovine α-lactalbumin. J. Biol. Chem. 245, 4570—4582.

Brignon, G., Ribadeau Dumas, B., Mercier, J.-C., Pelissier, J.-P., Das, B.C., 1977. The complete amino acid sequence of bovine α_{s2}-casein. FEBS Lett. 76, 105—158.

Brittan, H., Mudford, J.C., Norris, G.E., Kitson, T.M., Hill, J.P., 1997. Labelling the free sulfhydryl group in β-lactoglobulin A, B and C. In: International Dairy Federation, Special Issue 9702. International Dairy Federation, Brussels, pp. 200—203.

Brodbeck, U., Ebner, K.E., 1966. Resolution of a soluble lactose synthetase into two protein components and solubilization of microsomal lactose synthetase. J. Biol. Chem. 241, 762—764.

Brodbeck, U., Denton, W.L., Tanahashi, N., Ebner, K.E., 1967. The isolation and identification of lactose synthetase as α-lactalbumin. J. Biol. Chem. 242, 1391—1397.

Brownlow, S., Morais-Cabral, J.H., Cooper, R., Floer, D.R., Yewdall, S.J., Polikarpov, I., North, A.C.T., Sawyer, L., 1997. Bovine β-lacto-globulin at 1.8 A resolution — still an enigmatic lipocalin. Structure 5, 481—495.

Brumby, P.E., Welch, V.A., 1970. Fractionation of bovine serum lipo-proteins and their characterization by gradient gel electrophoresis. J. Dairy Res. 37, 121—128.

Carter, D.C., Ho, J.X., 1994. Structure of serum albumin. Adv. Protein Chem. 45, 153—203.

Chatterton, D.E.W., Smithers, G., Roupas, P., Brodkorb, A., 2006. Bioactivity of β-lactoglobulin and α-lactalbumin — technological implications for processing. Int. Dairy J. 16, 1229—1240.

Chin, S.F., Storkson, J.M., Albright, K.J., Cook, M.E., Pariza, M.W., 1992. Dietary sources of conjugated dienoic isomers of linoleic acid, a newly recognized class of anticarcinogens. J. Food Compos. Anal. 5, 185—197.

Christensen, H.N., 1975. Biological Transport. Benjamin, London.

Clark, J.H., Derrig, R.G., Davis, C.L., Spires, H.R., 1975. Metabolism of arginine and ornithine in the cow and rabbit mammary tissue. J. Dairy Sci. 58, 1808—1813.

Collomb, M., Schmid, A., Sieber, R., Wechsler, D., Ryhanen, E.-L., 2006. Conjugated linoleic acids in milk fat: variation and physiological effects. Int. Dairy J. 16, 1347—1361.

Coulson, E.J., Stevens, H., 1950. The serological relationship of bovine whey albumin to serum albumin. J. Biol. Chem. 187, 355—363.

Creamer, L.K., Berry, G.P., Mills, O.E., 1977. A study of the dissociation of β-casein of the bovine casein micelle at low temperature. N.Z. J. Dairy Sci. Technol. 12, 58—66.

Dalgleish, D.G., 1998. Casein micelles as colloids: surface structures and stabilities. J. Dairy Sci. 81, 3013—3018.

Dalgleish, D.G., Spagnuolo, P., Goff, H.D., 2004. A possible structure of the casein micelle based on high-resolution field-emission scanning electron microscopy. Int. Dairy J. 14, 1025—1031.

Danthine, S., Blecker, C., Paquot, M., Innocente, N., Deoanne, C., 2000. Progress in milk fat globule membrane research: a review. Lait 80, 209—222.

Davies, D.T., Law, A.J.R., 1980. The content and composition of protein in creamery milks in south-west Scotland. J. Dairy Res. 47, 83—90.

Deeth, H.C., 1997. The role of phospholipids in the stability of milk fat globules. Aust. J. Dairy Technol. 52, 44—46.

Delbecci, L., Ahnadl, C.E., Kennelly, J.J., Lacasse, P., 2001. Milk fatty acid composition and mammary lipid metabolism in Holstein cows fed protected or unprotected canola seeds. J. Dairy Sci. 84, 1375—1381.

Dewettinck, K., Rombaut, R., Thienpont, N., Trung Le, T., Messens, K., Van Camp, J., 2008. Nutritional and technological aspects of milk fat globule membrane material. Int. Dairy J. 18, 436—457.

Diaz, O., Gouldsworthy, A.M., Leaver, J., 1996. Identification of peptides from casein micelles by limited trypsinolysis. J. Agric. Food Chem. 44, 2517—2522.

Dileepan, K.N., Lin, C.Y., Smith, S., 1978. Release of two thioesterase domains from fatty acid synthetase by limited digestion with trypsin. Biochem. J. 175, 199—206.

Dong, C., Ng-Kwai-Hang, K.F., 1998. Characterization of a non-electrophoretic variant of β-casein by peptide mapping and mass spectrometric analysis. Int. Dairy J. 8, 967—972.

Donnelly, W.J., MacNeill, G.P., Buchheim, W., McGann, T.C.A., 1984. Comprehensive study of the relationship between size and protein composition in natural bovine casein micelles. Biochim. Biophys. Acta. 789, 136–143.

Eigel, W.N., 1977. Formation of γ1-A2, γ2-γ2 and γ3-A caseins by *in vitro* proteolysis of β-casein A2 with bovine plasmin. Int. J. Biochem. 8, 187–192.

Eigel, W.N., Butler, J.E., Ernstrom, C.A., Farrell Jr., H.M., Halwalkar, V.R., Jennes, R., Whitney, R., McL, 1984. Nomenclature of proteins of cow's milk. Fifth revision. J. Dairy Sci. 67, 1599–1631.

Erhardt, G., 1993. A new α_{s1}-casein allele in bovine milk and its occurrence in different breeds. Anim. Genet. 24, 65–66.

Evers, J.M., 2004. The milk fat globule membrane – compositional and structural changes post secretion by the mammary secretory cell. J. Food Sci. 14, 661–674.

Evers, J.M., Haverkamp, R.G., Holroyd, S.E., Jameson, G.B., Mackenzie, D.D.S., McCarthy, O.J., 2008. Heterogeneity of milk fat globule membrane structure and composition as observed using fluorescence microscopy techniques. Int. Dairy J. 18, 1081–1089.

Farrell Jr., H.M., Pessen, H., Brown, E.M., Kumosinski, T.F., 1990. Structural insights into the bovine casein micelle: small angle X-ray scattering studies and correlations with spectroscopy. J. Dairy Sci. 73, 3592–3601.

Farrell Jr., H.M., Jimenez-Flores, R., Bleck, G.T., Brown, E.M., Butler, J.E., Creamer, L.K., Hicks, C.L., Hollar, C.M., Ng-Kwai-Hang, K.F., Swaisgood, H.E., 2004. Nomenclature of the proteins of cow's milk – sixth revision. J. Dairy Sci. 87, 1641–1674.

Farrell, H.M., Malin, E.L., Brown, E.M., Qi, P.X., 2006. Casein micelle structure: what can be learned from milk synthesis and structural biology? Curr. Opin. Colloid Interface Sci. 11, 135–147.

Fox, P.F., 1981. Proteinases in dairy technology. Neth. Milk Dairy J. 35, 233–253.

Fox, P.F., Guiney, J., 1973. Casein micelle structure: susceptibility of various casein systems to proteolysis. J. Dairy Res. 40, 229–234.

Fox, P.F., McSweeney, P.L.H., 1998. Dairy Chemistry and Biochemistry. Blackie Academic and Professional (Chapman and Hall), New York.

Fox, P.F., McSweeney, P.L.H., 2003. Advanced Dairy Chemistry – 1. Proteins, (3rd ed.). Kluwer Academic, New York.

Fox, P.F., McSweeney, P.L.H. (Eds.), 2006. Advanced Dairy Chemistry – 2. Lipids, (3rd ed.). Kluwer Academic, New York.

Fox, P.F., McSweeney, P.L.H. (Eds.), 2009. Advanced Dairy Chemistry – 3. Lactose, Water and Minor Constituents, (3rd ed.). Springer, New York.

Fox, P.F., Mulvihill, D.M., 1982. Milk proteins, molecular, colloidal and functional properties. J. Dairy Res. 49, 679–693.

Garnier, J., Ribadeau-Dumas, B., 1970. Structure of the casein micelle. A proposed model. J. Dairy Res. 37, 493–505.

Garton, G.A., 1964. In: Dawson, R.M.C., Rhodes, D.N. (Eds.), Metabolism and Physiological Significance of Lipids. Wiley, New York, p. 335.

Glantz, M., Lindmark Mansson, H., Stalhammar, H., Barstrom, L.-O., Frojelin, M., Knutsson, A., Teluk, C., Paulsson, M., 2009. Effects of animal selection on milk composition and processability. J. Dairy Sci. 92, 4589–4603.

Goff, H.D., Griffiths, M.W., 2006. Major advances in fresh milk and milk products: fluid milk and frozen desserts. J. Dairy Sci. 89, 1163–1173.

Goldberg, I.J., Eckel, R.H., Abumrad, N.A., 2009. Regulation of fatty acid uptake into tissues: lipoprotein lipase- and CD36-mediated pathways. J. Lipid Res. 50, S86–S90.

Gordon, W.G., Groves, M.L., Greenberg, R., Jones, S.B., Kalan, E.B., Peterson, R.F., Townsend, R.E., 1972. Probable identification of γ, TS-, R- and S-caseins as fragments of β-casein. J. Dairy Sci. 55, 261–263.

Griinari, J.M., Chouinard, P.Y., Bauman, D.E., 1997. *Trans* fatty acid hypothesis of milk fat depression revised. In: Proceedings of the Cornell Nutrition Conference for Feed Manufacturers. Cornell University, Ithaca, NY, pp. 208–216.

Griinari, J.M., Cori, B.A., Lucy, S.H., Chouinard, P.Y., Nurmela, K.V.V., Bauman, D.E., 2000. Conjugated linoleic acid is synthesized endogenously in lactating dairy cows by Δ^9-desaturase. J. Nutr. 130, 2285–2291.

Groenen, M.A.M., Dijkhof, R.E.M., Verstege, A.J.M., van der Poel, J.J., 1993. The complete sequence of the gene coding bovine α_{s2}-casein. Gene 123, 5106–5114.

Grosclaude, F., Joudrier, P., Mahé, M.-F., 1966. Polmorphisime de la caséine α-bovine; étroite liason du locus α-Cn avec les loci déletion dans le variant α-CnD. Ann. Genet. Sel. Anim. 10, 313–327.

Grosclaude, F., Mahé, M.-F., Ribadeau-Dumas, B., 1973. Structure primaire de la caséine et de la caséine β-bovine correctif. Eur. J. Biochem. 40, 323–324.

Groves, M.L., Gordon, W.G., Kalan, E.B., Jones, S.B., 1973. TS-A2, TS-B, R- and S-caseins: their isolation, composition and relationship to the β- and α-casein polymorphs A2 and B. J. Dairy Sci. 56, 558–568.

Guigoz, Y., Solms, J., 1976. Bitter peptides, occurrence and structure. Chem. Senses Flavour. 2, 71–84.

Han, S.K., Shin, Y.C., Byun, H.D., 2000. Biochemical, molecular and physiological characterization of a new β-casein variant detected in Korean cattle. Anim. Genet. 31, 49–51.

Harfoot, C.G., Hazelwood, G.P., 1997. Lipid metabolism in the rumen. In: Hobson, P.M. (Ed.), The Rumen Microbial Ecosystem, (2nd ed.). Elsevier, New York, pp. 382–426.

Harrison, R., 2002. Structure and function of xanthine oxidoreductase: where are we now? Free Rad. Biol. Med. 33, 774–797.

Heck, J.M.L., van Valenberg, H.J.F., Dijkstra, J., van Hooijdonk, A.C.M., 2009. Seasonal variation in the Dutch bovine raw milk composition. J. Dairy Sci. 92, 4745–4755.

Heertje, I., Visser, J., Smits, P., 1985. Structure formation in acid milk gels. J. Ultrastruct. Res. 25, 193–213.

Hele, P., 1954. The acetate activating enzyme of beef heart. J. Biol. Chem. 206, 671–676.

von Hippel, P.H., Waugh, D.F., 1955. Casein, monomers and polymers. J. Am. Chem. Soc. 77, 4311.

Hill, J.P., Boland, J., Creamer, L.K., Anema, S.G., Otter, D.E., Paterson, G.R., 1996. Effect of bovine β-lactoglobulin phenotype on the properties of β-lactoglobulin, milk composition, and dairy products. In: Parris, N., Kato, A., Creamer, L.K., Pearce, R.J. (Eds.), Macromolecular Interactions in Food Technology, ACS Symposium Series 650. American Chemical Society, Washington, DC, pp. 281–294.

Hirayama, K., Akashi, S., Furuya, M., Fukuhara, K., 1990. Rapid confirmation and revision of the primary structure of bovine serum albumin by ESIMS and Frit-FAB LC/MS. Biochem. Biophys. Res. Commun. 173, 639–646.

Hoagland, P.D., Thompson, M.P., Kalan, E.B., 1971. Amino acid composition of α_{s3}-, α_{s4}-, and α_{sr}-caseins. J. Dairy Sci. 54, 1103.

Holt, C., Dalgleish, D.G., 1986. Electrophoretic and hydrodynamic properties of bovine casein micelles interpreted in terms of particles with an outer hairy layer. J. Colloid Interface Sci. 114, 513–524.

Holt, C., Horne, D.S., 1996. The hairy casein micelle: evolution of the concept and its implications for dairy technology. Neth. Milk Dairy J. 50, 85–111.

Holt, C., Timmins, P.A., Errington, N., Leaver, J., 1998. A coreshell model of calcium phosphate nanoclusters stabilized by β-casein phosphopeptides derived from sedimentation equilibrium and small-angle X-ray and neutron-scattering experiments. Eur. J. Biochem. 252, 73–78.

Holt, C., de Kruif, C.G., Tunier, R., Timmons, P.A., 2003. Substructure of bovine casein micelles by small angle X-ray and neutron scattering. Colloids Surfaces A 213, 275–284.

Horne, D.S., 1998. Casein interactions: casting light on the black boxes, the structure of dairy products. Int. Dairy J. 8, 171–177.

Horne, D.S., 2002. Casein structure, self assembly and gelation. Curr. Opin. Colloid Interface Sci. 7, 456–461.

Horne, D.S., 2006. Casein micelle structure: models and muddles. Curr. Opin. Colloid Interface Sci. 11, 148–153.

Jenness, R., 1974. The composition of milk. In: Larson, B.I., Smith, V.R. (Eds.), Lactation, Vol. III. Academic Press, New York, pp. 3–107.

Jenness, R., Larson, B.L., McMeekin, T.L., Swanson, C.H., Whitnah, C.H., Whitney, R., 1956. Nomenclature of the proteins of bovine milk. J. Dairy Sci. 39, 536–541.

Jollès, J., Fiat, A.-M., 1979. The carbohydrate portions of milk glyco-proteins. J. Dairy Res. 46, 187–191.

Jollès, J., Schoentgen, F., Alais, C., Jollès, P., 1972. Studies on the primary structure of cow κ-casein: the primary structure of cow para-κ-casein. Chimia 20, 148.

Kanno, C., 1990. Secretory membranes of the lactating mammary gland. Protoplasma 159, 184–208.

Karlsson, A.O., Ipsen, R., Ardo, Y., 2007. Observations of casein micelles to skim milk concentrate by transmission electron microscopy. Food Sci. Technol. 40, 1102–1107.

Keppler, C.R., Tove, S.B., 1967. Biohydrogenation of unsaturated fatty acids. 3. Purification and properties of a linoleate $\Delta 12$-cis, $\Delta 11$-trans isomerase from Butyrivibriofibrisolvens. J. Biol. Chem. 242, 5686–5692.

Kinsella, J.E., 1975. Coincident synthesis of fatty acids and secretory triglycerides in bovine mammary tissue. Int. J. Biochem. 6, 65–67.

Kitchen, B.J., 1974. A comparison of the properties of membranes isolated from bovine skim and cream. Biochim. Biophys. Acta. 356, 257–269.

Koczan, D., Hobom, G., Seyfert, H.-M., 1991. Genomic organization of the bovine α_{s1}-casein gene. Nucleic Acids Res. 19, 1590–1594.

Kraft, J., Collomb, M., Mockel, P., Sieber, R., Jahreis, G., 2003. Differences in CLA isomer distribution of cow's milk lipids. Lipids 38, 657–664.

Kruger, M.C., Plimmer, G.G., Schollum, L.M., Haggarty, N., Ram, S., Palmano, K., 2005. The effect of whey acidic protein fractions on bone loss in the ovariectomised rat. Br. J. Nutr. 93, 244–252.

de Kruif, C.G., 1998. Supra-aggregates of casein micelles as a prelude to coagulation. J. Dairy Sci. 81, 3019–3028.

de Kruif, C.G., Holt, C., 2003. Casein micelle structure, functions and interactions. In: Fox, P.F., McSweeney, P.L.H. (Eds.), Advanced Dairy Chemistry 1. Proteins, Part A. Kluwer Academic/Plenum, New York, pp. 233–270.

Kuhn, N.J., White, A., 1975. Milk glucose as an index of the intracellular glucose concentration of rat mammary gland. Biochem. J. 152, 153–155.

Kuhn, N.J., White, A., 1976. Evidence for specific transport of uridine diphosphate galactose across the Golgi membrane of rat mammary gland. Biochem. J. 154, 243.

Kuhn, N.J., White, A., 1977. The role of nucleoside diphosphatase galactose across the Golgi membrane of rat mammary gland. Biochem. J. 168, 423–433.

Kuhn, N.J., Carrick, D.T., Wilde, C.J., 1980. Lactose synthesis: the possibilities of regulation. J. Dairy Sci. 63, 328–336.

Kumar, S., Dorsey, J.A., Muesing, R.A., Porter, J.W., 1970. Comparative studies of the pigeon liver fatty acid synthetase complex and its subunits. J. Biol. Chem. 245, 4732–4744.

Lindstrom-Lang, K., Kodoma, S., 1925. Studies over kasein. C.R. Trav. Lab. Carlsberg 16, 1.

Lo, N.W., Shaper, J.H., Pevsner, J., Shaper, N.L., 1998. The expanding of β-4 galactosyltransferase gene family. Glycobiology 8, 517–526.

Loor, J.J., Herbein, J.H., Polan, C.E., 2002. Trans 18:1 and 18:2 isomers in blood plasma and milk fat of grazing cows fed a grain supplement containing solvent-extracted or mechanically extracted soybean meal. J. Dairy Sci. 85, 1197–1207.

McGann, T.C.A., Donnelly, W.J., Kearney, R.D., Buchheim, W., 1980. Composition and size distribution of bovine casein micelles. Biochim. Biophys. Acta. 630, 261–270.

McMahon, D.J., Brown, R.J., 1984. Composition, structure and integrity of casein micelles: a review. J. Dairy Sci. 67, 499–512.

McMahon, D.J., McManus, W.R., 1998. Rethinking casein micelle structure using electron microscopy. J. Dairy Sci. 81, 2985–2993.

McMahon, D.J., Oommen, B.S., 2008. Supramolecular structure of the casein micelle. J. Dairy Sci. 91, 1709–1721.

McPherson, A.V., Kitchen, B.J., 1983. Reviews of the progress of dairy science: the bovine milk fat globule membrane – its formation, composition, structure and behaviour in milk and dairy products. J. Dairy Res. 50, 107–133.

Mahe, M.F., Grosclaude, F., 1982. Polymorphisme de la caseine α_{s2}–des bovines. Characterization du variant C du yak (Bos grunniens). Ann. Genet. Sel. Anim. 14, 401–416.

Mahe, M.F., Miranda, G., Queral, R., Bado, A., Souvenir-Zafidrajaona, P., Grosclaude, F., 1999. Genetic polymorphism of milk proteins in African Bos taurus and Bos indicus populations characterization of variants α_{s1}-CN H and κ-CN. J. Genet. Sel. Evol. 31, 239–253.

Mansson, H.L., 2008. Fatty acids in bovine milk. Food Nutr. Res. 52, 1821–1828.

Marchin, S., Putaux, J.-L., Pignon, F., Leonil, J., 2007. Effects of the environmental factors in the casein micelle structure studied by cryo transmission electron microscopy and small angle x-ray scattering/ultrasmall-angle x-ray scattering. J. Chem. Phys. 126 art no. 045101.

Mariani, P., Summer, A., Anghinetti, A., Senese, C., Di Gregorio, P., Rando, P., Serventi, P., 1995. Effects of the c-CN G allele on the percentage distribution of caseins c1-, α_{s2}-, β- and κ- in Italian brown cows. Ind. Latte 31, 3–13.

Mather, I.H., 2000. A review of proposed nomenclature for major proteins of the milk-fat globule membrane. J. Dairy Sci. 83, 203–247.

Mather, I.H., Keenan, T.W., 1998. Origin and secretion of milk lipids. J. Mamm. Gland Biol. Neoplasia 3, 259–273.

Mazur, A., Rayssiguier, Y., 1988. Lipoprotein profile of the lactating cow. Ann. Rech. Vet. 19, 53–58.

Meister, A., Tate, S.S., Rose, L.L., 1976. Membrane bound γ-glutamyl transpeptidase. In: Martonsi, A. (Ed.), The Enzymes of Biological Membranes, Vol. III. Plenum Press, New York, p. 315.

Mellenberger, R.W., Bauman, D.E., Nelson, D.R., 1973. Metabolic adaptations during lactogenesis; fatty acid and lactose synthesis in cow mammary tissue. Biochem. J. 136, 741–748.

Mercier, J.-C., Grosclaude, F., Ribadeau-Dumas, B., 1971. Structure primaire de la caséine αs1-bovine. Séquence complète. Eur. J. Biochem. 23, 41–51.

Mercier, J.-C., Brignon, G., Ribadeau-Dumas, B., 1973. Structure primaire de la caséine κB bovine. Séquence complète. Eur. J. Biochem. 35, 222–235.

Michaelidou, A., Steijns, J., 2006. Nutritional and technological aspects of minor bioactive components in milk and whey: growth factors, vitamins and nucleotides. Int. Dairy J. 16, 1421–1426.

Moore, J.H., Christie, W.W., 1979. Lipid metabolism in the mammary gland of ruminant animals. Prog. Lipid Res. 17, 347–395.

Moore, J.H., Steele, W., Noble, R.C., 1969. The relationship between dietary fatty acids, plasma lipid composition and milk fat secretion in the cow. J. Dairy Res. 36, 383–392.

Morley, N.H., Kuksis, A., 1972. Positional specificity of lipoprotein lipase. J. Biol. Chem. 247, 6389–6393.

Morley, N.H., Kuksis, A., Buchna, D., Myher, J., 1975. Hydrolysis of diacylglycerols by lipoprotein lipase. J. Biol. Chem. 250, 3414–3418.

Mosley, E.E., Shafii Dagger, N., Moate, P.J., McGuire, M.A., 2006. Cis-9, trans-11 conjugated linoleic acid is synthesized directly from vaccenic acid in lactating dairy cattle. J. Nutr. 136, 570–575.

Nagao, M., Maki, R., Sasaki, R., Chuba, H., 1984. Isolation and characterization of bovine αs1-casein cDNA clone. Agric. Biol. Chem. 48, 1663–1667.

Neeling, J.M., 1964. Variants of κ-casein revealed by improved starch gel electrophoresis. J. Dairy Sci. 47, 506–509.

Ng-Kwai-Hang, K.F., Grosclaude, F., 2003. Genetic polymorphism of milk proteins. Part B. In: Fox, P.F., McSweeney, P.L.H. (Eds.) Advances in Dairy Chemistry, 3rd ed., Vol. 1. Kluwer Academic/Plenum, New York, pp. 739–816.

Norman, H.D., Kuck, A.L., Cassell, B.G., Dickinson, F.N., 1978. Effect of age and month-of-calving on solids-not-fat and protein yield for five dairy breeds. J. Dairy Sci. 61, 239–245.

Paltauf, F., Wagner, E., 1976. Stereospecificity of lipases. Enzymatic hydrolysis of enantiomeric alkyldiacyl and dialkylacylglycerols by lipoprotein lipase. Biochim. Biophys. Acta. 431, 359–362.

Paltauf, F., Esfandi, F., Holasek, A., 1974. Stereospecificity of lipases. Enzymic hydrolysis of enantiometric alkyl diacylglycerols by lipoprotein lipase, lingual lipase and pancreatic lipase. FEBS Lett. 40, 119–123.

Parodi, P.W., 1977. Conjugated octadecadienoic acids in milk fat. J. Dairy Sci. 60, 1550–1553.

Patton, S., Jensen, R.G., 1975. Lipid metabolism and membrane functions of the mammary gland. Prog. Chem. Fats Other Lipids 14, 167–277.

Payens, T.A.J., 1966. Association of caseins and their possible relation to the structure of the casein micelle. J. Dairy Sci. 49, 1317–1324.

Payens, T.A.J., 1979. Casein micelles, the colloid–chemical approach. J. Dairy Res. 46, 291–306.

Phillips, G.T., Nixon, J.E., Dorsey, J.A., Butterworth, P.H.W., Chesteron, C.J., Porter, J.W., 1970. The mechanism of synthesis of fatty acids by the pigeon liver enzyme system. Arch. Biochem. Biophys. 138, 380–391.

Piperova, L.S., Sampugna, J., Teter, B.B., Kalscheur, K.F., Yurawecz, M.P., Ku, Y., Morehouse, K.M., Erdman, R.A., 2002. Duodenal and milk trans octadecenoic acid and conjugated linoleic acid (CLA) isomers indicate that postabsorptive synthesis is the predominant source of cis-9 containing CLA in lactating dairy cows. J. Nutr. 132, 1235–1241.

Pocius, P.A., Baumrucker, C.R., 1980. Amino acid uptake by bovine mammary slices. J. Dairy Sci. 63, 746–749.

Pocius, P.A., Clark, J.H., Baumrucker, C.R., 1981. Glutathione in bovine blood: possible source of amino acids for milk protein synthesis. J. Dairy Sci. 64, 1551–1554.

Polis, B.D., Shmukler, H.W., Custer, J.H., 1950. Isolation of a crystalline albumin from milk. J. Biol. Chem. 187, 349–354.

Precht, D., Molkentin, J., 1995. Trans fatty acids: implications for health, analytical methods, incidence in edible fats and intake (a review). Nahrung 39, 343–374.

Quin, B.Y., Bewley, M.C., Creamer, L.K., Baker, H.M., Baker, E.N., Jameson, G.B., 1998a. Structural basis of the Tanford transition of bovine β-lactoglobulin. Biochemistry 37, 14014–14023.

Quin, B.Y., Ceamer, L.K., Baker, E.N., Jameson, G.B., 1998b. 12-Bromododecanoic acid binds inside the calyx of bovine β-lactoglobulin. FEBS Lett. 438, 272–278.

Quin, B.Y., Bewley, M.C., Creamer, L.K., Baker, Baker, E.N., Jameson, G.B., 1999. Functional implications of structural differences between variants A and B of bovine β-lactoglobulin. Protein Sci. 8, 74–83.

Ramakrishnan, B., Sha, P.S., Qasa, P.K., 2001. α-Lactalbumin (LA) stimulates milk β-1,4-galactosyltransferase I (β4Gal-T1) to transfer glucose from UDP–glucose to N-acetylglucosamine. J. Biol. Chem. 276, 37665–37671.

Ramakrishnan, B., Boeggeman, B., Qasba, P.K., 2002. β-1,4-galactosyltransferaseand lactose synthase: molecular mechanical devices. Biochem. Biophys. Res. Commun. 291, 1113–1118.

Rasmussen, J.T., Berglund, L., Pallesen, L.T., Petersen, T.E., 2002. Proteins from the milk fat globule membrane. Poster at the 26th IDF World Dairy Congress 24–27 September, Paris, France.

Reinhardt, T.A., Lippolis, J.D., 2008. Developmental changes in the milk fat globule membrane proteome during the transition from colostrums to milk. J. Dairy Sci. 91, 2307–2318.

Ribadeau-Dumas, B., Brignon, G., Grosclaude, F., Mercier, J.-C., 1972. Structure primaire de la caseine β bovine. Eur. J. Biochem. 25, 505–514.

Rose, D.J., Brunner, J.R., Kalan, E.B., Larson, B.L., Melchnychyn, P., Swaisgood, H.E., Waugh, D.F., 1970. Nomenclature of the proteins of cow's milk, third revision. J. Dairy Sci. 53, 1–17.

Sanchez-Juanes, F., Alonso, J.M., Zencada, L., Hueso, P., 2009. Distribution and fatty acid content of phospholipids from bovine milk and bovine milk fat globule. Int. Dairy J. 19, 173–278.

Saito, T., Itoh, T., 1992. Variations and distributions of O-glycosidically linked sugar chains in bovine κ-casein. J. Dairy Sci. 75, 1768–1775.

Schmidt, D.G., 1964. Starch gel electrophoresis of κ casein. Biochim. Biophys. Acta. 90, 411–414.

Schmidt, D.G., 1979. Properties of artificial casein micelles. J. Dairy Res. 46, 351–355.

Schmidt, D.G., 1980. Colloidal aspects of casein. Neth. Milk Dairy J. 34, 42–64.

Schmidt, D.G., Payens, T.A.J., 1976. Micellar aspects of casein. Surf. Colloid Sci. 9, 165–229.

Schmidt, D.G., Buchheim, W., 1970. An electron microscopical investigation of the sub-structure of the casein micelles in cow's milk. Milchwissenschaft 35, 596–600.

Schmidt, D.G., Buchheim, W., 1976. Particle size distribution in casein solutions. Neth. Milk Dairy J. 30, 17–28.

Scow, R.O., Blanchette-Mackie, E.J., Smith, L.C., 1976. Role of capillary endothelium in the clearance of chylomicrons: a model for lipid transport from blood by lateral diffusion in cell membranes. Circ. Res. 39, 149–162.

Sehat, N., Rickert, R., Mossoba, M.M., Yurawecz, M.P., Roach, J.A.G., Eulitz, K., Morehouse, K.M., Ku, Y., 1998. Identification of conjugated linoleic acid isomers in cheese by gas chromatography, silver ion high performance liquid chromatography and mass spectral reconstructed ion profiles. Comparison of chromatographic elution sequences. Lipids 33, 963–971.

Schanbacher, F.L., Goodman, R.E., Talhouk, R.S., 1993. Bovine mammary lactoferrin: implications from messenger ribonucleic acid (mRNA) sequence and regulation contrary to other milk proteins. J. Dairy Sci. 76, 3812–3831.

Singh, H., 2006. The milk fat globule membrane — a biophysical system for food applications. Curr. Opin. Colloid Interface Sci. 11, 154–163.

Slattery, C.W., 1976. Review: Casein micelle structure: an examination of models. J. Dairy Sci. 59, 1547–1556.

Smith, S., 1976. Structural and functional relationships of fatty acid synthetases from various tissues and species. In: Salton, M.R.J. (Ed.), Immunochemistry of Enzymes and their Antibodies. Wiley, New York, Chapter 5.

Smith, S., 1980. Mechanism of chain length determination in biosynthesis of milk fatty acids. J. Dairy Sci. 63, 337–352.

Smith, S., Agradi, E., Libertini, L., Dileepan, K.N., 1976. Specific release of the thioesterase component of the fatty acid synthetase complex by limited trypsinization. Proc. Natl. Acad. Sci. U.S.A. 73, 1184–1188.

Smithers, G.W., 2008. Whey and whey proteins — from 'gutter to gold'. Int. Dairy J. 18, 695–704.

Sommerfeldt, J.L., Baer, R.J., 1986. Variability of milk components in 1705 herds. J. Food Prot. 49, 729–733.

Spagnuolo, P., Dalgleish, D.G., Goff, H.D., Morris, E.R., 2005. Kappa-carrageenan interactions in systems containing casein micelles and polysaccharide stabilizers. Food Hydrocolloids 19, 371–377.

Spitsberg, V.L., 2005. Invited review: Bovine milk fat globule membrane as a potential nutraceutical. J. Dairy Sci. 88, 2289–2294.

Spitsberg, V.L., Gorewitt, R.C., 1998. Solubilization and purification of xanthine oxidase from bovine milk fat globule membrane. Protein Expr. Purif. 13, 229–234.

Stead, D., Welch, V.A., 1975. Lipid composition of bovine serum lipoproteins. J. Dairy Sci. 58, 122–127.

Stewart, A.F., Wills, I.M., Mackinlay, A.G., 1984. Nucleotide sequence of bovine α_{s1}- and κ-casein cDNAs. Nucleic Acids Res. 12, 3895–3907.

Stewart, A.F., Bonsing, J., Beattie, C.W., Shah, F., Willis, I.M., Mackinlay, A.G., 1987. Complete nucleotide sequences of bovine α_{s2}- and β-casein cDNAs: comparisons with related sequences of other species. Mol. Biol. Evol. 4, 231–241.

Stoop, M., Bovenhuis, H., van Arendonk, J., 2006. Genetic parameters for milk urea nitrogen in relation to milk production traits. J. Dairy Sci. 90, 1981–1989.

Stoop, M., van Arendonk, J., Hock, J.M.I., van Valenberg, H.J.F., Bovenhuis, H., 2008. Genetic parameters for major milk fatty acids and milk production traits of Dutch Holstein-Friesians. J. Dairy Sci. 91, 385–394.

Swaisgood, H.E., 1982. Chemistry of milk protein. In: Fox, P.F. (Ed.), Developments in Dairy Chemistry. I. Proteins. Applied Science Publishers, London, p. 1. Chapter 1.

Swaisgood, H.E., 1995. Nitrogenous components of milk. F. Protein and amino acid composition of bovine milk. In: Jensen, R.G. (Ed.), Handbook of Milk Composition. Academic Press, New York, pp. 464–468.

Thompson, M.P., Kiddy, C.A., Pepper, L., Zittle, Cs., A., 1962. Casein variants in the milk from individual cows. J. Dairy Sci. 45, 650.

Trieu-Cuot, P., Gripon, J.-C., 1981. Electrofocusing and two-dimensional electrophoresis of bovine caseins. J. Dairy Res. 48, 303–310.

US Department of Health and Human Services, 2011. Grade A Pasteurized Milk Ordinance. Food and Drug Administration, College Park, MD.

Van der Strate, B.W.A., Beljaars, L., Molema, G., Harmsen, M.C., Meijer, S.K.F., 2001. Antiviral activities of lactoferrin (Review). Antiviral Res. 52, 225–239.

Vanaman, T.C., Brew, K., Hill, R.L., 1970. The disulfide bonds of bovine α-lactalbumin. J. Biol. Chem. 245, 4583–4590.

Vanderghem, C., Blecker, C., Danthine, S., Deroanne, C., Haubruge, E., Guillonneau, F., De Pauw, E., Francis, F., 2008. Proteome analysis of the bovine milk fat globule: enhancement of membrane purification. Int. Dairy J. 18, 885–893.

Visser, S., Stangen, C.J., Van Dongen, D.F., Haverkamp, J., 1995. Identification of a new genetic variant of bovine β-casein using reversed-phase high-performance liquid chromatography and mass spectrometric analysis. J. Chromat. A 711, 141–150.

Walstra, P., 1974. High melting triglycerides in fat globule membrane-artifact. Neth. Milk Dairy J. 28, 3–9.

Walstra, P., 1979. The voluminosity of bovine casein micelles and some of its implications. J. Dairy Res. 46, 317–323.

Walstra, P., 1985. Some comments on the isolation of fat globule membrane material. J. Dairy Res. 52, 309–312.

Walstra, P., 1990. On the stability of casein micelles. J. Dairy Sci. 73, 1965–1979.

Walstra, P., 1999. Casein sub-micelles: do they exist? Int. Dairy J. 9, 189–192.

Walstra, P., Bloomfield, V.A., Wei, G.J., Jenness, R., 1981. Effect of chymosin action on the hydrodynamic character of casein micelles. Biochim. Biophys. Acta. 669, 258–259.

Walstra, P., Wouters, J.T.M., Guerts, T.J., 2006. Dairy Science and Technology, (2nd ed.). CRC Taylor & Francis, New York.

Ward, Q.J., Uribe-Luna, S., Conneeley, O.M., 2002. Lactoferrin and host defense. Biochem. Cell Biol. 80, 95–102.

Watkins, W.M., Hassid, W.Z., 1962. The synthesis of lactose by particulate enzyme preparation from guinea pig and bovine mammary glands. J. Biol. Chem. 237, 1432–1440.

Waugh, D.F., von Hippel, P.H., 1956. κ-Casein and the stabilization of casein micelles. J. Am. Chem. Soc. 78, 4576–4582.

Waugh, D.F., Noble Jr., R.W., Casein micelles. Formation and structure. II, 1965. J. Am. Chem. Soc. 87, 2246–2257.

Waugh, D.F., Creamer, L.K., Slattery, C.W., Dresdner, G.W., 1970. Core polymers of casein micelles. Biochemistry 9, 786–795.

West, C.E., Bickerstaffe, R., Annison, E.F., Linzell, J.L., 1972. Studies on the mode of uptake of blood triglycerides by the mammary gland of the lactating goat. Biochem. J. 126, 477–490.

Whitney, R., McL., Brunner, J.R., Ebner, K.E., Farrell Jr., M., Josephson, R.V., Morr, C.V., Swaisgood, H., 1976. Nomenclature of the proteins of cow's milk, fourth revision. J. Dairy Sci. 59, 795–815.

Wilcox, C.J., Gaunt, S.N., Farthing, B.R., 1971. Genetic interrelationship of milk composition and yield. South. Coop. Ser. Bull. 155 University of Florida, Gainesville, FL.

Wooding, F.B.P., Kemp, P., 1975. High melting triglycerides and the milk-fat globule membrane. J. Dairy Sci. 42, 419–426.

Woychik, J.H., 1964. Polymorphism in κ-casein of cow's milk. Biochem. Biophys. Res. Commun. 16, 267.

Zinder, O., Hamosh, M., Clary Fleck, T.R., Scow, R.O., 1974. Effect of prolactin of lipoprotein lipase in mammary gland and adipose tissue of rats. Am. J. Physiol. 226, 744–748.

Chapter 5

Egg Components in Food Systems

Yoshinori Mine and Hua Zhang

Department of Food Science, University of Guelph, Guelph, Ontario, Canada

Chapter Outline

I. INTRODUCTION

The avian egg has a long history of being recognized as an important food ingredient and nutrient source for humans. With advances in technology during the twentieth century, the functional properties and chemical composition of avian eggs have been intensively studied, and most of these studies have focused on the egg produced by the domestic chicken. It is vital and basic for food science and biotechnology to comprehensively understand the chemical composition of the egg, and several reviews have been written which describe in detail the chemical composition of eggs (Li-Chan and Nakai, 1989; Huopalahti *et al.*, 2007; Li-Chan and Kim, 2008). Likewise, the functionalities of eggs on a molecular basis have been studied by innovative research which is summarized by multiple scientific reviews (Li-Chan and Nakai, 1989; Mine, 1995, 2002; Campbell *et al.*, 2003; Lomakina and Mikova, 2006). Understanding the functional properties of eggs and the chemistry behind these properties is important for the application of hen's eggs in the food industry to develop novel food products and improve the quality of existing products. The food industry also endeavors to prolong the shelf-life of eggs, protect the nutritional value of eggs

Biochemistry of Foods. DOI: http://dx.doi.org/10.1016/B978-0-12-242352-9.00005-9

during the storage period, and incorporate egg products into other marketable products such as cake and ice cream via food processing (Burley and Vadehra, 1989c). However, chemical changes in egg components and their functional properties are the consequences of food processing. Some of these changes are desirable, but some have to be prevented. Thus, it is helpful to improve the quality of egg products by preventing the chemical changes in eggs caused by food processing.

The objective of this chapter is to provide detailed information on egg composition and functionality in order to improve the understanding of the changes that occur in eggs during food processing. Details of composition, biosynthesis, and processing-induced changes in eggshell, albumen, and yolk are also summarized.

II. STRUCTURE AND CHEMICAL COMPONENTS OF EGGS

A. Structure of Eggs

The three main constituents of eggs are the eggshell (9−11%), the albumen, also referred to as egg white (60−63%), and the egg yolk (28−29%), as indicated in Table 5.1 and Figure 5.1. The egg yolk is located in the center of the egg surrounded by albumen, and enclosed by the eggshell. There is also a layer of eggshell membrane in the interval between the albumen and eggshell. The structures of each of these egg constituents are described in this section.

1. Structure of Eggshell

The eggshell has a polycrystalline structure which includes a porous layer of cuticle, a calcite layer, and two shell membranes. The cuticle layer contains 7000−17,000 unevenly distributed pore canals used to exchange gases. The structure of the eggshell layer is presented in Figure 5.2. There are four main layers in the eggshell: (1) the cuticle, a 10−30 μm thin layer which contains mineralized and organic layers as well as most of the pigment; (2) the palisade region, which is a dense vertical crystal layer about 200 μm thick and is composed of a calcified spongy matrix with a crystalline structure; (3) the mammillary layer, also referred to as the inner calcified layer, which is located in the basal part of calcified columns and includes the calcium reserve assembly and crown region; and (4) the shell membranes including the inner (20 μm) and outer membranes (50 μm) which are located between the albumen and the mammillary layer and are formed by organic fibers and used to protect against penetration by microorganisms. The complex eggshell structure is formed through a sophisticated process. The organic cores in the mammillary layer are formed as a seeding site used to grow calcium carbonate crystals, which is eventually built up to form the cuticle. Owing to the growth of calcite crystallites being inhibited by the fiber component of shell membranes, the orientation of the crystal in the palisade layer is outward (Nys et al., 2004; Li-Chan and Kim, 2008).

2. Structure of Egg White

The egg white, or albumen, is comprised of four separate layers. About 23.3% of albumen is made up of a thin layer attached to the shell inner membrane, while the majority of albumen (57.3%) is composed of a viscous or thick white layer. Around 16.8% of albumen is composed of an inner thin white layer, and 2.7% is composed of a chalaziferous layer (Burley and Vadehra, 1989a; Li-Chan and Kim, 2008). The viscosity varies between thick and thin layers of egg white, because of the different contents of ovomucin. The proportions of egg white layers are affected by hen breed, environmental conditions, size of the egg, and rate of production (Li-Chan et al., 1995). In fresh eggs, thick albumen covers the inner thin albumen and the chalaziferous layer, holding the yolk in the center of the egg.

TABLE 5.1 Composition of Albumen, Yolk, and Whole Egg—Wet Basis

Egg Component	% Protein	% Lipid	% Carbohydrate	% Ash
Albumen	9.7−10.6	0.03	0.4−0.9	0.5−0.6
Yolk	15.7−16.6	32.0−35.0	0.2−1.0	1.1
Whole egg	12.8−13.4	10.5−11.8	0.3−1.0	0.8−1.0

Adapted from Li-Chan, 1995.

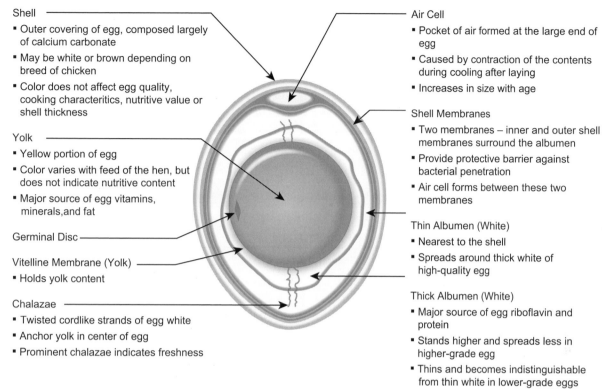

Shell
- Outer covering of egg, composed largely of calcium carbonate
- May be white or brown depending on breed of chicken
- Color does not affect egg quality, cooking characteritics, nutritive value or shell thickness

Yolk
- Yellow portion of egg
- Color varies with feed of the hen, but does not indicate nutritive content
- Major source of egg vitamins, minerals,and fat

Germinal Disc

Vitelline Membrane (Yolk)
- Holds yolk content

Chalazae
- Twisted cordlike strands of egg white
- Anchor yolk in center of egg
- Prominent chalazae indicates freshness

Air Cell
- Pocket of air formed at the large end of egg
- Caused by contraction of the contents during cooling after laying
- Increases in size with age

Shell Membranes
- Two membranes – inner and outer shell membranes surround the albumen
- Provide protective barrier against bacterial penetration
- Air cell forms between these two membranes

Thin Albumen (White)
- Nearest to the shell
- Spreads around thick white of high-quality egg

Thick Albumen (White)
- Major source of egg riboflavin and protein
- Stands higher and spreads less in higher-grade egg
- Thins and becomes indistinguishable from thin white in lower-grade eggs

FIGURE 5.1　Structure of the hen's egg. The egg is composed of a shell, shell membranes, an air cell, the chalazae, the albumen (egg white), and the yolk. The yolk is centered in the albumen and surrounded by the vitelline membrane, which is colorless. The germinal disk, where fertilization takes place, is attached to the yolk. On opposite sides of the yolk are two twisted cord-like strands known as chalazae. Their function is to support the yolk in the center of the albumen. Surrounding the albumen are two shell membranes and the shell itself. *(Adapted from American Egg Board, 1981, http://www.aeb.org).*

3. Structure of Egg Yolk

The macrostructure of egg yolk consists of vitelline membrane, yellow and white yolk, as shown in Figure 5.3. The vitelline membrane is a thin (about 10 μm) protein fiber membrane which contains three multiple layers (Mineki and Kobayashi, 1998). The yellow yolk is composed of a light yellow layer (0.25−0.45 mm thick) and a deep yellow layer (2 mm thick) of plasma which are mainly formed by lipid−protein particles. These particles have been classified as

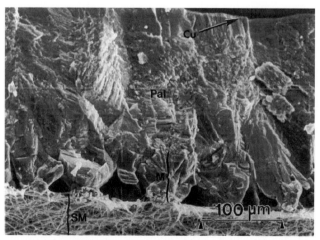

FIGURE 5.2　Scanning electron micrograph of a fractured eggshell showing a cross-section of the mineralized and non-mineralized zones. The shell membranes (SM) are a non-mineralized, collagen-based matrix interposed between the egg white and the mineralized shell. The mammillary zone (M) or cone region is a mineralized zone on the outer surface of the outer shell membrane and forms the base for the palisade region (Pal), which extends to the outermost portion of the eggshell, the cuticle (Cu). *(From Dennis et al., 1996.)*

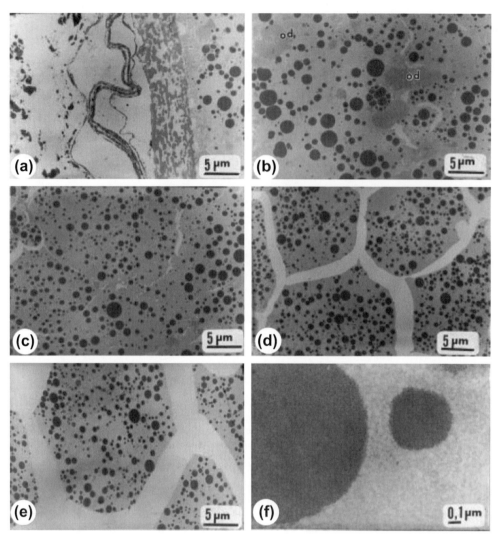

FIGURE 5.3 Transmission electron micrographs of fresh egg yolk. (a) Vitelline membrane; (b) cortical layer of yolk (od: oil droplet); (c) yolk sphere bordering cortical layer; (d) yolk sphere in outer layer; (e) yolk sphere in inner layer; (f) protein granules in yolk sphere. *(From Mineki and Kobayashi, 1997.)*

spheres (4–150 μm diameter), profiles (12–48 μm), or granules (0.3–2 μm) depending on their size. Egg yolk can be separated into two distinct fractions by dilution and centrifugation, resulting in a dark orange supernatant (plasma) and a pale pellet (granule) (Figure 5.4) (Anton, 2007). The white yolk accounts for only 2% of the total egg yolk by weight and contains several structures including the latebra, neck of latebra, nucleus of pander, and embryonic disk. The embryonic disk is 2–3 mm in diameter, is located in the nucleus of the pander, and is used by the developing embryo (Mineki and Kobayashi, 1998).

The microstructure of egg yolk has been analyzed using microscopy techniques including light microscopy (LM), transmission electron microscopy (TEM), and scanning electron microscopy (SEM), and the detailed structure of the yolk sphere and vitelline membrane has been shown (Mineki and Kobayashi, 1998). Mineki and Kobayashi used the frozen-section method as a novel approach to analyze egg yolk by fixing the egg yolk specimen at extremely low temperatures, followed by a second fixation step using chemicals. The result of their study is shown in Figure 5.3. The cortical layer of egg yolk (Figure 5.3b, c) was described as a distinct structure characterized by undeveloped yolk spheres with a shapeless membrane structure, small granules, possibly proteins, and oil spheres seen as larger granules. Furthermore, the yolk spheres observed in the outer layer (Figure 5.3d) were described as a round shape and smaller than the polyhedral spheres observed in the inner layer (Figure 5.3e). In the yolk spheres, protein granules with high electron density were shown to be highly dispersed (Figure 5.3f).

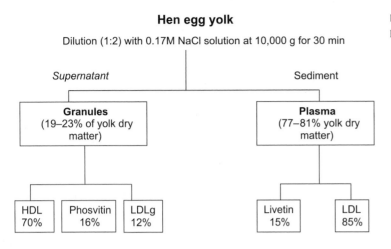

Hen egg yolk

Dilution (1:2) with 0.17M NaCl solution at 10,000 g for 30 min

Supernatant   Sediment

Granules
(19–23% of yolk dry matter)

Plasma
(77–81% yolk dry matter)

| HDL 70% | Phosvitin 16% | LDLg 12% |

| Livetin 15% | LDL 85% |

FIGURE 5.4 **Fractionation of egg yolk into granules and plasma.** *(From Anton, 2007.)*

B. Chemical Composition of Eggs

Eggs are made up of a variety of chemical components, including water, protein, fatty acids, minerals, vitamins, and pigments. Accordingly, the egg is recognized as a valuable food with high nutrient value. Eggs comprise 75% water, after which they are mainly made up of protein and lipid. Eggs also contain a smaller amount of carbohydrate, which includes glucose, sucrose, fructose, lactose, maltose, and galactose (Li-Chan and Kim, 2008). Sugino *et al.* (1997) indicated that the chemical composition of eggs is affected by the feed as well as by other factors including the species and age of the hen.

1. Chemical Composition of Eggshell

The eggshell consists of 95% minerals, primarily calcium and others including phosphorus and magnesium (Sugino *et al.*, 1997). About 3.5% of the organic composition of eggshells is protein, fatty acids, and polysaccharides rich in sulfated molecules (Nys *et al.*, 1999; Li-Chan and Kim, 2008). Li-Chan and Kim (2008) reviewed some organic acid components in the eggshell which are present in trace amounts. The cuticle layer, a thin organic layer covering the mineral crystal layer of the eggshell that provides protection from microbial infections, is composed of 90% insoluble protein, 5% carbohydrate, and 3% ash. This layer also contains a large proportion of the pigments which contribute to the different colors of eggs.

Following the cuticle layer, the vertical crystal layer is a thin monolayer on top of the palisade layer and is composed of calcium carbonate. As a foundation to the crystallization of calcium carbonate, there is a group of proteins called eggshell matrix proteins that belong to the basal part of the palisade layer and mammillary layer. Hunton (2005) reviewed the distribution of eggshell matrix proteins. Some eggshell matrix proteins are also widely expressed in egg white or in various chicken tissues, including ovalbumin, lysozyme, ovotransferrin, osteopontin, and clusterin. Osteopontin, a phosphorylated glycoprotein, is distributed in the basal part of the eggshell including membrane fibers, and the mammillary and palisade layers. The main function of osteopontin is to prevent calcium oxalate crystal precipitation extending over the basal part of the eggshell. Ovalbumin, ovotransferrin, and lysozyme are major egg white proteins distributed in both the mammillary layer and the shell membrane. The main role of these proteins in eggs is antimicrobial. Furthermore, ovotransferrin and lysozyme both have a dual role in the eggshell. These two proteins in eggshell membranes and the basal calcified layer influence the nucleation and crystallization of calcite and inhibit microorganism growth in egg albumen (Hincke *et al.*, 2000; Gautron *et al.*, 2001b).

Another group of eggshell matrix proteins is specific to the eggshell and identified only in domestic hens. These proteins include ovocleidin (OC)-17, -23, and -116, and ovocalyxin (OCX)-21, -25, -32, and -36. These specific eggshell matrix proteins have been successfully purified and identified as having particular functional properties which regulate the eggshell mineralization process (Dominguez-Vera *et al.*, 2000; Li-Chan and Kim, 2008). The mineralization develops in an acellular uterine fluid which contains the ionic and matrix precursors of the eggshell. These proteins, produced from uterine tubular gland cells, are widely distributed throughout the mammillary and palisade layers of the eggshell (Gautron *et al.*, 1997). OC-17 is a 142 amino acid phosphorylated protein with C-type lectin-like domains, which was the first matrix protein identified which regulates the precipitation of calcium carbonate in eggshell (Mann and Siedler, 1999; Reyes-Grajeda *et al.*, 2004). The glycosylated form of OC-17 protein is OC-23, which has a molecular mass of 23 kDa (Li-Chan and Kim, 2008). OC-116 is a 742 amino acid glycosylated protein with two disulfide bonds, which plays an important role in controlling calcite growth during eggshell calcification (Hincke *et al.*, 1999). OCX-32, identified by Gautron *et al.*, is secreted by the surface epithelial cells of the uterus during the phase of termination of eggshell formation (Gautron *et al.*, 2001a). Xing *et al.* (2007) recently determined that OCX-32 has the capacity to reinforce the antimicrobial activity of the eggshell. Gautron and Nys (2007) have also successfully cloned OCX-21, -25, and -36. OCX-36 is expressed abundantly during shell calcification and plays a role in natural defense mechanisms in eggs. However, the functional properties of OCX-21 and -25 have not been determined.

The inner and outer shell membranes consist of mainly protein and a smaller amount of ash and glucose based on the dry weight (Sugino *et al.*, 1997). The eggshell membranes are primarily constituted by collagens, keratin sulfate, and matrix proteins. The matrix proteins, including ovotransferrin, lysozyme, and ovalbumin, are distributed in the eggshell membranes to enhance the antimicrobial defense system of the egg. Dermatan, a proteoglycan substance, and keratin sulfate, are covered by these fibers to form a core (Dennis *et al.*, 1996). The mammillae are the vesicles of calcium reserve body containing the calcium binding molecule, which is made up of keratin sulfate (Li-Chan and Kim, 2008). The cross-linked collagen fiber system plays an important role in protecting the egg from contamination by microorganisms.

2. Chemical Composition of Egg White

About 60% of the total egg weight is composed of egg white. The major constituent of egg white is water, which makes up more than 80% of egg white (Li-Chan and Kim, 2008). The proteins in the egg white were first purified and fractionated by Rhodes *et al.* (1958). On a dry weight basis, proteins are the main component of egg white and predominantly include 54% ovalbumin, 12−13% ovotransferrin, 11% ovomucoid, 3.5% lysozyme, 2% G2 and G3 ovoglobulins, and 1.5−3% ovomucin. Some trace amounts of other proteins are also found in egg white, including ovostatin, ovoflavoproteins, avidin, and enzymes such as α-mannosidase, β-galactosidase, and catalase (Li-Chan and Nakai, 1989). The physicochemical and biological functions of the main egg white proteins are summarized in Table 5.2.

Ovalbumin is the most abundant constituent of egg white proteins. It is a phosphoglycoprotein with a molecular mass of 45 kDa and is composed of 386 amino acid residues (McReynolds *et al.*, 1978; Huntington and Stein, 2001; Lechevalier *et al.*, 2007). As a secretory protein, ovalbumin has a hydrophobic sequence between residues 21 and 47, used as an internal signal sequence involved in transmembrane location instead of a classical N-terminal leader sequence (Huntington and Stein, 2001). Half of the amino acids in ovalbumin are hydrophobic and one-third are negatively charged; consequently, the protein has an isoelectric point of 4.5 (Li-Chan and Nakai, 1989). Fothergill and Fothergill (1970) determined that ovalbumin contained six cysteine residues, but only two of them were involved in a disulfide bond and the other four were composed in a sulfhydryl (SH) group. Three of these SH groups are masked in the native state whereas the fourth is reactive only in the denatured protein (Lechevalier *et al.*, 2007). Apart from cysteine residues, ovalbumin also contains one carbohydrate unit and zero, one, or two residues of phosphoserine. Furthermore, the ovalbumin structure has four crystallographically independent molecule parts and the distance between the position of the helical reactive center loop and the protein core is 2−3 Å (Huntington and Stein, 2001). It also contains three β-sheets and nine α-helices (Stein *et al.*, 1991). As result of denaturation, ovalbumin can be transformed into a more heat-stable *S*-ovalbumin, which is an intermediate species (Lechevalier *et al.*, 2007). High pH and temperature both increase the rate of conversion. The appearance of *S*-ovalbumin can be enhanced by storage time of eggs. The *S*-ovalbumin level may reach 81% in an egg after 6 months' storage at low temperature (Vaderhra and Nath, 1973). About 2−5% loss of α-helices and increases in antiparallel β-sheets are attributed to the conformational changes in *S*-ovalbumin (Huntington and Stein, 2001). Ovalbumin belongs to the serpin (serine protease inhibitor) superfamily but lacks inhibitory activity; its biological function remains largely unknown (Huntington and Stein, 2001). For the application of egg ovalbumin in the food industry, the functionality of ovalbumin is mainly responsible for the gelling properties of egg white (Mine, 1995). Ovalbumin is also the major allergen found in egg white and can activate immunoglobulin E (IgE)-mediated allergic reactions in mammals. The structural and physicochemical properties of ovalbumin are reviewed in detail by Huntington and Stein (2001) and Lechevalier *et al.* (2007).

TABLE 5.2 Physicochemical Properties of Proteins Found in Egg White

Protein	% (w/w)	pI	M_w (kDa)	T_d(°C)	Cysteines	−SH	S−S
Ovalbumin	54	4.5−4.9	45	75−84	6	4	1
Ovotransferrin (conalbumin)	12−13	6.0−6.1	77.7	61−65 (76.5, Al^{3+})	30	−	15
Ovomucoid	11	4.1	28	77	18	−	9
Ovomucin	1.5−3.5	4.5−5.0	110, 5500−8300, 220−270,000		(2)	−	
Lysozyme	3.4−3.5	10.7	14.3−14.6	69−77	6		4
G2 ovoglobulin	1.0	4.9−5.5	47−49				
G3 ovoglobulin	1.0	4.8, 5.8	49−50				
Ovoflavoprotein	0.8	4.0	32−35, 80		5		2
Ovostatin	0.5	4.5−4.7	760−900				
Cystatin	0.05	5.1	12				
Avidin	0.05	10.0	55−68.3		2		1

From Mine (1995).

Egg white lysozyme, which is the *N*-acetylmuramic hydrolase used to hydrolyze peptidoglycan at the β1−4 glycosidic bond, has a molecular weight of 14.4 kDa and consists of 129 amino acid residues (Lechevalier *et al.*, 2007; Li-Chan and Kim, 2008). It is a basic protein with an isoelectric point of 10.7. Lysozyme is made up of two domains linked by a long α-helix, while also separated by a helix-loop-helix (HLH) (Young *et al.*, 1994). HLH has recently been shown to possess membrane-permeabilizing and antibacterial activity (Ibrahim *et al.*, 2001). As an enzyme, lysozyme has its polar groups outside and hydrophobic groups buried inside the molecule. The conformational transition in lysozyme, called 'hinge-bending', results in a relative movement of its two lobes and provides a site for catalysis to access the substrates. Ibrahim *et al.* (1997) demonstrated that the calcium cation was able to induce conformational changes in lysozyme to activate its antibacterial activity. Some of the antibacterial activity of lysozyme has been shown to be independent of its catalytic function via analysis of the activity of antibacterial peptides prepared from the enzymatic hydrolysis of lysozyme (Mine and Kovacs-Nolan, 2006).

Ovotransferrin, also referred to as conalbumin, is a glycoprotein found in egg white. It contains 686 amino acid residues and has a molecular mass of 78−80 kDa and an isoelectric point of 6.0. It has the capacity to bind with various ions, especially ferric Fe^{3+} ions, thereby transferring iron into host cells by membrane receptors (Mason *et al.*, 1996). Ovotransferrin has two lobes with four domains, including an N- and a C-terminal lobe. Each lobe consists of two distinct alpha/beta domains of similar size and with a single binding site as well as 15 disulfide bridges (Superti *et al.*, 2007; Li-Chan and Kim, 2008). Ovotransferrin has antimicrobial activity against various Gram-negative and Gram-positive bacteria, fungi, and viruses.

Another glycoprotein distributed in the egg white is ovomucoid, which is a thermally stable protein and makes up 11% of the egg white protein (Li-Chan and Nakai, 1989). Ovomucoid contains about a quarter of asparaginyl-linked carbohydrate moieties, and is made up of 186 amino acids with a molecular mass of 28 kDa (Kato *et al.*, 1987b; Li-Chan and Nakai, 1989). It is a trypsin inhibitor with three distinct domains cross-linked by intradomain disulfide bonds. In total, nine disulfide bonds are identified in ovomucoid, while no free sulfhydryl groups are found. The active site is found only in domain II, which is attributed to inhibiting trypsin activity in the chicken egg white (Li-Chan and Nakai, 1989). Ovomucoid purified from chicken egg white has allergenic potential to trigger IgE-mediated reactions in humans (Mine and Zhang, 2001; Mine and Rupa, 2004).

Ovomucin is a sulfated glycoprotein found in egg white with insoluble and soluble subtypes, which contains proteins and a smaller amount of carbohydrates. The carbohydrates found in ovomucin are in the form of oligosaccharides (Hiidenhovi, 2007). Insoluble ovomucin plays a key role in forming the gel-like insoluble fraction of the thick albumen in egg white, while soluble ovomucin is mainly distributed in the outer and inner albumen (Burley and

Vadehra, 1989d; Li-Chan and Kim, 2008). Ovomucin consists of two subunits, α- and β-ovomucin; these are found in both insoluble and soluble subtypes of ovomucin in different proportions (Li-Chan and Kim, 2008). α-Ovomucin contains 91% protein and 9% carbohydrate with 2087 amino acid residues and a molecular mass of 230–250 kDa (Watanabe *et al.*, 2004). β-Ovomucin consists of 872 amino acids and has a molecular mass of about 400–720 kDa (Itoh *et al.*, 1987; Hiidenhovi, 2007; Hammershøj *et al.*, 2008).

The other proteins in albumen include ovoglycoprotein, flavoprotein, avidin, protease, and minor proteins including lipocalins, clusterin, and Ch21 proteins. These are described in detail by Li-Chan and Kim (2008) and Huopalahti *et al.* (2007).

In addition to proteins, other types of chemical components are found in the egg white (Li-Chan and Kim, 2008). Carbohydrates are present in egg white in the form of both conjugated oligosaccharides and free glucose. A low content of lipid, 0.03% of egg albumen by weight, is also found in egg white, as are trace amounts of various minerals and water-soluble vitamins. Li-Chan and Kim (2008) provide a detailed review of the chemical composition of egg white.

3. Chemical Composition of Egg Yolk

Egg yolk contains 50% solids with a lipid:protein ratio of 2:1 (Li-Chan and Kim, 2008). The distribution of protein and lipid in the vitelline is different from the yolk itself. It contains 87% protein, 10% carbohydrate, and 3% lipid on a dry weight basis (Li-Chan and Kim, 2008). Egg yolk can be separated into two phases: plasma and granule. Plasma is composed of up to 80% liquid yolk with a higher content of lipid, while granules contain approximately three times more proteins than plasma (Li-Chan *et al.*, 1995). The components of egg yolk are shown in Figure 5.4 and the compositional analysis of fresh yolk is summarized in Table 5.3.

The protein content of liquid yolk is approximately 16%. Egg yolk protein constitutes mainly 16% high-density lipoproteins (HDL), 68% low-density lipoproteins (LDL), 10% livetins, 4% phosvitin and very low-density lipoproteins (VLDL) (McCully *et al.*, 1962). Approximately two-thirds of yolk solids are LDLs, which are spherical particles with a lipid core surrounded by a layer of phospholipid and protein (Anton *et al.*, 2003). The LDLs are composed of about 14% protein and 86% lipid, which includes 74% neutral lipid and 26% phospholipid (Martin *et al.*, 1964). Because LDLs consist of apoproteins and phospholipids, they have amphiphilic properties and can be dispersed at the oil–water interface. Thus, LDLs are the essential component responsible for the emulsifying properties of egg yolk. There are six types of apoprotein in the LDLs of egg yolk, with molecular weights between 15 and 130 kDa and a p*I* range from 6.3 to 7.5. Apoprotein I constitutes about 70% of total apoproteins, and is less soluble in water. Apoprotein II contains a high proportion of amphipathic α-helix chains, which contribute to dispersion at the oil–water interface (Anton *et al.*, 2003). VLDLs are precursors of egg yolk LDL and transferred from the hen's blood into the ovary.

HDLs in hen's egg yolk, also referred to as lipovitellins, are distributed in granules. They have a molecular mass of 400 kDa and contain 75–80% proteins and 20–25% lipid, comprising 65% phospholipids, 30% triglycerides, and 5% cholesterol (Cook and Martin, 1969; Anton *et al.*, 2003). Two subtypes of lipovitellin, α and β, are found in egg yolk in a ratio of 1:1.5 and contain different amino acid sequences as well as phosphorus and carbohydrate residues (Li-Chan and Kim, 2008). There are about five types of apoprotein identified in HDLs with molecular masses ranging from 35 to 110 kDa; these apoproteins are glycosylated by mannose, galactose, glucosamine, and sialic acid

TABLE 5.3 Compositional Analysis of Egg Yolk

	Fresh Yolk (%)	Dry Yolk (%)
Water	51.1	—
Lipids	3.6	62.5
Proteins	16.0	33.0
Carbohydrates	0.6	1.2
Minerals	1.7	3.5

From Li-Chan *et al.* (1995).

(Anton *et al.*, 2007). Furthermore, HDLs and phosvitins have the same precursor of vitellogenin which is synthesized in the hen's liver. A granular complex in egg yolk is formed by HDLs and phosvitin via phosphocalcic bridges (Wang *et al.*, 1983). Li-Chan and Kim (2008) indicated that egg yolk HDLs may have therapeutic potential including antioxidant and antimicrobial properties.

Phosvitin is a phosphoglycoprotein and makes up 4% of egg yolk dry matter (Anton *et al.*, 2007). Almost half of the amino acids in phosvitin are serine residues, with 90% phosphorylation, and form a central hydrophilic area surrounded by two hydrophobic areas at the N- and C-termini. Two subtypes of phosvitin, α- and β-phosvitin, are found in chicken egg yolk and account for 80% of the phosphorus binding proteins in yolk (Li-Chan and Kim, 2008). The higher content of phosphoserine residues and special conformational structure of phosvitin contribute to its resistance to heat denaturation and proteolytic cleavage (Juneja and Kim, 1997; Anton *et al.*, 2000). Phosvitin has strong chelating properties, and is a natural metal binding biomolecule.

Livetin accounts for 30% of the plasma proteins in egg yolk. Livetin contains 20% α-livetin, a serum albumin, 50% β-livetin, an α_2-glycoprotein, and 30% γ-livetin, immunoglobulin Y (IgY), which is similar to mammalian IgG (Schade and Chacana, 2007). The α-livetin has a molecular weight of 70 kDa and a p*I* value between 4.3 and 5.7, and has allergenic activity to induce type I hypersensitivity (Schade and Chacana, 2007; Williams, 1962). The β-livetin has a molecular weight of 45 kDa and contains 7% hexose. The molecular weight of γ-livetin (IgY) is approximately 167 kDa. The IgY antibody is made up of two light chains and two heavy chains, and is transferred to the yolk to passively protect the developing embryo.

The main constituents in dry egg yolk are lipids, which make up 62% of egg yolk powder (Juneja, 1997). Among those lipids, triglycerides, phospholipids, and cholesterol occupy 65%, 31%, and 4%, respectively. The fatty acid compositional analysis of these lipids is summarized in Table 5.4. Triglycerides and phospholipids are both glycerolipids with a glycerol backbone. Phospholipids have a glycerol—phosphate backbone. In addition, as amphiphilic molecules, phospholipids have both polar and non-polar groups with emulsification properties. The two major components of egg yolk phospholipids are phosphatidylcholine and phosphatidylethanolamine, with the remainder

TABLE 5.4 Proteins and Lipids in Egg Yolk

Constituent	Major Components	Relative %
Proteins[a]	Apovitellenin I—VI	37.3
	Lipovitellin apoproteins	
	α-Lipovitellin	26.7
	β-Lipovitellin	13.3
	Livetins	
	α-Livetin (serum albumin)	2.7
	β-Livetin (α_2-glycoprotein)	4.0
	γ-Livetin (γ-globulin)	2.7
	Phosvitin	13.3
	Biotin binding protein	Trace
Lipids[b]	Triglyceride	65
	Phosphatidylcholine	26
	Phosphatidylethanolamine	3.8
	Lysophosphatidylcholine	0.6
	Cholesterol	4
	Sphingomyelin	0.6

[a]*Modified from Burley and Vahedra (1989a).*
[b]*Adapted from Juneja (1997).*
From Li-Chan and Kim (2008).

consisting of lysophosphatidylcholine, lysophosphatidylethanolamine, and sphingomyelins (Li-Chan and Kim, 2008). The cholesterol content is about 1.6% in raw egg yolk. About 80% of total egg yolk cholesterol is free cholesterol and the remainder exists in the form of cholesterol esters.

There are smaller amounts of carbohydrate present, around 1% of the dry mass of egg yolk (Li-Chan and Kim, 2008). Glucose is a free carbohydrate and a major constituent found in egg yolk. The rest of the carbohydrates found in egg yolk are in a conjugated form, such as sialic acid bound to glycoproteins and glycolipids (Li-Chan and Kim, 2008). Egg yolk also contains a trace amount of vitamins, minerals, and pigments, which are listed in Table 5.5. The mineral content in egg yolk is about 1%. Phosphorus is the most abundant mineral owing to the high content of phospholipid (Li-Chan and Kim, 2008). Egg yolk contains various vitamins and is a good source of vitamins A, D, E, and B_{12}. Water-soluble vitamins including folic acid, riboflavin, and niacin are present in both egg white and yolk. As the precursor of vitamin A, carotenoids are also present in egg yolk and are its main

TABLE 5.5 Minerals and Vitamins Present in Whole Egg, Egg Albumen, and Egg Yolk

Constituent (units)	Whole Egg	Egg Albumen	Egg Yolk
Minerals (mg)			
Calcium	29.2	3.8	25.2
Chlorine	96.0	66.1	29.9
Cobalt	0.033	0.009	0.024
Iodine	0.026	0.001	0.024
Iron	1.08	0.053	1.02
Magnesium	6.33	4.15	2.15
Manganese	0.021	0.002	0.019
Phosphorus	111	8	102
Potassium	74	57	17
Sodium	71	63	9
Sulfur	90	62	28
Zinc	0.72	0.05	0.66
Vitamins			
Vitamin A (IU)	264	–	260
Vitamin D (IU)	27	–	27
Vitamin E (mg)	0.88	–	0.87
Vitamin B_{12} (µg)	0.48	–	0.48
Choline (mg)	11.0	2.58	8.35
Folic acid (mg)	237	0.46	238
Inositol (mg)	0.023	0.006	0.026
Niacin (mg)	5.94	1.52	4.35
Pantothenic acid (mg)	0.045	0.035	0.010
Pyroxidine (mg)	0.83	0.09	0.73
Riboflavin (mg)	0.065	0.008	0.057
Thiamine (mg)	0.18	0.11	0.07
	0.05	0.004	0.048

Adapted from Watkins (1995); Li-Chan and Kim (2008).

pigment. It should be noted that carotenoids cannot be synthesized by the hen; the carotenoids in egg yolk are obtained from the feed of laying hens.

III. BIOSYNTHESIS OF EGGS

A. Introduction to the Egg Formation Process

The chicken egg as a developed ovum is formed in the ovary, where it grows into a follicle, and is finally delivered to the uterus, where it can be further assembled (Burley and Vadehra, 1989c; Okubo et al., 1997). The constituents used for egg yolk development are supplied by the liver and delivered in the blood. First, the oocyte in the ovary begins to grow into a white follicle which is surrounded by the hen's veins. Later, a yellow follicle starts to develop at the base of a white follicle in the ovary approximately 7−10 days before ovulation, resulting in the deposition of yellow egg yolk (Okubo et al., 1997). Eventually, the encapsulated yellow follicle, known as the mature ovum, is ovulated into the oviduct, a process which takes about 24−27 hours. The oviduct in laying hens is about 60−80 cm long and is composed of five regions: the infundibulum, magnum, isthmus, uterus, and vagina (Burley and Vadehra, 1989a). The length and ovum-holding time of these five portions are distinguished by the particular functionality of each region (Okubo et al., 1997). The longest part of the oviduct is the magnum, which is also recognized as the albumen-secreting region. The eggs stay in this region for approximately 2−3 hours to accumulate albumen. The uterus has thick walls, and it is here that the eggshell is assembled. The eggs are held in the uterus for about 21 hours to complete the process of eggshell mineralization before the egg is laid.

B. Biosynthesis of Eggshell and Related Biochemical Changes

1. Regulation of Eggshell Biosynthesis

The eggshell is an essential part of all avian eggs and has a highly organized and porous structure to sustain water and gas exchanges, provide calcium to the embryo during development, and protect the egg from microbial infections. Eggshells are mainly made up of inorganic mineral and organic matrix components, the latter being composed of proteins, glycoproteins, and proteoglycans provided by uterine fluid (Hincke and Wellman-Labadie, 2008). The main inorganic component of eggshell is calcium ions, which are transported through the uterine mucosa by binding with calbindin. Subsequently, bicarbonate ions are produced from calcium and hydration of carbon dioxide by carbonic anhydrase catalysis (Gautron et al., 1997). The synthesis of the eggshell is regulated by the endocrine system of the laying hen, which includes estrogen, calcitonin, parathyroid hormone, and 1,25-dihydroxyvitamin D_3 (Burley and Vadehra, 1989b). The concentration of blood calcium and rhythms of laying eggs are controlled by hormones of the endocrine system. Formation of a whole eggshell is initiated by deposition of the shell membranes on the outside surface of the albumen in the isthmus region of the oviduct. The organic matrix of the eggshell membrane is mainly made up of a fibrillar protein formed by a disulfide cross-linking network, and collagens including types I, V, and X (Nys et al., 2004; Hincke and Wellman-Labadie, 2008). These components are synthesized and released by numerous tubular gland cells and other types of cell located in the isthmus. A lysine-derived cross-linked structure of eggshell membrane is produced with the assistance of a copper-containing enzyme (Burley and Vadehra, 1989c). This cross-linked fiberic structure plays a dual role in the control of eggshell mineralization by preventing calcification toward the inner membranes and forming nucleation sites on the outer surface of membranes.

2. Biomineralization of Eggshell

The biomineralization of eggshell is a distinct process compared with the mineralized tissues in other species, both vertebrate and invertebrate. In other species, the matrix usually contains collagenous or non-collagenous elements that directly interact with the mineral phase to form a biphasic calcified matrix and control the deposition of minerals (Addadi and Weiner, 1992; Belcher et al., 1996; Robey, 1996). For the avian shell, there is also a spatial separation in the eggshell between its organic framework and mineralized components, by which the eggshell membrane interacts with organic aggregates known as mammillary knobs (Hincke and Wellman-Labadie, 2008). The mammillary knobs are distributed on the outer surface of eggshell membrane and used as a nucleus for calcium carbonate aggregating into a polycrystalline structure. The calcification is a sequential process with a specific timeline related to ovulation. A variety of biochemical changes are associated with the biomineralization of eggshell, including changes in the protein matrix composition and formation of calcite in the uterine fluid. The acellular uterine fluid contains both organic and inorganic precursors of the shell matrix, thereby influencing the precipitation kinetics, morphology, and orientation of crystals. The protein profile collected from uterine fluid is different during the three stages of eggshell formation, i.e. the initial, growth, and terminal phases (Dominguez-Vera *et al.*, 2000). Recent studies have shown that the uterine fluid can enhance calcium carbonate precipitation in both the initial and growth phases, whereas it has inhibitory effects on mineralization (Gautron *et al.*, 1997; Dominguez-Vera *et al.*, 2000).

In the initial phase of mineralization of the eggshell, numerous proteins have been identified in uterine fluid that are involved in calcium carbonate precipitation and calcite crystalline formation, which mainly include ovocleidins, ovalbumin, albumen, osteopontin, and ovotransferrin. Ovotransferrin and ovalbumin are predominant in the uterine fluid in this phase and have calcium-binding properties, which may play a role in regulating the nucleation of calcite (Gautron *et al.*, 2001b). OC-17 and -116 are also prevalent in the uterine fluid during the initial phase. OC-17 may be involved in regulating calcite morphology as crystal growth of the eggshell (Nys *et al.*, 2004). OC-116, known as ovoglycan, which is a dermatan sulfate proteoglycan, is a unique eggshell matrix protein belonging to the secretory calcium-binding phosphoprotein (SCPP) family and produced from the granular cells of the uterine epithelium of the isthmus (Arias and Fernandez, 2001; Hincke and Wellman-Labadie, 2008). OC-116 may have the same functionality as a proteoglycan which plays a key role in promoting cartilage calcification and collagen mineralization (Hunter, 2001; Hincke and Wellman-Labadie, 2008).

In the growth phase, the uterine fluid still actively promotes precipitation kinetics of calcium carbonate. However, some protein constituents collected from uterine fluid in this phase have inhibitory effects on the precipitation of calcite and delay calcite growth by directly binding to the crystal (Gautron *et al.*, 1997; Nys *et al.*, 2004). OCX-32 and -36 are present in uterine fluid during the growth phase. Owing to its involvement in natural defense mechanisms, OCX-36 may enhance the development of the eggshell during this phase, but its activity in the mineralization of calcite is unknown (Hincke and Wellman-Labadie, 2008). OCX-32 is capable of inhibiting mineralization and therefore plays a role in controlling the deposition of calcium carbonate during the eggshell growth phase (Nys *et al.*, 2004; Hincke and Wellman-Labadie, 2008). Some proteins, including lysozyme, were identified with biphasic effects on regulating calcium carbonate precipitation relevant to their concentration in the uterine fluid (Hernandez-Hernandez *et al.*, 2003). At high concentration, these proteins can block the growth of calcite crystallization by binding to the crystal surface. The biphasic roles of these proteins may inhibit the growth of calcite crystals at the end of the eggshell formation process.

In the terminal phase, the uterine fluid mainly contains an organic matrix which is assumed to have inhibitory effects on the precipitation of calcite. The organic matrix contained in this phase delays the pH drop associated with calcium carbonate precipitation, thereby inhibiting the growth of calcite crystals (Gautron *et al.*, 1997). The proteins identified in the organic matrix collected from this stage are primarily OCX-32 and -36 (Nys *et al.*, 2004; Hincke and Wellman-Labadie, 2008). It takes about 1.5 hours to terminate mineralization and compose the organic cuticle layer surrounding the eggshell before egg ovulation. In the whole process of mineralization and formation of eggshells, the matrix components in the uterine fluid play an active role in the control of the calcite growth kinetics and crystal morphology.

C. Biosynthesis of Egg Albumen

The egg albumen proteins are primarily synthesized by tubular gland cells which are distributed along the oviduct wall. The egg albumen is assembled in the magnum after a developed ovum reaches this region. The biosynthesis of egg albumen proteins begins only in response to hormone stimulation (Schutz *et al.*, 1978; Burley and Vadehra, 1989d). The proportions of the three major egg albumen proteins synthesized from the tubular gland cells are 50—60% ovalbumin, 8% ovomucoid, and 2—3% lysozyme (Palmiter, 1972; Schutz *et al.*, 1978). Synthesis of ovalbumin results from primary stimulation by estrogen followed by withdrawal and secondary stimulation. The estrogen stimulation leads to an increase in ovalbumin messenger RNA expression and subsequently enhances the glycosylation of albumen in the chicken oviduct (Schutz *et al.*, 1978; Burley and Vadehra, 1989d). Other mechanisms, such as the signal transduction, are also involved in regulating the synthesis of egg albumen proteins. Cooney *et al.* (1993) identified that the upstream promoter transcriptional factor (COUP-TF) of chicken ovalbumin regulated the hormone expression via interaction with other transcriptional factors. The expression of hormones involved in regulating egg albumen protein secretions from the oviduct is under the control of the particular signals which, in turn, are stimulated by those secreted proteins; therefore, an integrated control system is established. Other egg albumen proteins produced by hens are mainly antimicrobial proteins, such as lysozyme and ovotransferrin, which are used to protect the embryo from microbial infections. The lysozyme in egg albumen is primarily produced by tubular gland cells in the oviduct, rather than being delivered from the hen's blood (Shawkey *et al.*, 2008). The biosynthesis of ovotransferrin is regulated by hormones in the oviduct including estrogen and progesterone (Lee *et al.*, 1978; Shawkey *et al.*, 2008). The activities of the antimicrobial proteins are enhanced during the hatching process owing to an increase in temperature. The production and deposition of antimicrobial proteins in eggs are a typical protective system for reproduction and genetics established during the evolutionary process.

The gel structure of egg albumen is primarily attributed to the interactions of glycoproteins. The main constituents of egg albumen are glycoproteins, including ovalbumin, lysozyme, and ovomucin. Glycoproteins are a group of macromolecules with carbohydrate moieties attached to the polypeptide chains through covalent bonds, such as *N*-acylglycosylamine linkages (Robinson, 1972). The variable degrees of modification in amino acid residues result in the generation of different protein moieties of the egg white glycoproteins; these modifications include N-acetylation of N-terminal residues, phosphorylation of serine and threonine residues, and methylation of lysine and arginine residues (Robinson, 1972). The particular physical properties of egg white are attributed to the interactions of chemical bonds in the egg glycoproteins. Moreover, both ovomucin and lysozyme contain substantial amounts of cysteine residues which can form disulfide bonds and subsequently form a lysozyme–ovomucin complex. This complex contributes to a rigid gel network being established in the egg white, whereas the disulfide bond of the lysozyme–ovomucin complex can be inhibited by existing divalent cations, such as magnesium and calcium (Robinson, 1972). An integrated regulation of the gel structure in the egg white depends on interaction among the glycoproteins and other constituents of the egg white.

D. Biosynthesis of Egg Yolk

The early formation of egg yolk is initiated after differentiation of the embryo. The egg yolk cells are assembled in the epithelium of the ovaries. Subsequently, the yolk cells divide rapidly and form oocytes, surrounded by continuous follicle cells (Burley and Vadehra, 1989b). The egg yolk develops inside the wall of the hen's ovarian follicle which is supplied with blood capillaries by which the egg yolk protein precursor, known as vitellogenin, is transferred and deposited into the yolk. Egg yolk proteins are primarily synthesized in the liver under hormonal control, and then transported into the developing oocyte via the blood (Burley and Vadehra, 1989b; Stevens, 2004). The activation of transcription and translation of genes encoding egg proteins is also involved in this process (Stevens, 2004). The whole process of egg yolk formation involves an increase in protein and lipid synthesis in the liver and is influenced by the hen's habitat and dietary nutrients (Stevens, 2004).

1. Formation of Vitellogenin

About 95% of the egg yolk protein is made up of LDL, lipovitellin, phosvitin, and livetin. After entering the yolk, vitellogenin is enzymatically cleaved into phosvitin and lipovitellin, which are continually incorporated into granules (Burley and Vadehra, 1989b). Vitellogenin is synthesized in liver cells in response to estrogen stimulation. Estrogen triggers signals that induce vitellogenin translation on the rough endoplasmic reticulum of the hepatocytes by binding with membrane receptors (Stevens, 2004). The completed vitellogenin polypeptide is formed by glycosylation and subsequent phosphorylation in the hepatocytes prior to secretion. The vitellogenin circulating in the blood is eventually picked up by the oocyte (Stevens, 2004).

2. Synthesis of Yolk Low-Density Lipoprotein

Egg yolk lipids are synthesized in the liver and transported to peripheral tissues in the form of VLDL, which is composed of protein and triglyceride, phospholipid, cholesterol, and cholesteryl esters. Apolipoprotein B (apoB), composed of apoVLDL I and II, is the primary protein constituent of VLDL and is produced in the liver following stimulation by estrogen (Burley and Vadehra, 1989b). The proteins apoB and apoVLDL are synthesized on the rough endoplasmic reticulum, whereas VLDL is assembled in the Golgi apparatus (Burley and Vadehra, 1989b). Estrogen and other transacting proteins, known as liver-enriched transcription factors, help to induce the transcriptional production of apoB (Beekman *et al.*, 1991; Stevens, 2004). Evans *et al.* (1987) suggested that five genes of yolk proteins are potentially regulated by estrogen in chicken liver, including *vtgI*, *vtgII*, *vtgIII*, *apoVLDL-II*, and *apoB*. These genes play a key role in controlling the biosynthesis of egg yolk lipoproteins.

3. Biosynthesis of Yolk Livetins

The livetins consist of soluble proteins in egg yolk and contain three subtypes, α, β, and γ. Unlike other egg yolk proteins, the biosynthesis of livetins is not controlled by estrogen (Burley and Vadehra, 1989b). The α- and β-livetins from egg yolk have similar characteristics to serum albumen and $α_2$-glycoprotein, respectively (Schade and Chacana, 2007). Furthermore, γ-livetin (IgY) is produced in the bone marrow, while the other two types of livetin are synthesized in the liver. γ-Livetin belongs to a group of chicken immunoglobulins, known as antibodies, which are transferred from blood into egg yolk, and IgY is the only type of antibody found in egg yolk (Burley and Vadehra, 1989b; Schade and Chacana, 2007). The IgY antibody is produced by hens and transferred into the egg yolk to protect their offspring from microbial infections (Schade and Chacana, 2007). Owing to its potential involvement in passive immunity, research on the biological functions and activities of egg yolk γ-livetin is increasing.

IV. CHANGES IN EGG COMPONENTS INDUCED BY FOOD PROCESSING

Eggs are an excellent source of nutrients and play a critical role in influencing the food consumer market. The nutrient components in egg include protein, lipid, various vitamins, and minerals. Because they are rich in protein and lipid, eggs have been associated with a variety of functionalities and are widely applied in the food processing industry. Moreover, egg processing technology has been improved to develop higher quality and more stable egg products to meet the increased demand for processed egg products (Froning, 2008). The main objectives of improving egg processing technology are to extend the shelf-life of egg products and to incorporate egg ingredients into other marketable products based on the chemical and physical properties of eggs. However, regardless of processing techniques, the processing itself can lead to chemical and physical changes in the egg components. In this section, processing-induced chemical changes in eggs will be discussed, as well as the chemical modifications of egg constituents that are generally applied in the food industry to improve their functionality.

A. Denaturation of Egg Proteins

The processing of egg products usually results in changes to the egg proteins. These changes are caused by the modification of protein structure, called denaturation. Under normal conditions, with constant pH and temperature, the protein molecule assumes one specific conformation, referred to as the native state (Boye *et al.*, 1997). In the native state, the protein molecule has a minimal free energy and is considered thermodynamically stable. Any changes to normal conditions cause changes in the thermodynamic homeostasis of the protein molecule, thereby altering the native structure of the molecule. The denaturation of proteins is defined as a process in which protein molecules lose their native structure and change into a more disordered arrangement through the spatial rearrangement of polypeptide chains within the molecule (Kauzmann, 1959). The conformational changes in protein molecules that occur during denaturation are more specific in their secondary, tertiary, and quaternary structural levels. The primary structure of protein molecules, composed of polypeptide sequences, is not affected by denaturation. Denaturation cannot affect the hydrolysis of peptide bonds in the protein.

To produce egg products or incorporate egg ingredients into other food products, denatured or partly denatured egg proteins are required, in order to improve their functional properties. The denaturation or partial denaturation of egg proteins is beneficial for egg foaming and emulsifying abilities and to enhance digestibility and palatability (Burley and Vadehra, 1989c). However, the denaturation of egg proteins needs to be avoided during the process of egg preservation as egg albumen aggregation increases with aging. Denaturation can be induced by a variety of physicochemical agents including heat, pH, salt, and surface effects. The process of heat-induced denaturation and formation of aggregates of ovalbumin is represented in Figures 5.5 and 5.6. In egg proteins, especially ovalbumin, there is an intermediate state known as the molten globule state, in which proteins are partly denatured and maintain their native compactness (Mine, 1995). The molten globule state is defined as a stable partially folded conformation that can be distinguished from both the native and completely denatured states.

B. Changes in Egg Proteins during Preservation

The composition of eggs undergoes a variety of chemical changes with increasing storage time. The major changes in the different parts of the egg as a result of aging have been identified and are as follows: (1) the pH of the albumen is increased by the loss of carbon dioxide; (2) the vitelline membrane becomes weak and eventually disappears;

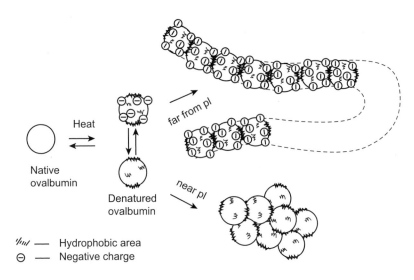

FIGURE 5.5 Model for heat denaturation and formation of aggregates of ovalbumin. *(From Doi and Kitabatake, 1989.)*

Native ovalbumin

Heat

Denatured ovalbumin

far from pI

near pI

𝖫𝗂𝗇𝗅 — Hydrophobic area
⊖ — Negative charge

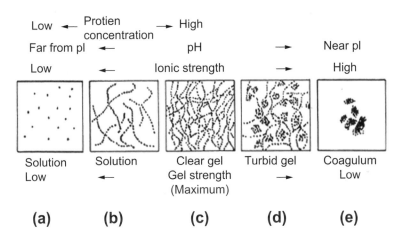

Low ← Protien → High
concentration
Far from pI ← pH → Near pI
Low ← Ionic strength → High

Solution Low | Solution | Clear gel Gel strength (Maximum) | Turbid gel | Coagulum Low

Low ← Gel strength →

(a) (b) (c) (d) (e)

FIGURE 5.6 Factors affecting the texture of heat-induced ovalbumin gels. (a) At pH values far from the p*I* and at low ionic strength, linear aggregates are formed. (b) With decreasing electrostatic repulsion at low ionic strength or at 7.0 > pH > p*I*, three-dimensional networks form a transparent gel. (c) At intermediate ionic strength of pH, both linear aggregates and random aggregates are formed. In this case, the linear aggregates form a cross-linked primary gel network and the random aggregates are interdispersed within this network. This mixed gel of linear and random aggregates has either a translucent or an opaque appearance depending on the relative amounts of the linear and random aggregates. (d) At high ionic strength or at pH values near the p*I*, proteins aggregate to form a turbid gel composed of random aggregates. Among these gel types, the transparent and the opaque/translucent gels exhibit higher gel strength and water-holding capacity than the others. *(From Doi and Kitabatake, et al., 1989.)*

(3) sulfhydryl (SH) groups in the egg yolk are cross-linked and lipids are oxidized; and (4) water is lost by evaporation from the eggshell (Burley and Vadehra, 1989c). The changes caused by the aging process can induce irreversible alterations to egg products which are attributed to the denaturation of egg proteins. The denaturation of egg proteins involves two parts: endothermic denaturation and exothermic aggregation. In the denaturation process, the hydrogen bonds and non-covalent bonds of egg proteins are extensively rearranged; therefore, the egg proteins aggregate and form a cross-linked network (Burley and Vadehra, 1989c).

1. Effects of Aging on Eggs

As eggs age, the gelling characteristics of albumen deteriorate and consequently albumen becomes watery. Meanwhile, the pH of the albumen progressively rises from an initial value of 7.6 to a final value of 9.5 (Fromm, 1967). The modification of ovalbumin among egg white proteins is significantly increased with the aging time. *S*-ovalbumin was detected in egg albumen after 34 days of aging (Rossi and Schiraldi, 1992). The formation of a relatively stable intermediate state in egg albumen during aging is attributed to the conversion of *n*-ovalbumin into *S*-ovalbumin. This transformation results in a cohesive film forming on the interface between air and water, thereby reducing foam stability (Lomakina and Mikova, 2006). The gelation and precipitation of egg albumen are initiated after denaturation, and subsequently increase the viscosity of egg albumen.

Mineki and Kobayashi (1998) determined that the storage duration and conditions can influence the microstructure of egg yolk. Longer storage time or increased temperature results in expansion of the interstitial spaces between yolk spheres and fusion of granules. The vitelline membrane of the egg yolk becomes more elastic and weaker, and eventually disintegrates. Kirunda and McKee (2000) indicated that the original integrity of the vitelline membrane dissipates as a result of aging. The deterioration of the vitelline membrane as a result of aging is influenced by the same factors causing degeneration of the gel structure in the egg albumen (Feeney *et al.*, 1952). The rate of deterioration of the vitelline membrane is enhanced by increasing storage temperature (Kirunda and McKee, 2000). Fromm (1967) identified that the weight of the vitelline membrane was decreased by a half after five days at 35°C, meanwhile the protein and hexosamine concentration of the vitelline membrane also decreased. The deterioration of the vitelline membrane is due to disruption of the cross-linked fibers which make up the structure of the vitelline membrane (Fromm, 1967). The degradation of the glycoprotein II structure and the disulfide bonds of the ovomucin is mainly attributed to the disruption of the fibrous structure of the vitelline membrane during aging (Kido *et al.*, 1976; Kato *et al.*, 1979). In addition, excess water penetrating into the yolk from the egg albumen results in a decrease in elasticity of the vitelline membrane during prolonged storage (Kirunda and McKee, 2000).

2. Effects of Heat on Eggs

Heat is the most common factor causing denaturation of proteins. Heat treatment of globular proteins leads to an increase in free energy and thermal motion of the whole protein molecule. Therefore, the thermodynamic homeostasis of the native proteins is disrupted, which is sustained by intermolecular and intramolecular bonds (Boye *et al.*, 1997). The thermal denaturation leads to disruption of the original bonds in the native proteins and the formation of a new three-dimensional network which results from two types of protein aggregations, thermal coagulation and thermal gelation (Boye *et al.*, 1997). Heat treatment is generally applied in commercial egg production to eliminate bacterial contamination or facilitate the production of desirable products. New technology is needed to minimize the effects of heat-induced denaturation on egg proteins. A flash heating method has been used to preserve eggs by placing eggs in boiling water for 5 seconds to form a thin layer next to the shell (Romanoff and Romanoff, 1944). The undesired effect of heat treatment on whole eggs is the development of a dark layer on the surface of the egg yolk, caused by the deposition of ferrous sulfide residues, which result from the reaction between hydrogen sulfide from the albumen and iron in the yolk granules (Tinkler and Soar, 1920; Burley and Vadehra, 1989c). The following subsections illustrate the effects of heating on individual egg components as well as mixed yolk and albumen.

a. Effects of Heat on Yolk

Heat treatment of egg yolk results in changes to yolk lipoproteins. A sharp increase in the viscosity of egg yolks was observed after heating above 65°C, and coagulation of the yolk occurred around 70°C (Denmat *et al.*, 1999). This is caused by heat-induced denaturation of LDL and lipovitellins in the egg yolk (Burley and Vadehra, 1989c). However, some egg yolk proteins, such as phosvitin and some livetins, are less sensitive to heat treatment. The denatured egg yolk proteins are able to form a cross-linked network which results in thermal gelation. Nguyen and Burley (1984) showed that sulfhydryl—disulfide interactions of denatured LDL are mostly responsible for the gelation of heated egg yolk, owing to LDL in egg yolk containing SH groups, which are used to attach to apoprotein. According to Denmat *et al.* (1999), egg yolk granules are less sensitive than egg yolk plasma to heat treatment. The different constituents of egg yolk have been shown to have different sensitivity to heat treatment. This may be caused by the distinct distribution of egg yolk proteins in the different constituents of egg yolk.

b. Effects of Heat on Albumen

Owing to the substantial role of egg white proteins in the food industry, the effects of heat on egg albumen have been extensively studied (Seideman *et al.*, 1963; Shimada and Matsushita, 1980; Egelandsdal, 1980; Mine *et al.*, 1990; Rossi and Schiraldi, 1992). The egg albumen undergoes coagulation and gelation upon heating. Therefore, heat treatment of egg albumen can facilitate the improvement of water solubility, foaming, and emulsification capacities of food products. The major protein constituents of egg albumen, including ovalbumin and conalbumin, have a high heat coagulability (Shimada and Matsushita, 1980). Payawal *et al.* (1946) identified a series of discontinuous changes in egg albumen induced by heating. Based on their results, the egg albumen liquid changes appearance from turbid to clear after heating from 63°C to 70°C. The egg albumen forms a clear gel on the surface at 63°C. Precipitation of the egg albumen mixture is initiated between 63°C and 66°C, but a clear gel appearance is detected at 66°C. Finally, egg albumen forms a white coagulum above 66°C and takes on a solid white appearance at even higher temperatures.

Using differential scanning calorimetry, Donovan *et al.* (1975) observed that heat-induced denaturation of the three main protein constituents of egg albumen was represented by endothermic peaks within 60–100°C. The contributions of these three proteins to the whole denaturation of egg albumen were 64.5% from ovalbumin, 16.8% from conalbumin, and 3.7% from lysozyme. Ovotransferrin is the least heat-stable protein, with a denaturation temperature of about 57°C. Globulins, ovalbumin, and lysozyme have denaturation temperatures around 72°C, 71.5°C, and 81.5°C, respectively (Cunningham, 1994). Ovomucoid and ovotransferrin contribute less to the denaturation of egg albumen owing to their lower heat coagulability.

The intermolecular interactions produce a continuous three-dimensional network exhibiting structural rigidity, which plays a control role in the thermocoagulation of proteins. Different chemical interactions, including hydrophobic, ionic, and intermolecular sulfhydryl–disulfide interactions, are involved in the heat-induced aggregation among the heterogeneous proteins in egg albumen (Shimada and Matsushita, 1980). The formation of a cross-linked gel network is mainly attributed to sulfhydryl–disulfide exchanges in egg albumen after heat treatment, while hydrophobic interactions are primarily responsible for the coagulation of egg albumen, as hydrophobic interactions are enhanced by increasing temperature (Shimada and Matsushita, 1980). The heat denaturation of egg white proteins is significantly increased by the exposure of hydrophobic residues on the molecular surface (Mine *et al.*, 1990). Mine *et al.* (1990) described the changes in secondary structure content in egg white proteins during heat denaturation. The β-sheet content was increased more than three-fold, while the helical content was decreased with increasing temperature. According to their results, the heat-denatured egg albumen contained a considerable amount of protein secondary structure which formed a cross-linked network by disulfide bonding, and this network of β-sheet structures is strengthened by the exposed hydrophobic residues. The heat-induced denaturation of egg albumen is represented by the aggregation of egg white proteins, resulting from the transformation of the protein from a native to a denatured state.

c. Pasteurization of Eggs by Heat

Pasteurization has been widely used in the egg production industry to destroy microbial contamination and has been generally recognized as a valid method since the twentieth century. Among pathogenic bacteria, salmonellae are of primary concern in the egg processing industry because of the high prevalence detected in eggs and egg products. The compositional differences in egg constituents account for the wide range of pasteurization conditions (Cunningham, 1994). Since salmonellae are most heat resistant at mild acidic pH or near-neutral pH, they are more viable in egg yolk (pH 6.0) than in egg albumen (pH 9.1). Owing to the differences in pH values, solids, and the nature of constituents of the whole egg, the heat resistance of salmonellae is different in egg albumen and yolk. Therefore, the pasteurization conditions applied to egg yolk are more severe than for egg albumen (Cunningham, 1994). The heat-induced denaturation of egg products is also affected by the mechanical conditions of the pasteurization process, including equipment design, flow rate, and temperature differential between the heating medium and egg products (Cunningham, 1994).

The physical and chemical changes induced by the heating of egg proteins can adversely affect the quality of resulting products. Pasteurization of egg albumen at 60°C results in a rise in viscosity and a decrease in foaming activity (Burley and Vadehra, 1989c). The reduction in the foaming capacity of egg albumen is caused by the denaturation of ovotransferrin during the pasteurization process and the irreversible denaturation of the ovomucin–lysozyme network (Lomakina and Mikova, 2006). Metal ions can be added to restore the foaming properties of egg albumen after pasteurization. During pasteurization of liquid egg albumen, chemical reagents are commonly added to stabilize conalbumin, which forms a heat-stable complex by binding with metal ions such as aluminum sulfate. Hydrogen peroxide, a well-known bactericidal agent, is also added to liquid egg albumen to eliminate microbial contamination so as to allow the pasteurization of egg albumen at relatively low temperatures. Pasteurization standards for egg processing have been established in different countries. The United States Department of Agriculture (USDA) requires the liquid whole egg to be heated for 3.5 minutes at 60°C (Burley and Vadehra, 1989c; Cunningham, 1994). Under these conditions, the pasteurized egg products are considered safe to be consumed.

3. Effects of Freezing on Eggs

The freezing process facilitates the preservation and production of egg products. Frozen egg products are normally produced from the liquid form of egg white, yolk, or other egg ingredient-containing food products. After freezing and thawing, the mixture of yolk—albumen becomes non-homogeneous and aggregated, and possibly undergoes gelation.

a. Effects of Freezing on Whole Egg Mixture

Torten and Eisenberg (1982) studied freezing-induced alterations in colloidal properties of the whole egg. Increases in viscosity and surface tension were detected in whole egg samples after freezing. Based on these results, it was concluded that the freezing process disrupts the three-dimensional system in eggs; thus, complexes are formed by random intermolecular contacts, which results in a lumpy appearance in the yolk—white mixture after thawing. Therefore, the freezing process can lead to major textural changes in egg products. The viscosity of the supernatant fraction from the thawed liquid of mixed whole egg is consistent with the unfrozen whole egg mixture. The precipitate fraction from the whole egg mixture, which contains the yolk granules, does not undergo gelation upon freezing (Cotterill, 1994). Accordingly, the gelation of the whole egg mixture depends on the interaction of components from both egg yolk and albumen. Freezing has no significant effects on the functional properties of whole egg mixture (Miller and Winter, 1950).

b. Effects of Freezing on Egg-Yolk Gelation

During freezing and storage of egg yolks below $-6°C$, viscosity is increased and gelation occurs. Below this temperature, the gelation of egg yolk is irreversible. However, the gelation of egg yolk does not occur when it is frozen at a supercooling or high freezing rate (Lopez et al., 1954). The freezing rate, temperature, and thawing conditions play an integral role in the control of egg yolk gelation. It is preferable to minimize the gelation of egg yolk when it is used as an ingredient in food products. The irreversible gelation of egg yolk can be controlled by adding solutes such as salt and sucrose (Burley and Vadehra, 1989c; Cotterill, 1994). One possible mechanism is that the solutes depress the freezing point of the products, thereby inhibiting protein denaturation caused by freezing (Burley and Vadehra, 1989c). LDL, as the major lipoprotein in egg yolk, is the primary component altered by freezing. Kamat et al. (1976) determined that solute addition inhibits the gelation of egg yolk lipoproteins by increasing solvation at interfacial regions to form stabilized layers. Wakamatu et al. (1983) studied the freezing-induced gelation of egg yolk LDL and the effects of salt on this process. The pH changes in the unfrozen phase and transformation of water into ice also affect LDL aggregation. The formation of ice crystals results in the denaturation of proteins and subsequent aggregation of yolk lipoproteins (Cotterill, 1994). A high concentration of salt in the unfrozen phase disrupts the egg yolk granules. In addition, pretreating yolk with proteolytic and lipolytic enzymes can help to inhibit freezing-induced gelation (Lopez et al., 1955).

Several intramolecular and intermolecular interactions are involved in the aggregation of egg yolk lipoproteins, and these are summarized by Cotterill (1994). The calcium and phosvitin in egg yolk may contribute towards the formation of intracellular bridges to combine proteins, as identified in the gelled fraction of yolk (Cotterill, 1994). Hydrophobic bonding, hydrogen bonding, and electrostatic forces all contribute to interactions between protein molecules in the gelled yolk after freezing-induced disruption in the egg yolk components (Palmer et al., 1970). Burley and Vadehra (1989c) observed that the electrophoretic behavior, chromatographic properties, and electron microscopic appearance are altered in egg yolk LDL as a result of freezing-induced gelation.

c. Effects of Cooling and Freezing on the Vitelline Membrane

Deformation of the vitelline membrane and loss of its elasticity occur during the cold storage progress. However, changes in the strength of the vitelline membrane are not noticeable under cold storage conditions. Jones and Musgrove (2005) showed that the vitelline membrane strength was consistent, with no significant decrease, when stored for up to 10 weeks at 4°C in a cool room at 80% relative humidity. In contrast, freezing causes changes in the strength of the vitelline membrane; it is reduced during extended cold storage (Jones et al., 2002). The water-holding capacity of the isolated membrane is increased by freezing at $-18°C$ (Cotterill, 1994). Thus, the properties of the vitelline membrane may be altered by freezing. During the slow freezing process, the formation of ice crystals can also cause physical damage by puncturing the vitelline membrane.

d. Effects of Freezing on Egg Albumen

The functional properties of egg albumen are affected by freezing. Frozen storage results in a decrease in the foaming ability of egg albumen. The amount of thick white of the total albumen is dramatically decreased when frozen at

−16°C and stored for 3 months at −3°C (Moran, 1925). One possible reason is that denaturation of proteins occurs, leading to coagulation of the thick white under frozen storage conditions (Cotterill, 1994). Wootton *et al.* (1981) observed that ovotransferrin was the most susceptible and, along with ovalbumin, was responsible for the freezing-induced changes in egg albumen. Davis *et al.* (1952) determined that the formation of ice crystals was a major factor causing damage to cooked egg albumen when it was frozen.

4. Effects of Irradiation on Eggs

Electromagnetic and ionizing radiation, at a dosage up to 3 kGy, has been approved for use in shell eggs by the US Food and Drug Administration (USDA, 2000). Irradiation has been regarded as an alternative method to heat pasteurization for shell eggs. The irradiation of the whole shell egg influences the physical properties of the egg by weakening the vitelline membrane and decreasing the viscosity of the albumen two-fold (Burley and Vadehra, 1989c). Loss of ovomucin, ovotransferrin, and ovalbumin was observed using gel electrophoresis in egg albumen following exposure to irradiation (Burley and Vadehra, 1989c). The viscosity of egg albumen is dramatically reduced after irradiation, because irradiation-induced denaturation of proteins results in the transformation of carbohydrate and protein moieties, and subsequently the disruption of ovomucin complexes (Ma *et al.*, 1990). Irradiation-induced changes in ovomucin may be attributed to cleavage of *O*-glycosides from the protein moiety (Wong and Kitts, 2002). Wong and Kitts (2002) reported that egg albumen irradiated with dosages between 2 and 4 kGy sustained a loss of thick albumen accompanied by an increase in free SH groups, which led to a reduction in the foam volume and gel hardness of albumen. In this study, the number of SH groups released from albumen did not increase with the elevated dosage of electron beam.

Since denaturation enhances the formation of a rigid film at the interface, the functional properties of fresh eggs, including foaming, emulsifying activity, gel rigidity, and angel cake volume, are improved by irradiation (Ma, 1996). However, Min *et al.* (2005) demonstrated that the foaming capacity and foam stability of egg albumen were reduced by increasing the dosage of irradiation. The deterioration in foaming properties of egg albumen is possibly caused by irradiation-induced oxidative changes in globulins, ovomucin, and lysozyme.

Irradiation-induced changes in egg yolk are less drastic than those in egg albumen. The major effect of irradiation on egg yolk is oxidation of polyunsaturated fatty acids, which results in the formation of hydroperoxides (Thakur and Singh, 1994). The emulsifying capacity and gelling ability of egg yolk are not affected by exposure to irradiation. No significant effects of irradiation on frozen and spray-dried egg have been detected. Ma *et al.* (1993) showed that no noticeable changes were induced in irradiation-treated frozen liquid egg white, by analyzing the scanning calorimetric profiles and electrophoretic patterns of egg protein components, as well as the functional properties. In a study of the effects of irradiation on the functionality of frozen liquid egg yolk, Huang *et al.* (1997) demonstrated that the emulsifying capacity of irradiated samples was significantly higher than that of non-irradiated samples during the frozen storage progress.

C. Changes in Egg Protein Functionality Induced by Processing

The unique functional properties of eggs have numerous benefits for their application as ingredients in a variety of food products. Egg products incorporated into other food products can improve the profits of the food industry owing to the high nutritional value of eggs. When used as food ingredients, egg products can contribute various functionalities including foaming, gelation, and emulsification properties (Table 5.6). The two main components of eggs directly responsible for their functional properties are proteins and lipids. In the food industry, there is increasing concern about the chemical and structural changes induced in egg ingredients by the physicochemical procedures used in food processing. These changes are associated with modifications in the functional properties of egg components. The major focus of this section is to improve understanding of the food processing-induced modifications in egg functional properties.

1. Effects of Dry-Heating on Gelation

A gel can be described as a continuous solid cross-linked system, which consists of a three-dimensional network embedded in an aqueous solvent (Smith, 1994). Gels have water-holding capacity in food products with rheological properties. Heat-induced gelation in food proteins generally happens during heat processing. Gel formation involves two main processes: (1) heat leads to the denaturation of proteins, which depends on the time,

TABLE 5.6 Functional Properties Attributed to Egg Proteins in Food Systems

Function	Underlying Mechanisms	Examples
Water binding	Hydrogen bonding and ionic hydration	Cakes and bread
Gelation	Water entrapment and immobilization, network formation	Gels, cakes, baked goods
Cohesion, adhesion	Hydrophobic, ionic, and hydrogen bonds	Pasta, baked goods
Emulsification	Adsorption and film formation at interface	Cakes, dressings
Foaming	Adsorption and film formation at interface	Whipped toppings, ice cream, cakes, desserts
Aroma − flavor binding	Hydrophobic bonds, entrapment	Low-fat bakery products, doughnuts

Reproduced from the American Egg Board (http://www.aeb.org).

temperature, and nature of the proteins; and (2) the unfolded proteins aggregate and form intermolecular interactions, which leads to the development of a coagulum or gel, depending on the conditions (Raikos *et al.*, 2007). The difference between a protein coagulum and a gel is that a gel has an ordered network system formed by the polymerization of protein molecules, whereas a coagulum consists of a disorganized aggregated structure (Hayakawa and Nakai, 1985).

Egg white proteins play an important role in improving the consistency of food products by forming heat-induced gels which provide a medium for holding flavor and a homogeneous texture (Sun and Hayakawa, 2002). The egg white proteins are mainly responsible for the gelation properties of eggs owing to their higher content, flexibility, thiol content, and ability to denature and form a cross-linked network. As the major constituent of egg albumen, ovalbumin plays a key role in gel formation. The factors affecting the textural changes in heat-induced ovalbumin gels are illustrated in Figure 5.6. The rigidity and turbidity of heat-induced ovalbumin gels are influenced by factors such as pH, ionic strength, and protein concentration (Hatta *et al.*, 1986). The heat-induced gelation of ovalbumin and other egg white proteins is attributed to intermolecular interactions of the denatured proteins. The aggregation of ovalbumin after heat treatment results from hydrophobic interaction and disulfide bonding, due to sulfyhydryl−disulfide exchange reactions and sulfyhydryl oxidation (Sun and Hayakawa, 2002).

Kato *et al.* (1989) first observed that heating in a dry state was an effective approach by which to improve the gelation properties of egg albumen. The dry-heating process can result in the formation of transparent and firm ovalbumin gels within a wide range of pH and ionic strengths (Matsudomi *et al.*, 1991). Matsudomi *et al.* (1991) suggested that the possible mechanism for this is that dry-heating provides a balance between attractive and repulsive forces, which facilitates the formation of soluble linear aggregates by partially unfolded ovalbumin. Mild alkaline treatment contributes to the development of firm and elastic ovalbumin gels during the dry-heating process (Mine, 1996, 1997). The gelation and polymerization of proteins are improved by the increasing hydrophobicity with the heating time and sulfyhydryl−disulfide interchange. Watanabe *et al.* (2000) demonstrated that dry-heated ovalbumin has inhibitory effects on the coagulation of ovotransferrin, because ovalbumin interferes with the interaction of ovotransferrin molecules via the formation of disulfide bonds with ovotransferrin.

2. Effects of Heat on Foaming

Foams are a complex system in which dispersed air bubbles are entrapped by interfacial films between air and liquid or solid continuous phases (Davis and Foegeding, 2007). The denaturation of egg proteins at the air−water interface occurs during blending or whipping processes, which lead to the introduction of air into the protein solution. As a consequence of the application of mechanical forces into a protein solution, which results in an increase in free energy, the proteins unfold and form biphasic films by exposing their hydrophobic groups to the air phase, while their hydrophilic groups remain in the liquid phase. Egg white proteins are optimal foaming agents in the food industry because of their ability to establish a complex of interactions between the various protein components. The hierarchy of the foaming capacity of egg white proteins, from low to high, is as follows: globulins; ovalbumin; ovotransferrin; lysozyme; ovomucoid; and ovomucin (Mine, 1995). Glycoproteins contain hydrophilic carbohydrate moieties, which increase the viscosity and foam stability by binding with water. The foaming properties of egg white proteins contribute to the spongy and foamy structure in various food products including angel food cakes, meringues, soufflés, and mousses. In these products, egg white proteins are the main surface-active agents that facilitate the stabilization of the dispersed gas phase. Foaming stability can be attributed to a variety of forces, including viscosity of the liquid phase and electrostatic and steric forces between proteins. Among the egg white proteins, globulins and ovalbumin contribute to the formation of foams, whereas ovomucin and lysozyme contribute to foam stability, and ovoglobulins to increasing viscosity (Yang and Baldwin, 1995).

Parameters affecting the foaming properties of egg white proteins include protein concentration, film thickness, ionic strength, pH, temperature, other constituents in the food product, and physicochemical properties of the proteins. Kim and Setser (1982) demonstrated that egg yolk lipids have detrimental effects on the foaming capacity of egg albumen. Recently, Wang and Wang (2009) also showed that egg yolk contamination has the most significant impact on heat-induced foaming properties of fresh egg albumen. Furthermore, intrinsic factors also affect the foaming quality of egg white proteins, including the ratio of thin albumen, storage conditions, aging, and the hen's genetic conditions. As mentioned previously, the increase in S-ovalbumin content in egg albumen during the prolonged storage process results in the reduction of foam stability (Lomakina and Mikova, 2006). Numerous studies have been carried out to improve the foaming properties of egg albumen. Kato et al. (1994) revealed that the dry-heating of egg albumen could enhance foaming capability and stability four-fold without loss of solubility. An increase in molecular flexibility and surface hydrophobicity after dry-heating facilitates the intermolecular interactions and formation of a cohesive interface film (Kato et al., 1994). Relkin et al. (1999) identified that the mild heat treatment of ovalbumin led to the formation of a molten globule state which was an intermediate stable unfolded state. The partial denaturation of ovalbumin in the molten globule state, which is reversible, improves foaming properties by increasing flexibility and surface hydrophobicity (Campbell et al., 2003).

3. Effects of Heat on Emulsification

Fluid emulsions are thermodynamically unstable mixtures of immiscible liquids such as lipid and water (Mangino, 1994). The surface-active molecules are dispersed in an immiscible liquid system and cover the oil−water interface to form emulsions. This process is facilitated by mechanical homogenization. Amphiphilic compounds which contain both polar and non-polar moieties normally have emulsifying activity. Egg yolk itself is an emulsion system in which lipid is dispersed into a continuous aqueous phase. Egg yolk contains substantial amounts of lipoprotein−cholesterol complexes and phospholipids that are efficient emulsifying agents. Accordingly, egg yolk is an essential ingredient in various foods, including mayonnaise and salad dressings, to stabilize emulsions. To study the roles of different components in egg yolk emulsion, the emulsifying abilities of yolk have been divided into two parts: plasma (LDL and livetins), and granules (HDL and phosvitin) (Denmat et al., 1999). The phospholipid−protein interaction and absorption behavior of egg yolk components, including HDL, LDL, phosvitin, and livetin, contributes to the emulsifying properties. The conditions for the emulsifying capacity of egg components are similar to their gelation properties, and mainly affect the formation of interfacial films. These conditions include pH, ionic concentration, and protein concentration. Mine (1998) studied the effects of pH on the emulsifying properties of egg yolk and showed that the size of the emulsion particles formed by egg yolk proteins decreased as pH levels increased. Excellent solubility of egg yolk plasma has been found in common pH ranges and salt concentrations (Sirvente et al., 2007). In addition, heat treatment has different effects on the emulsifying properties of egg yolk plasma and granules. Denmat et al. (1999) demonstrated that egg yolk granules were less sensitive than plasma to heat, because the HDL and phosvitin components were bound to form an insoluble complex by phosphocalcic bridges, which protected the protein molecules from thermal denaturation. Based on their results, egg yolk granules have better emulsifying activity and stability than plasma, especially during heat treatment over 72°C. The emulsifying property of plasma was not affected by heating up to 69°C (Denmat et al., 1999). However, after continuously increasing the heating temperature, the solubility of plasma was dramatically decreased, which led to a rise in viscosity due to aggregation of LDL induced by thermal denaturation.

Mine et al. (1991) showed that egg white proteins had effective emulsifying capacity to be used as surface-active agents. The emulsifying properties of egg white proteins are more heat resistant than those of the egg yolk components. Kato et al. (1989) demonstrated that dry-heating of egg albumen improved the emulsifying capacity and emulsion stability owing to the increased flexibility and surface hydrophobicity of egg white proteins. The thermal denaturation of egg white proteins results in an increase in absorbing ability at the oil−water interface caused by the exposure of hydrophobic residues at the oil phase (Campbell et al., 2003). The emulsifying capacity of egg white ovalbumin is particularly influenced by factors such as the oil-phase volume, presence of salts, and protein concentrations. Dry-heating of egg albumen is one application of dehydration in food production. Dehydration is a standard method used in food processing for the preservation of food components. The dry-heating process is indeed a promising approach to improve egg albumen functionality. The beneficial effects of dry-heating on the gelling, foaming, and emulsifying properties of egg albumen have been discussed and summarized in this section. The optimal conditions for dry-heating egg albumen to improve the functionalities of egg white proteins are heating at 80°C in a dry state with 7.5% moisture content (Kato *et al.*, 1989). In a recent study, the conformational changes in egg white proteins induced by dry-heating treatment were shown to be dependent on the moisture content (Plancken *et al.*, 2007). Based on the results of this study, the optimal moisture content was below 6.8% during dry-heating of egg albumen at 80°C.

D. Modification of Egg Protein Functionality

Research is being carried out to develop new approaches to modify the denaturation and aggregation of egg proteins in order to improve egg functionality with respect to the preparation of food products with stable qualities. Developments mainly focus on the application of chemical reagents and new processing techniques. Currently, the most successful approach is the use of the Maillard reaction to improve the functional properties of egg proteins.

1. Chemical and Physical Modifications of Eggs

Chemical modification using carboxylation and succinylation of spray-dried egg white solids has been developed to improve the foaming properties. However, Ma *et al.* (1986) demonstrated that carboxyl modification had moderate effects on improving foaming properties in spray-dried egg white solids, and succinylation caused a decrease in foaming ability and low heat coagulation. The enzymatic hydrolysis of egg albumen has been studied to improve egg white protein foaming ability and a variety of enzymes has been analyzed (Lomakina and Mikova, 2006). The application of pepsin in food proteins has been shown to have effects on increasing foaming stability owing to the hydrolysis of hydrophobic regions on the protein surface (Horiuchi and Fukushima, 1978). Hydrolysis of egg albumen with papain has also been found to have a promising effect on foaming ability (Lee and Chen, 2002). Phillips *et al.* (1987) indicated that adding copper ions to fresh egg albumen improved foaming stability through the formation of a copper–ovotransferrin complex, thereby preventing protein denaturation. Knorr *et al.* (2004) reported that a combination of ultrasound and high pressure contributed to an even distribution of protein and lipid in the whole egg liquid and increased its foaming ability.

The emulsifying properties of LDL have also been shown to be improved by high-pressure treatment combined with an alkaline pH. A significant decrease in droplet flocculation of LDL dispersions was detected under the above treatment conditions (Speroni *et al.*, 2005). Consequently, the aggregation and denaturation of proteins were shown to be enhanced without altering the capacity of LDL adsorption at the oil–water interface (Speroni *et al.*, 2005). Kato *et al.* (1987a) studied the modification of phosvitin emulsifying properties by chemical and enzymatic neutralization or the removal of phosphate anionic residues. These modifications resulted in a significant impact on emulsion stability and a decrease in the emulsifying properties of phosvitin. Kitabatake *et al.* (1989) reported that the application of freeze-drying and spray-drying improved the emulsifying properties of egg ovalbumin. Li *et al.* (2004) studied the effects of a combination of phosphorylation and dry-heating on the functional properties of egg white proteins. They showed that dry-heating of egg white proteins in the presence of pyrophosphate improved the emulsifying properties of egg white proteins by increasing the exposure of hydrophobic residues at the oil–water interface. The combination of phosphorylation and dry-heating also enhanced the gelling ability of egg white proteins and led to the formation of a transparent gel, which resulted from hydrophobic interaction and electrostatic-repulsive forces between phosphate residues added to proteins in the process.

2. Improvement of Functional Properties of Eggs using Maillard Reactions

The Maillard reaction is a chemical reaction by which the amino groups of proteins interact with the carbonyl groups of reducing polysaccharides to form covalent cross-links (Danehy, 1986), and it has been recognized as a safe and promising approach to modify egg functionality. The Maillard reaction, a non-enzymatic browning reaction widely applied in the food industry, plays a key role in the development of flavor and color in food products. Previously, the Maillard reaction was prevented during the preservation of fresh eggs by desugarization prior to pasteurization, because it results in browning reactions during the storage of eggs. However, it has been shown that the addition of carbohydrate moieties to egg protein molecules improves their functional properties, including emulsification and gelation. The carbohydrate moieties added to the protein molecules enhance the amphiphilic nature of the complex of conjugated molecules.

Kato *et al.* (1990) applied this approach to the preparation of ovalbumin—dextran conjugates to improve the functional properties of ovalbumin. The Maillard reaction in egg white proteins can facilitate solubility and improve heat stability owing to the decreased hydrophobicity of protein residues by covalent binding with hydrophilic sugar molecules (Campbell *et al.*, 2003). Handa and Kuroda (1999) reported that application of the Maillard reaction in dried egg albumen improved the gelling properties when carried out at 55°C and 35% relative humidity. In their study, the degree of polymerization was increased with increased heating time. The increase in polymerization may be attributed to disulfide bonding and non-disulfide covalent interactions. Furthermore, Matsudomi *et al.* (2002) found that galactomannan modification of dried egg albumen, using the Maillard reaction, could improve its gelling ability. Several studies have demonstrated that application of the Maillard reaction has a positive impact on the emulsifying properties of egg albumen (Kato *et al.*, 1993; Aoki *et al.*, 1999; Begum *et al.*, 2003). Aoki *et al.* (1999) showed that the emulsifying activity of ovalbumin was improved by conjugation with glucuronic acid through the Maillard reaction under certain conditions. Begum *et al.* (2003) identified that conjugating ovoinhibitor with galactomannan under controlled dry-heating conditions resulted in an improvement in emulsifying properties, with potential benefits for industrial applications. One possible explanation is that the amphiphilic conjugates formed by covalent interaction between sugars and proteins are absorbed better at the oil—water interface, with the hydrophobic residues being exposed to the oil phase, while the hydrophilic side-chains interact with water.

REFERENCES

Addadi, I., Weiner, S., 1992. Control and design principles in biological mineralization. Angew Chem. Int. Ed. Engl. 31, 153—169.

Anton, M., 2007. Composition and structure of hen egg yolk. In: Huopalahti, R., Lopez-Fandino, R., Anton, M., Schade, R. (Eds.), Bioactive Egg Compounds. Springer, Heidelberg, pp. 1—5.

Anton, M., Le Denmat, M., Deaumal, V., Pilet, P., 2000. Thermostability of hen egg yolk granules: contribution of native structure of granules. J. Food Sci. 65, 584.

Anton, M., Martinet, V., Dalgalarrondo, M., Beaumal, V., David-Briand, E., Rabesona, H., 2003. Chemical and structural characterization of low-density lipoproteins purified from hen egg yolk. Food Chem. 83, 175—183.

Anton, M., Castellani, O., Guerin-Dudiard, C., 2007. Phosvitin. In: Huopalahti, R., Lopez-Fandino, R., Anton, M., Schade, R. (Eds.), Bioactive Egg Compounds. Springer, Heidelberg, pp. 7—12.

Aoki, T., Hiidome, Y., Kitahata, K., Sugimoto, Y., Ibrahim, H.R., Kato, Y., 1999. Improvement of heat stability and emulsifying activity of ovalbumin by conjugation with glucuronic acid through the Maillard reaction. Food Res. Int. 32, 129—133.

Arias, J.L., Fernandez, M.S., 2001. Role of extracellular matrix molecules in shell formation and structure. World Poult. Sci. J. 57, 349—357.

Beekman, J.M., Wijnholds, J., Schippers, I.J., Pot, W., Gruber, M., Ab, G., 1991. Regulatory elements and DNA-binding proteins mediating transcription from the chicken very-low-density apolipoprotein II gene. Nucleic Acids Res. 19, 5371—5377.

Begum, S., Saito, A., Xu, X., Kato, A., 2003. Improved functional properties of the ovoinhibitor by conjugating with galactomannan. Biosci. Biotechnol. Biochem. 67, 1897—1902.

Belcher, A.M., Wu, X.H., Christenen, R.J., Hansma, P.K., Stuchy, G.D., Morse, D.E., 1996. Control of crystal phase switching and orientation by soluble mollusc-shell proteins. Lett. Nat. 381, 56—58.

Boye, J.I., Ma, C.Y., Harwalkar, V.R., 1997. Thermal denaturation and coagulation of proteins. In: Damodaran, S., Paraf, A. (Eds.), Food Proteins and their Applications. Marcel Dekker, New York, pp. 25—56.

Burley, R.W., Vadehra, D.V., 1989a. An outline of the physiology of avian egg formation. In: The Avian Egg Chemistry and Biology. John Wiley and Sons, New York, pp. 17—23.

Burley, R.W., Vadehra, D.V., 1989b. Egg yolk: biosynthesis and assembly. In: The Avian Egg Chemistry and Biology. John Wiley and Sons, New York, pp. 235—268.

Burley, R.W., Vadehra, D.V., 1989c. Processing-induced chemical and other changes to eggs. In: The Avian Egg Chemistry and Biology. John Wiley and Sons, New York, pp. 299—336.

Burley, R.W., Vadehra, D.V., 1989d. The albumen: biosynthesis. In: The Avian Egg Chemistry and Biology. John Wiley and Sons, New York, pp. 129—145.

Campbell, L., Raikos, V., Euston, S.R., 2003. Modification of functional properties of egg-white proteins. Nahrung/Food 47, 369—376.

Cook, W.H., Martin, W.G., 1969. Egg lipoproteins. In: Tria, E., Scanu, A.M. (Eds.), Structural and Functional Aspects of Lipoproteins in Living Systems. Academic Press, London, pp. 579—615.

Cooney, A.J., Leng, X., Tsai, S.Y., Malley, B.W., Tsai, M., 1993. Multiple mechanisms of chicken ovalbumin upstream promoter transcription factor-dependent repression of transactivation by the vitamin D, thyroid hormone, and retinoic acid receptors. J. Biol. Chem. 268, 4152—4160.

Cotterill, O.J., 1994. Freezing egg products. In: Stadelman, W.J., Cotterill, O.J. (Eds.), Egg Science and Technology. Food Products Press, New York, pp. 265—288.

Cunningham, F.E., 1994. Egg-product pasteurization. In: Stadelman, W.J., Cotteril, O.J. (Eds.), Egg Science and Technology. Food Products Press, New York, pp. 289—315.

Danehy, J.P., 1986. Maillard reactions: nonenzymatic browning in food systems with special reference to the development of flavor. In: Chichester, C.O. (Ed.), Advances in Food Research. Academic Press, New York, pp. 77—124.

Davis, J.G., Hanson, H.L., Lineweaver, H., 1952. Characterization of the effect of freezing on cooked egg white. J. Food Sci. 17, 393—401.

Davis, J.P., Foegeding, E.A., 2007. Comparisons of the foaming and interfacial properties of whey protein isolate and egg white proteins. Colloids Surf. B Biointerfaces 54, 200–210.

Denmat, M.L., Anton, M., Gandemer, G., 1999. Protein denaturation and emulsifying properties of plasma and granules of egg yolk as related to heat treatment. J. Food Sci. 64, 194–197.

Dennis, J.E., Xiao, S.-Q., Agarwal, M., Fink, D.J., Heuer, A.H., Caplan, A.I., 1996. Microstructure of matrix and mineral components of egg shells from white leghorn chickens (Gallus gallus). J. Morphol. 228, 287–306.

Doi, E., Kitabatake, N., 1989. Structure of gycinin and ovalbumin gels. Food Hydrocoll. 3, 327.

Dominguez-Vera, J.M., Gautron, J., Garcia-Ruiz, J.M., Nys, Y., 2000. The effect of avian uterine fluid on the growth behavior of calcite crystals. Poult. Sci. 79, 901–907.

Donovan, J.W., Mapes, C.J., Davis, J.G., Garibaldi, J.A., 1975. A differential scanning calorimetric study of the stability of egg white to heat denaturation. J. Sci. Food Agric. 26, 73–83.

Egelandsdal, B., 1980. Heat-induced gelling in solutions of ovalubumine. J. Food Sci. 45, 981–993.

Evans, M.I., O'Malley, P.J., Krust, A., Burch, J.B., 1987. Developmental regulation of the estrogen receptor and the estrogen responsiveness of five yolk protein genes in the avian liver. Proc. Natl. Acad. Sci. U.S.A. 84, 8493–8497.

Feeney, R.E., Weaver, J.M., Jones, J.R., Rhodes, M.B., 1952. Studies of the kinetics of yolk deterioration in shell eggs. Poult. Sci. 35, 1061–1066.

Fothergill, L.A., Fothergill, J.E., 1970. Thiol and disulphide contents of hen ovalbumin: C-terminal sequence and location of disulphide bond. Biochem. J. 116, 555–561.

Fromm, D., 1967. Some physical and chemical changes in the vitelline membrane of the hen's egg during storage. J. Food Sci. 32, 52–56.

Froning, G.W., 2008. Egg products industry and future perspectives. In: Mine, Y. (Ed.), Egg Bioscience and Biotechnology. John Wiley and Sons, Hoboken, NJ, pp. 307–320.

Gautron, J., Nys, Y., 2007. Eggshell matrix protein. In: Huopalahti, R., Lopez-Fandino, R., Anton, M., Schade, R. (Eds.), Bioactive Egg Compounds. Springer, Heidelberg, pp. 103–108.

Gautron, J., Hincke, M.T., Nys, Y., 1997. Precursor matrix proteins in the uterine fluid change with stages of eggshell formation in hens. Connect. Tissue Res. 36, 195–210.

Gautron, J., Hincke, M.T., Mann, K., Panheleux, M., Bain, M., McKee, M.D., Solomon, S.E., Nys, Y., 2001a. Ovocalyxin-32, a novel chicken eggshell matrix protein, isolation, amino acid sequencing, cloning, and immunocytochemical localization. J. Biol. Chem. 276, 39243–39252.

Gautron, J., Hincke, M.T., Panheleux, M., Garcia-Ruiz, J.M., Boldicke, T., Nys, Y., 2001b. Ovotransferrin is a matrix protein of the hen eggshell membranes and basal calcified layer. Connect. Tissue Res. 42, 255–267.

Hammershøj, M., Nebel, C., Carstens, J.H., 2008. Enzymatic hydrolysis of ovomucin and effect on foaming properties. Food Res. Int. 41, 522–531.

Handa, A., Kuroda, N., 1999. Functional improvements in dried egg white through the Maillard reaction. J. Agric. Food Chem. 47, 1845–1850.

Hatta, H., Kitabatake, N., Doi, E., 1986. Turbidity and hardness of a heat-induced gel of hen egg ovalbumin. Agric. Biol. Chem. 50, 2083–2089.

Hayakawa, S., Nakai, S., 1985. Contribution of hydrophobicity, net charge and sulfhydryl groups to thermal properties of ovalbumin. Can. Inst. Food Sci. Technol. J. 18, 290–295.

Hernandez-Hernandez, A., Navarro, A.R., Garcia-Ruiz, J.M., 2003. Influence of model proteins on the precipitation of calcium carbonate. In: Proceedings of the 16th European Symposium on the Quality of Poultry Meat and 10th European Symposium on the Quality of Eggs and Egg Products, Vol. III. Saint-Brieuc, France. 28–34.

Hiidenhovi, J., 2007. Ovomucin. In: Huopalahti, R., Lopez-Fandino, R., Anton, M., Schade, R. (Eds.), Bioactive Egg Compounds. Springer, Heidelberg, pp. 61–68.

Hincke, M.T., Wellman-Labadie, O., 2008. Biosynthesis and structural assembly of eggshell components. In: Mine, Y. (Ed.), Egg Bioscience and Biotechnology. John Wiley and Sons, New York, pp. 97–128.

Hincke, M.T., Gautron, J., Tsang, C.P., McKee, M.D., Nys, Y., 1999. Molecular cloning and ultrastructural localization of core protein of an eggshell matrix proteoglycan, ovocleidin-116. J. Biol. Chem. 12, 32915–32923.

Hincke, M.T., Gautron, J., Panheleux, M., Garcia-Ruiz, J., McKee, M.D., 2000. Identification and localization of lysozyme as a component of eggshell membranes and eggshell matrix. Matrix Biol. 19, 443–453.

Horiuchi, T., Fukushima, D., 1978. Studies on enzyme-modified proteins as foaming agents; effect of structure on foam stability. Food Chem. 3, 35–41.

Huang, S., Herald, T.J., Mueller, D.D., 1997. Effect of electron beam irradiation on physical, physicochemical and functional properties of liquid egg yolk during frozen storage. Poult. Sci. 76, 1607–1615.

Hunter, G.K., 2001. Role of proteoglycan in the provisional calcification of cartilage: a review and reinterpretation. Clin. Orthop. Relat. Res. 262, 256–280.

Huntington, J.A., Stein, P.E., 2001. Structure and properties of ovalbumin. J. Chromatogr. B. 756, 189–198.

Hunton, P., 2005. Research on eggshell structure and quality: an historical overview. Braz. J. Poult. Sci. 7, 67–71.

Huopalahti, R., Lopez-Fandino, R., Anton, M., Schade, R. (Eds.), 2007. Bioactive Egg Compounds. Springer, Heidelberg.

Ibrahim, H.R., Higashiguchi, S., Sugimoto, Y., Aoki, T., 1997. Role of divalent cations in the novel bactericidal activity of the partially unfolded lysozyme. J. Agric. Food Chem. 45, 89–94.

Ibrahim, H.R., Thomas, U., Pellegrini, A., 2001. A helix-loop-helix peptide at the upper lip of the active site cleft of lysozyme confers potent antimicrobial activity with membrane permeabilization action. J. Biol. Chem. 276, 43767–43774.

Itoh, T., Miyazaki, J., Sugawara, H., Adachi, S., 1987. Studies on the characterization of ovomucin and chalaza of the hen's egg. J. Food Sci. 52, 1518–1521.

Jones, D.R., Musgrove, M.T., 2005. Effects of extended storage on egg quality factors. Poult. Sci. 84, 1774–1777.

Jones, D.R., Tharrington, J.B., Curtis, P.A., Anderson, K.E., Keener, K.M., Jones, F.T., 2002. Effects of cryogenic cooling of shell eggs on egg quality. Poult. Sci. 81, 727–733.

Juneja, L.R., 1997. Egg yolk lipids. In: Yamamoto, T., Juneja, L.R., Hatta, H., Kim, M. (Eds.), Hen Eggs: Their Basic and Applied Science. CRC Press, New York, pp. 73–99.

Juneja, L.R., Kim, M., 1997. Egg yolk proteins. In: Yamamoto, T., Juneja, L.R., Hatta, H., Kim, M. (Eds.), Hen Eggs: Their Basic and Applied Science. CRC Press, New York, pp. 57–71.

Kamat, V., Graham, G., Barratt, M., Stubbs, M., 1976. Freeze−thaw gelation of hen's egg yolk low density lipoprotein. J. Sci. Food Agric. 27, 913−927.

Kato, A., Ogino, K., Kuramoto, Y., Kobayashi, K., 1979. Degradation of the O−glycosidically linked carbohydrate units of ovomucin during egg white thinning. J. Food Sci. 44, 1341−1344.

Kato, A., Miyazaki, S., Kawamoto, A., Kobayashi, K., 1987a. Effects of phosphate residues on the excellent emulsifying properties of phosphoglycoprotein phosvitin. Agric. Biol. Chem. 51, 2989−2994.

Kato, I., Schrode, J., Kohr, W.J., Laskowski, M.J., 1987b. Chicken ovomucoid: determination of its amino acid sequence, determination of the trypsin reactive site, and preparation of all three of its domains. Biochemistry 26, 193−201.

Kato, A., Ibrahim, H.R., Watanabe, H., Honma, K., Kobayashi, K., 1989. New approach to improve the gelling and surface functional properties of dried egg white by heating in dry state. J. Agric. Food Chem. 37, 433−437.

Kato, A., Sasaki, Y., Furuta, R., Kobayashi, K., 1990. Functional protein−polysaccharide conjugate prepared by controlled dry-heating of ovalbumin−dextran mixtures. Agric. Biol. Chem. 54, 107−112.

Kato, A., Minaki, K., Kobayashi, K., 1993. Improvement of emulsifying properties of egg white proteins by the attachment of polysaccharide through Maillard reaction in a dry state. J. Agric. Food Chem. 41, 540−542.

Kato, A., Ibrahim, H.R., Nakamura, S., Kobayashi, K., 1994. New methods for improving the functionality of egg white proteins. In: Sim, J.S., Nakai, S. (Eds.), Egg Uses and Processing Technologies: New Developments. CAB International, Wallingford, pp. 250−267.

Kauzmann, W., 1959. Some factors in the interpretation of protein denaturation. Adv. Prot. Chem. 14, 1−63.

Kido, S., Janado, M., Nunoura, H., 1976. Macromolecular components of the vitelline membrane of hens eggs I. Membrane structure and deterioration with age. J. Biochem. 79, 1351−1356.

Kim, K., Setser, C.S., 1982. Foaming properties of fresh and commercially dried eggs in the presence of stabilizers and surfactants. Poult. Sci. 61, 2194−2199.

Kirunda, D.F., McKee, S.R., 2000. Relating quality characteristics of aged eggs and fresh eggs to vitelline membrane strength as determined by a texture analyzer. Poult. Sci. 79, 1189−1193.

Kitabatake, N., Indo, K., Doi, E., 1989. Changes in interfacial properties of hen egg ovalbumin caused by freeze-drying and spray drying. J. Agric. Food Chem. 37, 905−910.

Knorr, D., Zenker, M., Heinz, V., Lee, D.U., 2004. Application and potential of ultrasonics in food processing. Trends Food Sci. Technol. 15, 261−266.

Koseki, T., Kitabatake, N., Doi, E., 1989. Irreversible thermal denaturation and formation of linear aggregates of ovalbumin. Food Hydrocoll. 3, 123.

Lechevalier, V., Croguennec, T., Nau, F., Guerin-Dubiard, C., 2007. Ovalbumin and gene-related proteins. In: Huopalahti, R., Lopez-Fandino, R., Anton, M., Schade, R. (Eds.), Bioactive Egg Compounds. Springer, Heidelberg, pp. 51−60.

Lee, D.C., McKnight, G.S., Palmiter, R.D., 1978. The action of estrogen and progesterone on the expression of the transferrin gene. J. Biol. Chem. 253, 3494−3503.

Lee, W.C., Chen, T.C., 2002. Functional characteristics of egg white solids obtained from papain treated albumen. J. Food Eng. 51, 263−266.

Li, C.P., Ibrahim, H.R., Sugimoto, Y., Hatta, H., Aoki, T., 2004. Improvement of functional properties of egg white protein through phosphorylation by dry-heating in the presence of pyrophosphate. J. Agric. Food Chem. 52, 5752−5758.

Li-Chan, E.C.Y., Kim, H.-O., 2008. Structure and chemical composition of eggs. In: Mine, Y. (Ed.), Egg Bioscience and Biotechnology. Wiley-Interscience, Hoboken, NJ, pp. 1−8.

Li-Chan, E., Nakai, S., 1989. Biochemical basis for the properties of egg white. Crit. Rev. Poult. Biol. 2, 21−58.

Li-Chan, E.C.Y., Powrie, W.D., Nakai, S., 1995. The chemistry of eggs and egg products. In: Stadelmen, W.J., Cotterill, O.J. (Eds.), Egg Science and Technology. Food Products Press, New York, pp. 105−175.

Lomakina, K., Mikova, K., 2006. A study of the factors affecting the foaming properties of egg white − a review. Czech. J. Food Sci. 24, 110−118.

Lopez, A., Fellers, C.R., Powrie, W.D., 1954. Some factors affecting gelation of frozen egg yolk. J. Milk Food Technol. 17, 334−339.

Lopez, A., Fellers, C.R., Powrie, W.D., 1955. Enzymic inhibition of gelation in frozen egg yolk. J. Milk Food Technol. 18, 77−80.

Ma, C.Y., 1996. Effects of gamma irradiation on physicochemical and functional properties of eggs and egg products. Radiat. Phys. Chem. 48, 375.

Ma, C.Y., Poste, L.M., Holme, J., 1986. Effects of chemical modifications on the physicochemical and cake-baking properties of egg white. Can. Inst. Food Sci. Technol. J. 19, 17−22.

Ma, C.Y., Sahasrabudhe, M.R., Poste, L.M., Harwalker, V.R., Chambers, J.R., 1990. Gamma irradiation of shell eggs: internal and sensory quality, physicochemical characteristics and functional properties. Can. Inst. Food Sci. Technol. J. 23, 226−232.

Ma, C.Y., Harwalkar, V.R., Poste, L.M., Sahasrabudhe, M.R., 1993. Effect of gamma irradiation on the physicochemical and functional properties of frozen liquid egg products. Food Res. Int. 26, 247−254.

McCully, K.A., Mok, C.C., Common, R.H., 1962. Paper electrophoresis characterization of proteins and lipoproteins of hen's yolk. Can. J. Biochem. Physiol. 40, 937−952.

McReynolds, L., O'Malley, B.W., Nisbet, A.D., Fothergill, J.D., Givol, D., Fields, S., Robertson, M., Brownlee, G.G., 1978. Sequence of chicken ovalbumin mRNA. Nature 273, 723−728.

Mangino, M.E., 1994. Protein interactions in emulsions: protein−lipid interactions. In: Hettiarachchy, N.S., Ziegler, G.R. (Eds.), Protein Functionality in Food Systems. Marcel Dekker, New York, pp. 147−180.

Mann, K., Siedler, F., 1999. The amino acid sequence of ovocleidin 17, a major protein of the avian eggshell calcified layer. Biochem. Mol. Biol. Int 47, 997−1007.

Martin, W.G., Augustyniak, J., Cook, W.H., 1964. Fractionation and characterization of the low-density lipoproteins of hen's egg yolk. Biochim. Biophys. Acta 84, 714−720.

Mason, A.B., Woodworth, R.C., Oliver, R.W., Green, B.N., Lin, L.N., Brandts, J.F., Savage, K.J., Tam, B.M., MacGillivray, R.T., 1996. Association of the two lobes of ovotransferrin is a prerequisite for receptor recognition: studies with recombinant ovotransferrins. Biochem. J. 15, 361−368.

Matsudomi, N., Ishimura, Y., Kato, A., 1991. Improvement of gelling properties of ovalbumin by heating in dry state. Agric. Biol. Chem. 55, 879−881.

Matsudomi, N., Nakano, K., Soma, A., Ochi, A., 2002. Improvement of gel properties of dried egg white by modification with galactomannan through the Maillard reaction. J. Agric. Food Chem. 50, 4113−4118.

Miller, C., Winter, A.R., 1950. The functional properties and bacterial content of pasteurized and frozen whole egg. Poult. Sci. 29, 88−97.

Min, B.R., Nam, K.C., Lee, E.J., Ko, G.Y., Trample, D.W., Ahn, D.U., 2005. Effect of irradiating shell eggs on quality attributes and functional properties of yolk and white. Poult. Sci. 84, 1791−1796.

Mine, Y., 1995. Recent advances in the understanding of egg white protein functionality. Trends Food Sci. Technol. 6, 225−232.

Mine, Y., 1996. Effect of pH during the dry heating on the gelling properties of egg white proteins. Food Res. Int. 29, 155−161.

Mine, Y., 1997. Effect of dry heat and mild alkaline treatment on functional properties of egg white proteins. J. Agric. Food Chem. 45, 2924−2928.

Mine, Y., 1998. Emulsifying characterization of hens egg yolk proteins in oil-in-water emulsions. Food Hydrocoll. 12, 409−415.

Mine, Y., 2002. Recent advances in egg protein functionality in the food system. World Poult. Sci. J. 58, 31−39.

Mine, Y., Kovacs-Nolan, J., 2006. New insights in biologically active proteins and peptides derived from hen eggs. World Poult. Sci. J. 62, 87−95.

Mine, Y., Rupa, P., 2004. Immunological and biochemical properties of egg allergens. World Poult. Sci. J. 60, 321−330.

Mine, Y., Zhang, J.W., 2001. The allergenicity of ovomucoid and the effect of its elimination from hen's egg white. J. Sci. Food Agric. 81, 1540−1546.

Mine, Y., Noutomi, T., Haga, N., 1990. Thermally induced changes in egg white proteins. J. Agric. Food Chem. 38, 2122−2125.

Mine, Y., Noutomi, T., Haga, N., 1991. Emulsifying and structural properties of ovalbumin. J. Agric. Food Chem. 39, 443−446.

Mineki, M., Kobayashi, M., 1997. Microstructure of yolk from fresh eggs by improved method. J. Food Sci. 62, 757−761.

Mineki, M., Kobayashi, M., 1998. Microstructural changes in stored hen egg yolk. Jpn. Poult. Sci. 35, 285−294.

Moran, T., 1925. Effect of low temperature on hen eggs. Proc. R. Soc. Lond. B 98, 436−456.

Nguyen, L.T., Burley, R.W., 1984. Studies on the apoproteins of the major lipoprotein of the yolk of hen's egg. Aust. J. Biol. Sci. 37, 7−16.

Nys, Y., Hincke, M., Arias, J.L., Garcia-Ruiz, J.M., Solomon, S., 1999. Avian eggshell mineralization. Poultry Avian Biol. Rev. 10, 143−166.

Nys, Y., Gautron, J., Garcia-Ruiz, J.M., Hincke, M.T., 2004. Avian eggshell mineralization: biochemical and functional characterization of matrix proteins. Comptes Rendus Palevol. 3, 549−562.

Okubo, T., Akachi, S., Hatta, H., 1997. Structure of hen eggs and physiology of egg laying. In: Yamamoto, T., Juneja, L.R., Hatta, H., Kim, M. (Eds.), Hen Eggs: Their Basic and Applied Science. CRC Press, New York, pp. 1−12.

Palmer, H.H., Ijichi, K., Roff, H., 1970. Partial thermal reversal of gelation in thawed egg yolk products. J. Food Sci. 35, 403−406.

Palmiter, R., 1972. Regulation of protein synthesis in chick oviduct. J. Biol. Chem. 247, 6450−6461.

Payawal, S.R., Lowe, B., Stewart, G.F., 1946. Pasteurization of liquid-egg products. II. effect of heat treatments on appearance and viscosity. Food Res. 11, 246−260.

Phillips, L.G., Haque, Z., Kinsella, J.E., 1987. A method for the measurement of foam formation and stability. J. Food Sci. 52, 1074−1077.

Plancken, I.V., Loey, A.V., Hendrick, M., 2007. Effect of moisture content during dry-heating on selected physicochemical and functional properties of dried egg white. J. Agric. Food Chem. 55, 127−135.

Raikos, V., Campbell, L., Euston, S.R., 2007. Rheology and texture of hen's egg protein heat-set gels as affected by pH and the addition of sugar and/or salt. Food Hydrocoll. 21, 237−244.

Relkin, P., Hagolle, N., Dalgleish, D.C., Laurey, B., 1999. Foam formation and stabilisation by pre-denatured ovalbumin. Colloid Surf. B. Biointerfaces 12, 409−416.

Reyes-Grajeda, J.P., Moreno, A., Romero, A., 2004. Crystal structure of ovocleidin-17, a major protein of the calcified Gallus gallus eggshell − implications in the calcite mineral growth pattern. J. Biol. Chem. 39, 40876−40881.

Rhodes, M.B., Azari, P.R., Feeney, R.E., 1958. Analysis, fractionation and purification of egg white proteins with cellulose-cation exchanger. J. Biol. Chem. 230, 399−408.

Robey, P.G., 1996. Vertebrate mineralized matrix proteins: structure and function. Connect. Tissue Res. 35, 131−136.

Robinson, D.S., 1972. Egg white glycoproteins and the physical properties of egg white. In: Freeman, B.M., Lake, P.E. (Eds.), Egg White Glycoproteins and the Physical Properties of Egg White. British Poultry Science, Edinburgh, pp. 65−86.

Romanoff, A.L., Romanoff, A.J., 1944. A study of preservation of eggs by flash heat treatment. Food Sci. 9, 358−366.

Rossi, M., Schiraldi, A., 1992. Thermal denaturation and aggregation of egg proteins. Thermochim. Acta. 199, 115−123.

Schade, R., Chacana, P.A., 2007. Livetin fractions (IgY). In: Huopalahti, R., Lopez-Fandino, R., Anton, M., Schade, R. (Eds.), Bioactive Egg Compounds. Springer, Heidelberg, pp. 25−32.

Schutz, G., Nguyen-Huu, M.C., Giesecke, K., Hynes, N.E., Groner, B., Wurtz, T., Sippekl, A.E., 1978. Hormonal control of egg white protein messenger RNA synthesis in the chicken oviduct. Cold Spring Harb. Symp. Quant. Biol. 24, 617−624.

Seideman, W.E., Cotterill, O.J., Funk, E.M., 1963. Factors affecting heat coagulation of egg white. Poult. Sci. 42, 406−417.

Shawkey, M.D., Kosciuch, K.L., Liu, M., Rohwer, F.C., Loos, E.R., Wang, J.M., Bessinger, S.R., 2008. Do birds differentially distribute antimicrobial proteins within clutches of egg? Behav. Ecol. 19, 920−927.

Shimada, K., Matsushita, S., 1980. Thermal coagulation of egg albumine. J. Agric. Food Chem. 28, 409−412.

Sirvente, H., Beaumal, V., Gaillard, C., Bialek, L., Hamm, D., Anton, M., 2007. Structuring and functionalization of dispersions containing egg yolk, plasma and granules induced by mechanical treatments. J. Agric. Food Chem. 5, 9537−9544.

Smith, D.M., 1994. Protein interactions in gels: protein−protein interactions. In: Hettiarachchy, N.S., Ziegler, G.R. (Eds.), Protein Functionality in Food System. Marcel Dekker, New York, pp. 209−224.

Speroni, F., Puppo, M.C., Chapleau, N., Lamballerie, M.D., Castellani, O., Anon, M.C., Anton, M., 2005. High-pressure induced physicochemical and functional modifications of low-density lipoproteins from hen egg yolk. J. Agric. Food Chem. 53, 5719−5725.

Stein, P.E., Leslie, A.G., Finch, J.T., Carrell, R.W., 1991. Crystal structure of uncleaved ovalbumine at 1.95 Å resolution. J. Mol. Biol. 221, 941−950.

Stevens, L., 2004. Metabolic adaptation in avian species. In: Stevens, L. (Ed.), Avian Biochemistry and Molecular Biology. Cambridge University Press, New York, pp. 82–99.

Sugino, H., Nitoda, T., Juneja, L.R., 1997. General chemical composition of hen eggs. In: Yamamoto, T., Juneja, L.R., Hatta, H., Kim, M. (Eds.), Hen Eggs: Their Basic and Applied Science. CRC Press, New York, pp. 13–24.

Sun, Y., Hayakawa, S., 2002. Heat-induced gels of egg white/ovalbumins from five avian species: thermal aggregation, molecular forces involved, and rheological properties. J. Agric. Food Chem. 50, 1636–1642.

Superti, F., Ammendolia, M.G., Berlutti, F., Valenti, P., 2007. Ovotransferrin. In: Huopalahti, R., Lopez-Fandino, R., Anton, M. (Eds.), Bioactive Egg Compounds. Springer, Heidelberg, pp. 43–50.

Thakur, B.R., Singh, R.K., 1994. Food irradiation chemistry and applications. Food Rev. Int. 10, 437–473.

Tinkler, C.K., Soar, M.C., 1920. The formation of ferrous sulfide in eggs during cooking. Biochem. J. 14, 114–119.

Torten, J., Eisenberg, H.K., 1982. Studies on colloidal properties of whole egg magma. J. Food Sci. 47, 1423–1428.

US Department of Agriculture, Food and Drug Administration, 2000. Irradiation in the production, processing and handling of food: shell eggs, fresh; safe use of ionizing radiation for salmonella reduction. Federal Register 65, 45280–45282.

Vaderhra, D.V., Nath, K.R., 1973. Eggs as a source of protein. Crit. Rev. Food Sci. Nutr. 4, 193–309.

Wakamatu, T., Sato, Y., Saito, Y., 1983. On sodium chloride action in gelation process of low density lipoprotein (LDL) from hen egg yolk. J. Food Sci. 48, 507–516.

Wang, G., Wang, T., 2009. Effects of yolk contamination, shearing, and heating on foaming properties of fresh egg white. J. Food Sci. 74, 147–156.

Wang, S., Smith, D.E., Williams, D.L., 1983. Purification of avian vitellogenin III: comparison with vitellogenins I and II. Biochem. 22, 6206–6212.

Watanabe, K., Nakamura, Y., Xu, J.Q., Shimoyamada, M., 2000. Inhibition against heat coagulation of ovotransferrin by ovalbumin dry-heated at 120°C. J. Agric. Food Chem. 48, 3965–3972.

Watanabe, K., Tsuge, Y., Shimoyamada, M., Onizuka, T., Niwa, M., Ido, T., Tsuge, Y., 2004. Amino acid sequence of alpha-ovomucin in hen egg white ovomucin deduced from cloned cDNA. DNA Seq. 15, 251–261.

Watkins, B.A., 1995. The nutritive value of the egg. In: Stadelman, W.J., Cotterill, O.J. (Eds.), Egg Science and Technology, (4th ed.). Food Products Press, New York, pp. 177–194.

Williams, J., 1962. Serum proteins and the livetins in hen's-egg yolk. Biochem. J. 83, 346–355.

Wong, P.Y., Kitts, D.D., 2002. Physicochemical and functional properties of shell eggs following electron beam irradiation. J. Sci. Food Agric. 83, 44–52.

Wootton, M., Hong, N.T., Thi, L.P., 1981. A study of the denaturation of egg white proteins during freezing using differential scanning calorimetry. J. Food Sci. 46, 1336–1338.

Xing, J., Wellman-Labadie, O., Gautron, J., Hincke, M.T., 2007. Recombinant eggshell ovocalyxin-32: expression, purification and biological activity of the glutathione S-transferase fusion protein. Comp. Biochem. Physiol. B 172, 172–177.

Yang, S.C., Baldwin, R.E., 1995. Functional properties of eggs in foods. In: Stadelman, W.J., Cotterill, O.J. (Eds.), Egg Science and Technology. Haworth Press, New York, pp. 405–464.

Young, A.C., Tilton, R.F., Dewan, J.C., 1994. Thermal expansion of hen egg-white lysozyme comparison of the 19 Å resolution structures of the tetragonal form of the enzyme at 100K and 298K. J. Mol. Biol. 235, 302–317.

Biochemistry of Food Processing

Chapter 6

Browning Reactions in Foods

N. A. Michael Eskin,* Chi-Tang Ho[†] and Fereidoon Shahidi**

*Department of Human Nutritional Sciences, Faculty of Human Ecology, University of Manitoba, Winnipeg, Canada,
[†]Food Science Department, Cook College, Rutgers University, New Brunswick, New Jersey, USA, **Department of Biochemistry,
Memorial University of Newfoundland, St. John's, Newfoundland, Canada

Chapter Outline

I. INTRODUCTION

Browning reactions in food are widespread phenomena which take place during processing and storage. These reactions occur during the manufacture of meat, fish, fruit, and vegetable products, as well as when fresh fruits and vegetables are subjected to mechanical injury. Browning affects the flavor, appearance, and nutritive value of the food products involved. However, for certain foods, browning is an important part of the preparation process. For example, in the manufacture of coffee, tea, beer, and maple syrup, and in the toasting of bread, it enhances the appearance and flavor of these products. Browning, to a limited degree, is considered desirable in apple juice, potato chips, and French fries. To control or inhibit these reactions it is important to understand the mechanisms involved.

Three browning mechanisms appear to be involved in foods, as shown in Table 6.1. Ascorbic acid browning can proceed either by the enzyme ascorbic acid oxidase or by direct atmospheric oxygen and oxidation of ascorbic acid.

Biochemistry of Foods. DOI: http://dx.doi.org/10.1016/B978-0-12-242352-9.00006-0

TABLE 6.1 Mechanisms of Browning Reactions

Mechanism	Requires Oxygen	Requires Amino Group in Initial Reaction	pH Optimum
Maillard	−	+	Alkaline
Caramelization	−	−	Alkaline, acid
Ascorbic acid oxidation	+	−	Slightly acid

II. NON-ENZYMATIC BROWNING

During the preparation and processing of foods, one soon becomes acquainted with the phenomenon of browning associated with heated and stored products. This phenomenon, referred to as non-enzymatic browning, distinguishes it from the enzyme-catalyzed reactions described in Chapter 10. The importance of this reaction in the production of foods is amply illustrated by its contribution to the flavor, color, and aroma of coffee, caramel, bread, and breakfast cereals. Careful control must be exercised to minimize excessive browning, which could lead to unpleasant changes in the food product. In recent years there has been considerable focus on the deleterious effects of non-enzymatic browning reactions in food (Nursten, 2005). Of particular concern is the toxicity and potential mutagenicity of some of the intermediates formed (Aeschbacher et al., 1981; Lee et al., 1982; Sugimura et al., 1988; Mottram et al., 2002; Tareke et al., 2002; Friedman, 2003). However, not all the intermediates formed are deleterious as some appear to exert considerable antioxidant activity (Yamaguchi and Fujimaki, 1974; Kawashima et al., 1977; Lingnert and Eriksson, 1980; Eichner, 1981; Lingnert and Hall, 1986; Yen and Hsieh, 1995; Yoshimura et al., 1997; Wagner et al., 2002, 2007; Tagliazucchi et al., 2010a, b).

Since the second edition of this book a large number of papers have been published on non-enzymatic browning systems. Nevertheless, our knowledge of this area remains fragmentary. Current evidence still supports the existence of three major pathways: Maillard reaction, caramelization, and ascorbic acid oxidation.

A. Maillard Reaction

The formation of brown pigments and melanoidins was first observed by the French chemist Louis Camille Maillard (1912) following the heating of a solution of glucose and lysine (Finot, 2005). This reaction was subsequently referred to as the Maillard reaction and essentially covers all those reactions involving compounds with amino groups and carbonyl groups present in foods. These include amines, amino acids, and proteins interacting with sugars, aldehydes, and ketones, as well as with products of lipid oxidation (Feeney et al., 1975; Hidalgo and Zamora, 2004a, b, 2005; Kwon et al., 1965; Montgomery and Day, 1965). The general mechanism of browning was first proposed by Hodge (1953) and subsequently reviewed by Ellis (1959), Heyns and Paulsen (1960), Reynolds (1963, 1965, 1969), Baltes (1973), Namiki (1988), and Ledl and Schleicher (1990). In spite of the volumes of research on this reaction, the original reaction sequence (Scheme 6.1) proposed by Hodge (1953) still remains valid.

The importance of this reaction in living systems has also been the subject of extensive studies over the past 30 years. This reaction can also occur *in vivo*, originally being referred to as non-enzymatic glycosylation, and later as glycation (Tessier, 2010). The random damage to extracellular proteins in the human body, as a result of glycation, has important implications for aging and chronic diseases such as diabetes (Brownlee, 2001).

1. Carbonylamino Reaction

The first step in the Maillard reaction involves condensation between the α-amino groups of amino acids or proteins and the carbonyl groups of reducing sugars: this defines the carbonylamino reaction. The initial product is an addition compound which rapidly loses water to form a Schiff base followed by cyclization to the corresponding N-substituted glycosylamine:

These reactions are all reversible as an equilibrium exists for these compounds in aqueous solution.

2. Mechanism of the Carbonylamino Reaction

The formation of the N-substituted glycosylamine involves condensation of the amine group of the amino acid with a carbonyl group of a reducing sugar. This reaction is not necessarily restricted to α-amino acids and can involve the participation of other amino groups found in peptides and proteins. This is facilitated when the pH of the medium is above the isoelectric point of the amino group, thus producing basic amino groups.

The protein molecules are composed of many amino acids joined covalently by peptide bonds, in which the amino acids are presumably unavailable for interaction. Harris and Mattil (1940) observed that lysine provided the majority of free amino groups in proteins, in the form of ε-amino groups, which were the main participant in this reaction. While this is true, other amino acids with additional amino groups could also participate, for example, arginine. As the temperature is increased, many more amino acids are rendered unavailable, which cannot be explained by the cleavage of the peptide bonds, a process which appears to be slight even at fairly high temperatures. Horn et al. (1968) found it difficult to explain the rapid and extensive destruction of amino acids in proteins in the presence of sugars simply on the basis of the free amino groups present. A common group such as the imide group of the peptide bond was suggested to be involved, in which the hydrogen of this group was replaced by a carbohydrate moiety. The resulting complex was thought to render the amino acid involved unavailable or prevent enzymatic hydrolysis of the peptide bond itself.

SCHEME 6.1 Non-enzymatic browning
(based on Hodge, 1952) (Nursten, 1981).

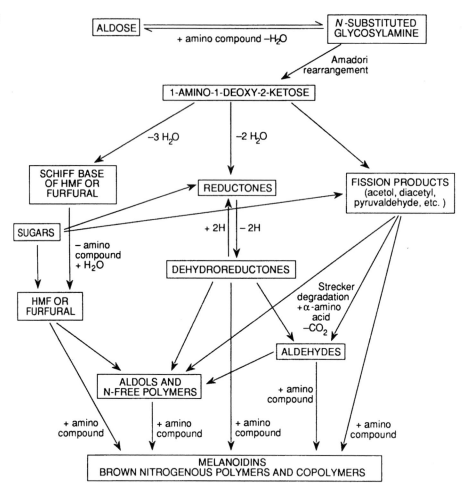

In addition to the mechanisms discussed here, the possibility of anhydride linkages between carboxylic and amino groups was suggested earlier by Harris and Mattil (1940). It was later proposed that these linkages might arise from the interaction between the ε-amino groups of lysine and free dicarboxylic acids in the protein chain. Patton *et al.* (1954) proposed that such linkages could involve aspartic and glutamic acids in the protein.

Besides proteins, oligopeptides have also been identified in a variety of natural and artificial protein hydrolysates such as seafood, coffee beans, soy, and wheat gluten (Aaslyng *et al.*, 1998; Ludwig *et al.*, 2000). Kim and Lee (2009) found that the Maillard reaction rate was affected by the length of the peptide chain. In particular, the reaction between glucose and diglycine was far greater than with glycine or triglycine. This suggested that the reaction rate may be greatly affected by the extent of hydrolysis of the peptide bond and its stability as the heating time is increased. On the other hand, it has been reported that although the diglycine—glucose mixtures had a higher degree of browning, followed by glycine—glucose and triglycine—glucose, the overall reactivity of peptides for flavor formation was determined to be glycine > triglycine > diglycine (Lu *et al.*, 2005). Thus, chain length could be an important factor affecting the reaction and the formation of Maillard reaction products (MRPs).

3. Amadori Rearrangement

The final condensation product in the carbonylamino reaction is the N-substituted glycosylamine. It soon became evident, however, that this compound was extremely unstable and underwent a series of rearrangements, which explained why the reducing power for a casein—glucose system was of the same order as the original glucose (Lea and Hannan, 1950). These changes involved isomerization of the N-substituted glycosylamine to the corresponding fructose—amino acid. The transition from an aldose to a ketose sugar derivative (Scheme 6.2) is referred to as the

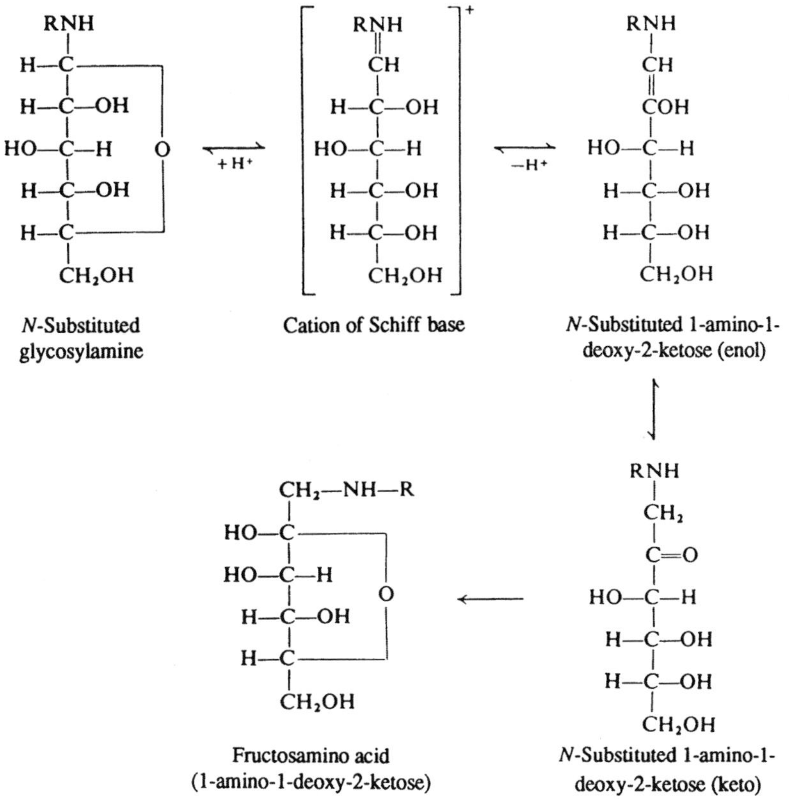

SCHEME 6.2 Amadori rearrangement.

RNH
|
H—C——————┐
|
H—C—OH
|
HO—C—H   O
|
H—C—OH
|
H—C——————┘
|
CH₂OH

N-Substituted glycosylamine

$\xrightarrow{+ H^+}$

$\left[\begin{array}{c}\text{RNH}\\\|\\\text{CH}\\|\\\text{H—C—OH}\\|\\\text{HO—C—H}\\|\\\text{H—C—OH}\\|\\\text{H—C—OH}\\|\\\text{CH}_2\text{OH}\end{array}\right]^+$

Cation of Schiff base

$\xleftarrow{- H^+}$

RNH
|
CH
‖
COH
|
HO—C—H
|
H—C—OH
|
H—C—OH
|
CH₂OH

N-Substituted 1-amino-1-deoxy-2-ketose (enol)

↕

RNH
|
CH₂
|
C=O
|
HO—C—H
|
H—C—OH
|
H—C—OH
|
CH₂OH

N-Substituted 1-amino-1-deoxy-2-ketose (keto)

←

CH₂—NH—R
|
HO—C——————┐
|
HO—C—H
| O
H—C—OH
|
H—C——————┘
|
CH₂OH

Fructosamino acid
(1-amino-1-deoxy-2-ketose)

Amadori rearrangement (Weygand, 1940) and involves protonation of nitrogen at carbon 1 (C1). In the case of ketones and amines, ketosylamines are formed which then undergo the Heyns rearrangement to form 2-amino-2-deoxy aldoses (Reynolds, 1965) by protonation of the oxygen at C6 (Kort, 1970).

The Amadori rearrangement has been demonstrated for a series of glucose—amino acid complexes synthesized by Abrams *et al.* (1955). The reactions leading up to the formation of 1-amino-1-deoxy-2 ketone are all reversible. In fact, these products are quite stable and have been identified in freeze-dried peaches and apricots (Anet and Reynolds, 1957), tomato powder (Eichner *et al.*, 1994), soy sauce (Hashiba, 1978), and milk (Finot *et al.*, 1968). Moll *et al.* (1982) isolated and purified a number of Amadori compounds from crude extracts of Maillard reaction systems using high-performance liquid chromatography (HPLC). These included Amadori compounds of alanine—fructose, leucine—fructose, hydroxyproline—fructose, and tryptophan—fructose, the structures of which are shown in Scheme 6.3. Lee *et al.* (1979) followed the development of MRPs during the processing of apricots and found that the level of Amadori compounds reached a maximum prior to the development of any brown color. In foods containing proteins and glucose, one of the Amadori compounds formed is ε-fructosyl—lysine or furosine (Olano and Martinez-Castro, 1996). The most significant consequence of this reaction is the loss of available lysine (Henle *et al.*, 1991a). Milk, with lactose as the major carbohydrate, forms lactulosyl—lysine when heated. Amadori compounds, such as furosine, are used to assess the quality of a variety of food products such as dairy, eggs, cereals, vegetables, tomatoes, infant cereals, honey, and dehydrated fruits (Molnar-Perl *et al.*, 1986; Resmini and Pellegrino, 1991; Hildago *et al.*, 1998; Guerra-Hernandez *et al.*, 1999; Sanz *et al.*, 2000, 2001, 2003). Gokmen *et al.* (2008) showed that the formation of furosine in cookies was highly correlated with the initial water content of the dough under the same baking conditions.

Yaylayan and Locas (2007) recently showed that 4-hydroxy-alkenals can be formed in a non-lipid system from 2-deoxyribose catalyzed by amino acids. These are important, including lipid peroxidation products that can form Schiff base adducts with nitrogen nucleophiles, such as amino acids and proteins, and then undergo vinylogous Amadori rearrangement. While these Amadori intermediates do not contribute directly to browning or flavor, they result in a loss of nutritional value due in large part to the unavailability of the ε-amino group of lysine (Mauron, 1970; Finot and Mauron, 1972; Hurrell and Carpenter, 1974, 1981; Dworschak, 1980; Friedman, 1982; Lee *et al.*, 1982; Plakas *et al.*, 1988).

Amadori compounds readily undergo acid hydrolysis, so several direct and indirect methods of analysis have been developed (Silvan *et al.*, 2006). Early methods for the direct analysis of Amadori compounds used column

SCHEME 6.3   Amadori compounds. *(Reprinted with permission from Moll et al., 1982. Copyright © by the American Chemical Society.)*

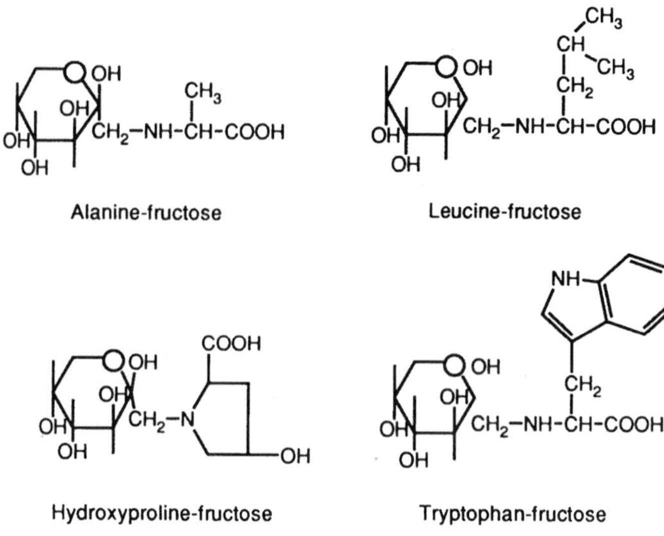

Alanine-fructose

Leucine-fructose

Hydroxyproline-fructose

Tryptophan-fructose

chromatography in an amino acid analyzer (Ellis, 1959; Reynolds, 1959; Moller *et al.*, 1977; Ciner-Doruk and Eichner, 1979; Henle *et al.*, 1991b). This method has since been replaced by HPLC coupled with electrochemical and/or diode array detector. Amadori compounds can also be measured by fast atom bombardment and tandem mass spectrometry (Staempfli *et al.*, 1994). Recently, Davidek *et al.* (2005) developed a high-performance cation-exchange chromatography coupled to tandem mass spectrometry or electrochemical detection as an efficient method for analyzing Amadori compounds derived from hexose and pentose sugars. Indirect measurement of Amadori compounds uses rapid colorimetric or fluorimetric methods (Silvan *et al.*, 2006).

4. Conditions for the Maillard Reaction

a. pH and Buffers

The carbonylamino reaction can develop in acidic or alkaline media, although it is favored under alkaline conditions, where the amine groups of the amino acids, peptides, and proteins are in the basic form. Increasing the pH also ensures that more of the hexoses are in the open chain or reducing form (Burton and McWeeney, 1963). A higher pH environment intensifies Maillard browning in most instances. More browning was observed in leucine–glucose (Renn and Sathe, 1977), lysine–glucose (Ajandouz and Puigserver, 1999), lysine–fructose (Ajandouz *et al.*, 2001), and proline–glucose (Blank *et al.*, 2003) systems. Similar effects were observed for microwave-irradiated samples (Yeo and Shibamato, 1991a; Zamora and Hidalgo, 1995). When extrusion cooking was used, the effects of pH were temperature dependent. Volatile compounds increased with decreasing pH at 180°C but increased with rising pH at 120°C (Ames *et al.*, 2001). The role of buffers in non-enzymatic reactions has been shown to increase the rate of browning for sugar–amino acid systems as a result of their influence on the ionic environment in which the reaction takes place. For example, Lee *et al.* (1984) monitored the formation of the Amadori compounds monofructosyllysine (MFL) and difructosyllysine (DFL) in glucose–lysine mixtures at different temperatures and pH. Their results, as shown in Figure 6.1, indicate pseudo-first-order plots for MFL and DFL formation, which increased from pH 4 to 8. This pattern was similar to that observed for pigment formation.

A plot of pigment formation as a function of pH, however, showed a parabolic curve with break points at pH 6 and 5 for systems heated at 100°C and 110°C, respectively (Figure 6.2). Using a glycine–glucose model system, Bell (1997) showed that phosphate buffer enhanced the browning rate and loss of amino acid at pH 7 and 25°C while citrate buffer had only a minimal effect. The catalytic ability of the phosphate buffer was attributed, in part, to its ability to act as a base catalyst for Amadori rearrangement.

b. Temperature

The temperature dependence of this reaction has been demonstrated in a number of quantitative studies, where increased rates were reported with rise in temperature. Lea and Hannan (1949) found that the decrease in free amino

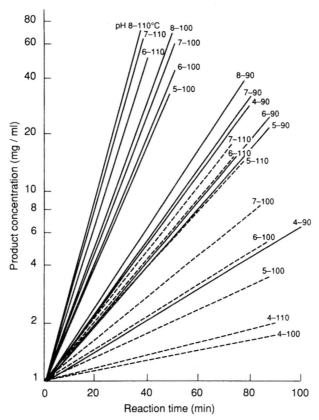

FIGURE 6.1 Formation of monofructosyllysine (MFL) (—) and difructosyllysine (DFL) (− −) as a function of reaction time at various temperatures and pH. All lines were drawn for best fit of the data points. *(Reprinted with permission from Lee* et al.*, 1984. Copyright © by the American Chemical Society.)*

nitrogen for a casein−glucose system conformed to the Arrhenius equation over a temperature range of 0−80°C, where a linear relationship existed between the rate of reaction over this range. The α-amino nitrogen loss possessed 29 cal/mole activation energy in the casein−glucose system, while an increase in activation energy from 26 to 36 cal/mole was noted by Hendel *et al.* (1955) during the browning of dehydrated potato products with increased humidity. Using the formation of hydroxymethylfurfural (HMF) as a measure of progress of the Maillard reaction, Dworschak and Hegedüs (1974) noted an increase in the activation energy for lysine from 29.4 to 34.8 cal/mole when milk powder was heated while increasing the humidity from 2.45 to 5.7%. The amount of HMF produced, however, decreased from 43.6% to 34.6%. Hurrell and Carpenter (1974) noted that the loss in ε-amino lysine groups in an albumin−glucose system at 37°C over 30 days was almost equivalent to that in the same system heated at 121°C for 15 minutes. In both cases the loss in ε-amino groups was 80%, thus emphasizing the importance of duration of storage as well as temperature. Horak and Kessler (1981) monitored the loss of lysine residues during the heating of milk at 160°C and described the loss as a second order reaction. A semilog plot of lysine retention during heating of the glucose−lysine model systems at 69°C was shown by Lee *et al.* (1984) to be linear during the first 2 hours, with an extrapolated decimal reduction time *D* (time required for a 90% reduction of lysine at 69°C) of 3.5 hours (Figure 6.3).

More volatile compounds were detected as xylose and glucose were reacted at increasing temperatures (Benzing-Purdie *et al.*, 1985). The high-molecular-weight products formed were characterized by different lengths of aliphatic carbons and fewer unsaturated carbons. A leucine and glucose system also exhibited a greater browning rate at 122°C than at 100°C (Renn and Sathe, 1997). Similarly, mixture of glucose and glycine generated more volatile compounds at elevated reaction temperatures from 100°C to 300°C (Tehrani *et al.*, 2002).

c. Moisture Content

The Maillard reaction proceeds rapidly in solution, although complete dehydration or excessive moisture levels inhibit this process (Wolfrom and Rooney, 1953). Lea and Hannan (1949, 1950) recorded the optimum moisture level

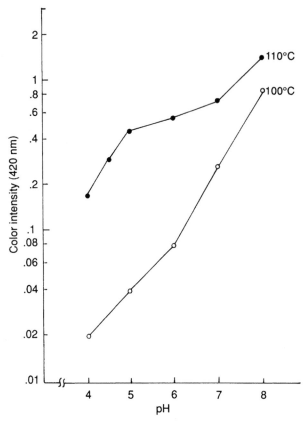

FIGURE 6.2   Changes in pigment formation as a function of pH. *(Reprinted with permission from Lee et al., 1984. Copyright © by the American Chemical Society.)*

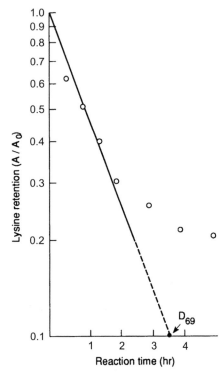

FIGURE 6.3   Lysine retention as a function of reaction time. D_{69}: time required for a 90% reduction in the concentration of lysine at 69°C. *(Reprinted with permission from Lee et al., 1984. Copyright © by the American Chemical Society.)*

for a casein–glucose system and found that the maximum loss of free amino groups occurred between 65% and 70% relative humidity, which corresponded to a level at which the reactants were still in a comparatively dry state. Loncin *et al.* (1968) monitored browning in milk powder at 40°C as a function of water activity (a_w) and lysine over a period of a day. Their results, shown in Figure 6.4, show that the loss of lysine paralleled the extent of browning with a maximum between a_w of 0.6 and 0.7. An increase in humidity was also shown by Dworschak and Hegedüs (1974) to cause an increase in the loss of lysine and tryptophan with the concomitant increase in HMF formation and browning. In general, it appears that this reaction is favored at an optimum moisture content corresponding to fairly low moisture levels (Danehy, 1986).

Investigations into the effect of water on volatile MRPs showed that both quantitative and qualitative changes occurred in the volatile profile of a meat flavor model system with different moisture levels (Hartman *et al.*, 1984a, b). The maximum amount of volatiles was observed at an a_W of 0.72. The kind of major volatiles also differed: sulfur-containing compounds were present in the high water system while dehydration-type products predominated in the low water system. In a glycine–glucose model system, the presence of water drastically increased the amount of volatiles, with furans as the predominant product (Ames *et al.*, 2001). Moisture content also affected the Maillard reaction when microwave heating was used (Yeo and Shibamoto, 1991b; Peterson *et al.*, 1994).

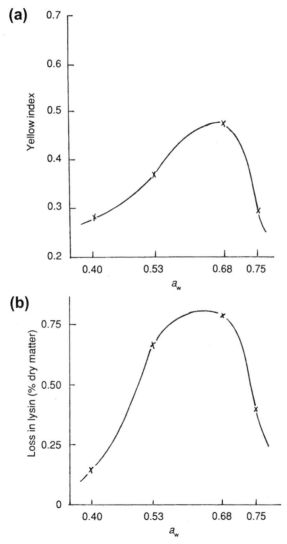

FIGURE 6.4 (A) Color change and (B) loss of free lysine of milk powder kept at 40°C for 10 days as a function of water activity (a_w). *(Loncin et al., 1968).*

d. High Pressure

High-pressure processing has gained much attention because of its minimal effect on food flavor and nutrient quality. However, in combination with other factors, high pressure may lead to some flavor variation.

According to Hill *et al.* (1996), the effect of high pressure on the Maillard reaction varied depending on the pH of the medium. In a glucose and lysine system with pH ranging from 7.0 to 7.5, high pressure had insignificant effects. However, at a higher pH, high pressure hastened the Maillard reaction, with the opposite effect observed at a lower pH. This suggested that pressure induced the ionization of the acid groups in the system, resulting in a lower pH and a subsequent rate reduction. This pH dependency on the effect of pressure on the browning reaction was confirmed by Moreno *et al.* (2003).

Pressure could also change the reaction profile of a system. For example, when glucose—lysine solutions (with initial pH 10.1 and incubated at 60°C) were subjected to atmospheric pressure or 600 MPa, significant quantitative differences in the reaction products were observed (Hill *et al.*, 1998). High pressure also enhanced the formation of tetramethylpyrazine under weak acidic conditions (Huang *et al.*, 1996). Moreno *et al.* (2003) indicated that Amadori arrangement products formed more rapidly and subsequently degraded under high pressure, resulting in an increase in intermediate and advanced reaction products.

e. Pulsed Electric Field and Ultrasound

Using an asparagine—glucose model system, Guan *et al.* (2010a) showed that applying a pulsed electric field at an intensity above 30 kV/cm resulted in a dramatic increase in the formation of MRPs. Guan *et al.* (2010b) also reported that ultrasonic treatment, at high intensities, can be used to promote the Maillard reaction in a glycine—maltose model system. However, the length of time used in their study (100 minutes) for the ultrasonic treatment was impractical for a commercial process, so further studies are needed for this to become a viable process.

f. Sugars

Reducing sugars are essential ingredients in these reactions, providing the carbonyl groups for interaction with the free amino groups of amino acids, peptides, and proteins. The initial rate of this reaction is dependent on the rate at which the sugar ring opens to the oxo or reducible form. Burton and McWeeny (1963) monitored the concentration of the oxo form of the sugar using polarography and found that it increased with increasing pH. The amount of the oxo form was much higher for pentoses than hexoses, thus explaining the greater reactivity of pentoses in browning systems. This was confirmed by Spark (1969), who found that the order of reactivity was greater for aldopentoses than aldohexoses while reducing disaccharides were considerably less reactive. Tu and Eskin (1973) noted that reducing sugars exerted an inhibitory effect on the hydrolysis of casein by trypsin because of the unavailability of certain essential amino acids resulting from non-enzymatic browning reactions. They found that xylose exerted the greatest inhibitory effect, followed by fructose and glucose. Katchalsky (1941) reported that fructose did not condense with amino acids in dilute solution, although scientists have since confirmed that a definite interaction does take place (Heyns and Breuer, 1958; Heyns and Noack, 1962, 1964). D-Fructose has also been reported by Shallenberger and Birch (1975) and Bobbio *et al.* (1973) to brown at a much faster rate than glucose during the initial stages of the browning reaction, but it then falls behind. This was confirmed by Reyes *et al.* (1982) using model systems containing glucose—glycine and fructose—glycine (1:1 molar ratio) at 60°C, pH 3.5, and held for 280 hours. The fructose system browned at a faster rate during the first 80 hour period but was subsequently taken over by the glucose system. The consumption pattern of glucose and fructose paralleled the rate of browning (Figure 6.5). The difference in losses was attributed to the greater polymerization of the glucose-derived melanoidins as measured by the formation of a haze after 240 hours of storage compared to the fructose system, which remained essentially clear. A similar haze formation was noted for a sucrose—glycine system (1:1 molar ratio) resulting from the hydrolysis of sucrose and the release of glucose.

Using a shrimp hydrolysate, Laroque *et al.* (2008) examined the effect of five reducing sugars (ribose, xylose, arabinose, glucose, and fructose) on the kinetics of the Maillard reaction at 55°C and pH 6.5. The pentose sugars were all more reactive than the corresponding hexose sugars, with ribose being the most active, followed by xylose and arabinose (Figure 6.6).

g. Metals

The formation of metal complexes with amino acids can influence the Maillard reaction. This reaction was catalyzed by copper and iron, and inhibited by manganese and tin (Ellis, 1959; Markuze, 1963). Inhibition of browning in

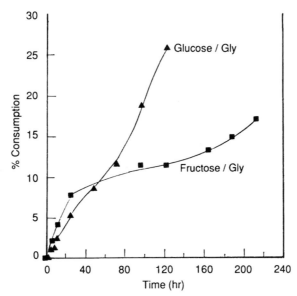

FIGURE 6.5 **Consumption of glucose and fructose in the glucose and fructose–glycine systems during storage at 60°C and pH 3.5.** % Consumption represents the percentage individual sugar lost (Reyes *et al.*, 1982). *(Copyright © by the Institute of Food Technologists.)*

glucose–glycine model systems by trace metals was reported by Bohart and Carson (1955). Using an ovalbumin–glucose mixture Kato *et al.* (1981) examined the effect of Na^+, Cu^{2+}, Fe^{2+}, and Fe^{3+} on the rate of browning at 50°C and 65% relative humidity. Figure 6.7 shows that an acceleration of browning occurred in the presence of Cu^{2+} and Fe^{3+}, while Na^+ had no effect. Fe^{3+} was more effective than Fe^{2+} in accelerating the browning reaction, which suggested that the first step was an oxidation activation resulting in a reduction of the metal. The more rapid browning of a dried egg white–solid glucose system was attributed by both Kato *et al.* (1978) and Watanabe *et al.* (1980) to the presence of trace metals in the egg white. In addition to the catalytic effect of iron on the browning reaction (Hashiba, 1979), it was shown that iron also participated as a chromophore of the pigment (Hashiba, 1986). The possible interaction of iron with hydroxypyridone and hydroxypyranone, both capable of chelating iron, in the melanoidin polymer may be responsible for color formation. The presence of these heterocyclic compounds was reported previously by Tsuchuda *et al.* (1976) after the pyrolysis of non-dialyzable melanoidins.

The ability of monovalent and divalent cations to prevent the formation of acrylamide in a fructose–asparagine model system was reported recently by Gokmen and Senyuva (2007). Divalent cations, such as Ca^{2+} and Mg^{2+}, were

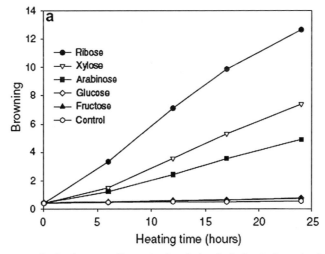

FIGURE 6.6 **Effect of different sugars on the development of browning in a shrimp hydrolysate.** Browning development is expressed as absorbance at 420 nm × dilution factor *(Laroque* et al., *2008).*

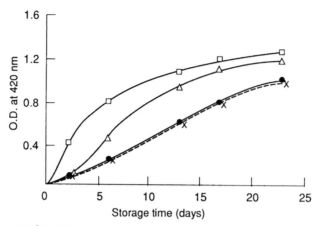

FIGURE 6.7   Effect of Na$^+$, Cu^{2+}, and Fe^{3+} additions on the browning color development in ovalbumin—glucose (OVG) mixtures. (●) OVG, (×) OVG—Na; (□) OVG—Cu; (△) OVG—Fe. *(Reprinted with permission of Kato et al., 1981. Copyright © by the American Chemical Society.)*

able to completely stop the formation of acrylamide compared to the monovalent cations, Na$^+$ and K$^+$, which only reduced acrylamide formation by half. One explanation was the ability of the cations to affect the rate of decomposition of one of the acrylamide precursors, for example the reducing sugars.

III. PIGMENT FORMATION

A. Via Amadori Compounds

The reactions involved in the conversion of 1-amino-1-deoxy-1-ketose derivatives to brown pigments or melanoidins are extremely complex and incompletely understood. Nevertheless, three distinct pathways have been proposed, two of which are directly involved in pigment formation (Scheme 6.4). They involve different labile intermediates, which are the enol forms of the Amadori compounds. In one pathway, enolization of 1-amino-1-deoxy-2-ketose occurs at 2 and 3 positions to irreversibly produce 2,3-enediol. This then undergoes a series of changes including the loss of the amine from C1 to form a methyl dicarbonyl intermediate (Hodge, 1953; Hodge *et al.*, 1963; Simon and Heubach, 1965). The second pathway involves formation of 1,2-eneaminol from the Amadori product in which a hydroxyl group is lost at C3, followed by deamination at C1 and addition of water to form 3-deoxyhexosulose (Anet, 1960, 1964; Kato, 1962). Both hexose derivatives, 1-deoxyhexosulose and 3-deoxyhexosulose, are very reactive and will undergo retro-aldolization to form α-dicarbonyls such as glyoxal, methylglyoxal, and 2,3-butanedione (Weenen and Apeldoorn, 1996). The subsequent reactions are complex and involve a series of aldol condensation and polymerization reactions. The final products are nitrogenous compounds which give rise to the dark brown pigmentation. A low pH favors the 1,2-eneaminol pathway while a high pH favors the pathway involving the conversion of 2,3-enediol to reductones and the subsequent fragmentation to furaneol and pyrones.

B. Alternative Pathways

The formation of free radicals in browning mixtures of carbonyl compounds and amines or amino acids was first reported in the 1970s by Namiki *et al.* (1973), Namiki and Hayashi (1975), and Hayashi *et al.* (1977). The generation of free radicals during the initial stages of the carbonylamino acid reaction for D-glucose—aminobutyric acid isomers was established by Milic *et al.* (1978, 1979, 1980). Namiki and Hayashi (1981) reported that model systems with alanine and arabinose gave rise to electron spin resonance spectra with 17 and 23 lines. These simple signals were attributed to the presence of *N,N*-dialkyl pyrazine cation radicals, which were detected prior to the formation of Amadori compounds. These researchers proposed the formation of a C$_2$ sugar fragment as the precursor of this radical, which was confirmed by isolation and identification of glyoxal dialkylimine. This pointed to an alternative pathway for browning in which the sugar moiety of the Schiff base was cleaved prior to the Amadori rearrangement, leading to the formation of glycolaldehyde alkylimine or its corresponding eneaminol (Scheme 6.5) (Namiki and Hayashi, 1983). Further research by Hayashi and Namiki (1986) confirmed the formation of methylglyoxal dialkylimine, a C$_{2n}$ compound, during the initial stages of the Maillard reaction. The formation of this C$_2$ compound was

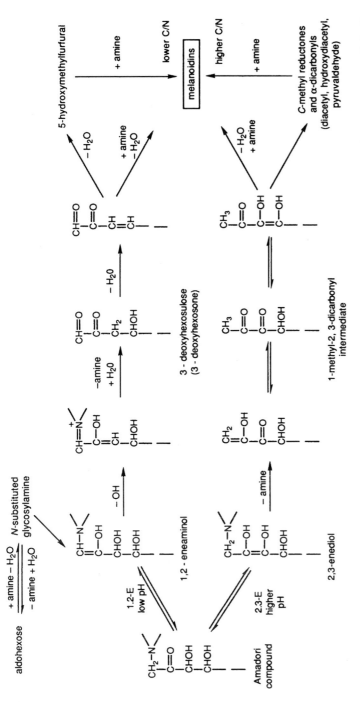

SCHEME 6.4 Maillard reaction: two major pathways from Amadori compounds to melanoidins *(based on Hodge, 1967) (Nursten, 1981).*

thought to arise directly from the Amadori rearrangement. Glycoaldehyde and methylglyoxal, which represented the C_2 and C_3 sugar fragments, exhibited much higher browning rates, corresponding to 2000 and 650 times faster than those of glucose, fructose, or xylose when heated with β-alanine (Table 6.2). Another C_3 compound, glyceraldehyde, also showed close to a 2000-fold increase in the rate of browning compared to the corresponding sugars. Hayashi and Nimiki (1986) presented a summary of the early stages of browning in which they concluded that, under acidic conditions, the traditionally accepted pathway involved osone formation via the Amadori rearrangement. Under alkaline conditions, however, they largely attributed the increase in browning to sugar fragmentation to C_2 and C_3 fragments (Scheme 6.6). Danehy (1986), however, suggested that this pathway be considered as a concomitant one occurring along with the established Maillard reaction scheme.

SCHEME 6.5 Alternative pathway for browning *(Namiki and Hayashi, 1983).*

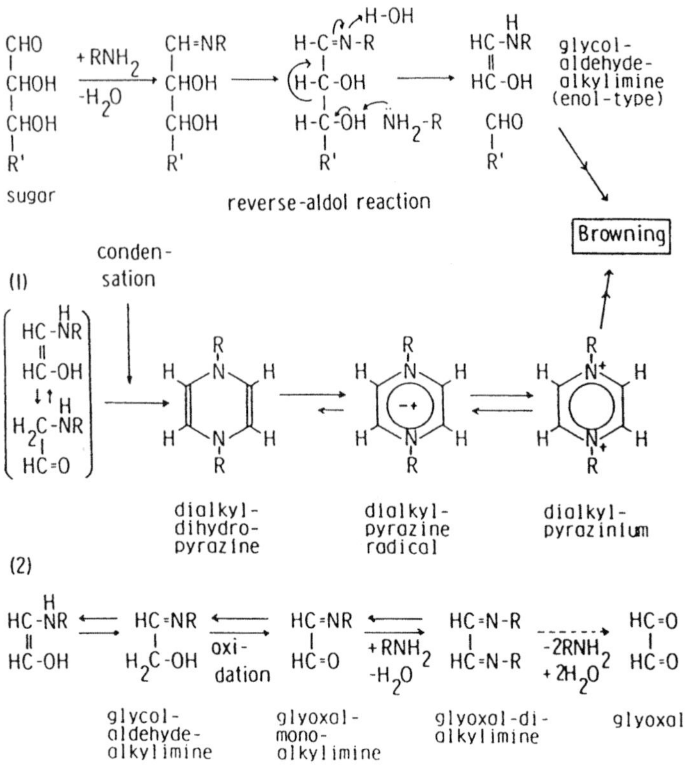

TABLE 6.2 Browning Rates of β-Alanine with Sugar or Carbonyl System

Sugar or Carbonyl Compound	Reaction Temperature (°C)	Browning Activity[a] (liters/min)	Relative Value
Glucose	95	0.019	1
Fructose	95	0.014	0.74
Xylose	95	0.166	8.74
Xylose	80	0.037	
Methylglyoxal	80	2.77	654.3
Glyceraldehyde	80	8.33	1967
Glyoxal	80	0.515	121.6
Glycoaldehyde	80	8.93	2109

[a]*Browning rate measured as change in absorbance at 420 nm.*
From Hayashi and Namiki (1986).

C. Strecker Degradation

The third pathway in the Maillard reaction is concerned with the oxidative degradation of amino acids in the presence of α-dicarbonyls or other conjugated dicarbonyl compounds formed from Amadori compounds. This reaction is based on the work originally carried out by Strecker over a century ago in which he observed the oxidation of alanine by alloxan, a weak oxidizing agent. The reaction, now referred to as the Strecker degradation, is not directly concerned with pigment formation but provides reducing compounds essential for its formation (Rizzi, 2008). The initial reaction involves the formation of a Schiff base with the amino acid. The tuatomeric

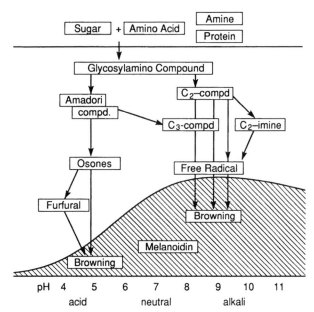

SCHEME 6.6 Different pathways for melanoidin formation depending on reaction pH *(Hayashi and Namiki, 1986)*.

end-form then decarboxylates to produce the eneaminol, which then undergoes hydrolysis to the corresponding aldehyde with one carbon less, together with a 1-amino-2-keto compound (Scheme 6.7). The aldehydes formed during the Strecker degradation reaction contribute to flavor. Some of these aldehydes are listed in Table 6.3 together with their flavor characteristics. At one time aldehydes were considered to be directly responsible for the flavor of roasted foods, although they are now recognized to have a contributory role as auxiliary flavor compounds (Reynolds, 1970; Hodge *et al.*, 1972). Van Praag *et al.* (1968) showed that it is the secondary products of Strecker degradation that are responsible for the strong aroma of cocoa.

Hodge *et al.* (1972) similarly noted that the contribution of the Strecker aldehydes isobutyric, isovaleric, and methional to roasted food aroma was only auxiliary. Condensation of the intermediates formed by Strecker degradation produced many heterocyclic compounds, pyrazines, pyrrolines, oxazoles, oxazolines, and thiazole derivatives responsible for the flavor of heated foods (Hodge *et al.*, 1972; Maga and Sizer, 1973; Maga, 1982).

The generation of flavor compounds by the Maillard reaction has been based primarily on mixtures of sugars and amino acids. Izzo and Ho (1992) first pointed out that the formation of aroma from amino acids bound in proteins and peptides has not been investigated, as in the absence of free amino acids, the Strecker degradation reaction cannot take place. As pointed out by van Boekel (2006), the situation has not changed, as research has still not generated peptide- or protein-specific flavor compounds.

1. Strecker Degradation Reaction and Acrylamide Formation

Tareke *et al.* (2002) first detected the neurotoxin and carcinogen acrylamide in heated foods. The consumption of high-acrylamide food was shown to be associated with a higher incidence of ovarian, endometrial, breast, and kidney cancers in humans (Hogervorst *et al.*, 2007, 2008; Olesen *et al.*, 2008). This led to extensive research to improve the accuracy of its analysis as well as to understand better its formation and subsequent reduction in foods (Locas and Yaylayan, 2008; Anese *et al.*, 2010; Knol *et al.*, 2010; Pedreschi *et al.*, 2010; Kotsiou *et al.*, 2011). Acrylamide appeared to be formed by the thermal degradation of asparagine in the presence of carbonyl compounds (Zamora *et al.*, 2011). Scheme 6.8 outlines the role that Strecker aldehydes and α-keto acids play in the conversion of asparagine to acrylamide. Asparagine is first decarboxylated to 3-alkylaminopropionamides or 3-amino-propionamide, which requires carbonyls (Yayalayan *et al.*, 2003), and then deaminated to form acrylamide. Fortunately, various methods have been developed which prevent the formation of acrylamide in heated foods (Amrein *et al.*, 2007; Yuan *et al.*, 2011).

SCHEME 6.7 Strecker degradation reaction *(Adapted from Schonberg and Moubacher, 1952).*

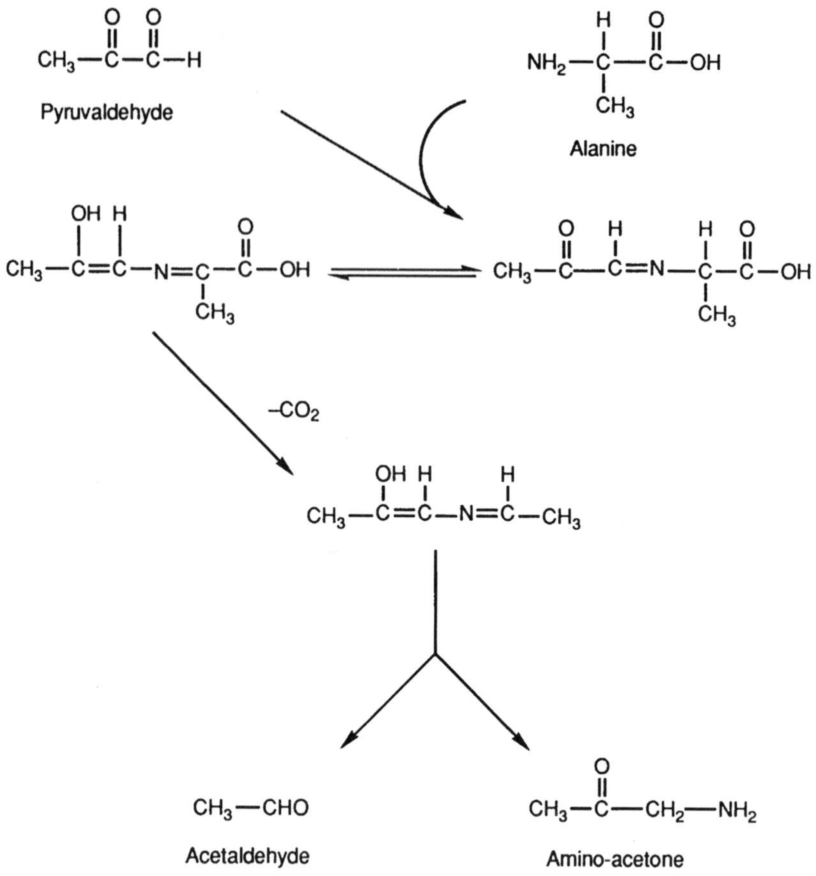

IV. HETEROCYCLIC COMPOUNDS

A. Pyrazine

Among the heterocyclic compounds formed from the Strecker degradation products are the pyrazines. These are very potent flavored compounds which have been identified in almost all processed foods, including beef products, soy products, processed cheese, coffee, potatoes, tea, and roasted pecans (Maga and Sizer, 1973; Maga, 1981, 1982). Dawes and Edwards (1960) identified a number of substituted pyrazines in sugar–amino acid model systems including 2,5-dimethylpyrazine and trimethylpyrazine.

TABLE 6.3 Aromas and Volatile Compounds Produced from L-Amino Acids in Maillard Reaction Systems

Amino Acid	Volatile Compound	Aroma
Alanine	Acetaldehyde	Roasted barley
Cysteine	Thiol, H$_2$S	Meaty
Valine	2-Methylpropanal	
Leucine	3-Methylbutanal	Cheesy
Lysine		Bread-like
Methionine	Methional	

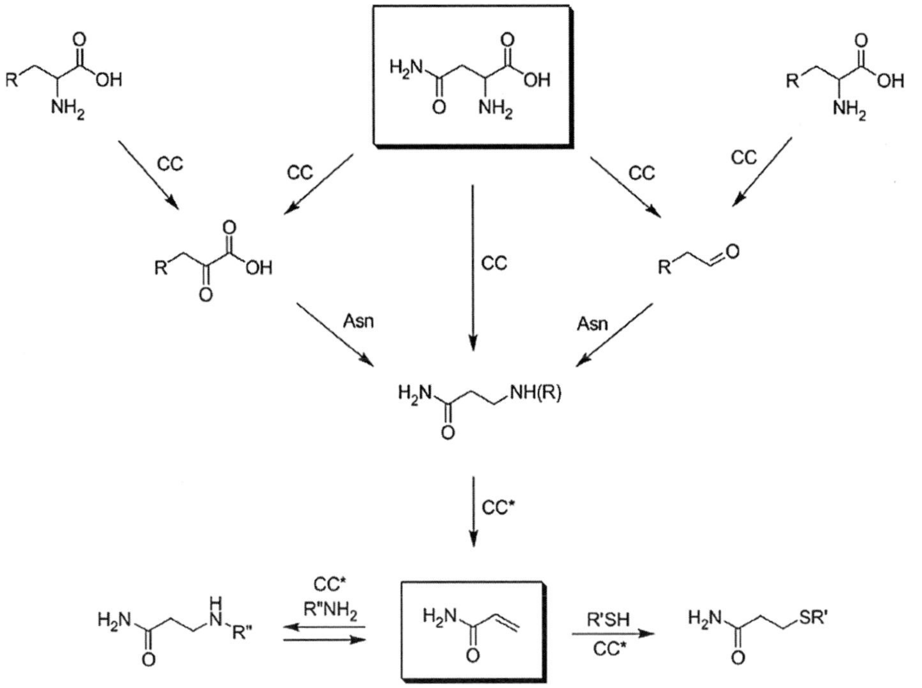

SCHEME 6.8 Schematic role of carbonyl compounds and amino acids in the formation and elimination reactions of acrylamide. CC means that the presence of a carbonyl compound is required; CC* means that carbonyls play a role in the reaction (*Zamora* et al., *2011*).

2,5-Dimethylpyrazine   Trimethylpyrazine

Koehler *et al.* (1969) showed that the C-ring in the substituted pyrazines was derived from the fragmentation of sugars. Koehler and Odell (1970) monitored the formation of methylpyrazines and dimethylpyrazines from sugar–asparagine systems. They noted that fructose gave the highest yields while arabinose gave the smallest yields of these compounds. This suggested that the yields and distribution patterns of pyrazine rings were determined by the nature of the sugar. Shibamato and Bernard (1977), using sugar–ammonia model systems, found similar distribution patterns of pyrazines for both pentose and aldose sugars. Only in the case of the aldose sugars was the level of unsubstituted pyrazines higher, although higher yields were obtained in the presence of the pentoses. One of the major pathways leading to the formation of pyrazines is illustrated in Scheme 6.9, involving condensation of amino-ketones.

Koehler *et al.* (1969) proposed an alternative pathway involving Strecker degradation in which the bound amino acid nitrogen was the main contributor to nitrogen in the pyrazine rings. Condensation of two 2-carbon sugar fragments with nitrogen produced pyrazine, while a similar reaction involving condensation of a 2-carbon fragment with a 3-carbon sugar fragment produced methylpyrazine. The formation of dimethylpyrazine was attributed to the

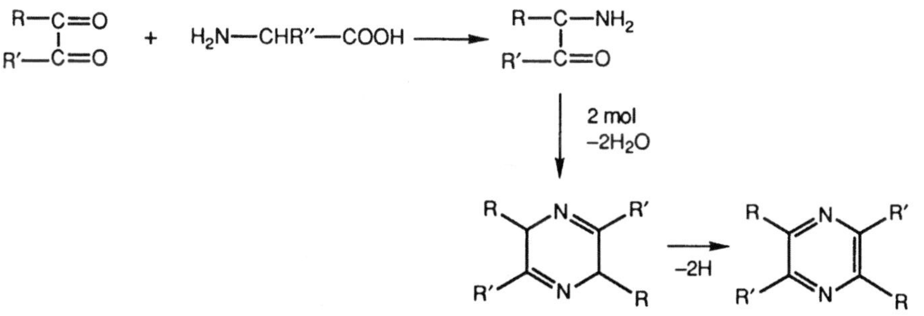

SCHEME 6.9 Formation of pyrazines (*Shibamato and Bernard, 1977*).

condensation of two 3-carbon sugar fragments and nitrogen. Wong and Bernhard (1988) examined five different nitrogen sources (ammonium hydroxide, ammonium acetate, ammonium formate, glycine, and monosodium glutamate) for the formation of pyrazines. They concluded that the nitrogen source had a marked effect on both the amount and types of pyrazines formed during non-enzymatic browning, as suggested earlier by Koehler and Odell (1970) and Koehler et al. (1969). Since there is more than one amino acid in foods, competition between amino acids to generate pyrazines can occur. To investigate this competition, a ^{15}N isotope-labeled glycine was used as a reference with another test amino acid (glutamine, glutamic acid, asparagines, aspartic acid, lysine, arginine, phenylalanine, or isoleucine) added to compete with glycine in the production of pyrazines (Hwang et al., 1995). They found that the reaction mixture containing lysine had the highest yield of pyrazines, with the reaction mixture containing arginine producing the lowest yields. These results indicated that lysine was able to increase the reactivity of glycine. Consequently, the variety and quantity of pyrazine formation appeared to depend on the reactivity and type of amino acids used.

B. Pyrroles

Pyrroles are an important group of heterocyclic compounds formed during the browning of foods (Hodge, 1953). One of the pathways leading to their formation involves cyclization of methyldicarbonyls to 2,4-dideoxypentulose-3-ene, which cyclizes to furfural or reacts with an amino acid at C2 to form a Schiff base, which then cyclizes to N-substituted pyrrole-2-aldehyde (Kato and Fujimaki, 1968). The formation of pyrrole derivatives has since been identified in a number of sugar—amino acid systems (Ferretti and Flanagan, 1971, 1973; Shigematsu et al., 1972; Rizzi, 1974). Shaw and Berry (1977) reported the formation of 2-acetylpyrrole and 5-methylpyrrole-2-carboxyaldehyde in fructose—alanine model systems. This pathway is shown in the diagram below, in which the 3-deoxyhexulose derivative underwent Strecker degradation with the amino acid leading to the formation of 1-amino-3-deoxy-2-ketose. Further changes included enolization and dehydration, resulting in the formation of 2-acetylpyrrole.

3,4-Dideoxypentosulos-3-ene

Furfural   N–Substituted 2-pyrrole aldehyde

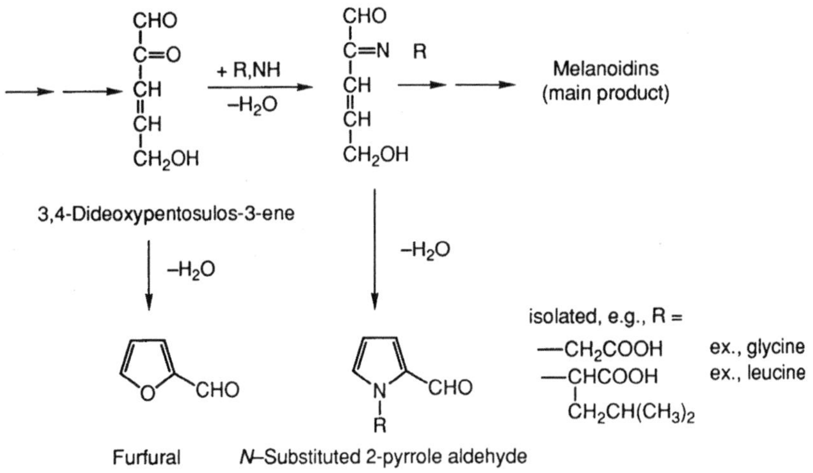

C. Oxazoles and Oxazolines

Oxazoles and oxazolines have been identified among the flavor volatiles of coffee (Stoeffelsman and Pypker, 1968), baked potato (Coleman et al., 1981), and roasted peanuts (Lee et al., 1981). The role of these compounds in the flavor of foods has been reviewed by Maga (1978, 1981). One such compound, 2,3,5-trimethyl-2-oxazole, was identified in the volatiles of boiled beef (Chang et al., 1968) and canned beef stew (Peterson et al., 1975). The latter researchers also reported the presence of the corresponding oxazoline (2,4,5-trimethyl-3-oxazoline). The role of the Strecker degradation reaction in the formation of these compounds was first suggested by Rizzi (1969), in which 2-isopropyl-4,5-dimethyl-3-oxazoline was formed from D-histidine and 2,3-butadione (diacyl). The formation of oxazoles and oxazolines by Strecker degradation was confirmed by Ho and Hartman (1982) to explain the presence of these compounds in the flavor volatiles of meat and roasted peanuts. These researchers proposed the pathway in Scheme 6.10 to explain the mechanism for the formation of 2,4,5-trimethyloxazole and 2,4,5-trimethyloxazoline

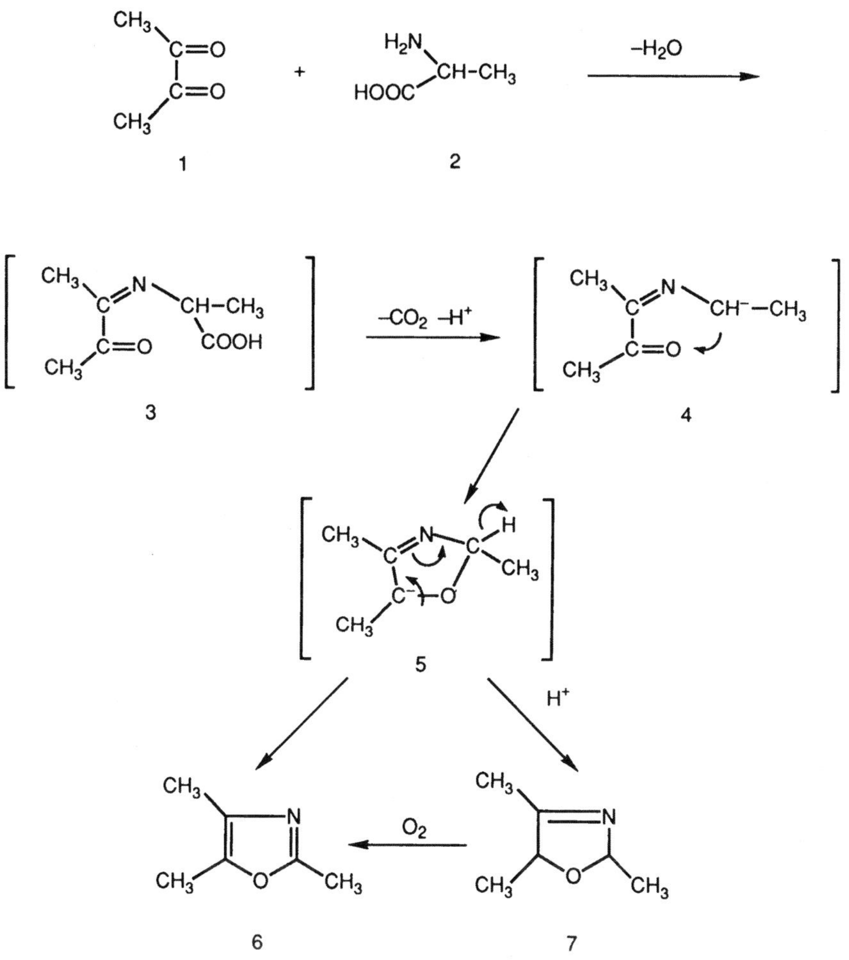

SCHEME 6.10 Formation of 2,4,5-trimethyloxazole and 2,4,5-trimethyl-3-oxazoline from the reaction of DL-alanine and 2,3-butanedione. *(Reprinted with permission from Ho and Hartman, 1982. Copyright © by the American Chemical Society.)*

from DL-alanine and butanedione. Elimination of water resulted in an unstable Schiff base which then underwent decarboxylation to the corresponding anion followed by cyclization to the 3-oxazolinide ion. Protonation or loss of hydride ion was thought to produce the 2,4,5-trimethyloxazoline. The corresponding oxazole was attributed to the oxidation of the oxazoline or loss of hydride.

D. Thiazoles

Thiazoles are formed from sulfur amino acids. These compounds have been identified in coffee, roasted peanuts, cooked beef, and potato chips (Stoll *et al.*, 1967a, b; Walradt *et al.*, 1971; Buttery and Ling, 1974; Buttery *et al.*, 1983). The presence of 2-acetyl-2-thiazoline in beef broth was attributed by Tonsbeek *et al.* (1971) to the Strecker degradation reaction between cysteine and methylglyoxal followed by cyclization.

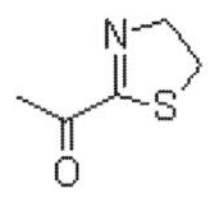

Hofmann and Schieberle (1995) later performed model experiments to gain a more detailed insight into the reaction mechanisms and intermediates governing the formation of 2-acetyl-2-thiazoline. The intermediates in the reaction path leading to 2-acetyl-2-thiazoline were identified as the odorless 2-(1-hydroxyethyl)-4,5-dihydrothiazole.

Formation of 3-thiazolines from an α-hydroxyketone involves the substitution of the hydroxyl group with a thiol group. This is followed by nucleophilic attack by the sulfur atom at the carbon atom of an imine intermediate formed by the reaction between ammonia and an aldehyde. Subsequent ring closure, with the elimination of a molecule of water, gives the 3-thiazolines. Oxidation of the thiazoline results in the formation of the corresponding thiazole (Schutte, 1974).

The Strecker degradation reaction plays a key role in the production of important flavor compounds by condensation and cyclization of the different aldehydes formed. In addition to the heterocyclic compounds discussed there are many others, including pyrrolidines and pyridines. The isolation and chemical synthesis of 2-acetyl-1-pyrroline, a key compound responsible for the characteristic smell of cooked rice, provides another example of the role of Strecker degradation in flavor genesis (Buttery *et al.*, 1982, 1983). It is now confirmed that 2-acetyl-1-pyrroline is a Strecker degradation product of proline and is considered as a character impact compound also giving roasty odors to foods such as bread and popcorn (Schieberle, 1990; Adams and De Kimpe, 2006).

2-Acetyl-1-pyroline

In addition to these compounds, other heterocyclic and carboxyclic compounds identified from heated sugar—amine systems include furanones, pyrrolinones, and cyclopentenones (Ledl and Fritsch, 1984).

V. MAILLARD REACTION—LIPID INTERACTIONS

Lipids are common food components present intrinsically or added as ingredients. Their widespread occurrence unavoidably causes them to affect the Maillard reaction and to interact with its products. Major flavor changes occur as a consequence. The mechanism of flavor variation due to such interactions is being studied in both model and food systems.

Degradation products from lipid oxidation and the Maillard reaction give rise to certain aromatic compounds. Lipid oxidation results in the formation of aldehydes, etones, and other compounds. On the other hand, Strecker degradation in the Maillard reaction scheme produces other reactive intermediates that can react with lipid degradation products. The volatiles formed from these interactions include heterocyclic compounds containing oxygen, nitrogen or sulfur, with long-chain *n*-alkyl substituents (Whitfield, 1992).

The Maillard reaction and lipid oxidation products react to form pyrazines. Chiu *et al.* (1990) demonstrated the synthesis of pentylpyrazines or hexylpyrazines in a system containing ammonium acetate, acetol, and pentanal or hexanal. These alkyl-substituted pyrazines were also detected in fried potatoes (Carlin *et al.*, 1986), corn-based systems (Bruechert *et al.*, 1988), and fried chicken (Tang *et al.*, 1983). Interaction between 2,4-decadienal, a linoleic oxidation product, and cyteine or glutathione yielded long-chain alkyl dithiazines and trithiolanes (Zhang and Ho, 1989; Zhang *et al.*, 1994). Zhang and Ho (1989) also observed the formation of a large amount of 2-pentylpyridine in the model system containing glutathione instead of cysteine. They proposed that 2,4-decadienal was involved in a direct Schiff base formation with the amino group of cysteine or glutathione, followed by electrocyclic reaction and aromatization to form pentylpyridine. The relatively stable glutathione, rather than cysteine, provided more primary free amino groups for Schiff base formation and was responsible for the production of the larger amount of 2-pentylpyridine. The latter compound has been identified in deep-fried foods (Tang *et al.*, 1983) and meat (Mottram, 1985).

Sulfur-containing heterocyclic compounds are important to meat flavor (Mottram, 1998). A number of alkylthiazoles has been reported in meat (Hartman *et al.*, 1983; Tang *et al.*, 1983), with many alkyl-3-thiozolines and alkylthiazoles identified in cooked beef (Elmore *et al.*, 1997). The formation of these compounds was thought to be derived from α-hydroxyketones or α-diones, hydrogen sulfide, ammonia, and Strecker or lipid-oxidation derived aldehydes (Elmore and Mottram, 1997).

Phospholipids are often part of a food matrix. They are rich in polyunsaturated fatty acids, making them extremely prone to oxidation. Their presence was shown to alter the configuration of volatile products in a cysteine—ribose system (Farmer *et al.*, 1989). In this system, phospholipids caused a decline in the amount of sulfur-containing heterocyclic compounds.

Edible oils play a role in pyrazine production. Negroni *et al.* (2001) observed the opposite effects of edible oils on unsubstituted and substituted pyrazines in lysine—xylose—glucose model systems. Decreasing amounts of unsubstituted pyrazines were reported in the presence of olive, canola, or sunflower oil. In contrast, increasing levels of 2-methylpyrazine, 2,5-methylpyrazine, and 2,3-dimethylpyrazine were observed with olive, canola, and sunflower oils. The sensitivity of the pyrazines was attributed to differences in the degree of unsaturation between the oils. However, this hypothesis remains to be tested using purified oils. Commercially available vegetable oils also contain different amounts and types of phenolic antioxidants which may influence the degree of lipid interactions in the Maillard reaction. As discussed earlier, Yayalan and Locas (2007) showed that important lipid peroxide products can form Schiff base adducts with nitrogen nucleophiles, such as amino acids and proteins; these can then undergo vinylogous Amadori rearrangements which can then cyclize and form pyrrole moieties. In the case of proline, however, this cyclization is prevented so that stable vinylogous Amadori rearrangement products (vARPs) are formed.

The Strecker degradation of amino acids by α-epoxyenals was reported by Hildago and Zamora (2004b), as part of their extensive studies on the interaction of lipid degradation products with amino acids and proteins. In their model studies with phenylalanine and 4,5-epoxy-2-decenal at 37°C overnight, products indicative of Strecker degradation occurring were identified by gas chromatography–mass spectrometry, including phenylacetaldehyde and 2-pentylpyridine.

Many foods exist as emulsions which affect the perception of aroma compounds. Van Ruth et al. (2002) reported that lower lipid and emulsifying fractions and greater particle diameter of oil–water emulsions enhanced the aroma release of alcohols, ketones, esters, aldehydes, terpenes, and sulfur compounds.

VI. EFFECT OF POLYPHENOLS ON THE MAILLARD POLYMERS

Polyphenols are secondary metabolites of plants which are used in their defense against severe environments such as ultraviolet radiation or attack by pathogens. These compounds are generally classified as flavonoids, phenolic acids, lignans, and stilbenes. They are widely distributed in plant-based food products; however, their effects on the Maillard reaction have only recently been studied.

Certain polyphenols were recently shown to have trapping activity for reactive carbonyl species, indicating a new chemical characteristic of polyphenols (Jiang and Peterson, 2010; Lo et al., 2006, 2011; Lv et al., 2011; Sang et al., 2007; Shao et al., 2008). Reactive carbonyl species glyoxal, methylglyoxal, and 3-deoxyglucosone, generated by the Maillard reaction, are important intermediates for flavor and color formation. Trapping reactive carbonyl species by polyphenols can alter Maillard reaction pathways. For example, some flavor compounds normally generated in the Maillard reaction, such as pyrazine, methylpyrazine, 2,5-dimethylpyrazine, and tri-methylpyrazine, were inhibited when epicatechin was added to a glucose–glycine model system (Totlani and Peterson, 2005). Using the glucose–glycine system, Noda and Peterson (2007) compared the trapping efficiency of flavan-3-ol epicatechin (EC), epigallocatechin (EGC), epigallocatechin gallate (EGCG), and several phenolic compounds (1,3,5-trihydroxybenzene, 1,2,3-trihydroxybenzene, and methyl gallate) on the formation of pyrazines. A similar significant reduction in pyrazine formation was observed for EC, ECG, and ECGC. Methyl gallate followed by 1,2,3-trihydroxybenzene was the least reactive, while 1,3,5-trihydroxybenzene proved to be the most effective. This suggested that the mechanism responsible for the inhibition was the carbonyl trapping activity on the A-ring of a flavonoid compound.

VII. MELANOIDIN–MAILLARD POLYMERS

The final products formed in the Maillard reaction are polymers or melanoidins. Unlike the flavor and aroma compounds discussed earlier, the origin and nature of these polymers are poorly understood. Numerous studies have attempted to examine melanoidins in model systems, including Barbetti and Chiappini (1976a, b), Ledl (1982a, b), Ledl and Severin (1982), Velisek and Davidek (1976a, b), Imasato et al. (1981), and Bobbio et al. (1981). A study by Feather and Nelson (1984) attempted to isolate the Maillard polymers produced in model systems composed of D-glucose/D-fructose/5-(hydroxymethyl)-2-furaldehyde and glycine and D-glucose/D-fructose with methionine. Increasing amounts of water-soluble, non-dialyzable polymers with molecular weights greater than 16,000 were obtained for both glycine and methionine systems as a function of time (Figures 6.8 and 6.9). Elemental analyses (carbon, hydrogen, and nitrogen) of polymers prepared from glycine model systems were similar, which suggested that the amino acid was incorporated into the polymer. The polymer isolated from D-glucose/D-fructose and glycine was composed of sugar and amino acid minus three molecules of water. Detection of sulfur and nitrogen in polymers obtained from D-glucose and methionine also pointed to incorporation of the amino acid intact. The binding of metal ions to these melanoidins was evident by their greater solubility in tap water than in distilled water.

The nuclear magnetic resonance (NMR) spectra of these polymers suggested that some aromaticity was present. Feather and Huang (1986) examined the ^{13}C-NMR spectra of water-soluble polymers produced from labeled

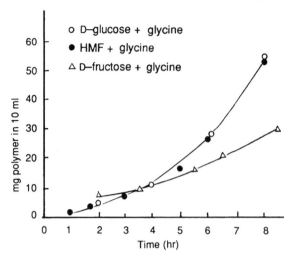

FIGURE 6.8 **Yields of non-dialyzable polymers as a function of time using glycine as the amino acid.** HMF: hydroxymethylfurfural. *(Reprinted with permission from Feather and Nelson, 1984. Copyright © by the American Chemical Society.)*

D-[1-^{13}C] glucose, L-[1-1^{13}C]alanine, and L-[2-1^{13}C]alanine (90 atom%). Polymers prepared with the C1-labeled carbon atom in L-alanine had lower activity compared to the C2-labeled amino acid. This could be due to its degradation to volatile aldehydes, suggesting a more direct role for the Strecker degradation in the Maillard reaction as proposed by Holtermand (1966) (Section III, B). Earlier work by Olsson *et al.* (1982) found that the aldehyde NMR spectrum was similar to that of an analogous Amadori compound. Based on this study and other NMR spectra, it appeared that the non-dialyzable polymer was formed by dehydration and polymerization of an Amadori compound, as suggested previously by Olsson *et al.* (1982).

Benzing-Purdie *et al.* (1985) examined the effect of temperature on the structure of the melanoidins formed in model systems composed of D-xylose and glycine. In the presence of equimolar amounts of the reactants, an increase in temperature (22, 68, and 100°C) was accompanied by an increase in the aromatic nature of both the low- and high-molecular-weight melanoidin products. These researchers also noted considerable differences in the nature of the melanoidins produced at 22°C compared to those formed at the higher temperatures, with different types of aliphatic carbons and fewer unsaturated carbons.

The non-dialyzable melanoidins produced from glucose–glucose systems heated at 95°C at pH 6.8 were examined by Kato *et al.* (1985). These melanoidins were composed of saturated aliphatic carbons together with smaller amounts of aromatic carbons. Milic (1987) examined the kinetics of melanoidin formation between D-glucose and 2-, 3-, and 4-aminobutanoic acid isomers using cross-polarization—magic angle spinning ^{13}C-nuclear

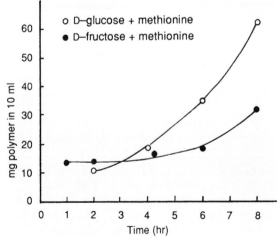

FIGURE 6.9 **Yields of non-dialyzable polymers as a function of time using methionine as the amino acid** *(Feather and Nelson, 1984).*

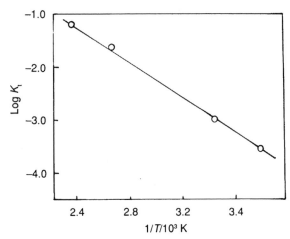

FIGURE 6.10 Temperature dependence of the approximate rate constant K_t for the D(+)-glucose—2-aminobutanoic acid model system *(Milic, 1987).*

magnetic resonance (CP-MAS ^{13}C-NMR) spectroscopy. The systems were heated in sealed quartz test-tubes at 313, 343, and 371 K for 1.0×10^5 to 3.60×10^5 seconds under alkaline conditions (pH 9.0) and the brown melanoidins were eluted on an ion-exchange Permutit ES resin with 5% NaCl. The purified melanoidins were then concentrated, dialyzed, and dried under vacuum before analysis by CP-MAS ^{13}C-NMR. Based on the CP-MAS spectra, an increase in unsaturation and/or aromaticity was observed at 100 and/or 180 ppm with increasing temperature and time, which leveled off when both glucose and aminobutanoic acid were depleted. Milic (1987) calculated the order of reaction of melanoidin formation, assuming that aromaticity, measured by CP-MAS ^{13}C-NMR spectroscopy, corresponded to the rate of browning temperature (K). The amount of amino acid (A_0) (peak area at 180 ppm) was plotted against the reaction temperature at a constant time of 2.34×10 seconds and the straight line obtained showed that melanoidin formation followed first order kinetics (Figure 6.10).

The formation of melanoidins appears to involve two separate pathways. One pathway leads to the formation of protein-based melanoidins from deoxysones, while the other is based on sugar degradation products (Kroh *et al.*, 2008). The latter reaction was shown by Hofmann (1998) to result in the formation of low-molecular-weight colored compounds with furan and pyrrolidine structures and high-molecular-weight brown compounds. α-Dicarbonyls were recently shown by Kroh *et al.* (2008) to be key intermediates in the formation of the carbohydrate-based melanoidins. These include the long-chain dicarbonyls, 1,3-deoxyosone and 1,4-dideoxyosones, and the short-chain dicarbonyls, methylglyoxal and glyoxal. They proposed the following carbohydrate-based melanoidin structure resulting from the aldol condensation of α-dicarbonyl compounds (Scheme 6.11). Colored melanoidins were formed in the presence of methylglyoxal, as under alkaline conditions it forms a carbanion on the third carbon atom, which then proceeds to form colored melanoidins by aldol condensation. In the case of glyoxal a similar reaction is not possible as it is unable to form a carbanion.

Extensive studies conducted on the formation of protein-based melanoidins used a D-xylose—glycine model system. Using this model system, Hayase *et al.* (1999) isolated a Maillard reaction-1 pigment (Blue-M1) that was

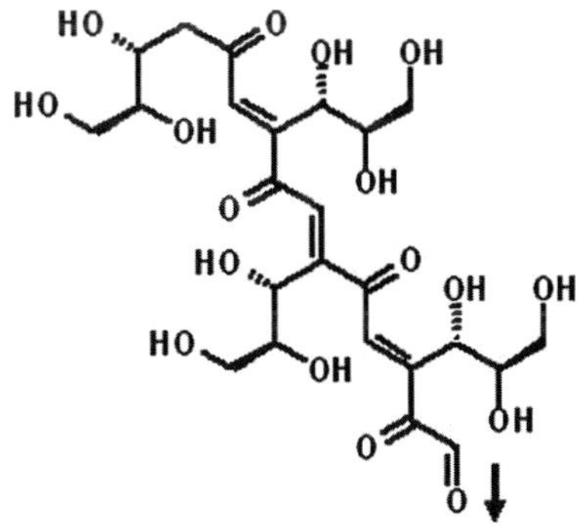

SCHEME 6.11 Proposal of a carbohydrate-based melanoidin structure *(Kroh et al., 2008).*

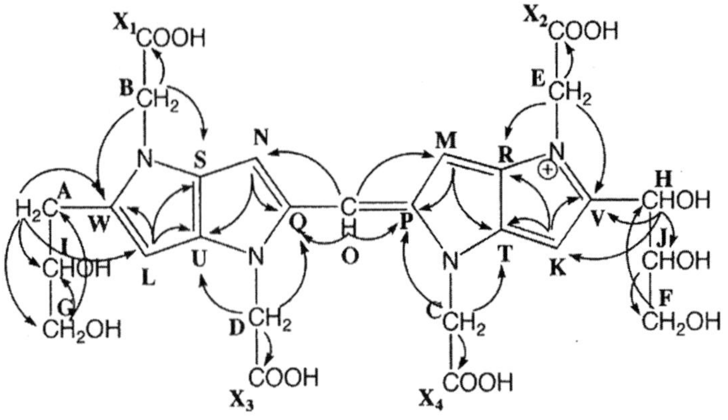

SCHEME 6.12   Identified structure of Blue-M1 *(Hayase* et al., *1999).*

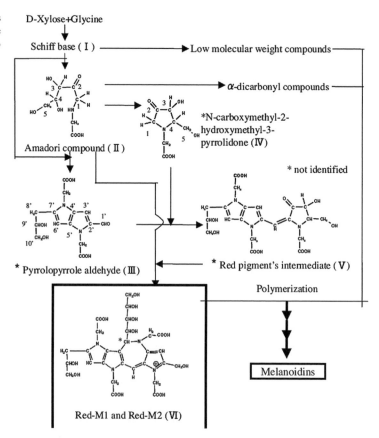

considered to be an oligomer intermediate in the formation of melanoidins. As seen in Scheme 6.12, it is made up of four molecules of D-xylose and glycine with a methine proton between two pyrrolpyrrole rings. Further work by Sasaki *et al.* (2006) identified a second blue pigment (Blue-M2) which had a higher molecular weight compound than Blue-M1 and was thought to arise from the interaction between Blue-M1 and di-D-xyuloseglycine. In addition to the blue pigments, Blue-M1 and Blue-M2, Shirahashi *et al.* (2009) recently identified several red pigments (Red-M1 and Red-M2) using the same D-xylose–glycine system. Once again, these red pigments were intermediate compounds in the formation of melanoidins. They proposed the pathway leading to the formation of melanoidins shown in Scheme 6.13.

The actual mechanism involved in the formation of melanoidins, advanced glycation end products from these intermediate pigments, was finally resolved by Hayase *et al.* (2008). Scheme 6.14 shows the formation of a Schiff base and Amadori compound from xylose and glycine, followed by the formation of 3-deoxyxylosone (3-DX). The latter

SCHEME 6.13   Proposed formation pathway for products from the reaction of D-xylose (1 M) incubated with glycine (0.1 M) and sodium hydrogen carbonate (0.1 M) in a 60% ethanol solution (pH 8.1) *(Shirahashi* et al., *2009).*

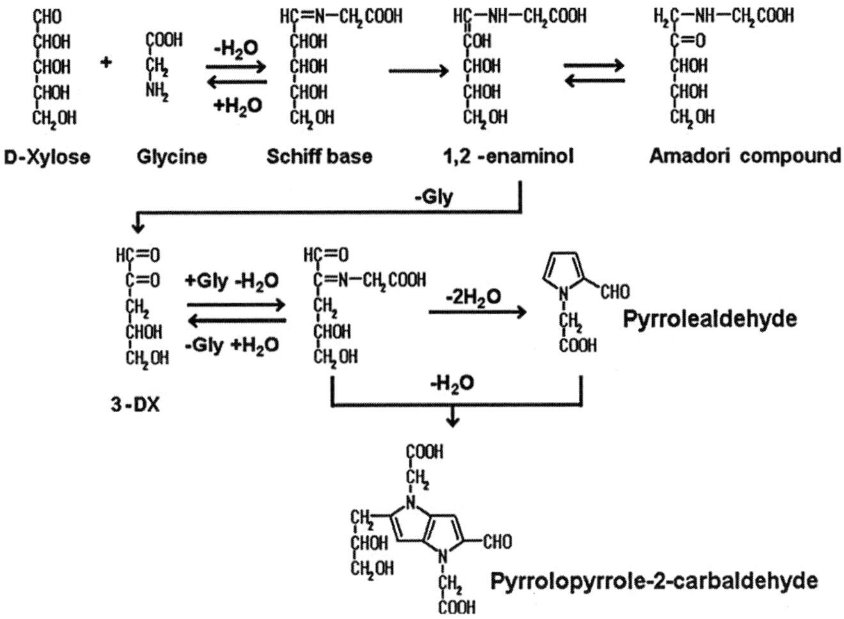

SCHEME 6.14 Proposed pathway for the formation of pyrrolopyrrole-2-carbaldehyde (PPA) *(Hayase et al., 2008).*

compound then reacts with glycine to form pyrrole-2-carbaldehyde, which then reacts with imine to form pyrrolopyrrole-2-carbaldehydes (PPA). PPA is the precursor of the blue pigment-M1, an intermediate pigment formed by decarboxylation of two molecules of PPA. The blue pigments then readily polymerize to the brown-colored melanoidins. Similar blue pigments were also detected in glucose—glycine and D-xylose—β-alanine model systems.

VIII. CARAMELIZATION

Caramelization is another example of non-enzymatic browning involving the degradation of sugars and generally proceeds simultaneously with the Maillard reaction (Ajandouz and Puigserver, 1999). Caramelization of sugars contributes markedly to the production of brown pigments and may lead to an overestimation of the Maillard reaction and its associated properties in foods (Ajandouz *et al.*, 2001). This was evident by the more rapid disappearance of fructose in a lysine-containing fructose solution compared to the amino acid. Caramelization of fructose accounted for 10—36% of total brown development between pH 4.0 and 7.0. When sugars are heated above their melting points they darken to a brown coloration under alkaline or acidic conditions. If this reaction is not carefully controlled it can lead to the production of unpleasant, burned, and bitter products. Consequently, it is important to control this reaction during food processing while still retaining the pleasant qualities of caramel. Caramel colors used for coloring foods vary in color from very dark brown to black, syrup-like liquids or powders. In 1987, von Smolnik classified caramel colors into four distinct groups based on differences in functional properties, as shown in Table 6.4. The different caramel colors are prepared by heating food-grade carbohydrates with ammonia, sulfite, and/or acid and alkaline reactants (Licht *et al.*, 1992a). Of these, caramel color IV accounts for approximately 70% of all caramel colors produced worldwide (Licht *et al.*, 1992b). This classification is recognized by both the Joint FAO/WHO Expert Committee on Food Additives (JECFA) and the US Food and Drug Administration definition of Caramel, Code of Federal Regulations. Caramel remains one of the most widely used coloring agents in food and pharmaceuticals.

The chemical composition of caramel is extremely complex and still poorly understood, although caramels produced from different sugars all show similarity in composition. Bryce and Greenwood (1963), using chromatographic techniques, found that pyrolysis of sucrose, glucose, and starch all produced caramels of similar composition. Heyns and Klier (1968), in a series of studies on a whole group of different monosaccharides, disaccharides, and polysaccharides, also found that the volatile products formed at high temperatures were almost identical. Studies by these and other researchers, as reviewed by Feather and Harris (1973), clearly indicated a common pathway for both the acidic and alkaline degradation of sugars. Later research identified new groups of compounds formed during this process (Popoff and Theander, 1976; Theander, 1981).

TABLE 6.4 Classification of Caramel Colors for Food Use

Classification	Food Use
I. Plain (alcohol) caramel	Spirits
II. Caustic sulfite caramel	Spirits
III. Ammonia caramel	Beer
IV. Sulfite ammonia caramel	Soft drinks

From von Smolnik (1987).

A. Acidic Degradation

The first step involves the stepwise conversion of D-glucose to D-fructose and D-mannose, referred to as the Lobry de Bruyn—Alberda van Eckenstein transformation (Scheme 6.15). These transformations can be mediated by organic acid catalysts over a pH range of 2.2—2.9 (Hodge and Osman, 1976). The interconversion of these sugars occurs primarily through the 1,2-enolic form and depends on the ease with which the ring opens. Since D-glucose is the most conformational stable form in both acid and alkaline media there is much less of the carbonyl (open-chain) form present in solution. This explains the presence of relatively high levels of glucose when fructose is heated at high temperatures over a pH range of 3.0—6.9, whereas only trace amounts of fructose are found when D-glucose is heated under identical conditions. Enolization takes place very slowly under acidic conditions, whereas the hydroxyl group next to the carbonyl group is rapidly removed.

The process of enolization via 1,2-enediol of sugars under acidic conditions was questioned when Ohno and Ward (1961) reported the presence of only small amounts of fructose when D-glucose was treated with 2.5% sulfuric acid, with no mention of mannose. Mawhinney *et al.* (1980) detected the presence of both fructose and mannose when D-glucose was isomerized in acidic solution. Table 6.5, taken from their data, shows that fructose levels off at around 0.8 µg, while mannose increased with reaction time. These data show the levels of sugars generated but give no information on their degradation by dehydration, which probably occurred at different

SCHEME 6.15 The Lobry de Bruyn—Alberda van Eckenstein transformation *(Eskin et al., 1971).*

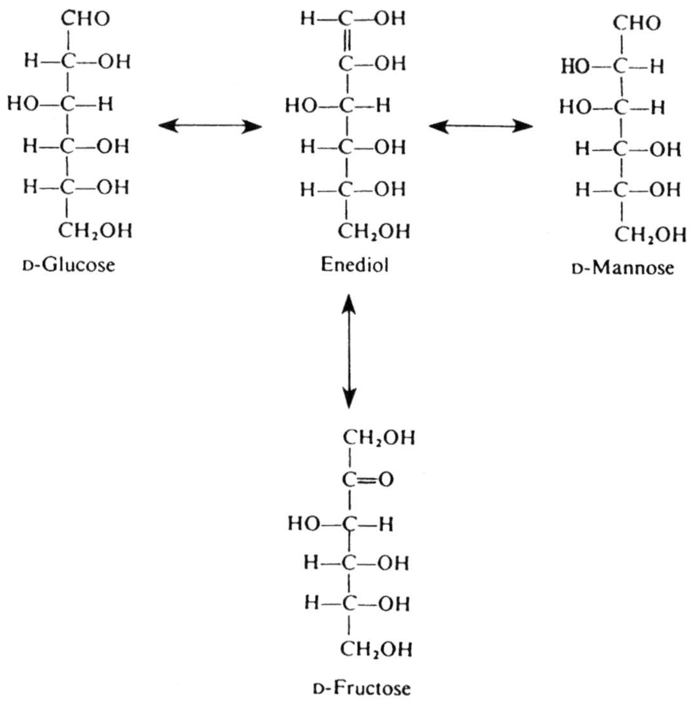

TABLE 6.5 Yield of Sugar Obtained from 50 mg of D-Glucose after Treatment with 2.5% Sulfuric Acid at 120°C

Reaction Time (hours)	Glucose (µg)	Mannose (µg)	Fructose (µg)
0	50.0	0	0
1.0	46.2	4.7	0.6
2.5	43.6	10.8	0.8
5.0	39.7	19.1	0.8
7.5	37.1	27.7	0.9
10.0	35.9	43.8	0.8

From Mawhinney et al. (1980).

reaction rates. The mechanism of this reaction resembled the corresponding isomerase enzyme reaction involving a C1 → C2 intramolecular hydrogen transfer in which D-glucose-2-H is converted to D-fructose-1-H (Harris and Feather, 1973, 1975):

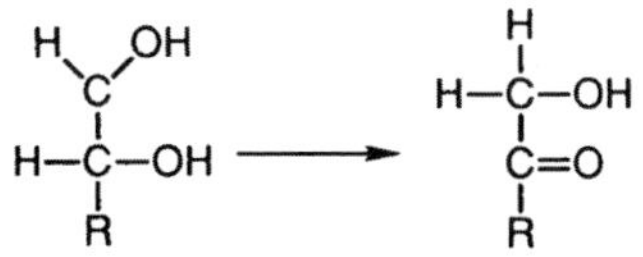

Continued heating results in the dehydration of sugars, leading to the formation of hydroxymethylfurfural, levulinic acid, and humin. This process is initiated by the removal of a hydroxyl group from the 1,2-enediol form located in the α position to the carbonyl group. The initial product, a dicarbonyl, undergoes further degradation. The postulated intermediates in this reaction were thought to be 3-deoxyaldos-2-ene,3-deoxyosulose, and osulos-3-ene (Isbell, 1944; Wolfrom et al., 1948). These compounds were isolated by Anet (1962) during the acidic degradation of fructose. If the initial sugar was a pentose, the final product was 2-furaldehyde. For example, D-xylose yielded approximately 93% 2-furaldehyde, although the yields from other pentoses were much lower. The dehydration rate of D-glucose was reported to be approximately one-fortieth of that observed for D-fructose, with considerably lower product yields (Kuster and Van der Bean, 1977). The mechanism of sugar dehydration from 1,2-enol to 5-(hydroxymethyl-2-furaldehyde) originally described by Anet (1964) has since been modified. This resulted from work by Feather et al. (1972) using isotope exchange in which D-xylose in tritiated water was found to be converted to 2-furaldehyde. By monitoring the proportion and amount of isotope converted into 2-furaldehyde they established the existence of an aldose—ketose 1,2-enediol equilibrium as incorporation at C1 of the sugar corresponded to the α-carbon of the 2-furaldehyde. Formation of 3-deoxyglyculose as an intermediate, however, should result in isotope incorporation at the C3 position in the furan ring. The absence of any isotope exchange eliminated any 1,2-enediol equilibrium during the reaction, thus making 1,2-enediol the rate-limiting step. It also eliminated 3-deoxyglyculose as an intermediate and supported the reaction sequence shown in Scheme 6.16.

Isomaltol and 2-(hydroxyacetyl)furan are formed during the acid treatment of D-fructose, suggesting that 2,3-enediol is the precursor:

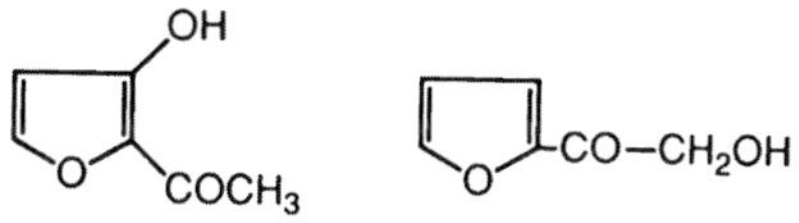

Isomaltol   2-Hydroxyacetyl furan

Their formation involved removal of a hydroxyl at C4 and C1 followed by dehydration of the furanone ring. Using tritium-labeled D-glucose, D-fructose, and D-mannose, Harris and Feather (1975) found that D-fructose underwent acid-catalyzed dehydration and degradation at a much faster rate than the aldose sugars. Among the major products detected were 5-(hydroxymethyl)-2-furaldehyde, 2-(hydroxyacetyl)furan, and levulinic acid. The

272

SCHEME 6.16 Mechanism of sugar dehydration from 1,2-enediol to 5-hydroxymethyl-2-furaldehyde. *(Adapted from Feather and Harris, 1973.)*

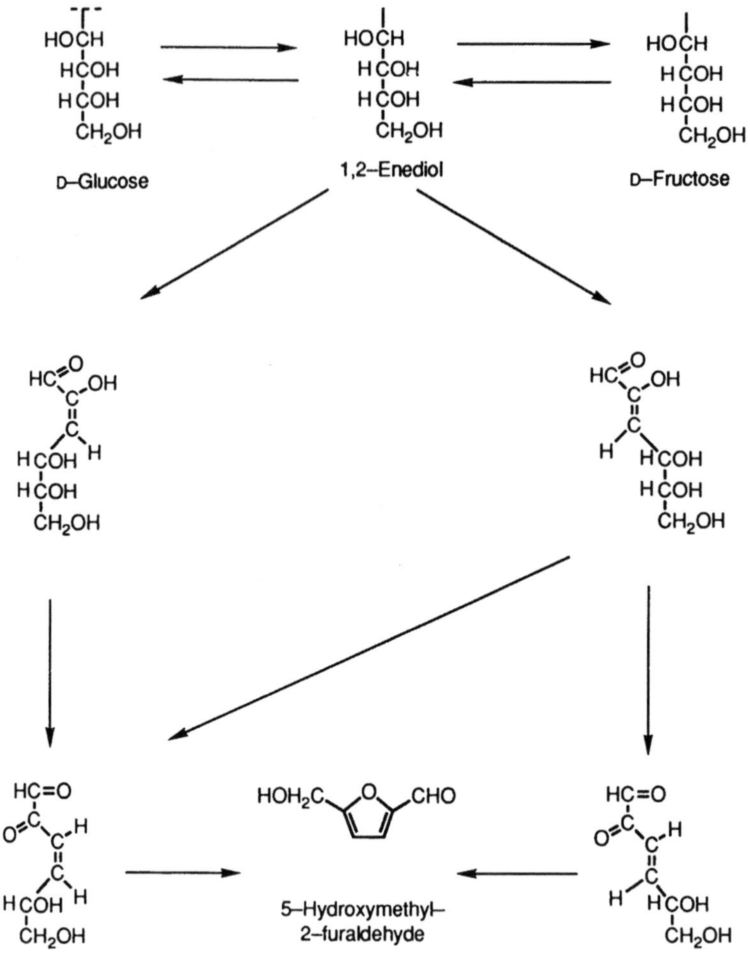

difference in rates of degradation explained, in part, why the levels of fructose observed by Mawhinney *et al.* (1980) remained low and constant during the isomerization of acidified D-glucose. Kuster and Temmink (1977) investigated the influence of pH and weak acid anions on the dehydration of D-fructose but were unable to detect 5-hydroxymethyl-2-furaldehyde (HMF) formation from D-fructose at pH > 3.9, while at pH > 2.7 no levulinic acid was formed. Isomerization of D-fructose to D-glucose was observed at pH above 4.5. It was apparent that the formation of HMF by dehydration of D-fructose and rehydration of HMF to levulinic and formic acids were catalyzed by acids, the latter requiring greater acidity. The formation of HME, one of the major caramelization products produced over a pH range of 6.0–6.7, is presumably a precursor of the pigment. In addition to HMF, a minor product, 2-(2-hydroxyacetyl)furan, also appeared to be formed by 2,3-enolization instead of 1,2-enolization of D-fructose.

The pyrolysis of sucrose was reported by Johnson *et al.* (1969) to produce maltol. The formation of maltol and isomaltol, together with ethyl lactate, furfural, 3-hydroxypropionic acid, 5-hydroxymethylfurfural, levulinic acid, and 2-furoic acid, was later detected by Ito (1977) when an aqueous solution of sucrose at pH 2.3 was heated to 120°C. The initial step in the acid-catalyzed browning of sucrose is hydrolytic cleavage with release of the constituent monosaccharides. Several studies have shown that this can occur in freeze-dried sucrose systems at 37°C, where the monolayer of absorbed water is involved in the hydrolysis (Karel and Labuza, 1968; Schoebel *et al.*, 1969). Flink (1983) monitored the development of non-enzymatic browning in sucrose-bound systems during freeze-drying and storage at room temperature. Hydrolysis of sucrose to glucose and fructose occurred following the primary sublimation stage of drying. The increased production of HMF, measured by monitoring absorbance at 280 nm, was evident at the end of the primary freeze-drying process. Storage of the freeze-dried samples over an a_w range of 0–0.40 showed a reduction in absorbance in the presence of increased levels of water due to the slowing down of the reaction. The reduction in weight at a_w of 0 was a clear indication that water can be produced in the browning reaction. Following production of HMF, a brown color developed which was monitored at 400 nm. The browning reaction was attributed to the increase in hydrogen ion concentration taking place with passage of ice interface during freeze-drying. Increase in temperature was accompanied by a rapid hydrolysis of sucrose to glucose and fructose in which fructose rapidly underwent dehydration. This study explained the stability of products during freeze-drying and the storage needs of high-acid foods containing sugar, such as fruit juices.

B. Alkaline Degradation

The initial reaction in the degradation of sugars under alkaline conditions follows the Lobry de Bruyn–Alberda van Eckenstein transformation via the 1,2- and 2,3-enediol. As discussed previously, enolization is a general reaction for carbonyl compounds with an α-hydrogen atom. Alkalis are much more effective catalysts than acids for the enolization of sugars (Pigman and Anet, 1972). Under mild alkaline conditions the series of reactions shown in Scheme 6.17 takes place. Under strong alkaline conditions, continuous enolization progresses along the carbon chain, resulting in a complex mixture of cleavage products including saccharinic acids. The formation of metasaccharinic acid is detailed in Scheme 6.17.

A particular feature of the alkaline degradation of hexoses is the extensive fragmentation which occurs, resulting in the production of 2- and 4-carbon fragments, including saccharinic acids, lactic acid, and 2,4-dihydroxybutyric acid (Harris, 1972; Feather and Harris, 1973). The recombination of some of these fragments accounts for the formation of a variety of compounds, including 2,4-dihydroxybutyric acid (Harris, 1972). A detailed discussion of fragmentation and recombination reactions of sugars under alkaline conditions is covered by Feather and Harris (1973).

C. Aromatic Compounds

A number of cyclic compounds were isolated among the products formed when aqueous solutions of D-glucose and D-fructose were heated to 160°C at pH 4.5. While these compounds were similar for both hexoses, the yields were

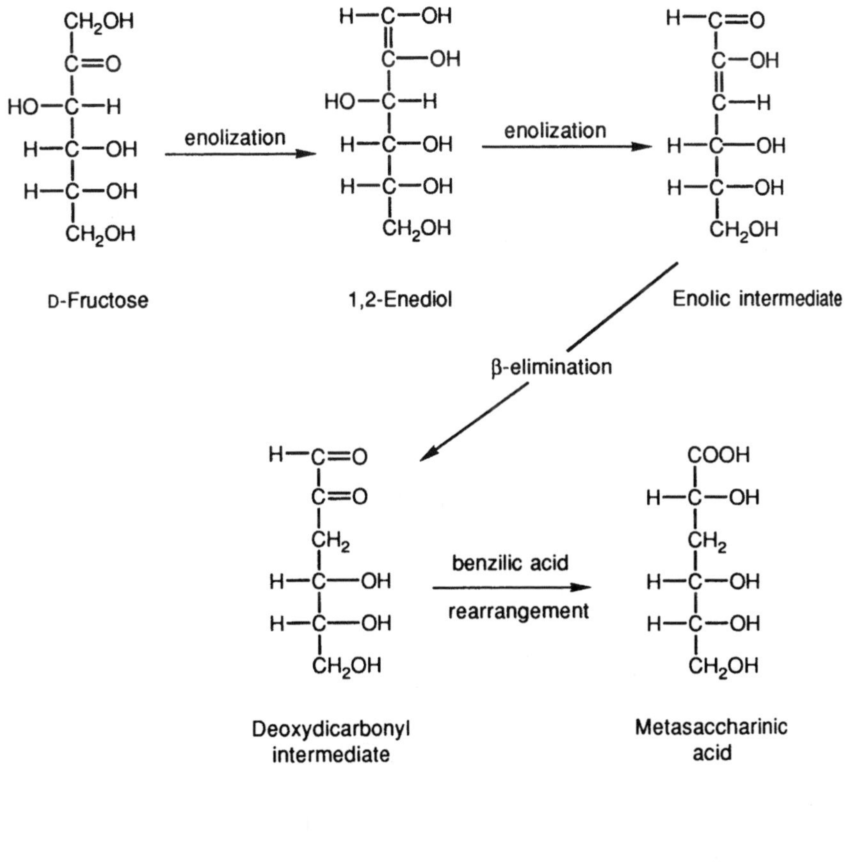

SCHEME 6.17   Alkaline degradation. *(Adapted from Feather and Harris, 1973.)*

much lower in the case of D-glucose. The major compound formed was hydroxymethylfurfural, although a large number of phenolic compounds were also produced.

The predominant phenol detected during the acid degradation of hexoses was isobenzene furanone, whereas chromone alginetin was the major product formed from pentoses and hexuronic acids (Theander, 1981):

Isobenzene furanone   Alginetin

This difference in specificity was not evident, however, when sugars were degraded under alkaline conditions. Forsskahl et al. (1976) noted a similarity in the pattern of phenolics formed under alkaline or neutral conditions. The compounds identified included a number of cyclic enols and phenols. The low yields obtained for these compounds reflected their instability under alkaline conditions. The formation of cyclopentones was reported earlier by Shaw et al. (1968) from the alkaline treatment of D-fructose. These compounds were isolated among the aroma components of roasted coffee by Gianturco et al. (1963) and had a strong caramel-like odor. The only common phenolic compounds identified from either the acid or alkaline treatments of glucose were catechol, 4-methyl-1,2-benzene diol, and 3,4-dihydroxy benzaldehyde (Popoff and Theander, 1976).

The development of color is extremely complex and involves a series of polymerization reactions. Theander (1981) reported that reductic acid and catechols were much more active color producers than furfurals.

D. Difructose Anhydrides: A Caramelization Tracer

The condensation of two fructose molecules via caramelization was reported to occur during the heating of sugars or food products rich in sugars, leading to the formation of a major non-volatile fraction containing difructose anhydrides (Defaye et al., 2000). Difructose anhydrides, which are pseudodisaccharides, were first reported by Tschiersky and Baltes (1989), with later structural studies by Defaye and Garcia Fernandez (1994, 1995). They are monitored as a tracer for caramelization in food and food additives as caramels, chicory, and dehydrated foods (Defaye et al., 2000). Montilla et al. (2006) reported that difructrose anhydrides could be used as an indicator of coffee quality as well as to test the authenticity of honey.

IX. ASCORBIC ACID OXIDATION

The browning of citrus juices and concentrates also involves Maillard-type reactions between amino acids and sugars present in citrus products. This was confirmed by Clegg (1969), who demonstrated improved color stability of lemon juice following the removal of amino nitrogen by cation-exchange resins. A patent was subsequently registered by Huffman in 1974 based on improved flavor stability when orange concentrate was treated with cation-exchange resins. The acceleration of browning by the addition of amino acids to model systems containing citrus confirmed their role in browning (Curl, 1949; Joslyn, 1957; Clegg, 1964). A review of citrus browning by Handwerk and Coleman (1988) suggested that the Maillard reaction was initiated in citrus juice by the formation of hexosamines from amino acids and sugars. The involvement of ascorbic and dehydroascorbic acids occurred at a later stage in this process via the formation of α-dicarbonyls, similar to that formed during the degradation of sugars. A recent paper by Schulz et al. (2007) used electrospray ionization mass spectrometry to simultaneously detect α-dicarbonyls formed from the thermal treatment of L-ascorbic acid. The compounds identified were glyoxal, methyl glyoxal, diacetyl, 3-deoxy-L-pentosone, and L-threosone. This section will focus on the degradation of ascorbic acid in citrus products in addition to their role, together with amino compounds, in the browning of dehydrated cabbage.

Ascorbic acid plays a central role in the browning of citrus juices and concentrates, for example, lemon and grapefruit. The reaction of ascorbic acid in fruit juices and concentrates is very much dependent on pH, as the browning process is inversely proportional to pH over a range of 2.0–3.5 (Braverman, 1963). Juices with a higher pH are much less susceptible to browning, for example, orange juice at a pH of 3.4. Below pH 4.0, browning is due primarily to decomposition of ascorbic acid to furfural (Huelin, 1953; Huelin et al., 1971).

The degradation of ascorbic acid was investigated by Herrmann and Andrae (1963), who identified 17 decomposition products, including dehydroascorbic acid and 2,3-diketogulonic and oxalic acids:

```
O=C┐              COOH
 |  |              |
 C=O │            C=O
 |   │O           |
 C=O │            C=O          COOH
 |   │            |            |
HC───┘            HCOH         COOH
 |                |
HOCH             HOCH
 |                |
CH₂OH            CH₂OH
```

Dehydroascorbic   2,3-Diketogulonic   Oxalic acid
acid   acid

Otsuka *et al.* (1986) identified a degradation product of 2,3-diketogulonic acid by preparative HPLC. The structure of this compound appeared to be the 3,4-enediol form of 2,3-diketogulono-δ-lactone. It was extremely unstable and developed intense brown coloration under mild temperature conditions. These researchers considered 3,4-enediol to be important in the browning of ascorbic acid:

```
    ┌C=O
    │ |
    │ C=O
    │ |
  O │ C─OH
    │ ‖
    │ C─OH
    │ |
    └CH
       |
      CH₂OH
```

3, 4 Enediol

Ascorbic acid degradation can occur under both aerobic and anaerobic conditions. While the level of air in juice is kept as low as possible by the use of vacuum deaeration and live steam injection, there is still some dissolved oxygen in the juice (0.05%) (Nagy, 1980). Only after the oxygen has been used does anaerobic degradation of vitamin C occur, but at a much slower rate. Tatum *et al.* (1967) reported degradation products of ascorbic acid, half of which were identical to the non-enzymatic browning products found in dehydrated orange and grapefruit powders, that is, instant juices. The aerobic and anaerobic degradation of ascorbic acid is outlined in Scheme 6.18 (Bauernfriend and Pinkert, 1970). The dependency of vitamin C degradation on headspace oxygen was recognized by Bauernfriend (1953). Kefford (1959) reported that the oxidative degradation of ascorbic acid in canned, pasteurized juice occurred during the first few days until the free oxygen was utilized (Nagy and Smoot, 1977). Following this, anaerobic breakdown of ascorbic acid proceeds, but at a tenth of the rate. Improved stability of vitamin C was found in juice sold in tin cans compared to enamel-lined cans as a result of oxygen reacting with the tin and competing with ascorbic acid (Riester *et al.*, 1945). A recent study on lemon juice concentrate by Mazin *et al.* (2007) found that an increase in concentration and temperature increased the rate of degradation of ascorbic acid. In addition, they developed a direct equation that estimated the shelf-life of stored lemon juice based on a first order loss of ascorbic acid at any specified temperature and degradation ratio.

The browning of citrus juices, as discussed earlier, is not due solely to ascorbic acid. Amino acids are also involved via the Maillard reaction, depending on the pH of the juice and basicity of the amine. This is illustrated by the fact that the main degradation product of juices with pH below 4.0 is furfural (Huelin, 1953; Huelin *et al.*, 1971). At pH above 4.0 this pathway is inoperative and explains the discoloration of dehydrated vegetables, which also involves ascorbic acid. Ranganna and Setty (1968) found that the discoloration of dehydrated cabbage was due to Strecker degradation between ascorbic acid and amino acid. This was facilitated by interactions between the oxidized products of ascorbic acid, dehydroascorbic or 2,3-diketogulonic acids, and amino acids during the final stages of the drying process.

The formation of pyrazines from ascorbic acid and amino acids was reported under dry-roasting conditions (Adams and De Kimpe, 2009). The reaction between L-ascorbic acid and L-threonine/L-serine resulted mainly in the formation of pyrazines. The alkylpyrazines formed were 2-methylpyrazine, 2,5-diethylpyraine, 2-ethylpyrazine,

SCHEME 6.18 Possible vitamin C (ascorbic acid) degradation pathways. AA: ascorbic acid; DHA: dehydroascorbic acid; DKA: diketogulonic acid; HF: hydroxyfurfural *(Nagy, 1980).*

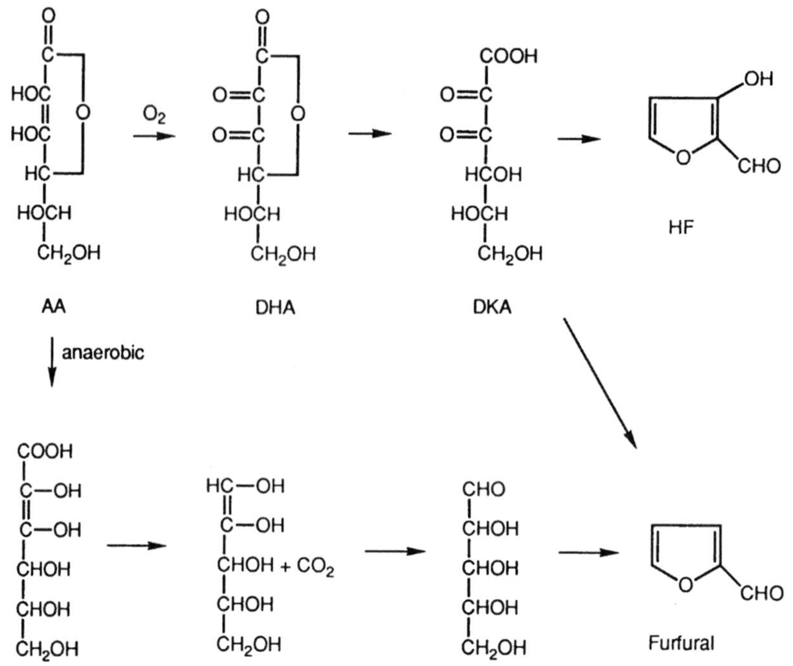

2-ethyl-6-methylpyrazine, 2-ethyl-5-methylpyrazine, 3-ethyl-2,5-dimethylpyrazine, 2,3-diethyl-5-methylpyrazine, and 3,5-diethyl-2-methylpyrazine (Yu and Zhang, 2010).

X. ANTIOXIDANT ACTIVITY OF NON-ENZYMATIC BROWNING PRODUCTS

The ability of MRPs to retard the development of rancidity has been reported by a number of researchers (Griffith and Johnson, 1957; Anderson *et al.*, 1963; Kato, 1973; Yamaguchi and Fujimaki, 1974; Lingnert, 1980). Most studies, including recent ones, have focused on the *in vitro* antioxidant effects of brown MRPs or melanoidins in model systems (Jing and Kitts, 2000; Morales and Jimenez-Perez, 2004; Hayase *et al.*, 2006; Gu *et al.*, 2009, 2010; Rao *et al.*, 2011). Antioxidant effects of melanoidins have also been reported in beer (Morales and Jimenes-Perez, 2004), roasted cocoa (Summa *et al.*, 2008), bread crusts (Michalska *et al.*, 2008), vinegar (Tagliazucchi *et al.*, 2010a), honey (Brudzynski and Miotto, 2011), and wine (Lopez de Lerma *et al.*, 2010). The antioxidant protective effects of melanoidins were attributed, in part, to their ability to chelate metals (Verzelloni *et al.*, 2010), while the anionic nature of melanoidins enabled them to chelate transition metals (Daglia *et al.*, 2008; Rufian-Henares and de la Cueva, 2009; Tagliazucchi *et al.*, 2010b). Cosovic *et al.* (2010) proposed that the nitrogen atoms in melanoidins may chelate Cu. In addition to chelating metal ions, the ability of melanoidins to scavenge free radicals appears to be another mechanism explaining their antioxidant properties. A strong linear relationship was reported between the color intensity of the melanoidin fraction and their ability to scavenge such free radicals as peroxyl (Gomez-Ruiz *et al.*, 2008) and 2,2′-azobis-2-methyl-propanimidamine, dichloride (AAH) (Brudzynski and Miotto, 2011). This is shown in Figure 6.11 by the linear relationship between browning (as expressed as a_{420}) and the ability of melanoidins to scavenge peroxyl radicals, expressed as R_{inh} (Morales and Jimenez-Perez, 2004). Some studies were unable to find any linear correlation between radical-scavenging activity and browning (Morales and Jimenez-Perez, 2004; Morales, 2005). However, this was attributed to the complex nature of the components in melanoidins, as well as the assay methods used to measure radical-scavenging activity (Wang *et al.*, 2011).

Using a coffee MRP model system, made up of arabinose and serine (C), Liu and Kitts (2011) showed the Maillard reaction to be the main contributor to the antioxidant activity in roasted coffee beans (RC). The formation of MRPs in roasted coffee beans, non-roasted coffee beans (NRC), and the arabinose—serine model system (C) was measured by fluorescence, ultraviolet—visible spectra and tristimulus Hunterlab measurements. The relative antioxidant activities of RC, NRC, and C were monitored by the oxygen radical absorbance capacity and Trolox equivalent antioxidant capacity methods and reducing power (Rp), while total chlorogenic acid content was

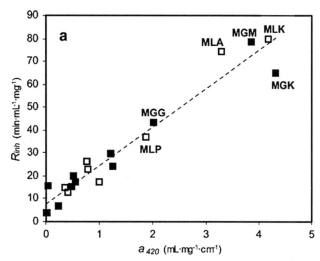

FIGURE 6.11 Relationship between peroxyl radical scavenging activity of melanoidins, expressed as R_{inh}, and browning (expressed as a_{420}), for glucose melanoidins (■) and lactose melanoidins (□). The dotted line shows the linear regression for melanoidins from model systems *(Morales and Jimenez-Perez, 2004).*

determined in RC and NRC extracts. The ultraviolet absorption (250–700 nm) spectrum for RC was similar to the model MRP extracts (C) but quite different from NRC (Figure 6.12). The NRC spectrum, however, was similar to the chlorogenic acid standard but its absence in RC confirmed its almost complete destruction during the roasting process (Budryn *et al.*, 2009). While the fluorescent spectrum for RC was similar to the control (C), no fluorescence was observed in the NRC extract, owing to the absence of a Maillard reaction and the lack of formation of MRPs. The high antioxidant activity associated with the NRC extract was attributed to the natural phenolics, which were substantially lost in the roasted sample. Consequently, any significant antioxidant activity in RC was attributed primarily to the MRPs formed during the roasting of the coffee beans and their similarity to the MRPs produced in the model system.

XI. INHIBITION OF NON-ENZYMATIC BROWNING

A major concern of food technologists is to control or minimize non-enzymatic browning reactions in food processing. The particular method used must be adapted to each food product. Various methods have been proposed for controlling these reactions.

A. Temperature

An increase in temperature or time of heat treatment accelerates the rate of these reactions (Labuza and Baisier, 1992). Thus, lowering the temperature during processing and storage can lengthen the lag phase, that is, the period needed for the formation of brown-colored products. Most foods will not brown below −10°C during normal storage (Nursten, 2005).

B. Moisture Content

The dependency of browning reactions on moisture content provides a convenient method for control. A reduction in moisture content in solid food products by dehydration reduces the mobility of reactive components (Loncin *et al.*, 1968; Eichner and Karel, 1972; Labuza and Saltmarch, 1981; Fox *et al.*, 1983). However, the rate of browning often exhibits a maximum with a moisture content of 5–30%, so that partial dehydration may make browning worse rather than better (Nursten, 2005). In the case of solutions, an increase in water activity will diminish the reaction velocity. Not only does this dilute the effect of the reactants but water also represents the first reaction product of the condensation step in the Maillard browning reaction. This is illustrated in Figure 6.13, where the browning rates of an avicel–glucose–glycine system were low at both high and low a_w (McWeeny, 1973).

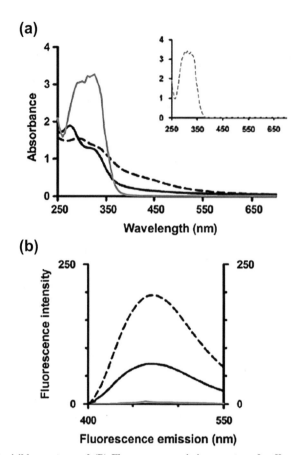

FIGURE 6.12   (A) Ultraviolet (UV) visible spectra and (B) Fluorescence emission spectra of coffee and model Maillard reaction products (MRPs). The UV spectra of 0.1 mg/ml chlorogenic acid (CGA) are also shown (dashed line; A insert). Samples were diluted to 0.5 mg/ml with distilled deionized water before UV spectrum measurement and diluted to 0.25 mg/ml with milli-Q water before fluorescence spectrum measurement. Non-roasted coffee beans (NRC) (solid gray line); roasted coffee beans (RC1) (solid black line); model Maillard reaction product (MRP-1) (dashed line). The spectra patterns of RC1 and RC2 are similar, so only the spectra of RC1 are shown in the figure *(Liu and Kitts, 2011)*.

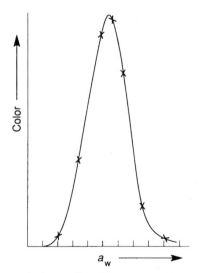

FIGURE 6.13   Browning of avicel−glucose−glycine as affected by water activity (a_w) after 8 days at 38°C *(McWeeney, 1973)*.

C. pH

The Maillard reaction is generally favored under more alkaline conditions, so lowering the pH provides a useful method of control (Fox *et al.*, 1983; Nursten, 2005). This method has been used in the production of dried egg powder in which acid is added prior to the dehydration process to lower the pH. The pH is restored by the addition of sodium bicarbonate to the reconstituted egg.

D. Gas Packing

Gas packing excludes oxygen by packing under an inert gas. This reduces the formation of lipid oxidation products capable of interacting with amino acids. The exclusion of oxygen is thought to affect those reactions involved in the browning process and not the initial carbonylamino reaction step.

E. Biochemical Agents

Removal or conversion of one of the reactants in the sugar—amino acid interactions forms the basis of the biochemical method. For instance, in the commercial production of egg white, glucose is removed by yeast fermentation prior to drying. The direct application of enzymes such as glucose oxidase and catalase mediates the conversion of glucose to gluconic acid, which is no longer capable of combining with amino acids. This enzyme has been used for many years to remove glucose from egg before spray-drying (Lightbody and Fevold, 1948). Glucose oxidase has the additional advantage of removing any residual oxygen and is used to reduce headspace oxygen during the production of bottled products.

F. Chemical Inhibitors

A variety of chemical inhibitors has been used to limit browning during the production and storage of a number of different food products. The most widely used are sulfur dioxide and sulfites, although thiols, calcium salts, and aspartic and glutamic acids have also been studied. The use of thiols as inhibitors is limited because of their unpleasant properties.

1. Sulfur Dioxide/Sulfites

Sulfur dioxide is unique in its ability to inhibit the Maillard reaction and can be applied as a gas or in solution as sulfite or bisulfite. Sulfur dioxide is not only capable of partially bleaching chromophores that have already formed but also inhibits color formation at the beginning of the reaction (McWeeny, 1984). The mechanism involves the binding of sulfur dioxide/sulfite with glucose to form hydroxy-sulfonate and other compounds from which sulfur dioxide/sulfite can be reversibly released. This results in the blocking of the carbonyl group of the sugar, rendering it unavailable for interaction in the typical Maillard reaction. As the reaction proceeds, sulfur dioxide/sulfite becomes irreversibly bound. This permits monitoring of the progress of browning by measuring the amount of sulfur dioxide or sulfite in the bound or free form. For example, color development for a glucose—glycine model system was not observed by McWeeny (1969) until all the free sulfite was depleted, while the ratio of bound to free sulfite increased as the reaction proceeded (Figure 6.14).

Inhibition of non-enzymatic browning by sulfite appears to involve the formation of stable sulfonates. In the case of ascorbic acid browning, 3-deoxy-4-sulfopentulose is formed with the corresponding 6-carbon compound, 6-deoxy-4-sulfohexulose, for the sulfited inhibition of the Maillard reaction (Knowles, 1971; Wedzicha and McWeeny, 1974a, b). Wedzicha and McWeeny (1975) monitored the formation of several organic sulfonates from sulfited foods. These compounds possess a dicarbonyl group, which makes them particularly reactive and at elevated temperatures can lead to the formation of sulfur compounds with other food components. McWeeny (1984) pointed out that future research on the role of sulfur dioxide and sulfites in inhibiting the Maillard reaction should focus on the nature of those precursors with which they react. In addition, further information is needed on the effect of time, temperature, pressure, pH, and additives on the quantitative and qualitative formation of compounds formed in foods to which sulfur dioxide or sulfite has been added. This work is particularly important in light of the current trend towards limiting intake of sulfur dioxide and sulfites in foods.

FIGURE 6.14 Color production and loss of sulfur dioxide (SO₂) from glucose–glycine–SO₂ during incubation at 55°C. (● ● ●) free SO₂; (● − ● −) bound SO₂; (− −) total SO₂; (—) absorbance at 490 nm (McWeeny, 1969). *(Copyright © by the Institute of Food Technologists.)*

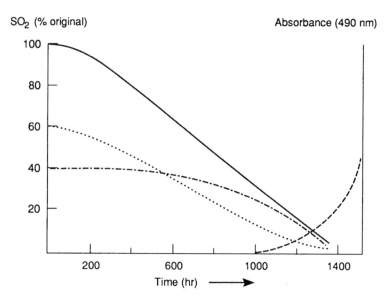

2. Aspartic and Glutamic Acids

A study by Nafisi and Markakis (1983) indicated the potential of aspartic and glutamic acids for inhibiting the Maillard browning reaction. Using model systems containing lysine–glucose/lysine–fructose (pH 8.0, 60°C for 58 hours), these researchers found that L-aspartic acid or L-glutamic acid decreased the rate of browning, as seen in Table 6.6. Dipping specially prepared potato chips into either aspartic or glutamic acid solutions before frying was accompanied by less darkening as measured by Hunter L values. This corresponded to a mean value of 38.4 for the untreated potato chips compared to 46.4 and 43.7 for the potato chips dipped in aspartic and glutamic acids, respectively.

While the various chemical inhibitors discussed can, with varying degrees of success, limit the progress of browning, the nutritional value of the food could still be reduced. For instance, the initial stage of the Maillard reaction, the carbonylamino step, could still render the amino acids unavailable without any visible browning during this stage. It is extremely difficult to ensure that this step is prevented.

TABLE 6.6 Hunter L Values of Control and Treated Potato Chips[a]

Replicate	Control	Dipped in Aspartic Acid	Dipped in Glutamic Acid
1	39.0[b]	47.6	44.2
2	38.3	44.6	44.7
3	38.2	45.3	43.0
4	39.0	47.0	43.0
5	37.9	46.8	43.7
6	36.7	46.6	43.9
7	39.5	47.2	43.6
Mean	38.4 ± 0.57[c]	46.4 ± 1.00	43.7 ± 0.57

[a] The treatment consisted of freeze-drying potato slices and dipping them in 0.04 M aspartic or glutamic acid before frying.
[b] Each L value is the average of three readings obtained by rotating the sample at 120° angles.
[c] The differences between control and treated samples are significant at the 99% probability level. The difference between treated samples is not significant.
Reproduced from Nafisi and Markarkis (1983).

REFERENCES

Aaslyng, M.D., Martens, M., Poll, L., Nielsen, P.M., Flyge, H., Larsen, L.M., 1998. Chemical and sensory characterization of hydrolyzed vegetable protein, a savory flavoring. J. Agric. Food Chem. 46, 481–489.

Abrams, A., Lowy, P.H., Borsook, H., 1955. Preparation of 1-amino-1-deoxy-2-keto hexoses from aldohexoses and amino acids. J. Am. Chem. Soc. 77, 4794–4796.

Adams, A., De Kimpe, N., 2006. Chemistry of 2-acetyl-1-pyrroline, 6-acetyl-1,2,3,4-tetrahydropyridine, 2-acetyl-2-thiazoline, and 5-acetyl-2,3-dihydro-4H-thiazine: Extraordinary Maillard flavor compounds. Chem. Rev. 106, 2299–2319.

Adams, A., De Kimpe, N., 2009. Formation of pyrazines from ascorbic acid and amino acids under dry-roasting conditions. Food Chem. 115, 1417–1423.

Aeschbacher, H.V., Chappus, C., Manganel, M., Aesbach, R., 1981. Investigation of Maillard products in bacterial mutagenicity test systems. Prog. Food Nutr. Sci. 5, 279–294.

Ajandouz, E.H., Puigserver, A., 1999. Nonenzymatic browning reaction of essential amino acids: effect of pH on caramelization and Maillard reaction kinetics. J. Agric. Food Chem. 47, 1786–1793.

Ajandouz, E.H., Tchiakpe, L.S., Dalle Ore, F., Benajiba, A., Puigserver, A., 2001. Effects of pH on caramelization and Maillard reaction kinetics in fructose–lysine model systems. J. Food Sci. 66, 926–931.

Ames, J.M., Guy, R.C.E., Kipling, G.J., 2001. Effect of pH, temperature, moisture on the formation of volatile compounds in glycine/glucose model systems. J. Agric. Food Chem. 49, 4315–4323.

Amrein, T.M., Andres, L., Escher, F., Amado, P., 2007. Occurrence of acrylamide in selected foods and mitigation options. Food Addit. Contam. 24, 13–25.

Anderson, R.H., Moran, D.H., Huntley, T.E., Holahan, J.L., 1963. Responses of cereals to antioxidants. Food Technol. 17, 115–120.

Anese, M., Suman, M., Nicoli, M.C., 2010. Acrylamide removal from heated foods. Food Chem. 119, 791–794.

Anet, E.F.L.J., 1960. Degradation of carbohydrates. I. Isolation of 3-deoxyhexosones. Aust. J. Chem. 13, 396–403.

Anet, E.F.L.J., 1962. Degradation of carbohydrates. III. Unsaturated hexosones. Aust. J. Chem. 15, 503–509.

Anet, E.F.L.J., 1964. 3-Deoxyglycosuloses (3-deoxyglycosones) and the degradation of carbohydrates. Adv. Carbohydr. Chem. 19, 181–218.

Anet, E.F.L.J., Reynolds, T.M., 1957. Chemistry of nonenzymic browning. 1. Reactions between amino acids, organic acids, and sugar in freeze-dried apricots and peaches. Aust. J. Chem. 10, 182–192.

Baltes, W., 1973. Nichtenzymatische Veraenderungen bei der Beund Verarbeitung von Lebens. Z. Ernaehrungswiss. Suppl. 16, 34–49.

Barbetti, P., Chiappini, I., 1976a. Fractionation and spectroscopic characterization of melanoidic pigments from a glucose–glycine non-enzymic browning system. Ann. Chim. (Rome) 66, 293.

Barbetti, P., Chiappini, I., 1976b. Identification of the furan group in the melanoidic pigments from a glucose–glycine Maillard reaction. Ann. Chim. (Rome) 66, 485.

Bauernfriend, J.C., 1953. The use of ascorbic acid in processing food. Adv. Food Res. 4, 359–431.

Bauernfriend, J.C., Pinkert, D.M., 1970. Food processing with ascorbic acid. Adv. Food Res. 18, 219–315.

Bell, L.N., 1997. Maillard reaction as influenced by buffer type and concentration. Food Chem. 59, 143–147.

Benzing-Purdie, L.M., Ripmeester, J.A., Ratcliffe, C.I., 1985. Effects of temperature on Maillard reaction products. J. Agric. Food Chem. 33, 31–33.

Blank, I., Devaud, S., Matthey-Doret, W., Robert, F., 2003. Formation of odorants in Maillard model systems based on L-proline as affected by pH. J. Agric. Food Chem. 51, 3643–3650.

Bobbio, P.A., Bobbio, F.O., Trevisan, L.M.V., 1973. Estudos sobre a meacao de Maillard. 1. Efeitos da temperaturia e do pH. An. Acad. Bras. Cienc. 45, 419.

Bobbio, P.A., Imasato, H., Leite, S.R., de, A., 1981. Maillard reaction. Preparation and characterization of melanoidins from glucose and fructose with glycine. An. Acad. Bras. Cienc. 53, 83.

Bohart, G.S., Carson, J.F., 1955. Effects of trace metals, oxygen, and light on the glucose–glycine browning reaction. Nature (London) 175, 470–471.

Braverman, J.B.S., 1963. Introduction to the Biochemistry of Foods. Elsevier, Amsterdam.

Brownlee, M., 2001. Biochemistry and molecular cell biology of diabetic complications. Nature (London) 414, 813–820.

Brudzynski, K., Miotto, D., 2011. The recognition that of high molecular melanoidins as the main components responsible for radical-scavenging capacity of unheated and heat-treated Canadian honeys. Food Chem. 125, 570–575.

Bruechert, L.J., Zhang, Y., Huang, T.C., Hartman, T.G., Rosen, R.T., Ho, C.T., 1988. Contribution of lipids to volatile generation in extruded corn-based model systems. J. Food Sci. 53, 1444–1447.

Bryce, D.J., Greenwood, C.T., 1963. Thermal degradation of starch. Staerke 15, 166–170.

Budryn, G., Nebesny, E., Podsedek, A., Zyzelewicz, D., Materska, M., Jankowski, S., Janda, B., 2009. Effect of different extraction methods on the recovery of chlorogenic acids, caffeine and Maillard reaction products in coffee beans. Eur. Food Res. Technol. 228, 913–922.

Buttery, R.G., Ling, L.C., 1974. Alkyl thiazoles in potato products. J. Agric. Food Chem. 22, 912–914.

Buttery, R.G., Ling, L.C., Juliano, B.O., 1982. 2-Acetyl-1-pyrroline: an important aroma component of cooked rice. Chem. Ind. (London) 12, 958–959.

Buttery, R.G., Ling, L.C., Juliano, B.O., Turnbaugh, J.G., 1983. Cooked rice aroma and 2-acetyl-1-pyrroline. J. Agric. Food Chem. 31, 823–826.

Carlin, J.T., Jin, Q.Z., Huang, T.C., Ho, C.T., Chang, S.S., 1986. Identification of alkyloxalzole in the volatile compounds from french-fried potatoes. J. Agric. Food Chem. 33, 621–623.

Chang, S.S., Hirai, C., Reddy, B.R., Herz, K.O., Kato, A., Sipma, G., 1968. Isolation and identification of 2,4,5-trimethyl-3-oxazoline and 3,5-dimethyl 1,2,4-trithiolane in the volatile flavour compounds of boiled beef. Chem. Ind. (London), 1639–1644.

Chiu, E.M., Kuo, M.C., Bruechert, L.J., Ho, C.T., 1990. Substitution of pyrazines by aldehydes in model systems. J. Agric. Food Chem. 89, 58–61.

Ciner-Doruk, M., Eichner, K.Z., 1979. Bildung und stabilitat von Amadori-Verbindungen in wasseramen lebensmitteln. Z. Lebensm.-Unters.-Forsch. 168, 9–20.

Clegg, K.M., 1964. Non-enzymic browning of lemon juice. J. Sci. Food Agric. 15, 878–885.

Clegg, K.M., 1969. Citric acid and the browning of solutions containing ascorbic acid. J. Sci. Food Agric. 17, 566.

Coleman, E.C., Ho, C.-T., Chang, S.S., 1981. Isolation and identification of volatile compounds from baked potatoes. J. Agric. Food Chem. 29, 42–48.

Cosovic, B., Vojvodic, V., Boskovic, N., Plavsic, M., Lee, C., 2010. Characterization of natural and synthetic humi substances (melanoidins) by chemical composition and adsorption measurements. Org. Geochem. 41, 200–205.

Curl, L.A., 1949. Ascorbic acid losses and darkening on storage at 69°C (120°F) of synthetic mixtures analogous to orange juice. Food Res. 14, 9–14.

Daglia, M., Papetti, A., Aceti, C., Sordelli, B., Gregotti, C., Gazzani, G., 2008. Isolation of high molecular weight components and contributions to the protective activity of coffee against lipid peroxidation in a rat liver microsome system. J. Agric. Food Chem. 56, 11653–11660.

Danehy, J.P., 1986. Maillard reaction: nonenzymic browning in food systems with special reference to the development of flavor. Adv. Food Res. 30, 77–138.

Dawes, I.W., Edwards, R.A., 1960. Methyl substituted pyrazines as volatile reaction products of heated aqueous aldose, amino acid mixtures. Chem. Ind. (London), 2203.

Defaye, J., Garcia Fernandez, J.M., 1994. Protonic thermal activation of sucrose and the oligosaccharide composition of caramel. Carbohydr. Res. 256, C1–C4.

Defaye, J., Garcia Fernandez, J.M., 1995. The oligosaccharide components of caramel. Zuckerindustrie 120, 700–704.

Defaye, J., Garcia Fernandez, J.M., Ratsimba, V., 2000. Les molécules de la caramélisation: structure et méthodologies détection et d'évaluation. L'Actualité Chim. 11, 24–27.

Dworschak, E., Hegedüs, M., 1974. Effect of heat treatment on the nutritive value of proteins in milk powder. Acta Aliment. Acad. Sci. Hung. 3, 337–347.

Dworshak, E., 1980. Nonenzymic browning and its effect on protein nutrition. CRC Crit. Rev. Food Sci. Nutr. 13, 1–40.

Eichner, D., 1981. Antioxidant effect of Maillard reaction intermediates. Prog. Food Nutr. Sci. 5, 441–451.

Eichner, K., Karel, M., 1972. The influence of water content and water activity on the sugar–amino browning reaction in model systems under various conditions. Food Chem. 20, 218–223.

Eichner, K., Ruetter, M., Wittman, R., 1994. In: Fiot, P.A., Aeschbacher, H.U., Hurrel, R.F., Liardon, R. (Eds.), Maillard Reaction in Food Processing, Human Nutrition and Physiology. Birkhauser, Basel, pp. 63–77.

Ellis, G.P., 1959. The Maillard reaction. Adv. Carbohydr. Chem. 14, 63–134.

Elmore, J.S., Mottram, D.S., 1997. Investigation of the reaction between ammonium sulfide, aldehydes, and α-hydroxyketones or α-dicarbonyls to form some lipid–Maillard interaction products found in cooked beef. J. Agric. Food Chem. 45, 3595–3602.

Elmore, J.S., Mottram, D.S., Enser, M.B., Wood, J.D., 1997. Novel thiazoles and 3-thiazolines in cooked beef aroma. J. Agric. Food Chem. 45, 3603–3607.

Eskin, N.A.M., Henderson, H.M., Townsend, R.J., 1971. Biochemistry of Foods, Academic Press, New York, p. 96.

Farmer, L.J., Mottram, D.S., Whitfield, F.B., 1989. Volatile compounds produced in Maillard reactions involving cysteine, ribose and phospholipid. J. Sci. Food Agric. 49, 347–368.

Feather, M.S., Harris, J.F., 1973. Dehydration reactions of carbohydrates. Adv. Carbohydr. Chem. 28, 161–224.

Feather, M.S., Huang, R.D., 1986. Some studies on a Maillard polymer derived from L-alanine and D-glucose. Dev. Food Sci. 13, 183–192.

Feather, M.S., Nelson, D., 1984. Maillard polymers derived from D-glucose, D-fructose, 5-(hydroxymethyl)-2-furaldehyde, and glycine and methionine. J. Agric. Food Chem. 32, 1428.

Feather, M.S., Harris, D.W., Nichols, S.B., 1972. Routes of conversion of D-xylose, hexuronic acids and L-ascorbic acid to 2-furaldehyde. J. Org. Chem. 37, 1606–1608.

Feeney, R.E., Blankenhorn, G., Dixon, H.B.F., 1975. Carbonylamine reactions in protein chemistry. Adv. Protein Chem. 29, 135–203.

Ferretti, A., Flanagan, C.P., 1973. Characterization of volatile constituents of an N-formyl-L-lysine–D-lactose browning system. J. Agric. Food Chem. 21, 35–37.

Ferretti, A., Flanagan, V.P., 1971. The lactose–casein (Maillard) browning system: volatile components. J. Agric. Food Chem. 19, 245–249.

Finot, P.A., Mauron, J., 1972. Le blocage de la lysine par la réaction de Maillard. II. Propriétés chimiques des dérivés N-(desoxy-1-D-fructosyl-1) et N-(desoxy-1-lactulosyl-1) de la lysine. Helv. Chim. Acta. 55, 1153–1164.

Finot, P.-A., 2005. Historical perspective of the Maillard reaction in food science. Ann. N. Y. Acad. Sci. 1043, 1–8.

Finot, P.-A., Bricout, J., Viani, R., Maurion, J., 1968. Identification of a new lysine derivative obtained upon acid hydrolysis of heated milk. Experientia 24, 1097–1099.

Flink, J.M., 1983. Nonenzymic browning of freeze-dried sucrose. J. Food Sci. 48, 539–542.

Forsskahl, I., Popoff, T., Theander, O., 1976. Reactions of D-xylose and D-glucose in alkaline aqueous solutions. Carbohydr. Res. 48, 13–21.

Fox, M., Loncin, M., Weiss, M., 1983. Investigations into the influence of water activity, pH and heat treatment on the velocity of the Maillard reactions in foods. J. Food Qual. 6, 103–118.

Friedman, M., 1982. Chemically reactive and unreactive lysine as an index of browning. Diabetes 31, 5–14.

Friedman, M., 2003. Chemistry, biochemistry, and safety of acrylamide. A review. J. Agric. Food Chem. 51, 4504–4526.

Gianturco, M., Giammarino, A.S., Pitcher, R.G., 1963. The structures of five cyclic diketones isolated from coffee. Tetrahedron 19, 2051–2059.

Gokmen, V., Senyuva, H.Z., 2007. Acrylamide formation is prevented by divalent cations during the Maillard reaction. Food Chem. 103, 196–203.

Gokmen, V., Serpen, A., Acar, O.C., Morales, F.J., 2008. Significance of furosine as heat-induced marker in cookies. J. Cereal Sci. 48, 843–847.

Gomez-Ruiz, J., Ames, J., Leake, D., 2008. Antioxidant activity and protective effects of green and dark coffee components against human low density lipoprotein oxidation. Eur. Food Res. Technol. 227, 1017–1024.

Griffith, T., Johnson, J.A., 1957. Relation of the browning reaction to storage stability of sugar cookies. Cereal Chem. 34, 159–169.

Gu, F.L., Kim, J.M., Abbas, S., Zhang, X.M., Xia, S.Q., Chen, Z.X., 2009. Characteristics and antioxidant activity of ultrafiltrated Maillard reaction products from a casein–glucose model system. Food Chem. 117, 48–54.

Gu, F.L., Kim, J.M., Abbas, S., Zhang, X.M., Xia, S.Q., Chen, Z.X., 2010. Structure and antioxidant activity of high molecular weight Maillard reaction products from casein–glucose. Food Chem. 120, 505–511.

Guan, Y.-G., Wang, J., Yu, S.-J., Zeng, X.-A., Han, Z., Liu, Y.-Y., 2010a. A pulsed electric field for promoting Maillard reaction in an asparagins–glucose model system. Int. J. Food Sci. Technol. 45, 1303–1309.

Guan, Y.-G., Wang, J., Yu, S.-Y., Xu, X.-B., Zhu, S.-M., 2010b. Effects of ultrasound intensities on a glycin—maltose model system — a means of promoting Maillard reaction. Int. J. Food. Sci. Technol. 45, 758—764.

Guerra-Hernandez, E., Corzo, N., Garcia-Vilanova, B., 1999. Maillard reaction evaluation by furosine determination during infant cereal processing. J. Cereal Sci. 29, 171—176.

Handwerk, R.L., Coleman, R.L., 1988. Approaches to the citrus browning problem: a review. J. Agric. Food Chem. 36, 231—236.

Harris, D.W., Feather, M.S., 1973. Evidence for a C-2→C-1 intramolecular hydrogen transfer during the acid-catalyzed isomerization of D-glucose to D-fructose. Carbohydr. Res. 30, 359.

Harris, D.W., Feather, M.S., 1975. Studies on the mechanism of the interconversion of D-glucose, D-mannose and D-fructose in acid solution. J. Am. Chem. Soc. 97, 178—181.

Harris, J.F., 1972. Alkaline decomposition of D-xylose-1-C, D-glucose-1-C and D-glucose-6-C. Carbohydr. Res. 23, 207—215.

Harris, R.L., Mattil, H.A., 1940. The effect of hot alcohol on purified animal proteins. J. Biol. Chem. 132, 477—485.

Hartman, G.J., Jin, Q.Z., Collins, G.J., Lee, K.N., Ho, C.-T., Chang, S.S., 1983. Nitrogen-containing heterocyclic compounds identified in the volatile flavor constituents of roast beef. J. Agric. Food Chem. 31, 1030—1033.

Hartman, G.J., Scheide, J.D., Ho, C.-T., 1984a. Effect of water activity on the major volatiles produced in a meat model system approximating cooked meat. J. Food Sci. 49 697—613.

Hartman, G.J., Scheide, J.D., Ho, C.-T., 1984b. Volatile products formed from a flavor model system at high and low moisture levels. Lebensm. Wiss. Technol. 17, 222—225.

Hashiba, H., 1978. Isolation and identification of Amadori compounds from soy sauce. Agric. Biol. Chem. 42, 763—768.

Hashiba, H., 1979. Oxidative browning of soy sauce. I. The roles of cationic fraction from soy sauce. Nippon Shoyu Kenkyusho Zasshi 5 (4), 169.

Hashiba, H., 1986. Oxidative browning of Amadori compounds color formation by iron with Maillard reaction products. Dev. Food Sci. 13, 155—164.

Hayase, F., Takahashi, Y., Tominaga, S., Miura, M., Gomyo, T., Kato, H., 1999. Identification of blue pigment in a D-xylose—glycine reaction system. Biosci. Biotechnol. Biochem. 63, 1512—1514.

Hayase, F., Usui, T., Watanabe, H., 2006. Chemistry and some biological effects of model melanoidins and pigments as Maillard intermediates. Mol. Nutr. Food Res. 50, 1171—1179.

Hayase, F., Usui, T., Shirahashi, Y., Machida, T., Ito, M., Nishitani, N., Shimohira, K., Watanabe, H., 2008. Formation mechanisms of melanoidin and fluorescent pyridinium compounds as advanced glycation end products. Ann. N. Y. Acad. Sci. 1126, 53—58.

Hayashi, T., Namiki, M., 1986. Role of sugar fragmentation: an early stage browning of amino—carbonyl reaction of sugar and amino acid. Agric. Biol. Chem. 50, 1965—1970.

Hayashi, T., Ohata, Y., Namiki, M., 1977. Electron spin resonance spectral study of the structure of the novel free radical products formed by the reactions of sugars with amino acids or amines. J. Agric. Food Chem. 25, 1282—1287.

Hendel, C.E., Silveira, V.G., Harrington, W.O., 1955. Rates of nonenzymatic browning of white potato during dehydration. Food Technol. 9, 433—438.

Henle, T., Walter, H., Krause, I., Klostermeyer, H., 1991a. Efficient determination of individual Maillard compounds in heat-treated milk products by amino acid analysis. Int. Dairy J. 1, 125—135.

Henle, T., Walter, H., Krause, I., Klostermeyer, H., 1991b. Evaluation of the extent of the early Maillard-reaction in milk products by direct measurement of the Amadori product lactulosylelysine. Z. Lebensm. Unters Forsch. 193, 119—122.

Herrmann, J., Andrae, W., 1963. Oxidative Abbauprodukte der L-Ascorbinsaure. I. Paper-chromatographischer Nachweis. Nahrung 7, 243.

Heyns, K., Breuer, H., 1958. Darstellung und Verhalten weiterer N-substituierter 2-Amino-2-desoxy-aldosen aus D-fructose und Aminosauren. Chem. Ber. 91, 2750—2762.

Heyns, K., Klier, M., 1968. Braunungsreaktionen und Fragmentierungen von kohlenhydaten. Carbohydr. Res. 6, 436—448.

Heyns, K., Noack, H., 1962. Die Umsetzung von D-Fructose mit L-Lysin und L-Histidine mit Hexosen. Chem. Ber. 95, 720—727.

Heyns, K., Noack, H., 1964. Die Umsetzung von L-Tryptophan und L-Histidine mit Hexosen. Chem. Ber. 97, 415—418.

Heyns, K., Paulsen, H., 1960. On the chemical basis of the Maillard reaction. Wiss. Veroff. Dtsch. Ges. Ernahr. 5, 15—44.

Hidalgo, F.J., Zamora, R., 2004a. Strecker-type degradation produced by the lipid oxidation product 4,5-(E)-heptenal. Chem. Res. Toxicol. 13, 501—508.

Hidalgo, F.J., Zamora, R., 2004b. Strecker-type degradation produced by the lipid oxidation products 4,5-epoxy-2-alkenals. J. Agric. Food Chem. 52, 7126—7131.

Hidalgo, F.J., Zamora, R., 2005. Interplay between Maillard reaction and lipid peroxidation in biochemical systems. Ann. N. Y. Acad. Sci. 1043, 319—326.

Hildago, A., Pompei, C., Zambuto, R., 1998. Heat damage evaluation during tomato products processing. J. Agric. Food Chem. 46, 4387—4390.

Hill, V.M., Ledward, D.A., Ames, J.M., 1996. Influence of high hydrostatic pressure and pH on the rate of browning in a glucose—lysine system. J. Agric. Food Chem. 44, 394—400.

Hill, V.N., Ames, J.M., Ledward, D.A., Royle, I., 1998. The use of capillary electrophoresis to investigate the effect of high hydrostatic pressure on the Maillard reaction. In: O'Brien, J., Nursten, H.E., Crabbe, M.J.C., Ames, J.M. (Eds.), The Maillard Reaction in Foods and Medicine. Royal Society of Chemistry, Cambridge, pp. 121—126.

Ho, C.T., Hartman, G.J., 1982. Formation of oxazolines and oxazoles in Strecker degradation of DL-alanine and L-cysteine with 2,3-butanedione. J. Agric. Food Chem. 30, 793—794.

Hodge, J.E., 1953. Chemistry of browning reactions in model systems. J. Agric. Food Chem. 1, 928—943.

Hodge, J.E., 1967. Origin of flavors in foods. Nonenzymic browning reactions. In: Schultz, H.W. (Ed.), Symposium on Foods: The Chemistry and Physiology of Flavors. Avi, Westport, CT, pp. 466—491.

Hodge, J.E., Osman, E.M., 1976. Carbohydrates. In: Fennema, O.R. (Ed.), Principles of Food Science. Part 1. Food Chemistry. Dekker, New York.

Hodge, J.E., Fisher, B.E., Nelson, E.C., 1963. Dicarbonyls, reductones and heterocyclics produced by the reactions of reducing sugars with secondary amine salts. Proc. Am. Soc. Brew. Chem., 84—92.

Hodge, J.E., Mill, F.D., Fisher, B.E., 1972. Compounds of browned flavor derived from sugar—amine reactions. Cereal Sci. Today. 17, 34—40.

Hoffmann, T., Schieberle, P., 1995. Studies in the formation and stability of the roast-flavor compound 2-acetyl-2-thiazoline. J. Agric. Food Chem. 43, 2946—2950.

Hofmann, T., 1998. 2-Alkylidene-2-imino-5-[4-alkylidene-5-oxo-1,3-imidazol-2-inyl]aza-methylidene-1,3-imidazolidine — a novel colored substructure in melanoidins formed by Maillard reactions of bound arginine with glyoxal and furan-2-carboaldehhyde. J. Agric. Food Chem. 46, 3896—3901.

Hogervorst, J.G., Schouten, L.J., Konings, E.J., Goldbohm, R.A., van den Brandt, P.A., 2007. A prospective study of the dietary acrylamide intake and the risk of endometrial, ovarian and breast cancer. Cancer Epidemiol. Biomarkers Prev. 16, 2304—2313.

Hogervorst, J.G., Schouten, L.J., Konings, E.J., Goldbohm, R.A., van den Brandt, P.A., 2008. Dietary acrylamide intake and risk of renal cell, bladder and prostate cancer. Am. J. Clin. Nutr. 87, 1428—1438.

Holtermand, A., 1966. The browning reaction. Staerke 18, 319—325.

Horak, F.P., Kessler, H.G., 1981. The influence of UHT heating sterilization on lysine in milk. Milchwissenschaft 36, 543—547.

Horn, J.M., Lichtenstein, H., Womack, M., 1968. Availability of amino acids. A methionine—fructose compound and its availability to microorganisms and rats. J. Agric. Food Chem. 16, 741—745.

Huang, T.C., Fu, H.Y., Ho, C.T., 1996. Mechanistic studies of tetramethyl pyrazine under weak acidic conditions and high hydrostatic pressure. J. Agric. Food Chem. 44, 240—246.

Huelin, P.E., 1953. Studies on the anaerobic decomposition of ascorbic acid. Food Res. 15, 78.

Huelin, P.E., Coggiola, I.M., Sidhu, G.S., Kennett, B.H., 1971. The anaerobic decomposition of ascorbic acid in the pH range of foods and in more acid solutions. J. Sci. Food Agric. 22, 540—542.

Huffman, C.F., 1974. Treatment of fruit juices with ion-exchange resins US Patent No. 3,801,717.

Hurrell, R.F., Carpenter, K.J., 1974. Mechanism of heat damage in proteins. IV. The reactive lysine content of heat-damaged material as measured in different ways. Br. J. Nutr. 32, 589—604.

Hurrell, R.F., Carpenter, K.J., 1981. The estimation of available lysine in foodstuffs after Maillard reaction. Prog. Food Nutr. Sci. 5, 159.

Hwang, H.-I., Hartman, T.G., Ho, C.-T., 1995. Relative reactivities of amino acids in the formation of pyridines, pyrroles, and oxazoles. J. Agric. Food Chem. 43, 2917—2921.

Imasato, H., Leite, S.R., de, A., Bobbio, P.A., 1981. Maillard reaction. VI. Structural determinations in melanoidin from fructose and glycine. An. Acad. Bras. Cienc. 53, 83—86.

Isbell, H.S., 1944. Interpretation of some reactions in the carbohydrate field in terms of consecutive electron displacement. J. Res. Natl. Bur. Stand. (U.S.) 32, 45.

Ito, H., 1977. The formation of maltol and isomaltol through degradation of sucrose. Agric. Biol. Chem. 41, 1307—1308.

Izzo, H.V., Ho, C.-T., 1992. Peptide-specific Maillard reaction products: a new pathway for flavour chemistry. Trends Food Sci. Technol. 3, 253—257.

Jiang, D., Peterson, D.G., 2010. Role of hydroxycinnamic acids in food flavors: a brief overview. Phytochem. Rev. 9, 187—193.

Jing, H., Kitts, D.D., 2000. Comparison of the antioxidative and cytotoxic properties of glucose—lysine and fructose—lysine Maillard reaction products. Food Res. Int. 33, 509—516.

Johnson, R.R., Alford, E.D., Kinzer, G.W., 1969. Formation of sucrose pyrolysis products. J. Agric. Food Chem. 17, 22—24.

Joslyn, M.A., 1957. Role of amino acids in the browning of orange juice. Food Res. 22, 1—14.

Karel, M., Labuza, T.P., 1968. Nonenzymatic browning in model systems containing sucrose. J. Agric. Food Chem. 16, 717—719.

Katchalsky, A., 1941. Interaction of aldoses with α-amino acids or peptides. Biochem. J. 35, 1024—1027.

Kato, H., 1962. Chemical studies on amino—carbonyl reaction. I. Isolation of 3-deoxypentosone and 3-deoxyhexosones formed by browning degradation of N-glycosides. Agric. Biol. Chem. 26, 187—192.

Kato, H., 1973. Antioxidative activity of amino-carbonyl reaction products. Shokuhin Eiseigaku Zasshi 14, 343—351.

Kato, H., Fujimaki, M., 1968. Formation of N-substituted pyrrole-2-aldehydes in the browning reaction between D-xylose and amino compounds. J. Food Sci. 33, 445—449.

Kato, K., Watanabe, K., Sato, Y., 1978. Effect of the Maillard reaction on the attributes of egg white proteins. Agric. Biol. Chem. 42, 2233—2237.

Kato, K., Watanabe, K., Sato, Y., 1981. Effect of metals on the Maillard reaction of ovalbumin. J. Agric. Food Chem. 29, 540—543.

Kato, Y., Matsua, T., Watanabe, K., Nakamura, R., 1985. Alteration of ovalbumin immunogenic activity by glycosylation through Maillard reaction. Agric. Biol. Chem. 49, 423—427.

Kawashima, K., Itoh, H., Chibata, I., 1977. Antioxidant activity of browning products prepared from low molecular carbonyl compounds and amino acids. J. Agric. Food Chem. 25, 202.

Kefford, J.F., 1959. The chemical constituents of fruits. Adv. Food Res. 9, 285—372.

Kim, J.-S., Lee, Y.-S., 2009. Study of Maillard reaction products from aqueous model systems with different peptide chain lengths. Food Chem. 116, 846—853.

Knol, J.J., Linssen, J.P.H., van Boekel, M.A.J.S., 2010. Unravelling the kinetics of the formation of acrylamide in the Maillard reaction of fructose and asparagine by multiresponse modeling. Food Chem. 120, 1047—1057.

Knowles, M.E., 1971. Inhibition of non-enzymic browning by sulphites: identification of sulphonated products. Chem. Ind. (London), 110—111.

Koehler, P.E., Odell, G.V., 1970. Factors affecting the formation of pyrazine compounds in sugar—amine reactions. J. Agric. Food Chem. 18, 895—898.

Koehler, P.E., Mason, M.E., Newell, J.A., 1969. Formation of pyrazine compounds in sugar—amine model systems. J. Agric. Food Chem. 17, 393—396.

Kort, M.J., 1970. Reactions of free sugars with aqueous ammonia. Adv. Carbohydr. Chem. Biochem. 25, 311—349.

Kotsuiou, K., Tasioula-Margari, M., Caicano, E., Fogliano, V., 2011. Effect of standard phenolic compounds and olive oil phenolic extract on acrylamide formation in an emulsion system. Food Chem. 124, 242—247.

Kroh, L.W., Fiedler, T., Wagner, J., 2008. Dicarbonyl compounds — key intermediates for the formation of carbohydrate-based melanoidins. Ann. N. Y. Acad. Sci. 1126, 210—215.

Kuster, B.F.M., Temmink, H.M.G., 1977. The influence of pH and weak-acid anions on the dehydration of D-fructose. Carbohydr. Res. 54, 185—191.

Kuster, B.F.M., Van der Bean, H.S., 1977. The influence of the initial and catalyst concentrations on the dehydration of D-fructose. Carbohydr. Res. 54, 165—176.

Kwon, T.W., Menzel, D.B., Olcott, H.S., 1965. Reactivity of malonaldehyde with food constituents. J. Food Sci. 30, 808—813.

Labuza, T.P., Baisier, W.M., 1992. The kinetics of nonenzymatic browning. In: Schwartzberz, H.G., Hartel, R.W. (Eds.), Physical Chemistry of Foods. Marcel Dekker, New York.

Labuza, T.P., Saltmarch, M., 1981. The nonenzymic browning reaction as affected by water in foods. In: Rockland, L.B., Stewart, B.S. (Eds.), Water Activity Influences Food Quality. Academic Press, New York, p. 605.

Laroque, D., Inisan, C., Berger, C., Vouland, E., Dufosse, L., Guerard, F., 2008. Kinetic study on the Maillard reaction. Consideration of sugar reactivity. Food Chem. 111, 1032–1042.

Lea, C.H., Hannan, R.S., 1949. Studies of the reaction between proteins and reducing sugars in the 'dry' state. I. The effect of activity of water, pH and of temperature on the primary reaction between casein and glucose. Biochim. Biophys. Acta 3, 313–325.

Lea, C.H., Hannan, R.S., 1950. Biochemical and nutritional significance of the reaction between proteins and reducing sugars. Nature (London) 165, 438–439.

Ledl, F., 1982a. Formation of coloured products in browning reactions: reaction of hydroxyacetone with furfural. Z. Lebensm.-Unters.-Forsch. 175, 203.

Ledl, F., 1982b. Formation of coloured products in browning reactions: reaction of dihydrox-yacetone with furfural. Z. Lebensm.-Unters.-Forsch. 175, 349.

Ledl, F., Fritsch, G., 1984. Formation of pyrroline reductones by heating hexoses with amino acids. Z. Lebensm.-Unters.-Forsch. 178, 41.

Ledl, F., Schleicher, E., 1990. New aspects of the Maillard reaction in foods and in the human body. Angew Chem. Int. Ed. Eng. 29, 565–594.

Ledl, F., Severin, T., 1982. Investigations of the Maillard reaction. XVI (1). Formation of coloured compounds from hexoses. Z. Lebensm.-Unters.-Forsch. 175, 262.

Lee, C.M., Lee, T.-C., Chichester, C.O., 1979. Kinetics of the production of biologically active Maillard browned products in apricot and glucose-L-tryptophan. J. Agric. Food Chem. 27, 478–482.

Lee, C.M., Sherr, B., Koh, Y.-N., 1984. Evaluation of kinetic parameters for a glucose–lysine Maillard reaction. J. Agric. Food Chem. 32, 379–382.

Lee, M.-H., Ho, C., Chang, S.S., 1981. Thiazoles, oxazoles and oxazolines identified in the volatile flavor of roasted peanuts. J. Agric. Food Chem. 26, 1049–1050.

Lee, T.-C., Pintauro, S.J., Chichester, C.O., 1982. Nutritional and toxicological effects of nonenzymatic browning. Diabetes 31, 37–46.

Licht, B.H., Shaw, K., Smith, C., Mendoza, M., Orr, J., Myers, D.V., 1992a. Characterization of caramel colours I, II, III. Food Chem. Toxicol. 30, 375–382.

Licht, B.H., Shaw, K., Smith, C., Mendoza, M., Orr, J., Myers, D.V., 1992b. Characterization of caramel colour IV. Food Chem. Toxicol. 30, 365–373.

Lightbody, H.D., Fevold, H.L., 1948. Biochemical factors influencing the shelf life of dried whole eggs and means of their control. Adv. Food Res. 1, 149–202.

Lingnert, N., 1980. Antioxidative Maillard reaction products. III. Application in cookies. J. Food Process. Preserv. 4, 219–233.

Lingnert, N., Eriksson, E., 1980. Antioxidative Maillard reaction products from sugars and free amino acids. J. Food Process. Preserv. 4, 161.

Lingnert, N., Hall, G., 1986. Formation of antioxidative Maillard reaction products during food processing. Dev. Food Sci. 13, 273.

Liu, Y., Kitts, D.D., 2011. Confirmation that the Maillard reaction is the principal contributor to the antioxidant capacity of coffee brews. Food Res. Int. 44, 2418–2424.

Lo, C.Y., Hsiao, W.T., Chen, X.Y., 2011. Efficiency of trapping methylglyoxal by phenols and phenolic acids. J. Food Sci. 76, H90–H96.

Lo, C.Y., Li, S., Tan, D., Pan, M.H., Sang, S., Ho, C.-T., 2006. Trapping reactions of reactive carbonyl species with tea polyphenols in simulated physiological conditions. Mol. Nutr. Food Res. 50, 1118–1128.

Locas, C.P., Yaylayan, V.A., 2008. Further insight into thermally and pH-induced generation of acrylamide from glucose/asparagine model systems. J. Agric. Food Chem. 56, 6069–6074.

Loncin, M., Bimbenet, J.J., Lenges, J., 1968. Influence of the activity of water on the spoilage of foodstuffs. J. Food Technol. 3, 131–142.

Lopez de Lerma, N., Peinado, J., Moreno, J., Peinado, R.A., 2010. Antioxidant activity, browning and volatile Maillard compounds in Pedro Ximenez sweet wines under accelerated oxidative aging. LWT-Food Sci. Technol. 43, 1557–1563.

Lu, C.Y., Hao, Z., Payne, R., Ho, C.T., 2005. Effects of water content on volatile generation and peptide hydrolysis in Maillard reaction of glycine, diglycine and triglycine. J. Agric. Food Chem. 53, 6443–6447.

Ludwig, E., Lipke, U., Raczek, U., Jager, A., 2000. Investigations of peptides and proteases in green coffee beans. Eur. Food Res. Technol. 211, 111–116.

Lv, L., Shao, X., Chen, H., Ho, C.-T., Sang, S., 2011. Genistein inhibits advanced glycation end product formation by trapping methylglyoxal. Chem. Res. Toxicol. 24, 579–586.

Maga, J.A., 1978. Oxazoles and oxazolines in food. J. Agric. Food Chem. 26, 1049–1050.

Maga, J.A., 1981. Pyrroles in foods. J. Agric. Food Chem. 29, 691.

Maga, J.A., 1982. Pyrazines in foods: an update. CRC Crit. Rev. Food Sci. Nutr. 16, 1–48.

Maga, J.A., Sizer, C.E., 1973. Pyrazines in foods. CRC Crit. Rev. Food Technol. 4, 39–115.

Maillard, L.C., 1912. Action des acides amines sur les sucres; formation des mélanoidines par voie méthodique. C. R. Hebol. Seances Acad. Sci. 154, 66–68.

Markuze, Z., 1963. Effects of traces of metals on the browning of glucose–lysine solutions. Rocz. Panstw. Zakl. Hig. 14, 65 (Cited Chem. Abstr. 59: 4980).

Mauron, J., 1970. Nutritional evaluation of proteins by enzymatic methods. Wenner-Gren Cent. Int. Symp. Ser. 14.

Mawhinney, T.P., Madson, M.A., Feather, M.S., 1980. The isomerization of D-glucose in acidic solution. Carbohydr. Res. 86, 147–150.

Mazin, M.I., Al-Zubaidy, Khalil, P.A., 2007. Kinetic and prediction studies of ascorbic acid in normal and concentrate local lemon juice during storage. Food Chem. 101, 254–259.

McWeeny, D.J., 1969. The Maillard reaction and its inhibition by sulfite. J. Food Sci. 34, 641–643.

McWeeny, D.J., 1973. The role of carbohydrates in non-enzymatic browning. In: Birch, G.G., Green, F. (Eds.), Molecular Structure and Functions of Food Carbohydrates. Applied Science Publishers, London.

McWeeny, D.J., 1984. Sulfur dioxide and the Maillard reaction in food. Proc. Food Nutr. Sci. 5, 395–404.

Michalska, A., Amigo-Benavent, M., Zielinski, H., del Castillo, M.D., 2008. Effect of bread making on formation of Maillard reaction products contributing to the overall antioxidant activity of rye bread. J. Cereal Sci. 48, 123–132.

Milic, B.L., 1987. CP-mass carbon-13 NMR spectral studies of the kinetics of melanoidin formation. Analyst 112, 783–785.

Milic, B.L., Piletic, M.V., Grujic-Injac, B., Premovic, P.I., 1978. A comparison of the chemical composition of boiled and roasted aromas of heated beef. J. Agric. Food Chem. 25, 113.

Milic, B.L., Piletic, M.V., Cembic, S.M., Odavic-Josic, J., 1979. Free radical formation kinetics in D(+)-glucose and amino acid model systems. Proceedings of the 27th IUPAC Congress 1979, 296.

Milic, B.L., Piletic, M.V., Cembic, S.M., Odavic-Josic, J., 1980. Kinetic behaviour of free radical formation on the nonenzymatic browning reaction. J. Food Process. Preserv. 4, 13–26.

Milic, B.L., Piletic, M.V., Grujic-Injac, B., Premovic, P.I., 1978. A comparison of the chemical composition of boiled and roasted aromas of heated beef. J. Agric. Food Chem. 25, 113.

Moll, N., Gross, B., That, V., Moll, M., 1982. A fully automated high-performance liquid chromatographic procedure for isolation and purification of Amadori compounds. J. Agric. Food Chem. 30, 782–786.

Moller, A.B., Andrews, A.T., Cheeseman, G.C., 1977. Chemical changes in ultra-heat-treated milk during storage: II. Lactuloselysine and fructoselysine formation in the Maillard reaction. J. Dairy Res. 44, 267–275.

Molnar-Perl, I., Pinter-Szakacs, M., Wittman, R., Reutter, M.M., Eichner, K., 1986. Optimum yield of pyridosine and furosine originating from Maillard reactions monitored by ion-exchange chromatography. J. Chromatogr. 361, 311–320.

Montgomery, M.W., Day, E.A., 1965. Aldehyde–amine condensation reaction, possible fate of carbonyls in foods. J. Food Sci. 30, 828–832.

Montilla, A., Ruiz-Matute, A.I., Sanz, M.L., Martinez-Castro, I., del Castillo, M.D., 2006. Difructose anhydrides as quality markers of honey and coffee. Food Res. Int. 39, 801–806.

Morales, F., Jimenez-Perez, S., 2004. Peroxyl radical scavenging activity of melanoidins in aqueous systems. Eur. Food Res. Technol. 218, 515–520.

Morales, F.J., 2005. Assessing the non-specific hydroxyl radical scavenging properties of melanoidins in a Fenton-type reaction system. Anal. Chim. Acta 534, 171–178.

Moreno, F.J., Molina, E., Olano, A., Lopez-Fandino, R., 2003. High pressure effects on the Maillard reaction between glucose and lysine. J. Agric. Food Chem. 51, 394–400.

Mottram, D.S., 1985. The effect of cooking conditions on the formation of volatile heterocyclic compounds in pork. J. Sci. Food Agric. 36, 337–382.

Mottram, D.S., 1998. Flavor formation in meat and meat products: a review. Food Chem. 62, 415–424.

Mottram, D.S., Wedzich, B.L., Dodson, A.T., 2002. Acrylamide is formed in the Maillard reaction. Nature 419, 448–449.

Nafisi, K., Markakis, P., 1983. Inhibition of sugar–amine browning by aspartic and glutamic acids. J. Agric. Food Chem. 31, 1115–1117.

Nagy, S., 1980. Vitamin C contents of citrus fruit and their products: a review. J. Agric. Food Chem. 28, 8–18.

Nagy, S., Smoot, J.M., 1977. Temperature and storage effects on percentage retention and percentage US recommended dietary allowance of vitamin C in canned single-strength orange juice. J. Agric. Food Chem. 25, 135–138.

Namiki, M., 1988. Chemistry of the Maillard reaction: recent studies on the browning reaction mechanism and the development of antioxidants and mutagens. Adv. Food Res. 32, 115–184.

Namiki, M., Hayashi, T., 1975. Development of novel free radicals during the amino–carbonyl reaction of sugars with amino acids. J. Agric. Food Chem. 23, 487–491.

Namiki, M., Hayashi, T., 1981. Formation of free radical products in an early stage of the Maillard reaction. In: Eriksson, C. (Ed.), Maillard Reactions in Food. Prog. Food Nutr. Sci., Vol. 5. Pergamon, Oxford.

Namiki, M., Hayashi, T., 1983. A new mechanism of the Maillard reaction involving sugar fragmentation and free radical formation. In: Walter, G.R., Feather, M.S. (Eds.), The Maillard Reaction in Foods and Nutrition. ACS Symposium Series, 215, p. 21.

Namiki, M., Hayashi, T., Kawakishi, S., 1973. Free radicals developed in the amino–carbonyl reaction of sugars with amino acids. Agric. Biol. Chem. 37, 2935–2936.

Negroni, M., D'Agostina, A., Arnoldi, A., 2001. Effects of olive, canola, and sunflower oils on the formation of volatiles from the Maillard reaction of lysine with xylose and glucose. J. Agric. Food Chem. 49, 439–445.

Noda, Y., Peterson, P.G., 2007. Structure–reactivity relationships of flavan-3-ols on product generation in aqueous glucose/glycine model system. J. Agric. Food Chem. 55, 3686–3691.

Nursten, H., 1986. Maillard browning reactions in dried foods. In: McCarthy, D. (Ed.), Concentration and Drying of Foods. Elsevier, London.

Nursten, H., 2005. The Maillard Reaction: Chemistry, Biochemistry and Implications. Royal Society of Chemistry, London.

Ohno, Y., Ward Jr., K., 1961. Acid epimerization of D-glucose. J. Org. Chem. 26, 3928–3931.

Olano, A., Martinez-Castro, I., 1996. Nonenzymatic browning. In: Nollet, L.M.L. (Ed.), Handbook of Food Analysis. Dekker, New York, pp. 1683–1721.

Olesen, P.T., Olsen, A., Frandsen, H., Fredericksen, K., Overvad, K., Tjonneland, A., 2008. Acrylamide exposure and incidence of breast cancer among postmenopausal women in the Danish Diet, Cancer and Health Study. Int. J. Cancer 122, 2094–2100.

Olsson, K., Peremalm, P.A., Theander, O., 1982. In: Eriksson, C. (Ed.), Maillard Reactions in Food. Prog. Food Nutr. Sci., Vol. 5. Pergamon, New York.

Otsuka, M., Kurata, T., Arakawa, N., 1986. Isolation and characterization of an intermediate product in the degradation of 2,3-diketo-L-gulonic acid. Agric. Biol. Chem. 50, 531–533.

Patton, A.R., Salander, R.C., Piano, M., 1954. Lysine destruction in casein–glucose interactions measured by quantitative paper chromatography. Food Res. 19, 444–450.

Pedreschi, F., Segtnan, V.H., Knutsen, S.H., 2010. On-line monitoring of fat, dry matter and acrylamide contents of potato chips using near infrared interactance and visual reflectance imaging. Food Chem. 121, 616–620.

Peterson, B.I., Tong, C.H., Ho, C.T., Welt, B.A., 1994. Effect of moisture content on Maillard browning kinetics of a model system during microwave heating. J. Agric. Food Chem. 42, 1884–1887.

Peterson, R.J., Izzo, H.J., Jungerman, E., Chang, S.S., 1975. Changes in volatile flavor compounds during the retorting of canned beef stew. J. Food Sci. 40, 948–954.

Pigman, W., Anet, E.F.L.J., 1972. Mutarotation and actions of acids and bases. In: Pigman, W., Horton, D. (Eds.)).The Carbohydrates, 2nd ed., Vol. 1A. Academic Press, New York, pp. 165–194.

Plakas, S.M., Lee, C.-T., Walke, K.E., 1988. Bioavailability of lysine in Maillard browned protein as determined by plasma response in rainbow trout (Salmo gairdneri). J. Nutr. 118, 19–22.

Popoff, T., Theander, O., 1976. Formation of aromatic compounds from carbohydrates. Part III. Reaction of D-glucose and D-fructose in slightly acidic, aqueous solution. Acta Chem. Scand. 30, 397–402.

Ranganna, S., Setty, L., 1968. Non-enzymatic discoloration in dried cabbage. Ascorbic acid–amino acid interaction. J. Agric. Food Chem. 16, 529–533.

Rao, M.S., Chawla, S.P., Chander, R., Shrma, A., 2011. Antioxidant potential of Maillard reaction products formed by irradiation of chitosan–glucose solution. Carbohydr. Polym. 83, 714–719.

Renn, R.T., Sathe, S.K., 1997. Effect of pH, temperature, and reactant molar ratio on L-leucine and D-glucose Maillard reaction in an aqueous system. J. Food Sci. 42, 38.

Resmini, P., Pellegrino, L., 1991. Analysis of food heat damage by direct HPLC of furosine. Int. Chromatogr. Lab. 6, 7–11.

Reyes, F.G.R., Poocharoen, B., Wrolstad, R.E., 1982. Maillard browning reaction of sugar–glycine model systems: changes in sugar concentration, color and appearance. J. Food Sci. 47, 1376–1377.

Reynolds, T.M., 1959. Chemistry of non-enzymatic browning III. Effect of bi-sulfite, phosphate and malate on the reaction of glycine and glucose. Aust. J. Chem. 12, 265–274.

Reynolds, T.M., 1963. Chemistry of nonenzymic browning. I. Adv. Food Res. 12, 1–52.

Reynolds, T.M., 1965. Chemistry of non-enzymatic browning. II. Adv. Food Res. 14, 167–183.

Reynolds, T.M., 1969. Nonenzymic browning. Sugar–amine interactions. In: Schultz, H.W., Cain, R.F., Wrolstadt, R.W. (Eds.), Foods, Carbohydrates and their Roles. Avi, Westport, CT.

Reynolds, T.M., 1970. Flavours from nonenzymic browning reactions. Food Technol. Austr. 22, 610–611.

Riester, D.W., Braun, O.G., Pearce, W.E., 1945. Why canned citrus juices deteriorate in storage. Food Ind. 17, 742–744.

Rizzi, G.P., 1969. The formation of tetraethyl pyrazine and 2-isopropyl-4,5-dimethyl-3-oxazoline in the Strecker degradation of DL-valine with 2,3-butanedione. J. Org. Chem. 34, 2002.

Rizzi, G.P., 1974. Formation of N-alkyl-2-acylpyrroles and aliphatic aldimines in model non-enzymic browning reactions. J. Agric. Food Chem. 22, 279–282.

Rizzi, G.P., 2008. The Strecker degradation of amino acid: newer avenues of flavor formation. Food Rev. Int. 24, 416–435.

Ruffian-Henares, J.A., de la Cueva, S.P., 2009. Antimicrobial activity of coffee melanoidins – a study of their metal-chelating properties. J. Agric. Food Chem. 57, 432–438.

Sang, S., Shao, X., Bai, N., Lo, C.Y., Yang, C.S., Ho, C.-T., 2007. Tea polyphenol (−)-epigallocatechin 3-gallate: a new trapping agent of reactive carbonyl species. Chem. Res. Toxicol. 20, 1862–1870.

Sanz, M.L., de Castillo, M.D., Corzo, N., Olano, A., 2000. Presence of 2-furoylmethyl derivatives in hydrolysates of processed tomato products. J. Agric. Food Chem. 48, 468–471.

Sanz, M.L., de Castillo, M.D., Corzo, N., Olano, A., 2001. Formation of Amadori compounds in dehydrated fruits. J. Agric. Food Chem. 49, 5228–5231.

Sanz, M.L., de Castillo, M.D., Corzo, N., Olano, A., 2003. 2-Furoylmethyl amino acids and hydroxymethylfurfural as indicators of honey quality. J. Agric. Food Chem. 51, 4278–4283.

Sasaki, S., Shirahashi, Y., Nishiyama, K., Watanabe, H., Havase, F., 2006. Identification of a novel blue pigment as a melanoid intermediate in the D-xylose–glycine reaction system. Biosci. Biotechnol. Biochem. 70, 2529–2531.

Schieberle, P., 1990. The role of free amino acids present in yeast as precursors of the odorants 2-acetyl-1-pyrroline and 2-acetyltetrahydropyridine in wheat bread crust. Z. Lebensm. Unters. Forsch. 191, 206–209.

Schoebel, T., Tannenbaum, S.R., Labuza, T.P., 1969. Reaction at limited water concentration. 1. Sucrose hydrolysis. J. Food Sci. 34, 324.

Schonberg, A., Moubacher, R., 1952. The Strecker degradation of α-amino acids. Chem. Rev. 50, 261–277.

Schultz, A., Trage, C., Schwarz, H., Kroh, L.W., 2007. Electrospray ionization mass spectrometric investigations of α-dicarbonyl compounds – probing intermediates formed in the course of the nonenzymatic browning reaction of l-ascorbic acid. Int. J. Mass Spectrom. 262, 169–173.

Schutte, L., 1974. Precursors of sulfur-containing flavor compounds. CRC Crit. Rev. Food Technol. 4, 457–505.

Shallenberger, R.S., Birch, C.G., 1975. Sugar Chemistry. Avi, Westport, CT, p. 189.

Shao, X., Bai, N., He, K., Ho, C.T., Yang, C.S., Sang, S., 2008. Apple polyphenols, phloretin and phloridzin: new trapping agents of reactive dicarbonyl species. Chem. Res. Toxicol. 21, 2042–2050.

Shaw, P.E., Berry, R.E., 1977. Hexose–amino acid degradation studies involving formation of pyrroles, furans and other low molecular weight products. J. Agric. Food Chem. 25, 641–644.

Shaw, P.E., Tatum, J.H., Berry, R.E., 1968. Base-catalyzed fructose degradation and its relation to nonenzymic browning. J. Agric. Food Chem. 16, 979–982.

Shibamoto, T., Bernard, R.A., 1977. Investigation of pyrazine formation pathways in sugar–ammonia model systems. J. Agric. Food Chem. 25, 609–614.

Shigematsu, H., Kurata, T., Kato, H., Fujimaki, M., 1972. Volatile compounds formed on roasting DL-α-alanine with d-glucose. Agric. Biol. Chem. 36, 1631–1637.

Shirahashi, Y., Watanabe, H., Hayase, F., 2009. Identification of red pigments formed in a d-xylose–glycine reaction system. Biosci. Biotechnol. Biochem. 73, 2287–2292.

Silvan, J.M., van de Lagemaat, J., Olano, A., Doores del Castillo, M., 2006. Analysis and biological properties of amino acid derivates formed by Maillard reaction in foods. J. Pharm. Biomed. Anal. 41, 1543–1551.

Simon, H., Heubach, G., 1965. Formation of alicyclic and open-chain nitrogenous reductones by reaction of secondary amine salts on monosaccharides. Chem. Ber. 98, 3703–3711.

von Smolnik, H.D., 1987. Herstellung und Andwebdung von Zucker-Kolor aus Starkeprodukten. Staerke 39, 28.

Spark, A.A., 1969. Role of amino acids in non-enzymic browning. J. Sci. Food Agric. 20, 308–316.

Staempfli, A.A., Blank, I., Fumeaux, R., Fay, L.B., 1994. Study on the decomposition of the amadori compound N-(1-deoxy-D-fructos-1-yl)-glycine in model systems: quantification by fast atom bombardment tandem mass spectrometry. Biol. Mass Spectrom. 23, 642–646.

Stoeffelsman, J., Pypker, J., 1968. Some new constituents of roasted coffee. Rec. Trav. Chim. Pays-Bas 87, 241–242.

Stoll, M., Dietrich, P., Sundt, E., Winter, M., 1967a. Sur l'aroma de café. I. Helv. Chim. Acta 50, 628–694.

Stoll, M., Winter, M., Gautschi, F., Filament, I., Willhalm, B., 1967b. Sur l'aroma du cacao. II. Helv. Chim. Acta 50, 2065–2067.

Sugimura, T., Sato, S., Wakabayashi, K., 1988. Mutagens/carcinogens in pyrolysates of amino acids and proteins and in cooked foods: heterocyclic aromatic amines. In: Woo, Y., Lai, D.Y., Aros, J.C., Argus, M.F. (Eds.), Chemical Induction of Cancer: Structural Bases and Biological Mechanism. CRC Press, Boca Raton, FL.

Summa, C., McCourt, J., Cammerer, B., Fiala, A., Probst, M., Kun, S., Anklam, E., Wagner, K.H., 2008. Radical scavenging activity, antibacterial and mutagenic effects of cocoa bean Maillard reaction products with degree of roasting. Mol. Nutr. Food Res. 52, 342−351.

Tagliazucchi, D., Verzelloni, E., Conte, A., 2010a. Contribution of melanoidins to the antioxidant activity of traditional balsamic vinegar. J. Food Biochem. 34, 1061−1078.

Tagliazucchi, D., Verzelloni, E., Conte, A., 2010b. Effect of dietary melanoidins on lipid peroxidation during simulated gastric digestion: their possible role in the prevention of oxidative damage. J. Agric. Food Chem. 58, 2513−2519.

Tang, J., Jin, Q.Z., Shen, G.H., Ho, C.T., Chang, S.S., 1983. Isolation and identification of volatile compounds in fried chicken. J. Agric. Food Chem. 31, 1287−1292.

Tareke, E., Rydberg, P., Karlsson, P., Eriksson, S., Tornqvist, M., 2002. Analysis of acrylamide, a carcinogen formed in heated foodstuffs. J. Agric. Food Chem. 50, 4998−5006.

Tareke, E., Rydberg, P., Karlsson, P., Eriksson, S., Tornqvist, M., 2002. Analysis of acrylamide, a carcinogen formed in heated foodstuffs. J. Agric. Food Chem. 50, 4998−5006.

Tatum, J.H., Shaw, P.E., Berry, R.E., 1967. Some compounds formed during nonenzymic browning of orange powder. J. Agric. Food Chem. 15, 773−775.

Tehrani, K.A., Kersiene, M., Adams, A., Venskutonis, R., de Kimpe, M., 2002. Thermal degradation studies of glucose/glycine melanoidins. J. Agric. Food Chem. 50, 4062−4068.

Tessier, F.J., 2010. The Maillard reaction in the human body: the main discoveries and factors that affect glycation. Pathol. Biol. 58, 214−218.

Theander, O., 1981. Novel developments in caramelization. Prog. Food. Nutr. Sci. 5, 471−476.

Tonsbeek, C.H.T., Copier, H., Plancken, A.J., 1971. Components contributing to beef flavor. Isolation of 2-acetyl-2-thiazoline from beef broth. J. Agric. Food Chem. 19, 1014−1016.

Totlani, V.M., Peterson, D.G., 2005. Reactivity of epicatechin in aqueous glycine and glucose reaction models: quenching of C2, C3 and C4 fragments. J. Agri. Food Chem. 53, 4130−4135.

Tschiersky, H., Baltes, W., 1989. Investigation of caramel Curie-point-pyrolysis of caramel syrups and other investigations of structure. Z. Lebensm.-Unters.-Forsch. 189, 132−137.

Tsuchuda, H., Tachibana, S., Kamoto, M., 1976. Identification of heterocyclic compounds produced by pyrolysis of the nondialyzable melanoidins. Agric. Biol. Chem. 40, 2051−2056.

Tu, A., Eskin, N.A.M., 1973. The inhibitory effect of reducing sugars on the hydrolysis of casein by trypsin. Can. Inst. Food Sci. Technol. J. 6, 50−53.

van Boekel, M.A.J.S., 2006. Formation of flavor compounds in the Maillard reaction. Biotechnol. Adv. 24, 230−233.

Van Praag, M., Stein, H.S., Tibbetts, M.S., 1968. Steam volatile aroma constituents of roasted cocoa beans. J. Agric. Food Chem. 16, 1005−1008.

Van Ruth, S.M., King, C., Giannouli, P., 2002. Influence of lipid fraction, emulsifier fraction, and mean particle diameter of oil-in-water emulsions on the release of 20 aroma compounds. J. Agric. Food Chem. 50, 2365−2371.

Velisek, J., Davidek, K., 1976a. Reactions of glyoxal with amino acids. I. Formation of brown pigments. Sb. Vys. Sk. Chem.-Technol. Praze Potraivin. E46, 35.

Velisek, J., Davidek, K., 1976b. Reactions of glyoxal with amino acids. II. Analysis of the brown pigments. Sb. Vys. Sk. Chem.-Technol. Praze Potraivin. E46, 51.

Verzelloni, E., Tagliazucchi, D., Conte, A., 2010. From balsamic to healthy: traditional balsamic vinegar melanoidins inhibit lipid peroxidation during simulated gastric digestion of meat. Food Chem. Toxicol. 48, 2097−2102.

Wagner, K.H., Derkits, S., Herr, M., Schuh, W., Elmadfa, I., 2002. Antioxidative potential of melanoidins isolated from a roasted glucose−glycine model. Food Chem. 78, 375−382.

Wagner, K.-H., Reichhold, S., Koschutnig, K., Cheriot, S., 2007. The potential antimutagenic and antioxidant affects of Maillard reaction products used as 'natural antibrowning' agents. Mol. Nutr. Food Res. 51, 498−504.

Walradt, J.P., Pittet, A.O., Kinlin, T.E., Muralidhara, R., Sanderson, A., 1971. Volatile components of roasted peanuts. J. Agric. Food Chem. 19, 972−979.

Wang, H.-Y., Qian, H., Yao, W.-R., 2011. Melanoidins produced by the Maillard reaction: structure and biological activity. Food Chem. 128, 573−584.

Watanabe, K., Kato, Y., Sato, Y., 1980. Chemical and conformational changes of ovalbumin due to the Maillard reaction. J. Food Process Preserv. 3, 263−274.

Wedzicha, B.L., McWeeny, D.J., 1974a. Non-enzymic browning reactions of ascorbic acid and their inhibition. The production of 3-deoxy-4-sulphopentulose in mixtures of ascorbic acid, glycine and bisulphite ion. J. Sci. Food Agric. 25, 577−587.

Wedzicha, B.L., McWeeny, D.J., 1974b. Non-enzymic browning reactions of ascorbic acid, glycine and bisulphite ion. The identification of 3-deoxy-4-sulphopentulose in dehydrated, sulphited cabbage after storage. J. Sci. Food Agric. 25, 589−593.

Wedzicha, B.L., McWeeny, D.J., 1975. Concentrations of some sulphonates derived from sulphite in certain foods. J. Sci. Food Agric. 26, 327−335.

Weenen, H., Apeldoorn, W., 1996. Carbohydrate cleavage in the Maillard reaction. In: Taylor, A.J., Mottram, D.S. (Eds.), Flavor Science − Recent Developments, Proceedings of the 8th Weurman Flavour Research Symposium. Royal Society of Chemistry, Reading.

Weygand, F., 1940. Uber N-Glykoside. II. AmadoriUmlagerungen. Ber. Dtsch. Chem. Ges. 73, 1259−1278.

Whitfield, F.B., 1992. Volatiles from interactions of Maillard reactions and lipids. Crit. Rev. Food Sci. Nutr. 31, 1−58.

Wolfrom, M.L., Rooney, C.C., 1953. Chemical interactions of amino compounds and sugars. VIII. Influence of water. J. Am. Chem. Soc. 75, 5435−5436.

Wolfrom, M.L., Schuetz, R.D., Calvalieri, L.F., 1948. Discoloration of sugar solutions and 5-(hydroxymethyl)furfural. J. Am. Chem. Soc. 70, 514−517.

Wong, J.M., Bernhard, R.A., 1988. Effect of nitrogen source on pyrazine formation. J. Agric. Food Chem. 36, 123−129.

Yamaguchi, N., Fujimaki, M., 1974. Studies on browning reaction products from reducing sugars and amino acids. XIV.

Antioxidative activities of purified melanoidins and their comparisons with those of legal antioxidants. Nippon Shokuhin Kogyo Gakkaishi 21, 6.

Yaylayan, V.A., Locas, C.P., 2007. Vinylogous Amadori rearrangement: implications in food and biological systems. Mol. Nutr. Food Res. 51, 437–444.

Yaylayan, V.A., Wnorowski, A., Locas, C.P., 2003. Why asparagine needs carbohydrates to generate acrylamide. J. Agric. Food Chem. 51, 1753–1757.

Yen, G.C., Hsieh, P.P., 1995. Antioxidative activity and scavenging effects on active oxygen of xylose–lysine Maillard reaction products. J. Sci. Food Agric. 67, 415–420.

Yeo, H.C.H., Shibamoto, T., 1991a. Microwave-induced volatiles of the Maillard model system under different pH conditions. J. Agric. Food Chem. 39, 370–373.

Yeo, H.C.H., Shibamoto, T., 1991b. Effects of moisture content on the Maillard browning model system upon microwave irradiation. J. Agric. Food Chem. 39, 1860–1862.

Yoshimura, Y., Iijima, T., Watanabe, T., Nakazawa, H., 1997. Anti-oxidative effect of Maillard reaction products using glucose-glycine model system. J. Agric. Food Chem. 45, 4106–4109.

Yu, A.N., Zhang, A.D., 2010. The effect of pH on the formation of aroma compounds produced by heating a model system containing L-ascorbic acid with threonine/serine. Food Chem. 119, 214–219.

Yuan, Y., Shu, C., Zhou, B., Qi, X., Xiang, J., 2011. Impact of selected additives on acrylamide formation in asparagine/sugar Maillard model systems. Food Res. Int. 44, 449–455.

Zamora, R.M.R., Hildago, F.J., 1995. Influence of irradiation time, pH, and lipid/amino acid ratio on pyrrole production during microwave heating of lysine/(E)-4. 5-epoxy-(E)-2-heptenal model systems. J. Agric. Food Chem. 43, 1029–1037.

Zamora, R.M.R., Delago, R.M., Hildago, F.J., 2011. Aminophospholipids and lecithins as mitigating agents for acrylamide in asparagine/glucose and asparagines/2,4-decadienal model systems. Food Chem. 126, 104–108.

Zhang, Y., Ho, C.-T., 1989. Volatile compounds formed from thermal interaction of 2,4-decadienal with cysteine and glutathione. J. Agric. Food Chem. 37, 1016–1020.

Zhang, Y., Ritter, W.J., Barker, C.C., Traci, P.A., Ho, C.T., 1994. Volatile formation of lipid-mediated Maillard reaction in model systems. In: Ho, C.T., Hartman, T.G. (Eds.), Lipids in Food Flavor. ACS Symposium Series, 558. American Chemical Society, Washington, DC.

Chapter 7

Biochemistry of Brewing

Graham G. Stewart

GGStewart Associates, 13 Heol Nant Castan, Rhiwbina, Cardiff, UK

I. INTRODUCTION

This chapter reviews the brewing of beer. The purpose of brewing is to hydrolyze starch from barley malt, together with wheat, maize, rice, sorghum, unmalted barley, and sugar/syrups into a sugary nitrogenous fermentable liquid called wort and to convert it into an alcoholic carbonated beverage using yeast. Although brewing is essentially a biochemical process it also involves a number of other disciplines. This chapter largely focuses on the enzyme reactions involved in beer brewing.

Saké (a high-alcohol rice-based beverage) is also brewed and similar to beer; however, it is not distilled and will not be considered here. Beer is the world's most widely consumed (Steiner, 2009) and probably the oldest alcoholic beverage, and is the third most popular drink overall, after water and tea (Priest and Stewart, 2006).

In 2009, 1800 million hl of beer was produced globally (1 hl ≡ 100 liters). The largest beer producer in 2009 was China (418 million hl), followed by the USA (234 million hl), Russia (110 million hl), Brazil (107 million hl), and Germany (98 million hl). Beer is produced by the brewing and fermentation of starches and sugar (sucrose), mainly derived from cereal grains, commonly malted barley together with wheat (sometimes malted), maize (corn), rice, sorghum (malted and unmalted), and unmalted barley. Most beer is flavored with hops, which add bitterness and aroma and can act as a preservative, although other flavorings such as herbs or fruit may occasionally be included. Beer brewing dates from 8000 BC and the earliest known chemical evidence dates from approximately 3500 BC from the site of Godin Tepe in the Zagros Mountains of western Iran.

The beer production process is outlined in the flow diagram in Figure 7.1. Malting, mashing, and fermentation/aging are essentially enzymatic processes. The main raw material, malt, contains extract components (starch, proteins, etc.) and enzymes (amylases, proteases, etc.). However, malt is an expensive raw material. Consequently, when microbial enzymes produced by fermentation became commercially available on a large scale in the 1960s, both brewers and enzyme producers began research into substituting malt with unmalted raw materials and exogenous enzymes. With some notable exceptions, this search was unsuccessful, although, as discussed later in this chapter, it has been reviewed

Biochemistry of Foods. DOI: http://dx.doi.org/10.1016/B978-0-12-242352-9.00007-2

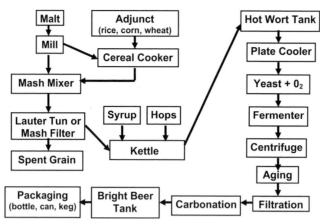

FIGURE 7.1 The brewing process.

and proposals are afoot to reintroduce 'barley brewing', as it has become known. This research has led to a range of applications for exogenous enzymes in brewing which has brought economic and technical benefits:

- improved control of the brewing process
- accelerated production resulting in increased brewing capacity
- potential for new and different beer types
- use of less expensive and unconventional raw materials.

However, there are still some unanswered questions when using exoenzymes in this context, including beer flavor match, product stability (in particular physical, flavor, and foam stability), and drinkability.

The purpose of brewing is to hydrolyze the starch/protein sources into a sugary nitrogenous fermentable liquid called wort, and to convert the wort into the alcoholic carbonated beverage known as beer in a fermentation process effected by yeast, followed by maturation. Brewing was one of the earliest biological processes to be undertaken on a commercial scale and it became one of the first processes to develop from a craft into a technology. Beer production is a unit process divided into five distinct, but related, interconnected stages:

- Malting is the germination of barley or other cereal and drying (or kilning) of the germinated cereal. Malt is made in malting plants which are usually, but not always, physically separate from breweries. The raw materials are specially selected brewing barley varieties, suitable for malting. The objective of malting from a brewing (and distilling) perspective is to permit the development of enzymes that will hydrolyze proteins and starch during the later stages of germination and subsequently during mashing.
- Mashing involves the hydrolysis of proteins/peptides, starch, and other materials from the ground malted barley and unmalted cereals (adjuncts) by a spectrum of enzymes (we will discuss this in more detail later, in Section III) to produce a water-soluble, largely fermentable extract which can be separated from the insoluble material (called spent grains). This unboiled, unhopped non-sterile liquid is called sweet wort.
- Wort boiling, with the inclusion of hops, or hop extracts, and sometimes sugar and/or syrups to produce a sterile medium called wort.
- Fermentation with yeast, followed by maturation and filtration.
- Packaging, used generally to mean kegging, bottling, and canning.

Brewing is essentially a biochemical process, but also involves many other disciplines such as microbiology and botany (both closely related to biochemistry), analytical, organic and inorganic chemistry, process engineering and control, structural engineering, separation systems including adsorbants and filtration, taste and flavor assessment, statistics, and physics (heat, refrigeration, etc.). Biochemistry is essentially a study of the biochemical processes in living organisms (in the case of brewing, plants and yeast). It deals with the structure and function of cellular components such as proteins, carbohydrates, lipids, nucleic acids, and other biomaterials. Perhaps the most important aspect of the biochemistry of brewing is that all biological reactions involve enzymes. Therefore, this chapter will focus on the enzyme reactions that are involved in the beer brewing process. There are many chemical reactions during brewing that do not involve enzymes, such as the isomerization of hops during wort boiling, the formation of coloring materials (particularly melanoidins, as discussed in Chapter 6) during green malt kilning and wort boiling, the development of skunky (light struck) off-flavors when beer is exposed to light, and the generation of haze and staling off-flavors when beer is stored for prolonged periods. These chemical reactions will not be discussed in detail in this chapter; however, further details can be found in Bamforth (2009).

The principal raw materials used in the brewing process are malted cereals (usually barley, but not always; sometimes wheat and sorghum), unmalted cereals (corn, wheat, rice, sorghum, oats, barley), sugar and syrups (usually called adjuncts), hops, water, and yeast (Figure 7.2). In addition, there are various additives (including some enzymes) and processing aids which will be discussed in this chapter if they involve biochemical reactions. Some processing aids and additives take part in chemical reactions, for example, silica gel and polyvinylpolypyrrolidone (PVPP) (Leiper *et al.*, 2003).

II. MALT AND MALTING

The objectives of malting are to develop a spectrum of enzymes which hydrolyze the constituents of barley malt (and other cereals) to develop a fermentable extract called wort. Wort is a medium that will support yeast growth and fermentation with beer as the end product. It is important that beer is drinkable (beer is not usually sipped, it is drunk!) and it exhibits a number of stability properties, i.e. flavor, physical, foam, and biological characteristics (Stewart, 2004). Malt contributes a large number of materials to wort (Briggs, 1998a). The principal components are free amino nitrogen (FAN) and fermentable sugars.

FAN is the sum of the individual wort amino acids, ammonium ions, and small peptides (dipeptides and tri-peptides). FAN is an important general measure of yeast nutrients which constitute the yeast assimilable nitrogen during brewery fermentations (Briggs, 1998b). One important aspect of barley and malt studies is the relationship between the level of nitrogen incorporated into barley during growth in the field and the level of enzyme activity developed in barley during malting (Jones and Pierce, 1965). During the malting of barley grains, large-molecular-weight components of the endosperm cell walls, the storage proteins, and the large and small starch granules in the endosperm are hydrolyzed (modified) enzymatically by, for example, amylases, proteases, and glucanases, rendering them more soluble in hot water during mashing.

The word 'malt' is derived from the Anglo-Saxon *mealt* and perhaps has the same root as melt, referring to grain softening that occurs during germination, or *malled* (mauled: broken or ground), as malts are milled before being used in brewing (and distilling) (Singer *et al.*, 1954). Malting is the limited germination of cereal grains, usually, but not always, barley. Sometimes malt is used 'green', meaning undried and not kilned. Although malt made from barley is by far the most important, it is also made from wheat, rye, oats, triticale, maize (corn), sorghum, various millets, and rice.

Malting is perhaps the oldest biotechnology. The cultivation of barley and wheat probably began around 10,000 BC (Briggs, 1998c) and wild grain must have been collected earlier. Malting is the controlled germination of cereals, followed by termination of this natural process by the application of heat to dry the grain (kilning). Further heat is then applied to kiln the grain in order to produce the required flavor and color. According to the Brewing and Malting Research Institute (BMBRI) based in Winnipeg, Canada, the following characteristics are required to produce superior malting barley:

- pure lot of an acceptable variety
- germination of 96% grains or higher

FIGURE 7.2 Principal brewing raw materials.

- no evidence of preharvest germination
- protein concentration of 11−12.5% on a dry weight basis
- maximum moisture content of 13%
- plump kernels of uniform size
- free from disease, mycotoxins such as deoxynivalenol (DON), and chemical residues
- free of frost damage and weathering
- less than 5% peeled or broken kernels
- free of insects, ergot, treated seeds, smut, and odor.

The stages of malting are usually divided into steeping, germination, and kilning. In reality, malting involves more than this and the divisions between these 'classical' stages are not clear-cut. Before the production process can begin, barley must be obtained by the maltster, either directly or indirectly from the farmer, and it must be clean and stored ready for use. Prior to this, farmers must have been persuaded to grow acceptable malting barley varieties that will meet malting quality standards. In turn, barley breeders must have developed these varieties. After the kilning process, malt must be cleaned ('dressed') and stored for a minimum period before blending and transportation to a brewer (or distiller). By-products of the process, including broken grains, malt sprouts (culms), and dust, are collected and sold for use as animal feed. Increasingly, maltsters blend culms and pelletize them. When grain is converted into malt, some malting losses are inevitably by-products.

The malting process is a blend of pure and applied science involving plant and microbial biochemistry, physiology, chemistry, physics, and engineering. The stages of a typical malting process are depicted in Figure 7.3. However, there are several variations on the basic procedure. The procedure employed is based on the principle that barley, or other grain, must be converted into malt of the best achievable quality, economically, in the shortest feasible time, with the best yield. The choice of malting procedure is guided by these considerations. However, it is beyond the scope of this chapter to discuss the detailed nuances of malting. The focus is the biochemistry of the process which includes the major metabolic pathways that operate during grain germination. These pathways include hundreds of compounds that are interconverted under the influence of a large number of enzymes and provide routes by which the carbon skeletons of carbohydrates, amino acids, and lipids can be interconverted. In addition, highly polymeric substrates, such as polysaccharides and proteins, can provide respiratory fuel and building blocks for polymeric substances, such as cell wall components. The grain tissues undergo changes as malting proceeds. The

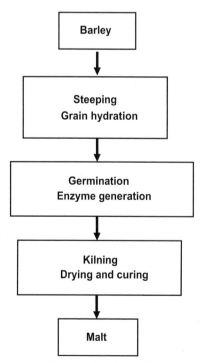

FIGURE 7.3 Typical barley malting process.

aleurone cells are partly depleted of their contents but are metabolically active. The contents of the dead starchy endosperms are partly degraded and depleted. The embryo metabolizes and grows, chiefly at the expense of hydrolysis products from the starchy endosperm. The hydrolysis and biosynthetic processes proceed simultaneously. There is a net breakdown of polymeric substances, such as starch, and a migration of substances from the aleurone layer and the endosperm to the embryo.

Although the changes that occur during malting are normally described in terms of physical modification of the grain and alteration in conventional malt analysis (Table 7.1), on a biochemical basis, the gross changes are the net result of the degradation of reserve substances, the interconversion of substances in the living embryo and aleurone layer, the net flow of substances to the embryo from the aleurone layer and starchy endosperm, the synthesis of new grain substances, and their incorporation into the new, growing tissues (the acrospire and rootlets) of the embryo. Allowances must be made for the losses of dry matter that occur during the conversion of grain into malt. These malting losses result from:

- leaching of substances from the grain during steeping
- fermentative processes and the respiratory oxidation of substances to carbon dioxide and water
- removal of the rootlets.

The total losses during malting are usually in the range 6—12% of the original dry weight. Losses can be much larger when cereals such as wheat and sorghum are malted.

The chemical and biochemical changes that take place during malting are complex. They can only be understood by appreciating the range of reactions that occur during the overlapping processes of steeping, germination, and kilning, and the effects of deculming and dressing (cleaning) the malt. During steeping, the grain is permitted to imbibe water so as to increase the grain moisture from 12—14% to 42—48%. This occurs by immersing the grain in water or by spraying with water, or usually a combination of both. The steep water becomes dirty and is replaced at least once to keep the grain fresh. During steeping the grain swells and softens, and the living tissues resume their metabolism, which had ceased during grain ripening and drying prior to harvesting and during storage awaiting malting. Sometimes air is blown through the grain—water mixture (aeration), or the grain can be air-rested — the water is drained away and air is sucked downwards through the grain. When the grain has achieved the correct moisture content, the steeping water is removed. Usually this steeped grain is transferred to a germination vessel. In some malting plants, steeping and germination, and occasionally kilning, take place in one container.

Each batch of grain being malted is referred to as a 'piece'. Following steeping, germination begins and the grain undergoes modification. Modification is an imprecise term that signifies all the desirable changes, both biochemical

TABLE 7.1 Typical Barley Malt Analysis	
Moisture (%)	3.8—4.2
Extract (%)	79.9—81.0
Wort color (°SRM)	1.4—1.7
Diastatic power (°ASBC)	120—145
α-Amylase (DM)	39—49
Malt protein (%)	10.8—12.3
Wort protein (%)	4.9—5.6
FAN (mg/l)	180—220
Wort viscosity (cP)	1.38—1.48
Wort β-glucan (mg/l)	25—150
Friabilimeter value (%)	70—86
Wort fermentability (%)	78—82

FAN: free amino nitrogen.
From Briggs (1998b).

and chemical, that occur when grain is converted to malt. Modification continues during the initial stages of kilning. The three major aspects of modification are:

- accumulation of hydrolytic enzymes (details discussed later, in Section III)
- a variety of chemical reactions occurring in the grains
- physical changes in the grain, which appear as weakening and softening.

Visible signs of germination include the initial appearance of a white chit at the end of the grain, followed by a tuft of rootlets or culms. At the same time, the acrospire (coleoptile or shoot) grows. It is covered by the husk in the barley but it grows freely in many other grains. The germinated grain is transferred to the kiln while it is still fresh (called green or undried, but not green colored). The moisture content of kilned barley malt is usually 4—4.5%, reduced from 42—48%.

Biochemical changes occur in malting grains' progress as modification in the starchy endosperm occurs. A partly malted grain retains some of its barley character. In each batch of grain there is heterogeneity in the sense that individual grains vary in size, shape, maturity, chemical composition, and potential for generating enzymes. Furthermore, individual grains modify at different rates and receive different treatments during steeping, germination, and kilning.

A. Grain Structure

Barley grains (Figure 7.4), except for naked (huskless) barley, contain husk, pericarps, testa, aleurone layer, starchy endosperm, and embryo (Briggs, 1998b). In terms of total dry weight of the barley grain, the husk is 10—12%, the pericarp and testa 2—3%, the aleurone layer 4—5%, the starchy endosperm 77—80%, and the embryo 2—3%.

The husk is composed of two leaf-like structures. The dorsal half is called the lemma and the ventral half is the palea. The husk protects the underlying structure of the grain, especially the embryo. Husk damage is regarded as unacceptable and barley samples are rejected if husk damage is beyond specification requirements. High damage implies embryo damage, uncontrollable embryo growth, and mash filter operational difficulties. The husk contains background levels of microorganisms such as fungi and bacteria. These microorganisms may have invaded the grains in the field before harvesting or during storage before malting (Flannigan, 1996). The pericarp is the fruit of the grain. Cereal grains are fruits and, strictly speaking, should not be referred to as seeds. Similar to the husk, the pericarp contains a waxy cuticle, and below this waxy layer is a compressed structure of cells. The pericarp is semipermeable, so certain chemicals will pass through it, while others, such as the plant hormone gibberellic acid, will not. Water can pass through the pericarp. Damage to the pericarp during abrasion (a process whereby 7—9% of a grain's weight is separated as hulls, mainly husk, with minimal reduction in germination or grain breakage) allows gibberellic acid to enter the aleurone layer directly, rather than via the germinated embryo, thereby improving modification of the starchy endosperm by improving the efficiency of the aleurone in producing endosperm-degrading enzymes (Sandegren and Beling, 1959). However, abrasion is not currently used to any great extent.

The testa comprises two lipid layers that enclose cellular material. The testa is permeable to gibberellic acid. Phenolic compounds such as anthocyanogens (proanthocyanogens) are associated with the aleurone and testa and can be clearly seen in some varieties of barley and sorghum. The small area of the pericarp—testa that lies over the coleorhiza (chit) is called the micropyle. The latter may facilitate the uptake of water and salts into the embryo during germination.

FIGURE 7.4 The barley grain. (*From Briggs, 1998c.*)

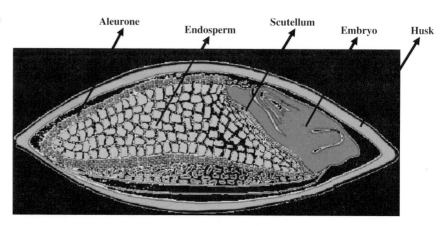

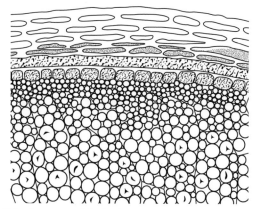

FIGURE 7.5 **Structure of the barley aleurone layer.** *(From Briggs, 1998c.)*

The aleurone layer is two or three cells deep over the starchy endosperm (Figure 7.5). It extends over the embryo as a single cell layer. Therefore, excised embryos contain alerone cells and produce endosperm-degrading enzymes. During malting, gibberellic acid from germinated embryos can induce alerone cells to produce endosperm-degrading enzymes such as α-amylase, endo-β-1,3:1,4-glucanases, limit dextrinases, endoproteases, and xylanases (pentosa-nases). It has been established that α-amylase, endoprotease, and limit dextrinase are produced *de novo* and aleurone layers stimulated by gibberellic acid.

III. MASHING AND BOILING

The purposes of mashing are:

- to extract starch, proteins, peptides and other components from the malt
- to render the extract fermentable by ensuring the necessary enzymatic hydrolysis of the above components to sugars, amino acid, small peptides, etc.

Before mashing, the malt is milled using a variety of milling procedures. This is a physical process and therefore beyond the scope of this chapter (Briggs, 1998a). The milled grain is wetted to stimulate enzyme activity which will hydrolyze the starch into fermentable sugars, the proteins into amino acids and small peptides, and the lipids into free fatty acids and sterols. In order to achieve this complex hydrolysis procedure, the mash is subjected to a series of heating and rest periods at set temperatures to realize the optimum catalytic conditions with reference to the type of beer being produced. The infusion mashing program is extensively used and is shown in Figure 7.6. The initial temperature of 52°C is used to stimulate both protease and glucanase activity (often called the protein rest) but the rates of activity of the two enzyme systems are unclear. After 15–20 minutes the temperature is increased gradually

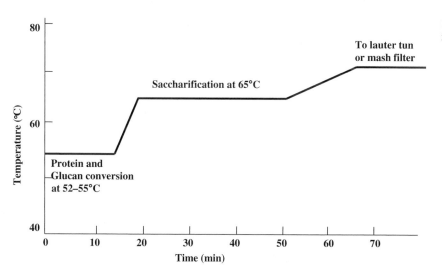

FIGURE 7.6 **Temperature profile of a typical infusion mashing program.**

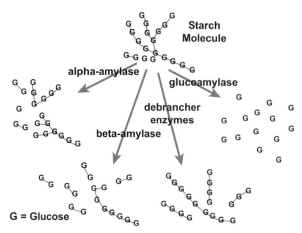

FIGURE 7.7   The enzymatic hydrolysis of starch.

to 65°C. This temperature is the saccharification temperature and is optimal for both α- and β-amylase activity. α-Amylase is an endoamylase that hydrolyzes 1,4-α-glucosidic linkages in amylose (contains $1 \rightarrow 4$ linkages) and amylopectin (contains both $1 \rightarrow 4$ and $1 \rightarrow 6$ linkages). This enzyme is virtually absent from mature barley unless it has pregerminated. However, considerable quantities are synthesized *de novo* in the embryo and aleurone layer and large proportions are generated into the starchy endosperm. β-Amylase is an exoenzyme that catalyzes the hydrolysis of the $1 \rightarrow 4$ linkages penultimate to the non-reducing chain ends (Figure 7.7), releasing the disaccharide maltose (Figure 7.8) and an oligosaccharide (also called dextrin) shortened by the removal of two glucose residues. Unlike α-amylase, this enzyme is present in unmalted barley. During malting, the level of free β-amylase may initially fall

FIGURE 7.8 Structure of maltose and maltotriose.

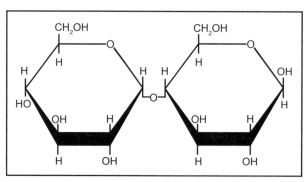

Maltose,
Molecular weight = 342

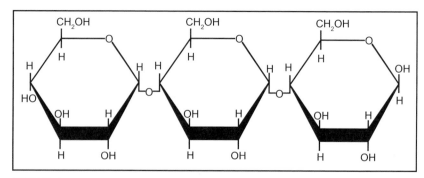

Maltotriose,
Molecular weight = 504

during steeping and subsequently, during germination, nearly all the β-amylase becomes free, and the bound form disappears. This enzyme will not attack $1 \rightarrow 6$ or $1 \rightarrow 4$ linkages immediately adjacent to each other.

Another important enzyme action during mashing is the debranching enzyme that hydrolyzes $1 \rightarrow 6$ linkages in amylopectin, dextrins, and other oligosaccharides. This enzyme is also known as limit dextrinase and as pullulanase because of its ability to degrade a particular bacterial polysaccharide, namely pullulan. In malt, this enzyme occurs in both freely soluble and bound forms. There exists a heat-stable protein in barley that inhibits malt limit dextrinase (Briggs, 1998c). The amount of inhibition increases during malting. However, there is evidence that this enzyme is synthesized *de novo* in the barley aleurone layer during germination.

After approximately 30 minutes at 65°C (saccharification rest) the temperature is raised to 78°C: the mash-off temperature. The principal purpose of this mash-off temperature is to inactivate most of the enzymes that were active in the mash. In addition, some final α-amylolysis will occur at this temperature, whereas β-amylase will be rapidly inactivated. Also, at this temperature, the viscosity of the sweet wort (due to β-glucans and arabinxylans) will be reduced and many, although not all, of the microorganisms contaminating the malt will be heat inactivated. This sweet wort is separated from the spent grains in a lauter tun or mash filter.

The sweet wort is boiled in a kettle (also called a copper), usually for 30−60 minutes with 4−6% evaporation. Boiling is needed to isomerize the hop α-acids (isomerized hop acids are bitter whereas non-isomerized hop α-acids are not), to strip out unwanted malt and hop volatiles, to denature proteins and coagulate proteins/polyphenols as hot break, and to fix the wort composition (see Section IV) by terminating all enzymatic and microbiological activity surviving the mashing process. As a consequence of boiling and evaporation, there will also be color development and an increase in wort gravity that need to be accommodated to achieve the finished target for a particular wort. Ideally, a kettle should be fitted with in-line probes to follow the isomerization of α-acids, the disappearance of dimethyl sulfide, the coagulation of colloidal size particles (0.1−10 μm) to form hot break (30−70 μm), and the wort gravity. Currently, this in-line control is not practical and the brewer has to rely on pragmatic experience to achieve these critical parameters.

IV. WORT COMPOSITION

Compared to other media used in the production of fermentation alcohol (both industrial and potable), wort is by far the most complicated. Therefore, when yeast is pitched (inoculated) into wort it is introduced into a complex environment because it consists of simple sugars, dextrins, amino acids, peptides, proteins, vitamins, ions, nucleic acids, and other constituents too numerous to mention. One of the major advances in brewing science during the past 40 years has been the elucidation of the mechanisms by which the yeast cell utilizes, in an orderly manner, the plethora of wort nutrients. Wort sugars (Stewart, 2006) and amino acids (Jones and Pierce, 1964), as discussed in Subsection D, Wort Free Amino Nitrogen, below, are removed in a distinct order at various points in the fermentation cycle.

A. Wort Sugars

Wort contains the sugars sucrose, fructose, glucose, maltose, and maltotriose, together with dextrin material. A typical percentage sugar spectrum of brewer's wort is shown in Table 7.2. In the normal situation, brewing yeasts are

TABLE 7.2 Typical Sugar Spectrum of Wort

Sugar	Composition (%)
Glucose	10−15
Fructose	1−2
Sucrose	1−2
Maltose	50−60
Maltotriose	15−20
Dextrins	20−30

From Stewart (2006).

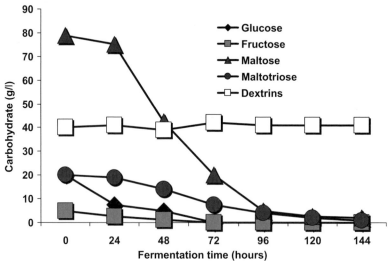

capable of utilizing sucrose, glucose, fructose, maltose, and maltotriose in this approximate sequence (or priority), although some degree of overlap does occur, leaving maltotetraose (G4) and the larger dextrins unfermented (Figure 7.9).

The objectives of wort fermentation are to consistently metabolize wort constituents into ethanol, carbon dioxide, and other fermentation products in order to produce beer with satisfactory quality and stability. Another objective is to produce yeast crops that can be confidently repitched into subsequent brews (Stewart and Russell, 1986).

Saccharomyces cerevisiae strains, including brewery yeast strains, are single-celled fungi (more details in Subsection IV, C, below). The requirements of an acceptable brewer's yeast strain are: 'In order to achieve a beer of high quality, it is axiomatic that not only the yeast be effective in removing the required nutrients from the growth/fermentation medium (wort), able to tolerate the prevailing environmental conditions (for example, ethanol tolerance), and impart the desired flavor to the beer, but the microorganisms themselves must be effectively removed from the wort by flocculation, centrifugation and/or filtration after they have fulfilled their metabolic role' (Stewart and Russell, 1986).

These strains have the ability to take up and ferment a wide range of sugars, including sucrose, glucose, fructose, galactose, mannose, maltose, and maltotriose. In addition, *Saccharomyces diastaticus* (a subspecies of *S. cerevisiae*) is able to utilize dextrin material due to the secretion of glucoamylase. The initial step in the utilization of any sugar is usually either its passage intact across the cell membrane, or its hydrolysis outside the cell membrane, followed by entry into the cell by some or all of the hydrolysis products (Figure 7.10). Maltose and maltotriose are examples of sugars that pass intact across the cell membrane, whereas sucrose and dextrins are hydrolyzed by extracellular enzymes [invertase for sucrose and glucoamylase (amyloglucosidase) for dextrins] and the hydrolysis products are

FIGURE 7.10 Uptake and metabolism of maltose and maltotriose by the yeast cell. *(From Stewart, 2006.)*

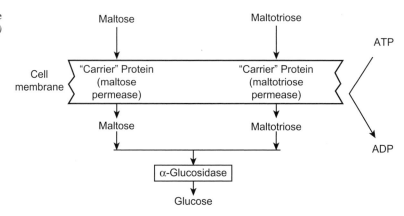

taken up into the cell. An important metabolic difference between the uptake of monosaccharides such as glucose and fructose disaccharides such as maltose or maltotriose uptake is that energy (ATP conversion to ADP) is required for maltose and maltotriose uptake (active transport) whereas glucose and fructose are taken up passively with no energy required (Bisson et al., 1993). As maltose and maltotriose (Figure 7.8) are the major sugars in brewer's wort, the ability of a brewing yeast to use these two sugars is vital and depends upon the correct genetic complement. Brewer's yeast possesses independent uptake mechanisms (maltose and maltotriose permeases) to transport the two sugars across the cell membrane into the cell. Once inside the cell, both sugars are hydrolyzed to glucose units by the α-glucosidase system (Figure 7.10). The transport, hydrolysis, and fermentation of maltose are particularly important in brewing, distilling, and baking since maltose is the major sugar component of brewing wort, spirit mash, and wheat dough. Maltose fermentation in brewing, distilling, and baking yeasts requires at least one of five unlinked *MAL* loci, each consisting of three genes encoding the structural gene for α-glucosidase (maltase) (*MALS*) maltose permease (*MALT*) and an activator whose product coordinately regulates the expression of the α-glucosidase and permease genes. The expression of *MALS* and *MALT* is regulated by maltose induction and repression by glucose. When glucose concentrations are high (> 10 g/l) the *MAL* genes are repressed and only when 40−50% of the glucose has been taken up from the wort will the uptake of maltose and maltotriose commence (Figure 7.9).

B. Effect of Osmotic Pressure and Ethanol on the Uptake of Wort Glucose, Maltose, and Maltotriose

High-gravity worts (> 16°Plato, where 1°Plato ≡ 1 g sucrose dissolved in 100 ml distilled water at 20°C) have been shown to exact a negative effect upon fermentation performance (Casey et al., 1984). High-gravity brewing is a procedure that employs wort at higher than normal concentration and thus requires dilution with water (usually specially treated, including being deoxygenated) at a later stage in processing. By this means, increased production demands can be met without expanding brewing, fermenting, and storage facilities. This process also enhances the sustainability of brewing because of a reduction in the amount of water and energy required in a brewhouse per unit of sales-gravity beer produced. In addition, high-gravity conditions can result in a reduction in labor, cleaning, and effluent costs (Stewart, 1999).

Although the high-gravity process has a number of advantages, inevitably there are disadvantages. Owing to the more concentrated mash (increased ratio of carbohydrate to water in order to produce high-gravity wort), there is a decrease in brewhouse material efficiency. This problem can be overcome by the use of modern mash filters (Andrews et al., 2011) instead of traditional lauter tuns and/or kettle syrups, where entry of carbohydrate into the process bypasses the process whereby the unboiled wort is separated from the spent grains. In addition, the more concentrated wort during boiling results in reduced isomerization of hops, leading to decreased hop utilization. This process disadvantage can be alleviated by the use of kettle and postfermentation hop extracts.

Another major disadvantage to the high-gravity brewing process is reduced foam stability (head retention) in the final diluted beer. This aspect will not be discussed in detail here, suffice to say that the major reason is greater loss of hydrophobic polypeptides during the brewing process. This is due to reduced extraction of these polypeptides during mashing and their hydrolysis by yeast proteases during fermentation (Cooper et al., 1998).

Another negative effect of high-gravity worts is their effect on overall yeast activity and performance. One of the negative influences is elevated osmotic pressure (Pratt et al., 2003). This was simulated by employing the non-metabolized sugar sorbitol in a synthetic medium. Fermentations were conducted with peptone−yeast extract containing 25% (w/v) sorbitol to which was added 2.5% (w/v) glucose, maltose, or maltotriose. The effect of the elevated osmotic pressure on the transport of these three sugars is shown in Figure 7.11 as a percentage of the untreated control that did not contain added sorbitol, with both an ale and a lager yeast strain. The sugar uptake rates of the three sugars were inhibited in both strains in the presence of sorbitol, but to a much greater extent with maltose and maltotriose than with glucose. Similar effects were obtained on the uptake of glucose, maltose, and maltotriose following treatment with 10% (v/v) ethanol (Stewart, 2006) (Figure 7.12). These effects are probably a reflection of the different transport mechanisms for glucose compared to maltose and maltotriose.

Yet another disadvantage of high-gravity brewing is that the flavor of the diluted beer is often not comparable to sales-gravity brewed beer. This is primarily due to disproportionate levels of beer acetate compounds such as ethyl acetate and isoamyl acetate (Table 7.3). Varying the wort sugar source (Younis and Stewart, 1999) has been reported to modify the levels of many metabolites, including esters, although reasons for these differences are unclear. Initially, 4% (v/v) glucose and maltose in a synthetic medium (yeast extract−peptone) were fermented separately with shaking at 21°C and the production of ethyl acetate and isoamyl acetate was monitored (Table 7.4). The

FIGURE 7.11 Effect of osmotic pressure [25% (w/v) sortibol*] on the uptake of glucose, maltose, and maltotriose by a lager strain (gray bars) and an ale strain (black bars). *(From Stewart, 2006.)*

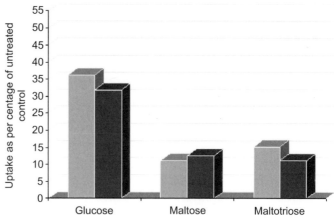

*Peptone – yeast extract – 2.5% (w.v) sugar medium

FIGURE 7.12 Effect of ethanol [10% (v/v) ethanol*] on the uptake of glucose, maltose and maltotriose by a lager strain (gray bars) and an ale strain (black bars). *(From Stewart, 2006.)*

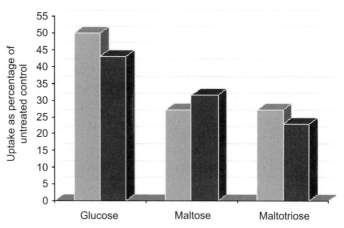

*Peptone – yeast extract – 2.5% (w.v) sugar medium

fermentation performances of the three ale and three lager brewing strains employed were similar. In all six strains studied, the strains cultured in maltose consistently produced lower levels of both esters when in the glucose medium.

The lower levels of the esters produced with maltose as the substrate compared to glucose could be due to a number of reasons. It is possible that fermentation with maltose inhibits the transport of esters out of the cell, perhaps by modifying the plasma membrane, thus giving the impression that fewer esters are produced. Another possibility is that maltose metabolism produces lower levels of acetyl-coenzyme A which, it has been suggested, results in fewer esters due to a lack of substrate. It has been proposed that ester production is linked to lipid

TABLE 7.3 Influence of Wort Gravity on Beer Ester Levels

	12°Plato	20°Plato
Ethanol (v/v)	5.1	5.0
Ethyl acetate (mg/l)	14.2	21.2
Isoamyl acetate (mg/l)	0.5	0.7

From Younis and Stewart (1999).

TABLE 7.4 Ethyl Acetate and Isoamyl Acetate Produced by Brewing Yeast Strains During Fermentation of Synthetic Media[a]

Strain	Ethyl Acetate (mg/l)		Isoamyl Acetate (mg/l)	
	Glucose	Maltose	Glucose	Maltose
Ale 1	4.13	2.79	0.14	0.14
Ale 2	2.97	2.59	0.06	0.04
Ale 3	3.13	2.71	0.05	0.03
Lager 1	6.00	5.22	0.22	0.21
Lager 2	3.75	3.28	0.26	0.22
Lager 3	4.13	3.51	0.23	0.17

[a]Peptone—yeast extract—4% sugar medium.
From Younis and Stewart (1999).

metabolism (Stewart, 2005). If this is the case, or if for some reason maltose metabolism produces fewer toxic fatty acids, it would be reasonable to assume that fewer fatty acids would also be produced.

It is generally agreed that a reduction in beer ester levels, particularly ethyl acetate and isoamyl acetate, from high-gravity brewed beers would be welcome. In order to study the influence of maltose and glucose levels in high-gravity worts, two 20°Plato worts were prepared, one containing 30% maltose syrup (MS) and the other containing 30% very high maltose syrup (VHMS). The sugar composition of the two brewing syrups is shown in Table 7.5. In addition, a 12°Plato wort containing 30% maltose syrup (MS) was prepared and used as a control. The sugar spectra of the three worts are shown in Figure 7.13. The maltose plus maltotriose concentration in the 20°Plato VHMS wort had increased compared to the 20°Plato MS wort, with a corresponding decrease in the concentration of glucose plus fructose.

The three worts were fermented in the 2 hl pilot brewery of the International Centre for Brewing and Distilling at Heriot-Watt University, Scotland, UK (Figure 7.14). A lager strain was used at 13°C and the concentration of ethyl acetate and isoamyl acetate determined throughout the fermentation (Figures 7.15 and 7.16). The profiles were similar with both esters. The concentration of both esters in the 20°Plato MS fermented wort was twice that in the 12°Plato MS fermented wort, as would be expected. However, the ester concentration in the 20°Plato VHMS wort was reduced by approximately 25% compared to the 20°Plato MS wort (Younis and Stewart, 1999). These results confirm the findings using synthetic media with simple sugars that maltose fermentations produce less ethyl acetate and isoamyl acetate than glucose fermentations (Younis and Stewart, 1998).

C. Uptake of Wort Maltose and Maltotriose: Differences between Ale and Lager Yeast Strains

The commercial worldwide production of ale has always been much lower than that of lager and over the years this difference has grown. The trends in ale compared with lager consumption in Ontario, Canada, and the UK are shown

TABLE 7.5 Sugar Composition of Brewing Syrups

	MS	VHMS
Glucose	15[a]	5
Maltose	55	70
Maltotriose	10	10
Dextrins	20	15

MS: maltose (55) syrup; VHMS: very high maltose (70) syrup.
[a]Percent composition.
From Priest and Stewart (2006).

304

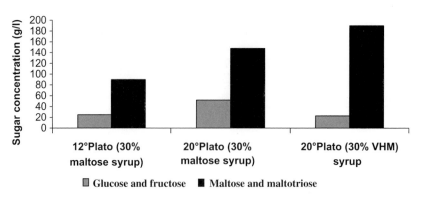

FIGURE 7.13 **Wort sugar profile.** *(From Younis and Stewart, 1998.)*

FIGURE 7.14 **Brewing pilot plant at Heriot-Watt University, Edinburgh, UK.** *(Source: author's own photograph.)*

in Tables 7.6 and 7.7, respectively. In the USA, currently 4.4% of beer produced is ale, largely due to the craft brewing sector.

There are several differences in the production of these two types of beer, one of the main ones being the characteristics of ale and lager yeast strains. Considerable research by many breweries and institutions on this topic has been conducted (Barnett, 1992; Pederson, 1995; Pulvirenti *et al.*, 2000; Rainieri *et al.*, 2003) and the typical differences between ale and lager yeast strains have been established (Table 7.8).

With the advent of molecular biology-based methodologies, gene sequencing of ale and lager brewing strains has shown that they are interspecies hybrids with homologous relationships to one another and also to *Saccharomyces*

FIGURE 7.15 **Ethyl acetate concentration in fermenting worts of differing gravities and sugar composition.** (*u*) 12°Plato (30% maltose syrup); (*v*) 20°Plato (30% maltose syrup); (σ) 20°Plato (30% very high maltose syrup). *(From Younis and Stewart, 1998.)*

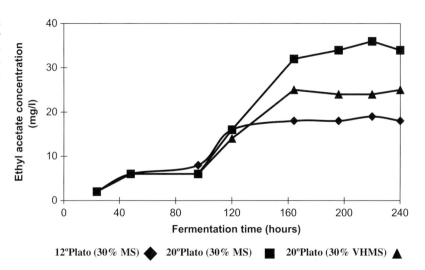

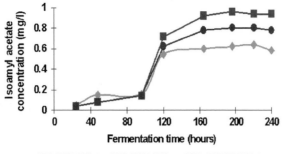

FIGURE 7.16 **Isoamylacetate concentration in fermenting worts of differing gravities and sugar composition.** *(From Younis and Stewart, 1998.)*

TABLE 7.6 Comparison of Ale versus Lager Consumption (%) Trends in Ontario, Canada (1970–2007)

Year	Ale	Lager
1970	60	40
1980	20	80
1990	10	90
2000	10	90
2005	15	85
2007	15	85

From Steiner (2009).

TABLE 7.7 Comparison of Ale versus Lager Consumption (%) Trends in the UK (1970–2007)

Year	Ale	Lager
1970	90	10
1980	70	30
1990	50	50
2000	40	60
2007	35	65

From Steiner (2009).

bayanus, a yeast species used in wine fermentation and identified as a wild yeast in brewing fermentation (Figure 7.17). The gene homology between *Saccharomyces pastorianus* and *S. bayanus* strains is high at 72%, whereas the homology between *S. pastorianus* and *S. cerevisiae* is much lower at 50% (Pederson, 1995).

Recently, a research group from Argentina, Portugal, and the USA published a paper entitled 'Microbe domestication and the identification of the wild genetic stock of lager-brewing yeast' (Libkinda *et al.*, 2011). This confirmed that *S. pastorianus* is a domesticated yeast species created by the fusion of *S. cerevisiae* with a previously unknown species that has now been designated *Saccharomyces eubayanus* because of its close relationship to *S. bayanus*. They also report that *S. eubayanus* exists in the forests of Patagonia and was not found in Europe until the advent of trans-Atlantic trade between Argentina and Europe. This paper (Libkinda *et al.*, 2011) contains a draft genome sequence of *S. eubayanus*; it is 99.5% identical to the non-*S. cerevisiae* portion of the *S. pastorianus* genome sequence and

TABLE 7.8 Differences Between Ale and Lager Yeast Strains

Ale Yeast	Lager Yeast
Saccharomyces cerevisiae (ale type)	*Saccharomyces uvarium* (*carlsbergensis*)
	Saccharomyces cerevisiae (lager type)
	Saccharomyces pastorianus: current taxonomic name
Fermentation temperature of 18–22°C	Fermentation temperature of 8–15°C
Cells can grow at 37°C or higher	Cells cannot grow at 37°C or higher
Cannot ferment the disaccharide melibiose	Ferments the disaccharide melibiose
'Top' fermenter	'Bottom' fermenter

suggests specific changes in wort sugar and sulfate metabolism compared to ale strains that are critical for determining lager beer characteristics.

Several ale and lager yeast strains have been used to explore the mechanisms of maltose and maltotriose uptake in wort. A 16°Plato all-malt wort was used in a 30 liter static fermenter at 15°C (Figure 7.18). Under these conditions, lager strains utilized maltotriose more efficiently than ale strains, whereas maltose utilization efficiency was not dependent on the type of brewing strain (Zheng *et al.*, 1994). This supports the proposal that maltotriose and maltose possess independent, but closely linked, uptake (permease) systems (Russell and Stewart, 1980a). In addition, this consistent difference between ale and lager strains supports the observation that ale strains appear to have greater difficulty than lager strains in completely fermenting wort, particularly in high-gravity wort (Stewart *et al.*, 1995).

In order to investigate the MAL gene cassettes further, a strain with two *MAL2* and two *MAL4* genes copies was constructed using hybridization techniques. The wort fermentation rate was compared to a strain containing only one copy of *MAL2*. As expected, the overall fermentation rate with the strain containing multiple *MAL* genes was considerably faster than the strain containing the single copy *MAL2* (Figure 7.19). The principal reason for this faster fermentation rate was due to an increased rate of maltose uptake and subsequent metabolism compared to the yeast strain containing the single *MAL2* copy (Figure 7.20).

As already discussed, wort contains unfermentable dextrins (Table 7.2). These dextrins remain in the finished beer and thus give it mouth-feel and contribute to its calorific value (Brenner, 1980). To produce a low-calorie beer the dextrins must be reduced, by any of a number of techniques. One method would be to employ a yeast strain that possesses an ability to metabolize wort dextrins. The fact that there is a grouping of yeast — *S. cerevisiae* var. *diastaticus* — that is taxonomically closely related to brewer's yeast strains has already been discussed in this chapter. These strains have the genetic ability to produce an extracellular glucoamylase that can hydrolyze the dextrins to glucose which will be taken up by the yeast during wort fermentation. These genes have been identified as

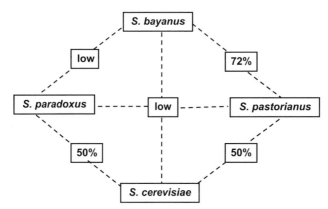

FIGURE 7.17   The *Saccharomyces sensu stricto* **group for ale and lager strains.** *(Adapted from Pederson 1995.)*

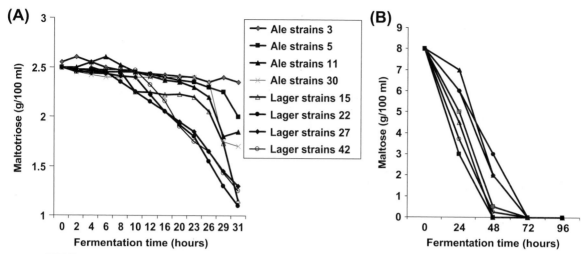

FIGURE 7.18 (A) Maltotriose and (B) maltose uptake profiles from a 16°Plato wort. *(From Zheng et al., 1994.)*

a polymeric set of three genes: *STA1/DEX1*, *STA2/DEX2*, and *STA3/DEX3* (Erratt and Stewart, 1978). A yeast strain incorporating these genes was constructed and its fermentation characteristics were assessed during a wort fermentation. The amylolytic yeast exhibited a faster fermentation rate and a lower final wort degree Plato than the control yeast that was unable to metabolize wort dextrins (Figure 7.21) (Erratt and Stewart, 1981).

The extracellular glucoamylase produced by this group of yeast is thermotolerant, probably because it is heavily glycosylated (it is a mannoprotein). Therefore, the glucoamylase was not inactivated during pasteurization (12 PU ≡ 12 minutes at 60°C) of the low-dextrin beer. This resultant beer produced with an amylolytic yeast contained increasing concentrations of beer and became sweeter and sweeter (Figure 7.22). Nevertheless, a low-dextrin beer, Nutfield Lyte, produced with a glucoamylase-containing yeast, has been produced on a semi-production scale as part of a collaborative project between the Brewing Research Foundation (now Campden BRI, Brewing Division) and Heriot-Watt University (Figure 7.23). This strain has been approved for use on a production basis by the UK's Novel Food Products and Processes (Baxter, 1995). However, it is not currently being used on a production basis for brewing. Indeed, brewers are still reluctant to use genetically manipulated yeast strains (Sofie *et al.*, 2010).

A second method of producing low-dextrin beer is to use bacterial amylases (exoenzymes) that are added either to the mash mixer or to the fermenter. Further details on the use of these exoenzymes will be discussed in Section V.

Several factors in the metabolism of wort sugars during brewing and distilling have to be considered. The effects of high-gravity brewing are particularly important and some aspects of this area will be reviewed in the next subsection.

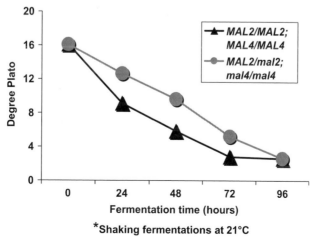

*Shaking fermentations at 21°C

FIGURE 7.19 Fermentation profile of a 16°Plato wort with a diploid yeast strain containing multiple maltose (*MAL*) genes. *(From Stewart, 2006.)*

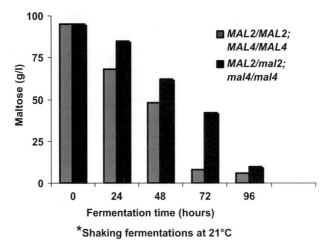

*Shaking fermentations at 21°C

FIGURE 7.20   Uptake of maltose from 16°Plato wort by a diploid yeast strain containing multiple (*MAL*) genes. *(From Stewart, 2006.)*

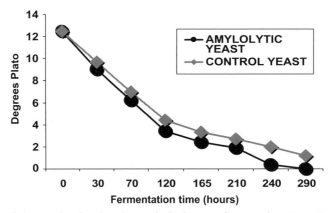

FIGURE 7.21   Effect of glucoamylase in a brewing strain during wort fermentation. *(From Erratt and Stewart, 1978.)*

D. Wort Free Amino Nitrogen

The complex composition of wort has already been discussed, with this medium containing many important ingredients. As well as the sugar spectrum, the level of FAN is important. Wort FAN consists of three components: amino acids, small peptides, and ammonia. FAN is a general measure of a yeast culture's assimilable nitrogen. It is a good index for yeast growth and therefore fermentation efficiency and sugar uptake (Inoue, 1992; Inoue and Kashihara, 1995). Wort FAN is essential for the formation of new yeast amino acids, the synthesis of new structural

FIGURE 7.22   Production of glucose in pasteurized beer during storage at 21°C. *(From Erratt and Stewart, 1978.)*

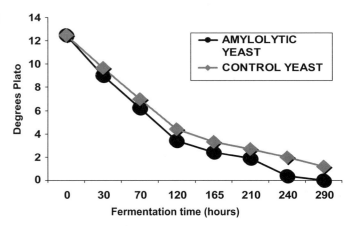

FIGURE 7.23 **Brewing Research Foundation International: Nutfield Lyte.** *(Source: author's own photograph.)*

and enzymatic proteins, cell proliferation, and cell viability and vitality. Lastly, FAN levels have a direct influence on flavor compounds in beer (for example, higher alcohols and esters).

There are differences between lager and ale yeast strains with respect to wort assimilable nitrogen uptake characteristics (O'Connor-Cox and Ingledew, 1989). Nevertheless, with all brewing strains the amount of wort FAN content required by yeast under normal brewery fermentation is directly proportional to yeast growth and also affects certain aspects of beer maturation, for example diacetyl management (Barton and Slaughter, 1992) (details described in the next subsection). There has been considerable polemic regarding the minimal FAN required to achieve satisfactory yeast growth and fermentation performance in conventional gravity (10−12°Plato) wort and it is regarded to be 130 mg FAN/l. For rapid attenuation of high-gravity wort (> 16°Plato), increased levels of FAN are required (Casey *et al.*, 1984). However, optimum wort FAN levels differ from fermentation to fermentation and from yeast strain to yeast strain. Furthermore, the optimum FAN values change with different wort sugar levels and type (Lekkas *et al.*, 2005).

During the 1960s, Margaret Jones and John Pierce, working in the Research Department of the Guinness Brewery in Park Royal, London, conducted notable studies on nitrogen metabolism during brewing, mashing, and fermentation. They reported that the absorption and utilization of exogenous nitrogenous wort compounds and their synthesis intracellularly are controlled by three main factors: (1) the total wort concentration of assimilable nitrogen; (2) the concentration of individual nitrogenous compounds and their ratio; and (3) the competitive inhibition of the uptake of these components (mainly amino acids) via various permease systems (Jones and Pierce, 1964).

Jones and Pierce (1964) established a unique classification of amino acids according to their rates of consumption during brewing wort fermentation (Table 7.9). There are four groups of amino acids. Three groups of wort amino acids are taken up at different stages of fermentation and the fourth consists of only one amino acid, proline (the

TABLE 7.9 Order of Wort Amino Acids and Ammonia Uptake During Fermentation

Group A	Group B	Group C	Group D
Fast absorption	*Intermediate absorption*	*Slow absorption*	*Little or no absorption*
Glutamic acid	Valine	Glycine	Proline
Aspartic acid	Methionine	Phenylalanine	
Asparagine	Leucine	Tyrosine	
Glutamine	Isoleucine	Tryptophan	
Serine	Histidine	Alanine	
Threonine		Ammonia	
Lysine			
Arginine			

From Jones and Pierce (1964).

largest concentration amino acid in wort), which is not taken up during brewing fermentations (Figure 7.24). This is because of the anaerobic conditions that prevail late in the fermentation. When this classification was developed, the methodology used (liquid chromatography for measuring individual amino acids, etc.) was iconic. Similar measurements today use automated computerized high-performance liquid chromatography (HPLC), and it is difficult to envisage the challenges that were overcome 50 years ago. The Jones and Pierce (1964) amino acid classification is the basis of today's understanding of the relative importance of individual wort amino acids during fermentation and manipulating wort nitrogen levels by the addition of yeast extract or specific amino acids during high-gravity brewing. However, this assimilation pattern is often specific to the conditions employed. The nutritional preferences of the yeast strain are perhaps the most significant. Because of the differences in malting barley varieties, brewing conditions, and yeast strains used worldwide in the brewing industry, a more detailed review is desirable.

Many studies (Lekkas *et al.*, 2007) have examined the absorption and utilization of wort amino acids and we now have a clear idea of their role during fermentation. However, approximately 30% of incorporated nitrogen compounds come from sources other than amino acids. Although the utilization of small peptides by brewing yeasts was confirmed before the 1950s, an understanding of their role in yeast nitrogen requirements is still limited. Small peptides can be used as nutritional sources of amino acids as carbon or nitrogen sources and precursors of cell wall peptides during yeast growth, although growth is much slower when they are the sole nitrogen source (Ingledew and Patterson, 1999). Polypeptides are also used as a substrate because yeasts can generate proteolytic enzymes extracellularly to provide additional assimilable nitrogen to the cells.

Most brewing yeast strains transport peptides with no more than three amino acid residues but this limit is strain dependent (Marder *et al.*, 1977; Nisbet and Payne, 1979). Nevertheless, small wort peptides are an important source

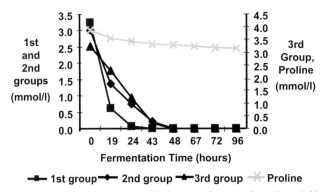

FIGURE 7.24   Amino acid absorption pattern during wort fermentation. *(From Lekkas et al., 2007.)*

of assimilable nitrogen and 20–40% of wort oligopeptides are used during fermentation. In a similar way to single amino acids, peptides probably contribute to beer character and flavor. Wort peptides are taken up in specific order dependent on the amino acid composition. The determination of small (two and three amino acid units) peptides has been an impediment to this area of research. Recently, a method to measure these small oligopeptides has been developed (Lekkas *et al.*, 2009). In brief, the sample is deproteinized, the supernatant ultrafiltered through a membrane with a 500 Da exclusion limit, the filtrate is acid and alkaline hydrolyzed, and the hydrolysate subjected to HPLC analysis.

During fermentation, the yeast secretes a number of proteinases into the fermented wort (mainly proteinase A). Therefore, hydrolysis of medium-sized peptides occurs to smaller peptides, which can be taken up by yeast, and this continues throughout fermentation (Lekkas *et al.*, 2007). This fact highlights a difference between the uptake of wort sugars and FAN during brewing. The spectrum of wort sugars at the beginning of fermentation is fixed because after wort boiling, all malt amylases have been inactivated, whereas because of proteinase secretion by yeast, the spectrum of nitrogen compounds is dynamic. It is worthy of note that stress effects on yeast can increase proteinase secretion. For example, this occurs to a greater extent during the fermentation of high-gravity worts (Cooper *et al.*, 2000).

E. Wort Fermentation

As previously discussed, wort is a very complex medium and fermentation occurs largely in static systems (except for the agitation engendered as a result of carbon dioxide evolution as part of the glycolytic pathway), usually in large cylindroconical vessels. The metabolism of wort sugars and FAN has already been discussed, but a plethora of other metabolic reactions also occurs; indeed, there are too many to discuss all of them in detail. Consequently, only three reactions will be considered: (1) yeast management, including the initial stages of fermentation, when the intracellular yeast glycogen is used in the presence of oxygen to synthesize unsaturated fatty acids (UFAs) and sterols, which are incorporated into the yeast membrane structure; (2) the later stages of fermentation, which overlap into maturation and involve the management of diacetyl and other vicinal diketones (VDK); and (3) the harvesting (cropping) of the yeast culture for reuse by flocculation and/or centrifugation.

Yeast management in the brewing context refers to yeast handling between fermentations and into a subsequent fermentation. The critical parameter for all stages of yeast management (cropping, storage, acid washing, propagation, and repitching) is to maintain the viability and vitality of cultures in order to ensure that when the yeast is pitched into wort the lag phase is kept to a minimum. At the beginning of fermentation, synthesis of UFAs and sterols, which are essential for cell membrane components, occurs at the expense of glycogen (which is the yeast culture's intracellular storage carbohydrate) (Figure 7.25) (Quain *et al.*, 1981). This must occur to ensure a normal growth pattern of the yeast population during the wort fermentation process. For many years it has been known that yeast cells are unable to synthesize UFAs and sterols under strictly anaerobic conditions (Andreason and Stier, 1953). Consequently, oxygen (by aeration or, increasingly, from the use of gaseous sterile oxygen) is supplied to the wort in the initial stages of fermentation and has been the subject of considerable polemic for many years (Kirsop, 1974). Indeed, with the advent of high-gravity brewing, the concentrated medium requires increasing concentrations of dissolved oxygen at the beginning of fermentation.

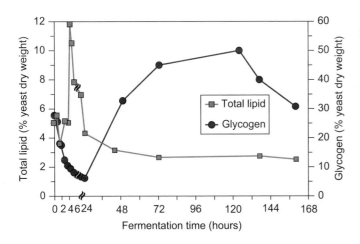

FIGURE 7.25 Intracellular concentration of glycogen and lipids in a lager yeast strain during fermentation of a 15°Plato wort. *(From Quain et al., 1981.)*

FIGURE 7.26   Fermentation of diacetyl and 2,3-pentanedione as by-products of pathways leading to the formation of valine and isoleucine. *(From Wainwright, 1973.)*

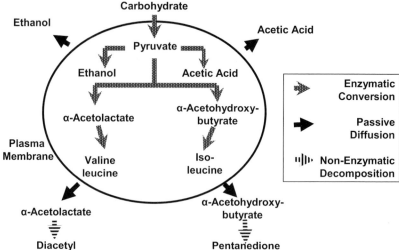

The VDKs diacetyl (2,3-butanedione) and 2,3-pentanedione impart an unwanted butterscotch or stale milk aroma to beer. Quantitatively, diacetyl is the most important since its flavor threshold is approximately 0.1 mg/l and is 10-fold lower than that of 2,3-pentanedione. The organoleptic properties of vicinal diketones contribute to the overall palate and aroma of some ales but in most lagers they impart an undesirable character. An exception to this is that in some lagers from the Czech Republic, VDKs can occur above threshold. A critical aspect of the management of lager fermentation and subsequent processing (maturation, aging, or lagering) is to ensure that the mature beer contains concentrations of VDKs lower than their flavor threshold.

Both VDKs arise in beer as by-products of the pathways leading to the formation of valine and isoleucine (Wainwright, 1973). The α-acetohydroxy acids, which are intermediates in these biosyntheses, are in part excreted into the fermenting medium. Here, they undergo spontaneous oxidative decarboxylation (Figure 7.26), giving rise to VDKs. Diacetyl is subsequently taken up by the yeast in suspension and reduced to acetoin and ultimately 2,3-butanediol (Figure 7.27) (Inoue, 1992). The flavor threshold concentration of the diols is relatively high and, therefore, the final reductive stages of VDK metabolism are critical in order to obtain a beer with acceptable organoleptic properties.

The reduction of VDKs in the later stages of fermentation and during maturation requires the presence of adequate yeast in suspension in the fermenting wort. Thus, where the yeast is particularly flocculent (this phenomenon will be discussed next), premature separation will be reflected by low rates of diacetyl reduction and potentially elevated levels in finished beer. Diacetyl removal is also affected by the physiological condition of the yeast. When the

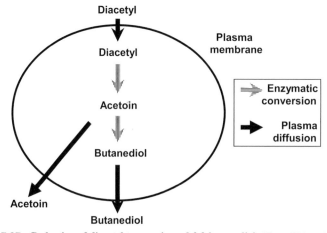

FIGURE 7.27   Reduction of diacetyl to acetoin and 2,3-butanediol. *(From Wainwright, 1973.)*

pitching yeast is in poor condition, such that the primary fermentation performance is suboptimal, the yeast present during the later stages will be stressed and the period of diacetyl reduction will be prolonged.

The flocculating properties of brewer's yeast strains have already been discussed and flocculation is one of the major factors when considering important characteristics during brewing fermentations (Stewart, 1973). Flocculation has many definitions (Stewart *et al.*, 1974). However, one that has been used for many years is 'the phenomenon wherein yeast cells adhere in clumps and either sediment from the medium in which they are suspended or rise to the medium's surface' (Stewart *et al.*, 1974). This definition excludes forms of 'clumpy growth' and 'chain formation' (Stewart *et al.*, 1973, 1974), which will not be discussed further.

The importance of flocculation to crop a yeast culture at the end of primary fermentation, so that it can be reused in a subsequent fermentation, cannot be overstated.

As flocculation is a cell surface phenomenon, the cell wall structure of the culture is critical. Therefore, appropriate findings on cell wall structure will be discussed. Genetic studies on yeast flocculation began over 50 years ago (Thorne, 1951) and Gilliland (1951) confirmed that this phenomenon was an inherited characteristic with flocculence being dominant over non-flocculence. The first flocculation gene (*FLO* gene) to be studied in detail was *FLO1*. Using traditional gene mapping techniques (mating, sporulation, micromanipulation, tetrad analysis, etc.) it has been shown that *FLO1* is located on chromosome I, 33 cM from the centromere on the right-hand side of the chromosome (Russell and Stewart, 1980a).

Novel genetic techniques have been developed, the principle of which is the sequencing of the *Saccharomyces* genome (Goffeau *et al.*, 1996). This sequenced laboratory strain contains five *FLO* genes, four located near the chromosome telomeres *FLO1*, *FLO5*, *FLO9*, and *FLO10*, and one neither at the centromere nor at the telomeres, *FLO11* (Querol and Bond, 2009). These genes encode lectin-like proteins which are also known as adhesins, zymolectins, or flocculins. The widely accepted model for yeast flocculation describes it as the result of the interaction between adhesins and mannans, polysaccharides built up of mannose residues, present on mannoproteins in the cell wall (Klis *et al.*, 2006). In most laboratory strains (haploid and diploid), added mannose will block adhesin-binding sites and thus inhibit flocculation by preventing the adhesions binding the mannose present on neighboring cells (Stratford, 1989). A similar adhesion of considerable industrial importance is responsible for the mannose-, glucose-, and maltose-sensitive 'new flo'-type of lager yeast strains (Teunissen and Steensma, 1995). In this case, competitive binding of such carbohydrates by this adhesion takes place and ensures that flocculation occurs only at the appropriate stage in the wort fermentation; namely, when all fermentable carbohydrates have been depleted and, hopefully, when VDK levels (particularly diacetyl) are under control.

It has been reported that yeast strains that exhibit ideal properties on day one of a fermentation can rapidly change and flocculate too early, or late flocculation onset occurs, or they may even lose the ability to flocculate altogether. A plethora of studies has described the presence of mutated sequences in the *FLO1* gene (Ogata *et al.*, 2008). Frequent intragenic recombination events will typically result in the net loss or gain of tandem repeat units. Expansion of the *FLO1* tandem repeat domain size results in stronger flocculation.

The genetic variability of flocculation genes has important consequences for studies and applications targeting these genes in brewing and distilling strains which have unknown genomes because of their polyploid and aneuploid characteristics. It has been revealed that there is considerable genetic variability in the chromosomal regions where flocculation genes are located. In future, knowledge of yeast flocculation mechanisms and control, particularly in the context of the brewing process, research, and industrial application, should be adapted to the flocculation gene families present in brewing yeast strains.

The other way to crop yeast for reuse and to clarify primary fermented wort is with the use of a disk stack centrifuge. As this process cannot be regarded as a biochemical process, it is beyond the scope of this chapter. The use of centrifuges in brewing has recently been reviewed (Chlup and Stewart, 2011). However, when a yeast is passed through a centrifuge it experiences mechanical and hydrodynamic shear stresses. This shear stress can cause a decrease in cell viability and flocculation, cell wall damage, increased extracellular proteinase A levels, hazier beers, and reduced foam stability.

V. EXOENZYMES IN BREWING

This chapter has endeavored to discuss many (but not all) of the important endoenzymes involved in the brewing process. As already described here, several enzymes of microbiological origin (exoenzymes) are being applied to the brewing process, many of which have been developed by genetic manipulation. The use of exoenzymes can be considered according to their use in the brewing process: adjunct liquefaction, malt and cereal enhancement (high-adjunct brewing), smooth brewing operations, attenuation control, maturation, stability enhancement, troubleshooting, and innovations. Many exoenzymes are used in brewing as troubleshooting palliatives to try to overcome problems. It is not the intention here to discuss in detail each commercially available enzyme, because of their proprietary characteristics.

A. Adjunct

Adjunct (unmalted starch) is usually converted to fermentable sugars using enzymes from malt. However, the use of bacterial α-amylase with high heat stability instead of malt can offer the brewer the possibility of a simpler and less expensive process. There are many brewers who maintain that with high-quality malt containing a good enzyme level, with high diastatic power (DP), it does not matter that some activity is lost during the adjunct liquefaction process at 100°C. When using other malted cereals such as sorghum (or as an adjunct), where the DP is considerably less than barley malt, high dosages of α-amylase and pentosanase can be used as exoenzymes.

High-adjunct brewing requires separate cooking and liquefaction stages and the proportion of the malt in the grist may not be high enough to provide sufficient amylase activity for efficient starch conversion to fermentable sugars. This problem can be overcome by adding a saccharifying fungal or bacterial α-amylase either to the mash tun or at the start of fermentation. In addition, when small malt proportions are used, there is a possibility that insufficient FAN will be dissolved in the wort to provide good yeast nutrition. To overcome this problem, a protease can be used to enhance the wort FAN level, assuming there is sufficient protein to be hydrolyzed.

B. Barley Brewing

Barley brewing is the ultimate in high-adjunct brewing without any barley malt being used. This process has been investigated for its economic advantages. In addition, a ban on the import of barley malt was introduced in Nigeria to conserve valuable foreign currency. This ban forced Nigerian breweries to develop methods for brewing lager beer entirely from locally grown raw materials such as sorghum, maize (corn), and sorghum malt, although very little of this malt was available at the time. Brewing with this type of raw material has several problems, such as a lack of enzymes from malt, low nitrogen with the use of proteases, wort separation (enhanced with a mash filter), and gelatinization temperature, which involves the need to cook large proportions of the grist at temperatures much higher than the optimum of the available saccharification enzymes. One way to solve this problem is to use very low water-to-grist ratios for the cooker mashes (thick mashes) and to use low-temperature water and very high ratios of water-to-grist (thin mashes) for mashing the part of the grist that does not need cooking.

VI. WORT SEPARATION AND BEER FILTRATION

Wort separation and beer filtration are two common bottlenecks in the brewing process. Poor lautering not only causes a loss in production capacity, but can also lead to loss of extract yield. Furthermore, a slow lautering process negatively affects the quality of the wort, which may lead to problems with beer filtration, flavor, and stability.

A very important factor influencing filtration rates is wort viscosity. A thorough breakdown of β-glucans during mashing is one method leading to fast wort separation. Undegraded, dissolved β-glucans lead to worts with high viscosity, resulting in slower diffusion through the spent grains and less efficient extraction of fermentable sugars. Undegraded wort β-glucans that are carried into the fermenter will also negatively influence beer filtration, reduce filter capacity, and increase the consumption of filter aids (kieselgur or diatomaceous earth). Given the raw material quality and composition, as well as milling procedures, the solution to these problems is enhanced mashing. Sufficient levels of β-glucanase and pentosanase should be present in the mash, and mashing conditions should permit these enzymes to operate effectively. A wide range of β-glucanase/pentosanase preparations is available for reducing wort viscosity caused by glucans and pentosans. An unconventional method of speeding up the lautering process by reducing wort viscosity is to mash-off and lauter at 95°C. However, high mashing temperatures can negatively influence beer flavor and stability.

The availability of industrially produced amylases provides the brewer with a means to control the degree of fermentation. A wort produced under normal brewing conditions will generally give an apparent attenuation of 80–85% and approximately 25% of residual extract will be present in the beer. As already discussed, this residual extract, mainly non-fermentable short-chain dextrin material, contributes to beer mouth-feel and its calorific value (Brenner, 1980). Sometimes the residual extract specifications are not achieved because of malt quality deficiencies and/or mashing

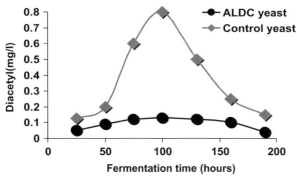

FIGURE 7.28 Effect of α-acetolactate decarboxylase (ALDC) expression in a brewing yeast strain on diacetyl metabolism during wort fermentation. *(Graphs plotted from data by Yamano et al., 1994.)*

process problems, so corrections have to be made. Alas, where non-traditional raw materials are used, there may also be a need to use external saccharifying amylases to achieve the required degree of fermentation. Another possibility is to create new products by increasing the degree of fermentation even further by reducing the residual unfermentable dextrins; light (lite) and dry beers fall into this category. The goal might be to make small adjustments or to completely remove all of the residual dextrins, resulting in what is known as a super-attenuated beer. Such a beer will have fewer calories and less mouth-feel than a beer of normal attenuation (assuming a similar alcohol content).

Traditional brewing methods are available to achieve these adjustments, usually by modifying the mashing regime and using high DP malts. However, a range of exoenzymes exists for the same purpose and includes a number of saccharifying enzymes such as α-amylase, glucoamylase, and pullulanase (debranching enzyme) that are useful, practical tools when the attenuation needs to be adjusted for one reason or another.

The management of diacetyl and other VDKs during the later stages of fermentation and during maturation has already been discussed in this chapter. An alternative way to manage diacetyl and related compounds is to use the enzyme α-acetolactate decarboxylase (ALDC). This enzyme can decarboxylate α-acetolactate into the flavor-inactive acetoin and bypasses the formation of diacetyl. ALDC is obtained from the generally regarded as safe (GRAS) Gram-negative microorganism *Acetobacter aceti*, which is also used to metabolize ethanol into acetic acid during vinegar manufacture. The availability of an ALDC (marketed under the tradename Maturex®) now offers the brewer a possibility of bypassing the conversion of α-acetolactate to diacetyl, which is a slow reaction. However, ALDC must be added to the wort at the start of fermentation.

A novel alternative approach was to clone the gene for ALDC into a brewing strain. The ALDC gene was obtained from a strain of *A. aceti*. Wort fermentation trials with the ALDC yeast were conducted and compared to the uncloned strain as control. The diacetyl production and reduction profile were profoundly different compared to the uncloned control culture (Yamano *et al.*, 1994) (Figure 7.28). Because of the presence of ALDC, the α-acetolactic acid was not spontaneously converted to diacetyl, but to acetoin instead. The overall fermentation performance of some cloned yeast strains can be adversely affected compared to the uncloned control strain. This was not the situation with the ALDC yeast (Figure 7.29).

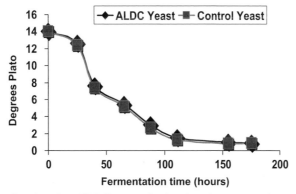

FIGURE 7.29 Effect of α-acetolactate decarboxylase (ALDC) expression in a brewing yeast strain on overall fermentation rate during wort fermentation. *(Graphs plotted from data by Yamano et al., 1994.)*

Beer stability can be enhanced by the use of exoenzymes. This is particularly true of physical stability, also called haze formation. With a few notable exceptions, consumers prefer their beer to be bright and free of particles. Stored beer, particularly at room temperature, has the potential to produce haze and its brightness will be compromised. This haze consists of an association of small polymerized polyphenols and polypeptides which are high in the amino acid proline (Siebert and Lynn, 1997). This permanent haze does not redissolve even when the beer is warmed to 30°C or higher. The balance between flavenoid polyphenols and 'sensitive' proline-containing hydrophilic proteins largely dictates the physical (colloidal) stability of beer. Haze will not form, or its formation will be retarded, when either of these components is removed or the factors promoting the interaction are largely excluded. A number of procedures can be employed to retard and/or prevent haze formation:

- Prevent the formation of large amounts of complex protein degradation products during beer production, particularly mashing.
- Remove some of the polyphenols or the sensitive proteins during brewing with specific adsorbants (polyvinylpolypyrrolidone or silica gel).
- Store maturing beer cold (0°C or less) to precipitate haze precursors — this is a time consuming process.
- Store packaged beer cold (2—4°C) to retard haze formation.
- Hydrolyze, with exoenzymes, the complex sensitive polypeptides. Broad-spectrum proteases, e.g. papain (from papaya), ficin (from figs), and biomelain (from pineapple) have been used for many years to enhance the physical stability of beer. However, these enzymes are non-specific; therefore, as well as hydrolyzing sensitive polypeptides, they will hydrolyze the hydrophobic polypeptides that stabilize beer foam, and foam-enhancing agents, such as polyglycol alginate, have to be added to put a foam head on a glass of beer. Recently, a proline-specific endopeptidase has been developed. This enzyme specifically hydrolyzes sensitive polypeptides because of their enhanced proline content and promotes the physical stability of beer with no effect on foam stability (Lopez and Edens, 2005).

VII. CONCLUSIONS

Brewing is largely, but not exclusively, a biochemical/enzymatic process. However, the current process also involves plant science, microbiology (closely related to biochemistry), chemistry, physics, engineering, process control, and flavor assessment. It is normally a batch multienzyme process that involves a number of clearly defined, but over-lapping, stages. The modern-day brewing process has two objectives. The first objective is efficiency, making maximum use of the available overheads, both fixed and variable. The second objective is to produce high-quality drinkable beers that possess enhanced stability. Brewing is a traditional process with a long history that focuses upon four major raw materials: barley malt; hops; water; and yeast. The advent of high-quality microbial enzymes, some of which have been developed with the use of genetic manipulation, has permitted the use of non-malt enzymes in the brewing process that will, under certain circumstances, assist the efficiency of the process and the quality and stability of the product. However, brewers are still reluctant to use such enzymes.

ACKNOWLEDGMENT

The author is grateful to Anne Anstruther for her invaluable assistance with the development of this manuscript.

REFERENCES

Andreason, A.A., Stier, T.J.B., 1953. Anaerobic nutrition of *Saccharomyces cerevisiae*. 1. Ergosterol requirements for growth in a defined medium. J. Cell. Comp. Physiol. 41, 23—26.

Andrews, J.M.H., Hancock, J.C., Ludford-Brooks, J., Murfin, I.J., Houldsworth, L., Phillips, M., 2011. 125th anniversary review: Some recent engineering advances in brewing and distilling. J. Inst. Brew. 117, 23—32.

Bamforth, C.W., 2009. Beer: A Quality Perspective. Academic Press, New York.

Barnett, J.A., 1992. The taxonomy of the genus *Saccharomyces Meyen ex Reese*: a short review for non-taxonomists. Yeast 8, 1—23.

Barton, S., Slaughter, J.C., 1992. Amino acids and vicinyl diketone concentration during fermentation. Tech. Q. Master Brew. Assoc. Am. 29, 60—63.

Baxter, E.D., 1995. The application of genetics in brewing. Fermentation 8, 307—311.

Bisson, L.F., Coors, D.M., Frankel, A.L., Lewis, D.A., 1993. Yeast sugar transporters. CRC Crit. Rev. Biochem. Mol. Biol. 284, 269—308.

Brenner, M.W., 1980. Beers of the future. Tech. Q. Master Brew. Assoc. Am. 17, 185—195.

Briggs, D.E., 1998a. The principles of mashing. In: Malts and Malting. Blackie Academic & Professional, London, pp. 229—244.

Briggs, D.E., 1998b. An introduction to malts and their uses. In: Malts and Malting. Blackie Academic & Professional, London, pp. 1−34.

Briggs, D.E., 1998c. Grains and pulses. In: Malts and Malting. Blackie Academic & Professional, London, pp. 35−78.

Casey, G.P., Magnus, C.A., Ingledew, W.M., 1984. High-gravity brewing: effects of nutrition on yeast composition, fermentative ability and alcohol production. Appl. Environ. Microbiol. 48, 639−646.

Chlup, P.H., Stewart, G.G., 2011. Centrifuges in brewing. Tech. Q. Master Brew. Assoc. Am. 48, 46−50.

Cooper, D.J., Stewart, G.G., Bryce, J.H., 1998. Hydrophobic polypeptide extraction during high gravity mashing − experimental approaches for its improvement. J. Inst. Brew. 104, 283−287.

Cooper, D.J., Stewart, G.G., Bryce, J.H., 2000. Yeast proteolytic activity during high and low gravity wort fermentation and its effect on head retention. J. Inst. Brew. 106, 197−201.

Erratt, J.A., Stewart, G.G., 1978. Genetic and biochemical studies on yeast strains able to utilize dextrins. J. Am. Soc. Brew. Chem. 36, 151−161.

Erratt, J.A., Stewart, G.G., 1981. Fermentation studies using Saccharomyces diastaticus yeast strains. Dev. Ind. Microbiol. 22, 577−586.

Flannigan, B., 1996. The microflora of barley and malt. In: Priest, F.G., Campbell, I. (Eds.), Brewing Microbiology. Chapman & Hall, London, pp. 83−126.

Gilliland, R.B., 1951. The flocculation characteristics of brewing yeasts during fermentation. In: Proceedings of the European Brewing Convention. Elsevier, Amsterdam, pp. 35−58.

Goffeau, A., Barrell, B.G., Bussey, H., Davis, R.W., Dujon, B., Feldmann, H., Galibert, F., Hoheisel, J.D., Jacq, C., Johnston, M., Louis, E.J., Mewes, H.W., Murakami, Y., Philippsen, P., Tettelin, H., Oliver, S.G., 1996. Life with 6000 genes. Science 274, 563−567.

Ingledew, W.M., Patterson, C.A., 1999. Effect of nitrogen source and concentration on the uptake of peptides by a lager yeast in continuous culture. J. Am. Soc. Brew. Chem. 47, 102−108.

Inoue, T., 1992. A review of diacetyl control technology. In: Proceedings of the Institute of Brewers, pp. 109−113. Australia/New Zealand Section, 23rd Convention, Melbourne.

Inoue, T., Kashihara, T., 1995. The importance of indices related to nitrogen metabolism in fermentation control. Tech. Q. Master Brew. Assoc. Am. 32, 109−113.

Jones, M., Pierce, J., 1964. Absorption of amino acids from wort by yeasts. J. Inst. Brew. 70, 307−315.

Jones, M., Pierce, J., 1965. Nitrogen requirements in wort − practical applications. In: Proceedings of the European Brewery Convention 10th Congress, pp. 182−194. Stockholm.

Kirsop, B., 1974. Oxygen in brewery fermentations. J. Inst. Brew. 80, 252−259.

Klis, F.M., Boorsma, A., De Groot, P.W., 2006. Cell wall construction of Saccharomyces cerevisiae. Yeast 23, 185−202.

Leiper, K.A., Stewart, G.G., McKeown, I.P., 2003. Beer polypeptides and silica gel. J. Inst. Brew. 109, 57−79.

Lekkas, C., Stewart, G.G., Hill, A., Taidi, B., Hodgson, J., 2005. The importance of free amino nitrogen in wort and beer. Tech. Q. Master Brew. Assoc. Am. 42, 113−116.

Lekkas, C., Stewart, G.G., Hill, A.E., Taidi, B., Hodgson, J., 2007. Elucidation of the role of nitrogenous wort components in yeast fermentation. J. Inst. Brew. 113, 183−191.

Lekkas, C., Hill, A.G., Taidi, B., Hodgson, J., Stewart, G.G., 2009. The role of small peptides in brewing fermentations. J. Inst. Brew. 115, 134−138.

Libkinda, D., Hittinger, C.T., Valério, E., Gonçalves, C., Dover, J., Johnston, M., Gonçalves, P., Sampaio, J.P., 2011. Microbe domestication and the identification of the wild genetic stock of lager-brewing yeast. Proc. Natl. Acad. Sci. U.S.A. 1105430108.

Lopez, M., Edens, L., 2005. Effective prevention of chill-haze in beer using an acid proline-specific endoprotease from Aspergillus niger. J. Agric. Food Chem. 53, 7944−7949.

Marder, R., Becker, J.M., Naider, F., 1977. Peptide transport in yeast: utilization of leucine and lysine containing peptides by Saccharomyces cerevisiae. J. Bacteriol. 131, 906−916.

Nisbet, T.M., Payne, J.W., 1979. Peptide uptake in Saccharomyces cerevisiae. Characteristics of transport systems shared by di- and tri-peptides. J. Gen. Microbiol. 105, 127−133.

O'Connor-Cox, E.S.C., Ingledew, W.M., 1989. Wort nitrogenous sources − their use in brewing yeasts: a review. J. Am. Soc. Brew. Chem. 47, 102−108.

Ogata, T., Isumikawa, M., Kohno, K., Shibata, K., 2008. Chromosomal location of Lg-FL01 in bottom-fermenting yeasts at the FL05 locus of industrial yeast. J. Appl. Microbiol. 105, 1186−1198.

Pederson, M.B., 1995. Recent views and methods for the classification of yeasts. Cerevisia − Belg. J. Brew. Biotechnol. 20, 28−33.

Pratt, P.L., Bryce, J.H., Stewart, G.G., 2003. The effects of osmotic pressure and ethanol on yeast viability and morphology. J. Inst. Brew. 109, 218−228.

Priest, F.G., Stewart, G.G., 2006. Handbook of Brewing. Taylor & Francis, Boca Raton, FL.

Pulvirenti, A., Nguyen, H.-V., Caggie, C., Giudici, P., Rainieri, S., Zambonelli, C., 2000. Saccharomyces uvarum, a proper species within Saccharomyces sensu stricto. FEMS Microbiol. Lett. 192, 191−196.

Quain, D.E., Thurston, P.A., Tubb, R.S., 1981. The structural and storage carbohydrates of Saccharomyces cerevisiae changes during fermentation of wort and a role for glycogen catabolism in lipid biosynthesis. J. Inst. Brew. 87, 108−111.

Querol, A., Bond, U., 2009. The complex and dynamic genomes of industrial yeasts. FEMS Microbiol. Lett. 293, 1−10.

Rainieri, S., Zambonelli, C., Kaneko, Y., 2003. Review Saccharomyces sensu stricto systematics, genetic diversity and evolution. J. Biosci. Bioeng. 96, 1−9.

Russell, I., Stewart, G.G., 1980a. Revised nomenclature of genes that control flocculation. J. Inst. Brew. 86, 120−121.

Russell, I., Stewart, G.G., 1980b. Transformation of maltotriose uptake ability into a haploid strain of Saccharomyces spp. J. Inst. Brew. 86, 55−59.

Sandegren, E., Beling, H., 1959. Gibberellic acid in malting and brewing. In: Proceedings of the European Brewery Convention 7th Congress, pp. 278−289. Rome.

Siebert, K.J., Lynn, P.Y., 1997. Mechanisms of beer colloidal stabilization. J. Am. Soc. Brew. Chem. 55, 73−78.

Singer, C., Holmyard, E.J., Hall, A.R. (Eds.), 1954. A History of Technology. Clarendon Press, Oxford, pp. 229−244.

Sofie, M.G., Saerens, C., Duong, T., Nevoigt, E., 2010. Genetic improvement of brewer's yeast: current state, perspectives and limits. Appl. Microbiol. Biotechnol. 86, 1195−1212.

Steiner, S. H. 2009. Guidelines for Hop Buying. Hopsteiner, Mainburg.

Stewart, G.G., 1973. Recent developments in the characterization of brewery yeast strains. Tech. Q. Master Brew. Assoc. Am. 9, 183−191.

Stewart, G.G., 1999. High gravity brewing. Brewers' Guardian 128, 31–37.

Stewart, G.G., 2004. The chemistry of beer instability. J. Chem. Educ. 81, 963–968.

Stewart, G.G., 2005. Esters – the most important group of flavour-active beer components. In: Proceedings of the European Brewery Convention 30th Congress, Prague, CD Paper No. 100.

Stewart, G.G., 2006. Studies on the uptake and metabolism of wort sugars during brewing fermentations. Tech. Q. Master Brew. Assoc. Am. 43, 264–269.

Stewart, G.G., Russell, I., 1986. One hundred years of yeast research and development in the brewing industry. J. Inst. Brew. 92, 537–558.

Stewart, G.G., Russell, I., Garrison, I.F., 1973. Further studies on flocculation and co-flocculation in brewer's yeast strains. Am. Soc. Brew. Chem. Proc. 31, 100–106.

Stewart, G.G., Russell, I., Garrison, I.F., 1974. Factors influencing the flocculation of brewer's yeast strains. Tech. Q. Master Brew. Assoc. Am. 11, 155–163.

Stewart, G.G., Zheng, X., Russell, I., 1995. Wort sugar uptake and metabolism – the influence of genetic and environmental factors. In: Proceedings of the European Brewery Convention 25th Congress, pp. 403–410, Brussels.

Stratford, M., 1989. Evidence for two mechanisms of flocculation in *Saccharomyces cerevisiae*. Yeast 5, 441–445.

Teunissen, A.W., Steensma, A.Y., 1995. Review: The dominant flocculation genes of *Saccharomyces cerevisiae* constitute a new sub-telomeric gene family. Yeast 11, 1001–1013.

Thorne, R.S.W., 1951. Some aspects of yeast flocculence. In: Proceedings of the European Brewery Convention 3rd Congress, Brighton, Elsevier, Amsterdam, pp. 21–34.

Wainwright, T., 1973. Diacetyl – a review. J. Inst. Brew. 79, 451–470.

Yamano, G., Tanaka, J., Inoue, T., 1994. Cloning and expression of the gene encoding alpha-acetolactate decarboxylase from *Acetobacter aceti ssp.* in brewer's yeast. J. Biotechnol. 32, 165–171.

Younis, O.S., Stewart, G.G., 1998. Sugar uptake and subsequent ester and alcohol production in *Saccharomyces cerevisiae*. J. Inst. Brew. 104, 255–264.

Younis, O., Stewart, G.G., 1999. The effect of malt wort, very high gravity malt wort and very high gravity adjunct wort on volatile production in *Saccharomyces cerevisiae*. J. Am. Soc. Brew. Chem. 52, 38–45.

Zheng, X., D'Amore, T., Russell, I., Stewart, G.G., 1994. Factors influencing maltotriose utilization during brewery wort fermentations. J. Am. Soc. Brew. Chem. 52, 41–47.

Chapter 8

Dairy Products: Cheese and Yogurt

Arthur R. Hill and Prashanti Kethireddipalli

Department of Food Science, University of Guelph, Guelph, Ontario, Canada

Chapter Outline

I. INTRODUCTION

The production of cheese began thousands of years ago in the Middle East (Scott *et al.*, 1998). Cheese making was subsequently introduced into Europe during the period of the Roman Empire, where it was produced either in monasteries or on farms. Factory production of cheese began in the middle of the nineteenth century in both Europe and the New World, and beginning in the 1930s, many established varieties were defined by national standards of

Biochemistry of Foods. DOI: http://dx.doi.org/10.1016/B978-0-12-242352-9.00008-4

identity, such as the Appellation d'Origine Contrôlées (AOC) system in France, the Denominazione di Origine Controllata (DOC) in Italy, and the international European Community Protected Designation of Origin (PDO) standards. Standards of identity were strengthened internationally by the Stresa Convention in 1951, and later updated by European Community rules (Echols, 2008). The Center for Dairy Research, Wisconsin, online cheese directory lists more than 1400 varieties of cheese, many differing only in shape, size, degree of ripening, type of milk, condiments used, packaging, and/or region of production. Of the main commodity cheeses the most popular globally is Mozzarella, mainly due to its use on pizza.

In addition to cheese, there are a large number of fermented milk products differentiated by manufacturing processes and starter microorganisms. The main cultured product consumed in Australia, Canada, the UK, and the USA is yogurt, while other varieties of cultured milk and cream products are more popular in the Scandinavian and Eastern European countries. Figure 8.1 illustrates the principal types of fermented milk products. This chapter will discuss the manufacturing procedures, the cultures used, and the biochemical changes that take place during the production of yogurt and cheese. Figure 8.2 illustrates the principal ingredients and processes involved in cheese making.

II. MILK COMPOSITION

Cheese and yogurt depend on the growth of bacteria to produce acidity, flavor compounds, and ripening enzymes. Milk is a good source of nutrients, including carbon, nitrogen, and macrominerals; many micronutrients such as vitamins and microminerals are also available (Jenness, 1988). However, milk is unique with respect to its sugar; lactose is naturally present only in milk. Most microorganisms lack the enzyme lactase, which is required to break down lactose into its component sugars, glucose, and galactose. Lactic acid bacteria (LAB), which possess lactase, readily break down lactose and use glucose as an energy source. Furthermore, some LAB are able to convert galactose to glucose. LAB, therefore, have a competitive advantage in milk.

Milk composition data for the most common dairy species are given in Table 8.1. Cheese- and yogurt-making principles are similar for milk of all species, with some modifications required to account for the higher milk solids in species such as buffalo and sheep and for differences in the properties of caseins and enzymes, among other factors. For example, goat's milk has smaller fat globules, which allows higher fat recovery (McSweeney *et al.*, 1993b), and relative to cow's milk, the milk of most goat breeds is low in α_{s1}-casein (Moatsou *et al.*, 2004). The practical effect is that goat's milk is more suitable than cow's milk for some varieties such as Feta, but less suitable for other varieties such as Cheddar, in which flavor development is strongly dependent on the breakdown of α_{s1}-casein by rennet.

FIGURE 8.1 Principal processes used to differentiate milk into fermented dairy products. *(Adapted from Robinson et al., 2002.)*

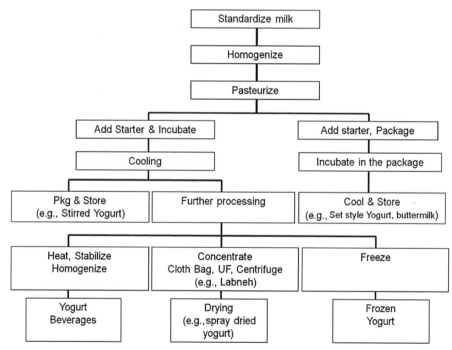

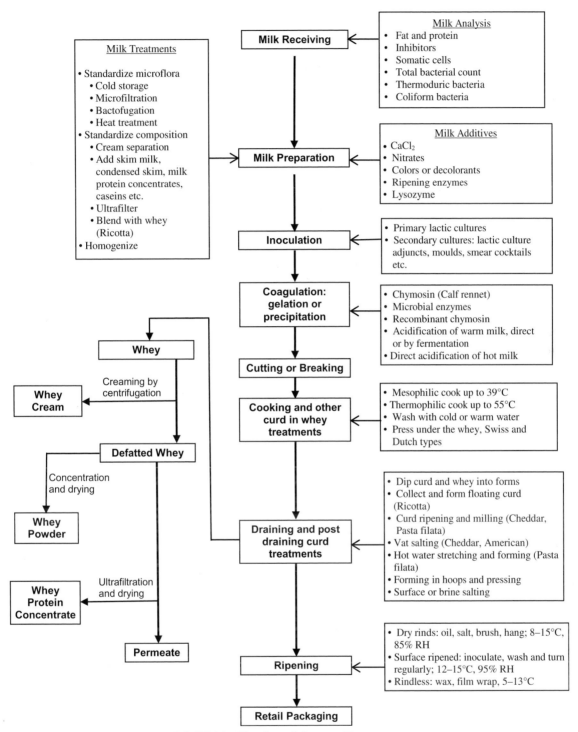

FIGURE 8.2   Flowchart of cheese-making processes.

Buffalo's milk has a higher buffering capacity than cow's milk, which means that the acidification process in the former has to be suitably modified (Ahmad *et al.*, 2008). Owing to differences in the composition of cow's and buffalo's milk, the changes to casein micelle and curd rheology during acidification were found to be qualitatively similar but quantitatively different (Ahmad *et al.*, 2008; Hussain *et al.*, 2011).

There are also important within-species differences. Throughout modern history, farmers have selectively bred dairy cattle to increase milk production or fat content or both. More recently, genetic selection has focused on other

TABLE 8.1 Typical Gross Composition (kg/100 kg) of Cow, Sheep, Water Buffalo, and Goat Milk

	Cow	Dairy Sheep	Water Buffalo	Goat
Fat	3.9	7.2	7.4	4.5
Total protein	3.2	4.6	3.8	3.2
Casein	2.6	3.9	3.2	2.6
Whey	0.6	0.7	0.6	0.6
Lactose	4.6	4.8	4.8	4.3
Ash	0.7	0.9	0.8	0.8
Total solids	12.4	17.5	16.83	12.8

Adapted from Wong *et al.* (1988).

milk properties, such as increasing the proportion of milk protein to fat (PF ratio). With respect to other solids, mineral content (mainly calcium, magnesium, and phosphorus) generally varies in proportion to protein content, and lactose content is relatively stable. Since lactose is largely a wasted component, increasing protein and fat by feed or genetic selection has economic advantages in terms of feed conversion, milk transportation costs, and waste handling.

Season is yet another factor that affects milk composition (Hill, 2011). In general, in the northern hemisphere, fat content reaches a minimum in August and a maximum in October. The protein content changes roughly in parallel with fat content, but because seasonal variations in protein are smaller, the result is a higher PF ratio during summer and a lower ratio in the winter. Casein number (casein as a percentage of total protein) is relatively constant over the seasons; that is, casein varies seasonally but mostly in proportion to the total protein. However, there is a significant positive correlation between total protein and casein number. The average value for casein number in Ontario, where Holstein Friesian is the dominant breed, was reported to be 77.0% ($N = 1067$, SD 1.07%, range 70.31−81.06%). Based on these data, casein can be predicted from total crude protein ($N \times 6.38$) as follows (Hill, unpublished):

$$\text{Casein} = (0.833 \times \text{total protein}) - 0.208$$

Milk fat is unique with respect to its diversity of fatty acids. As indicated in Table 8.2, some noteworthy lipid components include substantial amounts of short-chain fatty acids and rumenic acid (conjugated linoleic acid; C18:2, c9, t11). In its natural state, milk fat is found as globules surrounded by surface-active phospholipid membranes; the latter allow fat particles to remain dispersed in the aqueous milk medium. During the formation of milk gels, as described in Section IV, the fat globules occlude within the protein gel structure and mainly function as shortening agents that soften cheese texture. Short-chain fatty acids tend to be located in the *sn*1 or *sn*3 position (Villeneuve *et al.*, 1996) and for that reason are more subject to lipolysis with associated rancid flavor. (See also the review by Collins *et al.*, 2003a.)

The general aspects of milk proteins, which are briefly described below, have been reviewed by Fox (2003a). Apart from enzymes and minor proteins, the principal groups of proteins in milk are the caseins and whey proteins, making up roughly 80% and 20% of total lactoproteins, respectively. In addition to dispersed fat globules (1−10 μm diameter), milk's colloidal system contains casein−calcium phosphate particles called casein micelles (about 80−500 nm diameter) dispersed in a solution (serum) of lactose, whey proteins, and minerals. Bovine casein micelles are made up of four different casein fractions, the α_{s1}-, α_{s2}-, β-, and κ-caseins, in an approximate ratio of 4:1:3.5:1.5, respectively. The size of casein micelles, content of caseins, and relative proportions of individual caseins differ between species. The principal whey/serum proteins in bovine milk are β-lactoglobulin ($\approx$ 50%), α-lactalbumin ($\approx$ 20%), immunoglobulins ($\approx$ 3%; up to 10% in colostrum), bovine serum albumin (0.3−1%), and some lactoferrin (< 0.1%). β-Lactoglobulin is the most abundant whey protein in the milk of most mammals; exceptions are humans, rats, mice, guinea-pigs, and camels, in which the protein is absent. α-Lactalbumin is present in the milk of all mammals. Each molecule of β-lactoglobulin contains two intramolecular disulfide bonds and one cysteine residue. The sulfhydryl group is especially important in heated milk chemistry; following thermal denaturation, it participates

TABLE 8.2 Fatty Acid Composition of Milks from Cow, Goat, and Sheep

| Fatty Acid | Common Name | Composition (% w/w) | | | | | |
| | | Cow[a] | | Goat[b] | | Sheep[c] | |
		Typical	Range	Typical	Range	Typical	Range
C4:0	Butyric	3.9	3.1–4.4	2.2	2.0–2.4	3.5	3.1–3.9
C6:0	Caproic	2.5	1.8–2.7	2.4	2.0–2.7	2.9	2.7–3.4
C8:0	Caprylic	1.5	1.0–1.7	2.7	2.3–3.0	2.6	2.1–3.3
C10:0	Capric	3.2	2.2–3.8	10.0	8.9–11.0	7.8	5.5–9.7
C12:0	Lauric	3.6	2.6–4.2	5.0	3.9–6.2	4.4	3.5–4.9
C14:0	Myristic	11.1	9.1–11.9	9.8	7.7–11.2	10.4	9.9–10.7
C14:1	Myristoleic	0.8	0.5–1.1	0.18	0.17–0.2	0.3	0.2–0.5
C15:0		1.2	0.9–1.4	0.7	0.5–0.9	1.0	0.9–1.1
C16:0	Palmitic	27.9	23.6–31.4	28.2	23.2–34.8	26.0	22.5–28.2
C16:1	Palmitoleic	1.5	1.4–2.0	1.6	1.0–2.7	1.0	0.7–1.3
C18:0	Stearic	12.2	10.4–14.6	8.9	5.8–13.2	9.6	8.5–11.0
C18:1 cis	Oleic	17.2	14.9–22.0	–	–	–	–
C18:1 trans		3.9	–	–	–	–	–
C18:1 total		21.1	–	19.3	15.4–27.7	21.1	17.8–23.0
C18:2	Linoleic	1.4	1.2–1.7	3.2	2.5–4.3	3.2	2.9–3.6
C18:2 conj	Conjugated Linoleic	1.1	0.8–1.5	0.7	0.3–1.2	0.7	0.6–1.0
C18:3	α-Linolenic	1.0	0.9–1.2	0.4	0.2–0.9	0.8	0.5–1.0
	Minor acids	6.0	4.8–7.5	3.2	2.2–4.6	3.6	3.1–4.3

[a]Creamer and MacGibbon (1996); MacGibbon (1996)
[b]Alonso et al. (1999)
[c]Goudjil et al. (2004).

in thiol/disulfide exchange reactions with the intermolecular disulfide of κ-casein. As discussed in Section IV, this reaction severely impairs rennet coagulation and cheese-making properties of heated milk, but is essential for yogurt making. Some properties of caseins and whey proteins are listed in Tables 8.3 and 8.4, respectively, and will be discussed in the relevant sections; the structural features of the casein micelle and the mechanisms of its destabilization in cheese and yogurt making are briefly described in Section IV.

III. MILK QUALITY

Milk quality has a great influence on the quality of the resulting cheese and yogurt products. Here, some biochemical attributes of milk quality that are associated with the quality and safety of cheese and yogurt are described.

A. Types of Microorganisms and their Activity in Milk

Over a period of 1–4 days, warm raw milk (25–40°C) is first fermented by LAB to pH 4.8–4.2 depending on the acid tolerance of the LAB present and the rate of growth of other species (Figure 8.3). As the LAB approach their stationary growth phase, acid-tolerant yeasts and molds begin to utilize the lactic acid that is produced and also metabolize proteins, causing the pH to rise. The released peptides and amino acids further encourage the growth of LAB until they use up most of the lactose. When the lactose is gone, the pH begins to rise, which causes proteolytic

TABLE 8.3 Principal Caseins and Some Properties of Importance to Cheese Making

Name	Symbol	% of Casein	Properties
Alpha-s1 casein	α_{s1}	40	Binds Ca^{2+} strongly
			Readily hydrolyzed by chymosin during cheese ripening
			Less susceptible to natural milk protease, plasmin
Alpha-s2 casein	α_{s2}	10	Binds Ca^{2+} strongly
			Limited susceptibility to chymosin action
			Eight peptide bonds are susceptible to plasmin
Beta-casein	β	35	Binds Ca^{2+} weakly
			Partially soluble in cold milk
			Broken down by plasmin but not chymosin
			Peptide β-CN (f193–209) and related fragments are very bitter
Kappa-casein	κ	15	Does not bind calcium
			Stabilizes casein micelles; its hydrolysis by chymosin initiates milk coagulation
			Covalently binds to whey proteins when milk is heated
			Para-κ-casein is not broken down by chymosin or plasmin

bacteria to proliferate, and along with yeasts and molds they elevate the pH to neutral or slightly alkaline values. Physical changes at these respective stages include coagulation (pH 5.0–4.6) and increased translucence due to proteolytic breakdown and resulting dissociation of casein micelles. In the manufacture of dairy products, these natural biochemical/physical processes are directed to specified ends. The following subsections describe several types of bacteria that are typically present in raw milk, categorized according to how they change the properties of milk. Often these changes are negative (spoilage), but many of these bacteria, along with contaminants transferred to the milk and/or cheese during manufacture, are also important adjuncts to primary and secondary cultures, particularly in aged cheese varieties.

TABLE 8.4 The Principal Whey Proteins: Some Properties of Importance to Cheese Making

Name	% of Whey Protein	Properties
β-Lactoglobulin	40	Its heat-induced ($> 70°C$) binding with κ-casein interferes with rennet coagulation
		Principal component of Ricotta cheese
α-Lactalbumin	15	Principal protein of breast milk
		Used in infant formulae
Immunoglobulins	6	Present in high proportions in colostrum
Other heat-sensitive whey proteins	4.0	Mainly includes bovine serum albumin
Heat-stable whey proteins	14	Cannot be recovered by heat–acid precipitation as in Ricotta cheese manufacture
Non-protein nitrogen	21	Consists of amino acids, ammonia, urea, and small peptides

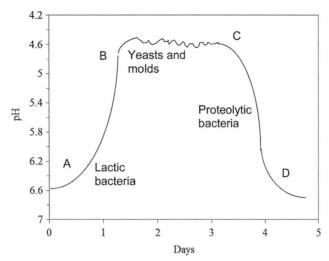

FIGURE 8.3 **Natural fermentation of raw milk.** A−B: At the natural pH of milk (6.6−6.8) and at temperatures > 20°C, lactic acid bacteria (LAB) rapidly ferment lactose to lactic acid. The lactic acid that is produced lowers pH and inhibits most bacteria, eventually even LAB. B−C: Acid-tolerant yeasts and molds begin to grow and utilize lactic acid, which permits further growth of LAB. This synergistic relation continues until all of the lactose is exhausted. C−D: Eventually, yeasts and molds are joined by proteolytic bacteria. Together they consume lactic acid and/or neutralize it with protein breakdown products. Note that all ripened cheese follows this pattern to a greater or lesser extent depending on the minimum pH reached during manufacture and the highest pH during ripening. The 'textbook' example of this pattern is Camembert cheese, which has a minimum pH of < 4.7 about 24 hours after inoculation and a ripened pH of about 7.5.

1. Lactic Acid Bacteria

LAB are: (1) non-motile, Gram positive, and catalase negative, so simple tests for catalase activity can be used to identify spoilage bacteria in lactic cultures, microaerophilic, or facultative anaerobes, which means that they create and thrive in anaerobic conditions in fermented dairy products; (2) mesophilic or thermophilic, which means that cold storage depletes their numbers and encourages the growth of psychrotrophic spoilage bacteria; and (3) with respect to morphology, cocci (spherical cells) 1 μm in diameter, or rod-shaped cells 1 μm wide and 2−3 μm long. As noted above, LAB have competitive advantages in milk because they are lactase positive, readily able to convert lactose into glucose and galactose. However, LAB prefer temperatures greater than 30°C, so depending on their initial counts psychrotrophic bacteria, including some coliforms and pseudomonads, may outgrow LAB at room temperatures. Lactic cultures are discussed further in Section V.

2. Psychrotrophic Bacteria

Psychrotrophic bacteria grow at less than 7°C. Common species in cold stored milk, which is the storage norm in most jurisdictions, are *Micrococcus*, *Bacillus*, *Staphylococcus*, *Pseudomonas*, *Flavobacterium*, and coliforms. *Pseudomonas* spp. are the most common and typically have the greatest impact on quality. *Pseudomonas fluorescens* is able to grow slowly at near 2°C and, along with other psychrotrophic species of this genus, it produces heat-stable lipases and proteases that remain active even after the bacterial cells have been killed by pasteurization (Lelievre *et al.*, 1978; Law *et al.*, 1979; Hicks *et al.*, 1982; Ellis and Marth, 1984; Cromie, 1992). Elaboration of heat-stable enzymes is maximal in the stationary phase, which most often occurs during the growth of *Pseudomonas* spp. in dirty equipment.

3. Gas Formers

Gas-producing microorganisms such as yeasts, propionibacteria (Todesco *et al.*, 2000), and heterofermentative LAB (Laleye *et al.*, 1987) cause undesirable openness in cheese. The floating curd defect in cottage cheese is also due to gassy LAB. The early gas defect and barney flavor in cheese can be caused by coliform bacteria such as *Enterobacter aerogenes* (Abo-Elnaga, 1971; Nieuwoudt and Bester, 1975; Bester, 1976; Melilli *et al.*, 2004), and late gas defect in Dutch and Swiss type cheese is usually due to *Clostridium tyrobutyricum* (Nieuwoudt and Bester, 1975).

Yeasts occur spontaneously in milk and are common contaminants during the cheese-making process. Yeasts commonly associated with dairy products include *Kluyveromyces lactis*, *Saccharomyces cerevisiae*, *Pichia anomala*, and *Debaryomyces hansenii* (Klein *et al.*, 2002). Some of their undesirable effects are the formation of yeast slits or holes, development of fruity flavor in many cheese varieties, and browning on and under the rind in white and blue mold-ripened cheeses (Nichol *et al.*, 1996). On the positive side, yeasts contribute peptidases (Klein *et al.*, 2002), which assist in interior ripening. In bloomy cheeses, yeasts importantly reduce acidity, encouraging the growth of the white mold *Penicillium candidum* or *P. camemberti*. In smear-ripened (washed rind) cheeses, yeasts not only reduce the acidity, encouraging growth of *Brevibacterium linens*, but also contribute peptidases to the ripening process (Bockelmann and Hoppe, 2001). They also influence the hue and intensity of color development by *B. linens*. Furthermore, yeasts are the dominant flora in some smear cheeses such as Portuguese sera cheese (Macedo *et al.*, 1993).

Coliform bacteria are usually present in raw milk and have been traditionally used as indicators of sanitation or good manufacturing practice (GMP). Alternative GMP markers are total counts for the family Enterobacteriaceae, which includes enteric pathogens commonly associated with milk such as pathogenic serotypes of *Escherichia coli* O157:H7 and various *Salmonella* spp. With the notable exception of *E. coli* O157:H7, which can survive in acidic milk products, most bacteria in this group do not tolerate low pH and compete poorly with LAB, so their numbers decrease in the presence of a growing lactic culture (Bester, 1976). Poor sanitation combined with poor acid development, however, can result in excessive early gas (spongy cheese) with associated barney flavors. Early gas can also be caused by *Propionibacterium* spp. (Todesco *et al.*, 2000) and heterofermentative LAB.

Clostridium tyrobutyricum is a thermoduric (survives pasteurization) spore-forming organism of legendary fame among cheese makers. It causes gas formation (carbon dioxide) during the later stages of ripening of Swiss- and Dutch-type cheeses, and sometimes in others such as hard Italian varieties. The resulting craters and cracks in the cheese are termed 'late gas defect'. European cheese makers frequently check raw milk for thermoduric and/or spore-forming bacteria. Five hundred spores per liter of milk are sufficient to cause late gas defect. Control strategies include: (1) added egg lysozyme, which requires an egg allergy declaration in many jurisdictions (see summary in Fox, 1993); (2) culture adjuncts that at least partly inhibit *Clostridium* spp.; these are commercially available and research in this area is continuing (Christiansen *et al.*, 2005; Martínez-Cuesta *et al.*, 2010); (3) spore removal by bactofugation or microfiltration; (4) addition of nitrate salts, the traditional remedy, which is falling out of favor owing to subsequent formation of nitrosamines when cheese is heated and a general consumer distaste for additives; and (5) for hard Italian cheeses, preripening at cool temperatures (Spolaor and Marangon, 1997).

4. Pathogenic Bacteria

This summary of pathogenic bacteria associated with fermented dairy products is adapted from Hill and Warriner (2011a, b). *Listeria monocytogenes*, *Salmonella* species, and enteropathogenic *E. coli* represent the greatest concern for cheese makers (Johnson *et al.*, 1990a–c). With respect to yogurt and most fermented dairy beverages, such as buttermilk, the risk associated with pathogenic bacteria is greatly reduced by high heat treatment, which is normally well in excess of required pasteurization levels, and closed processing systems that largely mitigate the risk associated with postpasteurization contamination. Sources of contamination in raw milk are shown in Figure 8.4.

Listeria monocytogenes probably represents the highest risk associated with cheese, because both the severity and probability of listeriosis are high relative to other cheese-borne diseases. On the severity side of the equation, the death rate associated with listeriosis is about 30%, and for those who survive, there may be serious complications such as spontaneous abortion. On the probability side of the equation, the rate of incidence of *L. monocytogenes* in raw milk is relatively high; its thermal tolerance is close to pasteurization (72°C, 16 seconds); it survives, if not thrives, under the conditions of cheese making; and it is able to grow at refrigeration temperatures (i.e. cheese ripening and storage temperatures). Similarly, *Salmonella* species and *E. coli* O157:H7 are relatively high risk in cheese because they are frequently present in raw milk, grow at, or at least tolerate, the cold temperatures and acidity of cheese and other fermented milk products, and may persist for several months during ripening.

Staphylococcus aureus is a frequent biological contaminant in cheese, originating from raw milk or from personnel, but is listed as low risk because both growth and toxin production are readily suppressed by competing lactic cultures and cheese acidity (Johnson *et al.*, 1990a, b). Similarly, *Campylobacter jejuni* is widespread in the environment and occurs in raw milk, but does not normally survive the cheese-making conditions. So, it is considered low risk for cheese making provided that either heat treatment or fermentation (reduced pH) is included in the manufacturing process. *Yersinia enterocolitica* apparently tolerates cheese-making conditions; however, recent studies suggest that virulent strains, especially in North America, do not occur frequently in milk (Kushal and Anand, 2001, 2006).

Bacillus cereus is a nuisance because it forms difficult-to-remove bacterial films on milk processing equipment, survives pasteurization, and grows well at refrigeration temperatures. However, it is considered low risk because

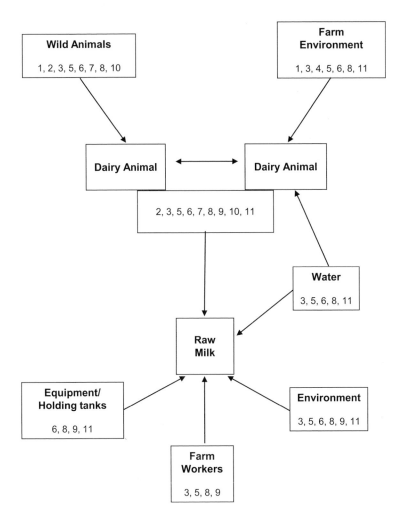

FIGURE 8.4 Sources of contamination in raw milk. 1: *Bacillus* spp.; 2: *Brucella*; 3: *Campylobacter jejuni*; 4: *Clostridium botulinum*; 5: *Escherichia coli* O157:H7; 6: *Listeria monocytogenes*; 7: *Mycobacterium tuberculosis*; 8: *Salmonella*; 9: *Staphylococcus aureus*; 10: *Coxiella burnetii* (Q fever); 11: *Yersinia enterocolitica*. (*Courtesy of K. Warriner, University of Guelph, Canada.*)

production of its diarrheal toxin in milk is associated with obvious spoilage and cell counts typically greater than 10 million/ml (Griffiths, 1990; Sutherland and Limond, 1993; Sutherland, 1993). It is also interesting that pasteurization of milk encourages germination of *B. cereus* spores, so its numbers may be higher in pasteurized milk than in raw milk (Stewart, 1975; Ahmed *et al.*, 1983).

In summary, it is critical to approach all raw milk processing with the assumption that pathogenic bacteria are present and must be eliminated or reduced to safe levels. Considering the average incidence rates in various reports, it seems likely that greater than 10% of cow's milk is contaminated with one or more pathogenic bacteria before it leaves the farm. A South Dakota study found that 26% of bulk tank samples were contaminated with one or more of *E. coli* O157:H7, *L. monocytogenes*, *Salmonella* spp., or *C. jejuni* (Jayarao and Henning, 2001). In Tennessee, 25% of bulk tanks were contaminated with one or more of *L. monocytogenes*, *C. jejuni*, *Y. enterocolitica*, or *Salmonella* spp. (Rohrbach *et al.*, 1992). The pooling process during milk transport and storage ensures that the proportion of contaminated milk reaching the cheese vat is usually much higher, and new rapid and more accurate molecular detection and quantification techniques will certainly increase the estimates of incidence rates. Furthermore, some of the traditional pathogens that provided the initial motivation for milk pasteurization, such as *Corynebacterium diphtheriae* and *Coxiella burnetii*, still occur in raw milk [EFSA Panel on Animal Health and Welfare (AHAW), 2010].

B. Antibiotics

Lactic cultures are very sensitive to the antibiotics used on dairy farms for therapeutic and prophylactic treatment of various diseases, especially mastitis. Common antibiotics used on dairy farms include β-lactams (penicillin), sulfonamides, streptomycin, tetracycline, and amphenicols. In most jurisdictions, increasing penalties have greatly reduced antibiotic residues in milk. Nevertheless, antibiotic testing of cheese milk before unloading raw milk for bulk milk trucks is still commonly done. In many jurisdictions one of several available growth inhibition assays is still used as the official test, usually based on the inhibition of *Bacillus stearothermophilus*. These tests have the advantage of a broad detection range but are non-specific, slow, and not very sensitive. Commonly used rapid and more sensitive test kits include: (1) the Penzyme Test Kit for β-lactam antibiotics, based on the binding of penicillin by DD-carboxypeptidase; (2) Charm II tests, which test a wide range of antibiotics based on *B. stearothermophilus*, which contains natural receptor sites for antibiotics on or within its cells; the radioactive labeled (^{14}C or ^{3}H) antibiotics are displaced by antibiotics present in milk; and (3) immunoassays, enzyme-linked assays, or enzyme-linked immunosorbent assays. High-performance liquid chromatography is typically used as the reference method for other assays. Several recent reviews on antibiotic testing are available on sulfonamides (Wang *et al.*, 2006; Zhang and Wang, 2009), β-lactams (Cui *et al.*, 2007; Samanidou *et al.*, 2007), tetracycline (Pastor-Navarro *et al.*, 2009), chloramphenicols (Santos and Ramos, 2006), and also an overview of veterinary residues in milk (Nag, 2010).

C. Mastitic Milk

Mastitis is an infection of the cow's udder that has negative impacts on milk quality. Pooling milk dilutes the effect of single infected cows and herds, but in most jurisdictions the cumulative effect of mastitis, especially subclinical mastitis, is significant. Olson, as cited in Eck and Gillis (2000), estimates a 1% loss in cheese yield if 10% of the milk is from cows with subclinical mastitis. Furthermore, as noted below, the quality effects of mastitic milk are probably of more economic importance than the yield effects. Causative organisms include human pathogens such as *E. coli* and *S. aureus*. Non-bacterial infections such as by *Prototheca* also cause high somatic cell counts (SCCs) and are associated with high bacterial counts; see University of California extension report (Kirk and Mellenberger, 2011).

Mastitis is indicated by an increased number of somatic cells in milk. SCC greater than 100,000/ml in cows and about one million in goats (Rupp *et al.*, 2011) indicates clinical or subclinical mastitis. The association of SCC with milk quality in goat's milk is not clear. In cow's milk, SCC of less than 500,000 cells/ml, perhaps as low as 100,000, is associated with reduced cheese yield and reduced quality of fluid milk and other dairy products (Barbano *et al.*, 1991; Klei *et al.*, 1998; Ma *et al.*, 2000; Santos *et al.*, 2003). When counts exceed 1,000,000 cells/ml, altered milk composition and reduced cheese yields are obvious. Principal changes include increased activity of plasmin (an alkaline protease indigenous to milk) and plasminogen, accelerated protein breakdown, elevated milk pH, increased mineral content, reduced casein content, and higher levels of immunoglobulins.

IV. CHEMISTRY OF MILK COAGULATION

The essential step in the making of cheese and yogurt is to induce a clotting reaction in milk, primarily involving the casein proteins. In most cheeses, coagulation of milk is brought about by an enzymatic method, i.e. by the addition of rennet. For thousands of years rennets were extracted from the inner stomach lining of young calves. Chymosin, the aspartic proteinase contained in crude rennet extracts, is now largely produced by fermentation methods using genetically modified (GM) microbes. Yogurt and some cheese varieties are solely or predominantly acid coagulated products in which gelation of caseins, sometimes together with denatured whey proteins, is induced by slow acidification of milk by lactic cultures. Enzymatic and acid coagulation of milk are discussed in separate sections below.

A. Enzymatic Coagulation of Milk

Casein proteins naturally exist in milk as highly hydrated supramolecular colloidal aggregates, the casein micelles, with an average diameter of about 200 nm. The structure and function of casein micelles have been extensively studied and reviewed (Dalgleish, 1998, 2007, 2011; Horne, 1998, 2002, 2006; Walstra, 1999; Holt *et al.*, 2003; de Kruif and Holt, 2003; Dalgleish *et al.*, 2004; Farrell *et al.*, 2006; Fox and Brodkorb, 2008). The casein micelle is made of several thousand individual casein molecules organized around nanodomains of inorganic calcium phosphate. Two-thirds of the calcium in milk exists as these nanoclusters of casein–calcium phosphate. Whereas the calcium-sensitive α_s- and β-caseins are assembled within the micelle's interior, the calcium-insensitive κ-casein largely resides on the micelle's surface. Micelle integrity is due to casein–calcium interactions, hydrogen bonding, and electrostatic and hydrophobic interactions. The κ-casein protein is the key to a stable casein colloidal system in milk; its negatively charged and flexible C-terminus (106–169) protrudes from the micelle's surface, forming a highly hydrated 'hairy' layer that provides electrostatic and (especially) steric stability to the micelle (Holt and Dalgleish, 1986; Horne, 1986, 2002; Walstra, 1990; de Kruif, 1992). At low temperatures, the hydrophilic N-terminal of β-casein may also contribute to micelle stability against aggregation (Roefs et al., 1990).

When rennet is added to milk, after a certain lag period, the milk rather abruptly begins to coagulate and becomes a three-dimensional gel. The primary phase is entirely proteolytic, in which chymosin hydrolyzes the C-terminal part of κ-casein. The hydrophilic peptide, referred to as caseinomacropeptide (CMP), diffuses away from the micelle (para-casein) into the serum phase of milk. Loss of the hairy layer causes micelles to become unstable and this initiates the secondary clotting phase, i.e. the aggregation of para-casein micelles into a three-dimensional gel. The chemistry of enzymatic coagulation of milk ends here, i.e. with the formation of the casein curd. However, in a subsequent phase, sometimes called the tertiary phase, the newly formed gel continues to become firmer and eventually begins to release whey (syneresis) (see reviews by Dalgleish, 1992, 1993; Hyslop, 2003; Horne and Banks, 2004).

The primary phase in the coagulation of milk by rennet can be approximated to a first order rate reaction (Dalgleish, 1992). The rate increases with enzyme concentration and temperature (so long as the enzyme is stable). The early part of the reaction, i.e. until about three-quarters of κ-casein is proteolyzed, was found to vary linearly with rennet concentration (van Hooydonk and Walstra, 1987; Fox et al., 2000).

In the secondary phase, the destabilized micelles aggregate due to a loss of steric hindrance and reduced electrostatic repulsion ($\approx$ 50% decrease in zeta potential). The onset of clotting requires that at least 85–90% of κ-casein is proteolyzed (Dalgleish, 1979; Chaplin and Green, 1980). Using diffusing-wave spectroscopy, Sandra et al. (2007) monitored rennet gelation in undiluted milk at its natural pH. They suggested that partially renneted micelles continue to diffuse freely until there is a significant breakdown of κ-casein 'hairs' and only when > 70% of the hairs have been cleaved do the micelles become restricted in motion (owing to decreased electrostatic repulsion), but they do not aggregate yet. The real onset of aggregation requires more extensive proteolysis ($\approx$ 90% CMP release) and a space-filling gel is formed only when about 95% of κ-casein has been cleaved. Several models have been proposed to explain the mechanism of micelle aggregation (Horne and Banks, 2004), but there is no clear consensus about the nature of attractive forces that drive the aggregation reaction; calcium bridges, hydrophobic interactions, and van der Waals' forces are likely to be involved. Aggregation of destabilized micelles is greatly inhibited in the absence of free Ca^{2+} and at temperatures below 15°C, but rennet can remain active (although greatly reduced) even at temperatures as low as 0°C. This temperature dependence of the clotting reaction suggests the role of hydrophobic interactions in para-casein aggregation.

The clotting of milk by rennet and the strength of the resulting gel are influenced by several factors. Optimum temperature and pH for rennet action and micelle aggregation are about 30°C and 6.5, respectively. Free Ca^{2+} is essential for the milk to clot and produce a firm gel; ionic calcium (as $CaCl_2$), up to 1 mM, is routinely added to cheese milk. Concentrations of ions such as Na^+ (as NaCl), content of casein, presence of homogenized fat, and denatured whey proteins are also important in the clotting process and influence the final texture of rennet gels. These factors have been discussed previously by several authors (Dalgleish, 1993; Green and Grandison, 1993; Hyslop, 2003; Horne and Banks, 2004; Janhøj and Qvist, 2010).

Heat treatment of milk is a routine practice in the dairy industry and is an essential step in the manufacture of yogurt and some soft cheeses. The classic casein/whey protein interaction that occurs when milk is heated above 75°C is the covalent binding (through thiol/disulfide exchange reactions) of denatured whey proteins with κ-casein on the micelle surfaces and the formation of soluble whey protein/κ-casein complexes in the serum (O'Connell and Fox, 2003). The heat-induced changes in casein micelle structure unfavorably alter its rennet functionality. Cheese produced from heated milk has poor texture and unacceptable organoleptic properties. The rennet curds are soggy and ragged in appearance, with poor stretching and melting properties (Singh and Waungana, 2001). Heat-treated milk is therefore seldom used in cheese making. It is now well established that the proteolytic action of rennet (rate of CMP release) is similar in unheated and heated milk (Vasbinder et al., 2003; Anema et al., 2007; Sandra and Dalgleish, 2007; Kethireddipalli et al., 2010) and is therefore not the reason for heat-impaired rennet clotting. However, heating milk severely impairs the aggregation of renneted micelles and this is routinely attributed to the binding of denatured whey proteins to the casein micelles and to the heat-induced losses of ionic calcium. Kethireddipalli et al. (2010, 2011) demonstrated that the poor clotting ability of heated milk can be attributed not only to the heat-induced association of denatured whey proteins with micelle surfaces, but also to a further binding in the course of renneting, of the soluble fraction of whey protein/κ-casein complexes to the micelles. It was suggested also that the heat-induced binding of denatured whey proteins impairs clotting to a greater extent than similar levels of rennet-induced micellar binding of serum protein complexes, possibly because of the distinct ways of protein binding, i.e. through covalent linkages in the former and by a largely hydrophobic interaction in the latter. In addition to heat-induced serum protein complexes, certain unknown serum components were implicated in clotting inhibition. Further, it was shown by these authors that decreased levels of ionic calcium in heated milk has no significant effect on clotting inhibition; heated milk samples, with or without restored ionic equilibrium, had similarly long clotting times and produced weak gels with comparable low elastic moduli. These new findings on heat-induced milk protein interactions may help researchers to develop novel cheese varieties from heat-impaired milk.

B. Acid Coagulation of Milk

Yogurt, fresh acid cheeses, and acidified milk drinks are popular fermented foods produced throughout the world. Commercial production of yogurt involves gradual acidification at 40–45°C of previously heat-treated milk (85°C, 10 minutes) by lactic cultures. Fermentation of lactose to lactic acid by the thermophilic starter bacteria, *Streptococcus thermophilus* and *Lactobacillus delbrueckii* subsp. *bulgaricus*, gradually causes the pH of the milk to drop from about 6.7 to below 5.0. Titration of negative charges on the surfaces of casein micelles collapses the κ-casein hairy layer, destabilizes the micelles, and induces clotting as the isoelectric point is approached (pH ≈ 4.6). As the pH is reduced, the colloidal calcium phosphate (CCP) is dissolved and this alters/weakens the micelle's internal structure. The chemistry of progressive acidification of milk can be observed in three distinct pH regions from pH 6.7 to 4.6 (Lucey, 2004a): (1) as pH is lowered from 6.7 to 6.0, there is a decrease in the net negative charge on the micelle surface and reduced electrostatic repulsions; CCP and micelle integrity are preserved; (2) a decrease in pH from 6.0 to 5.0 further neutralizes the micelle's surface and with the shrinkage/collapse of κ-casein hairs, micelles are sterically and electrostatically destabilized; CCP is completely dissolved when a pH around 5.0 is reached and this increases the internal flexibility of micelles (Donato *et al.*, 2007); and (3) at pH less than 5.0, as the destabilized micelles come closer, hydrophobic interactions come into play (Horne, 1998, 2001); gelation occurs at pH 4.9 in unheated milk, but at a higher pH (≈ 5.2) in heat-treated milk. Hydrophobic interactions are important in the initiation of clotting, but subsequent cooling/refrigeration causes the gels to swell, increasing the particle contact area, and thereby gel firmness/strength (storage modulus). Because hydrophobic interactions are weakened at low temperatures, this suggests that gel integrity may be due to other forces such as electrostatic and van der Waals' interactions. For reviews on acid gelation of milk, the reader may refer to Dalgleish *et al.* (2005), de Kruif *et al.* (1995), Lee and Lucey (2010), Lucey (2002, 2004a, b), Lucey *et al.* (1997), and van Vliet *et al.* (2004).

Heating milk (and subsequent cooling) prior to acidification is the key to producing yogurts with desired texture and consistency. Unheated milk forms very weak acid gels, probably because the interface between aggregating micelles is still dominated by κ-casein hairs, which although collapsed, could prevent attractive interactions owing to their hydrophilic tendency (Li and Dalgleish, 2006). In heated milk, the attachment of heat-denatured whey proteins to micelle surfaces and the presence of serum-dispersed whey protein/κ-casein particles facilitate extensive cross-linking of gels, greatly increase gel strength and stiffness, and reduce syneresis (Lucey *et al.*, 1998; Donato *et al.*, 2007). Heat-treated acid milk gels also set at a higher pH, around 5.2 compared to unheated milk gels (setting pH ≈ 4.9), mostly because of the higher isoelectric pH (≈ 5.3) of β-lactoglobulin than that of caseins (Lucey *et al.*, 1997; Guyomarc'h *et al.*, 2003).

In some fresh acid cheeses such as cottage cheese, quark, and fromage frais, a small amount of rennet may also be used as coagulant. Studies on milk gelation by simultaneous acidification and renneting have received greater attention during recent years (Lucey *et al.*, 2000; Gastaldi *et al.*, 2003; Li and Dalgleish, 2006; Cooper *et al.*, 2010; Tranchant *et al.*, 2001). This method produces gels that are firmer and also set earlier (at a higher pH) than pure acid gels. The enzymatic removal of less than 25% of the micellar hairy layer has a synergistic effect on micelle destabilization by acid and enzyme. As the pH drops, the charge at the micelle's surface decreases and the CMP layer collapses, and so the extent of κ-casein breakdown needed to cause aggregation becomes smaller, in contrast to the requirement of over 90% proteolysis in pure rennet gels. Or, the κ-casein that needs to be cleaved to produce a reactive 'hot spot' will be smaller (Li and Dalgleish, 2006). With weakened steric hindrances due to the collapse of κ-casein hairs, even a small gap in the surface layer will allow interaction between the inner surfaces of micelles. The casein aggregates that are formed are larger and have increased connectivity (Cooper *et al.*, 2010) and so the combination gels are stronger. Depending on the concentration of rennet used (10–25% proteolysis), combined gels set at pH values between 5.0 and 5.3, in contrast to the lower gelation pH (4.8–4.9) of acid gels (Li and Dalgleish, 2006; Cooper *et al.*, 2010). Addition of small amounts of rennet during acidification of preheated milk (≥ 85°C for 5 minutes) also has the effect of improving gel firmness and increasing the pH of the onset of gelation. The inter-micelle bonds in these gels are not entirely made up of cross-linked denatured whey proteins (as in acidified heated milk), but the inner surfaces of the partially renneted micelles may also be involved (Li and Dalgleish, 2006). This principle is exploited in the commercial manufacture of thermo quark (Schulz-Collins and Senge, 2004).

Heat—acid coagulation, a third type of coagulation, is primarily acid induced, but no fermentation is involved and the acid is added to hot milk at temperatures in the range of 75—100°C. Some of the thermally labile whey proteins are coagulated along with the casein and recovered in the cheese. Although total protein recovery increases by only a small amount (about 2—5%, compared to rennet coagulation), yield increases are largely due to the higher water-holding capacity of the denatured whey proteins. Increasing milk protein content also increases the proportion (percent recovery) of protein in the cheese (Hill *et al.*, 1982). Acid coagulation at high temperatures requires less acidification, so the final cheese pH is in the range of 5.2—6.0. Due to the interaction of denatured whey proteins with caseins, the cheese is rendered unmeltable, but has excellent frying/cooking properties; examples of such cheeses are Ricotta and Indian Paneer.

V. CULTURES FOR CHEESE AND YOGURT

A. Functions of Cheese Cultures

Culture refers to prepared inocula of bacteria, yeasts, and/or molds that are directly added to milk and/or incorporated into the fermented dairy product at a later stage during its manufacture. Specific LAB are selected as primary acid formers and are referred to as primary cultures; other cultures including yeasts, molds, certain bacteria, and other LAB may be added as secondary or adjunct cultures for specific purposes (see Cogan *et al.*, 2007, for a review). In broad terms, cultures have five purposes in fermented milk products: (1) to develop acidity; (2) to develop typical flavors such as the diacetyl and acetaldehyde flavor of fermented milks and certain cheeses; (3) to promote ripening of cheese (Section VII); (4) to mitigate specific quality or safety concerns such as the late gas defect in cheese caused by *Clostridium tyrobutyricum* or a food hazard such as listeriosis; and (5) for health-promoting properties such as from probiotics.

McKay and Baldwin (1975) were the first to demonstrate that many properties that are of technological importance to cheese making are encoded on the plasmids of lactococci. As listed by Callanan and Ross (2004) in their review of starter culture genetics, plasmid-encoded properties include the ability to contribute enzymes necessary for casein hydrolysis and for the transport and metabolism of lactose and citrate, the ability to produce bacteriocins, and the ability to combat phages. The rapidly growing area of LAB genetics is beyond the scope of this chapter; a search of *Food Science and Technology Abstracts* for LAB genetics or genomics returned more than 1500 papers during 2006—2010. For example, a review by Broadbent and Steele (2007a) illustrates the importance of genomics research in understanding and directing flavor development in cheese.

B. Classification of Lactic Acid Cultures

Table 8.5 lists the Latin names of some common lactic cultures. Both homofermentative and heterofermentative cultures produce lactic acid as the principal metabolite. However, heterofermentative cultures also produce substantial amounts of carbon dioxide (CO_2) and certain flavor compounds such as diacetyl (the principal flavor note in sour cream and buttermilk). Gas formation by heterofermentative cultures creates the desired open texture in some cheese types such as Gouda and Havarti.

The other major divider is mesophilic versus thermophilic cultures. The optimum growth range for mesophilic cultures is 30—35°C; acid production is slow or absent at temperatures less than 20°C and growth is inhibited at temperatures greater than 39°C. In general, cheeses that do not employ high temperatures (> 40°C) at the cooking stage utilize mesophilic cultures. These varieties include Cheddar, soft ripened cheese, most fresh cheese, and most washed cheeses.

Thermophilic cultures prefer temperatures in the approximate range of 40—45°C, but can survive temperatures of 55°C or higher, as required in cooked cheese varieties like Swiss and Italian types. These cheeses are generally cooked at temperatures exceeding 45°C before the separation of curd and whey. Cell counts decrease rapidly at temperatures less than 20°C, so bulk thermophilic cultures cannot be chilled to extend their shelf-life. The two most common thermophilic mixed cultures are *Streptococcus thermophilus* with *Lactobacillus delbrueckii* subsp. *bulgaricus*, and *S. thermophilus* with *Lb. helveticus* (Table 8.5).

Thermophilic starters grow together in an associative or symbiotic relationship, in which the growth rate and acid production of the mixed cultures are faster than for either culture on its own (Carrasco *et al.*, 2005). The proteolytic

TABLE 8.5 Properties and Functions of Some Lactic Cultures

Culture	Properties and Functions
Mesophilic Cultures	
Lactococcus lactis ssp. *cremoris; Lactococcus lactis* ssp. *lactis*	As a blend, these two form the most common mesophilic and homofermentative culture used in many low-temperature cheese varieties including fresh cheese, Cheddars, American Colby types, Dutch varieties, soft-ripened varieties, and others
Leuconostoc mesenteroides ssp. *cremoris; Lactococcus lactis*	Heterofermentative cultures. Often mixed with citrate fermenting *L. lactis* ssp. *cremoris/lactis* for cultured butter and buttermilk, and for cheese with small holes such as Havarti and Gouda
Thermophilic Cultures	
Streptococcus thermophilus; Lactobacillus delbrueckii ssp. *bulgaricus* or *Lactobacillus helveticus* or *Lactobacillus delbrueckii* ssp. *lactis*	Commonly used coccus/rod blend for high-temperature cheese varieties. *Lb. helveticus* is galactose positive and helps to reduce browning in pizza cheese; also, being proteolytic, it can accelerate ripening in Cheddar cheese. *Lb. delbrueckii* ssp. *lactis* is an alternative to *L. helveticus* and *L. bulgaricus* where less acid is preferred, as in 'stabilized' Brie cheese, and some mild and probiotic yogurts

activity of rods produces amino acids and peptides that stimulate the growth of cocci, and cocci in turn produce formic acid, which is required by rods (Galesloot *et al.*, 1968; Veringa *et al.*, 1968; Higashio *et al.*, 1977; Kikuchi *et al.*, 1984; Suzuki *et al.*, 1986). Cocci also have higher temperature and lower acid tolerances, so the rod/cocci balance is also influenced by temperature and pH. For example, in cooked cheese, cocci dominate the initial growth when the temperature and pH are high after curd separation; then, as the temperature and pH decrease, the rods flourish. Also of technological significance are the differences in galactose fermentation: *Lb. helveticus* readily metabolizes galactose, whereas *Lb. delbrueckii* subsp. *bulgaricus* does not (Turner and Martley, 1983; Matzdorf *et al.*, 1994; Mukherjee and Hutkins, 1994; Hassan, 2000; Baskaran and Sivakumar, 2003). This is important because faster metabolism of galactose reduces Maillard browning, so pizza color, for example, can be modified by culture selection.

C. Secondary and Adjunct Cultures

There are two general approaches to directed specified technological outcomes using cultures. The first is to select or engineer primary cultures to be used not only for acidification, but also for additional specified functionalities. For example, a heterofermentative mesophilic culture can function as a primary acid former, as well as produce an open cheese structure and characteristic flavors. The second approach is to select and/or engineer other bioagents with specific functionality. Some of these functional agents are described below.

Large hole (eye) formation in Swiss cheese and its cousins is caused by *Propionibacterium freudenreichii* subsp. *shermanii*. The growth and gas production of this bacterium are induced by warming young cheese to 20–25°C for 1–2 weeks. Cheese permeability and elasticity are important and are determined by a range of factors including cheese composition and the extent of proteolysis (Grappin *et al.*, 1993; Ekinci and Gurel, 2008). Eye formation occurs when the rate of gas production exceeds its rate of permeation through cheese. If the cheese is not sufficiently elastic, cracks rather than eyes will form.

Bloomy (white) rinds are formed mainly by *Penicillium camemberti* Thom, which is an amalgamation of two closely related phenotypic forms, namely, *P. album* (grey–green color) and *P. caseiocolum* or *P. candidum* (white) (see review by Chamba and Irlinger, 2004). According to these authors, other white molds found on cheese include *P. thonii, P. nalglovensis,* and *P. verrucosum*. The characteristic mushroom flavor of bloomy varieties is due to production of 1-octen-3-ol (Spinnler and Gripon, 2004).

With the exception of so-called stabilized Brie, soft ripened cheeses such as Camembert, Brie, blue, and Feta are initially acidic, reaching a minimum pH of less than 4.8 (Table 8.6). Stabilized Brie typically reaches a minimum pH of 5.2. The lag phase of white mold is reduced by 'contaminating' or adding yeasts that consume lactic acid and

TABLE 8.6 Some Properties of Cheese Categorized According to Type of Coagulation and Procedures used for pH and Moisture Control[a] (Hill, 2007)

Category	Examples	Coagulation[b]	MNFS[c]	pH[d]	Calcium[e]	Ripening
1. Acid coagulated	Fresh: cottage, quark, cream. ripened: Valençay, Harzer	Predominantly acid coagulation at pH 4.6–4.9; rennet: 0–1200 IMCU/1000 liters of milk	72–80%; a_w 0.980–0.997; controlled by cooking and washing treatments	4.3–4.9; inhibition of culture by low pH, high temperature cooking, rapid cooling and/or washing	50–350	Normally consumed fresh; or mold and/or smear ripened
2. Heat–acid coagulated	Cooking cheeses: Paneer and Channa, Ricotta, Requeson; cream: Mascarpone	Whey proteins coprecipitate with caseins and inhibit melting	75–84%; a_w 0.975–0.997; increases with whey protein content, decreases with cooking after acidification	5.0–5.8; amount of acidulant added; 3–6% lactose in cheese due to absence of fermentation		Consumed fresh, unless hot packed, pickled, or packed in sugar syrup
3. Fresh: rennet coagulated	Hispanic white frying cheeses; Italian fresh cheese	Rennet+; little or no culture; cutting pH 6.4–6.6; milk may be salted before renneting	60–80%; controlled by cooking, stirring out, milling before draining, vat salting; syneresis often occurs in the package	5.8–6.6; little or no culture; high pH prevents melting		Consumed fresh; high pH limits shelf-life; Hispanic varieties may develop yeasty flavor
4. Soft ripened	Feta, Camembert, blue	Rennet+++; culture+++; ripening time+++; cutting pH 6.5–6.4	60–70%; a_w 0.940–0.980; syneresis induced by acid development and by salting	4.5–4.8; acid inhibition of culture, salting and cooling	400–600	2–8 weeks
5. Mesophilic washed	Gouda, Edam, Colby, Havarti, Montasio, and many others	Rennet++; culture−++; ripening time++; cutting pH 6.6–6.5	55–65%; a_w 0.950–0.970; controlled by cooking, temperature of wash water, rate of acid development, curd handling, salting treatments	4.8–5.2; washing step to remove lactose	500–700	2 weeks– 12+ months
6. Mesophilic unwashed	Cheddar, Provolone	Rennet++; culture++; ripening time++; cutting pH 6.6–6.5	52–60%; a_w 0.940–0.960; controlled by cooking, curd ripening, rate of acid development, and salting	5.0–5.3; rate of acid development and moisture determines residual lactose; draining pH is critical	500–700	1–24+ months
7. Thermophilic	Swiss and hard Italian types	Rennet+; culture+; ripening+ or none; cutting pH > 6.6	39–52%; a_w 0.900–0.960; controlled by high-temperature cooking (52–55°C)	5.1–5.3; acidity and moisture determine residual lactose; draining pH is critical	600–800	1–24+ months

a_w : water activity.
[a]Representative data from various sources are given to define broad ranges and trends only.
[b]+ symbols indicate amounts of rennet and culture and ripening time relative to other categories.
[c]Moisture in non-fat substance.
[d]Minimum pH reached during manufacture or in the first days of ripening 5 mM/kg non-fat solids.
[e]Calcium content in mM per Kg of nonfat cheese solids.

reduce acidity. Proteolysis by white molds further reduces acidity, so the pH on the surface and interior ripened portions of bloomy cheese increases rapidly to 6.8−8.0. Combined with low calcium content and high moisture (Table 8.6), the increase in pH softens the casein matrix, creating the characteristic creamy texture of bloomy cheeses (Vassal *et al.*, 1986; Spinnler and Gripon, 2004). Increased pH also explains the relatively high incidence of food-borne illnesses associated with bloomy rinds, especially by *E. coli* O157:H7 and *Listeria monocytogenes*; these organisms survive the low initial pH and then multiply when the pH increases during ripening (Griffiths, 1989; El-Gazzar and Marth, 1991a; Ramsaran *et al.*, 1998).

Blue cheese is produced through interior veining by blue/green molds, mainly *Penicillium roqueforti*, and is usually of rennet-coagulated type. However, there are also some acid-coagulated varieties, mostly the Chevre type, which are surface ripened with blue mold. Blue molds produce a similar effect as described above for white molds (but to a lesser extent), i.e. they increase the pH and soften the texture. For surface-ripened Chevre-type cheese, proteolysis and increased surface pH tend to create a gelatinous layer under the rind.

Commercially available strains of *P. camemberti* and *P. roqueforti* vary in their lipolytic and proteolytic properties. β-Oxidation of free fatty acids during lipolysis produces methyl ketones and their secondary alcohols, some of which are important flavor notes (Collins *et al.*, 2004). Proteolysis includes all principal caseins. Some *P. roqueforti* strains are more lipolytic and produce the sharp rancid flavor notes typical of Danish blues; other proteolytic strains that are less lipolytic produce more mellow notes, such as in Stilton.

Smear/washed/mixed rinds are formed by complex mixtures of yeasts, various coryneform bacteria such as *Brevibacterium linens*, various species of *Micrococci* and *Staphylococci*, sometimes molds, and others (see review by Chamba and Irlinger, 2004). Similar to the bloomy varieties, the surface pH of smear/mixed rind varieties increases during ripening, but generally not to the same extent. There is evidence that *Listeria monocytogenes* is able to grow on some smear-ripened cheeses (El-Gazzar and Marth, 1991b; Farrag and Marth, 1992; Rudolf and Scherer, 2001). In a challenge study using 24 market cheeses (Genigeorgis *et al.*, 1991), *L. monocytogenes* was found to survive but did not grow on cheese surfaces with pH less than 5.5.

Ripening adjuncts include bacterial or yeast cultures added in addition to the regular lactic acid cultures, attenuated cultures that are not intended to grow but only to contribute their enzymes, and bacterial culture adjuncts, such as lactobacilli and pediococci, that are intended to grow during cheese ripening and also contribute enzymes. For example, in cooked cheese varieties, *Lb. bulgaricus* is the primary acid former, as well as a proteolytic ripening agent, but in mesophilic aged cheeses such as Cheddar, *Lb. bulgaricus* and sometimes other lactobacilli such as *Lb. casei* are mainly used as flavor adjuncts that contribute aminopeptidases (Simova and Beshkova, 2007; Slattery *et al.*, 2010). The search for ripening adjuncts is currently an active area of research, often employing non-starter bacteria and yeasts from traditional cheeses (see Section VII and reviews: Laleye *et al.*, 1990; Klein *et al.*, 2002; Broadbent and Steele, 2007b; Nieto-Arribas *et al.*, 2009; Law, 2010a; Milesi *et al.*, 2010; Morales *et al.*, 2010).

LAB bacteriocins are grouped as lantibiotics, small heat-stable non-lantibiotics, and large heat-stable bacteriocins (see summary by Parente and Cogan, 2004). Many lactococci produce bacteriocins which, when combined with other control factors, can help to control the growth of non-starter (Rea *et al.*, 2003) and pathogenic bacteria such as *L. monocytogenes* (Dal Bello *et al.*, 2010). In recent years this has been a very active area of investigation, with much effort focused on identifying new sources of bacteriocins associated with LAB and their potential application in various food products such as ready-to-eat meats. Examples include: (1) identification of bacteriocin-producing species of *Lactobacillus*, *Pediococcus*, and *Enterococcus* from various sources (Todorov, 2008, 2010; Tome *et al.*, 2009; Todorov *et al.*, 2010, 2011a−c); (2) use of nisin-producing lactococci strains to improve the safety of low-salt Domiati (Ayad, 2009); (3) description of the lantibiotic mutacin 1140 produced by *Streptococcus mutans* (Smith, 2002); (4) description of pentocin 31-1, an anti-listeria bacteriocin produced by *Lactobacillus pentosus* (Zhang *et al.*, 2009); (5) description of bacteriocins produced by *Lb. acidophilus* and their pH dependence (Soomro and Masud, 2008); (6) production of enterocin by *Enterococcus faecium* in Cheddar cheese (Rea *et al.*, 2003); (7) description of macedocin produced by *Streptococcus gallolyticus* subsp. *macedonicus* (Leroy and De Vuyst, 2010); (8) a 2010 PhD dissertation (Martin-Visscher, 2010) reporting that each of the four LAB bacteriocins (nisin, gallidermin, CclA, and enterocin 710C) killed one or more of *E. coli* DH5α, *Pseudomonas aeruginosa* ATCC 14207, and *Salmonella typhimurium* ATCC 23564; and (9) inhibition of late gas defect (*Clostridium beijerinckii* INIA 63) in Manchego cheese by nisin- and lacticin 481-producing *Lactococcus lactis* subsp. *lactis* INIA 415 (Garde *et al.*, 2011). Cultures imparting health functionalities will be discussed in Section VIII, B.

Exopolysaccharides produced by some non-starter and starter bacteria may be important health-promoting and texture-building agents in fermented milk products and particularly in improving the texture of low-fat cheese (see reviews by Vlahopoulou *et al.*, 2001; Zisu and Shah, 2005; Hassan, 2008; Robitaille *et al.*, 2009; Welman, 2009; Costa *et al.*, 2010; Badel *et al.*, 2011).

D. Culture Management

Provided that milk is not chilled and that the cheese maker has lots of patience, it is possible to make fermented dairy products without additional cultures, but the normal practice is to add domesticated cultures for the manufacture of cheese from both pasteurized and raw milk. In terms of increasing direction or definition, cultures are selected and maintained in three general ways: natural starters, mixed strain starters, and defined strain starters.

1. Natural Cultures

The practice of using natural cultures based on traditional culture handling techniques has been preserved, particularly among artisan cheese makers, and in some cases by national and/or international cheese identity standards such as the AOC. For both primary LAB cultures and secondary cultures, the simplest type of natural handling technique is back-slopping, in which a small portion of a fermented product is used to inoculate a new batch. For example, finished yogurt can be used as the source of LAB culture for a new batch of yogurt, and bloomy cheese surfaces can be inoculated by placing the young cheese in the curing area with older cheese. Similarly, smear cultures for washed rind cheeses can be transferred via wooden curing shelves or hoops, or by washing young cheese with brine that was previously used to wash older cheese. Another example of a type of back-slopping is the Sicilian Ragusano cheese, in which the PDO requires milk fermentation in a wooden vat called a tina. Lortal *et al.* (2009) found that most of the interior surface of the vat is covered in a biofilm, a matrix of exopolysaccharides, which is the source of the primary starter.

Natural primary starters are prepared from both milk and whey. The various techniques were reviewed by Limsowtin *et al.* (1996) and summarized by Parente and Cogan (2004). For both milk and whey natural starters selective pressure is applied by heat treatment, incubation temperature, and incubation time. For example, Parente and Cogan (2004) reported that a traditional milk culture composed mainly of *S. thermophilus*, but also containing a number of other LAB, can be prepared by thermization (62–65°C) followed by incubation at 37–45°C. Similarly, natural whey-based culture for Parmigiano Reggiano and Grana Padano, composed mainly of aciduric *Lb. helveticus* with lower counts of other thermophilic LAB (*S. thermophilus*), is prepared by incubating fresh whey at 45°C until the pH reaches a low of 3.3. Another example is the AOC-prescribed whey starter (called Recuite) for Comté cheese. According to Dr. Sébastien Roustel (personal communication), dried calf intestines are heated in fresh whey at 54–55°C for 34 minutes and fermented at 42°C overnight. Vinegar may be added to encourage aciduric LAB. The LAB counts in whey are usually 10^{6-7}/ml after cheese manufacture and 10^{9-10}/ml in the Recuite after fermentation.

Natural cultures have the disadvantage of inconsistent performance and the advantage of inherent phage resistance, although phages may still cause a slow vat by attaching to dominant strains. Natural starters, including spontaneous starters (starters that grow spontaneously in uncooled raw milk), are also used as co-starters or adjunct starters. One example is AOC Gruyere, which must be made from a blend of evening and morning milk not cooled to less than 15°C. The spontaneous culture is then supplemented with mixed cultures. Another example is the common practice of subpasteurization heat treatment of Cheddar cheese milk to reduce pathogens and select for thermoduric pediococci and lactobacilli which contribute aminopeptidases and other enzymes.

Natural cultures are also rich sources of LAB genetics. Rapid identification techniques such as the randomly amplified polymorphic DNA (RAPD) technique make it possible to screen large numbers of samples. A 2011 survey of fermented yak's, goat's, sheep's, and cow's milk products in Mongolia, on the basis of 16s RNA analysis, identified a total of 664 strains and 10 species of LAB, of which *S. thermophilus* and *Lb. helveticus* were the most abundant species (Yu *et al.*, 2011). Similarly, Morales *et al.* (2010) identified and characterized the lipolytic and proteolytic properties of 10 halotolerant/halophilic LAB in Mexican Cotija and doble crema cheeses. See also characterization of LAB in artisanal goat cheese (Colombo *et al.*, 2010) and an artisanal Corsican cheese (Casalta *et al.*, 2009).

2. Mixed Strain and Defined Strain Starters

Primary starters prepared under controlled conditions are mixed strain starters (MSS) or defined strain starters (DSS) (for reviews, see Limsowtin *et al.*, 1996, 1997; Parente and Cogan, 2004). In brief, MSS cultures are prepared by culturing from natural sources, but retain undefined strains that are diverse with respect to properties such as sugar metabolism, citrate metabolism, and bacteriocin production. With respect to phage resistance, MSS may contain both phage-resistant and lytic strains, and may harbor their own phages (Lodics and Steenson, 1993). So diversity provides some protection, but as with natural cultures, the principal strains may become susceptible. With respect to performance, strain ratios are constantly evolving, again similar to natural cultures, so properties such as rate of acid development and enzyme activities during ripening may not be consistent.

DSS are selected from mixed cultures by intensive subculturing to obtain more accurately directed and repeatable performance with respect to flavor and texture development. The original DSS starters in New Zealand were based on a single phage-resistant strain with daily rotations to prevent phage build-up. However, phages still occurred and the daily rotations were tedious. So, multiple strains were introduced. Current practices for DSS cultures rely less on rotation and more on improved phage resistance. Reduced rotation also improves consistency of performance.

In summary, it seems that the distinction between natural, MSS, and DSS cultures will become less important with the application of rapid techniques to isolate, accurately identify, and characterize functional properties of large numbers of microbial species and strains, and the opportunity to use that information to create complex but more defined culture blends that mimic natural cultures and more accurately account for other microbial fermentation and ripening agents. In other words, with increasing ability to intelligently target an overall balance of fermentative and ripening activities, culture strategies can become more holistic. A good example of this is the recent development of cultures for smear (washed rind) ripened cheese. It is well known that washed rinds may harbor *L. monocytogenes* (Rudolf and Scherer, 2001; Wagner *et al.*, 2006) and this is exacerbated by the common practice of using different types of back-slopping to transfer smear from old to young cheese. This has resulted in commercial cultures and ongoing research to prepare mixed or defined strain cultures to replace back-slopping (Bockelmann and Hoppe, 2001; Hannon *et al.*, 2004) or to otherwise mitigate the risk. A 2011 report that anti-listerial strains were isolated from the washed rind of raclette cheese (Roth *et al.*, 2011) suggests that it may be possible to improve safety by supplementing natural smear cultures with anti-listeria cultures. This could be quite important considering the large number of cheeses with PDO or otherwise specified natural smears and the growth of the artisan cheese industry in many countries.

3. Culture Handling in the Plant

Cultures are typically prepared for cheese milk and yogurt inoculation as bulk set cultures (one transfer and scale-up from the commercial culture) or direct-to-vat cultures (requiring no scale-up at the cheese plant). High cell counts for bulk set cultures can be obtained by internal pH control using buffers or by external pH control by titrating with sodium or ammonium hydroxide. Inclusion of phosphate salts in the culture media effectively prevents the proliferation of bacteriophages in bulk cultures (Suarez *et al.*, 2007). Industrial experience suggests that bulk mesophilic cultures can be cooled to near 4°C and stored for up to 7 days. Thermophilic cultures should not be cooled to less than 20°C and have a shelf-life of only about 24 hours.

4. Bacteriophages

Commercial experience suggests that bacteriophages continue to pose a significant challenge to cheese makers, especially for mesophilic cultures and for processing in open vats that are exposed to the environment. Common phages associated with dairy cultures are c2, 936, and p335 (Ahn *et al.*, 2009). Both lysogenic and lytic life cycles occur. With a short latent period of 30–50 minutes and a large burst size of 50–175, lytic phages can multiply quickly and destroy a culture within hours. Culture growth stops when phage levels reach $10^3–10^7$/ml. Culture failure due to phage can be recognized by an initial normal acid development, followed by a decrease or termination of culture growth at a later stage. This is different from inhibition due to antibiotics, which is characterized by no or slow initial growth; if antibiotic inhibition is not severe, culture growth and acid development by resistant strains or mutants may resume with time.

Means to prevent culture infection with bacteriophage include: (1) culture rotation, which is effective because bacteriophages are highly (although not completely) strain specific; (2) general good hygiene, especially with respect to whey spillage (removing or pasteurizing whey immediately after separation from the curd); (3) routine tests for normal acid development, when the culture currently in use is grown in skimmed milk containing a little whey from the most recent vat; and (4) selection of LAB with inherent phage resistance. To this end, there is a great deal of ongoing work to characterize LAB phages (see reviews by Limsowtin *et al.*, 1997; Martin *et al.*, 2008; Suarez *et al.*, 2008; Ahn *et al.*, 2009; Powell, 2010; Quiberoni *et al.*, 2010; Zinno *et al.*, 2010).

VI. CHEESE MANUFACTURE

The essential ingredients in cheese making are milk and a protein coagulant such as rennet and/or acid; the acid is normally produced by LAB. Rennet and acid alone, or in combination, cause milk proteins to aggregate and eventually transform fluid milk to a semi-firm gel (Section IV). When this gel is cut into small pieces (curds), the whey (mostly water and lactose) begins to separate from the curds. The curds are then differentiated by various cheese-making procedures to produce all of the many different cheese varieties. Exceptions to gelation are the heat-acid coagulated varieties (Table 8.6, Category 2) in which curd is recovered as falling or floating (precipitated) aggregates.

Properties of some representative cheese varieties are listed in Table 8.6, grouped into families according to basic manufacturing procedures. Detailed descriptions of the cheese families can be found in Hill (2007). Cheese recipes representing each of the families in Table 8.6 can also be found in Hill (2011). Using these categories, most cheese can be classified into technological groups, but the categories themselves cannot be applied rigidly. For example: (1) pasta filata varieties vary widely in composition, manufacturing techniques, and degree of ripening, so they do not fit well into any category; and (2) the manufacture of Cheshire types is similar to Cheddar up to the point of draining, but after that their high acid development is similar to Feta.

With respect to quality, the objectives of cheese making are: (1) to achieve an optimum composition with respect to moisture, acidity (pH), fat, protein, and minerals (especially calcium); (2) to establish the correct ultrastructure of cheese at the microscopic level; and (3) to ripen cheese to the desired flavor and texture. Objectives (1) and (2) are achieved by varying initial make procedures, and by doing so, objective (3) is achieved (Adda *et al.*, 1982; Green and Manning, 1982; Lawrence *et al.*, 1983, 1984). These variations in initial make procedures are mostly different means of controlling the rate and extent of acid development and the rate and extent of moisture release. Figure 8.2 is a flowchart illustrating the principal unit operations that may be applied to cheese making depending on the variety. A detailed discussion of these alternative operations is beyond the scope of this chapter; rather, the following paragraphs describe the principles that can be applied to control the most important cheese-making parameters, namely, pH history, moisture, calcium, texture, and flavor. Much of what is presented in this section is discussed in greater detail in Hill (2007).

A. Moisture Control

Cheese making is the process of removing moisture from a rennet or acid coagulum which is made of fat globules (unless the milk is skimmed) and water droplets trapped in a matrix of casein micelles. Cheese is, therefore, primarily a concentrate of milk protein (casein) and fat. The many cheese-making operations are directed to this process of removing water from the milk gel by inducing syneresis. Syneresis refers to the contraction of the protein network with resulting expulsion of water (whey) from the curd. Contraction is primarily due to increased hydrophobic interactions, which are strengthened by a rise in temperature up to a maximum at about 80°C. This means, for example, that thermophilic (cooked) cheese varieties have short make procedures because the higher cooking temperatures (up to 52°C) dry the curds quickly. Contraction is also encouraged by reduced electrostatic repulsion caused by decreasing the pH towards the isoelectric point of caseins (pH 4.6). The whey contains lactose, whey proteins, lactic acid, and some of the milk minerals. Moisture content, therefore, influences the final cheese pH because it determines the amount of residual/fermentable lactose in cheese; and the pH history, in turn, influences cheese moisture by affecting the rate and extent of syneresis. The final cheese moisture is also strongly influenced by packaging and ripening conditions (Section VII).

B. pH Control

Proper development of acidity, as indicated by the pH history, is the most important process control tool that determines cheese safety and quality. Critical process control points are pH at draining, the pH at salting for varieties that are salted before forming, the pH at demolding or at 24 hours after manufacturing, the minimum pH during $1-14$ days after manufacture, and the pH of the ripened cheese. Some typical pH profiles are presented in Table 8.7. With respect to minimum pH, which is usually reached within 7 days of ripening, most rennet-coagulated varieties can be categorized into three groups: (1) fresh varieties with little or no acid development and a minimum pH $> 5.8-6.5$

TABLE 8.7 pH versus Time Profiles for Several Cheese Varieties (Hill, 2007)

Operation	Swiss Type		Gouda		Cheddar MNFS 53%		Cheddar MNFS 57%		Feta		Cottage	
	Time	pH	Time	pH	Time	pH	Time	pH	Time	pH	Time	pH
Add starter	0	6.60	0	6.60	0	6.60	0	6.60	0	6.60	0	6.60
Add rennet	15 min	6.60	35 min		60 min	6.55	30 min	6.55	75 min	6.50	60 min	6.50
Cut	45 min	6.55	70 min	6.45	90 min	6.50	75 min	6.50	115 min		300 min	4.80
Drain or dip into forms	150 min	6.35	100 min		210 min	6.20	195 min	6.3	130 min	NA	360 min	
Milling	NA	NA	NA	NA	360 min	5.40	315 min	5.45	NA	NA	NA	NA
Pressing	165 min	6.35	130 min		420 min	5.35	390 min	5.40	NA	NA	NA	NA
Demolding	16 h	5.30	8 h	5.40	24 h	5.20	10	5.20	24 h		NA	NA
Minimum pH	1 wk	5.20	1 wk	5.20	1 wk	5.10	1 wk	5.10	1 wk		NA	NA
Retail	6 mo	5.6	6 mo	5.6	24 mo	5.50	4 mo	5.3	6 wk	4.4	2–14 d	5.2

MNFS: moisture as a percentage of the non-fat solids in the cheese.

(Table 8.6, Category 3); (2) soft ripened varieties with minimum pH < 5.0 (Table 8.6, Category 4); and (3) varieties with minimum pH > 5.0 (Table 8.6, Categories 5–7).

Increased acidity induces syneresis (due to reduced charge repulsion between casein micelles), causes solubilization of CCP, and disrupts casein micelles (Section IV). This alters curd texture and results in reduced lactose levels in the curd due to fermentation (conversion to lactic acid) and syneresis (lactose removed with the whey). Acid development takes place mainly within the curd because most bacteria, following coagulation, are trapped in the gel matrix. The minimum pH value usually occurs within 3–14 days after manufacture and is dependent on: (1) the initial extent of acid development; (2) the amount of fermentable lactose that remains during early curing; and (3) the activity and concentration of ripening agents that utilize lactic acid and/or cause proteolysis. Also important for pH control are salting, which reduces the rate of acid development, and the ability of the culture to ferment galactose.

C. Calcium Control

More calcium is retained at a higher draining pH. For example, the calcium content of Swiss cheese (draining pH 6.3–6.5) is greater than that of Cheddar (draining pH 6.0–6.2). The important parameter is the ratio of total calcium to casein or calcium to non-fat solids (NFS), which is easier to measure (see Table 8.6). Little calcium is retained in Feta cheese, a fact that needs some explanation. Feta curds are dipped into the forms early, while the pH is still quite high. However, the moisture is also high because no cooking has taken place. The decrease in pH while the cheese is in the forms combined with the high salt levels used in Feta greatly increase syneresis and associated moisture removal. The net result is that a great deal of moisture (whey) is removed at the low pH and high salt concentration and most of the calcium phosphate is also removed with it. This is also true for other soft ripened cheeses such as blue and Camembert. In addition, soft ripened cheese making procedures have a long fermentation time before rennet addition and an extended setting time. The acid development before cutting encourages the release of minerals into the whey. The degree of solubilization of micellar calcium phosphate determines the extent of casein micelle disruption which, in turn, determines the basic cheese structure.

D. Texture (Cheese Body)

When cheese graders refer to cheese texture they often mean the amount and type of openness or holes in the cheese. Here, texture refers to the sensory attributes of firmness, elasticity, brittleness, etc. A typical texture in young cheese is a strong indication of probable flavor defects later. Therefore, an important objective of cheese making is to develop the proper ultrastructure that will eventually lead to the desired cheese texture (see reviews by Olson *et al.*, 1996; Gunasekaran and Mehmet Ak, 2003; Lucey *et al.*, 2003). The following are the principal determinants of cheese texture.

Percent fat in dry matter (FDM) ranges from 60% for double cream varieties, through 50% for the so-called full-fat varieties, to 30% for part-skim varieties such as pizza cheese. Fat globules physically disrupt and weaken the casein matrix, so higher fat content softens cheese texture. FDM is mostly determined from the relative proportions of fat and protein in milk.

Percent moisture in non-fat substance (MNFS), ranging from about 40% to 80% in rennet-coagulated varieties (Table 8.6), also weakens the casein matrix and softens the cheese.

The pH profile is extremely important to cheese texture. Rennet-coagulated cheese with little acid development (pH > 5.8; Table 8.6, Category 3) is chewy and crumbly, and does not melt or stretch when heated. Hence it can be used in stir fries, for example. Rennet-coagulated cheese with a minimum pH in the range of 5.3–5.0 (Table 8.6, Categories 5 and 6) becomes stretchable or meltable when heated depending on the degree of ripening. As the cheese ages, the protein matrix is broken down so the caseins become more meltable and less stretchable. Finally, rennet-coagulated cheese with minimum pH < 5.0 (Table 8.6, Category 4) is brittle (e.g. Feta and Cheshire). Although low pH solubilizes calcium phosphate and disrupts casein micelles, the casein proteins remain intact and tightly packed owing to reduced charge repulsion. Therefore, while Feta remains brittle, Camembert becomes soft and smooth due to increased alkalinity from the ammonia released during ripening.

Calcium content, and the relative proportions of soluble and insoluble/colloidal calcium, combined with their interactive effects with pH are important determinants of cheese texture (Lee *et al.*, 2005; O'Mahony *et al.*, 2006a, b). The pH history up to the point of draining, along with postdraining syneresis, determine the calcium content of the curd, and the minimum curd pH determines the distribution of soluble and insoluble calcium in cheese. So, together with pH, calcium influences the strength of the casein matrix and, therefore, the texture. In brine-salted cheeses it is important that the Ca^{2+} concentration and the pH of brine are at levels similar to those in cheese. High pH and/or low Ca^{2+} induce more ion exchange (Ca^{2+} into the brine and Na^+ into the cheese) to produce more sodium caseinate, which holds much more water than calcium caseinate. The result is moist mushy rind (Genigeorgis *et al.*, 1991).

Physical and timing aspects of curd handling are diverse and strongly influence texture and flavor. A few of the many examples that could be cited follow:

- Cheese with eyes (Swiss) requires a smooth fused texture which is obtained by forming and/or pressing the curd under warm whey.
- Vat salting, in which the curd is salted before forming and pressing (Cheddar, Cheshire, and American type varieties,) creates a distinct granular structure that never fuses as smoothly or as completely as brine-salted varieties. Vat salting versus brine or surface salting is probably the most important difference between traditional European varieties and their Americanized imitations such as American Havarti and American Mozzarella.
- One of the more interesting curd handling processes is the stretching operation in the manufacture of pasta filata cheeses and its many ethnic cousins around the world. Traditional pasta filata is mesophilic cheese similar to Cheddar in the early stages of its manufacture. After 'cheddaring', Cheddar is milled, salted, formed, and pressed, whereas the traditional pasta filata types are milled, immersed in hot water, and stretched. The structure and texture of stretched cheese are dependent on all of the factors discussed above and also on other factors: moisture, fat content, plasmin activity, pH history, total and insoluble calcium, temperature and time of heating before stretching, and ripening time (Walsh *et al.*, 1998; Guinee *et al.*, 2001; Feeney *et al.*, 2002; Somers *et al.*, 2002; O'Mahony *et al.*, 2006b; Tunick and Van Hekken, 2006).

E. Flavor Control

In the broadest terms, directed flavor development involves retention and/or addition of ripening agents and controlling their activity over time. See Weimer's recent text (2007) on this topic. Milk heating and clarification treatments determine the number and type of non-starter bacteria present in the milk. A great deal of direction is obtained through selection of cultures, coagulants, and other additives such as lipases. Debittering cultures, for example, reduce bitter protein fragments to shorter, non-bitter peptides and are an important tool to reduce bitterness and extend shelf-life in high-moisture varieties, such as Monterey Jack, and low-fat varieties. See the discussion of ripening agents in Section VII.

All cooking and curd-handling procedures exert specific effects on the various cheese-ripening agents (bacteria and enzymes) and determine the types and numbers/levels of these biological agents that remain in the curd to ripen the cheese. For example, the milk enzyme plasmin prefers neutral to slightly alkaline pH and is inactivated by low pH. Therefore, varieties in which ripening pH is high (e.g. smear-ripened types) have higher retention and greater activity of plasmin. Plasmin activity is also increased by higher cooking temperatures, due to the activation of plasminogen, for example, in traditional Swiss and Italian varieties (Bastian and Brown, 1996; Bastian *et al.*, 1997). On the other hand, calf rennet is more soluble at higher pH, but more active at lower pH. Therefore, rennet retention is higher in varieties that are drained at lower pH and more so in varieties that are cut at lower pH. Rennet activity is also drastically reduced by higher cooking temperatures, as in traditional Swiss and Italian types; therefore, rennet is more active in mesophilic varieties.

Finally, a great deal of flavor and texture development and differentiation among varieties takes place when cheese ripens. See the discussion on cultures (Section V) and ripening (Section VII).

VII. BIOCHEMISTRY OF CHEESE RIPENING

Some varieties of cheese, mostly acid- or acid/heat-coagulated types, are consumed fresh, i.e. soon after the curds are processed. Most other cheeses, which are of rennet type, have to be ripened to the desired texture and flavor over a period ranging from 2 weeks (e.g. Mozzarella) to over 2 years (e.g. extra-mature Cheddar and Parmigiano Reggiano). Cheese is a biologically and biochemically dynamic food system in which the proteins, lactose, and milk fat undergo physicochemical changes that facilitate the conversion of the 'green' curd to mature cheese with its characteristic microstructure, texture, flavor, and desired functionality (Guinee, 2003). The biochemical processes during cheese ripening have been reviewed several times (Fox *et al.*, 1995, 1996, 2000, 2004; Fox and Wallace, 1997; McSweeney and Sousa, 2000; Fox and McSweeney, 2006). The complex process of cheese ripening is discussed under three main headings:

- metabolism of residual lactose and catabolism of lactate and citrate
- lipolysis and catabolism of fatty acids
- proteolysis and amino acid catabolism.

The products of these metabolic pathways undergo further biochemical and chemical modifications. Cheese texture and flavor are mostly the outcome of the above primary reactions, while flavor is also probably due to the modification of primary reaction products (Povolo *et al.*, 1999).

A. Metabolism of Residual Lactose and Catabolism of Lactate and Citrate

Cheese is a fermented dairy product and the metabolism of lactose to lactic acid is essential in the production of all varieties. These reactions were reviewed by Fox *et al.* (1990), McSweeney and Sousa (2000), and McSweeney and Fox (2004). The cheese microbes (starter and/or non-starter) are mainly responsible for the fermentation of lactose (in the vat stage of cheese making) and the breakdown of lactate and citrate. Fresh cheese curd contains only about 1–2% lactose depending on whether the curds are washed (Dutch-type cheeses; $\approx$ 1% lactic acid) or not (Emmental and Parmigiano Reggiano; $\approx$ 1.5% acid) and whether the curd is dry-salted (Cheddar type; $\approx$ 1.5% in young Cheddar). About 98% of lactose is lost in the drained whey (Huffman and Kristoffersen, 1984). Fermentation of the remaining lactose is critical to the manufacture of cheese curd. Unless it is completely metabolized, the residual sugar can lead to undesirable secondary fermentations during ripening (Fox *et al.*, 1990, 2000; McSweeney and Fox, 2004). Lactose trapped in the curd matrix is rapidly and completely fermented by the starter bacteria before the salt/moisture ratio reaches a level that inhibits starter activity (Fox and Kelly, 2003; McSweeney, 2007); that is, when the curd pH is about 6.2–6.4 at molding and prior to salting (McSweeney, 2011); in about 12 hours the pH reaches $\approx$ 5.0–5.3 (Fox *et al.*, 2000). In the production of Cheddar cheese, the low curd pH of nearly 5.4 and dry-salting before molding retard lactose metabolism (Povolo *et al.*, 1999; McSweeney, 2011), and lactose is slowly, in about a month, fermented by residual starter activity [or by non-starter lactic acid bacteria (NSLAB)], to L(+) lactic acid (McSweeney, 2004).

The catabolic pathway of lactic acid is characteristic of the cheese variety, and the range of reactions that take place has a positive or negative impact on cheese ripening and/or quality. In Swiss-type cheese, the secondary fermentation of lactate by *Propionibacterium freudenreichii* subsp. *shermanii* is of great importance. The organic acids (propionate, acetate) produced contribute to the nutty flavor and the released CO_2 collects at weak points in the curd and forms the large eyes characteristic of this cheese variety (Martley and Crow, 1996; McSweeney, 2004). Catabolism of lactate is also important in surface mold-ripened (e.g. Camembert and Brie) and smear-ripened (e.g. Tilsit or Limburger) cheeses. The oxidation of lactate to CO_2 and water by *Penicillium camemberti* greatly increases the surface pH of this cheese type, and causes a pH gradient from the surface (pH $\approx$ 7.5) to the core (pH $\approx$ 6.5) and the diffusion of lactate towards the surface (Bonaiti *et al.*, 2004; Spinnler and Gripon, 2004). The characteristic soft texture of this cheese comes from the precipitation of calcium phosphate at the high-pH surface and a concomitant migration of soluble calcium and phosphate to the surface (Vassal *et al.*, 1986; Walstra *et al.*, 2006). In surface smear-ripened cheeses, yeast activity deacidifies the surface and this favors the growth of coryneform bacteria. In Cheddar and Dutch-type cheeses, excessive formation of D/L-lactate (racemic mixture of L(+) and D(−) lactate) by adventitious NSLAB activity produces undesirable white specks of Ca-D-lactate crystals on the surface of mature cheese (Chou *et al.*, 2003; McSweeney and Fox, 2004; Swearingen *et al.*, 2004). Another undesirable, but minor, reaction in cheese (when the packaging material becomes more permeable to oxygen) is the oxidation of lactic acid by some members of NSLAB to acetic or formic acids and CO_2 (McSweeney and Fox, 2004). Late gas blowing and off-flavors are sometimes caused in hard and semi-hard cheese varieties by the anaerobic catabolism of lactate to butyrate, CO_2, and hydrogen by *Clostridium tyrobutyricum*. This defect can be avoided by minimizing the spore count in milk through good hygiene and by the removal of spores by bactofugation or microfiltration; increased levels of NaCl in cheese and lower ripening temperatures can also alleviate the problem (McSweeney, 2007).

Most ($\approx$ 94%) of the citrate in milk is soluble and is lost in the whey. Mostly, the colloidal citrate is metabolized by the citrate-positive strains of *Lactococcus lactis* subsp. *lactis* and/or *Leuconostoc* spp. to diacetyl, acetate/2,3-butanediol, acetoin, and CO_2. Diacetyl and sometimes acetate contribute to the cheese flavor. The CO_2 produced is responsible for the formation of small eyes in Dutch-type cheese, but gives an undesirable open texture in Cheddar cheese and the floating curd defect in cottage cheese (McSweeney and Fox, 2004).

B. Lipolysis and Catabolism of Fatty Acids

Milk fat is a critical component that determines cheese quality. Milk lipids affect cheese rheology and texture (Yoshida, 1989), influence flavor by acting as a source of fatty acids and their derivatives and by acting as a solvent for sapid compounds produced from lipids and other precursors. They also form the fat—water interface where many important reactions occur (Collins *et al.*, 2004).

Lipolysis during cheese ripening has been reviewed a number of times (Fox and Wallace, 1997; Fox *et al.*, 2000; McSweeney and Sousa, 2000; Collins *et al.*, 2003a; Collins *et al.*, 2004; McSweeney *et al.*, 2006; Wolf *et al.*, 2009). The breakdown of milk fat by lipases/esterases in cheese releases highly flavored short-chain (C_4—C_{10}) fatty acids. Low levels of these fatty acids are important contributors of flavor in many cheese varieties. Lipolysis is limited in most cheeses; even slight excess degradation of fat can cause rancidity or unbalanced flavor. Lipases in cheese possibly originate from milk, rennet paste, starter and non-starter LAB, secondary starter microbes, and exogenous lipase preparations (Fox and Wallace, 1997; McSweeney and Sousa, 2000; Collins *et al.*, 2004; Deeth and Fitz-Gerald, 2006). In varieties such as Cheddar and Gouda, the weakly lipolytic starter and sometimes the non-starter bacteria contribute to the slow and limited production of free fatty acids (Fox *et al.*, 2000; Collins *et al.*, 2003b) and in Swiss cheese, *Propionibacterium freudenreichii*, along with the thermophilic starter, plays this role (McSweeney, 2004; Thierry *et al.*, 2005). Extensive lipolysis is the major pathway to flavor generation in mold-ripened (blue and Camembert), surface bacterially ripened (e.g. Limburger), and some hard Italian cheeses manufactured from raw milk (e.g. Parmigiano-Reggiano, Grana Padano, and Provolone). Potent extracellular lipases are produced by *Penicillium roqueforti* and *P. camemberti* (Spinnler and Gripon, 2004), and the complex Gram-positive surface microflora, especially *Brevibacterium linens* of smear cheeses (Brennan *et al.*, 2004), also produce extracellular lipases. The pregastric esterase contained in rennet paste (e.g. Pecorino varieties, Provolone, and certain traditional Greek cheeses) (Hamosh, 1990; Fox, 2003a) and milk's indigenous lipoprotein lipases in raw milk cheeses (Olivecrona *et al.*, 2003) also possess strong lipolytic activity.

In addition to their direct impact on cheese flavor, the volatile short-chain fatty acids are important precursors in the series of reactions that lead to the production of different fatty acid flavor derivatives (Collins *et al.*, 2003a): ethyl esters (fruity aromatic notes, floral, goaty), thioesters (cheesy, cooked vegetable), branched chain keto acids (pungent cheesy), and unsaturated alcohols and ketones (mushroom-like) (Collomb *et al.*, 1998). *N*-Methyl ketones, produced by partial β-oxidation of fatty acids, are of particular importance to blue cheese flavor.

C. Proteolysis and Amino Acid Catabolism

Proteolysis is perhaps the most important of the three biochemical events that occur during the maturation of most cheese varieties (especially the internal bacterially ripened cheeses such as Cheddar, Swiss, or Gouda) and is definitely the most complex of the events. The subject has been extensively reviewed (Grappin *et al.*, 1985; Rank *et al.*, 1985; Fox, 1989; Fox and Law, 1991; Fox *et al.*, 1994; Fox and McSweeney, 1996, 2006; Fox and Wallace, 1997; Upadhyay *et al.*, 2004a; Mikulec *et al.*, 2010). Proteolysis contributes to: (1) the development of cheese texture through the hydrolysis of the para-casein matrix, by increasing the water binding capacity of the curd (the newly liberated ionized α-carboxyl and α-amino groups can bind water), and indirectly through an increase in pH due to the release of ammonia during amino acid catabolism; and (2) the flavor of cheese through the production of short—medium peptides (these contribute to the brothy background flavor of cheese; some hydrophobic peptides are bitter) and free amino acids, but most importantly, when these amino acids are catabolized to generate many important volatile flavor compounds, and also by the release of sapid compounds from the cheese matrix during mastication. The caseins are extensively proteolyzed by the action of a wide array of proteinases and peptidases that in cheese originate from six sources: the coagulant, milk's indigenous enzymes (mainly plasmin), starter LAB, NSLAB, secondary starter, and exogenous proteinases and peptidases.

1. Coagulant

In cheeses cooked to a temperature of less than 40°C, about 5—30% of rennet activity is retained depending on the enzyme type, pH at draining, and the moisture content of the cheese (McSweeney, 2011; Upadhyay *et al.*, 2004a). For example, Cheddar curd retains $\approx$ 6% of the added chymosin; the amount increases as the pH is reduced at whey draining (Holmes *et al.*, 1977; Creamer *et al.*, 1985). Rennet is extensively or completely denatured in varieties where the curd cooking temperatures exceed 55°C, as in Emmental, Parmigiano-Reggiano, and Mozzarella cheeses (Singh and Creamer, 1990; Boudjellab *et al.*, 1994; McSweeney, 2011). The coagulant is mainly responsible for primary proteolysis that generates large and intermediate-sized peptides, which are subsequently hydrolyzed by enzymes from the starter and non-starter microflora of the cheese (McSweeney, 2004). Apart from its specificity for the Phe_{105}—Met_{106} bond in κ-casein, chymosin is weakly proteolytic; α_{s1}-, α_{s2}-, and β-caseins are not hydrolyzed during milk coagulation, but may be slowly acted upon during cheese ripening (Upadhyay *et al.*, 2004a). The primary site of chymosin action on α_{s1}-casein is Phe_{23}—Phe_{24}, which is completely hydrolyzed in 4 months in Cheddar and related cheeses (Carles and Dumas, 1985; McSweeney *et al.*, 1993a; McSweeney, 2011). Two other chymosin-susceptible bonds in this protein are Leu_{101}—Lys_{102} and Trp_{164}—Tyr_{165}, both of which are extensively hydrolyzed in mature Cheddar. Chymosin appears to have limited action on α_{s2}-casein; there are not many studies in this area. Although the Leu_{192}—Tyr_{193} bond of β-casein is very susceptible to the action of chymosin, its hydrolysis is strongly inhibited by just 5% NaCl. This is likely to have significance for cheese flavor since the peptide β-CN (f193—209) and related fragments are very bitter.

The proteolytic action of rennet substitutes and pepsin (commonly found in commercial rennet preparations) differs from that of chymosin and has been extensively studied and reviewed (Hassan *et al.*, 1988; Broome *et al.*, 2006; Dervisolgu *et al.*, 2007; Jacob *et al.*, 2011). For example, fungal rennet substitutes have very different specificity from that of chymosin; the principal cleavage sites of *Rhizomucor miehei* proteinase in α_{s1}-casein in solution are Phe_{23}—Phe_{24}, Met_{123}—Lys_{124}, and Tyr_{165}—Tyr_{166}, and those in β-casein are Glu_{31}—Lys_{32}, Val_{58}—Val_{59}, Met_{93}—Gly_{94}, and Phe_{190}—Leu_{191}. Pepsins have specificity similar to chymosin, but it has not been established precisely; bovine pepsin has quite a rapid action on the Leu_{101}—Lys_{102} bond of α_{s1}-casein, which is relatively slowly hydrolyzed by chymosin.

2. Indigenous Milk Proteases

Of the many indigenous enzymes in milk, plasmin is the most significant in proteolysis during cheese ripening. Ismail and Nielsen (2010) recently reviewed the current knowledge on this protease and its relevance to the dairy industry. Bastian and Brown (1996) have also reviewed the subject. Plasmin is a serine protease derived from blood with pH and temperature optima of $\approx$ 7.5 and 37°C, respectively (Upadhyay *et al.*, 2004a). Plasminogen, the precursor, plasmin, and plasminogen activator (PA) are all associated with casein micelles and are incorporated into the enzyme-coagulated casein curd; inhibitors of both plasmin and PA are found in milk serum and so are lost in the whey (Bastian and Brown, 1996). Plasmin readily hydrolyzes β-casein to produce γ-caseins [(γ^1-CN (β-CN f29—209),

γ^2-CN (β-CN f106−209), γ^3-CN (β-CN f108−209)] and some proteose peptones [PP5 (β-CN f105/107), PP8 slow (β-CN f29105/107), and PP8 fast (β-CN f1−28)]. In α_{s2}-casein eight peptide bonds are rapidly hydrolyzed by the action of plasmin, but α_{s1}-casein is less susceptible to plasmin, and κ-casein is immune to its action (Bastian and Brown, 1996). Plasmin is mainly responsible for the limited proteolysis of β-casein in internal bacterially ripened cheeses; large and intermediate-sized peptides are the main products of its action. Being heat stable, plasmin plays a particularly important role in pasta filata and high-cook cheeses in which the coagulant is largely denatured by heat (Gobbetti, 2004). Its contribution is also vital in cheeses in which pH increases during ripening, such as Camembert-type (Spinnler and Gripon, 2004) and smear-ripened cheeses (O'Farrell *et al.*, 2002).

Exogenous plasminogen activators, urokinase and streptokinase, were successfully used to accelerate proteolysis in cheese by increasing the activity of plasmin. For example, urokinase was used in the manufacture of Cheddar (Bastian *et al.*, 1997; Barrett *et al.*, 1999) and ultrafiltered Havarti and Saint Paulin (Bastian *et al.*, 1991) cheeses. In addition to an exogenous streptokinase, Upadhyay *et al.* (2004b, 2006) successfully used a streptokinase-producing strain of *Lactococcus* to accelerate proteolysis in Cheddar.

3. Lactic Acid Bacteria

The auxotrophic LAB are weakly proteolytic, but possess a comprehensive proteolytic system which is essential to meet their complex amino acid requirements (Beresford and Williams, 2004; Upadhyay *et al.*, 2004a). Milk contains very low concentrations of these peptides and amino acids. While their enzymes degrade caseins into small peptides and amino acids, the LAB grow to high cell populations [10^9−10^{10} colony-forming units (cfu)/ml] in cheese milk and inadvertently contribute to the flavor of fermented dairy foods (Law and Mulholland, 1995; Steele, 1995, 1996).

Lactococcus and the thermophilic *Lactobacillus* spp. are economically important starters and their proteolytic systems, especially of the former, have been well characterized and extensively reviewed (Mulholland, 1995; Steele, 1995, 1996; Kunji *et al.*, 1996; Law and Haandrikman, 1997; Christensen *et al.*, 1999; Savijoki *et al.*, 2006). Their proteolytic system consists of a serine protease or cell envelope protease (CEP) anchored to the cell membrane and extending out of the cell wall, giving it ready access to extracellular substrates, peptide and amino acid transport systems, intracellular proteinases, and a number of intracellular peptidases (Upadhyay *et al.*, 2004a; McSweeney, 2011). The CEP is mainly responsible for the hydrolysis of the larger peptides produced from α_{s1}-casein and β-casein by chymosin and plasmin, respectively. The intracellular peptidases (aminopeptidases, dipeptidases, and tripeptidases) are responsible for the release of free amino acids after the cells have lysed. Research (Farkye *et al.*, 1990; Lane and Fox, 1997; Broadbent *et al.*, 2002) using CEP negative (Prt$^-$) and CEP positive (Prt$^+$) strains of the starter has shown that, although CEP is active in cheese during ripening, it is not essential; some differences in quality were found between Prt$^-$ and Prt$^+$ cheeses, but Prt$^-$ cheeses were also good. Other lactococcal enzymes or those from the NSLAB serve its function. The starter LAB, both *Lactococcus* and thermophilic *Lactobacillus*, die off rapidly after curd manufacture; following lysis, the dead cells release their intracellular enzymes.

Bitterness is a common problem in cheese, especially in Cheddar, Gouda, and other internal bacterially ripened cheeses. The hydrophobic peptides (molecular mass < 6 kDa and a mean hydrophobicity > 1400 cal per residue) derived from caseins, especially from the C-terminal region of β-casein (β-CN f193−209), are very bitter (Singh *et al.*, 2005; McSweeney, 2007, 2011). Bitterness in cheese develops due to incorrect patterns of proteolysis causing excessive production of bitter peptides, or due to peptidase activity that is insufficient to degrade the hydrophobic peptides to free amino acids (Lemieux and Simard, 1991; HabibiNajafi and Lee, 1996; McSweeney, 1997; Frister *et al.*, 2000). Recent studies (Broadbent *et al.*, 1998, 2002; Pillidge *et al.*, 2003; Broadbent and Steele, 2007b) suggest that both the lactococcal CEP and peptidase systems play a significant role with respect to bitterness and cheese quality in general. Besides, excessive chymosin activity can also cause bitterness, but plasmin is unlikely to produce bitter peptides. Changing the milk coagulant to a more suitable type, using a starter culture or adjunct with high peptidase activity, blending bitter and non-bitter strains, using a *Lactobacillus* adjunct culture with a bitter strain, and ensuring an adequate level of NaCl in cheese are some strategies that can be used to ameliorate bitterness in cheese (McSweeney, 2007, 2011).

The activity of NSLAB appears to supplement the proteolytic activity of the starter LAB. Several cheese studies with controlled microflora (McSweeney *et al.*, 1994; Rehman *et al.*, 2000) suggest that NSLAB are less significant than the starter in flavor development in Cheddar and other similar varieties. The initial numbers of NSLAB in the freshly produced Cheddar curd are very low (< 100 cfu/g); depending on the ripening temperature, they grow to reach 10^7−10^8 cfu/g within about 3 months (Beresford and Williams, 2004; McSweeney, 2011) and thereafter their numbers more or less remain constant, as in Cheddar (Fox *et al.*, 1998; Peterson and Marshall, 1990), or may decline, as in Swiss cheese (Beuvier *et al.*, 1997). So, the viable microflora of long ripened cheese is dominated by mesophilic lactobacilli during most of its ripening. The NSLAB are more numerous and more diverse in raw milk than in pasteurized milk cheese (McSweeney *et al.*, 1993b; Grappin and Beuvier, 1997; Albenzio *et al.*, 2001; de Angelis *et al.*, 2001; Mannu and Paba, 2002; Dasen *et al.*, 2003) and they contribute to the formation of small peptides and amino acids (Beresford and Williams, 2004). In general, raw milk cheeses are known to ripen more quickly and develop more intense flavor than those made from pasteurized milk (Beuvier and Buchin, 2004; McSweeney, 2011). This is primarily due to the presence of indigenous NSLAB in raw milk which are incorporated into the cheese; heat-induced inactivation of the indigenous lipoprotein lipase is also of significance. For health and safety reasons and for the production of cheese with consistence in quality, it is unlikely that large cheese-making firms will revert to the use of raw milk. Starter and NSLAB adjuncts, usually selected *Lactobacillus* strains, are being used to simulate the flavor characteristics of raw milk cheese in the pasteurized milk product (Law, 2010b). However, adjuncts added to pasteurized milk do not closely simulate the flavor of raw milk cheese, probably because they contain only a few strains of lactobacilli, whereas the NSLAB of raw milk cheese are quite heterogeneous. Development of improved adjuncts is likely in the future as more work on NSLAB is being carried out.

The secondary microflora added to, or encouraged, to grow in a number of cheese types have specific functions and possess proteolytic systems similar to LAB. For example, various strains of *Lactobacillus* are added to Cheddar to improve flavor or accelerate ripening. Proteolysis by *Brevibacterium linens* at the surface of smear-ripened cheeses is due to the extracellular proteinase and aminopeptidase, and a number of intracellular peptidases secreted by the bacterium (Rattray and Fox, 1999). Both *Penicillium roqueforti* and *P. camemberti* produce potent extracellular aspartyl and metalloproteinases and various peptidases that contribute to the extensive proteolysis in blue, Camembert and Brie-type cheeses, respectively (Cantor *et al.*, 2004; Spinnler and Gripon, 2004).

4. Amino Acid Catabolism

Amino acids are the end products of proteolysis, and together with small peptides they contribute directly to the savory background flavor of cheese and to its taste; amino acids may be sweet, bitter, or sour. Research has now made it apparent that the production of amino acids is not the rate-limiting step because accelerating proteolysis does not necessarily accelerate flavor development (Upadhyay and McSweeney, 2003). It is now generally agreed that amino acids in principle serve as precursors for a complex series of reactions that produce a wide range of sapid and aromatic compounds such as amines, acids, carbonyls, ammonia, and sulfur compounds. The catabolism of amino acids was reviewed by Ardo (2006), Curtin and McSweeney (2004), Ganesan and Weimer (2007), and Yvon and Rijnen (2001), and is believed to proceed via two major pathways:

- Transaminase action catalyzed by aminotransferases, in which the amino group of an amino acid (usually aromatic and branched chain amino acids and methionine) is transferred to an acceptor molecule (usually α-ketoglutarate) to produce the corresponding α-keto acid. α-Keto acids are unstable and they degrade to aldehydes, carboxylic acids, hydroxyacids, alcohols, and other compounds.
- Elimination reaction by amino acid lyases (side-chains of amino acids are cleaved), which is particularly important in the production of volatile sulfur compounds from the side-chain of methionine.

In addition, decarboxylases in certain strains of LAB remove the carboxylic group of amino acids and produce amines. Of particular concern are the biogenic amines (amines from His, Trp, and Tyr), which at high levels can bring about adverse physiological reactions in susceptible consumers. The action of deaminases, for example, in surface smear-ripened cheeses, removes the α-amino group of amino acids and results in the formation of carboxylic acids and ammonia.

D. Controlled/Accelerated Cheese Ripening and Cheese Flavor Technology

There is an ever-increasing consumer demand for cheeses with distinctive characteristics and interesting flavors. That an increasing proportion of cheese is made from pasteurized milk does not offer many possibilities to cheese manufacturers for flavor diversification. But, with advances in cheese ripening technologies, it is possible to make cheese both exciting and safe. Flavorful cheeses with characteristic body attributes also have premium prices and need to be ripened at a low temperature for months or even years, so there is considerable industrial interest in technologies to accelerate the ripening process. Some of the options available for the control of cheese ripening and flavor technology are listed below (Farkye, 2004):

- elevated ripening temperatures
- high-pressure processing
- addition of exogenous enzymes
- attenuated starter cultures
- use of culture adjuncts
- GM starters.

It is obvious that any technology that can accelerate ripening and shorten cheese storage time will increase the profit margin and/or offer more competitive pricing to the cheese manufacturer. Temperature control and enzyme technology are the methods of choice to hasten the ripening process; culture-based technology did not meet with much success (Law, 2010b). High-pressure treatment of cheese is known to accelerate proteolysis and has the potential to speed up cheese maturation. These techniques are discussed below.

1. Elevated Storage Temperatures

This is a relatively simple and low-cost technology that is frequently applied to accelerate cheese ripening. However, using elevated temperatures poses the risk of stimulating the growth of spoilage and pathogenic microbes (Fox *et al.*, 1996) and should only be applied to cheeses made from pasteurized milk using GMP. The practical choices for forced ripening are hard and semi-hard cheeses such as Cheddar, Gouda, and Edam, which are very stable and have relatively simple LAB microflora (Law, 2010b). Relatively low temperatures of less than 10°C are typically used in cheese stores. Temperature increases up to 12°C decrease the maturation time of Cheddar cheese by about 60% without adversely affecting its body or texture (Law, 2001). Hannon *et al.* (2002, 2004) used combinations and successions of ripening temperature (20°C for 1 week/12°C for 6 weeks, followed by the normal 8°C for 8 months) and gained 2 months' maturation in Cheddar cheese without a loss in flavor balance.

2. High-Pressure Processing

There is an increasing body of evidence (Yokoyoma *et al.*, 1992; Messens *et al.*, 2000, 2001; Saldo *et al.*, 2000, 2002; O'Reilly *et al.*, 2001, 2002) that the use of ultra-high-pressure technology in young cheese speeds up ripening. Very high pressures in the range of 100−1000 MPa are applied for a short span of time. High pressure breaks up the cells of the starter to release the intracellular enzymes, activates these enzymes, and thereby rapidly increases the proteolytic and possibly other flavor-generating reactions (Stewart *et al.*, 2006; Law, 2010b). At present, this technological option is limited to research laboratories, but the food industry already has an interest in this branch. With more studies on its industrial relevance and with the development of the required hardware, high-pressure processing can be extended to factory applications (Stewart *et al.*, 2006; Law, 2010b).

3. Enzyme-Based Cheese Ripening and Flavor Technology

The use of enzyme-modified cheese (EMC) as an ingredient in processed cheese and as food flavoring is well known. However, the application of enzymes to the controlled ripening and flavor development of cheeses meant for direct consumption is a relatively new concept (Collomb *et al.*, 1998; Law, 2010b). The main challenge to this method is to distribute gram quantities of the enzyme within tonnes of cheese; addition to cheese milk is not an option, as 95% of ripening enzymes are lost in the whey. In addition, currently there is a limited availability of approved commercial enzymes. The development of microencapsulation technology makes it possible to physically enmesh the liposome-entrapped enzymes into the curd matrix as it is being formed (Kirby *et al.*, 1987). Owing to the high cost of phospholipids this method is not economically viable for industrial applications, but is suitable for small-scale applications (Law, 2010b). Alternative approaches to proteinase encapsulation exist, for example, the use of food-grade gums (Kailasapathy and Lam, 2005), and these may be an option. In dry-salted varieties such as Cheddar, the enzymes can be granulated together with salt and distributed into the milled curd (Law and Wigmore, 1982, 1983). Based on the research findings on the role of starter lactococci and commercial proteinases in Cheddar cheese ripening (Law and Wigmore, 1982, 1983), the Accelase™ range of enzymes was developed for commercial use by Imperial Biotechnology Limited (IBT, now a part of Danisco).

Added to the ease of incorporation and efficiency in the distribution of enzymes within the curd, a wide range of flavor options is also available in the Accelase series. Accelase contains a mixture of endopeptidases and exopeptidases (also lipases) which promote a balanced breakdown of caseins into non-bitter peptides and flavor-generating amino acids. The Accelase enzymes are fabricated into a special formulation based on the type of cheese and coagulant, the starter culture regime used, and the market destination for the cheese (Law, 2010b). Unlike the dry-salted varieties, washed curd cheeses such as Gouda and Edam are particularly difficult to ripen with enzymes. Although the enzymes can be introduced into the cheese matrix with wash water or at the soft curd stage, both methods result in curd softening, yield reduction, and loss of enzymes into the wash water (Collomb *et al.*, 1998; Law, 2010b). Mechanical injection of enzymes into the finished cheese may be an option (Wilkinson and Kilcawley, 2005), but this technique has to be expanded to the market stage.

Applications for flavor diversification are mostly culture based and involve the addition of top notes to traditional flavor profiles using selected strains of lactobacilli (*Lactobacillus casei*, *Lb. plantarum*, and *Lb. helveticus*). The methods in use are briefly discussed in the following sections.

4. Attenuated Starter Cultures

Starter LAB function well as cheese-ripening agents, but because of their primary acidification role, it is not possible simply to add a bit more of the culture to enhance flavor. Attenuation or weakening of starter cultures by various means prevents or reduces their acidification function, at the same time retaining their intracellular enzymes to be subsequently released into the cheese matrix during ripening via lysis (Upadhyay and McSweeney, 2003). Petterson and Sjöström (1975) were the pioneers in the use of attenuated starters to accelerate the ripening of a Swedish semi-hard cheese, Svecia; attenuation was by heat treatment of cells. Attenuated bacterial cells can also be prepared by freeze shocking, spray drying, treatment with lysozymes under sublethal conditions, and selection of lactose negative (Lac^{-ve}) mutants (Klein and Lortal, 1999; Azarnia *et al.*, 2006; Law, 2010a, b). Madkor *et al.* (2000) have found that Cheddar cheese made with heat- or freeze-shocked (HS or FS) *Lb. helveticus* strain exhibited greatly enhanced rates of free amino group formation and lipolysis and the highest flavor and aroma scores were obtained for the FS *Lb. helveticus*-treated cheeses. Based on the work of Ardo and Pettersson (1988), a heat-shocked strain of *Lb. helveticus* was made by Medipharm (Sweden) and is commercially sold in Sweden and Finland as Enzobact™. Originally, Enzobact was developed to accelerate ripening in full-fat hard and semi-hard cheeses, but is now mostly used in the ripening of reduced-fat varieties of traditional Swedish hard cheese (Azarnia *et al.*, 2006; Law, 2010b).

Lysozyme is commercially extracted from hen egg white and is a relatively inexpensive enzyme. Although it is a promising tool for starter culture attenuation, its application is limited by the fact that the pretreatment process is laborious and not practical for routine use in cheese production (Law, 2010b). The Lac^{-ve} mutants occur spontaneously in all starter cultures and these natural LAB variants provide the most practical and economically feasible sources of attenuated cultures (Law, 2010a, b). Vindfeldt (1993) has described the technical development and scope for commercial application of these cultures. They are first selected based on their flavor profiles, easily isolated and grown on a glucose medium, packaged, and sold as frozen concentrates in amounts corresponding to their cheese-ripening power per liter of cheese milk or kilogram of cheese.

In 2007, Upadhyay *et al.* demonstrated that high-pressure treatment ($\approx 200\,\mathrm{MPa}$) can impair acid production in lactococcal strains without causing cell lysis and can offer a potential means for attenuation of LAB cultures. Smith (2005) developed a method to accelerate cheese ripening using a biological agent (not an attenuated bacterial starter culture; for example, *Brevibacterium linens*, *Kluyveromyces lactis*, *Staphylococcus xylosus*, *Arthrobacter nicotianae*, and *Geotrichum candidum*) that has been treated with a surface active agent such as sodium dodecyl sulfate to kill, incapacitate, or otherwise reduce cell viability (by acting on the cell membranes), but allowing the intracellular enzymes to access the substrate, as well as allowing the release of the reaction products. The treated biological agent is added to the cheese milk along with a starter culture possessing significant aminopeptidase activity.

Flavor Control CR™ is a range of natural non-acidifying *Lactococcus* cultures (selected from starter cultures) developed by Chr. Hansen A/S in Denmark for commercial application as a flavor-enhancing system. The culture enhances the overall flavor intensity of cheese by accentuating all important flavor notes. It enhances balanced, mellow, rounded, and clean flavors, and suppresses unwanted flavors such as sour, bitter, and flat notes. The CR 500 series and CR 213 ripening blends are especially well suited for low-fat Cheddar-type cheese, but can also bring out a special flavor profile in full-fat cheeses (Chr. Hansen, 2008; Law, 2010b). The effectiveness of these CR cultures is most pronounced in low-fat cheese at near to normal salt concentrations ($\approx 1.75\,\mathrm{g}/100\,\mathrm{g}$). CR cultures are priced similarly to quality enzyme blends, but are much easier to apply; they can be used 'off the shelf', can be thawed (sold as frozen concentrates) and added directly to cheese milk along with the primary starter (Law, 2010b). Furthermore, there are no regulatory restrictions on their use as is the case with GM LAB cultures.

5. Non-Starter Adjunct Cultures

Adjuncts are the secondary or non-starter microflora that are deliberately added to cheese milk (at levels < 0.01%) to bring out desired cheese flavor and functionality (Crow *et al.*, 2002). Examples of some traditional adjuncts are *Penicillium*, *Brevibacterium*, *Geotrichum*, yeasts, *Propionibacterium*, and *Leuconostoc*. The flavor control option provided by secondary microflora in Cheddar-type cheeses is based on adding distinctive flavor/aroma top notes (e.g. cooked, fermented, sweet, nutty) to basic flavor profiles. Their use as commercial adjunct flavor cultures, alone or in combination with attenuated starters, therefore provides complete control over flavor intensity and flavor character (Law, 2010b). Unlike attenuated cultures, adjunct technology also has the potential to harness metabolic flavor pathways of viable cells (Broome, 2007). Culture companies carefully select desired strains by screening out many potential defect-forming traits and, based on their ability to compete against adventitious microflora, produce them in quantities, and blend them reproducibly for sale and use in cheese factories (Crow *et al.*, 2002; Law, 2010b). Popular culture adjuncts are mostly NSLAB (*Lb. casei*, *Lb. plantarum*, *Lb. helveticus*), although surface-smear microflora (*Brevibacterium linens*, staphylococci, yeasts) are also available to give additional flavor notes. Cheddar cheese, with its relatively high pH and more solid matrix, has the potential to deliver probiotics (Playne, 2002). Selected strains of the probiotic cultures *Bifidobacterium* (McBrearty *et al.*, 2001), *Enterococcus* (Gardiner *et al.*, 1999), and *Lactobacillus* (Gardiner *et al.*, 1998) can also improve cheese flavor (McBrearty *et al.*, 2001). Inner-Balance is a brand of Cheddar cheese marketed in Australia (Mainland Dairies) and the UK and it contains the probiotic culture adjunct *Lb. rhamnosus* DR20™, which also provides flavor consistency and improves and accelerates flavor quality (Crow *et al.*, 2002). Enterococci, despite some of their negative effects on cheese flavor and some food safety concerns, have a great potential as cheese flavor adjuncts and probiotic cultures and are being tried in dairy products (Crow *et al.*, 2002; El-din *et al.*, 2002; Bulajic and Mijacevic, 2003; Abeijon *et al.*, 2006; Bhardwaj *et al.*, 2008).

In the case of mold cultures, the challenge is to keep a fine balance between mold growth and the development of flavor notes via the metabolism of living organisms; there is less scope to control the flavor and aroma of mold-ripened cheeses (Law, 2010b). Still, Chr. Hansen has developed the commercial SWING™ range of mold cultures by selecting from the inherent variability of enzyme and pigment production and growth rate/metabolic ratios within available cultures.

6. Genetically Modified Lactic Acid Bacteria

Over the past two decades, much research knowledge has accumulated in relation to the biochemical basis of the vital functions of dairy LAB, including acid production, flavor/aroma production, protein utilization, extracellular polysaccharide secretion, and bacteriophage resistance; this advanced knowledge base forms the basis of the many GM LAB that the culture companies now provide for industrial trials (Law, 2010b). The USA and the European Union require that the GM strains comply fully with all the federal/national regulations covering GM organisms for food applications; the GM LAB are constructed using food-grade cloning systems.

Three general strategies are used for the genetic modification of LAB to be applied to cheese-ripening and cheese flavor technology (Law, 2010b): (1) altered lactose metabolism; (2) increased peptidase production capacity; and (3) increased starter cell lysis in the matrix of young cheese. Mutated strains of *Lactococcus lactis* can be produced with altered metabolic routes from lactose to the key intermediates, α-acetolactate (precursor of diacetyl, which contributes to the buttery aroma) and acetate; the result is an accumulation of aroma compounds at higher than normal concentrations (de Vos, 1996; Swindell *et al.*, 1996). Such GM lactococcal cultures are now commercially available. Commercial strains of *L. lactis* spp. with genetically enhanced production of two general aminopeptidases have been developed and these cultures are now being used to increase the flavor quality and intensity in Cheddar and Dutch-type cheese (Law, 2010b). Peptidases prevent the accumulation of bitter peptides that may cause flavor defects. Lastly, starter cultures can be genetically altered to lyse quickly in cheese. It is now established that the quality and pace of cheese flavor development are positively correlated with the rate and extent of starter cell lysis and subsequent release of enzymes into young cheese (Crow *et al.*, 1995). Based on the original work of Gasson's group in the UK, lactococcal variants of commercial starters containing the gene for the bacteriophage lysin are constructed in a way that the gene expression is controllable by external stimuli such as changes in pH, salt concentration, and temperature (Law, 2010b). The NIZO food research company in the Netherlands adopted a similar approach and developed and patented the NICE system, in which microgram quantities of Nisin are added to trigger the cloned phage lysin. The latter technology is based on the research findings of de Ruyter *et al.* (1997).

VIII. YOGURT

A. Introduction to Fermented Dairy Foods

Cheese and yogurt are the most popular dairy products derived from fermented milk. Apart from these, cultured buttermilk (used as an ingredient in baking; uses *Lactococcus lactis* subsp. *cremoris* or *diacetylactis*), acidophilus milk (therapeutically benefits the gastrointestinal tract; uses *Lactobacillus acidophilus*), and sour cream (cultured cream; uses *L. lactis* subsp. *cremoris* or *diacetylactis*) are also consumed in North America and Europe. Other fermented dairy products enjoy regional popularity in certain parts of the world, for example kefir (an alcoholic fermented milk drink that uses kefir grains that contain a combination of bacteria and yeasts capable of producing kefiran, a ropy polysaccharide), kumiss (produced from mare's milk; uses a liquid culture of bacteria and yeasts), filmjölk (a Nordic dairy product similar to cultured buttermilk; uses *L. lactis* and *Leuconostoc mesenteroides*), långfil (an elastic variant of filmjölk containing ropy LAB that produce exopolysaccharides), and viili (also similar to filmjölk, but includes *Geotrichum candidum* and sometimes ropy LAB). For further information, the reader may refer to Robinson and Tamime (1990) and Tamime and Robinson (1999b).

The set-style, stirred, and drinking forms of yogurt are the most common; the first type is incubated and cooled in the final package and is characterized by a firm, gel-like structure, whereas in the latter two forms, the final coagulum is broken by stirring (stirred type) or by homogenizing to a low-viscosity drink (drinkable yogurt) before cooling and packing. Another popular American type is the frozen yogurt, which first originated in New England. Physically, it resembles ice cream but is characterized by the sharp acidic taste of yogurt, and is available in three categories: soft, hard, and mousse. Frozen yogurts contain high levels of sugar and stabilizers to maintain the air-bubble structure during freezing. Concentrated/strained yogurt (e.g. Eastern Mediterranean labneh, Egyptian laban, Icelandic skyr, Indian chakka, and shrikhand from buffalo milk) is produced by straining cold/unsweetened yogurt by application of pressure in a cloth bag (Berge system), centrifugation, or ultrafiltration. Finally, a powder form of dried yogurt is also produced and finds different applications in the food industry (e.g. baked goods and confectionery) (see Tamine and Robinson (1999b)).

B. Yogurt Manufacture

Yogurt making dates back thousands of years and is an ancient craft, but still today the commercial production of yogurt is a complex process and combines both art and science. The manufacturing process of yogurt was well described by Tamine and Robinson (1999a) and is also briefly discussed by Goff on the University of Guelph's Dairy Science and Technology Education website. When refrigerated milk arrives at the plant, the first step is to modify its composition to suit yogurt specifications. This process involves clarification of milk into cream and skim milk, followed by standardization to the desired fat and solids-not-fat (SNF) content. While the average fat content of milk ranges from 3.7% to 4.2% (w/w), the fat content of commercial yogurt can range anywhere from 0.1% to 10%. The Pearson's square method (Tamine and Robinson, 1999a) is a very convenient way to calculate the components required to standardize milk. To increase the SNF content, the industry most commonly adds milk powder (whole or skimmed, 3−4% is recommended) to the yogurt mixture, but one or more other ingredients may also be used: concentrated milk, high-protein milk powders, buttermilk powder, whey powder/concentrate, casein powder, and even non-milk proteins (e.g. proteins from soy, legumes, or sweet potato). Added solids improve yogurt viscosity/consistency, reduce syneresis, and impart a better mouth-feel, besides contributing to specific functionalities related to the material added (e.g. phospholipids from buttermilk possess emulsifying properties).

The yogurt milk base is usually also fortified with non-milk solids such as stabilizers/emulsifiers, and sweetening agents. Stabilizers are mostly gums/hydrocolloids which are added (at a level of 0.1−0.5%) as single compounds or as a blend of either natural (gelatin, pectin, guar and locust bean gums, cereal starches, alginates, and carrageenans) or modified natural/semi-synthetic (carboxymethyl cellulose, xanthan, low-methoxy pectin, and modified starches) gums. The hydrocolloids in yogurt function as gelling/thickening and stabilizing agents by binding water as water of hydration, and by reacting with and stabilizing milk protein molecules in the form of a network that retards free water movement (Tamine and Robinson, 1999a). Sweeteners are normally added to fruit and flavored yogurts and can be sugars such as sucrose, invert sugar, glucose, and fructose, or non-caloric high-intensity sweeteners such as aspartame and saccharin.

Following standardization and fortification of the base milk with a cocktail of milk and non-milk solids, the mixture is pasteurized or heat treated using a continuous plate heat exchanger for 30 minutes at 85°C or 10 minutes at 95°C. The higher heat treatments that are used not only achieve the destruction or elimination of pathogens and spoilage microbes, but also encourage production of factors capable of stimulating or inhibiting starter culture activity, and most importantly bring about very favorable changes in the physicochemical properties of milk through denaturation of whey proteins and their attachment to the surfaces of casein micelles (refer to Section IV).

The pasteurized mix (at 65–70°C) is then homogenized, usually by a single-stage homogenizer at pressures ranging from 15 to 20 MPa. Besides uniformly blending in all the ingredients, this process effectively reduces fat globule size, increases fat surface area, and coats the surface with mainly proteins; casein micelles cover nearly 25% of the surface. In effect, the homogenized fat globules act as large casein micelles and participate in acid precipitation reactions. Yogurt made from homogenized milk is firmer, smoother, and more stable (reduced creaming and wheying off) during storage.

The homogenized mixture is cooled to the incubation temperature (40–45°C) and pumped into jacketed fermentation tanks. The starter culture ($\approx$ 3% w/w of *Streptococcus thermophilus* and *Lactobacillus delbrueckii* subsp. *bulgaricus* in a 1:1 ratio) is directly metered into the mix while pumping, or added to the fermentation tanks. The actual fermentation can take place either in retail containers (set-style yogurt) or in bulk tanks (stirred yogurt). A temperature of typically 42°C is maintained for 2–2.5 hours under quiescent conditions; pH and/or titratable acidity (TA) are carefully monitored during this period. When pH reaches 4.6 (TA of 0.85–0.90%), i.e. when a continuous solid mass of gel is formed, the yogurt is immediately cooled to nearly 5°C, at which it is stored, or processed further to produce other yogurt forms. Cooling starts when the pH of yogurt is relatively high and the rate is carefully controlled so that the final product has the desired level of acidity and gel structure. While slow cooling can increase yogurt acidity, very rapid cooling may lead to whey separation, possibly due to excessive contraction of the protein matrix (Rasic and Kurmann, 1978).

Fruit and/or flavored yogurts, available in a great variety of forms, are the most popular types, especially in North America, Europe, and Australia. Before the 1950s, yogurt was virtually unknown outside the Middle East and Balkan region, but the addition of sweetening agents and fruits greatly increased its popularity and acceptability worldwide. Various flavoring agents (fruits, natural, and/or synthetic flavors) are added to yogurt. In set-style yogurt these are normally added to the mix before incubation, but in the stirred type they are often incorporated into the formed gel.

C. Yogurt Starter Cultures

Yogurt is made with live and active cultures of *Lactobacillus delbrueckii* subsp. *bulgaricus* and *Streptococcus thermophilus*, and the US Food and Drug Administration (FDA) requires that these two specific LAB be present in a product for it to be called yogurt (Van de Water and Naiyanetr, 2008). Although other cultures may be added to yogurt, they are not required. In the milk environment, *Lb. bulgaricus* and *S. thermophilus* coexist and interact beneficially in a stable associative relationship also known as protocooperation (Liu *et al.*, 2009). Protocooperation, previously defined as biochemical mutualism, involves the exchange of metabolites and/or stimulatory factors (Pette and Lolkema, 1950). Most strains of *S. thermophilus* have fewer nutritional requirements and hence grow preferentially in milk. In fact, during the first exponential growth of *S. thermophilus*, no growth of *Lb. bulgaricus* is observed. In the second phase, as the pH of milk begins to drop, growth of *S. thermophilus* (less acid tolerant) slows down and it provides several growth factors such as formate, pyruvate, folate, CO_2, and even some long-chain fatty acids that stimulate *Lb. bulgaricus* (more acid tolerant) to grow exponentially. The latter, in turn, releases cell wall proteases and cytoplasm peptidases that hydrolyze caseins into peptides, subsequently broken down to amino acids. Since *S. thermophilus* strains lack extracellular proteases, their growth is strongly stimulated in cocultures with *Lb. bulgaricus* strains serving as a source of these amino acids able to support a second exponential growth phase for *S. thermophilus*. The growth of *Lb. bulgaricus* continues in the third growth phase. (Please refer to the work of Sieuwerts *et al.*, 2008.)

D. Bioyogurt

The dairy sector is the largest (33%) functional food market and dairy products are the main vehicle for probiotic supplementation (Leatherhead Food International, 2006). Added to the fact that many consumers associate yogurt with good health, supplementation with probiotics has drastically increased yogurt consumption in recent years (Hekmat and Reid, 2006). As defined by FAO/WHO (2001), probiotics are 'live microorganisms which when administered in adequate amounts confer a health benefit on the host'. Some commonly used dairy probiotics include specific strains of LAB such as *Lactobacillus acidophilus*, *L. casei*, *L. rhamnosus*, *Bifidobacterium bifidum*, *B. animalis*, *B. longum*, and *B. infantis* (Granato *et al.*, 2010). Other LAB such as *Enterococcus faecalis* and

E. faecium, also known to possess probiotic properties, and *Propionibacterium freudenreichii* and *Saccharomyces boulardii*, are non-lactic probiotics (Granato *et al.*, 2010). Some examples of probiotic yogurt products available on the world market are acidophilus bifidus/AB yogurt (A + B + yogurt culture; A = *L. acidophilus*, B = *Bifidobacteria*), Bifidus yogurt/Biobest (*B. bifidum* or *B. longum* + yogurt culture), Bioghurt/ABT yogurt (A + B + *S. thermophilus*), Bifilak/Cultura/Biomild (A + B), and BA Bifidus active (*B. longum* + yogurt culture) (Lourens-Hattingh and Viljoen, 2001). In recent years, the health enhancement effects of dairy probiotics have been attributed to the metabolic release of physiologically active peptides during milk fermentation. The bioactive peptides may be released during food processing or after degradation by digestive enzymes and have been shown to possess opiate, antithrombotic, antihypertensive, immunomodulating, antibacterial, antigastric, and mineral carrier properties (Vinderola *et al.*, 2008). Some documented physiological/health benefits of probiotics (Van de Water and Naiyanetr, 2008; Granato *et al.*, 2010) are:

- enhancement of the immune system (of the gastrointestinal tract, and possibly the respiratory tract)
- treatment of gastrointestinal disturbances: primary or secondary lactose maldigestion, diarrhea caused by antibiotics, *Clostridium difficile*, or food allergies, infant diarrhea due to rotavirus enteritis, traveler's diarrhea, and possibly diarrhea from *Helicobacter pylori* infections, inflammatory bowel disease, and colon cancer
- regulation of gut motility (constipation, irritable bowel syndrome)
- improved absorption of nutrients, e.g. iron, calcium, and important B vitamins
- lowering of serum cholesterol (prevention of atherosclerosis) and hypertension
- detoxification of carcinogens (prevention of cancer and suppression of tumors) and reduction of catabolic products eliminated by the liver and kidney (prevention of urogenital infections).

The key challenge to making any probiotic food is to improve/maintain culture viability in the product during processing, storage, and transit through the stomach and small intestine, besides finding cost-effective ways to produce these foods (Prado *et al.*, 2008; Figueroa-Gonzalez *et al.*, 2011). In the USA, the National Yogurt Association (NYA) specifies a therapeutic minimum of 10^8 cfu/g of LAB (at the time of manufacture) as a prerequisite to using the NYA 'Live and Active Culture' logo on the product's containers. A dairy probiotic product must not only contain a minimum number of cells to confer health effects, but must also have good sensory properties to be acceptable by consumers. Several factors influence the survival of probiotic bacteria in fermented dairy bioproducts: the strains used, culture conditions, inoculation levels and interspecies interactions, the fermentation medium's composition, acidity, presence of nutrients/growth promoters/inhibitors, osmotic pressure, dissolved oxygen levels, and the temperature and time of incubation and storage (Lourens-Hattingh and Viljoen, 2001).

Besides having beneficial effects on human health, an ideal probiotic strain is able to withstand food processing and storage conditions, has good growth characteristics, is compatible with the starter cultures, is resistant to acid and bile, attaches to the human gut epithelial cells, colonizes in the human intestine, and produces antimicrobial substances such as bacteriocins (Granato *et al.*, 2010). Several selected strains of *Lactobacillus* and *Bifidobacterium* cultures are available commercially. *Lactobacillus acidophilus* is more acid tolerant (can tolerate pH 4.6) than bifidobacteria (growth is retarded at pH < 5.0) (Lankaputhra *et al.*, 1996), so for practical applications, the pH of the final product must be maintained above 4.6. GMP, use of starter *L. bulgaricus* cultures with reduced 'over-acidification' behavior, or use of ABT-yogurt starter cultures (*L. acidophilus*, *B. bifidum*, and *S. thermophilus*) can aid in pH control (Lourens-Hattingh and Viljoen, 2001). Other ways to prevent acid stress include previous exposure of a strain to a lower pH for a short time to induce acid tolerance (Sanz, 2007), applying heat shock (58°C for 5 minutes) to yogurt before the addition of probiotic culture (Marshall, 1992), storage at temperatures less than 3−4°C, addition of whey protein concentrate to increase the buffering capacity of yogurt (Kailasapathy and Rybka, 1997), and addition of sodium citrate or calcium carbonate to neutralize the lactic acid produced during fermentation (Zhao and Li, 2008). Probiotic bacteria prefer an anaerobic/microaerobic environment; oxidative stress can be minimized with active packaging (Miller *et al.*, 2003), addition of ascorbic acid (Dave and Shah, 1998; Zhao and Li, 2008), electroreduction of milk (Bolduc *et al.*, 2006), use of glucose oxidase (Cruz *et al.*, 2010), and microencapsulation (Talwalkar and Kailasapathy, 2003). Addition of prebiotics (selectively fermented ingredients that allow specific changes in the composition and/or activity of the gastrointestinal microbiota) (Roberfroid, 2007) such as chicory inulin and oligosaccharides (Figueroa-Gonzalez *et al.*, 2011) and growth-promoting substances such as casein hydrolysate, whey protein concentrate, cysteine, and tryptone can improve the growth and viability of probiotic bacteria (Lourens-Hattingh and Viljoen, 2001). A higher level of inoculum (10−20% recommended) and the use of concentrated or freeze-dried direct-to-vat cultures containing a minimum of 5×10^9 cfu/g (IDF, 1996) will ensure a high cell count at the end of incubation, during storage, and until consumption (Lourens-Hattingh and Viljoen, 2001).

E. Yogurt Flavor

Yogurt's popularity as food largely depends on its sensory characteristics, with aroma and taste being the most important. Yogurt is well liked for its delicate and low intense acidic flavor. So far, more than 90 flavor compounds have been identified in yogurt including carbohydrates, alcohols, aldehydes, ketones, acids, esters, lactones, sulfur compounds, pyrazines, and furan derivatives (Ott et al., 1997). But the flavor of yogurt is mainly attributed to the group of carbonyl compounds, acetaldehyde, diacetyl, acetoin, and acetone, which are present in relatively high concentrations (in decreasing order) (Imhof et al., 1994; Kaminarides et al., 2007). Of these, acetaldehyde is suggested as the indispensable component of the typical yogurt flavor; good flavored yogurt requires about 23–40 mg/kg (at least 8–10 mg/kg) of acetaldehyde to be present (Gaafar, 1992; Kneifel et al., 1992; Georgala et al., 1995). Acetaldehyde imparts to yogurt its characteristic green apple or nutty flavor (Bodyfelt et al., 1988). Production of acetaldehyde takes place via several pathways which use different compounds as precursors, such as glucose, catechol, glyceraldehydes, acetylene, threonine, glycine, and even DNA (Zourari and Desmazeaud, 1991; Chaves et al., 2002; Tamime and Robinson, 2007). Breakdown of threonine to acetaldehyde and glycine is reported as the major pathway and the reaction is catalyzed by the enzyme threonine aldolase, present in both Lb. bulgaricus and S. thermophilus. At the higher temperatures (40–45°C) used in yogurt manufacture, Lb. bulgaricus is the main contributor of the enzyme (Zourari and Desmazeaud, 1991) as the aldolase is inactivated in S. thermophilus at 30–42°C (Wilkins et al., 1986).

Diacetyl is the other major volatile compound that contributes to the buttery notes in yogurt aroma; the typical concentration in yogurt ranges from 0.2 to 3.0 mg/kg (Cheng, 2010). When present with acetaldehyde, it is reported to contribute to the well-rounded delicate flavor of yogurt; overproduction of acetaldehyde in relation to diacetyl can result in harsh flavors. There are different views on the ideal ratio of acetaldehyde to diacetyl that may be required to result in the desired fullness in yogurt flavor (Zourari and Desmazeaud, 1991; Panagiotidis and Tzia, 2001; Boelrijk et al., 2003). The diketone is derived from the fermentation of citrate in milk (Vedamuthu, 2006) or, according to Nilsson (2008), from both lactose and citrate. Acetoin is readily produced from diacetyl by the enzyme diacetyl reductase (Collins, 1972); about 1.2–28.2 mg/kg of acetoin is typically present in milk (Beshkova et al., 1998; Alonso and Fraga, 2001). Diacetyl and acetoin together impart the mild, pleasant, and buttery taste, and their combination is critical for the rich flavor perception of yogurt (Cheng, 2010). Two other volatile components of minor importance are acetone and 2-butanone, both of which contribute to the sweet, fruity aroma of yogurt (Carcoba et al., 2000; Gallardo-Escamilla et al., 2005).

Acetaldehyde, diacetyl, and the other volatiles described above are the most important volatile flavor compounds present in yogurt, but are not the only flavor components. The overall flavor of yogurt comes from lactic acid and a complex and well-balanced mixture of several aromatic and taste components, which include volatiles and non-volatiles (e.g. lactic, pyruvic, acetic, and formic acids) already present in milk and specific compounds produced during fermentation (Imhof et al., 1994; Ott et al., 1997). Factors such as the source of milk (cow, sheep, or goat), processing techniques, added components (stabilizers, fruits, flavorings, probiotics, and prebiotics), packaging materials, and storage conditions all have an impact on the final taste and aroma of yogurt (Routray and Mishra, 2011). Further, undesired odorants can be produced during the storage of yogurt, and as in any other fat-rich dairy product, lipid oxidation is the major contributor to these off-flavors.

REFERENCES

Abeijon, M.C., Medina, R.B., Katz, M.B., Gonzalez, S.N., 2006. Technological properties of Enterococcus faecium isolated from ewe's milk and cheese with importance for flavour development. Can. J. Microbiol. 52, 237–245.

Abo-Elnaga, I.G., 1971. Early blowing of white pickled cheese. Milchwissenschaft 26, 747–750.

Adda, J., Gripon, J.C., Vassal, I., 1982. The chemistry of flavour and texture generation in cheese. Food Chem. 9, 115–129.

Ahmad, S., Gaucher, I., Rousseau, F., Beaucher, E., Piot, M., Grongnet, J.F., Gaucheron, F., 2008. Effects of acidification on physico-chemical characteristics of buffalo milk: a comparison with cow's milk. Food Chem. 106, 11–17.

Ahmed, A.A.H., Moustafa, M.K., Marth, E.H., 1983. Incidence of Bacillus cereus in milk and some milk products. J. Food Prot. 46, 126–128.

Ahn, S., Azzouny, R.A., Huyen, T.T.T., Kwak, H., 2009. The characteristics, detection and control of bacteriophage in fermented dairy products. Korean J. Food Sci. Anim. Resour. 29, 1–14.

Albenzio, M., Corbo, M., Rehman, S., Fox, P., De Angelis, M., Corsetti, A., Sevi, A., Gobbetti, M., 2001. Microbiological and biochemical characteristics of Canestrato Pugliese cheese made from raw milk, pasteurized milk or by heating the curd in hot whey. Int. J. Food Microbiol. 67, 35–48.

Alonso, L., Fraga, M.J., 2001. Simple and rapid analysis for quantitation of the most important volatile flavor compounds in yogurt by headspace gas chromatography—mass spectrometry. J. Chromatogr. Sci. 39, 297—300.

Alonso, L., Fontecha, J., Lozada, L., Fraga, M.J., Juarez, M., 1999. Fatty acid composition of caprine milk: major, branched chain, and *trans* fatty acids. J. Dairy Sci. 82, 878—884.

Anema, S.G., Lee, S.K., Klostermeyer, H., 2007. Effect of pH at heat treatment on the hydrolysis of kappa-casein and the gelation of skim milk by chymosin. LWT-Food Sci. Technol. 40, 99—106.

de Angelis, M., Corsetti, A., Tosti, N., Rossi, J., Corbo, M., Gobbetti, M., 2001. Characterization of non-starter lactic acid bacteria from Italian ewe cheeses based on phenotypic, genotypic, and cell wall protein analyses. Appl. Environ. Microbiol. 67, 2011—2020.

Ardo, Y., 2006. Flavour formation by amino acid catabolism. Biotechnol. Adv. 24, 238—242.

Ardo, Y., Pettersson, H.E., 1988. Accelerated cheese ripening with heat-treated cells of *Lactobacillus helveticus* and a commercial proteo-lytic-enzyme. J. Dairy Res. 55, 239—245.

Ayad, E.H.E., 2009. Starter culture development for improving safety and quality of Domiati cheese. Food Microbiol. 26, 533—541.

Azarnia, S., Robert, N., Lee, B., 2006. Biotechnological methods to accelerate Cheddar cheese ripening. Crit. Rev. Biotechnol. 26, 121—143.

Badel, S., Bernardi, T., Michaud, P., 2011. New perspectives for lacto-bacilli exopolysaccharides. Biotechnol. Adv. 29, 54—66.

Barbano, D.M., Rasmussen, R.R., Lynch, J.M., 1991. Influence of milk somatic-cell count and milk age on cheese yield. J. Dairy Sci. 74, 369—388.

Barrett, F.M., Kelly, A.L., McSweeney, P.L.H., Fox, P.F., 1999. Use of exogenous urokinase to accelerate proteolysis in Cheddar cheese during ripening. Int. Dairy J. 9, 421—427.

Baskaran, D., Sivakumar, S., 2003. Galactose concentration in pizza cheese prepared by three different culture techniques. Int. J. Dairy Technol. 56, 229—232.

Bastian, E.D., Brown, R.J., 1996. Plasmin in milk and dairy products: an update. Int. Dairy J. 6, 435—457.

Bastian, E.D., Hansen, K.G., Brown, R.J., 1991. Activation of plasmin with urokinase in ultrafiltered milk for cheese manufacture. J. Dairy Sci. 74, 3669—3676.

Bastian, E.D., Lo, C.G., David, K.M.M., 1997. Plasminogen activation in cheese milk: influence on Swiss cheese ripening. J. Dairy Sci. 80, 245—251.

Beresford, T., Williams, A., 2004. The microbiology of cheese ripening. In: Fox, P.F., McSweeney, P.L.H., Cogan, T.M., Guinee, T.P. (Eds.), Cheese: Chemistry, Physics and Microbiology, Vol. 1. General Aspects, (3rd ed.), Elsevier, Amsterdam, pp. 287—317.

Beshkova, D., Simova, E., Frengova, G., Simov, Z., 1998. Production of flavour compounds by yogurt starter cultures. J. Ind. Microbiol. Biotechnol. 20, 180—186.

Bester, B.H., 1976. Some aspects of gas production by coliform bacteria in cheese. S. Afr. J. Dairy Technol. 8, 51—55.

Beuvier, E., Buchin, S., 2004. Raw milk cheeses. In: Fox, P.F., McSweeney, P.L.H., Cogan, T.M., Guinee, T.P. (Eds.), Cheese: Chemistry, Physics and Microbiology, Vol. 1. General Aspects, (3rd ed.), Elsevier, Amsterdam, pp. 319—345.

Beuvier, E., Berthaud, K., Cegarra, S., Dasen, A., Pochet, S., Buchin, S., Duboz, G., 1997. Ripening and quality of Swiss-type cheese made from raw, pasteurized or microfiltered milk. Int. Dairy J. 7, 311—323.

Bhardwaj, A., Malik, R.K., Chauhan, P., 2008. Functional and safety aspects of enterococci in dairy foods. Indian J. Microbiol. 48, 317—325.

Bockelmann, W., Hoppe, S.T., 2001. The surface flora of bacterial smear-ripened cheeses from cow's and goat's milk. Int. Dairy J. 11, 307—314.

Bodyfelt, F.W., Tobias, J., Trout, G.M., 1988. Sensory Evaluation of Dairy Products. Van Nostrand Reinhold, New York.

Boelrijk, A.E.M., de Jong, C., Smit, G., 2003. Flavor generation in dairy products. In: Smit, G. (Ed.), Dairy Processing — Improving Quality. CRC Press, Boca Raton, FL, pp. 130—154.

Bolduc, M.P., Raymond, Y., Fustier, P., Champagne, C.P., Vuillemard, J.C., 2006. Sensitivity of bifidobacteria to oxygen and redox potential in non-fermented pasteurized milk. Int. Dairy J. 16, 1038—1048.

Bonaiti, C., Leclercq-Perlat, M.N., Latrille, E., Corrieu, G., 2004. Deacidification by *Debaryomyces hansenii* of smear soft cheeses ripened under controlled conditions: relative humidity and tempera-ture influences. J. Dairy Sci. 87, 3976—3988.

Boudjellab, N., Roletrepecaud, O., Collin, J., 1994. Detection of residual chymosin in cheese by an enzyme-linked-immunosorbent-assay. J. Dairy Res. 61, 101—109.

Brennan, N.M., Cogan, T.M., Loessner, M., Scherer, S., 2004. Bacterial-surface ripened cheeses. In: Fox, P.F., McSweeney, P.L.H., Cogan, T.M., Guinee, T.P. (Eds.), Cheese: Chemistry, Physics and Microbiology, *Vol. 2*, Major Cheese Groups, (3rd ed.), Elsevier, Amsterdam, pp. 199—225.

Broadbent, J.R., Steele, J.L., 2007a. Biochemistry of cheese flavor development: insights from genomic studies of lactic acid bacteria. In: Cadwallader, K.R., Drake, M.A., McGorrin, R.J. (Eds.), Flavor of Dairy Products. ACS Symposium Series, 971. American Chemical Society, Washington, DC, pp. 177—192.

Broadbent, J.R., Steele, J.L., 2007b. Proteolytic enzymes of lactic acid bacteria and their influence on bitterness in bacterial-ripened cheeses. In: Cadwallader, K.R., Drake, M.A., McGorrin, R.J. (Eds.), Flavor of Dairy Products. ACS Symposium Series, 971. American Chemical Society, Washington, DC, pp. 193—203.

Broadbent, J.R., Strickland, M., Weimer, B.C., Johnson, M.E., Steele, J.L., 1998. Peptide accumulation and bitterness in Cheddar cheese made using single strain *Lactococcus lactis* starters with distinct proteinase specificities. J. Dairy Sci. 81, 327—337.

Broadbent, J.R., Barnes, M., Brennand, C., Strickland, M., Houck, K., Johnson, M.E., Steele, J.L., 2002. Contribution of *Lactococcus lactis* cell envelope proteinase specificity to peptide accumulation and bitterness in reduced-fat cheddar cheese. Appl. Environ. Microbiol. 68, 1778—1785.

Broome, M.C., 2007. Adjunct culture metabolism and cheese flavour. In: Weimer, B.C. (Ed.), Improving the Flavor of Cheese. CRC Press, Boca Raton, FL, pp. 177—198.

Broome, M.C., Xu, X., Mayes, J.J., 2006. Proteolysis in Cheddar cheese made with alternative coagulants. Aust. J. Dairy Technol. 61, 85—87.

Bulajic, S., Mijacevic, Z., 2003. The technological acceptability of enterococci isolated from cheeses. In: FEMS Congress of European Microbiologists, Abstract Book, Elsevier106.

Callanan, M.J., Ross, R.P., 2004. Starter cultures: genetics. In: Fox, P.F., McSweeney, P.L.H., Cogan, T.M., Guinee, T.P. (Eds.), Cheese: Chemistry, Physics and Microbiology, *Vol. 1*, General Aspects, (3rd ed.), Elsevier, Amsterdam, pp. 191—206.

Cantor, M.D., van den Tempel, T., Hansen, T.K., Ardo, Y., 2004. Blue cheese. In: Fox, P.F., McSweeney, P.L.H., Cogan, T.M., Guinee, T.P. (Eds.), Cheese: Chemistry, Physics and Microbiology, Vol. 2, Major Cheese Groups, (3rd ed.), Elsevier, Amsterdam, pp. 175−198.

Carcoba, R., Delgado, T., Rodriguez, A., 2000. Comparative performance of a mixed strain starter in cow's milk, ewe's milk and mixtures of these milks. Eur. Food Res. Technol. 211, 141−146.

Carles, C., Dumas, B., 1985. Kinetics of the action of chymosin (rennin) on a peptide-bond of bovine alpha-S1-casein − comparison of the behavior of this substrate with that of beta-caseins and kappa-caseins. FEBS Lett. 185, 282−286.

Carrasco, M.S., Scarinci, H.E., Simonetta, A.C., 2005. Associative growth of lactic acid bacteria for cheese starters: acidifying and proteolytic activities and redox potential development. J. Food Agric. Environ. 3, 116−119.

Casalta, E., Sorba, J., Aigle, M., Ogier, J., 2009. Diversity and dynamics of the microbial community during the manufacture of Calenzana, an artisanal Corsican cheese. Int. J. Food Microbiol. 133, 243−251.

Chamba, J., Irlinger, F., 2004. Secondary and adjunct cultures. In: Fox, P.F., McSweeney, P.L.H., Cogan, T.M., Guinee, T.P. (Eds.), Cheese: Chemistry, Physics and Microbiology, Vol. 1, General Aspects, (3rd ed.), Elsevier, Amsterdam, pp. 191−206.

Chaplin, B., Green, M.L., 1980. Determination of the proportion of kappa-casein hydrolyzed by rennet on coagulation of skim milk. J. Dairy Res. 47, 351−358.

Chaves, A.C.S.D., Fernandez, M., Lerayer, A.L.S., Mierau, I., Kleerebezem, M., Hugenholtz, J., 2002. Metabolic engineering of acetaldehyde production by Streptococcus thermophilus. Appl. Environ. Microbiol. 68, 5656−5662.

Cheng, H., 2010. Volatile flavor compounds in yogurt: a review. Crit. Rev. Food Sci. Nutr. 50, 938−950.

Chou, Y.E., Edwards, C.G., Luedecke, L.O., Bates, M.P., Clark, S., 2003. Nonstarter lactic acid bacteria and aging temperature affect calcium lactate crystallization in Cheddar cheese. J. Dairy Sci. 86, 2516−2524.

Chr. Hansen, N.C., 2008. Improved low fat cheese with Chr. Hansen's FLAVOR CONTROL (TM) ripening cultures. Available at http://newhope360.com/improved-low-fat-cheese-chr-hansen-s-flavor-controltm-ripening-cultures.

Christensen, J., Dudley, E., Pederson, J., Steele, J., 1999. Peptidases and amino acid catabolism in lactic acid bacteria. Anton. Leeuw. Int. J.G. 76, 217−246.

Christiansen, P., Petersen, M.H., Kask, S., Moller, P.L., Petersen, M., Nielsen, E.W., Vogensen, F.K., Ardo, Y., 2005. Anticlostridial activity of Lactobacillus isolated from semi-hard cheeses. Int. Dairy J. 15, 901−909.

Cogan, T.M., Beresford, T.P., Steele, J., Broadbent, J., Shah, N.P., Ustunol, Z., 2007. Advances in starter cultures and cultured foods. J. Dairy Sci. 90, 4005−4021.

Collins, E.B., 1972. Biosynthesis of flavor compounds by microorganisms. J. Dairy Sci. 55, 1022−1028.

Collins, Y.F., McSweeney, P.L.H., Wilkinson, M.G., 2003a. Lipolysis and free fatty acid catabolism in cheese: a review of current knowledge. Int. Dairy J. 13, 841−866.

Collins, Y.F., McSweeney, P.L.H., Wilkinson, M.G., 2003b. Evidence of a relationship between autolysis of starter bacteria and lipolysis in Cheddar cheese during ripening. J. Dairy Res. 70, 105−113.

Collins, Y.F., McSweeney, P.L.H., Wilkinson, M.G., 2004. Lipolysis and catabolism of fatty acids in cheese. In: Fox, P.F., McSweeney, P.L.H.,

Cogan, T.M., Guinee, T.P. (Eds.), Cheese: Chemistry, Physics and Microbiology, Vol. 1, General Aspects, (3rd ed.), Elsevier, Amsterdam, pp. 373−389.

Collomb, M., Spahni, M., Buehler, T., 1998. Analysis of triglycerides. II. Validation of a multilinear regression method for the determination of foreign fats in milk fat. Mitt. Geb. Lebensmittelunters. Hyg. 89, 75−83.

Colombo, E., Franzetti, L., Frusca, M., Scarpellini, M., 2010. Phenotypic and genotypic characterization of lactic acid bacteria isolated from artisanal Italian goat cheese. J. Food Prot. 73, 657−662.

Cooper, C., Corredig, M., Alexander, M., 2010. Investigation of the colloidal interactions at play in combined acidification and rennet of different heat-treated milks. J. Agric. Food Chem. 58, 4915−4922.

Costa, N.E., Hannon, J.A., Guinee, T.P., Auty, M.A.E., McSweeney, P.L.H., Beresford, T.P., 2010. Effect of exopolysaccharide produced by isogenic strains of Lactococcus lactis on half-fat Cheddar cheese. J. Dairy Sci. 93, 3469−3486.

Creamer, L.K., MacGibbon, A.K.H., 1996. Some recent advances in the basic chemistry of milk proteins and lipids. Int. Dairy J. 6, 539−568.

Creamer, L., Lawrence, R., Gilles, J., 1985. Effect of acidification of cheese milk on the resultant Cheddar cheese. N.Z.J. Dairy Sci. Tech. 20, 185−203.

Cromie, S., 1992. Psychrotrophs and their enzyme residues in cheese milk. Aust. J. Dairy Technol. 47, 96−100.

Crow, V.L., Coolbear, T., Gopal, P.K., Martley, F.G., McKay, L.L., Riepe, H., 1995. The role of autolysis of lactic acid bacteria in the ripening of cheese. Int. Dairy J. 5, 855−875.

Crow, V., Curry, B., Christison, M., Hellier, K., Holland, R., Liu, S.Q., 2002. Raw milk flora and NSLAB as adjuncts. Aust. J. Dairy Technol. 57, 99−105.

Cruz, A.G., Feria, J.A.F., Walter, E.H.M., Andrade, R.R., Cavalcanti, R.N., Oliveira, C.A.F., Granato, D., 2010. Processing optimization of probiotic yogurt containing glucose oxidase using response surface methodology. J. Dairy Sci. 93, 5059−5068.

Cui, S., Li, J., Hu, C., Jin, S., Ma, Y., 2007. Development of a method for the detection of beta-lactamases in milk samples. J. AOAC Int. 90, 1128−1132.

Curtin, A.C., McSweeney, P.L.H., 2004. Catabolism of amino acids in cheese during ripening. In: Fox, P.F., McSweeney, P.L.H., Cogan, T.M., Guinee, T.P. (Eds.), Cheese: Chemistry, Physics and Microbiology, Vol. 1, General Aspects, (3rd ed.), Elsevier, Amsterdam, pp. 435−454.

Dal Bello, B., Rantsiou, K., Bellio, A., Zeppa, G., Ambrosoli, R., Civera, T., Cocolin, L., 2010. Microbial ecology of artisanal products from north west of Italy and antimicrobial activity of the autochthonous populations. LWT-Food Sci. Technol. 43, 1151−1159.

Dalgleish, D.G., 1979. Proteolysis and aggregation of casein micelles treated with immobilized or soluble chymosin. J. Dairy Res. 46, 653−661.

Dalgleish, D.G., 1992. The enzymatic coagulation of milk. In: Fox, P.F. (Ed.), Advanced Dairy Chemistry, Vol. 1, Proteins, (2nd ed.), Elsevier, Essex, pp. 579−619.

Dalgleish, D.G., 1993. The enzymatic coagulation of milk. In: Fox, P.F. (Ed.), Cheese: Chemistry, Physics and Microbiology, Vol. 1, General Aspects, (2nd ed.), Chapman and Hall, London, pp. 69−100.

Dalgleish, D.G., 1998. Casein micelles as colloids: surface structures and stabilities. J. Dairy Sci. 81, 3013−3018.

Dalgleish, D.G., 2007. The casein micelle and its reactivity. Lait 87, 385−387.

Dalgleish, D.G., 2011. On the structural models of bovine casein micelles — review and possible improvements. Soft Matter 7, 2265–2272.

Dalgleish, D.G., Spagnuolo, P.A., Goff, H.D., 2004. A possible structure of the casein micelle based on high-resolution field-emission scanning electron microscopy. Int. Dairy J. 14, 1025–1031.

Dalgleish, D.G., Alexander, M., Corredig, M., 2005. Mechanism of acid coagulation of milk studied by a multi-technique approach. In: Dickinson, E. (Ed.), Royal Society of Chemistry Special Publications, Vol. 298, Royal Society of Chemistry, Cambridge, pp. 16–25.

Dasen, A., Berthier, F., Grappin, R., Williams, A., Banks, J., 2003. Genotypic and phenotypic characterization of the dynamics of the lactic acid bacterial population of adjunct-containing Cheddar cheese manufactured from raw and microfiltered pasteurised milk. J. Appl. Microbiol. 94, 595–607.

Dave, R.I., Shah, N.P., 1998. Ingredient supplementation effects on viability of probiotic bacteria in yogurt. J. Dairy Sci. 81, 2804–2816.

Deeth, H.C., Fitz-Gerald, C.H., 2006. Lipolytic enzymes and hydrolytic rancidity. In: Fox, P.F., McSweeney, P.L.H. (Eds.), Advanced Dairy Chemistry, Vol. 2, Lipids, (3rd ed.). Springer, New York, pp. 481–556.

Dervisolgu, M., Aydemir, O., Yazici, F., 2007. Coagulating enzymes utilized in cheesemaking and their effects on casein fractions. Gida 32, 241–249.

Donato, L., Alexander, M., Dalgleish, D.G., 2007. Acid gelation in heated and unheated milks: interactions between serum protein complexes and the surfaces of casein micelles. J. Agric. Food Chem. 55, 4160–4168.

Echols, M.A., 2008. Indications to the origin prior to the TRIPS agreement. In: Geographical Indications for Food Products: International, Legal, and Regulatory Perspectives. Kluwer Law International, The Netherlands, pp. 33–60.

Eck, A., Gillis, J.-C., 2000. Cheesemaking from Science to Quality Assurance, (2nd ed.). Lavoisier, Paris.

EFSA Panel on Animal Health and Welfare (AHAW), 2010. Scientific opinion on Q fever. EFSA J. 8, 1595 (114 pp.).

Ekinci, F.Y., Gurel, M., 2008. Effect of using propionic acid bacteria as an adjunct culture in yogurt production. J. Dairy Sci. 91, 892–899.

El-din, B.B., El-Soda, M., Ezzat, N., 2002. Proteolytic, lipolytic and autolytic activities of enterococci strains isolated from Egyptian dairy products. Lait 82, 289–304.

El-Gazzar, F.E., Marth, E.H., 1991a. Listeria monocytogenes and listeriosis related to milk, milk products and dairy ingredients — a review. I. Listeria monocytogenes, listeriosis and responses of the pathogen to environmental conditions. Milchwissenschaft 46, 14–19.

El-Gazzar, F.E., Marth, E.H., 1991b. Listeria monocytogenes and listeriosis related to milk, milk products and dairy ingredients: a review. II. Listeria monocytogenes and dairy technology. Milchwissenschaft 46, 82–86.

Ellis, B.R., Marth, E.H., 1984. Growth of Pseudomonas or Flavobacterium in milk — reduced yield of Cheddar cheese. J. Food Prot. 47, 713–716.

Farkye, N.Y., 2004. Cheese technology. Int. J. Dairy. Technol. 57, 91–98.

Farkye, N., Fox, P., Fitzgerald, G., Daly, C., 1990. Proteolysis and flavor development in Cheddar cheese made exclusively with single strain proteinase positive or proteinase negative starters. J. Dairy Sci. 73, 874–880.

Farrag, S.A., Marth, E.H., 1992. Interactions between Listeria monocytogenes and other psychrotrophic bacteria in dairy foods — a review. Food Aust. 44, 281–286.

Farrell Jr., H.M., Malin, E.L., Brown, E.M., Qi, P.X., 2006. Casein micelle structure: what can be learned from milk synthesis and structural biology? Curr. Opin. Colloid Interface Sci. 11, 135–147.

Feeney, E.P., Guinee, T.P., Fox, P.F., 2002. Effect of pH and calcium concentration on proteolysis in Mozzarella cheese. J. Dairy Sci. 85, 1646–1654.

Figueroa-Gonzalez, I., Quijano, G., Ramirez, G., Cruz-Guerrero, A., 2011. Probiotics and prebiotics — perspectives and challenges. J. Sci. Food Agric. 91, 1341–1348.

Food and Agriculture Organization of the United Nations/World Health Organization. 2001. Joint FAO/WHO Expert Consultation on Evaluation of Health and Nutritional Properties of Probiotics in Food Including Powder Milk with Live Lactic Acid Bacteria. Córdoba, Spain, 34 pp.

Fox, P.F., 1989. Proteolysis during cheese manufacture and ripening. J. Dairy Sci. 72, 1379–1400.

Fox, P.F., 1993. Exogenous enzymes in dairy technology — a review. J. Food Biochem. 17, 173–199.

Fox, P.F., 2003a. Milk proteins: general and historical aspects. In: Fox, P.F., McSweeney, P.L.H. (Eds.), Advanced Dairy Chemistry, Vol. 1, Proteins Part A, (3rd ed.). Kluwer/Plenum, New York, pp. 1–48.

Fox, P.F., Brodkorb, A., 2008. The casein micelle: historical aspects, current concepts and significance. Int. Dairy J. 18, 677–684.

Fox, P.F., Kelly, A.L., 2003. Developments in the chemistry and technology of milk proteins. 3. Product related aspects of milk proteins. Food Aust. 55, 337–340.

Fox, P.F., Law, J., 1991. Enzymology of cheese ripening. Food Biotechnol. 5, 239–262.

Fox, P.F., McSweeney, P.L.H., 1996. Proteolysis in cheese during ripening. Food Rev. Int. 12, 457–509.

Fox, P.F., McSweeney, P.L.H., 2006. Chemistry and biochemistry of cheese manufacture and ripening. Food Sci. Technol. Today 20, 28–32.

Fox, P.F., Wallace, J.M., 1997. Formation of flavor compounds in cheese. Adv. Appl. Microbiol. 45, 17–85.

Fox, P.F., Lucey, J.A., Cogan, T.M., 1990. Glycolysis and related reactions during cheese manufacture and ripening. Crit. Rev. Food Sci. Nutr. 29, 237–253.

Fox, P.F., Singh, T.K., McSweeney, P.L.H., 1994. Proteolysis in cheese during ripening. In: Andrews, A.T., Varley, J. (Eds.), Biochemistry of Milk Products. Royal Society of Chemistry, Cambridge, pp. 1–31.

Fox, P.F., Singh, T.K., McSweeney, P.L.H., 1995. Biogenesis of flavor compounds in cheese. In: Malin, E.L., Tunick, M.H. (Eds.), Chemistry of Structure/Function Relationships in Cheese. Plenum Press, London, pp. 59–98.

Fox, P.F., O'Connor, T.P., McSweeney, P.L.H., Guinee, T.P., O'Brien, N.M., 1996. Cheese: physical, biochemical and nutritional aspects. Adv. Food Nutr. Res. 39, 163–328.

Fox, P., McSweeney, P., Lynch, C., 1998. Significance of non-starter lactic acid bacteria in Cheddar cheese. Aust. J. Dairy Technol. 53, 83–89.

Fox, P.F., Guinee, T.P., Cogan, T.M., McSweeney, P.L.H., 2000. Fundamentals of Cheese Science. Aspen, Gaithersburg, MD.

Fox, P.F., McSweeney, P.L.H., Cogan, T.M., Guinee, T.P., 2004. Cheese: Chemistry, Physics and Microbiology, (3rd ed.), Elsevier, Amsterdam.

Frister, H., Michaelis, M., Schwerdtfeger, T., Folkenberg, D.M., Sorensen, N.K., 2000. Evaluation of bitterness in Cheddar cheese. Milchwissenschaft 55, 691–695.

Gaafar, A.M., 1992. Volatile flavor compounds of yogurt. Int. J. Food Sci. Technol. 27, 87–91.

Galesloot, Th., E., Hassing, F., Veringa, H.A., 1968. Symbiosis in yoghurt. I. Stimulation of *Lactobacillus bulgaricus* by a factor produced by *Streptococcus thermophilus*. Neth. Milk Dairy J. 22, 50–63.

Gallardo-Escamilla, F.J., Kelly, A.L., Delahunty, C.M., 2005. Influence of starter culture on flavor and headspace volatile profiles of fermented whey and whey produced from fermented milk. J. Dairy Sci. 88, 3745–3753.

Ganesan, B., Weimer, B.C., 2007. Amino acid catabolism in relation to cheese flavor development. In: Weimer, B.C. (Ed.), Improving the Flavour of Cheese. CRC Press, Boca Raton, FL, pp. 70–100.

Garde, S.S., Avila, M.M., Arias, R.R., Gaya, P.P., Nuñez, M.M., 2011. Outgrowth inhibition of Clostridium beijerinckii spores by a bacteriocin-producing lactic culture in ovine milk cheese. Int. J. Food Microbiol. 150, 59–65.

Gardiner, G., Ross, R.P., Collins, J.K., Fitzgerald, G., Stanton, C., 1998. Development of a probiotic cheddar cheese containing human derived *Lactobacillus paracasei* strains. Appl. Environ. Microbiol. 64, 2192–2199.

Gardiner, G.E., Ross, R.P., Wallace, J.M., Scanlan, F.P., Jagers, P.P.J.M., Fitzgerald, G.F., Collins, J.K., Stanton, C., 1999. Influence of a probiotic adjunct culture of *Enterococcus faecium* on the quality of Cheddar cheese. J. Agric. Food Chem. 47, 4907–4916.

Gastaldi, E., Trial, N., Guillaume, C., Bourret, E., Gontard, N., Cuq, J.L., 2003. Effect of controlled kappa-casein hydrolysis on rheological properties of acid milk gels. J. Dairy Sci. 86, 704–711.

Genigeorgis, C., Carniciu, M., Dutulescu, D., Farver, T.B., 1991. Growth and survival of *Listeria monocytogenes* in market cheeses stored at 4 to 30°C. J. Food Prot. 54, 662–668.

Georgala, A.I.K., Tsakalidou, E., Kandarakis, I., Kalantzopoulos, G., 1995. Flavor production in ewe's milk and ewe's milk yogurt, by single strains and combinations of *Streptococcus thermophilus* and *Lactobacillus delbrueckii* subsp. *bulgaricus*, isolated from traditional Greek yogurt. Lait 75, 271–283.

Gobbetti, M., 2004. Extra-hard varieties. In: Fox, P.F., McSweeney, P.L.H., Cogan, T.M., Guinee, T.P. (Eds.), Cheese: Physics, Chemistry and Microbiology, *Vol. 2*, Major Cheese Groups, (3rd ed.). Elsevier, Amsterdam, pp. 51–70.

Goff, H. D. Dairy Science and Technology Education website, University of Guelph, Canada. www.foodsci.uoguelph.ca/dairyedu/home.html

Goudjil, H., Fontecha, J., Luna, P., de la Fuente, M.A., Alonso, L., Juarez, M., 2004. Quantitative characterization of unsaturated and *trans* fatty acids in ewe's milk fat. Lait 84, 473–482.

Granato, D., Branco, G.F., Cruz, A.G., Fonseca Faria, J.d. A., Shah, N.P., 2010. Probiotic dairy products as functional foods. Compr. Rev. Food Sci. Food Safety 9, 455–470.

Grappin, R., Beuvier, E., 1997. Possible implications of milk pasteurization on the manufacture and sensory quality of ripened cheese. Int. Dairy J. 7, 751–761.

Grappin, R., Rank, T.C., Olson, N.F., 1985. Primary proteolysis of cheese proteins during ripening: a review. J. Dairy Sci. 68, 531–540.

Grappin, R., Lefier, D., Dasen, A., Pochet, S., 1993. Characterizing ripening of Gruyere de Comte: influence of time × temperature and salting conditions on eye and slit formation. Int. Dairy J. 3, 313–328.

Green, M.L., Grandison, A.S., 1993. Secondary (non-enzymatic) phase of rennet coagulation and post-coagulation phenomena. In: Fox, P.F. (Ed.), Cheese: Chemistry, Physics and Microbiology, *Vol. 1*, General Aspects, (2nd ed.). Aspen, Gaithersburg, MD, pp. 101–140.

Green, M.L., Manning, D.J., 1982. Development of texture and flavour in cheese and other fermented products. J. Dairy Res. 49, 737–748.

Griffiths, M.W., 1989. *Listeria monocytogenes*: its importance in the dairy industry. J. Sci. Food Agric. 47, 133–158.

Griffiths, M.W., 1990. Toxin production by psychrotrophic *Bacillus* spp. present in milk. J. Food Prot. 53, 790–792.

Guinee, T.P., 2003. Role of protein in cheese and cheese products. In: Fox, P.F., McSweeney, P.L.H. (Eds.), Advanced Dairy Chemistry, *Vol. 1*, Proteins, Part B, (3rd ed.). Kluwer/Plenum, New York, pp. 1083–1174.

Guinee, T.P., Feeney, E.P., Fox, P.F., 2001. Effect of ripening temperature on low moisture Mozzarella cheese: 2. Texture and functionality. Lait 81, 475–485.

Gunasekaran, S., Mehmet Ak, M., 2003. Cheese Rheology and Texture. CRC Press, Boca Raton, FL.

Guyomarc'h, F., Queguiner, C., Law, A.J.R., Horne, D.S., Dalgleish, D.G., 2003. Role of the soluble and micelle-bound heat-induced protein aggregates on network formation in acid skim milk gels. J. Agric. Food Chem. 51, 7743–7750.

HabibiNajafi, M.B., Lee, B.H., 1996. Bitterness in cheese: a review. Crit. Rev. Food Sci. Nutr. 36, 397–411.

Hamosh, M., 1990. Lingual and Gastric Lipases: Their Role in Fat Digestion. CRC Press, Boca Raton, FL.

Hannon, J.A., Wilkinson, M., Delahunty, C.M., Morrissey, P.A., Beresford, T.P., 2002. The use of novel starter systems to accelerate Cheddar cheese ripening. Farm Food 12, 48–51.

Hannon, J.A., Sousa, M., Lillevang, S., Sepulchre, A., Bockelmann, W., McSweeney, P., 2004. Effect of defined-strain surface starters on the ripening of Tilsit cheese. Int. Dairy J. 14, 871–880.

Hassan, A.N., 2008. Possibilities and challenges of exopolysaccharide producing lactic cultures in dairy foods. J. Dairy Sci. 91, 1282–1298.

Hassan, F.A.M., 2000. Effect of mixed coccus to rod starter ratio on Mozzarella cheese quality. Egypt. J. Dairy Sci. 28, 219–229.

Hassan, H.N., El-Deeb, S.A., Mashaly, R.I., 1988. Action of rennet and rennet substitutes on casein fractions in polyacrylamide gel electrophoresis. Indian J. Dairy Sci. 41, 485–490.

Hekmat, S., Reid, G., 2006. Sensory properties of probiotic yogurt is comparable to standard yogurt. Nutr. Res. 26, 163–166.

Hicks, C.L., Allauddin, M., Langlois, B.E., O'Leary, J., 1982. Psychrotrophic bacteria reduce cheese yield. J. Food Prot. 45, 331–334.

Higashio, K., Yoshioka, Y., Kikuchi, T., 1977. Isolation and identification of growth factor of *Lactobacillus bulgaricus* produced by *Streptococcus thermophilus*. 2. Studies on symbiosis in yogurt cultures. J. Agric. Chem. Soc. Jpn 51, 209–215.

Hill, A.R., 2007. Physical factors affecting the flavor of cheese. In: Weimer, B.C. (Ed.), Improving the Flavour of Cheese. Woodhead, Cambridge, pp. 252–283.

Hill, A.R., 2011. Welcome to our cheese site. Available at http://www.foodsci.uoguelph.ca/cheese/welcom.htm.

Hill, A.R., Warriner, K., 2011a. Cheese safety 101. Part 2: I have seen the enemy and he speaks Latin. Ontario Cheese Soc. Newslett. 8 August 2011.

Hill, A.R., Warriner, K., 2011b. Cheese safety 101. Part 3: Where do they come from? Ontario Cheese Soc. Newslett. 8 August 2011.

Hill, A.R., Bullock, D.H., Irvine, D.M., 1982. Manufacturing parameters of Queso Blanco made from milk and recombined milk. Can. Inst. Food Sci. Technol. J. 15, 47–53.

Holmes, D., Duersch, J., Ernstrom, C., 1977. Distribution of milk clotting enzymes between curd and whey and their survival during Cheddar cheese making. J. Dairy Sci. 60, 862–869.

Holt, C., Dalgleish, D.G., 1986. Electrophoretic and hydrodynamic properties of bovine casein micelles interpreted in terms of particles with an outer hairy layer. J. Colloid Interface Sci. 114, 513–524.

Holt, C., de Kruif, C.G., Tuinier, R., Timmins, P.A., 2003. Substructure of bovine casein micelles by small-angle X-ray and neutron scattering. Colloid Surf. A. 213, 275–284.

van Hooydonk, A.C.M., Walstra, P., 1987. Interpretation of the kinetics of the renneting reaction in milk. Neth. Milk Dairy J. 41, 19–48.

Horne, D.S., 1986. Steric stabilization and casein micelle stability. J. Colloid Interface Sci. 111, 250–260.

Horne, D.S., 1998. Casein interactions: casting light on the black boxes, the structure in dairy products. Int. Dairy J. 8, 171–177.

Horne, D.S., 2001. Factors influencing acid-induced gelation of skim milk. In: Dickinson, E., Miller, R. (Eds.), Food Hydrocolloids: Fundamentals of Formulation. Royal Society of Chemistry, Cambridge, pp. 345–351.

Horne, D.S., 2002. Casein structure, self-assembly and gelation. Curr. Opin. Colloid Interface Sci. 7, 456–461.

Horne, D.S., 2006. Casein micelle structure: models and muddles. Curr. Opin. Colloid Interface Sci. 11, 148–153.

Horne, D.S., Banks, J.M., 2004. Rennet-induced coagulation of milk. In: Fox, P.F., McSweeney, P.L.H., Cogan, T.M., Guinee, T.P. (Eds.), Cheese: Chemistry, Physics and Microbiology, *Vol. 1*, General Aspects, (3rd ed.). Elsevier, Amsterdam, pp. 47–70.

Huffman, L.M., Kristoffersen, T., 1984. Role of lactose in Cheddar cheese manufacturing and ripening. N.Z.J. Dairy Sci. Technol. 19, 151–162.

Hussain, I., Bell, A.E., Grandison, A.S., 2011. Comparison of the rheology of Mozzarella type curd made from buffalo and cows' milk. Food Chem. 128, 500–504.

Hyslop, D.B., 2003. Enzymatic coagulation of milk. In: Fox, P.F., McSweeney, P.L.H. (Eds.), Advanced Dairy Chemistry, *Vol. 1*, Proteins, Part B, (3rd ed.). Kluwer/Plenum, New York, pp. 839–878.

IDF, 1996. Dairy starter cultures of lactic acid bacteria (LAB) – standard of identity. IDF International Standard 149. International Dairy Federation, Brussels.

Imhof, R., Glattli, H., Bosset, J.O., 1994. Volatile organic aroma compounds produced by thermophilic and mesophilic mixed strain dairy starter cultures. Lebensm. Wiss. Technol. 27, 442–449.

Ismail, B., Nielsen, S.S., 2010. Invited review: plasmin protease in milk: current knowledge and relevance to dairy industry. J. Dairy Sci. 93, 4999–5009.

Jacob, M., Jaros, D., Rohm, H., 2011. Recent advances in milk clotting enzymes. Int. J. Dairy Technol. 64, 14–33.

Janhøj, T., Qvist, K.B., 2010. The formation of cheese curd. In: Law, B.A., Tamime, A.Y. (Eds.), Technology of Cheesemaking, pp. 130–165. Blackwell, Sussex.

Jayarao, B.M., Henning, D.R., 2001. Prevalence of foodborne pathogens in bulk tank milk. J. Dairy Sci. 84, 2157–2162.

Jenness, R., 1988. Composition of milk. In: Wong, N.P., Jenness, R., Keeney, M., Marth, E.M. (Eds.), Fundamentals of Dairy Chemistry, (3rd ed.). Van Nostrand Reinhold, New York, p. 779.

Johnson, E.A., Nelson, J.H., Johnson, M., 1990a. Microbiological safety of cheese made from heat-treated milk. I. Executive summary, introduction and history. J. Food Prot. 53, 441–452.

Johnson, E.A., Nelson, J.H., Johnson, M., 1990b. Microbiological safety of cheese made from heat-treated milk. II. Microbiology. J. Food Prot. 53, 519–540.

Johnson, E.A., Nelson, J.H., Johnson, M., 1990c. Microbiological safety of cheese made from heat-treated milk. III. Technology, discussion, recommendations, bibliography. J. Food Prot. 53, 610–623.

Kailasapathy, K., Lam, S.H., 2005. Application of encapsulated enzymes to accelerate cheese ripening. Int. Dairy J. 15, 929–939.

Kailasapathy, K., Rybka, S., 1997. L. acidophilus and Bifidobacterium spp. – their therapeutic potential and survival in yogurt. Aust. J. Dairy Technol. 52, 28–35.

Kaminarides, S., Stamou, P., Massouras, T., 2007. Comparison of the characteristics of set type yoghurt made from ovine milk of different fat content. Int. J. Food Sci. Technol. 42, 1019–1028.

Kethireddipalli, P., Hill, A.R., Dalgleish, D.G., 2010. Protein interactions in heat-treated milk and effect on rennet coagulation. Int. Dairy J. 20, 838–843.

Kethireddipalli, P., Hill, A.R., Dalgleish, D.G., 2011. Interaction between casein micelles and whey protein/kappa-casein complexes during renneting of heat-treated reconstituted skim milk powder and casein micelle/serum mixtures. J. Agric. Food Chem. 59, 1442–1448.

Kikuchi, M., Yamaguchi, K., Matsui, Y., 1984. The influence of heating milk on the symbiotic relationships of *Lactobacillus bulgaricus* and *Streptococcus thermophilus*. J. Coll. Dairy. (Ebetsu) 10, 349–358.

Kirby, C.J., Brooker, B.E., Law, B.A., 1987. Accelerated ripening of cheese using liposome encapsulated enzyme. Int. J. Food Sci. Technol. 22, 355–375.

Kirk, J., Mellenberger, R., 2011. Mastitis control program for *Prototheca* mastitis in dairy cows. Available at: http://www.uwex.edu/milkquality/PDF/prototheca.pdf.

Klei, L., Yun, J., Sapru, A., Lynch, J., Barbano, D., Sears, P., Galton, D., 1998. Effects of milk somatic cell count on cottage cheese yield and quality. J. Dairy Sci. 81, 1205–1213.

Klein, N., Lortal, S., 1999. Attenuated starters: an efficient means to influence cheese ripening – a review. Int. Dairy J. 9, 751–762.

Klein, N., Zourari, A., Lortal, S., 2002. Peptidase activity of four yeast species frequently encountered in dairy products – comparison with several dairy bacteria. Int. Dairy J. 12, 853–861.

Kneifel, W., Ulberth, F., Erhard, F., Jaros, D., 1992. Aroma profiles and sensory properties of yogurt and yogurt-related products. 1. Screening of commercially available starter cultures. Milchwissenschaft 47, 362–365.

de Kruif, C.G., 1992. Casein micelles: diffusivity as a function of renneting time. Langmuir 8, 2932–2937.

de Kruif, C.G., Holt, C., 2003. Casein micelle structure, functions and interactions. In: Fox, P.F., McSweeney, P.L.H. (Eds.), Advanced Dairy Chemistry, *Vol. 1*, Proteins, (3rd ed.). Kluwer/Plenum, New York, pp. 233–276. Part A.

de Kruif, K.G., Hoffmann, M.A.M., Van Marle, M.E., Van Mil, P.J.J.M., Roefs, S.P.F.M., Verheul, M., Zoon, N., 1995. Gelation of proteins from milk. Faraday Discuss. 101, 185–200.

Kunji, E., Mierau, I., Hagting, A., Poolman, B., Konings, W., 1996. The proteolytic systems of lactic acid bacteria. Anton. Leeuw. Int. J.G. 70, 187–221.

Kushal, R., Anand, S.K., 2001. A comparison of different virulence markers of *Yersinia enterocolitica*. J. Food Sci. Technol. Mysore 38, 254–256.

Kushal, R., Anand, S.K., 2006. Evaluation of the virulence potential of *Yersinia enterocolitica* isolates from milk by cell invasion-inhibition assay. Lait 86, 171–176.

Laleye, L.C., Simard, R.E., Lee, B.H., Holley, R.A., Giroux, R.N., 1987. Involvement of heterofermentative lactobacilli in development of open texture in cheeses. J. Food Prot. 50, 1009–1012.

Laleye, L.C., Simard, R.E., Lee, B.H., Holley, R.A., 1990. Quality attributes of Cheddar cheese containing added lactobacilli. J. Food Sci. 55, 114–118.

Lane, C., Fox, P., 1997. Role of starter enzymes during ripening of cheddar cheese made from pasteurized milk under controlled microbiological conditions. Int. Dairy J. 7, 55–63.

Lankaputhra, W.E.V., Shah, N.P., Britz, M.L., 1996. Evaluation of media for selective enumeration of *Lactobacillus acidophilus* and *Bifidobacterium* species. Food Aust. 48, 113–118.

Law, B.A., 2001. Controlled and accelerated cheese ripening: the research base for new technologies. Int. Dairy J. 11, 383–398.

Law, B., 2010a. Cheese adjunct cultures. Aust. J. Dairy Technol. 65, 45–49.

Law, B.A., 2010b. Cheese ripening and cheese flavor technology. In: Law, B.A., Tamime, A.Y. (Eds.), Technology of Cheesemaking, (2nd ed.). Sussex, Blackwell, pp. 231–259.

Law, B., Mulholland, F., 1995. Enzymology of lactococci in relation to flavour development from milk proteins. Int. Dairy J. 5, 833–854.

Law, B.A., Wigmore, A., 1982. Accelerated cheese ripening with food grade proteinases. J. Dairy Res. 49, 137–146.

Law, B.A., Wigmore, A.S., 1983. Accelerated ripening of Cheddar cheese with a commercial proteinase and intracellular enzymes from starter streptococci. J. Dairy Res. 50, 519–525.

Law, B.A., Andrews, A.T., Cliffe, A.J., Sharpe, M.E., Chapman, H.R., 1979. Effect of proteolytic raw milk psychrotrophs on Cheddar cheese making with stored milk. J. Dairy Res. 46, 497–509.

Law, J., Haandrikman, A., 1997. Proteolytic enzymes of lactic acid bacteria. Int. Dairy J. 7, 1–11.

Lawrence, R.C., Gilles, J., Creamer, L.K., 1983. The relationship between cheese texture and flavour. N.Z.J. Dairy Sci. Technol. 18, 175–190.

Lawrence, R.C., Heap, H.A., Gilles, J., 1984. A controlled approach to cheese technology. J. Dairy Sci. 67, 1632–1645.

Leatherhead Food International, 2006. The international market for functional foods. In: Functional Food Market Report. Leatherhead Food International, London.

Lee, M.R., Johnson, M.E., Lucey, J.A., 2005. Impact of modifications in acid development on the insoluble calcium content and rheological properties of Cheddar cheese. J. Dairy Sci. 88, 3798–3909.

Lee, W.J., Lucey, J.A., 2010. Formation and physical properties of yogurt [report]. Asian Austral. J. Anim. Sci. 23, 1127–1136.

Lelievre, J., Kelso, E.A., Stewart, D.B., 1978. Effects of *Pseudomonas* species on the syneresis of renneted milk gels. Proceedings of the XX International Dairy Congress, 760–762. Paris, Vol. E.

Lemieux, L., Simard, R.E., 1991. Bitter flavor in dairy products. 1. A review of the factors likely to influence its development, mainly in cheese manufacture. Lait 71, 599–636.

Leroy, F., De Vuyst, L., 2010. Bacteriocins of lactic acid bacteria to combat undesirable bacteria in dairy products. Aust. J. Dairy Technol. 65, 143–149.

Li, J., Dalgleish, D.G., 2006. Controlled proteolysis and the properties of milk gels. J. Agric. Food Chem. 54, 4687–4695.

Limsowtin, G.K.Y., Powell, I.B., Parente, E., 1996. Types of starters. In: Cogan, T.M., Accolas, J.-P. (Eds.), Dairy Starter Cultures. Wiley-VCH, Weinheim, pp. 101–129.

Limsowtin, G.K.Y., Bruinenberg, P.G., Powell, I.B., 1997. A strategy for cheese starter culture management in Australia. J. Microbiol. Biotechnol. 7, 1–7.

Liu, M., Siezen, R.J., Nauta, A., 2009. *In silico* prediction of horizontal gene transfer events in *Lactobacillus bulgaricus* and *Streptococcus thermophilus* reveals protocooperation in yogurt manufacturing. Appl. Environ. Microbiol. 75, 4120–4129.

Lodics, T.A., Steenson, L.R., 1993. Phage–host interactions in commercial mixed-strain dairy starter cultures: practical significance – a review. J. Dairy Sci. 76, 2380–2391.

Lortal, S., Di Blasi, A., Pediliggieri, C., Tuminello, L., Fauquant, J., Lecuona, Y., Campo, P., Carpino, S., Licitra, G., 2009. Tina wooden vat biofilm: a safe and highly efficient lactic acid bacteria delivering system in PDO Ragusano cheese making. Int. J. Food Microbiol. 132, 1–8.

Lourens-Hattingh, A., Viljoen, B.C., 2001. Yogurt as probiotic carrier food. Int. Dairy J. 11, 1–17.

Lucey, J.A., 2002. Formation and physical properties of milk protein gels. J. Dairy Sci. 85, 281–294.

Lucey, J.A., 2004a. Formation, structure, properties and rheology of acid-coagulated milk gels. In: Fox, P.F., McSweeney, P.L.H., Cogan, T.M., Guinee, T.P. (Eds.), Cheese: Chemistry, Physics and Microbiology, *Vol. 1*, General Aspects, (3rd ed.). Elsevier, Amsterdam, pp. 105–122.

Lucey, J.A., 2004b. Cultured dairy products: an overview of their gelation and texture properties. Int. J. Dairy Technol. 57, 77–84.

Lucey, J.A., Teo, C.T., Munro, P.A., Singh, H., 1997. Rheological properties at small (dynamic) and large (yield) deformations of acid gels made from heated milk. J. Dairy Res. 64, 591–600.

Lucey, J.A., Tamehana, M., Singh, H., Munro, P.A., 1998. Effect of interactions between denatured whey proteins and casein micelles on the formation and rheological properties of acid skim milk gels. J. Dairy Res. 65, 555–567.

Lucey, J.A., Tamehana, M., Singh, H., Munro, P.A., 2000. Rheological properties of milk gels formed by a combination of rennet and glucono-delta-lactone. J. Dairy Res. 67, 415–427.

Lucey, J.A., Johnson, M.E., Horne, D.S., 2003. Invited review: perspectives on the basis of the rheology and texture properties of cheese. J. Dairy Sci. 86, 2725–2745.

Ma, Y., Ryan, C., Barbano, D.M., Galton, D.M., Rudan, M.A., Boor, K.J., 2000. Effects of somatic cell count on quality and shelf-life of pasteurized fluid milk. J. Dairy Sci. 83, 264–274.

Macedo, A.C., Xavier, M.F., Oliveira, J.C., 1993. The technology, chemistry, and microbiology of Serra cheese: a review. J. Dairy Sci. 76, 1725–1739.

McBrearty, S., Ross, R.P., Fitzgerald, G.F., Collins, J.K., Wallace, J.M., Stanton, C., 2001. Influence of two commercially available bifidobacteria cultures on Cheddar cheese quality. Int. Dairy J. 11, 599–610.

McKay, L.L., Baldwin, K.A., 1975. Plasmid distribution and evidence for a proteinase plasmid in *Streptococcus lactis* C2. Appl. Microbiol. 29, 546–548.

McSweeney, P.L.H., 1997. The flavour of milk and dairy products: III. Cheese: taste. Int. J. Dairy Technol. 50, 123–128.

McSweeney, P.L.H., 2004. Biochemistry of cheese ripening: introduction and overview. In: Fox, P.F., McSweeney, P.L.H., Cogan, T.M.,

Guinee, T.P. (Eds.), In Cheese: Chemistry, Physics and Microbiology, *Vol. 1*, General Aspects, (3rd ed.). Elsevier, Amsterdam, pp. 347–360.

McSweeney, P.L.H., 2007. Flavor, texture, and flavor defects in hard and semi-hard cheeses. In: McSweeney, P.L.H. (Ed.), Cheese Problems Solved. Woodhead, Cambridge, pp. 189–201.

McSweeney, P.L.H., 2011. Biochemistry of cheese ripening. In: Fuquay, J.W., Fox, P.F., McSweeney, P.L.H. (Eds.). Encyclopedia of Dairy Sciences, (2nd ed.), Vol. 1. Elsevier, Amsterdam, pp. 667–674.

McSweeney, P.L.H., Fox, P.F., 2004. Metabolism of residual lactose and of lactate and citrate. In: Fox, P.F., McSweeney, P.L.H., Cogan, T.M., Guinee, T.P. (Eds.), Cheese: Chemistry, Physics and Microbiology, *Vol. 1*, General Aspects, (3rd ed.). Elsevier, Amsterdam, pp. 361–371.

McSweeney, P.L.H., Sousa, M.J., 2000. Biochemical pathways for the production of flavour compounds in cheeses during ripening: a review. Lait 80, 293–324.

McSweeney, P.L.H., Olson, N.F., Fox, P.F., Healy, A., Hojrup, P., 1993a. Proteolytic specificity of chymosin on bovine alpha-S1-casein. J. Dairy Res. 60, 401–412.

McSweeney, P.L.H., Fox, P.F., Lucey, J.A., Jordan, K.N., Cogan, T.M., 1993b. Contribution of the indigenous microflora to the maturation of Cheddar cheese. Int. Dairy J. 3, 613–634.

McSweeney, P., Walsh, E., Fox, P., Cogan, T., Drinan, F., Castelogonzalez, M., 1994. A procedure for the manufacture of Cheddar cheese under controlled bacteriological conditions and the effect of adjunct lactobacilli on cheese quality. Ir. J. Agric. Food Res. 33, 183–192.

McSweeney, P.L.H., Hayaloglu, A.A., O'Mahony, J.A., Bansal, N., 2006. Perspectives on cheese ripening. Aust. J. Dairy Technol. 61, 69–77.

MacGibbon, A.K.H., 1996. Herd-to-herd variations in the properties of milkfat. Proc. N.Z. Soc. Anim. Prod. 56, 224–227.

Madkor, S.A., Tong, P.S., El Soda, M., 2000. Ripening of cheddar cheese with added attenuated adjunct cultures of lactobacilli. J. Dairy Sci. 83, 1684–1691.

Mannu, L., Paba, A., 2002. Genetic diversity of lactococci and enterococci isolated from home-made Pecorino Sardo ewe's milk cheese. J. Appl. Microbiol. 92, 55–62.

Marshall, V.M., 1992. Inoculated ecosystems in a milk environment. J. Appl. Bacteriol. 73, S127–S135.

Martin, M.C., del Rio, B., Martinez, N., Magadan, A.H., Alvarez, M.A., 2008. Fast real-time polymerase chain reaction for quantitative detection of *Lactobacillus delbrueckii* bacteriophages in milk. Food Microbiol. 25, 978–982.

Martin-Visscher, L.A., 2010. Examining the structure, function and mode of action of bacteriocins from lactic acid bacteria. ProQuest Dissertations and Theses.

Martínez-Cuesta, M.C., Requena, T., Peláez, C., 2010. Use of a bacteriocin-producing transconjugant as starter in acceleration of cheese ripening. Int. Dairy J. 20, 18–24.

Martley, F.G., Crow, V.L., 1996. Open texture in cheese: the contributions of gas production by microorganisms and cheese manufacturing practices. J. Dairy Res. 63, 489–507.

Matzdorf, B., Cuppett, S.L., Keeler, L., Hutkins, R.W., 1994. Browning of Mozzarella cheese during high temperature pizza baking. J. Dairy Sci. 77, 2850–2853.

Melilli, C., Barbano, D.M., Caccamo, M., Calvo, M.A., Schembari, G., Licitra, G., 2004. Influence of brine concentration, brine temperature, and presalting on early gas defects in raw milk pasta filata cheese. J. Dairy Sci. 87, 3648–3657.

Messens, W., Foubert, I., Dewettinck, K., Huyghebaert, A., 2000. Proteolysis of a high pressure treated smear-ripened cheese. Milchwissenschaft 55, 328–332.

Messens, W., Foubert, I., Dewettinck, K., Huyghebaert, A., 2001. Proteolysis of a high pressure treated mould-ripened cheese. Milchwissenschaft 56, 201–204.

Mikulec, N., Habus, I., Antunac, N., Vitale, L., Havranek, J., 2010. Influence of peptide and amino acids on the formation of cheese flavour. Mljekarstvo 60, 219–227.

Milesi, M.M., Wolf, I.V., Bergamini, C.V., Hynes, E.R., 2010. Two strains of nonstarter lactobacilli increased the production of flavor compounds in soft cheeses. J. Dairy Sci. 93, 5020–5031.

Miller, C.W., Nguyen, M.H., Rooney, M., Kailasapathy, K., 2003. The control of dissolved oxygen content in probiotic yoghurts by alternative packaging materials. Packag. Technol. Sci. 16, 61–67.

Moatsou, G., Samolada, M., Panagiotou, P., Anifantakis, E., 2004. Casein fraction of bulk milks from different caprine breeds. Food Chem. 87, 75–81.

Morales, P., Calzada, J., Juez, C., Nunez, M., 2010. Volatile compounds in cheeses made with Micrococcus sp. INIA 528 milk culture or high enzymatic activity curd. Int. J. Dairy Technol. 63, 538–543.

Mukherjee, K.K., Hutkins, R.W., 1994. Isolation of galactose-fermenting thermophilic cultures and their use in the manufacture of low browning Mozzarella cheese. J. Dairy Sci. 77, 2839–2849.

Mulholland, F., 1995. The peptidases of lactic acid bacteria. Food Technol. Int. Eur. 64, 61–62.

Nag, S.K., 2010. Pesticides, veterinary residues and other contaminants in milk. In: Griffiths, M.W. (Ed.), Improving the Safety and Quality of Milk, *Vol. 1*, Milk Production and Processing. Woodhead, Cambridge, p. 113.

Nichol, A.W., Harden, T.J., Tuckett, H., 1996. Browning defects in mould ripened cheese. Food Aust. 48, 136–138.

Nieto-Arribas, P., Poveda, J.M., Sesena, S., Palop, L., Cabezas, L., 2009. Technological characterization of *Lactobacillus* isolates from traditional Manchego cheese for potential use as adjunct starter cultures. Food Control 20, 1092–1098.

Nieuwoudt, J.A., Bester, B.H., 1975. The effect of different methods of salting on gas development in Gouda cheese. S. Afr. J. Dairy Technol. 7, 247–252.

Nilsson, D., 2008. Metabolically engineered lactic acid bacteria and their use. US Patent No. US 7 465 575 B2.

O'Connell, J.E., Fox, P.F., 2003. Heat-induced coagulation of milk. In: Fox, P.F., McSweeney, P.L.H. (Eds.), Advanced Dairy Chemistry: Proteins, *Vol. 1*, Proteins, (3rd ed.). Kluwer/Plenum, New York, pp. 879–945. Part A.

O'Farrell, I., Sheehan, J., Wilkinson, M., Harrington, D., Kelly, A., 2002. Influence of addition of plasmin or mastitic milk to cheesemilk on quality of smear-ripened cheese. Lait 82, 305–316.

O'Mahony, J.A., McSweeney, P.L.H., Lucey, J.A., 2006a. A model system for studying the effects of colloidal calcium phosphate concentration on the rheological properties of Cheddar cheese. J. Dairy Sci. 89, 892–904.

O'Mahony, J.A., Mulholland, E.O., Guinee, T.P., 2006b. Modifying the functionality of reduced-fat Mozzarella cheese by reduction of calcium level or by the addition of emulsifying salts during curd plasticization. J. Anim. Sci. 84, 313–314.

O'Reilly, C.E., Kelly, A.L., Murphy, P.M., Beresford, T.P., 2001. High pressure treatment: applications in cheese manufacture and ripening. Trends Food Sci. Technol. 12, 51–59.

O'Reilly, C.E., O'Connor, P.M., Murphy, P.M., Kelly, A.L., Beresford, T.P., 2002. Effects of high pressure treatment on viability and autolysis of starter bacteria and proteolysis in Cheddar cheese. Int. Dairy J. 12, 915−922.

Olivecrona, T., Vilaro, S., Olivecrona, G., 2003. Lipases in milk. In: Fox, P.F., McSweeney, P.L.H. (Eds.), Advanced Dairy Chemistry, Vol. 1, Proteins, (3rd ed.). Kluwer/Plenum, New York, pp. 473−488. Part A.

Olson, N.F., Gunasekaran, S., Bogenrief, D.D., 1996. Chemical and physical properties of cheese and their interactions. Neth. Milk Dairy J. 50, 279−294.

Ott, A., Fay, L.B., Chaintreau, A., 1997. Determination and origin of the aroma impact compounds of yogurt flavor. J. Agric. Food Chem. 45, 850−858.

Panagiotidis, P., Tzia, C., 2001. Effect of milk composition and heating on flavor and aroma of yogurt. In: Spanier, A.M., Shahidi, F., Parliment, T.H., Mussinan, C., Ho, C.-T., Contis, E.T. (Eds.), Food Flavors and Chemistry: Advances of the New Millennium. RSC Special Publication No. 274. Royal Society of Chemistry, Cambridge, pp. 160−167.

Parente, E., Cogan, T.M., 2004. Starter cultures: general aspects. In: Fox, P.F., McSweeney, P.L.H., Cogan, T.M., Guinee, T.P. (Eds.), Cheese: Chemistry, Physics and Microbiology, Vol. 1, Major Cheese Groups, (3rd ed.). Elsevier, Amsterdam, pp. 123−147.

Pastor-Navarro, N., Maquieira, A., Puchades, R., 2009. Review on immunoanalytical determination of tetracycline and sulfonamide residues in edible products. Anal. Bioanal. Chem. 395, 907−920.

Peterson, S., Marshall, R., 1990. Nonstarter lactobacilli in Cheddar cheese − a review. J Dairy Sci. 73, 1395−1410.

Pette, J.W., Lolkema, H., 1950. Yogurt. I. Symbiosis and antibiosis of mixed cultures of Lactobacillus bulgaricus and Streptococcus thermophilus. Neth. Milk Dairy J. 4, 197−208.

Petterson, H.E., Sjöström, G., 1975. Accelerated cheese ripening: a method for increasing the number of lactic starter bacteria in cheese without detrimental effect to the cheese-making process, and its effect on the cheese ripening. J. Dairy Res. 42, 313−326.

Pillidge, C.J., Crow, V.L., Coolbear, T., Reid, J.R., 2003. Exchanging lactocepin plasmids in lactococcal starters to study bitterness development in Gouda cheese: a preliminary investigation. Int. Dairy J. 13, 345−354.

Playne, M., 2002. Researching, developing and commercializing probiotic cheese. Aust. Dairy Foods, February, 28−30.

Povolo, M., Bonfitto, E., Contarini, G., Toppino, P.M., 1999. Study on the performance of three different capillary gas chromatographic analyses in the evaluation of milk fat purity. J. High Resolut. Chromatogr. 22, 97−102.

Powell, I.B., 2010. Issues in cheese starter culture microbiology. Aust. J. Dairy Technol. 65, 40−44.

Prado, F.C., Parada, J.L., Pandey, A., Soccol, C.R., 2008. Trends in nondairy probiotic beverages. Food Res. Int. 41, 111−123.

Quiberoni, A., Moineau, S., Rousseau, G.M., Reinheimer, J., Ackermann, H., 2010. Streptococcus thermophilus bacteriophages. Int. Dairy J. 20, 657−664.

Ramsaran, H., Chen, J., Brunke, B., Hill, A., Griffiths, M.W., 1998. Survival of bioluminescent Listeria monocytogenes and Escherichia coli O157:H7 in soft cheeses. J. Dairy Sci. 81, 1810−1817.

Rank, T.C., Grappin, R., Olson, N.F., 1985. Secondary proteolysis of cheese during ripening: a review. J. Dairy Sci. 68, 801−805.

Rasic, J.L., Kurmann, J.A., 1978. Yoghurt − Scientific Grounds, Technology, Manufacture and Preparations. Technical Dairy Publishing House, Copenhagen.

Rattray, F., Fox, P., 1999. Aspects of enzymology and biochemical properties of Brevibacterium linens relevant to cheese ripening: a review. J. Dairy Sci. 82, 891−909.

Rea, M.C., Cogan, T.M., Vuyst, L.D., 2003. Applicability of a bacteriocin-producing Enterococcus faecium as a co-culture in Cheddar cheese manufacture. Int. J. Food Microbiol. 81, 73−84.

Rehman, S., Banks, J., McSweeney, P., Fox, P., 2000. Effect of ripening temperature on the growth and significance of non-starter lactic acid bacteria in Cheddar cheese made from raw or pasteurised milk. Int. Dairy J. 10, 45−53.

Roberfroid, M., 2007. Prebiotics: the concept revisited. J. Nutr. 137, 830S−837S.

Robinson, R.K., Tamime, A.Y., 1990. Microbiology of fermented milks. In: Robinson, R.K. (Ed.), Dairy Microbiology, Vol. 2, The Microbiology of Milk Products, (2nd ed.). Elsevier Applied Science, London, pp. 291−344.

Robinson, R.K., Tamine, A.Y., Wszolek, M., 2002. Microbiology of fermented milks. In: Robinson, R.K. (Ed.), Dairy Microbiology Handbook : the Microbiology of Milk and Milk products, (3rd ed.). John Wiley & Sons Inc, New York, pp. 367−421.

Robitaille, G., Tremblay, A., Moineau, S., St-Gelais, D., Vadeboncoeur, C., Britten, M., 2009. Fat-free yogurt made using a galactose positive exopolysaccharide producing recombinant strain of Streptococcus thermophilus. J. Dairy Sci. 92, 477−482.

Roefs, S.P.F.M., Vanvliet, T., Vandenbijgaart, H.J.C.M., Degrootmostert, A.E.A., Walstra, P., 1990. Structure of casein gels made by combined acidification and rennet action. Neth. Milk Dairy J. 44, 159−188.

Rohrbach, B.W., Draughon, F.A., Davidson, P.M., Oliver, S.P., 1992. Prevalence of Listeria monocytogenes, Campylobacter jejuni, Yersinia enterocolitica, and Salmonella in bulk tank milk: risk factors and risk of human exposure. J. Food Prot. 55, 93−97.

Roth, E., Schwenninger, S.M., Eugster-Meier, E., Lacroix, C., 2011. Facultative anaerobic halophilic and alkaliphilic bacteria isolated from a natural smear ecosystem inhibit Listeria growth in early ripening stages. Int. J. Food Microbiol. 147, 26−32.

Routray, W., Mishra, H.N., 2011. Scientific and technical aspects of yogurt aroma and taste: a review. Compr. Rev. Food Sci. F 10, 208−220.

Rudolf, M., Scherer, S., 2001. High incidence of Listeria monocytogenes in European Red smear cheese. Int. J. Food Microbiol. 63, 91−98.

Rupp, R., Clement, V., Piacere, A., Robert-Granie, C., Manfredi, E., 2011. Genetic parameters for milk somatic cell score and relationship with production and udder type traits in dairy Alpine and Saanen primiparous goats. J. Dairy Sci. 94, 3629−3634.

de Ruyter, P.G.G.A., Kuipers, O.P., Meijer, W.C., de Vos, W.M., 1997. Food-grade controlled lysis of Lactococcus lactis for accelerated cheese ripening. Nat. Biotechnol. 15, 976−979.

Saldo, J., McSweeney, P.L.H., Sendra, E., Kelly, A.L., Guamis, B., 2000. Changes in curd acidification caused by high pressure treatment. Ir. J. Agric. Food Res. 39, 169.

Saldo, J., McSweeney, P.L.H., Sendra, E., Kelly, A.L., Guamis, B., 2002. Proteolysis in caprine milk cheese treated by high pressure to accelerate cheese ripening. Int. Dairy J. 12, 35−44.

Samanidou, V.F., Nisyriou, S.A., Papadoyannis, I.N., 2007. Residue analysis of Penicillins in food products of animal origin by HPLC: a review. J. Liq. Chromatogr. Rel. Technol. 30, 1145−1204.

Sandra, S., Dalgleish, D.G., 2007. The effect of ultra high pressure homogenization (UHPH) on rennet coagulation properties of unheated and heated fresh skimmed milk. Int. Dairy J. 17, 1043–1052.

Sandra, S., Alexander, M., Dalgleish, D.G., 2007. The rennet coagulation mechanism of skim milk as observed by transmission diffusing wave spectroscopy. J. Colloid Interface Sci. 308, 364–373.

Santos, M.V., Ma, Y., Barbano, D.M., 2003. Effect of somatic cell count on proteolysis and lipolysis in pasteurized fluid milk during shelf life storage. J. Dairy Sci. 86, 2491–2503.

Santos, U., Ramos, F., 2006. Analytical methodologies for chloramphenicol residues determination in food matrixes: a brief review. Curr. Pharmaceut. Anal. 2, 53–57.

Sanz, Y., 2007. Ecological and functional implications of the acid adaptation ability of *Bifidobacterium*: a way of selecting improved probiotic strains. Int. Dairy J. 17, 1284–1289.

Savijoki, K., Ingmer, H., Varmanen, P., 2006. Proteolytic systems of lactic acid bacteria. Appl. Microbiol. Biotechnol. 71, 394–406.

Schulz-Collins, D., Senge, B., 2004. Acid- and acid/rennet-curd cheeses – Part A: Quark, cream cheese and related varieties. In: Fox, P.F., McSweeney, P.L.H., Cogan, T.M., Guinee, T.P. (Eds.), Cheese: Chemistry, Physics and Microbiology, *Vol. 2*, Major Cheese Groups, (3rd ed.), Elsevier, Amsterdam, pp. 301–328.

Scott, R., Robinson, R.K., Wilbey, R.A., 1998. Cheesemaking Practice, (3rd ed.). Aspen, Gaithersburg, ML.

Sieuwerts, S., de Bok, F.A.M., Hugenholtz, J., Vlieg, J.E.T.v.H., 2008. Unraveling microbial interactions in food fermentations: from classical to genomics approaches. Appl. Environ. Microbiol. 74, 4997–5007.

Simova, E., Beshkova, D., 2007. Effect of growth phase and growth medium on peptidase activities of starter lactic acid bacteria. Lait 87, 555–573.

Singh, H., Creamer, L., 1990. A sensitive quantitative assay for milk coagulants in cheese and whey products. J. Dairy Sci. 73, 1158–1165.

Singh, H., Waungana, A., 2001. Influence of heat treatment of milk on cheesemaking properties. Int. Dairy J. 11, 543–551.

Singh, T.K., Young, N.D., Drake, M., Cadwallader, K.R., 2005. Production and sensory characterization of a bitter peptide from beta-casein. J. Agric. Food Chem. 53, 1185–1189.

Slattery, L., O'Callaghan, J., Fitzgerald, G.F., Beresford, T., Ross, R.P., 2010. Invited review: *Lactobacillus helveticus* – a thermophilic dairy starter related to gut bacteria. J. Dairy Sci. 93, 4435–4454.

Smith, J.L., 2002. Structural and functional characterization of the lantibiotic mutacin 1140. ProQuest Dissertations and Theses.

Smith, M.R., 2005. Cheese ripening process. Official Gazette of the United States Patent and Trademark Office Patents.

Somers, J.M., Guinee, T.P., Kelly, A.L., 2002. The effect of plasmin activity and cold storage of cheese milk on the composition, ripening and functionality of Mozzarella-type cheese. Int. J. Dairy Technol. 55, 5–11.

Soomro, A.H., Masud, T., 2008. Partial characterisation of bacteriocin produced by Lactobacillus acidophilus J1 isolated from fermented milk product Dahi. Aust. J. Dairy Technol. 63, 8–14.

Spinnler, H., Gripon, J., 2004. Surface mould-ripened cheeses. In: Fox, P.F., McSweeney, P.L.H., Cogan, T.M., Guinee, T.P. (Eds.), Cheese: Physics, Chemistry and Microbiology, *Vol. 2*, Major Cheese Groups, (3rd ed.). Elsevier, Amsterdam, pp. 157–174.

Spolaor, D., Marangon, M., 1997. Pre-ripening at low temperature of hard cheeses to prevent late blowing: effect on the ripening. Scienza e Tecnica Lattiero Casearia 48, 43–52.

Steele, J.L., 1995. Contribution of lactic acid bacteria to cheese ripening. In: Malin, E.L., Tunick, M.H. (Eds.), Chemistry of Structure–Function Relationships in Cheese. Plenum Press, New York, pp. 209–220.

Steele, J.L., 1996. Proteolytic enzyme systems of lactic acid bacteria and their importance in cheese flavor development. In: Book of Abstracts, IFT Annual Meeting 1996. New Orleans, USA. Institute of Food Technologists, Chicago, IL.

Stewart, D.B., 1975. Factors influencing the incidence of *B. cereus* spores in milk. J. Soc. Dairy Technol. 28, 80–85.

Stewart, D.I., Kelly, A.L., Gulinee, T.P., Beresford, T.P., 2006. High pressure processing: review of application to cheese manufacture and ripening. Aust. J. Dairy Technol. 61, 170–178.

Suarez, V.B., Capra, M.L., Rivera, M., Reinheimer, J.A., 2007. Inactivation of calcium dependent lactic acid bacteria phages by phosphates. J. Food Prot. 70, 1518–1522.

Suarez, V.B., Zago, M., Quiberoni, A., Carminati, D., Giraffa, G., Reinheimer, J., 2008. Lysogeny in *Lactobacillus delbrueckii* strains and characterization of two new temperate prolate-headed bacteriophages. J. Appl. Microbiol. 105, 1402–1411.

Sutherland, A.D., 1993. Toxin production by *Bacillus cereus* in dairy products. J. Dairy Res. 60, 569–574.

Sutherland, A.D., Limond, A.M., 1993. Influence of pH and sugars on the growth and production of diarrhoeagenic toxin by *Bacillus cereus*. J. Dairy Res. 60, 575–580.

Suzuki, I., Kato, S., Kitada, T., Yano, N., Morichi, T., 1986. Growth of *Lactobacillus bulgaricus* in milk. 1. Cell elongation and the role of formic acid in boiled milk. J. Dairy Sci. 69, 311–320.

Swearingen, P.A., Adams, D.E., Lensmire, T.L., 2004. Factors affecting calcium lactate and liquid expulsion defects in Cheddar cheese. J. Dairy Sci. 87, 574–582.

Swindell, S.R., Benson, K.H., Griffin, H.G., Renault, P., Ehrlich, S.D., Gasson, M.J., 1996. Genetic manipulation of the pathway for diacetyl metabolism in *Lactococcus lactis*. Appl. Environ. Microbiol. 62, 2641–2643.

Talwalkar, A., Kailasapathy, K., 2003. Effect of microencapsulation on oxygen toxicity in probiotic bacteria. Aust. J. Dairy Technol. 58, 36–39.

Tamime, A.Y., Robinson, R.K., 1999a. Background to manufacturing practice. In: Yoghurt Science and Technology, (2nd ed.). Woodhead, Cambridge, pp. 11–128.

Tamime, A.Y., Robinson, R.K., 1999b. Traditional and recent developments in yoghurt production and related products. In: Yoghurt Science and Technology, (2nd ed.). Woodhead, Cambridge, pp. 306–388.

Tamime, A.Y., Robinson, R.K., 2007. Biochemistry of fermentation. In: Yoghurt Science and Technology, (3rd ed.). Woodhead, Cambridge, pp. 535–607.

Thierry, A., Maillard, M.B., Richoux, R., Kerjean, J.R., Lortal, S., 2005. *Propionibacterium freudenreichii* strains quantitatively affect production of volatile compounds in Swiss cheese. Lait 85, 57–74.

Todesco, R., Carcano, M., Lodi, R., Crepaldi, P., 2000. Indirect conductimetry in the study of propionibacteria inhibition. Lait 80, 337–346.

Todorov, S.D., 2008. Bacteriocin production by *Lactobacillus plantarum* AMA-K isolated from Amasi, a Zimbabwean fermented milk product and study of the adsorption of bacteriocin AMA-K to *Listeria* spp. Braz. J. Microbiol. 39, 178–187.

Todorov, S.D., 2010. Diversity of bacteriocinogenic lactic acid bacteria isolated from boza, a cereal based fermented beverage from Bulgaria. Food Control 21, 1011–1021.

Todorov, S.D., Ho, P., Vaz-Velho, M., Dicks, L.M.T., 2010. Characterization of bacteriocins produced by two strains of Lactobacillus plantarum isolated from Beloura and Chourico, traditional pork products from Portugal. Meat Sci. 84, 334–343.

Todorov, S.D., Furtado, D.N., Saad, S.M.I., Tome, E., Franco, B.D.G.M., 2011a. Potential beneficial properties of bacteriocin producing lactic acid bacteria isolated from smoked Salmon. J. Appl. Microbiol. 110, 971–986.

Todorov, S.D., Prevost, H., Lebois, M., Dousset, X., LeBlanc, J.G., Franco, B.D.G.M., 2011b. Bacteriocinogenic Lactobacillus plantarum ST16Pa isolated from papaya (Carica papaya) from isolation to application: characterization of a bacteriocin. Food Res. Int. 44, 1351–1363.

Todorov, S.D., Rachman, C., Fourrier, A., Dicks, L.M.T., van Reenen, C.A., Prevost, N., Dousset, X., 2011c. Characterization of a bacteriocin produced by Lactobacillus sakei R1333 isolated from smoked Salmon. Anaerobe 17, 23–31.

Tome, E., Todorov, S.D., Gibbs, P.A., Teixeira, P.C., 2009. Partial characterization of nine bacteriocins produced by lactic acid bacteria isolated from cold-smoked salmon with activity against Listeria monocytogenes. Food Biotechnol. 23, 50–73.

Tranchant, C.C., Dalgleish, D.G., Hill, A.R., 2001. Different coagulation behaviour of bacteriologically acidified and renneted milk: the importance of fine tuning acid production and rennet action. Int. Dairy J. 11, 483–494.

Tunick, M.H., van Hekken, D.L., 2006. Chemistry and rheology of cheese. In: Shahidi, F., Weenan, H. (Eds.), Food Lipids: Chemistry, Flavor, and Texture, (2nd ed.). American Chemical Society, Washington, DC, pp. 133–141.

Turner, K.W., Martley, F.G., 1983. Galactose fermentation and classification of thermophilic lactobacilli. Appl. Environ. Microbiol. 45, 1932–1934.

Upadhyay, V.K., McSweeney, P.L.H., 2003. Acceleration of cheese ripening. In: Smit, G. (Ed.), Dairy Processing: Maximizing Quality. Woodhead, Cambridge, pp. 419–447.

Upadhyay, V.K., McSweeney, P.L.H., Magboul, A.A.A., Fox, P.F., 2004a. Proteolysis in cheese during ripening. In: Fox, P.F., McSweeney, P.L.H., Cogan, T.M., Guinee, T.P. (Eds.), Cheese: Chemistry, Physics and Microbiology, Vol. 1, General Aspects, (3rd ed.). Elsevier, Amsterdam, pp. 391–433.

Upadhyay, V.K., Sousa, M.J., Ravn, P., Israelsen, H., Kelly, A.L., McSweeney, P.L.H., 2004b. Use of exogenous streptokinase to accelerate proteolysis in Cheddar cheese during ripening. Lait 84, 527–538.

Upadhyay, V.K., Ravn, P., Israelsen, H., Sousa, M.J., Kelly, A.L., McSweeney, P.L.H., 2006. Acceleration of proteolysis during ripening of Cheddar type cheese using a streptokinase-producing strain of Lactococcus. J. Dairy Res. 73, 70–73.

Upadhyay, V.K., Huppertz, T., Kelly, A.L., McSweeney, P.L.H., 2007. Use of high pressure treatment to attenuate starter bacteria for use as adjuncts for Cheddar cheese manufacture. Innov. Food Sci. Emerg. Technol. 8, 485–492.

Van de Water, J., Naiyanetr, P., 2008. Yogurt and immunity: the health benefits of fermented milk products that contain lactic acid bacteria. In: Farnworth, E.R. (Ed.), Handbook of Fermented Functional Foods, (2nd ed.). CRC Press, Boca Raton, FL, pp. 129–164.

Vasbinder, A.J., Rollema, H.S., de Kruif, C.G., 2003. Impaired rennetability of heated milk; study of enzymatic hydrolysis and gelation kinetics. J. Dairy Sci. 86, 1548–1555.

Vassal, L., Monnet, V., Lebars, D., Roux, C., Gripon, J.C., 1986. Relation between pH, chemical composition and texture of Camembert cheese. Lait 66, 341–351.

Vedamuthu, E.R., 2006. Starter cultures for yogurt and fermented milk. In: Chandan, R.C., White, C., Kilara, A., Hui, Y.H. (Eds.), Manufacturing Yogurt and Fermented Milks. Blackwell Publishing Professional, Ames, IA, pp. 89–116.

Veringa, H.A., Galesloot, Th., E., Davelaar, H., 1968. Symbiosis in yoghurt. 2. Isolation and identification of a growth factor for Lactobacillus bulgaricus produced by Streptococcus thermophilus. Neth. Milk Dairy J. 22, 114–120.

Villeneuve, P., Pina, M., Graille, J., 1996. Determination of pregastric lipase specificity in young ruminants. Chem. Phys. Lipids 83, 161–168.

Vinderola, G., LeBlanc, A.M., Perdigón, G., Matar, C., 2008. Biologically active peptides released in fermented milk: role and functions. In: Farnworth, E.R. (Ed.), Handbook of Fermented Functional Foods, (2nd ed.). CRC Press, Boca Raton, FL, pp. 209–241.

Vindfeldt, K., 1993. A new concept for improving the quality and flavour of cheese. Scand. Dairy Inf. 7, 34–35.

Vlahopoulou, I., Bell, A.E., Wilbey, R.A., 2001. Effects of starter culture and its exopolysaccharides on the gelation of glucono-delta-lactone acidified bovine and caprine milk. Int. J. Dairy Technol. 54, 135–140.

van Vliet, T., Lakemond, C.M.M., Visschers, R.W., 2004. Rheology and structure of milk protein gels. Curr. Opin. Colloid Interface Sci. 9, 298–304.

de Vos, W.M., 1996. Metabolic engineering of sugar metabolism in lactic acid bacteria. Antonie Van Leeuwenhoek 70, 223–242.

Wagner, M., Eliskases-Lechner, F., Rieck, P., Hein, I., Allerberger, F., 2006. Characterization of Listeria monocytogenes isolates from 50 small scale Austrian cheese factories. J. Food Prot. 69, 1297–1303.

Walsh, C.D., Guinee, T.P., Harrington, D., Mehra, R., Murphy, J., Fitzgerald, R.J., 1998. Cheesemaking, compositional and functional characteristics of low moisture part skim Mozzarella cheese from bovine milks containing kappa-casein AA, AB or BB genetic variants. J. Dairy Res. 65, 307–315.

Walstra, P., 1990. On the stability of casein micelles. J. Dairy Sci. 73, 1965–1979.

Walstra, P., 1999. Casein sub-micelles: do they exist? Int. Dairy J. 9, 189–192.

Walstra, P., Wouters, J.T.M., Geurts, T.J., 2006. Cheese ripening and properties. In: Walstra, P., Wouters, J.T.M., Geurts, T.J. (Eds.), Dairy Science and Technology. CRC Press, Boca Raton, FL, pp. 641–675.

Wang, S., Zhang, H.Y., Wang, L., Duan, Z.J., Kennedy, I., 2006. Analysis of sulphonamide residues in edible animal products: a review. Food Addit. Contam. 23, 362–384.

Weimer, B.C., 2007. Improving the Flavour of Cheese. Woodhead, Cambridge.

Welman, A.D., 2009. Exploitation of exopolysaccharides from lactic acid bacteria: nutritional and functional benefits. In: Ullrich, M. (Ed.), Bacterial Polysaccharides: Current Innovations and Future Trends. Caister Academic Press, Norfolk, pp. 331–344.

Wilkins, D.W., Schmidt, R.H., Shireman, R.B., Smith, K.L., Jezeski, J.J., 1986. Evaluating acetaldehyde synthesis from [C-14] U)l-]threonine by Streptococcus thermophilus and Lactobacillus bulgaricus. J. Dairy Sci. 69, 1219–1224.

Wilkinson, M.G., Kilcawley, K.N., 2005. Mechanisms of incorporation and release of enzymes into cheese during ripening. Int. Dairy J. 15, 817–830.

Wolf, V.I., Meinardi, A.C., Zalazar, A.C., 2009. Production of flavour compounds from fat during cheese ripening by action of lipases and esterases. Protein Pept. Lett. 16, 1235–1243.

Wong, N.P., Jenness, R., Keeney, M., Marth, E.H., 1988. Fundamentals of Dairy Chemistry, (3rd ed.). Van Nostrand Reinhold, New York, p. 21.

Yokoyoma, H., Sawamura, N., Motobayashi, N., 1992. Method for accelerating cheese ripening. Patent Application, EP 0 469 857 A1. Fuji Oil Company, Japan.

Yoshida, S., 1989. Preparation of lactoferrin by hydrophobic interaction chromatography from milk acid whey. J. Dairy Sci. 72, 1446–1450.

Yu, J., Wang, W.H., Menghe, B.L.G., Jiri, M.T., Wang, H.M., Liu, W.J., Bao, Q.H., Lu, Q., Zhang, J.C., Wang, F., Xu, H.Y., Sun, T.S., Zhang, H.P., 2011. Diversity of lactic acid bacteria associated with traditional fermented dairy products in Mongolia. J. Dairy Sci. 94, 3229–3241.

Yvon, M., Rijnen, L., 2001. Cheese flavour formation by amino acid catabolism. Int. Dairy J. 11, 185–201.

Zhang, H., Wang, S., 2009. Review on enzyme-linked immunosorbent assays for sulfonamide residues in edible animal products. J. Immunol. Methods 350, 1–13.

Zhang, J., Liu, G., Shang, N., Cheng, W., Chen, S., Li, P., 2009. Purification and partial amino acid sequence of Pentocin 31-1, an anti-Listeria bacteriocin produced by *Lactobacillus pentosus* 31-1. J. Food Prot. 72, 2524–2529.

Zhao, X., Li, D., 2008. A new approach to eliminate stress for two probiotics with chemicals *in vitro*. Eur. Food Res. Technol. 227, 1569–1574.

Zinno, P., Janzen, T., Bennedsen, M., Ercolini, D., Mauriello, G., 2010. Characterization of Streptococcus thermophilus lytic bacteriophages from Mozzarella cheese plants. Int. J. Food Microbiol. 138, 137–144.

Zisu, B., Shah, N.P., 2005. Low-fat Mozzarella as influenced by microbial exopolysaccharides, preacidification, and whey protein concentrate. J. Dairy Sci. 88, 1973–1985.

Zourari, A., Desmazeaud, M.J., 1991. Characterization of lactic acid bacteria isolated from Greek yogurts. 2. Strains of *Lactobacillus delbrueckii* subsp. *bulgaricus* and mixed cultures with *Streptococcus salivarius* subsp. *thermophilus*. Lait 71, 463–482.

Chapter 9

Oilseed Processing and Fat Modification

Fereidoon Shahidi

Department of Biochemistry, Memorial University of Newfoundland, St. John's, Newfoundland, Canada

Chapter Outline

I. INTRODUCTION

Edible oils originate from seeds (oilseeds), animal tissues (lard, tallow, or fish), fruits (e.g. palm), processing by-products of cereals (e.g. rice bran oil), and fisheries industries (fish oil, cod liver oil, etc.), among others. In general, cleaning, followed by oil recovery are essential first steps. Thus, seed crushing and refining are the major operations in the procurement of edible oils.

If the oil comes from seeds, it must be cleaned before extraction to remove unwanted solid particles. The seeds are then subjected to drying followed by heating to temper them, thus helping to deactivate enzymes that may lead to degradation of oil or other matters in the seed and better release of the oil from oil compartments (Meshehdani *et al.*, 1990).

In the intact seeds, the enzymes are separated from the oil compartments within the cells. However, if seeds are damaged the enzymes present may come into contact with the oil or other sensitive material, such as glucosinolates in canola. This would lead to adverse reactions, to different degrees, depending on the temperature, moisture, and the extent of damage. In some cases, oilseeds may subsequently undergo dehulling, such as in soybean. In this way, not only may a better oil result following dehulling, but the resultant deoiled soybean meal would contain a higher content of proteins, e.g. 48% versus 44%. Procurement of a better oil may also be achieved in cases where the hulls contain less desirable components such as waxes in sunflower or corn which would otherwise need to be removed at a later stage. However, dehulling must be carried out carefully to avoid rupture of oil cells and their adsorption by the hulls.

Depending on the seeds, they may be heated before crushing and pressing by a screw, which generally releases about one-third of the oil. The screw press provides a continuous operation that allows ejection and separation of the oil by draining. The solid material that still contains two-thirds of the oil is often a sticky solid mass, known as the 'cake'. The large pieces of cake may then be subjected to flaking for size reduction before extraction of the remaining oil by appropriate solvents (Kemper, 2005).

Biochemistry of Foods. DOI: http://dx.doi.org/10.1016/B978-0-12-242352-9.00009-6

Oil extraction is often achieved in crown extractors using a counter-current flow which reduces the amount of solvent required. In this process, the solvent flows over the surface of the particles and diffuses through the miscella during the percolation process (Kemper, 2005).

The solvents used for oil extraction are hexanes, in a mixture of about 60% *n*-hexane and other hexane isomers. Since hexane is flammable, the use of other solvents has been considered, but none of these is yet in commercial use because of the lower yield of oil and other considerations. Following solvent extraction, the oil is separated and the resultant solids are passed to the desolventizer-toaster to remove the remaining hexanes. The residual oil in the resultant meal is usually less than 1%. The toasting process is needed when the meal is to be used for food or feed. The nearly desolventized flakes are removed from the system using a vapor-tight cyclone or by vapor desolventizing with steam, leading to the production of a desolventized meal. This meal may then be ground to the desired size depending on its end use. Crude oil is also obtained following the removal of hexanes by distillation and mixing of the pressed oil with the solvent-extracted oil.

In the case of fruit oils, such as those of the palm, oil may be procured from both the fruit and the kernel. The processing of olive fruit and palm differs somewhat, but details will not be provided here. Bunches of palm fruits are harvested at the desired stage of maturity and transported to the mill, where they are sterilized by heating with pressurized steam for about an hour to inactivate the enzymes present so that the oil is not hydrolyzed.

Following sterilization, the fruits are stripped from stalks, and then transferred to a digester where the material is reheated to about 100°C for a few minutes to loosen the pericarp from the nuts and to break and release the oil from oil-bearing cells. This material is then subjected to continuous screw-pressing. The press liquor contains nearly two-thirds oil along with water and some solids, while the press cake contains the nuts and the flesh fiber. Addition of water to the press liquor facilitates the settling of the solids upon filtration and the resultant crude oil may be separated by decanting or centrifugation. The oil is then cooled and stored while the solids (press cake) are transferred to an aspirator, which allows separation of the nuts (kernels) from the fiber. The resultant fiber is often burned to provide heat for the steam boiler. The seed may then be conditioned by drying to loosen the kernels from the shells. The nuts may then be cooled and stored or cracked and the kernels separated from the shells based on density differences. Kernels are subsequently screw-pressed or solvent-extracted to produce palm kernel oil and palm kernel meal (Fairhurst and Mutert, 1999).

For the recovery of oil of animal origin, both wet and dry rendering may be practiced. This subject is beyond the scope of this chapter, but in wet rendering the material is cooked by steam and this eventually leads to three phases, with precipitated denatured proteins as solids, the water phase, and the fat layer on the top which can be separated by decanting using centrifuges (Henry, 2009). The crude oil thus obtained may be subjected to further processing, as required.

Cold-pressing is another method used for producing oils that may otherwise be sensitive to oxidation or to procure the prime quality of oil that is often used in the crude form. In this case the oil is extracted by traditional methods or commonly by screw-pressing, but usually in a batch processor. The pressed oil is released through small holes or slots. The cold-pressed oil has the advantage of retaining minor components of the oil without many changes. Following oil extraction and refining, other possible steps may be carried out to prepare products. As an example, the oils may be cooled slowly to about 4°C in a process known as winterization, which removes saturated fats that could precipitate out during refrigerated storage. The oils may also be subjected to the introduction of water to prepare margarines. In all these cases, oils could also be partially hydrogenated, a process that has raised concern due to the production of *trans* fats. Nonetheless, this process is often necessary to produce more solid-like fats that provide mouth-feel and texture to the end products. However, more recent efforts have concentrated on producing formulations that make use of palm and other more saturated oils to provide texture and mouth-feel without introducing *trans* fats into products.

Finally, the stability of edible oils is of much interest to the industry and consumers alike because adequate shelf-life of products is necessary, especially when products are used in frying or high-temperature operations. In such cases, the use of antioxidants, especially for more unsaturated oils, is practiced. Nonetheless, the oilseed industry has moved gradually towards producing high-oleic products, such as high-oleic sunflower oil, to obtain more stable oils with extended shelf-life characteristics.

II. PROCESSING OF EDIBLE OILS

The crude oil obtained following pressing and extraction may further be subjected to a series of processing steps known as refining. Specialty oils and lard, as well as tallow, are usually consumed without further refining.

The process of refining is carried out to produce a bland oil, mainly triacylglycerols (TAGs), commonly known as triglycerides. The non-TAG components present in the crude oil may include small amounts of a number of products and other solids as well as phospholipids, free fatty acids (FFAs), monoacylglycerols, diacylglycerols, pigments, tocopherols and/or tocotrienols, phytosterols, waxes, and possibly sulfur-containing compounds, along with hydroperoxides, secondary oxidation products, and possibly squalene and other hydrocarbons, including carotenoids. Phenolic compounds other than tocols may also be present in certain oils, such as sesame seed and olive oils (Pokorny, 1991).

During the refining process, oils are subjected to degumming, refining, bleaching, deodorization and, in certain cases, winterization. Antioxidants are sometimes also added to the oil following deodorization to enhance the oxidative stability of the oil. Some unit operations lead to the removal of certain beneficial components, such as tocophenols and sterols, from the oil.

A. Removal of Phospholipids: The Degumming Process

Phospholipids present in the oils are generally more unsaturated in nature and darken in color upon storage as a result of oxidation. These phospholipids exist in hydratable (HPL) and non-hydratable (NHPL) forms. The HPLs (phosphatides) may be removed by water washing as they are precipitated out from the oil. The sludge of phospholipids is referred to as 'gums' and hence the process is known as degumming. The NHPLs therefore require acidification to remove them from the oil. Thus, industrially, the oil may be subjected to acidification with 0.05–0.2% of phosphoric acid at a temperature of 70–80°C over a period of 5–30 minutes along with stirring (Zufarov et al., 2008). This process may also help to remove some of the chlorophyll in the oil that may otherwise act as a photosensitizer, thus speeding up oil deterioration via photooxidation. More recently, the use of citric or malic acid has been considered for the degumming process. After acidification, the oil is subjected to water washing and removal of phospholipids. The phospholipids removed may be used in different applications. For example, phospholipids from soybean oil processing, known as soy lecithin, may be purified and used as a dietary supplement or in other applications (Ceci et al., 2008).

The first refining step involves the removal of phospholipids using the degumming process. Degumming purifies the seed oils, which normally contain impurities in the colloidal state or in solution (Bernardini, 1985). Crude vegetable oil is degummed to produce an oil substantially free of materials that settle down during transportation or storage. Solvent-extracted vegetable oils contain considerable amounts of phosphatides and other mucilaginous materials which form deposits in the storage tanks. It is vital to remove the phosphatides from the crude oil because their presence would impart undesirable flavor and color to the oil, and shorten its shelf-life. They also lead to increased refining losses by emulsifying considerable amounts of neutral oil, which is lost in the soap stock. The recovered phosphatides can be further processed to produce lecithin for use as an emulsifier in products such as margarine, chocolate, and emulsion paints, or as a dietary supplement.

The common oilseeds such as soybean, cottonseed, sunflower, and rapeseed are rich sources of phospholipids (Indira et al., 2000; Willem and Mabel, 2008), which are either hydratable (HPLs) or non-hydratable (NHPLs). Most of the phospholipids in crude sunflower and rapeseed oils are hydratable and can be removed by water degumming (Zufarov et al., 2008). The NHPLs cannot swell and form gels or precipitate from the oil (Szydlowska-Czerniak, 2007), hence their removal requires a more complex process at increased temperatures with the use of phosphoric acid, citric acid, malic acid, or other degumming agents.

The resulting insoluble, hydrated gum, following acid treatment, is separated as a sludge by centrifugal action and, when dried, affords crude lecithin. The degummed oil can be dried and pumped to storage containers or can proceed to the refining step. Excessive amounts of phosphoric acid should be avoided, as it may increase the content of phosphorus in the medium, which is difficult to remove and may contribute to further refining problems. Different degumming processes, namely dry degumming, water degumming, acid degumming (Andersen, 1962), and total degumming (Dijkstra and Opstal, 1989) have been described in detail. More recently, enzymatic degumming, simultaneous degumming/dewaxing, and membrane degumming have received attention for their role in reducing the refining loss and color intensity of the finished oil.

The dry degumming process involves the removal of gums through precipitation by acid conditioning and via filtration during the bleaching process, not via centrifugal separation. This process is used for low-phosphatide oils such as palm oil, lauric oils, and edible tallow, and is suitable for preparing oils for subsequent physical refining. Water degumming consists of treating the natural oil with a small amount of water, followed by centrifugal separation. The process is applied to many oils that contain phospholipids in significant amounts. Phosphatidylcholine and phosphatidylinositol are completely hydratable. Phosphatidylethanolamine and phosphatidic acid are only partially hydratable or non-hydratable, when they form salts with divalent cations (mainly calcium and magnesium) or when they are not in the dissociated form. The phosphatide—metal complexes can be decomposed by the addition of acid or a complexing agent, followed by hydration with water. Partial neutralization of acid is used to avoid the migration of phosphatides back to the oil phase (Kovari, 2004).

For physical refining of vegetable oils, water degumming is not sufficient. In the acid degumming process, gums are precipitated by an acid-conditioning process (using e.g. phosphoric, citric, malic, or tartaric acid) and subsequently removed by centrifugal separation. The total degumming process, known by its Dutch acronym TOP, is designed to further treat the oil that has already been water degummed. The process has two variations to cater for different needs (Dijkstra and Opstal, 1987). In the first approach a dilute acid is finely dispersed into the oil. After a sufficient contact time a base is added and mixed into the acid-in-oil dispersion. The base can be sodium hydroxide, sodium carbonate, or sodium silicate. During the process the acid initially decomposes metal—phosphatidic acid complexes into insoluble metal salts and phosphatidic acid (in acid form). Phosphatidic acid is then hydrated by partial neutralization with the base added, and removed from the oil by centrifugation. The second approach uses a combination of two centrifuges to remove the hydrated phospholipids with high efficiency and minimal loss. The first one removes the bulk of the gum phase. Clearly, the quality of the water-degummed oil is critical to TOP degumming. When the water-degummed oil has a higher content of calcium and magnesium, the TOP process becomes less effective (Cleenewerck and Dijkstra, 1992). The removal of phospholipids from vegetable oils by membrane technologies is a relatively new development (Ochoa et al., 2001).

The traditional degumming processes, including water degumming, TOP degumming, acid treatment, and others, cannot guarantee the low phosphorus contents required for physical refining, and they are not always optimally suited for all oil qualities because of the high content of NHPLs (Copeland and Belcher, 2005).

The latest degumming processes are soft degumming and enzymatic degumming. The soft degumming process involves complete elimination of phospholipids by a chelating agent, such as ethylenediaminetetraacetic acid (EDTA), in the presence of an emulsifying agent. Different kinds of crude oil are degummed by the soft degumming method; the content of phospholipids in the treated oil samples is lowered to approximately 5 ppm. However, the high cost of EDTA does not make this process industrially attractive (Choukri et al., 2001).

The first enzymatic degumming process that was used industrially was the EnzyMax1 process, which launched in 1992 and used porcine phospholipase A_2. Subsequently, various microbial phospholipases with different specificities have been developed. They have the advantage of being kosher and halal, and of having an unlimited availability and a low cost. Enzymatic degumming is probably the best process available for reducing the phosphorus content of crude oils below the 5 ppm level (Chakrabarti and Rao, 2004). The enzyme, phospholipase A_2, catalyzes the non-hydratable phosphatides into hydratable lysophospholipids, which are then removed by centrifugation, yielding oil that is low in phosphorus. Owing to the low temperature used, this process produces no color deterioration in the degummed oil compared to conventional phosphoric acid degumming. Moreover, the oil content of the gums from enzymatic degumming is only 25—30% compared to 50—60% in the conventional method for rice bran oil. The oryzanol present in crude rice bran oil remains almost intact during the enzymatic process. Processes for simultaneous dewaxing and degumming use water and an aqueous solution of calcium chloride, followed by centrifugation or low-temperature (20°C) crystallization, which facilitates the precipitation of hydratable and non-hydratable phosphatides along with wax, and the phosphorus content decreases to below 5 ppm (Kaimal et al., 2000; Rajam et al., 2005).

The economic feasibility of these processes is quite good owing to the elimination of one step from the entire process. The membrane process is usually used in extraction plants. According to Lin et al. (1997), membrane separation is primarily a size-exclusion-based, pressure-driven process. It separates different components according to their molecular weight or particle size and shape and is dependent on their interactions with membrane surfaces and other components of the mixture. Phospholipids can be separated from TAGs in the miscella stage using appropriate membranes. The membrane-based crude oil degumming produces a permeate and a retentate containing TAGs and phospholipids, respectively. The majority of the coloring bodies and some of the FFAs and other impurities are included in phospholipid micelles and also removed (Lin et al., 1997).

B. Refining

The degumming step, prior to chemical refining, is not always necessary. The best known and the most widely used chemical refining procedure is the caustic soda process. Refining of oil is practiced as a purifying treatment designed to remove FFAs. Usually, the refined oils are neutral (i.e. neither acidic nor alkaline), free of materials that separate on heating (break material), lighter in color, less viscous, and more susceptible to oxidation.

The refining process may be carried out in either a batch or a continuous system. In batch refining, the aqueous emulsion of soaps formed from FFAs, along with other impurities (soapstock), settles to the bottom and is drawn off. In the continuous system the emulsion is separated by centrifugation. Essentially all soybean oils in the USA are refined by the continuous process. Based on the FFA content of the oil, caustic with an excess of 0.10−0.13% is proportioned into the crude oil and mixed in a high-shear in-line mixer. The soap−oil mixture is heated to 75−80°C and fed to a pressure or hermetic type centrifuge for separation into light- and heavy-density phases. Light-phase discharge consists of the refined oil containing traces of moisture and soap; the heavy phase is primarily soap, insoluble material, free caustic, phosphatides, and a small quantity of neutral oil. Refined oil is washed with 10−20% by weight of soft water at 90°C. The water washing process removes about 90% of the soap content in the refined oil; the remainder of the soap is removed in the bleaching process (Mounts and Khym, 1980). Soapstock and wash water are combined and treated with sulfuric acid to convert the soap into crude fatty acids. Most of the acidified soapstock is used as a high-energy ingredient in animal feed. Depending on market demand, acidified soapstock may be sold to fatty acid producers who recover the crude fatty acids by distillation as a valuable by-product of oil refining (Mounts, 1981). The residue from this distillation is a good source of sterols.

Conventional chemical refining is time consuming and has several disadvantages. It has substantial energy requirements and the by-products formed (soapstock and deodorizer distillate) are neither environmentally friendly nor commercially valuable. Furthermore, the chemical process leads to considerable oil loss; soapstock can hold as much as 50% of its weight of neutral oil. Despite having several disadvantages, it is still used in many industries because of the successful reduction of FFA content to an acceptable level. The refining of high FFA oil has been accomplished by miscella refining as it provides three immediate benefits: a lower refining loss; a lighter colored refined oil without bleaching; and elimination of the need for water washing of the refined oil or miscella (Canavag, 1976). The miscella refining process has been commercialized in Japan for many years. It is a simple process that is carried out in an explosion-proof system. However, the cost of the equipment is somewhat higher than that of an ordinary refining plant and control of the process is more difficult. Miscella deacidification involves slight modification of the chemical refining process, in which the oil is mixed with hexanes to create miscella. The mixed solvent process using hexane as the main solvent and ethanol or iso-propanol as the second solvent has been carried out for refining of high FFA rice bran oil (Ghosh, 2007). The miscella are mixed with sodium hydroxide in a neutralization step and then reacted with phosphatides. This process also induces decolorization. Soapstock is removed by centrifugation, which results in minimal loss of neutral oil; however, it is very expensive and solvent removal requires several steps. Miscella deacidification is only used for the refining of cottonseed oil because a lighter colored final product is obtained compared to using the classical methods (Bhosle and Subramanian, 2005).

Physical refining processes use steam stripping under vacuum to avoid chemical neutralization. This is a simplified operation that removes FFAs, unsaponifiable matter, and pungent compounds, and also reduces the amount of oil lost. Physical refining is also known as deacidification (deodorization) by steam distillation in which FFAs and other volatile components are distilled off from the oil using an effective stripping agent, which is usually steamed under suitable processing conditions (Ceriani and Meirelles, 2006). It consumes less steam, water, and power, and requires less capital investment than the chemical refining process (Cvengros, 1995). Physical refining of crude vegetable oil has several advantages over the traditional alkali refining process. For example, there are improvements with regard to simplicity of the procedure, product yield, energy conservation, and reduced generation of environmental pollutants. There are also many drawbacks as not all types of oils are suitable for this process. The use of high temperature and high vacuum often results in the formation of side products such as polymers and *trans* isomers (Sengupta and Bhattacharyya, 1992). Steam refining of certain high-FFA oils has been carried out in Europe for many years. The economics of deacidification by steam refining versus caustic refining normally favor steam refining only when high-FFA oils are processed (Sullivan, 1976).

A new supercritical fluid-based process has been developed which permits the counter-current refining (treatment) of extracted soybean oil to produce a refined feedstock suitable for direct deodorization. The process makes use of a packed vessel that facilitates interfacial contact between the high-pressure carbon dioxide and a liquid soybean oil in a counter-current mode (List *et al.*, 1993).

C. Bleaching

Bleaching of alkali-refined oils removes entrained soaps and reduces color bodies in the oil as well as decomposing hydroperoxides to secondary oxidation products; it is more appropriately referred to as adsorption treatment. Carotenoids, chlorophylls, residual soap, phospholipids, metals, and oxidized products are removed by bleaching. Bleaching often reduces the resistance of oils to rancidity, because some natural antioxidants are also removed with impurities. Three types of bleaching method can be used in the edible oil industry: adsorption bleaching; heat bleaching; and chemical oxidation.

The fat may be treated with various bleaching agents. Heated oils are treated with fuller's earth (a natural earthy material that will decolorize oils), activated carbon, or activated clays. Effective adsorption requires a large surface and highly specific surface area and the use of a very porous adsorbent. The channels by which molecules reach this surface depend on the type of molecule involved. The nature of the process must allow acceptably firm bonds, chemical or physical, between the bleaching clay and the adsorbate (Patterson, 1992). Therefore, to attain maximum bleaching performance, an efficient bleaching earth is required which has surfaces of the correct chemical composition and a pore distribution that is selectively attractive to the detrimental components present in crude TAG oils. A few types of bleaching agent are used in the vegetable oil industry, such as acid-activated bleaching earth, natural bleaching earth, activated carbon, synthetic silicates, and synthetic resins. Many impurities, including chlorophyll and carotenoid pigments, are adsorbed onto such agents and removed by filtration. Trace metal complexes, such as those of iron and copper, phosphatides, and oxidation products are also removed by the adsorptive effect of the bleaching earth and any residues of phosphoric acid are also removed during this stage. Usually, bleaching earth does not remove all the color-producing materials; many of these are removed by thermal destruction during the subsequent deodorization process.

Although batch atmospheric bleaching is still used to some extent in the USA, batch or continuous vacuum bleaching is generally practiced. Bleaching generally improves oil quality with respect to color, initial and aged flavor, and oxidative stability, but the process also has other less obvious effects, some of which are desirable and some undesirable. Several factors affect the degree of bleaching of an oil. Adsorption of pigments by the adsorbent and a reduction in color through oxidation of certain pigments are two favorable factors. Color increases brought about by oxidation of other pigments and stabilization of oxidized pigments against adsorption are unfavorable factors. Bleaching earth has been shown to catalyze such oxidation reactions. Vacuum bleaching minimizes these unfavorable events. Such reactions and color changes are complex, for example, oxidation and heat can bleach carotenoids; and these conditions may also encourage the formation of new pigments (Mounts, 1981). Natural and activated earths that have little or no acidity will produce little or no change in the FFA content of the neutralized oil. Some of the more acidic activated earths may increase FFA content by 0.05–0.10%, especially if the contact time is long or if moisture or soap is present.

Some pigments, such as the carotenes, become colorless if heated sufficiently. When many oils are heated to above 175°C, a phenomenon known as heat bleaching takes place. Apparently, heat decomposes some pigments, such as the carotenoids, and converts them to colorless compounds. However, this will leave the pigment molecules in the oil and may have adverse effects on oil quality. According to Gunstone and Norris (1983), if this oil comes into contact with air, colored degradation products such as chroman-5,6-quinones from γ-tocopherol may be formed and these are very difficult to remove. In addition, oxidation of carotenoids invariably affects the acylglycerols and may destroy natural antioxidants present in the oil. Consequently, oxidation-mediated bleaching is never used for edible oil, but is restricted to oils for technical purposes, such as soap making.

De-Smet and Alfa-Laval now offer counter-current bleaching and steam-agitated bleaching plants for crude rice bran oil processing. Industrial grade rice bran oil is often bleached by conventional chlorate bleaching and used in soap production.

Conjugation of oxidized polyunsaturated fatty acids during bleaching is known to occur. Oxidation of the oil before or during bleaching will promote conjugation; therefore, procedures such as deaerating both the initial oil and the adsorbent and vacuum bleaching would help to suppress the conjugation reaction; trans-isomerization has been shown to occur with acid-activated earths, but only at temperatures of 150°C and above, which far exceed those normally employed in the bleaching process (Mounts, 1981).

D. Deodorization

In the edible oil refinery, deodorization is the last process step used to improve the taste, odor, color, and stability of the oil by the removal of undesirable substances. All commercial deodorization, whether in continuous, semicontinuous, or batch units, is essentially a steam stripping of the oil for the removal of FFA and other volatile compounds.

During the process, peroxide decomposition products, color bodies, and their decomposition products are eliminated, and the contents of sterols, sterol esters, and tocopherols are reduced. The goal of the deodorization is to produce a finished oil that has a bland flavor, an FFA content of less than 0.05%, and a zero peroxide value. The modern commercial deodorizers are equipped with a pollution control system that consists of three steps: the deodorizer distillate recovery system; the closed circuit condensing water system; and the vapor scrubbing system. The distillate recovery system removes 80–90% of the distillate from the steam train before it reaches the condenser. The short-chain fatty acid (SCFA) fractions pass through the recovery unit and are recovered from the vapor scrubbing system. The deodorizer distillate is a concentrate of tocopherols and sterols and is a valuable source of these materials that could be used as value-added components in different applications.

Deodorization is primarily a high-temperature, high-vacuum, steam-distillation process. To produce a high-quality finished product, each deodorizer installation must deaerate the oil, heat the oil, steam strip the oil, and cool the oil, all with zero exposure to air. Steam deodorization is feasible because the flavor and odor compounds that are to be removed have appreciably greater volatility than TAGs present. Operation at high temperatures increases the volatility of these odoriferous compounds, and the introduction of stripping steam into the deodorizer greatly increases the rate at which these compounds are volatilized. Reduced pressure operation further aids in the removal of the odoriferous compounds and protects the oil from atmospheric oxidation. The deodorization process consists of blowing steam through heated oil held under a high vacuum. Small quantities of volatile components, responsible for taste and odor, distill, leaving a neutral, virtually odorless oil that is suitable for the manufacture of bland shortening or delicately flavored margarine. Originally, deodorization was a batch process, but increasingly, continuous systems are being used in which hot oil flows through an evacuated column counter-current to the upward passage of steam. In Europe, a deodorization temperature of 175–205°C is common, but in the USA higher temperatures of 235–250°C are usually employed. About 0.01% of citric acid is commonly added to deodorized oils to inactivate trace-metal contaminants such as soluble iron or copper compounds that would otherwise promote oxidation and the development of rancidity. The oil has to be given an adequate residence time for proper deodorization and the destruction of heat-labile pigments. The deodorized oil is counter-currently cooled first by the incoming oil and then by water to around 50°C. The cooled, deodorized oil is then passed through a polishing press to give it a transparent look.

Olive oil is invariably marketed in the undeodorized form. The natural flavor is an important asset, and olive oil commands a premium price in the market because of its distinctive flavor. The common cooking oils of Asia — soybean, rapeseed, peanut, sesame, and coconut — are consumed in their crude form as expressed from oilseeds. In contrast, deodorized oils are in particular demand in the USA and Europe. For many years the only important vegetable oil consumed in the USA was cottonseed oil, which in its crude form has such a strong and unpleasant flavor that further processing is necessary to render it suitable for consumption. Because of the widespread sale of neutrally flavored cottonseed oil products over many years, a general preference was developed for odorless and tasteless oils.

Another reason for the practice of deodorizing edible oils in Europe and the USA relates to differences in oil quality by Western and Eastern extraction techniques. In China and Southeast Asia, edible oils have been produced principally by small, relatively crude equipment. The yield of oil is relatively low, and a minimum amount of non-acylglycerol substances is expressed from the seed, with the result that the flavor of the oil is fairly mild. In Europe and the USA, oil extraction is carried out in large factories that operate on an extremely competitive basis. Very-high-pressure expression or solvent extraction is used, and to improve yields the seeds are heat-treated before extraction. Oils obtained in high yield under such conditions are stronger in flavor than oils prepared by low-pressure expression, and the refining and deodorizing steps are required to improve palatability. The improvement in yields more than compensates for the added costs of refining and deodorizing.

E. Winterization and Hydrogenation

Cloud formation during storage under low temperature is a concern for many edible oils. A winterization process is usually carried out to separate the crystallized part (waxes or triacylglycerols) from the oil and thus prevent turbidity of the oil in the winter or during cold storage. Waxes in general are esters of long-chain aliphatic fatty alcohols and long-chain fatty acids (LCFAs) that have 36–60 carbon atoms. They can be divided into soluble waxes, with chain lengths lower than C40; partially soluble waxes, with lengths between C40 and C42; and crystallizable waxes, with lengths of more than C44, which are responsible for the turbidity or sediment formation during oil storage and need to be removed by winterization. Winterization operations in the processing of edible oils are basically the separation of oils into two or more fractions with different melting points. In the winterization process, the oils are cooled in a simple way and kept at a low temperature for some time. The liquid and the solid fractions are generally separated by filtration. To separate the waxes from the oil by filtration, it is necessary to mix the winterized oil with a filter aid. This is to form a layer of the filtering material on the filter mats, giving a suitable drainage system. The wax-containing oil can then be pumped through the filter to remove waxes and the filter aid, as performed in winterization of sunflower oil (Raβ *et al.*, 2008). Winterization has a broad application in edible oil technology, including the production of cocoa butter equivalents from palm oil, palm-kernel oil, and shea fat, and from hydrogenated soybean and cottonseed oils (Kreulen, 1976).

Winterization can be difficult for certain oils such as palm olein because of the high content of the higher melting point crystallized part and the high viscosity of the oil that may hinder filtration (Leiboritz and Ruckenstein, 1984). In such cases, solvent winterization, with hexane or acetone, is used to reduce viscosity and facilitate crystallization of the waxes. Additives may also be used to help separate waxes from the oil. Calcium chloride and sodium lauryl sulfate solutions have been used in the winterization of rice bran oil to facilitate crystallization. The crystals that become dispersed in the water phase are separated by centrifugation (Ghosh, 2007).

When different degrees of hardness are required, the liquid oils may be subjected to a hydrogenation process to produce stable semi-solid plastic fats such as margarine and shortening. Processing conditions for hydrogenation can be modified to create a range of fats and oils with different melting and other characteristics. One such example is that of soybean oil, where relatively unstable linolenic acid (C18:3) is converted to more stable fatty acids. The hydrogenation process converts unsaturated fatty acids into their saturated or less unsaturated counterparts by the addition of hydrogen to the double bonds. Two types of process are involved in hydrogenation, namely saturation and isomerization. Saturation is where a molecule of hydrogen is added directly across a double bond in the presence of catalysts to give a saturated single bond. Hydrogenation is usually accompanied by the generation of *trans* fatty acids through the *cis* bond isomerization. *Trans* fatty acids formed during hydrogenation have been implicated in an increased risk of coronary heart disease, and have compelled consumers, health authorities, and manufacturers to reconsider the process. The process may also be accompanied by a shift in the double-bond location, leading to the formation of conjugated dienes and trienes.

The hydrogenation of edible oils is complex. Conventional hydrogenation uses a nickel catalyst and hydrogen gas at high temperatures of 140−230°C. It is a three-phase process with hydrogen in the gas phase, the liquid oil, and the solid catalyst, usually nickel. To achieve high reaction rates, an active catalyst as well as good mass transfer conditions between the gas and liquid and between the liquid and the catalyst are required. The high temperature produces high levels of *trans* fatty acids in the final oils. Low-temperature electrocatalytic hydrogenation may serve as an alternative method for the production of hydrogenated edible oils low in *trans* fatty acids. Low-temperature electrocatalytic hydrogenation uses an electrically conducting catalyst such as Raney nickel or platinum black as a cathode. Electrocatalytic hydrogenation has been used to produce a variety of organic compounds such as aromatic compounds, phenols, ketones, nitrocompounds, dinitriles, and glucose (Jang *et al.*, 2005). However, different catalysts are required for low-temperature hydrogenation since nickel catalysts are not very active below 120°C. Nickel can be replaced by precious metal catalysts, which are active at the low temperature of 70°C. Palladium, platinum, and ruthenium were the most potential precious metal catalysts for the hydrogenation of vegetable oils. Each metal catalyst has different characteristics in selectivity, reactivity, and *cis−trans* isomerization during hydrogenation. It has generally been accepted that platinum catalysts produce the least amount of *trans* fatty acids during hydrogenation (Jang *et al.*, 2005). Another hydrogenation method is supercritical fluid state hydrogenation, which improves the mass transfer of multiphase systems and reduces the formation of *trans* fat. A supercritical fluid state improves hydrogen transfer to the catalyst surface during hydrogenation by providing a good homogeneous phase. The efficacy of hydrogenation of oleochemicals and vegetable oils using supercritical carbon dioxide or propane as the solvent has been investigated (Macher *et al.*, 1999; Macher and Holmqvist, 2001; King *et al.*, 2001).

III. FAT MODIFICATION

Lipids have long been recognized for the richness they impart to foods as well as their satiety value in the diet. Lipid is an important component of the diet, because it provides both energy and essential fatty acids (EFAs). It is the most concentrated energy source in the diet, with an average energy value of 9 kcal/g compared to 4 kcal/g for carbohydrates and proteins. Lipid is an essential constituent of the membranes of every cell in the body. In addition to its role as a structural component of cells, lipid plays other roles in the body that include serving as an energy reserve, a regulator of body functions, and an insulator against heat loss. The role of dietary lipids in health and disease (notably, coronary heart disease, obesity, hyperlipidemia, and cancer) is one of the most active areas of research in modern food science, nutrition, and biochemistry. In this respect, the role of structured lipids should also be considered.

Structured lipids are TAGs or phospholipids containing SCFAs and/or medium-chain fatty acids (MCFAs) along with LCFAs located in the same glycerol molecule, and are produced by a chemical or an enzymatic process (Senanayake and Shahidi, 2000). These specialty lipids may be synthesized via direct esterification, acidolysis, alcoholysis, or interesterification reactions. However, the common methods reported in the literature for the synthesis of structured lipids are based on the reactions between two TAG molecules (interesterification) or between a TAG and an acid (acidolysis). These specialty lipids have been developed to fully optimize the benefit of various fatty acid moieties. Structured lipids have been reported to have beneficial effects by affecting a range of metabolic parameters including immune function, nitrogen balance, and improved lipid clearance from the bloodstream (Quinlan and Moore, 1993). Structured lipids are also produced to improve or change the physical and/or chemical properties of TAG. Research on structured lipids remains an interesting area that holds great promise for the future.

Nutraceutical is a term used to describe commodities procured from foods, but used in the medicinal form; they provide demonstrated physiological and health benefits beyond those ascribed to their nutritional value (Scott and Lee, 1996). These products may be incorporated into products that have the usual appearance of food to provide specific health benefits (Scott and Lee, 1996). Structured lipids can be designed for use as medical, nutraceutical, or functional food ingredients, depending on the form of use.

Lipids, mainly TAGs, can be modified to incorporate specific fatty acids of interest to achieve desired functionalities. Structured lipids may be synthesized via the hydrolysis of fatty acyl groups from a mixture of TAG followed by random re-esterification onto the glycerol backbone (Babayan, 1987). Various fatty acids, including different classes of saturated, monounsaturated, and n-3 and n-6 polyunsaturated fatty acids (PUFAs) or their mixtures may be used in this process, depending on the desired metabolic effect. Structured lipids containing MCFAs and LCFAs exhibit changes in their absorption rates because MCFAs are rapidly oxidized for energy while LCFAs are oxidized very slowly. These specialty lipids are structurally and metabolically different from the simple physical mixtures of medium-chain triacylglycerols (MCTs) and long-chain triacylglycerols (LCTs) (Akoh and Moussata, 1998).

A. Structured Lipids and their Fatty Acid Constituents

The unsaturated fatty acids from both the n-3 and n-6 families, as well as those from the n-9 family, may be included in structured lipids to promote health. The clinical advantages of these specialty lipids are derived from the combined effect of the short-, medium-, and long-chain fatty acids and the uniqueness of the structured lipid molecule itself. Many of these effects are due to the existing differences in the metabolic fate of the various fatty acids involved. Here, a brief description of SCFAs, MCFAs, and EFAs or PUFAs belonging to the n-3 and n-6 families is provided.

SCFAs are saturated fatty acids with two to four carbon atoms and include acetic acid (2:0), propionic acid (3:0), and butyric acid (4:0). These fatty acids are volatile and produced in the human gastrointestinal tract via bacterial fermentation of dietary carbohydrates (Stein, 1999). SCFAs are present in the diet in small amounts, for example, acetic acid in vinegar and butyric acid in bovine milk and butter. They may also be present in fermented foods. In humans, SCFAs contribute to 3% of total energy expenditure (Hashim and Babayan, 1978) and these are more easily absorbed in the stomach and provide fewer calories than MCFAs and LCFAs. Thus, acetic, propionic, and butyric acids have caloric values of 3.5, 5.0, and 6.0 kcal/g, respectively.

In certain nutritional applications, the use of SCFAs as alternatives to their MCFA and LCFA counterparts has been of interest. SCFAs are easily hydrolyzed from TAG and rapidly absorbed by the intestinal mucosa (Ruppin et al., 1980). These fatty acids go directly into the portal vein for transportation to the liver, where they are broken down to acetate via β-oxidation. The acetate can then be metabolized for energy or use in new fatty acid synthesis. SCFAs affect gastrointestinal function by stimulating pancreatic enzyme secretion (Harada and Kato, 1983) and increasing sodium and water absorption in the gut (Roediger and Rae, 1982).

MCFAs are saturated fatty acids with six to 12 carbon atoms (Senanayake and Shahidi, 2000) and are commonly found in tropical fruit oils such as those of coconut and palm kernel (Bell et al., 1991). For example, coconut oil naturally contains some 65% MCFA (Young, 1983). One of the first medical foods developed, as an alternative to conventional lipids, was based on MCTs. MCTs serve as an excellent source of MCFAs for the production of structured and specialty lipids. Pure MCTs have an energy value of 8.3 kcal/g. However, they do not provide EFAs (Heird et al., 1986; Lee and Hastilow, 1999). MCFAs are more hydrophilic than their LCFA counterparts, and hence solubilization as micelles is not a prerequisite for their absorption (Ikeda et al., 1991).

MCTs can also be directly incorporated into mucosal cells without hydrolysis and may be readily oxidized in the cell. MCTs pass directly into the portal vein and are readily oxidized in the liver to serve as an energy source. Thus, they are less likely to be deposited in the adipose tissues (Megremis, 1991) and are more susceptible to oxidation in tissues (Mascioli *et al.*, 1987).

MCTs are liquid or solid products at room temperature. They have a smaller molecular size, lower melting point, and greater solubility than their LCFA counterparts. These characteristics account for their easy absorption, transport, and metabolism compared to LCT (Babayan, 1987). MCTs are hydrolyzed by pancreatic lipase more rapidly and completely than are LCTs (Bell *et al.*, 1991). They may be directly absorbed by the intestinal mucosa with minimum pancreatic or biliary function. They are transported predominantly by the portal vein to the liver for oxidation (Heydinger and Nakhasi, 1996) rather than through the intestinal lymphatics. In addition, MCFAs are more rapidly oxidized to produce acetyl-coenzyme A and ketone bodies, and are independent of carnitine for entry into mitochondria.

MCTs need to be used with LCTs to provide a balanced nutrition in enteral and parenteral products (Ulrich *et al.*, 1996; Haumann, 1997a). In many medical foods, a mixture of MCTs and LCTs is used to provide both rapidly metabolized and slowly metabolized fuel as well as EFAs. Clinical nutritionists have taken advantage of the simpler digestion of MCTs to nourish individuals who cannot utilize LCTs owing to fat malabsorption. Thus, patients with certain diseases (Crohn's disease, cystic fibrosis, colitis, enteritis, etc.) have shown improvement when MCTs are included in their diet (Kennedy, 1991). MCTs are also increasingly used to feed critically ill or septic patients who presumably gain benefits in the setting of associated intestinal dysfunction. MCTs may be used in confectionery and in other functional foods as carriers for flavors, colors, and vitamins (Megremis, 1991). MCTs have clinical applications in the treatment of lipid malabsorption, maldigestion, obesity, and deficiency of the carnitine system (Bach and Babayan, 1982).

The EFAs are PUFAs belonging to the *n*-3 or the *n*-6 fatty acid families. The biological activity of PUFAs depends on the position of the double bond in the molecule nearest to the methyl end of the chain, being *n*-3 when the double bond is located between the third and fourth carbon atoms and *n*-6 when it is between the sixth and seventh carbon atoms. The parent compounds of the *n*-3 and *n*-6 groups of fatty acids are linoleic acid and α-linolenic acid (ALA), respectively. Within the body, these parent compounds are metabolized by a series of alternating desaturation (in which an extra double bond is inserted by removing two hydrogen atoms) and elongation (in which two carbon atoms are added) (Figure 9.1), but only to a limited extent of up to 4–5% for producing long-chain PUFAs (Plourde and Cunnane, 2007). The enzymes metabolizing linoleic acid and ALA are thought to be identical (Horrobin, 1990).

The role of EFAs as precursors of a wide variety of short-lived hormone-like substances called eicosanoids has received much attention. These 20-carbon endogenous biomedical mediators are derived from EFAs, notably arachidonic acid and dihomo-γ-linolenic acid (DGLA) from the *n*-6 family, and eicosapentaenoic acid (EPA) from the *n*-3 family (Branden and Carroll, 1986). Eicosanoids include prostaglandins, prostacyclins, thromboxanes, leukotrienes, and hydroxy fatty acids, which play a role in regulating the cell-to-cell communication involved in cardiovascular, reproductive, respiratory, renal, endocrine, skin, nervous, and immune system actions. Arachidonic acid is derived from linoleic acid, which gives rise to series-2 prostaglandins, series-2 prostacyclins, series-2 thromboxanes, and series-4 leukotrienes. These end products of *n*-6 fatty acid metabolism induce inflammation and immunosuppression. Prostanoids (the collective name for prostaglandins, prostacyclins, and thromboxanes) of series-1 and leukotrienes of series-3 are produced from DGLA. When *n*-3 fatty acids are processed in the eicosanoid cascade, series-3 prostaglandins, series-3 prostacyclins, series-3 thromboxanes, and series-5 leukotrienes are formed.

The biological activities of the eicosanoids derived from *n*-3 fatty acids differ from those produced from *n*-6 fatty acids. For example, series-2 prostaglandins formed from arachidonic acid may impair the immune function while series-3 prostaglandins produced from EPA ameliorate immunodysfunction. Thromboxane A_2 produced from arachidonic acid is a potent vasoconstrictor and platelet aggregator, whereas thromboxane A_3 synthesized from EPA is a mild vasoconstrictor and has shown antiaggregatory properties (von Schacky, 2003). Furthermore, *n*-3 fatty acids competitively inhibit the formation of eicosanoids derived from the *n*-6 family of fatty acids.

The *n*-3 fatty acids, such as ALA, EPA, and docosahexaenoic acid (DHA), have a myriad of health benefits related to cardiovascular disease, inflammation, allergies, cancer, and immune and renal disorders. Bang and Dyerberg (1972, 1986) suggested that the relatively high *n*-3 fatty acid content (especially EPA and DHA) of the diet of Inuits was related to their lower incidence of cardiovascular disease. Research has shown that DHA is essential for the proper function of the central nervous system and visual acuity of infants. The *n*-3 fatty acids are essential for normal growth and development throughout the life cycle of humans and therefore should be included in the diet. Fish and

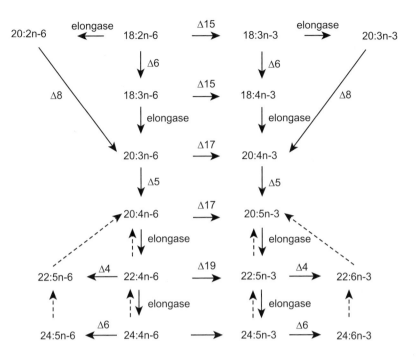

FIGURE 9.1 Biosynthesis of *n*-3 fatty acids.

marine oils are rich sources of *n*-3 fatty acids, especially EPA and DHA. Most fish oils, such as cod liver, menhaden, and sardine oils, contain approximately 30% EPA and DHA.

The *n*-6 fatty acids exhibit various physiological functions in the human body. The main functions of these fatty acids are related to their roles in the membrane structure and in the biosynthesis of short-lived derivatives (eicosanoids) which regulate many aspects of cellular activity. The *n*-6 fatty acids are involved in maintaining the integrity of the water-impermeable barrier of the skin. They are also involved in the regulation of cholesterol transport in the body.

γ-Linolenic acid (GLA) has shown therapeutic benefits in a number of diseases and syndromes, notably atopic eczema, cyclic mastalgia, premenstrual syndrome, cardiovascular disease, inflammation, diabetes, and cancer (Horrobin, 1990). While arachidonic acid is found in meats, egg yolk, and human milk, GLA is present in oats, barley, and human milk. GLA is also found in higher amounts in plant seed oils such as those from borage, evening primrose, and blackcurrant (Gunstone, 1992). Algae such as *Spirulina* and various species of fungi also seem to be desirable sources of GLA (Carter, 1988).

B. Production of Structured and Specialty Lipids

Production of specialty lipids used for confectionary fat formulations and nutritional applications may be carried out using enzyme-catalyzed reactions. In the area of confectionary fats, interesterification of high oleic sunflower oil and stearic acid using immobilized *Rhizomucor miehei* lipase produces mainly 1,3-distearoyl-2-monoolein (StOSt) (Macrae, 1983). Other reactants may also be used for production of specialty confectionary fats. In particular, there are many reports on the enzymatic interesterification of mixtures of palm oil fractions and stearic acid or its esters to produce fats containing high concentrations of StOSt and 1-palmitoyl-2-oleoyl-3-stearoylglycerol (POSt) (Macrae, 1983). These products are the main components of cocoa butter, and hence enzymatic interesterification may produce fats with similar compositions and physical properties to cocoa butter (Macrae, 1985).

Enzyme-catalyzed reactions have been used to synthesize a human milk fat substitute for use in infant formula (Quinlan and Moore, 1993). An acidolysis reaction of a mixture of tripalmitin and unsaturated fatty acids, using an *sn*-1,3-specific lipase as a biocatalyst, afforded TAG derived entirely from vegetable oils rich in palmitate in the 2-position with unsaturated fatty acyl groups in the *sn*-1 and *sn*-3 positions. These TAG closely mimic the fatty acid distribution found in human milk fat, and when they are used in infant formula instead of conventional fats, the presence of palmitate in the *sn*-2 position of the TAG has been shown to improve digestibility of the fat and absorption of other important nutrients such as calcium. Hamam and Shahidi (2008) investigated the effect of chain length, number of double bonds, location and geometry of double bonds, reaction conditions, and reactivity of different lipases on the incorporation of selected LCFAs into TAGs, such as trilinolein (tri C18:2) and trilinolenin (tri C18:3). The conditions for the synthesis of structured lipids were also optimized using response surface methodology (RSM) (Hamam and Shahidi, 2006a). Structured lipids having LCFAs at the *sn*-2 and medium-chain caprylic acid (8:0) at their *sn*-1,3-positions from corn oil were also prepared and the effects of substrate mole ratio, amount of enzyme, and reaction time on the incorporation of caprylic acid into corn oil were optimized (Ozturk *et al.*, 2010).

Enzyme-assisted reactions may also be used for the production of common lipid commodities such as margarine hardstocks and cooking oils. When non-specific lipases such as *Candida cylindraceae* and *Candida antarctica* are used as biocatalysts for the interesterification of oil blends, the TAG products are very similar to those obtained by chemical interesterification (Macrae, 1983), but the process is not economically attractive. However, enzymatic reactions may best be employed for the production of fats and oils containing nutritionally important PUFAs, such as EPA and DHA. For example, various vegetable and fish oils have been enriched with EPA and DHA by enzyme-catalyzed reactions (Senanayake and Shahidi, 1999a, b, 2001). The use of this technique to produce structured lipids with MCFAs and PUFAs located specifically in either the *sn*-2 or *sn*-1,3 position of the TAG has been described. Enzymatic processes are particularly suitable for the production and modification of lipids containing PUFAs, owing to the instability of the fatty acids involved and their susceptibility to oxidation under harsh chemical processing conditions. Wang and Shahidi (2010) examined the effect of chemical randomization on the stability of menhaden and seal blubber oils.

Interesterification of blends of palm and fully hydrogenated canola oils, and cottonseed and hydrogenated soybean oils using *sn*-1,3-specific lipases as catalysts, gave fats with a low *trans* fatty acid content that were effective as margarine hardstock (Mohamed and Larsson, 1994). Reaction of mixtures of palm stearine and lauric fats, using immobilized *R. miehei* as a catalyst, also produced fats that were functional as margarine hardstocks (Posorske *et al.*, 1988). With these enzymatically interesterified fats, margarine could be formulated without using hydrogenated fats.

In addition to the food applications, the process of interesterification or acidolysis could provide medical or nutritional benefits. For example, structured lipids may be used to improve the nutrition profile of certain TAGs (Osborn and Akoh, 2002). Saturated fatty acids at the *sn*-2 position are beneficial in terms of providing increased caloric intake through infant formula and enteral supplements (Decker, 1996). Structured lipids also have beneficial effects on a range of metabolic parameters including immune function, nitrogen balance, and improved lipid clearance from the bloodstream (Senanayake and Shahidi, 2005).

C. Structured Lipids Containing *n*-3 Polyunsaturated Fatty Acids

Plant seed oils such as those from borage, evening primrose, and blackcurrant are predominant sources of GLA (18:3*n*-6). GLA has been used in the treatment of certain skin disorders as well as a variety of other pathological conditions. The *n*-3 PUFAs have potential in the prevention of cardiovascular disease, arthritis, hypertension, immune and renal disorders, diabetes, and cancer (Senanayake and Shahidi, 2000). Structured lipids containing both GLA and *n*-3 PUFAs may be of interest because of their desired health benefits. Enzymatic synthesis of structured lipids containing GLA, EPA, and DHA in the same glycerol molecule, using selected oils such as those of borage and evening primrose as the main substrates, has been reported (Senanayake and Shahidi, 1999a, 2001). In these studies, three microbial enzymes, namely lipases from *C. antarctica* (Novozym-435), *Mucor miehei* (Lipozyme-IM), and *Pseudomonas* sp. (Lipase PS-30) were used as biocatalysts with free EPA and DHA as acyl donors. A higher incorporation of EPA + DHA (34.1%) in borage oil was obtained with *Pseudomonas* sp. lipase, compared to 20.7% and 22.8% EPA + DHA, respectively, with *C. antarctica* and *M. miehei* lipases (Table 9.1). Similarly, in evening primrose oil *Pseudomonas* sp. lipase afforded the highest degree of EPA + DHA incorporation (31.4%), followed by lipases from *M. miehei* (22.8%) and *C. antarctica* (17.0%) (Table 9.2). The modified borage and evening primrose oils thus obtained may be useful in the treatment of certain clinical disorders.

The incorporation of *n*-3 fatty acids into the acylglycerols of borage oil was also reported by Ju *et al.* (1998). In this work, borage oil was selectively hydrolyzed using immobilized *Candida rugosa* lipase and the resultant product was then used with *n*-3 fatty acids in the acidolysis reaction. The total content of *n*-3 and *n*-6 fatty acids in acyl-glycerols was 72.8% following acidolysis. The contents of GLA, EPA, and DHA in the structured lipid so prepared were 26.5%, 19.8%, and 18.1%, respectively. The *n*-3/*n*-6 ratio was increased from 0 to 1.09 following the acidolysis reaction. In another study, the fatty acid composition of borage oil was modified using EPA ethyl ester with an immobilized lipase from *C. antarctica* (Akoh and Sista, 1995). The highest incorporation (31%) was obtained with 20% *C. antarctica* lipase. At a substrate mole ratio of 1:3, the ratio of *n*-3 to *n*-6 fatty acids was 0.64. Under similar

TABLE 9.1 Fatty Acid Composition of Borage Oil Triacylglycerols Before and After Modification with EPA and DHA

Fatty Acid (wt%)	Before Modification	After Modification		
		Lipozyme-IM	Novozym-435	Lipase PS-30
16:0	9.8	6.9	7.3	5.5
18:0	3.1	2.6	2.8	2.0
18:1n-9	15.2	11.6	13.3	10.4
18:2n-6	38.4	28.8	31.2	26.1
18:3n-6	24.4	22.3	19.3	17.6
20:1n-9	4.1	3.1	3.3	2.6
20:5n-3	—	11.8	8.4	24.1
22:1n-11	2.5	1.8	2.1	1.7
24:1	1.5	—	—	—
22:6n-3	—	11.0	12.3	10.0

EPA: eicosapentaenoic acid; DHA: docosahexaenoic acid.

conditions, the n-3 fatty acid content (up to 43%) of evening primrose oil was increased, with a corresponding increase in the n-3/n-6 ratio from 0.01 to 0.6. Sridhar and Lakshminarayana (1992) modified the fatty acid composition of groundnut oil by incorporating EPA and DHA using an sn-1,3-specific lipase from M. miehei as the biocatalyst. The modified groundnut oil had 9.5% EPA and 8.0% DHA.

Incorporation of EPA and capric acid (10:0) into borage oil using two immobilized lipases, SP435 from C. antarctica and IM60 from R. miehei, as biocatalysts has also been reported (Akoh and Moussata, 1998). Higher incorporation of EPA (10.2%) and 10:0 (26.3%) was obtained with IM60 lipase, compared to 8.8% and 15.5%, respectively, with SP435 lipase. In another study, Huang et al. (1994) incorporated EPA into crude melon seed oil by two immobilized lipases, IM60 from M. miehei and SP435 from C. antarctica, as biocatalysts. Higher EPA incorporation was obtained using EPA ethyl ester than using EPA itself for both enzyme-catalyzed reactions. Furthermore, immobilized lipases, IM60 from M. miehei and SP435 from C. antarctica, were used to modify the fatty acid composition of soybean oil by incorporation of n-3 fatty acids (Huang and Akoh, 1994). The transesterification reaction was carried out with FFA and ethyl esters of EPA and DHA as acyl donors. With free EPA as the acyl donor,

TABLE 9.2 Fatty Acid Composition of Evening Primrose Oil Triacylglycerols Before and After Modification with EPA and DHA

Fatty Acid (wt%)	Before Modification	After Modification		
		Lipozyme-IM	Novozym-435	Lipase PS-30
16:0	6.2	4.5	4.9	3.6
18:0	1.7	1.3	1.4	1.0
18:1n-9	8.7	6.4	7.1	6.2
18:2n-6	73.6	55.1	60.6	49.7
18:3n-6	9.9	9.7	9.0	7.5
20:5n-3	—	12.3	8.4	22.0
22:6n-3	—	10.5	8.6	9.4

EPA: eicosapentaenoic acid; DHA: docosahexaenoic acid.

M. miehei lipase gave a higher incorporation of EPA than *C. antarctica* lipase. However, when ethyl esters of EPA and DHA were the acyl donors, *C. antarctica* lipase gave a higher incorporation of EPA and DHA than *M. miehei* lipase.

Marine oils can also be modified to incorporate MCFAs to improve their nutritional properties. Specific structured lipids with PUFA residues at the *sn*-2 position and MCFA residues at the *sn*-1,3 positions have high potential in biomedical and nutraceutical applications. In this form, the PUFA residues are protected against oxidation by the two saturated MCFA residues. In addition, PUFAs are better absorbed in the intestinal tract as 2-monoacylglycerol (2-MAG) upon hydrolysis by pancreatic lipase. MCFAs are absorbed efficiently and are a quick source of energy without being deposited in the adipose tissues. An immobilized *sn*-1,3-specific lipase from *M. miehei* was used to incorporate capric acid (10:0; an MCFA) into seal blubber oil. Upon enzymatic reaction, the fatty acid composition of seal blubber oil was modified and under optimum reaction conditions, a structured lipid containing 27.1% capric acid, 2.3% EPA, and 7.6% DHA was obtained (Table 9.3). Positional distribution of fatty acids in the structured lipid revealed that *M. miehei* lipase incorporated capric acid predominantly at the *sn*-1,3-positions of TAG molecules (Senanayake and Shahidi, 2002). In another study, structured lipids were successfully produced using seal blubber oil or menhaden oil with GLA and using lipase PS-30 from *Pseudomonas* sp. For the acidolysis reaction, optimum reaction conditions were 1:3 mole ratio of oil to GLA, reaction temperature of 40°C, reaction time of 24 hours, and enzyme concentration of 500 enzyme activity units/g oil. Under these conditions, incorporation of GLA into seal blubber oil was 37.1%, and 39.6% incorporation was achieved when menhaden oil was used (Spurvey *et al.*, 2001). Incorporation of capric acid (10:0) into fish oil TAG using immobilized lipase from *R. miehei* (IM 60) has also been reported (Jennings and Akoh, 1999). The fish oil concentrate used contained 40.9% EPA and 33.0% DHA. After a 24-hour incubation in hexane, there was an average of 43% incorporation of capric acid into fish oil, while the content of EPA and DHA decreased to 27.8% and 23.5%, respectively. Furthermore, capric acid (10:0) and EPA were incorporated into borage oil using lipase from *C. antarctica* and *R. miehei* as biocatalysts. Higher incorporation of EPA (10.2%) and 10:0 (26.3%) was obtained with *R. miehei* lipase, compared to 8.8% and 15.5%, respectively, with *C. antarctica* lipase.

A single-cell oil (produced by a marine microorganism, *Schizochytrium* sp.) containing docosapentaenoic acid (DPA; 22:5*n*-6) and DHA and caprylic acid (8:0) were used to produce a structured lipid using lipases from *R. miehei* and *Pseudomonas* sp. (Iwasaki *et al.*, 1999). The products contained caprylic acid at the *sn*-1 and *sn*-3 positions and DHA or DPA at the *sn*-2 position of glycerol. When *Pseudomonas* sp. was used, more than 60% of total fatty acids (palmitic, myristic, pentadecanoic, stearic, DHA, and DPA) in single-cell oil was exchanged with caprylic acid. The amount of TAG containing two caprylic acid and one DHA or DPA residue was 36%. With *R. miehei* lipase, the

TABLE 9.3 Enzymatic Modification of Seal Blubber Oil with Capric Acid to Produce Structured Lipids

Fatty Acid (wt%)	Before Modification	After Modification
10:0	—	27.1
14:0	3.4	2.7
14:1	1.0	0.8
16:0	5.0	3.7
16:1*n*-7	15.1	11.9
18:1*n*-9 and *n*-11	26.4	19.3
18:2*n*-6	1.3	1.7
20:1*n*-9	15.0	9.1
20:5*n*-3	5.4	2.3
22:1*n*-11	3.6	1.9
22:5*n*-3	4.9	3.0
22:6*n*-3	7.9	7.6

incorporation of caprylic acid was only 23%. A large amount of DHA and DPA remained unexchanged with this enzyme, so that the resulting oil was rich in TAG species containing two or three DHA or DPA residues (46%). The difference in the degree of acidolysis by the two enzymes was suggested to be due to their different selectivity towards DPA and DHA, as well as the difference in their positional specificities.

Hamam and Shahidi (2005, 2006b) prepared a number of structured lipids containing long-chain n-3 PUFAs and studied their oxidative stability. They found that a loss of endogenous tocopherols in the oil during the enzymatic esterification was responsible for the compromised oxidative stability of the structured lipids. It was further demonstrated that tocopherols were esterified with the FFAs in the medium and hence provided support for the mechanism by which tocopherols were lost during the process (Hamam and Shahidi, 2006b).

D. Production of *n*-3 Polyunsaturated Fatty Acid Concentrates from Marine and Algal Oils

Concentrates of n-3 fatty acids from marine and algal oils may be obtained using a number of techniques. The resultant concentrates may be in the form of FFAs, simple alkyl esters, or acylglycerols. Techniques that may be used include urea complexation, low-temperature crystallization, chromatography, distillation, supercritical fluid extraction, and enzyme-assisted reactions (Shahidi and Wanasundara, 1998).

For the preparation of PUFA concentrates on a large scale, each of the above physical and chemical methods has some disadvantages in terms of low yield, a requirement for large volumes of solvent or sophisticated equipment, a risk of structural changes in the fatty acid products, or high operation costs. Lipases work under mild temperature and pH conditions (Gandhi, 1997), and hence their potential use for the enrichment of PUFAs in oils is of much interest. Lipases (EC 3.1.1.3) are enzymes that catalyze the hydrolysis, esterification, interesterification, acidolysis, and alcoholysis reactions. The common feature among lipases is that they are activated by an interface. Lipases have been used for many years to modify the structure and composition of food lipids. Lipases act on neutral lipids and generally hydrolyze the esters of PUFAs at a slower rate than those of more saturated fatty acids (Villeneuve and Foglia, 1997). This relative substrate specificity has been used to increase the concentration of n-3 PUFAs in seal blubber and menhaden oils by subjecting them to hydrolysis by a number of microbial lipases (Wanasundara and Shahidi, 1998). The concentration of n-3 fatty acids by enzyme-assisted reactions involves mild reaction conditions and provides an alternative to the traditional concentration methods such as distillation and chromatographic separation. Furthermore, concentration via enzymatic means may also produce n-3 fatty acids in the acylglycerol form, which is nutritionally preferred.

In general, PUFAs in TAG molecules are resistant to *in vitro* hydrolysis by pancreatic enzymes. Microbial lipases have also been found to discriminate against PUFAs in enzyme-catalyzed reactions. Therefore, it is possible to concentrate n-3 fatty acids of marine oils. Preparation of n-3 PUFA-enriched acylglycerols from seal blubber and menhaden oils has been attempted via enzymatic hydrolysis (Wanasundara and Shahidi, 1998). Several microbial enzymes, namely lipases from *C. cylindraceae*, *Rhizopus oryzae*, *Pseudomonas* sp., *Chromobacterium viscosum*, *Geotrichum candidum*, *Rhizopus niveus*, *M. miehei*, and *Aspergillus niger*, were screened to enrich n-3 PUFAs in both oils. All microbial lipases tested were able to hydrolyze fatty acids in both oils, but at different rates (Table 9.4). Among the enzymes examined, *C. cylindraceae* was found to be the most effective biocatalyst for the production of n-3 fatty acid concentrates from seal blubber oil. However, in menhaden oil, lipase from *R. oryzae* gave the highest degree of hydrolysis. Other lipases studied gave lower degrees of hydrolysis than *C. cylindraceae* and *Rhizopus oryzae* lipases in both oils. At a given hydrolysis time, all lipases had considerably higher degrees of hydrolysis in seal blubber oil than in menhaden oil (data not shown). This difference may be due to the presence of higher amounts of PUFAs, especially EPA and DHA, in menhaden oil than in seal blubber oil, which exhibit resistance to enzymatic hydrolysis.

Among the enzymes tested, lipase from *C. cylindraceae* appeared to be the most active biocatalyst in increasing the contents of total n-3 fatty acids in the non-hydrolyzed fraction of both seal blubber and menhaden oils (Table 9.4). In menhaden oil, the total n-3 fatty acid content increased from 30% (original oil) to 46.4% and 46.1% after 75 hours of hydrolysis by lipases from *R. oryzae* and *C. cylindraceae*, respectively. In seal blubber oil the maximum increase in total n-3 fatty acids, from 20.2% to 45.0%, was reached when lipase from *C. cylindraceae* was used under similar experimental conditions (Table 9.4). The use of enzymes to produce n-3 fatty acid concentrates has an advantage over traditional methods of concentration (chromatographic separation, molecular distillation, etc.) since such methods involve extremes of pH and high temperatures which may partially destroy the natural all-*cis* n-3 PUFAs by oxidation and by *cis*—*trans* isomerization or double-bond

TABLE 9.4 Enzymatic Hydrolysis of Seal Blubber and Menhaden Oils

	Seal Blubber Oil		Menhaden Oil	
Enzyme Source	Hydrolysis (%)[a]	Total n-3 Fatty Acid Content (%)	Hydrolysis (%)[a]	Total n-3 Fatty Acid Content (%)
Candida cylindraceae	84	45	60	46
Rhizopus oryzae	77	34	70	46
Pseudomonas sp.	70	26	53	39
Mucor miehei	59	29	51	40
Chromobacterium viscosum	52	25	50	38
Geotrichum candidum	40	30	33	43
Rhizopus niveus	38	25	25	37
Aspergillus niger	26	23	5	35

[a]After 75 hours of hydrolysis.

migration. Therefore, mild conditions used in enzymatic hydrolysis provide a promising alternative that could also save energy and increase product selectivity.

The mechanism of resistivity of lipases towards the long-chain n-3 PUFAs in marine oils has been demonstrated by Bottino *et al.* (1967). The presence of carbon—carbon *cis*-double bonds in the fatty acids results in bending of the chains. Therefore, the terminal methyl group of the fatty acid lies close to the ester bond, which may cause a steric hindrance effect on lipases. The high bending effect of EPA and DHA is due to the presence of five and six double bonds, respectively, which enhance the steric hindrance effect; therefore, enzymes cannot reach the ester linkage of these fatty acids and glycerol. However, saturated and monounsaturated fatty acids do not present any barriers to lipase approach and could be easily hydrolyzed. Therefore, the fatty acid selectivity of a lipase for EPA and DHA allows separation of these fatty acids from the remaining fatty acids in marine oils, which is very important in the production of n-3 fatty acid concentrates. In addition, lipases have frequently been used to discriminate against EPA and DHA in concentrates containing both of these fatty acids; this enables the preparation of both EPA- and DHA-enriched concentrates.

Much attention has also been paid to using microbial lipases to produce n-3 fatty acid concentrates either by hydrolysis or by transesterification reactions of marine oils. Tanaka *et al.* (1992) used a number of microbial lipases (from *C. cylindraceae*, *R. oryzae*, *Pseudomonas* sp., *Ch. viscosum*, and *A. niger*) to hydrolyze tuna oil and found that *C. cylindraceae* lipase was the most effective biocatalyst in increasing the DHA content in the concentrates. This enzyme was able to increase the DHA content in the non-hydrolyzed fraction to three times that present in the original tuna oil; however, other lipases did not increase the DHA content in the oil. Shimada *et al.* (1994) reported that hydrolysis of tuna oil by lipase from *G. candidum* increased the content of both EPA and DHA from 32.1% to 57.5%. In another study, Hoshino *et al.* (1990) used several lipases for selective hydrolysis of cod liver and sardine oils. The best hydrolysis results were obtained for the non-specific *C. cylindraceae* lipase and *sn*-1,3 specific *A. niger* lipase, but none of the enzymes were able to raise the EPA content of the acylglycerols to any great extent. However, over 50% of total fatty acids were produced when these two enzymes were used.

The n-3 PUFA concentrates obtained using urea complexation and/or other concentration methods (Shahidi and Wanasundara, 1998) may be reacted with glycerol by enzymatic means to produce concentrates in the acylglycerol form. This form of n-3 fatty acid is preferred over FFAs and their alkyl esters. While TAG could be prepared, the presence of partial acylglycerols cannot be easily avoided. Various enzymes are known to catalyze the formation of acylglycerols. Osada *et al.* (1992) used lipases from *Ch. viscosum* and *C. cylindraceae* for direct esterification of glycerol with individual FFAs, including EPA and DHA. The lipase from *Ch. viscosum* was superior to *C. cylindraceae* lipase and resulted in 89—95% incorporation. With the latter lipase, 71—75% incorporation was obtained for all fatty acids, except for DHA, which reached 63% incorporation.

E. Low-Calorie Structured and Specialty Lipids

The high incidence of obesity due to overconsumption of lipids or high-calorie foods has encouraged researchers to develop products that are only partially digestible and provide fewer calories than regular fats and oils while being similar to conventional lipids in other aspects. Thus, the production of low-calorie lipids, which are characterized by a combination of SCFAs and/or MCFAs and LCFAs in the same glycerol backbone, is an active area of research and development in the field of structured and specialty lipids. Interest in these types of products emerged from the fact that they contain 5–7 kcal/g compared to the 9 kcal/g of conventional fats and oils, because of the lower caloric content of SCFAs or MCFAs compared to their long-chain counterparts. Reduced-calorie specialty lipids are designed for use in baking chips, coatings, dips, bakery and dairy products, or as cocoa butter substitutes. Currently, such products are synthesized by random chemical interesterification between a short-chain triacylglycerol (SCT) and an LCT, typically a hydrogenated vegetable oil such as soybean or canola oil (Smith *et al.*, 1994). Examples of commercially available low-calorie lipids include Caprenin®, Salatrim®, and Neobee®, as briefly reviewed here.

Caprenin is composed of one molecule of a very long-chain saturated fatty acid, behenic acid (C22:0), and two molecules of medium-chain saturated fatty acids, caprylic acid (C8:0) and capric acid (C10:0), and is a commercially available reduced calorie structured lipid. It provides 5 kcal/g compared to 9 kcal/g of conventional fats and oils. This product was first produced by Procter & Gamble Company. The constituent fatty acids for Caprenin synthesis are obtained from natural food sources. For example, caprylic and capric acids are obtained by fractionation of palm kernel and coconut oils while behenic acid is produced from rapeseed oil. Behenic acid, being a very long-chain saturated fatty acid, is poorly absorbed regardless of its position on the glycerol moiety. The MCFAs provide fewer calories than absorbable LCFAs. Caprenin displays functional characteristics similar to cocoa butter and can be used as a cocoa butter substitute in selected confectionary products. It is digested, absorbed, and metabolized by the same pathway as other TAGs (Artz and Hansen, 1996). Caprenin is a liquid or semi-solid product at room temperature, has a bland taste and is fairly heat stable. A petition by Proctor & Gamble to the United States Food and Drug Administration (FDA) for Caprenin for use as generally recognized as safe (GRAS) in soft candy bars and in confectionary coatings for nuts, fruits, and cookies was made, but subsequently withdrawn.

Salatrim, another reduced calorie structured lipid, is composed of a mixture of very short-chain fatty acids (C2:0–C4:0) and LCFAs (predominantly C18:0) (Smith *et al.*, 1994). The SCFAs are chemically transesterified with vegetable oils such as highly hydrogenated canola or soybean oil. The very short-chain fatty acids reduce the caloric value to approximately 5 kcal/g and LCFAs provide lipid functionality. Salatrim was developed by Nabisco Foods Group and is now marketed under the brand name Benefat™ by Cultor Food Science, Inc. It has the taste, texture, and functional characteristics of conventional fats. It may display different melting profiles depending on the amounts of SCFA and LCFA used in its chemical synthesis. Reduced fat baking chips are one of the products in the market that contain Salatrim and were introduced to the market in 1995 by Hershey Food Corporation. Salatrim received FDA GRAS status in 1994 and can also be used as a cocoa butter substitute. It was intended for use in chocolate-flavored coatings, chips, caramel, fillings for confectionery and baked goods, peanut spreads, savory dressings, dips and sauces, and dairy products (Kosmark, 1996).

Neobee, another caloric reduced fat, is composed of capric and caprylic acids and produced by Stepan Company. This class of specialty lipids includes different products. For example, Neobee 1053 and Neobee M-5 contain both capric and caprylic acids, while Neobee 1095 is made up of only capric acid (Heydinger and Nakhasi, 1996). Neobee 1095 is a solid product. Therefore, this product may be suitable in certain applications which require solid fats. Neobee 1814 is an MCT derivative made by interesterification of MCT with butter oil (Babayan *et al.*, 1990); it contains half of the long-chain saturated fatty acids found in conventional butter oil and is suitable to replace butter oil in a variety of applications. Neobee 1814 may serve as a flavor carrier and functions as a textural component for low-fat food products (Heydinger and Nakhasi, 1996).

F. Modified Lipids in Health Promotion and Disease Risk Reduction

Various fatty acids (*n*-3 and *n*-6 series) may be incorporated into structured and modified lipids to promote health and nutrition. These fatty acids have shown health benefits related to various disease conditions such as arthritis, thrombosis, cardiovascular disease, diabetes, and cancer (Horrobin, 1990; Vartak *et al.*, 1997; Senanayake and Shahidi, 2000).

Diets rich in *n*-3 and *n*-6 fatty acids have shown beneficial effects in arthritic patients. Arthritic patients demonstrated a significant improvement in morning stiffness and number of tender joints when consuming EPA supplements compared to placebo in a double-blinded, crossover study (Kremer *et al.*, 1987).

Thrombosis is the formation of blood clots. Blood clotting involves the clumping together of platelets into large aggregates and is triggered when endothelial cells lining the artery walls are damaged. If the platelet membranes are rich in long-chain *n*-3 PUFAs, formation of certain eicosanoids such as prostacyclin I_3 and thromboxane A_3 is promoted. These do not trigger platelet aggregation as much as the corresponding eicosanoids, prostacyclin I_2 and thromboxane A_2, that are formed from *n*-6 PUFA. Therefore, long-chain *n*-3 PUFAs may help to reduce the tendency for blood to clot (Groom, 1993).

Inuits in Greenland have been shown to suffer from a lower incidence of cardiovascular disease compared to their Danish counterparts (Bang and Dyerberg, 1972, 1986). Dyerberg *et al.* (1975) suggested that the relatively high dietary *n*-3 PUFA intake of Inuits was related to their lower incidence of cardiovascular disease. Mori *et al.* (1997) suggested that *n*-3 fatty acid intake from fish consumption in conjunction with a low-fat diet was most beneficial in terms of reducing cardiovascular disease. Recent studies indicate that the *n*-3 fatty acids, especially EPA and DHA, may be effective in reducing the clinical risk of cardiovascular disease by favorably altering lipid and hemostatic factors such as bleeding time and platelet aggregation (Hornstra, 1989). Dietary supplementation of *n*-3 fatty acids has also been recommended for lowering the risk of cardiovascular disease and improving the overall health of humans, mainly due to lowering both the TAG levels in the plasma and the incidence of arrhythmia, among others.

Possible effects of *n*-3 PUFA on diabetic patients have been studied. The development of insulin resistance in normal rats fed a high-fat, safflower oil diet was found to be prevented by partial replacement of linoleic acid with EPA and DHA from fish oil (Storlien *et al.*, 1987). From human studies on diabetes, it is clear that *n*-3 PUFAs exert beneficial effects on lipid metabolism and may decrease the severity of cardiac disorder and hence lower the incidence of coronary artery disease (Bhathena, 1992).

Mitsuyoshi *et al.* (1992) studied the effect of structured lipid, containing caprylic and linoleic acids, as an energy substrate after hepatic resection in diabetic rats. The lipid sources used in this study were MCT, LCT, a simple physical mixture of MCT/LCT, and structured lipid. The blood ketone body ratio (acetoacetate/β-hydroxybutyrate) and the cumulative excretion of $^{14}CO_2$ in expired breath after [^{14}C]glucose administration were significantly higher in the structured lipid group than in the other groups. These findings suggest that structured lipids may be a superior energy substrate compared to other TAG preparations during the critical period after hepatectomy in diabetic patients.

Ling *et al.* (1991) demonstrated that tumor growth in mice was decreased when they were fed with a structured lipid made from fish oil and MCT. In another study, the tumor growth rate was reduced in rats fed with SL containing MCFAs and fish oil (Mendez *et al.*, 1992). In contrast to the tumor-promoting effects of diets high in fat, diets high in fish oil failed to promote tumor development in rats (Branden and Carroll, 1986). Reddy and Maruyama (1986) also showed that diets containing high levels of fish oil inhibit or suppress tumor growth in animal models. Dietary intake of fish oils was effective in destroying some cancer cells, but it is not known whether such results are reproducible with humans, or what potential side-effects exist (Haumann, 1997b). Although it is known that *n*-3 PUFAs play an important role in the growth of certain cells in the human body, the mechanisms involved in their effect on cancer treatment remains somewhat elusive.

Impact® (Novartis Nutrition) is another example of a structured lipid, produced by interesterifying a high-lauric oil with a high-linoleic acid oil. This product has been used for patients suffering from trauma, surgery, sepsis, or cancer (Haumann, 1997a). A structured lipid containing MCFA and linoleic acid is more effective in cystic fibrosis patients than safflower oil, which has about twice as much linoleic acid (McKenna *et al.*, 1985). The structured lipid diet, Impact, containing low levels of linoleic acid, resulted in decreased infection and decreased length of hospital stay compared to other enteral formulae. Bower *et al.* (1995) also demonstrated a decreased length of hospital stay and infection rate when using diets with a low level of linoleic acid and added fish oil.

G. New Developments in Production of Fatty Acid Conjugates and their Potential Health Effects

Recent studies have demonstrated unexpected potential benefits when conjugating *n*-3 fatty acids such as stearic acid, EPA, and DHA with other bioactive molecules such as the green tea polyphenol, epigallocatechin gallate (EGCG), and phytosterols. The EGCG—DHA products thus obtained displayed excellent bioactivities including antioxidant, anti-inflammatory, antiviral, and anticancer properties, some of which were greater than that of the EGCG or DHA alone (Shahidi and Zhong, 2010). The EGCG—fatty acid conjugates acted as radical scavengers against 1,1-diphenyl-2-picrylhydrazyl (DPPH) and peroxyl radicals and metal ion chelators which were more potent than or comparable to EGCG. The conjugates were able to inhibit lipid oxidation in food model systems (bulk oil, oil-in-water emulsion, and meat), suggesting their potential use as antioxidant preservatives in the food industry. They have also shown effectiveness in protecting various biological model systems from oxidative damage, including copper-induced low-density lipoprotein-cholesterol oxidation, radical-induced DNA scission, and ultraviolet-induced liposome photooxidation. These suggest their antiatherosclerotic, antimutagenic, and membrane protection properties, and hence their potential in preventing/treating cardiovascular diseases, cancer, and skin disorders, among other oxidation-mediated diseases. In addition, the EGCG—fatty acid conjugates exhibited antiviral activity by inhibiting NS3/4A protease and α-glucosidase, which are important enzymes for the infectivity of hepatitis C virus and human immunodeficiency virus, respectively. The antiviral effect was not found for EGCG itself. Moreover, the DPA conjugates of EGCG played an anti-inflammatory role in lipo-polysaccharide-stimulated murine macrophages by suppressing the gene expression of inducible nitric oxide synthase and cyclooxygenase-2. The EGCG—DHA conjugates were found to be effective in inhibiting azoxy-methane-induced colon tumorigenesis. Meanwhile, the docosahexaenoate esters of phytosterol effectively reduced plasma cholesterol levels and atherosclerotic lesions in mice (Tan *et al.*, 2012). Thus, conjugation of polyphenols or phytosterols with fatty acids, especially long-chain *n*-3 PUFAs, may be useful in developing novel bioactives with health-promoting properties as potential functional food ingredients and as natural health products.

REFERENCES

Akoh, C.C., Moussata, C.O., 1998. Lipase-catalyzed modification of borage oil: incorporation of capric and eicosapentaenoic acids to form structured lipids. J. Am. Oil Chem. Soc. 75, 697—701.

Akoh, C.C., Sista, R.V., 1995. Enzymatic modification of borage oil: incorporation of eicosapentaenoic acid. J. Food Lipids 2, 231—238.

Andersen, A.J.C., 1962. Refining of fats and oils for edible purposes. In: Williams, P.W. (Ed.), (2nd ed.), Pergamon, Oxford, p. 40.

Artz, W.E., Hansen, S.L., 1996. Current developments in fat replacers. In: McDonald, R.E., Min, D.B. (Eds.), Food Lipids and Health. Marcel Dekker, New York, pp. 385—415.

Babayan, V.K., 1987. Medium-chain triglycerides and structured lipids. Lipids 22, 417—420.

Babayan, V.K., Blackburn, G.L. and Bistrian, B. R. 1990. Structured lipid containing dairy fat. US Patent No. 4,952,606.

Bach, A.C., Babayan, V.K., 1982. Medium-chain triglycerides: an update. Am. J. Clin. Nutr. 36, 950—962.

Bang, H.O., Dyerberg, J., 1972. Plasma lipids and lipoproteins in Greenlandic West-coast Eskimos. Acta Med. Scand. 192, 85—94.

Bang, H.O., Dyerberg, J., 1986. Lipid metabolism and ischemic heart disease in Greenland Eskimos. Adv. Nutr. Res. 3, 1—21.

Bell, S.J., Mascioli, E.A., Bistrian, B.R., Babayan, V.K., Blackburn, G.L., 1991. Alternative lipid sources for enteral and parenteral nutrition: long- and medium-chain triglycerides, structured triglycerides, and fish oils. J. Am. Diet. Assoc. 91, 74—78.

Bernardini, E., 1985. Vegetable Oils and Fats Processing. Vol. II. Inter-stampa, Rome.

Bhathena, S.J., 1992. Fatty acids and diabetes. In: Chow, C.K. (Ed.), Fatty Acids in Foods and their Health Implications. Marcel Dekker, New York, pp. 823—855.

Bhosle, B.M., Subramanian, R., 2005. New approaches in the deacidifi-cation of edible oils — a review. J. Food Eng. 69, 481—494.

Bottino, N.R., Vandenburg, G.A., Reiser, R., 1967. Resistance of certain long-chain polyunsaturated fatty acids of marine oils to pancreatic lipase hydrolysis. Lipids 2, 489—493.

Bower, R.H., Daly, J.M., Lieberman, M.D., Goldfine, J., Shou, J., Weintraub, F., Rosato, E.F., Lavin, P., 1995. Early enteral administra-tion of a formula (Impact®) supplemented with arginine, nucleotides, and fish oil in intensive care unit patients: results of a multicenter prospective, randomized clinical trial. Crit. Care Med. 23, 436—449.

Branden, L.M., Carroll, K.K., 1986. Dietary polyunsaturated fats in relation to mammary carcinogenesis in rats. Lipids 21, 285—288.

Canavag, G.C., 1976. Miscella refining. J. Am. Oil. Chem. Soc. 53, 361—363.

Carter, J.P., 1988. Gamma-linolenic acid as a nutrient. Food Technol. 42 (72), 74—82.

Ceci, L.N., Constenla, D.T., Capiste, G.H., 2008. Oil recovery and leci-thin production using water degumming sludge of crude soybean oils. J. Sci. Food Agric. 88, 2460—2466.

Ceriani, R., Meirelles, A.J.A., 2006. Simulation of continuous physical refiners for edible oil deacidification. J. Food Eng. 76, 261—271.

Chakrabarti, P.P., Rao, B.V.S.K., 2004. Process for the pre-treatment of vegetable oils for physical refining. US Patent No. 005,399.

Choukri, A., Kinany, M.A., Gibon, V., Tirtiaux, A.J., Jamil, S., 2001. Improved oil treatment conditions for soft gumming. J. Am. Oil Chem. Soc. 78, 1157—1160.

Cleenewerck, B., Dijkstra, A., 1992. The total degumming process — theory and industrial application in refining and hydrogenation. Eur. J. Lipid Sci. Technol. 94, 317—322.

Copeland, D., Belcher, M.W., 2005. Vegetable oil refining Int.Cl.C11B3/00. US Patent No. 6,844,458.

Cvengros, J., 1995. Physical refining of edible oils. J. Am. Oil Chem. Soc. 72, 1193—1196.

Decker, E.A., 1996. The role of stereospecific saturated fatty acid posi-tions on lipid nutrition. Nutr. Rev. 54, 108—110.

Dijkstra, A.J., Opstal, M.W., 1987. Process for producing degummed vegetable oils and gums of high phosphatidic acid content Int.Cl.C11B3/00. US Patent No. 4,698,185.

Dijkstra, A.J., Opstal, M.V., 1989. The total degumming process (TOP). In: Erickson, D.R. (Ed.), Proceedings of World Conference on

Edible Fats and Oils Processing. AOCS Press, Champaign, IL, pp. 176–177.

Dyerberg, J., Bang, H., Hjorne, N., 1975. Fatty acid composition of plasma lipids in Greenland Eskimos. Am. J. Clin. Nutr. 28, 958–966.

Fairhurst, T.H., Mutert, E., 1999. Introduction to oil palm production. Better Crops Int. 13, 3–6.

Gandhi, N.N., 1997. Applications of lipase. J. Am. Oil Chem. Soc. 74, 621–634.

Ghosh, M., 2007. Review on recent trends in rice bran oil processing. J. Am. Oil Chem. Soc. 84, 315–324.

Groom, H., 1993. Oil-rich fish. Nutr. Food Sci. Nov.–Dec., 4–8.

Gunstone, F.D., 1992. Gamma linolenic acid – occurrence and physical and chemical properties. Prog. Lipid Res. 31, 145–161.

Gunstone, F.D., Norris, F.A., 1983. Lipids in Foods: Chemistry, Biochemistry and Technology. Pergamon Press, Oxford.

Hamam, F., Shahidi, F., 2005. Enzymatic incorporation of capric acid into a single cell oil rich in docosahexaenoic acid and docosapentaenoic acid and oxidative stability of the resultant structured lipid. Food Chem. 91, 583–591.

Hamam, F., Shahidi, F., 2006a. Synthesis of structured lipids containing medium-chain and omega-3 fatty acids. J. Agric. Food Chem. 54, 4390–4396.

Hamam, F., Shahidi, F., 2006b. Acidolysis reactions lead to esterification of endogenous tocopherols and compromised oxidative stability of modified oils. J. Agric. Food Chem. 54, 7319–7323.

Hamam, F., Shahidi, F., 2008. Incorporation of selected long-chain fatty acids into trilinolein and trilinolenin. Food Chem. 106, 33–39.

Harada, E., Kato, S., 1983. Effect of short-chain acids on the secretory response of the ovine exocrine pancrease. Am J. Physiol. 244, G284–G290.

Hashim, A., Babayan, V.K., 1978. Studies in man of partially absorbed dietary fats. Am. J. Clin. Nutr. 31, 5273–5276.

Haumann, B.F., 1997a. Structured lipids allow fat tailoring. INFORM 8, 1004–1011.

Haumann, B.F., 1997b. Nutritional aspects of n-3 fatty acids. INFORM 8, 428–447.

Heird, W.C., Grundy, S.M., Hubbard, V.S., 1986. Structured lipids and their use in clinical nutrition. Am. J. Clin. Nutr. 43, 320–324.

Henry, J., 2009. Processing, manufacturing, uses and labelling of fats in the food supply. Ann. Nutr. Metabol. 55, 273–300.

Heydinger, J.A., Nakhasi, D.K., 1996. Medium chain triacylglycerols. J. Food Lipids 3, 251–257.

Hornstra, G., 1989. Effects of dietary lipids on some aspects of the cardiovascular risk profile. In: Ziant, G. (Ed.), Lipids and Health. Elsevier Applied Science, New York, pp. 39–42.

Horrobin, D.F., 1990. Gamma linolenic acid: an intermediate in essential fatty acid metabolism with potential as an ethical pharmaceutical and as a food. Rev. Contemp. Pharmacother 1, 1–41.

Hoshino, T., Yamane, T., Shimuzu, S., 1990. Selective hydrolysis of fish oil by lipase to concentrate w3 polyunsaturated fatty acids. Agric. Biol. Chem. 54, 1459–1467.

Huang, K., Akoh, C.C., 1994. Lipase-catalyzed incorporation of n-3 polyunsaturated fatty acids into vegetable oils. J. Am. Oil Chem. Soc. 71, 1277–1280.

Huang, K., Akoh, C.C., Erickson, M.C., 1994. Enzymatic modification of melon seed oil: incorporation of eicosapentaenoic acid. J. Agric. Food Chem. 42, 2646–2648.

Ikeda, I., Tomari, Y., Sugano, M., Watanabe, S., Nagata, J., 1991. Lymphatic absorption of structured glycerolipids containing medium-chain fatty acids and linoleic acid, and their effect on cholesterol absorption in rats. Lipids 26, 369–373.

Indira, T.N., Hemavathy, J., Khatoon, S., Gopala Krisna, A.G., Bhattacharya, S., 2000. Water degumming of rice bran oil: a response surface approach. J. Food Eng. 43, 83–90.

Iwasaki, Y., Han, J.J., Narita, M., Rosu, R., Yamane, T., 1999. Enzymatic synthesis of structured lipids from single cell oil of high docosahexaenoic acid content. J. Am. Oil Chem. Soc. 76, 563–569.

Jang, E.S., Jung, M.Y., Min, D.B., 2005. Hydrogenation for low *trans* and high conjugated fatty acids. Comp. Rev. Food Sci. Food Safety 4, 22–30.

Jennings, B.H., Akoh, C.C., 1999. Enzymatic modification of triacylglycerols of high eicosapentaenoic and docosahexaenoic acids content to produce structured lipids. J. Am. Oil Chem. Soc. 76, 1133–1137.

Ju, Y., Huang, F., Fang, C., 1998. The incorporation of n-3 polyunsaturated fatty acids into acylglycerols of borage oil via lipase-catalyzed reactions. J. Am. Oil Chem. Soc. 175, 961–965.

Kaimal, T.N.B., Vali, S.R., Rao, B.V.S.K., Turaga, V., Rao, C., Bhaerao, U.T., 2000. A process for the preparation of purified rice bran oil by simultaneous dewaxing and degumming. Indian Patent INP 183,639.

Kemper, T.G., 2005. Oil extraction. In: Shahidi, F. (Ed.), Bailey's Industrial Oil and Fat Products, 6th ed., Vol. 5. John Wiley & Sons, Hoboken, NJ, pp. 57–98.

Kennedy, J.P., 1991. Structured lipids: fats for the future. Food Technol. 11, 76–83.

King, J.W., Holliday, R.L., List, G.R., Snyder, J.M., 2001. Hydrogenation of vegetable oils using mixtures of supercritical carbon dioxide and hydrogen. J. Am. Oil Chem. Soc. 78, 107–113.

Kosmark, R., 1996. Salatrim: properties and applications. Food Technol. 50, 98–101.

Kovari, K., 2004. Recent developments, new trends in seed crushing and oil refining. Oléagineux Corps. Gras Lipides 11, 381–387.

Kremer, J.M., Jubiz, W., Michalek, A., Rynes, R.I., Bartholomew, L.E., Bigaouette, J., Timchalk, M., Beeler, D., Lininger, L., 1987. Fish oil fatty acid supplementation in active rheumatoid arthritis. Ann. Intern. Med. 106, 497–502.

Kreulen, H.P., 1976. Fractionation and winterization of edible fats and oils. J. Am. Oil Chem. Soc. 53, 393–396.

Lee, T.W., Hastilow, C.I., 1999. Quantitative determination of triacylglycerol profile of structured lipid by capillary supercritical fluid chromatography and high-temperature gas chromatography. J. Am. Oil Chem. Soc. 76, 1405–1413.

Leiboritz, Z., Ruckenstein, C., 1984. Winterization of sunflower oil. J. Am. Oil Chem. Soc. 61, 870–872.

Lin, L., Rhee, K.C., Koseoglu, S.S., 1997. Bench-scale membrane degumming of crude vegetable oil: process optimization. J. Membr. Sci. 134, 101–108.

Ling, P.R., Istfan, N.W., Lopes, S.M., Babayan, V.K., Blackburn, G.L., Bistrian, B.R., 1991. Structured lipid made from fish oil and medium chain triglyceride alters tumor and host metabolism in Yoshida sarcoma-bearing rats. Am. J. Clin. Nutr. 53, 1177–1184.

List, G.R., King, J.W., Johnson, J.H., Warner, K., Mounts, T.L., 1993. Supercritical CO2 degumming and physical refining of soybean oil. J. Am. Oil Chem. Soc. 70, 473–476.

Macher, M., Holmquivst, A., 2001. Hydrogenation of palm oil in near-critical and supercritical propane. Eur. J. Lipid Sci. Technol. 101, 81–84.

Macher, M., Hogberg, J., Moller, P., Harrod, M., 1999. Partial hydrogenation of fatty acid methyl esters at supercritical conditions. Fette/Lipid 8, 301–305.

McKenna, M.C., Hubbard, V.S., Pieri, J.G., 1985. Linoleic acid absorption in patients with cystic fibrosis with pancreatic insufficiency and in control subjects. J. Pediatr. Gastroenterol. Nutr. 4, 45–48.

Macrae, A.R., 1983. Lipase-catalyzed interesterification of oils and fats. J. Am. Oil Chem. Soc. 60, 291–294.

Macrae, A.R., 1985. Interesterification of fats and oils. In: Tramper, J., van der Plas, H.C., Linko, P. (Eds.), Biocatalysis in Organic Syntheses. Elsevier Applied Science, Amsterdam, pp. 195–208.

Mascioli, E.A., Bistrian, B.R., Babayan, V.K., Blackburn, G.L., 1987. Medium-chain triglycerides and structured lipids as unique nonglucose energy sources in hyperalimentation. Lipids 22, 421–423.

Megremis, C.L., 1991. Medium-chain triglycerides: a nonconventional fat. Food Technol. 45, 108–110.

Mendez, B., Ling, P.R., Istfan, N.W., Babayan, V.K., Bistrain, B.R., 1992. Effects of different lipid sources in total parenteral nutrition on whole body protein kinetics and tumor growth. J. Parenter. Enteral Nutr. 16, 545–551.

Meshehdani, T., Pokorny, J., Davidek, J., Panek, J., 1990. Deactivation of lipoxygenases during rapeseed processing. Corps Gras. 37, 23–27.

Mitsuyoshi, K., Hiramatsu, Y., Nakagawa, M., Yamamura, M., Hioki, K., Yamamoto, M., 1992. Effect of structured lipids as energy substrate after hepatectomy in rats with streptozocin-induced diabetes. Nutrition 8, 41–46.

Mohamed, H.M.A., Larsson, K., 1994. Modification of fats by lipase interesterification. 2. Effect on crystallisation behaviour and functional properties. Fat Sci. Technol. 96, 56–59.

Mori, T.A., Beilin, L.J., Burke, V., Morris, J., Ritchie, J., 1997. Interactions between dietary fat, fish and fish oils and their effects on platelet-function in men at risk of cardiovascular disease. Arterioscler. Thromb. Vasc. Biol. 17, 279–286.

Mounts, T.L., 1981. Chemical and physical effects of processing fats and oils. J. Am. Oil Chem. Soc. 58, 51A–54A.

Mounts, T.L., Khym, F.P., 1980. Refining. In: Erickson, D.R., Pryde, E.H., Brekke, O.L., Mounts, T.L., Falb, R.A. (Eds.), Handbook of Soybean Oil Processing Technology. American Soybean Association, St. Louis, MO, pp. 89–103.

Ochoa, N., Pagliero, C., Marchese, J., Mattea, M., 2001. Ultrafiltration of vegetable oils degumming by polymeric membranes. Separ. Purif. Technol. 22–23, 417–422.

Osada, K., Nakamura, M., Nonaka, M., Hatano, M., 1992. Esterification of glycerol with EPA and DHA by Chromobacterium viscosum and Candida cylindracea lipases. J. Jpn. Oil Chem. Soc. 41, 39–43.

Osborn, H.T., Akoh, C.C., 2002. Structured lipids – novel fats with medical, nutraceutical, and food applications. Comp. Rev. Food Sci. Food Safety 1, 110–120.

Ozturk, T., Ustun, G., Aksoy, H.A., 2010. Production of medium-chain triacylglycerols from corn oil: optimization by response surface methodology. Bioresour. Technol. 101, 7456–7461.

Patterson, H.W.B., 1992. Bleaching and Purifying Fats and Oils. Theory and Practice. AOCS Press, Champaign, IL.

Plourde, M., Cunnane, S.C., 2007. Extremely limited synthesis of long chain polyunsaturates in adults: implications for their dietary essentiality and use as supplements. Appl. Physiol. Nutr. Metab. 32, 619–634.

Pokorny, J., 1991. Natural antioxidants for food use. Trends Food Sci. Technol. 2, 223–227.

Posorske, L.H., LeFebvre, G.K., Miller, C.A., Hansen, T.T., Glenvig, B.L., 1988. Process considerations of continuous fat modification with an immobilised lipase. J. Am. Oil Chem. Soc. 65, 922–926.

Quinlan, P., Moore, S., 1993. Modification of triglycerides by lipases: process technology and its application to the production of nutritionally improved fats. INFORM 14, 580–585.

Rajam, L., Kumar, D.R.S., Sundarsan, A., Arumugham, C., 2005. A novel process for physically refined rice bran oil through simultaneous degumming and dewaxing. J. Am. Oil Chem. Soc. 82, 213–220.

Raß, M., Schein, C., Matthäus, B., 2008. Virgin sunflower oil. Eur. J. Lipid Sci. Technol. 110, 618–624.

Reddy, B.S., Maruyama, H., 1986. Effect of dietary fish oil on azoxymethane-induced colon carcinogenesis in male F344 rats. Cancer Res. 46, 3367–3370.

Roediger, W.E.W., Rae, D.A., 1982. Trophic effect of short-chain fatty acids on mucosal handling of ions by the defunctioned colon. Br. J. Surg. 69, 23–25.

Ruppin, H., Bar-Meir, S., Soergel, K.H., Wood, C.M., Schmitt, M.G., 1980. Absorption of short chain fatty acids by the colon. Gastroenterology 78, 1500–1507.

von Schacky, C., 2003. The role of omega-3 fatty acids in cardiovascular disease. Curr. Atheroscler. Rep. 5, 139–145.

Scott, F.W., Lee, N.S., 1996. Bureau of Nutritional Science Committee on Functional Foods. Food Directorate. Health Protection Branch, Ottawa, ON.

Senanayake, S.P.J.N., Shahidi, F., 1999a. Enzyme-assisted acidolysis of borage (Borago officinalis L.) and evening primrose (Oenothera biennis L.) oils: incorporation of omega-3 polyunsaturated fatty acids. J. Agric. Food Chem. 47, 3105–3112.

Senanayake, S.P.J.N., Shahidi, F., 1999b. Enzymatic incorporation of docosahexaenoic acid into borage oil. J. Am. Oil Chem. Soc. 76, 1009–1015.

Senanayake, S.P.J.N., Shahidi, F., 2000. Structured lipids containing long-chain omega-3 polyunsaturated fatty acids. In: Shahidi, F., F. (Eds.), Seafood in Health and Nutrition. Transformation in Fisheries and Aquaculture: Global Perspectives. ScienceTech, St. John's, NF, Canada, pp. 29–44.

Senanayake, S.P.J.N., Shahidi, F., 2001. Modified oils containing highly unsaturated fatty acids and their stability. In: Shahidi, F., Finley, J.W. (Eds.), Omega-3 Fatty Acids. Chemistry, Nutrition and Health Effects. ACS Symposium Series, 788. American Chemical Society, Washington, DC, pp. 162–173.

Senanayake, S.P.J.N., Shahidi, F., 2002. Enzyme-catalyzed synthesis of structured lipids via acidolysis of seal blubber oil with capric acid. J. Am. Oil Chem. Soc. 35, 745–752.

Senanayake, S.P.J.N., Shahidi, F., 2005. Modification of fats and oils via chemical and enzymatic methods. In: Shahidi, F. (Ed.), Bailey's Industrial Oil and Fat Products, 6th ed., Vol. 3. John Wiley & Sons, Hoboken, NJ, pp. 555–584.

Sengupta, R., Bhattacharyya, D.K., 1992. A comparative study between biorefining combined with other processes and physical refining of high acid mohua oil. J. Am. Oil Chem. Soc. 69, 1146–1149.

Shahidi, F., Wanasundara, U.N., 1998. Omega-3 fatty acid concentrates: nutritional aspects and production technologies. Trends Food Sci. Technol. 9, 230–240.

Shahidi, F. and Zhong, Y. 2010. US Provisional Patent. Application No. 61/322,004.

Shahidi, F., Zhong, Y., Tan, Z., 2010. Food bioactives and enhancement of their beneficial health effects by structure modification. Book of Abstracts, TCH-251, #333. In: Chemistry, Safety, Quality and Regulations Aspects of Functional Food Ingredients, Nutraceuticals and Natural Health Products. International Chemical Congress of Pacific Basin Societies, Honolulu, HI 15–20 December.

Shimada, Y., Murayama, K., Okazaki, S., Nakamura, M., Sugihara, A., Tominaga, Y., 1994. Enrichment of polyunsaturated fatty acids with *Geotrichum candidum* lipase. J. Am. Oil Chem. Soc. 71, 951–954.

Smith, R.E., Finley, J.W., Leveille, G.A., 1994. Overview of Salatrim, a family of low-calorie fats. J. Agric. Food Chem. 42, 432–434.

Spurvey, S.A., Senanayake, S.P.J.N., Shahidi, F., 2001. Enzyme-assisted acidolysis of menhaden and seal blubber oils with gamma-linolenic acid. J. Am. Oil Chem. Soc. 78, 1105–1112.

Sridhar, R., Lakshminarayana, G., 1992. Incorporation of eicosapentaenoic and docosahexaenoic acids into groundnut oil by lipase-catalyzed ester interchange. J. Am. Oil Chem. Soc. 69, 1041–1042.

Stein, J., 1999. Chemically defined structured lipids: current status and future directions in gastrointestinal diseases. Int. J. Colorect. Dis. 14, 79–85.

Storlien, L.H., Kraegen, E.W., Chisholm, D.J., Ford, G.L., Bruce, D.G., Pascoe, W.S., 1987. Fish oil prevents insulin resistance induced by high-fat feeding rats. Science 237, 885–888.

Sullivan, F.E., 1976. Steam refining. J. Am. Oil Chem. Soc. 53, 358–361.

Szydlowska-Czerniak, A., 2007. MIR spectroscopy and partial least-squares regression for determination of phospholipids in rapeseed oils at various stages of technological process. Food Chem. 105, 1179–1187.

Tan, Z., Le, K., Moghadasian, M., Shahidi, F., 2012. Enzymatic síntesis of phytosteryl docosahexaenoates and their evaluation of anti-atherogenic effects in apo-E deficient mice. Food Chem. 134, 2097–2104.

Tanaka, Y., Hirano, J., Funada, T., 1992. Concentration of docosahexaenoic acid in glyceride by hydrolysis of fish oil with *Candida cylindracea* lipase. J. Am. Oil Chem. Soc. 69, 1210–1214.

Ulrich, H., Pastores, S.M., Katz, D.P., Kvetan, V., 1996. Parenteral use of medium-chain triglycerides: a reappraisal. Nutrition 112, 231–238.

Vartak, S., Robbins, M.E.C., Spector, A.A., 1997. Polyunsaturated fatty acids increase the sensitivity of 36B10 rat astrocytoma cells to radiation-induced cell kill. Lipids 32, 283–292.

Villeneuve, P., Foglia, T.A., 1997. Lipase specificities: potential application in lipid bioconversions. INFORM 8, 640–650.

Wanasundara, U.N., Shahidi, F., 1998. Lipase-assisted concentration of *n*-3 polyunsaturated fatty acids in acylglycerols from marine oils. J. Am. Oil Chem. Soc. 75, 945–951.

Wang, J., Shahidi, F., 2010. Stability characteristics of omega-3 oil and their randomized counterparts. In: Ho, C.T., Mussinan, C.J., Shahidi, F., Contis, T. (Eds.), Recent Advances in Food and Flavour Chemistry. RSC Publishing, Cambridge, pp. 297–307.

Willem, V.N., Mabel, C.T., 2008. Update on vegetable lecithin and phospholipid technologies. Eur. J. Lipid Sci. Technol. 110, 472–486.

Young, F.V.K., 1983. Palm kernel and coconut oils: analytical characteristics, process technology and uses. J. Am. Oil Chem. Soc. 60, 374–379.

Zufarov, O., Schmidt, S., Sekretár, S., 2008. Degumming of rapeseed and sunflower oils. Acta Chim. Slovac. 1, 321–328.

PART III

Biochemistry of Food Spoilage

Chapter 10

Enzymatic Browning

Vera Lúcia Valente Mesquita and Christiane Queiroz

Departamento de Nutrição Básica e Experimental, Instituto de Nutrição, Josué de Castro, University of Rio de Janeiro, Brazil

Chapter Outline

I. INTRODUCTION

Enzymatic browning is a phenomenon which occurs in many fruits and vegetables, such as potatoes, mushrooms, apples, and bananas. When the tissue is bruised, cut, peeled, diseased, or exposed to any abnormal conditions, it rapidly darkens on exposure to air as a result of the conversion of phenolic compounds to brown melanins (Figure 10.1).

The international nomenclature for the enzymes involved in the browning reaction has changed. The first enzyme, monophenol monooxygenase or tyrosinase (EC 1.14.18.1), initiates the browning reaction, which later involves diphenol oxidase or catechol oxidase (EC 1.10.3.2) and laccase (EC 1.10.3.1). Catecholase oxidase will be referred to as 'polyphenol oxidase' (PPO) in this chapter. This enzyme requires the presence of both a copper prosthetic group and oxygen. The copper is believed to be monovalent in the case of the mushroom PPO and divalent in the case of the potato enzyme (Bendall and Gregory, 1963). PPO is classified as an oxidoreductase, with oxygen functioning as the hydrogen acceptor. The enzyme is widely distributed in higher plants and fungal and animal tissue and was reviewed by Swain (1962), Mathew and Parpia (1971), Mayer and Harel (1979), Vamos-Vigyazo (1981), Mayer (1987), and (2006).

A. Historical Aspects of Polyphenol Oxidase

The earliest work is attributed to Lindet, who in 1895 recognized the enzymatic nature of browning while working on cider. At the same time Bourquelot and Bertrand commenced studies on mushroom tyrosine oxidase. Subsequently, Onslow in 1920 showed that enzymatic browning of plant tissue in air was attributed to the presence of *o*-diphenolic

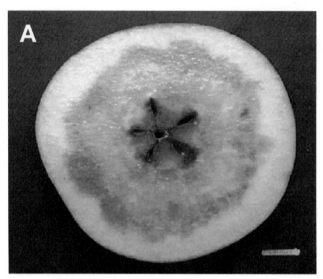

FIGURE 10.1   Browning disorder in 'Conference' pears after 4 months in browning-inducing storage conditions (no cooling period, 1% O_2, 10% CO_2, $-1°C$) *(Reprinted from Franck et al., 2007). © 2007 with permission form Elsevier.*

compounds, such as catechol, protocatechuic acid, and caffeic acid, plus the appropriate enzymes (oxygenases). The product of this reaction was thought to be a peroxide which reacted with a 'chromogen' to form a brown pigment. Many fruits and vegetables, including apples, pears, apricots, and potatoes, were all found to be rich in phenolic compounds and oxygenases. Other fruits, such as citrus, pineapples, and redcurrants, lacked these substances and were thus termed 'peroxidase plants'.

This distinction was eliminated when it was shown that peroxidase and catalase were present in both groups of plants and in fact were ubiquitous in plant tissues. The term 'oxygenase' was subsequently replaced by 'phenolase' or 'polyphenol oxidase'. In 1937, Kubowitz demonstrated that PPO was a copper-containing enzyme.

II. CHARACTERISTICS OF POLYPHENOL OXIDASE

A. Structure and Sequence

In order to understand the physiological role of PPO, much research has been conducted to identify gene expression and protein sequences of the PPO of different plants. It is known that plant PPOs are synthesized as preproteins and contain putative plastid transit peptides at the N-terminal region, which target the enzyme into chloroplasts and thykaloid lumen (Marusek *et al.*, 2006). Molecular studies have described PPO as a single genome or as a multiple gene family. For example, the cherimoya PPO gene was shown to be present in one copy of the genome and had a very different nucleotide sequence from other published sequences. Despite these divergences, it presented conserved proteins in the proposed active site and also basic and strategic amino acids related to active site accessibility, protein structure, and stability (Prieto *et al.*, 2007). In Fuji apple plants the expression of two PPO genes (APO5 and MD-PPO2) was observed during vegetative and reproductive development and in response to wounding. However, they were not expressed at the same moment or by the same stress stimulus. The maximum level of APO5 messenger RNA (mRNA) was observed in damaged fruits after 24 hours, while the expression of MD-PPO2 was not increased, suggesting that the selective activation of individual genes under different conditions may reflect the existence of distinct signal transduction pathways to turn on the different PPO genes (J. Y. Kim *et al.*, 2001). In addition, seven genes were identified from tomato (Newmann *et al.*, 1993) and five distinct PPO cDNAs were isolated from potato plants (Hunt *et al.*, 1993).

The existence of multiple genes leads to the presence of PPO isozymes and the variability on protein sequences observed in the studies explains the differences among enzymes extracted from distinct sources. Two izoenzymes were present in persimmon, Fuji apple (J. Y. Kim *et al.*, 2001), coffee (Mazzafera and Robinson, 2000), and artichoke (Aydemir, 2004), while medlar PPO showed four isoforms (Dincer *et al.*, 2002). PPO purified from field bean is a tetramer with 120 kDa and exists as a single isoform in the seed (Paul and Gowda, 2000), while pineapple PPO has three isoenzymes, with the major isoform being a tetramer of identical subunits of 25 kDa (Das *et al.*, 1997).

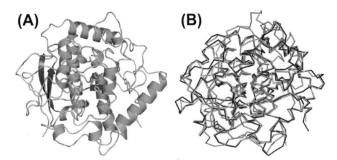

FIGURE 10.2 X-ray structure of polyphenol oxidase (PPO) from *Vitis vinifera*. (A) Ribbon model showing the overall ellipsoidal shape, two β-sheets, and the dicopper center within four-helix bundle. (B) Cα representation of *V. vinifera* PPO (blue) overlapping with those PPO from sweet potato (yellow) and *Streptomyces castaneoglobisporus* (green) See online for a colour version of this figure *(Virador et al., 2010). Reprinted with permission, © 2010 American Chemical Society.*

The molecular mass of PPO from other species has been reported as follows: mulberry, 65 kDa (Arslan *et al.*, 2004); Henry chestnuts, 69 kDa (Xu *et al.*, 2004); and pulp pine, 90 kDa (Lima *et al.*, 2001).

The crystal structure of PPO has been elicited for sweet potato, grape, and *Neurospora crassa*. The overall structure of the PPO from *Vitis vinifera* is a monomeric protein of 38.4 kDa. It is an ellipsoid with dimensions $56.7 \times 48.0 \times 48.3$ Å^3 (Figure 10.2A, B). The secondary structure is primarily R-helical with the core of the protein formed by a four-helix bundle composed of R-helices R4, R5, R12, and R14 (Figure 10.2A). Figure 10.2(B) shows overlapping images of the three existing PPO structures. There are a few areas where the folding is slightly different. For the most part, these differences occur on the surface and do not involve changes in R-helices or β-strands (Virador *et al.*, 2010).

B. Mechanism of Reaction

The mechanism of action proposed for PPO is based on its capacity to oxidize phenolic compounds. When the tissue is damaged the rupture of plastids, the cellular compartment where PPO is located, leads to the enzyme coming into contact with the phenolic compounds released by rupture of the vacuole, the main storage organelle of these compounds (Mayer and Harel, 1979).

PPO catalyzes two types of reaction: cresolase activity, in which monophenols are hydroxylated to *o*-diphenols, and catecholase activity. The catecholase or diphenolase type of reaction is best illustrated by the oxidation of catechol, an *o*-diphenol which is a commonly used laboratory substrate:

$$\text{Catechol} + E\text{–}2Cu^{2+} + \tfrac{1}{2}O_2 \xrightarrow{-2e^-} o\text{-Benzoquinone} + E\text{–}2Cu^+ + H_2O$$

The cresolase or monophenolase activity involves hydroxylation of monophenols to *o*-diphenols, as shown by the oxidation of ʟ-tyrosine to 3,4-dihydroxyphenylalanine, which occurs in potatoes (Schwimmer and Burr, 1967):

$$\text{L-Tyrosine} + E\text{–}2Cu^+ + O_2 + 2H^+ \xrightarrow{+2e^-} \text{3,4-Dihydroxyphenylalanine} + E\text{–}Cu^{2+} + H_2O$$

The active site of PPO consists of two copper atoms coordinated with six histidines, and exists in three oxidation states: *deoxy* (E_d), *met* (E_m), and *oxy* (E_o) (Figure 10.3). One of the Cu^{2+} atoms is bound to monophenols while

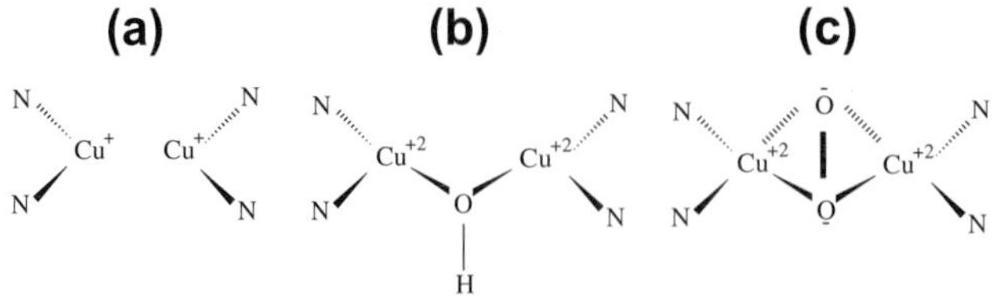

FIGURE 10.3 Polyphenol oxidase structure: (a) *deoxyPPO*, (b) *metPPO*, and (c) *oxyPPO*. *(Adapted with permission from Espín et al., 1998.)* © 2010 American Chemical Society.

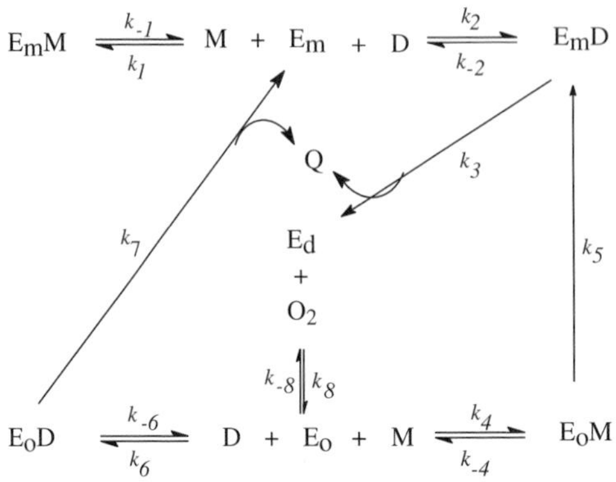

SCHEME 10.1 Kinetic reaction mechanism of polyphenol oxidase (PPO) on monophenols and *o*-diphenols. Em: *metPPO*; Ed: *deoxyPPO*; Eo: *oxyPPO*; M: monophenol; D: diphenol; Q: *o*-quinone. *(Reprinted with permission, Espín et al., 1998).* © 1998 American Chemical Society.

diphenols bind to both of them. As shown in the reaction mechanisms, monophenolase activity is intimately coupled to diphenolase activity and produces two electrons. These are required to incorporate one oxygen atom into the monophenol substrate.

The structural and kinetic mechanisms for the hydroxylation of monophenols (M) and the oxidation of *o*-diphenols (D) to *o*-quinones (Q) catalyzed by PPO have been established (Fenoll *et al.*, 2004; Espín *et al.*, 1998; Rodriguez-Lopes *et al.*, 1992). Monophenolic substrate initially coordinates to an axial position of one of the coppers of E_o that leads to the hydroxylation of monophenol by the bound peroxide, loss of water and formation of the enzyme—diphenol (E_mD) complex. This complex can either render free diphenol or undergo oxidation of the diphenolate intermediate that bound to the active site, giving a free quinone and a reduced binuclear copper on the enzyme site (E_d). *OxyPPO* is then regenerated after the binding of molecular oxygen to E_d (Rodriguez-Lopez *et al.*, 1992). When a diphenol is present in the medium, this substrate binds to both E_o and E_m to give E_oD and E_mD intermediates, which, in turn, give rise to two quinones (Orenes-Piñero *et al.*, 2005) (Scheme 10.1).

The presence of a dead-end complex, E_mM, is related to the occurrence of a lag period in the monophenolase activity of PPO (Rodriguez-Lopez *et al.*, 1992) that can be reversed by the addition of small quantities of reducing agents or *o*-diphenols as cosubstrates.

Quinone formation is both enzyme and oxygen dependent. Once this has taken place, the subsequent reactions occur spontaneously and no longer depend on the presence of PPO or oxygen. Joslyn and Ponting (1951) summarized those chemical reactions which may account for the formation of brown melanins.

The first reaction is thought to be a secondary hydroxylation of the *o*-quinone or of excess *o*-diphenol:

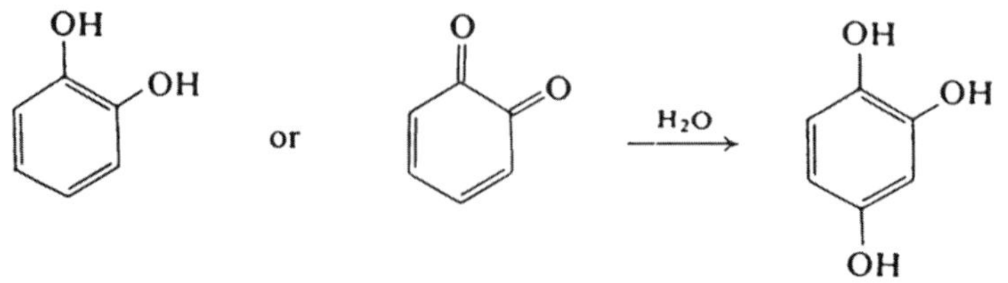

The resultant compound (triphenolic trihydroxybenzene) interacts with *o*-quinone to form hydroxyquinones:

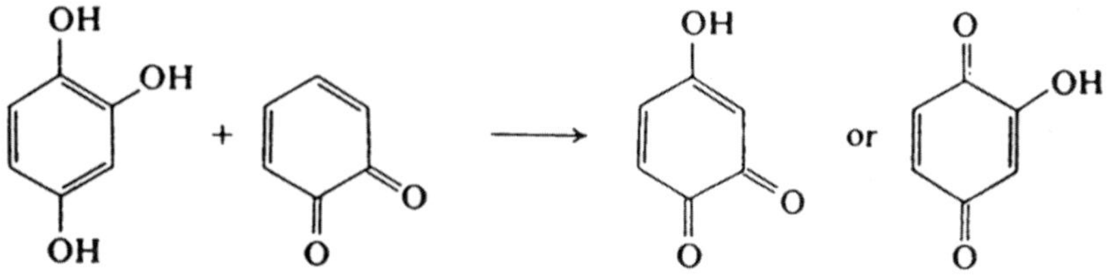

Hydroxyquinones undergo polymerization and are progressively converted to red and red–brown polymers, and finally to the brown melanins which appear at the site of plant tissue injury (Matheis and Whitaker, 1984; Whitaker, 1972).

The ratio of diphenolase to monophenolase activity depends on the plant source and can vary from 1:10 to 1:40 (Vamos-Vigyazo, 1981). Sanchez-Ferrer *et al.* (1988) partially purified PPO from Monastrel grapes and identified both cresolase and catecholase activity. Rocha and Morais (2001) also found cresolase and catecholase activity in Jonagorad apple PPO. Cresolase activity was described for strawberry and pear by Espín *et al.* (1997a, b). Several plants, however, have been found to be devoid of monophenolase activity, including coffee (Mazzafera and Robinson, 2000), broccoli florets (Gawlik-Dziki *et al.*, 2007), butter lettuce (Gawlik-Dziki *et al.*, 2008), and cashew apple (Queiroz *et al.*, 2011).

C. Biological Significance of Polyphenol Oxidase in Plants

The role of PPO in the living intact cell has until recently remained somewhat obscure. Early studies suggested its involvement as a terminal oxidase in respiration (James, 1953) or in the biosynthesis of lignin (Mason *et al.*, 1955). This was later discounted in studies by Nakamura (1967), who examined the role of three enzymes isolated from the latex of the Japanese lacquer tree (*Rhis vermicifera*): phenolase; peroxidase; and laccase. Of these enzymes only peroxidase was involved in lignification. Recent studies have demonstrated the role of laccase in lignification (Srebotnik and Hammel, 2000; Arora *et al.*, 2002; Shleev *et al.*, 2006). The mechanism is discussed in Section II, E.

Another possible role for PPO is in the biosynthesis of betalain. Red beet (*Beta vulgaris*) and common portulaca (*Portulaca portiflora*) present PPO that hydroxylates tyrosine, forming 3,4-dihydroxylphenylalanine (DOPA), and oxidizes DOPA to dopaquinone (Steiner *et al.*, 1996, 1999). Enzymatic activity is complemented by dioxygenase activity, leading to betalain formation. However, more evidence is needed to confirm this function.

PPO is restricted to the plastids. The enzyme appears to be in a latent form and bound to the thylakoid membrane, where the photochemical reactions of photosynthesis occur. The latent form can be activated in the presence of fatty acids (Siegenthaler and Vaucher-Boniour, 1971), detergents (Sellés-Marchart *et al.*, 2006), or trypsin (Tolbert, 1973). Pinto *et al.* (2008) isolated soluble and insoluble fractions of PPO from cowpea plants and reported activation by sodium dodecyl sulfate, a detergent, only in the soluble fraction, suggesting that insoluble PPO was purified in the active form.

In intact cells PPO seems to have little activity towards phenolics, which are located in the vacuole somewhat isolated from the plastid. Since the enzyme functions normally only when the cells are damaged or undergo senescence, it may also have a protected role, as proposed originally by Craft and Audia (1962). PPO is involved in phenolic metabolism when the plastid and vacuole contents are mixed. This occurs during senescence, when the integrity of the cell is disrupted, and the activation was reported by Goldbeck and Cammarata (1981). However, recent studies have demonstrated higher activity during development of the plant. Yu *et al.* (2010) evaluated PPO activity in longan fruit pericarp collected from 30 days after full bloom to mature fruit. They observed the highest enzymatic activity at first stage, followed by a sharp decrease, and then a slow increase in the final stage. These data are also in agreement with those described previously by Yang *et al.* (2000) and Sun *et al.* (2009). Ayaz *et al.* (2008) investigated PPO activity in medlar fruits during ripening and overripening and showed that, in mature fruit, the reaction velocity was higher than in overripe fruit. This could indicate that the enzyme was more active than in ripe and overripe fruits. A molecular study by J. Y. Kim *et al.* (2001) showed differential expression of two PPO genes in apple. The authors investigated the mRNA of PPO extracted from flower, fruit, and leaf in different stages and blotting analyses showed higher expression in the earlier stages (Figure 10.4).

Enzymatic browning also results following injury to the fruit or vegetable, causing disruption of the plastid, activation of latent PPO, and catalysis of phenolics released from the vacuole. Activation of PPO by biotic or abiotic stress had been demonstrated by several studies. Apple PPO showed maximum activity after 24 hours of

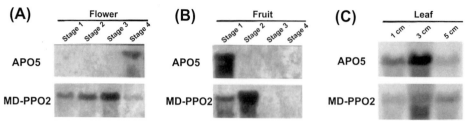

FIGURE 10.4 Tissue-specific expression of two Fuji apple polyphenol oxidase (PPO) mRNAs. Total RNAs were isolated from the different developmental stages of (A) flower, (B) fruit, and (C) leaf tissues as indicated *(Reprinted from J. Y. Kim et al., 2001). © 2001, with permission from Elsevier.*

wounding, whereas in cowpea PPO, the activity reached the maximum after 48 hours of wounding (J. Y. Kim *et al.*, 2001; Pinto *et al.*, 2008). Temperature is an important factor in determining the level of activation since PPO is thermosensitive. Queiroz *et al.* (2011) studied the effect of mechanical injury and storage for 24 hours at different temperatures on cashew apple PPO (Table 10.1). Five-fold higher activation was observed in fruits kept at 2°C and 27°C, but low activity at the higher temperature studied (40°C). Environmental stress can also modulate PPO activity. Stress situations are related to enhanced activity of phenylalanine ammonia-lyase, which regulates the synthesis of phenolic compounds, increasing the content of PPO substrates in plant cell (Dixon and Paiva, 1995). Tegelberg *et al.* (2008) found higher enzymatic activity in leaves of *Betula pendula* after exposure to elevated carbon dioxide (700 ppm) associated with elevated temperature (2.5°C higher than ambient temperature) and elevated ultraviolet (UV)-B radiation (7.95 kJ/m^2/day). On the other hand, Thipyapong *et al.* (2004b) reported higher tolerance to water stress in transgenic tomato with suppressed PPO than in non-transformed plants and in overexpressing PPO. However, in these conditions the authors also observed upregulation of two PPO genes, probably related to stress resistance. Rivero *et al.* (2001) submitted watermelon and tomato to cold and heat stress and observed that PPO activity was decreased and phenylalanine ammonia-lyase activity increased, leading to an accumulation of polyphenols in stressed plants. These findings demonstrate the different ways that plants use to defend against an undesirable situation.

The enzyme appears to play an important role in the resistance of plants to infection by viruses, bacteria, and fungi (Tyagi *et al.*, 2000; Mohammadi and Kazemi, 2002; Wang and Constabel, 2004). Under these conditions the activity of the enzyme increases with the production of insoluble polymers which serve as a barrier to the spread of infection in the plant. Alternatively, some of the intermediates in the oxidative polymerization of polyphenols prevent or reduce infection by inactivating or binding some labile plant enzymes or viruses. Mahanil *et al.* (2008) observed increased resistance to the common cutworm [*Spodoptera litura* (F.)] in transgenic tomato plants overexpressing PPO (Figure 10.5). The study also showed that increased PPO activity led to higher larval mortality. According to the authors, the efficiency of conversion of both ingested food and digested food of third instars was found to be significantly different among tomato genotypes with differing PPO activity levels, suggesting that PPO activity rendered foliage less nutritious. Other studies in tomato have shown the role of PPO in disease resistance; antisense suppression of PPO increases susceptibility, and PPO overexpression increases resistance, to *Pseudomonas syringae* pv. *tomato* (Li and Steffens, 2002; Thipyapong *et al.*, 2004a).

TABLE 10.1 Polyphenol Oxidase Activity by Wounded Cashew Apple

Enzyme Extract	Specific Activity (U/min/mg Protein)	Activation
0 hour (control)	0.62	—
2°C/24 hours	2.92	4.8
27°C/24 hours	3.33	5.4
40°C/24 hours	1.07	1.7

Reprinted from Queiroz *et al.* (2011). © 2011, with permission from Elsevier.

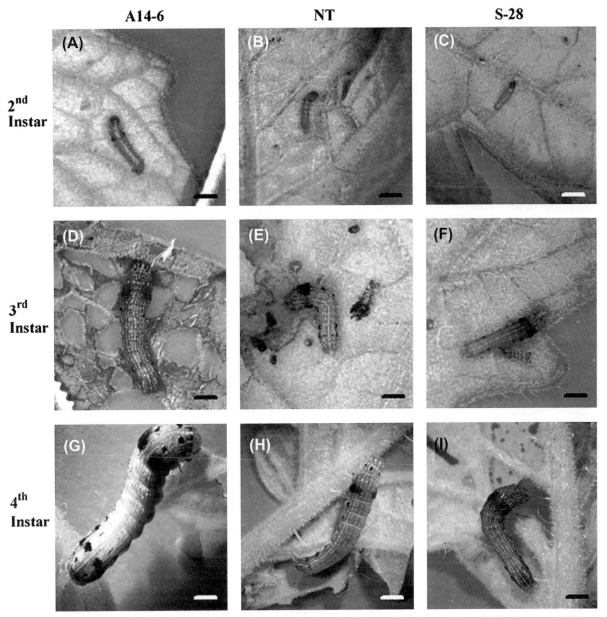

FIGURE 10.5 (A−C) Second instar, (D−F) third instar and (G−I) fourth instar larvae of *Spodoptera litura* (F.) feeding on node 8 leaves of tomato plants with varied polyphenol oxidase (PPO) activity levels. A14-6: transgenic line with suppressed PPO activity; NT: non-transformed control; S-28: transgenic line overexpressing PPO activity. Scale bars = 2.5 mm *(Reprinted from Mahanil et al., 2008). © 2008, with permission from Elsevier.*

D. Phenolic Compounds in Food Material

Phenolic compounds are natural substances that contribute to the sensorial properties (color, taste, aroma, and texture) associated with fruit quality (Marshall *et al.*, 2000). They form one of the main classes of secondary metabolites, with a large range of structures and functions, but generally possessing an aromatic ring bearing one or more hydroxy substituents. The phenolic composition of fruits is determined by genetic and environmental factors but may be modified by oxidative reactions. They are synthesized during plant development, but their synthesis is stimulated in stress conditions by activation of phenylalanine ammonia-lyase. Thus, these compounds play a role in defense and adaptive mechanisms of plants. The phenolic compounds which occur in food material and which participate in browning may be classified into four groups, namely, simple phenols, phenolic acids, cinnamic acid derivatives, and flavonoids.

1. Simple Phenols

The simple phenols include monophenols such as L-tyrosine and *o*-diphenols such as catechol, resorcinol, and hydroquinone. However, among these compounds only catechol can be oxidized by PPO, because hydroxyl is at the *ortho* position. Catechol was identified in *Diospyro kaki* roots by Jeong *et al.* (2009) and PPO showed high affinity towards this phenolic compound (Özen *et al.*, 2004).

catechol   resorcinol   hydroquinone

2. Phenolic Acids

This class comprises the acids synthesized from precursor benzoic acid and they are widely distributed in plants. Gallic acid is present in an esterified form in tea flavonoids. Gallic and protocatechuic acids were found in cashew apple (Michodjehoun-Mestres *et al.*, 2009; Queiroz *et al.*, 2011) and protocatechuic was the main free phenolic acid identified in medlar fruits (Gruz *et al.*, 2011).

Benzoic Acids

Gallic acid $R_1=R_2=R_3=OH$
Protocatechuic acid $R_1=H$, $R_2=R_3=OH$
Vanillic acid $R_1=H$, $R_2=OH$, $R_3=OCH_3$
Syringic acid $R_2=OH$, $R_1=R_3=OCH_3$

3. Cinnamic Acid Derivatives

The most important member of this group of compounds in food material is chlorogenic acid, which is the key substrate for enzymatic browning, particularly in apples and pears (Gauillard and Forget, 1997; Song *et al.*, 2007). Although potato is rich in chlorogenic acid, it is not a determinant factor for blackspot development. Two potato cultivars (cv. 'Bildtstar' and cv. 'Lady Rosetta') were bruised and the pigments identified. Quinic acid was detectable in hydrolysates of the pigments from 'Bildtstar', but not in those of 'Lady Rosetta', which indicated that chlorogenic acid may take part in blackspot formation, but is not essential for the discoloration (Stevens and Davelaar, 1996). These data were confirmed by Lærke *et al.* (2002), who found a correlation between black end products and free tyrosine, but none between black color and chlorogenic and caffeic acids in potato cultivars (cv. 'Dali' and cv. 'Oleva'). Discoloration derived from chlorogenic acid is attributed to the oxidation of complexes formed between iron and caffeic and chlorogenic acids.

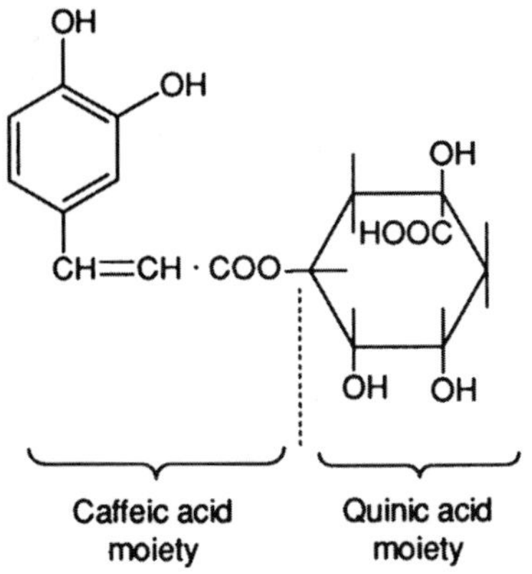

Caffeic acid moiety　　Quinic acid moiety

Other members of this group of compounds include *p*-coumaric, caffeic, ferulic, and sinapic acids. The universal distribution and high concentration of cinnamic acids in fruits may be due to their function as precursors in the biosynthesis pathway of more complex polyphenols.

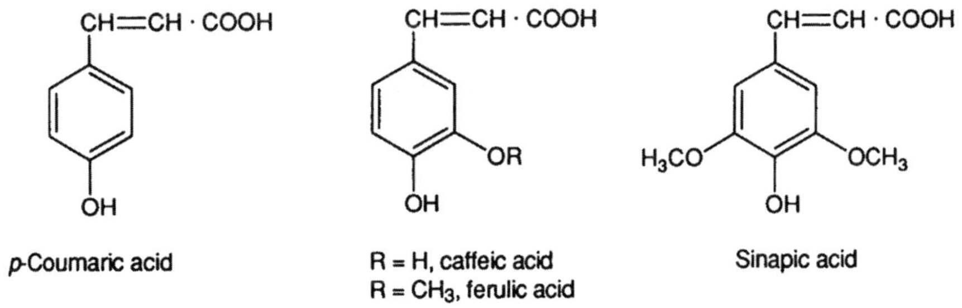

p-Coumaric acid

R = H, caffeic acid
R = CH₃, ferulic acid

Sinapic acid

4. Flavonoids

This group is the most widespread and structurally diverse among polyphenols. All members of this group of compounds are structurally related to flavone:

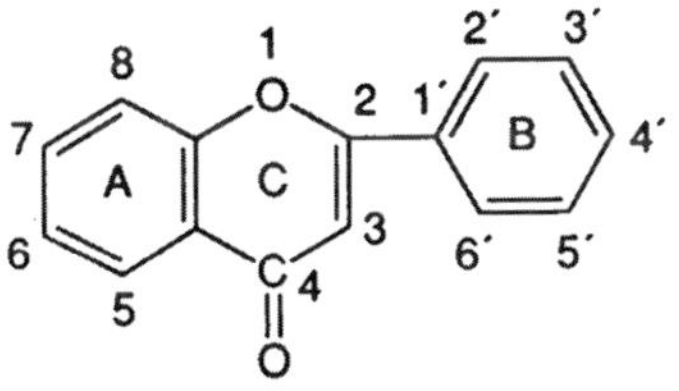

In food material, the important flavonoids are the catechins, anthocyanins, and flavonols. Catechin has the following structure:

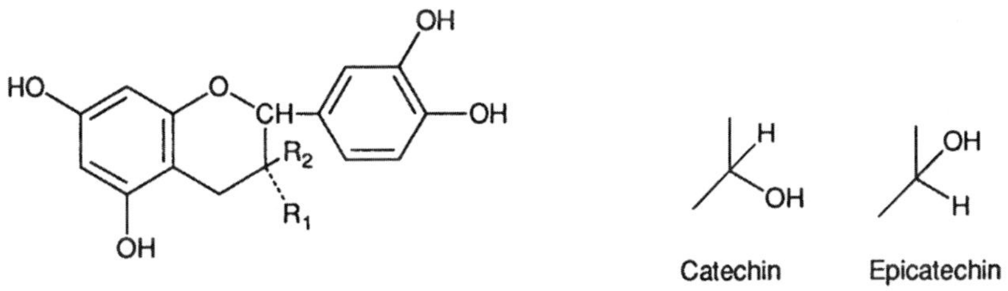

Catechin　　Epicatechin

(+)–Catechin (R$_1$ = H; R$_2$ = OH)

(−)–Epicatechin (R$_1$ = OH; R$_2$ = H)

An extra hydroxyl group attached to the 5' position on the B ring of catechin and epicatechin gives rise to gallocatechin and epigallocatechin, respectively. The catechin gallates are esters of catechins and gallic acid, the ester linkage being formed from the carboxyl group of gallic acid and the hydroxyl group attached to position 3 of the catechin C ring. An example is (−)-epigallocatechin gallate, which is the major polyphenol in dried tea leaves.

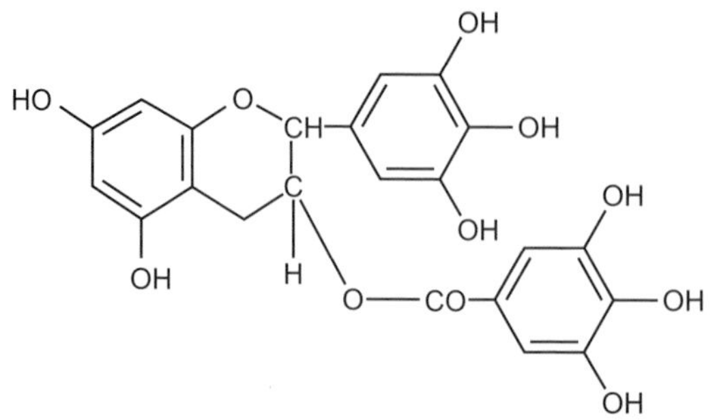

Such compounds are not the major phenols in fruits; however, they are important constituents of fruits in oligomeric or polymeric forms as proanthocyanidins or condensed tannins. Procyanidin is a dimer and is present in apple, grape, and cherry (Robards *et al.*, 1999).

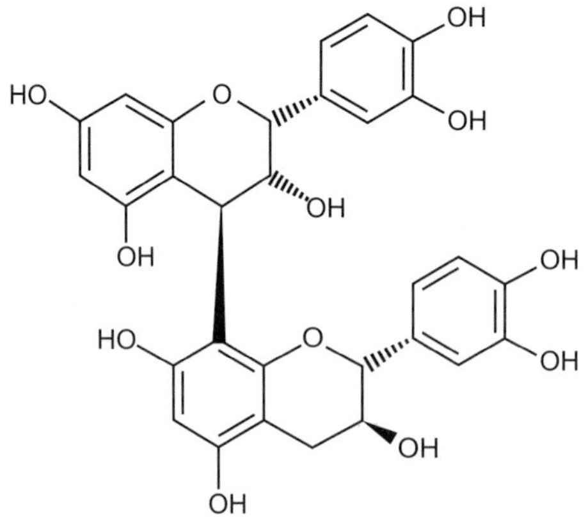

Procyanidin B-1

The flavonols also participate in browning reactions and are widely distributed in plant tissues. The most commonly occurring flavonols are kaempferol, quercetin, and myricetin:

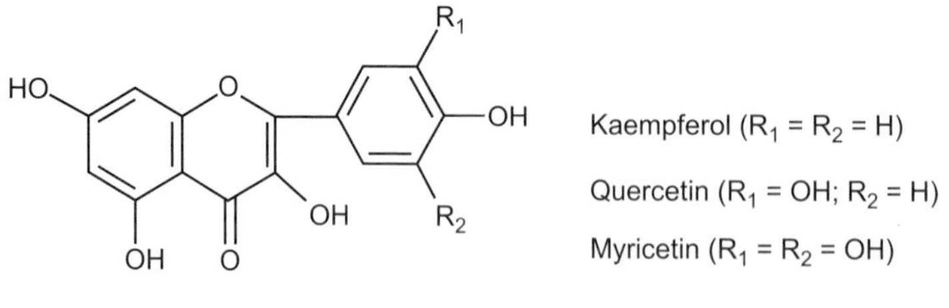

Kaempferol ($R_1 = R_2 = H$)

Quercetin ($R_1 = OH; R_2 = H$)

Myricetin ($R_1 = R_2 = OH$)

Flavonols possess a light yellow color and are particularly important in fruits and vegetables in terms of the astringency they impart to the particular food. They occur naturally as glycosides, examples of which are rutin and the glycosides of quercetin, the latter occurring in tea leaves and apple skins (Hulme, 1958).

All of the compounds discussed so far are substrates for PPO. Such oxidation reactions are important in tea fermentation, in the browning of cling peaches (Luh *et al.*, 1967), and in the drying stage of the curing of fresh cacao seeds (Roelofsen, 1958). It appears to be an important step in the development of the final color, flavor, and aroma of cacao and chocolate. PPO plays a beneficial role in the case of tea and cacao fermentation, in contrast to its role in the browning of fruits and vegetables.

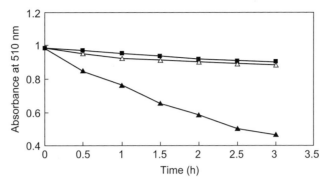

FIGURE 10.6 Changes in absorbance of bayberry anthocyanin at 510 nm during its degradation by bayberry polyphenol oxidase (PPO) in the presence (▲) and absence (△) of gallic acid and after thermal denaturation of the PPO in the presence of gallic acid (■) *(Reprinted from Fang et al., 2007).© 2007, with permission from Elsevier.*

Several studies have been conducted to identify other endogenous substrates for PPO in fruits. Epicatechin is the endogenous substrate for litchi (lychee) and longan PPOs (Sun *et al.*, 2006; Shi *et al.*, 2008). Other flavonoids, such as eriodictyol, myricetin, and fisetin, can also be oxidized by PPO (Jiménez *et al.*, 1998; Jiménez and García-Carmona, 1999; Jiménez-Atiénzar *et al.*, 2005a, b). Anthocyanins are not directly oxidized by PPO or are poor substrates (Mathew and Parpia, 1971), but they could be degradated by a coupled oxidation mechanism. Ruenroengklin *et al.* (2009) suggested that litchi PPO directly oxidized epicatechin, then oxidative products of epicatechin, in turn, catalyzed litchi anthocyanin degradation, and finally resulted in the browning reaction, which can account for pericarp browning of postharvest litchi fruit. Similar results were found by Fang *et al.* (2007) which demonstrated that the rate of anthocyanin degradation in bayberry was stimulated by the addition of gallic acid. In the absence of gallic acid, the rate of cyaniding 3-glucoside discoloration was almost the same as when the enzymatic extract was inactivated by heating and gallic acid was present (Figure 10.6).

E. Laccase

Early studies on enzymatic browning of phenolic compounds identified two types of activity which were initially termed tyrosinase and laccase. The respective systematic names formerly used were *o*-diphenol: oxygen oxidoreductase (EC 1.10.3.1) and *p*-diphenol: oxygen oxidoreductase (EC 1.10.3.2), which are now combined as monophenol monooxygenase (EC 1.14.18.1). The diagnostic feature of laccase is its ability to oxidize *p*-diphenols, a property not possessed by tyrosinase or PPO (Mayer and Harel, 1979). It is present in many plants, fungi, and microorganisms; however, most of the known laccases are of fungal origin, in particular from the white rot fungi

(Thurston, 1994; Revankar and Lele, 2006; Fonseca *et al.*, 2010). The basic laccase-catalyzed reaction responsible for the oxidation of *p*-diphenol is:

Laccase is responsible for the oxidation of flavonoids and browning during *Arabidopsis* seed development (Figure 10.7) and is one of the major enzymes responsible for oxidation of lignin, an amorphous polymer which functions as a cementing material in wood cells. The enzyme acts in cooperation with other ligninolytic (lignin-degrading) fungal enzymes such as lignin peroxidase, manganese peroxidase, and versatile peroxidase (Arora *et al.*, 2002). Laccase alone can only oxidize the phenolic lignin units. Therefore, laccase is often applied with an oxidation mediator, a small molecule able to extend the effect of laccase to non-phenolic lignin units and to overcome the accessibility problem (Srebotnik and Hammel, 2000; Shleev *et al.*, 2006). The mediator is first oxidized by laccase and then diffuses into the cell wall, oxidizing lignin inaccessible to laccase. The use of these mediators in laccase reactions enables the application of the enzyme in the forest industry, removing pitch, phenolic contaminants, and dyes from wood-based materials and water; laccase technology is applicable to virtually the entire production chain of paper products from pulping to recovery of secondary fibers and effluent treatment (Widsten and Kandelbauer, 2008).

F. Specificity of Polyphenol Oxidase

As discussed previously, PPO catalyzes two different reactions, either the hydroxylation of monophenols to *o*-dihydroxyphenols or the oxidation of *o*-dihydoxyphenols to *o*-quinones. The most appropriate chemical structure with respect to PPO activity, when the reaction rate is at a maximum, appears to correspond to the *o*-dihydroxy structure as evident in such compounds as catechol, caffeic acid, and the catechins (Rocha and Morais, 2001; Gawlik-Dziki *et al.*, 2007). Oxidation of *o*-diphenols to the corresponding *o*-quinones is a general reaction of all known PPO, irrespective of whether the source material is potato, sweet potato (Hyodo and Uritani, 1965), lettuce (Gawlik-Dziki *et al.*, 2008), apple (Harel *et al.*, 1966), tomato (Hobson, 1967), banana (Ünal, 2007), artichoke (Aydemir, 2004), tobacco (Shi *et al.*, 2002), cashew apple (Queiroz *et al.*, 2011), litchi (Ruenroengklin *et al.*, 2009; Yue-Ming *et al.*, 1997), or green olives (Segovia-Bravo *et al.*, 2009). Monophenols

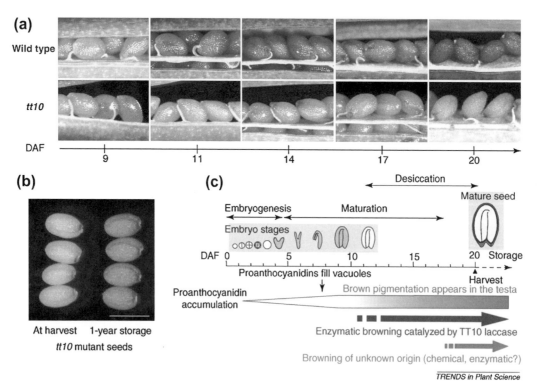

FIGURE 10.7   Seed coat pigmentation in *Arabidopsis*: illustration of a browning process. (a) Photographs showing the appearance of a brown pigmentation in the testa of the wild-type genotype during seed desiccation. The brown pigment is absent from the transparent testa 10 (tt10) mutant defective in a laccase enzyme. (b) Mutant tt10 seeds slowly become brown after harvest and eventually resemble wild-type seeds. Scale bar = 550 mm. (c) Schematic drawing indicating the occurrence of brown pigmentation during *Arabidopsis* seed development. DAF: days after flowering (*Reprint from Pourcel* et al., *2006*).© *2006, with permission from Elsevier.*

are more slowly acting substrates as they have to be hydroxylated before their oxidation to the corresponding *o*-quinones. The oxidation of monophenols is less widespread than that of diphenols, being catalyzed, for example, by potato and mushroom enzyme preparations. The relationship between cresolase and catecholase activities is not yet fully understood. It appears that many PPOs are specific to a high degree in that they attack only *o*-diphenols. Aydemir (2004), in a study on artichoke PPO, found that it exhibited the highest specificity towards catechol, followed by 4-methycatechol and pyrogallol. According to Espín *et al.* (1998), the best substrates are those with low molecular size substituent side-chain and high electron donor capacity. Erat *et al.* (2006), studying PPO activity in *Ferula* sp., found the best results using catechol and (−)-epicatechin with enzyme extracted from leaf and stem, respectively. Sellés-Marchart *et al.* (2006) found chlorogenic acid to be the most efficient substrate, followed by 4-methylcatechol, 4-*tert*-butylcatechol, epicatechin, catechol, and isoproterenol, for loquat fruit (*Eriobotrya japonica* Lindl.). PPO from Henry chestnuts catalyzed the oxidation of catechol and pyrogallic acid, but had no effect on cresol or tyrosine (Xu *et al.*, 2004). The substrate with the highest activity, for Amasya apple (*Mallus sylvestris* Miller cv. Amasya) PPO, was catechol, followed by 4-methylcatechol, pyrogallol, and 3,4-dihydroxylphenylalanine (L-DOPA) (Oktay *et al.*, 1995). Other kinetic parameters found by researchers are described in Table 10.2.

TABLE 10.2 Characteristics of Polyphenol Oxidases from Plant Sources

Plant	Substrate	K_m (mM)	Optimum temperature (°C)	Optimum pH
Apple[a]	Catechol	34	15	7.0
	4-Methylcatechol	3.1		
Artichoke[b]	Catechol	10.2	25	6.0
	4-Methylcatechol	12.4		
Banana[c]	Catechol	8.5	30	7.0
Broccoli[d]	Catechol	12.3	−	5.7
	4-Methylcatechol	21.0		
Cashew apple[e]	Catechol	18.8	−	6.5
Grape[f]	Chlorogenic acid	3.2	25	5.0
	Catechin	4.3		
Mango[g]	Catechol	6.3	30	7.0
Melon[h]	DOPAC[l]	7.2	60	7.0
Strawberry[i]	Catechol	5.9	25	5.0
Vanilla[j]	Catechol	85.0	37	3.4
	4-Methylcatechol	10.6	37	3.0
Yacon root[k]	Caffeic acid	0.2	30	6.6
	Chlorogenic acid	1.1		

[a]*Oktay* et al. *(1995)*
[b]*Aydemir* et al. *(2004)*
[c]*Ünal (2007)*
[d]*Gawlik-Dziki* et al. *(2007)*
[e]*Queiroz* et al. *(2011)*
[f]*Rapeanu* et al. *(2006)*
[g]*Wang* et al. *(2007)*
[h]*Chisari* et al. *(2008)*
[i]*Dalmadi* et al. *(2006)*
[j]*Waliszewski* et al. *(2009)*
[k]*Neves and Silva (2007)*
[l]*3,4-dihydroxyphenylacetic acid.*
Adapted from Queiroz *et al.* (2008).

III. POLYPHENOL OXIDASE IN FOODS AND FOOD PROCESSING

A. Role in Tea Fermentation

The production of black tea is dependent on the oxidative changes that *Camellia sinensis* leaf polyphenols undergo during processing. Such changes are particularly important for the development of color as well as the reduction of the bitter taste associated with unoxidized tannin (polyphenol compound). Several types of tea, namely white, green, oolong, and black tea, come from the same plant, *C. sinensis*. It is the processing that the tea leaves go through which determines what type they will become and will also alter the phenolic composition of the tea.

The main tea leaf polyphenols, determined by partition chromatography, include (+)-catechin, (−)-epicatechin, (+)-gallocatechin, (−)-epigallocatechin, (−)-epicatechin gallate, and (−)-epigallocatechin gallate. Of these compounds, (−)-epigallocatechin gallate is the major component in the tea shoot. During tea processing, Muthumani and Kumar (2007) observed that considerable quantities of epigallocatequin gallate, epigallocatechin, and epicatechin gallate were oxidized to form theaflavins and their gallates. Changes in catechin and epicatechin contents were not observed, probably owing to the formation of free gallic acid and catechin from the other catechin fractions such as epigallocatequin gallate, epigallocatechin, and epicatechin gallate, by oxidative degallation. Different results were published by Munoz-Munoz *et al.* (2008). These authors found that the best substrates for PPO in tea leaves are epicatechin followed by catechin, owing to better access of enzyme to hydroxyl radicals.

The production of black tea, the most popular form of the beverage, is carried out in four stages. The first stage is called withering, when the shoots from the tea plant are allowed to dry out. This is followed by rolling with a roller, which disrupts the tea leaf tissue and causes cell damage, providing the necessary conditions for the development of the oxidative processes. The next step is the fermentation of the fragmented tea leaves, which are held at room temperature in a humid atmosphere with a continuous supply of oxygen. These conditions are optimal for PPO action on the tea leaf catechins, which, in addition to reducing astringency, converts the green color of the rolled tea leaves to give coppery-red and brown pigments. Fermentation is terminated by firing, where the tea is dried at 90−95°C and the moisture reduced to 3−4%.

The critical biochemical reaction during tea fermentation is oxidation of catechins by PPO to the corresponding *o*-quinones. These quinones are intermediate compounds which are subjected to secondary oxidation leading to the production of theaflavin and theaflavin gallate, the yellow−orange pigments in black tea, and to a group of compounds referred to as thearubigins. These thearubigins are dark brown and the main contributors to the familiar color of black tea, and they are the oxidative products of theaflavins. A simplified scheme for the oxidative reactions occurring during tea fermentation is outlined in Scheme 10.2. The theaflavin content of tea was shown by Hilton and Ellis (1972) to correlate with the tea taster's evaluation. This was consistent with earlier studies by Roberts (1952) and Sanderson (1964), who noted a positive correlation between tea quality and PPO activity. The enzyme was later purified by Hilton (1972). The oxidative degradation of phloroglucinol rings of the theaflavins by peroxidase caused a loss of theaflavins and a decline in tea quality (Cloughley, 1980a, b). Consequently, the presence of both these enzymes affects the quality of tea. Van Lelyveld and de Rooster (1986) examined the browning potential of black tea clones and seedlings. They found a much higher level of PPO in a high-quality hybrid clone (MT12) compared to a low-quality seedling tea (Table 10.3). The reverse was true for peroxidase, in which lower quality tea had more than double the activity of peroxidase. This suggested that the combination of higher theaflavin levels and PPO activity was responsible for the better quality associated with the MT12 clone. Subramanian *et al.* (1999) studied the role of PPO and peroxidase in the formation of theaflavins. They demonstrated that PPO generates hydrogen peroxide (H_2O_2) during oxidation of catechins and peroxidase utilizes the formed H_2O_2 for subsequent oxidation of the products of PPO-catalyzed reactions, decreasing the content of theaflavins. The fermentation time is also important to the formation of theaflavins and thearubigins. Muthumani and Kumar (2007) reported an optimum time of 45 minutes, when theaflavins reached their maximum content; after this time their content declined to 20%, leading to a lower quality tea.

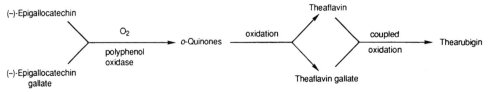

SCHEME 10.2   Oxidative transformations of (−)-epigallocatechin and its gallate during tea fermentation.

TABLE 10.3 Specific Activity (Δ OD/min/mg protein) of MT12 and Seedling Tea Leaves

Clone	Peroxidase[a]	Polyphenol Oxidase[b]
M12	1.821	0.055
Seedling	0.735	0.020

[a]Significant at $p < 0.05$
[b]Significant at $p < 0.01$.
From Van Lelyveld and de Rooster (1986).

Green tea is particularly popular in Oriental countries, such as Japan. It is an unfermented tea with a light color and a characteristic degree of astringency owing to a high content of catechins. This is achieved by the application of heat during the early stages of tea manufacture, which inhibits or prevents oxidation. Red and yellow teas are intermediate between black and green teas and are semifermented products (partially fermented before firing). An example of the latter is the Chinese variety oolong. Tea leaf catechins generate theaflavins and thearubigins during the longer processing time. White and green teas have higher levels of catechin, whereas oolong and black teas are rich in theaflavins and thearubigins (Table 10.4).

B. Shrimps and Crustaceans

Enzymatic browning, having been studied extensively in fruits and vegetables, has also been implicated in the discoloration, called melanosis or blackspot, of shrimps and other crustaceans (Zamorano *et al.*, 2009; Giménez *et al.*, 2010), which renders these products unattractive to the consumer and lowers their market value. In crustaceans, PPO is mainly located in the cuticle, specifically on the internal surface, inside chromatophores. During iced storage of crustaceans, the inactive PPO stored in homocytes and digestive glands can also be activated by the action of proteolytic enzymes leaching from the digestive tract. Moreover, protein hydrolysis by these proteases provides substrates for active PPO (Ali *et al.*, 1994).

A recent study by Zamorano *et al.* (2009) evaluated PPO activity *in situ* and in partially purified extracts from different organs of deepwater pink shrimp (*Parapenaeus longirostris*). They found higher enzyme activity in carapace extracts, but marked melanosis developed on the cephalothorax and head after 1 day at 4°C (Figure 10.8). Even after 7 days at 4°C, there was no melanosis in the carapace, confirming that the development of melanosis in the different tissues depends on another factor in addition to PPO levels. In native non-denaturing conditions, a band of 500 kDa was identified which was capable of oxidizing dihydroxyphenylalanine. Table 10.5 shows some biochemical characteristics of PPO extracted from marine sources.

TABLE 10.4 Flavonoid Composition of Tea (Percent by Dry Weight)

Component	Green Tea	Black Tea
Total flavonoids	15−25	15−25
Total catechins	12−18	2−3
(−)-Epicatechin	1−3	< 1
(−)-Epicatechin gallate	3−6	< 1
(−)-Epigallocatechin	3−6	< 1
(−)-Epigallocatechin gallate	9−13	1−2
Flavonols	2−3	1−2
Theaflavins	< 1	4
Other polyphenols	2−4	7−15

Reprinted from Hodgson and Croft (2010). © 2010, with permission from Elsevier.

FIGURE 10.8 Melanosis appearance in different anatomical parts of deepwater pink shrimp during 0, 1, 4, and 7 days of storage at 4°C. (A) Whole shrimp; (B) Whole shrimp with the carapace removed; (C) Head (carapace + cephalothorax + pereopods and maxillipeds); (D) Carapace; (E) Head with the carapace removed; (F) Pereopods + maxillipeds; (G) Whole abdomen (including the exoskeleton, muscle, pleopods, and telson); (H) Abdomen with the muscle removed (left) and corresponding muscle (right); and (I) Telson *(Reprinted from Zamorano et al., 2009). © 2009, with permission from Elsevier.*

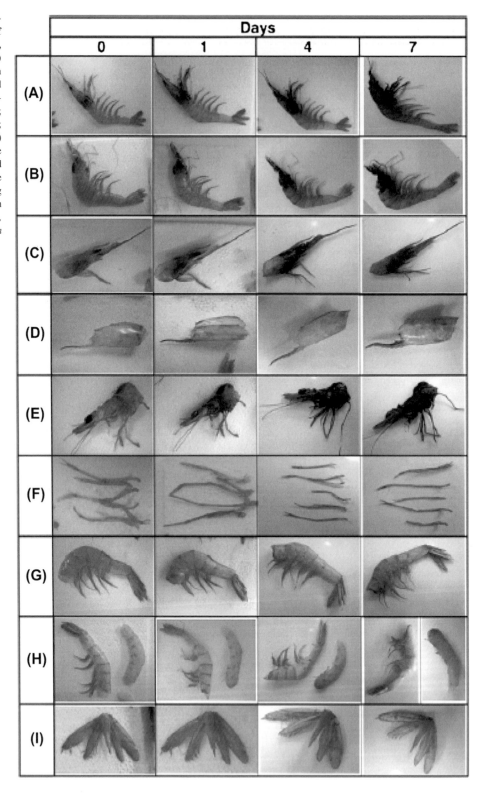

Melanosis inhibitors are used in marine food to prevent browning and extend the shelf-life of products. Traditional methods include the use of sulfite-based formulations, but sulfite causes adverse reactions. Therefore, other compounds, including those of natural origin, have been studied as alternatives to sulfur compounds. Martínez-Alvarez et al. (2007) sprayed 4-hexylresorcinol-based formulations on Norway lobsters. After 12 days of storage they

TABLE 10.5 Characteristics of Polyphenol Oxidases from Marine Sources

Crustacean	Substrate	K_m	Optimum temperature (°C)	Optimum pH
Deepwater pink shrimp[a]	DOPA	1.85	15–60	4.5
Prawn[b]	DOPA	–	40–60	5.0 and 8.0
Norway lobster[c] (carapace)	Catechol	19.40	60	7.0
Norway lobster[c] (viscera)	Catechol	5.90	60	8.0

DOPA: 3,4-dihydroxylphenylalanine.
[a]*Zamorano* et al. *(2009)*
[b]*Montero* et al. *(2001)*
[c]*Giménez* et al. *(2010)*.

observed lower PPO activity and melanosis and improvement in the appearance of the lobsters. Cystein and glutathione inhibited the oxidation of L-DOPA catalyzed by kuruma prawn PPO (Benjakul *et al.*, 2006). Among natural substances, recent studies have demonstrated the potential use of phenolic compounds inhibiting browning. Shrimps treated with 2% ferulic acid had, after 10 days of storage, a lower melanosis score and higher scores for color, flavor, and overall likability, compared with the control and sodium metabisulfite-treated shrimps (Nirmal and Benjakul, 2009). Treatment combining catechin and ferulic acid before freeze–thawing was efficient in retarding melanosis in Pacific white shrimp. Phenolics also acted to decrease the growth of psychrophilic bacteria and retard lipid oxidation during the subsequent refrigerated storage (Nirmal and Benjakul, 2010).

Novel technology such as high hydrostatic pressure was studied by Montero *et al.* (2001), achieving partial inactivation of prawn PPO after treatment at 300–400 MPa for 10 minutes. The mechanism of enzyme inactivation by high pressure is discussed in Section IV, D.

IV. CONTROL OR INHIBITION OF ENZYMATIC BROWNING

The inactivation of PPO is required to minimize product losses caused by browning. Several methods and technologies have been studied. Heat treatment and addition of antibrowning agents are usually applied, but several researchers have proposed the application of other methods as alternatives to thermal processing for PPO inactivation (Queiroz *et al.*, 2008).

Inhibitors of PPO activity can be based on their mode of action, for example, the exclusion of reactants such as oxygen, denaturation of enzyme protein, interaction with the copper prosthetic group, and interaction with phenolic substrates or quinones.

A. Exclusion of Oxygen

The simplest method of controlling enzymatic browning is by immersing the peeled product, such as potato, in water before cooking. This can be done very easily in the home to limit access of oxygen to the cut potato tissue. This procedure has been used for a long time on a large scale for the production of potato chips and French fries (Talburt and Smith, 1967). The method is limited as the fruit or vegetable will brown on re-exposure to air or via the oxygen occurring naturally in the plant tissues. The removal of oxygen from fruit or vegetable tissue could lead to anaerobiosis if it is stored for extended periods, which in turn could lead to abnormal metabolites and tissue breakdown. In the case of frozen sliced peaches, the surfaces are treated with excess ascorbic acid to use up the surface oxygen.

More recently, modified atmospheres have also been used to prevent enzymatic browning. Since browning is an oxidative reaction, it can be retarded by eliminating oxygen from the cut surface of the vegetables (Limbo and Piergiovanni, 2006). Modified atmosphere packaging has the potential to extend the shelf-life of fruits and vegetables, mainly by limiting the oxidation processes. Low-oxygen and elevated carbon dioxide atmospheres can reduce surface browning. This decrease in the browning phenomenon is accompanied by several physiological effects such as a decrease in respiration rate and a delay in the climacteric onset of the rise in ethylene (Solomos, 1997). Day (1996) hypothesized that high oxygen concentrations may cause substrate inhibition of PPO or, alternatively, high levels of colorless quinones formed may cause feedback inhibition of the enzyme. Gorny *et al.* (2002) found that low-oxygen (0.25 or 0.5 kPa), elevated carbon dioxide (air enriched with 5, 10, or 20 kPa CO_2), or high-oxygen (40, 60, or 80 kPa) active atmospheres alone did not effectively prevent surface browning of fresh-cut pear slices. Studies on the influence of modified atmosphere packaging and the use of antioxidant compounds on the shelf-life of fresh-cut pears have mainly focused on the sensory qualities of the commodity (Sapers and Miller, 1998; Gorny *et al.*, 2002; Soliva-Fortuny *et al.*, 2002, 2004). According to Oms-Oliu *et al.* (2008), knowledge about the impact of dipping treatments and packaging conditions on antioxidant properties of fresh-cut pears is still incomplete, especially in regard to high-oxygen active packaging. Y. Kim *et al.* (2007) evaluated the effects of hot water treatment in combination with controlled atmosphere (CA) storage and the resultant changes in quality and antioxidant polyphenolics in mango. They observed that the external quality of mango fruit during ripening was greatly improved by low-oxygen and/or high-carbon dioxide storage conditions and hot water immersion treatments. Carbon dioxide storage delayed fruit ripening, as evidenced by physicochemical changes, and the hot water treatment plus CA storage was an effective treatment combination to extend the postharvest life of mangoes, without adversely changing the nutritional profile of the fruit.

Limbo and Piergiovanni (2006) studied the effects of high oxygen partial pressures in combination with ascorbic and citric acid on the development of the enzymatic browning of peeled and cut potatoes ('Primura' variety), packaged in flexible pouches and stored at 5°C for 10 days. They observed that the treatments with high oxygen partial pressures had some positive effects on enzymatic browning only if the initial atmosphere was close to 100 kPa and the dipping acid concentrations were chosen with care. As Wszelaki and Mitcham (2000) have shown, near 100 kPa oxygen atmospheres could be difficult to maintain either in a package or on a larger scale, as well as being dangerous owing to flammability. A study conducted by Saxena *et al.* (2008) showed that the synergistic effect of antibrowning and antimicrobial compounds with reduced oxygen and elevated carbon dioxide atmosphere could enhance the keeping quality of fresh-cut jackfruit bulbs by minimizing deteriorative changes in physiological, sensorial, and microbial attributes. Martínez-Sánchez *et al.* (2011), studying the influence of oxygen partial pressures (pO_2) and light exposure during storage on the shelf-life of fresh-cut Romaine lettuce, noticed that the consumption of oxygen in samples exposed to 24-hour light differed significantly from that of those stored in a 12-hour light/12-hour dark photoperiod or in 24-hour darkness (10.6 ± 7.0, 18.3 ± 3.5, and 25.8 ± 8.6 nmol O_2/kg/second, respectively). Packages exposed to light showed higher pO_2 than packages stored in darkness, while those exposed to the photoperiod had intermediate values. This study showed that under light conditions respiration activity was compensated by photosynthesis, resulting in a higher pO_2. Thus, browning of fresh-cut Romaine lettuce can be promoted by light exposure during storage as it increases headspace pO_2. In fact, uncontrolled light conditions during storage and commercial distribution can contribute to the induction of the browning process, which can cause the product to be unsaleable.

B. Chemical Inhibitors of Polyphenol Oxidase

Different effectors can control enzymatic browning and these compounds are classified, based on the inhibition mechanism, as reducing agents, chelating agents, acidulants, enzyme inhibitors, enzyme treatments, and complexing agents (Özoğlu and Bayındırlı, 2002).

Several compounds possess chemical structures closely related to *o*-diphenols but do not function as substrates of PPO. These include methyl-substituted derivatives such as guaiacol and ferulic acid (Finkle and Nelson, 1963; Nirmal and Benjakul, 2009), and *m*-diphenols such as resorcinol and phloroglucinol, which have been shown to have an inhibitory effect on PPO activity, either in a competitive way (Montero *et al.*, 2001; Waliszewski *et al.*, 2009) or by a slow-binding inhibition mechanism (Jiménez and García-Carmona, 1997), depending on the substrate.

Guaiacol   Resorcinol   Phloroglucinol

Since PPO is a metalloprotein in which copper is the prosthetic group, it is inhibited by a variety of chelating agents, including sodium diethyldithiocarbamate (DIECA), sodium azide, potassium ethylxanthate, and ethylenediaminetetraacetate (EDTA). While sodium azide and EDTA inhibit PPO, they appear to be less specific chelators than DIECA or potassium ethylxanthate (Kavrayan and Aydemir, 2001; Aydemir, 2004; Gawlik-Dziki et al., 2007; Pinto et al., 2008; Gao et al., 2009). The latter compounds were reported by Anderson (1968) to combine with the quinones produced by PPO in addition to chelating copper.

Sulfites are strong inhibitors and have been used for a long time in the food industry. However, an excess of sulfite-based formulations can cause adverse reactions, mainly in asthmatic people. Since it was reported by the 51st meeting of the Joint Expert Committee on Food Additives that the intake of sulfites could be above safe limits, their use as food additives is not recommended. Researchers have been carrying out studies to find potent antibrowning agents which cause no damage to the organism. L-Cysteine has been shown to be an effective inhibitor of PPO by acting as a quinone coupler as well as a reducing agent. It reacts with o-quinone intermediates to produce stable and colorless products (İyidoğan and Bayındırlı, 2004). Since Kahn (1985) reported L-cysteine to be the most effective inhibitor of avocado, banana, and mushroom PPO, many studies have shown its effect on the PPO activity of artichoke, apple juice, mango purée, grape, and lettuce (İyidoğan and Bayındırlı, 2004; Guerrero-Beltrán et al., 2005; Rapeanu et al., 2006; Altunkaya and Gökmen, 2009). However, effective concentrations of cysteine produce an undesirable odor, limiting its use in food processing. Altunkaya and Gökmen (2009), trying to select the most effective antibrowning compound on fresh lettuce, evaluated the effectiveness of ascorbic acid, cysteine, citric acid, and oxalic acid. They found that lettuce treated with oxalic acid and ascorbic acid maintained a higher level of phenolic compounds than citric acid and cysteine. Interestingly, cysteine had no positive effect on the prevention of oxidation of phenolic compounds even though it prevented browning in lettuce.

The application of acids to control enzymatic browning is used extensively. The acids employed are those found naturally in plant tissues, including citric, malic, phosphoric, and ascorbic acids. This method is based on the fact that lowering the tissue pH will reduce or retard the development of enzymatic browning. The optimum pH of most PPO lies between pH 4.0 and 7.0, with little activity below pH 3.0, as illustrated in Table 10.2. Muneta (1977) examined the effect of pH on the development of melanins during enzymatic browning. Although, as discussed earlier, the initial reaction involving the formation of quinone is enzyme catalyzed, polymerization of these quinones to the brown or brown–black melanins is essentially non-enzymatic. Since both are pH dependent, Muneta (1977) studied the effect of pH on the non-enzymatic reactions leading to the formation of melanins in potatoes. The formation of dopachrome from dopaquinone was monitored in phosphate buffers at pH 5.0, 6.0, and 7.0. Very rapid melanin development occurred at pH 7.0, compared to a rather slow process at pH 5.0. This observation could be important to the processor, particularly if the lye-peeled potatoes are inadequately washed, resulting in a high surface pH that facilitates enzymatic browning and melanin formation.

Citric acid has been used in conjunction with ascorbic acid or sodium sulfite as a chemical inhibitor of enzymatic browning (Limbo and Piergiovanni, 2006; Ducamp-Collin et al., 2008; Hiranvarachat et al., 2011; Queiroz et al., 2011). Cut fruit or vegetables are often immersed in dilute solutions of these acids just before processing. This is particularly important in the case of lye-peeled cling fruits, where the acid dip counteracts the effect that any residual lye might have on enzymatic browning. Citric acid also inhibits PPO by chelating the copper moiety of the enzyme. Compared to citric acid, however, malic acid, the principal acid in apple juice, is a much more effective inhibitor of enzymatic browning.

A particularly effective inhibitor of PPO is ascorbic acid (Özoğlu and Bayındırlı, 2002). It does not have a detectable flavor at the level used to inhibit this enzyme, nor does it have a corrosive action on metals. This vitamin acts as an antioxidant because it reduces the quinone produced before it undergoes secondary reactions that lead to browning, and also decreases the pH for the enzyme activity (Guerrero-Beltrán et al., 2005). The mode of action of ascorbic acid is outlined in Scheme 10.3 (Walker, 1976). Ascorbic acid can inhibit the enzyme by decreasing pH and acting as an antioxidant. Thus, it reduces quinones to diphenols, reversing the reaction and decreasing brown pigments. Degl'Innocenti et al. (2007) evaluated lettuce, escarole, and rocket salad and their sensitivity to enzymatic browning. They observed that the resistance of rocket salad to browning was associated with the highest ascorbic acid content compared to other species. A study by Hsu et al. (1988) compared the effectiveness of several ascorbic acid derivatives, including dehydroascorbic acid, isoascorbic acid, ascorbic acid-2-phosphate, and ascorbic acid-2-sulfate, with that of ascorbic acid. On the basis of kinetic studies they found ascorbic acid and isoascorbic acid to be the most effective inhibitors of mushroom PPO, followed by dehydroascorbic acid. According to Özoğlu and Bayındırlı (2002), ascorbic acid has been shown to

SCHEME 10.3 Reduction by ascorbic acid of the primary quinone oxidation products of enzymatic browning *(Walker, 1976)*.

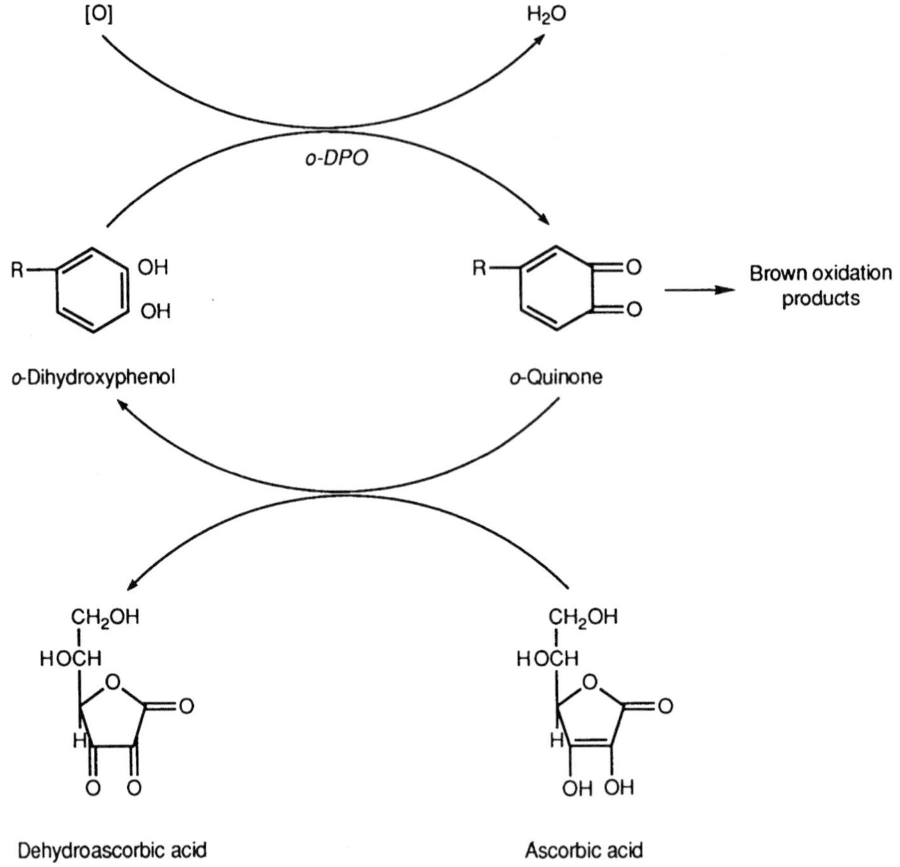

Dehydroascorbic acid   Ascorbic acid

be more effective than its isomer isoascorbic acid. The effect of ascorbic acid can be considered temporary because it is oxidized irreversibly by reaction with intermediates, such as pigments, endogenous enzymes, and metals such as copper. In apple juice, an inhibitory effect of ascorbic acid at a concentration of 1.8 mM was observed for 4 hours.

An adequate amount of ascorbic acid must be added to food material to delay enzymatic browning. In fruit juices treated with ascorbic acid, autooxidation of ascorbic acid, or natural ascorbic acid oxidase activity, will use up any dissolved oxygen in the fruit juice. Thus, oxygen would become the limiting factor determining the rate of enzymatic browning. The addition of ascorbic acid at a concentration of 300 mg per pound (0.454 kg) of fruit was shown by Hope (1961) to control browning as well as reduce headspace oxygen in canned apple halves. This method was particularly effective in controlling enzymatic browning in spite of the thickness of the apple tissue and its comparatively high oxygen content.

Muneta (1977) suggested sodium acid pyrophosphate as an alternative inhibitor of enzymatic browning. It has several advantages over the organic acids, being much less sour than citric acid, as well as minimizing after-cooking blackening of potatoes by complexing iron. An additional benefit of chelating iron is to limit its role in the catalysis of rancidity in fried or dehydrated potatoes.

Kojic acid is found in many fermented Japanese foods. Although it is present in certain foods as a natural fermentation product, the use of kojic acid in the food industry may be restricted by the difficulties of large-scale production and its high cost. İyidoğan and Bayındırlı (2004) showed a significant antibrowning effect of kojic acid in concentrations ranging from 1 to 4 mM. This agent inhibited PPO and bleached melanin through chemical reduction of the browning pigment to colorless compounds.

Benzoic and cinnamic acids, aromatic carboxylic acids, are PPO inhibitors owing to their structural similarities with the phenolic substrates. Undissociated forms of these acids are able to inhibit PPO through complexation with copper at the active site of the enzyme (Marshall *et al.*, 2000). Rapeanu *et al.* (2006) reported a weak inhibition at 0.5 mM (23%) when benzoic acid was tested in grapefruit.

Cyclodextrins are a family of cyclic oligosaccharides composed of α-1,4-linked glucopyranose subunits. Cyclodextrins are produced from starch by enzymatic degradation. These macrocyclic carbohydrates with apolar internal cavities can form complexes with and solubilize many normally water-insoluble compounds (Singh et al., 2002). β-Cyclodextrins bind substrate in its hydrophobic core. The most important functional property of cyclodextrins is their ability to form inclusion complexes with a wide range of organic guest molecules, including PPO substrates (Irwin et al., 1994). Because they can form inclusion complexes the properties of the materials with which they complex can be modified significantly. As a result of molecular complexation phenomena cyclodextrins are widely used in many industrial products, technologies, and analytical methods (Martin Del Valle, 2004). Özoğlu and Bayındırlı (2002), using concentrations ranging from 0.3 to 1.8 mM, showed no effect of β-cyclodextrin, indicating that a higher amount of the compound was necessary to achieve inhibition. In addition, PPO activity in apple juice was affected by the type of cyclodextrin (López-Nicolás et al., 2007). To achieve efficiency in PPO inhibition, food industries may pay attention to the major phenolic compounds present in the product and the type of β-cyclodextrins used.

Sodium chloride is a strong oxidizing agent which can generate chlorine dioxide under acidic conditions. Below pH values of about 5 a strongly pH-dependent inhibitory effect is observed and the degree of inhibition increases with the acidity of the reaction medium. In contrast, an activation effect is observed at higher pH values (Valero and García-Carmona, 1998; M. H. Fan et al., 2005). Severini et al. (2003) evaluated the effect of sodium chloride on the prevention of enzymatic browning in sliced potatoes and observed that all the considered blanching treatments allowed PPO inactivation. With regard to color, the use of calcium chloride, at low concentrations, would seem better than the use of sodium chloride.

These differences in the mechanisms allow the use of combinations of antibrowning agents that may result in enhancement of inhibition. A combination of ascorbic acid, cysteine, and cinnamic acid showed a synergistic effect compared to the individual compounds in cloudy apple juice (Özoğlu and Bayındırlı, 2002). İyidoğan and Bayındırlı (2004) obtained 89% of enzymatic inhibition in Amasya apple juice treated with 3.96 mM L-cysteine, 2.78 mM, kojic acid, and 2.34 mM 4-hexylresorcinol (4-HR). Also in apple juice, a synergic effect was observed between β-cyclodextrin and 4-HR which was not observed for the combination of β-cyclodextrin and methyl jasmonate. In the first combination, β-cyclodextrin reduced the concentration of free substrate that could be oxidized, while 4-HR interacted directly with the enzyme by a competitive mechanism. In the second case, β-cyclodextrin complexed with the substrate and methyl jasmonate, increasing the amount of free substrate and decreasing the concentration of methyl jasmonate that could inhibit PPO activity (Alvarez-Parrilla et al., 2007). The inhibition by methyl jasmonate observed in this study might be an exception, since some authors have described PPO activation by this compound (Koussevitzky et al., 2004; Melo et al., 2006).

Some natural agents have an antibrowning effect. Honey contains a number of components that act as preservatives, such as α-tocopherol, ascorbic acid, flavonoids, and other phenolic compounds. Honey from different floral sources reduced PPO activity by 2–45% in fruit and vegetable homogenates. When combined with ascorbic acid, this inhibitory effect is enhanced (Chen et al., 2000). It was also shown that native procyanidins, flavanol polymers that occur naturally in plants, inhibited PPO activity in cider apple juices and the inhibition intensity increased with degree of polymerization of the procyanidins. The mechanism is probably due to the binding of the polyphenol to the protein, affecting the catalytic activity of the enzyme or by forming an inactive enzyme–polyphenol–substrate complex (Le Bourvellec et al., 2004). Lee and Park (2005) tested Maillard reaction products, synthesized from various amino acids and sugar solutions, on potato PPO activity and observed an inhibitory effect which was dependent on the amino acid (arginine > cysteine > histidine > lysine) and the type of sugar used (monosaccharides > disaccharides).

Proteins, peptides, and amino acids can affect PPO activities by reacting with the o-quinones and by chelating the copper at the active site of PPO. Girelli et al. (2004) measured PPO activity in the presence of various glycyl-dipeptides and they found that these compounds can affect PPO activities by reacting with o-quinones and by chelating the copper at the active site of PPO. Glycylaspartic acid, glycylphenilalanine, glycylglycine, glycyllysine, glycyltyrosine, and glycylhistidine affected the formation of quinone at all concentrations used, varying from 1 to 50 mM. In a study conducted by Shi et al. (2005), different inhibition types between cinnamic acid and its derivatives were demonstrated. Strength and inhibition types were: cinnamic acid (non-competitive) > 4-hydroxycinnamic acid (competitive) > 4-methoxycinnamic acid (non-competitive). No inhibitory effect of 2-hydroxycinnamic acid on the diphenolase activity was found. According to the authors, these inhibitors were attached to a different region from the active site and hindered the binding of substrate to the enzyme through steric hindrance or by changing the protein conformation. Chen *et al.* (2005) demonstrated that PPO was inhibited by several *p*-alkoxybenzoic acids and that *p*-methoxybenzoic acid was the most potent inhibitor. Song *et al.* (2006) studied the inhibitory effects of *cis*- and *trans*-isomers of 3,5-dihydroxystilbene. Although both compounds inhibited PPO activity, the *cis*-form had higher inhibitory ability because it can bind more tightly to the active site of the enzyme than its isomer.

Oms-Oliu *et al.* (2010) reviewed some recent advances in the maintenance of fresh-cut fruit quality with respect to the use of chemical compounds, including plant natural antimicrobials and antioxidants, as well as calcium salts for maintaining texture. They focused on the use of natural preservatives, which are of increasing interest because of the toxicity and allergenicity of some traditional food preservatives, and on the difficulties in the application of these substances to fresh-cut fruit without adversely affecting the sensory characteristics of the product. Edible coatings are presented as an excellent way to carry additives since they have been shown to maintain high concentrations of preservatives on the food surface, reducing the impact of such chemicals on overall consumer acceptability of fresh-cut fruit.

C. Thermal Processing

Heat treatment is the most widely used method for stabilizing foods because of its capacity to destroy microorganisms and to inactivate enzymes. Blanching is the most common method used to inactivate vegetable enzymes (Marshall *et al.*, 2000). It causes denaturation and therefore inactivation of the enzymes but also causes destruction of thermosensitive nutrients, and it is rarely used for soft fruits (Lado and Yousef, 2002) because it results in losses in vitamins, flavor, color, texture, carbohydrates, and other water-soluble components.

The inactivation of PPO as well as other spoilage enzymes can be achieved by subjecting the food article to high temperatures for an adequate length of time to denature the protein. In general, exposure of PPO to temperatures of 70–90°C destroys their catalytic activity, but the time required for inactivation depends on the product (Chutintrasri and Noomhorm, 2006). Fortea *et al.* (2009), studying thermal inactivation of grape PPO and peroxidase, described that they showed similar thermostability, losing over 90% of relative activity after only 5 minutes of incubation at 78°C and 75°C, respectively. Khandelwal *et al.* (2010) estimated the concentrations of polyphenol in different cultivars of four pulses commonly consumed in India and examined the effects of domestic processing. They demonstrated that processing reduced the concentrations of phenolic compounds by 19–59%. Chutintrasri and Noomhorm (2006), studying thermal inactivation of pineapple PPO, described that the enzyme activity reduced by approximately 60% after exposure to 40–60°C for 30 minutes. Denaturation increased rapidly above 75°C. Thus, residual activity was about 7% after 5 minutes at 85°C and 1.2% after 5 minutes at 90°C. In another study, to determine the effect of heating conditions on enzymatic browning, Krapfenbauer *et al.* (2006) evaluated heated apple juice from eight different varieties of apples at high temperature (60–90°C) and short time (20–100 seconds) (HTST) combinations. The results showed that HTST treatment at 80°C already inactivated PPO, whereas pectinesterase activity was reduced to half and could not be inactivated completely, even at 90°C. The highest residual pectinesterase activity was found at 60°C. Heating at 70°C caused stable pectinesterase activity and even a slight increase for 50 and 100 second heating times. In Victoria grape PPO (Rapeanu *et al.*, 2006), complete inactivation was reported after 10 minutes at 70°C (Figure 10.9). The activity of PPO from *Castanea henryi* nuts after incubation at 70°C for 30 minutes was 8% (Xu *et al.*, 2004).

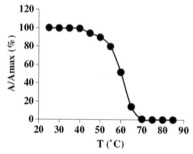

FIGURE 10.9   **Thermal stability of Victoria grape polyphenol oxidase extract. Residual activity was measured after a 10-minute treatment at different temperatures.** *(Reprinted from Rapeanu* et al.*, 2006). © 2006, with permission from Elsevier.*

The heat inactivation of enzymes in foods is not only dependent on time but is also affected by pH. The optimum temperature for PPO varies considerably for different plant sources as well as among cultivars, as shown in Table 10.2, and with the substrate used in the assay.

D. High-Pressure Processing

High hydrostatic pressure (HHP) treatment of fruit and vegetable products offers a natural, environmentally friendly alternative to pasteurization and the chance to produce food of high quality, greater safety, and increased shelf-life (Welti-Chanes et al., 2005). HHP reduces microbial counts and inactivates enzymes (Bayındırlı et al., 2006). The HHP treatment is expected to be less detrimental than thermal processes to low-molecular-mass food compounds such as flavoring agents, pigments, and vitamins, as covalent bonds are not affected by pressure (Butz et al., 2003). In cashew apple juice processed by HHP at 250 or 400 MPa for 3, 5, and 7 minutes, there was no change in pH, acidity, total soluble solids, ascorbic acid, or hydrolyzable polyphenol content. These data demonstrate that HHP can be used in the food industry, generating products with higher nutritional quality (Queiroz et al., 2010). Perera et al. (2010), studied cubes of two different varieties of apple, which were vacuum packed in barrier bags with 0–50% (v/v) pineapple juice and subjected to HHP at 600 MPa for 1–5 minutes (22°C). They observed no visible color changes during 4 weeks of storage in these bags. However, they also noticed that the texture and the in-pack total color change after 5 hours' air exposure were significantly affected by the apple variety, HHP time, and percentage of pineapple juice used. The combined treatment significantly reduced residual PPO activity, whereas pectinmethylesterase activity was not affected in either variety.

HHP can affect protein conformation and lead to protein denaturation, aggregation, or gelation, depending on the protein system, the applied pressure, the temperature, and the duration of the pressure treatment. Protein denaturation is associated with conformational changes and it can change the functionality of the enzyme by increasing or losing biological activity or changing substrate specificity. The effectiveness of treatment depends on the type of enzyme, pH, medium composition, temperature, time, and pressure level applied (Hendrickx et al., 1998).

For several PPOs, it has been reported that pressure-induced inactivation proceeds more rapidly at lower pH and that it is influenced by the addition of salts, sugars, or chemical antibrowning effectors (Rapeanu et al., 2005). It is known that PPO is more resistant to pressure than thermal treatment, when compared to other enzymes (Y.-S. Kim et al., 2001). In general, HHP is more effective at pressures above 600 MPa (Figure 10.10), but the combination of these methods increases the effectiveness of inactivation. The best results from studies aiming to decrease PPO activity in banana purée (Palou et al., 1999), litchi (Phunchaisri and Apichartsrangkoon, 2005), and carrot juice (Y.-S. Kim et al., 2001) were obtained at pressures higher than 400 MPa combined with mild heat (50°C). In contrast, low pressure (up to 400 MPa) induced PPO activation in pear (200–400 MPa, 25°C, 10 minutes) (Asaka and Hayashi, 1991) and apple juice (100 MPa, 1 minute) (Anese et al., 1995).

HHP can be used to create new products (new texture or new taste) or to obtain analogue products with minimal effects on flavor, color, and nutritional value, and without any thermal degradation (Messens et al., 1997). Pressure can also influence biochemical reactions by reducing molecular spacing and increasing interchain reactions (Marshall et al., 2000).

E. Gamma Irradiation

Irradiation is a physical treatment involving direct exposure to electrons or electromagnetic rays, for food preservation and improvement of safety and quality. For these reasons, it can be used to extend the shelf-life of fruits and vegetables. Radiation inactivates microorganisms, guaranteeing disinfection and delaying the ripening process and senescence (Lacroix and Ouattara, 2000). In addition, retention of ascorbic acid and synthesis of polyphenols during storage of irradiated food have been reported (Moussaid et al., 2000; Zhang et al., 2006).

Low-dose γ-irradiation is commonly applied to fruit and vegetable products to extend shelf-life. Prakash et al. (2000) noticed that treatment with a dose of 0.35 kGy decreased 1.5 and 1 log of total aerobic microorganisms and yeasts and molds in cut Romaine lettuce, respectively, and that this dose did not adversely affect sensory attributes, such as visual quality or off-flavor development. Latorre et al. (2010) observed the effect of low doses of γ-irradiation (1 or 2 kGy) on PPO activity of fresh-cut red beetroot and concluded that it produced biochemical changes in cellular contents as well as in the cell wall constitutive networks which could not necessarily be sensed by consumers. They also mentioned changes involving an increase in the antioxidant capacity of red beetroot tissue, showing that the studied doses could be used in the frame of a combined technique for red beet processing. D. Kim et al. (2007) related

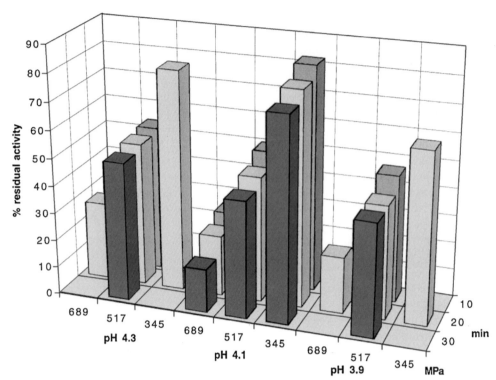

FIGURE 10.10   Effect of high hydrostatic pressure treatments and initial pH on residual polyphenol oxidase activity of avocado purée. *(Reprinted from Lopez-Malo* et al.*, 1998). © 1998, with permission from Elsevier.*

that total aerobic bacteria in fresh kale juice, prepared by a general kitchen process, were detected in the range of 10^6 cfu/ml, and about 10^2 cfu/ml of the bacteria survived in the juice in spite of γ-irradiation treatment with a dose of 5 kGy. In the inoculation test, the growth of the surviving *B. megaterium* and *E. acetylicum* in the 3–5 kGy-irradiated kale juice retarded and/or decreased significantly during a 3 day postirradiation storage period. However, there were no significant differences in the residual PPO activity and browning index between the non-irradiated control and the irradiated kale juice during the postirradiation period. Lu *et al.* (2005) showed a decrease in PPO activity of 73% in fresh-cut celery treated with 1.0 kGy (0.5 kGy/hour) after 3 days under refrigeration (4°C). In 9 days, the enzyme showed an activity around 25% lower than the control sample.

The use of irradiation in combination with other methods, or chemical antibrowning inhibitors, can result in products that are microbiologically safe and of high quality (X. Fan *et al.*, 2005).

F. Pulsed Electric Field

Pulsed electric field (PEF) is a non-thermal food preservation technology, used as an alternative to other conventional techniques, which consists of introducing the food into a chamber containing two electrodes that apply high-voltage pulses (20–80 kV) for a short time (microseconds).

In the past few years, several studies have demonstrated the ability of intense treatments to obtain safe and shelf-stable liquid foods. Novel applications such as the improvement of mass transfer processes or the generation of bioactive compounds using moderate field strengths are currently under development (Soliva-Fortuny *et al.*, 2009).

This emerging food processing technology is being investigated owing to its ability to inactivate undesirable microorganisms and enzymes, with limited increase in food temperature. As a result, more stable foods with fewer changes in composition, physicochemical properties, and sensory attributes can be obtained (Martín-Belloso and Elez-Martínez, 2005). The PEF applied to the food causes electroporation, that is, the irreversible loss of cell membrane functionality, leading to inactivation of the microbial cell (Zhong *et al.*, 2005; Cserhalmi *et al.*, 2006), and it also induces changes in the secondary structure of some enzymes. After PEF treatment, the loss of the relative α-helix fractions of PPO and peroxidase were calculated and these results showed that PPO was more susceptible to the treatment than peroxidase (Zhong *et al.*, 2007).

There are many studies in the literature focusing on the effect of PEF in the inactivation of microorganisms (García *et al.*, 2003, 2005; Evrendilek *et al.*, 2004; Li and Zhang, 2004), but few about the effect of this technology on PPO activity (Giner *et al.*, 2001, 2002; Zhong *et al.*, 2007). The inactivation of commercial PPO depends on the electric field strength and treatment time. The greatest reduction in PPO activity was 76.2% at 25 kV/minute for 744 μs (Zhong *et al.*, 2007). Giner *et al.* (2001, 2002) reported a decrease of 97% on the PPO activity in apple extract at 24.6 kV/cm for 6000 μs, 72% in pear at 22.3 kV/cm for the same treatment time, and 70% in peach PPO at 24.3 kV/cm for 5000 μs. Riener *et al.* (2008) observed a decrease of 71% in PPO activity using a combination of pre-heating to 50°C and a PEF treatment time of 100 μs at 40 kV/cm in freshly prepared apple juice. This level of inactivation was significantly higher ($p < 0.05$) than that recorded in juice processed by conventional mild pasteurization, where the activity of PPO decreased by 46%.

According to Yang *et al.* (2004) the sensitivity to PEF treatment varies from enzyme to enzyme. The sequence of sensitivity to PEF of five enzymes tested was: pepsin > PPO > peroxidase > chymotrypsin and lysozyme. The inactivation effect of PEF on enzymes was affected by electric field strength, electrical conductivity, and pH.

G. Other Technologies

Although most studies have focused on HHP, irradiation, and PEF treatments, other emerging techniques may hold promise for food technology.

Ohmic heating is defined as a process in which electric currents are passed through foods in order to heat them. The heating method affects the temperature distribution inside a food, and directly modifies the time–temperature relationship for enzyme deactivation. Ohmic heating is distinguished from other electrical heating methods by the presence of electrodes contacting the foods, the frequency, and the waveform of the electric field imposed between the electrodes. Icier *et al.* (2008) evaluated fresh grape juice ohmically heated at different voltage gradients (20, 30, and 40 V/cm) from 20°C to 60, 70, 80, or 90°C and the activity of PPO. The critical inactivation temperatures were found to be 60°C or lower for 40 V/cm, and 70°C for 20 and 30 V/cm. The activation energy of PPO inactivation for the temperature range 70–90°C was 83.5 kJ/mol. According to Castro *et al.* (2004), the inactivation kinetics of PPO were first order for conventional and ohmic heating treatments. These authors demonstrated that, using this technology, the time needed for enzyme inactivation was reduced. Icier *et al.* (2006) showed that ohmic heating can also be used for blanching food products. After treatment by ohmic blanching, pea purée peroxidase was inactivated in a shorter processing time than by conventional water blanching.

Supercritical carbon dioxide is a non-thermal technology in which a pressurization step ensures that the applied gas penetrates the microbial cells and subsequent explosive decompression results in rapid gas expansion within the cells, physically destroying them (Corwin and Shellhammer, 2002). Supercritical carbon dioxide has an effect on enzyme inactivation due to conformational changes caused by gas in the secondary and tertiary structure (Gui *et al.*, 2007). According to a study conducted by Gui *et al.* (2007), cloudy apple juice exposed to supercritical carbon dioxide at 30 MPa and 55°C for 60 minutes presented a reduction in PPO activity of over 60%, compared to a 27.9% reduction obtained under atmospheric conditions at 55°C, indicating that the combined effects of pressure, temperature, and time occurred after carbon dioxide treatment. Corwin and Shellhammer (2002) demonstrated the inhibition of PPO activity, with a residual activity of 57.6% (500 MPa, 3 minutes at 50°C), even when the carbon dioxide was added before HHP. In this study, addition of carbon dioxide decreased the PPO activity significantly at all pressures (0, 500, and 800 MPa) and temperatures (25 and 50°C) beyond the effects of pressure alone. Niu *et al.* (2010) evaluated the qualities of cloudy apple juices from apple slices treated with high pressure carbon dioxide and mild heat, and observed that PPO was completely inactivated and its minimal residual activity at 65°C was 38.6%.

Ultrasound is a non-thermal technology; it causes enzyme inactivation by cell lysis using vibration energy, which produces cavitation bubbles and temporarily generates spots of extremely high pressure and temperature when imploded (Morris *et al.*, 2007). Ultrasonication has been found to be more effective in inhibiting enzyme activity when combined with other processes, such as high pressure and/or heat, contrary to the minimal inhibitory effects of individual application. The effects of ultrasound and ascorbic acid on changes in activity of PPO and peroxidase of fresh-cut apple during storage were investigated by Jang and Moon (2011) and Jang *et al.* (2009). They found that the combined treatment inactivated monophenolase, diphenolase, and peroxidase, whereas individual treatment with ultrasound or ascorbic acid had an inverse and limited inhibitory effect on the enzymes. This investigation revealed that simultaneous treatment with ultrasound and ascorbic acid had synergistic inhibitory effects on several enzymes related to enzymatic browning.

Blanching in hot water or steam has been widely used for thermal inactivation of undesirable enzymes, including PPO. Because thermal treatments are also responsible for undesirable changes in tissue texture, several other methods have been tried to inhibit PPO activity and avoid color changes in fruits (Lamikanra, 2002). Microwave heating is an alternative method for liquid food pasteurization and, compared to the conventional method, microwaves can heat products internally, and have greater penetration depth and faster heating rates that could potentially improve the retention of thermolabile constituents in food (Heddleson and Doores, 1994; Deng et al., 2003). According to Cañumir et al. (2002), microwave energy induces thermal effects in microorganisms and enzymes similar to those of conventional heating mechanisms. Matsui et al. (2007) submitted solutions simulating the chemical constituents of coconut water to a batch process in a microwave oven, and observed that PPO activity in water and in sugar solution reduced after treatment. In salt solution, the stability of PPO was significantly affected and the contact between salt and enzyme promoted a drastic reduction in the initial activity. The authors found that, at temperatures above 90°C, the combined effects of salts and microwave energy reduced enzymatic activity to undetectable levels. However, at 90°C, the inactivation effect could be due to temperature alone. In a further study, Matsui et al. (2008) showed that thermal inactivation of PPO during microwave processing of green coconut water was significantly faster than conventional processes reported in the literature.

Short-wave ultraviolet light (UV-C) radiation is an alternative and inexpensive method to reduce the number of microorganisms on the surface of fresh and cut fruits and vegetables (Yaun et al., 2004; Fonseca and Rushing, 2006; Shama, 2006). The potential of UV-C light for commercial use in minimal processing of fruits depends on its ability to contribute to food safety without causing undesirable quality changes. Few studies have been conducted on the influence of UV-C on produce quality (color, texture, taste, and aroma) during storage. The reported effects are quite diverse depending on the type of produce and the dosages applied (Shama, 2006). Although it has been claimed that UV-C treatment does not produce undesirable by-products that could change taste, odor, and color, skin discoloration in tomatoes, browning of calyxes in strawberries, and increasing susceptibility to brown rot in peaches have been reported (Shama and Alderson, 2005). Erkan et al. (2001) found that high doses of UV-C caused a slight reddish-brown discoloration on the surface of zucchini squash. Gómez et al. (2010) examined the effect of UV-C irradiation at different doses, associated with some pretreatments (hot water blanching, dipping into a solution containing ascorbic acid and calcium chloride) in order to minimize browning on the surface color of fresh-cut apple disks. They found that the color and compression parameters were dependent on UV-C dose, storage time, and type of pretreatment. Samples exposed to only UV-C light turned darker and less green compared to fresh-cut apple slices or to samples on day 0, and this effect was more pronounced at the greatest UV-C dose. Light microscopic images showed the breakage of cellular membranes in UV-C-treated samples, which may explain the increase in browning of irradiated apples.

REFERENCES

Ali, M.T., Gleeson, R.A., Wei, C.I., Marshall, M.R., 1994. Activation mechanisms of pro-phenoloxidase on melanosis development in Florida spiny lobster (Panulirus argus) cuticle. J. Food Sci. 59, 1024–1030.

Altunkaya, A., Gökmen, V., 2009. Effect of various anti-browning agents on phenolic compounds profile of fresh lettuce (L. sativa). Food Chem. 117, 122–126.

Alvarez-Parrilla, E., de la Rosa, L.A., Rodrigo-García, J., Escobedo-González, R., Mercado-Mercado, G., Moyers-Montoya, E., Vázquez-Flores, A., González-Aguilar, G.A., 2007. Dual effect of b-cyclodextrin (b-CD) on the inhibition of apple polyphenol oxidase by 4-hexylresorcinol (HR) and methyl jasmonate (MJ). Food Chem. 101, 1346–1356.

Anderson, J.W., 1968. Extraction of enzyme and subcellular organelles from plant tissues. Phytochemistry 7, 1973–1988.

Anese, M., Nicoli, M.C., Dall'Aglio, G., Lerici, C.R., 1995. Effect of high pressure treatments on peroxidase and polyphenoloxidase activities. J. Food Biochem. 18, 285–293.

Arora, D.S., Chander, M., Gill, P.K., 2002. Involvement of lignin peroxidase, manganese peroxidase and laccase in degradation and selective ligninolysis of wheat straw. Int. Biodeterior. Biodegrad. 50, 115–120.

Arslan, O., Erzengin, E., Sinan, S., Ozensoy, O., 2004. Purification of mulberry (Morus alba L.) polyphenol oxidase by affinity chromatography and investigation of its kinetic and electrophoretic properties. Food Chem. 88, 479–484.

Asaka, M., Hayashi, R., 1991. Activation of polyphenoloxidase in pear fruits by high pressure treatment. Agric. Biol. Chem. 55, 2439–2440.

Ayaz, F.A., Demir, O., Torun, H., Kolcuoglu, Y., Colak, A., 2008. Characterization of polyphenoloxidase (PPO) and total phenolic contents in medlar (Mespilus germanica L.) fruit during ripening and over ripening. Food Chem. 106, 291–298.

Aydemir, T., 2004. Partial purification and characterization of polyphenol oxidase from artichoke (Cynara scolymus L.) heads. Food Chem. 87, 59–67.

Bayındırlı, A., Alpas, H., Bozoğlu, F., Hizal, M., 2006. Efficiency of high pressure treatment on inactivation of pathogenic microorganisms and enzymes in apple, orange, apricot and sour cherry juices. Food Control 17, 52–58.

Bendall, D.S., Gregory, R.P.F., 1963. Purification of phenol oxidases. In: Pridham, J.B. (Ed.), Enzyme Chemistry of Phenolic Compounds. MacMillan, New York, pp. 7—24.

Benjakul, S., Visessanguan, W., Tanaka, M., 2006. Inhibitory effect of cysteine and glutathione on phenoloxidase from kuruma prawn (*Penaeus japonicus*). Food Chem. 98, 158—163.

Butz, P., García, A.F., Lindauer, R., Dieterich, S., Bognár, A., Tausher, B., 2003. Influence of ultra high pressure processing on fruit and vegetable products. J. Food Eng. 56, 233—236.

Cañumir, J.A., Celis, J.E., de Bruijn, J., Vidal, L.V., 2002. Pasteurization of apple juice by using microwaves. Food Sci. Technol. 35, 389—392.

Castro, I., Macedo, B., Teixeira, J.A., Vicente, A.A., 2004. The effect of electric field on important food-processing enzymes: comparison of inactivation kinetics under conventional and ohmic heating. J. Food Sci. 69, C696—C701.

Chen, L., Mehta, A., Berenbaum, M., Zangeri, A.R., Engeseth, N.J., 2000. Honeys from different floral sources as inhibitors of enzymatic browning in fruit and vegetable homogenates. J. Agric. Food Chem. 48, 4997—5000.

Chen, Q.-X., Song, K.-K., Qiu, L., Liu, X.-D., Huang, H., Guo, H.-Y., 2005. Inhibitory effects on mushroom tyrosinase by *p*-alkoxybenzoic acids. Food Chem. 91, 269—274.

Chisari, M., Barbagallo, R.N., Spagna, G., 2008. Characterization and role of polyphenol oxidase and peroxidase in browning of fresh-cut melon. J. Agric. Food Chem. 56, 132—138.

Chutintrasri, B., Noomhorm, A., 2006. Thermal inactivation of polyphenoloxidase in pineapple puree. LWT — Food Sci. Technol 39, 492—495.

Cloughley, J.B., 1980a. The effect of fermentation on the quality parameters and price evaluation of Central African black teas. J. Sci. Food Agric. 31, 911—919.

Cloughley, J.B., 1980b. The effect of temperature on enzyme activity during the fermentation phase of black tea manufacture. J. Sci. Food Agric. 31, 920—923.

Corwin, H., Shellhammer, T.H., 2002. Combined carbon dioxide and high pressure inactivation of pectin methylesterase, polyphenol oxidase, *Lactobacillus plantarum* and *Escherichia coli*. J. Food Sci. 67, 697—701.

Craft, C.C., Audia, W.V., 1962. Phenolic substances associated with wound-barrier formation in vegetables. Bot. Gaz. (Chicago) 123, 211—219.

Cserhalmi, Z., Sass-Kiss, Á., Tóth-Markus, M., Lechner, N., 2006. Study of pulsed electric field treated citrus juices. Innov. Food Sci. Emerg. Technol. 7, 49—54.

Dalmadi, I., Rapeanu, G., Van Loey, A., Smout, C., Hendrickx, M., 2006. Characterization and inactivation by thermal and pressure processing of strawberry (*Fragaria ananassa*) polyphenol oxidase: a kinetic study. J. Food Biochem 30, 56—76.

Das, J.R., Boht, S.G., Eowda, L.R., 1997. Purification and characterization of a polyphenol oxidase from the Kew cultivar of Indian pineapple fruit. J. Agric. Food Chem. 45, 2031—2035.

Day, B.P.F., 1996. High oxygen modified atmosphere packaging for fresh prepared produce. Postharvest News Inf. 7, 31—34.

Degl'Innocenti, E., Pardossi, A., Tognoni, F., Guidi, L., 2007. Physiological basis of sensitivity to enzymatic browning in 'lettuce', 'escarole' and 'rocket salad' when stored as fresh-cut products. Food Chem. 104, 209—215.

Deng, Y., Singh, R.K., Lee, J.H., 2003. Estimation of temperature profiles in microwaved particulates using enzyme and vision system. LWT — Food Sci. Technol 36, 331—338.

Dincer, B., Colak, A., Aydin, N., Kadioglu, A., Güner, S., 2002. Characterization of polyphenoloxidase from medlar fruits (*Mespilus germanica* L. Rosaceae). Food Chem. 77, 1—7.

Dixon, R.A., Paiva, N.L., 1995. Stress-induced phenylpropanoid metabolism. Plant Cell 7, 1085—1097.

Ducamp-Collin, M.-N., Ramarson, H., Lebrun, M., Self, G., Reyne, M., 2008. Effect of citric acid and chitosan on maintaining red colouration of litchi fruit pericarp. Postharvest Biol. Technol. 49, 241—246.

Erat, M., Sakiroglu, H., Kufrevioglu, O.I., 2006. Purification and characterization of polyphenol oxidase from *Ferula sp*. Food Chem. 95, 503—508.

Erkan, M., Wang, C.Y., Krizek, D.T., 2001. UV-C irradiation reduces microbial population and deterioration in *Cucurbita pepo* fruit tissue. Environ. Exp. Bot. 45, 1—9.

Espín, J.C., Morales, M., Varón, R., Tudela, J., García-Canovas, F., 1997a. Monophenolase activity of polyphenol oxidase from blanquilla pear. Phytochemistry 44, 17—22.

Espín, J.C., Ochoa, M., Tudela, J., García-Canovas, F., 1997b. Monophenolase activity of strawberry polyphenol oxidase. Phytochemistry 45, 667—670.

Espín, J.C., García-Ruiz, P.A., Tudela, J., Varón, R., García-Canovas, F., 1998. Monophenolase and diphenolase reaction mechanisms of apple and pear polyphenol oxidases. J. Agric. Food Chem. 46, 2968—2975.

Evrendilek, G.A., Li, S., Dantzer, W.R., Zhang, Q.H., 2004. Pulsed electric field processing of beer: microbial, sensory, and quality analyses. J. Food Sci. 69, M228—M232.

Fan, M.H., Wang, M., Zou, P., 2005. Effect of sodium chloride on the activity and stability of polyphenol oxidase from Fuji apple. J. Food Biochem. 29, 221—230.

Fan, X., Niemera, B.A., Mattheis, J.P., Zhuang, H., Olson, D.W., 2005. Quality of fresh-cut apple slices as affected by low-dose ionizing radiation and calcium ascorbate treatment. J. Food Sci. 70, S143—S148.

Fang, Z., Zhang, M., Sun, Y., Sun, J., 2007. Polyphenol oxidase from bayberry (*Myrica rubra* Sieb. et Zucc.) and its role in anthocyanin degradation. Food Chem. 103, 268—273.

Fenoll, L.G., Peñalver, M.J., Rodríguez-López, J.N., Varón, R., García-Cánovas, F., Tudela, J., 2004. Tyrosinase kinetics: discrimination between two models to explain the oxidation mechanism of monophenol and diphenol substrates. Int. J. Biochem. Cell Biol. 36, 235—246.

Finkle, B.J., Nelson, R.F., 1963. Enzyme reactions with phenolic compounds: effects of *o*-methyltransferase on a natural substrate of fruit polyphenol oxidase. Nature (London) 197, 902—903.

Fonseca, J.M., Rushing, J.W., 2006. Effect of ultraviolet-C light on quality and microbial population of fresh-cut watermelon. Postharvest Biol. Technol. 40, 256—261.

Fonseca, M.I., Shimizu, E., Zapata, P.D., Villalba, L.L., 2010. Copper inducing effect on laccase production of white rot fungi native from Misiones (Argentina). Enzyme Microb. Technol. 46, 534—539.

Fortea, M.I., López-Miranda, S., Serrano-Martínez, A., Carreño, J., Núñez-Delicado, E., 2009. Kinetic characterisation and thermal inactivation study of polyphenol oxidase and peroxidase from table grape (Crimson Seedless). Food Chem. 113, 1008—1014.

Franck, C., Lammertyn, J., Quang Tri Ho, Q.T., Verboven, P., Verlinden, B., Nicola, B.T., 2007. Browning disorders in pear fruit. Postharvest Biol. Technol. 43, 1—13.

Gao, Z.-J., Han, X.-H., Xiao, X.-G., 2009. Purification and characterisation of polyphenol oxidase from red Swiss chard (*Beta vulgaris* subspecies *cicla*) leaves. Food Chem. 117, 342—348.

García, D., Gómez, N., Condón, S., Raso, J., Pagán, R., 2003. Pulsed electric fields cause sublethal injury in *Escherichia coli*. Lett. Appl. Microbiol. 36, 140–144.

García, D., Hassani, M., Manas, P., Condon, S., Pagan, R., 2005. Inactivation of *Escherichia coli* O157:H7 during the storage under refrigeration of apple juice treated by pulsed electric fields. J. Food Saf. 25, 30–42.

Gauillard, F., Forget, R.F., 1997. Polyphenoloxidases from Williams pear (*Pyrus communis* L. cv. Williams): activation, purification and some properties. J. Sci. Food Agric. 74, 49–56.

Gawlik-Dziki, U., Szymanowska, U., Baraniak, B., 2007. Characterization of polyphenol oxidase from broccoli (*Brassica oleracea* var. botrytis italica) florets. Food Chem. 105, 1047–1053.

Gawlik-Dziki, U., Złotek, U., Świeca, M., 2008. Characterization of polyphenol oxidase from butter lettuce (*Lactuca sativa* var. capitata L.). Food Chem. 107, 129–135.

Giménez, B., Martínez-Alvarez, O., Montero, P., Gómez-Guillén, M.C., 2010. Characterization of phenoloxidase activity of carapace and viscera from cephalothorax of Norway lobster (*Nephrops norvegicus*). LWT − Food Sci. Technol. 43, 1240–1245.

Giner, J., Gimeno, V., Barbosa-Cánovas, G.V., Martin, O., 2001. Effects of pulsed electric field processing on apple and pear polyphenoloxidases. Food Sci. Technol. Int 7, 339–345.

Giner, J., Ortega, M., Mesegue, M., Gimeno, V., Barbosa-Cánovas, G.V., Martin, O., 2002. Inactivation of peach polyphenoloxidase by exposure to pulsed electric fields. J. Food Sci. 67, 1467–1472.

Girelli, A.M., Mattei, E., Messina, A., Tarola, A.M., 2004. Inhibition of polyphenol oxidases activity by various dipeptides. J. Agric. Food Chem. 52, 2741–2745.

Goldbeck, J.H., Cammarata, K.V., 1981. Spinach thylakoid polyphenoloxidase. Isolation, activation, and properties of the native chloroplast enzyme. Plant Physiol. 67, 977–984.

Gómez, P.L., Alzamora, S.M., Castro, M.A., Salvatori, D.M., 2010. Effect of ultraviolet-C light dose on quality of cut-apple: microorganism, color and compression behavior. J. Food Eng. 98, 60–70.

Gorny, J.R., Hess-Pierce, B., Cifuente, R.A., Kader, A.A., 2002. Quality changes in fresh-cut pear slices as affected by controlled atmospheres and chemical preservatives. Postharvest Biol. Technol. 24, 271–278.

Gruz, J., Ayaz, F.A., Torun, H., Strnad, M., 2011. Phenolic acid content and radical scavenging activity of extracts from medlar (*Mespilus germanica* L.) fruit at different stages of ripening. Food Chem. 124, 271–277.

Guerrero-Beltrán, J.A., Swanson, B.G., Barbosa-Cánovas, G.V., 2005. Inhibition of polyphenoloxidase in mango puree with 4-hexylresorcinol, cysteine and ascorbic acid. LWT − Food Sci. Technol 38, 625–630.

Gui, F., Wu, J., Chen, F., Liao, X., Hu, X., Zhang, Z., Wang, Z., 2007. Inactivation of polyphenol oxidases in cloudy apple juice exposed to supercritical carbon dioxide. Food Chem. 100, 1678–1685.

Harel, E., Mayer, A.M., Shain, Y., 1966. Catechol oxidases, endogenous substrates and browning in developing apples. J. Sci. Food Agric. 17, 389–392.

Heddleson, R.A., Doores, S., 1994. Factors affecting microwave heating of foods and microwave induced destruction of foodborne pathogens − a review. J. Food Protec. 57, 1025–1037.

Hendrickx, M., Ludikhuyze, L., Van Den Broeck, I., Weemaes, C., 1998. Effects of high pressure on enzymes related to food quality. Trends Food Sci. Technol. 9, 197–203.

Hilton, P.J., 1972. *In vitro* oxidation of flavonols from tea leaf. Phytochemistry 11, 1243–1248.

Hilton, P.J., Ellis, R.T., 1972. Estimation of the market value of Central African tea by theaflavin analysis. J. Sci. Food Agric. 23, 227–232.

Hiranvarachat, B., Devahastin, S., Chiewchan, N., 2011. Effects of acid pretreatments on some physicochemical properties of carrot undergoing hot air drying. Food Bioprod. Process. 89, 116–127.

Hobson, G.E., 1967. Phenolase activity in tomato fruit in relation to growth and to various ripening disorders. J. Sci. Food Agric. 18, 523–526.

Hodgson, J.M., Croft, K.D., 2010. Tea flavonoids and cardiovascular health. Mol. Aspects Med. 312, 478–481.

Hope, G.W., 1961. The use of antioxidants in canning apple halves. Food Technol. 15, 548–556.

Hsu, A.F., Shieh, J.J., Bills, D.D., White, K., 1988. Inhibition of mushroom polyphenoloxidase by ascorbic acid derivatives. J. Food Sci. 53, 765–767.

Hulme, A.C., 1958. Some aspects of the biochemistry of apple and pear fruits. Adv. Food Res. 8, 297–413.

Hunt, M.D., Eannetta, N.T., Yu, H., Newmann, S.M., Steffens, J.C., 1993. cDNA cloning and expression of potato polyphenol oxidase. Plant Mol. Biol. 21, 59–68.

Hyodo, H., Uritani, I., 1965. Purification and properties of o-diphenol oxidases in sweet potato. J. Biochem. (Tokyo) 58, 388–395.

Icier, F., Yildiz, H., Baysal, T., 2006. Peroxidase inactivation and colour changes during ohmic blanching of pea puree. J. Food Eng. 74, 424–429.

Icier, F., Yildiz, H., Baysal, T., 2008. Polyphenoloxidase deactivation kinetics during ohmic heating of grape juice. J. Food Eng. 85, 410–417.

Irwin, P.L., Pfeffer, P.E., Doner, L.W., Sapers, G.M., Brewster, J.D., Nagahashi, G., Hicks, K.B., 1994. Binding geometry, stoichiometry, and thermodynamics of cyclomalto-oligosaccharide (cyclodextrin) inclusion complex formation with chlorogenic acid, the major substrate of apple polyphenol oxidase. Carbohydr. Res. 256, 13–27.

İyidoğan, N.F., Bayındırlı, A., 2004. Effect of l-cysteine, kojic acid and 4-hexylresorcinol combination on inhibition of enzymatic browning in Amasya apple juice. J. Food Eng 62, 299–304.

James, W.O., 1953. The terminal oxidases of plant respiration. Biol. Rev. Cambridge Philos. Soc. 28, 245–260.

Jang, J.-H., Moon, K.-D., 2011. Inhibition of polyphenol oxidase and peroxidase activities on fresh-cut apple by simultaneous treatment of ultrasound and ascorbic acid. Food Chem. 124, 444–449.

Jang, J.-H., Kim, S.-T., Moon, K.-D., 2009. Inhibitory effects of ultrasound in combination with ascorbic acid on browning and polyphenol oxidase activity of fresh-cut apples. Food Sci. Biotechnol. 18, 1417–1422.

Jeong, E.-Y., Jeon, J.-H., Lee, C.-H., Lee, H.-S., 2009. Antimicrobial activity of catechol isolated from *Diospyros kaki* Thunb. roots and its derivatives toward intestinal bacteria. Food Chem. 115, 1006–1010.

Jiménez, M., García-Carmona, F., 1997. 4-Substituted resorcinols (sulfite alternatives) as slow-binding inhibitors of tyrosinase chatecholase activity. J. Agric. Food Chem. 45, 2061–2065.

Jiménez, M., García-Carmona, F., 1999. Myricetin, an antioxidant flavonol, is a substrate of polyphenol oxidase. J. Sci. Food Agric. 79, 1993–2000.

Jiménez, M., Escribano-Cebrián, J., García-Carmona, F., 1998. Oxidation of the flavonol fisetin by polyphenol oxidase. Biochim. Biophys. Acta 1425, 534–542.

Jiménez-Atiénzar, M., Escribano, J., Cabanes, J., Gandía-Herrero, F., García-Carmona, F., 2005a. The flavonoid eriodictyol as substrate of peach polyphenol oxidase. J. Food Sci. 70, C540–C544.

Jiménez-Atiénzar, M., Josefa Escribano, J., Cabanes, J., Gandía-Herrero, F., García-Carmona, F., 2005b. Oxidation of the flavonoid eriodictyol by tyrosinase. Plant Physiol. Biochem. 43, 866–873.

Joslyn, M.A., Ponting, J.D., 1951. Enzyme-catalyzed oxidative browning of fruit products. Adv. Food Res. 3, 1.

Kahn, V., 1985. Effect of protein hydrolyzates and amino acids on o-dihydroxyphenolase activity of polyphenol oxidase of mushroom, avocado, and banana. J. Food Sci. 50, 111–115.

Kavrayan, D., Aydemir, T., 2001. Partial purification and characterization of polyphenoloxidase from peppermint (Mentha piperita). Food Chem. 74, 147–154.

Khandelwal, S., Udipi, S.A., Ghugre, P., 2010. Polyphenols and tannins in Indian pulses: effect of soaking, germination and pressure cooking. Food Res. Int 43, 526–530.

Kim, D., Song, H., Lim, S., Yun, H., Chung, J., 2007. Effects of gamma irradiation on the radiation-resistant bacteria and polyphenol oxidase activity in fresh kale juice. Radiat. Phys. Chem. 76, 1213–1217.

Kim, J.Y., Seo, Y.S., Kim, J.E., Sung, S.K., Song, K.J., An, G., Kim, W.T., 2001. Two polyphenol oxidases are differentially expressed during vegetative and reproductive development and in response to wounding in the Fuji apple. Plant Sci. 161, 1145–1152.

Kim, Y., Brecht, J.K., Talcott, S.T., 2007. Antioxidant phytochemical and fruit quality changes in mango (Mangifera indica L.) following hot water immersion and controlled atmosphere storage. Food Chem. 105, 1327–1334.

Kim, Y.-S., Park, S.-J., Cho, Y.-H., Park, J., 2001. Effects of combined treatment of high hydrostatic pressure and mild heat on the quality of carrot juice. J. Food Sci. 66, 1355–1360.

Koussevitzky, S., Ne'eman, E., Harel, E., 2004. Import of polyphenol oxidase by chloroplasts is enhanced by methyl jasmonate. Planta 219, 412–416.

Krapfenbauer, G., Kinner, M., Gossinger, M., Schonlechner, R., Berghofer, E., 2006. Effect of thermal treatment on the quality of cloudy apple juice. J. Agric. Food Chem. 54, 5453–5460.

Lacroix, M., Ouattara, B., 2000. Combined industrial process with irradiation to assure innocuity and preservation of food products — a review. Food Res. Int. 33, 719–724.

Lado, B.H., Yousef, A.E., 2002. Alternative food-preservation technologies: efficacy and mechanisms. Microbes Infect. 4, 433–440.

Lærke, P.E., Christiansen, J., Veierskov, B., 2002. Colour of blackspot bruises in potato tubers during growth and storage compared to their discolouration potential. Postharvest Biol. Technol. 26, 99–111.

Lamikanra, G., 2002. Enzymatic effects on flavor and texture of fresh-cut fruits and vegetables. In: Lamikanra, G. (Ed.), Fresh-Cut Fruits and Vegetables. Science, Technology and Market. CRC Press, Boca Raton, FL, pp. 125–185.

Latorre, M.E., Narvaiz, P., Rojas, A.M., Gerschenson, L.N., 2010. Effects of gamma irradiation on bio-chemical and physico-chemical parameters of fresh-cut red beet (Beta vulgaris L. var. conditiva) root. J. Food Eng. 98, 178–191.

Le Bourvellec, C., Le Quéré, J.-M., Sanoner, P., Drilleau, J.-F., Guyot, S., 2004. Inhibition of apple polyphenol oxidase activity by procyanidins and polyphenol oxidation products. J. Agric. Food Chem. 52, 122–130.

Lee, M.-K., Park, I., 2005. Inhibition of potato polyphenol oxidase by Maillard reaction products. Food Chem. 91, 57–61.

Li, L., Steffens, J.C., 2002. Overexpression of polyphenol oxidase in transgenic tomato plants results in enhanced bacterial disease resistance. Planta 215, 239–247.

Li, S.-Q., Zhang, Q.H., 2004. Inactivation of E. coli 8739 in enriched soymilk using pulsed electric fields. J. Food Sci. 69, 169–174.

Lima, E.D.P.A., Pastore, G.M., Lima, C.A.A., 2001. Purificação da enzima polifenoloxidase (PFO) de polpa de pinha (Annona squamosa L.) madura. Ciênc. Tecnol. Aliment. 21, 98–104.

Limbo, S., Piergiovanni, L., 2006. Shelf life of minimally processed potatoes: Part 1. Effects of high oxygen partial pressures in combination with ascorbic and citric acids on enzymatic browning. Postharvest Biol. Technol. 39, 254–264.

Lopez-Malo, A., Palou, E., Barbosa-Cánovas, G.V., Welti-Chanes, J., Swanson, B.G., 1998. Polyphenoloxidase activity and color changes during storage of high hydrostatic pressure treated avocado puree. Food Res. Int. 31, 549–556.

López-Nicolás, J.M., Núñez-Delicado, E., Sánchez-Ferrer, A., García-Carmona, F., 2007. Kinetic model of apple juice enzymatic browning in the presence of cyclodextrins: the use of maltosyl-β-cyclodextrin as secondary antioxidant. Food Chem. 101, 1164–1171.

Lu, Z., Yu, Z., Gao, X., Lu, F., Zhang, L., 2005. Preservation of gamma irradiation on fresh-cut celery. J. Food Eng. 67, 347–351.

Luh, B.S., Hsu, E.T., Stachowicz, K., 1967. Polyphenolic compounds in canned cling peaches. J. Food Sci. 32, 251–258.

Mahanil, S., Attajarusit, J., Stout, M.J., Thipyapong, P., 2008. Overexpression of tomato polyphenol oxidase increases resistance to common cutworm. Plant Sci. 174, 456–466.

Marshall, M.R., Kim, J., Wei, C.I., 2000. Enzymatic browning in fruits, vegetables and seafoods. Available at http://www.fao.org/ag/ags/agsi/ENZYMEFINAL/Enzymatic%20Browning.html (accessed November 2010).

Martin Del Valle, E.M., 2004. Cyclodextrins and their uses: a review. Process Biochem. 39, 1033–1046.

Martín-Belloso, O., Elez-Martínez, P., 2005. Enzymatic inactivation by pulsed electric fields. In: Sun, D.-W. (Ed.), Emerging Technologies for Food Processing. Elsevier, London, p. 155.

Martínez-Alvarez, O., López-Caballero, M.E., Montero, P., Gómez-Guillén, M.C., 2007. Spraying of 4-hexylresorcinol based formulations to prevent enzymatic browning in Norway lobsters (Nephrops norvegicus) during chilled storage. Food Chem. 100, 147–155.

Martínez-Sánchez, A., Tudela, J.A., Luna, C., Allende, A., Gil, M.I., 2011. Low oxygen levels and light exposure affect quality of fresh-cut Romaine lettuce. Postharvest Biol. Technol. 59, 34–42.

Marusek, C.M., Trobaugh, N.M., Flurkey, W.H., Inlow, J.K., 2006. Comparative analysis of polyphenol oxidase from plant and fungal species. J. Inorg. Biochem. 100, 108–123.

Mason, H.S., Fowlks, W.L., Peterson, E., 1955. Oxygen transfer and electron transport by the phenolase complex. J. Am. Chem. Soc. 77, 2914–2915.

Matheis, G., Whitaker, J.R., 1984. Peroxidase-catalyzed crosslinking of proteins. J. Protein Chem. 3, 35–48.

Mathew, A.G., Parpia, H.A.B., 1971. Food browning as a polyphenol reaction. Adv. Food Res. 19, 75–145.

Matsui, K.N., Granado, L.M., Oliveira, P.V., Tadini, C.C., 2007. Peroxidase and polyphenol oxidase thermal inactivation by microwaves in green coconut water simulated solutions. LWT — Food Sci. Technol. 40, 852–859.

Matsui, K.N., Gut, J.A.W., Oliveira, P.V., Tadini, C.C., 2008. Inactivation kinetics of polyphenol oxidase and peroxidase in green coconut water by microwave processing. J. Food Eng. 88, 169–176.

Mayer, A.M., 1987. Polyphenol oxidases in plants – recent progress. Phytochemistry 26, 11–20.

Mayer, A.M., 2006. Polyphenol oxidases in plants and fungi: going places? A review. Phytochemistry 67, 2318–2331.

Mayer, A.M., Harel, E., 1979. Polyphenol oxidase in plants (review). Phytochemistry 18, 193–215.

Mazzafera, P., Robinson, S.P., 2000. Characterization of polyphenol oxidase in coffee. Phytochemistry 55, 285–296.

Melo, G.A., Shimizu, M.M., Mazzafera, P., 2006. Polyphenoloxidase activity in coffee leaves and its role in resistance against the coffee leaf miner and coffee leaf rust. Phytochemistry 67, 277–285.

Messens, W., Camp., J.V., Huyghebaert, A., 1997. The use of high pressure to modify the functionality of food proteins. Trends Food Sci. Technol. 8, 107–112.

Michodjehoun-Mestres, L., Souquet, J.-M., Hélène Fulcrand, H., Bouchut, C., Reynes, M., Brillouet, J.-M., 2009. Monomeric phenols of cashew apple (*Anacardium occidentale* L.). Food Chem. 112, 851–857.

Mohammadi, M., Kazemi, H., 2002. Changes in peroxidase and polyphenol oxidase activities in susceptible and resistant wheat heads inoculated with *Fusarium graminearum* and induced resistance. Plant Sci. 162, 491.

Montero, P., Ávalos, A., Pérez-Mateosa, M., 2001. Characterization of polyphenoloxidase of prawns (*Penaeus japonicus*). Alternatives to inhibition: additives and high-pressure treatment. Food Chem. 75, 317–324.

Morris, C., Brody, A.L., Wicker, L., 2007. Non-thermal food processing/preservation technologies: a review with packaging implications. Pack. Technol. Sci. 20, 275–286.

Moussaid, M., Lacroix, M., Nketsia-Tabiri, J., Boubekri, C., 2000. Phenolic compounds and the colour of oranges subjected to a combination treatment of waxing and irradiation. Radiat. Phys. Chem. 57, 273–275.

Muneta, P., 1977. Enzymatic blackening in potatoes: influence of pH on dopachrome oxidation. Am. Potato J. 54, 387–393.

Munoz-Munoz, J.L., García-Molina, F., Molina-Alarcón, M., Tudela, J., García-Cánovas, F., Rodríguez-López, J.N., 2008. Kinetic characterization of the enzymatic and chemical oxidation of the catechins in green tea. J. Agric. Food Chem. 56, 9215–9224.

Muthumani, T., Kumar, R.S.S., 2007. Influence of fermentation time on the development of compounds responsible for quality in black tea. Food Chem. 101, 98–102.

Nakamura, W., 1967. Studies on the biosynthesis of lignin. I. Disproof against the catalytic activity of laccase in the oxidation of coniferyl alcohol. J. Biochem. (Tokyo) 62, 54–61.

Neves, V.A., Silva, M.A., 2007. Polyphenol oxidase from yacon roots (*Smallanthus sonchifolius*). J. Agric. Food Chem. 55, 2424–2430.

Newmann, S.M., Eannetta, N.T., Yu, H., Prince, J.P., De Vicente, C.M., Tanksley, S.D., Steffens, J.C., 1993. Organization of the tomato polyphenol oxidase gene family. Plant Mol. Biol. 21, 1035–1051.

Nirmal, N.P., Benjakul, S., 2009. Effect of ferulic acid on inhibition of polyphenoloxidase and quality changes of Pacific white shrimp (*Litopenaeus vannamei*) during iced storage. Food Chem. 116, 323–331.

Nirmal, N.P., Benjakul, S., 2010. Effect of catechin and ferulic acid on melanosis and quality of Pacific white shrimp subjected to prior freeze–thawing during refrigerated storage. Food Control 21, 1263–1271.

Niu, S., Xu, Z., Fang, Y., Zhang, L., Yang, Y., Liao, X., Hu, X., 2010. Comparative study on cloudy apple juice qualities from apple slices treated by high pressure carbon dioxide and mild heat. Innov. Food Sci. Emerg. Technol. 11, 91–97.

Oktay, M., Küfrevioğlu, I., Kocaçaliskan, I., Sakiroğlu, H., 1995. Polyphenoloxidase from Amasya apple. J. Food Sci. 60, 494–496.

Oms-Oliu, G., Odriozola-Serrano, I., Soliva-Fortuny, R., Martín-Belloso, O., 2008. Antioxidant content of fresh-cut pears stored in high-O_2 active packages compared with conventional low-O_2 active and passive modified atmosphere packaging. J. Agric. Food Chem. 56, 932–940.

Oms-Oliu, G., Rojas-Graü, M.A., González, L.A., Varela, P., Soliva-Fortuny, R., Hernando, M.I.H., Munuera, I.P., Fiszman, S., Martín-Belloso, O., 2010. Recent approaches using chemical treatments to preserve quality of fresh-cut fruit: A review. Postharvest Biol. Technol. 57, 139–148.

Orenes-Piñero, E., García-Carmona, F., Sánchez-Ferrer, A., 2005. A kinetic study of *p*-cresol oxidation by quince fruit polyphenol oxidase. J. Agric. Food Chem. 53, 1196–1200.

Özen, A., Colak, A., Dincer, B., Güner, S., 2004. A diphenolase from persimmon fruits (*Diospyros kaki* L., Ebenaceae). Food Chem. 85, 431–437.

Özoğlu, H., Bayındırlı, A., 2002. Inhibition of enzymatic browning in cloudy apple juice with selected antibrowning agents. Food Control 13, 213–221.

Palou, E., López-Malo, A., Barbosa-Cánovas, G.V., Welti-Chanes, J., Swanson, B.G., 1999. Polyphenoloxidase activity and color of blanched and high hydrostatic pressure treated banana puree. J. Food Sci. 64, 42–45.

Paul, B., Gowda, L.R., 2000. Purification and characterization of a polyphenol oxidase from the seeds of field bean (*Dolichos lablab*). J. Agric. Food Chem. 88, 3839–3846.

Perera, N., Gamage, T.V., Wakeling, L., Gamlath, G.G.S., Versteeg, V., 2010. Colour and texture of apples high pressure processed in pineapple juice. Innov. Food Sci. Emerg. Technol. 11, 39–46.

Phunchaisri, C., Apichartsrangkoon, A., 2005. Effects of ultra-high pressure on biochemical and physical modification of lychee (*Litchi chinensis* Sonn.). Food Chem. 93, 57–64.

Pinto, M.S.T., Siqueira, F.P., Oliveira, A.E.A., Fernandes, K.V.S., 2008. A wounding-induced PPO from cowpea (*Vigna unguiculata*) seedlings. Phytochemistry 69, 2297–2302.

Pourcel, L., Routaboul, J.-M., Cheynier, V., Lepiniec, L., Debeaujon, I., 2006. Flavonoid oxidation in plants: from biochemical properties to physiological functions. Trends Plant Sci. 12, 29–36.

Prakash, A., Guner, A.R., Caporaso, F., Foley, D.M., 2000. Effects of low-dose gamma irradiation on the shelflife and quality characteristics of cut romaine lettuce packaged under modified atmosphere. J. Food Sci. 65, 549–553.

Prieto, H., Utz, D., Castro, A., Aguirre, C., González-Agüero, M., Valdés, H., Cifuentes, N., Defilippi, B.G., Zamora, P., Zúñiga, G., Campos-Vargas, R., 2007. Browning in *Annona cherimola* fruit: role of polyphenol oxidase and characterization of a coding sequence of the enzyme. J. Agric. Food Chem. 55, 9208–9218.

Queiroz, C., Lopes, M.L.M., Fialho, E., Valente-Mesquita, V.L., 2008. Polyphenol oxidase: characteristics and mechanisms of browning control. Food Rev. Int. 24, 361–375.

Queiroz, C., Moreira, C.F.F., Lavinas, F.C., Lopes, M.L.M., Fialho, E., Valente-Mesquita, V.L., 2010. Effect of high hydrostatic pressure on the phenolic compounds, ascorbic acid and antioxidant activity in cashew apple juice. High Press. Res. 30, 507–513.

Queiroz, C., Silva, A.J.R., Lopes, M.L.M., Fialho, E., Valente-Mesquita, V.L., 2011. Polyphenol oxidase, phenolic acid composition and browning in cashew apple (Anacardium occidentale, L.) after processing. Food Chem. 125, 128–132.

Rapeanu, G., Loey, A.V., Smout, C., Hendrickx, M., 2005. Effect of pH on thermal and/or pressure inactivation of Victoria grape (Vitis vinifera sativa) polyphenol oxidase: a kinetic study. J. Food Sci. 70, E301–E307.

Rapeanu, G., Loey, A.V., Smout, C., Hendrickx, M., 2006. Biochemical characterization and process stability of polyphenoloxidase extracted from Victoria grape (Vitis vinifera ssp. sativa). Food Chem. 94, 253–261.

Revankar, M.S., Lele, S.S., 2006. Enhanced production of laccase using a new isolate of white rot fungus. WR-1. Process Biochem. 41, 581–588.

Riener, J., Noci, F., Cronin, D.A., Morgan, D.J., Lyng, J.G., 2008. Combined effect of temperature and pulsed electric fields on apple juice peroxidase and polyphenoloxidase inactivation. Food Chem. 109, 402–407.

Rivero, R.M., Ruiz, J.M., García, P.C., López-Lefebre, L.R., Sánchez, E., Romero, L., 2001. Resistance to cold and heat stress: accumulation of phenolic compounds in tomato and watermelon plants. Plant Sci. 160, 315–321.

Robards, K., Prenzler, P.D., Tucker, G., Swatsitang, P., Glover, W., 1999. Phenolic compounds and their role in oxidative processes in fruits. Food Chem. 66, 401–436.

Roberts, E.A.H., 1952. Chemistry of tea fermentation. J. Sci. Food Agric. 23, 227–232.

Rocha, A.M.C.N., Morais, A.M.M.B., 2001. Characterization of polyphenoloxidase (PPO) extracted from Jonagored apple. Food Control 12, 85–90.

Rodriguez-Lopez, J.N., Tudela, J., Varon, R., Garcia-Carmona, F., Garcia-Canovas, F., 1992. Analysis of a kinetic model for melanin biosynthesis pathway. J. Biol. Chem. 267, 3801–3810.

Roelofsen, P.A., 1958. Fermentation, drying and storage of cocoa beans. Adv. Food Res. 8, 225–296.

Ruenroengklin, N., Sun, J., Shi, J., Xue, S.J., Jiang, Y., 2009. Role of endogenous and exogenous phenolics in litchi anthocyanin degradation caused by polyphenol oxidase. Food Chem. 115, 1253–1256.

Sanchez-Ferrer, A., Bou, R., Cabanes, J., Garcia-Carmona, F., 1988. Characterization of catecholase and cresolase activities of Manartrell grape polyphenol oxidase. Phytochemistry 27, 319–320.

Sanderson, G.W., 1964. The chemical composition of fresh tea flush as affected by clone and climate. Tea Q. 35, 101–109.

Sapers, G.M., Miller, R.L., 1998. Browning inhibition in fresh-cut pears. J. Food Sci. 63, 342–346.

Saxena, A., Bawa, A.S., Raju, P.S., 2008. Use of modified atmosphere packaging to extend shelf-life of minimally processed jackfruit (Arotocarpus heterophyllus L.) bulbs. J. Food Eng. 87, 455–466.

Schwimmer, S., Burr, H.K., 1967. Structure and chemical composition of the potato tuber. In: Talburt, W.F., Smith, O. (Eds.), Potato Processing. Avi, Westport, CT, p. 12.

Segovia-Bravo, K.A., Jarén-Galán, M., García-García, P., Garrido-Fernández, A., 2009. Browning reactions in olives: mechanism and polyphenols involved. Food Chem. 114, 1380–1385.

Sellés-Marchart, S., Casado-Vela, J., Bru-Martínez, R., 2006. Isolation of a latent polyphenol oxidase from loquat fruit (Eriobotrya japonica Lindl.): kinetic characterization and comparison with the active form. Arch. Biochem. Biophys. 446, 175–185.

Severini, C., Baiano, A., De Pilli, T., Romaniello, R., Derossi, A., 2003. Prevention of enzymatic browning in sliced potatoes by blanching in boiling saline solutions. LWT – Food Sci. Technol. 36, 657–665.

Shama, G., 2006. Ultraviolet light. In: Hui, Y.H. (Ed.), Handbook of Food Science, Technology and Engineering. CRC/Taylor and Francis, Boca Raton, FL 122–1–122-14.

Shama, G., Alderson, P., 2005. UV hormesis in fruits: a concept ripe for commercialization. Trends Food Sci. Technol. 16, 128–136.

Shi, C., Dai, Y., Xu, X., Xie, Y., Liu, Q., 2002. The purification of polyphenol oxidase from tobacco. Protein Expr. Purif. 24, 51–55.

Shi, J., Sun, J., Wei, X., Shi, J., Cheng, G., Zhao, M., Wang, J., Yang, B., Jiang, Y., 2008. Identification of (−)-epicatechin as the direct substrate for polyphenol oxidase from longan fruit pericarp. LWT – Food Sci. Technol. 41, 1742–1747.

Shi, Y., Chen, Q.-X., Wang, Q., Song, K.-K., Qiu, L., 2005. Inhibitory effects of cinnamic acid and its derivates on the diphenolase activity of mushroom (Agaricus bisporus) tyrosinase. Food Chem. 92, 707–712.

Shleev, S., Persson, P., Shumakovich, G., Mazhugo, Y., Yaropolov, A., Ruzgas, T., Gorton, L., 2006. Interaction of fungal laccases and laccase-mediator system with lignin. Enzyme Microb. Technol. 39, 841–847.

Siegenthaler, P.-A., Vaucher-Bonjour, P., 1971. Vieillessement de l'appareil photosynthétique. III. Variations et caractéristiques de l'activité o-diphenyloxydase (polyphenoloxydase) au cours du vieillissement in vitro de chloroplastes isolés dépinard. Planta 100, 106–123.

Singh, M., Sharma, R., Banerjee, U.C., 2002. Biotechnological applications of cyclodextrins. Biotechnol. Adv. 20, 341–359.

Soliva-Fortuny, R.C., Biosca-Biosca, M., Grigelmo-Miguel, N., Martín-Belloso, O., 2002. Browning, polyphenol oxidase activity and head-space gas composition during storage of fresh-cut pears using modified atmosphere packaging. J. Sci. Food Agric. 82, 1490–1496.

Soliva-Fortuny, R.C., Alós-Saiz, N., Espachs-Barroso, A., Martín-Belloso, O., 2004. Influence of maturity at processing on quality attributes of fresh-cut 'Conference' pears. J. Food Sci. 69, 290–294.

Soliva-Fortuny, R., Balasa, A., Knorr, D., Martín-Belloso, O., 2009. Effects of pulsed electric fields on bioactive compounds in foods: a review. Trends Food Sci. Technol. 20, 544–556.

Solomos, T., 1997. Principles underlying modified atmosphere packaging. In: Wiley, R.C. (Ed.), Minimally Processed Refrigerated Fruits and Vegetables. Chapman and Hall, New York, pp. 183–225.

Song, K.-K., Huang, H., Han, P., Zhang, C.-L., Shi, Y., Chen, Q.-X., 2006. Inhibitory effects of cis- and trans-isomers of 3,5-dihydroxystilbene on the activity of mushroom tyrosinase. Biochem. Biophys. Res. Commun. 342, 1147–1151.

Song, Y., Yao, Y.-X., Zhang, H., Du, Y.-P., Chen, F., Wei, S.-W., 2007. Polyphenolic compound and degree of browning in processing apple varieties. Agric. Sci. China 6, 607–612.

Srebotnik, E., Hammel, K.E., 2000. Degradation of nonphenolic lignin by the laccase/1-hydroxybenzotriazole system. J. Biotechnol. 81, 179–188.

Steiner, U., Schliemann, W., Strack, D., 1996. Assay for tyrosine hydroxylation activity of tyrosinase from betalain-forming plants and cell cultures. Anal. Biochem. 238, 72–75.

Steiner, U., Schliemann, W., Böhm, H., Strack, D., 1999. Tyrosinase involved in betalain biosynthesis of higher plants. Planta 208, 114–124.

Stevens, L.H., Davelaar, E., 1996. Isolation and characterization of blackspot pigments from potato tubers. Phytochemistry 42, 941–947.

Subramanian, N., Venkatesh, P., Ganguli, S., Sinkar, V.P., 1999. Role of polyphenol oxidase and peroxidase in the generation of black tea theaflavins. J. Agric. Food Chem. 47, 2571–2578.

Sun, J., Jiang, Y., Wei, X., Shi, J., You, Y., Liu, H., Kakuda, Y., Zhao, M., 2006. Identification of (−)-epicatechin as the direct substrate for polyphenol oxidase isolated from litchi pericarp. Food Res. Int. 39, 864–870.

Sun, J., Xiang, X., Yu, C., Shi, J., Peng, H., Yang, Y., Yang, S., Yang, E., Jiang, Y., 2009. Variations in contents of browning substrates and activities of some related enzymes during litchi fruit development. Sci. Hortic. 120, 555–559.

Swain, T., 1962. Economic importance of flavonoid compounds. Foodstuffs. In: Geissman, T.A. (Ed.), The Chemistry of Flavonoid Compounds. Pergamon, Oxford, pp. 513–552.

Talburt, W.F., Smith, O. (Eds.), 1967. Potato Processing. Avi, Westport, CT.

Tegelberg, R., Julkunen-Tiitto, R., Vartiainen, M., Paunonen, R., Rousi, M., Kellomäki, S., 2008. Exposures to elevated CO_2, elevated temperature and enhanced UV-B radiation modify activities of polyphenol oxidase and guaiacol peroxidase and concentrations of chlorophylls, polyamines and soluble proteins in the leaves of *Betula pendula* seedlings. Environ. Exp. Bot. 62, 308–315.

Thipyapong, P., Hunt, M.D., Steffens, J.C., 2004a. Antisense downregulation of polyphenol oxidase results in enhanced disease susceptibility. Planta 220, 105–117.

Thipyapong, P., Melkonian, J., Wolfe, D.W., Steffens, J.C., 2004b. Suppression of polyphenol oxidases increases stress tolerance in tomato. Plant Sci. 167, 693–703.

Thurston, C.F., 1994. The structure and function of fungal laccases. Microbiology 140, 19–26.

Tolbert, N.E., 1973. Activation of polyphenol oxidase of chloroplasts. Plant Physiol. 51, 234–244.

Tyagi, M., Kayastha, A.M., Sinha, B., 2000. The role of peroxidase and polyphenol oxidase isozymes in wheat resistance to *Alternaria triticina*. Biol. Plant 43, 559–562.

Ünal, M.U., 2007. Properties of polyphenol oxidase from Anamur banana (*Musa cavendishii*). Food Chem. 100, 909–913.

Valero, E., García-Carmona, F., 1998. pH-dependent effect of sodium chloride on latent grape polyphenol oxidase. J. Agric. Food Chem. 46, 2447–2451.

Vamos-Vigyazo, L., 1981. Polyphenol oxidase and peroxidase in fruits and vegetables. CRC Crit. Rev. Food Sci. Nutr. 15, 49–127.

Van Lelyveld, L.J., de Rooster, K., 1986. Browning potential of tea clones and seedlings. J. Hortic. Sci. 61, 545–548.

Virador, V.M., Reyes-Grajeda, J.P., Blanco-Labra, A., Mendiola-Olaya, E., Smith, G.M., Moreno, A., Whitaker, J.R., 2010. Cloning, sequencing, purification, and crystal structure of grenache (*Vitis vinifera*) polyphenol oxidase. J. Agric. Food Chem. 58, 1189–1201.

Waliszewski, K.N., Márquez, O., Pardio, V.T., 2009. Quantification and characterisation of polyphenol oxidase from vanilla bean. Food Chem. 117, 196–203.

Walker, J.R.L., 1976. The control of enzymic browning in fruit juices by cinnamic acids. J. Food Technol. 11, 341–345.

Wang, J., Constabel, C.P., 2004. Polyphenol oxidase overexpression in transgenic *Populus* enhances resistance to herbivory by forest tent caterpillar (*Malacosoma disstria*). Planta 220, 87–96.

Wang, J., Jiang, W., Wang, B., Liu, S., Gong, Z., Luo, Y., 2007. Partial properties of polyphenol oxidase in mango (*Mangifera indica* L. cv. Tainong) pulp. J. Food Biochem. 31, 45–55.

Welti-Chanes, J., López-Malo, A., Palou, E., Bermúdez, D., Guerrero-Beltrán, J.A., Barbosa-Cánovas, G.V., 2005. Fundamentals and applications of high pressure processing of foods. In: Barbosa-Cánovas, G.V., Tapia, M.S., Cano, M.P. (Eds.), Novel Food Processing Technologies. CRC Press, New York, pp. 157–182.

Whitaker, J.R., 1972. Principles of Enzymology for the Food Sciences. Dekker, New York, pp. 571–582.

Widsten, P., Kandelbauer, A., 2008. Laccase applications in the forest products industry: a review. Enz. Microb. Technol. 42, 293–307.

Wszelaki, A.L., Mitcham, E.J., 2000. Effects of superatmospheric oxygen on strawberry fruit quality and decay. Postharvest Biol. Technol. 20, 125–133.

Xu, J., Zheng, T., Meguro, S., Kawachi, S., 2004. Purification and characterization of polyphenol oxidase from Henry chestnuts (*Castanea henryi*). J. Wood Sci. 50, 260–265.

Yang, C.P., Fujita, S., Ashrafuzzaman, M.D., Nakamura, N., Hayashi, N., 2000. Purification and characterization of polyphenol oxidase from banana (*Musa sapientum* L.) pulp. J. Agric. Food Chem. 48, 2732–2735.

Yang, R.-J., Li, S.-Q., Zhang, Q.H., 2004. Effects of pulsed electric fields on the activity of enzymes in aqueous solution. J. Food Sci. 69, FCT241–FCT248.

Yaun, B.R., Summer, S.S., Eifert, J.D., Marcy, J.E., 2004. Inhibition of pathogens on fresh produce by ultraviolet energy. Int. J. Food Microbiol. 90, 1–8.

Yu, C., Sun, J., Xiang, X., Yang, B., Jiang, Y., 2010. Variations in contents of (−)-epicatechin and activities of phenylalanineammonialyase and polyphenol oxidase of longan fruit during development. Sci. Hortic. 125, 230–232.

Yue-Ming, J., Zauberman, G., Fuchs, Y., 1997. Partial purification and some properties of polyphenol oxidase extracted from litchi fruit pericarp. Postharvest Biol. Technol. 10, 221–228.

Zamorano, J.-P., Martínez-Álvarez, O., Montero, P., Gómez-Guillén, M.C., 2009. Characterisation and tissue distribution of polyphenol oxidase of deepwater pink shrimp (*Parapenaeus longirostis*). Food Chem. 112, 104–111.

Zhang, L., Lu, Z., Lu, F., Bie, X., 2006. Effect of γ irradiation on quality-maintaining of fresh-cut lettuce. Food Control 17, 225–228.

Zhong, K., Chen, F., Wu, J., Wang, Z., Liao, X., Hu, X., Zhang, Z., 2005. Kinetics of inactivation of *Escherichia coli* in carrot juice by pulsed electric field. J. Food Process. Eng. 28, 595–609.

Zhong, K., Wu, J., Wang, Z., Chen, F., Liao, X., Hu, X., Zhang, Z., 2007. Inactivation kinetics and secondary structural change of PEF-treated POD and PPO. Food Chem. 100, 115–123.

Chapter 11

Lipid Oxidation

Karen M. Schaich,* Fereidoon Shahidi,[†] Ying Zhong[†] and N. A. Michael Eskin**

*Department of Food Science, Rutgers University, New Brunswick, New Jersey, USA, [†]Department of Biochemistry, Memorial University of Newfoundland, St. John's, Newfoundland, Canada, **Department of Human Nutritional Sciences, Faculty of Human Ecology, University of Manitoba, Winnipeg, Manitoba, Canada

Chapter Outline

Biochemistry of Foods. DOI: http://dx.doi.org/10.1016/B978-0-12-242352-9.00011-4

I. INTRODUCTION

In all foods, the first mode of spoilage is microbial. However, after microbes have been controlled by processing, oxidation becomes the set of chemical reactions most limiting shelf-life and degrading the quality of foods. Consumers commonly recognize lipid oxidation by characteristic painty or oily 'rancid' off-flavors and odors, but the impact of the reaction is far greater than this. As the most sensitive chemical functional groups in biological molecules, unsaturated fatty acids essentially harvest oxidizing potential from the atmosphere and transform it into highly reactive chemical species. Once initiated, the free radical chains of lipid oxidation are responsible for a cascade of oxidations that affect structural proteins and enzymes, nucleic acids, polysaccharides, vitamins, and lipids. The overall effect is to alter physical properties and degrade molecular functionality as well as to destroy the palatability of foods.

Lipid oxidation is not limited merely to foods, as radicals, hydroperoxides, epoxides, and aldehydes are involved in normal physiology and pathological processes in living tissues, including aging (Pryor, 1985), cancer (McBrien and Slater, 1982), atherosclerosis (Uchida, 2000), Alzheimer's disease (Sayre *et al.*, 1997) and other dementias, inflammatory bowel disease (Kruidenier and Verspaget, 2002), and macular degeneration (Gu *et al.*, 2003; Ebrahem *et al.*, 2006).

Most of the time, lipid oxidation is considered to be a toxic process leading to decomposition of membranes, inactivation of enzymes, adduct formation and strand scission in DNA, and impairment of cell functions (Logani and Davies, 1980; Borg *et al.*, 1981; McBrien and Slater, 1982; Borg and Schaich, 1983, 1984; Fadeel *et al.*, 2007). However, there are some positive impacts, as low levels of oxidation products play important roles in signal transduction in tissues, such as regulating responses to environmental oxygen (Suzuki *et al.*, 1997; Gutierrez *et al.*, 2006). One example of this is that when *Phanerochaete chrysosporium* (white rot fungi) are grown in high oxygen, lipid oxidation in membranes upregulates production of lignan peroxidases that utilize the excess oxygen and of antioxidant enzymes (catalase and glutathione peroxidase) that reduce oxidation products (C. Frenkel and K. M. Schaich, unpublished data, 1992). Hence, lipid oxidation has moved beyond the traditional purview of food science and has become a hot research topic in biology and medicine as well.

Whatever the field of application, new information is showing lipid oxidation to be a complex series of reactions with some fascinating chemistry and effects that move far beyond the neighboring lipid molecule or release of rancid odors. This chapter provides an introduction to lipid oxidation as a dynamic chemical process. The main focus is on lipid oxidation in foods rather than medicine, although the principles discussed are broadly applicable to all living tissues, both plant and animal.

II. WHERE DOES LIPID OXIDATION OCCUR?

Lipid oxidation primarily involves the reaction of oxygen with unsaturated fatty acids (fatty acids with double bonds), although secondary reactions with saturated fatty acids (no double bonds) cannot be ruled out. Lipid oxidation in foods is most commonly associated with fats and oils in bulk or in oil phases of emulsions, and certainly that is where the reaction is most obvious to consumers. However, it is important to recognize that oxidation occurs wherever unsaturated fatty acids are found, and that means in:

- triacylglycerols: in adipose tissue and marbling fats (Watts, 1954; Ladikos and Lougovois, 1990), oil bodies in grains, essential oils in fruits, chocolate (Rossi-Olson, 2011), bulk oils (Schaich, 2005a), and oil phases of emulsions (Sun *et al.*, 2011)
- phospholipids: in membranes (particularly muscle foods, organ meats, vegetables), bran layers, egg yolks, and natural and synthetic emulsifiers (Corliss and Dugan, 1971; Igene *et al.*, 1980; Mead, 1980; Yamamoto *et al.*, 1984; Porter and Wagner, 1986)

- free fatty acids: hydrolysis products of triacylglycerols and phospholipids resulting from acids, bases, heat with and without water, and lipase action (Heaton and Uri, 1961; deGroot *et al.*, 1973; Campbell *et al.*, 1974; Miyashita and Takagi, 1986); found especially in foods that are heated, have microbial contamination, or contain dairy or tropical lipids
- fatty acids esterified to other molecules such as sterols and alcohols (as in waxes).

This means that lipid oxidation can be rather pervasive in foods, even in low-lipid formulations.

III. HOW DOES LIPID OXIDATION OCCUR?

A. Lipid Oxidation Mechanisms: Traditional Free Radical Chain

Lipid oxidation has long been recognized as a free radical chain reaction which occurs in three stages: initiation, propagation, and termination (Figure 11.1) (Farmer *et al.*, 1943; Farmer and Sutton, 1943; Bolland, 1945, 1949; Swern, 1961). The radical chain reaction is responsible for several unique kinetic characteristics that present distinct challenges in measuring and controlling lipid oxidation, and are part of the reason why lipid oxidation is a major problem in storage stability of foods:

- Lipid oxidation is autocatalytic: once started, the reaction is self-propagating and self-accelerating.
- Many more than one lipid molecule is oxidized and more than one LOOH is formed per initiation. Chain lengths of up to several hundred lipid molecules have been measured (Hyde and Verdin, 1968; Cosgrove *et al.*, 1987).
- Very small amounts of pro-oxidants or antioxidants cause large rate changes.
- The reaction produces multiple intermediates and products that change with reaction conditions and time.

1. Initiation (LH → L•)

Because lipid oxidation occurs so readily and is found so ubiquitously, it is often referred to as a spontaneous process (Anonymous, 1981). However, lipid oxidation is not thermodynamically spontaneous, i.e. it cannot happen on its own. Normal oxygen is in a triplet spin state (odd electrons parallel) while double bonds are in singlet spin states (electrons with opposite spin), so atmospheric oxygen cannot react directly with lipid double bonds:

$$\underset{\text{Triplet}}{\overset{\bullet\uparrow \quad \bullet\uparrow}{O-O}} \quad + \quad \underset{\text{Singlet}}{\overset{\uparrow\bullet\bullet\downarrow}{-C=C-}} \quad \xrightarrow{\quad\times\quad} \quad ROOH \tag{11.1}$$

Thus, lipid oxidation always *requires* an initiator or catalyst to remove an electron from either the lipid or oxygen, creating radicals, or to change the electron spin of the oxygen so that it can add to the double bond directly to form hydroperoxides that break down to radicals. Whatever the initiator, the final result is formation of initial lipid alkyl radicals that react with oxygen to start the oxidation process. Reactions of the most common initiators are described in Section IV, C.

2. Propagation and Branching

a. Basic reactions

Propagation is the heart of the oxidation process (Kochi, 1973a). In it, oxygen adds at diffusion controlled rates (almost instantaneously) to relatively unreactive lipid alkyl radicals, L•, converting them to reactive peroxyl radicals, LOO• (Reaction 2 in Figure 11.1), that establish the free radical chain and keep it going (Ingold, 1969a). Peroxyl radicals abstract hydrogens from adjacent lipid molecules to form hydroperoxides, LOOH, and generate new L• radicals in the process (Reaction 3, Figure 11.1). Each new L• radical in turn adds oxygen, forms a peroxyl radical, abstracts a hydrogen from another lipid, forms another hydroperoxide, and generates a new L• radical to provide the driving force in the chain reaction (Reaction 4, Figure 11.1). The process continues indefinitely until no hydrogen source is available or the chain is intercepted.

Peroxyl radicals are the main chain carriers in early oxidation. Their abstractions are rather slow and specific ($k = 36-62$ l/mol/s) (Gaddis *et al.*, 1961; Howard and Ingold, 1967; Gebicki and Bielski, 1981), which contributes to an initial slow period in which lipid oxidation may or may not be detected. The chain continues one abstraction at a time from the initiation point. Without forces that decompose hydroperoxides, this process can continue indefinitely at a slow rate. However, reactions accelerate when hydroperoxides accumulate and then are decomposed to alkoxyl radicals, peroxyl radicals, and hydroxyl radicals by metals, heat, and ultraviolet (UV) light (Reactions 5, 6, and 7 in Figure 11.1). An important distinction in hydroperoxide decompositions is that metal reactions are heterolytic, yielding one radical and an ion, while heat and UV light induce homolytic scission that generates two radicals, alkoxyl ($LO^\bullet$) and hydroxyl ($HO^\bullet$), both of which react much more rapidly and more generally than $LOO^\bullet$. Once formed, these radicals greatly increase the rate of recycling in the chain reaction, and they attack more sites on lipids.

FIGURE 11.1 Classical free radical chain reaction of lipid oxidation as traditionally understood. *(From Schaich, 2005b; used with permission.)*

CLASSICAL FREE RADICAL CHAIN REACTION MECHANISM OF LIPID OXIDATION

Initiation *(formation of ab initio lipid free radical)*

$$L_1H \xrightarrow{k_i} L_1^\bullet \tag{1}$$

Propagation

Free radical chain reaction established

$$L_1^\bullet + O_2 \underset{k_\beta}{\overset{k_o}{\rightleftharpoons}} L_1OO^\bullet \tag{2}$$

$$L_1OO^\bullet + L_2H \xrightarrow{k_{p1}} L_1OOH + L_2^\bullet \tag{3}$$

$$L_2OO^\bullet + L_3H \xrightarrow{k_{p1}} L_2OOH + L_3^\bullet \text{ etc.} \dashrightarrow L_nOOH \tag{4}$$

Free radical chain branching (initiation of new chains)

$$L_nOOH \xrightarrow{k_{d1}} L_nO^\bullet + OH^- \text{ (reducing metals)} \tag{5}$$

$$L_nOOH \xrightarrow{k_{d2}} L_nOO^\bullet + H^+ \text{ (oxidizing metals)} \tag{6}$$

$$L_nOOH \xrightarrow{k_{d3}} L_nO^\bullet + {}^\bullet OH \text{ (heat and uv)} \tag{7}$$

$$\left.\begin{array}{c} L_nO^\bullet \\ L_nOO^\bullet \\ HO^\bullet \end{array}\right\} + L_4H \begin{array}{c} \xrightarrow{k_{p2}} \\ \xrightarrow{k_{p1}} \\ \xrightarrow{k_{p3}} \end{array} \left.\begin{array}{c} L_nOH \\ L_nOOH \\ HOH \end{array}\right\} + L_4^\bullet \begin{array}{c} \text{(8a)} \\ \text{(8b)} \\ \text{(8c)} \end{array}$$

$$L_1OO^\bullet + L_nOOH \xrightarrow{k_{p4}} L_1OOH + L_nOO^\bullet \tag{9}$$

$$L_1O^\bullet + L_nOOH \xrightarrow{k_{p5}} L_1OH + L_nOO^\bullet \tag{10}$$

Termination *(formation of non-radical products)*

$$\left.\begin{array}{c} L_n^\bullet \\ L_nO^\bullet \\ L_nOO^\bullet \end{array}\right\} + \left.\begin{array}{c} L_n^\bullet \\ L_nO^\bullet \\ L_nOO^\bullet \end{array}\right\} \quad \begin{array}{l} \text{Radical recombinations} \\ \xrightarrow{k_{t1}} \text{ polymers, non-radical monomer products} \\ \xrightarrow{k_{t2}} \text{ (ketones, ethers, alkanes, aldehydes, etc.)} \\ \xrightarrow{k_{t3}} \end{array} \begin{array}{l} \text{(11a)} \\ \text{(11b)} \\ \text{(11c)} \end{array}$$

$$\left.\begin{array}{c} LOO^\bullet \\ LO^\bullet \end{array}\right\} \quad \begin{array}{l} \text{Radical scissions} \\ \xrightarrow{k_{ts1}} \text{ non-radical products} \\ \xrightarrow{k_{ts2}} \text{ (aldehydes, ketones, alcohols, alkanes, etc.)} \end{array} \begin{array}{l} \text{(12a)} \\ \text{(12b)} \end{array}$$

i - initiation; o-oxygenation; β-O_2 scission; p-propagation; d-dissociation; t-termination; ts-termination/scission

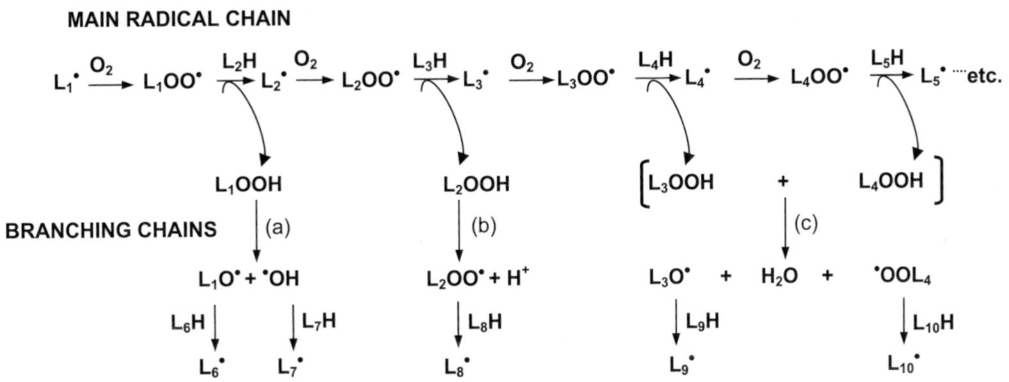

MAIN RADICAL CHAIN

FIGURE 11.2 Expansion of the lipid oxidation chain reaction by chain branching. Propagation extends the original chain from the first radical by hydrogen abstraction (top chain). Chain branching occurs when hydroperoxide products of the original chain decompose by (a) reduction, (b) oxidation, or (c) bimolecular dismutation to multiple radicals that all initiate new chains. Propagation rates are faster on chains with LO• as the chain carrier. *(Adapted from Schaich, 2005b; used with permission.)*

The change in propagation rate and abstraction specificity marks progression into a second stage of propagation called branching, in which the radical chain reaction expands, establishing new chains at faster rates. The effect of branching is shown diagrammatically in Figure 11.2. Lipid oxidation gathers steam, increasing in rate and extent as LO• becomes the dominant, faster chain carrier ($k = 10^6 - 10^7$ l/mol/s) (Pryor, 1986; Erben-Russ *et al.*, 1987) and secondary chains dramatically amplify and broadcast lipid oxidation beyond the initial radical chain. In this manner, a single initiating event can lead to sequential oxidation of literally hundreds of molecules in the primary chain and in secondary branching chains (Hyde and Verdin, 1968; Cosgrove *et al.*, 1987).

One final point about propagation must be made before moving on. In the early stages of lipid oxidation when hydroperoxides are in low concentrations, hydroperoxide decompositions occur monomolecularly (one at a time) as shown in Reactions 5, 6, and 7 of Figure 11.1. However, as oxidation progresses and hydroperoxides accumulate, decomposition shifts to bimolecular mechanisms in which two hydroperoxides interact to induce decomposition. The traditional explanation proposes that the hydroperoxides hydrogen bond and then undergo concerted hydrolysis to yield two radicals (Reaction 11.2); these go on to initiate branching reactions and accelerate the overall oxidation rate (Sliwiok *et al.*, 1974; Hiatt and McCarrick, 1975):

$$2\,LOOH \longrightarrow LOOH...HOOL \longrightarrow LO\bullet + H_2O + \bullet OOL \qquad (11.2)$$

An alternative explanation suggests that the dramatic increase in oxidation is kinetically more likely when one slowly reacting radical reacts with a non-propagating hydroperoxide to generate a powerful cascade of *three* very reactive radicals: LO•, epoxy-LO•, and •OH (Elson *et al.*, 1975):

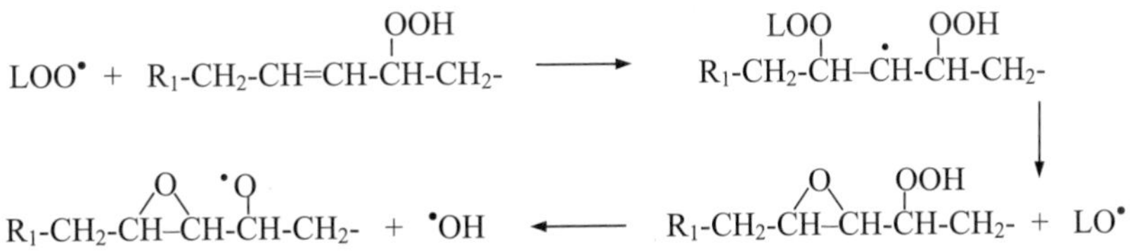

$$(11.3)$$

$$(11.3a)$$

Whichever mechanism is operative, the initial effect of bimolecular decomposition is to dramatically accelerate lipid oxidation. However, as oxidation progresses, eventually the hydroperoxide breakdowns and termination reactions, such as radical recombination or alkoxyl radical scission, become faster than the initiation of new chains. Oxidation then slows as stable secondary products form and off-flavors and odors become detectable.

b. Sites of Hydrogen Abstraction and L•/LOOH Formation in Unsaturated Fatty Acids

During propagation, lipid oxyl free radicals abstract hydrogens from carbon positions with the weakest bonding. The lowest C—H bond energies in unsaturated fatty acids are at the allylic hydrogens (next to double bonds) (Kerr, 1966) so these become the preferred sites for H removal and formation of a free radical. The —CH$_2$— groups between two double bonds in lipids (called doubly allylic) are doubly activated so C—H bond energies drop tremendously, as shown in the structure below (Scott, 1965; Kerr, 1966).

Thus, the order of preference for hydrogen abstractions in fatty acids is H between two double bonds (which partially explains why oxidizability of fatty acids increases with the number of double bonds) > singly allylic H next to double bonds > > H adjacent to the —COOH group > H on methylene groups further down the acyl chains (Patterson and Hasegawa, 1978).

When a hydrogen is abstracted between two double bonds, the free electron remaining (i.e. the radical) becomes distributed across a resonance stabilized double bond system (Reaction 11.4). The highest electron density concentrates in the center (weakest C—H bond), so the outside positions become relatively electron deficient, providing enhanced targets for oxygen addition and hydroperoxide formation.

$$(11.4)$$

Linoleic acid thus forms hydroperoxides almost exclusively at external positions carbons 9 and 13 during autoxidation. In polyunsaturated fatty acids with more than two double bonds and multiple 1,4-diene structures, the dominant hydroperoxides of autoxidizing fatty acids also are found at the external positions (Figure 11.3), regardless of the number of double bonds.

Two notable exceptions to this pattern are: (1) hydrogens are abstracted (and hydroperoxides are formed) equally at both double bond carbons and both neighboring carbons in isolated double bonds, such as in oleic acid; and (2) internal hydroperoxides, e.g. carbons 10 and 12 in linoleic acid, are formed during photosensitized oxidation by singlet oxygen, as will be discussed further in Section IV, D, 1.

Another important point to recognize about lipid oxidation is that hydrogen abstraction does not break the double bonds: when radicals are formed, the double bonds migrate to the next carbon and invert from *cis* to *trans*, even in isolated double bonds such as oleic acid (Reaction 11.5) (Farmer *et al.*, 1943; Porter, 1990; Porter *et al.*, 1995). In the

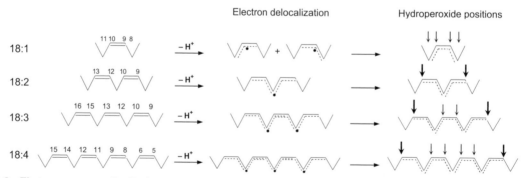

FIGURE 11.3 Electron resonance distributions and corresponding locations of hydroperoxides formed when hydrogens are abstracted from unsaturated fatty acids. Heavy arrows denote dominant positions for hydroperoxide formation. (*From Schaich, 2005b; used with permission.*)

1,4-diene systems of the polyunsaturated fatty acids, migration of the double bond generates a conjugated diene (Reaction 11.6).

$$RCH_2CH_2\text{-}CH\text{=}CH\text{-}CH_2(CH_2)_6\,COOH \longrightarrow RCH_2CH\text{=}CH\text{-}CH\text{-}CH_2(CH_2)_6\,COOH$$
$$+$$
$$RCH_2CH_2\overset{\bullet}{C}H\text{-}CH\text{=}CH(CH_2)_6\,COOH \tag{11.5}$$

$$\text{-}CH_2CH\text{=}CH\text{-}CH_2\text{-}CH\text{=}CHCH_2\text{-} \longrightarrow \text{-}CH_2CH\text{=}CH\text{-}CH\text{=}CH\overset{\bullet}{\text{-}}CHCH_2\text{-}$$
$$\overset{\bullet}{+}$$
$$\text{-}CH_2\overset{\bullet}{C}H\text{-}CH\text{=}CH\text{-}CH\text{=}CHCH_2\text{-} \tag{11.6}$$

Conjugated dienes have two important consequences in lipid oxidation. They are the first detectable chemical change in lipid oxidation, so are important intermediates for following early stages of lipid oxidation by their UV absorption at 234 nm (Parr and Swoboda, 1976). In addition, the conjugated system primes the chain for further reactions (see Section III, B), and converts the acyl chain from bent and flexible to straight and rigid. In membranes, therefore, even early oxidation alters membrane fluidity and disrupts function.

3. Termination Processes

In lipid oxidation, 'termination' is almost a misnomer since lipid oxidation never fully stops. A specific radical may be terminated and form some product, but usually there is another radical left behind, so the chain reaction continues. Net oxidation slows down when radical quenching processes exceed the rate of new chain production, and the momentum of oxidation shifts from radical propagation and chain expansion to generation of stable products. Thus, 'termination' in the discussion below refers to converting an individual lipid radical to a product, not stopping the overall reaction. The cumulative effect on a food system is determined by the number of radical chains being terminated.

Lipid free radicals terminate to form non-radical products by four major mechanisms (Schaich, 2005b):

- radical recombinations
- α and β scission reactions of alkoxyl radicals when proton sources (e.g. water) are present to stabilize the products
- co-oxidation of non-lipid molecules such as proteins
- group eliminations or dismutation.

The mechanisms dominating in a given food are influenced by the nature and concentration of the radicals, the temperature and oxygen pressure, and the solvent. Basic aspects of the reaction chemistry will be covered here; effects of co-oxidation reactions will be covered in more detail in Section V.

a. Radical Recombinations

Radicals formed from oxidizing lipids can recombine in limitless combinations to generate a broad range of oxidation products. Nevertheless, recombinations are not random, and distinct patterns of favored recombinations have been identified. Some of the most important recombinations responsible for major lipid oxidation products are:

alkyl radicals:
$$R_1^{\bullet} + R_2^{\bullet} \longrightarrow R_1\text{-}R_2 \qquad \text{alkane polymers} \tag{11.7}$$

peroxyl radicals:
$$2\ \underset{\displaystyle R_1}{\overset{\displaystyle OO^{\bullet}}{R_1CHR_2}} \longrightarrow \underset{\displaystyle R_2}{\overset{\displaystyle R_1}{HC\text{-}OOOO\text{-}}}\underset{\displaystyle R_2}{\overset{\displaystyle R_1}{CH}} \longrightarrow \underset{\displaystyle }{\overset{\displaystyle OH}{R_1CHR_2}} + \underset{\displaystyle \text{alcohols and ketones}}{\overset{\displaystyle O}{R_1CR_2}} + O_2 \tag{11.8}$$

$$ROO^{\bullet} \overset{O_2}{\underset{ROO^{\bullet}}{\longrightarrow}} R^{\bullet} \longrightarrow ROOR \qquad \text{alkyl peroxides} \tag{11.9}$$

alkoxyl radicals:
$$R_1O^{\bullet} + R_2^{\bullet} \longrightarrow R_1OR_2 \qquad \text{ethers} \tag{11.10}$$

$$R_1O^{\bullet} \ + \ R_2O^{\bullet} \ \longrightarrow \ R_1OOR_2 \qquad \text{peroxides} \qquad (11.11)$$

$$\underset{\underset{O^{\bullet}}{|}}{R_1\text{--CH--}R_2} \ + \ R^{\bullet} \ \longrightarrow \ \underset{\underset{O}{\|}}{R_1\text{--C--}R_2} \ + \ RH \qquad \text{ketones, alkanes} \qquad (11.12)$$

$$\underset{\underset{O^{\bullet}}{|}}{R_1\text{--CH--}R_2} \ + \ RO^{\bullet} \ \longrightarrow \ \underset{\underset{O}{\|}}{R_1\text{--C--}R_2} \ + \ ROH \qquad \text{ketones, alcohols} \qquad (11.13)$$

Temperature and oxygen pressure are key determinants of radical recombination pathways. L$^{\bullet}$ recombinations dominate under low oxygen pressures ($pO_2 = 1$ to about $80-100$ mmHg) and high temperatures (reduced oxygen solubility) (Figure 11.4) (Lundberg and Chipault, 1947; Labuza, 1971). High oxygen favors LOO$^{\bullet}$ reactions, but then addition to double bonds becomes competitive with combination. LO$^{\bullet}$ contributions to the product mix dominate in secondary stages of oxidation and at moderate temperatures and oxygen pressures when LOOH or LOO$^{\bullet}$ decompositions are faster than their formation (Bolland, 1949). Radical recombinations increase in importance relative to other products as oxidation increases and intermediates accumulate. For example, alkoxyl radicals from hydroperoxyepidioxides heated at 40°C generate $> 90\%$ dimers (Neff *et al.*, 1988).

Radical recombinations are responsible for many of the characteristics of oxidized oils or lipids. Recombinations are diffusion and concentration controlled (Erben-Russ *et al.*, 1987; Tsentalovich *et al.*, 1998), so they occur most readily when viscosity and radical concentrations are both high, i.e. in neat oils after extensive oxidation; recombinations decrease with lipid dilution and are probably unimportant in polar solvents. Primary alkyl radical recombinations generate the dimers and polymers that account for the increased viscosity in oxidized oils. Perhaps most importantly, recombinations of alkoxyl radicals and their fragmentation products (see discussion below) generate low levels of volatile compounds and flavor components that augment those produced in scission reactions and provide the undertones and secondary notes that round out flavors (Grosch, 1987). Ketones and dialkyl peroxides, in particular, result uniquely from recombination reactions.

b. Scission Reactions of LO$^{\bullet}$

Scission of alkoxyl radicals probably has the greatest practical consequences in lipid oxidation because the products generated are responsible for the distinctive volatile off-odors and off-flavors that are the strong markers of rancidity (Grosch, 1987). Lipid alkoxyl radicals undergo scission of the C—C bond on either side of the alkoxyl group to yield a mixture of carbonyl final products (typically aldehydes and oxo-esters from the initial alkoxyl radicals) and alkyl free radicals that can continue the chain reaction:

$$\underset{\underset{O^{\bullet}}{|}}{\overset{\beta \quad \alpha}{R_1 \{CH\} R_2}} \ \longrightarrow \ R_1^{\bullet} \ + \ \overset{\beta}{\underset{\underset{O}{\|}}{CH\text{--}R_2}} \quad \textbf{OR} \quad \overset{\alpha}{\underset{\underset{O}{\|}}{R_1CH}} \ + \ ^{\bullet}R_2 \qquad (11.14)$$

Unsaturated radical fragments oxidize further and then undergo secondary scissions to produce carbonyls and alkanes of shorter chain length. Consequently, product mixtures that accumulate in oxidized lipids can become quite complex.

Scission of alkoxyl radicals requires a strong proton donor (Russell, 1959). Hydrogen atoms hydrogen bond to both intermediate transition states and final polar cleavage products, reducing the activation energy for bond rupture

FIGURE 11.4 Effects of oxygen and temperature on termination processes in lipid oxidation. More oxygenated products are favored by high oxygen and low temperatures; alkyl reactions and dimerizations are favored by low oxygen and elevated temperatures (Schaich, 2005b). *(Redrawn from Labuza, 1971; used with permission.)*

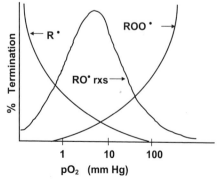

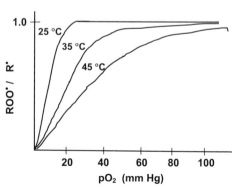

(Walling and Padwa, 1963; Walling and Wagner, 1963; Kochi, 1973b; Avila *et al.*, 1993; Tsentalovich *et al.*, 1998); H^+ from the solvent then adds immediately to the scission radicals to provide the driving force for the reaction (Schauenstein, 1967). Thus, alkoxyl radical scissions are quite rapid in the presence of water (e.g. in emulsions) and indeed account for about half the alkoxyl radical reactions in polar media, especially in dilute solutions of lipids where there is reduced competition from hydrogen abstraction (Bors *et al.*, 1984). However, the greatest contribution of this termination (and propagation) reaction is at elevated temperatures (Kochi, 1973b) because LO$^\bullet$ scission has a large E_a and log A (Arrhenius factor) (Horner *et al.*, 2000). Heat accelerates alkoxyl radical scissions in all solvents, although the pattern of cleavage may change as temperature increases. Scission is a minor process in neat lipids at room temperature.

Figures 11.5–11.7 show some of the multiple scissions that occur leading to complex mixtures of products in oxidizing oleic acid, linoleic acid, and linolenic acids, respectively. Further details of the scission reactions responsible for the hundreds of volatile products in lipid oxidation may be found in reviews by Frankel and by Grosch, pioneers in this field (Frankel, 1982, 1984, 1985, 1987; Grosch, 1987).

Discussion of scission reactions would be incomplete without mention of malondialdehyde (MDA), one of the most notorious products of lipid oxidation. MDA arises from multiple scissions of cyclic internal hydroperoxides formed in fatty acids with three or more double bonds (linolenic and higher) (Reaction 11.15) (Dahle *et al.*, 1962; Pryor *et al.*, 1976).

(11.15)

MDA is very commonly used – and misused – in assays of lipid oxidation because it can be detected without lipid extraction. However, it is not a universal product and not appropriate for assaying oxidation fatty acids with fewer than three double bonds. MDA formation is facilitated in pure lipid phases, by low lipid and oxygen, and by photosensitized oxidation (Porter *et al.*, 1984); it also requires mild heat and acid for endoperoxide cleavage (Frankel and Neff, 1983). In autoxidizing lipids, yields of authentic MDA are usually less than 0.1% (Shamberger *et al.*, 1977; Frankel and Neff, 1983), although in photosensitized fatty acids where internal hydroperoxides are formed in high concentrations MDA concentrations can reach 5% or higher (Frankel and Neff, 1983). Hence, this assay is not the best choice for lipid oxidation analysis in most food systems.

c. Co-oxidation of Non-lipid Molecules

While hydrogen transfer is needed to quench a particular lipid radical, the hydrogen atom can come from any molecule that has abstractable (loosely bonded) hydrogens and is located close to the lipid. In foods and in cells, lipids are closely associated with proteins, carotenoids and other pigments, starches, antioxidants, and vitamins, and radical transfer to any of these molecules via hydrogen abstraction or addition of LOO$^\bullet$ or LO$^\bullet$ to double bonds can lead to co-oxidation of these molecules.

Co-oxidations are a process in which the interception of lipid free radicals by non-lipid molecules stops propagation and forms lipid products on one side, while transferring radicals and oxidizing potential to proteins and other biomolecules (Schaich, 2008). This reaction is similar to that of antioxidants but differs critically in that the radicals formed are not stable. Most of these non-lipid radicals add oxygen to form peroxyl radicals that also abstract H from other molecules and lead to oxidative degradation of the molecular target. In this way, lipids serve to 'broadcast' oxidation damage to other molecules which then provide footprints of lipid oxidation in foods and biological systems (Pryor, 1978, 1989; Schaich, 1980a, 2008; Borg and Schaich, 1984).

Generic co-oxidation reactions of target molecules are shown in Reactions 11.16 and 11.17. TH is any target molecule and RH is any molecule with an abstractable hydrogen, either lipid or non-lipid. The conjugated double bonds in Reaction 11.17 can be in any molecule but are particularly numerous in carotenoids.

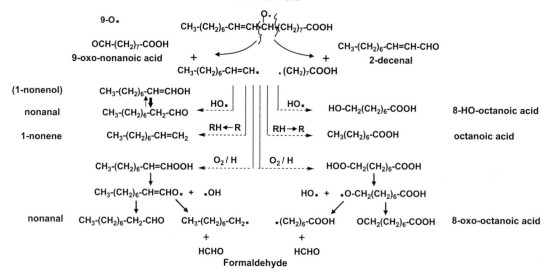

OLEIC ACID

Following the same fragmentation pattern –

β-scission		α-scission	
8–O$^\bullet$ → 8-oxo-octanoic acid +	decanal	2-undecenal +	7-HO-heptanoic acid
	1-decene		heptanoic acid
	⌈nonanol⌉		7-oxo-heptanoic acid
	nonane		⌈6-HO-hexanoic acid⌉
	nonanal		hexanoic acid
	⌊formaldehyde⌋		6-oxo-hexanoic acid
			⌊formaldehyde⌋
10–O$^\bullet$ → 10-oxo-8-decenoic acid +	octanol	nonanal +	9-oxo-nonanoic acid
	octane		8-nonenoic acid
	octanal		⌈octanol⌉
	⌈heptanol⌉		octane
	heptane		octanal
	⌊heptanal⌋		⌊formaldehyde⌋
	formaldehyde		
11–O$^\bullet$ → 11-oxo-9-undecenoic acid +	heptanol	octanal +	10-oxo-decanoic acid
	heptane		9-decenoic acid
	heptanal		⌈nonanol⌉
	⌈hexanol⌉		nonane
	hexane		nonanal
	hexanal		⌊formaldehyde⌋
	⌊formaldehyde⌋		

FIGURE 11.5   Scission pathways for oleic acid. (*From Schaich, 2005b; used with permission.*)

$$LOO^\bullet + TH \longrightarrow LOOH + T^\bullet \xrightarrow{O_2} TOO^\bullet \xrightarrow{RH} TOOH \tag{11.16}$$

$$LOO^\bullet + \text{-C=C-C=C-} \longrightarrow \text{-C=C-}\overset{\bullet}{\text{C}}\text{-C-} \longrightarrow \text{additional oxidation} \tag{11.17}$$
$$\underset{OOL}{}$$

Perhaps the most noticeable co-oxidations involve proteins. Abstractable hydrogens are available on side-chain amino (Reaction 11.18) and thiol groups (Reaction 11.19), so histidine, lysine, arginine, and cysteine are prime

LINOLEIC ACID

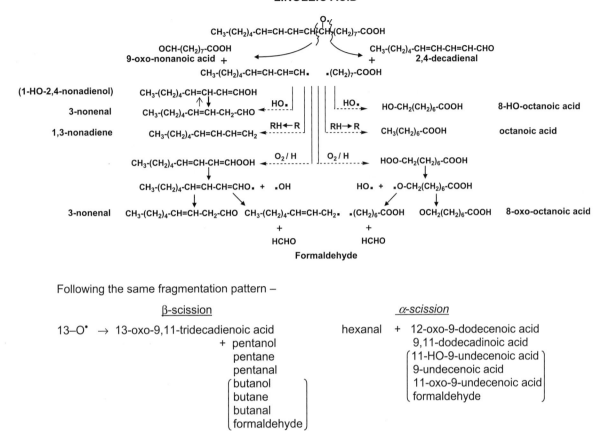

Following the same fragmentation pattern –

β-scission	α-scission
13–O• → 13-oxo-9,11-tridecadienoic acid	hexanal + 12-oxo-9-dodecenoic acid
+ pentanol	9,11-dodecadinoic acid
pentane	11-HO-9-undecenoic acid
pentanal	9-undecenoic acid
butanol	11-oxo-9-undecenoic acid
butane	formaldehyde
butanal	
formaldehyde	

FIGURE 11.6 Scission pathways for linoleic acid. *(From Schaich, 2005b; used with permission.)*

targets for hydrogen abstraction from proteins (Karel, 1975; Schaich and Karel, 1975, 1976; Yong and Karel, 1978, 1979; Schaich, 1980a). Radical addition reactions have also been reported (Reactions 11.18a and 11.19a) (Gardner and Weisleder, 1976; Gardner *et al.*, 1977, 1985).

$$\left.\begin{array}{c} LOO^\bullet \\ LO^\bullet \end{array}\right\} + \left\{\begin{array}{c} -NH_2 \\ >NH \end{array}\right. \longrightarrow \left\{\begin{array}{c} -NH^\bullet \\ >N^\bullet \end{array}\right. + \begin{array}{c} LOOH \\ LOH \end{array} \tag{11.18}$$

$$\begin{array}{c} -N\text{-OOL, -N-OL adducts} \\ >N\text{-OOL, }>N\text{-OL adducts} \end{array} \tag{11.18a}$$

$$\left.\begin{array}{c} LOO^\bullet \\ LO^\bullet \end{array}\right\} + RSH \longrightarrow RS^\bullet \left\{\begin{array}{c} LOOH \\ LOH \end{array}\right. \tag{11.19}$$

$$RS\text{-OOL, RS-OL, RS-epoxy-L adducts} \tag{11.19a}$$

Co-oxidation reactions should not be ignored in consideration of lipid oxidation kinetics, mechanisms, and overall effects in foods and biological systems. A critical issue is that while co-oxidation reactions terminate lipid oxidation chains they act as antioxidants at the same time, and lipid oxidation measured by hydroperoxides or downstream products slows. Co-oxidation products also limit extractability of lipids for analysis and often remove lipids from

LINOLENIC ACID

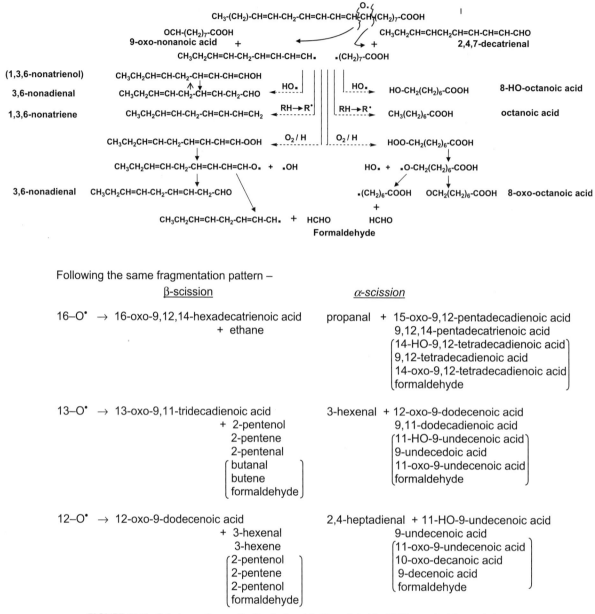

FIGURE 11.7   Scission pathways for linolenic acid. *(From Schaich, 2005b; used with permission.)*

product streams normally analyzed. As a result, lipid oxidation in complex systems is probably always under-estimated — sometimes severely — unless co-oxidation products are also measured. In foods, for example, that means monitoring at least oxidation in proteins and pigment bleaching. Sensitive target molecules need to be identified for each food product to tailor analyses and fully account for lipid oxidation.

Effects of co-oxidations will be considered in more detail in Section V.

d. Group Eliminations

Group eliminations are a minor form of termination but it is important to recognize the reaction because it accounts for some lipid oxidation products that are not easily formed by other mechanisms. The most common eliminations are HO^- and HOO^-, which can be eliminated from LOOH, yielding an internal carbonyl (ketone) (a) and a desaturated product with an additional double bond (b) (Reaction 11.20) (Terao *et al.*, 1975; Bothe *et al.*, 1978). The specific product formed is determined by the hydroperoxide position on the fatty acid chain.

$$R_1CH=CH-CH=CH-\overset{\overset{\displaystyle OOH}{|}}{CH}-CH_2-R_2$$

$$\overset{-OH}{\swarrow} \qquad \overset{-OOH}{\searrow}$$

$$R_1CH=CH-CH=CH-\overset{\overset{\displaystyle O}{||}}{C}-CH_2-R_2 \qquad R_1CH=CH-CH=CH-CH=CH-R_2$$

(11.20)

Formation of the ketone, the dominant product, prevents hydroperoxide decomposition to reactive alkoxyl radicals. Addition of a double bond to a polyunsaturated fatty acid forms a very reactive conjugated triene which is highly susceptible to radical addition and a variety of other secondary oxidations. Thus, while terminating a single radical chain, this product may still contribute to additional chains, adding to the complexity of lipid oxidation.

B. Reaction Mechanisms: Multiple Reaction Pathways

The chain reaction scheme shown in Figure 11.1 has been the accepted explanation of lipid oxidation for more than 50 years, yet it does not accurately account for the wide array of products observed in oxidizing lipids or for kinetics of the total process. Indeed, many published results are inconsistent with the chain reaction as written. In particular, the reaction scheme as drawn predicts that products should be generated and observed in strict sequence, i.e. hydroperoxides form first and their decomposition leads to a cascade of unspecified products. In some studies, products follow this pattern very generally, but under many conditions the predicted products form simultaneously with hydroperoxides or do not occur at all. Strategies for accurately analyzing lipid oxidation and for designing antioxidants require a clear understanding and tracking of all the oxidation processes active in a food. Thus, it is important to begin considering the multiple alternative pathways of both peroxyl and alkoxyl radicals that can be active in addition to and in competition with classical hydrogen abstraction in lipid oxidation. Figure 11.8 shows an integrated scheme for lipid oxidation which incorporates the classical free radical chain, shown flowing vertically down the center, but adds important side-reactions of LOO•, LOOH, LO•, and secondary radicals that compete with hydrogen abstraction in propagation and alter termination processes (Schaich, 2005b). This scheme has been

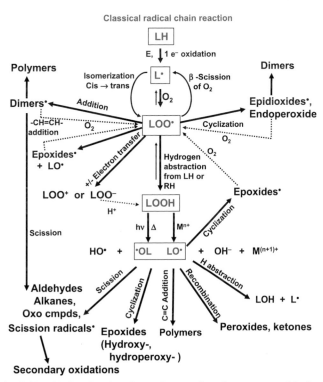

FIGURE 11.8 Integrated scheme for lipid oxidation showing alternative reactions that compete with classical hydrogen abstraction (vertical reactions in center). *(From Schaich, 2005b; used with permission.)*

compiled from extensive data in the literature and has not yet been proven per se, but it shows how thinking about lipid oxidation needs to be broadened in order to account for kinetics and products observed under different oxidation conditions and to meet the challenges of stabilizing foods with high polyunsaturated fatty acids, including *n*-3 fatty acids.

The major consequences of multiple reaction pathways are:

- Multiple LOO• and LO• reaction pathways occur simultaneously and in competition with each other. Mixtures of products can be formed at all stages, not just at the end of chains. Competing alternative reactions of LOO• and LO• alter propagation sequences and thus change kinetics, dominant products, detectability, and effects of lipid oxidation dramatically and in complex ways.

- Traditional lipid oxidation assays which monitor only one product may greatly underestimate lipid oxidation or miss it altogether if alternative pathways are more active than the one measured. This is particularly true for hydroperoxides and single volatile compounds such as hexanal. A combination of product analyses is needed to fingerprint all alternative pathways and accurately measure the full extent of lipid oxidation.

- All changes caused by lipid oxidation (flavors, odors, browning, texture alterations, color loss, etc.) derive from specific lipid oxidation products. Shifts away from H abstraction mean that aldehydes, epoxides, dimers, or other products are formed in place of alcohols, and this alters food qualities and stability patterns markedly. Understanding how lipid oxidation affects food properties thus requires analysis of a full range of lipid oxidation products.

- Toxicity and safety issues: shifts from volatile scission products to non-volatile internal rearrangement and addition products reduce off-odors, which is desirable. However, off-odors are the consumer's first clue that foods are rancid and should not be eaten. Epoxides are more toxic and react more rapidly with DNA and proteins, while dimers have reduced digestibility and some evidence of toxicity. These products have no sensory warning signals for consumers.

A brief description of propagation processes and products formed in alternative pathways and the conditions under which specific pathways are favored is provided as follows.

1. Hydrogen Abstraction

Hydrogen abstraction is the best known chain carrying reaction cited in all lipid oxidation schemes and is the ultimate reaction even with alternative pathways. Hydrogen abstraction by peroxyl radicals yields hydroperoxides (Reactions 3 and 4, Figure 11.1), the first stable product of lipid oxidation and the product most often measured to determine the extent of lipid oxidation. Hydrogen abstraction by alkoxyl radicals (Reactions 8a and 10, Figure 11.1) produces alcohols or hydroxylipids. However, alcohols are usually minor products, while they should be the only products if hydrogen abstraction is the only reaction occurring. Thus, there must be other reaction competing pathways that produce the aldehydes, epoxides, and other products most commonly observed in lipid oxidation.

Hydrogen abstractions from other lipid molecules by LOO• and LO• are favored under conditions providing close contact between lipid chains without competition from other H sources, e.g. in pure lipids, in highly polyunsaturated fatty acids that have bent chain conformations that align multiple bis-allylic hydrogens, in the lipid interior of membranes, in aprotic solvents that do not compete with lipids, in low-viscosity media that facilitate molecular movement, and at elevated temperatures that contribute activation energy (Ingold, 1969a; Kim *et al.*, 1993; Schaich, 2005b). In solvents, H abstraction is favored at moderate lipid concentrations where enough substrate is present to supply hydrogens. However, at low lipid concentrations cyclization or scission dominates, while at high concentrations radical additions and recombinations become more important (Schaich, 2005b). These differences are important to understand when designing model systems for experiments or when interpreting lipid oxidation patterns in different food systems.

When abstractable hydrogens are limited in quantity or accessibility, the following alternative reactions divert the radical stream.

2. Internal Rearrangement, or Cyclization

Lipid free radicals are very short lived. If a compound with abstractable hydrogen is not nearby, the free electron looks for another avenue to find a pairing electron. One option is to add to the first carbon of an adjacent double bond, generating epidioxides with peroxyl radical rearrangements (Reaction 11.21) and epoxides with alkoxyl radical cyclization (Reaction 11.22) and leaving a new free radical on the distal carbon of the double bond. Note that, with two oxygen atoms, peroxyl radicals add to a double bond two carbons away while alkoxyl radicals add to an immediately adjacent double bond.

$$(11.21)$$

$$(11.22)$$

As with all lipid alkyl radicals, addition of oxygen generates new peroxyl radicals now at a new position, and these go on to abstract hydrogens from neighboring lipid molecules. This propagates the radical chain while forming hydroperoxy epidioxide and epoxide products (Chan et al., 1980).

Although internal rearrangement can occur from any position, for both LOO$^\bullet$ and LO$^\bullet$ there is a stronger tendency to cyclize from internal positions, probably because the $-O^\bullet$ is oriented more advantageously. Internal hydroperoxides are found in higher concentrations in fatty acids with four or more double bonds and in photosensitized oxidations (see Section IV, D, 1), so cyclic products epidioxides and epoxides are correspondingly higher. Indeed, the levels and positional distribution of internal cyclic products are characteristic markers distinguishing autoxidation from photosensitized oxidation.

Internal cyclization reactions are reasonably fast (Kochi, 1962; Porter, 1990). Their reaction rates — k(LOO$^\bullet$) ~10^3 s^{-1} and k(LO$^\bullet$)~10^5 s^{-1} — are comparable to hydrogen abstraction, especially in fatty acids with three or more double bonds, so reaction conditions will determine the dominant pathway in oils and foods.

Cyclization is favored when oxygen is limited (Van Sickle et al., 1967) and abstractable hydrogens are not available, e.g. in aprotic solvents (Kochi, 1962) and at low lipid concentrations (Van Sickle et al., 1967). When lipids are dissolved in aprotic solvents such as chloroform or acetonitrile at low concentrations and stored at room temperature or moderately elevated temperatures, as in typical lipid oxidation experiments, external abstractable hydrogens are absent or limited so cyclization can be the only active reaction (Haynes and Vonwiller, 1990). In oils or pure lipids, abstractable hydrogens are both readily available and close by, so abstraction competes with cyclization to generate mixed products whose proportions vary with temperature and extent of oxidation. Cyclization percentages ranging from 30 to 100% in oils or fatty acids have been reported (Schaich, 2005b). Cyclization is also favored by moderately elevated temperatures (e.g. 40°C) (Bors et al., 1984) and lipid chain orientation (Wu et al., 1977). Indeed, epoxides have been reported as the dominant lipid oxidation products in surface monolayers (e.g. oil sprayed or coated onto food surfaces) and in membrane interiors. Propagation via internal rearrangements decreases with temperature because epidioxy and epoxy peroxyl radicals have an increasing tendency to dimerize rather than abstract hydrogens (Schaich, 2005b).

One final comment about cyclic lipid oxidation products: they are highly reactive and thus are often difficult to detect. Epidioxide-OO$^\bullet$ radicals are particularly prone to dimerization with even moderate heat (Neff et al., 1988) and the dimers often decompose still further. Epoxides react extremely rapidly with proteins in complex systems, and thus disappear from lipid extracts. Thus, it is often necessary to look for footprints of epidioxides and epoxides to determine their contributions to lipid oxidation in foods or biological tissues.

3. Addition

Lipid peroxyl radicals have a strong attraction to double bonds and add to them with relative ease, transferring the unpaired electron to the second carbon of the double bond:

$$(11.23)$$

This third propagation pathway has two possible endpoints. The most obvious is formation of dimers and polymers when the peroxyl group remains intact. This reaction is favored by addition to conjugated double bonds (Reaction series 11.24); the chain carrier is a new peroxyl radical formed at the second carbon of the double bond. This peroxyl radical can abstract hydrogens, forming new hydroperoxides (Reaction 11.24a), or can add to additional double bonds (Reaction 11.24b), leading eventually to development of long polymers and increased viscosity characteristic of very oxidized oils (Witting *et al.*, 1957; Privett, 1959; Sims and Hoffman, 1962). Contributions of LOO$^\bullet$ dimerizations increase with oxidation because doubly allylic abstractable hydrogens have been removed during initial reactions and the double bond system has been shifted to a conjugated system.

$$LOO^\bullet + -CH_2CH=CH-CH=CH-CH_2- \longrightarrow \overset{\overset{\textstyle OOL}{\textstyle |}}{-CH_2CH=CH-\overset{\bullet}{CH}-CHCH_2-} \tag{11.24}$$

$$L^\bullet + \overset{\overset{\textstyle HOO \quad OOL}{\textstyle | \qquad |}}{-CH_2CH=CH-CH-CHCH_2-} \overset{LH}{\longleftarrow} \underset{O_2\downarrow}{\overset{\overset{\textstyle \bullet OO \quad OOL}{\textstyle | \qquad |}}{-CH_2CH=CH-CH-CHCH_2-}} \tag{11.24a}$$

$$\downarrow \text{Conjugated diene}$$

$$\overset{\overset{\textstyle OOL}{\textstyle |}}{\underset{\overset{\textstyle |}{\textstyle O}}{-CH_2CH=CH-CH-CHCH_2-}}$$

$$-CH_2CH=CH-\overset{\bullet}{CH}-CHCH_2- \tag{11.24b}$$

The reaction is different when LOO$^\bullet$ adds to isolated or non-conjugated double bonds, as in oleic acid or unoxidized linoleic acid, respectively (Reaction 11.25) (Schaich, 2005b). In this case, the peroxyl adduct decomposes, forming an epoxide and allylic radical and releasing an alkoxyl radical (Reaction 11.25a). The chain is propagated by both the alkoxyl radical and the new epoxyperoxyl radical formed in the presence of oxygen (Reaction 11.25a). This is a true branching reaction since two new propagating radicals (LO$^\bullet$ and epoxyOO$^\bullet$) with increased reactivities are generated from the initial LOO$^\bullet$.

$$LOO^\bullet + R_1-CH_2CH=CH-R_2 \longrightarrow \underset{O_2\downarrow}{\overset{\overset{\textstyle OOL}{\textstyle |}}{R_1-CH_2\overset{\bullet}{CH}-CH-R_2}} \tag{11.25}$$

$$L_2(epoxy)OO^\bullet \longleftarrow R_1-\overset{\bullet}{HC}-\underset{O}{\underset{\diagdown\diagup}{CH-CH}}-R_2 + L_1O^\bullet \tag{11.25a}$$

LO$^\bullet$ radicals do add to double bonds (Reaction 11.26), but not readily, because they are strong hydrogen abstractors and very rapidly cyclize to epoxides (Schaich, 2005b). Addition of LO$^\bullet$ is favored by absence of allylic hydrogens and by conjugation, conditions which only hold after oxidation is reasonably well established. Hence, propagation by LO$^\bullet$ addition is most active in catalyzing chain branching in secondary stages of oxidation. In addition, LO$^\bullet$ adds only to *cis* double bonds (Walling and Thaler, 1961), in contrast to LOO$^\bullet$, which can add to either *cis* or *trans* double bonds.

$$LO^\bullet + \quad\diagup\!\!\diagdown\!\!=\!\!\diagup\!\!\diagdown\!\!\diagup\!\!\diagdown \quad \longrightarrow \quad \overset{\overset{\textstyle LO}{\textstyle |}}{\diagup\!\!\diagdown\!\!\overset{\bullet}{\diagup}\!\!\diagdown\!\!\diagup\!\!\diagdown} \tag{11.26}$$

4. Scission

Scission of alkoxyl radicals has already been introduced as a termination process (Reaction 11.14, Section III, A, 3, b), but scission reactions are also important in propagation. Some of the radicals arising from α and β scissions rearrange internally to non-radical products (mostly aldehydes), but most of them add oxygen to form peroxyl radicals and then abstract hydrogens to propagate the radical chain (Reaction 11.27). Note that the peroxyl radicals and resulting hydroperoxides in this case are in terminal positions rather than the midchain products formed via hydrogen abstraction. In addition, unsaturated fragments, particularly those containing conjugated dienes, are still susceptible to oxidation and their subsequent reactions also contribute to chain branching.

$$
\begin{array}{c}
\overset{\beta}{}\overset{\alpha}{} \\
R_1\text{-CH-}R_2 \\
| \\
O^\bullet
\end{array}
$$

$$R_1OO^\bullet \xleftarrow{O_2} R_1^\bullet + \underset{O}{CH\text{-}R_2} \qquad \underset{O}{R_1CH} + {}^\bullet R_2 \xrightarrow{O_2} R_2OO^\bullet \tag{11.27}$$

$$\downarrow L_3H \qquad\qquad\qquad\qquad\qquad\qquad \downarrow L_4H$$

$$R_1OOH + L_3^\bullet \qquad\qquad\qquad\qquad\qquad R_2OOH + L_4^\bullet$$

Scission is favored over H abstraction in the presence of water or strong proton donors (e.g. polar protic solvents such as alcohols) that provide the protons necessary to stabilize the scission products. However, too much water shifts propagation to termination because protons are drawn from non-lipid sources, stopping radical chains, and increased hydrolysis yields tertiary lipid oxidation products. Scission also increases markedly with temperature since thermal energy facilitates covalent bond rupture.

5. Disproportionation

Disproportionation occurs when two identical radicals join, then decompose to produce two or more different radicals. In lipid oxidation, disproportionation occurs when peroxyl radicals accumulate to high enough concentrations that they begin to react, forming tetroxide intermediates. These then decompose to form two alkoxyl radicals that are 1000 times more reactive as hydrogen abstractors than the original peroxyl radicals (Reaction 11.28) (Thomas, 1965; Lindsay *et al.*, 1973), and subsequent reactions of RO$^\bullet$ lead to greatly accelerated oxidation in true propagation reactions (Traylor and Russell, 1965). In an alternative slower reaction, the tetroxide decomposes to release oxygen and form a peroxide (R_1OOR_2) as a termination reaction (Reaction 11.28a).

$$R_1OO^\bullet + R_2OO^\bullet \longrightarrow [R_1OOOOR_2] \longrightarrow R_1O^\bullet + {}^\bullet OOOR \longrightarrow R_1O^\bullet + O_2 + {}^\bullet OR_2 \tag{11.28}$$

$$\longrightarrow R_1OOR_2 + O_2 \tag{11.28a}$$

Note that in both reactions, oxygen is released back into the headspace or sample to provide fuel to maintain the radical chain. Hence, it can be an important means of prolonging lipid oxidation even when oxygen becomes limiting in samples. Paradoxically, release of oxygen can also confound analyses of lipid oxidation by oxygen consumption, making it appear that oxidation has slowed when it is just shifting gears from LOO$^\bullet$ to LO$^\bullet$ mediation. Recycling of oxygen may explain in part why oxygen consumption appears to slow as oxidation progresses, as will be discussed in Subsection C, next.

Favored by high oxygen that increases ROO$^\bullet$, disproportionation is most important in secondary stages of oxidation or during thermal oxidation. It occurs only in lipids oxidized as pure oils or in aprotic solvents. In polar solvents such as alcohols, or aqueous systems such as emulsions, ROO$^\bullet$ decomposition increases dramatically, the preferred reaction of ROO$^\bullet$ shifts to direct release of oxygen (ROO$^\bullet \rightarrow$ R$^\bullet$), and disproportionation becomes a termination, rather than a propagation process (Walling *et al.*, 1970; Heijman *et al.*, 1985). Increasing solvent or system viscosity also favors termination over propagation by slowing radical movement and inhibiting Reaction 11.19 (Hiatt and Traylor, 1965).

C. Progression and Kinetics of Lipid Oxidation

1. Progression of Lipid Oxidation

One of the most challenging aspects of lipid oxidation is that it is such an ill-defined reaction. Most chemical reactions have a known beginning and fixed end products. However, the precursors of lipid oxidation are always present in foods and rates of reaction as well as reaction pathways and products change over time. Hence, lipid oxidation is a cumulative mix of reactions that create a dynamic and constantly changing process.

Hydroperoxides (LOOH) are the first stable products and they break down to keep the oxidation going, so lipid oxidation is often described in terms of hydroperoxide reactions. Three rate periods are customarily defined (Labuza, 1971):

- Induction period: very low levels of oxidation, undetectable formation of LOOH (and other products).
- Monomolecular rate period: initial stages of oxidation up to ~1% oxidation, LOOH accumulate slowly and decompose as single, isolated molecules (Reaction 11.29); LOO• is the main chain carrier in early stages; both LOO• and LO• become chain carriers as oxidation becomes established.

$$LOOH \longrightarrow LO^{\bullet} + {}^{\bullet}OH \text{ (UV light, heat) or } LO^{\bullet} + {}^{-}OH \text{ (metals)} \tag{11.29}$$

- Bimolecular rate period: later stages of lipid oxidation up to ~7–15% oxidation, LOOH accumulate rapidly to high levels and begin to decompose via LOOH or LOO• + LOOH interactions, as was shown in Reactions 11.2 and 11.3, respectively. However, LO• reactions are much faster and more specific, so LO• becomes the dominant chain carrier and controls the directions of reactions throughout the bimolecular rate periods.

These three rate periods are integrated conceptually with the progression of individual reactions in initiation, propagation, and termination of lipid oxidation in Figure 11.9. Included in the figure are the intermediates or products used to measure the extent of lipid oxidation in foods and biological materials.

Oxygen consumption detects the earliest events in lipid oxidation and reflects the rate of initiation; just as importantly, it is independent of products and pathways so is a true reflection of early oxidation. Oxygen consumption continues throughout oxidation, but its practical use ends when oxygen becomes limiting in the headspace or when peroxyl radicals accumulate to levels that disproportionately release oxygen back into the headspace.

Conjugated dienes, the first chemical change in lipids, and the corresponding hydroperoxides are also early indicators of lipid oxidation and they remain useful indicators at least until hydroperoxides begin to decompose bimolecularly. Because oxygen can add to and come off fatty acids repeatedly during the induction period until a hydrogen donor is available, development of hydroperoxides may lag behind conjugated dienes; and conjugated dienes may last longer than hydroperoxides because they remain intact in many products. However, both conjugated dienes and hydroperoxides decompose as lipid oxidation progresses, so while they are useful to follow the progress of

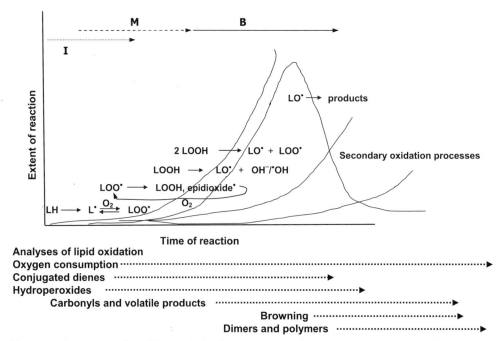

FIGURE 11.9 **Diagrammatic representation of changes in dominant reactions and products over the course of lipid oxidation as traditionally described.** Three separate rate periods are usually designated: induction period (I), monomolecular rate period (M), and bimolecular rate period (B). Shown below the graph are intermediates and products used to measure lipid oxidation at different stages. However, expectations that secondary products develop only after hydroperoxide decomposition are not valid in all systems. Multiple products should be measured at all points for accurate assessment of oxidation extent.

reactions in the early stages and continuously over time, neither of these measures can accurately measure the extent of oxidation in later stages or in isolated samples.

Secondary products such as aldehydes, epoxides, other volatiles, browning, and dimers/polymers were traditionally expected to develop only after decomposition of hydroperoxides, as shown in Figure 11.9. Indeed, many papers have reported this pattern. However, late detection may have resulted from insensitive methodologies and long times between sampling. Recent studies using more sensitive contemporary gas and high-pressure liquid chromatographies with mass spectrometry detection, as well as new chemical analyses accurate at nanomolar levels, have found secondary products formed in parallel with hydroperoxides, as would be predicted from the integrated reaction scheme in Figure 11.5. Thus, the graph in Figure 11.9 should be taken as a useful guide only. More research with multiple simultaneous product analyses will be needed to determine whether the graph is accurate as drawn, and how the various product curves change with different foods and different reaction conditions. The safest approach is always to follow lipid oxidation by simultaneously assaying multiple products from different stages and pathways to provide an accurate picture of the extent and progress of oxidation.

2. Kinetics of Lipid Oxidation

Some consideration of the kinetics of lipid oxidation needs to be included to provide a basis for understanding the driving forces underlying lipid oxidation as a process. Rate equations must be developed to account for conditions in each individual system, so they can become quite complicated and are different in every reference. To illustrate the principles, only very general approaches will be presented here.

At low pO_2 and during early oxidation (monomolecular LOOH decomposition), the reaction is directly dependent on oxygen and the rate of initiation. Thus, the overall rate of lipid oxidation may be expressed simplistically as:

$$\text{Rate of oxidation (low } pO_2) = k_o(k_i/2k_{t1})^{1/2}[LH]\,[O_2] \tag{11.30}$$

where rate constants refer to the corresponding oxygenation, initiation, and termination reactions in Figure 11.1. The overall rate thus reflects the balance between initiation reactions (k_i) and alkyl radical ($L^\bullet$) termination (k_t), primarily by recombination. Assuming that initiators are ubiquitous so $L^\bullet$ are constantly being generated, the rate-limiting step at low oxygen concentrations is formation of the peroxyl radicals necessary for propagation; this process depends on the presence and diffusion of oxygen to radical sites.

At moderate to high pO_2, sufficient oxygen is present to add to every $L^\bullet$ formed, so the rate dependence shifts from oxygen to the concentration of available oxidizable lipids and propagation processes involving LOOH:

$$\text{Rate of oxidation (non-limiting } pO_2) = k_p(k_i/2k_{t3})^{1/2}[LOOH]^{1/2}\,[LH] \tag{11.31}$$

Under these conditions, the overall rate is proportional to the available lipid (total amount and degree of unsaturation) plus propagating LOOH concentrations, and it is controlled by the balance between initiation, propagation, and peroxyl radical recombination. The rate-limiting step is hydrogen abstraction to form and accumulate LOOH. Thus, although they are not directly accounted for in rate equations or by measurement, factors such as light, heat, and metal contaminants that increase decomposition of LOOH and enhance chain branching, or conversely interfere with chain propagation (e.g. antioxidants or co-oxidation targets such as proteins), also strongly influence reaction rates.

In the monomolecular rate period, oxidation kinetics can be described by Equation 11.30 or 11.31, or by combining rate constants, more generally as:

$$\text{Oxidation rate} = -dO_2/dt = K_M\,[ROOH]^{1/2} \tag{11.32}$$

where K_M is the monomolecular rate constant. K_M can readily be determined by plotting the square root of the peroxide values versus time in early oxidation; the slope is $K_M/2$. Typical values are $[10^{-2}$ to 10^{-3} moles oxidized/mole lipid$]^{1/2}$ per hour (Labuza, 1971).

However, in the bimolecular rate period, lipids are consumed rapidly to form hydroperoxides that subsequently decompose in pairs, and oxidation now becomes more complicated. Adding bimolecular LOOH decomposition, the generalized rate now becomes proportional to $[LOOH]^2$:

$$\text{Rate of oxidation} = \underbrace{\frac{k_p k_i^{1/2}}{(2k_{t1\text{-}3})^{1/2}}}_{\text{overall bimolecular rate constant } K_B}\,[LH]\,[LOOH] \tag{11.33}$$

In this case, both LH and LOOH are rapidly changing. A semilog plot of [LOOH]/[unreacted LH] versus reaction time will generate a curve in which the final slope equals K_B. The early and final slopes then differentiate mono-molecular from bimolecular rate periods for a given system.

IV. WHICH FACTORS CONTROL LIPID OXIDATION?

The discussion of mechanisms above remains only a collection of theoretical reactions until they are connected to real systems. Foods and biological systems are much more complicated in their oxidation than the reactions described above because they are structurally complex with many catalysts, oxidation targets, and antioxidants; and all of these factors interact to influence the final kinetics, pathways, and extent of lipid oxidation. This chapter will outline the most important factors that influence lipid oxidation as a guide to evaluating each food or biological system and identifying what pro-oxidant and antioxidant factors are active. Effects of lipid composition, physical structure of the product; presence of catalysts, antioxidants, and sensitive co-oxidation targets; and environmental factors must all be considered when deciding how to formulate and process a product for stability, analyze for lipid oxidation, and finally to develop effective stabilization strategies. The latter will be discussed in Section VI.

Factors affecting lipid oxidation can be divided into five general categories (Table 11.1): nature of the lipids, surfaces, presence of other pro-oxidant and antioxidant compounds, and the environment and solvent system. Only major effects will be discussed here as an introduction.

A. Nature of Lipids

1. Degree of Unsaturation

The number of highly abstractable doubly allylic hydrogens increases with the number of double bonds, so it is not surprising that oxidizability of fatty acids in oils increases the degree of unsaturation (Table 11.2). In particular, the presence of fatty acids with three or more double bonds — linolenic in vegetable oils, arachidonic acid in animal fats, and docosahexaenoic and eicosapentaenoic acids in fish oils — markedly sensitizes oils to rapid oxidation. However, the degree of increase is not proportional to the number of double bonds because each double bond puts a bend in a fatty acid chain. Difficulty in aligning chains so that allylic hydrogens are in close juxtaposition decreases the rate of hydrogen abstraction. Also, with four or more double bonds, the acyl chains bend back on themselves, so that opposite halves of the molecule align and facilitate intramolecular hydrogen abstractions in competition with intermolecular radical transfers. Each additional double bond thus adds progressively less to the oxidation rate because multiple oxidations within molecules increase, reducing the total number of lipid molecules affected.

2. Free Fatty Acids versus Esters versus Triacylglycerols

Both pro-oxidant and antioxidant effects of fatty acids have been reported, mediated primarily through the acid group, so specific effects of free fatty acids will probably reflect the composition and environment of the food or reaction system where they are produced.

It has long been recognized that free fatty acids often oxidize more slowly than their esters (Table 11.2). This is due in large part to non-radical decomposition of hydroperoxides by the acid groups (Figure 11.10). The acidic H^+ contributed from ionized fatty acids adds to the −OOH of the hydroperoxide and induces nucleophilic rearrangement resulting in scission of the chain on either side of the peroxyl carbon. Products can be two aldehydes, as shown in Figure 11.10, or an aldehyde plus an alcohol (Uri, 1961; Scott, 1965). LOOH decomposition can also be induced by hydrogen bonding between undissociated acids (−COOH) and −OOH (Sliwiok et al., 1974). These reactions prevent chain branching and slow overall oxidation.

Carboxylic acids are also excellent metal complexers and they are surface active. These two activities can be complementary or antagonistic depending on the system. Complexation blocks electron transfer orbitals in metals and usually reduces redox potential, which should decrease oxidation rates. Fatty acids can move metals to interfaces, away from double bond targets in bulk oils, but if hydroperoxides and other metal targets concentrate at the interface, oxidation can be greatly accelerated. Moving in the opposite direction, fatty acids can pick up metals at interfaces with aqueous phases and carry them into bulk oils where their polarity attracts metals strongly to polar double bonds to initiate oxidation. Because the effects of free fatty acids are unpredictable, the safest guideline is to limit their addition or prevent their formation as much as possible.

TABLE 11.1 Factors Influencing the Rate and Course of Lipid Oxidation

Nature of lipids	Degree of unsaturation	
	Free fatty acids vs esters vs TAGs	
	trans vs *cis*	
	Conjugation	
	Phospholipids	
Surface effects	Monolayer dispersion	
	Emulsions	
	Bulk lipids	
Presence of catalysts	Pro-oxidants	Preformed hydroperoxides
		Metals
		Porphyrins, e.g. chlorophyll
		Heme compounds
	Enzymes	Lipoxygenase, cyclooxygenase, xanthine oxidase
	Amino acids	
	Ascorbic acid (low concentrations)	
Presence of inhibitors	Antioxidants (endogenous and added)	
	Polyphenols	
	Metal chelators and complexers	
	Synergists	
	Amino acids	
	Enzymes	Glutathione peroxidase
	Interceptors	Proteins
		DNA
		Vitamins
		Pigments
Environment and solvent system	Temperature	
	Light	
	Oxygen pressures	
	Water	
	pH	
	Solvent	
	Packaging	

TAGs: triacylglycerols.
Modified from Schaich (2005a).

Without a free carboxyl group to induce LOOH decomposition, oxidation increases with the number of double bonds in esters and acylglycerols, and with the degree of esterification in acylglycerols (Table 11.3) (Cosgrove *et al.*, 1987). Triacylglycerols oxidize more rapidly than free fatty acids, but considerably more slowly than free esters, because the irregular organization and orientation of fatty acyl chains in triacylglycerols prevent close alignment and thus

TABLE 11.2 Effects of Degree of Unsaturation on Relative Oxidizability of Fatty Acids

Fatty Acid	Relative Rate of Oxidation	
	Fatty Acid	Ester
18:1	1	1
18:2	28	41
18:3	77	98
20:4		195

Extracted from Schaich (2005b), data from Scott (1963); used with permission.

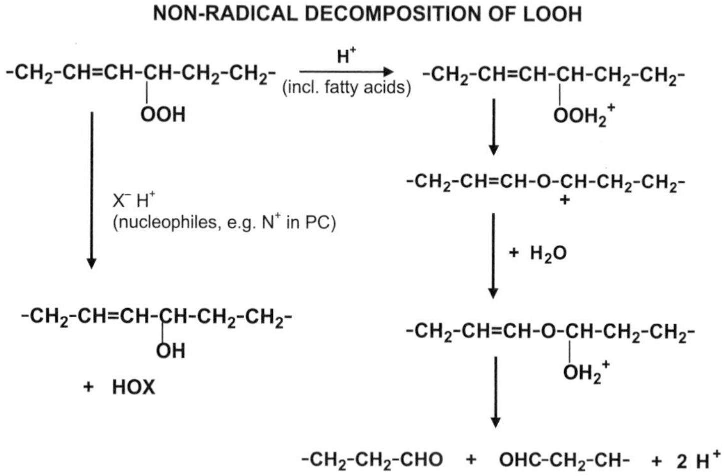

FIGURE 11.10 Non-radical decomposition of lipid hydroperoxides by acids and nucleophiles. These pathways account in part for lower rates of oxidation for fatty acids and phospholipids, particularly phosphatidylcholine, respectively. *(Figure from Schaich, 2005a; redrawn from O'Brien, 1969; used with permission.)*

NON-RADICAL DECOMPOSITION OF LOOH

TABLE 11.3 Lipid Oxidizability

	Oxidizability
	$(M^{-1/2}\ s^{-1/2} \times 10^2)$
Free fatty acids	2.03
Monoacylglycerol	2.83
Diacylglycerol	5.89
Triacylglycerol	7.98

inhibit hydrogen atom transfer between chains (Carless and Nixon, 1960), and limit access of initiators to double bonds. The arrangement of fatty acids in triacylglycerol crystals also influences oxidation: concentrating unsaturated fatty acids at *sn*-2 stabilizes triacylglycerols, while unsaturation at *sn*-1 and 3 enhances oxidation and randomization of triacylglycerols decreases oxidation (Raghuveer and Hammond, 1967). A symmetrical SUS or USU structure favors β-crystal structures, where the close association between chains thus facilitates radical transfers. In contrast, randomization shifts chains to an α-structure with less organization, and this interferes with radical transfer.

3. Trans *versus* cis *Isomers*

Lessons learned in organic chemistry apply equally well to lipid oxidation. Allylic hydrogens in *cis* double bonds are more exposed and accessible for abstraction in *cis* fatty acids than in their *trans* isomers. Thus, *cis* fatty acids oxidize more readily than their *trans* counterparts (Sargis and Subbaiah, 2003). This reactivity pattern explains in part why lipid oxidation is initially rapid then slows over time, i.e. reactive *cis* double bonds that migrate when a lipid radical is formed convert to the less reactive *trans* isomers, as was explained in Section III, A, 2, b. Decreased *trans* reactivity also has distinct consequences for secondary oxidations and for the *in vivo* action of dietary *trans* fatty acids (Pokorny *et al.*, 1976b).

4. Conjugation

Conjugation of fatty acids has very interesting effects on lipid oxidation. Normal non-conjugated double bonds tend to disrupt acyl chain associations. In contrast, conjugation of double bonds straightens the acyl chain and allows closer association of that region between chains which should facilitate hydrogen abstractions. However, the requisite doubly allylic hydrogens are missing in conjugated systems so they react more slowly in initial oxidation. On the other hand, as was considered in the sections on alternative reactions, conjugated systems facilitate both internal rearrangements and radical additions, so can exert critical directing effects on reaction pathways and product distributions in secondary stages of oxidation. Effects of conjugation can be missed or completely misinterpreted if only peroxide values are measured.

5. Phospholipids

Although the acyl chains of phospholipids follow the same oxidation pattern as triacylglycerols and fatty acids/esters, i.e. increasing oxidation with increasing unsaturation of component fatty acids, the oxidative behavior of phospholipids in oils as contaminants or as emulsifiers in multiphase products is complicated. Other phospholipid components participate in the reaction, and the fatty acid composition of phospholipids relative to other system lipids also affects reactivity. Hence, phospholipids can be either pro-oxidants or antioxidants, depending on the system and concentration (Nwosu *et al.*, 1997).

a. Pro-oxidant Effects

Fatty acids in phospholipids are highly unsaturated, often containing arachidonic acid in animal tissues, and the fatty acid acyl chains of phospholipids are highly oriented in membranes and at interfaces, which facilitates electron transfer via the aligned double bonds (Weenan and Porter, 1982). Both of these factors contribute to very rapid oxidation of phospholipids, especially phosphatidylethanolamine, which has the highest unsaturation. In mixed systems, phospholipids are preferentially oxidized and the radicals generated can seed oxidation of triacylglycerols (Nwosu *et al.*, 1997). Phospholipids bind large amounts of water (e.g. each molecule of phosphatidylcholine binds 37 molecules of water), and their hydration water allows them to mobilize catalysts. Phospholipids also complex and activate metals (Nwosu *et al.*, 1997), moving them from aqueous phases into closer contact with lipid phases.

b. Antioxidant Effects

In mixed systems, phospholipids exert a 'sparing' effect on triacylglycerols in oil phases. Exposed to catalysts at interfaces (Bishov *et al.*, 1960), phospholipids are preferentially oxidized in order of unsaturation (Sugino *et al.*, 1997). This could be pro-oxidant, but the association of phospholipids into bilayers or mesophases isolates lipid radicals and limits interaction with the oil phase. High viscosity in bilayers slows migration of catalysts, radical transfers, and chain propagation, while it accelerates terminations (Barclay and Ingold, 1981). Phospholipids also act as co-solvents for antioxidants, increasing the accessibility of tocopherol and other antioxidants to chain-initiating radicals in aqueous microenvironments (Koga and Terao, 1995).

However, the major antioxidant effect of phospholipids by far is from non-radical decomposition of LOOH by the nucleophilic choline group [$-N^+(CH_3)_3$] of lecithin (Figure 11.11, left reaction), which interrupts the radical chain and blocks further reactions (O'Brien, 1969; Corliss and Dugan, 1971). This action does not stop initiation, though, so there can be a build-up of conjugated dienes or products from alternative pathways that are not detected if lipid oxidation in systems containing phospholipids is measured only by peroxide values.

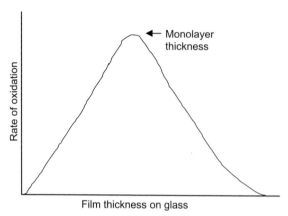

FIGURE 11.11   Effects of film thickness on oxidation of lipids spread on a surface. Maximum oxidation occurs at film thickness approximating a molecular monolayer of lipids. *(Figure from Schaich, 2005a; redrawn from Koch, 1962; used with permission.)*

Such mixed effects argue again that lipid oxidation should always be followed by assays of multiple products from different stages and pathways to provide an accurate picture of the extent and progress of oxidation.

c. Mixed Effects

Because phospholipids bind both metals and water, they can be either (or both) pro-oxidants or antioxidants depending on the system. Although effects on oxidation have been measured, few details about metal binding by phospholipids are yet available. Speculative explanations are that when the metal ligand orbitals are filled by phospholipids, when the redox potential is decreased out of the range of any reducing agents present, or when the phospholipids move metals away from reactive sites, the actions inhibit oxidation (Chen and Nawar, 1991; Yoshida *et al.*, 1991). However, metal binding can accelerate oxidation if the metal redox potential is lowered in a system with LOOH and the metal and LOOH are both concentrated at interfaces. The metal then catalyzes a cascade of LOOH decomposition and generation of LO$^{\bullet}$. Water binding by the phosphate group of phospholipids may be protective by stabilizing LOOH and metals (preventing decomposition and electron transfer, respectively) or may be destructive by activating electron transfer from metals or facilitating contact between metals and LOOH. The specific effects are determined ultimately by the specific nature of the food or reaction system, which is one reason why so many different effects of phospholipids have been reported.

B. Surface Effects

Lipid oxidation is obvious in bulk oils, but foods are complex with multiple phases and complex structure. How does the form of lipid and its organization in a food matrix affect oxidation? All food matrices have surfaces, both solid and liquid, which govern lipid exposure and molecular organization as well as contact with pro-oxidants and antioxidants. Some authorities even argue that all oxidation takes place at surfaces or interfaces, although this view is controversial. It is well established that lipid oxidation rates increase with surface area (Bishov *et al.*, 1960). For lipids dispersed on a solid surface such as a dehydrated food matrix or surfaces of crackers or snack foods, maximum oxidation occurs with dispersal at just about a molecular monolayer of lipid over the solid matrix (Koch, 1962) (Figure 11.11). Monolayer coverage is the point of maximum surface area and oxygen exposure for a lipid film. Lower coverage separates lipid molecules so radical transfers do not easily occur, while multiple layers force lipid molecules to compete for radical transfers, enhance side-reactions that reduce efficiency of the primary radical chain, and increase the thickness through which oxygen must diffuse. As bulk oil volume increases further, oxygen diffusion and interfacial character become the controlling factors, and the oxidation rate declines (Mikula and Khayat, 1985).

Films and surfaces alter oxidation pathways as well as kinetics. Spreading lipid in films in low concentrations orients chains but restricts migrational mobility of hydrogens, so abstraction reactions are greatly impeded. This favors radical recombination over hydrogen abstraction as the initial reaction, and rearrangements of LOO$^{\bullet}$ and LOOH to hydroxyepoxides, endoperoxides, and similar products (Reaction 11.34) rather than decomposition to LO$^{\bullet}$ as secondary reactions (Slawson *et al.*, 1973). As a result, fewer products are found than in emulsions or bulk oil systems, and they are mostly internal cyclization products (Leermakers *et al.*, 1966; Porter *et al.*, 1971; Wu *et al.*, 1978).

$$(11.34)$$

Lipid oxidation is arguably considered to be fastest in emulsions which have the greatest surface area of lipids, greatest diffusion of oxygen, and greatest interface with the aqueous phase containing multiple sources of protons and catalysts (Table 11.4). A categorical statement about whether emulsions oxidize more quickly than spread surface films cannot be made since individual systems vary tremendously and relative oxidation rates are affected by many factors in addition to surfaces. Certainly, there is more lipid to sustain oxidation in emulsions, while LH supply in films is limited, so secondary oxidation can be more extensive in emulsions. Even so, initial oxidations in some films can be faster than in emulsions. For the purpose of this discussion, it is sufficient to note that any system with high surface exposure, whether solid or liquid, has a very high potential for rapid lipid oxidation.

Lipid oxidation in emulsions differs from dry film systems in that modulating effects of water, emulsifying agents, and the nature of the interface compete with dispersion and surface area enhancement of lipid oxidation. Hence, the design and even preparation method for the emulsion can have marked pro-oxidant and antioxidant effects on its oxidative stability. In general, decreasing droplet size (increased surface area) and increasing proportions of water and concentrations of emulsifying agents (Hyde, 1968) enhance initiation at the oil droplet surface (Labuza, 1971). However, the type of emulsifier has been recently shown to exert a major control over oxidation by altering or facilitating specific molecular associations. With lecithin as an emulsifier, for example, apparent oxidation rates in emulsions can be lower than on spread surfaces owing to non-radical decomposition of LOOH as described above. The reverse may be true if a negatively charged emulsifier is used and binds redox metals at oil—water interfaces. Also, emulsifiers, like antioxidants, appear to exhibit a kind of 'polar paradox' where polar and non-polar agents preferentially stabilize phases of the opposite polarity.

Contrary to what might be expected, lipid oxidation is slowest in bulk oil phases where oxygen solubility is high but diffusion limited, and surface area is usually low relative to volume. In the triacylglycerols of oils, lipid hydrocarbon chains are only loosely associated without orientation, making radical transfer relatively inefficient. Hydroperoxides that are necessary for propagation migrate to entrained water microdroplets and less hydrophobic interfaces or self-associate within the oil phase. When the hydroperoxides decompose, propagating LO• are released away from target lipid chains with abstractable hydrogens. When high local concentrations of radicals are generated in oil phases, extensive polymerization is favored in preference to hydrogen abstraction and chain propagation (Treibs, 1948). This is one reason why viscosity increases occur in oxidizing food oils almost as rapidly as production of off-odors and flavors.

TABLE 11.4 Effects of Surface on Overall Rate Constants for Lipid Oxidation During the Monomolecular (K_M) and Bimolecular (K_B) Rate Periods

Methyl Linoleate System	K_M ($\times 10^{-3}$)	K_B ($\times 10^{-2}$)
Bulk	13	6.4
Filter paper — wet	20	6.8
Filter paper — dry	7	4.6
Emulsion	32	8.5
TAG dispersed on cellulose	1.7	0.8

TAG: triacylglycerol.
Data from Labuza (1971); extracted from Schaich (2006); used with permission.

C. Initiators and Catalysts

1. Preformed Free Radicals

Lipid oxidation is so constantly present in foods that it often seems to initiate itself. However, it must be remembered that, in foods and biological systems, many reactions occur that produce radicals. Even in trace quantities these radicals can initiate lipid oxidation by abstracting hydrogens and generating the first $L^\bullet$:

$$LH + R^\bullet, RO^\bullet, ROO^\bullet \longrightarrow L^\bullet + RH, ROH, ROOH \qquad (11.35)$$

Free radicals that most commonly initiate lipid oxidation chains and their reactivity with oleic, linoleic, and linolenic acids are shown in Table 11.5. Hydroxyl radical ($HO^\bullet$), the most strongly oxidizing radical known, is produced by reduction of hydrogen peroxide that is produced by enzymes in living tissues and by metals in foods. It is so reactive that it abstracts hydrogens rather non-selectively from acyl chains (Patterson and Hasegawa, 1978; Heijman et al., 1985). However, also because it is so reactive, it does not migrate, so must be generated very close to its target.

The perhydroxyl radical, $HOO^\bullet$, is also generated by decomposition of hydroperoxide, as well as from protonation of superoxide anion, $O_2^{-\bullet}$, the reduction product of oxygen. Its reactions are slower than $HO^\bullet$, but competitive with organic peroxyl radicals (Bielski et al., 1985). Particularly when generated in enzyme reactions, $HOO^\bullet$ can be an effective initiator of lipid oxidation; $O_2^{-\bullet}$, however, reacts only with lipid hydroperoxides and does not react with unoxidized lipids (Bielski et al., 1983) (Table 11.5).

As has been noted in previous sections, organic peroxyl radicals abstract hydrogens several orders of magnitude more slowly than alkoxyl radicals ($1-200$ vs 10^6-10^7 l/mol/s), but even so, $ROO^\bullet$ and $RO^\bullet$ arising from oxidation of non-lipid molecules in the reaction system very effectively start oxidation chain reactions in lipids.

One final comment: remember that these radicals are needed only to start the lipid oxidation chain reaction, so only trace levels are needed. Once started, the lipid oxidation chain is autocatalytic. The presence of any non-lipid radicals then competes with lipid radicals and may either augment lipid oxidation or have no effect.

2. Metals

Transition metals such as iron, copper, and cobalt are perhaps the most active initiators of lipid oxidation in foods because they are so ubiquitous; metals are also important, but less dominant, in intact plant or animal tissues where

TABLE 11.5 Hydrogen Abstraction Rates of Common Radicals that Initiate Lipid Oxidation

Fatty Acid	Rate Constant (l/mol/s)				Radical Half-life (s)	References
	18:1	18:2	18:3	20:4		
Radical						
$HO^{\bullet\bullet}$ (allylic H)	$\sim 10^9$	9.0×10^9	7.3×10^9	$\sim 10^{10}$	10^{-9}	Patterson and Hasegawa (1978), Hasegawa and Patterson (1978)
(non-allylic H)	4×10^2	3.4×10^3	7.0×10^3	1.0×10^4		Patterson and Hasegawa (1978), Hasegawa and Patterson (1978)
$RO^\bullet$	3.3×10^6	8.8×10^6	1.3×10^7	2.0×10^7	10^{-6}	Erben-Russ et al. (1987)
$ROO^\bullet$	1.1	6×10^1	1.2×10^2	1.8×10^2	10	Hasegawa and Patterson (1978)
$O_2^{-\bullet}$	No reaction	No reaction	<1	<1		Bielski et al. (1985), Gebicki and Bielski (1981)
(MLOOH)		7.4×10^3				Bielski et al. (1983)
$HOO^\bullet$	No reaction	1.1×10^3	1.7×10^3	3.1×10^3		Bielski et al. (1985)
$O^{-\bullet}$	7.5×10^2	9.7×10^3	1.2×10^4	1.9×10^4		Hasegawa and Patterson (1978)

Adapted from Schaich (2005b).

compartmentalization and metal complexers in cells slow the access of metals to lipids. Both metal valences are active (Schaich, 1992). Oxidized metals are stronger direct initiators, forming initial radicals by withdrawing an electron from the double bonds (Reaction 11.36). In contrast, reduced metals tend to be indirect initiators; they react preferentially with oxygen to form complexes (Reaction 11.37) or reduced oxygen species (Reactions 11.38 and 11.39) that can then react more efficiently with lipids. Reducing metals also decompose trace levels of unreactive hydroperoxides to reactive alkoxyl radicals that initiate free radical chains (Reactions 11.40 and 11.41).

Direct:
$$RCH{=}CHR + M^{(n+1)+} \longrightarrow \overset{+}{R}CH{-}\overset{\bullet}{C}HR + M^{n+} \xrightarrow{LH} \left[L^{\bullet} + RH \right] \tag{11.36}$$

Indirect:

(a) formation of active metal–oxygen complexes:

$$M^{n+} + O_2 \longrightarrow [M^{(n+1)+}...O_2^{-\bullet}] \xrightarrow{LH} L^{\bullet} + M^{n+} + HO_2^{\bullet} \atop \quad\quad\quad L'H \downarrow \atop \quad\quad\quad L'^{\bullet} + H_2O_2 \tag{11.37}$$

(b) autoxidation of reduced metals (active at moderate to high pO_2):

$$Fe^{2+} + O_2 \longrightarrow Fe^{3+} + O_2^{-\bullet} \underset{}{\overset{H}{\rightleftarrows}} HOO^{\bullet} \xrightarrow{L} L^{\bullet} + H_2O_2 \tag{11.38}$$

$$O_2^{-\bullet} + HOO^{\bullet} \longrightarrow H_2O_2 \xrightarrow{Fe^{2+}} Fe^{3+} + HO^- + {}^{\bullet}OH \xrightarrow{LH} H_2O + L^{\bullet} \tag{11.39}$$

(c) reduction or oxidation of hydroperoxides ($ROOH$, H_2O_2, $LOOH$) (dominates with low metal, substrate, and oxygen concentrations):

$$ROOH + M^{n+} \xrightarrow{fast} RO^{\bullet} + {}^-OH \tag{11.40}$$

$$ROOH + M^{(n+1)+} \xrightarrow{slow} ROO^{\bullet} + H^+ \tag{11.41}$$

Decomposition of hydrogen peroxide and miscellaneous organic peroxides by metals provides a major source of radicals for initiating for lipid oxidation. However, hydroperoxide decomposition by metals is just as important in accelerating chain propagation and branching in later stages of oxidation when higher concentrations of LOOH accumulate. This critical influence on both initiation and propagation of lipid oxidation provides solid arguments for why metals are generally considered the most important lipid oxidation catalysts in foods and also in biological systems.

Although oxidized and reduced metals are each active alone, their catalytic effectiveness is amplified tremendously when redox cycling occurs (Schaich, 1992). The combination of oxidized and reduced states of metals, particularly iron, creates a powerful factory for continuous cycling of oxygen and active catalysts, while generating new lipid radicals, as shown in Reaction Series 11.42. This can involve cycling either hydroperoxide oxidation and reduction (Reaction 11.42), or combining oxidation of double bonds to generate $L^{\bullet}$ with reduction of LOOH to generate $LO^{\bullet}$ (Reaction 11.42a). Rates of cycling can be accelerated by several orders of magnitude in the presence of reducing agents, particularly ascorbic acid, that maintain the faster reacting lower valence state in metals.

Cycling:

$$\text{Cycling}\quad Fe^{2+} + \begin{Bmatrix} ROOH \\ LOOH \end{Bmatrix} \xrightarrow{fast} Fe^{3+} + \begin{Bmatrix} RO^{\bullet} + OH^- \\ LO^{\bullet} + OH^- \end{Bmatrix} \xrightarrow{LH} \begin{Bmatrix} ROH + L^{\bullet} \\ LOH + L^{\bullet} \end{Bmatrix} \atop \downarrow ROOH \atop Fe^{2+} + ROO^{\bullet} + H^+ \xrightarrow{LH} ROOH + L^{\bullet} \tag{11.42}$$

$$\begin{aligned} LH + M^{(n+1)+} &\longrightarrow L^{\bullet} + H^+ + M^{n+} \xrightarrow[LH]{O_2} LOOH + L^{\bullet} \\ LOOH + M^{n+} &\longrightarrow LO^{\bullet} + OH^- + M^{(n+1)} \\ &\quad\quad\quad AH \end{aligned} \tag{11.42a}$$

3. Enzymes

Enzymes are a major class of lipid oxidation initiators in unprocessed tissues, including postharvest plants and grains and postslaughter meats. Normally separated from their substrates in plant and animal cells, enzymes are brought into contact with their lipid substrates when food preparation and processing disrupts cell organization. Lipoxygenase reacts directly with lipids to form hydroperoxides without radicals, while other enzymes catalyze oxidations that generate radicals capable of initiating lipid oxidation. Their basic reactions are described below.

a. Lipoxygenase

Lipoxygenases are iron metallo enzymes that catalyze the aerobic oxidation of fatty acids with *cis*-1,4-pentadiene structures to generate conjugated LOOHs without releasing lipid free radicals (Reaction 11.43). To do this, the enzyme binds the diene section of the fatty acid in its active site and, via the iron also bound there (Aoshima *et al.*, 1977), removes the bisallylic hydrogen (Egmond *et al.*, 1973; deGroot *et al.*, 1975). The acyl chain rotates, oxygen bound to a separate site on the enzyme reacts with the free radical, and H^+ is donated from histidine residues in the active site to complete the LOOH before it is released. The resulting conjugated dienes are always *trans, cis* relative to the hydroperoxide (Egmond *et al.*, 1972).

$$(11.43)$$

Lipoxygenase action by itself does not initiate lipid oxidation. However, lipoxygenase acts as a kind of engine for producing reservoirs of LOOH that later decompose to generate initiating $LO^\bullet$ and $^\bullet OH$ radicals by light and heat, $LO^\bullet/LOO^\bullet$ by metals, or $LO^\bullet$ in secondary reactions of the enzyme itself (Aoshima *et al.*, 1981). Very low LOOH levels produced in plant or animal tissues may provide the 'invisible' initiators that make lipid oxidation sometimes appear spontaneous. Under some conditions (e.g. cold and dark, as in frozen unblanched materials), LOOH produced by lipoxygenases can accumulate to relatively high levels, then lead to a cascade of rapid oxidation when LOOH decomposes.

The rate and specificity of lipoxygenase oxidations depend on the type and concentration of fatty acids, oxygen concentration, pH, and enzyme source and isozyme form. The preferred substrate is linoleic acid; linolenic and arachidonic acids are also oxidized but more slowly (Gardner, 1986). There are three different isozyme forms of lipoxygenase that differ in their specificity for C9 and C13 on linoleic acid, their pH optima, and their phase preference for reactions:

- LPOx-1: optimum pH = 9, prefers charged substrates, (e.g. $LOO^\bullet$), has no phase specificity, so reacts equally in organic and aqueous phases
- LPOx-2: optimum pH ~6.8; prefers un-ionized acids, esters, and organic phases; reactions are depressed in emulsions; strong co-oxidation capabilities
- LPOx-3: optimum pH ~6.8, prefers emulsions; strong co-oxidation capabilities.

In addition to these general preferences, isozymes from different materials add their own specificities or eccentricities. Examples are shown in Table 11.6. It should also be noted that at pH ~6, lipoxygenases have some hydroperoxidase activity wherein they catalyze the decomposition of LOOH to LOH (Reaction 11.44). This action contributes to formation of some characteristic flavor compounds in plant tissues (e.g. cucumber).

$$\text{Hydroperoxidase activity of LPOx:}\quad \text{LOOH} \xrightarrow{\text{LPOx}} \text{LOH} + H_2O \qquad (11.44)$$

Lipoxygenases are found in low levels in all biological materials. Relatively high lipoxygenase levels are found in grains, peanuts, tomatoes, potatoes, soybeans and other legumes, beans, peas, and muscle tissue.

TABLE 11.6 Reaction Characteristics of Lipoxygenases from Various Plant Sources

Plant	pH Optimum	Position Preference	Reference
Soybean (1)	8.5—9.0	13-OOH	Hamberg and Samuelsson (1967)
	7.0	13-OOH:9-OOH 2:9	
	6.5	9-OOH	
Soybean (2)	6.1	13-OOH:9-OOH 80:20	Fukushige et al. (2005)
Soybean (3)	6.5	13-OOH:9-OOH 40:60	
Corn	6.6	Mostly 9-OOH (10t, 12c)	Gardner and Weisleder (1970)
		Small amount of 13-OOH (9t, 11t)	
	9.0	13-OOH:9-OOH 85:15	Veldink et al. (1972)
Wheat (1)	5.7—6.5	13-OOH:9-OOH 15:85	Grosch et al. (1976)
Peanut (1)	6.0	100% 13-OOH	Belitz and Grosch (1986)
	6.2	84% 13-OOH (9c, 11t)	Pattee and Singleton (1979)
		16% 9-OOH (10t, 12c)	
	8.3	10% as much product formed	
		13-OOH:9-OOH 2:1	
		Less 13-OOH at low pO_2	
Tomato (1)	5.5	13-OOH:9-OOH 95:5	Belitz and Grosch (1986)
Peas (2)	6.5	13-OOH:9-OOH 50:50	Belitz and Grosch (1986)

Numbers in parentheses refer to the isozyme form.

b. Xanthine oxidase

Xanthine oxidase is an enzyme with two flavin adenine dinucleotide molecules, two molybdenum atoms, and eight iron atoms in its active site. It catalyzes the oxidation of hypoxanthine to xanthine, with the simultaneous reduction of oxygen to superoxide anion, $O_2^{-\bullet}$. $O_2^{-\bullet}$ dismutates to form hydrogen peroxide, which decomposes to reactive hydroxyl radicals that initiate lipid oxidation (Kellogg and Fridovich, 1975; Porter and Lehman, 1982; Thomas et al., 1982). This reaction is particularly important in initiating lipid oxidation in dairy products.

$$\text{hypoxanthine} + O_2 \longleftrightarrow \text{xanthine} + O_2^{-\bullet} \tag{11.45}$$

$$\text{xanthine} + O_2 \longleftrightarrow \text{uric acid} + O_2^{-\bullet} \tag{11.45a}$$

$$2\, O_2^{-\bullet} + H_2O \longrightarrow 2\, O_2^{-\bullet}/HOO^{\bullet} \longrightarrow 2\, H_2O_2 \xrightarrow{M^+} 2\, HO^{\bullet} + 2\, HO^{-} \tag{11.45b}$$

M^+ is a reducing metal.

4. Hemes

The ability of heme proteins such as hematin, hemoglobin, myoglobin and metmyoglobin, catalases and peroxidases to catalyze lipid oxidation much more rapidly than free iron was first reported nearly 90 years ago. The pioneering work of Watts (Watts, 1954, 1962; Younathon and Watts, 1959) and Tappel (1953a, b, 1955, 1961, 1962, 1995) in the

1950s and 1960s established the effects clearly, but the mechanisms of heme action remained controversial. Several characteristics or requirements for heme catalysis, however, were identified (Schaich, 1980a, 1992):

- the porphyrin–Fe structure is required (not just iron)
- not all heme proteins show the same catalytic activity, due in part to differences in degree of exposure and/or accessibility of the hematin structure
- Fe^{3+}-hemes are most active even without oxygen; Fe^{2+}-hemes require oxygen for catalysis
- no change of heme iron valence is involved in catalysis
- catalysis reverses to inhibition at high heme levels.

From these observations, three fundamental mechanisms were proposed: (1) fundamental electron transfer from ferric hemes (Watts, 1954, 1962; Younathon and Watts, 1959); (2) hemes form complexes with hydroperoxides, and alkoxyl radicals generated in subsequent decomposition of the complex then increase rates of chain propagation (Tappel, 1953a, b, 1955, 1961, 1962, 1995); and (3) free inorganic iron released from hemes, rather than the hemes themselves, catalyzed lipid oxidation (Love and Pearson, 1974). While each of these theories addresses some characteristics, none of them is completely consistent with kinetics, structure–activity patterns, product mixes, and solvent effects of heme catalysis (Schaich, 1980a).

Modern analytical techniques have now revealed that heme catalysis involves generation of ferryl iron – Fe(IV)=O or Fe(IV)-OH – in complexes of the hemes with hydroperoxides or oxygen (Rao $et\ al.$, 1994; Nam $et\ al.$, 2000a, 2003). Active proteins bind fatty acids in a heme pocket, with the hydrocarbon end inside and the double bonds oriented towards the heme iron (Rao $et\ al.$, 1994; Nam $et\ al.$, 2000a, 2003). Ferric heme iron combines with a hydroperoxide or oxygen to form Fe(IV), which is a very strong oxidant, equivalent to $HO^{\bullet}$ in reaction rates but more selective owing to a lower redox potential (Traylor and Xu, 1990). Ferryl iron then rapidly abstracts H from the doubly allylic C11 of linoleic acid (Rao $et\ al.$, 1994) or even more rapidly from hydroperoxides (Traylor and Xu, 1990), in contrast to the very slow oxidation of hydroperoxides by non-heme Fe^{3+}:

$$(P^{\bullet+})Fe^{4+}(O) \ + LOOH \ \xrightarrow{\text{very fast}} \ LOO^{\bullet} \ + \ Fe^{3+}OH \ + \ H_2O \qquad (11.46)$$

Fe^{4+} states are maintained by shuttling redox electrons between states in the apoprotein without involving the iron center.

$$(P^{\bullet+})Fe^{4+}(O) \ \rightleftharpoons \ (P)Fe^{4+}(OH) \qquad (11.47)$$

Thus, electrons can be shuttled easily between two reactive states without the loss of oxidizing power or reduction of Fe^{4+} to less reactive Fe^{3+} (Nam $et\ al.$, 2000b). In this way, the hemes become a kind of engine for continuously producing cascades of radicals and rapidly catalyzing lipid oxidation.

Figure 11.12 shows a general representation of the current understanding of ferric heme-mediated formation and reaction of Fe^{4+} (Schaich, 2005b). The three reaction streams support three unique characteristics differentiating heme catalysis from other oxidants: a cascade of radical propagations and extremely rapid oxidation rates plus high proportions of epoxides and hydroxylated products (Dix and Marnett, 1983, 1985; Nam $et\ al.$, 2000a, b).

Fe^{2+}-hemes generate ferryl complexes but more slowly than ferric hemes. They either use hydroperoxides as oxygen sources or react directly with oxygen to form ferryl iron.

$$(P)Fe^{2+}(H_2O) \ + \ \tfrac{1}{2}O_2 \ \longrightarrow \ Fe^{4+}(O) \ + \ H_2O \qquad (11.48)$$

This ability to use oxygen directly provides constant replenishment of $Fe^{4+}(O)$ in longer reactions. Fe^{2+} hemes may also contribute to delayed catalysis by regenerating more rapidly reacting Fe^{3+} hemes after some peroxyl radicals have been formed (Bruice $et\ al.$, 1988).

$$ROO^{\bullet} \ + \ Mb(Fe^{II})\,O_2 \ \longrightarrow \ ROOH \ + \ Mb(Fe^{III}) \ + \ O_2 \qquad (11.49)$$

Altogether, these characteristics make Fe^{2+}-hemes more important in maintaining oxidation over long periods during storage, in contrast to the rapid immediate action of Fe^{3+}-hemes. Mechanisms of heme catalyses of lipid oxidation have been reviewed in detail by Schaich (2005b); heme catalyses of lipid oxidation in foods has been reviewed by Baron and Andersen (2002).

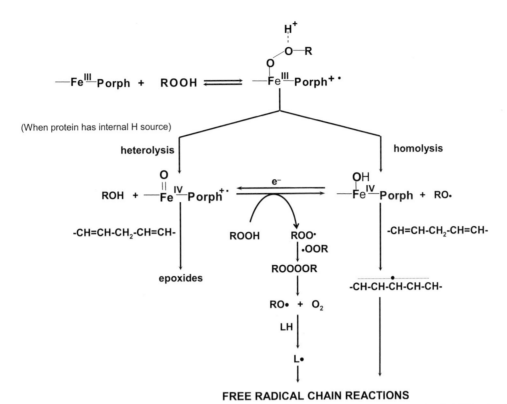

FIGURE 11.12   Formation and reaction pathways of ferryl iron in catalysis of lipid oxidation. *(Figure from Schaich, 2005a; data from Nam et al., 2000a, b; used with permission.)*

5. Chlorophyll and Other Pigments

With extended conjugated double-bond systems and/or activated carbonyls, chlorophyll and many other pigments act as light gatherers in plants. They retain that ability even outside the plant system, so when present in foods or as contaminants in oils, they act as powerful photosensitizers. Photosensitizers absorb low-energy visible light which is normally inert, are bumped into excited states, then transform the excitation energy to reactive species, either free radicals or singlet oxygen, $^1O_2$, which then react rapidly with lipids. This process will be explained in detail in the next section.

Chlorophyll is probably the most notorious oxidation catalyst among pigments because it can oxidize lipids by both free radicals and singlet oxygen (Figure 11.13); even traces of chlorophyll remaining in oils or lipid extracts

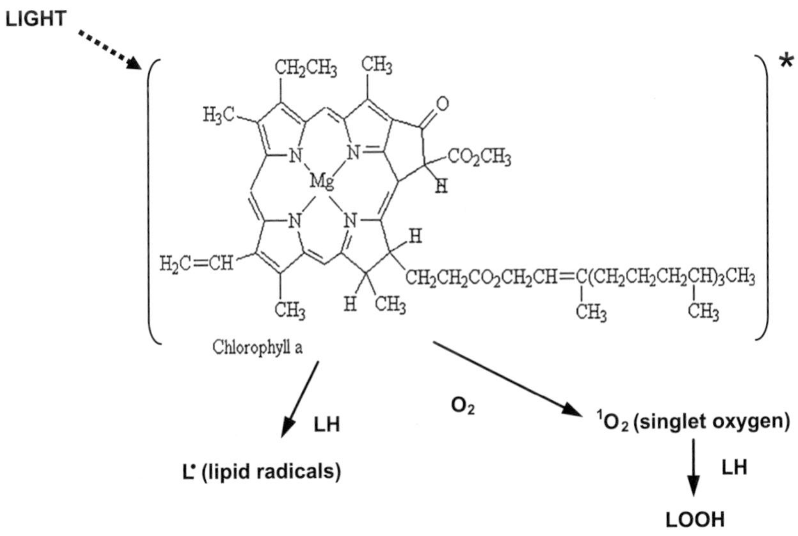

FIGURE 11.13   Multiple modes of photosensitization active with chlorophyll.

rapidly set off lipid oxidation. Many synthetic food colors are known to be active photosensitizers (Umehara *et al.*, 1979), and it is likely that other natural pigments similarly influence lipid oxidation.

One interesting point about pigments is that they have the potential to be both pro-oxidants and antioxidants. With extended regions of conjugated double bonds, carotenoids act as $^1O_2$ quenchers, offering multiple double bonds for formation of hydroperoxides in preference to reactions with lipids. Thus, at low concentrations carotenoids, especially β-carotene, are antioxidants (see also Section VI). However, carotenoid alkoxyl and peroxyl radicals are reactive, and at high concentrations these pigments become pro-oxidants. It is likely that other pigments act in the same manner, so the safest approach is to limit pigment concentrations in lipid-containing foods.

D. Environment and Solvent System

1. Light

a. Ultraviolet Light

Direct initiation of lipid oxidation by light is not a favorable reaction because light energy at wavelengths above 200 nm is insufficient to break bonds and also provide activation energy for ionization and then push the molecular fragments apart (Table 11.7).

$$LH \xrightarrow{\ h\nu\ } L^\bullet + {}^\bullet H \quad \text{or} \quad L_a^\bullet + {}^\bullet L_b \qquad (11.50)$$

The principal UV-absorbing groups of lipids are carbonyls, double bonds, and peroxide O—O bonds. Of these, only O—O bonds have bond energies accessible to UV light. Thus, catalysis of lipid oxidation by UV light is mediated through homolytic scission of any preformed hydroperoxides to generate the true initiators — $LO^\bullet$, $HO^\bullet$, and $RO^\bullet$ — that abstract hydrogens from lipid molecules to begin the free radical chain reaction. What makes UV light so important as an initiator is that it generates two reactive radicals and consequently two lipid oxidation chains for each hydroperoxide decomposition, and perhaps more importantly, it generates a very strongly oxidizing hydroxyl radical, $HO^\bullet$, instead of a hydroxide ion.

$$
\left.
\begin{array}{l}
ROOH \\
HOOH \\
LOOH
\end{array}
\right\}
\xrightarrow{h\nu\ (uv)}
\left.
\begin{array}{l}
RO^\bullet + {}^\bullet OH \\
HO^\bullet + {}^\bullet OH \\
LO^\bullet + {}^\bullet OH
\end{array}
\right\}
\xrightarrow{2\,L_2H}
\begin{array}{ll}
ROH + H_2O & (11.51) \\
2\,L_2^\bullet + 2\,H_2O & (11.51a) \\
LOH + H_2O & (11.51b)
\end{array}
$$

TABLE 11.7 Photon Energies (E_p) Available in Light at Various Wavelengths in Comparison to Bond Dissociation Energies (BDEs) of Important Functional Groups

Bond	BDE (kJ/mol)[a]	Wavelength (nm)	E_p (kJ)[b]
C=C	612	200	596
O—H	463	230	518
C—H	412	260	458
C—O	360	290	411
C—C	348	320	372
C—N	305	350	341
O—O	157	410	291

[a]Hecht (2000)
[b]Atkins and Beran (1990), Kamiya et al. (1963).

ROOH is any organic hydroperoxide, HOOH is hydrogen peroxide, LOOH is a lipid hydroperoxide.

Light is always around unless deliberately excluded in dark or packaging. Altogether, these factors make UV light one of the most powerful catalysts of lipid oxidation. Hence, consideration of light effects is critical when handling and processing lipids in any form: as reagents, in oils, or in foods.

b. Visible Light

Visible light ($> 400\,nm$) lacks the energy to produce radicals directly, but it can initiate lipid oxidation indirectly through photosensitizers (Table 11.8), molecules that absorb low-level light energy, and transform it to chemical energy sufficient to drive reactions. In this process, called photosensitization, a sensitizer molecule ^1S absorbs light and is excited to a high energy state $^3S^*$, then transfers the excitation energy either to molecular bonds in lipids to form free radicals directly (Type 1) (Foote, 1976) or to oxygen to form singlet oxygen $^1O_2$, which then adds to double bonds of unsaturated fatty acids (Type 2) (Murray, 1979):

Type 1 sensitization (free radical) $\longrightarrow L^\bullet$ (e^- transfer reaction)

$$^1S \xrightarrow{h\nu} {}^3S^* \Bigg\langle \begin{array}{l} \xrightarrow[-H^+]{LH} {}^\bullet SH + {}^\bullet L \xrightarrow{O_2} {}^0S + LOO^\bullet \xrightarrow{H} LOOH \\[2mm] \xrightarrow[-e^-]{L} (S^{-\bullet} + L^{+\bullet}) \text{ or } (S^{+\bullet} + L^{-\bullet}) \xrightarrow[H^+]{O_2} {}^0S + LOOH \end{array} \tag{11.52}$$

Type 2 sensitization ($^1O_2$, singlet oxygen) $\longrightarrow$ LOOH (no free radicals produced)

$$^1S \xrightarrow{h\nu} {}^3S^* \Bigg\langle \begin{array}{l} \xrightarrow{{}^3O_2} {}^1S + {}^1O_2 \xrightarrow{LH} {}^0S + LOOH \\[2mm] \xrightarrow{{}^3O_2} S\text{-}O_2^{*-} \left[\begin{array}{l} \longrightarrow {}^1S + {}^1O_2 \xrightarrow{LH} LOOH \\ \longrightarrow {}^0S + LOO^\bullet \xrightarrow{H^+} LOOH \end{array} \right. \end{array} \tag{11.53}$$

The dominant reaction depends on the photosensitizer. Some sensitizers (such as chlorophyll) can act by either radical or singlet oxygen reactions (Sastry and Lakshminarayana, 1971; Terao and Matsushita, 1977b) with the dominant mechanisms determined by reaction conditions and solvent; other sensitizers act by only one mechanism, radical or singlet oxygen (Murray, 1979) (Table 11.8).

TABLE 11.8 Photosensitizers Found in Foods or Used Commonly in Food Chemistry Research

Type 1 — Free Radical	Type 2 — Singlet Oxygen
Chlorophyll, pheophytin	Chlorophyll
Hemes — protoporphyrin	Hemes
— myoglobin	Erythrosine
— hemoglobin	Rose bengal
— flavins (especially riboflavin)	Flavins
Xanthenes	Methylene blue
Anthracenes	Proflavine
Anthroquinones	Eosin
Crystal violet	Food dyes (?)
Food dyes	

Type 1 photosensitization generates radical chain reactions that are indistinguishable from normal autoxidations, while Type 2 photosensitization is direct oxygenation in which singlet oxygen forms hydroperoxides 1500 times faster than with normal triplet oxygen (Rawls and Van Santen, 1970). In a concerted 'ene' reaction, $^1O_2$ attaches to either carbon of a double bond and adds an allylic proton to form a hydroperoxide directly in a cage reaction without generating radicals:

$$\hspace{10cm} (11.54)$$

An important distinguishing characteristic of singlet oxygen oxidation is that it shows no preference for a particular carbon in the double bond, so yields comparable amounts of hydroperoxides at both internal and external positions for isolated or non-conjugated double bonds in unsaturated fatty acids (Terao and Matsushita, 1977a). (Remember that autoxidation generates hydroperoxides primarily at external positions.) $^1O_2$ addition to conjugated double bonds (e.g. natural conjugated linoleic acid or oxidized linoleic acid) yields cyclic endoperoxides in high concentrations:

$$\hspace{10cm} (11.55)$$

When exposed to UV light, heat, or metals, these hydroperoxides then decompose in a cascade of radicals that can lead to explosive oxidation. What makes singlet oxygen sensitization so damaging as an initiator of lipid oxidation is that the rate of hydroperoxide formation depends primarily on the concentration of the sensitizer, not oxygen, and the sensitizers can act in parallel rather than in series. Consequently, high concentrations of hydroperoxides can be generated in a short time.

Light catalysis of lipid oxidation can easily be detected in kinetic patterns and product distributions (Table 11.9). In early stages of autoxidation, formation of LOO$^\bullet$ depends on oxygen, and an induction period with very low or no hydroperoxides is characteristic. Antioxidant strategies are designed to extend the induction period as long as possible. In contrast, photosensitized oxidations depend on concentrations of sensitizers, are independent of pO_2 except under very limited oxygen, and show no induction period, accumulating high levels of hydroperoxides immediately. Eliminating light during handling and by packaging thus has dramatic immediate effects on photosensitized oxidations, but will only affect autoxidations after LOOH accumulation. Similarly, as was discussed in Section IV, A, 1, oxidation rates increase geometrically with numbers of double bonds in autoxidation owing to abstraction sensitivity of doubly allylic hydrogens. However, photosensitized oxidation rates are directly proportional to the number of double bonds since $^1O_2$ shows no preference in bond addition (Terao and Matsushita, 1977a).

Increased proportions of non-conjugated and internal 10- and 12-hydroperoxides, as well as cyclic products, can also be used diagnostically to differentiate photosensitization from autoxidation (Table 11.9). Finding these products is a clear indication of light effects, so protection of the product from light during processing and with packaging will be critical for stabilization. It is important to note also that internal and cyclic hydroperoxides decompose into different secondary products than those resulting from autoxidation, so 'light-struck' oils and foods will have altered flavors, odors, and co-oxidations of other molecules compared to 'rancid' products. Cyclic products, in particular, are thought to be important precursors of light-induced off-flavors (Neff et al., 1982). Flavor reversion in soybean oil is a light-induced oxidation (Chang et al., 1966).

2. Heat

When considering heat effects on lipid oxidation, storage temperatures are always the first targets. However, heat-induced changes are much more pronounced at the high temperatures used in processing than at typical transportation and storage temperatures. Even relatively small increases in processing or storage temperatures can dramatically

TABLE 11.9 Comparison of Kinetics and Products of Free Radical Autoxidation and Type 1 Photosensitization versus Singlet Oxygen Type 2 Photosensitized Oxidation

	Free Radical — Type 1 Autoxidation	$^1O_2$ — Type 2 Photosensitization
Mechanism	Conventional free radical chain reaction	'ene' reaction, concerted addition of $^1O_2$, no free radicals involved
		Radicals generated by LOOH decomposition
Kinetics	Induction period present	No induction period
	Dependent on pO_2	Independent of O_2 (when O_2 not limiting)
		Dependent on sensitizer concentration
	Rate not proportional to number of double bonds	Rate directly proportional to number of double bonds
Relative reactivity	18:1 ~1	18:1 ~1
	18:2 17	18:2 2
	18:3 25	18:3 3
Products	Conjugated LOOHs	Both non-conjugated and conjugated LOOHs
	External 9, 13, 16 positions dominant	LOOHs at all positions
	Scission products all from external LOOHs	High proportion of cyclics, endoperoxides, epoxides, dihydroperoxides, and trihydroperoxides at internal positions
		Scission products from internal LOOHs, altered scission position preferences

Data from Foote (1968, 1976), Kearns and Khan (1969), Foote *et al.* (1970), Murray (1979), Neff (1980), Neff and Frankel (1984), Neff *et al.* (1982), Frankel *et al.* (1982), Frankel (1985, 1987), Pryor (1978), Umehara *et al.* (1979). Adapted from Schaich (2005a); used with permission.

shorten shelf-life. Thus, control of temperature effects on lipid oxidation must be addressed at every stage of food production, handling, and storage.

Heat has three major effects on lipid oxidation:

- Especially at low to moderate temperatures, heat accelerates lipid oxidation by increasing the rate of LOOH decomposition and initiation of secondary chains, so faster oxidation kinetics is where heat effects are most clearly observed (Labuza, 1971; Marcuse and Fredriksson, 1968).
- Heat mediates shifts in dominant reaction pathways and resulting products:
    - hydrogen abstraction from LOH and LOOH in preference to LH (Frankel, 1982)
    - less scission, altered scission patterns (e.g. hexanal to decadienal from linoleic acid), and increased di 90% dimers and polymers (Pokorny *et al.*, 1976a)
    - increased dimerization and polymerization, changes in types of dimers formed: more C—O—C and C—C dimers and fewer C—O—O—C cross-links (Frankel, 1984, 1991)
    - increased *trans* isomers (Piretti *et al.*, 1978).
- At high temperatures, e.g. frying processes, > 150°C, thermal energy induces scissions of acyl chains to provide a base of radicals that underlie all molecular degradation and initiate chains of autoxidation when air is present (Nawar, 1969, 1986).

In the interest of space, only the first of these effects will be discussed here.

a. Thermal Decomposition of Hydroperoxides at Low to Moderate Temperatures

The effects of low to moderate heat on LOOH decomposition are rather straightforward. In any system, heat gives the greatest boost to reactions that have high activation energies. Of all the steps in lipid oxidation, only

TABLE 11.10 Activation Energies for the Various Reaction Steps in Lipid Autoxidation

Reaction	Activation Energies (E_a) (kcal/mole)
($L^\bullet + O_2$)	0
k_p ($LOO^\bullet + LH$)	~5–15
k_t ($2\ ROO^\bullet$)	~4
k_t ($2\ R^\bullet$)	5
k_t ($R^\bullet + ROO^\bullet$)	1
* k_{d3} (monomolecular) LOOH →	31
* k_{d3} (bimolecular) 2 LOOH →	50 (uncatalyzed system)

Data from Labuza (1971) and Marcuse and Fredriksson (1968).

hydroperoxide decomposition has energy hurdles, requiring 31 and 50 for monomolecular and bimolecular LOOH decomposition, respectively (Table 11.10). Thus, at the low to moderate temperatures used in most handling, storage, and some processing, heat acts primarily by breaking the O—O bonds in traces of organic or lipid hydroperoxides preformed by metals, lipoxygenases, photosensitizers, or other reactions. The reactive $RO^\bullet$, $LO^\bullet$, and $^\bullet OH$ radicals generated then abstract hydrogens from neighboring lipids to form $L^\bullet$ and initiate radical chains. As oxidation progresses and appreciable levels of authentic LOOH accumulate, the major effect of heat shifts to acceleration of propagation rather than initiation (Marcuse and Fredriksson, 1968; Labuza, 1971).

3. Water

No discussion of lipid oxidation is complete without consideration of moisture effects. As shown in the classical graph of degradation processes versus water activity (Figure 11.14), lipid oxidation is the only degradation reaction in foods that cannot be stopped by sufficient removal or binding of water. Lipid oxidation is rapid at high moistures and water activity, as are all reactions (region D), but unlike other degradation reactions, lipid oxidation is just as rapid in very dry systems (region A). Lipid oxidation is lowest (region B) when only a monolayer of water molecules is bound to macromolecules in foods.

FIGURE 11.14   Classical graph of effects of water activity on chemical reactions in foods. *(Adapted from Karel, 1980.)*

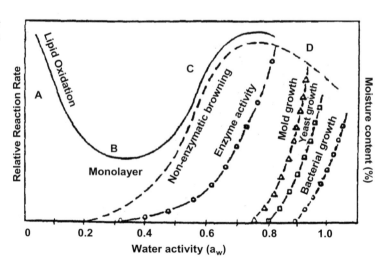

Water has both pro-oxidant and antioxidant effects on lipid oxidation, as shown by the rate changes between regions on the graph. This biphasic action may be explained as follows (Labuza, 1971; Karel, 1980):

- A, Very dry systems — high oxidation: molecular sites of oxidation are bare, providing facile access of oxygen to lipids, metals are unhydrated and reactive, hydroperoxides are uncomplexed and decomposed readily, the food matrix is open and porous, allowing free flow of oxygen.
- A → B, Water binding to molecules provides surface protection, decreasing oxidation: water hydrogen bonding to LOOH retards decomposition, hydration of metals inhibits electron transfers and shifts redox potentials, water binding to surfaces of the macromolecular matrix retards oxygen diffusion, water attachment to potentially reactive sites excludes oxygen.
- B, Monolayer value — lowest oxidation: conceptual single layer of water molecules bound to molecular surface protects reactive sites from oxygen and catalysts but is insufficient for mobilization.
- B → C, Intermediate moisture region — increasing oxidation: hydration of molecules increases molecular mobility and reactivity; multilayers of water accumulating on molecular surfaces dissolve catalysts and increase their diffusion and interaction with lipids, and mobilize and activate metals.
- C → D, Fluid water region — high but decreasing oxidation: water dilutes metals and other catalysts and reactants, emulsion formation separates lipid phases from most catalysts, high moisture promotes non-enzymatic browning, products are antioxidants.

Figure 11.14 shows clearly that the moisture content and water activity (a_w) strongly influence the rate of oxidation of products with unsaturated fatty acids, over and above all other catalysts present. Although foods seldom move from one extreme moisture level to the other during storage, they do constantly dry out or absorb moisture to moderate extents during storage, and each moisture change can significantly alter the kinetics and products of lipid oxidation. For example, a dry product stabilized at the monolayer a_w can rapidly oxidize when it either loses or gains moisture and changes a_w. Intermediate moisture foods are moderately reactive in oxidation, but they can be stabilized by dehydration or their shelf-life can be severely shortened by gaining moisture and increasing a_w. Consequently, moisture content and degree of water binding by system molecules (a_w) must be carefully controlled by formulation and packaging in order to stabilize foods and other materials against lipid oxidation.

4. Oxygen

Oxygen is obviously a major player in lipid oxidation, forming peroxyl radicals by addition to $L^\bullet$ radicals, so it is logical to expect that oxidation is dependent on and proportional to the amount of oxygen available. However, oxygen addition is diffusion controlled and almost instantaneous, so it is not the rate-limiting step in lipid oxidation. The key issue in sustaining oxidation is that enough oxygen must be present to react with all the $L^\bullet$ formed. When $O_2 < LH$, oxidation does not progress at full steam, and the rate of oxidation increases rapidly with increasing pO_2 as long as oxygen is limiting. For most lipid oxidations, formation of $LOO^\bullet$ depends on O_2 only up to ~10 mm pO_2. However, when $O_2 > LH$, the excess oxygen is wasted and has no additional effect on kinetics.

There are several important points to recognize about oxygen contributions to lipid oxidation:

- Oxidation rate is determined by the rate of $L^\bullet$ formation by initiators, not by oxygen concentration (De Groot and Noll, 1987). Consequently, stopping initiation can be just as important as limiting oxygen for control of lipid oxidation.
- Oxygen often influences formation of $L^\bullet$ as much as oxygen addition to these radicals. For example, Fe^{2+} requires oxygen to oxidize to Fe^{3+} and produce $O_2^{-\bullet}$ and radicals deriving from it, so the presence or absence of a metal chelator can dramatically affect apparent oxygen dependence of subsequent lipid oxidation over and above oxygen reactions with fatty acids.
- At higher pO_2, oxygen has little effect on oxidation rate, but it controls the course of oxidation and degree of oxygenation of products. Oxygen alters the balance between alternative reaction paths (e.g. internal rearrangement versus abstraction) and determines the proportion of oxygenated recombination products formed, as was discussed in Termination Processes (Section III, A, 3).
- Oxygen has different effects on initiation and propagation, so the oxygen dependence observed varies with the products measured.

- Oxygen catalysis of lipid oxidation is proportional to exposed surface area in both liquid and solid foods. Reaction of lipids with oxygen in the atmosphere is faster than oxygen diffusion in oils.
- The effects of oxygen on oxidation decrease at elevated temperatures where increased thermal initiation can become greater than the solubility of oxygen.

V. WHAT ARE THE CONSEQUENCES OF LIPID OXIDATION?

It is unlikely that lipid oxidation would have received so much attention as a reaction if it only produced off-odors and flavors. The problem with lipid oxidation is that even at extremely low levels, it broadcasts oxidation far beyond the lipid molecules immediately involved, so that nearly every aspect of chemistry and quality is affected. Some of the most important consequences are shown in Figure 11.15. All except those involving lipids alone result from co-oxidations of other macromolecules.

A. Direct Effects due to Changes in Lipids

1. Production of Off-Odors and Flavors, Flavor Fade

Lipid oxidation is known among non-scientists as 'rancidity' due to the characteristic sharp, oily off-odors that consumers smell when they open a jar of old peanut butter, a packet of potato chips, a bar of chocolate, a bottle of salad oil, or a bag of pet food. The sensory perception of rancidity also includes a variety of pungent off-flavors. These odors and flavors arise mostly from secondary lipid oxidation products, particularly aldehydes and ketones, which humans can detect at trace concentrations (parts per billion). The odors and flavors rapidly become objectionable as concentrations increase, so that most foods are rejected at less than 1% oxidation.

Some of the flavors and odors associated with oxidized lipids, along with the product responsible and levels of detection, are listed in Tables 11.11 and 11.12. One important point to note is that oxidation products are 10−10,000 times more detectable in water than in oil because the compounds become more volatile and have less competition for sensors on the tongue. In practical terms, this means that salad oil in a bottle may smell and taste rather innocuous but become inedible when combined with water and vinegar in salad dressings. Indeed, any food with appreciable moisture content will develop recognizable 'rancidity' faster than dry or oil-based products.

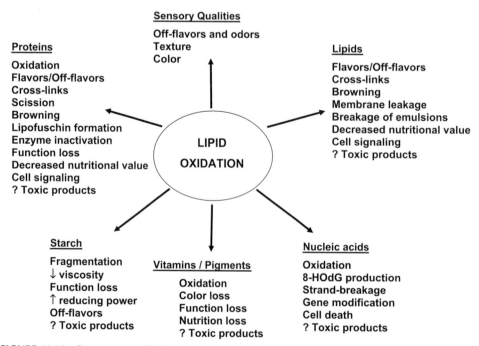

FIGURE 11.15   Consequences of lipid oxidation to sensory characteristics and chemical properties of foods.

TABLE 11.11 Off-flavors Produced by Various Lipid Oxidation Products

Compound[a]	Off-Flavor	Threshold Value (ppm)		In Water
		In Oil		
		Odor	Taste	Odor
Aldehydes				
Pentanal	Sharp, bitter almond	0.24	0.15	0.012
Hexanal	Green–fruity, bitter almond	0.32	0.08	0.008
Octanal	Fatty, soapy–fruity	0.32	0.04	0.0007
Nonanal	Tallowy, soapy–fruity	13.5	0.2	0.001
Decanal	Orange peels	6.7	0.7	0.0001
Nonenal (3c)	Green cucumber	0.25	0.03	
Nonenal (2t)	Tallowy, starch–glue	3.5	0.04	0.0008
Nonadienal (2t,4t)	Fatty, oily	2.5	0.46	
Nonadienal (2t,6c)	Cucumbers	0.01	0.0015	
Nonadienal (2t,6t)	Tallowy, green	0.21	0.018	
Decadienal (2t,4c)	Frying odor		0.02	
Decadienal (2t,4t)	Deep-fried	2.15	0.1	
Ketones and furans				
1-Pentene-3-one	Sharp, fishy		0.003	
1-Octen-3-one	Metallic			
1-Octen-3-ol	Moldy, mushroomy	0.077	0.0001	0.00009
2-Pentylfuran	Buttery, beany	2		
2-(1-Pentenyl)furan (cis and trans)	Licorice	2 to 6		

Data from Belitz and Grosch (1986).

A paradoxical flipside of this issue is that carbonyl products from lipid oxidation enter into Strecker degradation reactions with amino acids to generate some desirable and characteristic flavor products. Peanut butter, for example, tastes flat when fresh and unoxidized.

A phenomenon referred to as flavor fade also occurs. As lipid oxidation progresses, some of the characteristic fresh flavors are lost. It is not clear whether this occurs because flavor compounds are oxidized or because increasing levels of lipid aldehydes and other lipid oxidation products merely overwhelm the perception of native flavors.

2. Degradation of Membranes and Alteration of Lipid Structures

Lipid physical structures in foods depend largely on hydrophobic associations between acyl chains, but these are disrupted when lipids oxidize and either accumulate more polar products or cross-link and lose structural organization. Membranes become leaky or stiffen, either way losing functionality when component phospholipids oxidize in living tissues. In foods, the same changes in membranes lead to loss of turgor and degradation of texture. Oxidation of lipid phases in emulsions produces polar products that move to the interface and change surface tensions. At high enough product levels, emulsions will break. Oxidation affects lipid crystal structures as well. Introduction of polar groups disrupts chain packing in crystals and forces reorganization of the triacylglycerols to reduce repulsions. When this occurs in shortenings, margarines, or chocolate, invariably a shift to more dense β-crystal forms occurs, resulting

TABLE 11.12 Threshold Values of Major Volatiles in Oxidized Edible Oils

Compounds	Threshold Value Range (ppm)
Hydrocarbons	90–2150
Furans	2–27
Vinyl alcohols	0.5–3
1-Alkenes	0.02–9
2-Alkenals	0.04–2.5
Alkanals	0.04–1
2,4-Alkadienals	0.04–0.3
trans,cis-Alkadienals	0.02–0.06
Vinyl ketones	0.00002–0.007

Data from Frankel (1985).

in hardening, grittiness, and in the case of chocolate, an increased tendency to bloom (form white crystals on the surface).

B. Effects due to Interactions of Lipid Oxidation Intermediates and Products with Other Molecules, with Focus on Proteins

Although rancidity is a sensory condition generally attributed to lipids alone, most of the deleterious effects of lipid oxidation in foods listed in Figure 11.15 result not from lipid changes per se, but from reactions of lipid oxidation intermediate radicals and secondary products with other molecules. Many off-flavors attributed to lipids, e.g. warmed-over flavor in meats and some stale flavors, instead arise from co-oxidations. Thus, when tracking and measuring lipid oxidation in foods, it is just as important to follow the footprints of lipid oxidation in other molecules as to measure direct lipid oxidation products themselves. Unfortunately, these reactions are too often overlooked because they are not familiar and they are more difficult to analyze. Even more problematic is that, when co-oxidations occur, they have an antioxidant effect, removing lipid oxidation products from the analytical stream. As a result, lipid peroxide, carbonyl, and especially epoxide levels (and hence overall oxidation) may appear to be low or even negligible when the oxidation has actually occurred, perhaps extensively, and been transferred to other molecules.

One important factor that complicates sorting out co-oxidations is that nearly all lipid oxidation intermediates and products — especially radicals, hydroperoxides, aldehydes, and epoxides — react with proteins (Schaich, 2008), starches (Ishii *et al.*, 1976; Kawakishi *et al.*, 1983), pigments, DNA (Blair, 2001; Yang, 1993; Yang and Schaich, 1996), and other molecules. Radical reaction products are particularly difficult to track because they are usually oxidations rather than adducts, and they often transform or degrade further. New analytical methodologies that make it easier to detect individual modified amino acids may change this situation in the future. Carbonyls and epoxides form adducts that are more persistent and easier to detect, so co-oxidation has more often been attributed to reactions of secondary products than to radicals. The issue is not which reaction is right or wrong, because they all occur. Rather, it is critical to recognize that a continuum of damage caused by different lipid oxidation species develops and changes as oxidation (and co-oxidation) progresses. In early oxidation, radical co-oxidations dominate; with more extensive oxidation, reactions of epoxides and aldehydes become more important.

A general overview of how various lipid oxidation intermediates and products cause co-oxidation of proteins is presented here as a small introduction to an aspect of lipid oxidation that could fill an entire separate chapter.

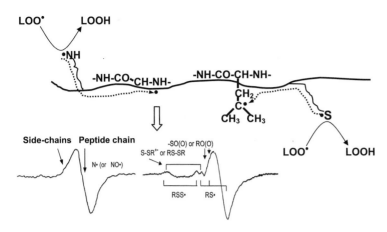

FIGURE 11.16 **Free radical transfer from lipid peroxyl radicals to proteins.** Hydrogen abstraction from His, Lys, Trp, Arg, and Cys side-chains generates N• and S• radicals. Overlapping unresolved spectra of multiple radical sites are evident in broad envelopes of electron paramagnetic resonance spectra. Radicals also become delocalized on nitrogens along the peptide backbone and can migrate to the side-chains of hydrophobic amino acids such as valine and leucine.

1. Reactions of Lipid Free Radicals

As noted above, proteins react with nearly every intermediate and product class of lipid oxidation — radicals, hydroperoxides, carbonyls, and epoxides — with variable consequences. Radical transfer from oxidizing lipids to proteins is the dominant process in the early stages of oxidation. The targets are any amino acid residues with readily abstractable hydrogens that can substitute for and compete with lipid acyl chains, i.e. those with amino or sulfhydryl groups on their side-chains (tryptophan, histidine, arginine, lysine, and cysteine), and the results are protein free radicals (Schaich and Karel, 1976; Schaich, 1980b). Free electrons that stay localized on side-chains create nitrogen- or sulfur-centered free radicals that give characteristic broad wings in electron paramagnetic resonance (EPR) spectra (Figure 11.16). Free electrons also migrate and become delocalized along the peptide backbone or on the β-carbon of hydrophobic amino acids such as valine. These radicals appear as narrower singlet EPR spectra. All these types of radicals may be present simultaneously. They remain quite stable in dry materials, often persisting for months to years.

Reactions of protein radicals parallel reactions of lipid radicals: addition of oxygen to form peroxyl radicals (POO•), abstraction of hydrogens to generate hydroperoxides (POOH), and decomposition of the POOH to yield protein alkoxyl radicals (POO•), and subsequent protein carbonyls (P—C=O) (Figure 11.17). Originally, protein radicals and hydroperoxides were thought to be relatively inactive. Current evidence, though, shows that they are indeed reactive, transferring radicals to other proteins (Soszylqski *et al.*, 1996), DNA (Gebicki and Gebicki, 1999), lipids (Gardner and Weisleder, 1976; Gardner *et al.*, 1977; Avdulov *et al.*, 1997), and potentially other molecules to broadcast and perpetuate oxidative damage.

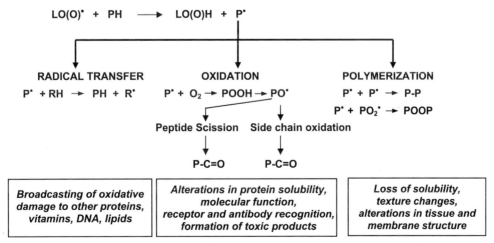

FIGURE 11.17 **Pathways and consequences of free radical production in proteins.** P• is a protein radical on any α-carbon of the main peptide backbone or on an amino acid side-chain, and RH is any molecule with abstractable hydrogens. *(From Schaich, 2008; used with permission.)*

The formation of protein free radicals, hydroperoxides, and carbonyls sets the stage for a wide range of subsequent damaging reactions in foods, including alterations in protein conformation and solubility, loss of enzyme and other biological activity, scission and cross-linking that lead to marked texture degradation and loss of nutritional value. Amino acid targets are oxidized, transformed, and degraded. Many of the essential amino acids, as listed below, are among those most sensitive to damage by radical transfer from peroxidizing lipids (Schaich, 1980a, 2008):

Cysteine	Serine	Glycine
Tryptophan	Threonine	Alanine
Histidine		Valine
Lysine		Proline
Arginine		Leucine
Tyrosine		Isoleucine
Methionine		

In vivo, these same changes have disastrous consequences to physiological function and are intimately involved in the pathological processes that were noted in the Introduction.

2. Reactions of Lipid Hydroperoxides

Lipid hydroperoxides are not directly reactive themselves, but their ability to hydrogen bond to amine groups on proteins creates a reaction cage within which there is induced decomposition of LOOH and direct reaction of the resulting LO(O)$^\bullet$ with amino acid targets (Karel *et al.*, 1975; Schaich and Karel, 1976). The process, called molecule assisted homolysis, is very fast and may be metal independent or involve metals bound to proteins.

Without metals, lipid hydroperoxides bind at the surface of proteins, usually to amino or sulfhydryl groups on side-chains, and the nucleophilic atom induces hydroperoxide homolysis and radical transfer from lipid to protein:

$$\text{LOOH} + \text{PH} \longrightarrow [\,\text{LO-OH}...\text{H-P}\,] \longrightarrow \text{LO}^\bullet + \text{P}^\bullet + \text{H}_2\text{O} \qquad (11.56)$$

$$\text{LOOH} + \text{R}_3\text{N} \longrightarrow [\text{R}_3\text{N}...\text{HOOR'}] \longrightarrow \text{R}_3\text{NO} + \text{LO}^\bullet \qquad (11.57)$$

$$\text{LOOH} + 2\,\text{RSH} \longrightarrow \text{RSSR} + \text{H}_2\text{O} + \text{LO}^\bullet \qquad (11.58)$$

This concerted reaction may contribute to the sensitivity of histidine, arginine, lysine, tryptophan, cysteine, serine, and threonine to LOOH — all of which contain hydrogen bonding amino, carboxylic acid, and hydroxyl groups on their side-chains (Gardner, 1979; Schaich, 1980b). It also may be enhanced in lipid-bonding proteins such as bovine serum albumin where hydrophobic side-chains facilitate associations with lipids and bring reactive residues into close proximity with LOOH.

Metals bound to proteins can be even more reactive with lipids and more damaging overall than metals free in solution. In metalloproteins, LOOH binds at the metal ligand site and is reduced by the metal in a reaction cage; the LO$^\bullet$ generated oxidizes nearby amino acids, particularly histidine (Kowalik-Jankowska *et al.*, 2004) (Figure 11.18,

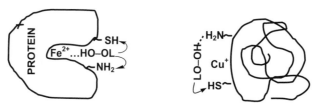

FIGURE 11.18   Metal-mediated radical transfer from LOOH to proteins. Lipid hydroperoxides bound in close proximity to metals in active sites or on protein surfaces are reduced in cage reactions and the resulting LO$^\bullet$ then abstracts a hydrogen from a susceptible amino acid nearby before release. *(From Schaich, 2008; used with permission.)*

left). In non-metalloproteins, metal-binding sites such as histidine, glutamic acid, or aspartic acid side-chains on protein surfaces also provide foci for metal-catalyzed reduction of LOOH in cage reactions (Figure 11.18, right).

Why is induced decomposition of LOOH important? With normal radical transfer by hydrogen abstraction, protein radicals and LOOH are generated simultaneously so there is a direct, measurable connection between initiation processes and lipid and protein oxidation products. Both LOOH and protein radicals appear at rates proportional to $L^\bullet$ and $LOO^\bullet$. However, since proteins supply the abstractable hydrogens for LOOH formation, the overall process is antioxidative: lipid free radical chains are stopped and fewer lipid molecules are oxidized.

$$LH \longrightarrow L^\bullet \xrightarrow{O_2} LOO^\bullet \xrightarrow{PH} \mathbf{LOOH} + \mathbf{P^\bullet} \tag{11.59}$$

In contrast, in concerted cage reactions, lipid molecules provide the abstractable hydrogens for LOOH formation, so LOOH is continually generated but then also removed as protein radicals form.

$$LH \longrightarrow L^\bullet \xrightarrow{O_2} LOO^\bullet \xrightarrow{LH} LOOH + LH \qquad P^\bullet \tag{11.60}$$

$$LOOH + PH \longrightarrow [\, LO\text{-}OH\ldots\overbrace{H\text{-}P}\,] \longrightarrow \mathbf{LO^\bullet} + \mathbf{P^\bullet} + H_2O \tag{11.61}$$

This is a pro-oxidant process since lipid radical chains are not intercepted, LOOH decomposition provides rapid, direct radical transfer to proteins, and $LO^\bullet$ are still released to propagate more radical chains. Of particular concern is that when lipid oxidation is followed by peroxide values alone, it can look like lipids are not oxidizing even while proteins are being extensively degraded, so lipids will be missed as the source of damage.

Concerted LOOH reactions with proteins have been reported in buffered solutions of methyl linoleate bound to β-lactoglobulin (Hildago and Kinsella, 1989; Yuan *et al.*, 2007), lyophilized emulsions of methyl linoleate and lysozyme and other proteins (Schaich and Karel, 1975; Schaich, 1980a), HDL apoA1 and A2 reacted with cholesterol and phospholipid hydroperoxides (Garner *et al.*, 1998), low-density lipoprotein (LDL) and cytochrome *c* incubated with LOOH or phospholipid (PE) hydroperoxides (Fruebis *et al.*, 1992), and lupine conglutins reacted with LOOH (Fruebis *et al.*, 1992; Lqari *et al.*, 2003). However, it is likely that these reactions occur in most systems but are not recognized.

3. Reactions of Lipid Epoxides

Epoxides are probably the most underrated and understudied lipid-derived oxidant, in part because they are so reactive that it is not easy to track them analytically. It is especially difficult to connect epoxides to protein co-oxidations because epoxides react with proteins about 1000 times more rapidly than do aldehydes. This greatly reduces or even eliminates epoxides from the mix of products in oxidizing lipids, so protein damage detected is attributed to other lipid oxidation products. Thus, following footprints of epoxides in oxidized proteins may be more revealing than analyzing lipid epoxides alone.

Epoxide reactions with proteins are strongly dependent on reaction conditions and the position of the epoxy group on the lipid acyl chain. When epoxy groups are next to double bonds, they readily hydrolyze (Lederer, 1996) and also undergo nucleophilic attack from amine groups on proteins (Ingold, 1969b). The fundamental process as it occurs between a linoleic acid epoxide and lysine, histidine, and cysteine (left to right) is shown in Reaction 11.62. The products, called aminols, have two key features: (a) a hydroxyl group at the β-carbon of the original epoxide, and (b) protein covalently linked to the α-carbon of the epoxide as shown below, or to the α-carbon on the opposite side of the double bond.

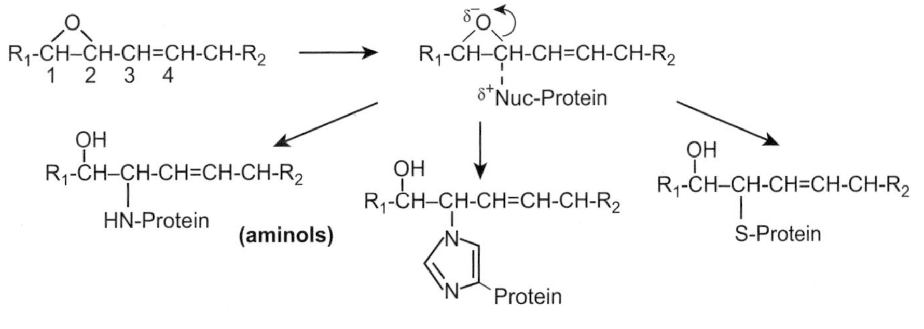

$$\tag{11.62}$$

When water is present, the epoxide rings open but hydrolysis prevents addition of the protein (Lederer, 1996). Thus, these epoxide reactions with proteins are most important under anhydrous conditions, e.g. in dry foods and in hydrophobic interior regions of biomembranes and blood lipoproteins.

As lipid oxidation progresses, epoxide co-oxidations become more complex. Most importantly, the presence of both aldehyde and epoxide functional groups on the same lipid provides two sites for reaction and sets the stage for cyclization to form hydroxyalkyl pyrroles and furans (Reactions 11.63 and 11.64, respectively) (Zamora and Hidalgo, 1995, 2003, 2005; Zamora et al., 1999; Hidalgo and Zamora, 2000) and also alkylpyridines (Reaction 11.65) (Zamora and Hidalgo, 2005; Zamora et al., 2006). The reaction occurring in a given system is determined by the solvent, pH, temperature, and the nature of amine groups on the protein.

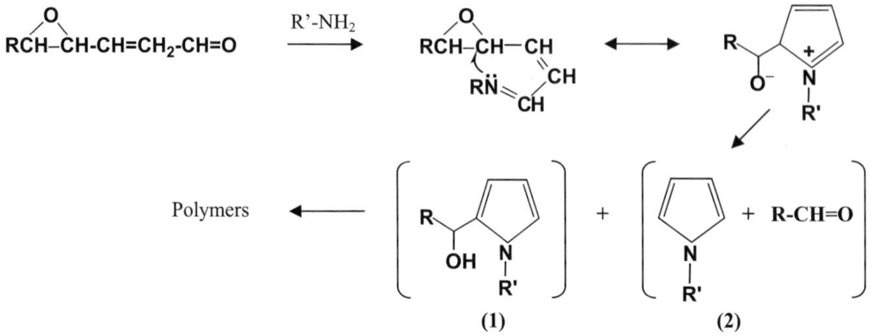

(11.63)

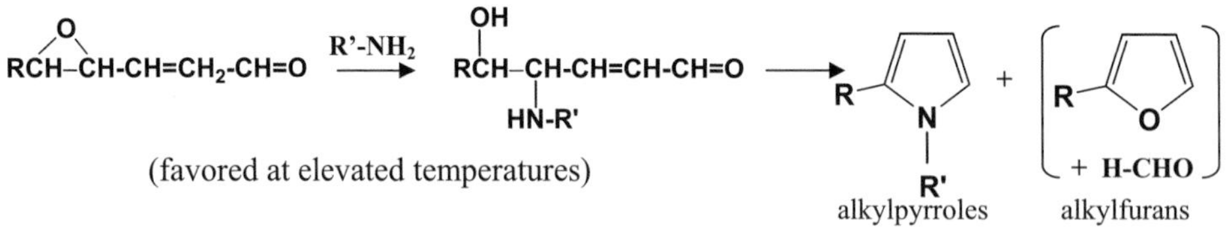

(11.64)

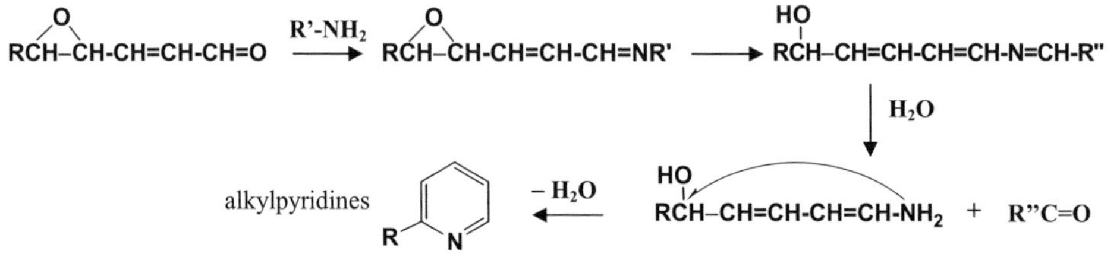

(11.65)

Detection of aminols is clearly diagnostic of epoxide co-oxidation of proteins. However, pyrroles and pyridines are also the major products of aldehyde co-oxidations of proteins, as will be explained in the next section, so it can be exceedingly difficult to distinguish epoxides from aldehyde reactions, except perhaps by kinetics. As a result, it is quite likely that reactions of epoxides with proteins have been overlooked and misinterpreted as aldehyde-mediated damage.

4. Reactions of Lipid Aldehydes

In the progression of co-oxidation, reactions of secondary products of lipid oxidation, particularly aldehydes, develop in later stages of lipid oxidation and introduce different types of protein oxidation products. Aldehydes form covalent adducts with proteins, whereas lipid radicals generally do not, and this has critical consequences for the properties that are modified in proteins. Except for cross-linking, damage done by free radicals remains largely 'invisible' at the molecular level. In contrast, reactions of aldehydes cause clearly perceivable changes such as production of flavors and odors, browning, and textural aggregation. Reactions of lipid aldehydes and other secondary products are responsible for the degradation that causes consumers to reject food products as inedible.

Aldehydes react with nucleophilic groups on proteins via Schiff bases, Michael addition, or a combination of both to form adducts, with three general outcomes:

- Linear adducts that change surface chemistry and protein recognition and potentially contribute to browning. This is the initial step for all aldehydes and is the dominant reaction between single, isolated carbonyl and amine groups.
- Cyclic products, especially dihydropyridines and pyrroles, as adducts or in cross-links. These contribute to browning, flavor and odor production, and loss of protein functionality. Cyclic products are the dominant products when multiple reactive groups are available, e.g. two amino groups on a terminal amino acid or two functional groups on an aldehyde (e.g. hydroxyl + aldehyde, two aldehydes, aldehydes in great excess over available amine).
- Protein cross-links, both intramolecular and intermolecular, that toughen textures, decrease solubility, and lead to phase separations. Cross-links require multiple reactive groups in the aldehyde, e.g. dialdehydes such as malonaldehyde and glutaraldehyde, or hydroxylalkenals such as 4-hydroxynonenal.

The exact reaction pathways that occur or dominate in a given system are influenced by the nature of the protein, the type of lipid oxidation products present, relative protein–aldehyde concentrations, pH, phase or solvent, oxygen tension, and other factors. The reactions described below provide some examples of the various products possible with different aldehydes. In real food systems, the products become much more complex than these simplified reactions.

a. Saturated Aldehydes

α and β scissions of lipid alkoxyl radicals release a variety of saturated aldehydes such as hexanal, nonanal, and 8-oxo-methyloctanoate. Although not very reactive, these aldehydes do add selectively to terminal amino acids via standard Schiff base reactions (McMurray, 2000):

$$R(CH_2)_n\text{-}C(H){=}O \ + \ H_2N\text{-}CH(R)COOH \ \longrightarrow \ R\text{-}(CH_2)_n\text{-}C(H){=}N\text{-}CH(R)COOH \qquad (11.66)$$

Since both peptide amine groups and side-chain amine and thiol groups are available, Reaction 11.66 is the first step and the remaining lone pair of electrons on the aldehydic oxygen undergoes a second Schiff base reaction with side-chain nucleophilic groups to generate heterocyclic products (Fenaille et al., 2003):

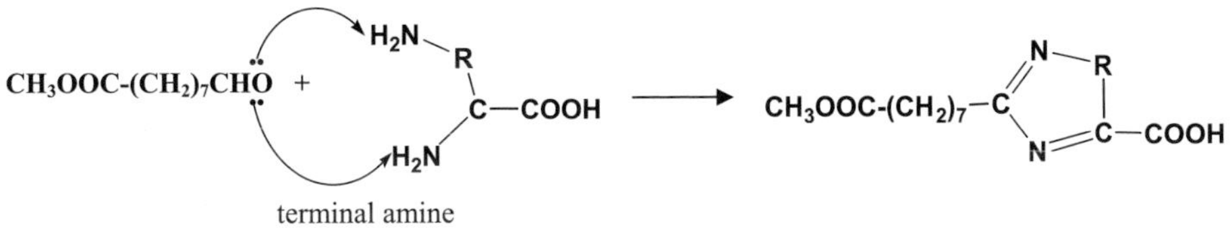

terminal amine (11.67)

terminal thiol　　**Thiazolidine carboxylic acid derivative** (11.68)

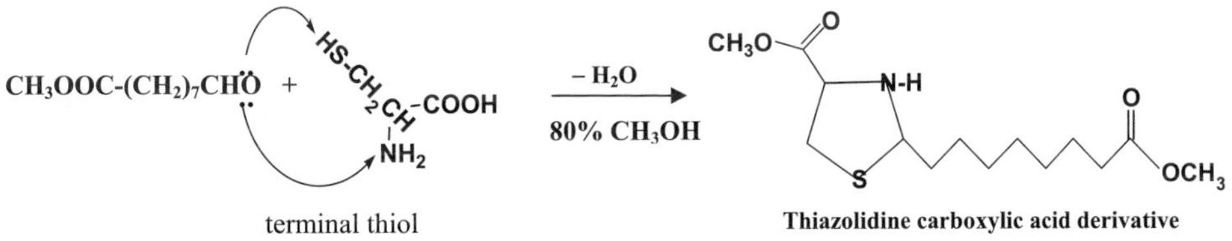

Dialdehydes such as malonaldehyde (MDA) are produced in downstream secondary oxidations of poly-unsaturated fatty acids with three or more double bonds. Two aldehyde groups provide two sites for reaction with proteins. At low MDA levels, the results are usually cross-links:

$$\text{Protein } {-}NH_2 \leftarrow [O{=}C\text{-}CH_2\text{-}C{=}O] \rightarrow H_2N\text{-Protein} \qquad (11.69)$$

Under conditions of high lipid oxidation, two molecules of MDA combine with another aldehyde and an amino group to form pyridine derivatives (ring adducts on the protein (Freeman et al., 2005; Kikugawa and Ido, 1984; Nair et al., 1988):

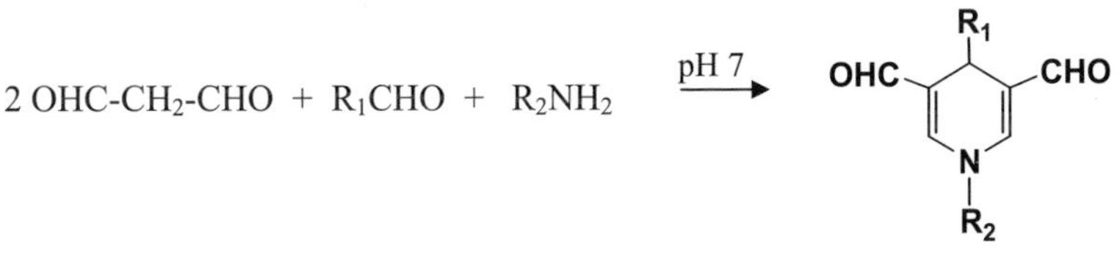

$$2\,OHC\text{-}CH_2\text{-}CHO \ + \ R_1CHO \ + \ R_2NH_2 \ \xrightarrow{pH\ 7} \quad$$

(11.70)

(1,4-dihydropyridine-3,5-dicarbaldehydes)

Dihydropyridines are important flavor precursors in foods (Buttery *et al.*, 1977; Suyama and Adachi, 1980; Maga, 1981).

b. Unsaturated Aldehydes

Unsaturated aldehydes, particularly 2-enals, are extraordinarily reactive compounds because they have three tautomeric forms in equilibrium,

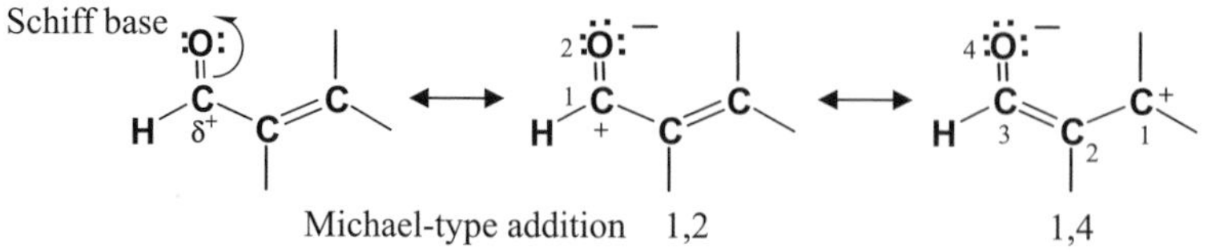

(11.71)

and thus three potential reaction sites: Schiff base formation at the carbonyl and Michael-type 1,2 and 1,4 additions at the carbocations (Esterbauer *et al.*, 1991; Ege, 1999; McMurray, 2000). Thiols of cysteine, ε-amino groups of lysine, and imidazole nitrogens of histidine are the main protein targets (Esterbauer *et al.*, 1991; Petersen and Doorn, 2004).

Direct (1,2) addition is comparable to the first step in Maillard browning reactions. The amine (or thiol) adds to the carbonyl carbon, and generates a carbinolamine intermediate that rearranges and dehydrates to a Schiff base as the final product (Ege, 1999; McMurray, 2000):

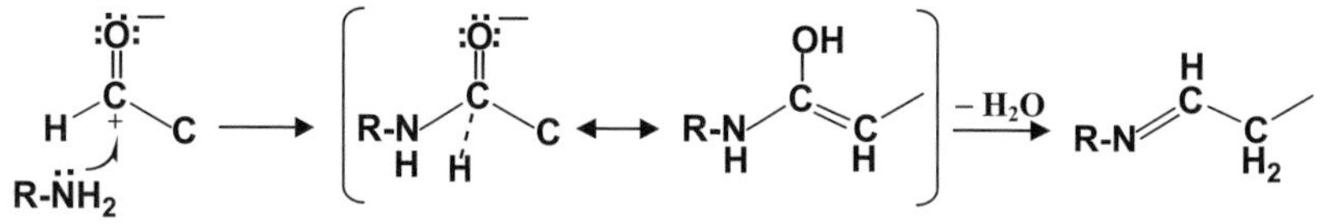

(11.72)

This is the main reaction in organic solvents and hydrophobic microenvironments; it is reversible and catalyzed by acid (Esterbauer *et al.*, 1991).

In aqueous solution or emulsions, the dominant reaction of lipid alkenals with amino acids is conjugated (1,4) addition. In this case, the amine adds to the carbocation formed at the β-carbon of the double bond initially in conjugation with the carbonyl:

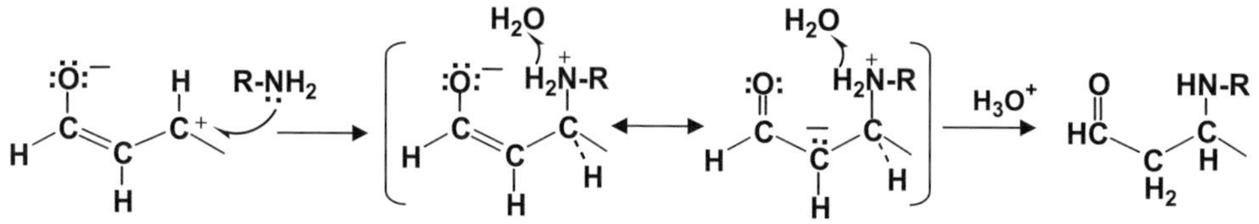

(11.73)

A key difference here is that the carbonyl group remains intact. Now complexed to the protein, this carbonyl is detectable in dinitrophenylhydrazine (DNPH) or antibody analyses for protein carbonyls (oxidation products). It can also form a Schiff base by reacting with an additional amine. This Michael addition—Schiff base protein—lipid aldehyde—protein complex is an important source of protein—lipid cross-links in more oxidized systems, and is the initial sequence found repeatedly in the formation of many complex adducts. Some of these will be described below.

Acrolein. Acrolein ($CH_2=CH-CHO$) is an unsaturated aldehyde formed by the oxidation of glycerol during frying, and it has been identified among secondary lipid oxidation products of arachidonic acid and *n*-3 polyunsaturated fatty acids *in vivo* (Uchida *et al.*, 1998a, b; Uchida, 1999; Kehrer and Biswal, 2000). Recently, this compound has received considerable attention as a carcinogen.

Acrolein is highly electrophilic so has very strong reactivity with nucleophiles such as thiols, the imidazole groups of histidine, and the ε-amino group of lysine via both Schiff base and Michael additions, and all these reactions contribute to its toxicity (Uchida *et al.*, 1998a; Esterbauer *et al.*, 1991). Michael addition is usually the preferred reaction because it is faster and the products are more stable (Esterbauer *et al.*, 1991; Uchida and Stadtman, 1992; Uchida *et al.*, 1998a, b). When aldehyde concentrations are low, the products are generally substituted propanals, as shown in Reactions 11.74 and 11.75.

$$\text{(imidazole-NH}^\tau\text{)} + CH_2=CH\text{-}CHO \longrightarrow \text{(imidazole-N}^\tau\text{-CH}_2CH_2\text{-C=O)} \tag{11.74}$$

$$\text{R-NH}_2 + CH_2=CH\text{-}CHO \longrightarrow \text{R-NH-CH}_3CH\text{-}CHO \tag{11.75}$$

However, lysine adducts have a strong tendency to cyclize. When aldehydes are present in excess of the amines, lysine undergoes multiple additions with acrolein to generate the cyclic pyrrole derivatives shown in Reaction 11.76.

$$\text{R-NH}_2 + 2\,CH_2=CH\text{-}CHO \longrightarrow \text{R-(CH}_2)_4\text{-N}\cdots\text{CH=O} \tag{11.76}$$

With long incubation time or high aldehyde concentrations, accumulated Michael adducts of acrolein cyclize, yielding N^ε-(2,5-dimethyl-3-formyl-3,4-dehydropiperidino) (FDP) structures by 1,4 addition (Ichihashi *et al.*, 2001).

$$\tag{11.77}$$

Formyl dehydropiperidino (FDP) -lysine

FDP adducts retain a carbonyl function and thus are detected in carbonyl assays of oxidized proteins. Antibodies to FDP-lysine have detected acrolein adducts in oxidized LDL (Ichihashi *et al.*, 2001) and in bovine serum albumin co-oxidized with methyl arachidonate (Uchida *et al.*, 1998b).

4-Hydroxy-2-alkenals and 4-oxo-2 alkenals. Oxidized alkenals such as 4-hydroxy-2-nonenal (HNE) and 4-oxo-2-nonenal (ONE) are the aldehydes most reactive with proteins (Esterbauer *et al.*, 1991).

HO-2-alkenals   4-oxo-2-alkenals (isoketals)

4-Oxo-2-alkenals, also called γ-ketoaldehydes, may be formed independently but are actually tautomers of 4-HO-2-alkenals:

$$\underset{\text{OH}}{\text{R-C-CH=CHCH=O}} \longleftrightarrow \underset{\text{O}}{\text{R-C-CH}_2CH_2CH=O} \tag{11.78}$$

Under oxidizing conditions, reactions of these two oxidants are often difficult to distinguish except by kinetics (ONE reactions are orders of magnitude faster) (Lin *et al.*, 2005) and balance between pathways. Since both of these compounds have two reactive functional groups, their major reaction sequence is the same as that described above for dialdehydes. The first step is usually (1,4) Michael addition to nucleophilic amino acid side-chains (Uchida and Stadtman, 1992; Bruenner *et al.*, 1995; Schaur, 2003). HNE reacts with cysteine, histidine, or lysine; ONE reacts with these amino acids plus arginine. Subsequent reactions vary with the aldehyde, the specific amino acids present, relative concentrations of aldehyde and protein, and the reaction conditions.

The simplest reaction is cyclization of the aldehyde to form 2-hydroxy-furan adducts, shown below for cysteine and lysine (Uchida *et al.*, 1994, 1997; Uchida, 2003):

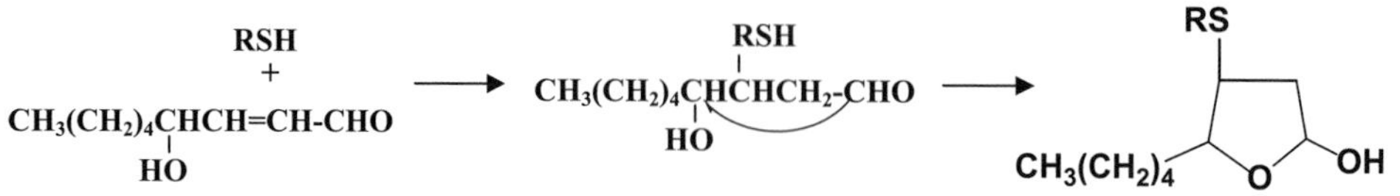

$$(11.79)$$

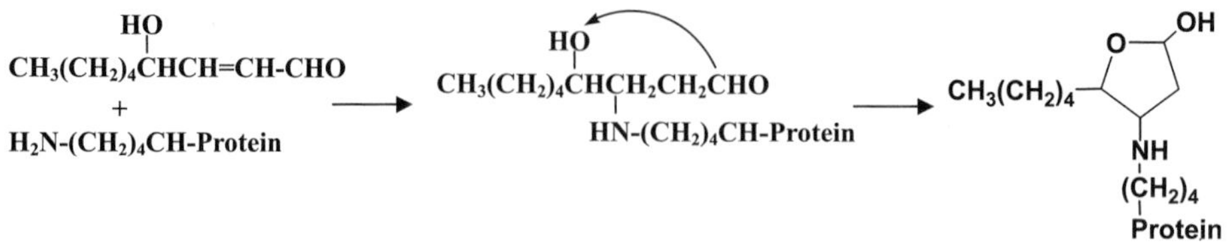

$$(11.80)$$

This reaction occurs when the amines and aldehydes are approximately equal in concentration, but the products hydrolyze readily so are not always found in product mixes.

When amines are present in molar excess over aldehydes (e.g. in early oxidation), the reaction sequence changes. The first step here is Schiff base formation. The unreacted HO-then oxidizes to generate a second carbonyl (for HNE), which undergoes Michael addition to a second amine (usually lysine); the complex then oxidizes and cyclizes to form stable dihydropyrrol-iminium complexes (Sayre *et al.*, 1993, 1997; Xu *et al.*, 1999; Schaur, 2003; Zhang *et al.*, 2003):

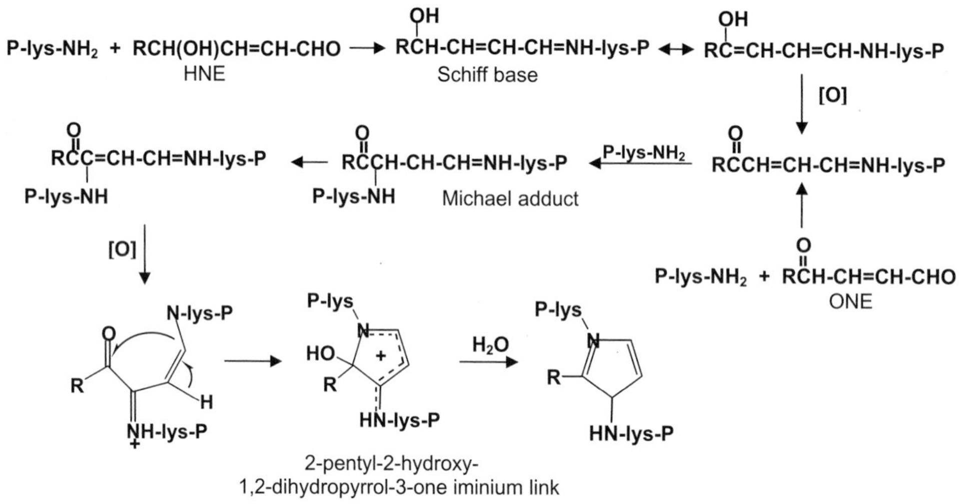

$$(11.81)$$

2-pentyl-2-hydroxy-
1,2-dihydropyrrol-3-one iminium link

The complexes formed by both HNE and ONE are very interesting products. The adducts themselves are fluorescent, and probably contributed to the fluorescent products originally detected by Tappel and co-workers (Chio and Tappel, 1969; Fletcher and Tappel, 1970; Fletcher *et al.*, 1973). The pyrroles are hydrolysable and the pyrrole -OH can react further, e.g. with another lysine, to form cyclic or acyclic mixed aminals; thus, the pyrroles may serve as precursors for development of flavor compounds during food processing and storage. Finally, the entire complex creates still another kind of peptide cross-link that contributes to hardening and toughening of textures in foods. These complexes also increase in Alzheimer's disease and may be involved in the tangling of β-amyloid chains (Sayre *et al.*, 1997).

5. Connecting Co-oxidation Reactions to Protein Damage

All lipid oxidation products react in some way with proteins, so there is really no time at which proteins are safe from the effects of co-oxidation. Rather, a continuum of co-oxidation damage develops in parallel with lipid oxidation as new lipid products are generated and shift the types of reactions that dominate. This section has focused on chemical details of the reactions to shift thinking about the processes from the global behaviors usually reported to the molecular level chemistry responsible. It is only by understanding the detailed chemistry that we can ever gain the ability to manipulate and control it to protect food quality.

Table 11.13 summarizes the connections between specific lipid oxidation products, their protein co-oxidation reactions, and the global protein behaviors affected. The overlap between pathway effects is clear, so measuring a single effect alone cannot determine what caused it. Neither is it sufficient to correlate protein changes with lipid oxidation products since, as has been noted several times, lipid oxidation products reacting with proteins are removed

TABLE 11.13 Protein Co-oxidation Products of Lipid Oxidation Species and Associated Effects on Protein Properties

Lipid Oxidation Product	Reaction	Products	Global Behaviors Affected
LOO• and LO• radicals	Radical	Radicals	Solubility changes, loss of enzyme activity
	Transfer	Amino acid oxidation products	Fragmentation, surface properties, texture
		Protein hydroperoxides, carbonyls	Amino acid destruction, loss of nutritional value
		Protein dimers	Cross-linking, solubility, antibody recognition, texture
LOOH	Cage reaction → radical transfer, LO• released	Protein radicals, hydroperoxides, carbonyls	All effects of radicals but with enhanced oxidation rates
		Disulfide cross-links	
Epoxides	Nucleophilic	(Low oxidation) Aminol adducts	Altered surface properties, loss of antibody recognition
	Addition	(High oxidation) Hydroxylalkyl pyrroles	Loss of enzyme activity
		Hydroxylalkyl furans	Nutritional value, functional properties
		Alkylpyridines	
Saturated aldehydes	Schiff base	(Low oxidation) Schiff base linear adducts	Browning, loss of nutritional value, age pigments
		(High oxidation) Heterocyclic adducts	Flavors, functional properties, solubility, surface properties
Saturated dialdehydes	Schiff base, Michael addition	(Low oxidation) Schiff base cross-links	Browning, cross-linking, enzyme activity, nutritional value
		(High oxidation) Dihydropyridine adducts and cross-links	Texture, flavors, odors, functional properties
α,β-Unsaturated aldehydes	Michael addition, Schiff base	(Low oxidation) β-substituted alkanals, pyrroles	Browning, flavors, surface properties, age pigments
		(High oxidation) Methylpyridinium, dehydropiperidino adducts	Cross-linking, texture, enzyme activity, biological activity
Oxidized aldehydes, e.g.	Michael addition, Schiff base	Linear adducts, furans, pyrroles	Browning, fluorescence, texture hardening, flavors
4,5-Epoxy-2-alkenals		(High amine) Dihydropyrrol—iminium adducts, aminals	Cross-linking, texture, surface and functional properties
4-Hydroxy-2-alkenals			Enzyme activity, antibody recognition, nutritional value
Epoxyoxo acids, oxoacids		(High aldehyde) Epoxides; pyrrole, pyrrolidone, thiazolidine adducts	Browning, fluorescence, flavors, functional properties

from the analytical stream; these would then appear as negative correlations at best. In the past, direct determination of product structures was exceedingly difficult or impossible. However, modern mass spectrometry should now make identifying complex products almost routine. The reactions and product structures outlined in this section can thus provide guides for tracking footprints of lipid oxidation among protein co-oxidation products.

VI. HOW CAN LIPID OXIDATION BE CONTROLLED?

Since lipid oxidation causes so many problems, the overarching concern with lipid oxidation in foods is its prevention. Antioxidant approaches fall into three categories, all of which are necessary to stabilize products against oxidation. These can be applied individually but are usually combined to most effectively limit lipid oxidation:

- Type 1 antioxidants: agents that prevent initiation
- Type 2 antioxidants: compounds that quench radicals
- Type 3 antioxidants: environmental and processing factors.

A. Type 1 Antioxidants: Agents that Prevent Free Radical Production

1. Metal Chelators and Complexers

For the most part, these are metals chelators (the most important of which is EDTA, ethylenediamine tetraacetic acid, shown below) and metal complexers (e.g. citric acid, polyphosphates, diamines, some amino acids, and to a lesser extent, ascorbic acid). Other common chelating groups include polycarboxylic acids, hydroxamates, and vicinal diphenols.

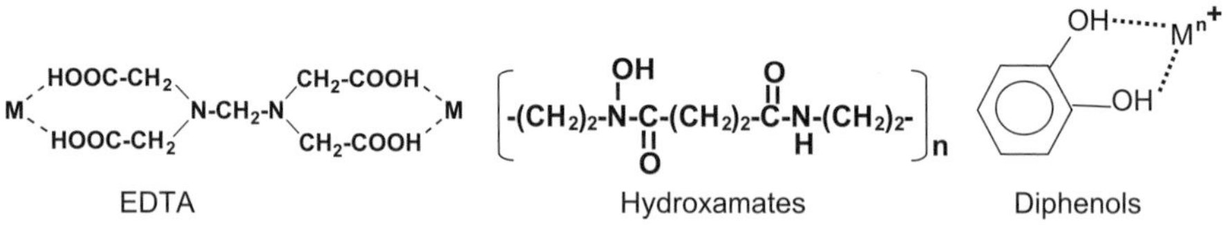

Two points need to be stressed about the actions of metal chelators and complexers. First, to be effective, chelators and complexers must have a structure that wraps around the metal and completely blocks all metal orbitals, preventing electron transfer, or be present at high enough concentrations that multiple molecules can act in concert to do the same.

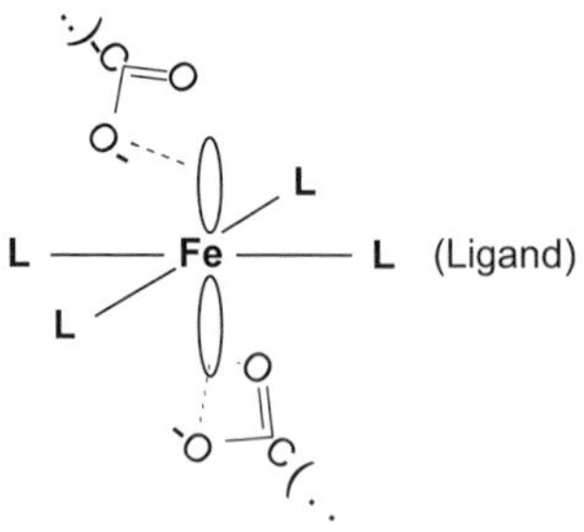

EDTA is an example of the first action, and citric acid is an example of the second. Compounds that incompletely block metal orbitals may act as synergists and slow electron flow but cannot stop initiation reactions. Most organic acids and phosphates are examples of synergistic action.

Secondly, chelators and complexers do not physically remove the metals from reaction; they merely block orbitals and reduce redox potential, making them less active than the parent metal as oxidizing agents. However, the complexes may become paradoxically more reactive in the total system if reducing agents with lower redox potential are also present to cycle the metal. An oxidized metal with low E° is a poor catalyst, but reducing that metal generates a much more powerful reducing agent that will react very rapidly with oxygen and all hydroperoxides and be even more damaging than the parent metal. Thus, attention must be given to potential reducing agents in food systems when adding chelators.

2. Singlet Oxygen Scavengers, Particularly Carotenoids

As described in Section IV, D, 1, singlet oxygen adds across double bonds in lipids, particularly conjugated double bonds. With extended conjugated double-bond systems, carotenoids provide many sites that preferentially react with $^1O_2$ and spare fatty acids (Stahl and Sies, 2005). Only a few are noted in the structure below, but all are reactive with $^1O_2$.

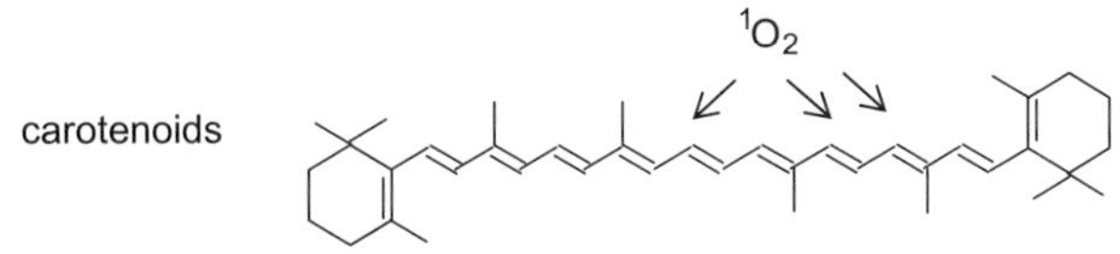

Pyridoxine (vitamin B_6) and its derivatives are also efficient singlet oxygen scavengers (Bilski *et al.*, 2000). Initiation inhibitors are frequently overlooked as antioxidants, yet they serve a very critical role in controlling the radical load that must be overcome by free radical scavengers. The radical load is the cumulative total of radicals from all sources that must be quenched to effectively limit progression of lipid oxidation. The radical load determines the type, amount, and reactivity of antioxidant necessary to stabilize a given system. If the radical load is too high, no amount of radical chain quencher (classical antioxidants) can prevent lipid oxidation. However, neither can initiation inhibitors work alone. Some radicals from various sources are always produced and these must be quenched to prevent establishment of cycling chain reactions. Thus, Type 1 antioxidants are usually combined with Type 2 antioxidants for most efficient action.

B. Type 2 Antioxidants: Compounds that Quench Radicals

These are classical or primary antioxidants. Antioxidants (AH) quench free radicals by several mechanisms.

1. Hydrogen Atom Transfer

Hydrogen atom transfer (HAT) to quench radicals is the best known antioxidant action (Leopoldini *et al.*, 2004):

$$ROO^\bullet + AH \longrightarrow ROOH + A^\bullet \qquad (11.82)$$

$$ROO^\bullet + AH_2 \longrightarrow ROOH + AH^\bullet \xrightarrow{R^\bullet} A + RH \qquad (11.83)$$

For a compound to be an antioxidant, its radical $A^\bullet$ must be stable and unreactive or yield non-radical products so that it does not propagate radical chains. The most important antioxidant structures with H-transferring activity are phenolic compounds in which the $A^\bullet$ reactivity is reduced by resonance delocalization of the free electron over the aromatic ring.

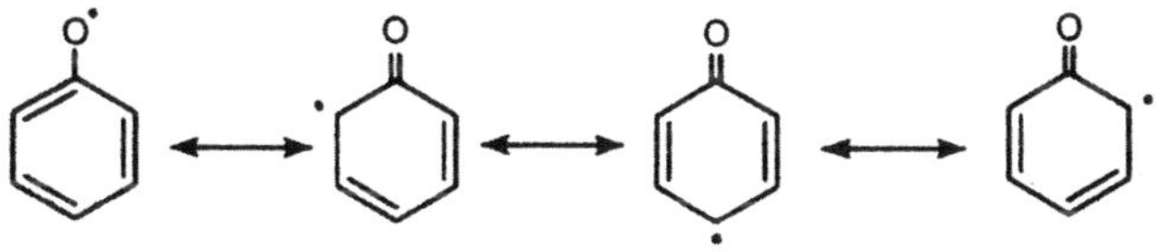

The synthetic antioxidants BHA, BHT, propyl gallate, and TBHQ in commercial use in foods are all phenolic compounds, as are natural antioxidant tocopherols, flavonoids in fruits, and many herb and spice components, to mention just a few. The structures of some common phenolic antioxidants are shown in Figure 11.19.

There is currently great interest in replacing synthetic antioxidants with natural compounds. Some natural antioxidants are monophenols, while many have more than one phenolic group per ring (polyphenols). In general, antioxidant activity increases with the number of phenolic groups, although the position of the groups is also important. As noted above, diphenols and triphenols also complex metals (Afanas'ev *et al.*, 1989), so they have multiple mechanisms for inhibiting lipid oxidation and thus can be very powerful antioxidants. Many polyphenols in rosemary, oregano, and cranberries, for example, have this structure.

Additional compounds that inhibit lipid oxidation by HAT include aromatic amines, hydroxyl amines, thiophenols, aminophenols, ascorbic acid, glutathione, uric acid, carotenoids, amino acids, and proteins.

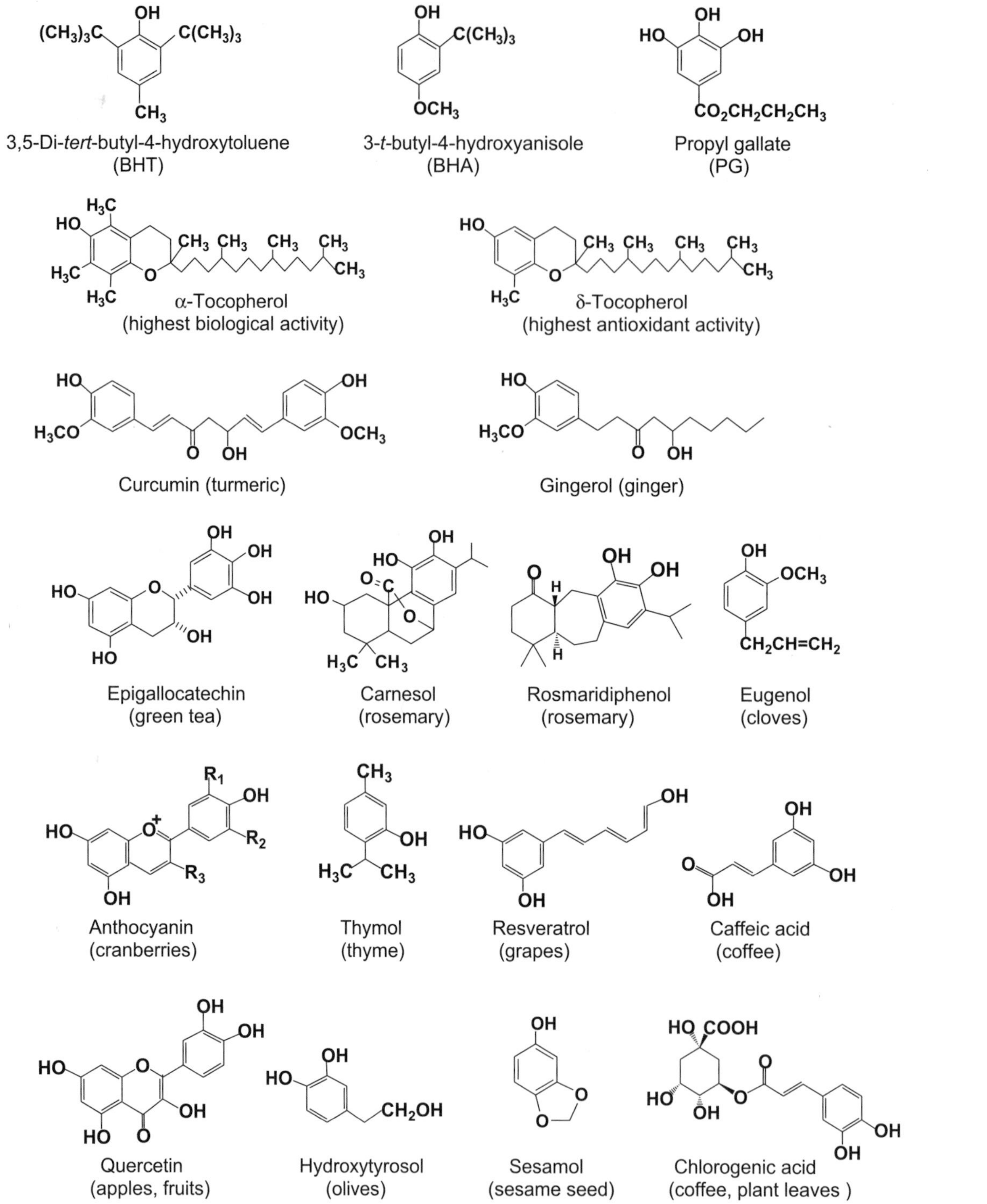

FIGURE 11.19 Structures of some common antioxidants. The top row shows synthetic antioxidants; the rest are antioxidants from natural sources (source in parentheses).

2. Reduction or Oxidation

Reactive radicals can be reduced or oxidized to unreactive ions (single electron transfer):

$$ROO^{\bullet} + ArOH \xrightarrow{+e^-} ROO^- + ArOH^{\bullet+} \tag{11.84}$$

$$ROO^{\bullet} + A \xrightarrow{-e^-} ROO^+ + A^- \tag{11.85}$$

Compounds with this capability include ascorbic acid and other reducing agents, some phenols and quinones, and high metal concentrations (Ganapathi et al., 2000; Yen et al., 2002; Leopoldini et al., 2004). This antioxidant effect of high metal concentrations is frequently ignored or unrecognized.

3. Termination of Chains

Termination of chains can occur through direct reaction with $R^\bullet$, e.g. quinones, nitro compounds, and quinone imines:

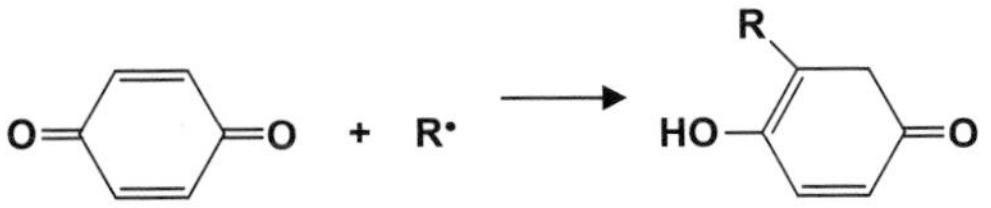

$$(11.86)$$

4. Decomposition of Radicals or ROOH

Radicals or ROOH may be decomposed without generating new radicals. Compounds with this action include sulfides and disulfides (Kulich and Shelton, 1991), phosphates and thiophosphates, carboxylic acids (including phenolic acids and free fatty acids), enzymes (superoxide dismutase, glutathione peroxidase, catalase), amines (Fueno et al., 1959), and phosphatidylcholines (O'Brien, 1969). In the reactions shown below, AcH denotes an acid and Ar denotes an aryl group.

Sulfides and disulfides:

$$LOOH \ + \ R_1SR_2 \longrightarrow LOH \ + \ R_1S(O)R_2 \xrightarrow{\text{heat}} R_1SOH \ + \ R_3CH{=}CH_2 \qquad (11.87)$$

$$2\,LOOH \ + \ R_1SOH \longrightarrow 2\,LOH \ + \ R_1SO_3H \qquad (11.88)$$

$$LOOH \ + \ R_1SSR_2 \longrightarrow LOH \ + \ R_1S(O)R_2 \longrightarrow R_1SSOH \ + \ R_3CH{=}CH_2 \qquad (11.89)$$

$$2\,LOOH \ + \ R_1SSOH \longrightarrow 2\,LOH \ + \ R_1SSO_3H \qquad (11.90)$$

$$LOO^\bullet \ + \ R_1SOH \longrightarrow LOOH \ + \ RSO^\bullet \longrightarrow \text{inactive products} \qquad (11.91)$$

Acids and nucleophiles (e.g. N^+ in choline):

$$Ar\text{-}COOH \ + \ LOOH \longrightarrow Ar\text{-}CHO \ + \ LOH \qquad (11.92)$$

$$AcH...HOOL \longrightarrow [Ac^-H^+\underset{\overset{|}{H}}{OOL}] \longrightarrow AcOH \ + \ LOH \qquad (11.93)$$

$$X^-H^+ \ + \ LOOH \longrightarrow LOOH_2^+ \longrightarrow \beta\text{-scission aldehydes} \qquad (11.94)$$

$$RH_2N{:} \ + \ LOO^\bullet \longrightarrow (LOO{:})^- \ + \ RH_2N^{\bullet+} \qquad (11.95)$$

$$Ar\text{-}SOH \ + \ 2\,ROOH \longrightarrow 2\,ROH \ + \ ArSO_3H \qquad (11.96)$$

C. Type 3 Antioxidants: Environmental Factors that Limit Oxidation

Macroenvironmental factors such as refrigeration, blanching, dark, inert gas or vacuum, moisture content, and pH cannot stop lipid oxidation themselves but provide the microenvironments in which lipid oxidation must occur. When managed appropriately, they certainly slow the rate of lipid oxidation and they can also shift reaction pathways. Thus, they provide the very important functions of reducing the radical load so that primary antioxidants can be more effective, and controlling products to minimize detrimental effects to food quality. Indeed, if these environmental factors are not controlled, no amount of chelator or radical quencher can successfully hold lipid oxidation at levels low enough to maintain sensory acceptability and safety of food products. Control of environmental factors is particularly important when multiple reaction mechanisms are active because they provide a means of directing the oxidation to or from specific pathways.

Environmental factors affect products from formulation through processing to storage. Formulation factors include moisture levels and pro-oxidant enzymes such as lipoxygenase and lipase that require blanching to inactivate. Processing environment factors that are important encompass everything from potential metal contamination from equipment and processing water to temperatures and light, and oxygen exposure during processing. Packaging and storage environments include the pO_2 inside the package and surrounding it, vacuum versus inert atmosphere versus air headspace, moisture inside the package and migrating into or out of it, temperature and light exposures, and so on.

Simplistically, the easiest approach to stabilizing foods against lipid oxidation might be to vacuum pack all foods in light-impermeable packages and then to freeze them. However, other considerations such as cost, protection of other food qualities, and storage requirements must be balanced against requirements for limiting lipid oxidation in each food product. In addition, real foods always have more than one catalyst and complex matrices so the catalytic factor effects described in this chapter are never as straightforward as when written in an equation on a page. Thus, in actual practice, it is usually necessary to creatively combine multiple antioxidant approaches and even multiple antioxidants to effectively block lipid oxidation in foods.

It is hoped that the information presented in this chapter will enable the reader to evaluate more fully how lipids are oxidized in individual systems, which factors are driving the oxidation, and which products are causing the most trouble. This information can then be applied to tailor combinations of antioxidant approaches to effectively and economically maintain sensory food quality while retaining the highest possible levels of nutritional quality and safety.

REFERENCES

Afanas'ev, I.B., Dorozhko, A.I., Brodskii, A.V., Kostyuk, V.A., Potapovitch, A.I., 1989. Chelating and free radical scavenging mechanisms of inhibitory action of rutin and quercetin in lipid oxidation. Biochem. Pharmacol. 38, 1763–1769.

Anonymous, 1981. Fatty acids' spontaneous oxidation clarified. Chem. Eng. News 26 (Oct.), 18–20.

Aoshima, H., Kajiwara, T., Hatanaka, A., 1981. Decomposition of lipid hydroperoxide by soybean lipoxygenase-1 under aerobic conditions studied by high performance liquid chromatography and the spin trapping method. Agric. Biol. Chem. 45, 2245–2251.

Aoshima, H., Kajiwara, T., Hatanaka, A., Hatano, H., 1977. Electron spin resonance studies on the lipoxygenase reaction by spin trapping and spin labelling methods. J. Biochem. 82, 1559–1565.

Atkins, P.W., Beran, J.A., 1990. General Chemistry. Scientific American Books, W.H. Freeman & Co., New York.

Avdulov, N.A., Chochina, S.V., Igbavboa, U., O'Hare, E.O., Schroeder, F., Cleary, J.P., Wood, W.G., 1997. Amyloid beta-peptides increase annular and bulk fluidity and induce lipid peroxidation in brain synaptic plasma membranes. J. Neurochem. 68, 2086–2091.

Avila, D.V., Brown, C.E., Ingold, K.U., Lusztyk, J., 1993. Solvent effects on the competitive β-scission and hydrogen atom abstraction reactions of the cumyloxyl radical. Resolution of a long-standing problem. J. Am. Chem. Soc. 115, 466–470.

Barclay, L.R.C., Ingold, K.U., 1981. Autoxidation of biological membranes. 2. The autoxidation of a model membrane. A comparison of the autoxidation of egg lecithin phosphatidylcholine in water and chlorobenzene. J. Am. Chem. Soc. 103, 6478–6485.

Baron, C.P., Andersen, H.J., 2002. Myoglobin-induced lipid oxidation. A review. J. Agric. Food Chem. 50, 3887–3897.

Belitz, H.-D., Grosch, W., 1986. Food Chemistry. Springer, Berlin, p. 130.

Bielski, B.H.J., Arudi, R.L., Sutherland, M.W., 1983. A study of the reactivity of HO_2/O_2^- ion with unsaturated fatty acids. J. Biol. Chem. 258, 4759–4761.

Bielski, B.H.J., Cabelli, D.E., Arudi, R.L., Ross, A.B., 1985. Reactivity of HO_2/O_2^- radicals in aqueous solution. J. Phys. Chem. Ref. Data 14, 1041–1100.

Bilski, P., Li, M.Y., Ehrenshaft, M., Daub, M.E., Chignell, C.F., 2000. Vitamin B_6 (pyridoxine) and its derivatives are efficient singlet oxygen quenchers and potential fungal antioxidants. Photochem. Photobiol. 71, 129–134.

Bishov, S.J., Henick, A.S., Koch, R.B., 1960. Oxidation of fat in model systems related to dehydrated foods. Food Res. 25, 174–181.

Blair, I.A., 2001. Lipid hydroperoxide-mediated DNA damage. Exp. Gerontol. 36, 1473–1481.

Bolland, J.L., 1945. The course of autoxidation reactions in polyisoprenes and allied compounds. Part IX. The primary thermal oxidation products of ethyl linoleate. J. Chem. Soc., 445–447.

Bolland, J.L., 1949. Kinetics of olefin oxidation. Q. Rev. 3, 1–21.

Borg, D.C., Schaich, K.M., 1983. Reactions connecting autoxidation with oxy radical production, lipid peroxidation, and cytotoxicity. In: Cohen, G., Greenwald, R.A. (Eds.), Oxy Radicals and their Scavenger Systems, Vol. I: Molecular Aspects. Elsevier, New York, pp. 122–129.

Borg, D.C., Schaich, K.M., 1984. Cytotoxicity from coupled redox cycling of autoxidizing xenobiotics and metals. Isr. J. Chem. 24, 38–53.

Borg, D.C., Schaich, K.M., Elmore Jr., J.J., 1981. Autoxidation and cytotoxicity. In: Rodgers, M.A.J., Powers, E.L. (Eds.), Oxygen and Oxy-radicals in Chemistry and Biology. Academic Press, New York, pp. 177–186.

Bors, W., Tait, D., Michel, C., Saran, M., Erben-Russ, M., 1984. Reactions of alkoxyl radicals in aqueous solutions. Isr. J. Chem. 24, 17—24.

Bothe, E., Schuchmann, M.N., Schulte-Frohlinde, D., von Sonntag, C., 1978. HO$_2\bullet$ elimination from α-hydroxyalkylperoxyl radicals in aqueous solution. Photochem. Photobiol. 28, 639—644.

Bruenner, B.A., Jones, A.D., German, J.B., 1995. Direct characterization of protein adducts of the lipid peroxidation product 4-hydroxy-2-nonenal using electrospray mass spectrometry. Chem. Res. Toxicol. 8, 552—559.

Bruice, T.C., Balasubramanian, P.N., Lee, R.W., Lindsay Smith, J.R., 1988. The mechanism of hydroperoxide O—O bond scission on reaction of hydroperoxides with iron(III) prophyrins. J. Am. Chem. Soc. 110, 7890—7892.

Buttery, R.G., Ling, L.C., Taeranishi, R., Mon, T.R., 1977. Roasted fat: basic volatile components. J. Agric. Food Chem. 25, 1227—1229.

Campbell, I.M., Caton, R.B., Crozier, D.N., 1974. Complex formation and reversible oxygenation of free fatty acids. Lipids 9, 916—920.

Carless, J.E., Nixon, J.R., 1960. The oxidation of solubilized and emulsified oils. Part III. The oxidation of methyl linoleate in potassium laurate and cetomacrogol dispersions. J. Pharm. Pharmacol. 12, 348—359.

Chan, H.W.-S., Matthew, J.A., Coxon, D.T., 1980. A hydroperoxy-epidioxide from the autoxidation of a hydroperoxide of methyl linolenate. J. Chem. Soc. Chem. Commun. 235—236.

Chang, S.S., Smouse, T.H., Krishnamurthy, R.G., Mookherjee, R.D., Reddy, R.B., 1966. Isolation and identification of 2-pentyl-furan as contributing to the flavor reversion of soyabean oil. Chem. Ind. 1926—1927.

Chen, Z.Y., Nawar, W.W., 1991. The role of amino acids in the autoxidation of milk fat. J. Am. Oil Chem. Soc. 68, 47—50.

Chio, K.S., Tappel, A.L., 1969. Synthesis and characterization of the fluorescent products derived from malondialdehyde and amino acids. Biochemistry 8, 2821—2827.

Corliss, G.A., Dugan Jr., L.R., 1971. Phospholipid oxidation in emulsions. Lipids 5, 846—853.

Cosgrove, J.P., Church, D.F., Pryor, W.A., 1987. The kinetics of the autoxidation of polyunsaturated fatty acids. Lipids 22, 299—304.

Dahle, L.K., Hill, E.G., Holman, R.T., 1962. The thiobarbituric acid reaction and the autoxidation of polyunsaturated fatty acid methyl esters. Arch. Biochem. Biophys. 98, 253—261.

De Groot, H., Noll, T., 1987. The role of physiological oxygen partial pressures in lipid peroxidation. Theoretical considerations and experimental evidence. Chem. Phys. Lipids 44, 209—226.

deGroot, J.J.M.C., Garssen, G.J., Vliegenthart, J.F.G., Boldingh, J., 1973. Detection of linoleic acid radicals in the anaerobic reaction of lipoxygenase. Biochim. Biophys. Acta 326, 279—284.

deGroot, J.J.M.C., Garssen, G.J., Vliegenthart, J.F.G., Boldingh, J., 1975. Demonstration by EPR spectroscopy of the functional role of iron in soybean 1-lipoxygenase. Biochim. Biophys. Acta 377, 71—79.

Dix, T.A., Marnett, L.J., 1983. Hematin-catalyzed rearrangement of hydroperoxylinoleic acid to epoxy alcohols via an oxygen rebound. J. Am. Chem. Soc. 105, 7001—7002.

Dix, T.A., Marnett, L.J., 1985. Conversion of linoleic acid hydroperoxide to hydroxy, keto, epoxyhydroxy, and trihydroxy fatty acids by hematin. J. Biol. Chem. 260, 5351—5357.

Ebrahem, Q., Ranganathan, K., Sears, J., Vasanji, A., Gu, X., Lu, L., Salomon, R.G., Crabb, J.W., Anand-Apte, B., 2006. Carboxyethylpyrrole oxidative protein modifications stimulate neovascularization: implications for age-related macular degeneration. Proc. Natl. Acad. Sci. U.S.A. 103, 13480—13484.

Ege, S.N., 1999. Organic Chemistry: Structure and Reactivity. Houghton Mifflin, Boston, MA, pp. 517—518, 631—654, 692—693, 725—733.

Egmond, M.R., Veldink, G.A., Vliegenthart, J.F.G., Boldingh, J., 1973. C-11 H abstraction from linoleic acid, the rate limiting step in lipoxygenase reaction. Biochem. Biophys. Res. Commun. 54, 1178—1184.

Egmond, M.R., Vliegenthart, J.F.G., Boldingh, J., 1972. Stereospecificity of the hydrogen abstraction in carbon atom n-8 in the oxygenation of linoleic acid by lipoxygenases from corn germs and soya beans. Biochim. Biophys. Res. Commun. 48, 1055—1060.

Elson, I.H., Mao, S.W., Kochi, J.K., 1975. Electron spin resonance study of addition of alkoxy radicals to olefins. J. Am. Chem. Soc. 97, 335—341.

Erben-Russ, M., Michael, C., Bors, W., Saran, M., 1987. Absolute rate constants of alkoxyl radical reactions in aqueous solution. J. Phys. Chem. 91, 2362—2365.

Esterbauer, H., Schaur, R.J., Zollner, H., 1991. Chemistry and biochemistry of 4-hydroxynonenal, malonaldehyde, and related aldehydes. Free Radic. Biol. Med. 11, 81—128.

Fadeel, B., Quinn, P., Xue, D., Kagan, V., 2007. Fat(al) attraction: oxidized lipids act as 'eat-me' signals. HFSP J. 1, 225—229.

Farmer, E.H., Sutton, D.A., 1943. Course of autoxidation reactions in polyisoprenes and allied compounds. IV. The isolation and constitution of photochemically-formed methyl oleate peroxide. J. Chem. Soc., 119—122.

Farmer, E.H., Koch, H.P., Sutton, D.A., 1943. The course of autoxidation reactions in polyisoprenes and allied compounds. VII. Rearrangement of double bonds during autoxidation. J. Chem. Soc., 541—547.

Fenaille, F., Guy, P.A., Tabet, J.-C., 2003. Study of protein modification by 4-hydroxy-2-nonenal and other short chain aldehydes analyzed by electrospray ionization tandem mass spectrometry. J. Am. Soc. Mass Spectrom 14, 215—226.

Fletcher, B.L., Tappel, A.L., 1970. Fluorescent modification to serum albumin by lipid peroxidation. Lipids 6, 172—175.

Fletcher, B.L., Dillard, C.J., Tappel, A.L., 1973. Measurement of fluorescent lipid peroxidation products in biological systems and tissues. Anal. Biochem. 52, 1—9.

Foote, C.S., 1968. Photosensitized oxygenation and the role of singlet oxygen. Acc. Chem. Res. 1, 104—110.

Foote, C.S., 1976. Photosensitized oxidation and singlet oxygen: consequences in biological systems. In: Pryor, W.A. (Ed.), Free Radicals in Biology. Academic Press, New York, pp. 85—134.

Foote, C.S., Denny, R.W., Weaver, L., Chang, Y., Peters, J., 1970. Quenching of singlet oxygen. Ann. N.Y. Acad. Sci. 171, 139—148.

Frankel, E.N., 1982. Volatile lipid oxidation products. Prog. Lipid Res. 22, 1—33.

Frankel, E.N., 1984. Lipid oxidation: mechanisms, products, and biological significance. J. Am. Oil Chem. Soc. 61, 1908—1917.

Frankel, E.N., 1985. Chemistry of free radical and singlet oxidation of lipids. Prog. Lipid Res. 23, 197—221.

Frankel, E.N., 1987. Secondary products of lipid oxidation. Chem. Phys. Lipids 44, 73—85.

Frankel, E.N., 1991. Recent advances in lipid oxidation. J. Sci. Food Agric. 54, 495—511.

Frankel, E.N., Neff, W.E., 1983. Formation of malonaldehyde from lipid oxidation products. Biochim. Biophys. Acta 754, 264—270.

Frankel, E.N., Neff, W.E., Selke, E., Weisleder, D., 1982. Photosensitized oxidation of methyl linoleate: secondary and volatile thermal decomposition products. Lipids 17, 11–18.

Freeman, T.L., Haver, A., Duryee, M.J., Tuma, D.J., Klassen, L.W., Hamel, F.G., White, R.L., Rennard, S.I., Thiele, G.M., 2005. Aldehydes in cigarette smoke react with the lipid peroxidation product malonaldehyde to form fluorescent protein adducts on lysines. Chem. Res. Toxicol. 18, 817–824.

Fruebis, J., Parsasarathy, S., Steinberg, D., 1992. Evidence for a concerted reaction between lipid hydroperoxides and polypeptides. Proc. Natl. Acad. Sci. U.S.A. 89, 10588–10592.

Fueno, T., Ree, T., Eyring, H., 1959. Quantum-mechanical studies on oxidation potentials and antioxidizing action of phenolic compounds. J. Phys. Chem. 63, 1940–1948.

Fukushige, H., Wang, C., Simpson, T.D., Gardner, H.W., Hildebrand, D.F., 2005. Purification and identification of linoleic acid hydroperoxides generated by soybean seed lipoxygenases 2 and 3. J. Agric. Food Chem. 53, 5691–5694.

Gaddis, A.M., Ellis, R., Currie, G.T., 1961. Carbonyls in oxidizing fats. V. The composition of neutral volatile monocarbonyl compounds from autooxidized oleate, linoleate, linolenate esters and fats. J. Am. Oil Chem. Soc. 38, 371–375.

Ganapathi, M.R., Hermann, R., Naumov, S., Brede, O., 2000. Free electron transfer from several phenols to radical cations of non-polar solvents. Phys. Chem. Chem. Phys. 2, 4947–4955.

Gardner, H.W., 1979. Lipid hydroperoxide reactivity with proteins and amino acids: a review. J. Agric. Food Chem. 27, 220–229.

Gardner, H.W., 1986. Lipoxygenase pathway in cereals. In: Pomeranz, Y. (Ed.), Advances in Cereal Science and Technology. American Association of Cereal Chemists, St. Paul, MN, pp. 161–215.

Gardner, H.W., Weisleder, D., 1970. Lipoxygenase from Zea mays: 9-d-hydroperoxy- trans-10, cis-12-octadecadienoic acid from linoleic acid. Lipids 5, 678–683.

Gardner, H.W., Weisleder, D., 1976. Addition of N-acetylcysteine to linoleic acid hydroperoxide. Lipids 11, 127–134.

Gardner, H.W., Kleiman, R., Weisleder, D., Inglett, G.E., 1977. Cysteine adds to lipid hydroperoxide. Lipids 12, 655–660.

Gardner, H.W., Plattner, R.D., Weisleder, D., 1985. The epoxyallylic radical from homolysis and rearrangement of methyl linoleate hydroperoxide combines with the thiyl radical of N-acetylcysteine. Biochim. Biophys. Acta 834, 65–74.

Garner, B., Waldeck, A.R., Witting, P.K., Rye, K.-A., Stocker, R., 1998. Oxidation of high density lipoproteins. II. Evidence for direct reduction of lipid hydroperoxides by methionine residues of apolipoproteins A1 and A2. J. Biol. Chem. 273, 6088–6095.

Gebicki, J.M., Bielski, B.H.J., 1981. Comparison of the capacities of the perhydroxyl and the superoxide radicals to initiate chain oxidation of linoleic acid. J. Am. Chem. Soc. 103, 7020–7022.

Gebicki, S., Gebicki, J.M., 1999. Crosslinking of DNA and proteins induced by protein hydroperoxides. Biochem. J. 338, 629–636.

Grosch, W., 1987. Reactions of hydroperoxides – products of low molecular weight. In: Chan, H.W.-S. (Ed.), Autoxidation of Unsaturated Lipids. Academic Press, London, pp. 95–139.

Grosch, W., Laskawy, G., Weber, F., 1976. Formation of volatile carbonyl compounds and co-oxidation of β-carotene by lipoxygenase from wheat, potato, flax, and beans. J. Agric. Food Chem. 24, 456–459.

Gu, X., Meer, S.G., Miyagi, M., Rayborn, M.E., Hollyfield, J.G., Crabb, J.W., Salomon, R.G., 2003. Carboxyethylpyrrole protein adducts and autoantibodies, biomarkers for age-related macular degeneration. J. Biol. Chem. 278, 42027–42035.

Gutierrez, J., Ballinger, S.W., Darley-Usmar, V.M., Landar, A., 2006. Free radicals, mitochondria, and oxidized lipids: the emerging role in signal transduction. Circ. Res. 99, 924–932.

Hamberg, M., Samuelsson, B., 1967. On the specificity of the oxygenation of unsaturated fatty acids catalyzed by soybean lipoxidase. J. Biol. Chem. 242, 5329–5335.

Hasegawa, K., Patterson, L.K., 1978. Pulse radiolysis studies in model lipid systems: formation and behavior of peroxy radicals in fatty acids. Photochem. Photobiol. 28, 817–823.

Haynes, R.K., Vonwiller, S.C., 1990. Iron(III) and copper(II) catalysed transformation of fatty acid hydroperoxides: efficient generation of peroxy radicals with copper(II) trifluoromethane sulphonate. J. Chem. Soc. Chem. Commun., 1102–1104.

Heaton, F.W., Uri, N., 1961. The aerobic oxidation of unsaturated fatty acids and their esters: cobalt stearate-catalyzed oxidation of linoleic acid. J. Lipid Res. 2, 152–160.

Hecht, E., 2000. Physics: Calculus. Brooks/Cole, Pacific Grove, CA.

Heijman, M.G.J., Nauta, H., Levine, Y.K., 1985. A pulse radiolysis study of the dienyl radical in oxygen-free linoleate solutions: time and linoleate concentration dependence. Radiat. Phys. Chem. 26, 73–82.

Hiatt, R., McCarrick, T., 1975. On 'bimolecular initiation' by hydroperoxides. J. Am. Chem. Soc. 97, 5234–5237.

Hiatt, R., Traylor, T.G., 1965. Cage recombination of t-butoxy radicals. J. Am. Chem. Soc. 87, 3766–3768.

Hidalgo, F.J., Kinsella, J.E., 1989. Changes in β-lactoglobulin B following interactions with linoleic acid 13-hydroperoxide. J. Agric. Food Chem. 37, 860–866.

Hidalgo, F.J., Zamora, R., 2000. Modification of bovine serum albumin structure following reaction with 4,5(E)-epoxy-2-(E)-heptenal. Chem. Res. Toxicol. 13, 501–508.

Horner, J.H., Choi, S.-Y., Newcomb, M., 2000. Laser flash photolysis studies of alkoxyl radical kinetics using 4-nitrobenzenesulfenate esters as radical precursors. Org. Lett. 2, 3369–3372.

Howard, J.A., Ingold, K.U., 1967. Absolute rate constants for hydrocarbon autoxidation. VI. Alkyl aromatic and olefinic hydrocarbons. Can. J. Chem. 45, 793–802.

Hyde, S.M., 1968. Oxidation of methyl oleate induced by cobalt-60 γ-radiation. II. Emulsions of methyl oleate in water. Trans. Faraday Soc. 64, 155–162.

Hyde, S.M., Verdin, D., 1968. Oxidation of methyl oleate induced by cobalt-60 γ-irradiation. I. Pure methyl oleate. Trans. Faraday Soc. 64, 144–154.

Ichihashi, K., Osawa, T., Toyokuni, S., Uchida, K., 2001. Endogenous formation of protein adducts with carcinogenic aldehydes. J. Biol. Chem. 276, 23903–23913.

Igene, J.O., Pearson, A.M., Dugan Jr., L.R., Price, J.F., 1980. Role of triglycerides and phospholipids on development of rancidity in model meat systems during frozen storage. Food Chem. 5, 263–276.

Ingold, K.U., 1969a. Peroxy radicals. Acc. Chem. Res. 2, 1–9.

Ingold, K.U., 1969b. Structure and Mechanism in Organic Chemistry, 2nd ed. Cornell University Press, Ithaca, NY.

Ishii, K., Shimada, A., Yoshimatsu, J., 1976. Effects of the oxidized linseed oil on the physico-chemical properties of starch. J. Jpn. Soc. Starch Sci. 23, 82–90.

Kamiya, Y., Beaton, S., Lafortune, A., Ingold, K.U., 1963. The metal-catalyzed autoxidation of tetralin. I. Introduction. The cobalt-catalyzed autoxidation in acetic acid. Can. J. Chem. 41, 2020–2032.

Karel, M., 1980. Lipid oxidation, secondary reactions, and water activity of foods. In: Simic, M.G., Karel, M. (Eds.), Autoxidation in Food and Biological Systems. Plenum Press, New York, pp. 191–206.

Karel, M., Schaich, K.M., Roy, R.B., 1975. Interaction of peroxidizing methyl linoleate with some proteins and amino acids. J. Agric. Food Chem. 23, 159–164.

Kawakishi, S., Satake, A., Komiya, T., Namiki, M., 1983. Oxidative degradation of β-cyclodextrin induced by lipid peroxidation. Starch/Starke 35, 54–57.

Kearns, D.R., Khan, A.U., 1969. Sensitized photooxygenation reactions and the role of singlet oxygen. Photochem. Photobiol. 10, 193–210.

Kearns, D.R., Hollins, R.A., Khan, A.U., Radlick, P., 1967. Evidence for the participation of $^1sigma_{g+}$ and $^1delta_g$ oxygen in dye-sensitized photooxygenation reactions. II. J. Am. Chem. Soc. 89, 5456–5457.

Kehrer, J.P., Biswal, S.S., 2000. The molecular effects of acrolein. Toxicol. Sci. 57, 6–15.

Kellogg, E.W., Fridovich, I., 1975. Superoxide, hydrogen peroxide, and singlet oxygen in lipid peroxidation by a xanthine oxidase system. J. Biol. Chem. 250, 8812–8817.

Kerr, J.A., 1966. Bond dissociation energies by kinetic methods. Chem. Rev. 66, 465–500.

Kikugawa, K., Ido, Y., 1984. Studies on peroxidized lipids. V. Formation and characterization of 1,4-dihydropyridine-3,5-dicarbaldehydes as model of fluorescent components in lipofuschin. Lipids 19, 600–608.

Kim, S.S., Kim, S.Y., Ryou, S.S., Lee, C.S., Yoo, K.H., 1993. Solvent effects in the hydrogen abstractions by tert-butoxy radical: veracity of the reactivity/selectivity principle. J. Org. Chem. 58, 192–196.

Koch, R.B., 1962. Dehydrated foods and model systems. In: Schultz, H., Day, E.A., Sinnhuber, R.O. (Eds.), Symposium on Foods: Lipids and their Oxidation. Avi, Westport, CT, pp. 230–249.

Kochi, J.K., 1962. Chemistry of alkoxyl radicals. J. Am. Chem. Soc. 84, 1193–1197.

Kochi, J.K., 1973a. Oxygen radicals. In: Kochi, J.K. (Ed.), Free Radicals, Vol. 2. John Wiley, New York, pp. 665–710.

Kochi, J.K., 1973b. Free Radicals Vol. 1 John Wiley and Sons, New York.

Koga, T., Terao, J., 1995. Phospholipids increase radical-scavenging activity of vitamin E in a bulk oil model system. J. Agric. Food Chem. 43, 1450–1454.

Kowalik-Jankowska, T., Ruta, M., Winiewska, K., Lankiewicz, L., Dyba, M., 2004. Products of Cu(II)-catalyzed oxidation in the presence of hydrogen peroxide of the 1–10, 1–16 fragments of human and mouse β-amyloid peptide. J. Inorg. Biochem. 98, 940–950.

Kruidenier, L., Verspaget, H.W., 2002. Oxidative stress as a pathogenic factor in inflammatory bowel disease – radicals or ridiculous? Aliment. Pharmacol. Ther. 16, 1997–2015.

Kulich, D.M., Shelton, J.R., 1991. Organosulfur antioxidants: mechanisms of action. Polymer Degrad. Stabil. 3, 397–410.

Labuza, T.P., 1971. Kinetics of lipid oxidation in foods. CRC Crit. Rev. Food Sci. Nutr. 2, 355–405.

Ladikos, D., Lougovois, V., 1990. Lipid oxidation in muscle foods: a review. Food Chem. 35, 295–314.

Lederer, M.O., 1996. Reactivity of lysine moieties toward g-hydroxy-a, b-unsaturated epoxides: a model study on protein–lipid oxidation product interaction. J. Agric. Food Chem. 44, 2531–2537.

Leermakers, P.A., Thomas, H.T., Weis, L.D., James, P.C., 1966. Spectra and photochemistry of molecules adsorbed on silica gel. IV. J. Am. Chem. Soc. 88, 5075–5083.

Leopoldini, M., Marino, T., Russo, N., Toscano, M., 2004. Antioxidant properties of phenolic compounds: H-atom versus electron transfer mechanism. J. Phys. Chem. 108, 4916–4922.

Lin, D., Lee, H., Liu, Q., Perry, G., Smith, M.A., Sayre, L.M., 2005. 4-Oxo-2-nonenal is both more neurotoxic and more protein reactive than 4-hydroxy-2-nonenal. Chem. Res. Toxicol. 18, 1219–1231.

Lindsay, D.A., Howard, J.A., Horswill, E.C., Iton, L., Ingold, K.U., Cobbley, T., Li, A., 1973. The bimolecular self-reaction of secondary peroxy radicals, product studies. Can. J. Chem. 51, 870–880.

Logani, M.K., Davies, R.E., 1980. Lipid oxidation: biological effects and antioxidants – a review. Lipids 15, 485–495.

Love, J.D., Pearson, A.M., 1974. Metmyoglobin and non-heme iron as prooxidants in cooked meat. J. Agric. Food Chem. 22, 1031–1034.

Lqari, H., Pedroche, J., Giron-Calle, J., VIoque, J., Millan, F., 2003. Interaction of Lupinus angustifolius L. a and g conglutins with 13-hydroperoxide-11,9-octadienoic acid. Food Chem. 80, 517–523.

Lundberg, W.O., Chipault, J.R., 1947. The oxidation of methyl linoleate at various temperatures. J. Am. Chem. Soc. 69, 833–836.

Maga, J.A., 1981. Pyridines in foods. J. Agric. Food Chem. 29, 895–898.

Marcuse, R., Fredriksson, P.-O., 1968. Fat oxidation at low oxygen pressure. I. Kinetic studies on the rate of fat oxidation in emulsions. J. Am. Oil Chem. Soc. 45, 400–407.

McBrien, D.C.H., Slater, T.F. (Eds.), 1982. Free Radicals, Lipid Peroxidation, and Cancer. Academic Press, New York.

McMurray, J., 2000. Organic Chemistry. Brooks/Cole, Pacific Grove, CA.

Mead, J.F., 1980. Membrane lipid peroxidation and its prevention. J. Am. Oil Chem. Soc. 57, 393–397.

Mikula, M., Khayat, A., 1985. Reaction conditions for measuring oxidative stability of oils by thermogravimetric analysis. J. Am. Oil Chem. Soc. 62, 1694–1698.

Miyashita, K., Takagi, T., 1986. Study on the oxidative rate and prooxidant activity of free fatty acids. J. Am. Oil Chem. Soc. 63, 1380–1384.

Murray, 1979. Chemical sources of $^1O_2$. In: Wasserman, H., Murray, R.W. (Eds.), Singlet Oxygen. Academic Press, New York, Chapter 3.

Nair, V., Offerman, R.J., Turner, G.A., Pryor, A.N., Baenziger, N.C., 1988. Fluorescent 1,4-dihydropyridines – the malondialdehyde connection. Tetrahedron 44, 2793–2803.

Nam, W., Han, H.J., Oh, S.-Y., Lee, Y.J., Choi, M.-H., Han, S.-Y., Kim, C., Woo, S.K., Shin, W., 2000a. New insights into the mechanism of O–O bond cleavage of hydrogen peroxide and tert-alkyl hydroperoxides by iron(III) porphyrin complexes. J. Am. Chem. Soc. 122, 8677–8684.

Nam, W., Lim, M.H., Lee, H.J., Kim, C., 2000b. Evidence for the participation of two distinct reactive intermediates in iron(III) porphyrin complex-catalyzed epoxidation reactions. J. Am. Chem. Soc. 122, 6641–6647.

Nam, W., Park, S.-E., Lim, I.K., Lim, M.H., Hong, J., Kim, J., 2003. First direct evidence for stereospecific olefin epoxidation and alkane hydroxylation by an oxoiron(IV) porphyrin complex. J. Am. Chem. Soc. 125, 14674–14675.

Nawar, W.W., 1969. Thermal degradation of lipids. J. Agric. Food Chem. 17, 18–21.

Nawar, W.W., 1986. Lipids. In: Fennema, O.R. (Ed.), Food Chemistry. Academic Press, New York, pp. 225–320.

Neff, W.E., 1980. Quantitative analyses of hydroxystearate isomers from hydroperoxides by high pressure liquid chromatography of autoxidized and photosensitized–oxidized fatty esters. Lipids 15, 587–590.

Neff, W.E., Frankel, E.N., 1984. Photosensitized oxidation of methyl linolenate monohydroperoxides: hydroperoxy cyclic peroxides, dihydroperoxides and hydroperoxy bis-cyclic peroxides. Lipids 19, 952–957.

Neff, W.E., Frankel, E.N., Fujimoto, K., 1988. Autoxidative dimerization of methyl linolenate and its monohydroperoxides, hydroperoxy epidioxides and dihydroperoxides. J. Am. Oil Chem. Soc. 65, 616–623.

Neff, W.E., Frankel, E.N., Weisleder, D., 1982. Photosensitized oxidation of methyl linolenate. Secondary products. Lipids 17, 780–790.

Nwosu, C.V., Boyd, L.C., Sheldon, B., 1997. Effect of fatty acid composition of phospholipids on their antioxidant properties and activity index. J. Am. Oil Chem. Soc. 74, 293–297.

O'Brien, P.J., 1969. Intracellular mechanisms for the decomposition of a lipid peroxide. I. Decomposition of a lipid peroxide by metal ions, heme compounds, and nucleophiles. Can. J. Biochem. 47, 485–493.

Parr, L.J., Swoboda, P.A.T., 1976. The assay of conjugable oxidation products applied to lipid deterioration in stored foods. J. Food Technol. 11, 1–12.

Pattee, H.E., Singleton, J.A., 1979. Evidence of enzymic production of 9-hydroperoxy- *trans*-12, *cis*-12-octadecadienoic acid by peanut lipoxygenase. J. Agric. Food Chem. 27, 216–220.

Patterson, L.K., Hasegawa, K., 1978. Pulse radiolysis studies in model lipid systems. The influence of aggregation on kinetic behavior of OH induced radicals in aqueous sodium linoleate. Ber. Bunsenges. Phys. Chem. 82, 951–956.

Petersen, D.R., Doorn, J.A., 2004. Reactions of 4-hydroxynonenal with proteins and cellular targets. Free Radic. Biol. Med. 37, 937–945.

Piretti, M.V., Cavani, C., Zeli, F., 1978. Mechanism of the formation of hydroperoxides from methyl oleate. Rev. Franc. Corps Gras. 25, 73–78.

Pokorny, J., Kundu, M.K., Pokorny, S., Bleha, M., Coupek, J., 1976a. Lipid oxidation. 4. Products of thermooxidative polymerization of vegetable oils. Nahrung 20, 157–163.

Pokorny, J., Rzepa, J., Janicek, G., 1976b. Lipid oxidation. Part 1. Effect of free carboxyl group on the decomposition of lipid hydroperoxide. Nahrung 20, 1–6.

Porter, N.A., 1990. Autoxidation of polyunsaturated fatty acids: initiation, propagation, and product distribution (basic chemistry). In: Membrane Lipid Oxidation, Vol. I. C. Vigo-Pelfrey. CRC Press, Boca Raton, FL, pp. 33–62.

Porter, N.A., Lehman, L.S., 1982. Xanthine oxidase initiated oxidation of model membranes. Effect of position of abstractable hydrogen atoms in the bilayer on the distribution of products. J. Am. Chem. Soc. 104, 4731–4732.

Porter, N.A., Wagner, C.R., 1986. Phospholipid autoxidation. Adv. Free Radical Biol. Med. 2, 283–323.

Porter, N.A., Caldwell, S.E., Mills, K.A., 1995. Mechanisms of free radical oxidation of unsaturated lipids. Lipids 30, 277–290.

Porter, N.A., Lehman, L.S., Wujek, D.G., 1984. Oxidation mechanisms of poly-unsaturated fatty acids. In: Bors, W., Saran, M., Tait, D. (Eds.), Oxygen Radicals in Chemistry and Biology. D. Walter de Gruyter, Berlin, pp. 235–237.

Porter, W.L., Levasseur, L.A., Jeffers, J.I., Henick, A.S., 1971. UV spectrophotometry of autoxidized lipid monolayers while on silica gel. Lipids 6, 16–25.

Privett, O.S., 1959. Autoxidation and autoxidative polymerization. J. Am. Oil Chem. Soc. 36, 507–512.

Pryor, W.A., 1978. The formation of free radicals and the consequences of their reactions in vivo. Photochem. Photobiol. 28, 787–801.

Pryor, W.A., 1985. Free radical involvement in chronic diseases and aging: the toxicity of lipid hydroperoxides and their decomposition products. In: Finley, J.W., Schwass, D.E. (Eds.), Xenobiotic Metabolism: Nutritional Effects. American Chemical Society, Washington, DC, pp. 76–96.

Pryor, W.A., 1986. Oxy-radicals and related species: their formation, lifetimes and reactions. Annu. Rev. Physiol. 48, 657–667.

Pryor, W.A., 1989. On the detection of lipid hydroperoxides in biological samples. Free Radic. Biol. Med. 7, 177–178.

Pryor, W.A., Stanley, J.P., Blair, E., 1976. Autoxidation of poly-unsaturated fatty acids. Part II. A suggested mechanism for the formation of TBA-reactive materials from prostaglandin-like endoperoxides. Lipids 11, 370–379.

Raghuveer, K.G., Hammond, E.G., 1967. The influence of glyceride structure on the rate of autoxidation. J. Am. Oil Chem. Soc. 44, 239–243.

Rao, S.I., Wilks, A., Hamberg, M., Ortiz de Montellano, P.R., 1994. The lipoxygenase activity of myoglobin. Oxidation of linoleic acid by the ferryl oxygen rather than the protein radical. J. Biol. Chem. 269, 7210–7216.

Rawls, H.R., Van Santen, P.J., 1970. A possible role for singlet oxygen in the initiation of fatty acid autoxidation. J. Am. Oil Chem. Soc. 47, 121–125.

Rossi-Olson, A., 2011. The impact of nut inclusions on properties and stability of dark chocolate. Dissertation, Food Science. Rutgers University, New Brunswick, NJ.

Russell, C.A., 1959. Solvent effects in the reaction of free radicals and atoms. V. Effects of solvents on the reactivity of *t*-butoxy radicals. J. Am. Chem. Soc. 21, 300–302.

Sargis, R.M., Subbaiah, P.V., 2003. *Trans* unsaturated fatty acids are less oxidizable than *cis* unsaturated fatty acids and protect endogenous lipids from oxidation in lipoproteins and lipid bilayers. Biochemistry 42, 11533–11543.

Sastry, Y.S.R., Lakshminarayana, G., 1971. Chlorophyll-sensitized peroxidation of saturated fatty acid esters. J. Am. Oil Chem. Soc. 48, 452–454.

Sayre, L.M., Arora, P.K., Iyer, R.S., Salomon, R.G., 1993. Pyrrole formation from 4-hydroxynonenal and primary amines. Chem. Res. Toxicol. 6, 19–22.

Sayre, L.M., Zelasko, D.A., Harris, P.L., Perry, G., Salomon, R.G., Smith, M.A., 1997. 4-Hydroxynonenal-derived advanced lipid peroxidation end products are increased in Alzheimer's disease. J. Neurochem. 68, 2092–2097.

Schaich, K.M., 1980a. Free radical initiation in proteins and amino acids by ionizing and ultraviolet radiations and lipid oxidation. Part III: Free radical transfer from oxidizing lipids. CRC Crit. Rev. Food Sci. Nutr. 13, 189–244.

Schaich, K.M., 1980b. Free radical initiation in proteins and amino acids by ionizing and ultraviolet radiations and lipid oxidation. Part III: Free radical transfer from oxidizing lipids. CRC Crit. Rev. Food Sci. Nutr. 13, 189–244.

Schaich, K.M., 1992. Metals and lipid oxidation: contemporary issues. Lipids 27, 209–218.

Schaich, K.M., 2006. Lipid oxidation in specialty oils. In: Shahidi, F. (Ed.), Nutraceutical and Specialty Lipids and Their Co-Products. CRC Press/Taylor & Francis, Boca Raton, FL, pp. 401–448.

Schaich, K.M., 2005b. Lipid oxidation in fats and oils: an integrated view. In: Shahidi, F. (Ed.), Bailey's Industrial Fats and Oils, 6th ed John Wiley, New York, Vol.1, pp. 269–355.

Schaich, K.M., 2008. Co-oxidations of oxidizing lipids: reactions with proteins. In: Kamal-Eldin, A., Min, D.B. (Eds.), Lipid Oxidation Pathways, Vol. 2. CRC Press, Boca Raton, FL, pp. 183–274.

Schaich, K.M., Karel, M., 1975. Free radicals in lysozyme reacted with peroxidizing methyl linoleate. J. Food Sci. 40, 456–459.

Schaich, K.M., Karel, M., 1976. Free radical reactions of peroxidizing lipids with amino acids and proteins: an ESR study. Lipids 11, 392–400.

Schauenstein, E., 1967. Autoxidation of polyunsaturated esters in water: chemical structure and biological activity of the products. J. Lipid Res. 8, 417–428.

Schaur, R.J., 2003. Basic aspects of the biochemical reactivity of 4-hydroxynonenal. Mol. Aspects Med. 24, 149–159.

Scott, G., 1965. Atmospheric Oxidation and Antioxidants. Elsevier, London.

Shamberger, R.J., Shamberger, B.A., Willis, C.E., 1977. Malonaldehyde content of food. J. Nutr. 107, 1404–1409.

Sims, R.P., Hoffman, W.H., 1962. Oxidative polymerization. In: Lundberg, W.O. (Ed.), Autoxidation and Antioxidants, Vol. II. Interscience, London, pp. 629–694.

Slawson, V., Adamson, A.W., Mead, J.F., 1973. Autoxidation of polyunsaturated fatty esters on silica. Lipids 8, 129–134.

Sliwiok, J., Kowalska, T., Kowalski, W., Biernat, A., 1974. The influence of hydrogen-bond association on the destruction of hydroperoxides in the autoxidation process of oleyl alcohol, oleic acid, and methyl oleate. Microchem. J. 19, 362–372.

Soszylqski, M., Filipiak, A., Bartosz, G., Gebicki, J.M., 1996. Effect of amino acid peroxides on the erythrocyte. Free Radic. Biol. Med. 20, 45–51.

Stahl, W., Sies, H., 2005. Bioactivity and protective effects of natural carotenoids. Biochim. Biophys. Acta 1740, 101–107.

Sugino, H., Ishikawa, M., Nitoda, T., Koketsu, M., Juneja, L.R., Kim, M., Yamamoto, T., 1997. Antioxidative activity of egg yolk phospholipids. J. Agric. Food Chem. 45, 551–554.

Sun, Y.-E., Wang, W.-D., Chen, H.-W., Li, C., 2011. Autoxidation of unsaturated lipids in food emulsions. Crit. Rev. Food Sci. Nutr. 51, 453–466.

Suyama, K., Adachi, A., 1980. Origin of alkyl-substituted pyridines in food flavor: formation of the pyridines from the reaction of alkanals with amino acids. J. Agric. Food Chem. 28, 546–549.

Suzuki, Y.J., Forman, H.J., Sevanian, A., 1997. Oxidants as stimulators of signal transduction. Free Radic. Biol. Med. 22, 269–285.

Swern, D., 1961. Primary products of olefinic autoxidation. In: Lundberg, W.O. (Ed.), Autoxidation and Antioxidants, Vol. 1. Interscience, New York, pp. 1–54.

Tappel, A.L., 1953a. Oxidative rancidity in food products. I. Linoleate oxidation catalyzed by hemin, hemoglobin, and cytochrome c. Food Res. 18, 560–573.

Tappel, A.L., 1953b. The mechanism of the oxidation of unsaturated fatty acids catalyzed by hematin compounds. Arch. Biochem. Biophys. 44, 378–394.

Tappel, A.L., 1955. Unsaturated lipide oxidation catalyzed by hematin compounds. J. Biol. Chem. 217, 721–733.

Tappel, A.L., 1961. Biocatalysts: lipoxidase and hematin compounds. In: Lundberg, W.O. (Ed.), Autoxidation and Antioxidants, Vol. 1. Interscience, New York, pp. 326–366.

Tappel, A.L., 1962. Hematin compounds and lipoxidase as biocatalysts. In: Schultz, H., Day, E.D., Sinnhuber, W.O. (Eds.), Symposium on Foods: Lipids and their Oxidation. Avi, Westport, CT, pp. 122–138.

Tappel, A.L., 1995. Unsaturated lipide oxidation catalyzed by hematin compounds. J. Biol. Chem. 217, 721–733.

Terao, J., Matsushita, S., 1977a. Products formed by photosensitized oxidation of unsaturated fatty acid esters. J. Am. Oil Chem. Soc. 54, 234–238.

Terao, J., Matsushita, S., 1977b. Structures of monohydroperoxides produced from chlorophyll sensitized photooxidation of methyl linoleate. Agric. Biol. Chem. 41, 2467–2468.

Terao, J., Ogawa, T., Matsushita, S., 1975. Degradation process of autoxidized methyl linoleate. Agric. Biol. Chem. 39, 397–402.

Thomas, J.R., 1965. The self-reaction of t-butylperoxy radicals. J. Am. Chem. Soc. 87, 3935–3940.

Thomas, M.J., Mehl, K.S., Pryor, W.A., 1982. The role of superoxide in xanthine oxidase-induced autoxidation of linoleic acid. J. Biol. Chem. 257, 8343–8347.

Traylor, T.G., Russell, C.A., 1965. Mechanism of autoxidation. Terminating radicals in cumene autoxidation. J. Am. Chem. Soc. 87, 3698–3706.

Traylor, T.G., Xu, F., 1990. Mechanism of reactions of iron(III) porphyrins with hydrogen peroxide and hydroperoxides: solvent and solvent isotope effects. J. Am. Chem. Soc. 112, 178–186.

Treibs, W., 1948. Bei der Autoxydation yon Lind- und Linolensureestern werden durch 2 Mol Sauerstoff. Chem. Ber. 81, 472–477.

Tsentalovich, Y.P., Kulik, L.V., Gritsan, N.P., Yurkovskaya, A.V., 1998. Solvent effect on the rate of β-scission of the $tert$-butoxyl radical. J. Phys. Chem. A. 102, 7975–7980.

Uchida, K., 1999. Current status of acrolein as a lipid peroxidation product. Trends Cardiovasc. Med. 9, 109–113.

Uchida, K., 2000. Role of reactive aldehyde in cardiovascular diseases. Free Radic. Biol. Med. 28, 1685–1696.

Uchida, K., Stadtman, E.R., 1992. Modification of histidine residues in proteins by reaction with 4-hydroxynonenal. Proc. Natl. Acad. Sci. U.S.A. 89, 4544–4548.

Uchida, K., Hasui, Y., Osawa, T., 1997. Covalent attachment of 4-hydroxy-2-nonenal to erythrocyte proteins. J. Biochem. (Tokyo) 122, 1246–1251.

Uchida, K., Kanematsu, M., Morimutsu, Y., Osawa, T., Noguchi, N., Niki, E., 1998a. Acrolein is a product of lipid peroxidation reaction. Formation of free acrolein and its conjugate with lysine residues in oxidized low density lipoproteins. J. Biol. Chem. 273, 16058–16066.

Uchida, K., Kanematsu, M., Sakai, K., Matsuda, T., Hattori, N., Mizuno, Y., Suzuki, D., Miyata, T., Noguchi, N., Niki, E., Osawa, T., 1998b. Protein-bound acrolein: potential markers for oxidative stress. Proc. Natl. Acad. Sci. U.S.A. 95, 4882–4887.

Uchida, K., Toyokuni, S., Nishikawa, K., Kawakishi, S., Oda, H., Hiai, H., Stadtman, E.R., 1994. Michael addition-type 4-hydroxy-2-nonenal adducts in modified low-density lipoproteins: markers for atherosclerosis. Biochemistry 33, 12487–12494.

Umehara, T., Terao, J., Matsushita, S., 1979. Photosensitized oxidation of oils with food colors. J. Agric. Chem. Soc. Jpn 53, 51–56.

Uri, N., 1961. Physico-chemical aspects of autoxidation. In: Lundberg, W.O. (Ed.), Autoxidation and Antioxidants, Vol. 1. Interscience, New York, pp. 55–106.

Van Sickle, D.E., Mayo, F.R., Gould, E.S., Arluck, R.M., 1967. Effects of experimental variables in oxidations of alkenes. J. Am. Chem. Soc. 89, 977–984.

Veldink, G.A., Garssen, G.J., Vliegenthart, J.F.G., Boldingh, J., 1972. Positional specificity of corn germ lipoxygenase as a function of pH. Biochem. Biophys. Res. Commun. 47, 22−26.

Walling, C., Padwa, A., 1963. Positive halogen compounds. VI. Effects of structure and medium on the B-scission of alkoxyl radicals. J. Am. Chem. Soc. 85, 1593−1597.

Walling, C., Thaler, W., 1961. Positive halogen compounds, III. Allyl chlorination with *t*-butyl hypochlorite. The stereochemistry of allylic radicals. J. Am. Chem. Soc. 83, 3877−3884.

Walling, C., Wagner, P.J., 1963. Effects of solvents on transition states in the reactions of *t*-butoxy radicals. J. Am. Chem. Soc. 85, 2333−2334.

Walling, C., Waits, H.P., Milanovic, J., Pappiaonnou, C.G., 1970. Polar and radical paths in the decomposition of diacyl peroxides. J. Am. Chem. Soc. 92, 4927−4932.

Watts, B.M., 1954. Oxidative rancidity and discoloration in meat. Adv. Food Res. 1, 1−52.

Watts, B.M., 1962. Meat products. In: Schultz, H., Day, E.D., Sinnhuber, W.O. (Eds.), Symposium on Foods: Lipids and their Oxidation. Avi, Westport, CT, pp. 202−214.

Weenan, H., Porter, N.A., 1982. Autoxidation of model membrane systems: cooxidation of polyunsaturated lecithins with steroids, fatty acids, and α-tocopherol. J. Am. Chem. Soc. 104, 5216−5221.

Witting, I.A., Chang, S.S., Kummerow, F.A., 1957. The isolation and characterization of the polymers formed during the autoxidation of ethyl linoleate. J. Am. Oil Chem. Soc. 34, 470−473.

Wu, G.-S., Stein, R.A., Mead, J.F., 1977. Autoxidation of fatty acid monolayers adsorbed on silica gel: II. Rates and products. Lipids 12, 971−978.

Wu, G.-S., Stein, R.A., Mead, J.F., 1978. Autoxidation of fatty acid monolayers adsorbed on silica gel: III. Effects of saturated fatty acids and cholesterol. Lipids 13, 517−524.

Xu, G., Liu, Y., Sayre, L.M., 1999. Independent synthesis, solution behavior, and studies on the mechanism of formation of the primary amine-derived fluorophore representing cross-linking of proteins by (*E*)-4-hydroxy-2-nonenal. J. Org. Chem. 64, 5732−5745.

Yamamoto, Y., Niki, E., Kamiya, Y., Shimasaki, H., 1984. Oxidation of lipids. 7. Oxidation of phosphatidylcholines in homogeneous solution and in water dispersion. Biochim. Biophys. Acta 795, 332−340.

Yang, M.-H., 1993. Damage of DNA and its Constituents by Oxidizing Lipids. Rutgers University, New Brunswick, NJ.

Yang, M.-H., Schaich, K.M., 1996. Factors affecting DNA damage caused by lipid hydroperoxides and aldehydes. Free Radic. Biol. Med. 20, 225−236.

Yen, G.-C., Duh, P.-D., Tsai, H.-L., 2002. Antioxidant and pro-oxidant properties of ascorbic acid and gallic acid. Food Chem. 79, 307−313.

Yong, S.H., Karel, M., 1978. Reaction of histidine with methyl linoleate: characterization of the histidine degradation products. J. Am. Oil Chem. Soc. 55, 352−357.

Yong, S.H., Karel, M., 1979. Cleavage of the imidazole ring in histidyl residue analogs reacted with peroxidizing lipids. J. Food Sci. 22, 568−574.

Yoshida, K., Terao, J., Suzuki, T., Takama, K., 1991. Inhibitory effect of phosphatidylserine on iron-dependent lipid peroxidation. Biochem. Biophys. Res. Commun. 179, 1077−1081.

Younathon, M.T., Watts, B.M., 1959. Relation of meat pigments to lipide oxidation. Food Res. 24, 728−734.

Yuan, Q., Zhu, X., Sayre, L.M., 2007. Chemical nature of stochastic generation of protein-based carbonyls: metal-catalyzed oxidation versus modification by products of lipid oxidation. Chem. Res. Toxicol. 20, 129−139.

Zamora, R., Hidalgo, F.J., 1995. Linoleic acid oxidation in the presence of amino compounds produces pyrroles by carbonylamine reactions. Biochim. Biophys. Acta 1258, 319−327.

Zamora, R., Hidalgo, F.J., 2003. Phosphatidylethanolamine modification by oxidative stress product 4,5(*E*)-epoxy-2(*E*)-heptenal. Chem. Res. Toxicol. 16, 1632−1641.

Zamora, R., Hidalgo, F.J., 2005. 2-Alkylpyrrole formation from 4,5-epoxy-2-alkenals. Chem. Res. Toxicol. 18, 342−348.

Zamora, R., Alaiz, M., Hidalgo, F.J., 1999. Modification of histidine residues by 4,5-epoxy-2-alkenals. Chem. Res. Toxicol. 12, 654−660.

Zamora, R., Gallardo, E., Hidalgo, F.J., 2006. Amine degradation by 4,5-epoxy-2-decenal in model systems. J. Agric. Food Chem. 54, 2398−2404.

Zhang, W.-H., Liu, J., Xu, G., Yuan, Q., Sayre, L.M., 2003. Model studies on protein side chain modification by 4-oxo-2-nonenal. Chem. Res. Toxicol. 16, 512−523.

Chapter 12

Off-Flavors in Milk

Juan He,* Pedro Vazquez-Landaverde,† Michael C. Qian* and N. A. Michael Eskin**

*Department of Food Science and Technology, Oregon State University, Corvallis, Oregon, USA, †CICATA-IPN Unidad Queretaro, Colonia Colinas del Cimatario, Queretaro, Mexico, **Department of Human Nutritional Sciences, Faculty of Human Ecology, University of Manitoba, Winnipeg, Manitoba, Canada

I. INTRODUCTION

The nutritional benefits of cow's milk, described in Chapter 4, make it an important source of nutrients for infants and young adults in their daily diets. Besides milk fulfilling nutritional requirements, consumers enjoy the delicate flavor of milk and other milk-derived dairy products. The flavor of milk becomes a key parameter of product quality as acceptance is largely dependent on flavor (Drake et al., 2006). Fresh milk is a rather bland product: it has a pleasant, slightly sweet aroma and flavor, and a pleasant mouth-feel and aftertaste. Since fresh milk has a very delicate flavor, any off-balance of the flavor profile can emerge into 'off-flavor' which can be easily detected by the consumer. The flavor of milk is influenced by a variety of factors involved in milk production, including the genetics of the cow, the physical and physiological condition of the cow, the type of feed consumed by the cow, the environment around the cow and the milking area, and biological, chemical and enzymatic changes in milk during production and distribution (Franklin, 1951). The flavor composition of milk is complex; at least 400 volatile compounds have been reported in milk, covering a wide range of chemical classes including lactones, acids, esters, ketones, aldehydes, alcohols, furans, carbonyls, pyrazines, sulfur compounds, and aliphatic and aromatic hydrocarbons (Moio et al., 1994). The off-balance of these volatile compounds in milk, as well as the generation of some new off-flavor compounds, can cause off-flavor in milk. According to the Committee on Flavor Nomenclature and Reference Standards of the American Dairy Science Association, off-flavors in milk can be categorized into heated, light-induced, lipolyzed, microbial, oxidized, transmitted, and miscellaneous, as shown in Table 12.1 (Shipe et al., 1978). The most common off-flavor issues in dairy industry, and the possible methods to minimize or eliminate these off-flavors, will be addressed in this chapter.

Biochemistry of Foods. DOI: http://dx.doi.org/10.1016/B978-0-12-242352-9.00012-6

TABLE 12.1 Categories of Off-Flavors in Milk

Cause	Descriptive or Associated Terms
Heated	Cooked, caramelized, scorched
Light-induced	Light, sunlight, activated
Lipolyzed	Rancid, butyric, bitter, goaty[a]
Microbial	Acid, bitter, fruity, malty, putrid, unclean
Oxidized	Papery, cardboard, metallic, oily, fishy
Transmitted	Feed, weed, cowy, barny
Miscellaneous	Flat, chemical, foreign, lacks freshness, salty

[a]*Bitter flavor may arise from a number of different causes. If a specific cause is unknown it should be classified under miscellaneous.*
Reprinted from Shipe *et al.* (1978) with permission from Elsevier.

II. OFF-FLAVORS IN MILK

A. Lipolyzed Flavors

Lipolyzed flavors, one of the most common types of off-flavor in milk and dairy products, are produced by the enzymatic hydrolysis of milk fat triglycerides. This results in the accumulation of free fatty acids (FFAs) as major degradation products, mono glycerides and diglycerides, and possibly glycerol (Figure 12.1). At one time this type of flavor defect was described as 'rancid', which caused considerable confusion because of the term's association with lipid oxidation. This was eventually resolved by differentiating hydrolytic rancidity from oxidative rancidity, which more closely described 'oxidized flavors'. Heat-resistant lipases from psychrotrophic bacteria, predominantly *Pseudomonas* species, have been associated with lipolyzed flavors (Saxby, 1992).

FFAs can be derived from lipolysis, proteolysis, and lactose fermentation during cheese ripening. Both esterase and lipase have lipolytic activities, and can hydrolyze milk lipids to FFAs. Most of the FFAs with carbon chain

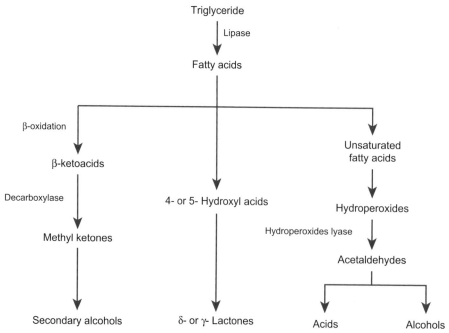

FIGURE 12.1   General pathways for the metabolism of milk triglycerides and fatty acids. (*From Singh* et al., *2003. With kind permission from Springer Science and Business Media.*)

lengths from C4 to C20 come from triglyceride hydrolysis by milk and microbial lipases during cheese aging. A lower proportion of FFAs with carbon chain length from C2 to C6 can come from lactose fermentation and amino acid degradation. Lactose fermentation produces acetic acid, propionic acid, and butanoic acid. Amino acid degradation, mainly catalytic deamination, can generate branched chain fatty acids such as isobutanoic and isovaleric acid (Kuzdzal-Savoie, 1980). A small amount of FFAs can also be generated from the oxidation of alcohols, aldehydes, ketones, and esters.

1. Methods for Determination of Free Fatty Acids

The determination of lipolysis is very important for evaluating milk quality. The degree of lipolysis can be measured by several methods. The acid degree value (ADV) has been widely used to monitor FFA liberation in milk, cream, and cheese (Richardson, 1985; Lin and Jeon, 1987; Ikins et al., 1988). However, the method measures only the total liberated FFA by titration. It does not provide information about the concentration of individual FFAs. The ability of ADV to accurately predict rancidity in milk was questioned by a number of researchers (Duncan et al., 1990, 1991) as milk with high ADV values was not always found to be rancid. This discrepancy was attributed to the difference in solubility of the fatty acids in the milk (Duncan et al., 1990). ADV methods only partially recover medium-chain fat-soluble fatty acids (C10–C16) as the shorter chain fatty acids implicated in rancid flavor (C4–C12) are more hydrophilic and remain in the aqueous phase of the milk. Thus, the determination of individual fatty acid concentrations is important, particularly short-chain FFAs (between C4:0 and C12:0), as they have strong, often undesirable flavors, and are the major contributors to lipolyzed flavor in milk.

Individual FFAs can be accurately determined by gas chromatography (GC) with or without derivatization. The determination of individual FFAs by GC consists of lipid extraction, FFA isolation, and quantification (Deeth et al., 1983). High extraction efficiency was achieved with acidified diethyl ether when the sample pH was brought to 1.5 with sulfuric acid (Needs et al., 1983). Before the individual FFAs can be analyzed by GC, the extracted FFAs need to be separated out from fat. The separation technique needs to provide quantitative separation of FFAs from fat and have no contamination from the fat.

Among many methods used to isolate FFAs, the alkaline arrest silicic acid column method was widely used in early work (Woo and Lindsay, 1982). In this method, both FFAs and lipids were dissolved in petroleum ether–diethyl ether (80:20), and the triglycerides were passed through the KOH–silicic acid column. The FFAs were trapped by the column as the potassium salts, and eluted from the column with 2% formic acid in diethyl ether. However, not only is the preparation of KOH–silicic acid column tedious, but also the column performance lacks reproducibility. The potential for channeling in the column can greatly reduce the separation efficiency. In addition, the prolonged fat contact with KOH column can greatly increase fat hydrolysis and cause variability in the analysis.

An alumina column has been used to separate FFA from triglyceride (Deeth et al., 1983; De Jong and Badings, 1990). The neutral lipids were eluted from a deactivated alumina column with diethyl ether–heptane (1:1), while FFAs were extracted with diethyl ether containing 3% formic acid. The recovery of FFA in cheese samples was high (De Jong and Badings, 1990).

Strong anion-exchange resins have also been used to isolate FFAs in milk (Needs et al., 1983; Spangelo et al., 1986). After the triglycerides were eluted out, the absorbed FFAs were methylated by mixing the dried resin with HCl–methanol. The methyl esters of fatty acids were extracted with diethyl ether and analyzed by GC. However, the prolonged contact with the strong basic resin can cause triglyceride hydrolysis (Needs et al., 1983). The isolated FFAs are typically methylated for GC analysis. The methylation of FFAs is catalyzed by acid or base.

FFAs can be methylated and analyzed without being separated from triglycerides. Dowex strong anion-exchange resin has been used as a heterogeneous catalyst to methylate FFAs directly in milk extract (Spangelo et al., 1986). The methylation was carried out in dimethylformamide, CH_3I, and pyridine at 40°C. The methyl esters were then extracted with hexane and analyzed by GC. The catalyst allows complete methylation of FFAs without potential hydrolysis of triglycerides and transesterification with other lipids. However, the reproducibility of butyric acid is very poor, owing to potential evaporation loss of methyl butyrate during methylation steps.

Tetramethylammonium hydroxide (TMAH) has been directly added to a lipid extract to convert FFAs to the tetramethylammonium soaps, the latter being transformed to methyl esters in the GC injector and analyzed by GC (Metcalfe and Wang, 1981; Martinez-Castro et al., 1986; Martin-Hernandez et al., 1988; Chavarri et al., 1997). The methyl esters from transesterification of triglycerides were in the top ether layer, while the FFA ammonium soaps stayed in the methanol layer. After pH adjustment, the methanol layer was injected into the GC and the pyrolysis methylation was carried out in the injection port of a gas chromatograph. The method does not require FFA isolation from triglyceride. However, the short-chain fatty acid esters from glycerol can be dissolved in the lower metholic phase and gave higher results (Chavarri *et al.*, 1997). In addition, the pyrolysis process generates excess large amounts of trimethylamine, which flashes off short-chain fatty acid esters from the splitter and alters the results (Martin-Hernandez *et al.*, 1988).

An aminopropyl weak anion-exchange column has been successfully used to isolate FFAs from lipid extract (De Jong and Badings, 1990; Chavarri *et al.*, 1997). The neutral lipids were rinsed from the column with chloroform−isopropanol (2:1), and then the FFAs were eluted with diethyl ether containing 2% formic acid. The FFAs were directly injected onto a gas chromatograph and separated by an FFAP column. This method is simple and quick, and nearly 100% recoveries were achieved with most FFAs. Hydrolysis of triglycerides and lactic acid contamination were not observed with this method. All fatty acids can be analyzed, with good repeatability. The method has also been used to analyze FFAs in cheeses (Qian and Reineccius, 2002)

Liquid chromatographic methods were developed from a protocol utilizing FFA derivatization (Garcia *et al.*, 1990). Recently, several new approaches have been reported that allow rapid quantitative analysis of short-chain FFAs. Capillary electrophoresis and indirect ultraviolet (UV) absorption (Vallejo-Cordoba *et al.*, 1998) were used for the quantification of FFAs in lipolyzed cream. Solid-phase microextraction (SPME) and GC were used to quantitatively determine short-chain FFAs in milk (Gonzalez-Cordova and Vallejo-Cordoba, 2001). In this method, FFA extraction consisted of placing 40 ml of milk containing 28% NaCl at pH 1.5 in a sealed vial and equilibrating for 30 minutes at 70°C. The FFAs were then extracted with an SPME fiber and thermally desorbed for GC analysis. Using multiple regression analysis, Gonzalez-Cordova and Vallejo-Cordoba (2003) reported a highly significant ($p < 0.001$) correlation coefficient (R^2) of 0.84 between their SPME and gas chromatographic method for detecting short-chain FFAs and rancidity scores determined by sensory evaluation in 19 commercial milks. Using this method they could detect and predict hydrolytic rancidity in milk based on the formation of short-chain fatty acids.

2. Sensory Properties of Free Fatty Acids

The sensory detection thresholds of FFAs have been investigated in both water and oil by several authors (Patton, 1964; Siek *et al.*, 1969; Urbach *et al.*, 1972; Brennand, 1989), and the reported values vary widely owing to different methods used in the studies (Table 12.2). Short-chain FFAs have higher sensory threshold values in water than in oil, possibly due to their higher solubility (and thus lower vapor pressure) in water. Milk is an emulsion or a colloid of butterfat globules, which can be considered a water-based fluid. Butanoic (C4), hexanoic (C6), and octanoic (C8) acids are often described as rancid, sweaty (body odor), goat-like, and generally unpleasant, while decanoic (C10) and dodecanoic (C12) acids are described as having soapy and waxy aromas. From butanoic acid (C4:0) to octanoic acid (C8:0), sensory thresholds in water generally increase with increasing chain length. In ordinary fresh milk, the concentrations of butanoic (C4), hexanoic (C6), octanoic (C8), decanoic (C10), dodecanoic (C12) acids are typically below their sensory thresholds (De Jong and Badings, 1990; Gonzalez-Cordova and Vallejo-Cordoba, 2001). Thus, these compounds may not contribute significantly to the overall aroma of ordinary fresh milk. However, high lipolytic activity in milk can increase the short-chain fatty acid concentration to extents that exceed their sensory thresholds, and cause lipolyzed off-flavor in milk.

B. Lipases in Milk

Bovine milk contains very high lipolytic activity, predominantly from β-type esterases. These include glycerol tricarboxyl esterases, aliphatic esterases, diesterases, and lipases (EC 3.1.1.3), with a pH optimum of 8−9. The majority of lipase activity is associated with casein, of which 70% is bound to micellar casein (Downey and Andrews 1966). This association is largely electrostatic as the enzyme can be released from the micelle by sodium chloride or heparin. The remainder of lipase is present as a soluble casein−enzyme complex in the milk serum (Hoynes and Downey, 1973; Anderson, 1982).

One particular lipase in milk is lipoprotein lipase (LPL) (EC 3.1.1.34), a glycoprotein with two N-linked oligosaccharides, which appear to be necessary for its activity (Egelrud and Olivecrona, 1972). It accounts for most, if not all, of the lipolytic activity in bovine milk. It exists as a homodimer with a molecular mass of approximately 100 kDa (Kinnunen *et al.*, 1976). LPL plays an important role in removing lipids from the blood to the mammary gland and its presence may be due to leakage from the tissue (Shirley *et al.*, 1973; Mendelson *et al.*, 1977). However, bovine milk LPL does not serve any known biological purpose in milk. LPL is relatively unstable to heat and acid, and would normally be deactivated in the stomach. High-temperature, short-time (HTST) pasteurization

TABLE 12.2 Aroma Attributes and Sensory Thresholds of the Volatile Free Fatty Acids

Compound	Aroma Attributes	Threshold (ppm)	Medium	References
Acetic acid	Vinegar, sour, pungent	22–100	Water	Patton (1964), Siek *et al.* (1971), Manning and Robinson (1973)
		0.12–7	Oil	Patton (1964), Siek *et al.* (1971), Urbach *et al.* (1972), Reiners and Grosch (1998)
Propanoic acid	Sour, pungent	20–40	Water	Amoore *et al.* (1968), Salo (1970)
Butyric acid	Rancid, cheesy, sharp	0.3–6.8	Water	Patton (1964), Amoore *et al.* (1968), Siek *et al.* (1971), Baldwin *et al.* (1973)
		0.14–3	Oil	Patton (1964), Siek *et al.* (1971), Urbach *et al.* (1972), Schieberle *et al.* (1993)
2-Methylpropanoic acid	Cheesy, rancid, caramel	0.05–8.1	Water	Salo (1970), Brennand (1989), Larsen and Poll (1992)
Pentanoic acid	Cheesy, sour, meaty, sweaty	1.1–6.5	Water	Amoore *et al.* (1968), Brennand (1989)
2-Methylbutanoic acid	Cheesy, sour, rancid, sweaty	0.07	Water	Brennand (1989)
		0.02	Oil	Reiners and Grosch (1998)
Hexanoic acid	Cheesy, goaty, sharp	0.29–27	Water	Amoore *et al.* (1968), Siek *et al.* (1971), Baldwin *et al.* (1973), Buttery (1993)
		2.5–10	Oil	Patton (1964), Siek *et al.* (1971), Urbach *et al.* (1972), Schieberle *et al.* (1993)
Heptanoic acid	Cheesy, goaty, rancid	0.28–10.4	Water	Amoore *et al.* (1968), Brennand (1989)
Octanoic acid	Cheesy, sweaty	3–19	Water	Patton (1964), Amoore *et al.* (1968), Baldwin *et al.* (1973), Buttery (1993)
		10–350	Oil	Patton (1964), Urbach *et al.* (1972)
Nonanoic acid	Fatty, green	2.4–8.8	Water	Amoore *et al.* (1968), Brennand (1989)
Decanoic acid	Soapy, waxy	1.4–10	Water	Patton (1964), Amoore *et al.* (1968), Baldwin *et al.* (1973), Buttery (1993)
		5–200	Oil	Patton (1964), Urbach *et al.* (1972)
Dodecanoic acid	Soapy, metallic	2.2–16	Water	Brennand (1989)
		700	Oil	Patton (1964), Siek *et al.* (1971), Urbach *et al.* (1972)

(72°C held for 15 seconds) can deactivate most of the enzymes in milk, hydrolyzing long- and short-chain triacylglycerols, partial glycerides, and phospholipids (Egelrud and Olivecrona, 1973; Scow and Egelrud, 1976).

Another lipase source in milk is from psychrotrophic bacteria such as *Pseudomonas*, which have a major effect on the lipolysis of milk and dairy products (Sørhaug and Stepaniak, 1997). These bacterial lipases have different characteristics from LPL. The major difference is their ability to pass through the milk fat globule membrane (MFGM) into the intact fat globules (Fitz-Gerald and Deeth, 1983). Another significant difference is that bacterial lipases are stable to HTST, and even to ultra-high-temperature (UHT) treatment (~140°C for 4 seconds) (Christen *et al.*, 1986).

1. Lipolysis of Milk

Intrinsic milk enzymes are present in sufficient amounts to cause extensive fat hydrolysis with concomitant flavor impairment (Herrington, 1954). The level of enzymes is not the critical factor in determining the susceptibility of milk to lipolysis, but rather the MFGM, which protects the micelle triacylglycerols from lipolytic attack. In freshly secreted milk, this biological membrane is intact and forms an effective barrier around the fat. However, this protection is reduced or completely eliminated in certain situations, such as physical damage to the membrane in raw milk. Lipolysis in milk can be broadly categorized into two types: spontaneous and induced. Spontaneous lipolysis is initiated by the simple act of cooling raw milk below 10°C soon after secretion. By contrast, induced lipolysis is initiated by physical damage to the MFGM, which allows lipase access to the fat substrate (Deeth, 2006). Both spontaneous and induced lipolysis progress during storage, mostly on the first day of refrigeration (Ouattara *et al.*, 2004).

2. Microbial Generated Off-Flavors

Milk is an ideal medium for microbial growth. Thus, it is particularly important to utilize the most thorough sanitization procedures and proper cooling and holding temperatures to optimize raw milk quality on the farm. Off-flavors generally develop in processed milk when the bacterial population rises to $\geq 10^7$ cfu/ml (Schroder *et al.*, 1982). Off-flavors develop in three stages: loss of freshness; increased perception of staleness; and generation of rancid, fruity, and bitter flavors. Bitter flavors usually accompany protein degradation. Soapy and rancid flavors are generally a result of lipid breakdown (Cousin, 1982). Most microbial spoilage and associated off-flavors are due to postpasteurization contamination, generally involving psychrotrophic bacteria. Psychrotrophs are bacteria capable of growth at temperatures at $\leq 7°C$ (44.6°F). Rapid cooling and refrigerated storage of raw milk have favored the growth of psychrotrophic bacteria. During cold storage these bacteria dominate the flora and produce extracellular enzymes (mainly proteases and lipases), which are the major contributors to dairy spoilage (Sørhaug and Stepaniak, 1997). For example, digestion of casein by proteases can cause a bitter flavor and the gelation of milk. Lipases hydrolyze milk fat to produce FFAs, which cause milk to taste rancid, bitter, unclean, and soapy. Lecithinase degrades MFGMs and increases the susceptibility of milk fat to the action of lipases. A better understanding of bacterial-induced spoilage is needed.

Psychrotrophic bacteria from numerous genera have been isolated from milk, both Gram-negative (e.g. *Pseudomonas, Aeromonas, Serratia, Acinetobacter, Alcaligenes, Achromobacter, Enterobacter*, and *Flavobacterium* spp.) and Gram-positive (e.g. *Bacillus* (Meer *et al.*, 1991), *Clostridium, Corynebacterium, Microbacterium, Micrococcus, Streptococcus, Staphylococcus*, and *Lactobacillus* spp.) (Champagne *et al.*, 1994; Shah, 1994). Of these, *Pseudomonas* is the most frequently reported psychrotroph in raw milk. The growth of *Pseudomonas* strains and their production of proteases were reported as the cause of the release of plasmin and plasminogen from the casein micelle into the whey fraction (Fajardo-Lira and Nielsen, 1998; Nielsen, 2002). Eight unique milk spoilage aromas were used to differentiate milk spoiled by Pseudomonas strains: rotten; barn; shrimpy; medicinal; fruity; cheesy; cooked; and overall spoilage aroma. *Ps. fragi* was confirmed to produce fruity aromas in milk, while *Ps. putida* produced fruity, cheesy, rotten, and barn aromas. Spoilage aroma characteristics were found to be influenced not only by species, but also by fat level and time (Hayes *et al.* 2002). This suggested that the extracellular enzyme activity patterns among *Pseudomonas* isolates appeared to be associated with ribotypes (Dogan and Boor, 2003). A potential early detection system of microbial spoilage, which utilizes an electronic nose unit and 14 conducting polymer sensors, has been studied with bacteria (*Pseudomonas aureofaciens, P. fluorescens, Bacillus cereus*) and yeasts (*Candida pseudotropicalis, Kluyveromyces lactis*) (Magan *et al.*, 2001).

3. Proteolyzed Off-Flavors

Proteolytic enzymes degrade proteins and release a range of nitrogenous compounds. Those proteases, which attack either casein or whey proteins, result in the coagulation of milk and production of bitter flavors. While bacteria may be killed by heat treatment, certain heat-resistant enzymes produced in raw milk by psychrotrophic bacteria may cause proteolytic and lipolytic action. Proteolysis can be measured as the increase in trichloroacetic-acid-soluble free amino groups, and subsequently determined by trinitrobenzene sulfonic acid (TNBS) via colorimetric analysis (Cogan, 1977). The relationship between proteolysis measured with TNBS and off-flavor development was investigated by McKellar (1981). The study showed that this method may be used as an indicator of shelf-life as proteolysis can be detected prior to off-flavor development. Proteolysis can also be measured by the determination of tyrosine value, which has been reported to be associated with high somatic cell counts in milk (Senyk *et al.*, 1985). The reported proteolytic-derived aromatic products include *p*-cresol, methional, phenethanol, phenylacetaldehyde, 3/2-methylbutanal, and 2-methylpropanal (Dunn and Lindsay, 1985). Off-flavor associated with proteolysis can be affected by proteolytic activity, via species differences (Stead, 1986) or environmental conditions (Matselis and Roussis, 1998).

C. Oxidized Off-Flavors

Lipid oxidation influences the quality of food products through flavor and taste deterioration, as well as reduction in nutritive value. The flavors produced by oxidation of dairy products have been described as oxidized, cardboard-like, beany, green, metallic, oily, fishy, bitter, fruity, soapy, painty, rancid, grassy, buttery, and tallow-like. 'Oxidized flavor' was recommended as the generic term to describe all of these flavors (Shipe *et al.*, 1978). Lipid oxidation typically involves the reaction of molecular oxygen with unsaturated fatty acids via a free radical mechanism or light-induced oxidation, influenced by factors such as degree of unsaturation of fatty acids, presence of transition metal ions, and antioxidant tocopherols and carotenoids. The initial products of lipid oxidation, lipid hydroperoxides, are quite unstable; they can break down rapidly to produce short-chain volatile compounds such as hydrocarbons, acids, alcohols, aldehydes, and ketones, which elicit undesirable flavors.

Sensitivity to oxidation can be monitored also by measuring the antioxidative capacity, because oxidation can only occur in the case of imbalance between reactive oxidants and the antioxidant defense (Halliwell, 1996). A review by Antolovich *et al.* (2002) gave a comprehensive summary on the method for measuring the oxidative stability on antioxidative capacity, such as peroxide value, diene conjugation, thiobarbituric acid reactive substances, hexanal formation, total radical trapping antioxidant parameter, and electron spin resonance (ESR) spin-trap test. Recently, the ferric reducing antioxidant power (FRAP) and diphenyl picryl hydrazyl (DPPH) methods have been used to measure antioxidative capacity in order to monitor oxidation sensitivity in milk (Smet *et al.*, 2008). The methods showed higher sensitivity to oxidation than the more conventional peroxide value method during the first hours and days of storage.

Polyunsaturated acids, including oleic, linoleic, linolenic, and arachidonic acid are major precursors for the formation of aldehyde compounds owing to their prevalence in milk products. Table 12.3 lists the possible origins of aldehydes produced by autoxidation. In the autoxidation process, oxygen reacts with a methylene group adjacent to a double bond under catalysis of trace metals (such as copper) and enzymes (Forss, 1979), leading to the formation of hydroperoxides. These further decompose to straight-chain aldehydes.

Removal of oxygen can effectively prevent autoxidation. Flushing the sample headspace with nitrogen reduces lipid oxidation. Oxygen-scavenging packaging can lower the dissolved oxygen content in UHT milk, and thus reduce the formation of stale flavor volatiles, including methyl ketones and aldehydes, compared with milk packaged without oxygen-scavenging film (Perkins *et al.*, 2007).

Autoxidation also generates methyl ketones and hydrocarbons, although the majority of methylketones are generated through thermal breakdown of fatty acids. Methyl ketones with an odd number of carbon atoms (C7, C9, C11, C12), such as 2-pentanone, 2-heptanone, 2-nonanone, and 2-undecanone, are major ketones found in milk. Such are formed during heat treatment from the oxidation of FFAs into β-ketoacids, and the subsequent decarboxylation of these into methyl ketones (Moio *et al.*, 1993). A proposed pathway for the formation of ketones is presented in Figure 12.1, with their thresholds listed in Table 12.4.

D. Fishy Off-Flavors

The development of a fishy off-flavor, reminiscent of rotting fish, was reported in bulk milk from Red and White dairy breeds in Sweden (Lunden *et al.*, 2002a). This phenomenon, often confused with oxidized flavor, and was incorrectly thought to be associated with the formation of trimethylamine (TMA) oxide. The connection between fish odor and TMA is well established, with the olfactory threshold for detection being around 1−2 ppm (Mehta *et al.*, 1974;

TABLE 12.3 Possible Origins of Aldehyde Obtained from Specific Unsaturated Fatty Acids

Unsaturated Fatty Acid	Aldehyde Obtained
Oleic acid	Octanal, nonanal, decanal, 2-decenal, 2-undecenal
Linoleic acid	Hexanal, 2-octenal, 3-nonenal, 2,4-decadienal
Linolenic acid	Propanal, 3-hexenal, 2,4-heptadienal, 3,6-nonadienal, 2,4,7-decatrienal
Arachidonic acid	Hexanal, 2-octenal, 3-nonenal, 2,4-decadienal, 2,5-undecadienal, 2,5,8-tridecatrienal

TABLE 12.4 Aroma Attributes and Sensory Thresholds of Some Volatile Ketones

Compound	Aroma Attributes	Threshold (ppm)	Medium	References
Acetone	Acetone-like, pungent	500	Water	Manning and Robinson (1973)
		125	Oil	Siek et al. (1969)
2-Butanone	Acetone-like	50	Water	Wick (1966)
		30	Oil	Siek et al. (1969)
2-Pentanone	Floral, fruity, wine, acetone-like	2.3	Water	Siek et al. (1971)
		61	Butter	Siek et al. (1969)
2-Hexanone	Floral, fruity	0.93	Water	Siek et al. (1971)
2-Heptanone	Blue cheese, fruity, sweet	0.14	Water	Buttery et al. (1988)
		1.5—15	Butter	Siek et al. (1969), Preininger and Grosch (1994)
2-Octanone	Fruity, musty, unripe apple, green	2.5—3.4	Butter	Siek et al. (1969)
2-Nonanone	Fruity, musty, rose, tea-like	0.2	Water	Buttery et al. (1988)
		7.7	Cheese	Siek et al. (1969)
2-Decanone	Fruity, musty	0.19	Water	Siek et al. (1971)
2-Undecanone	Floral, herbaceous, fruity	0.007—5.4	Water	Karahadian et al. (1985), Buttery et al. (1988)
		3.4	Oil	Kubicková and Grosch (1998)

von Gunten et al., 1976). TMA is oxidized by the liver enzyme, flavin-containing monooxygenase (FMO), to TMA oxide, however, it is tasteless and colorless (Hlavica and Kehl, 1977). It is the impaired oxidation of TMA, however, that results in the fishy odor phenomenon (Pearson et al., 1979; Spellacy et al., 1979). Feeding wheat pasture was particularly associated with the development of this fish odor/flavor problem in milk (Mehta et al., 1974; von Gunten et al., 1976; Kim et al., 1980). Using dynamic headspace GC, Lunden et al. (2002a) showed that milk samples with a fish taint had > 1 mg TMA/kg of milk compared to normal milk in which TMA was not detected (Figure 12.2). There appeared to be a dose-dependent relationship between TMA levels and the development of a fishy off-flavor score.

Fishy odor or trimethylaminuria is an autosomal recessive inborn error of metabolism in humans in which there is an abnormal secretion of TMA in breath, urine, sweat, saliva, and vaginal secretions. This phenomenon appears to be due to impaired oxidation of TMA resulting from loss-of-function mutations in the *FMO3* gene encoding the isoform of flavin-containing monooxygenase (Dolphin et al., 1997; Treacy et al., 1998; Ackerman et al., 1999; Basarab et al., 1999; Forrest et al., 2001). Lunden et al. (2002b) showed that this phenomenon in cow's milk was due to the nonsense mutation (R238X) in bovine *FM03* gene ortholog. The R238X substitution was not found in Swedish Holstein, Polled, or Jersey cows, but was surprisingly common in the Swedish Red and White breeds.

E. Light-Induced Off-Flavors

The primary factor responsible for the development of light-induced—oxidized flavor in milk was shown by Aurand et al. (1966) to be riboflavin, with ascorbic acid playing a secondary role. Riboflavin acts as a photosensitizer in milk by accelerating the oxidation of amino acids, DNAs, and unsaturated fatty acids (Choe et al., 2005). The use of fluorescent lights to illuminate dairy display cases is responsible for the flavor deterioration and loss of nutrient quality in milk (Bradley, 1980; Dimick, 1982; Hoskin and Dimick, 1979; Sattar and deMan, 1975). In particular, the effects of 'white' fluorescent light, in widespread use in supermarkets, with a spectral output of 350—750 nm and peaks at 470 and 600 nm, are shown in Figure 12.3. The radiant energy emitted by the fluorescent light is absorbed by and interacts with milk components such as riboflavin (Dunkley et al., 1962). When exposed to light, riboflavin forms singlet oxygen and superoxide anions from triplet oxygen (Jernigan, 1985; Bradley and Min, 1992;

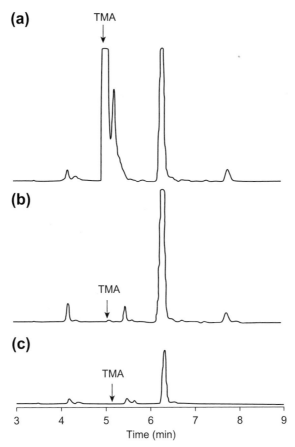

FIGURE 12.2 **Gas chromatography–flame ionization detection chromatogram of: (a) a milk sample with strong fishy off-flavor; (b) normal Swedish milk; and (c) Swiss UHT reference milk.** Sample treatment using a purge-and-trap system (dynamic headspace analysis) and gas chromatography with flame ionization detection. TMA: trimethylamine *(Lunden et al., 2002a).*

Naseem *et al.*, 1993). The mechanism of activation has been extensively studied and involves several mechanisms, referred to as type I and type II (de la Rochette *et al.*, 2003). Excitation by light leads to the formation of a riboflavin triplet active state, a diradical (Choe *et al.*, 2005). The type I mechanism involves the formation of free radicals by hydrogen or electron transfer between the riboflavin triplet activated state (^{1}RF*) and substrates (Edwards and

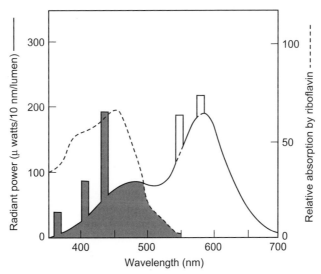

FIGURE 12.3 **Emission spectra of a cool white fluorescent lamp compared with absorption of riboflavin** *(Dunkley* et al., *1962).*

Silva, 2001). Type II involves the formation of singlet oxygen ($^1O_2^*$) by energy transfer from $^1RF^*$ to molecular oxygen (Boff and Min, 2002). Both of these mechanisms are outlined in Figure 12.4 (Choe *et al.*, 2005).

Riboflavin radicals are very strong oxidizing species, so that light-induced off-flavor (LIOF) development in milk is dependent on the availability of oxygen and ultraviolet light. LIOFs are created through photosensitization of lipids and amino acid sulfur groups, dependent on the wavelength and intensity of the light, exposure time, product temperature, and light-transmission properties of the container. The two major distinctive off-flavors in milk induced by irradiation of light energy are sunlight and cardboard flavor. Sunlight flavor refers to a burnt and oxidized odor in milk after light exposure for more than 2 days. Methionine was first implicated in the formation of LIOF by Patton and Josephson (1953) by the formation of dimethyl disulfide (DMDS) and methional. The mechanism involved in the formation of LIOF is somewhat controversial. Jung *et al.* (1998) reported that singlet oxygen, formed from triplet oxygen under sunlight in the presence of riboflavin in milk, reacts with methional, forming the hydroperoxide. Figure 12.5 shows that the hydroperoxide decomposes to form methional and thiomethyl radicals, with the latter producing dimethyl sulfide (DMS) (Choe *et al.*, 2005). The formation of DMDS was highly correlated with the LIOF sensory scores (Jung *et al.*, 1998). Because ascorbic acid is a good quencher of singlet oxygen, its presence reduces the formation of DMDS. Cardboard-like or metallic flavors, which develop in milk through prolonged light exposure, are caused by secondary lipid oxidation products including aldehydes, ketones, alcohols, and hydrocarbons

FIGURE 12.4   **Photosensitization of riboflavin and type I and type II mechanisms** *(Choe* et al.*, 2005).*

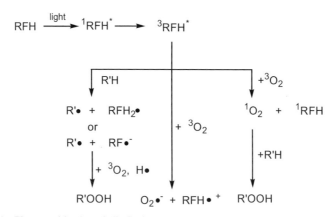

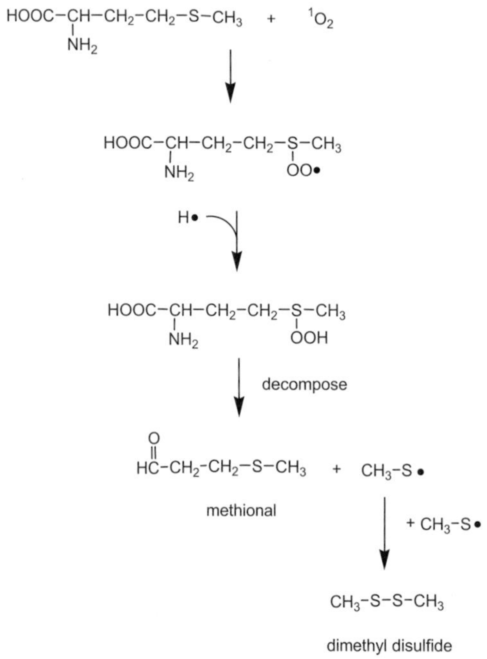

FIGURE 12.5   **Oxidation of methionine by singlet oxygen** *(Choe* et al.*, 2005).*

(Gaafar and Gaber 1992). Dynamic headspace analysis has been used to evaluate the volatile compounds hexanal, pentanal, DMDS, 2-butanone, and 2-propanol (Kim and Morr, 1996).

Packaging can directly prevent the development of LIOF by protecting the product from both light and oxygen. All plastic containers, such as polyethylene terephthalate (PET) and high-density polyethylene (HDPE), can provide very good convenience through easy opening and reclosing, minimizing recontamination. Pigmented PET bottles have excellent mechanical properties and offer good oxygen and light protection. Pigmented HDPE bottles, both monolayer and multilayer, with a greater thickness than current PET, are more suitable for the fresh milk packaging market (Cladman *et al.*, 1998). Chemical and sensorial aspects were studied in low-fat pasteurized milk bottled by various packaging materials (Moyssiadi *et al.* 2004). Multilayer TiO_2-pigmented HDPE, monolayer HDPE, clear PET, and pigmented PET were compared with paperboard cartons for a period of 7 days. The result showed that all packaging materials with regard to microbiological and chemical parameters provided good protection of milk flavor over the test period. The multilayer provided the best overall protection for the product, followed by the monolayer HDPE bottle. The degradation of volatiles by chemical reactions could explain these results, such as oxidation caused by oxygen diffusion and light transmission into the packaging. In the authors' unpublished work, samples of 2% pasteurized milk were packaged in three different treated bottles: translucent HDPE bottles stored in the dark, translucent 7HDPE bottles stored under fluorescent light, and HDPE bottles coated with light-blocking pigment stored under fluorescent light, in order to determine the cause of oxidation. Dynamic headspace analysis clearly demonstrated that translucent HDPE bottled milk stored under fluorescent light had much higher concentrations of hexanal, heptanal, and octanal than the milk stored under dark or in UV-blocking bottles (authors' unpublished data). The hexanal level in translucent HDPE bottled milk was five times higher than in those stored in dark or light-blocking bottles (Figure 12.6). This trend was also observed in cream cheese. The surface of cream cheese had much higher aldehyde formation than the center of the cream cheese (Figure 12.7) owing to poor packaging protection (authors' unpublished data). A similar study was conducted to assess the sensory differences in milk packaged with different materials (Boccacci Mariani *et al.*, 2006). No off-flavor was found in milk packaged in the paperboard during the storage period studied; however, a taint off-flavor was found in the PET-bottled milk (due to light-induced oxidative changes) after 1–2 days of storage (as assessed by the trained panel) and 2–3 days of storage (as assessed

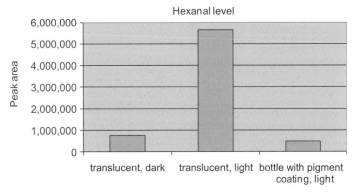

FIGURE 12.6 Hexanal level in milk stored in different bottles *(Qian* et al., *unpublished data).*

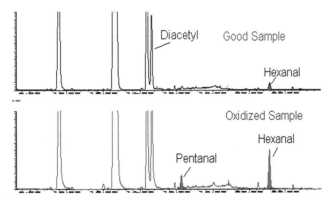

FIGURE 12.7 Gas chromatogram showing the oxidation of cream cheese *(Qian* et al., *unpublished data).*

by consumers). In addition to offering microbial and light protection, the packaging material can absorb aroma or off-aroma from the food matrix; this scalping property may help to maintain product quality, depending on the product. Polyethylene packaging has much stronger adsorption to lactones, aldehydes, and FFAs in UHT milk compared with glass bottles (Czerny and Schieberle, 2007).

Many consumers, however, prefer packaging material that allows them to see the food they are purchasing (Sattar *et al.*, 1976; Rosenthal, 1992; Cladman *et al.*, 1998; Doyle, 2004). Therefore, milk is still packaged in HDPE or PET containers which transmit between 62% and 85% of light with wavelengths between 300 and 700 nm. UV absorbers, such as iridescent films, can be added to polymer packaging materials to block UV wavelengths without affecting the clarity of the packaging material. Webster *et al.* (2009) demonstrated the ability of these films to protect against the development of LIOF in 2% milk exposed to visible excitation wavelengths of riboflavin (400, 446, and 570 nm). It was evident from their research, however, that blocking the transmission of all riboflavin excitation wavelengths was insufficient to completely protect against the development of LIOF. This suggested the presence of other components in milk, such as chlorins and porphyrins, identified previously by Wold *et al.* (2005) in the photooxidation of cheese.

F. Heat-Induced Off-Flavors

Thermal treatment can destroy spoilage bacteria and deactivate enzymes, increasing the shelf-life of milk. The most popular thermal treatments, pasteurization and UHT, develop new volatile compounds which bring desirable and undesirable flavors, affecting taste and texture. The main volatile compounds contributing to heated flavor are Maillard reaction products. Thermal degradation of lipids generates stale or oxidized flavor during storage. Thermal reactions involving amino acid side-chains generate sulfur compounds which are responsible for off-flavor in UHT milk. Other thermal-induced reactions such as hydrolysis of peptide bonds, dephosphorylation of proteins, and the interaction of lipid oxidation and Maillard reaction, may also affect the flavor to some extent. Heat treatments, particularly UHT processing, can promote the development of thermally derived off-flavor compounds such as aldehydes, methyl ketones, and various sulfur compounds (Scanlan *et al.*, 1968; Jeon *et al.*, 1978; Moio *et al.*, 1994; Contarini *et al.*, 1997; Contarini and Povolo 2002).

Contarini and Povolo (2002) studied the effect of heat treatments on volatile compounds in commercially processed milk samples using headspace SPME and GC. They identified 11 compounds, five of which (2-pentanone, 2-heptanone, 2-nonanone, benzaldehyde, and 2-undecanone) exhibited a correlation with the severity of the heat treatment. Vazquez-Landaverde *et al.* (2005) quantified some volatile flavor compounds in milk subjected to different heat treatments. Concentrations of ketones were not different in raw and pasteurized milk samples; however, their concentrations were markedly higher in UHT milk. At the same fat level, UHT milk contained approximately 12 times the amount present in raw and pasteurized milk. The major contributors were 2-heptanone and 2-nonanone, followed by 2,3-butanedione, 2-pentanone, and 2-undecanone. The concentrations of 2-heptanone and 2-nonanone in UHT milk were 34 and 52 times higher, respectively, than in raw and pasteurized samples. The odor activity values (OAVs; ratio of concentration to sensory threshold) for 2-heptanone and 2-nonanone were less than 1 in raw and pasteurized milk, indicating that they were not important aroma contributors. However, the OAVs in UHT milk were in the range of 4–10, suggesting that these compounds could be very important contributors to the aroma of heated milk. Contarini *et al.* (1997) and Contarini and Povolo (2002) reported that the concentrations of 2-pentanone, 2-hexanone, 2-heptanone, 2-nonanone, and 2-undecanone increased in direct proportion to the severity of the heat treatment and were associated with the development of stale-heated flavor in UHT milk. Moio *et al.* (1994) identified 2-heptanone and 2-nonanone as the most abundant volatile flavor compounds in UHT milk.

Although methyl ketones are naturally present in raw milk, they can be formed during heat treatment by β-oxidation of saturated fatty acids followed by decarboxylation (Nawar, 1996). Milk fat contains 10% (w/w) of C6, C8, C10, and C12 fatty acids, which are precursors for odd-carbon-numbered C5, C7, C9, and C11 methyl ketones produced during heat treatment. Fat content seems to have an impact on the concentration of methyl ketones in UHT milk, where their concentration in 3% fat milk was found to be almost double that found in 1% fat milk (Vazquez-Landaverde *et al.*, 2005). Methyl ketones can also be formed through direct decarboxylation of β-ketoacids present in raw milk. Milk fat contains approximately 1% lipids in which oxo fatty acids of various chain lengths are esterified to glycerol. These oxo fatty acids can be liberated as β-ketoacids and decarboxylated to C6–C16 methyl ketones when the fat is heated in the presence of water (Grosch, 1982; Jensen *et al.*, 1995).

Vazquez-Landaverde *et al.* (2005) found that the concentration of 2,3-butanedione in UHT milk was higher than in raw milk, while its concentration varied widely in pasteurized milk. The OAV was higher than 1 for UHT and pasteurized milk, suggesting that 2,3-butanedione is contributing to the aroma of heated milk. 2,3-Butanedione has been reported as a very important flavorant contributing to the rich 'heated' note in UHT milk, giving a buttery, pastry-like aroma (Scanlan *et al.*, 1968). Although its formation has been suggested to be heat induced (Scanlan *et al.*, 1968), it is also attributed to microbial activity in milk (Badis *et al.*, 2004), therefore being an ambiguous indicator for the heat treatment.

Aldehyde concentrations are also affected by heat treatment. The aroma contributors and sensory thresholds are summarized in Table 12.5. All these volatiles are associated with green, grass-like odors; the shorter chain aldehydes are also considered to be pungent and malt-like, and the longer chain aldehydes have more fatty notes. According to Vazquez-Landaverde *et al.* (2005), UHT milk had higher concentrations of total aldehydes than raw and pasteurized milk. Hexanal, octanal, and nonanal showed a higher concentration in 3% fat UHT milk while 2-methylpropanal, 3-methylbutanal, 2-furaldehyde, heptanal, and decanal concentrations were higher for both 1% and 3% fat UHT milk. The total aldehyde concentration was not different between raw and pasteurized milk. Based on

TABLE 12.5 Aroma Attributes and Sensory Thresholds of Some Volatile Aldehydes

Compound	Aroma Attributes	Threshold (ppm)	Medium	References
Acetaldehyde	Pungent, fruity penetrating	0.025	Water	Guth and Grosch (1994)
		0.0002	Oil	Buttery *et al.* (1995)
Propanal	Pungent, acrid, solvent	0.037	Water	Boelens and Van Gemert (1987)
		0.009	Oil	Reiners and Grosch (1998)
2-Methylpropanal	Malty, cocoa, green, pungent	0.002	Water	Amoore *et al.* (1976)
2-Methylbutanal	Cocoa, coffee, almond, malty	0.002–0.14	Oil	Guadagni *et al.* (1972), Reiners and Grosch (1998)
3-Methylbutanal	Malty, cocoa	0.013	Oil	Guadagni *et al.* (1972), Kubicková and Grosch (1998)
Butanal	Pungent, malty, green	0.018	Water	Boelens and Van Gemert (1987)
Pentanal	Malty, apple, green	0.012–0.07	Water	Siek *et al.* (1971), Buttery *et al.* (1988)
Hexanal	Grassy, green, tallow	0.009–0.05	Water	Ahmed *et al.* (1978), Larsen and Poll (1992)
		0.19–0.3	Oil	Siek *et al.* (1971), Guth and Grosch (1990)
(E)-2-Hexenal	Green, sweet, leafy, apple	0.017–0.05	Water	Ahmed *et al.* (1978), Buttery *et al.* (1988), Larsen and Poll (1992)
		0.42	Oil	Reiners and Grosch (1998)
Heptanal	Fatty, green, woody, fruity	0.031–0.25	Oil	Siek *et al.* (1971), Guadagni *et al.* (1972)
Octanal	Fatty, citrus	0.001	Water	Ahmed *et al.* (1978), Buttery *et al.* (1988)
		0.056	Oil	Reiners and Grosch (1998)
Nonanal	Citrus, green, fatty, floral	0.002	Water	Ahmed *et al.* (1978)
		1	Oil	Siek *et al.* (1969)
Decanal	Waxy, floral, citrus	0.002	Water	Ahmed *et al.* (1978), Boelens and Van Gemert (1987)
Dodecanal	Citrus, powerful	0.0005–0.002	Water	Ahmed *et al.* (1978), Boelens and Van Gemert (1987)
Furfural	Sweet, almond, penetrating	3	Water	Guadagni *et al.* (1972)
Phenyl acetaldehyde	Floral, hyacinth, green	0.002	Water	Whetstine *et al.* (2005)

their OAVs, nonanal and decanal appeared to be important contributors to the aroma of raw, pasteurized and UHT milk, while octanal, hexanal, 2-methylbutanal, 3-methylbutanal, and 2-methylpropanal were important only for UHT milk aroma. Contarini and Povolo (2002) found that 3-methylbutanal increased with the heat treatment severity, while hexanal and heptanal did not. The presence of 2-methylpropanal, 2-methylbutanal, and 3-methyl-butanal in heated milk is due to the Strecker degradation of amino acids during Maillard reactions (Damodaran, 1996). Hexanal, heptanal, octanal, nonanal, and decanal result from the autoxidation of unsaturated fatty acids (C18:1 and C18:2) and also the spontaneous decomposition of hydroperoxides promoted by heat (Grosch, 1982). Hexanal can also be transferred to milk from cow's feed (Scanlan et al., 1968) or originate from light-induced lipid oxidation (Marsili, 1999). Rerkrai et al. (1987) stated that the increase in C2 and C7−9 saturated aldehyde concentration is the main cause of the stale flavor in UHT milk, due to their low flavor thresholds. 2-Furaldehyde has been found in UHT milk (Vazquez-Landaverde et al., 2005), but the OAV was too low to be considered an important contributor to milk aroma. However, it is considered a good indicator of the heat treatment because it is the precursor of melanoidins in Maillard reactions between sugars and the free amino group of milk proteins or amino acids (BeMiller and Whistler 1996).

Ethyl acetate has been found to increase in concentration up to 10 times in UHT milk compared to raw milk (Vazquez-Landaverde et al., 2005). It has been reported that ethyl acetate is formed by esterification of ethanol and acetic acid via the Fischer reaction, catalyzed by heat (Hart, 1991). However, its very low OAV in the samples analyzed suggest that this compound is not an important contributor to the aroma of milk.

Thermal processing can generate a cooked, sulfurous, cabbage-like off-flavor in milk (Shipe, 1980). Researchers have identified that hydrogen sulfide (H_2S), methanethiol (MeSH), carbon disulfide (CS_2), dimethyl sulfide (DMS), dimethyl disulfide (DMDS), and dimethyl trisulfide (DMTS) are related to this cooked off-flavor defect (Shipe, 1980; Christensen and Reineccius, 1992; Simon and Hansen, 2001; Datta et al., 2002). Several other sulfur compounds, including benzothiazole, dimethyl sulfoxide (DMSO), dimethyl sulfone (Me_2SO_2), carbonyl sulfide (COS), sulfur dioxide, butanethiol, and thiophene, have been found in heated milk, but their roles in milk flavor have not been well elucidated (Scanlan et al., 1968; Shibamoto et al., 1980; Shipe, 1980; Moio et al., 1994).

The concentration of sulfur compounds in milk is related to the severity of heat treatment of the milk. Vazquez-Landaverde et al. (2006a) found that UHT milk contained significantly higher concentrations of H_2S, MeSH, CS_2, DMTS, and DMSO than raw and pasteurized milk. H_2S was the sulfur compound with the highest increase in concentration, up to seven times. MeSH increased about five times, and the DMSO concentration increased almost three times more in UHT than in raw milk (Vazquez-Landaverde et al., 2005).

Sulfur compounds have a very low sensory threshold. Calculated OAVs suggest that many of the sulfur compounds are very important contributors to the flavor of both heated and fresh milk. OAV values for MeSH and DMTS in UHT milk samples were much greater than 1, thus contributing to the aroma. According to the magnitude of its OAVs, MeSH could be the most important sulfur-containing contributor to the aroma of UHT milk, in which its concentration was 80−119 times higher than its reported threshold. It has a strong and unpleasant cabbage, sulfur-like aroma (Rychlik et al., 1998). A correlation has been found between the increase in concentration of this compound and the increase in 'cooked' flavor defect due to the heat treatment of milk (Badings et al., 1981; Christensen and Reineccius, 1992; Simon and Hansen, 2001). MeSH is thought to be liberated during heat treatment from methionine, by breakdown of the sulfur-bearing side-chain (Damodaran, 1996), but the actual pathway has not been well elucidated. Despite the importance of MeSH to the flavor of milk and dairy products, its study has been limited owing to its high reactivity and volatility. Only a few researchers have reported reliable quantification techniques for this compound (Burbank and Qian, 2005; Fang and Qian, 2005).

The concentration of H_2S in milk increases linearly with the intensity of heating (Hutton and Patton, 1952; Christensen and Reineccius, 1992). In addition, the log of the concentration of H_2S has a strong linear relationship with the heated flavor intensity of milk (Badings, 1978). This compound is also indirectly respon-sible for the formation of other sulfur compounds (Zheng and Ho, 1994). H_2S is produced mainly from temperature-activated sulfhydryl groups of sulfur-containing amino acids (cysteine) in β-lactoglobulin (Badings et al., 1981; Damodaran, 1996), which are then oxidized, forming H_2S. It has been suggested that H_2S could be the most important contributor to the 'cooked' flavor of milk because it is the major sulfur compound formed in heated milks (Badings et al., 1978; Jaddou et al., 1978; Badings et al., 1981; Rerkrai et al., 1987; Christensen and Reineccius, 1992) and it also has a characteristic sulfur-like aroma (Rychlik et al., 1998). However, Vazquez-Landaverde et al. (2006a) showed that H_2S concentration in UHT milk was only slightly higher than its reported threshold, and calculated OAV values indicate that H_2S could be less important to the aroma of heated milk than previously thought.

Although DMS is present naturally in raw milk (Toso *et al.*, 2002), it can also be formed from the sulfhydryl group of milk proteins subjected to thermal denaturation (Datta *et al.*, 2002). The formation of DMDS is probably due to the oxidation of MeSH (Ferreti, 1973; Chin and Lindsay, 1994). Jaddou *et al.* (1978) reported that DMDS concentration increased in UHT milk, but decreased in sterilized samples. DMDS has a sulfur- and cabbage-like aroma (Rychlik *et al.*, 1998), with low threshold values. DMTS has a low concentration in milk, but it also has a very low sensory threshold, indicating that this compound could also be a contributor to the sulfurous aroma in heated milk, although it may not be as important as MeSH.

Although its origin has not been well elucidated, carbon disulfide (CS_2) has been identified as a product of the breakdown of other sulfur compounds (Urbach, 1993). Since there is correlation between CS_2 and heat treatments (Vazquez-Landaverde *et al.*, 2006a), this compound could be a good indicator of the heat treatment. It has a sweet, ethereal, slightly green, sulfur-like aroma (Rychlik *et al.*, 1998). Because of its high sensory threshold, it probably will not contribute to milk flavor.

It has been proposed that dimethyl sulfone ($DMSO_2$) is produced in milk by the heat-induced oxidation of DMS via DMSO as the intermediate (Shibamoto *et al.*, 1980). Under oxidant conditions, methionine is easily oxidized to methionine sulfoxide and methionine sulfone, which eventually breakdown to yield DMSO and $DMSO_2$, respectively (Damodaran, 1996). Shibamoto *et al.* (1980) found that $DMSO_2$ concentration decreases when milk is subjected to treatments between 60 and 90°C, but it starts to increase considerably at temperatures above 90°C. Moio *et al.* (1994) found that $DMSO_2$ concentration was lower for UHT milk than that for raw and pasteurized samples. $DMSO_2$ has an aroma defined as being like hot milk, leather, and bovine sweat (Rychlik *et al.*, 1998).

A general trend has been observed that the concentration of H_2S, MeSH, and DMTS in heated milk increases with the fat content in milk (Vazquez-Landaverde *et al.*, 2006a), suggesting that the heat-induced formation of sulfur compounds in milk is affected by fat level, but the mechanism is not clear. It was proposed by de Koning *et al.* (1990) that the membrane proteins of the milk fat globules contributed to the formation of sulfides.

Lactones could be important contributors to the flavor of heated milk. Lactones are cyclic esters that usually have pronounced fruity aromas associated with peaches, apricots, and coconut (Table 12.6). From γ-hexalactone to γ-dodecalactone, or δ-hexalactone to δ-dodecalactone, their sensory thresholds in water generally decrease with increasing chain length. Lactones can be formed in the ruminant mammary gland from the hydrolysis of saturated fatty acids and subsequent cyclization of free hydroxyacids (Dumont and Adda, 1978), and are therefore present in very small amounts in fresh, unheated milk. These compounds can be formed during heat treatments from the thermal breakdown of γ- and δ-hydroxyacids through intramolecular esterification of hydroxyacids, where the loss of water results in ring formation (Fox *et al.*, 2000). Recent work by the authors of this chapter demonstrated that lactone concentrations are much higher in UHT milk than in pasteurized milk at the same fat content (Figure 12.8).

G. Non-Thermal Processing and Off-Flavor Formation

New processing technology is needed to increase the shelf-life of milk without compromising its natural flavor. Thermal processing is the prevailing method to achieve microbial safety and shelf-life stability of milk. Although HTST pasteurization of milk (typically at 72°C for 15 seconds) is acceptable to most consumers, the process does impart a slight cooked, sulfurous note, and the final product shelf-life is only 20 days at refrigeration temperatures. UHT processing (135−150°C for 3−5 seconds) produces a product that is stable at room temperature for up to 6 months; however, this process can induce strong 'cooked' off-aroma notes in milk (Shipe, 1980), thus limiting its marketing in the USA and many other countries (Steely, 1994).

Promising non-thermal methods including membrane filtration, high-pressure processing (HPP), and pulsed electric field treatment are used to achieve a microbial shelf-life similar to that of UHT milk while minimizing the generation of off-flavor compounds. To retain the 'fresh' milk flavor, HPP has been studied as an alternative to the pasteurization of milk. A similar microbiological reduction to that of pasteurized milk has been achieved using pressure treatments of 400 MPa for 15 minutes or 500 MPa for 3 minutes at room temperature (Rademacher and Kessler, 1996). At moderate temperature (55°C), HPP (586 MPa for 5 minutes) can significantly extend the shelf-life of milk to 45 days, which is beyond that of pasteurized milk (Tovar-Hernandez *et al.*, 2005). Although it is generally assumed that HPP at low temperature will not change the aroma or flavor of the product (Cheftel, 1995; Berlin *et al.*, 1999; Velazquez *et al.*, 2002), HPP under certain conditions has been reported to change the concentration of some important flavor compounds. Hofmann *et al.* (2005) reported that HPP could influence the formation of Maillard-derived compounds in a sugar−amino acid model solution. In another study using milk, Vazquez-Landaverde *et al.*

TABLE 12.6 Aroma Attributes and Sensory Thresholds of Some Major Lactones

Compound	Aroma Attributes	Threshold (ppm)	Medium	References
γ-Hexalactone	Coconut, fruity, sweet	1.6–13	Water	Siek et al. (1971), Engel et al. (1988)
		8	Oil	Siek et al. (1971)
γ-Heptalactone	Coconut, fruity, nutty	0.52	Water	Siek et al. (1971)
		3.4	Oil	Siek et al. (1971)
δ-Octalactone	Coconut, animal	0.4–0.57	Water	Siek et al. (1971), Engel et al. (1988)
		0.1–3	Oil	Siek et al. (1971), Urbach et al. (1972)
γ-Octalactone	Coconut, fruity	0.095	Water	Siek et al. (1971)
		3.5	Oil	Siek et al. (1971)
γ-Nonalactone	Coconut, peach	0.065	Water	Siek et al. (1971)
		2.4	Oil	Siek et al. (1971)
δ-Decalactone	Coconut, apricot	0.1–0.16	Water	Siek et al. (1971), Urbach et al. (1972), Engel et al. (1988)
		0.4–1.4	Oil	Siek et al. (1971), Preininger et al. (1994)
γ-Decalatone	Coconut, apricot, fatty	0.005–0.09	Water	Siek et al. (1971), Engel et al. (1988), Larsen et al. (1992)
		1	Oil	Siek et al. (1971)
δ-Dodecalactone	Fresh fruit, peach	0.1–1	Water	Siek et al. (1971)
		0.12–10	Oil	Siek et al. (1971), Schieberle et al. (1993)
γ-Dodecalactone	Peach, butter, sweet, floral	0.007	Water	Engel et al. (1988)
		1	Oil	Urbach et al. (1972)

(2006b) found that pressure, temperature, and time, as well as their interactions, all had significant effects ($p < 0.001$) on off-flavor generation in milk. Pressure and time effects were greatest at 60°C, while their effects were almost negligible at 25°C. It was observed that the off-flavor generation of pressure-heated samples at 60°C was different from that of heated-alone samples. Heat treatment at 60°C tended to promote mostly the formation of methanethiol, H_2S, and methyl ketones, while high-pressure treatment at the same temperature mostly formed H_2S and aldehydes such as hexanal and octanal. The results demonstrated that the off-flavor generation at high pressure and moderate temperature was different from that under atmospheric pressure conditions.

Although the actual formation mechanisms under high pressure are not known, oxygen becomes more soluble under high pressure, therefore potentially increasing the formation of hydroperoxides and leading to more aldehyde generation. It is also possible that high pressure affects the kinetics of volatile formation. According to Le Chatelier's principle (Galazka and Ledward, 1996), if the formation of hydroperoxides from oxygen and fatty acids involves equilibrium reactions with a volume reduction, high pressure will favor this reaction and thus lead to more aldehyde generation. Another highly likely possibility is that the hydrostatic pressure affects the rate of formation according to its reaction activation volume (ΔV^*) defined as the difference between the partial molar volume of the transition or activated state and that of the reactant at the same temperature and pressure (McNaught and Wilkinson, 1997). When pressure is applied, $\Delta V^* < 0$ leads to an increase in reaction rate, whereas $\Delta V > 0$ has the opposite effect. The sensitivity of a chemical reaction to pressure will increase with the absolute value of ΔV^* (Mussa and Ramaswamy, 1997). The formation of H_2S seems to be affected by both pressure and holding time (Vazquez-Landaverde et al., 2006b). A dramatic increase in H_2S was observed under high-pressure treatments even at 25°C. The concentration of MeSH also increased at 25°C under high pressure. However, when the pressure increased to 620 MPa, the concentration of MeSH decreased (Vazquez-Landaverde et al., 2006b). Although

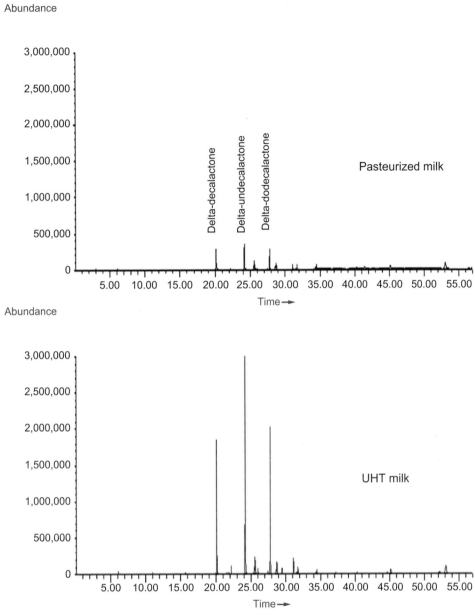

FIGURE 12.8 **Gas chromatography—mass spectrometry chromatograms (selected ion monitoring, 99 *m/z*) of UHT milk and pasteurized milk by stir bar sorptive extraction** *(Qian* et al., *unpublished data).*

methanethiol formation appeared to be inhibited under pressure, it is also possible that it was converted to other compounds. In addition, the formation and conversion of methanethiol could be pH dependent, due to pressure-induced pH shifts.

A kinetic study conducted by Vazquez-Landaverde *et al.* (2007) provided an improved understanding of the formation mechanisms of volatile compounds in milk subjected to high hydrostatic pressure. Hexanal, heptanal, octanal, nonanal, and decanal formation followed first order kinetics with rate constants increasing with pressure and temperature. Activation energies for these five straight-chain aldehydes decreased with pressure, suggesting that pressure has a catalytic effect on their formation reactions in milk. Formation of 2-methylpropanal, 2,3-butanedione, and H_2S followed zero-order kinetics with rate constants increasing with temperature but with an unclear pressure effect. Activation energies for 2-methylpropanal and 2,3-butanedione increased with pressure, whereas the values for H_2S remained constant in the pressure range studied. The concentrations of other off-flavor compounds studied, including the powerful off-flavor compound methanethiol, remained unchanged in all pressure-treated samples. In the case of the methyl ketones 2-pentanone, 2-hexanone, 2-heptanone, 2-octanone, 2-nonanone, 2-decanone, and 2-undecanone, their concentration did not depend on time and pressure. The results supported a previous suggestion (Vazquez-Landaverde et al., 2007) that high hydrostatic pressure affects the formation kinetics of off-flavor compounds in milk differently, inhibiting some and promoting others. Conventional heat treatment of food products will produce different responses depending on the free energy, ΔG, or activation free energy, ΔG^*, of the various reactions. But the reactions that are commonly observed during heat treatment will not be observed during HPP unless they have an optional pathway of reaction that involves the application of a mechanical reduction of the volume, which has been defined as the main difference between high pressure and thermal processing (Galazka and Ledward, 1996). Although temperature has to be controlled when applying HPP treatments so as not to change the flavor of milk, Vazquez-Landaverde and Qian (2007) suggested that the combination of high pressure, heat, and antioxidants could be used to develop a commercial product that is much more shelf stable while possibly reducing or completely eliminating cooked off-flavor.

REFERENCES

Ackerman, B.R., Forrest, S., Chow, L., Youil, R., Knight, M., Treacy, E.P., 1999. Two novel mutations of the *FM03* gene in a proband with trimethyl-aminuria. Hum. Mutat. 13, 376–379.

Ahmed, E., Dennison, P.A., Dougherty, R.H., Shaw, I.E., 1978. Flavor and odor thresholds in water of selected orange juice components. J. Agric. Food Chem. 26, 187–191.

Amoore, J.E., Forrester, L.J., Pelosi, P., 1976. Specific anosmia to isobutyraldehyde – malty primary odor. Chem. Senses Flavour 2, 17–25.

Amoore, J., Venstrom, D., Davis, A.R., 1968. Measurement of specific anosmia. Percept. Mot. Skills 26, 143–164.

Anderson, M., 1982. Factors Affecting the distribution of lipoprotein–lipase activity between serum and casein micelles in bovine-milk. J. Dairy Res. 49, 51–59.

Antolovich, M., Prenzler, P., Patsalides, E., McDonald, S., Robards, K., 2002. Methods for testing antioxidant activity. Analyst 127, 183–198.

Aurand, L.W., Singleton, J.A., Noble, B.W., 1966. Photooxidation reactions in milk. J. Dairy Sci. 49, 138–143.

Badings, H.T., 1978. Reduction of cooked flavor in heated milk and milk products. In: Land, D.G., Nursten, H.E. (Eds.), Progress in Flavor Research. Applied Science Publishers, London, pp. 263–265.

Badings, H.T., Neeter, R., Van der Pol, J.J.G., 1978. Reduction of heated flavor in heated milk and milk products by l-cystine. Lebensm. Wiss. Technol. 11, 237–242.

Badings, H.T., Van der Pol, J.J.G., Neeter, R., 1981. Aroma compounds which contribute to the difference in flavor between pasteurized milk and UHT milk. In: Schreider, P. (Ed.), Flavor '81. Walter de Gruyter, Berlin, pp. 683–692.

Badis, A., Guetarni, D., Moussa-Boudjemaa, B., Henni, D.E., Tornadijo, M.G., Kihal, M., 2004. Identification of cultivable lactic acid bacteria isolated from Algerian raw goat's milk and evaluation of their technological properties. Food Microbiol. 21, 343–349.

Baldwin, R., Cloninger, M., Lindsay, R.C., 1973. Flavor thresholds for fatty acids in buffered solutions. J. Food Sci. 38, 528–530.

Basarab, T., Ashton, G.H., Menage, H.P., McGrath, J.A., 1999. Sequence variations in the flavin-containing mono-oxygenase 3 gene (*FM03*) in fish odour syndrome. Br. J. Dermatol. 140, 164–167.

BeMiller, J.M., Whistler, R.L., 1996. Carbohydrates. In: Fennema, O.R. (Ed.), Food Chemistry. Marcel Dekker, New York.

Berlin, D.L., Herson, D.S., Hicks, D.T., Hoover, D.G., 1999. Response of pathogenic *Vibrio* species to high hydrostatic pressure. Appl. Environ. Microbiol. 65, 2776–2780.

Boccacci Mariani, M., Chiaccierini, E., Bucarelli, F.M., Quaglia, G.B., Menesatti, C., 2006. Comparative study of milk packaging materials. Note 2. Sensorial quality change in fresh milk during storage. Ind. Aliment. 45, 6–10.

Boelens, M., Van Gemert, L., 1987. Organoleptic properties of aliphatic aldehydes. Perfum. Flavor. 12, 31–43.

Boff, J.M., Min, D.B., 2002. Chemistry and reaction of singlet oxygen in foods. Compr. Rev. Food Sci. Food Saf. 1, 58–72.

Bradley, D.G., Min, D.B., 1982. Singlet oxygen detection in skim milk by electron spin resonance spectroscopy. J. Food Sci. 68, 491–494.

Bradley, R.L., 1980. Effect of light on alteration of nutritional value and flavor of milk. A review. J. Food Prot. 43, 314–320.

Brennand, H., 1989. Aroma properties and thresholds of some branched-chain and other minor volatile fatty acids occurring in milk fat and meat lipids. J. Sens. Stud. 4, 105–120.

Burbank, H.M., Qian, M., 2005. Quantification of volatile sulfur compounds in Cheddar cheeses of various age. In: IFT Annual Meeting, New Orleans, LA.

Buttery, R., 1993. Quantitative and sensory aspects of flavor of tomato and other vegetables and fruits. In: Acree, T.E., Teranishi, R. (Eds.), Flavor Science: Sensible Principles and Techniques. American Chemical Society, Washington, DC, pp. 259–286.

Buttery, R., Takeoka, G., Ling, C.L., 1995. Furaneol: odor threshold and importance to tomato aroma. J. Agric. Food Chem. 43, 1638–1640.

Buttery, R., Turnbaugh, J., Ling, C.L., 1988. Contribution of volatiles to rice aroma. J. Agric. Food Chem. 36, 1006–1009.

Champagne, C., Laing, R., Roy, D., Mafu, A.A., Griffiths, M.W., White, C., 1994. Psychrotrophs in dairy products: their effects and their control. Crit. Rev. Food Sci. Nutr. 34, 1–30.

Chavarri, F., Virto, M., Martin, C., Najera, A.I., Santisteban, A., Barron, L.J.R., de Renobales, M., 1997. Determination of free fatty acids in cheese: comparison of two analytical methods. J. Dairy Res. 64, 445–452.

Cheftel, J.C., 1995. High pressure, microbial inactivation and food preservation. C.R. Acad. Sci. Agric. France 81, 13–38.

Chin, H.-W., Lindsay, R.C., 1994. Ascorbate and transition-metal mediation of methanethiol oxidation to dimethyl disulfide and dimethyl trisulfide. Food Chem. 49, 387–392.

Choe, E., Huang, R., Min, D.B., 2005. Chemical reactions and stability of riboflavin in foods. J. Food Sci. 70, R28–R36.

Christen, G., Wang, W., Ren, T.-J., 1986. Comparison of the heat resistance of bacterial lipases and proteases and the effect on ultra-high temperature milk quality. J. Dairy Sci. 69, 2769–2778.

Christensen, K.R., Reineccius, G.A., 1992. Gas chromatographic analysis of volatile sulfur compounds from heated milk using static headspace sampling. J. Dairy Sci. 75, 2098–2104.

Cladman, W., Scheffer, S., Goodrich, N., Griffiths, W., 1998. Shelf-life of milk packaged in plastic containers with and without treatment to reduce light transmission. Int. Dairy J. 8, 629–636.

Cogan, T., 1977. A review of heat resistant lipases and proteinases and the quality of dairy products. Ir. J. Food Sci. Technol. 1, 95–105.

Contarini, G., Povolo, M., 2002. Volatile fraction of milk: comparison between purge and trap and solid phase microextraction techniques. J. Agric. Food Chem. 50, 7350–7355.

Contarini, G., Povolo, M., Leardi, R., Toppino, P.M., 1997. Influence of heat treatment on the volatile compounds of milk. J. Agric. Food Chem. 45, 3171–3177.

Cousin, M.A., 1982. Presence and activity of psychrotrophic microorganisms in milk and dairy-products — a review. J. Food Prot. 45, 172–207.

Czerny, M., Schieberle, P., 2007. Influence of the polyethylene packaging on the adsorption of odour-active compounds from UHT-milk. Eur. Food Res. Technol. 225, 215–223.

Damodaran, S., 1996. Amino acids, peptides, and proteins. In: Fennema, O.R. (Ed.), Food Chemistry. Marcel Dekker, New York, pp. 412–413.

Datta, N., Elliot, A.J., Perkins, M.L., Deeth, H.C., 2002. Ultra-high-temperature (UHT) treatment of milk: comparison of direct and indirect modes of heating. Aust. J. Dairy Technol. 57, 211–227.

De Jong, C.D., Badings, H.T., 1990. Determination of free fatty acids in milk and cheese: procedures for extraction, clean up, and capillary gas chromatographic analysis. J. High Resolut. Chromatogr. 13, 94–98.

de Koning, P.J., Badings, H.T., van der Pol, J.G., Kaper, J., vos-Klampmaker, E.A.J., 1990. Effect of heat treatment and fat content on UHT milk. Voedingsmiddelentechnologie 23, 11–14.

de la Rochette, A., Birlouez-Argon, I., Silva, E., Morliere, P., 2003. Advanced glycation endproducts as UVA photosensitizers of tryptophan and ascorbic acid: consequences for the lens. Biochim. Biophys. Acta 1621, 235–241.

Deeth, H.C., 2006. Lipoprotein lipase and lipolysis in milk. Int. Dairy J. 16, 555–562.

Deeth, H.C., Fitz-Gerald, C.H., Snow, A.G., 1983. A gas chromatographic method for the quantitative determination of free fatty acids in milk and milk products. N.Z.J. Dairy Sci. Technol. 18, 13–20.

Dimick, P.S., 1982. Photochemical effects on flavor and nutrients of fluid milk. Can. Inst. Food Sci. Technol. J. 15, 247–256.

Dogan, B., Boor, K., 2003. Genetic diversity and spoilage potentials among Pseudomonas spp. isolated from fluid milk products and dairy processing plants. Appl. Environ. Microbiol. 69, 130–138.

Dolphin, C.T., Janmohamed, A., Smith, R.L., Shepard, E.A., Phillips, J.R., 1997. Missense mutation of flavin-containing monooxygenase 3-gene, FM03, underlies fish-odor syndrome. Nat. Genet. 17, 491–494.

Downey, W., Andrews, P., 1966. Studies on the properties of cow's-milk tributyrinases and their interaction with milk proteins. Biochem. J. 101, 651–660.

Doyle, M., 2004. Consumers have a long list of packaging wishes and pet peeves. Food Drug Pack. 68, 24–28.

Drake, M.A., Miracle, R.E., Caudle, A.D., Cadwallader, K.R., 2006. Relating sensory and instrumental analyses. In: Marsaluis, R. (Ed.), Sensory-Directed Flavor Analysis. CRC Press, Taylor and Francis, Boca Raton, FL, pp. 23–55.

Dumont, J.-P., Adda, J., 1978. Flavour formation in dairy products. In: Land, D.G., Nursten, H.E. (Eds.), Progress in Flavour Research. Applied Science Publishers, London, pp. 245–262.

Duncan, S.E., Christen, G.L., Penfield, M.P., 1990. Acid degree value — does it really predict rancid flavor in milk? Dairy Food Environ. Sanit. 10, 715–718.

Duncan, S.E., Christen, G.L., Penfield, M.P., 1991. Rancid flavor of milk — relationship of acid degree value, fatty acids, and sensory perception. J. Food Sci. 36, 394–397.

Dunkley, W.L., Franklin, J.D., Pangborn, R.M., 1962. Effects of fluorescent light on flavor, ascorbic acid and riboflavin of milk. Food Technol. 16, 112–118.

Dunn, H., Lindsay, R., 1985. Evaluation of the role of microbial Strecker-derived aroma compounds in unclean-type flavors of Cheddar cheese. J. Dairy Sci. 68, 2859–2874.

Edwards, A.M., Silva, E., 2001. Effect of visible light on selected enzymes, vitamins and amino acids. J. Photochem. Photobiol. B 63, 126–131.

Egelrud, T., Olivecrona, T., 1972. The purification of a lipoprotein lipase from bovine skim milk. J. Biol. Chem. 247, 6212–6217.

Egelrud, T., Olivecrona, T., 1973. Purified bovine milk (lipoprotein) lipase: activity against lipid substrates in the absence of exogenous serum factors. Biochim. Biophys. Acta 306, 115–127.

Engel, K.H., Flath, R.A., Buttery, R.G., Man, T.R., Ramming, D.W., Teranishi, R., 1988. Investigation of volatile constituents in nectarines. 1. Analytical and sensory characterization of aroma components in some nectarine cultivars. J. Agric. Food Chem. 36, 549–553.

Fajardo-Lira, C., Nielsen, S., 1998. Effect of psychrotrophic microorganisms on the plasmin system in milk. J. Dairy Sci. 81, 901–908.

Fang, Y., Qian, M., 2005. Sensitive quantification of sulfur compounds in wine by headspace solid-phase microextraction technique. J. Chromatogr. A 1080, 177–185.

Ferreti, A., 1973. Inhibition of cooked flavor in heated milk by use of additives. J. Agric. Food Chem. 21, 939–942.

Fitz-Gerald, C.H., Deeth, H.C., 1983. Factors influencing lipolysis by skim milk cultures of some psychrotrophic microorganisms. Aust. J. Dairy Technol. 38, 97–103.

Forrest, S.M., Knight, M., Ackerman, B.R., Cashman, J.R., Treacy, E.P., 2001. A novel deletion in the flavin-containing monooxygenase gene (FM03) in a Greek patient with trimethylaminuria. Pharmacogenetics 11, 169–174.

Forss, D.A., 1979. Review of the progress of dairy science — mechanisms of formation of aroma compounds in milk and milk-products. J. Dairy Res. 46, 691–706.

Fox, P.F., Guinee, T.P., Cogan, T.M., McSweeney, P.L.H., 2000. Fundamentals of Cheese Science, Aspen, Gaithersburg, MD.

Franklin, U., 1951. Some factors affecting the flavor of milk and milk products. J. Med. Assoc. State Alabama 20, 383–384.

Gaafar, A., Gaber, F., 1992. Volatile flavour compounds of sunlight-exposed milk. Egypt. J. Dairy Sci. 20, 111–115.

Galazka, V.B., Ledward, D.A., 1996. Effects of high pressure on protein polysaccharide interactions. In: Parris, N., Kato, A., Creamer, L.K., Pearce, J. (Eds.), Macromolecular Interactions in Food Technology. American Chemical Society, Washington DC, pp. 113–123.

Garcia, H.S., Reyes, H.R., Malcata, F.X., Hill, C.G., Amundson, C.H., 1990. Determination of the major free fatty acids in milkfat using a three-component mobile phase for HPLC analysis. Milchwissenschaft 45, 747–759.

Gonzalez-Cordova, A., Vallejo-Cordoba, B., 2001. Quantitative determination of short-chain free fatty acids in milk using solid-phase microextraction and gas chromatography. J. Agric. Food Chem. 49, 4603–4608.

Gonzalez-Cordova, A., Vallejo-Cordoba, B., 2003. Detection and prediction of hydrolytic rancidity in milk by multiple regression analysis of short-chain free fatty acids determined by solid phase microextraction gas chromatography and quantitative flavor intensity assessment. J. Agric. Food Chem. 51, 7127–7131.

Grosch, W., 1982. Lipid degradation products and flavour. In: Morton, I.D., Macleod, A.J. (Eds.), Food Flavours. Part A. Introduction. Elsevier, Oxford, pp. 325–385.

Guadagni, D.G., Buttery, R.G., Turnbaugh, J.G., 1972. Odour thresholds and similarity ratings of some potato chip components. J. Sci. Food Agric. 23, 1435–1444.

von Gunten, R.L., Bush, L.J., Odell, G.V., Wells, M.E., Adams, G.D., 1976. Factors related to the occurrence of trimethylamine in milk. J. Milk Food Technol. 39, 526–529.

Guth, H., Grosch, W., 1990. Deterioration of soya-bean oil: quantification of primary flavour compounds using a stable isotope dilution assay. Lebensm. Wiss. Technol. 23, 513–522.

Guth, H., Grosch, W., 1994. Identification of the character impact odorants of stewed beef juice by instrumental analyses and sensory studies. J. Agric. Food Chem. 42, 2862–2866.

Halliwell, B., 1996. Antioxidants in human health and disease [review]. Annu. Rev. Nutr. 16, 33–50.

Hart, H., 1991. Organic Chemistry: A Short Course. Houghton Mifflin, Boston, MA.

Hayes, W., White, C.H., Drake, M.A., 2002. Sensory aroma characteristics of milk spoilage by Pseudomonas species. J. Food Sci. 67, 861–867.

Herrington, B., 1954. Lipase: a review. J. Dairy Sci. 37, 775–789.

Hlavica, P., Kehl, M., 1977. Studies on the mechanism of hepatic microsomal N-oxide formation, the role of cytochrome P450 and mixed function amine oxidase in the N-oxidation of N, N-dimethylamine. Biochem. J 164, 487–496.

Hofmann, T., Deters, F., Heberle, I., Schieberle, P., 2005. Influence of high hydrostatic pressure on the formation of Maillard-derived key odorants and chromophores. Ann. N.Y. Acad. Sci. 1043, 893.

Hoskin, J.C., Dimick, P.S., 1979. Evaluation of fluorescent light on flavor and riboflavin content of milk held in gallon returnable containers. J. Food Prot. 42, 105–109.

Hoynes, M., Downey, W., 1973. Relationship of the lipase and lipoprotein lipase activities of bovine milk. Biochem. Soc. Trans. 1, 256–259.

Hutton, J.T., Patton, S., 1952. The origin of sulfhydryl groups in milk proteins and their contributions to 'cooked' flavor. J. Dairy Sci. 35, 699–705.

Ikins, W.G., Kwak, H.S., Zink, G.S., Jean, I.J., 1988. Comparison of methods for quantitation of free fatty acids in cheese. J. Food Sci. 53, 1915–1916.

Jaddou, H.A., Pavey, J.A., Manning, D.J., 1978. Chemical analysis of flavor volatiles in heat-treated milks. J. Dairy Res. 45, 391–403.

Jensen, R.G., Bitman, J., Carlson, S.E., Couch, S.C., Hamosh, M., Newberg, D.S., 1995. Milk lipids. In: Jensen, R.G. (Ed.), Handbook of Milk Composition. Academic Press, San Diego, CA, pp. 571–572.

Jeon, I.J., Thomas, E.L., Reinneccius, C.A., 1978. Production of volatile flavor compounds in ultrahigh-temperature processed milk during aseptic storage. J. Agric. Food Chem. 26, 1183–1188.

Jernigan, H.M.J., 1985. Role of hydrogen peroxide in riboflavin-sensitized photodynamic damage to cultured rat lenses. Exp. Eye Res. 41, 121–129.

Jung, M., Yoon, S., Lee, H., Min, D., 1998. Singlet oxygen and ascorbic acid effects on dimethyl disulfide and off-flavor in skim milk exposed to light. J. Food Sci. 63, 408–412.

Karahadian, C., Josephson, D.B., Lindsay, R.C., 1985. Contribution of Penicillium sp. to the flavors of Brie and Camembert cheese. J. Dairy Sci. 68, 1865–1877.

Kim, H.S., Gilliland, S.E., von Gunten, R.L., Morrison, R.D., 1980. Chemical test for detecting wheat pasture flavor in cow's milk. J. Dairy Sci. 63, 368–374.

Kim, Y., Morr, C., 1996. Dynamic headspace analysis of light activated flavor in milk. Int. Dairy J. 6, 185–193.

Kinnunen, P., Huttunen, J.K., Ehnholm, C., 1976. Properties of purified bovine milk lipoprotein lipase. Biochim. Biophys. Acta 450, 342–351.

Kubicková, J., Grosch, W., 1998. Quantification of potent odorants in Camembert cheese and calculation of their odour activity values. Int. Dairy J. 8, 17–23.

Kuzdzal-Savoie, S., 1980. Determination of free fatty acids in milk and milk products. Int. Dairy Fed. Bull. 118, 53–66.

Larsen, M., Poll, L., 1992. Odour thresholds of some important aroma compounds in strawberries. Z. Lebensmitteluntersuch. Forsch. A 195, 120–123.

Lin, J.C.C., Jeon, I.J., 1987. Effect of commercial food grade enzymes on free fatty acid profiles in granular Cheddar cheese. J. Food Sci. 52, 78–87.

Lunden, A., Gustafsson, V., Imhof, M., Gauch, R., Bosset, J.-O., 2002a. High trimethamine concentration in milk from cows on standard diets is expressed as fishy off flavor. J. Dairy Res. 69, 383–390.

Lunden, A., Marklund, S., Gustafsson, V., Andersson, L., 2002b. A nonsense mutation in the FM03 gene underlies fishy off-flavor in cow's milk. Genome Res. 12, 1885–1888.

Magan, N., Pavlou, A., Chrysantakis, I., 2001. Milk-sense: a volatile sensing system recognises spoilage bacteria and yeasts in milk. Sens. Actuators B: Chem. 72, 28–34.

Manning, D.J., Robinson, H.M., 1973. Analysis of volatile substances associated with cheddar-cheese aroma. J. Dairy Res. 40, 63–75.

Marsili, R.T., 1999. Comparison of solid-phase microextraction and dynamic headspace methods for the gas chromatographic–mass spectrometric analysis of light-induced lipid oxidation products in milk. J. Chromatogr. Sci. 37, 17–23.

Martinez-Castro, I., Alonso, L., Juarez, M., 1986. Gas chromatographic analysis of free fatty acids and glycerides of milk fat using tetramethylammonium hydroxide as catalyst. Chromatographia 21, 37–40.

Martin-Hernandez, M.C., Alonso, L., Jubrez, N., Fontecha, J., 1988. Gas chromatographic method for determining free fatty acids in cheese. Chromatographia 25, 87–90.

Matselis, E., Roussis, I., 1998. Proteinase and lipase production by Pseudomonas fluorescens. Proteolysis and lipolysis in thermized ewe's milk. Food Control 9, 251–259.

McKellar, R., 1981. Development of off-flavors in ultra-high temperature and pasteurized milk as a function of proteolysis. J. Dairy Sci. 64, 2138–2145.

McNaught, A.D., Wilkinson, A., 1997. Compendium of Chemical Terminology, the Gold Book, (2nd ed.) Blackwell Science, Oxford.

Meer, R., Baker, J., Bodyfelt, F.W., Griffiths, M.W., 1991. Psychrotrophic *Bacillus* spp. in fluid milk products: a review. J. Food Prot. 54, 969–979.

Mehta, R.S., Bassette, R., Ward, G., 1974. Trimethylamine responsible for fishy flavor in milk from cows on wheat pasture. J. Dairy Sci. 57, 285–289.

Mendelson, C., Zinder, O., Blanchette-Mackie, E.J., Chemick, S.S., Scow, C.O., 1977. Lipoprotein lipase and lipid metabolism in mammary gland. J. Dairy Sci. 60, 666–676.

Metcalfe, L.D., Wang, C.N., 1981. Rapid preparation of fatty acid methyl esters using organic base-catalyzed transesterification. J. Chromatogr. Sci. 19, 530–535.

Moio, L., Dekimpe, J., Etievant, P., Addeo, F., 1993. Neutral volatile compounds in the raw milks from different species. J. Dairy Res. 60, 199–213.

Moio, L., Etievant, P., Langlois, D., Dekimpe, J., Addeo, F., 1994. Detection of powerful odorants in heated milk by use of extract dilution sniffing analysis. J. Dairy Res. 61, 385–394.

Moyssiadi, T., Badeka, A., Kondyli, E., VaKirtzi, T., Sawaidis, T., Kontominas, M.G., 2004. Effect of light transmittance and oxygen permeability of various packaging materials on keeping quality of low fat pasteurized milk: chemical and sensorial aspects. Int. Dairy J. 14, 429–436.

Mussa, D., Ramaswamy, H., 1997. Ultra high pressure pasteurization of milk: kinetics of microbial destruction and changes in physico-chemical characteristics. Lebensm. Wiss. Technol. 30, 551–557.

Naseem, J., Ahmed, M.S., Bhat, R., Hadi, S.M., 1993. Copper(II)-dependent degradation of DNA by riboflavin. Food Chem. Toxicol. 31, 589–597.

Nawar, W.W., 1996. Lipids. In: Fennema, O.R. (Ed.), Food Chemistry. Marcel Dekker, New York, pp. 290–291.

Needs, E.C., Ford, G.D., Owen, A.J., Tuckley, B., Anderson, M., 1983. A method for the quantitative determination of individual free fatty acids in milk by ion exchange resin adsorption and gas–liquid chromatography. J. Dairy Res. 50, 321–329.

Nielsen, S., 2002. Plasmin system and microbial proteases in milk: characteristics, roles, and relationship. J. Agric. Food Chem. 50, 6628–6634.

Ouattara, G.C., Jeon, I.J., Hart-Thakur, P., Schmidt, K., 2004. Fatty acids released from milk fat by lipoprotein lipase and lipolytic psychrotrophs. J. Food Sci. 69, C659–C664.

Patton, S., 1964. Flavor thresholds of volatile fatty acids. J. Food Sci. 29, 679–680.

Patton, S., Josephson, D.V., 1953. Methionine – origin of sunlight flavor in milk. Science 118, 211.

Pearson, A.W., Butler, E.J., Curtis, R.F., Fenwick, G.R., Hobson-Frohock, A., Land, D.G., 1979. Effect of rapeseed meal on trimethylamine metabolism in the domestic fowl in relation to egg taint. J. Sci. Food Agric. 30, 799–804.

Perkins, M.L., Zerdin, K., Rooney, M.L., D'Arcy, B.R., Deeth, H.C., 2007. Active packaging of UHT milk to prevent the development of stale flavour during storage. Pack. Technol. Sci. 20, 137–146.

Preininger, M., Grosch, W., 1994. Evaluation of key odorants of the neutral volatiles of Emmentaler cheese by the calculation of odour activity values. Lebens. Wiss. Technol. 27, 237–244.

Qian, M., Reineccius, G.A., 2002. Importance of free fatty acids in Parmesan cheese. In: Reineccius, G.A., Reineccius, T.A. (Eds.),

Heteroatomic Aroma Compounds, ACS Symposium Series 826. American Chemical Society, Washington, DC, pp. 243–256.

Rademacher, B., Kessler, H.G., 1996. High pressure inactivation of microorganisms and enzymes in milk and milk products. In: European High Pressure Research Conference. Leuven, Belgium.

Reiners, J., Grosch, W., 1998. Odorants of virgin olive oils with different flavor profiles. J. Agric. Food Chem. 46, 2754–2763.

Rerkrai, S., Jeon, I.J., Bassette, R., 1987. Effect of various direct ultra-high temperature heat treatments on flavor of commercially prepared milks. J. Dairy Sci. 70, 2046–2054.

Richardson, G.H. (Ed.), 1985. Standard Methods for the Examination of Dairy Products, (15th ed.). American Public Health Association, Washington, DC.

Rosenthal, I., 1992. Electromagnetic Radiations in Food Science. Springer, Berlin.

Rychlik, M., Schieberle, P., Grosch, W., 1998. Compilation of Odor Thresholds, Odor Qualities and Retention Indices of Key Food Odorants. Deutsche Forschungsanstalt fur Lebensmittelchemie and Institut fur Lebensmittelchemie der Technischen Universitat Munchen, Garching, Germany.

Salo, P., 1970. Variability of odour thresholds for some compounds in alcoholic beverages. J. Sci. Food Agric. 21, 597–600.

Sattar, A., deMan, J.M., 1975. Photooxidation of milk and milk products: a review. CRC Crit. Rev. Food Sci. Nutr. 7, 13–35.

Sattar, A., deMan, J.M., Alexander, J.C., 1976. Light-induced oxidation of edible oils and fats. Lebensm. Wiss. Technol. 9, 149–152.

Saxby, M.J., 1992. A survey of chemicals causing taints and off-flavours in food. In: Food Taints and Off-Flavours. Blackie, London, pp. 23–25.

Scanlan, R.A., Lindsay, R., Libby, L.M., Day, E.A., 1968. Heat-induced volatile compounds in milk. J. Dairy Sci. 51, 1001–1007.

Schieberle, P., Gassenmeier, K., Guth, H., Sen, A., Grosch, W., 1993. Character impact odour compounds of different kinds of butter. Lebensm. Wiss. Technol. 26, 347–356.

Schroder, M.J.A., Cousins, C.M., McKinnon, C.H., 1982. Effect of psychrotrophic post-pasteurization contamination on the keeping quality at 11 and 5 degrees C of HTST-pasteurized milk in the UK. J. Dairy Res. 49, 619–630.

Scow, R.O., Egelrud, T., 1976. Hydrolysis of chylomicron phosphatidylcholine *in vitro* by lipoprotein lipase, phospholipase A_2 and phospholipase C. Biochim. Biophys. Acta 431, 538–549.

Senyk, G., Barbano, D., Shipe, W., 1985. Proteolysis in milk associated with increasing somatic cell counts. J. Dairy Sci. 68, 2189–2194.

Shah, N.P., 1994. Psychrotrophs in milk – a review. Milchwissenschaft. Milk Sci. Int. 49, 432–437.

Shibamoto, T., Mihara, S., Nishimura, O., Kamiya, Y., Aitoku, A., Hayashi, J., 1980. Flavor volatiles formed by heated milk. In: Charalambous, G. (Ed.), The Analysis and Control of Less Desirable Flavors in Foods and Beverages. Academic Press, New York, pp. 260–263.

Shipe, W.F., 1980. Analysis and control of milk flavor. In: Charalambous, G. (Ed.), The Analysis and Control of Less Desirable Flavors in Foods and Beverages. Academic Press, New York, pp. 201–239.

Shipe, W.F., Bassette, R., Deane, D.D., Dunkley, W.L., Hammond, E.G., Harper, W.J., Kleyn, D.H., Morgan, M.E., Nelson, J.H., Scanlan, R.A., 1978. Off flavors of milk: nomenclature, standards, and bibliography. J. Dairy Sci. 61, 855–869.

Shirley, J., Emery, R., Convey, E., Oxender, W., 1973. Enzymic changes in bovine adipose and mammary tissue, serum and mammary tissue hormonal changes with initiation of lactation. J. Dairy Sci. 56, 569–574.

Siek, T., Albin, I.A., Sather, L.A., Lindsay, R.C., 1969. Taste thresholds of butter volatiles in deodorized butteroil medium. J. Food Sci. 34, 265–267.

Siek, T., Albin, I.A., Sather, L.A., Lindsay, R.C., 1971. Comparison of flavor thresholds of aliphatic lactones with those of fatty acids, esters, aldehydes, alcohols, and ketones. J. Dairy Sci. 54, 1–4.

Simon, M., Hansen, A.P., 2001. Effect of various dairy packaging materials on the shelf life and flavor of ultrapasteurized milk. J. Dairy Sci. 84, 784–791.

Singh, T.K., Drake, M.A., Cadwallader, K.R., 2003. Flavor of cheddar cheese: A chemical and sensory perspective. Comp. Rev. Food Sci. Food Safety. J. Dairy Sci. 2, 166–189.

Smet, K., Raes, K., De Block, J., Herman, L., Dewerrinck, K., Condijzer, K., 2008. A change in antioxidative capacity as a measure of onset to oxidation in pasteurized milk. Int. Dairy J. 18, 520–530.

Sørhaug, T., Stepaniak, L., 1997. Psychrotrophs and their enzymes in milk and dairy products: quality aspects. Trends Food Sci. Technol. 8, 35–41.

Spangelo, A., Karijord, O., Svensen, A., Abrahamsen, R.K., 1986. Determination of individual free fatty acids in milk by strong anion-exchange resin and gas chromatography. J. Dairy Sci. 69, 1787–1792.

Spellacy, E., Watts, R.W.E., Gollamali, S.K., 1979. Trimethylaminuria. J. Inherit. Metabol. Dis. 2, 85–88.

Stead, D., 1986. Microbial lipases – their characteristics, role in food spoilage and industrial uses. J. Dairy Res. 53, 481–505.

Steely, J.S., 1994. Chemiluminiscence detection of sulfur compounds in cooked milk. In: Mussinan, C.J., Keelan, M.E. (Eds.), Sulfur Compounds in Foods. American Chemical Society, Washington, DC, pp. 22–35.

Toso, B., Procida, G., Stafanon, B., 2002. Determination of volatile compounds in cows' milk using headspace GC-MS. J. Dairy Res. 69, 569–577.

Tovar-Hernandez, G., Peña, H.R.V., Velasquez, G., Ramirez, J.A., Torres, J.A., 2005. Effect of combined thermal and high pressure processing on the microbial stability of milk during refrigerated storage. In: IFT Annual Meeting, New Orleans, LA.

Treacy, E.P., Akerman, B.R., Chow, L.M.L., Youil, R., Bibeau, C., Lin, J., Bruce, A.G., Knight, M., Danks, D.M., Cashman, J., 1998. Mutations of the flavin-containing monooxygenase gene (FMO3) cause trimethylaminuria, a defect in detoxication. Hum. Mol. Genet. 7, 839–845.

Urbach, G., 1993. Relations between cheese flavor and chemical composition. Int. Dairy J. 3, 389–422.

Urbach, G., Stark, W., Forss, D.A., 1972. Volatile compounds in butter oil. 11. Flavor and flavor thresholds of lactones, fatty-acids, phenols, indole and skatole in deodorized synthetic butter. J. Dairy Res. 39, 35–47.

Vallejo-Cordoba, B., Mazorra-Manzano, M.A., Gonzalez-Cordova, A.F., 1998. Determination of short-chain free fatty acids in lipolyzed milk fat by capillary electrophoresis. J. Capill. Electrophor. 5, 111–114.

Vazquez-Landaverde, P.A., Qian, M.C., 2007. Antioxidant impacts on volatile formation in high-pressure-processed milk. J. Agric. Food Chem. 55, 9183–9188.

Vazquez-Landaverde, P.A., Qian, M.C., Torres, J.A., 2007. Kinetic analysis of volatile formation in milk subjected to pressure-assisted thermal treatments. J. Food Sci. 72, 389–398.

Vazquez-Landaverde, P.A., Torres, J.A., Qian, M.C., 2006a. Quantification of trace-volatile sulfur compounds in milk by solid-phase microextraction and gas chromatography-pulsed flame photometric detection. J. Dairy Sci. 89, 2919–2927.

Vazquez-Landaverde, P.A., Torres, J.A., Qian, M.C., 2006b. Effect of high pressure-moderate temperature processing on the volatile profile of milk. J. Agric. Food Chem. 54, 9184–9192.

Vazquez-Landaverde, P.A., Velazquez, G., Torres, J.A., Qian, M.C., 2005. Quantitative determination of thermally derived volatile compounds in milk using solid-phase microextraction and gas chromatography. J. Dairy Sci. 88, 3764–3772.

Velazquez, G., Gandhi, K., Torres, J.A., 2002. High hydrostatic pressure: a review. Biotam 12, 71–78.

Webster, J., Duncan, S., Marcy, J., O'Keefe, S., 2009. Controlling light oxidation flavor in milk by blocking riboflavin excitation wavelengths by interference. J. Food Sci. 74 (9), S390–S398.

Whetstine, M.E.C., Cadwallader, K.R., Drake, M.A., 2005. Characterization of aroma compounds responsible for the rosy/floral flavor in Cheddar cheese. J. Agric. Food Chem. 53, 3126–3132.

Wick, E., 1966. Flavor update: one opinion. Food Technol. 20, 1549–1554.

Wold, J.P., Veberg, A., Nilsen, A., Iani, V., Juzenas, P., Moan, J., 2005. The role of naturally occurring chlorophyll and porphyrins in light-induced oxidation of dairy products. A study based on fluorescence spectroscopy and sensory analysis. Int. Dairy J. 15, 343–353.

Woo, A.H., Lindsay, R.C., 1982. Rapid method for quantitative analysis of individual free fatty acids in cheddar cheese. J. Dairy Sci. 65, 1102–1109.

Zheng, Y., Ho, C., 1994. Kinetics of the release of hydrogen sulfide from cysteine and glutathione during thermal treatment. In: Mussinan, C.J., Keelan, M.E. (Eds.), Sulfur Compounds in Foods. American Chemical Society, Washington, DC, pp. 138–146.

PART IV

Biotechnology

Chapter 13

Recombinant DNA Technologies in Food

Peter Eck

Department of Human Nutritional Sciences, University of Manitoba, Winnipeg, Canada

Chapter Outline

Biochemistry of Foods. DOI: http://dx.doi.org/10.1016/B978-0-12-242352-9.00013-8

I. INTRODUCTION

Genetic modification has been used by humans for at least 10,000 years through selective breeding methods of crops, animals, and microorganisms to achieve higher disease resistance, yields, and food quality. During the past 25–30 years, developments in biotechnology have enabled selected traits to be transferred within a species or from one species to another in a much shorter time, using molecular biology techniques manipulating the blueprint of life, deoxyribonucleic acid (DNA).

DNA encodes all the information needed to create an organism. It is the genetic material found mainly in the nucleus of eukaryotic cells. DNA contains four nitrogen bases: adenine (A), thymine (T), guanine (G), and cytosine (C):

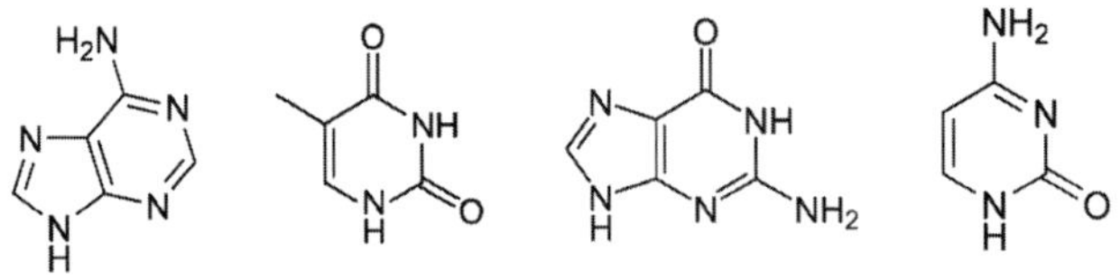

adenine thymine guanine cytosine

All DNA is made up of a backbone of deoxyribose molecules, which are connected by phosphate groups and have nitrogen bases attached to them. The nitrogen bases are found in pairs, with A&T and G&C paired together:

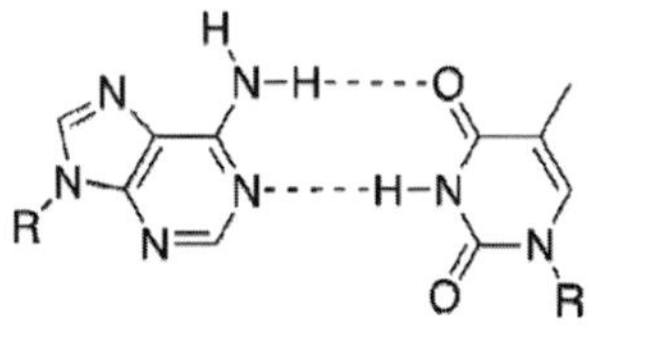

Adenine **Thymine**

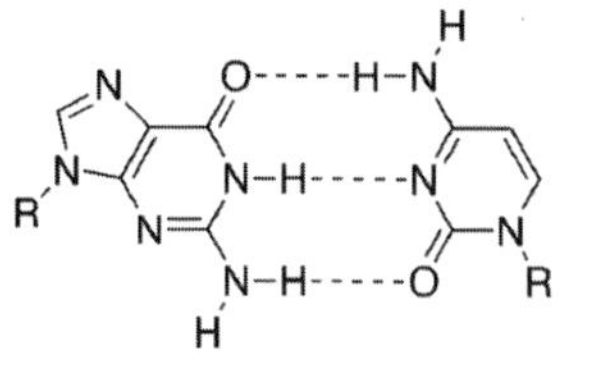

Guanine **Cytosine**

The nitrogen bases are identical in all organisms, but the arrangement of sequence and number of bases creates enormous diversity. Each base is also attached to a sugar molecule and a phosphate molecule. Together, a base, sugar, and phosphate are called a nucleotide. Through the backbone of deoxyribose molecules linked by phosphate groups, DNA forms a 'double-helix' structure (Figure 13.1).

Each DNA sequence that occupies a specific location on a chromosome and determines a particular characteristic in an organism is defined as a gene. The size of a gene may vary greatly, ranging from about 1000 bases to 1 million bases in humans. The DNA codes for the amino acid sequence in proteins or other non-protein coding sequences essential for the regulation of the genome. The DNA is transcribed into messenger ribonucleic acid (mRNA), which is

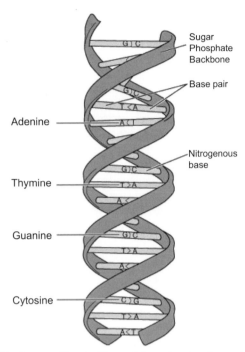

FIGURE 13.1 **Double-helix structure of DNA.** Supported by a backbone of alternating phosphate and sugar residues, the complementary nucleobases adenine-thymine or guanine-cytosine bind to form a helical structure. *(Source: US National Library of Medicine.)*

a single-stranded molecule of RNA that is synthesized in the cell's nucleus using the genomic DNA as template and then exported to the cytoplasm, where its genetic code is translated into the amino acid sequences constituting the proteins. In RNA the base thymine is substituted with uracil (U).

Recombinant DNA (rDNA) is a single molecule which has been combined artificially from two or more sources outside the living organisms. This recombinant DNA is intended to be introduced into a new host organism to transfer one or more traits for desirable characteristics, or remove genes for undesirable characteristics. There are many reasons to create recombinant DNA, such as the scientific study of the gene products or insertion of foreign genes into other organisms to alter commercially important traits.

Genetically modified organisms (GMOs) and genetically modified microorganisms (GMMs) can be defined as organisms in which the genetic material has been altered in a way that does not occur naturally by mating or natural recombination. The resulting organism is said to be genetically modified, genetically engineered, or transgenic. The technology is often called modern biotechnology, genetic engineering, or recombinant DNA technology. The techniques to produce recombinant DNA are routine laboratory procedures with highly specific outcomes. Modern recombinant genetic engineering techniques can be used to transfer genes from microorganisms, plants, or animals into cells from each of these living forms.

Genetic modifications of food have seen a rapid increase in the past decade. The advent of recombinant DNA technology and the possibility of gene transfer between organisms of distinct species, or even distinct phylogenic kingdoms, opened a wide range of recombinant DNA application in food technology. A lot of the food that we eat today contains genetically modified ingredients or is produced with the help of genetic engineering. Foods can consist of or contain living/viable transgenic organisms, e.g. maize, soybeans, and yeast cultures. Foods can contain ingredients derived from GMOs, e.g. flour, food protein products, or oil from genetically modified crops. Single ingredients used in food processing or as food additives can be produced by GMMs, e.g. enzymes and proteins.

The main rationale for the application of transgenic technology in food materials is to overcome the limitations of conventional breeding and selection. Whereas methods of cross-breeding and selection have significantly increased crop, livestock, and even microbial yields over many centuries, the future potential of these methods is constrained by the limitations of sexual-compatibility boundaries between species. Recombinant DNA technology overcomes the limitation of species incompatibility by selectively transferring DNA sequences encoding specific genetic traits (e.g. insect resistance, enhanced growth, better enzyme activity) into a recipient organism so that it expresses this trait. These genetically engineered products have a substantial presence in the food chain where, specifically, genetically engineered microorganisms and plants have a significant commercial value. Therefore, this chapter will focus on the commercialized applications of genetically engineered crops and microorganisms found in food. Owing to the lack of commercialized genetically modified livestock products, these will not be discussed.

II. GENETICALLY MODIFIED CROPS

Commercial transgenic crops were first planted in 1996 and are now soundly integrated into the agricultural system. The growing area expanded from 1.7 million hectares in 1996 to 148 million hectares worldwide in 2010. The accumulated hectarage from 1996 to 2010 exceeds one billion hectares, indicating the broad acceptance of the new crop traits. These so-called biotech crops are contributing significantly to food security, self-sufficiency, sustainability, alleviation of poverty and hunger, and mitigation of climate change; and the potential of biotech crops for the future is regarded as significant. Biotech crops are now legally planted in 29 countries, where they have an impact on more than half the world's population; hence, they are a significant element of the food chain. This section will give a brief introduction to transgenic crop technology and summarize the genetics and biochemistry for the most important commercially available transgenic crops (James, 2010).

A. Genetic Transformation of Crop Plants

For the process of transgenic plant production several components need to be established:

- A reliable tissue culture and regeneration systems must exist.
- Effective genetic constructs must be created using suitable vectors.
- Efficient transformation techniques for crop plants can be applied.
- Transgenic plants can be recovered and multiplied.
- Stable gene expression can be characterized.
- Trans-genes can be transferred into elite cultivars by conventional breeding.
- Transgenic plants will be evaluated for their effectiveness and environmental hazard (Birch, 1997; Sharma et al., 2004).

Although several approaches have been tried successfully for recombinant DNA integration into crops (Potrykus, 1991), only four approaches are used widely to introduce genes into crop plants (Dale et al., 1993):

- *Agrobacterium*-mediated gene transfer
- microprojectile bombardment with DNA (biolistics)
- microinjection of DNA into the cell nucleus
- direct DNA transfer into isolated protoplasts.

Agrobacterium tumefaciens-mediated gene transfer has been used widely for transformations of crop plants. *Agrobacterium tumefaciens* is a soil-inhabiting bacterium that has been implicated in gall formation at the wound sites in many dicotyledonous plants. This tumor-inducing capability is due to the presence of a large *Ti* (tumor-inducing) plasmid in virulent strains of *Agrobacterium*. Likewise, *Ri* (root-inducing) megaplasmids are found in virulent strains of *Agrobacterium rhizogenes*, the causative agent of 'hairy root' disease. The *Ti* and *Ri* plasmids, and the molecular biology of crown gall and hairy root induction, have been studied in great detail (for further reference see Zambryski et al., 1983; Zambryski, 1992). *Agrobacterium*-mediated transformation is brought about by the incorporation of genes of interest from an independently replicating *Ti* plasmid within the *A. tumefaciens* cell, which then infects the plant cell and transfers the *Ti*-DNA containing the gene of interest into the chromosomes of the actively dividing cells of the host plant (Sharma et al., 2004).

In the particle bombardment (biolistics) method, tungsten or gold particle microprojectiles are coated with the DNA to be inserted, and bombarded into cells or tissues capable of subsequent plant regeneration. Acceleration of heavy microprojectiles (0.5–5.0 μm diameter tungsten or gold particles) coated with DNA carries genes into virtually every type of cell and tissue (Klein et al., 1987; Sanford, 1990). The DNA-coated particles enter the plant cells, the DNA is incorporated in a small proportion of the treated cells, and the transformed cells are selected for plant regeneration (Sharma et al., 2004).

Genetically engineered DNA can also be directly injected into nuclei of embryogenic single cells, which can be induced to regenerate plants in cell culture (Neuhaus et al., 1987). This requires micromanipulation of single cells or small colonies of cells under the microscope, and precise injection of small amounts of DNA solution with a thin glass micropipette. Injected cells or clumps of cells are subsequently raised in *in vitro* culture systems and regenerated into plants (Sharma *et al.*, 2004).

In the protoplast transformation, the cell wall of the target cells is removed by enzymatic treatment, and the cells are bounded by a plasma membrane (Zhang and Wu, 1988). The DNA can be added into cell suspension, which can be introduced by affecting the plasma membrane with polyethylene glycol or by passing an electric current through the protoplast suspension. The DNA is incorporated into the genome of a few cells. A suitable marker should also be inserted to select the transformed protoplasts and the cell colonies that develop from them (Sharma *et al.*, 2004).

B. Effective Gene Expression in Crop Plants

Efficient genetic engineering relies on the ability to generate a specific gene product at the desired level of expression, in the appropriate tissues, at the right time. This can be accomplished by creating genetic expression constructs (also called expression cassettes) that include promoters and/or transcription regulation elements that control the level, location, and timing of gene expression. A typical expression cassette might include 5' and 3' flanking regions homologous to genetic host sequences, enabling integration of the cassette by homologous recombination. A promoter will determine the strength, and temporal and spatial distribution of the open reading frames' expression. The open reading frame contains the code for the heterologous protein. A terminator element regulates proper termination of the RNA's transcription:

| 5'flank | Promoter | Open Reading Frame | Terminator | 3'flank |

A major constraint in the development of effective transgenic products has been the lack of promoters that can offer a high level of gene expression at a high degree of specificity in the crop species of interest. Traditionally, transgene expression has been driven by strong constitutive promoters such as cauliflower mosaic virus *35S* promoter (*CaMV35S*) (Benfey and Chua, 1989, 1990) and *actin 1* (McElroy *et al.*, 1990). Although *CaMV35S* has been widely used in a number of dicotyledonous plant transformation systems, it has low activity in monocotyledonous systems (Wilmink *et al.*, 1995). Moreover, the pattern of *CaMV35S* promoter activity in different tissues of transgenic plants is difficult to predict (Benfey and Chua, 1990). In general, it has been found that monocot promoters are more active in monocot tissues than in dicot tissues (Wilmink *et al.*, 1995; Sharma *et al.*, 2004).

More recently, tissue-specific promoters have been successfully used for driving transgene expression solely in pith tissue. Phosphoenolpyruvate carboxylase (*PEPC*) regulative elements from maize can be used to direct gene expression specifically to green tissue (Hudspeth and Grula, 1989). From the perspective of potential crop yield, many transgenes should be expressed only in those organs where they are needed. For example, insect-resistant traits should be expressed only in those organs likely to be attacked by the insects. Otherwise, plants may be highly resistant, yet the metabolic cost may substantially reduce the crop yield. Tissue-specific expression also reduces the probability of unexpected negative effects on non-target organisms. It seems to be impossible to extrapolate results on gene expression levels from one species to another, and each crop needs be tested with a set of promoters to optimize results. Although the constitutive promoters such as *CaMV35S* are effective in providing high levels of gene expression, such expressions in some cases are not only unnecessary, but could also have unanticipated negative consequences towards non-target organisms. A more targeted expression of insecticidal genes by using tissue- and organ-specific promoters can form an important component for developing transgenic plants with resistance to insects (Wong *et al.*, 1992; Svab and Maliga, 1993; McBride *et al.*, 1995; Sharma *et al.*, 2004).

C. Genetically Modified Crops in the Food Chain

As of September 2011, there were 194 transgenic events for plants/crops approved worldwide for planting, or for use in animal feed or human foods. Twenty-four species, most of them suitable for consumption, were genetically modified to express a variety of traits, mainly herbicide tolerance and insect resistance (ISAAA GM Approval Database). Genetically engineered plants, also called biotech crops, are common commodities in today's agriculture and therefore an essential component of the food chain. In 2010, 148 million ha of biotech crops were grown in 29 countries, representing 10% of all 1.5 billion hectares of cropland in the world. The global worth of this seed alone was valued at US $11.2 billion in 2010, with commercial biotech maize, soybean grain, and cotton valued at approximately US $150 billion per year (James, 2010; Gatehouse *et al.*, 2011).

The advances in transgenic crop modification are often referred to as the 'green revolution'. A significant number of genetically modified food crop varieties gain higher yield and resistance to diseases and pests in many developed and developing countries. However, the resulting intensification and expansion of agriculture bring new forms of health and environmental challenges, for example, increased use of agrochemicals and intensified cultivation resulting in soil erosion. This highlights the fact that we are at the beginning of this green revolution and every genetic modification should be thoroughly evaluated for sustainability through systematic research (James, 2010; Gatehouse *et al.*, 2011).

Insect resistance and herbicide tolerance are the transgenic traits with the highest impact and acceptance on agriculture to date and account for over 99% of global GM crop area (James, 2010; Gatehouse *et al.*, 2011). In addition, there are limited varieties exhibiting virus resistance, delayed ripening, altered nutrient contents, medical applications, improved processing, and altered plant fertility. In the process of genetic engineering, traits establishing resistance to the antibiotics ampicillin, kanamycin, and streptomycin have been cointroduced to allow for the selection of agrobio traits.

D. Insect-Resistant Genetically Modified Crops

1. Bacillus thuringiensis *Toxin-Expressing Crops*

Crops containing *Bacillus thuringiensis* (*Bt*) toxins are currently the only commercialized insect-resistant genetically modified crops. Plants producing *B. thuringiensis* toxin are one of the dominant genetically engineered crops grown on a large scale and in many regions of the world (James, 2009; Then, 2010). Their development progressed rapidly from early genetically engineered tobacco plants in 1985 (Höfte *et al.*, 1986; Vaeck *et al.*, 1987). In 1995, potato plants producing *B. thuringiensis* toxin were approved as safe by the US Environmental Protection Agency (EPA), making them the first pesticide-producing crop to be approved in the USA. Since 1996, *Bt* maize, *Bt* potato, and *Bt* cotton have been grown by farmers in the USA (James, 2010).

Bacillus thuringiensis is a spore-forming pathogenic bacterium distinguished from other members of the *Bacillus* group by its production of crystalline inclusions known as Cry δ-endotoxins. The insect-resistant transgenic crops first commercialized in the mid-1990s expressed all genes encoding entomocidal δ-endotoxins from *B. thuringiensis*, also known as Cry proteins (Soberón *et al.*, 2010; Zhang *et al.*, 2011; Gatehouse *et al.*, 2011). The concept of using Cry proteins was not novel as *B. thuringiensis* spray formulations have been used commercially for approximately four decades to control insect pests, and in particular Lepidoptera (Cannon, 1996). Individual *Bt* toxins have defined insecticidal activity, usually restricted to a few species in one particular order of Lepidoptera (butterflies and moths), Diptera (flies and mosquitoes), Coleoptera (beetles and weevils), Hymenoptera (wasps and bees), and nematodes (de Maagd *et al.*, 2001). Early commercial varieties of insect-resistant transgenic crops expressed single Cry proteins with specific activity against lepidopteran pests, such as cotton expressing Cry1Ac or maize expressing Cry1Ab. Subsequently, other lepidopteran-active *Bt* toxins, such as Cry1F and Cry2Ab2, were introduced and often presented as pyramided genes in a single variety (cotton expressing both Cry1F + Cry1Ac, or cotton expressing Cry1Ac + Cry2Ab2). Cry3 toxins with activity against coleopteran pests are also being used in commercial transgenic crops, particularly maize, to protect against chrysomelid rootworms (e.g. maize expressing Cry3Bb1, maize expressing Cry34Ab1 and Cry35Ab1, and maize expressing a modified version of Cry3A). More recently released transgenic maize varieties express genes encoding Cry proteins active against Lepidoptera and Coleoptera. In China, *Bt* cotton cultivars expressing Cry1Ac together with a modified cowpea trypsin inhibitor (CpTI), which enhances effectiveness, were commercially released in 2000 (Gatehouse *et al.*, 2011).

2. Cry Proteins: Mode of Action

Even though Cry toxins have been extensively used commercially, the specifics of their mode of action are still controversial; however, a multistep toxicity process is highly probable (Figure 13.2). The toxicity of *Bt* toxins in target organisms depends on very specific factors such as intestinal pH, proteases, and receptors (Oppert, 1999; de Maagd *et al.*, 2001).

In the spores of *B. thuringiensis* the Bt toxins are produced as inactive crystalline protoxins. Upon ingestion by a susceptible insect larva, the alkaline midgut environment promotes solubilization of crystalline inclusions, releasing the protoxins. Subsequent cleavage by gut proteases at the amino- and carboxyl-terminal ends generates a 65−70 kDa truncated protein, the active δ-endotoxin (Höfte and Whiteley, 1989).

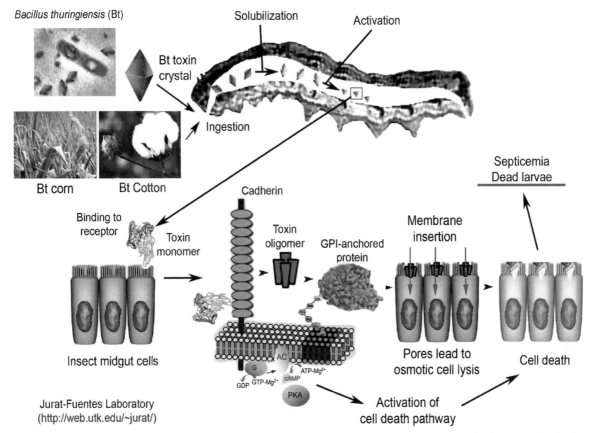

FIGURE 13.2 **Suggested mode of action of Cry proteins.** The multistep process of Cry toxicity includes ingestion by the insect, solubilization of the proteins, activation by peptide cleavage, binding to the specific cellular receptor catherin and GPI-anchored proteins, which results in activation of cell death pathways and/or oligomerization and pore formation. The pores will allow the release of intracellular contents resulting in cell lysis. *(Source: Jurat-Fuentes Laboratory: http://web.utk.edu/~jurat/.)*

The proteolytically activated toxin binds to receptors (some cadherins have been identified) located on the apical microvillus membrane of epithelial midgut cells (de Maagd *et al.*, 2001; Soberón *et al.*, 2010). Toxin binding to cadherin proteins results in activation of an oncotic cell death pathway and/or formation of toxin oligomers that bind to Glycosylphosphatidylinositol (GPI)-anchored proteins in regions of the cell membrane called lipid rafts (Figure 13.2). Receptor binding causes a conformation change, allowing the proteins' insertion into the cell membrane (Vadlamudi *et al.*, 1995). In the membrane environment several proteins oligomerize to form a pore that leads to osmotic cell lysis followed by death of the insect (de Maagd *et al.*, 2001; Likitvivatanavong *et al.*, 2011; Bravo *et al.*, 2011). Whether oncosis, pore formation, or both mechanisms are ultimately responsible for enterocyte death is still controversial (Figure 13.2). For more information on specific binding refer to Soberón *et al.* (2010) and Likitvivatanavong *et al.* (2011).

In most commercial crop varieties, the Cry proteins are expressed in the active form and as such differ from those used in biopesticide formulations where the Cry proteins are present as protoxins (Gatehouse *et al.*, 2011).

The significant future potential of these proteins is demonstrated by the fact that 634 different variations of toxins have so far been identified in diverse strains of *B. thuringiensis* (Crickmore, 2011).

3. Bacillus thuringiensis *Toxins: Environmental and Health Impact*

Owing to the specific intrinsic factors, such as pH, specific proteases, and specific receptor binding required for activation, the effects of *Bt* toxins are limited to the target insects, with little or no impact on humans, wildlife, pollinators, and most other beneficial insects (Gill *et al.*, 1992; Oppert, 1999; de Maagd *et al.*, 2001). The currently commercialized *Bt* toxins do not exhibit acute toxicity in mammals. Because no toxic effects of *Bt* toxins in humans have been detected in its 70 years of use, it is now considered an acceptable pest-control measure for the organic food industry (Whalon and Wingerd, 2003).

4. Commercial Bacillus thuringiensis Crops

Bacillus thuringiensis Cry proteins are being integrated in an ever-increasing variety of commercial corps, as demonstrated in Table 13.1.

TABLE 13.1 Selection of Commercial Transgenic Crops

Common Name of Food Product	Common Name of Parent Organism and OECD Unique Identifier	Details of Modified Trait	Gene(s) Responsible for Modified Trait
Corn			
Corn oil, flour, sugar or syrup	Yieldgard corn (MON810)	Insect protected	*cry 1A(b)* gene from *B. thuringiensis* subsp. *kurstaki*
	MON-00810-6		
Corn oil, flour, sugar or syrup	Bt-176 corn	Insect protected	*cry1A(b)* gene from *B. thuringiensis* subsp. *kurstaki*
	SYN-EV176-9		
Corn oil, flour, sugar or syrup	Bt-11 corn	Insect protected and glufosinate-ammonium tolerant	*cry1A(b)* gene from *B. thuringiensis*, *pat* gene derived from the bacteria *Streptomyces viridochromogenes*
	SYN-BT011-1		
Corn oil, flour, sugar or syrup	Bt – Liberty Link Corn line DBT418	Insect protected and glufosinate-ammonium tolerant	Gene for CryIAc from *B. thuringiensis*. Gene for Bar is derived from *Streptomyces hygroscopicus*
	DKB-89614-9		
Corn oil, flour, sugar or syrup	Herculex Insect Protection	Protected from major lepidopteran insect pests, including the European corn borer, and glufosinate-ammonium tolerant	*cry1F* gene is a synthetic version of a gene from *B. thuringiensis*. The *pat* gene is derived from *Streptomyces viridochromogenes*
	DAS-01507-1		
Corn oil, flour, sugar or syrup	Insect Resistant MON863 corn	Protected against corn rootworm	*cry3Bb1* variant derived from *B. thuringiensis*
	MON-00863-5		
Corn oil, flour, sugar or syrup	Bt Cry34/35Ab1 corn	Corn rootworm protection and glufosinate-ammonium tolerant	*cry34Ab1* and *cry35Ab1* genes from *B. thuringiensis* and the *pat* gene from *Streptomyces viridochromogenes*
	DAS-59122-7		
Foods derived from corn, e.g. corn syrup, flour, and oil	MIR604 corn	Corn rootworm protected	A modified *Cry3A* gene from *B. thuringiensis*. Also contains the *pmi* gene from *E. coli* (encodes phosphomannose isomerase, a selectable marker)
	SYN-IR604-5		
Foods derived from corn, e.g. corn syrup, flour, and oil	Roundup ready, insect protected corn MON88017	Glyphosate tolerance and corn rootworm protected	*Cry3Bb1* from *B. thuringiensis* and *cp4 EPSPS*, derived from *Agrobacterium* species strain *CP4*
	MON-88017-3		
Cotton			
Cotton oils and linters	Ingard® cotton	Insect protected	*cry1Ac* gene from *B. thuringiensis*

TABLE 13.1 Selection of Commercial Transgenic Crops—cont'd

Common Name of Food Product	Common Name of Parent Organism and OECD Unique Identifier	Details of Modified Trait	Gene(s) Responsible for Modified Trait
	MON-00531-6		
	MON-00757-7		
	MON-89924-2		
Cotton oils and linters	Bollgard II® cotton	Insect protected	*cry1Ac* and *cry2Ab* genes from *B. thuringiensis*
	MON-15985-7		
Cottonseed oils and linters	COT102 cotton	Insect protected	*vip3a* gene from *B. thuringiensis* subsp. *kurstaki* and *hph* gene (hygromycin resistance) from *E. coli*
	SYN-IR102-7		
Cottonseed oils and linters	MXB-13 cotton	Insect-protected and glufosinate-ammonium tolerant	*cry1Ac* and *cry1F* genes from *B. thuringiensis* subsp. *kurstaki* and *aizawai*, respectively, and the *pat* gene from *Streptomyces viridochromogenes*
	DAS-24236-5 × DAS-21023-5		
Potatoes			
Potatoes	New Leaf® potatoes	Protected against a range of insects, including the Colorado potato beetle	*cry3Aa* gene bacterium *B. thuringiensis* subsp. *tenebrionis* (*B.t.t.*)
	NMK-89812-3		
	NMK-89170-9		
	NMK-89879-1		
	NMK-89576-1		
Potatoes	New Leaf® Plus potatoes	Protection against CPB and PLRV	*cry3Aa* gene from the soil bacterium *B. thuringiensis* subsp. *tenebrionis* (*B.t.t.*) and *PLRVrep* gene from PLRV
	NMK-89185-6		
	NMK-89684-1		
	NMK-89896-6		
Potatoes	New Leaf® Y potatoes	Protected against a range of insects, including CPB, and protected against potato virus Y	*cry3Aa* gene from the soil bacterium *B. thuringiensis* subsp. *tenebrionis* (*B.t.t.*) and coat protein gene (*PVYcp*) from potato virus Y
	NMK-89653-6		
	NMK-89935-9		
	NMK-89930-4		
Soybean			
Soy foods, soybean oil or protein meal	MON-87701-2	Insect protected from lepidopteran larvae	*cry1Ac* gene from *B. thuringiensis* subsp. *kurstaki*

CPB: Colorado potato beetle; PLRV: potato leafroll virus; *B. thuringiensis*: *Bacillus thuringiensis*.

E. Herbicide-Tolerant Crops

From 1996 to 2010, herbicide-tolerant crops consistently occupied the largest planting area of biotech crops. In 2010 alone, herbicide-tolerant crops occupied 89.3 million hectares or 61% of the 148 million hectares of biotech crops planted globally. In 2010, 93% of all soybeans grown in the USA were herbicide resistant, as were 78% of all cotton and 70% of all maize varieties (USDA, 2011). The most common are the glyphosate- and glufosinate-tolerant varieties.

Many herbicides kill plants by interfering with enzyme functions in the plant. Most herbicides exert their effect on a single enzyme which catalyzes a key metabolic reaction in the plant. In general, plants exhibit a range of sensitivities to the herbicides used in agriculture, with some species exhibiting considerable tolerance to a single herbicide. There are several mechanisms by which plants can tolerate exposure to herbicides:

- The plant produces an enzyme which detoxifies the herbicide.
- The plant produces an altered target enzyme which is not affected by the herbicide.
- The plant produces physical or physiological barriers to uptake of the herbicide into the plant tissues and cells (OECD, 1999; Devine and Shukla, 2000).

Strategies to engineer herbicide tolerance into crops focus on the first two mechanisms. Transgenes are introduced to express enzymes detoxifying the herbicide or to replace the intrinsic herbicide-target enzyme with a variant resistant to the inhibition.

This section will summarize the information available on the source of the genes that have been used to construct herbicide-tolerant transgenic plants and the nature of the enzymes they encode and their role in the plants' metabolism. It will not discuss the wealth of information available on the herbicides or the uses of the herbicide in agricultural and other applications. Food safety aspects of the use of herbicide-tolerant transgenic plants will not be discussed extensively. Additional information on aspects not discussed is available from other sources, including the respective governmental organizations that regulate the use of the herbicides.

1. Tolerance to the Herbicide Glyphosate

Glyphosate (*N*-phosphonomethyl-glycine) is the most widely used broad-spectrum weed control agent with minimal human and environmental toxicity (Duke and Powles, 2008; Pollegioni *et al.*, 2011). It is not a natural product — it is derived by chemical synthesis — and it is the active ingredient of the herbicide Roundup® (Monsanto):

$$HO - \overset{\overset{\displaystyle O}{\|}}{\underset{\underset{\displaystyle OH}{|}}{P}} - CH_2 - NH - CH_2 - \overset{\overset{\displaystyle O}{\|}}{C} - OH$$

Glyphosate is a highly specific reversible competitive inhibitor of the enzyme 5-enolpyruvyl-3-phosphoshikimic acid synthase (EPSPS, EC 2.5.1.19) which catalyzes the transfer of the enolpyruvyl moiety of phosphoenolpyruvate (PEP) to the 5-hydroxyl of shikimate 3-phosphate (S3P) to produce 5-enolpyruvyl shikimate 3-phosphate (EPSP) and inorganic phosphate (KEGG R03460):

Phosphoenolpyruvate + Shikimate 3-phosphate <=> Orthophosphate + 5-*O*-(1-Carboxyvinyl)-3-phosphoshikimate

In this reaction glyphosate competes with phosphoenolpyruvic acid, but it does not inhibit any other phosphoenolpyruvic acid-dependent enzymatic reactions. This reaction forms the sixth step in the shikimate pathway leading to the synthesis of aromatic amino acids and other aromatic compounds in plants, fungi, bacteria, and apicomplexan parasites. The following (Figure 13.3) graph illustrates the shikimate pathway that leads to the biosynthesis of aromatic amino acids, and the mode of action of glyphosate on the reaction catalyzed by EPSPS (Pollegioni *et al.*, 2011):

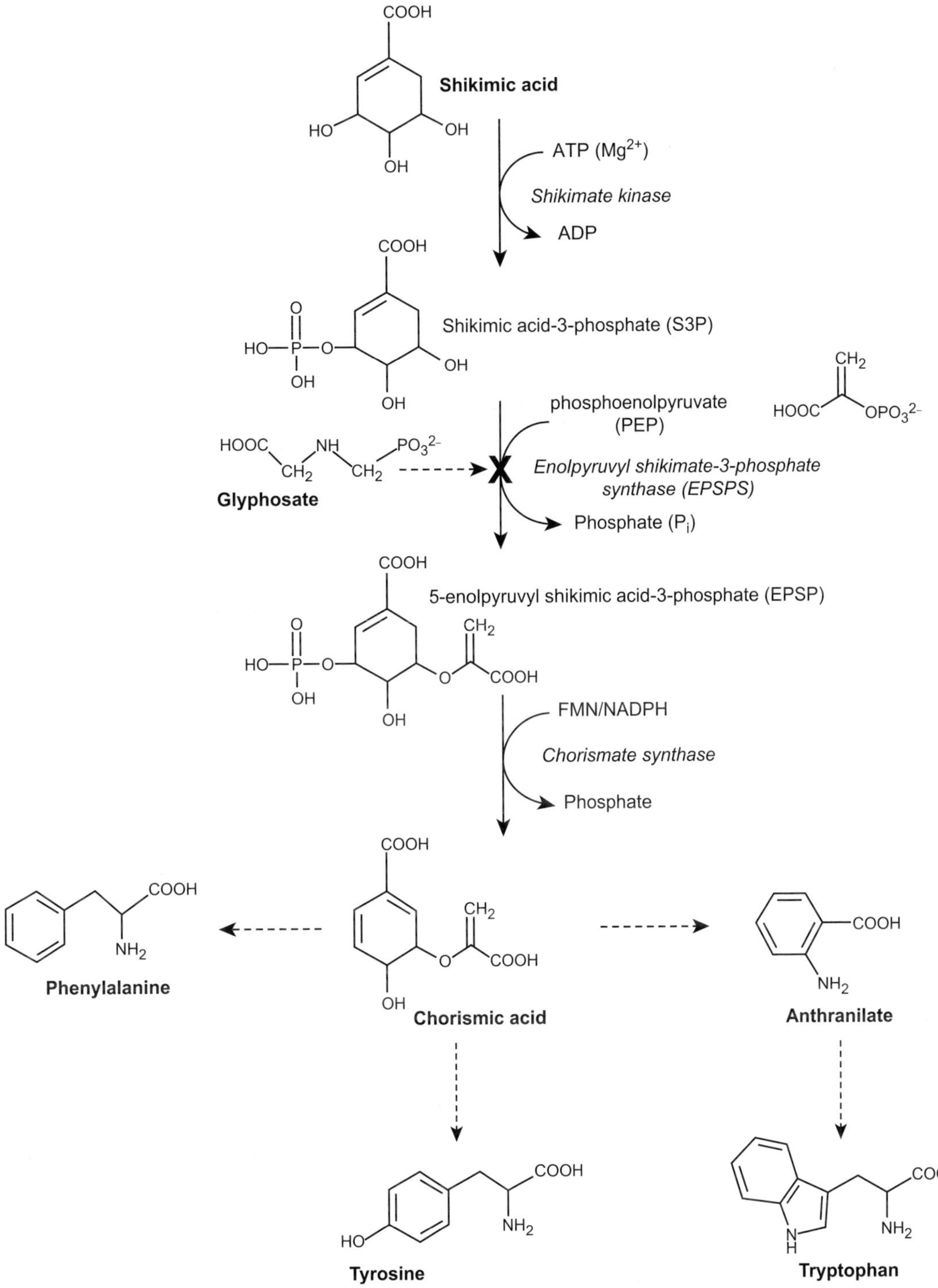

FIGURE 13.3   The shikimate pathway of aromatic amino acids biosynthesis and its inhibition by glyphosphate, which acis on the enzyme enolpyruvylshikimate-3-phosphate synthase *(Pollegiotti et al., 2011).*

As a consequence of the inhibition of aromatic amino acid biosynthesis, protein synthesis is disrupted, resulting in the plant's death (Kishore and Shah, 1988). No orthologs of the *EPSPS* gene are found in animals (Steinrucken and Amrhein, 1980).

The high sensitivity of crop plants to glyphosate has limited its use as a pre-crop emergence herbicide in no-till management strategies, and as a herbicide and crop desiccant when applied shortly before crop harvest. Traditional mutagenesis and selection techniques have failed to produce a useful level of tolerance in crop plants (OECD, 2010). With the development of genetically engineered crop plants that are resistant to glyphosate, this herbicide can be applied after both crops and weeds have emerged, with little or no damage to the crop (James, 2009).

Based on the knowledge of the mode of action of glyphosate, two major transgenic strategies have proven successful to confer herbicide tolerance, i.e. to produce glyphosate-tolerant plants: introduction of a glyphosate-tolerant EPSPS enzyme and introduction of the glyphosate-inactivating enzyme, glyphosate oxidoreductase (GOX). Recombinant DNA techniques have been used to express genes that encode glyphosate-tolerant EPSPS enzyme alone or a combination of EPSPS and GOX genes in susceptible plants (Nida *et al.*, 1996; Padgette *et al.*, 1995, 1996).

2. Glyphosate-Tolerant 5-Enolpyruvyl-3-Phosphoshikimic Acid Synthase Transgenes

The revolutionary new glyphosate use pattern commenced in 1996 with the introduction of a transgenic glyphosate-resistant soybean, launched and marketed under the Roundup Ready brand in the USA. Two EPSPS genes which provide field-level tolerance to glyphosate have been introduced into commercial cultivars.

Agrobacterium sp. CP4, isolated from a waste-fed column at a glyphosate production facility, yielded a glyphosate-resistant, kinetically efficient EPSPS (the so-called CP4 EPSPS) that is suitable for the production of transgenic, glyphosate-tolerant crops. The CP4 enzyme has unexpected kinetic and structural properties that make it unique among the known EPSPSs, and it is therefore considered to be a prototypic class II EPSPS (for further information see Pollegioni *et al.*, 2011). An intriguing feature is the strong dependence of the catalytic activity on monovalent cations, namely K^+ and NH_4^+. The lack of inhibitory potential ($K_i > 6$ mM) is primarily attributed to alanine (Ala)[100] and leucine (Leu)[105] in place of the conserved plant residues glycine (Gly)[96] and proline (Pro)[101] (Figure 13.3). The presence of Ala[100] in the CP4 enzyme is of no consequence for the binding of phosphoenolpyruvic acid, but glyphosate can only bind in a condensed, high-energy, and non-inhibitory conformation. Glyphosate sensitivity is partly restored by mutation of Ala[100] to glycine, allowing glyphosate to bind in its extended, inhibitory conformation (Pollegioni *et al.*, 2011).

When the *Agrobacterium* EPSPS is present in transgenic plants it realizes the aromatic amino acid needs of the plant in the presence of glyphosate, whereas the plant version of this enzyme (ubiquitous in nature) is sensitive to glyphosate. *Agrobacterium* spp. are not human or animal pathogens, but some species are pathogenic to plants (Holt, 1984; Croon, 1996).

Recently, the EPSPS gene from corn (*Zea mays*) has been mutated *in vitro* to obtain a glyphosate-tolerant enzyme. In the tolerant version of the enzyme, called mEPSPS, threonine (Thr)[102] is replaced by isoleucine (Ile) and proline (Pro)[106] is replaced by serine (Ser) (Monsanto, 1997). The location of the changed amino acids resembles the situation in the *Agrobacterium* isoform and therefore confers glyphosate tolerance by the mechanism described in Figure 13.4.

EPSPS enzyme is synthesized in the cytoplasm and then transported to the chloroplast (Kishore and Shah, 1988). The translocation of the protein to the chloroplast is carried out by an N-terminal protein sequence called the chloroplast transit peptide. Chloroplast transit peptides are typically cleaved from a mature protein and degraded following delivery to the plastid (Della-Cioppa *et al.*, 1986). A plant-derived coding sequence expressing a chloroplast transit peptide is often linked with each of the genes imparting glyphosate tolerance. This peptide facilitates the import of the newly translated enzymes into the chloroplasts, the site of both the shikimate pathway and glyphosate mode of action (OECD, 2010).

The constitution of the expression cassette for the CP4 EPSPS integrated into the sugar beet line H7-1 illustrates the different genetic elements necessary for optimal expression and targeting of the transgene. It contains a region of T-DNA that is delineated by left and right border sequences, and contains a single *cp4 epsps* gene with essential regulatory elements necessary for expression in the chloroplasts of the sugar beet plants. The organization of the T-DNA, corresponding to approximately 3.4 kb, is depicted in Figure 13.5 (FSANZ A525). The size and function of each of the genetic elements present in the expression cassette are described in Table 13.2 (FSANZ A525).

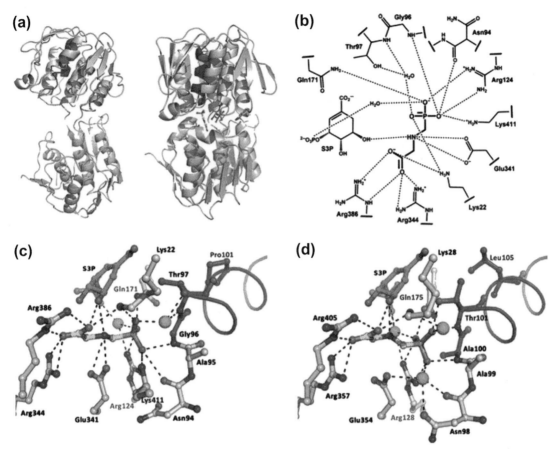

FIGURE 13.4 Molecular mode of action of glyphosate and the structural basis for glyphosate resistance. (a) In its ligand-free state, 5-enolpyruvyl-3-phoshoshikimic acid synthase (EPSPS) exists in the open conformation (left; PDB: 1eps). Binding of S3P induces a large conformational change in the enzyme to the closed state, to which glyphosate or the substrate PEP binds (right, PDB: 1g6s). The respective crystal structures of the E. coli enzyme are shown. (b) Schematic representation of potential hydrogen-bonding and electrostatic interactions between glyphosate and active site residues including bridging water molecules in EPSPS from E. coli (PDB: 1g6s). (c) The glyphosate-binding site in EPSPS from E. coli (PDB: 1g6s). Water molecules are shown as spheres, and the residues known to confer glyphosate resistance upon mutation are Pro[101], Thr[97], and Gly[96]. (d) The glyphosate-binding site in CP4 EPSPS (PDB: 2gga). The spatial arrangement of the highly conserved active site residues is almost identical for class I (E. coli) and class II (CP4) enzymes, with the exception of an alanine at position 100 (Gly[96] in E. coli). Another significant difference is the replacement of Pro[101] (E. coli) by a leucine (Leu[105]) in the CP4 enzyme. Note the markedly different, condensed conformation of glyphosate as a result of the reduced space provided for binding in the CP4 enzyme. *(Modified from Pollegioni et al., 2011.)*

In order to achieve efficient gene expression and protein translation of bacterial genes within plants the codon usage needs to be optimized. This has been done for the codons of the *Agrobacterium* glyphosate-tolerant EPSPS gene by chemical DNA synthesis for codon optimization in plants. The amino acid sequence of the resulting enzymes is not changed. In expression cassettes, the gene coding sequences are usually linked to regulatory sequences like promoters, terminators, enhancers, and introns, which are also optimized for plant expression (Parrish *et al.*, 1995).

During the life cycle of any herbicide-tolerant plant, the plant is exposed only rarely to the herbicide. Except for the production of the enzyme(s) encoding glyphosate tolerance, there should be no other changes in plant metabolism. After glyphosate application, the enzyme activities expressed by the transgenes enable the plant to survive herbicide exposure. In the case of introduced EPSPS, no new metabolic products are formed since the only difference from the native enzyme is its insensitivity to glyphosate. However, in situations of very high expression the levels of downstream metabolites might change (OECD, 1999).

Expression of glyphosate-tolerant EPSPS is not detrimental to plant growth, since such crops have agronomic performance similar to their parents. Governmental regulatory agencies in the USA (US Department of Agriculture, 1994, 1995, 1997), Canada (Agriculture and Agrifood Canada, 1995, 1996), Japan (Ministry of Agriculture, Forestry and Fisheries, 1996), and the European Union (European Commission, 1998a, b) have made decisions that the

FIGURE 13.5 Organization of the T-DNA cassette for the CP4 EPSPS gene integrated into the sugar beet line H7-1.

Right Border   Left Border

| P-FMV | ctp2 | Cp4 epsps | E93' |

TABLE 13.2 Size and Function of Genetic Elements Present in the Expression Cassette for the cp4 epsps Gene

Genetic Element	Size (kb)	Description and Reference
Right border	0.025	A 21−25 bp nucleotide sequence that acts as the initial point of DNA transfer into plant cells, originally isolated from *Agrobacterium tumefaciens* plasmid pTiT37
P-FMV	0.672	The 35S gene promoter from a modified figwort mosaic virus (FMV)
ctp2	0.31	The N-terminal chloroplast transit peptide sequence from the *Arabidopsis thaliana epsps* coding region
cp4 epsps	1.363	The 5-enolpyruvylshikimate-3-phosphate synthase (EPSPS) coding region from *Agrobacterium* sp. strain CP4
E93'	0.63	The 3' end of the *Pisum sativum rbc*S E9 gene, containing polyadenylation sites that direct mRNA processing and polyadenylation
Left border	0.025	A 21−25 bp nucleotide sequence that delimits the T-DNA transfer into plant cells, originally isolated from *A. tumefaciens* plasmid pTi 15955, a derivative of the octopine type plasmid, pTiA6

presence of the EPSPS protein in plants does not result in plants that are unsafe in their environments. Several lines of evidence support the conclusion that these enzymes show low mammalian toxicity:

- Neither enzyme shows amino acid homology to known allergens or mammalian toxins.
- Data from acute oral toxicity tests at high concentration of enzymes showed no toxicity (Harrison *et al.*, 1996). In acute oral toxicity tests of bacterially derived CP4 EPSPS protein, no adverse effects occurred at a dose of 572 mg/kg body weight of the test animals.
- The enzyme is readily inactivated by heat or mild acidic conditions and is readily degraded in an *in vitro* digestibility assay, which is consistent with the lack of oral toxicity (US EPA, 1996, 1997). This is consistent with the observation that most enzymes are not considered toxic to vertebrates (Kessler *et al.*, 1992).

3. Glyphosate Oxidoreductase

Glyphosate oxidoreductase (GOX) catalyzes the oxidative cleavage of the carbon−nitrogen bond on the carboxyl side of glyphosate, resulting in the formation of aminomethylphosphonic acid (AMPA) and glyoxylate, effectively deactivating the substrate as a herbicide (Pollegioni *et al.*, 2011):

AMPA pathway

B

$$HO-P(=O)(OH)-CH_2-NH-CH_2-C(=O)(OH) + E\text{-}FAD \longrightarrow HO-P(=O)(OH)-CH_2-NH_2 + O=CH-C(=O)(OH) + E\text{-}FADH_2$$

Glyphosate   [Glycine oxidase]   AMPA   Glyoxylate

$$E\text{-}FADH_2 + O_2 \longrightarrow E\text{-}FAD + H_2O_2$$

The gene that encodes for the glyphosate oxidoreductase was isolated from *Achromobacter* strain LBAA, a soil bacterium ubiquitous in nature (Barry and Kishore, 1997). GOX requires flavin adenine dinucleotide (FAD) and magnesium for activity; therefore, it is more appropriately designated an apoenzyme.

To facilitate the expression of GOX in plants, the gene sequence was redesigned to eliminate stretches of G and C of five or greater, A + T-rich regions that could function as polyadenylation sites or potential RNA-destabilizing regions, and codons not frequently found in plant genes. When this gene is transfected into plants, expression of GOX results in glyphosate tolerance (Pollegioni *et al.*, 2011).

The presence of GOX is not expected to have an impact on the plant's metabolome, since it only converts glyphosate to AMPA and glyoxylate when glyphosate herbicide is applied. Since aminomethylphosphonic is a naturally occurring plant metabolite involved in carbon cycling, it will be further metabolized to provide intermediates for the Krebs cycle. GOX is highly specific for its substrate, glyphosate; therefore, in the absence of glyphosate no metabolites are expected. The US EPA has decided that only glyphosate residues are to be regulated in plant and animal commodities, and that the major metabolite AMPA is not of toxicological concern regardless of its level in food (US EPA, 1997). Expression of GOX is not detrimental to plant growth, since such crops have agronomic performance similar to their parents. The presence of the GOX protein in plants does not result in plants that are unsafe in their environments or exhibit mammalian toxicity (US EPA, 1997).

4. Tolerance to the Herbicide Glufosinate

Glufosinate [phosphinothricin; DL-homoalanin-4-yl(methyl)phosphinic acid] is a racemic phosphinico amino acid (Hoerlein, 1994):

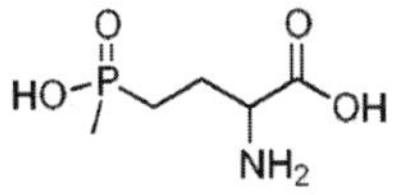

Its ammonium salt (glufosinate-ammonium) is widely used as a non-selective herbicide and is the active ingredient of many commercial herbicide formulations. The L-isomer of glufosinate is a structural analog of glutamate and, therefore, is a competitive inhibitor of the enzyme glutamine synthetase (GS) of bacteria and plants. The D-isomer is not a glutamine synthetase inhibitor and is not herbicidally active (OECD, 2002).

Owing to the inhibition of glutamine synthetase, non-tolerant plant cells accumulate large amounts of toxic ammonia produced by nitrate assimilation and photorespiration (Tachibana *et al.*, 1986) and the level of available glutamine drops (Sauer *et al.*, 1987). Damage to cell membranes and inhibition of photosynthesis are followed by plant cell death. The action of glufosinate is dependent on environmental conditions. Temperatures below 10°C, as well as drought stress, reduce its efficacy because of the limited metabolic activity of the plant (OECD, 2002).

Glufosinate is a contact herbicide and is taken up by the plant primarily through the leaves. There is no uptake from the soil through the roots, presumably because of the rapid degradation of glufosinate by soil microorganisms. There is limited translocation of glufosinate within the plant. After application of L-glufosinate, *N*-acetyl-L-glufosinate and further metabolites on distinct leaves, a preferential transport into the upper leaves and a low level of translocation into the lower plant parts were observed in both genetically modified and unmodified tobacco plants (Dröge-Laser *et al.*, 1994).

Glufosinate has a wide spectrum of activity encompassing monocotyledonous and dicotyledonous species. Because of its limited systemic action, there is no enduring effect on perennial weeds. Weeds emerging after herbicide application are not affected (OECD, 2002).

5. Phosphinothricin Acetyltransferase

Phosphinothricin acetyltransferase (EC 2.3.1.183) catalyzes the acetylation of L-phosphinothricin (KEGG R08938):

Acetyl-CoA + Glufosinate <=> CoA + *N*-Acetylphosphinothricin

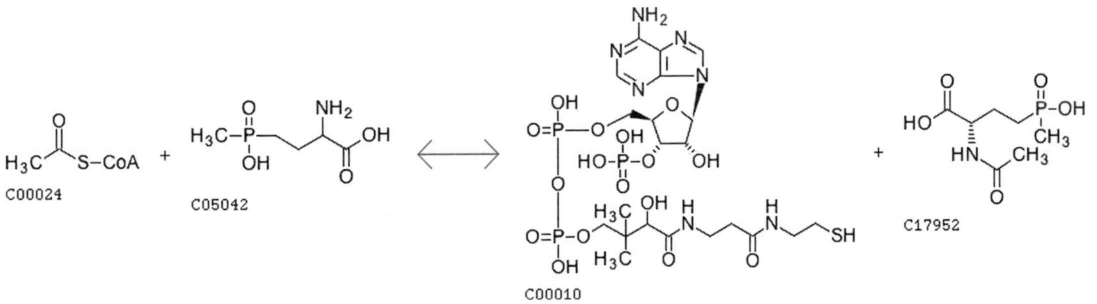

The phosphinothricin acetyltransferase (PAT) proteins, which are encoded by the *bar* coding sequence from *Streptomyces hygroscopicus* or the *pat* coding sequence from *Streptomyces viridochromogenes*, are present in glufosinate-ammonium-tolerant plant varieties of various crops such as corn, cotton, rice, oilseed rape, and soybean. The PAT enzymes acetylate L-phosphinothricin, the active isomer of the glufosinate-ammonium herbicide, resulting in tolerance of transgenic plants to postemergent application of the non-selective herbicide (Hérouet *et al.*, 2005).

The native *S. hygroscopicus pat* gene has been resynthesized to modify codon usage for improved protein expression in plant cells. At the nucleotide sequence level, the synthetic gene demonstrates 70% homology with the native *pat* gene from *S. viridochromogenes*. The amino acid sequences of the PAT enzyme encoded by the native and synthetic genes are identical.

Using recombinant DNA technologies, the *bar* or the *pat* coding sequences are each fused to an appropriate promoter and terminator for plant gene expression and introduced into the plant genome.

The *bar* gene is often under the control of the plant promoter (*Pssu-Ara*), which directs expression of PAT predominantly in the green tissues (leaves and stems). Alternatively, in constructs involving the synthetic *pat* gene, the plant viral promoter *P35S* has been used for constitutive expression of the PAT protein in all tissues of the plant.

In genetically modified glufosinate-tolerant plants expressing the PAT enzyme, it appears that two metabolic routes compete:

- The deamination of glufosinate and subsequent conversion of 4-methylphosphinico-2-oxo-butanoic acid (PPO) to 3-methylphosphinico-propionic acid (MPP) or to 4-methylphosphinico-2-hydroxy-butanoic acid.
- The N-acetylation of L-glufosinate by PAT (Dröge-Laser *et al.*, 1994). The second of these two routes predominates when PAT specific activity is relatively high (OECD, 2002).

In all glufosinate-tolerant crops, the principal metabolic residues are *N*-acetyl-L-glufosinate, glufosinate-ammonium, and 3-methylphosphinico-propionic acid. Besides these principal residues, trace levels of other metabolites were also identified in soybean including 2-methylphosphinico-acetic acid (MPA) and 4-methylphosphinico-butanoic acid (MPB). The herbicidally inactive D-glufosinate appears to be stable in plants owing to the L-specific acetylation activity of the PAT enzyme (Dröge-Laser *et al.*, 1994).

The *pat* and the *bar* coding sequences as well as their respective PAT proteins are not toxic and do not possess any of the characteristics associated with food allergens. There is a reasonable certainty of no harm resulting from the inclusion of the PAT proteins in human food or in animal feed (Hérouet *et al.*, 2005).

6. Tolerance to the Herbicide Bromoxynil

Bromoxynil (3,5-dibromo-4-hydroxybenzonitrile) is a systemic herbicide, which is widely used for post-emergence control of annual broadleaved weeds, especially young seedlings of cereals, ryegrass-seed crops, turf, and non-crop land:

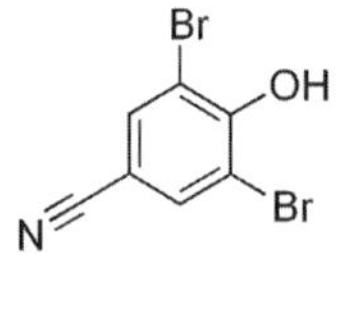

It is often used in combination with other herbicides to extend the spectrum of weed control. Bromoxynil inhibits photosynthetic electron transport and also uncouples oxidative phosphorylation in mitochondria, thereby stopping energy production and negatively affecting plant respiration (EPA738-R-98-013). Inhibition of electron transport causes superoxide production, resulting in the destruction of cell membranes and an inhibition of chlorophyll formation, leading to plant death. Bromoxynil octanoic and heptanoic acid esters are also applied as herbicides.

Bromoxynil nitrilase (EC 3.5.5.6) hydrolyzes carbon—nitrogen bonds, other than peptide bonds (KEGG R04349):

Bromoxynil + 2 H$_2$O <=> 3,5-Dibromo-4-hydroxybenzoate + NH$_3$

This degradation effectively inactivates the herbicide and enables the normally bromoxynil-sensitive plant to survive and grow when treated with applications of the herbicide.

Bromoxynil nitrilase is encoded by the *oxy* gene of the soil bacterium *Klebsiella pneumoniae* subsp. *ozaenae* (Stalker and McBride, 1987). When integrated in an expression cassette and transferred into plants, the gene, through its encoded protein, confers tolerance to the oxynil family of herbicides including bromoxynil and ioxynil.

In a typical expression cassette the *oxy* gene is flanked by specific sequences, as found in pBrx75, responsible for the trait in BXN cotton lines (FSANZ A379) (Figure 13.6, Table 13.3).

The 35s cauliflower mosaic virus promoter region ensures strong initiation of translation. The 3′ non-translated region of the *tml* gene from *A. tumefaciens* contains signals for termination of transcription and directs polyadenylation. The gene *nptII* codes for neomycin phosphotransferase II from Tn5 in *Escherichia coli* and confers resistance to the antibiotics kanamycin and neomycin. It is used as a selectable marker for plant transformation (Table 13.3).

F. Male Sterility: The *Barstar* and *Barnase* Gene System

Sterility of transgenic crops can be regulated through the selective expression of two genes called *barnase* and *barstar*, which control the development of pollen.

The *barnase* gene encodes the enzyme barnase, a ribonuclease derived from *Bacillus amyloliquefaciens*, which disrupts the production of RNA. When the *barnase* gene is selectively expressed in the tapetal cell layer of the anther, a cell layer that plays a vital nutritive role during pollen formation, it destroys the tapetal cell layer, rendering the anthers incapable of producing viable pollen grains. This inability to produce viable pollen grains renders the plant male sterile and provides reliable pollination control (FDA BNF No. 000031).

Fertility is restored by the expression of the *barstar* gene, also originating from *B. amyloliquefaciens*, coding for an inhibitor of the barnase protein. When the barnase protein is expressed in the tapetum cells it leads to the restoration of fertility by tightly binding and inactivating the barnase. The coexpression is achieved through traditional breeding of two separate crop lines, each carrying one of the two genes. Plant lines expressing *barstar* are referred to as fertility restorers.

Thus, the hybrid system consists of crossing a male sterile line (female parent) with a specific fertility restorer line, giving rise to progeny that are fully fertile. The primary objective of these modifications is the production of a range of parental lines with superior agronomic performance that are to be used in a breeding system for producing hybrids yielding significantly more seed (FSANZ A372).

G. Pathogen-Derived Virus-Resistant Crops

Plant viruses cause significant crop losses, specifically demonstrated by the effects of a few single viruses for a specific host in a particular geographical region (for reviews, see Bos, 1982; Waterworth and Hadidi, 1998). For

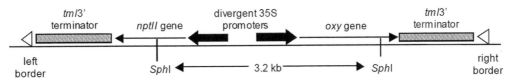

FIGURE 13.6 Expression cassette pBrx75 conferring bromoxynil resistance.

TABLE 13.3 Description of the Elements of Expression Cassettes pBrx75 Conferring Bromoxynil Resistance

Genetic Element	Source	Function
35S promoter	The cauliflower mosaic virus (CaMV) 35S promoter region	A promoter for high level constitutive expression in plant tissues
oxy	Gene isolated from *Klebsiella pneumoniae* subspecies *ozaenae* encoding the enzyme nitrilase	Inactivates the herbicide bromoxynil and confers bromoxynil tolerance when expressed in plants
tml 3.	The 3. non-translated region of the *tml* gene from *Agrobacterium tumefaciens* plasmid pTiA6	Contains signals for termination of transcription and directs polyadenylation
nptII	The gene coding for neomycin phosphotransferase II from Tn5 in *Escherichia coli*	Confers resistance to the antibiotics kanamycin and neomycin. Used as a selectable marker for plant transformation

instance, the annual global impact of *Tomato spotted wilt virus* (TSWV) and the viruses causing African cassava mosaic disease, which are of exceptional agronomic importance, has been estimated to be approximately US $1 billion each (Taylor *et al.*, 2004).

The concept of pathogen-derived resistance (PDR) offers a novel approach to developing virus-resistant crop plants. Resistance is engineered in otherwise susceptible hosts by introducing coding or non-coding DNA sequences from the pathogen's genome (Sanford and Johnston, 1985). For plant viruses, the concept of pathogen-derived resistance was first validated with the development of tobacco expressing the coat protein of *Tobacco mosaic virus* (TMV), which led to diminished or abolished infection (Powell Abel *et al.*, 1986). This breakthrough discovery paved the way for the creation of numerous virus-resistant transgenic plants, including horticultural crops. Crop plants expressing viral genetic elements have been tested successfully in the field and a few have been commercialized. The deployment of virus-resistant transgenic plants has become an important strategy for effective and sustainable control of major virus diseases (Gottula and Fuchs, 2009).

1. Mechanism of Pathogen-Derived Virus Resistance

Resistance to numerous plant viruses was initially engineered by using viral coat protein genes, following the discovery that transgenic tobacco expressing the coat protein gene of TMV exhibited resistance to infection (Powell Abel *et al.*, 1986; Register and Beachy, 1988; Prins *et al.*, 2008; Tepfer, 2002). However, other viral sequences, such as the RNA-dependent RNA polymerase readthrough domain of TMV, were also shown to induce resistance (Golembowski *et al.*, 1990), as well as the movement of protein (Malyshenko *et al.*, 1993), proteinase (Maiti *et al.*, 1993; Vardi *et al.*, 1993), satellite RNA (Gerlach *et al.*, 1987; Harrison *et al.*, 1987), defective interfering RNA (Kollar *et al.*, 1993), and 5′ (Stanley *et al.*, 1990; Nelson *et al.*, 1987) and 3′ (Zaccomer *et al.*, 1993) non-coding regions. It has become apparent that almost any viral genetic element could be used to confer resistance to virus infection in plants (Gottula and Fuchs, 2009).

The viral transgene protein product is not needed for engineered resistance but there is an inverse correlation between transgene expression and resistance to virus infection (Dougherty *et al.*, 1994). Therefore, the mRNA sequence is responsible for the resistance phenotype rather than the protein itself (Gottula and Fuchs, 2009), and the majority of pathogen-derived resistance phenomena seem to work through RNA-mediated mechanisms (Baulcombe, 2007; Eamens *et al.*, 2008; Prins *et al.*, 2008; Voinnet, 2008). Plant RNA-dependent RNA polymerase and double-stranded RNAase activities were proposed to be part of the mechanism of resistance by producing short RNA of 10—20 nucleotides in length complementary in sequence to the RNA to be degraded from the transgene RNAs (Lindbo *et al.*, 1993). These short RNAs would target specific RNAs for degradation by a dsRNase activity (Dougherty and Parks, 1995). These findings paved the way to the discovery of RNA silencing as a potent defense mechanism against plant viruses (Baulcombe, 2004, 2007; Eamens *et al.*, 2008; Lin *et al.*, 2007; Prins *et al.*, 2008; Voinnet, 2001, 2005, 2008; Waterhouse *et al.*, 1999, 2001).

RNA silencing is initiated by double-stranded RNA (dsRNA) structures that are identical to the RNA to be degraded (Waterhouse *et al.*, 1998). Silencing is associated with the production of 21—25 nucleotide duplexes called small interfering RNAs (siRNAs) (Hamilton and Baulcombe, 1999; Hamilton *et al.*, 2002). The siRNAs are produced from dsRNA precursors by an endonuclease known as Dicer and become incorporated and converted to single-stranded RNAs (ssRNAs) in an Argonaute-containing ribonuclease complex (RISC) that targets RNA for cleavage (Deleris *et al.*, 2006; Hannon, 2002; Obbard *et al.*, 2009; Voinnet, 2001, 2005, 2008). The pioneering work by Baulcombe and Waterhouse and their respective colleagues showed that RNA silencing is an innate and potent plant response to virus infection and a natural example of the concept of pathogen-derived virus resistance (Gottula and Fuchs, 2009).

2. Commercialized Virus-Resistant Transgenic Crops

a. Virus-Resistant Summer Squash

Summer squash expressing the coat protein gene of *Zucchini yellow mosaic virus* and *Watermelon mosaic virus* received exemption status in the USA in 1994 and was released thereafter. This was the first disease-resistant transgenic crop to be commercialized in the USA. Plants of line ZW-20 remain vigorous following exposure to aphid-mediated transmission of *Zucchini yellow mosaic virus* and *watermelon mosaic virus* and produce marketable fruits unlike conventional squash. Summer squash expressing the coat protein gene of *Cucumber mosaic virus*, *Watermelon mosaic virus*, and *Zucchini yellow mosaic virus* was deregulated and commercialized in 1996. Subsequently, numerous squash types and cultivars have been developed by crosses and backcrosses with the two initially deregulated lines. Currently there are five zucchini and six straightneck or crookneck yellow squash cultivars for which combinations of resistance to *Zucchini yellow mosaic virus* and *Watermelon mosaic virus* or resistance to all three viruses are available (Gottula and Fuchs, 2009).

The adoption of virus-resistant summer squash cultivars has increased steadily since 1996. In 2006, the adoption rate was estimated to be 22% (3250 ha) across the USA. The benefit to growers was estimated to be $24 million in 2006 (Gottula and Fuchs, 2009).

b. Virus-Resistant Papaya

Papaya expressing the coat protein gene of *Papaya ringspot virus* was deregulated in 1998 and commercialized in Hawaii. *Papaya ringspot virus* is a major limiting factor to papaya production in Hawaii and around the world. After extensive experimental testing, *Papaya ringspot virus*-resistant papaya was released in 1998 as devastation caused by the virus reached record proportions in the archipelago's main production region (Gonsalves, 1998). The impact of *Papaya ringspot virus*-resistant papaya on the papaya industry in Hawaii is evidenced by its rapid adoption rate. In 2000, the first wave of transgenic papaya bore fruit on more than 42% of the total acreage (Johnson *et al.*, 2007). Resumption of fruitful harvests put papaya packing houses back in business and provided a $4.3 million impact over a 6-year period (Gottula and Fuchs, 2009). By 2006, transgenic papaya cultivars were planted on more than 90% of the total papaya land in Hawaii (780 of a total 866 ha) (Johnson *et al.*, 2007).

Another PRSV-resistant papaya has been deregulated by two of the three US biotechnology regulatory authorities. Line X17-2 differs from the previously deregulated Hawaiian papaya in that it expresses the coat protein gene of a Florida isolate of *Papaya ringspot virus* and is suitable for cultivation in Florida (Davis, 2004). The US Department of Agriculture (USDA) Animal and Public Health Inspection Service (APHIS) and the FDA have granted X17-2 deregulated status (Shea, 2009).

c. Virus-Resistant Potato

Two virus-resistant potato lines were deregulated in 1998 and 2000 in the USA. After failed attempts to create a potato line resistant to *Potato leafroll virus* (PLRV) by coat protein gene expression, lines expressing a PLRV replicase gene were created, field tested, deregulated, and commercialized (Kaniewski and Thomas, 2004). Later, this resistance was stacked with a synthetic *Bt* gene that conferred resistance to Colorado potato beetle. Another potato cultivar was developed by adding the coat protein gene of PLRV. Although many growers in the Pacific Northwest, Midwest USA, and Canada were growing transgenic potato, and no resistance breakage or any detrimental impact on the environmental or human health was reported, virus-resistant potatoes were withdrawn from the market after the 2001 season owing to the reluctance of several large processors and exporters to adopt these products (Kaniewski and Thomas, 2004).

The PLRV-resistant line, called New Leaf Plus Potatoes, has been genetically engineered to express the Cry3a protein from *B. thuringiensis* subsp. *tenebrionis* (*B.t.t.*) and the *orf1/orf2* gene from PLRV. The mechanism of resistance provided by the *orf1/orf2* gene has not been elucidated at this time. Possible scenarios responsible for the inhibition of viral replication include:

- protein-driven inhibition (viral replication inhibited by the three proteins that could be encoded by the *orf1/orf2* gene)
- RNA-driven inhibition.

Protein-driven inhibition as a mechanism of action of the *orf1/orf2* gene is questionable owing to the lack of conclusive evidence that any of the potential proteins encoded for by the *orf1/orf2* gene are expressed.

d. Virus-Resistant Plum

On 7 May 2010, the Environmental Protection Agency (EPA) registered the pesticide product C5 HoneySweet Plum, which contains the new plant-incorporated protectant (PIP) active ingredient Coat Protein Gene of Plum Pox Virus (CPG-PPV) (EPA 2010 PC Code: 006354).

Plum pox virus (PPV) is a plant virus that reduces the quality of stone fruits, and eventually renders infected trees incapable of producing fruit. It was first described in Europe in 1915, where it was considered to be the most devastating viral disease of stone fruit. PPV is also present in the USA and Canada. Recent outbreaks in New York and Michigan underscore that PPV is becoming endemic despite containment efforts (bulldozing and disposal of infected vegetation, moratoria on the movement/transport of infected plant materials, and control of insect vectors).

PPV is an agricultural pest that causes significant economic losses to the stone fruit industry. The primary effects of the infection are reduction of fruit quality and crop yield. Stone fruits (including native or wild *Prunus* species) that are affected by PPV include plums, peaches, almonds, nectarines, apricots, and sweet and sour cherries. Various other non-*Prunus* dicotyledonous plants have been infected under experimental conditions.

When PPV infects a plant, its genetic material (a single strand of RNA) is inserted into the plant cells. This strand of RNA contains the genes needed to make new virions. One of these genes codes for the PPV coat protein: CPG-PPV. The infected cell makes viral coat protein in a similar manner as it produces its own plant proteins. RNA coding for the CPG-PPV are translated into the amino acid sequences that make up the protein. During virus replication, segments of dsRNA are produced but, ultimately, exact copies of the original single-stranded virus RNA chromosome are formed and packaged together with the coat proteins into new PPV virions. Small segments of dsRNA are formed during this process, and recognized by a defense mechanism within the host plant, post-transcriptional gene silencing. Post-transcriptional gene silencing blocks the transcription as well as the production of viral proteins and RNA. This sequence of events results in the development of natural resistance to further PPV infections, but not before fruit degradation, leaf chlorosis, and other serious damage caused by the virus have occurred (EPA 2010 PC Code: 006354).

Uninfected plum trees can be genetically engineered to express the CPG-PPV. But, because the CPG-PPV is responsible for only one component needed for the production of new virions, these engineered plum trees cannot produce the virus. The USDA Agricultural Research Service, Appalachian Fruit Research Station, has developed a genetically engineered plum tree, called the C5 HoneySweet Plum (C5 or C5 plum), which expresses the CPG-PPV and is resistant to PPV infection. To create the C5 plum, the CPG-PPV is isolated and inserted into the plum genome as a transgene. During the plant's naturally occurring cellular processes, the transgenic CPG-PPV gene is transcribed. The mRNA copied from the inserted viral coat proteins genes forms abnormal regions of dsRNA, and the pathogen-derived virus-resistant mechanism recognizes the abnormality and destroys segments with the same sequence. This process establishes the ability of the plant to respond quickly to a PPV infection, blocking the production of new virions and the spread of the disease (EPA 2010 PC Code: 006354).

The active ingredient of the inserted cassette is a reverse transcription derivation of the virus coat protein RNA, inserted with a 3′ untranslated region with fusion of a start codon and short leader sequence and an *A. tumefaciens* *NOS* terminator under direction of a *CaMV 35s* promoter. There is one complete copy of the PPV-CP gene, a small fragment of the PPV-CP *35s* promoter on another insert, and a third insert that is a 3′−3′ tail-to-tail copy of the PPV-CP with the *35S* promoter for each copy and a portion of GUS sequence flanking each PPV-CP copy. The insert is flanked by plum DNA.

e. Potential Food Safety Issues of Virus-Resistant Transgenic Plants

Since nucleic acids are universal in living organisms, the only novel molecules that are expected to exist in virus-resistant transgenic plants that could have a food safety impact are the proteins synthesized from the transgenes. From a practical perspective, this primarily concerns viral coat proteins, since the vast majority of virus-resistant transgenic plants developed so far, and all but one of the virus-resistant transgenic plants that have been deregulated or authorized for large-scale release, are ones that express coat protein transgenes. It has been argued that there is a long history of consumption of viral coat proteins in virus-infected non-transgenic plant foods that we consume regularly. Thus, although they may be transgene encoded, coat proteins are in fact very much familiar parts of human and animal diets (Prins *et al.*, 2008) and there is no evidence for adverse health effects of the consumption of the currently deregulated virus-resistant squash and papaya cultivars (Fuchs and Gonsalves, 2007).

Transgenes that express no protein, as is the case for those used in post-transcriptional gene silencing and artificial microRNA approaches, should in theory not pose this sort of question (Fuchs and Gonsalves, 2007).

The presence of antibiotic resistance markers in virus-resistant transgenic plants seems to pose no negative health risk. Papaya ringspot virus-resistant transgenic papaya contain two of the selection marker genes for antibiotic resistance, *npt II* and *gus*, and the consumption of millions of kilograms of transgenic papaya fruits over the past years has not provided any evidence of compromising safety. It is postulated that the usefulness of the *npt II* gene far outweighs any risks that might evolve from its presence in transgenic plants (Fuchs and Gonsalves, 2007).

3. Altered Nutrient Profiles in Transgenic Crops

a. High-Lysine Corn

Corn does not contain a significant amount of lysine compared to other dietary sources. Conventional corn—soy-based diets for swine and chicken are characteristically deficient in lysine and require the addition of supplemental lysine for optimal animal growth and performance. Therefore, a genetically modified corn line was developed to have higher levels of lysine in the grain. This corn line is used for feed and food use.

Dihydrodipicolinate synthase is the key enzyme in lysine biosynthesis (Figure 13.7). Dihydrodipicolinate synthase (DHDPS, EC 4.2.1.52) cleaves carbon oxygen bonds in the following reaction (KEGG R02292):

L-Aspartate 4-semialdehyde + Pyruvate <=> L-2,3-Dihydrodipicolinate + 2 H_2O

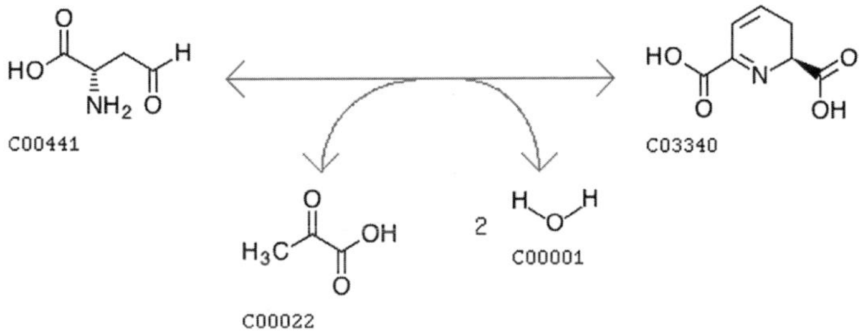

The *cordapA* gene from *Corynebacterium glutamicum*, which encodes the bacterial enzyme dihydrodipicolinate synthase, was incorporated into maize. The derived corn line is referred to as corn line LY038 or 'MAVERA HVC with Lysine', where HVC stands for High-Value Corn. The bacterial DHDPS enzyme, unlike the plant DHDPS enzyme, is not sensitive to lysine feedback inhibition, so lysine biosynthesis will continue in the presence of high levels of free lysine (FSANZ A549).

The genetic modification did indeed elevate the lysine levels in the corn; however, four amino acids are significantly reduced in LY038 corn compared to its control, and with the exception of glutamic acid, all are essential amino acids (histidine, isoleucine, and phenylalanine). However, when compared to conventional corn data, the reductions in these amino acids in LY038 corn are shown to remain within the normal variation observed in corn grain.

The high-lysine corn line LY038 is primarily intended for use as field corn for animal feed. Some types of food products that could contain ingredients derived from corn line LY038 in the case of inadvertent co-mingling are: margarine, cooking oil and baking and frying fats; various sweeteners including high fructose, dextrose, and maltodextrins; corn grain used as an additive; flaking grits used almost exclusively in the manufacture of corn flakes; fine grits utilized by the snack, breakfast cereal, and brewing industries; coarse grits eaten as a breakfast food; corn flour; dried-milled corn products used as a substrate for brewing beer; and corn grits and whole kernels used to produce many distilled hard liquors. Owing to the change in lysine levels, there is also the possibility of the occurrence of different Maillard reactions, which can make lysine unavailable by forming complexes.

The *cordapA* gene is under the control of the *Zea mays globulin 1* (*Glb1*) promoter, which in wild-type maize directs expression of the most abundant embryo-specific protein in maize grain. The utilization of the Glb1 promoter for *cordapA* transcription results in the expression of cDHDPS and the accumulation of lysine predominantly in the germ portion of the grain. Following the promoter is an intron sequence derived from the rice *actin-1* gene, the

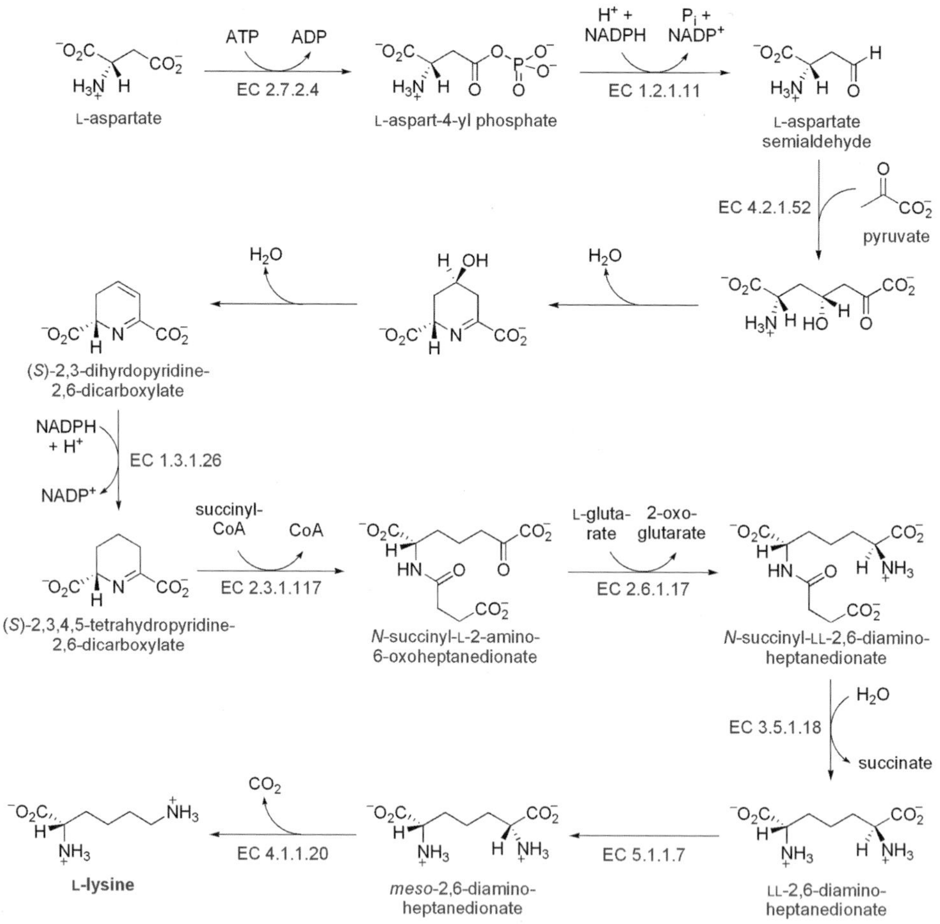

FIGURE 13.7   The lysine biosynthesis pathway. In corn, a key enzyme of the pathway, dihydrodipicolinate synthase (EC 4.2.1.52) is inhibited by lysine feedback. When an additional nonsensitive dihydrodipicolinate synthase from *Corynebacterium glutamicum* is expressed in corn, the lysine content is elevated *(Source: http://en.wikipedia.org/wiki/File:Lysine_Biosynthesis.png).*

5'flank   3'flank

FIGURE 13.8   Organization of the T-DNA cassette for the *cordapA* gene from *Corynebacterium glutamicum*, which encodes the bacterial enzyme dihydrodipicolinate synthase. This cassette is integrated into the genome of maize to enhance lysine biosynthesis. Glb1 promoter: *Zea mays* globulin 1 promoter; Act1 intron: rice actin-1 intron; DHDPS CTP: *Zea mays* dihydrodipicolinate synthase chloroplast transit peptide directing the gene product to the plastid; *cordapA*: gene encoding *Corynebacterium glutamicum* dihydrodipicolinate synthase; *Glb1* 3'UTR: *Zea mays* globulin 1 3' untranslated region to terminate transcription.

rAct1 intron, which enhances DNA transcription. The *cordapA* gene is preceded by the *Zea mays* dihydrodipicolinate synthase chloroplast transit peptide (mDHDPS CTP), to translocate cDHDPS to the plastid, where the majority of amino acid biosynthesis occurs. The 3' non-translated region of the globulin 1 gene follows the *cordapA* gene and contains the polyadenylation signal that directs the termination and maturation of the *cordapA* transcript (FSANZ A549) (Figure 13.8).

III. GENETICALLY MODIFIED MICROORGANISMS AND DERIVED PRODUCTS INTENDED FOR FOOD USE

Genetically modified microorganisms (GMMs) are found in food, involved in the production of food or additives. These microorganisms can be archaea, bacteria, and eukarya. Eukarya include filamentous fungi, yeasts, protozoa, and microalgae. Owing to the ease of genetic manipulation in microorganisms a high diversity of GMMs and their products exists in food. The European Food Safety Authority (EFSA, 2011) defines four categories:

- Category 1: Chemically defined purified compounds and their mixtures in which both GMMs and newly introduced genes have been removed (e.g. amino acids, vitamins).
- Category 2: Complex products in which both GMMs and newly introduced genes are no longer present (e.g. cell extracts, most enzyme preparations).
- Category 3: Products derived from GMMs in which GMMs capable of multiplication or of transferring genes are not present, but in which newly introduced genes are still present (e.g. heat-inactivated starter cultures).
- Category 4: Products consisting of or containing GMMs capable of multiplication or of transferring genes (e.g. live starter cultures for fermented foods and feed).

The different categories demonstrate the complex nature of products made with recombinant technology, ranging from a single isolated compound produced by a GMM to pure cultures of viable GMMs. Amino acids or vitamins that have been purified by crystallization would represent examples at one end of this spectrum and microbial food cultures such as probiotics or dairy starters at the other. Since GMMs are the 'workhorses' in molecular biology, an enormous body of information is available on heterologous expression systems with a potential application in food, which it is not possible to address here. Therefore, this section will focus on viable GMMs in food and the major protein products derived from heterologous systems, which are currently generally recognized as safe by major regulatory bodies, such as the US FDA, the EFSA and Food Standards Australia New Zealand (FSANZ) (Table 13.4). It is assumed that if a product is approved it is applied to foods.

A. Products Consisting of or Containing Genetically Modified Microorganisms Capable of Multiplication

1. Modified Yeast to Reduce Hydrogen Sulfide in Fermented Foods and Beverages

The production of volatile sulfur compounds such as hydrogen sulfide (H_2S) during alcoholic fermentation is a problem that affects the brewing, wine-making, and sake industries. Hydrogen sulfide is an undesirable by-product of the sulfate assimilation pathway formed in *Saccharomyces cerevisiae* during fermentation (Figure 13.9).

The undesirable characteristics of H_2S include introducing a rotten egg odor to the fermented beverage, which can render the product unsaleable. It also has the potential to form mercaptans and thiols which not only are dangerously reactive compounds, but also impart onion and canned vegetable aromas. The removal of these undesirable sulfur compounds is technically difficult and strips the wine of desirable flavor compounds. These sensory characteristics are extremely important to the wine producer. The current method used in the wine industry to remove sulfides from the wine is to add copper in attempts to chelate the sulfur. Copper can lead to the catalysis of deleterious compositional changes and increase the amount of waste produced by wineries. The use of copper as a fining agent may also lead to high residual copper levels in wine, which is allowed up to 0.5 mg/l for wine (FDA-GRN 350).

The commercial *S. cerevisiae* strain P1Y0, which does not produce H_2S as a by-product of fermentation, was developed by replacement of its native *MET10* gene with the 'low-H_2S' *MET10* allele from the *S. cerevisiae* wine yeast UCD932. The *MET10* gene codes for subunit α of assimilatory sulfite reductase (EC 1.8.1.2), which catalyzes the six-electron reduction of sulfite into sulfide (KEGG R00858):

$$\text{Hydrogen sulfide} + 3\ \text{NADP+} + 3\ H_2O <=> \text{Sulfite} + 3\ \text{NADPH} + 3\ H^+$$

TABLE 13.4 Recombinant Microorganisms and Enzymes Regulated by Major International Agencies such as the US Food and Drug Administration (FDA), European Food Safety Authority (EFSA), and Food Standards Australia New Zealand (FSANZ)

Source Microorganism	Enzyme	Reference and Trade name
Aspergillus niger	Asparaginase/*Aspergillus niger*	FDA-GRN 214, FSANZ A1003, PreventASe™
	Carboxypeptidase/*Aspergillus niger*	FDA-GRN 345, Accelerzyme® CPG
	Chymosin B/*Bos taurus*	FDA-GRN 230, 21 CFR 184.1685
	Lipase (triacylglycerol)/synthetic construct of various *Fusarium* species	FDA-GRN 296, Panamore™
	Lipase (triacylglycerol)/*Candida antarctica*	FDA-GRN 158, Lipozyme®
	Phospholipase A_2/*Sus scrofa* (porcine)	FDA-GRN 183, Bakezyme, Cakezyme, Maxapal A2
Aspergillus oryzae	Asparaginase/*Aspergillus oryzae*	FDA-GRN 201, Acrylaway®
	Aspartic proteinase/*Rhizomucor miehei*	FDA-GRN 34, NovoCarne™ Tender
	Glucose oxidase/*Aspergillus niger*	FDA-GRN 106, Gluzyme®
	Laccase/*Myceliophthora themophila*	FDA-GRN 122
	Lipase (triacylglycerol)/*Thermomyces lanuginosus*	FDA-GRN 43
	Lipase (triacylglycerol)/*Fusarium oxysporum*	FDA-GRN 75
	Lipase (triacylglycerol)/hybrid lipase from *Thermomyces lanuginosud/fusarium oxysporum*	FDA-GRN 103
	Lipase (triacylglycerol)/*Rhizomucor miehei*	FSANZ A402, palatase
	Pectin esterase/*Aspergillus aculeatus*	FDA-GRN 8
	Phospholipase A_1/*Fusarium venenatum*	FDA-GRN 142, Novozyme® 46016
	Phytase/*Peniophora lycii*	FSANZ A371
	Xylanase/*Thermomyces lanuginosus*	FDA-GRN 54
Bacillus licheniformis	α-Amylase/*Bacillus stearothemophilus*	FDA-GRN 24
	α-Amylase/*Bacillus licheniformis* + *Bacillus amyloliquefaciens* modified gene	
	Termamyl® LC	FDA-GRN 22
	α-Amylase/*Bacillus licheniformis* modified gene	FDA-GRN 79, Novozym® 28035
	Exo-maltotetraohydrolase, G4 amylase/*Pseudomonas saccharophila stutzeri*	FDA-GRN 277
	Glycerophospholipid cholesterol acyltransferase/*Aeromonas salmonicida*	FDA-GRN 265, FoodPro™, LysoMaxa Oil
	Pullulanase/*Bacillus deramificans*	FDA-GRN 72
Bacillus subtilis	α-Acetolactate decarboxylase/*Bacillus brevis*	21 CFR 173.115
	Maltogenic amylase/*Bacillus stearothermophilus*	GRASP 7G0326
	Pullulanase/*Bacillus naganoensis*	FDA-GRN 20, GRASP 4G0293
	Pullulanase/*Bacillus acidopullulyticus*	FDA-GRN 205
	Branching glycosyltransferase/*Rhodothermus obamensis*	FDA-GRN 274

TABLE 13.4 Recombinant Microorganisms and Enzymes Regulated by Major International Agencies such as the US Food and Drug Administration (FDA), European Food Safety Authority (EFSA), and Food Standards Australia New Zealand (FSANZ)—cont'd

Source Microorganism	Enzyme	Reference and Trade name
Escherichia coli K-12	Chymosin	21 CFR 184.1685
	Cyclodextrin glucanotransferase/*Klebsiella oxytoca*	FDA-GRN 155, FDA-GRN 64
Fusarium venenatum	Xylanase/*Thermomyces lanuginosus*	FDA-GRN 54
Hansenula polymorpha yeast	Hexose oxidase/*Chondrus crispus*	FSANZ A475, FDA-GRN 238
	Lipase (triacylglycerol)/*Fusarium heterosuorum*	FDA-GRN 238, FSANZ A569, GRINDAMYL™ POWERBake
Kluyveromyces marxianus var. *lactis*	Chymosin B/*Bos taurus*	21 CFR 184.1685
Myceliophthora thermophila	Cellulase/*Myceliophthora thermophila*	FDA-GRN 292
Pichia pastoris	Phospholipase C/DNA fragment from undetermined soil organism	FDA-GRN 204
Pseudomonas fluorescens Biovar I	α-Amylase/*Thermococcales hybrid modified gene*	FDA-GRN 126
Saccharomyces cerevisiae	Ice structuring protein/*Macrozoarces americanus* (ocean pout)	FDA-GRN 117
	Urea amidolyase/*Saccharomyces cerevisiae*	FDA-GRN 175
	Subunit α of assimilatory sulfite reductase, 'low-H_2S' *MET10*/*Saccharomyces cerevisiae*	FDA-GRN 350
	Malolactic enzyme/*Oenococcus oeni*	
	+ Malate permease/*Schizosaccharomyces pombe*	FDA-GRN 120
Streptomyces violaceruber	Phospholipase A_2/*Streptomyces violaceruber*	FDA-GRN 212, PLA2 Nagase
Trichoderma reesei	Aspergillopepsin I/*Trichoderma reesei*	FDA-GRN 333
	Chymosin B/*Bos taurus*	FDA-GRN 230, Chymostar Supreme
	Glucoamylase/*Trichoderma reesei*	FDA-GRN 372
	Pectin lyase/*Aspergillus niger*	FDA-GRN 32
	Transglucosidase/*Aspergillus niger*	FDA-GRN 315

The low-H_2S *MET10* allele differs by three single-nucleotide polymorphisms from the wild-type allele (C404A, G1278A, and C1985A). The single-nucleotide polymorphisms found at positions 404 and 1278 are common to a number of strains and the one found at position 1985 is unique to the *MET10* coding sequence of UCD932 (and now the *MET10* coding sequence of P1YO). Only two of these polymorphisms result in amino acid changes since the change at nucleotide position 1278 is a silent mutation. The two copies of the wild-type *MET10* gene were completely removed from the genome of the final low-H_2S commercial strain by knocking out the genes with markers *kanMX* and *hphMX*, via homologous recombination with the non-coding upstream and downstream *MET10* flanking sequences. Subsequently, the markers were replaced with the low-H_2S *MET10* allele of UCD932. The low-H_2S *MET10* allele is under the control of its intrinsic 5′ and 3′ flanking sequences, as seen in its expression cassette (Figure 13.10) (FDA-GRN 350).

The *MET10* allele swap results in minimal H_2S leakage from the yeast cell. The P1Y0 active dry yeast is used as a starter culture for alcoholic beverage fermentation such as grape must, brewing wort, and rice fermentations, and is applied to wine, champagne, sherry, sake, beer, brandy, cognac, whiskey, and rum.

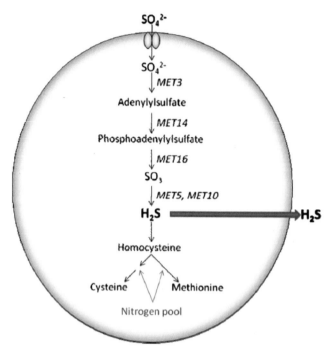

FIGURE 13.9   Pathway of the production of hydrogen sulfite during fermentation. The MET10 enzyme catalyzes the last step in hydrogen sulfite production. The replacement of the native enzyme with a 'low-H2S' transgenic isoform results in low sulfite production. *(Sweiegers and Pretorius, 2007)*

2. Modified Yeast for Reduction of Ethyl Carbamate in Fermented Beverages

Certain fermented foods and alcoholic beverages contain ethyl carbamate (Urethane, CAS No. 5-179-6) (see Zimmerli and Schlatter, 1991, for review), which is a known carcinogen in animals and a suspected carcinogen in humans (12th Report on Carcinogens 2011 by the US National Toxicology Program). Ethyl carbamate forms in a reaction of ethanol with cyanate or urea. In *S. cerevisiae* urea is formed as a breakdown product of L-arginine, and the reaction of ethanol and urea is known to be the primary source of ethyl carbamate in grape-based and rice-based fermented products (FDA-GRN 175):

H₃C—OH	+	H₂N—NH₂	→ H₂N—O—CH₃ + NH₃
ethanol		urea	Ethyl carbamate

Urea is normally metabolized in *S. cerevisiae* by the enzyme urea amidolyase, exhibiting urea carboxylase (EC 6.3.4.6) as well as allophanate hydrolase (EC 3.5.1.54) activity:

Urea carboxylase activity (EC 6.3.4.6, KEGG R00774):

ATP + Urea + HCO₃⁻ = ADP + Phosphate + Urea-1-carboxylate

$$ATP + Urea + HCO_3^- = ADP + Phosphate + Urea\text{-}1\text{-}carboxylate$$

C00086   C01010

C00002   C00288   C00008   C00009

Allophanate hydrolase activity (EC 3.5.1.54, KEGG R00005):

$$Urea\text{-}1\text{-}carboxylate + H_2O = 2\ CO_2 + 2\ NH_3$$

Urea amidolyase is encoded by the *S. cerevisiae DUR1,2* gene (systematic name YBR208C). During fermentation, in the presence of good nitrogen sources the *DUR1,2* gene is only weakly expressed and therefore urea is poorly metabolized. As a result, urea accumulates in the yeast cell and releases into the wine, where it reacts with ethanol to form ethyl carbamate. Heating and aging of the wine accelerate the production of ethyl carbamate, and therefore the levels are much higher in sherry and sake/rice wine (Zimmerli and Schlatter, 1991, FDA-GRN 175).

To reduce the formation of ethyl carbamate the genetically modified *S. cerevisiae* strain ECMoO1 was developed, where a *DUR1,2* containing expression cassette is integrated into the *Ura3* locus by homologous recombination of flanking sequences (*URA3* 5'- and *URA3* 3'-flanks). The *DUR1,2* gene expression is controlled by the *S. cerevisiae PGK1* promoter (*PGK1*p) and terminator (*PGK1*t) sequences (FDA-GRN 175):

| URA3 5' flank | PGK1p | DUR1,2 | PGK1t | URA3 3' flank |

As a result, the expression of the *DUR1,2* gene is increased by 17-fold in the recombinant strain ECMoO1, causing a more effective degradation of urea during alcoholic fermentation, resulting in 89% reduced ethyl carbamate levels (Figure 13.11).

The ECMoO1 active dry yeast is used as a yeast starter culture for alcoholic beverage fermentation such as grape must and rice fermentations (FDA-GRN 175).

3. Modified Yeast for Malolactic Fermentation

Alcoholic fermentation and malolactic fermentation are the two main biotechnological processes in wine making. After the alcoholic fermentation, most red wines and some white wines are subjected to malolactic fermentation. This secondary fermentation, usually carried out by lactic acid bacteria, is important for deacidification, flavor modification, and microbial stability of the wine. Wine deacidification takes place through the conversion of L-malate (dicarboxylic acid) to L-lactate (monocarboxylic acid). *Oenococcus oeni* is the major lactic acid bacterium responsible for malolactic fermentation of wines, but most wine lactic acid bacteria possess the malolactic enzyme which catalyzes the L-malate to L-lactate conversion:

malic acid → (malolactic enzyme) → lactic acid

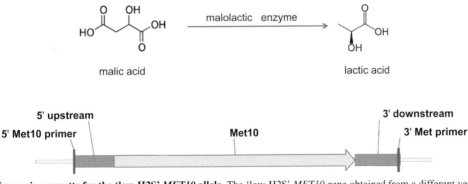

FIGURE 13.10 Expression cassette for the 'low-H2S' *MET10* allele. The 'low-H2S' *MET10* gene obtained from a different yeast strain is flanked by intrinsic 5' upstream and 3' downstream sequences that were used for homologous recombination into the *S. cerevisiae* wild-type *MET10* gene locus.

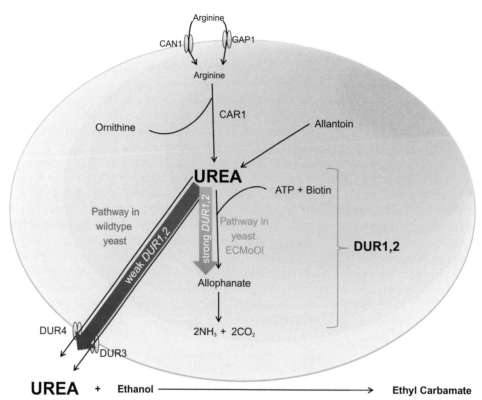

FIGURE 13.11   **Degradation of urea by DUR1,2 urea amidolase.** High DUR1,2 activity degrades urea effectively. However, in high nitrogen conditions activity is low resulting in urea accumulation in the wine and detrimental downstream reactions. The recombinant *ECMoOl* yeast with constitutive high DUR1,2 expression was developed to prevent urea excretion. *(Coulon et al., 2006)*

In contrast to the desired effects of malolactic fermentation, some lactic acid bacteria strains also show specific metabolic pathways which can lead to the formation of undesirable compounds. In addition, due to the sensitivity of lactic acid bacteria to nutritional conditions, pH, temperature, sulfur dioxide content, and ethanol concentration of the wine, malolactic fermentation is often unpredictable and difficult to achieve, even with the use of commercial starter cultures of lactic acid bacteria (FDA-GRN 120).

Saccharomyces cerevisiae is unable to efficiently degrade L-malate, but the yeast *Schizosaccharormyces pombe* is able to do so; however, it is unsuitable for the production of wine. Therefore, the genetically enhanced strain ML01 of *S. cerevisiae* was created to aid malolactic fermentation. To overcome the lack of malolactic fermentation, as well as transport of L-malate into the *S. cerevisiae* cell, two genes were transferred into the ML01 strain: the *mleA* gene encoding the malolactic enzyme from *O. oeni* and the *mae1* gene encoding malate permease from *S. pombe* (FDA-GRN 120) (Husnik *et al.*, 2007).

A linear expression cassette containing the tandem *mae1* and the *mleA* genes under the control of the *S. cerevisiae* *PGK1* promoter (*PGK1*p) and terminator (*PGK1*t) was integrated into the *URA3* (recombination with flanking ura3 sequences) locus of *S. cerevisiae* strain S92 (Husnik *et al.*, 2007) (Figure 13.12).

The resulting ML01 strain is capable of decarboxylating up to 9.2 g/l of malate to equimolar amounts of lactate in Chardonnay grape must during the alcoholic fermentation. The presence of the malolactic cassette in the genome does not affect growth, ethanol production, fermentation kinetics, or metabolism of ML01, is substantially equivalent to the parental industrial wine yeast (Husnik *et al.*, 2007), and is therefore generally regarded as safe by the FDA (FDA-GRN 120).

FIGURE 13.12   **The expression cassette containing the *mae1* gene from *S pombe* (encoding malate permease) and the *mleA* gene from *O oeni* (encoding the malolactic enzyme).** Expression is controlled by the *S. cerevisiae* *PGK1* promoter (*PGK1p*) and terminator (*PGK1t*). The ura3 flanking sequence was used for homologous recombination to create the ML01 yeast capable of malolactic fermentation.

The ML01 wine active dry yeast is used as a yeast starter culture for grape must fermentation to perform alcoholic fermentation. However, in contrast to traditional yeast strains, alcoholic and malolactic fermentation occur simultaneously and solely in the yeast. The use of ML01 wine yeast depends on the wine maker's choice and relies on whether the malolactic fermentation is required for deacidification, flavor modification, or microbial stability, or a combination of these characteristics.

If deacidification is the major contribution required from malolactic fermentation, the use of the ML01 wine active dry yeast is highly recommended. The bioconversion of a dicarboxylic acid (L-malate) into a monocarboxylic acid (L-lactate) decreases the total acidity of the wine and softens its mouth-feel. In some wines, such as red wines with aging potential, malolactic fermentation by bacteria contributes not only to deacidification, but also to flavor modification through bacterial secondary metabolism. In this case, it is often preferable not to use the ML01 yeast and to carry out malolactic fermentation with lactic acid bacteria. In contrast, the ML01 yeast is applicable when wine microbial stability is required without the development of buttery flavors (acetoyn, diacetyl) due to lactic acid bacteria secondary metabolism, as is often the case in white wines. The ML01 wine yeast strain can also be used in cases where wine stability is required as soon as possible, since wine microbial stability is enhanced once the malolactic fermentation is completed by removal of L-malate. Moreover, the sooner the malolactic fermentation is completed, the sooner the wine can be sulfited and bottled with a low risk of bacterial contamination (FDA-GRN 120).

IV. COMMERCIAL ENZYMES AND PROTEINS IN THE FOOD INDUSTRY DERIVED FROM GENETICALLY MODIFIED MICROORGANISMS

Enzymes are commonly used in food processing and in the production of food ingredients. Intrinsic enzymes are traditionally isolated from culturable microorganisms, plants, and mammalian tissues. Therefore, their production and isolation can be challenging and the available resources might be limiting. Moreover, enzymes often originate from microorganisms that cannot be easily cultured under industrial conditions or have toxic by-products. By judicious selection of host microorganisms, recombinant production strains can be constructed to allow efficient production of enzymes that are substantially free of undesirable by-products or other microbial metabolites. An ever-increasing number of enzymes is produced using heterologous expression systems, where transgenic technology is used to engineer and introduce transgenes into microbial host strains, which can themselves be genetically modified to optimize the biotechnological process. For example, several microbial strains recently developed for enzyme production have been engineered to increase enzyme yield by deleting native genes encoding extracellular proteases. Moreover, certain fungal production strains have been modified to reduce or eliminate their potential for production of toxic secondary metabolites (Olempska-Beer *et al.*, 2006).

Native enzymes are often not well adapted to the conditions used in modern food production. The increasing sophistication of food processing creates a demand for a broad variety of food-processing enzymes with characteristics compatible with food-processing conditions. Therefore, enzymes are specifically engineered for improved properties in industrial processes through DNA changes which result in an altered amino acid sequence. For example, commonly used sweeteners such as glucose or fructose syrups are typically produced from corn starch using hydrolytic enzymes. In the first step of starch hydrolysis, starch is liquefied with α-amylase by heating at 105°C for 2—5 minutes followed by 1—2 hours at 90—100°C. Thus, α-amylases with increased heat stability and improved compatibility with other parameters of the liquefaction process have been engineered (Olempska-Beer *et al.*, 2006).

The following paragraphs will focus on the biochemical properties as well as the genetic origin and modifications of recombinant enzymes produced by GMMs which are currently used in the production of food (Table 13.4). We will not discuss safety-related characteristics. The safety evaluation of food processing enzymes from recombinant microorganisms has been extensively discussed in the literature (Jonas *et al.*, 1996; Pariza and Johnson, 2001) and in guidance documents issued by regulatory authorities and international organizations, for example, by the Scientific Committee for Food (SCF, 1992) or the EFSA guidance document on the risk assessment of products derived from GMMs (EFSA, 2011). In principle, the same safety considerations apply to enzymes derived from native and recombinant microorganisms. The key component in evaluating enzyme safety is the safety assessment of the production strain, in particular, its pathogenic and toxigenic potential (Pariza and Johnson, 2001). Although neither pathogenic nor toxigenic microorganisms are intentionally used in the production of food-processing enzymes, certain fungi traditionally used as sources of enzymes have been found to produce low levels of toxic secondary metabolites under fermentation conditions conducive to the synthesis of these compounds. Some of these micro-organisms are now used as sources of recombinant enzymes (Olempska-Beer *et al.*, 2006).

A. Acetolactate Decarboxylase

Acetolactate decarboxylase (EC 4.1.1.5) reduces the undesirable butter aroma occurring as a by-product of beer brewing by converting diacetyl (acetolactate) into the neutral-tasting acetoin (KEGG R02948):

(S)-2-Hydroxy-2-methyl-3-oxobutanoate $\rightleftharpoons$ (R)-2-Acetoin + CO_2

During storage of beer, the diacetyl aroma is slowly converted into a neutral taste note. In order to shorten the conversion time, diacetyl is eliminated by addition of acetolactate decarboxylase.

Transgenic acetolactate-decarboxylase is currently derived from a modified *Bacillus subtilis* strain that contains the unmodified gene coding for the *Bacillus brevis* α-acetolactate decarboxylase (21 CFR 173.115).

To date, two biotechnologically produced acetolactate-decarboxylase specimens are marketed in Europe. They are produced with the help of GM brewer's yeast, *S. cerevisae*. In the USA an enzyme preparation from a modified *B. subtilis* strain that contains the gene coding for α-acetolactate decarboxylase from *B. brevis* is permitted in food for human consumption (21 CFR 173.115).

B. Aminopeptidase

Aminopeptidases (EC 3.4.11) are enzymes produced and secreted by glands of the small intestine where they participate in the digestion of proteins, but are widely distributed throughout the animal kingdom and also found in all microorganisms (KEGG database). An aminopeptidase cleaves single amino acids from the N-terminus of protein or peptide substrates. For example, aminopeptidase EC 3.4.11.1 can mediate the cleavage of an N-terminal glycine (KEGG R04951):

R-S-Cysteinylglycine + H_2O <=> S-Substituted L-cysteine + Glycine

These cleavages will change the flavors of the proteins in a food matrix. Therefore, aminopeptidases are used in the production of cheese, beverages, flavorings, meats, and milk products. They are typically meant to enhance and optimize aroma and flavor. The addition of aminopeptidases can accelerate the maturing of cheese. They are also used in animal feeds to improve the utilization of proteins in feed for piglets and poultry. The cleavage of an N-terminal glycine is described here to give an example of the mechanism of aminopeptidases.

Various aminopeptidases have been produced biotechnically for some time utilizing various fungal and bacterial cultures (such as *Aspergillus*, *Lactococcus*, and *Trichoderma*). Among the recombinant peptidases produced are the aminopeptidase (PepN) from *Lactobacillus rhamnosus* S93, *Lactococcus lactis* (FDA 21CFR184.1985), and *Aspergillus sojae* (European Patent 0967286).

C. Amylases

Amylases are amylolytic enzymes that break down starches into sugars and are very widespread throughout the animal, plant, and microorganism kingdoms (OrthoDB EC 3.2.1). Various amylases exist which cleave the branches of starch molecules with a particular specificity. Depending on the type of amylase, the resulting compounds may be simple sugars such as glucose or fructose, compound sugars such as maltose, malt sugar, or special forms of starch such as dextrins. All amylases are glycoside hydrolases which act upon the α-(1,4)-and/or α-(1,6)-linkages of starch polymers (Goesaert *et al.*, 2009). Figure 13.13 presents an overview of the modes of action of amylolytic enzymes.

Amylases are predominantly used to modify raw materials containing starch in the food-processing industry. The most important area of application of amylases is the production of sugars from starch (glucose syrup, fructose syrup), which later become ingredients in a wide variety of food products, e.g. sweets, baked goods, ice cream, or tomato ketchup. Amylases are naturally present in many raw materials, such as in cereals or yeasts. However, these

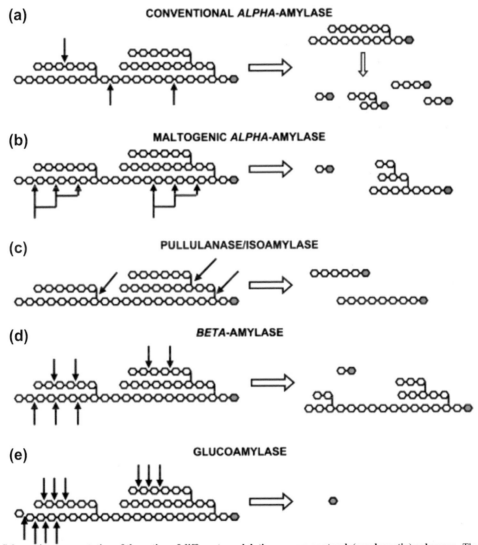

FIGURE 13.13 **Schematic representation of the action of different amylolytic enzymes on starch (amylopectin) polymers.** The gray ring structure represents a reducing glucose residue. (a) Endo-type action of α-amylase, yielding branched and linear low-molecular-weight dextrins; (b) mainly exo-type action of maltogenic α-amylase, yielding mainly maltose; (c) debranching enzyme action, yielding linear dextrins; (d) purely exo-type action of β-amylase, yielding maltose and β-limit dextrins; (e) purely exo-type action of glucoamylase, yielding glucose. *(From Goesaert et al., 2009.)*

naturally occurring amylases are often either insufficient or too slow in effect. In order to steer or accelerate the degradation of starch, industrially produced amylases are commonly added. These preparations mostly contain a mixture of several types of amylase (GMO Compass).

Amylases are routinely used in baking and bread making for flour standardization and as anti-staling agents. In sound wheat flour, α-amylases are virtually absent, while β-amylases are abundantly present, but the latter have little if any activity on undamaged, native starch granules and are inactivated before starch gelatinization. Therefore, amylase activity in wheat flour is often optimized by adding fungal α-amylases. The added amylases increase the level of fermentable and reducing sugars in flour and dough, thus promoting yeast fermentation and the formation of Maillard reaction products, which, in their turn, intensify bread flavor and crust color. However, amylase functionality may also be related to the reduction of dough viscosity during starch gelatinization, thus prolonging oven rise and resulting in an increased loaf volume. Certain amylases can delay crumb firming and hence function as anti-staling agents. Possible mechanisms or modes of action whereby these enzymes retard the staling/firming process are discussed below. Typical amylase-containing anti-staling products mainly consist of bacterial or fungal α-amylases with intermediate thermostability (Goesaert *et al.*, 2009).

During the production of fermented alcoholic drinks, the starch in the raw materials needs to be broken down into small sugar molecules, to allow the yeast to transform it into alcohol. Enzymes carry out this process in two stages: liquefaction and saccharification. Traditionally, enzymes have been provided by adding malt, which is now being increasingly replaced in distilling operations by the use of isolated enzyme preparations. A comparatively small quantity of industrial enzymes can replace significant amounts of malt.

In fruit juices, amylases eliminate lees that contain starch. This is especially important if the fruits are picked while unripe and stored for relatively long periods at low temperature. Under these conditions fruit pulp contains starch in sufficient amounts to cause turbidity or even gelatinize during processing, which makes productive procedures difficult. The addition of amylolytic enzymes counteracts these processes. As feed additives, amylases increase the breakdown of starches and thereby contribute to the better use of plant-based feed.

It has been possible for a long time to produce amylases with a variety of non-genetically modified fungal and bacterial cultures. As a rule, bacterial amylases are more stable in regard to temperature than are amylases derived from fungal cultures. Bacterial amylases now are produced predominantly with GMMs (various types of *Bacillus*). In the production of amylases using fungi, cultures are most often used that are not regarded as genetically modified (GMO Compass).

1. α-Amylases

α-Amylases (EC 3.2.1.1) are enzymes that catalyze the hydrolysis of internal α-1,4-glycosidic linkages in polysaccharides into low-molecular-weight products, such as glucose, maltose, and maltotriose units (de Souza and Magalhaes, 2010). α-Amylases act on starch, glycogen, and related polysaccharides or oligosaccharides in a random manner; reducing groups are liberated in the α-configuration (KEGG R02108):

$$Starch + H_2O \iff Dextrin + Starch$$

The term 'α' relates to the initial anomeric configuration of the free sugar group released and not to the configuration of the linkage hydrolyzed. A variety of recombinant α-amylase mutants engineered for specific applications are currently produced in *Bacillus subtilis*, *Bacillus licheniformis*, and *Pseudomonas fluorescens* (Olempska-Beer *et al.*, 2006). The *B. licheniformis* α-amylase (*amyl*) gene is commonly used as a transgene and template for modifications (FDA-GRN 79). In addition, various thermostable α-amylases have been isolated from thermophilic microorganisms and expressed in heterologous production strains. Treatment of starch with α-amylase may be followed by treatment with other suitable enzymes such as glucoamylase, pullulanase, and glucose isomerase.

Termamyl® SC is a preparation of heat-stable α-amylase from *Bacillus stearothemophilus* produced in *B. licheniformis* (GRASP 0G0363, FDA-GRN 24). The sequence of the *B. stearothemophilus* α-amylase *amyS* gene has been modified to increase enzyme stability at low calcium concentrations and low pH. The modified enzyme has a deletion of two amino acids (positions 181 and 182) and a single substitution (position 193) compared to the original *B. stearothemophilus amyS* gene. Termamyl® SC is applied in the starch industry for the continuous liquefaction at temperatures of up to 105–110°C, in the alcohol industry for thinning starch in distilling mashes, in brewing to liquefy starch added to wort, and in the sugar industry to break down the starch in cane juice (FDA-GRASP 0G0363, FDA-GRN 24).

Termamyl® LC is a preparation of heat-stable α-amylase from *B. licheniformis* produced in *B. licheniformis* (FDA-GRN 22). The sequence of the *B. licheniformis* α-amylase *amyL* gene has been modified to enable operation at lower pH and lower calcium levels than conventional thermostable α-amylases. The template for the Termamyl® LC enzyme was the *B. lichenifomis* α-amylase *amyS* gene used in the production of Termamyl® SC (FDA-GRASP 3G002612). The Termamyl® LC enzyme was constructed by splicing in the N-terminal region of the *B. amyloliquefaciens* α-amylase sequence and changing the DNA coding sequence for other specific amino acid residues to increase stability at low pH, low calcium concentration, and high temperature (FDA-GRN 22). The Termamyl® enzymes have been developed from gene bank nucleotide E01158 through an intermediate molecule named M38570. The enzyme preparations are used in the food industry as a processing aid in the liquefaction of starch to produce syrup or in the alcohol industry to thin starch in distilling mashes.

Based on the Termamyl® LC enzyme, another modified α-amylase, called Novozym® 28035, was developed for further increased stability at low pH, low calcium concentration, and high temperature. The Novozym® 28035 α-amylase enzyme has substitutions at four additional amino acid positions compared to its precursor, Termamyl® LC α-amylase, and its applications are equivalent (FDA-GRN 79).

Gene sequences from *Thermococcales archaebacteria* strains were used to create α-amylase BD5088 (FDA-GRN 126). The α-amylase designated BD5088 is a hybrid enzyme derived from three wild-type archaeal α-amylases. The hybridized enzyme was selected for commercialization owing to its stability and activity at low pH, high temperature, and low calcium concentration. Alignments of the thermophilic archaeal α-amylase amino acid sequences demonstrate at least 85% homology to BD5088 α-amylase. α-Amylase BD5088 was developed to be used in the hydrolysis of edible starch to produce starch hydrolysis products, including glucose syrup, high-fructose corn syrup (HFCS), and crystalline glucose (dextrose), and distilled ethanol for food and beverage use (FDA-GRN 126).

2. β-Amylases

β-Amylases (EC 3.2.1.2) or maltogenic-amylases act on the non-reducing end of starch, glycogen, and related polysaccharides and oligosaccharides to cleave β-maltose (two glucose units) by inversion. β-Amylase is intrinsic to bacteria and plants, where during the ripening of fruit, β-amylase breaks starch into maltose, resulting in the sweet flavor of ripe fruit. The term 'beta' relates to the initial anomeric configuration of the free sugar group released and not to the configuration of the linkage hydrolyzed (KEGG R02112):

Starch <=> Maltodextrin + Maltose

C00369   C01935   C00208

3. Maltogenic Amylase

Maltogenic amylase (EC 3.2.1.133) is an exo-acting enzyme catalyzing the hydrolysis of α-1,4-glucosidic linkages in amylose, amylopectin, and related glucose polymers. α-Maltose residues are successively removed from the non-reducing end of the starch or related polysaccharides and oligosaccharides polymers until the molecule is degraded or, in the case of amylopectin, a branch-point is reached (WHO Food Additives Series 40). Maltogenic amylase can be produced by submerged fermentation of a non-pathogenic and non-toxicogenic strain of *Bacillus subtilis*, which contains the *amyM* gene from *Bacillus stearothermophilus* coding for maltogenic amylase (WHO Food Additives Series 40; Diderichsen and Christiansen, 1988) (FDA-GRASP 7G0326).

4. γ-Amylases

γ-Amylases (EC 3.2.1.3) hydrolyze terminal 1,4-α-D-glucosidic bonds successively from non-reducing ends of the chains, releasing β-D-glucose. Most forms of the enzymes can rapidly hydrolyze 1,6-α-D-glucosidic bonds when the next bond in the sequence is 1,4, and some preparations of this enzyme hydrolyze 1,6- and 1,3-α-D-glucosidic bonds in other polysaccharides (KEGG R01790):

Starch + H₂O <=> alpha-D-Glucose + Starch

γ-Amylases or glucoamylases are only found in fungi (KEGG, OrthoDB), and unlike the other forms of amylase, γ-amylase is most efficient in acidic environments and has an optimum pH of 3.

A glucoamylase from *Trichoderma reesei* is overexpressed in the same organism, *T. reesei*. The production strain has been genetically modified by inactivation of several cellulase genes to facilitate the overexpression of the native *T. reesei TrGA* gene for the production and secretion of recombinant glucoamylase enzyme.

Two expression cassettes containing the *T. reesei cbh1* promoter and terminator flanking the *glaI* DNA encoding the *T. reesei* glucoamylase are integrated into the genome of the final production strain, designated as 70H2-TrGA #32-9 (FDA-GRN 372).

The derived glucoamylase is used to saccharify liquefied corn starch for the manufacture of corn sweeteners such as HFCS. It maximizes the conversion of starchy substrate to fermentable carbohydrates. It enhances the extraction and saccharification of starch (mashing) from malted cereal, cereal, and other plant sources (including barley, corn, wheat, rye, milo, rice, tapioca, and potatoes). The resultant process liquors (worts) are fermented, typically by a yeast, to produce ethanol (and sometimes organic acids). *Trichoderma reesei* glucoamylase is also utilized in lactic acid manufacture processes similar to those used in potable alcohol manufacture (FDA-GRN 372).

5. Pullulanases

Pullulanases (EC 3.2.1.41) are specific kinds of glucanases, amylolytic exoenzymes, that degrade pullulan, amylopectin, glycogen, and α- and β-limit dextrins of amylopectin and glycogen. They are produced as extracellular, cell surface-anchored lipoproteins by Gram-negative bacteria. Type I pullulanases specifically attack α-1,6 linkages, while type II pullulanases are also able to hydrolyze α-1,4 linkages. Pullulanases cleave the branches of amylopectin molecules in starch to produce chains of amylase and are therefore called debranching enzymes (Lee and Whelan, 1972). Maltose is the smallest sugar that they can release from an α-1,6 linkage:

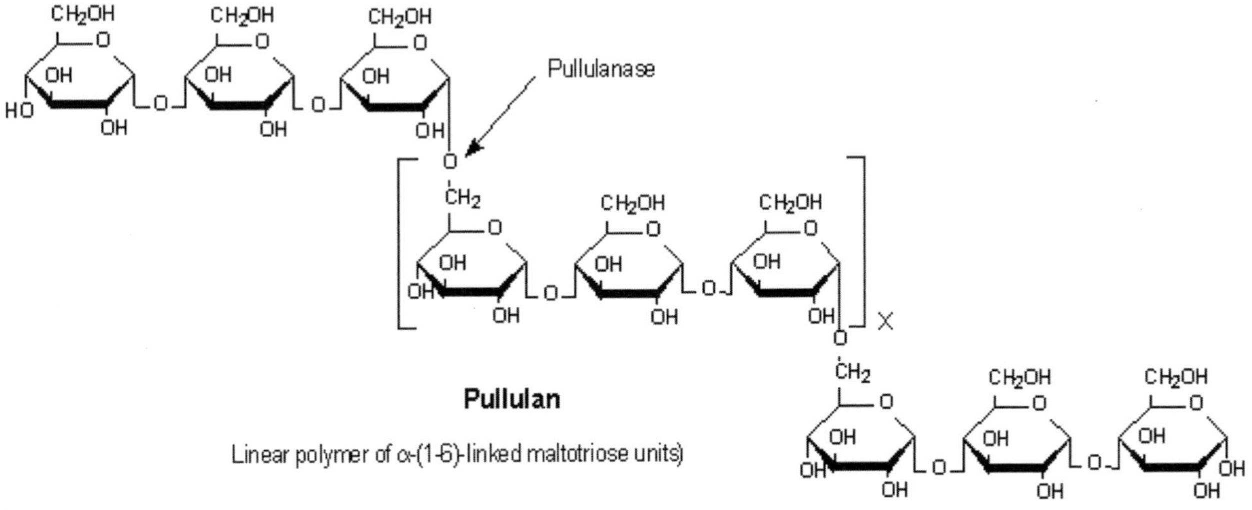

Pullulan

Linear polymer of α-(1-6)-linked maltotriose units

(Singh et al. 2008)

Pullulanases are used predominantly in conjunction with other enzymes to enhance the saccharification of starch into glucose and glucose syrup. They are occasionally applied as a baking enzyme or in the alcoholic beverage industry, where they increase the amount of fermentable sugars available for conversion into alcohol (FDA-GRN 20). Pullulanases are produced through fermentation, utilizing cultures of bacteria (*Bacillus* and *Klebsiella*) or fungi (*Trichoderma*).

One specific recombinant *Bacillus naganoensis* pullulanase is expressed in *Bacillus subtilis* (FDA FDA-GRN 20). *Bacillus subtilis* is also utilized to express *Bacillus acidopullulyticus* pullulanase (FDA FDA-GRN 205). A third pullulanase derived from *Bacillus deramificans* is expressed in *B. licheniformis* (FDA FDA-GRN 72).

6. Maltotetraose-Forming Amylases

Maltotetraose-forming amylases (EC 3.2.1.60) are used to delay the staling process in bread and other baked goods and thereby extend the period for which the products have an acceptable eating quality. One maltotetraose-forming amylase, called G4-amylase, is applied as a processing aid in the starch industry. Heterologous G4-amylase is produced in *B. licheniformis* strain GICC03279, which is expressing the *mta* gene from *Pseudomonas saccharophila* (FDA-GRN 277). In order to improve thermostability during baking, the wild-type DNA sequence of the *P. saccharophila PS4* gene was altered by removing the C-terminal starch-binding domain and changing 16 out of the remaining 429 amino acids of the catalytic core (FDA-GRN 277). By introducing these mutations, the temperature stability and baking performance of the enzyme are increased, making this G4-amylase far better suited to anti-staling applications than the wild-type maltotetraohydrolase enzyme from *P. saccharophila*.

7. A Branching Glycosyltransferase

A branching glycosyltransferase (EC 2.4.1.18) from *Rhodothermus obamensis* produced in *B. subtilis* is used as an enzyme in the starch industry to obtain dextrins with improved physical properties, such as higher solubility, lower viscosity, and reduced retrogradation (FDA-GRN 274). The improved dextrin preparation can be used in soups, sauces, dried instant food, low-fat products, and soft drinks. The branching glycosyltransferase catalyzes the transfer of a segment of a 1,4-α-D-glucan chain to a primary hydroxyl-group in a similar glucan chain to create 1,6-α-linkages and increase the number of branched points. It converts amylose into amylopectin (KEGG R02110):

D. Asparaginase

Asparaginase (EC 3.5.1.1) converts the amino acid asparagine into aspartic acid (KEGG R00485):

$$\text{L-Asparagine} + H_2O <=> \text{L-Aspartate} + NH_3$$

Asparagine is a precursor of acrylamide, a suspected carcinogen. In the process of browning during the baking, frying, and grilling of starchy food, acrylamides are formed at temperatures higher than 100°C through the Maillard reaction. Through the addition of asparaginase, asparagine becomes unavailable and acrylamide contents may be reduced by up to 90% in processed food. The nutritional value, taste, and browning of the product remain unaffected (Pruser and Flyn, 2011).

Asparaginase preparations are applied to L-asparagine- and carbohydrate-containing foods that are heated above 120°C, such as bread (e.g. tin bread, buns and rolls, French sticks, variety breads like multigrain types of bread, raisin bread, biscuits, and crackers), other cereal-based products (e.g. cakes, Swiss rolls, Dutch honey cake, and breakfast cereals), potato-based products (e.g. French fries and potato chips), and reaction flavors (FDA-GRN 214).

Asparaginase preparations are available under the brand names *Acrylaway*® and *PreventASe*™. *PreventASe* is manufactured with the aid of the genetically modified mold *Aspergillus niger* transgenic for its own asparaginase *aspA* gene under the regulation of a *glaA* promoter for enhanced expression of the 378 amino acid enzyme. In the recipient *A. niger* strain GAM-53 seven glucoamylase gene loci were removed by replacement with plug sites. The major protease *pepA* gene, as well as the major amylases *amyA* and *amyB* were deleted, enhancing the strain's capacity to secrete proteins (FDA-GRN 214).

Acrylaway® is produced in the mold *Aspergillus oryzae*, carrying its own transgenic asparaginase gene under the control of the *Pna2/tpi* promoter for enhanced expression (FDA-GRN 201). The *A. oryzae* host strain BECh2 was modified to eliminate TAKA-amylase secretion and neutral metalloproteinase I by genetic deletion. In addition, cyclopiazonic acid and Kojic acid synthesis was eliminated by gamma- and ultraviolet-irradiation (FDA-GRN 201).

E. Aspartic Proteinase

Aspartic proteinase or mucorpepsin (EC 3.4.23.23) hydrolyzes proteins, favoring hydrophobic residues at P1 and P1'. It does clot milk but does not accept Lysin at P1, and hence does not activate trypsinogen.

Aspartic proteinase cloned from *Rhizomucor miehei* and produced in *A. oryzae* is used as a milk clotting enzyme and approved for use in cheese production (21CFR 173.1 50). An aspartic proteinase subcloned from *R. miehei* and produced in *A. oryzae* strain IFO 4177 is used as a tenderizing agent under the trademark NovoCarne™. The recipient *A. oryzae* strain IFO 4177 is the well-known genetically modified industrial production strain from the Institute for Fermentation, Osaka, Japan. In this expression plasmid the aspartic proteinase is under the control of the *A. oryzae* TAKA amylase gene promoter (FDA-GRN 34). The enzyme is applied in liquid form by injection into the meat followed by tumbling to enhance distribution. It can also be mixed with the marinade for tumbling. NovoCarne Tender has a narrow specificity acting only on the myofibrillar proteins, not the connective tissue proteins. NovoCarne Tender also expresses self-limiting hydrolysis of myosin. These factors prevent NovoCarne Tender from overtenderizing the meat (FDA-GRN 34).

F. Aspergillopepsin I/Acid Fungal Protease

Acid fungal protease, synonymous with aspergillopepsin I (EC 3.4.23.18), is a protease hydrolyzing protein with broad specificity.

The gene for *Trichoderma reesei* acid fungal protease is overexpressed in the same microorganism, *T. reesei*. The DNA encoding the *T. reesei* mature secreted acid fungal protease was fused to the DNA encoding the *T. reesei* CBHI signal peptide, which enhances yield and secretion. This open reading frame is flanked by the promoter and terminator sequences of the *T. reesei* cellobiohydrolase I gene, to optimize overexpression. The isolated acid fungal protease is used in corn steeping, alcoholic beverage manufacture to increase free amino acids nitrogen or reduce foaming, dehazing apple juice, and degumming of membranes during orange juice manufacture (FDA-GRN 333).

G. Carboxypeptidase

Carboxypeptidases belong to the subclass of peptidases hydrolyzing polypeptide chains at the C-terminal end. They are also called exopeptidases (EC 3.4.11.x—3.4.19.x):

Carboxypeptidases

The technological function of carboxypeptidase is to release C-terminal amino acids from proteins and peptides present in various foods such as milk (casein, whey) and meat, in order to aid in and/or speed up the development of flavors during ripening.

A serine-type carboxypeptidase enzyme preparation is derived from genetically modified *Aspergillus niger* cultures and will be marketed under the trade name Accelerzyme® CPG. The *A. niger* host strain ISO-528 was derived from strain GAM-53 by deletion of seven loci of the glucoamylase gene *glaA*, which were replaced by so-called 'plug sites'. These *ΔglaA* plug sites allow for the site-directed insertion of expression cassettes. In addition, the *pepA* gene coding for the major protease was inactivated and the *amyA* and *amyB* genes encoding the major amylases were deleted. The strain also has an improved capacity to secrete proteins, which were selected by classical mutation breeding (FDA-GRN 345).

The carboxypeptidase expression cassette contains *A. niger*-derived DNA, starting with the *glaA* promoter, followed by the entire *pepG* genomic sequence encoding the carboxypeptidase protein, and 3′ flanked by the *gluA* terminator sequence. The *gluA* terminator sequence ensures efficient termination of *pepG* gene transcription and targeting of the expression unit to the *ΔglaA* loci (FDA-GRN 345).

This Accelerzyme CPG carboxypeptidase preparation is to be used in cheese, enzyme-modified cheese, and fermented meat. In cheese production the enzyme is added to milk, together with the lactic acid bacteria, enabling the enzyme to act on the proteins present throughout the cheese and to release the amino acids as precursors for flavor components. In this way, the same types of flavor components are developed as during regular ripening, but the speed of the flavor development is accelerated, resulting in a shorter ripening period of the cheese, as well as a debittering during the ripening process. In fermented meat, carboxypeptidase can also be used to speed up the development of flavor during ripening. Fermented dry meat is produced by cutting and mixing fresh pieces of meat and bacon together with lactic acid bacteria. The mixture is filled in foil or natural pig-intestine, and left to dry and ferment in a temperature- and moisture-controlled ripening cell. During the ripening period of 3 weeks, the moisture decreases by evaporation, the pH decreases because the lactic acid bacteria form lactic acid, and the taste develops because of flavor components formed by the lactic acid bacteria. By adding carboxypeptidase together with the lactic acid bacteria during cutting and mixing of the meat and bacon, amino acids are released that stimulate the lactic acid bacteria in their formation of flavor components. As a result, the meat reaches the desired taste after 22 instead of 28 days (FDA-GRN 345).

H. Cellulase

Cellulase (EC 3.2.1.4) is an enzyme responsible for the endohydrolysis of 1,4-P-D-glucosidic linkages in cellulose, lichenin, and cereal P-D-glucans. In addition, cellulases will hydrolyze the 1,4-linkages in P-D-glucans also containing 1,3-1inkages (KEGG R02886):

$$\text{Cellulose} + H_2O <=> \text{Cellulose} + \text{Cellobiose}$$

Cellulase enzyme preparations can be used to break down cellulose in a wide variety of cellulose-containing food products, such as the breakdown of cellulose in citrus products or other fruits used in juice and wine production, removal of fiber from edible oil press cakes, increase in starch recovery from potatoes and other starch sources, extraction of proteins from leaves and grasses, tenderizing fruits and vegetables prior to cooking, extraction of essential oils and flavoring material from plant materials, treatment of distiller's mash, extraction of green tea components, and other uses (FDA-GRN 292).

A cellulase enzyme preparation is derived from a genetically modified fungus. *Myceliophthora thermophila.* Additional copies of the *M. thermophila* cellulase gene *eg5* are incorporated into the chromosome of the *M. thermophila* recipient. These additional copies are under the control of the *cbh1* gene promoter, resulting in a hyperproducing cellulase strain (FDA-GRN 292).

I. Cyclodextrin Glucanotransferase

Cyclodextrin glucanotransferase (CGTase; EC 2.4.1.19) is a unique enzyme capable of converting starch or starch derivatives into cyclodextrins via the cyclization reaction. It cyclizes part of a 1,4-α-D-glucan chain by formation of a 1,4-α-D-glucosidic bond (KEGG EC 2.4.1.19).

The enzyme first cleaves the α-1,4-glycosidic bond between the residues bound at subsites +1 and −1, resulting in a covalent intermediate. The linear chain of the intermediate assumes a cyclic conformation, which is the circularization step. Subsequently, an α-1,4-glycosidic bond is reformed with the terminal 4-hydroxyl group of the intermediate. The catalytic residues involved in bond cleavage are Asp[229] and Glu[257] (numbering in *B. circulans* cyclodextrin glucanotransferase) (Figure 13.14) (Li *et al.*, 2007).

As a result, cyclodextrins (Schardinger dextrins) of various sizes are formed reversibly from starch and similar substrates. Linear maltodextrins can also be disproportionated without cyclizing. Cyclodextrins are cyclic α-1,4-glucans composed of between six to over 100 glucose-unit cyclodextrins, the most common forms being α-, β-, and γ-cyclodextrins. The enzymatic production of cyclodextrin from starch mediated by cyclodextrin glycosyltransferase results in a mixture of cyclodextrins α-, β-, and γ-, consisting of six, seven, and eight glucose units, respectively (Figure 13.15) (Li *et al.*, 2007).

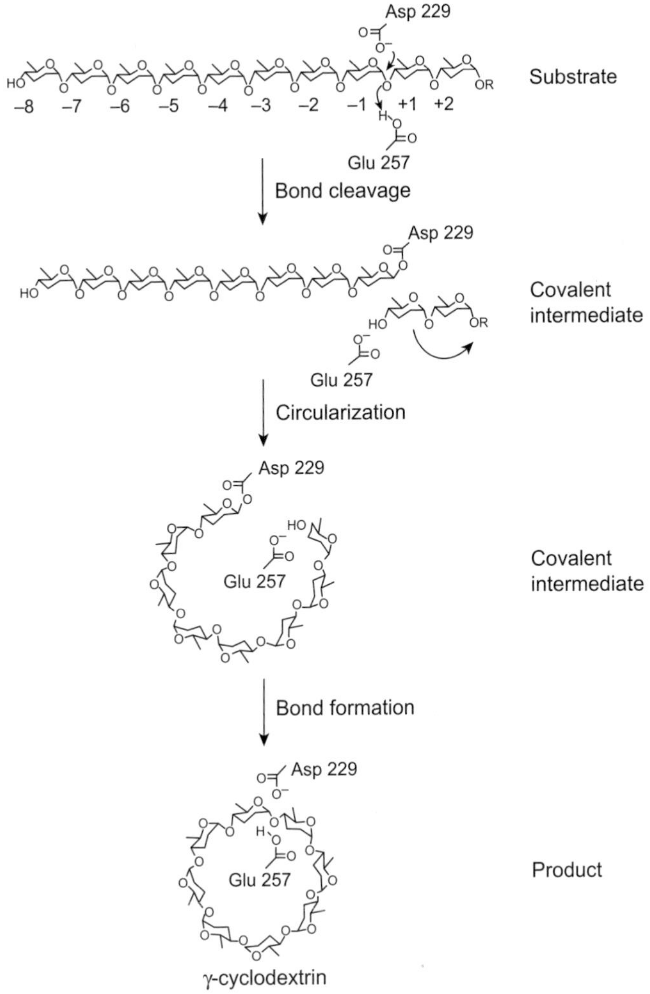

FIGURE 13.14   Circularization of starch derivatives into cyclodextrins catalyzed by cyclodextrin glucanotransferase (CGTase; EC 2.4.1.19). *(From Li et al., 2007.)*

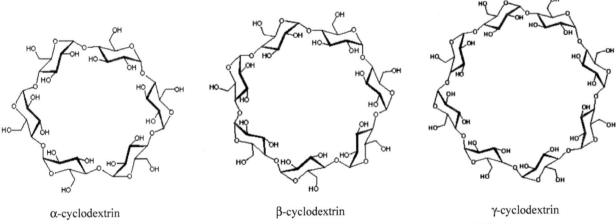

α-cyclodextrin   β-cyclodextrin   γ-cyclodextrin

FIGURE 13.15   The most common forms of circular cyclodextrins. *(From Li et al., 2007.)*

The circular molecules of α-, β- and γ-cyclodextrin are shaped like a hollow truncated cone or torus. Because the hydrogen atoms and the oxygen atoms of the glycosidic bonds are facing the inner side of the torus, while the hydroxyl groups are located on the outer side, cyclodextrins have a hydrophobic cavity and, at the same time, a hydrophilic outer surface which makes them water soluble. The hydrophobic cavity enables cyclodextrins to form inclusion complexes with a variety of organic compounds. The diameter of the cavity provides for a certain selectivity of the complexation of 'guest' molecules, i.e. the bigger ring of the eight-membered γ-cyclodextrin can accommodate a wider variety of guest molecules than the smaller rings of α- and β-cyclodextrins. Large guest molecules may complex with more than one cyclodextrin molecule (Tsoucaris *et al.*, 1987). The formation of an inclusion complex with a guest molecule is the basis for many applications of cyclodextrins in food, cosmetics, and pharmaceutical preparations (Li *et al.*, 2007).

Cyclodextrins can be used to increase the dietary fiber content of solid, semiliquid and liquid foods. Owing to their ability to form complexes with certain organic molecules, cyclodextrins can fulfill certain food-technological functions, e.g. as a flavor adjuvant; coloring adjunct; or a carrier, stabilizer, or solubilizer of certain vitamins (e.g. retinol acetate, vitamins K_1 and C, riboflavin, and β-carotene) and fatty acids (FDA-GRN 155). They are generally regarded as stabilizers, emulsifiers, carriers, and formulation aids in foods (FDA-GRN 64).

Cyclodextrins are produced using cyclodextrin glucanotransferase preparations. In the USA these preparations are produced in a recombinant strain of *E. coli* K12, which expresses the gene coding for the α-CGTase from *Klebsiella oxytoca* (FDA-GRN 155). The enzyme is completely removed from the final cyclodextrin products (FDA-GRN 155, FDA-GRN 64).

J. Chymosin

Chymosin (EC 3.4.23.4) is a peptidase with broad substrate specificity. It is specifically known for its ability to clot milk by cleaving at the Phe[105]−Met[106] bond in κ-casein (Gilliland *et al.*, 1991) and is therefore essential in the production of firm cheeses. Heterologously produced chymosin substitutes for the traditional preparations from calf stomach.

Bovine (*Bos taurus*) chymosin B is produced in submerged fermentation of *Trichoderma reesei*. Specifically, DNA encoding the prochymosin portion of the chymosin protein was synthesized using the preferred codon usage for *T. reesei* without changing the encoded amino acid sequence. The synthetic chymosin B gene is under the regulation of a native *T. reesei* promoter and terminator. A second expression cassette in the transforming DNA encodes for a *T. reesei* chaperonin that resides in the endoplasmic reticulum and is involved in folding of nascent secreted proteins fused to a different *T. reesei* promoter and terminator to control expression (FDA-GRN 230). *Trichoderma reesei* strain GICC03278 hosts the expression construct and the resulting preparation is known under the trade name Chymostar Supreme. The *T. reesei* strain GICC03278 is optimized by several gene deletions for the production of enzyme proteins.

Bovine chymosin is produced as a fusion protein using the *A. niger* glucoamylase gene as a carrier molecule. A synthetic copy of the bovine chymosine having optimized codon usage and a change of serine-351 to threonine was introduced into *A. niger* var. *awamori*. Chymosin is released from the glucoamylase−prochymosin fusion protein by autocatalytic cleavage of the propeptide (Kappeler *et al.*, 2006).

Recombinant bovine chymosin transcribed from the synthetic bovine prochymosin gene is also produced in *E. coli*, *A. niger* var. *awamori* (Chy-Max®), and the yeast *Kluyveromyces lactis* (Maxiren®) (FDA 21 CFR 184.1685).

K. Glucose Oxidase

Glucose oxidase (EC 1.1.3.4) is an oxidoreductase catalyzing the oxidation of glucose to hydrogen peroxide (H_2O_2) and D-glucono-1,5-lactone, acting on CH−OH groups with oxygen as the acceptor (KEGG R01522):

beta-D-Glucose + Oxygen <=> D-Glucono-1,5-lactone + Hydrogen peroxide

(From Li et al., 2007.)

Glucose oxidase is used in the baking industry as a processing aid to strengthen gluten in dough systems. It contributes to the oxidation of free sulfhydryl units in gluten, whereby disulfide linkages are formed. This results in stronger, more elastic dough with greater resistance to mechanical shock, as well as better oven spring and larger loaf volume (FDA-GRN 106). Glucose oxidase is also utilized to remove D-glucose from egg white to prevent browning. It is used to eliminate oxygen from food packaging, such as the headspace above bottled and canned drinks, as well as reducing non-enzymatic browning in wines and mayonnaises.

Recombinant glucose oxidase is produced in the genetically modified *Aspergillus oryzae* strain BECh 2, which is amylase negative, alkaline protease (*alp*) negative, neutral metalloprotease I (*Npl*) negative, cyclopiazonic acid deficient, and kojic acid deficient. This microorganism carries the gene coding for glucose oxidase from *A. niger* under the control of the neutral amylase I (*Pna2ltpI*) promoter. Moreover, the 5' non-translated part of this promoter has been replaced with the 5' non-translated part of the *Aspergillus nidulans* triose phosphate isomerase (*TPI*) promoter (FDA-GRN 106). The resulting enzyme preparation is registered under the trade name Gluzyme®.

L. Glycerophospholipid Cholesterol Acyltransferase

Glycerophospholipid cholesterol acyltransferases (EC 2.3.1.43) catalyze the transfer of acyl residues, such as palmitoyl, oleoyl, and linoleoyl residues, to a variety of sterols (KEGG R02114):

Phosphatidylcholine + Sterol <=> 1-Acyl-*sn*-glycero-3-phosphocholine + Steryl-ester

A glycerophospholipid cholesterol acyltransferase from *Aeromonas salmonicida* is heterologously expressed in *Bacillus licheniformis*. The host organism is *B. licheniformis* Bra7, which was modified through deletion of several enzyme activities (proteases, amylase), a sporulation gene, and the native chloramphenicol resistance genes to make it suitable for expression of heterologous proteins. The gene coding for the *A. salmonicida* glycerophospholipid cholesterol acyltransferase was synthesized and codon optimized. One amino acid was changed and the synthetic gene is under the regulation of a native *B. licheniformis* promoter and terminator. Integration was achieved through homologous recombination into the *B. licheniformis* chloramphenicol resistance gene (*catH*), which functions for selection, chromosomal integration, and cassette amplification. The *catH* genetic sequences surround the expression cassette with upstream and downstream sequences. Part of the upstream *catH* sequence (called 5' repeat) is present twice on the plasmid to allow amplification of the expression cassette on the chromosome (FDA-GRN 265).

The effectiveness of the *B. licheniformis* glycerophospholipid cholesterol acyltransferase is based on its effects on the cell membrane by transferring acyl groups from phospholipids and glycolipids to acceptors such as sterols, fatty alcohols, and other smaller primary alcohols. The acyl groups transferred are mainly C14 to C18 fatty acids: myristic, palmitic, stearic, oleic, linoleic, and linolenic acids. Cholesterol and other sterols accept the transferred acyl groups to become cholesterol-esters and sterol-esters. Fatty alcohols (defined as C12 and larger alcohols) can also be esterified. The reaction products formed depend on the substrate(s), but generally consist of lysophospholipids, cholesterol ester of C14 to C20 fatty acids and sterol esters of C14 to C20 fatty acids (campesterol, stigmasterol, betasitosterol, 5-avenasterol, and 7-stigmasterol).

The enzyme preparation is used in egg yolk and whole eggs, in processed meats, in degumming of vegetable oils, in milk products such as cheese, and in bakery products containing eggs, such as cake products. The trade name is FoodPro™ LysoMaxa Oil (FDA-GRN 265).

Egg yolk is well used in the food industry owing to its emulsification properties. Approximately 30% of the lipid in egg yolk is phospholipid, which contributes to egg yolk's emulsification properties. In many foods including mayonnaise, sauces, dressings, and cakes, the emulsification properties of egg yolk are exploited. For some food applications, however, the emulsification properties of egg yolk are not sufficient to obtain a homogeneous product without separation. In mayonnaise, for instance, pasteurization of the product at high temperatures causes the product to separate. The enzyme preparation will be used to modify phospholipid to lysophospholipid and cholesterol-ester in egg yolk. Product separation at high-temperature pasteurization can be avoided using enzyme-modified egg yolk for production of mayonnaise. In cakes, enzymatically modified egg yolk gives a softer and more tender crumb.

The enzyme preparation is used in processed meat products. It improves the emulsification of processed meat products and contributes to better consistency and reduced cooking loss. The enzyme preparation added to processed meat converts meat phospholipids to lysophospholipids.

Crude vegetable oils such as soybean oil contain around 12% phospholipids, which are removed from the oil during the refining process, in order to improve the quality and prevent sedimentation in the oil. The removal of phospholipids is conducted by a degumming process during the oil-refining process. The degumming can be conducted by chemical or enzymatic means. In the degumming process the enzyme converts phospholipids to lysophospholipids which are more water soluble and can be removed from the oil by washing with water. Enzymatic hydrolysis of phospholipids is a gentler process than chemical degumming, which needs more acids and alkalis. Furthermore, degumming with the enzyme will produce fewer effluents. During the degumming process the enzyme catalyzes the transfer of fatty acids form phospholipids to phytosterols in the oil during the formation of phytosterol esters. Phytosterols are normally removed by deodorization during oil refining, but when the enzyme preparation is used, the phytosterols are converted to phytosterol esters; these esters are not removed during the refining process because of the lower volatility of the sterol esters. Phytosterol esters are not unknown constituents in vegetable oil because a smaller part of phytosterol exists naturally in the form of esters.

The enzyme preparation in milk products contributes to increased yield during cheese production. The enzyme added to milk converts milk phospholipids to lysophospholipids, enhancing emulsification properties and increasing cheese yield by entrapping more lipid in the cheese curd (FDA-GRN 265).

M. Hexose Oxidase

Hexose oxidase (EC 1.1.3.5) catalyzes the oxidation of various monosaccharides and oligosaccharides (principally glucose, but also maltose, lactose, D-galactose, D-mannose, and cellobiose) with oxygen to produce lactones and hydrogen peroxide (KEGG R01522):

beta-D-Glucose + Oxygen <=> D-Glucono-1,5-lactone + Hydrogen peroxide

The main application of hexose oxidase is in bread making, to increase dough strength and bread volume. The enzyme acts in a similar way to glucose oxidase for this purpose, where the produced hydrogen peroxide acts as an oxidant with other food components. However, it has added advantages, since it acts on a wider range of substrates. Other applications in the food industries are in cheese and tofu manufacture, where it aids curd formation, limiting undesirable browning by limiting Maillard reactions in food, and as an oxygen scavenger during the production of dressings and sauces. It can also be used in the pasta and noodle manufacturing industries, where its use strengthens the structure of the dough, resulting in reduced loss of starch and protein on cooking and a firmer bite and better texture (FSANZ A475).

It has been known for some time that hexose oxidase can be extracted from a number of red algae. However, because of the difficulty in recovering the small amounts of the enzyme from such algae, little use has been made of its properties. This has been overcome in recent times by using recombinant DNA technologies to produce greater quantities of enzyme to enable more complete characterization. Such techniques have allowed industrial processes to be used to produce commercially viable enzyme preparations.

The gene encoding for hexose oxidase has been isolated from the alga *Chondrus crispus* and inserted into the yeast host *Hansenula polymorpha*. The plasmid used in this construction was created from the *E. coli* plasmid pBR322 by insertion of the *URA3* gene (encoding oritidine-5′-phosphate decarboxylase) from *S. cerevisiae*, a native *H. polymorphu* promoter and terminator, and the gene encoding for hexose oxidase from *C. crispus*. Furthermore, the genes coding for ampicillin resistance (*Apr*) and tetracycline resistance (*TCr*) in the original pBR322 have been removed. The subsequent organism produces the enzyme in commercial quantities during a submerged fermentation process (FSANZ A475, FDA-GRN 238).

N. Ice Structuring Protein

Ice structuring proteins are widely distributed in nature, for example in coldwater fish, vegetables, grains, lichens, and bacteria. Ice structuring proteins bind to ice and help organisms to cope with very cold environments, both by lowering the temperature at which ice crystals form and by modifying the size and shape of the ice crystals so that the ice is less damaging to tissues (Hall-Manning *et al.*, 2004; EFSA 768, 2008).

Ice structuring protein type III was originally isolated from the blood of *Macrozoarces americanus*, the ocean pout. The ice structuring protein from this fish consists of 12 isoforms that can be separated by high-performance liquid chromatography (HPLC). Isoform HPLC-12 gives the largest peak and is the most functionally active in *in vitro* studies on ice structuring (EFSA768, 2008). This protein is specifically identified by accession number P19614 in the Swiss Prot database. It has a molecular weight of 7.027 kDa, is not glycosylated, is heat stable, is stable over a pH range of 2−12, and consists of the following 66 amino acids:

NQASVVANQL IPINTALTLV MMRSEVVTPV GIPAEDIPRL VSMQVNRAVP LGTTLMPDMV KGYPPA

This *M. americanus* ice structuring protein isoform is prepared by fermentation of a genetically modified strain of food-grade baker's yeast *S. cerevisiae* in which a synthetic gene for the ice structuring protein has been inserted into the yeast's genome. The ice structuring protein expressed by the yeast has the same amino acid sequence as that from the ocean pout but the nucleotide sequence was constructed to favor codon usage in yeast to maximize expression. The protein is expressed and secreted into the growth medium (EFSA768, 2008, FDA-GRN 117). The ice structuring protein produced in the recombinant yeast strain is a mixture of 60% unglycosylated and 40% glycosylated protein. However, only the unglycosylated ice structuring protein is able to bind to ice crystals and alter the ice structure.

Ice structuring protein is used in the production of edible ice products, e.g. ice cream, milk ice, water ice, fruit ice, sorbets, frozen desserts, and any similar products such as iced smoothies. It allows for the production of products lower in fat, sugar, and calories. From 2003 to 2007 more than 470 million ice structuring protein-containing edible ice products were sold in the USA and 47 thousand liters of ice structuring protein containing ice cream was sold in Australia and New Zealand (EFSA768, 2008).

O. Laccase

Laccases (EC 1.10.3.2) are a group of multicopper oxidoreductases acting on both *o*- and *p*-quinols, and often acting also on aminophenols and phenylenediamine (KEGG R00083):

Oxygen + 4-Benzenediol <=> 4-Semiquinone + 2 H$_2$O

Alternatively:

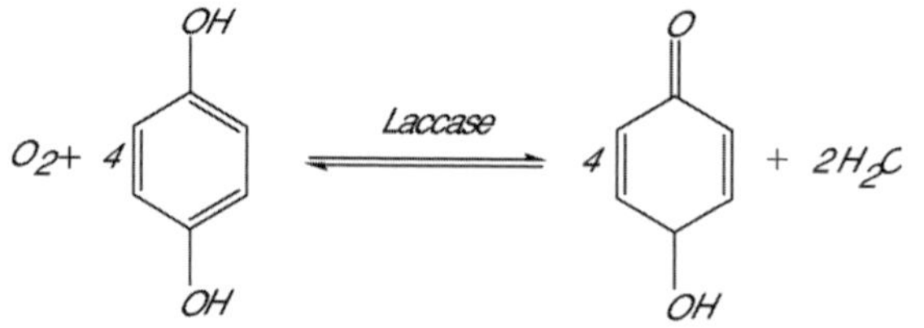

The semiquinone may react further either enzymatically or non-enzymatically. Laccases have a wide range of phenols and other substrates with the concomitant reduction of oxygen to water. Reactions between laccases and phenols are commonly found in nature. One prominent example is the oxidation of phenols in fruits and vegetables by laccases to form brown polymers, which is known as the bruising or enzymatic browning of fruits and vegetables.

In breath-freshening products such as breath mints, chewing gum, toothpaste, and mouthwash, laccase facilitates a reaction of naturally occurring polyphenolic compounds in food. The resulting semiquinones react with odor-causing volatile sulfur compounds in the oral cavity to deodorize them, therefore removing bad breath (FDA-GRN 122):

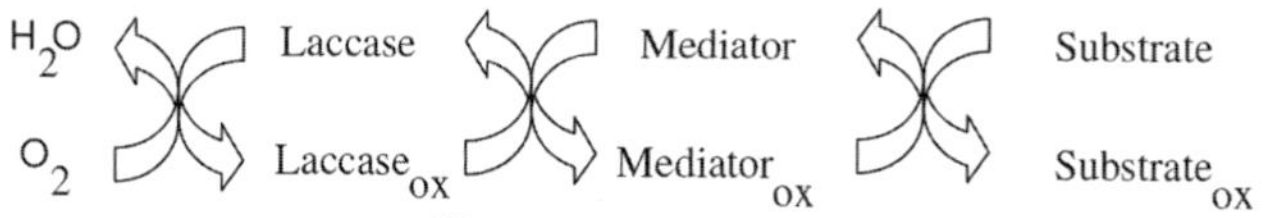

(Kunamneni et al., 2008)

Since laccases have the capability of catalyzing the oxidation and the polymerization of phenols and other substances, they are used in various applications such as in wine clarification, improved storage of beer, stabilization of fruit juices, dough improvement in bread making, and ethanol production. Fungal laccases are used to speed up the polymerization of polyphenols in the production of beverages such as beer, wine, and apple juice. In wine clarification, laccase is used to remove phenols from white grape must. Another potential application is oxygen removal during processing of foodstuffs, preventing oxidation of sensitive substances while preserving aromatic compounds or preventing the formation of off-flavors (FDA-GRN 122).

A transgenic laccase is produced by submerged fermentation of *Aspergillus oryzae* strain *How B711*, where the three resident TAKA amylase genes were removed by site-directed gene disruption and the expression plasmid for the encoding of a laccase from *Myceliophthora themophila* was inserted. Here, the laccase gene is under transcriptional control of the *A. oryzae* TAKA-amylase promoter and the *A. niger* glucoamylase terminator.

P. Lipases

1. Phospholipase A₁

Phospholipase A$_1$ (EC 3.1.1.32) hydrolyzes the *sn*-ester bond of diacylphospholipids to form 2-acyl-lysophospholipid and free fatty acid (KEGG R01314):

Phosphatidylcholine + H$_2$O <=> 2-Acyl-*sn*-glycero-3-phosphocholine + Carboxylate

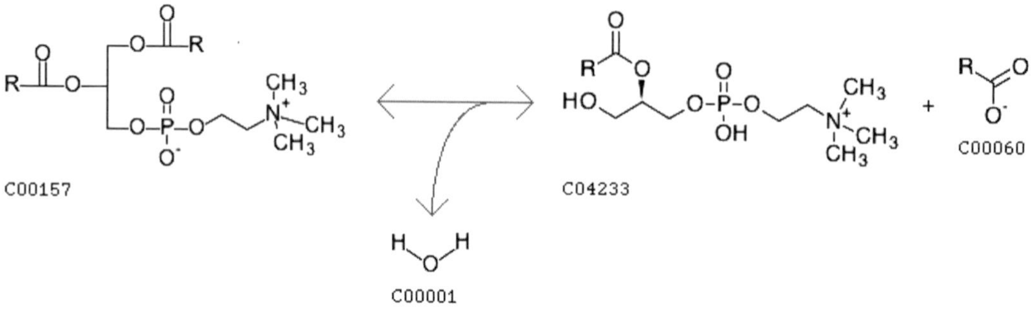

Phosphatidylcholine, often referred to by the name lecithin, is composed of a glycerol backbone esterified to phosphocholine and two fatty acids, and is one of the major components of the phospholipid portion of the cell membrane.

The *FvPLAl* gene from the fungus *Fusarium venenatum* is used for heterologous production of phospholipase A$_1$ in *A. oryzae* strain BECh 2. The expression cassette contains a double promoter element constructed from the *Pna2* TATA less neutral amylase II promoter from *A. niger*, fused to the Pna2/TPI unit, which is the neutral amylase II promoter from *A. niger* where the 5' non-translated part has been replaced with the 5' non-translated part of the *A. nidulans* triose phosphate isomerase (TPI) promoter. The *F. venenatum* phospholipase gene *FvPLAl* is flanked in the 3' by the amyloglycosidase terminator of *A. niger* (FDA-GRN 142).

The enzyme preparation has the trade name Novozyme® 46016 and specifically acts on the fatty acid in position 1 of phospholipids. It is used in the dairy industry as a processing aid during cheese production, where it is added to the cheesemilk. The modified phospholipids in the cheesemilk have improved emulsifying properties and as such result in better production efficiency by keeping more of the original milk components in the cheese and reducing the loss of fat and other solids into the whey stream (FDA-GRN 142).

2. Phospholipase A$_2$

Phospholipase A$_2$ (EC 3.1.1.4) catalyzes the hydrolysis of 3-*sn*-phospholipids exclusively at the 2-position, giving rise to the formation of 1-acyl-3-*sn*-lysophospholipids and free fatty acids (KEGG R01313):

Phosphatidylcholine + H$_2$O <=> 1-Acyl-*sn*-glycero-3-phosphocholine + Carboxylate

Although phospholipase A$_2$ is able to hydrolyze these phospholipids in their monomeric form, the enzyme is much more active when these substrates are organized in the form of micelles or lipid membranes. The presence of calcium ions is essential for the enzymatic activity of phospholipase A$_2$.

The genetically modified glucoamylase and protease negative *A. niger* strain ISO-502 is used as the host strain to express porcine phospholipase A$_2$. The coding sequence for the porcine phospholipase A$_2$ was derived from a pancreatic tissue cDNA library and integrated into a fusion protein. The coding sequences of the 498 N-terminal amino acids of the *A. niger* glucoamylase (*glaA*) gene are linked to the porcine phospholipase A$_2$ gene via a synthetic KEX2 proteolytic site. The expression of this construct is regulated by the *glaA* promoter and effective termination of the RNA is ensured by the 3'-flanking *glaA* terminator, both derived from the parental *A. niger* strain. The production strain, designated PLA-54, harbors multiple copies of the expression cassette, resulting in commercially attractive expression levels of the phospholipase A$_2$ enzyme (FDA-GRN 183).

The expression unit is translated into a glucoamylase—pro-phospholipase A$_2$ fusion protein. During secretion of this protein by the microbial cell, the endogenous kexine protease separates the pro-phospholipase A$_2$ from the glucoamylase part at the KEX2 proteolytic site. Also, the pro-part of the phospholipase A$_2$ is split off, resulting in the mature, active enzyme. The truncated secreted glucoamylase still has full enzymatic activity (FDA-GRN 183).

The heterologous porcine phospholipase A$_2$ hydrolyzes natural phospholipids present in foodstuffs, resulting in the formation of lysophospholipids. Lysophospholipids are surface-active agents with emulsifying properties to mimic the effects of chemical emulsifiers in food. During mixing and subsequent fermentation of bread dough, phospholipase A$_2$ hydrolyzes the phospholipids present in wheat flour mixes for bread making. The resulting lysophospholipids strengthen the dough to improve the dough mixing tolerance and machinability, as well as gas retention, to increase the loaf volume, crust crispiness, and crumb texture and softness. In egg-yolk-based fine bakery wares phospholipase A$_2$ hydrolyzes the phospholipids naturally present in the eggs that are part of the batter recipe. This contributes to the structure of batter-derived products such as sponge and pound cakes through efficient air incorporation and retention.

The addition of phospholipase A_2 to egg-yolk-based sauces and dressings significantly enhances the emulsifying properties, heat stability, and viscosity (FDA-GRN 183). The trade names for the spray-dried products are Bakezyme and Cakezyme (depending on the application) and the liquid product is sold under the trade name Maxapal A_2.

Another heterologous phospholipase A_2 is produced in the *Streptomyces violaceruber* strain AS-IO, carrying an expression cassette containing the gene encoding a phospholipase A_2 enzyme from *S. violaceruber* regulated by the *Streptomyces cinnamoneum* phospholipase D promoter and terminator. The enzyme preparation, called PLA2 Nagase, is used as a processing aid in egg yolk treatment or in hydrolyzing lecithin. The treated egg yolk or hydrolyzed lecithin has improved emulsifying properties and results in improved taste when added in food such as mayonnaise and/or a longer shelf-life of the product (FDA-GRN 212).

3. Phospholipase C

Phospholipases C (EC 3.1.4.3) catalyze the ester hydrolysis of a phospholipid to a diglyceride and choline phosphate (KEGG R01312):

Phosphatidylcholine + H_2O <=> 1,2-Diacyl-*sn*-glycerol + Choline phosphate

Bacterial Phospholipases C also act on sphingomyelin and phosphatidylinositol; that from seminal plasma does not act on phosphatidylinositol.

A phospholipases C encoding genetic fragment, designated as a BD1649 PLC gene, was isolated from a soil environmental DNA library and transferred into *Pichia pastoris* to create the strain DVSA-PLC-004. The specific DNA sequences used in the strain's construction include the phospholipases C encoding gene from an environmental library; linked to the *S. cerevisiae* α-mating factor secretion signal sequence; a fragment of the *P. pastoris* alcohol oxidase gene (3' *P.p. AOXI*); a *Pichia pastoris* selectable marker gene, *HIS4* (histidinol dehydrogenase); and the well-characterized, non-coding *P. pastoris* regulatory sequences comprising the alcohol oxidase promoter (*AOXI*) and the transcriptional terminator from *AUXI* (FDA-GRN 204).

The enzyme preparation is used for the degumming of oils, the first step in this refining process designed to remove contaminating phospholipids or phosphatides that otherwise interfere with the processing of high-phosphorus oils such as soybean, canola, corn, and sunflower oils. BD 16449 phospholipase C catalyzes the hydrolysis of the phosphodiester bond linking the glycerol and phosphate moieties at the *sn*-3 position of glycerophospholids, such as phosphatidylcholine, phosphatidylethanolamine, phosphatidylserine, and phosphatidic acid; for example:

Phosphatidylcholine + H_2O < = > 1, 2-Diacyl-*sn*-glycerol + Choline phosphate

Phosphatidylethanolamine + H_2O < = > 1, 2-Diacyl-*sn*-glycerol + Ethanolamine phosphate

The products of the hydrolysis are diacylglycerol and water-soluble phosphate esters, resulting in the following benefits:

- reduction in the amount of gum phospholipids and overall gum mass
- reduction in the total phosphorus contained in the oil
- reduction in the total mass of neutral oil entrained in the phospholipid gum
- increase in the quantity of diacylglycerol contained in the oil

Therefore, the oil yields are higher, and less bleaching is needed, resulting in lower usage of water and bleaching earth, and overall reduced use of environmental resources.

4. Triacylglycerol Lipases

Triacylglycerol lipases (EC 3.1.1.3) catalyze the hydrolysis of triglycerides, as well as phospholipids and galactolipids by acting on ester bonds (KEGG R01369):

$$\text{Triacylglycerol} + H_2O \rightarrow \text{Diacylglycerol} + \text{Carboxylate (fatty acid)}$$

Triglyceride lipases hydrolyze esters in an aqueous solution; however, they are also known to catalyze esterification of fatty acids and alcohols or rearrange fatty acids in glycerides under certain conditions where water content is low (FDA-GRN 43). Hence, the specificity of a lipase is not only determined by the species, but also depends on the reactants and the reaction conditions. For example, in some reactions a lipase shows 1,3 specificity, whereas in other reactions it functions as a non-positional lipase.

Rearrangement of fatty acids in glycerides can improve the physical and nutritional characteristics of the glyceride products. Lipases are applied in the fruit juice industry, baked goods, vegetable fermentation, and dairy products. Lipases have traditionally been used in the oils and fats industry, where they are utilized primarily to catalyze the cleavage of fatty acids from triglycerides. Lipases are used for degumming of edible oils by removing phospholipids. They can also be used to improve the emulsifying properties of ingredients (such as lecithin and egg yolk) during food processing. In bread making the enzymes improve dough stability and dough handling properties and subsequently bread volume and crumb homogeneity (FSANZ A569). There is a variety of heterologously produced triacylglycerol lipases in the food chain today and in the following section the most prominent will be discussed.

One recombinant triacylglycerol lipase enzyme preparation is produced by submerged fermentation using a selected strain of the yeast *Hansenula polymorpha* that carries the gene coding for a triacylglycerols lipase from the mold *Fusarium heterosporum*. The native gene was resynthesized with codon usage optimized for expression in yeast. In the expression cassette the synthetic gene encodes the same amino acid sequence as the native gene from *F. heterosporum* and it is under the control of a native *H. polymorphu* promoter and terminator (FDA-GRN 238, FSANZ A569). This specific enzyme preparation has the marketing name of GRINDAMYL™ POWERBake.

Transgenic triacylglycerol lipases produced in *Aspergillus niger* are marketed under the trade name Panamore™ for their ability to improve baking properties. In this specific case the *A. niger* host carries a gene coding for a pre-pro-lipase *lfs* gene which was constructed synthetically based on sequences of various *Fusarium* species. Expression is controlled by the glucoamylase *glaA* promoter and the 3′ flanking *glaA* terminator sequence from the parental *A. niger* (FDA-GRN 296).

A *Candida antarctica*-derived lipase is produced in the *A. niger* strain *MBinl18* which has been optimized for expression in heterologous proteins by deletion in the *pyrG* gene, interruption of the glucoamylase, acid-stable amylase, protease regulator, neutral amylase II, and neutral amylase I gene. The *C. antarctica* lipase is under the control of *Pna2/tpi*, the neutral amylase II promoter from *A. niger*, in which the 5′ non-translated part has been replaced with the 5′ non-translated part of the *Aspergillus nidulans* triose phosphate isomerase promoter. The amyloglycosidase terminator of *A. niger* completes this expression cassette. The enzyme preparation is marketed as Lipozyme® and is designated for the oils and fats industry to improve the physical and/or the nutritional properties of triglyceride products (FDA-GRN 158).

Cultures of *Aspergillus oryzae* are utilized to produce a variety of transgenic triacylglycerol lipases originating from the fungi *Fusarium oxysporum*, *Rhizomucor miehei*, and *Thermomyces lanuginosus* (previously known as *Humicola lanuginose*) (FSANZ A569, FDA FDA-GRN 43, FSANZ A402).

Cultures of the genetically modified *A. oryzae* strain H-1-52/c (synonym AI-11) are expressing the *T. lanuginosus* triacylglycerol lipase gene under the control of the *A. oryzae* TAKA amylase promoter and the *A. niger* glucoamylase terminator sequence. Two enzyme preparations derived from these cultures are marketed; Lipozyme TL IM, for interesterification of bulk fats, or production of frying fats, shortenings, and margarine components; and Novozym 677 BG, for increased dough stability and larger volume of baked goods, improved crumb softness, as well as structure and whiter crumb in bread making (FDA-GRN 43).

A triacylglycerol lipase from *F. oxysporum* is used for heterologous expression in *A. oryzae* strain *MStrl15*, which was genetically modified to be negative for amylase, alkaline protease, and neutral metalloprotease I. This expression cassette is under the control of the *A. niger* neutral amylase II gene promoter, which is fused to the 5' non-translated leader of the *A. nidulans* triose phosphate isomerase gene. Termination of transcription is ensured by the terminator of the *A. niger* amyloglycosidase gene. Resulting enzyme preparations are marketed as Lipopan F for baking applications and Lecitase® Novo in the fats and oils industry for degumming of vegetable oil as well as for the hydrolysis of lecithin for altered emulsifying properties and modification of egg yolk (FDA-GRN 75).

The *R. miehei* triacylglycerol lipase expressed in *A. niger* is marketed under the trade name Palatase®, but unfortunately no information on the nature of the genetic constructs seems to be available in the public records (FSANZ A402).

A hybrid genetic construct was derived from the two fungi *T. lanuginosud* and *F. oxysporum* to yield a triacylglycerol lipase specifically acting on the fatty acid in position 1 in both triglyceride substrates and phospholipids (FDA-GRN 103). The hybrid protein is constituted of the 284 N-terminal amino acids from the *T. lanuginosud* lipase *Tl1* gene fused to the 54 C-terminal amino acids of the *F. oxsporurn* lipase *FoL* gene. In addition, the sequence of the *T. lanuginosud* lipase was changed at three specific amino acid residues: glycine[113] → alanine, aspartic acid[118] → tryptophan, and glutamic acid[121] → lysine. The preparations' trade names are Lipopan® H and Lecitase® Ultra, used for baking applications and in the fats and oils industry, respectively. In the baking applications, the activity towards both triglyceride and phospholipid substrates is utilized, while in the fat and oil applications, the activity towards the phospholipids (phosphatides, lecithins) will predominate as a result of specific processing conditions. The specificity of the lipase towards long-chain fatty acids in position 1 is utilized in baking applications, where the lower tendency to release short-chain fatty acids from primary ester bonds in triglycerides and phospholipids decreases the risk of off-flavor generation in baking formulae that contain butterfat. In fats and oils applications this enzyme has increased activity towards phospholipids and high efficacy when used for the modification of egg yolk or whole egg, hydrolysis of lecithin for altered emulsifying properties, and vegetable oil degumming (FDA-GRN 103).

Q. Pectinesterase

Pectinesterases (EC 3.1.1.11) catalyze the de-esterification of pectin into pectate and methanol. They hydrolyze the ester linkage between methanol and galacturonic acid in esterified pectin (KEGG R02362):

Pectin + n H$_2$O <=> n Methanol + Pectate

Pectin is one of the main components of the plant cell wall. In plants, pectinesterases play important roles in cell wall metabolism during fruit ripening. The enzymatic de-esterification of pectins results in a low methoxylated pectin, which in the presence of calcium ions forms a strong gel. The enzymatic conversion of high methoxylated to low methoxylated pectin makes a gel formation possible and may, in products such as jam and ketchup, render further addition of thickening agents unnecessary. Pectin esterase also firms fruit and vegetables after it is infused into the tissue (FDA-GRN 8).

Heterologous pectin esterase is produced by *Aspergillus oryzae* strain IF0 4177 carrying the gene coding for pectin esterase from *Aspergillus aculeatus*. The expression cassette consists of the *A. oryzae* TAKA amylase gene promoter, followed by the *A. aculeatus* pectin esterase gene and the *A. niger* glucoamylase gene terminator sequence. The resulting enzyme preparation is marketed under the trade name Rheozyme™, used for the gelation of plant materials, thickening of plant preparations, hardening of fruits and vegetables, and controlled demethylation of high methoxylated pectins. It can be found in fruit preparations, compote, cider, jam, and tomatoes (FDA-GRN 8).

R. Pectin Lyase

Pectin is a heteropolysaccharide found in fruits with galacturonic acid and methanol as the main components, and some neutral sugars attached, such as D-galactose, L-arabinose, and D-xylose. Pectin molecules are formed by α-1,4-glycosidic linkages between the pyranose ring of D-galacturonic acid units. These galacturonan chains are interrupted at intervals by the insertion of 1,2-linked α-L-rhamnopyranosyl residues. The carboxyl groups of pectin are partially esterified with methanol, and the hydroxyl groups are sometimes partially acetylated.

Pectin lyases (EC 4.2.2.10) cleave 1,4-α-D-galacturonan methyl ester to oligosaccharides with 4-deoxy-6-*O*-methyl-α-D-galact-4-enuronosyl groups at their non-reducing ends. Hence, the enzyme cleaves the methylated polygalacturonic acid backbone of esterified pectin by a transelimination reaction and the cleaved oligomers contain a 4,5-dehydrogalacturonic acid group at their non-reducing end.

A pectin lyase from *Trichoderma reesei* is produced in *Aspergillus niger*. The *T. reesei* pectin lyase gene (*pe/D*) in the recombinant *T. reesei* strain is under the control of the *cbhl* promoter. This pectin lyase is a highly specific endo-enzyme with the ability to cleave highly esterified polygalacturonic acid. However, only those glycosidic bonds adjacent to a methyl ester group are split. Pectin lyase therefore shows high activity only towards highly esterified pectins. The degree of esterification in water-soluble pectin is 65−98%. The cleavage of only a small number of glycosidic linkages causes a sharp decrease in viscosity. The decrease in viscosity has a favorable influence both on the rate of filtration and on the heat transmission coefficient during concentrate manufacture. This specific heterologous pectin lyase enzyme preparation is used for applications in the processing of fruits and vegetables where a fast reduction of viscosity is needed. In the processing of juices the yield and the coloring matter will increase (FDA-GRN 32).

S. Phytases

A phytase (*myo*-inositol hexakisphosphate phosphohydrolase) is any type of phosphatase enzyme that catalyzes the hydrolysis of phytic acid (*myo*-inositol hexakisphosphate), an undigestible organic form of phosphorus that is found in grains and oil seeds, and releases a usable form of inorganic phosphorus (Mullaney *et al.*, 2000a, b). 4-Phytase, also named 6-phytase (EC 3.1.3.26), mediates the following reaction (KEGG R03372):

$$myo\text{-Inositol hexakisphosphate} + H_2O <=> \text{Inositol 1,2,3,5,6-pentakisphosphate} + \text{Orthophosphate}$$

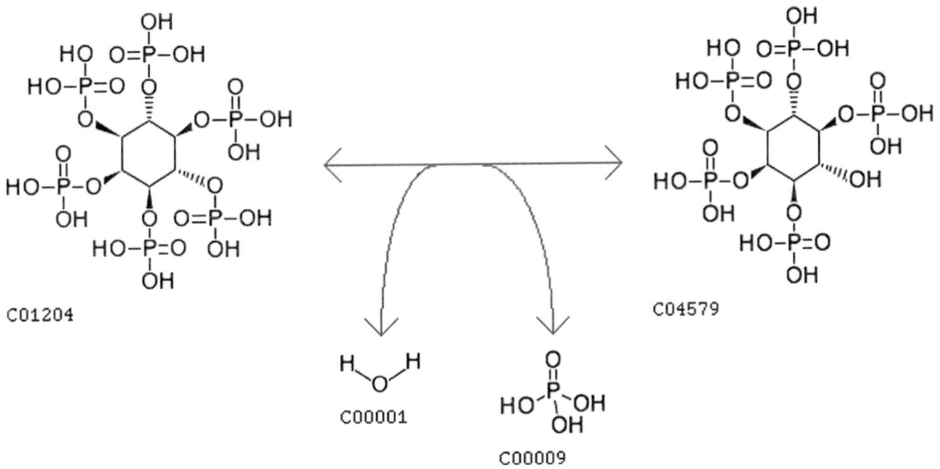

A heterologous 6-phytase from *Peniophora lycii* is heterologously produced in *Aspergillus oryza* strain Pz-3 (FSANZ A371) (EFSA, 2004). Expression is under the regulation of the Pna2/TPI promoter, constructed of the neutral amylase II promoter from *A. niger*, where the 5′ non-translated part has been replaced with the 5′ non-translated part of the *Aspergillus nidulans* triose phosphate isomerase (TPI) promoter. In Australia and New Zealand this enzyme preparation is applied as a processing aid for starch (FSANZ A371). Otherwise, it is known as a feed additive for fattening of monogastric livestock, improving phosphorus utilization in animals fed cereal-based diets (trade name Bio-Feed®).

T. Transglucosidase

Transglucosidases (EC 2.4.1.24) transfer an α-D-glucosyl residue in a 1,4-α-D-glucan to the primary hydroxy group of glucose, free or combined in a 1,4-α-D-glucan (KEGG).

The transglucosidase gene from *Aspergillus niger* is heterologously expressed in *Trichoderma reesei* strain RL-P37, which has been genetically modified by deletion of several cellulase genes. The expression cassette contains the DNA encoding *A. niger* mature secreted transglucosidase protein fused to the *T. reesei* CBHI signal peptide, for enhanced secretion. This open reading frame is flanked by the promoter and terminator sequences of the *T. reesei* cellobiohydrolase 1 (*cbhl*) gene (FDA-GRN 315).

Transglucosidase from *A. niger* only acts on oligosaccharides with a low degree of polymerization (Goffin *et al.*, 2010). This transglucosidase catalyzes both hydrolysis and transfer of α-D-glucooligosaccharides, resulting in a variety of products (Goffin *et al.*, 2010). Transfer occurs most frequently to HO-6, producing isomaltose from D-glucose, and panose from maltose. Transglucosidase can also transfer to the HO-2 or HO-3 of D-glucose to form kojibiose or nigerose, or back to HO-4 to form maltose. The action on maltose produces an equimolar concentration of panose and glucose. As a result of transglucosidase catalysis, malto-oligosaccharides are converted to isomalto-oligosaccharides containing high proportions of glucosyl residues linked by an α-D-1,6 linkage from the non-reducing end. Therefore, non-fermentable sugars including raffinose and stachyose are converted to sucrose, galactose, glucose, and fructose, which can then be fermented into alcohol (FDA-GRN 315). Therefore, the preparation is used as a processing aid in the production of isomalto-oligosaccharide syrup from starch and potable alcohol from molasses.

U. Xylanase

Xylanases (EC 3.2.1.8) hydrolyze 1,4-β-D-xylosidic linkages in the arabinoxylan backbone:

Xylanase Specificity

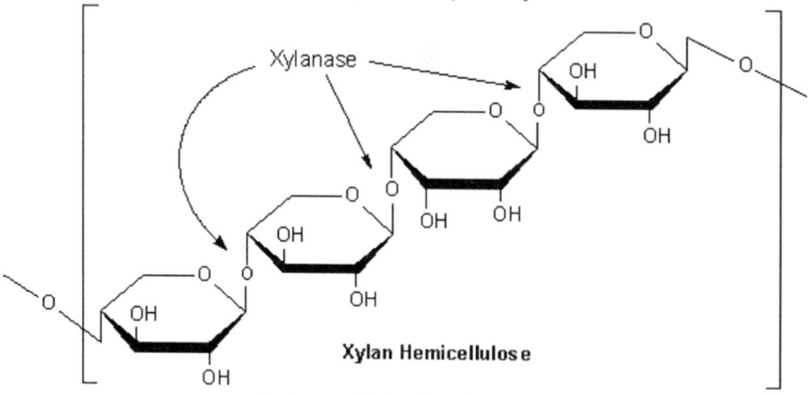

Xylan Hemicellulose

Polymer of β-(1-4)-D-xylopyranosyl units

(Source: sigma.com)

The linear polysaccharide 1,4-β-D-xylan is a component of hemicelluloses, the main constituents of plant cell walls. Arabinoxylans are highly branched xylans that are found in various cereals, and they exist in both a soluble and an insoluble form.

The endo-1,4-β-xylanase gene from *Thermomyces lanuginosus* has been transferred into a selected strain of *Fusarium venenatum* (trade name NOVOZYM® 899) as well as *Aspergillus oryzae* (FDA-GRN 54). The xylanase gene from *T. lanuginosus* is under the control of the *F. oxysporum* trypsin gene promoter and terminator. The derived heterologous xylanase enzyme preparations are used in the food industry as a processing aid in baking. The hydrolysis of the xylosidic linkages in an arabinoxylan backbone results in depolymerization of the arabinoxylan into smaller oligosaccharides. This increases the elasticity of the gluten network, improving handling of the dough.

REFERENCES

Agriculture and Agri-food Canada (as of 1 April 1997, the Canadian Food Inspection Agency), 1995. Monsanto Canada Inc.'s Glyphosate-tolerant soybean (*Glycine max* L.) Line GTS 40-3-2. DD95-05. (Available electronically at http://www.cfia-acia.ca/english/food/pbo/bhome.html.)

Agriculture and Agri-food Canada, Food Production and Inspection Branch (as of 1 April 1997, the Canadian Food Inspection Agency), 1996. Monsanto Canada Inc.'s Roundup Herbicide-Tolerant Brassica napus Canola line GT200. DD96-07. (Available electronically at http://www.cfia-acia.ca/english/food/pbo/bhome.html.)

Barry, G.F., Kishore G.M., 1997. Glyphosate tolerant plants. U.S. patent no. 5776760.

Baulcombe, D., 2004. RNA silencing in plants. Nature 431, 356–363.

Baulcombe, D.C., 2007. Amplified silencing. Science 315, 199–200.

Benfey, P.N., Chua, N.H., 1989. Regulated genes in transgenic plants. Science 244, 174–181.

Benfey, P.N., Chua, N.H., 1990. The cauliflower mosaic virus-35s promoter – combinatorial regulation of transcription in plants. Science 250, 959–966.

Birch, R.G., 1997. Plant transformation: problems and strategies for practical application. Annu. Rev. Plant Physiol. Mol. Biol. 48, 297–326.

Bos, L., 1982. Crop losses caused by viruses. Crop Protect. 1, 263–282.

Bravo, A., Likitvivatanavong, S., Gill, S.S., Soberon, M., 2011. *Bacillus thuringiensis*: a story of a successful bioinsecticide. Insect Biochem. Mol. Biol. 41, 423–431.

Cannon, R.J.C., 1996. *Bacillus thuringiensis* use in agriculture: a molecular perspective. Biol. Rev. 71, 561–636.

Coulon, J., Husnik, J.I., Inglis, D.L., van der Merwe, G.K., Lonvaud, A., Erasmus, D.J., van Vuuren, H.J.J., 2006. Metabolic engineering of *Saccharomyces cerevisiae* to minimize the production of ethyl carbamate in wine. Am. J. Enol. Vitic. 57, 113–124.

Crickmore, N., 2011. *Bacillus thuringiensis* toxin nomenclature. Available at. http://www.lifesci.sussex.ac.uk/home/Neil_Crickmore/Bt/.

Croon, K.S., 1996. Petition for Determination of Nonregulated Status: Insect-protected Roundup Ready Corn Line MON 802. USDA Petition No. 96-317-01P.

Dale, P.J., Irwin, J.A., Scheffler, J.A., 1993. The experimental and commercial release of transgenic crop plants. Plant Breeding 111, 1–22.

Davis, M.J., Ying, Z., 2004. Development of papaya breeding lines with transgenic resistance to Papaya ringspot virus. Plant Dis. 88, 352–358.

Deleris, A., Gallego-Bartolome, J., Bao, J.S., Kasschau, K.D., Carrington, J.C., Voinnet, O., 2006. Hierarchical action and inhibition of plant Dicer-like proteins in antiviral defense. Science 313, 68–71.

Della-Cioppa, G., Bauer, S.C., Klein, B.K., Shah, D.M., Fraley, R.T., Kishore, G.M., 1986. Translocation of the precursor of 5-enolpyruvylshikimate-3-phosphate synthase into chloroplasts of higher-plants *in vitro*. Proc. Natl. Acad. Sci. U.S.A. 83, 6873–6877.

Devine, M.D., Shukla, A., 2000. Altered target sites as a mechanism of herbicide resistance. Crop Protect 19, 881–889.

Diderichsen, B., Christiansen, L., 1988. Cloning of a maltogenic alpha-amylase from *Bacillus stearothermophilus*. FEMS Microbiol. Lett. 56, 53–59.

Dougherty, W.G., Parks, T.D., 1995. Transgenes and gene suppression – telling us something new. Curr. Opin. Cell Biol. 7, 399–405.

Dougherty, W.G., Lindbo, J.A., Smith, H.A., Parks, T.D., Swaney, S., Proebsting, W.M., 1994. RNA-mediated virus-resistance in transgenic plants – exploitation of a cellular pathway possibly involved in RNA degradation. Mol. Plant Microbe Interact. 7, 544–552.

Dröge-Laser, W., Siemeling, U., Puhler, A., Boer, I., 1994. Herbicide l-phosphinothricin (glufosinate) – identification, stability, and mobility in transgenic, herbicide-resistant, and untransformed plants. Plant Physiol. 105, 159–166.

Duke, S.O., Powles, S.B., 2008. Glyphosate: a once-in-a-century herbicide. Pest. Manag. Sci. 64, 319–325.

Eamens, A., Wang, M.B., Smith, N.A., Waterhouse, P.M., 2008. RNA silencing in plants: yesterday, today, and tomorrow. Plant Physiol. 147, 456–468.

EFSA, 2011. Guidance on the risk assessment of genetically modified microorganisms and their products intended for food and feed use. EFSA J. 9 (6), 20.

EFSA, 2004. Opinion of the Scientific Panel on additives and products or substances used in animal feed (FEEDAP) on the safety of the change of strain of the producing micro-organism of the enzyme preparation Bio-Feed Phytase. EFSA J. doi:10.2903/j.efsa.2004.66.

European Commission, Directorate General XXIV, Policy and Consumer Health Protection: Scientific Committee on Plants, 1998a. Opinion of the Scientific Committee on Plants regarding submission for placing on the market of fodder beet tolerant to glyphosate notified by DLF-Trifolium, Monsanto and Danisco Seed (notification C/K/97/01) (Opinion expressed by SCP on 23 June 1998). (See: http://europa.eu.int/comm/dg24/health/sc/scp/out16_en.htm.)

European Commission, Directorate General XXIV, Policy and Consumer Health Protection: Scientific Committee on Plants, 1998b. Opinion of the Scientific Committee on Plants regarding the genetically modified cotton tolerant to glyphosate herbicide notified by the Monsanto Company (notification C/ES/97/01) (Opinion expressed by the SCP on 14 July 1998). (See: http://europa.eu.int/comm/dg24/health/sc/scp/out17_en.htm.)

Fuchs, M., Gonsalves, D., 2007. Safety of virus-resistant transgenic plants two decades after their introduction: lessons from realistic field risk assessment studies. Annu. Rev. Phytopathol. 45, 173–202.

Gatehouse, A.M.R., Ferry, N., Edwards, M.G., Bell, H.A., 2011. Insect-resistant biotech crops and their impacts on beneficial arthropods. Philos. Trans. R. Soc. B Biol. Sci. 366, 1438–1452.

Gerlach, W.L., Llewellyn, D., Haseloff, J., 1987. Construction of a plant-disease resistance gene from the satellite RNA of tobacco ringspot virus. Nature 328, 802–805.

Gill, S.S., Cowles, E.A., Pietrantonio, P.V., 1992. The mode of action of *Bacillus thuringiensis* endotoxins. Annu. Rev. Entomol. 37, 615–636.

Gilliland, G.L., Oliva, M.T., Dill, J., 1991. Functional implications of the three-dimensional structure of bovine chymosin. Adv. Exp. Med. Biol. 306, 23–37.

GMO Compass. Available at www.gmo-compass.org/

Goesaert, H., Slade, L., Levine, H., Delcour, J.A., 2009. Amylases and bread firming – an integrated view. J. Cereal Sci. 50, 345–352.

Goffin, D., Wathelet, B., Blecker, C., Deroanne, C., Malmendier, Y., Paquot, M., 2010. Comparison of the glucooligosaccharide profiles produced from maltose by two different transglucosidases from *Aspergillus niger*, Biotechnology, Agronomy and Society and Environment 14 (4), 607–616.

Golemboski, D.B., Lomonossoff, G.P., Zaitlin, M., 1990. Plants transformed with a tobacco mosaic-virus nonstructural gene sequence are resistant to the virus. Proc. Natl. Acad. Sci. U.S.A. 87, 6311—6315.

Gonsalves, D., 1998. Control of papaya ringspot virus in papaya: a case study. Annu. Rev. Phytopathol. 36, 415—437.

Gottula, J., Fuchs, M., 2009. Toward a quarter century of pathogen-derived resistance and practical approaches to plant virus disease control. Adv. Virus Res. 75, 161—183.

Hall-Manning, T., Spurgeon, M., Wolfreys, A.M., Baldrick, A.P., 2004. Safety evaluation of ice-structuring protein (ISP) type IIIHPLC 12 preparation. Lack of genotoxicity and subchronic toxicity. Food Chem. Toxicol. 42, 321—333.

Hamilton, A.J., Baulcombe, D.C., 1999. A species of small antisense RNA in posttranscriptional gene silencing in plants. Science 286, 950—952.

Hamilton, A., Voinnet, O., Chappell, L., Baulcombe, D., 2002. Two classes of short interfering RNA in RNA silencing. EMBO J. 21, 4671—4679.

Hannon, G.J., 2002. RNA interference. Nature 418, 244—251.

Harrison, B.D., Mayo, M.A., Baulcombe, D.C., 1987. Virus-resistance in transgenic plants that express cucumber mosaic-virus satellite RNA. Nature 328, 799—802.

Harrison, L.A., Bailey, M.R., Naylor, M.W., Ream, J.E., Hammond, B.G., Nida, D.L., Burnette, B.L., Nickson, T.E., Mitsky, T.A., Taylor, M.L., Fuchs, R.L., Padgette, S.R., 1996. The expressed protein in glyphosate-tolerant soybean, 5-enolypyruvylshikimate-3-phosphate synthase from Agrobacterium sp. strain CP4, is rapidly digested in vitro and is not toxic to acutely gavaged mice. J. Nutr. 126, 728—740.

Hérouet, C., Esdaile, D.J., Mallyon, B.A., Debruyne, E., Schulz, A., Currier, T., Hendrickx, K., Van Der Klis, R.J., Rouan, D., 2005. Safety evaluation of the phosphinothricin acetyltransferase proteins encoded by the pat and bar sequences that confer tolerance to glufosinate-ammonium herbicide in transgenic plants. Regul. Toxicol. Pharmacol. 41, 134—149.

Hoerlein, G., 1994. Glufosinate (phosphinothricin), a natural amino-acid with unexpected herbicidal properties. Rev. Environ. Contam. Toxicol. 138, 73—145.

Höfte, H., de Greve, H., Seurinck Jansen, J.S., Mahillon, J., Ampe, C., Vanderkerchove, J., Vanderbruggen, H., Montagu, M.V., Zabeau, M., Vaek, M., 1986. Structural and functional analysis of a cloned delta-endotoxin of Bacillus thuringiensis. Eur. J. Biochem. 161, 273—280.

Hofte, H., Whiteley, H.R., 1989. Insecticidal crystal proteins of Bacillus thuringiensis. Microbiol. Rev. 53, 242—255.

Holt Editor-in-chief, J.G., 1984. Bergy's Manual of Systematic Bacteriology. William and Wilkins, Baltimore.

Hudspeth, R.L., Grula, J.W., 1989. Structure and expression of the maize gene encoding the phosphoenolpyruvate carboxylase isozyme involved in C-4 photosynthesis. Plant Mol. Biol. 12, 579—589.

Husnik, J.I., Delaquis, P.J., Cliff, M.A., Van Vuuren, H.J.J., 2007. Functional analyses of the malolactic wine yeast ML01. Am. J. Enol. Vitic. 58, 42—52.

ISAAA GM Approval Database. Available at http://www.isaaa.org/gmapprovaldatabase/default.asp

James, C., 2009. Executive summary. Global Status of Commercialized Biotech/GM Crops: 2009. ISAAA Brief No. 41. ISAAA, Ithaca, NY.

James, C., 2010. Global Status of Commercialized Biotech/GM Crops: 2010. ISAAA Brief No. 42. ISAAA, Ithaca, NY.

Johnson, S.R., Strom, S., Grillo, K., 2007. Quantication of the Impacts on US Agriculture and Biotechnology-Derived Crops in 2006. National Center for Food and Agriculture Policy, Washington, DC.

Jonas, D.A., Antignac, E., Antoine, J.M., Classen, H.G., Huggett, A., Knudsen, I., Mahler, J., Ockhuizen, T., Smith, M., Teuber, M., Walker, R., De Vogel, P., 1996. The safety assessment of novel foods. Guidelines prepared by ILSI Europe Novel Food Task Force. Food Chem. Toxicol. 34, 931—940.

Kaniewski, W.K., Thomas, P.E., 2004. The potato story. AgBioForum 7, 41—46.

Kappeler, S.R., Van Den Brink, H.M., Rahbek-Nielsen, H., Farah, Z., Puhan, Z., Hansen, E.B., Johansen, E., 2006. Characterization of recombinant camel chymosin reveals superior properties for the coagulation of bovine and camel milk. Biochem. Biophys. Res. Commun. 342, 647—654.

Kessler, H.A., Pottage, J.C., Bick, J.A., Benson, C.A., 1992. Aids Part II. Dis. Mon. 38 691—674.

Kishore, G.M., Shah, D.M., 1988. Amino-acid biosynthesis inhibitors as herbicides. Annu. Rev. Biochem. 57, 627—663.

Klein, T.M., Wolf, E.D., Wu, R., Sanford, J.C., 1987. High-velocity microprojectiles for delivering nucleic-acids into living cells. Nature 327, 70—73.

Kollar, A., Dalmay, T., Burgyan, J., 1993. Defective interfering RNA-mediated resistance against cymbidium ringspot tombusvirus in transgenic plants. Virology 193, 313—318.

Kunamneni, A., Camarero, S., García-Burgos, C., Plou, F.J., Ballesteros, A., Alcalde, M., 2008. Engineering and applications of fungal laccases for organic synthesis. Microb. Cell Fact. 7, 32.

Lee, E.Y.C., Whelan, W.J., 1972. Glycogen and starch debranching enzymes. In: Boyer, P. (Ed.), The Enzymes, vol 3. Academic Press, New York, pp. 191—234.

Li, Z., Wang, M., Wang, F., Gu, Z., Du, G., Wu, J., Chen, J., 2007. γ-Cyclodextrin: a review on enzymatic production and applications. Appl. Microbiol. Biotechnol. 77, 245—255.

Likitvivatanavong, S., Chen, J.W., Evans, A.M., Bravo, A., Soberon, M., Gill, S.S., 2011. Multiple receptors as targets of cry toxins in mosquitoes. J. Agric. Food Chem. 59, 2829—2838.

Lin, S.S., Henriques, R., Wu, H.W., Niu, Q.W., Yeh, S.D., Chua, N.H., 2007. Strategies and mechanisms of plant virus resistance. Plant Biotechnol. Rep. 1, 125—134.

Lindbo, J.A., Silvarosales, L., Proebsting, W.M., Dougherty, W.G., 1993. Induction of a highly specific antiviral state in transgenic plants — implications for regulation of gene-expression and virus-resistance. Plant Cell. 5, 1749—1759.

de Maagd, R.A., Bravo, A., Crickmore, N., 2001. How Bacillus thuringiensis has evolved specific toxins to colonize the insect world. Trends Genet. 17, 193—199.

McBride, K.E., Svab, Z., Schaaf, D.J., Hogan, P.S., Stalker, D.M., Maliga, P., 1995. Amplification of a chimeric bacillus gene in chloroplasts leads to an extraordinary level of an insecticidal protein in tobacco. Bio-Technology 13, 362—365.

McElroy, D., Rothenberg, M., Wu, R., 1990. Structural characterization of a rice actin gene. Plant Mol. Biol. 14, 163—171.

Maiti, I.B., Murphy, J.F., Shaw, J.G., Hunt, A.G., 1993. Plants that express a potyvirus proteinase gene are resistant to virus infection. Proc. Natl. Acad. Sci. U.S.A. 90, 6110—6114.

Malyshenko, S.I., Kondakova, O.A., Nazarova, J.V., Kaplan, I.B., Taliansky, M.E., Atabekov, J.G., 1993. Reduction of tobacco mosaic-virus accumulation in transgenic plants producing nonfunctional viral transport proteins. J. Gen. Virol. 74, 1149—1156.

Ministry of Agriculture, Forestry and Fisheries of Japan, 1996. Current Status of Commercialization of Transgenic Crop Plants in Japan. (Available electronically at http://ss.s.affrc.go.jp/docs/sentan/index.htm.)

Monsanto, 1997. Petition for determination of nonregulated status: Roundup Ready Corn Line GA21 (submitted to United States Department of Agriculture, Petition No. 97-099-01P). (Copies of the documents are available from USDA-APHIS, Unit 147, 4700 River Road, Riverdale, Maryland 20737.)

Mullaney, E.J., Daly, C.B., Sethumadhavan, K., Rodriquez, E., Lei, X.G., Ullah, A.H.J., 2000a. Phytase activity in Aspergillus fumigatus isolates. Biochem. Biophys. Res. Commun. 275, 759−763.

Mullaney, E.J., Daly, C.B., Ullah, A.H.J., 2000b. Advances in phytase research. Adv. Appl. Microbiol. 47, 157−199.

Nelson, R.S., Abel, P.P., Beachy, R.N., 1987. Lesions and virus accumulation in inoculated transgenic tobacco plants expressing the coat protein gene of tobacco mosaic-virus. Virology 158, 126−132.

Neuhaus, G., Spangenberg, G., Scheid, O.M., Schweiger, H.G., 1987. Transgenic rapeseed plants obtained by the microinjection of DNA into microspore-derived embryoids. Theor. Appl. Genet. 75, 30−36.

Nida, D.L., Kolacz, K.H., Buehler, R.E., Deaton, W.R., Schuler, W.R., Armstrong, T.A., Taylor, M.L., Ebert, C.C., Rogan, G.J., Padgette, S.R., Fuchs, R.L., 1996. Glyphosate-tolerant cotton: genetic characterization and protein expression. J. Agric. Food Chem. 44, 1960−1966.

Obbard, D.J., Gordon, K.H.J., Buck, A.H., Jiggins, F.M., 2009. The evolution of RNAi as a defence against viruses and transposable elements. Philos. Trans. R. Soc. B Biol. Sci. 364, 99−115.

OECD, 1999. Consensus document on general information concerning the genes and their enzymes that confer tolerance to glyphosate herbicide. In: OECD, Safety Assessment of Transgenic Organisms, OECD Consensus Documents, vol. 1. OECD Publishing (2006).

OECD, 2002. ENV/JM/MONO(2002)14. Available at www.oecd.org/dataoecd/17/39/46815748.pdf

OECD, 2010. Section 2, Molecular characterisation of plants derived from modern biotechnology. In: OECD, Safety Assessment of Transgenic Organisms, OECD Consensus Documents, vol. 3. OECD Publishing.

Olempska-Beer, Z.S., Merker, R.I., Ditto, M.D., DiNovi, M.J., 2006. Food-processing enzymes from recombinant microorganisms − a review. Reg. Toxicol. Pharmacol. 45, 144−158.

Oppert, B., 1999. Protease interactions with Bacillus thuringiensis insecticidal toxins. Arch. Insect Biochem. Physiol. 42, 1−12.

Padgette, S.R., Kolacz, K.H., Delannay, X., Re, D.B., Lavallee, B.J., Tinius, C.N., Rhodes, W.K., Otero, Y.I., Barry, G.F., Eichholtz, D.A., Peschke, V.M., Nida, D.L., Taylor, N.B., Kishore, G.M., 1995. Development, identification, and characterization of a glyphosate-tolerant soybean line. Crop Sci. 35, 1451−1461.

Padgette, S.R., Re, D.B., Barry, G.F., Eichholtz, D.E., Delannay, X., Fuchs, R.L., Kishore, G.M., Fraley, R.T., 1996. New weed control opportunities: Development of soybeans with a Roundup Ready gene. In: Duke, S.O. (Ed.), Herbicide-resistant Crops: Agricultural, Environmental, Economic, Regulatory, and Technical Aspects. CRC Press Inc., Boca Raton, Florida, and London, England, pp. 53−84.

Pariza, M.W., Johnson, E.A., 2001. Evaluating the safety of microbial enzyme preparations used in food processing: update for a new century. Regul. Toxicol. Pharmacol. 33, 173−186.

Parrish, S.K., Kaufmann, J.E., Croon, K.A., Ishida, Y., Ohta, K., Itoh, S., 1995. MON 37500: a new selective herbicide to control annual and perennial weeds in wheat. In: Brighton Crop Protection Conference, Weeds − 1995, Vols. 1−3 57−63.

Pollegioni, L., Schonbrunn, E., Siehl, D., 2011. Molecular basis of glyphosate resistance − different approaches through protein engineering. FEBS J. 278, 2753−2766.

Potrykus, I., 1991. Gene transfer to plants: assessment of published approaches and results. Annu. Rev. Plant Physiol. Plant Mol. Biol. 42, 205−225.

Powell-Abel, P., Nelson, R.S., De, B., Hoffmann, N., Rogers, S.G., Fraley, R.T., Beachy, R.N., 1986. Delay of disease development in transgenic plants that express the tobacco mosaic-virus coat protein gene. Science 232, 738−743.

Prins, M., Laimer, M., Noris, E., Schubert, J., Wassenegger, M., Tepfer, M., 2008. Strategies for antiviral resistance in transgenic plants. Mol. Plant Pathol. 9, 73−83.

Pruser, K.N., Flynn, N.E., 2011. Acrylamide in health and disease. Front. Biosci. S3, 41−51.

Register, J.C., Beachy, R.N., 1988. Resistance to TMV in transgenic plants results from interference with an early event in infection. Virology 166, 524−532.

Sanford, J.C., 1990. Biolistic plant transformation. Physiol. Plant. 79, 206−209.

Sanford, J.C., Johnston, S.A., 1985. The concept of parasite-derived resistance: deriving resistance genes from the parasite's own genome. J. Theoret. Biol. 113, 395−405.

Sauer, H., Wild, A., Ruhle, W., 1987. The effect of phosphinothricin (glufosinate) on photosynthesis. 2. The causes of inhibition of photosynthesis. Z. Naturforsch. C. 42, 270−278.

Scientific Committee on Food, 1992. Reports of the Scientific Committee on Foods (Twenty-sixth series). Food Science and Techniques. European Commission, Luxembourg.

Sharma, H.C., Sharma, K.K., Crouch, J.H., 2004. Genetic transformation of crops for insect resistance: potential and limitations. Crit. Rev. Plant Sci. 23, 47−72.

Shea, K., 2009. Determination of Nonregulated Status for Papaya Genetically Engineered for Resistance to the Papaya Ringspot Virus. US Federal Register Volume 74 (168), 45163−45164.

Singh, R.S., Sini, G.K., Kennedy, J.F., 2008. Pullulan: Micobial sources, production and applications. Carbohydr. Polymers. 73, 515−531.

Soberon, M., Pardo, L., Munoz-Garay, C., Sanchez, J., Gomez, I., Porta, H., Bravo, A., 2010. Pore formation by cry toxins. Adv. Exp. Med. Biol. 677, 127−142.

de Souza, P.M., Magalhães, P.O., 2010. Application of microbial α-amylase in industry − a review. Braz. J. Microbiol. 41, 850−861.

Stalker, D.M., McBride, K.E., 1987. Cloning and expression in Escherichia coli of a Klebsiella ozaenae plasmid-borne gene encoding a nitrilase specific for the herbicide bromoxynil. J. Bacteriol. 169, 955−960.

Stanley, J., Frischmuth, T., Ellwood, S., 1990. Defective viral-DNA ameliorates symptoms of geminivirus infection in transgenic plants. Proc. Natl. Acad. Sci. U.S.A. 87, 6291−6295.

Steinrucken, H.C., Amrhein, N., 1980. The herbicide glyphosate is a potent inhibitor of 5-enolpyruvyl-shikimic-acid 3-phosphate synthase. Biochem. Biophys. Res. Commun. 94, 1207−1212.

Svab, Z., Maliga, P., 1993. High-frequency plastid transformation in tobacco by selection for a chimeric aada gene. Proc. Natl. Acad. Sci. U.S.A. 90, 913−917.

Swiegers, J.H., Pretorius, I.S., 2007. Modulation of volatile sulfur compounds by wine yeast. Appl. Microbiol. Biotechnol. 74, 954–960.

Tachibana, K., Watanabe, T., Sekizawa, Y., Takematsu, T., 1986. Action mechanism of Bialaphos. 1. Inhibition of glutamine-synthetase and quantitative changes of free amino-acids in shoots of Bialaphos-treated Japanese barnyard millet. J. Pest. Sci. 11, 27–31.

Taylor, N., Chavarriaga, P., Raemakers, K., Siritunga, D., Zhang, P., 2004. Development and application of transgenic technologies in cassava. Plant Mol. Biol. 56, 671–688.

Tepfer, M., 2002. Risk assessment of virus-resistant transgenic plants. Annu. Rev. Phytopathol. 40, 467–491.

Then, C., 2010. Risk assessment of toxins derived from *Bacillus thuringiensis* — synergism, efficacy, and selectivity. Environ. Sci. Pollut. Res. 17, 791–797.

Tsoucaris, G., Le Bas, G., Rysanek, N., Villain, F., 1987. Conformational and enantiomeric discrimination in cyclodextrin inclusion compounds. J. Inclus. Phenom. 5, 77–84.

USDA, 1994. Petition under 7CFR Part 340. Petition Number 93-258-01p. Monsanto, Soybean, Herbicide Tolerance, Glyphosate-tolerant. Approved 5/19/94. (See: http://www.aphis.usda.gov/bbep/bp/petday.html.)

USDA, 1995. Petition under 7CFR Part 340. Petition Number 95-045-01p, Monsanto, Cotton, Herbicide Tolerance, Glyphosate-tolerant. Approved 7/11/95. (See: http://www.aphis.usda.gov/bbep/bp/petday.html.)

USDA, 1997. Petition under 7CFR Part 340. Petition Number 96-317-01p, Monsanto, Cotton, Herbicide Tolerance, Glyphosate-tolerant Approved 5/27/97. (See: http://www.aphis.usda.gov/bbep/bp/petday.html.)

USDA, 2011. Adoption of genetically engineered crops in the US. Available at. http://www.ers.usda.gov/Data/BiotechCrops/.

US EPA, 1996. Plant pesticide inert ingredient CP4 enolpyruvlshikimate-3-3 and the genetic material necessary for its production in all plants. Federal Register, vol. 61, No. 150, pp. 40338-40340. (Available electronically at http://www.access.gpo.gov.)

US EPA, 1997. Phosphinothricin acetyltransferase and the genetic material necessary for its production in all plants — exemption from the requirement of a tolerance on all raw agricultural commodities. Federal Register 70, 17717–17720.

Vadlamudi, R.K., Weber, E., Ji, I.H., Ji, T.H., Bulla, L.A., 1995. Cloning and expression of a receptor for an insecticidal toxin of *Bacillus thuringiensis*. J. Biol. Chem. 270, 5490–5494.

Vaeck, M., Reynaerts, A., Hofte, H., Jansens, S., Debeuckeleer, M., Dean, C., Zabeau, M., Vanmontagu, M., Leemans, J., 1987. Transgenic plants protected from insect attack. Nature 328, 33–37.

Vardi, E., Sela, I., Edelbaum, O., Livneh, O., Kuznetsova, L., Stram, Y., 1993. Plants transformed with a cistron of a potato virus-Y protease (Nia) are resistant to virus-infection. Proc. Natl. Acad. Sci. U.S.A. 90, 7513–7517.

Voinnet, O., 2001. RNA silencing as a plant immune system against viruses. Trends Genet. 17, 449–459.

Voinnet, O., 2005. Induction and suppression of RNA silencing: insights from viral infections. Nat. Rev. Genet. 6, 206–220.

Voinnet, O., 2008. Post-transcriptional RNA silencing in plant–microbe interactions: a touch of robustness and versatility. Curr. Opin. Plant Biol. 11, 464–470.

Waterhouse, P.M., Smith, N.A., Wang, M.B., 1999. Virus resistance and gene silencing: killing the messenger. Trends Plant Sci. 4, 452–457.

Waterhouse, P.M., Wang, M.B., Lough, T., 2001. Gene silencing as an adaptive defence against viruses. Nature 411, 834–842.

Waterworth, H., Hadidi, A., 1998. Economic losses due to plant viruses. In: Hadidi, A., Khetarpa, R.K., Koganezawa, H. (Eds.), Plant Virus Disease Control, APS Press, p. 13.

Whalon, M.E., Wingerd, B.A., 2003. Bt: mode of action and use. Arch. Insect Biochem. Physiol. 54, 200–211.

WHO Food Additives Series 40, 1998. World Health Organization International Program on Chemical Safety, Committee on Food Additives (JECFA). Safety evaluation of certain food additives and contaminants: Maltogenic Amylase. Report available under http://www.inchem.org/documents/jecfa/jecmono/v040je06.htm.

Wilmink, A., Vandeven, B.C.E., Dons, J.J.M., 1995. Activity of constitutive promoters in various species from the Liliaceae. Plant Mol. Biol. 28, 949–955.

Wong, E.Y., Hironaka, C.M., Fischhoff, D.A., 1992. *Arabidopsis thaliana* small subunit leader and transit peptide enhance the expression of *Bacillus thuringiensis* proteins in transgenic plants. Plant Mol. Biol. 20, 81–93.

Zaccomer, B., Cellier, F., Boyer, J.C., Haenni, A.L., Tepfer, M., 1993. Transgenic plants that express genes including the 3′ untranslated region of the turnip yellow mosaic-virus (Tymv) genome are partially protected against Tymv infection. Gene 136, 87–94.

Zambryski, P.C., 1992. Chronicles from the agrobacterium–plant cell–DNA transfer story. Annu. Rev. Plant Physiol. Plant Mol. Biol. 43, 465–490.

Zambryski, P., Joos, H., Genetello, C., Leemans, J., Vanmontagu, M., Schell, J., 1983. Ti-plasmid vector for the introduction of DNA into plant-cells without alteration of their normal regeneration capacity. EMBO J. 2, 2143–2150.

Zhang, B.Y., Chen, M., Zhang, X.F., Luan, H.H., Tian, Y.C., Su, X.H., 2011. Expression of Bt-Cry3A in transgenic *Populus alba* × *P. glandulosa* and its effects on target and non-target pests and the arthropod community. Transgen. Res. 20, 523–532.

Zhang, W., Wu, R., 1988. Efficient regeneration of transgenic plants from rice protoplasts and correctly regulated expression of the foreign gene in the plants. Theor. Appl. Genet. 76, 835–840.

Zimmerli, B., Schlatter, J., 1991. Ethyl carbamate: analytical methodology, occurrence, formation, biological activity and risk assessment. Mutat. Res. 259, 325–350.

Page references followed by "f" indicate figure, and "t" indicate table.